Encyclopedia of Applied Physics

SPONSORS

AMERICAN INSTITUTE OF PHYSICS
DEUTSCHE PHYSIKALISCHE GESELLSCHAFT
JAPAN SOCIETY OF APPLIED PHYSICS
PHYSICAL SOCIETY OF JAPAN

DEUTSCHE
PHYSIKALISCHE
GESELLSCHAFT

JAPAN SOCIETY
OF APPLIED PHYSICS

PHYSICAL SOCIETY
OF JAPAN

ENCYCLOPEDIA OF APPLIED PHYSICS

VOLUME 11
Mössbauer Effect to Nuclear Structure

Edited by
GEORGE L. TRIGG

Associate Editors
EDUARDO S. VERA
WALTER GREULICH

Managing Editor
EDMUND H. IMMERGUT

Assistant Managing Editor
CHRISTOPHER THOMAS MORAN

Editorial Assistant
HILARY GRANSON

George L. Trigg
275 Beaver Dam Road
Brookhaven, New York 11719

Walter Greulich
Ziegeleiweg 30
D-69488 Birkenau
Federal Republic of Germany

Eduardo S. Vera
NTT Advanced Technology Corporation
NTT Interdisciplinary Research Laboratories
Musashino-shi, Tokyo 180, Japan.

Edmund H. Immergut
Christopher Thomas Moran
Hilary Granson
2 Sidney Place
Brooklyn, New York 11201

Information Center on Science and Technology
University of Chile
Beaucheff 850
Santiago, Chile

Library of Congress Cataloging-in-Publication Data

Encyclopedia of applied physics.
(Revised for Vol. 11)

"Sponsors, American Institute of Physics . . . [et al]."
Includes bibliographical references and index.
1. Physics—Encyclopedias. 2. Engineering—Encyclopedias. I. Trigg, George L. II. Vera, Eduardo S. III. Greulich, Walter. IV. American Institute of Physics.
QC5.E543 1991 530'.05 91-8738
ISBN 1-56081-070-X (VCH Publishers)

British Library Cataloguing in Publication Data

Encyclopedia of applied physics.
Vol. 11
I. Trigg, George L. (George Lockwood) II. Vera, Eduardo S. III. Greulich, Walter
621
ISBN 1-56081-058-0 (set)
ISBN 1-56081-070-X (vol. 11)

Printed in the United States of America.

ISBN 3-527-28133-9 (Volume 11) VCH Verlagsgesellschaft mbH
ISBN 3-527-26841-3 (set) VCH Verlagsgesellschaft mbH

Printing History:
10 9 8 7 6 5 4 3 2 1

Published jointly by:

VCH Publishers, Inc.
220 East 23rd Street
New York, NY 10010

VCH Verlagsgesellschaft mbH
P.O. Box 10 11 61
69451 Weinheim
Federal Republic of Germany

VCH Publishers (UK) Ltd.
8 Wellington Court
Cambridge CB1 1HZ
United Kingdom

ADVISORY BOARD

EDITORIAL CONSULTANTS

MAIN ENTRIES

The subject matter in the *Encyclopedia of Applied Physics* is presented in approximately 500 individual articles, arranged alphabetically. The topics can be classified into 20 sections, similar to the AIP Physics and Astronomy Classification Scheme (PACS):

01	General Aspects: Mathematical, Computational, and Information Techniques	11	Condensed Matter B: Thermal, Acoustic, and Quantum Properties
02	Measurement Science, General Devices and/or Methods	12	Condensed Matter C: Electronic Properties
03	Nuclear and Elementary Particle Physics	13	Condensed Matter D: Magnetic Properties
04	Atomic and Molecular Physics	14	Condensed Matter E: Dielectrical and Optical Properties
05	Electricity and Magnetism	15	Condensed Matter F: Surfaces and Interfaces
06	Optics (classical and quantum)	16	Materials Science
07	Acoustics	17	Physical Chemistry
08	Thermodynamics and Properties of Gases	18	Energy Research and Environmental Physics
09	Fluids and Plasma Physics	19	Biophysics and Medical Physics
10	Condensed Matter A: Structure and Mechanical Properties	20	Geophysics, Meteorology, Space Physics, and Aeronautics

Each article has been assigned a code number consisting of two digits which denotes the section, and a letter which gives the type of article. There are six types: A = Devices, Equipment; B = Materials; C = Methods, Processes; D = Phenomena, Effects; E = Scientific or Technological Fields; F = Institutions, Companies, Societies and other organizations.

CONTRIBUTORS

Henry D. I. Abarbanel, Department of Physics and Marine Physical Laboratory, Scripps Institution of Oceanography, University of California–San Diego, La Jolla, CA 92093, U.S.A.
Nonlinear Systems

Jess H. Brewer, Canadian Institute of Advanced Research and Department of Physics, University of British Columbia, Vancouver, B.C., Canada V6T 1Z1
Muon Spin Rotation/Relaxation/Resonance

Allen G. Croff, Oak Ridge National Laboratory, Oak Ridge, TN 37831, U.S.A.
Nuclear Fuels and Isotopes

J. P. Davidson, Department of Physics and Astronomy, University of Kansas, Lawrence, KS 66045, U.S.A.
Nuclear Structure

Herman Feshbach, Center for Theoretical Physics, Laboratory for Nuclear Science, and Department of Physics, Massachusetts Institute of Technology, Cambridge, MA 02139, U.S.A.
Nuclear Reactions

Robert J. Finley, Bureau of Economic Geology, The University of Texas at Austin, Austin, TX 78713, U.S.A.
Natural Gas

Ehud Gavron, Aces Research, Inc., P.O. Box 14546, Tucson, AZ 85732, U.S.A.
Networks, Computer

F. J. Hartmann, Physik-Department, E-18, Technische Universität München, D-85747 Garching, Germany
Muonic, Mesonic, and Baryonic Atoms and Molecules

H. G. E. Hentschel, Department of Physics, Emory University, Atlanta, GA 30322, U.S.A.
Nonhomogeneous Flows

Ilko G. Iliev, Department of Orthopaedic Surgery, Louisiana State University Medical Center, Shreveport, LA 71130, U.S.A.
Neurobiophysics

Yasuo Ito, Research Center for Nuclear Science and Technology, University of Tokyo, Tokai, Ibaraki 319-11, Japan
Muonium and Positronium

L. K. Johnson, The Aerospace Corporation, Los Angeles, CA 90009, U.S.A.
Neutral Atomic and Molecular Collision Processes

Edwin C. Jones, Jr., Electrical Engineering and Computer Engineering Department, Iowa State University, Ames, IA 50011, U.S.A.
Network Theory, Electrical

Walter Y. Kato, Brookhaven National Laboratory, Upton, NY 11973, U.S.A.
Nuclear Energy, Fission

M. Kimura, Argonne National Laboratory, Argonne, IL 60439, U.S.A., and Department of Physics, Rice University, Houston, TX 77251, U.S.A.
Neutral Atomic and Molecular Collision Processes

Michał Kopcewicz, Institute of Electronic Materials Technology, 01-919 Warszawa, Poland
Mössbauer Effect

Andrew A. Marino, Department of Orthopaedic Surgery and Department of Cellular Biology and Anatomy, Louisiana State University Medical Center, Shreveport, LA 71130, U.S.A.
Neurobiophysics

Robert Moog, Big Briar, Inc., Asheville, NC 28801, U.S.A.
Music, Electronic

Thomas D. Rossing, Department of Physics, Northern Illinois University, DeKalb, IL 60115, U.S.A.
Musical Instruments

Thomas J. Ruth, UBC/TRIUMF PET Program, University of British Columbia and TRIUMF, Vancouver, B.C., Canada V6T 2B5
Nuclear Medicine

Richard W. Siegel, Materials Science Division, Argonne National Laboratory, Argonne, IL 60439, U.S.A.
Nanophase Materials

Harold G. Smith, 103 Walton Lane, Oak Ridge, TN 37830, U.S.A.
Neutron Scattering

Tommaso Toffoli, Massachusetts Institute of Technology, Laboratory for Computer Science, Cambridge, MA 02139, U.S.A.
Neural Networks

Thomas Vogt, Physics Department, Brookhaven National Laboratory, Upton, NY 11973, U.S.A.
Neutron Diffraction

Toshimitsu Yamazaki, Institute for Nuclear Study, University of Tokyo, Tanashi, Tokyo 188, Japan
Muonium and Positronium

MOSFET (METAL-OXIDE SEMICONDUCTOR FET)

See TRANSISTORS, FIELD EFFECT

MÖSSBAUER EFFECT

MICHAŁ KOPCEWICZ, *Institute of Electronic Materials Technology, Warszawa, Poland*

INTRODUCTION

The effect to be discussed here is the recoil-free nuclear gamma resonance commonly known as the "Mössbauer effect" in recognition of Rudolf L. Mössbauer who discovered the effect in 1958. The path to this discovery led through conventional investigations of the resonant absorption of nuclear gamma radiation. In the pre-Mössbauer period, the emission and absorption of nuclear gamma radiation were treated in a classical way. The energy and momentum conservation in the process of emission and absorption of gamma radiation was considered for a free nucleus. The forces binding the nucleus in a solid were ignored. Thus, a nuclear transition with energy E_0 was associated with the recoil of the nucleus, which decreased the energy of the emitted gamma radiation. The recoil energy E_R being large as compared with the width of the emission line, together with the shift of the emission and absorption lines due to recoil, causes that the emission and absorption lines usually do not overlap and the resonance does not occur. In the pre-Mössbauer experiments, it was possible to observe the gamma resonance effect either by compensating the recoil energy or by thermal broadening of the emission and absorption lines. In the 1950s, Mössbauer studied the conventional gamma resonance effect for ^{191}Ir and observed an unexpected increase of the resonance effect at low temperature, which was contradictory to the classic model. Mössbauer interpreted this effect in terms of the Lamb model of capture of neutrons by atoms in the crystal (Lamb, 1939). The basis of this discovery constitutes grounds for a fundamentally new concept of the recoil-free emission of nuclear gamma radiation from nuclei bound in a solid and its resonant recoil-free absorption introduced by Mössbauer in 1958 (Mössbauer, 1958a,b). The nuclide ^{191}Ir did not make this effect famous. A breakthrough was made when the recoil-free resonant gamma absorption for ^{57}Fe was observed (Schiffer and Marshall, 1959; Hanna

3-527-28133-9/94/$5.00 + .50

et al., 1960a). It was recognized that the unique properties of the Mössbauer effect, such as a very narrow resonance line and excellent energy resolution, allow the study of hyperfine interactions. In 1960, the isomer shift, the quadrupole interaction, and magnetic hyperfine structure in α-Fe_2O_3 (Kistner and Sunyar, 1960) and in ferromagnetic α-iron (Hanna, *et al.*, 1960b) were observed. In 1961, Mössbauer was awarded the Nobel Prize.

Since its discovery, the Mössbauer effect has been applied in many disciplines of natural science: physics, chemistry, biology, and mineralogy. Now it is one of the most sophisticated experimental methods, which allows us to measure with unrivaled accuracy the changes of energy of gamma radiation.

The Mössbauer effect and its applications have been described in detail in many books (see Further Reading). All references related to the Mössbauer studies are compiled in the Mössbauer Effect Data Index, which is published regularly (10 times a year) by the Mössbauer Effect Data Center (University of North Carolina, Asheville, NC).

In the present article, the principles, basic properties, and methodology of the Mössbauer effect are discussed. The introduction to the hyperfine interactions in solids is presented. The applications of the Mössbauer spectroscopy in solid state and nuclear physics, chemistry, biology, geology, materials science, and technology are described briefly. This article serves only as an introduction to Mössbauer spectroscopy. For more rigorous theoretical treatment and a more complete description of experimental results, the reader is referred to the books suggested for Further Reading at the end of this article.

1. THEORY

1.1 Emission of a Gamma Ray from a Free Nucleus

Let us consider the process of emission of a gamma ray from a free nucleus at rest. Let E_0 be the energy of the excited state of the nucleus, which de-excites to the ground state by emitting a gamma ray. The gamma emission event is associated with the recoil of the nucleus (to conserve momentum), and so the energy of the gamma ray is $E_\gamma = E_0 - E_R$. From energy and momentum conservation, the recoil energy of the nucleus can be calculated as

$$E_R = p^2/2M = E_\gamma^2/2Mc^2 \tag{1}$$

where M is the mass of the nucleus and p is the momentum transferred to the nucleus, equivalent to the momentum of the gamma photon. Similarly, when a gamma photon is absorbed, it loses the energy E_R because of the recoil imparted to the absorbing nucleus.

In order to find whether resonant gamma ray absorption will occur, the recoil energy must be compared to the linewidth Γ of the emission (absorption) line, which is related to the mean lifetime τ of the excited state:

$$\Gamma\tau = \hbar. \tag{2}$$

As can be seen from Eq. (1), the recoil energy depends strongly on the energy of the gamma rays. For example, for a nucleus with M = 100 and $E_\gamma \cong E_0 \cong 1 \times 10^4$ eV, the recoil energy $E_R \cong 5 \times 10^{-4}$ eV. However, if the energy of the gamma rays is 5×10^4 eV, then the recoil energy E_R increases already to about 1.3×10^{-2} eV. Since the typical natural linewidth Γ is of the order of 10^{-8} eV for $\tau \cong 100$ ns, the emission and absorption lines will be separated by $2E_R \gg \Gamma$ and resonance will not occur. In order to observe the resonant absorption, the emission and absorption lines must at least partially overlap (Fig. 1). Now one can easily see why nuclear gamma resonance is very unlikely ($2E_R \gg \Gamma$) and optical resonance is easy to observe. For optical transitions with photon energies of a few electronvolts, $E_R \sim 10^{-11}$ eV for M = 100 and $\Gamma \sim 10^{-7}$–10^{-8} eV, so that $2E_R \ll \Gamma$.

When gamma rays are emitted and absorbed by free nuclei (e.g., in a gas), which move isotropically in thermal motion, the width of the emission and absorption lines will be increased by the Doppler effect. The Doppler broadening at temperature T is

$$D \cong 2(k_B T E_R)^{1/2}, \tag{3}$$

where k_B is the Boltzmann constant. This broadening at room temperature is of the order of 10^{-3} eV, which is usually comparable to or smaller than E_R. Thus, even the Doppler broadening is in most cases insufficient to make the overlapping of the emission and absorption lines significant. A large increase

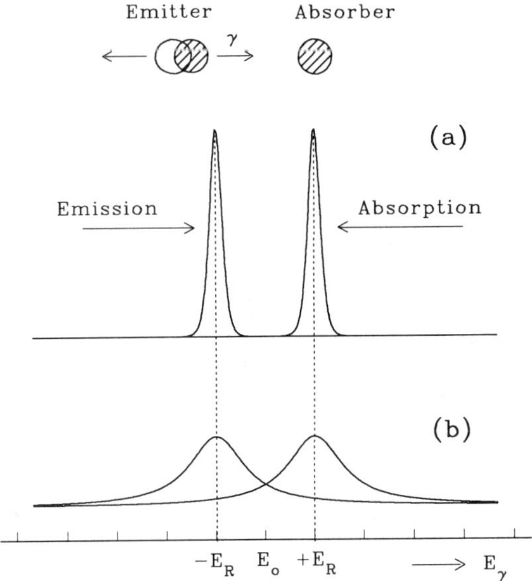

FIG. 1. Conditions for the occurrence of the nuclear gamma resonance for free nuclei: (a) case when $\Gamma < 2E_R$; (b) case when thermally broadened lines partially overlap.

of temperature will increase the overlap, but the lines become very broad, the cross section in the overlap region becomes small, and the resonance effect is of little value because of the very poor resolution.

Various methods have been used to compensate the recoil in order to make observation of the gamma resonance possible. Beside making use of the Doppler broadening, the recoil energy has been compensated by

1. the Doppler shift associated with the motion of the emitting nuclei (source) toward the absorber with a sufficiently high velocity [in the experiment by Moon (1950), the velocity of 8×10^4 cm/s was required] or
2. employing the recoil from the preceding nuclear decay or reaction.

In all these techniques, the recoil energy is *compensated,* in clear distinction to the Mössbauer effect discussed below in which the recoil energy is *eliminated* and compensation is not required.

1.2 Gamma Emission from the Nucleus Bound in a Solid—the Mössbauer Effect

In 1958, Mössbauer discovered that, when the emitting or absorbing nuclei are not free but bound in a solid, a certain fraction of such events may occur with negligible energy loss due to recoil. Mössbauer made this discovery when studying the scattering of 129-keV gamma rays from ^{191}Ir by Ir and Pt at low temperatures (Mössbauer, 1958a,b). In distinction from the existing models for the occurrence of the nuclear gamma resonance discussed above, Mössbauer observed an unexpected increase in the scattering when the temperature was decreased. The new idea of recoil-free gamma emission and absorption can be understood in terms of a simple phenomenological model. It can be shown that the mean energy transferred to the solid because of emission of a gamma quantum from a nucleus bound in the solid is equal to the recoil energy of the free nucleus E_R, Eq. (1). Three cases can be distinguished:

1. If the free-atom recoil energy E_R is larger than the binding energy of the atom in the solid (which typically is of the order of 10 eV), then the emitting atom (nucleus) will be displaced from its lattice site and the situation will be similar to emission from the free atom. Such a situation is typical for a gamma-ray energy of the order of 1 MeV or more. In such a case, the nuclear gamma resonance for lines with natural width will not occur.
2. If the free-atom recoil energy E_R is smaller than the binding energy but larger than the characteristic energy of lattice vibrations (phonon energy) defined in terms of the Debye frequency (ω_D) or Einstein frequency (ω_E), the atom will remain in its lattice position but will dissipate the recoil energy by creation of phonons. This situation is typical for gamma-ray energies of several hundred keV. In this case, the energy of the emitted gamma ray will be smaller than the resonance energy by $n\hbar\omega_D$ or $n\hbar\omega_E$, and resonance will not occur.
3. If the recoil energy E_R is smaller than the phonon energy $\hbar\omega_E$, a new effect arises, because the solid as a quantum system cannot be excited in an arbitrary way. The recoil energy is not dissipated either by displacing the nucleus from its lattice site or by heating the lattice (phonons). The nucleus behaves as if it were rigidly bound to the solid and the recoil is taken by the entire solid. Such a situation may occur for low gamma-ray energies of about 10–

150 keV. In this case, in Eq. (1), M in the denominator represents the mass of the entire solid, making E_R negligible as compared with Γ. When this occurs in the source and the absorber, then the conditions for nuclear gamma resonance are fulfilled and a large resonance for emission and absorption lines *with natural width* is observed. The recoil-free (zero-phonon) emission, absorption, and scattering of nuclear gamma radiation is called the *Mössbauer effect*.

1.3 Probability of the Mössbauer Effect

In the simple case discussed above as case **3** ($E_R \ll \hbar\omega_E$) and when we neglect multiphonon processes, the probability for recoil-free gamma transition (emission or absorption) is

$$f = 1 - E_R/\hbar\omega_E. \tag{4}$$

Since

$$E_R = \frac{E_\gamma^2}{2Mc^2} = \frac{p^2}{2M} = \frac{\hbar^2\kappa^2}{2M},$$

where κ is the magnitude of the wave vector of the gamma ray and $\hbar = h/2\pi$ (h is Planck's constant), we have

$$f = 1 - \hbar^2\kappa^2/2\hbar\omega_E M. \tag{5}$$

A general expression for the fraction of zero-phonon (recoil-free) processes is given by the same Debye–Waller (W) factor that is applicable to the scattering of x rays or neutrons by atoms:

$$\begin{aligned} f &= \exp(-2W) = \exp(-4\pi^2\langle x^2\rangle/\lambda^2) \\ &= \exp[-\kappa^2\langle x^2\rangle], \end{aligned} \tag{6}$$

where λ is the wavelength of the gamma quantum:

$$\kappa = 2\pi/\lambda = E/hc,$$

and $\langle x^2\rangle$ is the component of the mean square vibrational amplitude of the emitting nucleus in the direction of the gamma ray. A large probability for the Mössbauer effect (f close to 1) occurs when $\kappa^2 \langle x^2\rangle \ll 1$, i.e., when the rms displacement of the nucleus is small as compared to the wavelength λ. This is why the Mössbauer effect is not observed in gases and nonviscous liquids, where $\langle x^2\rangle$ is not limited.

Since $\langle x^2\rangle$ depends on temperature, the study of the temperature dependence of the Mössbauer effect provides useful information concerning lattice dynamics. The probability of the Mössbauer effect can be calculated in terms of the Debye model of lattice dynamics. Only $\langle x^2\rangle$ must be calculated, because κ is constant for a given gamma transition. The Debye model yields the following expression for the probability of the recoil-free fraction (see, e.g., Greenwood and Gibb, 1971):

$$\begin{aligned} f &= \exp\left[-\frac{6E_R}{k_B\Theta_D}\left(\frac{1}{4} + \frac{T^2}{\Theta_D^2}\int_0^{\Theta_D/T}\frac{x\,dx}{e^x - 1}\right)\right] \\ &= \exp[-2W], \end{aligned} \tag{7}$$

where Θ_D is the Debye temperature, k_B is the Bolzmann constant, T is the temperature, and $x = \hbar\omega/k_BT$.

The recoil-free fraction f can be evaluated in the high- and low-temperature limits with respect to Θ_D. At low temperatures ($T \ll \Theta_D$),

$$f = \exp\left[-\frac{E_R}{k_B\Theta_D}\left(\frac{3}{2} + \frac{\pi^2T^2}{\Theta_D^2}\right)\right]. \tag{8}$$

At absolute zero ($T = 0$ K),

$$f = \exp[-3E_R/2k_B\Theta_D]. \tag{9}$$

In the high-temperature range ($T \gg \Theta_D/2$),

$$f = \exp[-6E_RT/k_B\Theta_D^2]. \tag{10}$$

Equation (10) yields the linear dependence of the Debye–Waller factor W on T at high temperatures. This is true only when the harmonic approximation (Debye model) holds. In many experimental cases, when anharmonicity plays a significant role, a deviation from the proportionality of W to T is observed.

Concluding, the probability of the Mössbauer effect depends on three factors:

1. the free-atom recoil energy, related to the gamma transition energy [Eq. (1)];
2. the properties of the solid (i.e., the Debye temperature); and
3. the temperature of the solid (T).

A large probability is observed for low-energy gamma transitions in a solid with high Debye temperature when the measurements are performed at low temperature.

1.4 Shape of the Resonance Line

The gamma radiation emitted without recoil from the source has a Heisenberg width at half height [Eq. (2)], Γ_s, and the distribution of energies around the energy E_γ is given by the Breit–Wigner formula. The resulting shape of the line is Lorentzian:

$$w(E) = \frac{(\Gamma_s/2)^2}{(E - E_\gamma)^2 + (\Gamma_s/2)^2}. \qquad (11)$$

The cross section for resonance absorption (line shape of the absorption line with width Γ_a) is also given by the Breit–Wigner formula:

$$\sigma(E) = \sigma_0 \frac{(\Gamma_a/2)^2}{(E - E_\gamma)^2 + (\Gamma_a/2)^2}. \qquad (12)$$

The maximum cross section is given by

$$\sigma_0 = 2\pi\lambda\!\!\!^{-2} \frac{2I_e + 1}{2I_g + 1} \frac{1}{1 + \alpha}, \qquad (13)$$

where I_e and I_g are the nuclear spins of the excited and ground states and α is the internal conversion coefficient.

In a resonance experiment, an emission line of width Γ_s is moved over an absorption line of width Γ_a. The experimentally observed shape of the resonance line is

$$\sigma_{\exp}(E) = \int_{-\infty}^{\infty} w(e)\sigma(E - e)de. \qquad (14)$$

In the thin absorber approximation limit,

$$\sigma_{\exp}(E) = \sigma_0 \frac{(\Gamma_s + \Gamma_a)^2/4}{(E - E_\gamma)^2 + (\Gamma_s + \Gamma_a)^2/4}. \qquad (15)$$

If $\Gamma_s \cong \Gamma_a \cong \Gamma$, where Γ is the natural linewidth, then

$$\sigma_{\exp}(E) \cong \sigma_0 \frac{\Gamma^2}{(E - E_\gamma)^2 + \Gamma^2}. \qquad (16)$$

Since the theoretical line shape is well known [Eq. (16)], the Mössbauer effect allows us to measure changes of the gamma radiation energy of the order of a fraction of the natural linewidth Γ. The energy resolution, defined as

$$\Delta E/E_\gamma = \Gamma/E_\gamma, \qquad (17)$$

is of the order of 10^{-12}–10^{-15} depending on the Mössbauer isotope. The Mössbauer effect provides the most accurate method for measuring the changes of energy of electromagnetic radiation. Because of this feature, the Mössbauer effect is the most powerful technique for studying hyperfine interactions (see Sec. 3).

1.5 Parameters of the Mössbauer Line

The maximum resonance absorption occurs when the energies E_γ in the source (E_γ^s) and absorber (E_γ^a) coincide. The effective E_γ value can be changed by moving the source with respect to the absorber with velocity v, i.e., by using an externally applied Doppler effect. The change of the energy E_γ due to such motion is

$$\Delta E = (v/c)\, E_\gamma. \qquad (18)$$

Usually, absorption experiments are performed, and so the maximum resonance absorption corresponds to the minimum counting rate at the velocity v_r for which $E_\gamma^s = E_\gamma^a$ (Fig. 2). At any higher or lower velocity, the resonance absorption decreases until it finally vanishes at velocities far from v_r. The shape of the absorption line is Lorentzian [see Eq. (15)] with the full width at half maximum $\Gamma_r = \Gamma_s + \Gamma_a$, and the counting rate at $\pm v_\infty$ gives the nonresonant background. The nuclei in the absorber excited by resonance absorption re-emit gamma rays within $\tau \cong 10^{-7}$ s. However, when the internal conversion coefficient is high, relatively few gamma rays will be re-emitted. The re-emission process is not directional but occurs over the 4π solid angle. Thus, the number of counts in the detector corresponding to re-emitted gamma rays in a collimated, strongly directional transmission experiment is usually negligible.

Concluding, the parameters of a simple single-line Mössbauer spectrum (Fig. 2) are as follows:

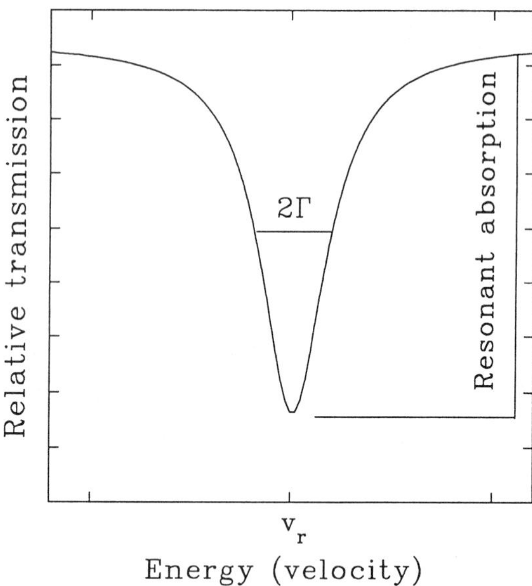

FIG. 2. Parameters of the Mössbauer transmission spectrum.

1. position of the resonance v_r;
2. full width at half maximum of the resonance line $\Gamma_r = \Gamma_s + \Gamma_a \cong 2\Gamma$;
3. depth of the resonance line, expressed as

$$D_r = \frac{\text{count rate at resonance} - \text{count rate at } \pm v_\infty}{\text{count rate at } \pm v_\infty}; \qquad (19)$$

4. nonresonant background count rate at $\pm v_\infty$; and
5. area of the resonance line,

$$A_r = \tfrac{1}{2}\pi \Gamma_r D_r. \qquad (20)$$

1.6 Mössbauer Isotopes

In order to observe the Mössbauer effect, a proper isotope must be found and favorable conditions for the solid matrix in which the isotope will be bound must be fulfilled. The Mössbauer nucleus should fulfill the following basic requirements:

1. low gamma transition energy (typically 10–150 keV) between the first excited and the stable ground state, and nucleus of mass $M > 50$ to minimize the recoil energy E_R;
2. mean lifetime of the excited level $\tau \cong 10^{-7}$–10^{-9} s; a longer lifetime results in a very narrow linewidth that makes the resonance difficult to observe experimentally, while a shorter lifetime results in a wide resonance line with poor resolution;
3. preferably, multipolarity of the gamma transition M1 or E2;
4. isotope production in the excited state as a product of natural radioactive decay of the parent isotope; the lifetime of the parent isotope should be reasonable, preferably of the order of 1 year or more. Mössbauer isotopes with radioactive metastable ground state also exist. The most important one is ^{129}I in which the ground state is radioactive with a half-life of 1.7×10^7 years. Other examples are ^{133}Ba (half-life of the ground state of 7.2 yr), ^{147}Pm (17.7 yr), ^{231}Pa (3.25×10^4 yr), ^{243}Am (7.95×10^3 yr); and
5. large maximum resonance cross section σ_0.

In view of the applications of the Mössbauer effect, the electric quadrupole moment and the nuclear magnetic moments of the levels involved in gamma transition should not be zero.

The solid matrix in which the isotope is bound should have a high Debye temperature Θ_D to allow observation of the Mössbauer effect with reasonable probability in a wide temperature range.

By far the best combination of the requirements listed above is found in ^{57}Fe, which is the most convenient and most popular Mössbauer isotope. However, the Mössbauer effect has been observed for 109 transitions in 90 isotopes of 47 elements.

1.7 ^{57}Fe Mössbauer Spectroscopy

Most Mössbauer investigations involve the isotope ^{57}Fe. Over 75% of all papers on experimental Mössbauer spectroscopy are concerned with the 14.4-keV transition in ^{57}Fe. It is fortunate that an element of great commercial importance and so abundant in nature, which forms a great variety of chemical compounds and alloys, has an isotope with an almost ideal combination of properties for Mössbauer spectroscopy. The gamma transition for which the Mössbauer effect is observed has very low energy (14.39 keV), and hence, the recoil energy is small (1.95×10^{-3} eV); an almost purely magnetic dipole (M1) transition (E2 admixture is negligible) occurs between the first excited state ($I_e = \frac{3}{2}^-$) and the ground state ($I_g = \frac{1}{2}^-$). The lifetime ($\tau_{1/2}$) of this level is about 100 ns, and hence, the natural linewidth is only 0.45×10^{-8} eV and produces the linewidth of the Mössbauer line of $2\Gamma = 0.194$ mm/s. The small amount of ^{57}Fe in natural iron (2.19%) is compensated by the large cross section for resonant absorption, $\sigma_0 \cong 257 \times 10^{-20}$ cm^2. The 14.4-keV transition has an internal conversion factor $\alpha \cong 10$, which allows us to use the conversion electrons for detecting the Mössbauer transition. This leads to a special technique: conversion-electron Mössbauer spectroscopy (CEMS). The conversion electrons (among which 7.3-keV ones emitted from the K shell are most abundant) emerge only from the depth equivalent to their range in a given material. Thus, the CEMS technique is very useful for surface studies, probing the depth of about 150 nm. ^{57}Fe is produced by β^+ (electron capture) decay from the parent ^{57}Co isotope whose lifetime is 270 days. ^{57}Co decay populates the second excited state in ^{57}Fe of the energy of 136 keV, which decays 90% through a cascade to the 14.4-keV level and then to the ground state, and 10% directly to the ground state. The Mössbauer effect has also been detected for the 136-keV transition, but its importance for the applications of the Mössbauer spectroscopy is negligible. The decay scheme of ^{57}Fe is shown in Fig. 3.

The recoil energy being small, the probability of the recoil-free emission and absorption of 14.4-keV gamma rays is large even at high temperatures (well above room temperature). Thus, the Mössbauer effect can be employed to study hyperfine interactions in a wide temperature range.

By using the Mössbauer effect, it is possible to measure changes of the energy of gamma radiation with astonishing accuracy. The energy resolution of the ^{57}Fe resonance is $\Gamma/E_\gamma \cong 3 \times 10^{-13}$.

1.8 Other Isotopes

The second most popular Mössbauer isotope is ^{119}Sn. The decay scheme and the most relevant parameters are shown in Fig. 4. The most convenient way of populating the 23.87-keV excited level used for the observation of the Mössbauer effect is to use metastable ^{119m}Sn, which can be produced by neutron capture in enriched ^{118}Sn. The metastable ^{119m}Sn has a lifetime of 250 days. The energy resolution is, however, poorer than for ^{57}Fe (the natural linewidth $2\Gamma = 0.626$ mm/s). This isotope has found fairly wide applications in chemistry. A range of tin-containing inorganic and organic compounds and alloys has been studied. It is of particular interest to study materials containing two Mössbauer isotopes, e.g., FeSn alloys, where both isotopes can be used for studying the hyperfine interactions.

The lightest nuclide for which the

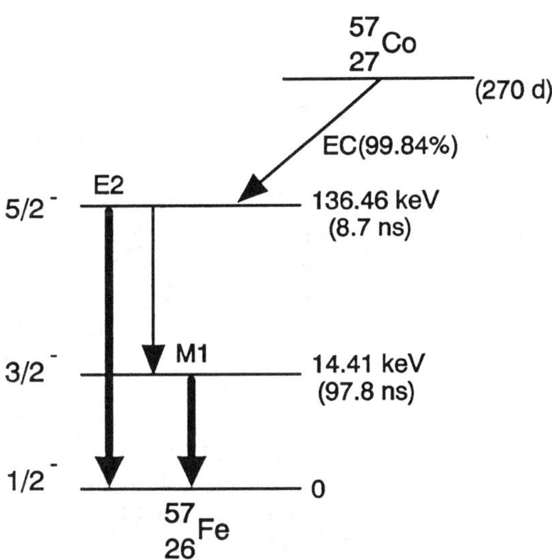

FIG. 3. Decay scheme for ^{57}Fe.

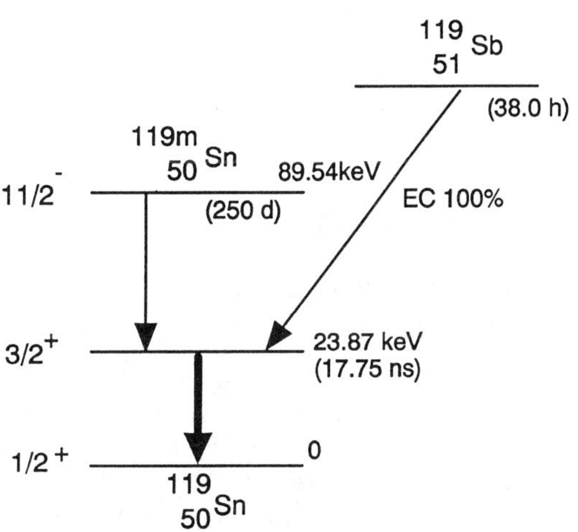

FIG. 4. Decay scheme for ^{119}Sn.

Mössbauer effect was observed is ^{40}K. However, the 29.4-keV Mössbauer level is not populated by any radioactive parent. This level is accessible only during a nuclear reaction [e.g., (d,p) or (n,γ)].

The isotope that reveals the highest energy resolution, of the order of 10^{-15}, is ^{67}Zn. The 93.26-keV excited state has a long lifetime of 9400 ns. The resulting linewidth of the Mössbauer line is 3.12×10^{-4} mm/s. Experiments with ^{67}Zn are very difficult, since the normal level of mechanical vibrations in the laboratory may easily disturb the resonance. Nevertheless, experiments with ^{67}Zn have been performed successfully in several laboratories.

Other isotopes that have found applications mostly in chemistry are ^{129}I, ^{127}I, and ^{125}Te. For the studies of magnetism of rare earth elements, Mössbauer isotopes such as ^{151}Eu, ^{153}Eu, and ^{161}Dy are available. The unique element is gadolinium, which has eight Mössbauer transitions in six isotopes. Unfortunately none of them is easy to observe; the experiments require expensive instrumentation, and even then the hyperfine effects cannot be fully resolved.

The Mössbauer isotopes of actinides (^{237}Np, ^{237}U, ^{241}Au, ^{232}Th) are interesting for measuring nuclear constants. The experiments are difficult because in most of these isotopes the Mössbauer nuclear level can be populated only in nuclear reaction or via Coulomb excitation. There is also high background due to other nuclear gamma transitions, which obscure the Mössbauer line.

An extensive list of isotopes with all the relevant data for which the Mössbauer effect was observed can be found in, e.g., Gonser (1975), Greenwood and Gibb (1971), and Shenoy and Wagner (1978).

2. INSTRUMENTATION

2.1 Basic Principles of the Mössbauer Experiment

The most typical Mössbauer experiment involves a radioactive source containing the Mössbauer isotope in an excited state and an absorber consisting of the material to be investigated that contains the same isotope in the ground state. Here, we will present the most common case of experiments with the use of the isotope ^{57}Fe. In transmission geometry, the gamma rays emitted from the source pass through the absorber in which resonant absorption occurs, and then they are detected by a suitable detector. In order to investigate the energy levels of the Mössbauer nucleus in the absorber, which may be shifted with respect to the energy levels of the same isotope in the source material or split by hyperfine interactions (see Sec. 3), it is necessary to modulate the energy of the emitted gamma radiation so that it can match its energy to the resonance. The most common method of energy modulation is Doppler modulation, consisting in a relative movement of the source with respect to the absorber. The energy change δE is given by the simple formula

$$\delta E = (v/c)\, E_\gamma, \tag{21}$$

where v is the velocity of the relative motion of the source with respect to the absorber, E_γ is the energy of the Mössbauer gamma radiation, and c is the velocity of light. For the 14.4-keV transition in ^{57}Fe, the velocity of 1 mm/s corresponds to 4.8×10^{-8} eV or 11.6 MHz. The motion of the source is usually oscillatory to provide an energy scan.

The block diagram of a simple Mössbauer spectrometer is shown in Fig. 5. The typical velocity transducer is an electromechanical device consisting of two coils fixed to a rod supported by springs and placed in a static magnetic field (double loudspeaker system). The drive signal delivered from the drive generator to one of the coils induces motion of the rod, to which the source is attached. As a result of this motion, a signal is induced in the pick-up coil that is used for correcting the shape of the drive signal via a negative-feedback loop. The shape of the drive signal is usually such that the motion occurs with constant acceleration (linear velocity scale). Sometimes a sinusoidal drive signal is used.

The detector, preamplifier, amplifier, and single-channel analyzer (SCA) form a conventional gamma spectrometer. The detector must have a sufficiently good energy resolution to ensure clear resolution of the Mössbauer gamma radiation from other gamma or x-ray lines and the background. For ^{57}Fe Mössbauer spectroscopy, usually proportional gas counters or scintillation counters do this job. All pulses from the detector, re-

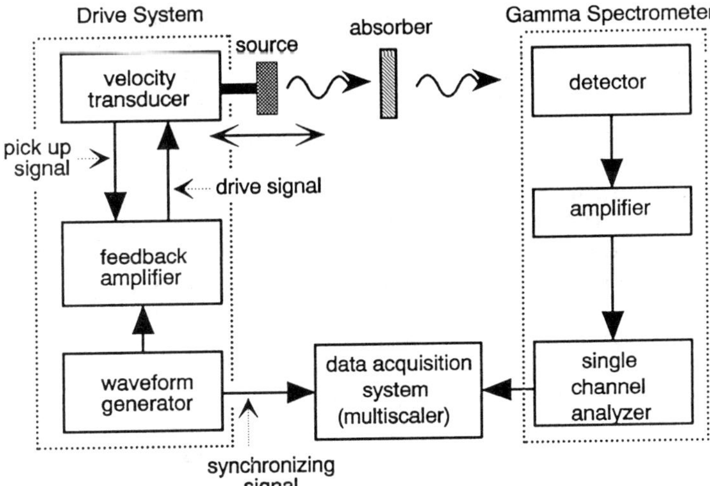

FIG. 5. Block diagram of a typical Mössbauer spectrometer.

gardless of their energy, are amplified. The SCA is used for selecting from the whole energy spectrum only the pulses corresponding to the Mössbauer gamma radiation. These pulses are delivered to the acquisition system working in the multiscaler mode (MSC). The address of the MSC is triggered by a starting pulse from the drive generator, and so the counting is synchronized with the motion of the source in such a way that each channel in the multiscaler stores the counts corresponding to a given velocity. In this way, a Mössbauer spectrum, i.e., counting rate vs velocity, is obtained.

The Mössbauer source must fulfill the following conditions. The radioactive isotope, e.g., ^{57}Co, must be embedded into a solid matrix that has a high Debye temperature to assure high probability for the recoil-free gamma emission. The preferred matrix material must be a nonmagnetic cubic metal (e.g., Rh, Cr, Pt, Pd, Cu) to avoid hyperfine splitting and ensure obtaining a single narrow emission line. The commercially available sources, such as ^{57}Co in Rh, have a probability for emission of the Mössbauer gamma radiation of about 0.75.

The absorber must be a solid (single crystal, polycrystalline powder or foil, amorphous, frozen solution, etc.) containing the resonant isotope. Its thickness should be small enough to allow transmission of gamma radiation but high enough to ensure measurable resonant absorption. The effective absorber thickness is

$$t_a = \sigma_o f_a n_a a_a d_a, \tag{22}$$

where σ_o is the maximum absorption cross section, f_a is the recoil-free fraction for the absorber material, n_a is the number of atoms/cm^3 of the particular element, a_a is the isotopic abundance of the resonance isotope, and d_a is the physical thickness of the absorber (in centimeters). In usual experimental conditions, $t_a \cong 1$.

It is possible to carry out Mössbauer experiments in which the material studied forms the source. In such a case, a single-line absorber should be used.

The experiments are performed not only in transmission geometry (which is most common) but also in a scattering geometry. The scattering geometry is useful when a thin absorber cannot be prepared, and it allows examination of a thin surface layer of the bulk materials. In the scattering geometry, the Mössbauer effect can be monitored by detecting either the Mössbauer gamma radiation or conversion electrons (conversion electron Mössbauer spectroscopy—CEMS), or else by x rays (conversion x-ray Mössbauer spectroscopy—CXMS) associated with the gamma absorption process.

Many experiments are carried out in a wide temperature range, and so cryostats or ovens are used. Investigations of magnetic properties require in many cases measurements with external static magnetic fields produced by superconducting coils. It is also possible to apply pressure to the material investigated, but such experiments are very difficult to perform.

In the case of ^{57}Fe Mössbauer spectroscopy, it is possible to produce polarized gamma rays either from a magnetically split six-line source of ^{57}Fe in ferromagnetic α-Fe or from a paramagnetic single-crystalline source exhibiting a quadrupole-split spectrum. The Mössbauer polarimetry technique is useful for studying the magnetic properties of solids. For the details concerning this method, the reader is referred to Gonser (1975).

3. HYPERFINE INTERACTIONS

The great accuracy of the Mössbauer effect of measuring the changes of the energy of electromagnetic (nuclear gamma) radiation is the most important feature of this technique. Its importance is enhanced by the fact that the typical energy resolution for the Mössbauer effect, related to the linewidth of the resonance line, is of the order of 10^{-12}–10^{-15}; this means that for 10–100-keV gamma radiation, energy changes of the order of 10^{-8} eV can be measured. Because of this feature, the Mössbauer effect is the most powerful tool for studying the hyperfine interactions. It allows us to observe directly the shifts and splittings of the Mössbauer lines resulting from the interactions of the nucleus with electrons. The hyperfine coupling mechanisms yield information regarding electron and spin density distributions, thus providing information concerning chemical, structural, and magnetic properties of solids.

The hyperfine interaction Hamiltonian for the atom contains terms related to interactions between the nucleus and its environment (electrons):

$$\mathcal{H} = E_0 + M_1 + E_2 + \cdots, \tag{23}$$

where E_0 refers to the Coulomb interaction between the nucleus and electrons at the nuclear site, M_1 is the interaction between the nuclear magnetic dipole moment and the effective magnetic field at the nucleus (magnetic dipole hyperfine interaction), and E_2 is the interaction between the nuclear quadrupole moment and the electric field gradient at the nucleus (electric quadrupole interaction). These are the most important interactions that determine the shape of the Mössbauer spectra. The term E_0 describes the isomer shift, M_1 is the nuclear Zeeman effect, which is responsible for the magnetic hyperfine structure, and E_2 causes the quadrupole splitting. Interactions of higher order (M_3, E_4, etc.) can be neglected because their energies are by several orders of magnitude smaller than E_0, M_1, and E_2, and the electric dipole interaction (E_1) is parity forbidden.

3.1 Isomer Shift

The presence of electrons affects the energy levels of the nucleus. The nuclear levels are shifted by the interaction of the nuclear charge with the electron charge density at the nucleus. In a given nuclear state, the nucleus, considered as a uniformly charged sphere of radius R (R_g or R_e for the ground and excited states engaged in the gamma transition, respectively), interacts with the nonzero electron charge density (s electrons) within the nuclear volume. During a gamma transition, the effective nuclear volume changes, thereby affecting the nucleus-electron interaction energy. The energy of the nuclear gamma transition is shifted as compared to that for a hypothetical point nucleus of the same charge by

$$\Delta E = (2\pi/5)Ze^2|\psi(0)|^2(R_e^2 - R_g^2), \tag{24}$$

where $-e|\psi(0)|^2$ is the electron charge density at the nucleus site.

Since in the Mössbauer effect the source and absorber are involved, for which the chemical environments of the resonant nucleus can be different, the isomer shift δ (or IS) measured in the experiment is

$$\begin{aligned}\delta = \mathrm{IS} &= \Delta E_a - \Delta E_s \\ &= (2\pi/5)Ze^2[|\psi_a(0)|^2 - |\psi_s(0)|^2](R_e^2 - R_g^2) \\ &= (4\pi/5)Ze^2\mathrm{R}^2(\delta R/R)[|\psi_a(0)|^2 - |\psi_s(0)|^2],\end{aligned} \tag{25}$$

where subscripts a and s denote the absorber and the source, respectively, and $\delta R = R_e - R_g$. Figure 6(a) schematically presents the shifts of the nuclear levels causing the isomer shift in the Mössbauer spectrum, a simple example of which is shown in Fig. 6(b).

As can be seen from Eq. (25), the isomer

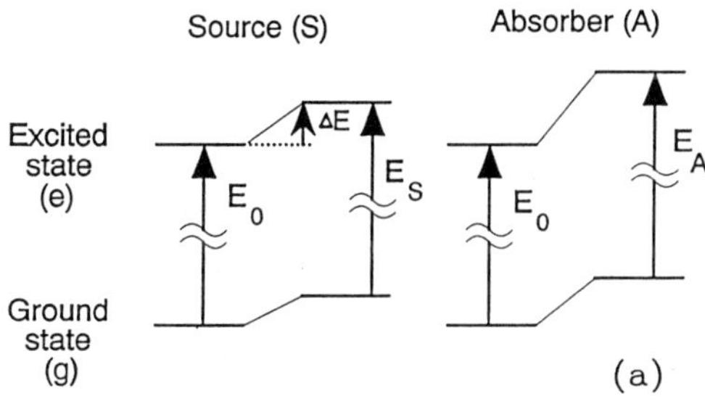

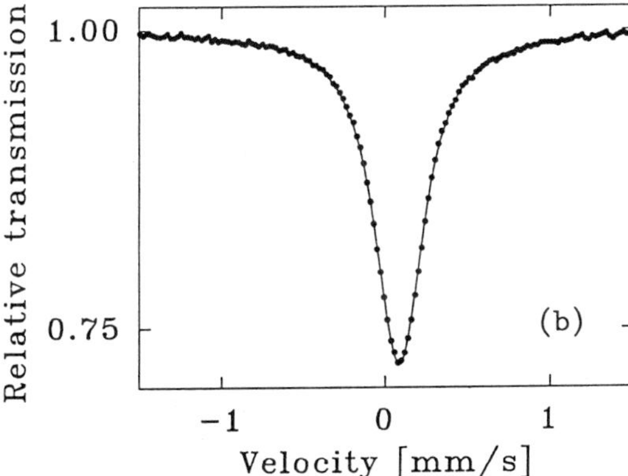

FIG. 6. The isomer shift: (a) the shift of the nuclear levels; (b) the Mössbauer spectrum of $K_4Fe(CN)_6$ enriched in ^{57}Fe revealing the isomer shift.

shift depends both on the nuclear properties (the change of the nuclear radius due to gamma transition) and on electronic properties, the latter related to the chemical properties of the material in which emitting or absorbing nuclei are contained. It is not possible to separate these contributions in a single Mössbauer experiment. Usually, the nuclear part, constant for a given gamma transition, is established first, or measured by another method, and the isomer shift is used to determine the electronic term, which allows study of the chemical properties via the changes of the electron density for various materials.

From the direction of the shift of the resonance line with increasing *s*-electron density, it is inferred that δR is negative for ^{57}Fe, which means that the ^{57}Fe nucleus is larger in the ground state than in the excited state. More typical is the opposite trend ($\delta R > 0$) as observed, e.g., for another typical Mössbauer isotope—^{119}Sn.

Despite the isomer shift being directly related to the *s*-electron density, it depends also on the *d* electrons through screening effects.

The isomer shift is measured relatively: the isomer shift of the absorber is measured with respect to the source host material. Various reference standards have been chosen for various isotopes. For the ^{57}Fe isotope, the reference point with respect to which the isomer shifts are measured is the center of the spectrum of ferromagnetic α-Fe at room temperature. For a detailed discussion of isomer shifts for various isotopes, the reader is referred to Shenoy and Wagner (1978).

The line position in the spectrum is determined not only by the isomer shift [Eq. (25)] but also by a relativistic temperature-dependent contribution (second-order Doppler shift) of the form

$$\delta E_\gamma / E_\gamma = -\langle v^2 \rangle / 2c^2, \tag{26}$$

where $\langle v^2 \rangle$ is the mean square velocity of the resonant nucleus related to the thermal motion and can be calculated in terms of the Einstein or Debye model.

3.2 Electric Quadrupole Interaction

The electric quadrupole interaction between the quadrupole moment of the nucleus in a state with spin I and the electric field gradient (EFG) at the site of the nucleus created by the asymmetry of the electronic or lattice charge distribution is described by the Hamiltonian

$$\mathcal{H}_Q = \hat{Q}\nabla E = \frac{e^2 Qq}{4I(2I-1)} \times \left[3I_z^2 - I(I+1) + \frac{\eta}{2}(I_+^2 + I_-^2) \right], \tag{27}$$

where $eq = V_{zz}$ and $\eta = (V_{xx} - V_{yy})/V_{zz}$ are the principal component and the asymmetry parameter of the diagonal EFG tensor, respectively, in such a reference system that $|V_{zz}| \geq |V_{xx}| \geq |V_{yy}|$ and $V_{zz} + V_{yy} + V_{xx} = 0$; thus, $0 \leq \eta \leq 1$; I_+ and I_- are the raising and lowering (shift) operators, respectively.

As a result of this interaction, the energy levels with $|Q| > 0$ ($I > \frac{1}{2}$) split into sublevels with energies given by the eigenvalues of

Eq. (27):

$$E_Q = \frac{e^2qQ}{4I(2I-1)}[3m_I^2 - I(I+1)]\left(1+\frac{\eta^2}{3}\right)^{1/2}, \tag{28}$$

where $m_I = I, I-1, \ldots, -I$ is the magnetic quantum number. The electric quadrupole interaction lifts only partially the degeneracy of the nuclear states, and the states that differ only in sign of m_I remain degenerate.

In the case of ^{57}Fe, for which the spins of the ground and excited states are $\frac{1}{2}$ and $\frac{3}{2}$ respectively, the quadrupole interaction causes the splitting of the excited state into two sublevels with $m_I = \pm\frac{3}{2}$ and $\pm\frac{1}{2}$, leaving the ground state unsplit. The Mössbauer spectrum, which corresponds to the transitions between two sets of hyperfine sublevels, splits into a two-line pattern (quadrupole-split doublet) (Fig. 7). The line intensities are determined by the angular dependence of the electric quadrupole interaction. In the case of an axially symmetric EFG ($\eta = 0$), this dependence is

$\frac{3}{2}(1+\cos^2\Theta)$ for $\pm\frac{3}{2}\rightarrow\pm\frac{1}{2}$ transition

$1+\frac{3}{2}\sin^2\Theta$ for $\pm\frac{1}{2}\rightarrow\pm\frac{1}{2}$ transition

where Θ is the angle between the direction of the principal axis of the EFG tensor and the propagation direction of the gamma radiation. For randomly oriented polycrystalline materials, the relative line intensities are 1:1. However, if the recoil-free fraction is anisotropic because of anisotropy of lattice vibrations (i.e., the f factor depends on the angle Θ measured with respect to the crystallographic axis of highest symmetry), then the angular dependence of the electric quadrupole interaction is modified:

$\frac{3}{2}(1+\cos^2\Theta)f(\theta)$ for $\pm\frac{3}{2}\rightarrow\pm\frac{1}{2}$ transition;

$[1+\frac{3}{2}\sin^2\Theta]f(\Theta)$ for $\pm\frac{1}{2}\rightarrow\pm\frac{1}{2}$ transition.

In such a case, averaging of Θ over a sphere yields, even for polycrystalline material, unequal line intensities of the quadrupole doublet. This effect is called the Goldanskii–Karyagin effect (Goldanskii, *et al.*, 1963; Karyagin, 1963).

As for the case of isomer shift, the measurements of quadrupole splitting give only the product of the nuclear parameter (Q) and the electric field gradient at the nucleus. To obtain a value of the quadrupole moment, which can serve as a valuable test of nuclear models, it is necessary to evaluate independently the EFG tensor. The nuclear parameter is, however, constant for a given nuclear level, and it is the details of the EFG, which can be derived from the Mössbauer spectrum, that are of primary interest.

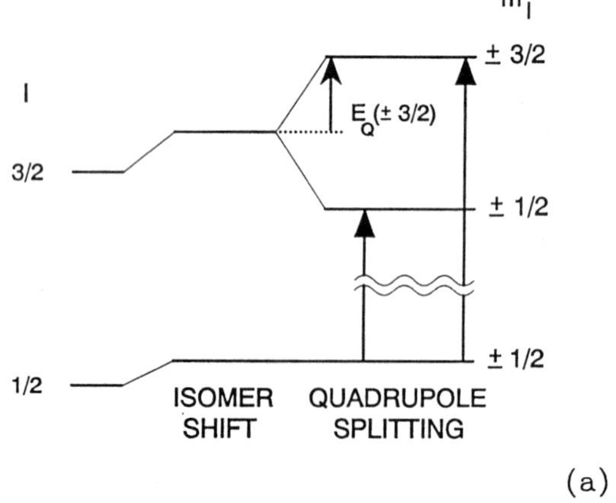

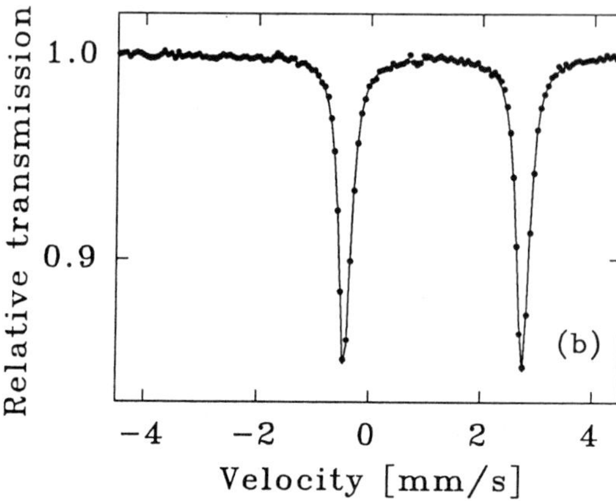

FIG. 7. The quadrupole splitting for a nucleus with $I_e = \frac{3}{2}$ and $I_g = \frac{1}{2}$ (e.g., ^{57}Fe): (a) splitting of the nuclear levels: (b) the Mössbauer spectrum of $FeSO_4 \cdot H_2O$ revealing the quadrupole splitting and isomer shift.

The EFG is determined by a number of different contributions. The most important are the asymmetry of the electronic charge distribution of the Mössbauer atom itself (due to the partially filled electronic shells occupied by the valence electrons) and the contribution from the lattice due to the asym-

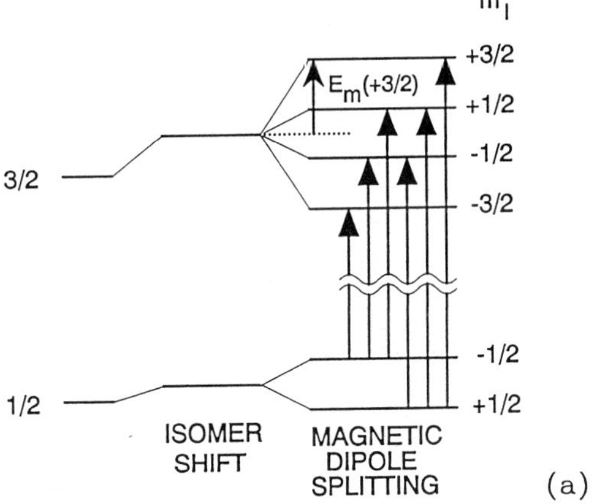

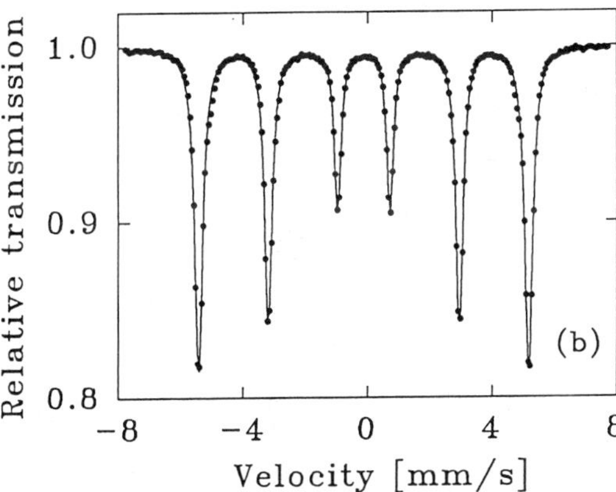

FIG. 8. The magnetic hyperfine splitting for a nucleus with $I_e = \frac{3}{2}$ and $I_g = \frac{1}{2}$ (e.g., ^{57}Fe): (a) the splitting of the nuclear levels and allowed gamma transitions ($\Delta m_I = 0, \pm1$); (b) the Mössbauer spectrum of ferromagnetic α-Fe revealing the magnetic hyperfine splitting.

metric arrangement of the ligand atoms in a noncubic lattice.

The quadrupole splitting is completely determined by the symmetry of the bonding environment and the local structure in the vicinity of the Mössbauer atom. Thus, the quadrupole interaction is of great significance, particularly in chemistry and solid-state physics.

3.3 Magnetic Dipole Interaction

The magnetic hyperfine interaction couples the nuclear magnetic moment $\boldsymbol{\mu}$ to the magnetic field $\mathbf{H}$ acting at the site of the nucleus. The Hamiltonian is

$$\mathcal{H}_M = -\boldsymbol{\mu} \cdot \mathbf{H} = -g\mu_N \mathbf{I} \cdot \mathbf{H}, \tag{29}$$

where g is the nuclear g factor ($g = \mu/I\mu_N$) and μ_N is the nuclear Bohr magneton ($\mu_N = e\hbar/2Mc$). This interaction lifts the degeneracy of the nuclear levels and splits the nuclear state with spin I ($I > 0$) into $2I + 1$ equally spaced sublevels with energies

$$E_m = -\mu H m_I/I = -g\mu_N H m_I, \tag{30}$$

where m_I is the magnetic quantum number ($m_I = I, I - 1, \ldots, -I$).

The magnetic field H is the total magnetic field at the nuclear site, created by the atom's own electrons and any externally applied field. In the case of ^{57}Fe, the magnetic hyperfine interaction splits a single Mössbauer line into a six-line pattern (Fig. 8). Only the transitions with $\Delta m_I = 0, \pm1$ are allowed. The relative line intensities are given by the angular dependence of each spectral line (Table 1).

From the line intensities, it is possible to determine the orientation of the magnetic field in the sample. In the general case, the line intensity ratio in the six-line spectrum is

$$3:\alpha:1:1:\alpha:3,$$

where

$$\alpha = \frac{4 \sin^2 \Theta_m}{1 + \cos^2 \Theta_m}.$$

The parameter α can vary from 0 to 4. The relative line intensities are 3:0:1:1:0:3 for $\Theta_m = 0$ (H parallel to the gamma-ray direction) and 3:4:1:1:4:3 for $\Theta_m = 90°$ (H per-

Table 1. Angular dependence of the allowed transitions in a Zeeman sextet of ^{57}Fe.

Transition	Δm_I	Angular dependence
$\pm\frac{3}{2} \rightarrow \pm\frac{1}{2}$	±1	$(\frac{3}{4})(1 + \cos^2 \Theta_m)$
$\pm\frac{1}{2} \rightarrow \pm\frac{1}{2}$	0	$\sin^2 \Theta_m$
$\mp\frac{1}{2} \rightarrow \pm\frac{1}{2}$	∓1	$(\frac{1}{4})(1 + \cos^2 \Theta_m)$

Θ_m is the angle between the direction of the hyperfine field H and the propagation direction of the gamma rays.

pendicular to the gamma-ray direction). For randomly oriented fields, we have 3:2:1:1:2:3.

The magnetic hyperfine interaction Hamiltonian (29) contains a nuclear parameter (μ) and an atomic parameter (H), which cannot be separated experimentally. The situation is more favorable here than in the case of the isomer shift and quadrupole splitting because both nuclear states are split, and it is possible to determine in one experiment two of the three parameters that describe the interaction: μ_g, μ_e, and H. Usually, the magnetic moment of the ground state, μ_g, is known, or determined from other than Mössbauer experiments (e.g., microwave resonance, atomic beam experiments), so that μ_e and H can be found. One can also apply a sufficiently strong external magnetic field and observe the resulting splitting, whereas it is not possible to apply a sufficiently high external electric field gradient in the case of quadrupole interaction.

The magnetic hyperfine splitting in the spectra yields an effective magnetic field, at the site of the nucleus, originating from various sources that are difficult to separate. The total magnetic field experienced by the nucleus is a vector sum of the following contributions:

$$H_{\text{eff}} = H_0 - DM + \tfrac{4}{3}\pi M + H_s + H_L + H_D, \tag{31}$$

where H_0 is the magnetic field at the nucleus generated by the externally applied field (in the absence of the external field $H_0 = 0$), $-DM$ is the demagnetizing field, and $\frac{4}{3}\pi M$ is the Lorenz field. The largest contribution to H_{eff} results from the Fermi contact interaction (H_s). Its origin either is related to the intrinsic impairing of s electrons or, indirectly, results from polarization effects on the filled s orbitals due to unpaired d or f orbitals:

$$H_S = -(16/3)\pi\beta\langle\Sigma(s_\uparrow - s_\downarrow)\rangle, \tag{32}$$

where $s_\uparrow$ and $s_\downarrow$ are the s-electron spin densities at the nucleus with spin up and spin down, respectively, and β is the Bohr magneton.

The nonzero orbital magnetic moment of the atom gives rise to the field

$$H_L = -2\beta\langle 1/r^3\rangle\langle L\rangle = -2\beta\langle 1/r^3\rangle(g-2)\langle S\rangle, \tag{33}$$

where $\langle L\rangle$ and $\langle S\rangle$ are the expectation values of the orbital and spin angular momentum, respectively, and g is the Lande factor. Evidently, $H_L = 0$ if $L = 0$.

The dipolar interaction of the nucleus with the spin of the atom creates the field

$$H_D = -2\beta\langle 3\mathbf{r}(\mathbf{S}\cdot\mathbf{r})/r^5 - \mathbf{S}/r^3\rangle. \tag{34}$$

In cubic metals, in the absence of spin-orbit coupling, $H_D = 0$.

The terms H_s, H_L, and H_D can all be of the order of 1 to 10 T (10^4–10^5 G), and their sum is usually called the "internal magnetic field."

3.4 Combined Magnetic Dipole and Electric Quadrupole Interactions

In the presence of the combined hyperfine interactions, the complete Hamiltonian is

$$\begin{aligned}\mathcal{H} &= \mathcal{H}(E_0) + \mathcal{H}_Q + \mathcal{H}_M = \\ &= \mathcal{H}(E_0) - g\mu_N \hat{I}'_z H + \frac{e^2qQ}{4I(2I-1)} \\ &\quad \times [3\hat{I}_z^2 - I(I+1) + \eta(\hat{I}_x^2 - \hat{I}_y^2)],\end{aligned} \tag{35}$$

where $\mathcal{H}(E_0)$ corresponds to the isomer shift and results in a shift of all spectral lines: I'_z and $\hat{I}_z$ are defined with respect to two different axis systems related to the orientations of the magnetic field or EFG axis relative to the gamma-ray direction. There is no analytical solution of the full Hamiltonian, and so a complete numerical analysis is necessary. In general, additional normally "forbidden" transitions appear. However, usually one interaction is small compared to the other. In the case of magnetically ordered systems, the electric quadrupole interaction is much weaker than the magnetic dipole one and causes a perturbation of the magnetic interaction. In the case of paramagnetic or diamagnetic systems, the applied external magnetic field causes a perturbation of the quadrupole interaction.

The Hamiltonian (35) has simple solutions in the following cases of ^{57}Fe Mössbauer spectroscopy.

1. Axially symmetric EFG tensor with symmetry axis parallel to the magnetic field H. For $I = \frac{3}{2}$, the energies of the magnetically split nuclear levels are shifted by the same amount as compared with the pure

magnetic splitting:

$$E = -g\mu_N H m_I + (-1)^{|m_I|+1/2}\, e^2qQ/4. \tag{36}$$

2. Axially symmetric EFG tensor with symmetry axis at angle Θ with respect to H and $e^2qQ \ll \mu H$ and $I = \frac{3}{2}$:

$$E = -g\mu_N H m_I + (-1)^{|m_I|+1/2} \times \frac{e^2qQ}{4}\,\frac{3\cos^2\Theta - 1}{2}. \tag{37}$$

The angle Θ cannot be determined from the spectrum, and so e^2qQ cannot be calculated unless the direction of H relative to the symmetry axis of the solid is measured independently.

3. Axially nonsymmetric EFG tensor with H along one of its principal axes ($I = \frac{3}{2}$):

$$E = \begin{cases} \dfrac{1}{2}\, g\mu_N H \pm \dfrac{e^2qQ}{4}\left[\left(1 + \dfrac{4g\mu_N H}{e^2qQ}\right)^2 + \dfrac{\eta^2}{3}\right]^{1/2}, \\ -\dfrac{1}{2}\, g\mu_N H \pm \dfrac{e^2qQ}{4}\left[\left(1 - \dfrac{4g\mu_N H}{e^2qQ}\right)^2 + \dfrac{\eta^2}{3}\right]^{1/2}. \end{cases} \tag{38}$$

The combined hyperfine interactions allow us to determine the sign of the quadrupole coupling, which in favorable conditions may allow the determination of the sign of the nuclear quadrupole moment.

3.5 Information Concerning the Properties of Solids from the Mössbauer Effect

The Mössbauer effect yields most information from the hyperfine interactions. As discussed above, the hyperfine interactions include nuclear and atomic parameters that are difficult to separate. However, for the most important isotopes, such as ^{57}Fe and ^{119}Sn, the nuclear parameters are well known and can be simply put into the fitting Hamiltonian. Most of the applications of the Mössbauer effect concern properties of solids, the nuclear parameters being of much less interest, since for a given gamma transition, they are constant. The isomer shift provides valuable chemical information. It reflects the changes in the electron density at the nucleus due to changes in the valence orbital population of the Mössbauer atom. The electron density is related to the type of chemical bonding, covalency effect, oxidation and reduction processes, and differences in the electronegativity of ligands coordinated to the Mössbauer atom. The study of a single compound is of little value for obtaining useful information unless data for other compounds are available for comparison. Only then is it possible to correlate isomer-shift data with the oxidation and spin state of the atom in a complex compound, the coordination between the Mössbauer atom and bonded ligands, etc. The richest data have been collected for iron compounds. The isomer shift is sensitive enough to the spin state of iron in a complex compound to allow distinguishing between high-spin and low-spin divalent and trivalent iron. The isomer-shift data alone make the distinction between low-spin iron(II) and iron(III) difficult, but with the help of quadrupole splitting [which is usually substantially larger for iron(III) compounds], they can be readily identified. High-spin iron compounds can be distinguished by isomer-shift data alone.

The quadrupole interaction provides information about the electric field gradient at the nucleus site, which can originate either from the valence electrons of the Mössbauer atom, when it is associated with asymmetry in the electronic structure, or from a nonspherical charge distribution in the ligand sphere and/or surrounding lattice with symmetry lower than cubic. Molecular orbitals can also contribute to the EFG. The effects of these contributions at the nucleus site are modified by the polarization of the core electrons of the Mössbauer atom, which may reduce or enhance the EFG. The quadrupole splitting observed in Mössbauer spectra of a given solid reflects the symmetry of the bonding environment and the local structure in the vicinity of the Mössbauer atom. Quadrupole splitting data are especially useful when combined with isomer-shift data. They yield chemical information concerning the electronic population of various orbitals, ligand structure, and structural information about the local atomic arrangement in both crystalline and amorphous solids. Quadrupole splitting is a particularly sensitive probe of short-range order in metallic glasses.

The magnetic hyperfine interaction yields information on the magnetic properties of solids, such as magnetic ordering and structure of magnetically ordered systems, the nature of the magnetic interaction (ferro- or antiferro-), and the magnitude of the magnetic moment on a particular atom. This information results from studying the magnetic hyperfine fields at the nucleus site. However, in order to compare such results with those obtained by conventional methods (e.g., magnetization measurements), phenomenological formulas have to be used to convert the information at the site of the nucleus to those at the site of the atom. Thus, the estimation of the magnetic moment of the atom is indirect. The temperature dependence of the hyperfine field is fortunately the same as that of magnetization; hence, the onset of magnetic ordering and magnetic phase transitions can be studied directly by the Mössbauer technique. Also, the shapes of the $H_{hf}(T)$ and $M(T)$ curves are identical and allow the study of deviation of experimental $H_{hf}(T)$ curves from the Brillouin function. Combined electric quadrupole and magnetic dipole interactions allow study of the direction of the magnetic field with respect to the crystallographic axes. The application of an external magnetic field to a system with no unpaired spins, which has no magnetic hyperfine field, leads to a magnetic splitting in addition to the quadrupole splitting and provides information on the geometry of the EFG at the nucleus.

From the complex spectra revealing a poorly resolved magnetic hyperfine structure of ferromagnetic amorphous alloys, it is possible to extract the distributions of hyperfine fields $P(H)$, which provide valuable information on the magnetic properties and structure of such materials.

The Mössbauer effect allows us to study not only static magnetic fields but also time-dependent relaxation effects. Every spectroscopy has a characteristic observation time of interaction of the radiation with the solid. The observation time of Mössbauer spectroscopy corresponds to the mean lifetime (τ_N) of the nuclear resonant level. For hyperfine interaction measurements, with the characteristic energy ΔE, it corresponds to the Larmor precession period $\tau_L = \hbar/\Delta E$. The times τ_N and τ_L are typically of the order of 10^{-7}–10^{-8} s. In the presence of time-dependent changes in the environment of the resonant nucleus, the experimental observation (shape of the Mössbauer spectrum and the hyperfine parameters) depends on the relative order of magnitude of the observation time (τ_N or τ_L) as compared with the residence time of the fluctuating environment, τ_R (relaxation time). When the observation time is much longer than the relaxation time (τ_N, $\tau_L \gg \tau_R$, slow relaxation), the quasistatic situation is observed. If the opposite relation occurs (τ_N, $\tau_L \ll \tau_R$, fast relaxation), a fully time-averaged situation occurs. For intermediate relaxation rates, we get complex spectra.

The time dependence can be introduced into the hyperfine interactions by using an external high-frequency magnetic field. When the frequency of such a field is higher than the Larmor frequency, the magnetic hyperfine splitting collapses and the Mössbauer spectrum consists of a single line or a quadrupole doublet instead of a Zeeman sextet. The shape of the spectra depends on the relation between the frequency of the field and the Larmor frequency. The relaxation effect induced by the radio-frequency field is fully controlled by the field frequency and intensity (Pfeiffer, 1972; Kopcewicz, 1989). The rf collapse of the magnetic hyperfine structure can be induced only in soft ferromagnetic materials.

4. APPLICATIONS OF THE MÖSSBAUER EFFECT IN VARIOUS FIELDS

The applications of the Mössbauer effect are so extensive in various branches of science and technology that only a few will be very briefly mentioned here. The interdisciplinary nature of the effect is demonstrated by applications as different as the theory of relativity, magnetism, metal physics, biology, nuclear physics, mineralogy, atmospheric physics, and even art. Most important, however, are the applications in solid state physics and chemistry. Some of the most rewarding applications of the Mössbauer spectroscopy were made in the field of magnetism (Thomas and Johnson, 1986, and, e.g., Long, 1987, and Long and Grandjean, 1989). Even though the various contributions to the effective magnetic field at the nucleus are difficult to separate experimentally and the magnetic moment associated with the atom is not necessarily proportional to the hyperfine field, valuable information on the type of magnetism, the temperature dependence of magnetization, and magnetic phase transitions was obtained. More exotic effects, such as spin-glass behavior, low-dimensional magnetism (Johnson, 1984), and superparamagnetism of ultrafine particles (Mørup, 1987), also have been studied. Magnetic properties of solids were investigated as a function of temperature (e.g., Thosar *et al.*, 1983), pressure (Williamson, 1978), and external static (e.g., Long, 1984; Long and Grandjean, 1989, 1993) and high-frequency magnetic fields (Kopcewicz, 1991b). A great variety of magnetic materials, such as alloys, intermetallic compounds, and metallic glasses, have received a lot of attention (see, e.g., Cranshaw, 1983). The Mössbauer effect greatly contributed to better understanding of Invar alloys, rare-earth magnetism (e.g., Thosar *et al.*, 1983; Long and Grandjean, 1993), and amorphous transition-metal–metalloid and metal-metal alloys (Longworth, 1987; Gonser, 1981). The thermal stability, crystallization, magnetic properties, and structure (short-range order) were extensively studied, and Mössbauer spectroscopy proved to be a powerful technique for such investigations. Recently, the possibility to form nanocrystalline phases of exceptional soft magnetic properties by controlled crystallization of some amorphous alloys was recognized, and the Mössbauer effect significantly contributed to the identification of the various nanocrystalline and microcrystalline phases formed.

The time-dependent effects readily observable by Mössbauer spectroscopy include electronic, spin, or lattice relaxation effects, the spin-flip processes in paramagnetic ions related to the spin-spin interaction with spins of the neighboring ions and spin-lattice interaction between electronic spins and phonons, valence fluctuations, after-effects of nuclear transitions, structural changes both macroscopic (diffusion) and microscopic (rotation within the molecule), and fluctuations in electronic spin orientation. The magnetic relaxation effects (time dependence of the magnitude and direction of the magnetic hyperfine field experienced by the nucleus) and the superparamagnetic relaxation in ultrafine particles were extensively studied. The fast magnetization reversal induced by an external magnetic radio-frequency field, causing a collapse of the magnetic hyperfine structure, can be observed in the Mössbauer spectra of soft crystalline and amorphous ferromagnets. The rf-collapse effect provides a unique possibility to separate the magnetic dipole and electric quadrupole interactions in the ferromagnetic state. The quadrupole splitting distributions can be extracted from the collapsed spectra, which allows a direct study of the short-range order in ferromagnetic amorphous alloys (Kopcewicz, 1991b).

The surface, thin-film, and multilayer structures were successfully studied by using CEMS and conversion x-ray Mössbauer spectroscopy (CXMS). The CEMS technique opened a wide class of applications to ion-beam modification of materials. The effects of ion implantation (especially nitrogen, carbon, boron, oxygen) into iron and steels are important for both science and technology (Williamson, 1993). Formation of new materials as a result of ion implantation or ion-beam mixing of multilayered structures, and the ion-induced transformation from crystalline to amorphous state, have been characterized by CEMS in various metallic systems (e.g., B-implanted α-Fe, ion-beam mixing of multilayers). Phenomena such as radiation damage, defects, consequences of the nuclear transformations in solids, and after-effects of radioactive decay affect the Mössbauer spectra. Ion implantation of radioactive Mössbauer isotopes into metal hosts allowed study of the migration of vacancies and interstitials, radiation damage cascades, cluster formation and dissolution, and migration of implanted impurities. The phenomena of internal oxidation and helium decoration were investigated (De Waard and Niesen, 1987). Ion implantation in semiconductors was also studied (Langouche, 1991).

The sophisticated technique of in-beam Mössbauer spectroscopy, developed already in the 1960s, makes use of Mössbauer nuclei excited directly by Coulomb excitation that simultaneously causes recoil implantation. In this way, a Mössbauer source with lifetime of the excited Mössbauer state is formed. No parent radioactive isotope is involved. Recoil implantation can be performed into any host material. In-beam Mössbauer spectroscopy yielded valuable information concerning radiation damage and the behavior of the implanted Mössbauer atoms in various materials. Pioneering studies were reviewed by Sprouce and Kalvius (1968) and Obenshain

(1968), and more recent results were presented by Sielemann (1991).

The surface magnetism of ultrathin iron films was investigated by conversion-electron and transmission Mössbauer spectroscopy. The dependence of the magnetic hyperfine field vs distance from the surface was studied by suspending one or only a few monolayers of ^{57}Fe in the ^{56}Fe layer at a well-defined depth; this yielded a good depth sensitivity. The experimentally determined hyperfine field as a function of depth was compared with theoretical predictions. Recent results are reviewed by Przybylski *et al.* (1991).

The Mössbauer effect was successfully applied for studying semiconductors (Si, GaAs) and solar cell materials (Williamson *et al.*, 1992).

When a new class of materials, such as high-T_C superconductors, appeared, the Mössbauer effect yielded useful structural information. Also, conventional superconductors have been studied (Shenoy, 1983).

Applications of the Mössbauer effect in metallurgy relate to steel research, phase formation, and phase transformation with respect to steel processing (Fujita, 1975).

The applications of the Mössbauer effect in magnetism are extensively reviewed in all books listed in Further Reading.

Chemical applications of the Mössbauer effect exploit its sensitivity to changes in electron density at the nucleus. The main fields of interest concern the electronic structure of the Mössbauer atom, oxidation states, spin states, bond properties (e.g., σ and π interaction of the Mössbauer atom with ligands and ionicity and covalency of bonds), structural properties (molecular geometry, polymerization), and dynamic processes (such as electron hopping in spinels, spin-spin and spin-lattice relaxation). The analytical potential of Mössbauer spectroscopy, especially involving the use of the isotope ^{57}Fe, was widely used for identification of various phases containing iron even in complex systems. It is possible to determine the iron content in various phases; however, the measurements of the relative concentrations are much more reliable than the determination of the iron content in absolute units. Reactions in solids, thermal- and radiation-induced decomposition, surface studies (e.g. corrosion process, adsorption, and surface reactions), and frozen solutions have been investigated. In the field of catalysis, Mössbauer spectroscopy has allowed *in situ* study of solids that catalyze gaseous reactions (see Gütlich *et al.*, 1978; Thosar *et al.*, 1983; Long, 1984; Long and Grandjean, 1987, 1989; Goldanskii and Herber, 1968; Greenwood and Gibb, 1971).

The Mössbauer effect has contributed also to nuclear physics. It is useful in the determination of nuclear parameters, especially of the excited states involved in the Mössbauer gamma transitions. Comparison of the experimentally determined parameters such as lifetime of the excited state, nuclear spin and magnetic and quadrupole moments, change of the nuclear radius, and internal conversion coefficient with the predictions of theoretical models is used for evaluation of the validity of the nuclear models. Especially interesting for nuclear physics are the Mössbauer transitions in rare-earth isotopes because of the great number of gamma transitions available, allowing the study of the effects of successive proton or neutron addition over a range of nuclei (consult Goldanskii and Herber, 1968; Greenwood and Gibb, 1971).

Mössbauer spectroscopy has contributed significantly to the better understanding of some biological systems (Johnson, 1975). As many biological molecules contain iron, the Mössbauer technique yields information on the local environment of iron in large and complex molecules in which iron atoms are at the active centers where biological changes occur. The most studied biological systems are iron proteins, such as hemoproteins (hemoglobin, myoglobin, cytochromes, catalases), iron sulfur proteins, and iron storage and transport proteins (ferritin). The structural dynamics of proteins has been investigated (Parak, *et al.* 1988). The Mössbauer effect has also found medical applications involving studies of tissue and blood samples that lead to identification of pathological factors.

In mineralogy, the Mössbauer effect has been used for studying chemical properties of various minerals, e.g., silicate minerals such as pyroxene and olivine (Bancroft, 1973). Phase identification and information concerning site population of Fe^{2+} and Fe^{3+} ions are obtained from the Mössbauer measurements. Many oxides (hematite, magnetite, goethite, etc.) and ores have been investigated. Studies of meteorites have been performed in view of their importance for reconstruction of the early history of the solar system. Even lunar samples delivered to the Earth by spacecraft were investigated by Mössbauer spectroscopy (Hafner, 1975). The Mössbauer effect has been applied for studying iron-containing atmospheric aerosol. Not only the iron concentration in the atmospheric air, but also the chemical compounds in which iron is contained and even the size of iron-containing particles of atmospheric aerosol can be evaluated. Such studies provide information concerning the sources of iron in the atmosphere, both natural and related to human activity (Kopcewicz and Kopcewicz, 1991).

The Mössbauer effect has been applied to archaeological problems. The study of ancient pottery and other archaeological artifacts provides information about the manufacturing technology and has revealed the level of technology in ancient times. The Mössbauer technique has been applied to the study of ancient bronzes and casting and firing techniques of ancient metallurgy (Zheng and Hsia, 1991).

The above list of applications is by far not complete but already shows the exceptional power of Mössbauer spectroscopy as a technique that can provide valuable information in many areas of science and technology.

5. EXOTIC MÖSSBAUER STUDIES

One of the most spectacular Mössbauer experiments performed in the early years of the "Mössbauer era" concerned the verification of the gravitational red shift predicted by the Einstein theory of relativity. Before the discovery of the Mössbauer effect, there was no experimental method that could measure the change of energy (or frequency) of electromagnetic radiation with sufficient precision to allow the measurement of the gravitational red shift in the laboratory. The relevant parameter is the high energy resolution of the Mössbauer effect ($\Gamma/E_\gamma \cong 3 \times 10^{-13}$ for ^{57}Fe or 5.2×10^{-16} for ^{67}Zn). The gravitational red shift can be considered as the change of the energy of a photon as it moves from one place to another of different gravitational potential. Let us consider a Mössbauer experiment in which a source is placed at a height h above the absorber; then the gamma photon travels a distance h through a gravitational field and interacts with it as though it had a mass of E_γ/c^2. As a result of this interaction, the energy of the photon changes by a fraction equal to

$$\delta E/E_\gamma = gh/c^2, \tag{39}$$

where g is the acceleration of the gamma photon in the gravitational field. This energy change is the gravitational red shift, which shifts the position of the Mössbauer resonance line by a factor of 1.1×10^{-16} m^{-1}. The experimental verification of this relativistic effect was performed by Pound and Rebka (1960) with the use of ^{57}Fe 14.4-keV resonance with a height separation between the source and absorber of 22.5 m. The experimentally measured shift was 0.94 ± 0.10 of that predicted. Later, the experiments gave an even more accurate value of 0.997 ± 0.0076 (Pound and Snider, 1965). In these studies, special care was taken to eliminate other than gravitational sources of the line shift, e.g., those due to the difference of the temperature at the sites of the source and absorber (second-order Doppler shift). Although the energy resolution of isotope ^{67}Zn is much better than that of ^{57}Fe, which makes the experiment with ^{67}Zn appear relatively simpler, the technical difficulties associated with the ^{67}Zn resonance prevented the use of this isotope and made the isotope ^{57}Fe a more attractive choice.

Experiments related to the red shift have also been performed in which the gravitational acceleration was replaced by a kinematic acceleration by means of a high-speed rotor with the source in the center and the absorber at the periphery (transverse Doppler effect); however, the results were not conclusive.

Another relativistic effect influencing the position of the Mössbauer resonance line is the thermal red shift (second-order Doppler shift) discussed already in Sec. 3.1, Eq. (26).

In 1984, new exotic experiments with the Mössbauer effect were performed. Synchrotron radiation was used as a source of the Mössbauer radiation (Gerdau, *et al.* 1985). The property that makes the synchrotron radiation especially attractive for Mössbauer studies is its exceptional brilliance, which exceeds the maximum brilliance of a conventional ^{57}Fe source by 1–2 orders of magnitude.

Mössbauer experiments with synchrotron radiation combine the unique properties of the Mössbauer effect, namely, its exceptional energy resolution, with the unique properties of synchrotron radiation such as outstanding brilliance, a high degree of linear polarization, and the pulsed structure of the beam. In order to use synchrotron radiation as a source of the Mössbauer radiation, the broad energy band coming from the storage ring must be monochromatized. Electronic monochromators as well as a nuclear monochromator, making use of the pure nuclear Bragg reflections that select a very narrow band corresponding to the width of the nuclear level Γ (GIAR films), were used (see, e.g., Rüffer, *et al.* 1990).

The experiments with synchrotron radiation are usually performed not on the conventional velocity (energy) scale but on a time scale. Synchrotron radiation opened new fields for Mössbauer studies, such as gamma optics and soft solid state excitations. Nuclear resonant scattering was studied in the spatially coherent decay channels in the Bragg and forward directions. Time evolution of the nuclear Bragg scattering was observed. The brilliance of the synchrotron radiation source allows experiments that are difficult in conventional Mössbauer spectroscopy, e.g., high-pressure and small-crystal investigations. The hyperfine parameters, especially the isomer shift, can be determined with much higher accuracy because of the quantum beat effect.

At present, there are several synchrotron radiation sources available for Mössbauer experiments: the European Synchrotron Radiation Facility in Grenoble (France), DESY in Hamburg (Germany), Stanford Synchrotron Radiation Laboratory in Stanford (U.S.A.), the Advanced Photon Source at Argonne National Laboratory in Argonne (U.S.A.), and the TRISTAN Accumulation Ring at KEK (Japan).

Since 1984, the technique of time-differential Mössbauer spectroscopy has developed and matured rapidly. Beginning as an exotic experiment, it is now on the way to becoming a widely used tool of Mössbauer spectroscopy.

6. PROSPECTS FOR THE FUTURE

Thirty-five years have elapsed since the discovery of the Mössbauer effect. During this time, the effect developed from an experimental curiosity to a commonly used spectroscopy that is finding wide applications in most disciplines of natural science. Nature was kind to provide a perfect isotope for Mössbauer spectroscopy—^{57}Fe, thanks to which a great variety of materials is accessible to Mössbauer studies. Mössbauer spectroscopy was quick to enter any new field that appeared in the last 35 years. Typical examples can be found in materials science. When new materials, such as amorphous alloys, high-T_C superconductors, nanocrystals, quasicrystals, and fullerenes appeared, the Mössbauer effect was immediately and successfully applied for the study of microstructure, chemical, and magnetic properties. Mössbauer spectroscopy was also combined with new experimental facilities of nuclear physics. Synchrotron radiation was used as a Mössbauer source opening new horizons. Recently, the possibility of using the Mössbauer effect to investigate the interaction of high-energy heavy ions with solids has attracted considerable attention. It seems that in the near future the Mössbauer experiments will involve combination with large installations, such as synchrotrons, cyclotrons, and other accelerators, to explore new frontiers of nuclear and solid-state physics. Mössbauer spectroscopy will certainly enter any new field of natural science that may appear in the future. The practical and technological applications, which still lag behind purely scientific research, offer wide and unexploited possibilities for future developments.

In his Nobel Laureate lecture in December 1961, Rudolf Mössbauer concluded, "We may therefore hope that this young branch of physics stands only at its threshold and that it will be developed in future, not only to extend the application of existing knowledge but to make possible new advances in the exciting world of unknown phenomena and effects." A lot has been done already. The effect has developed and matured, but Mössbauer's remark did not lose much of its novelty. Thirty-three years later Mössbauer spectroscopy has enormous and still unexploited possibilities for future applications.

GLOSSARY

Debye–Waller Factor: Originally derived in the Bragg x-ray scattering theory, describes the effect of temperature on the intensity of x-rays scattered elastically from the lattice. The Debye–Waller factor describes also the probability of the recoil-free gamma transition.

Doppler Broadening: Broadening of the lines emitted and absorbed by the atoms in a gas due to the thermal motion of atoms.

EFG: Electric field gradient.

***f* Factor:** See **Debye–Waller Factor**.

Isomer Shift: Shift of a whole Mössbauer spectrum relative to the reference standard due to the shift of the nuclear levels as a result of the interaction of nuclear charge with the electron charge density at the nucleus.

Lorentzian Line Shape: The shape of a spectral line in which the variation of intensity with energy is given by the reciprocal of a quadratic expression.

Magnetic Hyperfine Structure: Splitting of the nuclear levels due to the interaction between the nuclear magnetic moment with the effective magnetic field at the nucleus site; observed in the Mössbauer spectra as a splitting of the Mössbauer lines.

Mössbauer Effect: Recoil-free resonance absorption or scattering of nuclear gamma radiation occurring for atoms bound in solids.

Natural Linewidth: The width of the gamma emission line, related to the mean lifetime of the nuclear state, resulting from the Heisenberg uncertainty principle: $\Gamma\tau = h$.

Nuclear Zeeman Effect: See **Magnetic Hyperfine Structure.**

Quadrupole Splitting: Splitting of the nuclear levels due to the interaction of the nuclear quadrupole moment with the electric field gradient at the nucleus site; observed in the Mössbauer spectra as a splitting of the spectral lines.

Recoil Energy: The energy transferred to the nucleus due to the emission of gamma radiation from a free nucleus.

Second-Order Doppler Shift: The relativistic effect, related to thermal motion of atoms, that causes a temperature-dependent shift of the Mössbauer spectrum.

Velocity Transducer: An electromechanical device (essential part of the spectrometer) in which the source is moved with respect to the absorber (or vice versa); this provides the Doppler modulation of the Mössbauer gamma radiation and allows energy scan.

Works Cited

Bancroft, G. M. (1973), *Mössbauer Spectroscopy*, London: McGraw-Hill.

Cranshaw, T. E. (1983), in: B. V. Thosar, P. K. Iyengar, J. K. Srivastava, S. G. Bhargava (Eds.), *Advances in Mössbauer Spectroscopy*, Amsterdam: Elsevier, pp. 271–272.

De Waard, H., Niesen, L. (1987), in: G. J. Long (Ed.), *Mössbauer Spectroscopy Applied to Inorganic Chemistry*, Vol. II, New York: Plenum, pp. 1–86.

Fujita, F. E. (1975), in: U. Gonser (Ed.), *Mössbauer Spectroscopy*, Berlin: Springer, pp. 201–236.

Gerdau, E., Rüffer, R., Winkler, H., Tolksdorf, W., Klages, C. P., Hannon, J. P. (1985), *Phys. Rev. Lett.* **54**, 835–838.

Goldanskii, V. I., Makarov, E. F., Khrapov, V. V. (1963), *Phys. Lett.* **3**, 344–346.

Goldanskii, V. I., Herber, R. H. (1968), *Chemical Applications of Mössbauer Spectroscopy*, New York: Academic Press.

Gonser, U. (1975), *Mössbauer Spectroscopy*, Berlin: Springer.

Gonser, U. (1981), *At. Energy Rev.*, Suppl. No. 1, 302–228.

Greenwood, N. N., Gibb, T. C. (1971), *Mössbauer Spectroscopy*, London: Chapman and Hall.

Gütlich, P., Link, R., Trautwein, A. (1978), *Mössbauer Spectroscopy and Transition Metal Chemistry*, Berlin: Springer.

Hafner, S. S. (1975), in: U. Gonser (Ed.), *Mössbauer Spectroscopy*, Berlin: Springer, pp. 167–200.

Hanna, S. S., Heberle, J., Littlejohn, C., Perlow, G. J., Preston, R. S., Vincent, D. H. (1960a), *Phys. Rev. Lett.* **4**, 28–29.

Hanna, S. S., Heberle, J., Littlejohn, C., Perlow, G. J., Preston, R. S., Vincent, D. H. (1960b), *Phys. Rev. Lett.* **4**, 177–180.

Johnson, C. E. (1975), in: U. Gonser (Ed.), *Mössbauer Spectroscopy*, Berlin: Springer, pp. 139–166.

Johnson, C. E. (1984), in: G. J. Long (Ed.), *Mössbauer Spectroscopy Applied to Inorganic Chemistry*, Vol. I, New York: Plenum, pp. 619–640.

Kistner, O. C., Sunyar, A. W., (1960), *Phys. Rev. Lett.* **4**, 412–415.

Karyagin, S. V. (1963), *Dokl. Acad. Nauk SSSR*, 148, 1102–1105.

Kopcewicz, M. (1989), in: G. J. Long and F. Grandjean (Eds.), *Mössbauer Spectroscopy Applied to Inorganic Chemistry*, Vol. III, New York: Plenum, pp. 243–287.

Kopcewicz, M. (1991), *Struct. Chem.* **2**, 313–342.

Kopcewicz, B., Kopcewicz, M. (1991), *Struct. Chem.* **2**, 303–312.

Lamb, W. E., Jr. (1939), *Phys. Rev.* **55**, 190–197.

Langouche, G. (1991), *Hyperfine Int.* **68**, 95–106.

Long, G. J. (Ed.) (1984), *Mössbauer Spectroscopy Applied to Inorganic Chemistry*, Vol. I; (1987), Vol. II, New York: Plenum.

Long, G. J., Grandjean, F. (Eds.) (1989), *Mössbauer Spectroscopy Applied to Inorganic Chemistry*, Vol. III, New York: Plenum.

Longworth, G. (1987), in: G. J. Long (Ed.), *Mössbauer Spectroscopy Applied to Inorganic Chemistry*, Vol. II, New York: Plenum, pp. 289–342.

Moon, P. B. (1950), *Proc. Phys. Soc.* **63**, 1189–1196.

Mössbauer, R. L. (1958a), *Z. Phys.* **151**, 124–143.

Mössbauer, R. L. (1958b), *Naturwissenschaften* **45**, 538–539.

Mørup, S. (1987), in: G. J. Long (Ed.), *Mössbauer Spectroscopy Applied to Inorganic Chemistry*, Vol. II, New York: Plenum, pp. 89–123.

Obenshain, F. E. (1968), in: I. J. Gruvermann (Ed.), *Mössbauer Effect Methodology*, Vol. 4, New York: Plenum, pp. 61–74.

Parak, F., Heidemeier, J., Nienhaus, G. U. (1988), *Hyperfine Int.* **40**, 147–158.

Pfeiffer, L. (1972), in: I. J. Gruvermann (Ed.), *Mössbauer Effect Methodology*, Vol. 7, New York: Plenum, pp. 263–298.

Pound, R. V., Rebka, G. A., Jr. (1960), *Phys. Rev. Lett.* **4**, 337–341.

Pound, R. V., Snider, J. L. (1985), *Phys. Rev.* **140**, B788–B803.

Przybylski, M., Korecki, J., Gradmann, U. (1991), *Appl. Phys.* **A52**, 33–47.

Rüffer, R., Gerdau, E., Grote, M., Hollatz, R., Röhlsberger, R., Rüter, H. D., Sturhahn, W. (1990), *Hyperfine Int.* **61**, 1279–1294.

Schiffer, J., Marshall, W. (1959), *Phys. Rev. Lett.* **3**, 556–557.

Sielemann, R., Yoshida, Y. (1991), *Hyperfine Int.* **68**, 119–130.

Sprouce, G . D., Kalvius, G. M. (1968),. in: I. J. Gruvermann (Ed.), *Mössbauer Effect Methodology*, Vol. 4, New York: Plenum, pp. 37–59.

Shenoy, G. K. (1983), in: B. V. Thosar, P. K. Iyengar, J. K. Srivastava, S. G. Bhargava (Eds.), *Advances in Mössbauer Spectroscopy*, Amsterdam: Elsevier, pp. 561–585.

Shenoy, G. K., Wagner, F. E. (Eds.) (1978), *Mössbauer Isomer Shifts*, Amsterdam: North Holland.

Thomas, M. F., Johnson, C. E. (1986), in: D. P. E. Dickson, F. J. Berry (Eds.), *Mössbauer Spectroscopy*, Cambridge: Cambridge University Press, pp. 143–197.

Thosar, B. V., Iyengar, P. K., Srivastava, J. K., Bhargava, S. G. (Eds.), (1983), *Advances in Mössbauer Spectroscopy*, Amsterdam: Elsevier.

Williamson, D. L. (1978), in: G. K. Shenoy, F. E. Wagner (Eds.), *Mössbauer Isomer Shifts*, Amsterdam: North-Holland, pp. 317–360.

Williamson, D. L., Niesen, L., Weyer, G., Sielemann, R., Langouche, G. (1992), in: G. Langouche (Ed.), *Hyperfine Interaction of Defects in Semiconductors*, Amsterdam: Elsevier, pp. 3–75.

Williamson, D. L. (1993), *Nucl. Instr. Methods* **B76**, 262–267.

Zheng, Y. F., Hsia, Y. F. (1991), *Hyperfine Int.* **68**, 131–142.

Further Reading

Cohen, R. L. (Ed.) (1976), *Application of Mössbauer Spectroscopy*, Vol. I, New York: Academic Press.

Cohen, R. L. (Ed.) (1980), *Application of Mössbauer Spectroscopy*, Vol. II, New York: Academic Press.

Dickson, D. P. E., Berry, F. J. (1986), *Mössbauer Spectroscopy*, Cambridge: Cambridge University Press.

Goldanskii, V. I., Herber, R. H. (1968), *Chemical Applications of Mössbauer Spectroscopy*, New York: Academic Press.

Gonser, U. (Ed.) (1975), *Mössbauer Spectroscopy*, Berlin: Springer.

Gonser, U. (Ed.) (1981), *Mössbauer Spectroscopy II, The Exotic Side of the Effect*, Berlin: Springer.

Greenwood, N. N., Gibb, T. C. (1971), *Mössbauer Spectroscopy*, London: Chapman and Hall.

Gruvermann, I. J. (Ed.) (1965–1975), *Mössbauer Effect Methodology*, Vols. 1–10, New York: Plenum.

Gütlich, P., Link, R., Trautwein, A. (1978), *Mössbauer Spectroscopy and Transition Metal Chemistry*, Berlin: Springer.

Long, G. J. (Ed.) (1984), *Mössbauer Spectroscopy Applied to Inorganic Chemistry*, Vol. I; (1987), Vol II, New York: Plenum.

Long, G. J., Grandjean, F. (Eds.) (1989), *Mössbauer Spectroscopy Applied to Inorganic Chemistry*, Vol. III, New York: Plenum.

Long, G. J., Grandjean, F. (Eds.) (1993), *Mössbauer Applied to Magnetism and Materials Science*, New York: Plenum.

Shenoy, G. K., Wagner, F. E. (Eds.) (1978), *Mössbauer Isomer Shifts*, Amsterdam: North-Holland.

Thosar, B. V., Iyengar, P. K., Srivastava, J. K., Bhargava, S. G. (Eds.) (1983), *Advances in Mössbauer Spectroscopy*, Amsterdam: Elsevier.

Vertes, A., Korecz, L., Burger, K. (1979), *Mössbauer Spectroscopy*, Amsterdam: Elsevier.

Werthelm, G. K. (1964), *Mössbauer Effect: Principles and Applications*, New York: Academic Press.

MRI

See MAGNETIC RESONANCE IMAGING

MUON SPIN ROTATION/RELAXATION/RESONANCE

JESS H. BREWER, *Canadian Institute of Advanced Research and Department of Physics, University of British Columbia, Vancouver, British Columbia, Canada*

3-527-28133-9/94/$5.00 + .50

INTRODUCTION

The acronym "μSR" was coined in 1974 (*μSR Newsletter*, 1974); the definition and explanation offered on that occasion are still apt.

> μSR stands for Muon Spin Relaxation, Rotation, Resonance, Research or what have you. The intention of the mnemonic acronym is to draw attention to the analogy with NMR and ESR, the range of whose applications is well known. Any study of the interactions of the muon spin by virtue of the asymmetric decay is considered μSR, but this definition is not intended to exclude any peripherally related phenomena, especially if relevant to the use of the muon's magnetic moment as a delicate probe of matter.

Although muons were used as probes of magnetism in matter as early as 1944 (Rasetti, 1944), the essential property of the weak interaction that makes μSR possible—violation of parity (*P*) symmetry—was posited by Lee and Yang (1956) to explain anomalies in kaon decay experiments (Fitch, 1981; Cronin, 1981). Of the famous measurements (Garwin *et al.*, 1957; Friedman and Telegdi, 1957; Wu *et al.*, 1957) confirming their hypothesis (Lee and Yang, 1957), one also suggested that *P* nonconservation in $\pi \rightarrow \mu \rightarrow e$ decay might furnish a sensitive general-purpose probe of matter. The history of μSR began with that experiment (Garwin *et al.*, 1957), which used an experimental method similar to the most common and familiar of modern μSR techniques: transverse-field (TF)-μSR.

Later sections will treat TF-μSR and other μSR techniques in some detail, but it is useful to begin with a qualitative phenomenological description just to establish some terminology. A crude apparatus for TF-μ^+SR (μSR using positive muons) is pictured schematically in Fig. 1. The μ^+ arising from π^+ decay at rest is perfectly spin polarized as it enters the sample; later, it decays asymmetrically, with the decay positron emitted preferentially along the muon spin direction.

After stopping ~10^6 muons in the target sample, one obtains a time spectrum like that shown in Fig. 2 (top), which ideally has the following form:

$$N(t) = B + N_0 \exp(-t/\tau_\mu)[1 + A(t)], \tag{1}$$

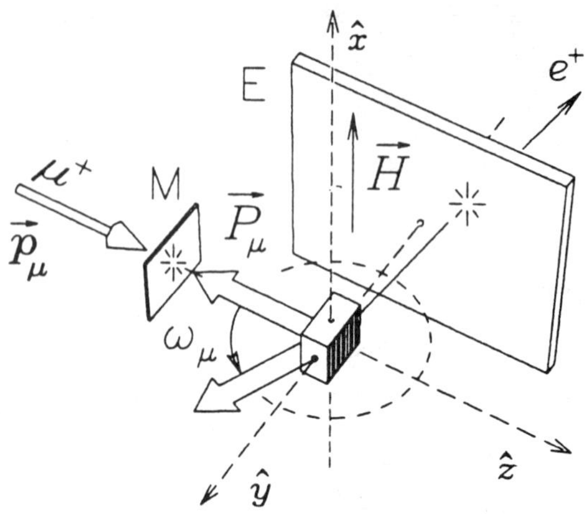

FIG. 1. A simple transverse-field (TF)-μ^+SR experiment: The μ^+ beam enters from the left with its polarization antiparallel to its momentum. A magnetic field **H** is applied vertically, causing the μ^+ spins to precess at the Larmor frequency $\omega_\mu = \gamma_\mu H$, where $\gamma_\mu/2\pi = 0.013\,553\,42$ MHz/Oe. An incoming μ^+ triggers the *M* counter, generating a start pulse for a fast time digitizer ("clock"), and an outgoing decay positron later stops the clock with a pulse from the *E* counter; for each such event, the time interval is digitized, and the corresponding bin in a discrete time spectrum is incremented.

where N_0 is an overall normalization, *B* is a time-independent background, $\tau_\mu = 2.197\ \mu$s, and $A(t)$ is the corresponding *asymmetry spectrum* shown in Fig. 2 (bottom), which can be extracted numerically from $N(t)$ as*

$$A(t) = \left[\frac{N(t) - B}{N_0}\right] e^{+t/\tau_\mu} - 1. \tag{2}$$

*In order to convert $N(t)$ to $A(t)$, one must know both *B* and N_0; fortunately, both constants can be extracted numerically from $N(t)$ in cases where the period of μ^+ Larmor precession is a negligible fraction of the muon lifetime. Treating $N(t)$ as a continuous function (the actual discrete sums can easily be deduced from the integrals below), we may write

$$N_0 = \frac{S\tau_\mu E_+ - \mathrm{RT}}{\tau_\mu^2 E_+ E_- - T^2}, \qquad B = \frac{R\tau_\mu E_- - ST}{\tau_\mu^2 E_+ E_- - T^2},$$

where

$$S \equiv \int_{t_i}^{t_f} N(t)dt, \qquad R \equiv \int_{t_i}^{t_f} N(t)\exp(t/\tau_\mu)dt,$$

$$E_\pm \equiv \pm[\exp(\pm t_f/\tau_\mu) - \exp(\pm t_f/\tau_\mu)],$$

and $T \equiv t_f - t_i$, the time interval over which the integrals are evaluated.

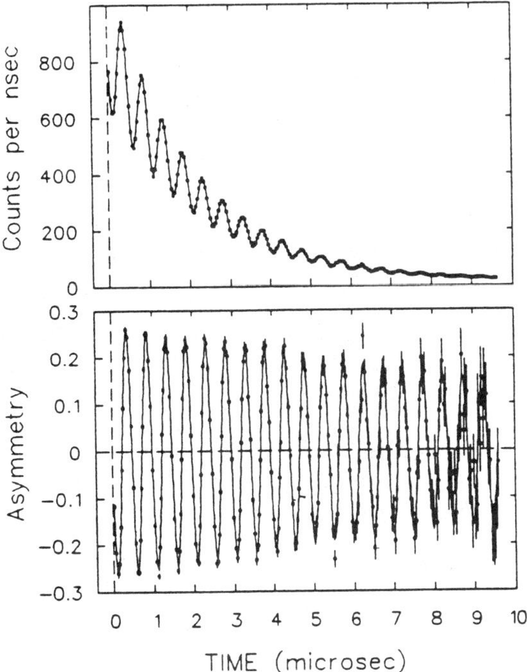

FIG. 2. Top: "Raw" time spectrum from a simple TF-μ^+SR experiment: The overall exponential decay reflects the muon lifetime, and the precession of the μ^+ spins is manifest in the superimposed oscillations as the muon polarization sweeps past the positron detector. Bottom: TF-μ^+SR asymmetry spectrum obtained from the raw time spectrum by subtracting any time-independent background and dividing out the exponential distribution of muon decay times.

Except for an empirical multiplicative constant, $A(t)$ represents the time evolution of the muon polarization, much like a free-induction decay (FID) signal in NMR.

For the next decade or so after 1957, μSR was developed primarily in the guise of a series of experiments using muons to test the predictions of quantum electrodynamics (QED) with unprecedented accuracy (Combley *et al.*, 1981, Hughes, 1988). Meanwhile, the muon's weak-interaction properties were being precisely determined in an ongoing (Stoker *et al.*, 1985) sequence of measurements of the *Michel parameters* (Michel, 1950) of normal muon decay. Many of today's areas of application of μSR began as peripheral problems in those fundamental physics experiments.

With the advent of the "meson factories" in the 1970s came a hundredfold to thousandfold increase in the intensity of muon beams, dramatically accelerating the development of new experimental techniques and the discovery of new applications for μSR. When the potential of μSR became apparent, most meson factories invested in the upgrading of muon beamlines and μSR facilities that came on line in the early 1980s. Since then, the techniques of μSR have been discovered by the chemistry and solid-state physics communities, and what was once an esoteric oddity has become one of the fastest-growing areas of intermediate-energy science.

Today's μSR is a standard magnetic-resonance tool, almost exclusively devoted to disciplines rarely associated with subatomic physics—primarily chemistry and condensed-matter physics. A list of some of these areas of application will appear later in this article, but first it is necessary to describe the fundamental physics that makes μSR possible and to explain the basic techniques of its use.

1. BASIC MUON PHYSICS

1.1 Muon Production

Although muons are produced in a variety of high-energy processes and elementary-particle decays such as $K \rightarrow \mu + \nu$ decay (Yamanaka *et al.*, 1984), μSR requires low-energy muons in order to stop the beam in samples of convenient thickness ($\lesssim$1 cm), and these are available in the required intensities only from the ordinary two-body decay of charged pions:

$$\pi^+ \rightarrow \mu^+ + \nu_\mu \quad \text{or} \; \pi^- \rightarrow \mu^- + \bar{\nu}_\mu, \tag{3}$$

from which the muon emerges (in the rest frame of the pion) with a momentum of 29.79 MeV/c and a kinetic energy of 4.119 MeV. The lifetime of a free charged pion is τ_π = 26.03 ns. Because the neutrino is only produced with negative helicity (spin antiparallel to momentum) and the antineutrino only with positive helicity—a succinct description of P nonconservation adequate for our purposes here—the simultaneous conservation of linear and angular momentum forces the μ^+ also to have negative helicity (and the μ^-, positive helicity) in the rest frame of the pion. Thus, muons emitted from pion decay at rest are also 100% spin polarized opposite to (for μ^+) or along (for μ^-) the direction of their momenta. This is the greatest advantage of μSR as a magnetic-resonance technique: whereas NMR and ESR rely upon a thermal-equilibrium spin polarization, usually achieved at low temperatures in strong magnetic fields, μSR begins with a *perfectly* polarized probe, regardless of conditions in the medium to be studied. It also implies that muon spin degrees of freedom usually start their evolution as far from thermal equilibrium as conceivable.

Most μ^+ beams today are literally emitted from π^+ decay at rest in the surface layer of the primary target, where the pions themselves are produced by collisions of high-energy protons with target nuclei—hence, the common mnemonic name, *surface muons* (Bowen, 1985). Unfortunately, this mode is not available for negative muons because a π^- stopping in the production target almost always undergoes nuclear capture from low-lying orbitals of pionic atoms before it has a chance to decay. This problem was actually solved long before the surface-muon beam was invented, when many fundamental physics experiments with muons were primarily concerned with the "heavy electron" behavior of the μ^-. In so-called "conventional" muon channels, pions are allowed to decay *in flight* down a relatively long straight section where the decay muons are collected by axial or alternating-gradient magnetic fields; the muons emitted "backward" in the pion rest frame have quite different momenta from those emitted forward (or from the pions themselves), and can thus be selectively extracted by a bending magnet. The disadvantages of such *backward muon beams* are their relatively higher momentum (usually ~ 50–100 MeV/c), their larger momentum spread (and, therefore, lower stopping density), and their much larger phase space (and, therefore, lower luminosity). Today, they are rarely used for μ^+SR, but for μ^-SR there is no alternative.

1.2 Muon Decay

Books have been written on the subject of this section (Primakoff, 1975); there follows only a synopsis of those aspects of normal muon decay ($\mu^+ \rightarrow e^\pm + \nu_{e,\mu} + \bar{\nu}_{\mu,e}$) that are essential to a qualitative understanding of μSR. The foremost of these is, of course, the propensity of the muon-decay positron (electron) to be emitted along (opposite to) the spin of the μ^+ (μ^-). This second example of P nonconservation in the weak interaction is what allows us to read out the information encoded in the evolution of an initially polarized muon spin ensemble. The information is delivered to the experimenter in the form of rather high-energy (up to 52 MeV) positrons or electrons, which readily penetrate sample holders, cryostats, or ovens and the detectors used to establish the time and direction of the muon decay.

The decay probability of the muon, illustrated in Fig. 3, depends upon the $e^\pm$ energy $x \equiv \epsilon_e/\epsilon_{\max}$ (where $\epsilon_{\max} = 52.83$ MeV is the maximum possible total relativistic energy of the $e^\pm$) and the angle θ between the muon spin direction and the direction and the direction of $e^\pm$ emission as

$$dP(x,\theta) = E(x)[1 + a(x)\cos\theta]dxd(\cos\theta). \tag{4}$$

The *asymmetry* factor a depends upon the $e^\pm$ energy as

$$a(x) = \pm(2x - 1)/(3 - 2x), \tag{5}$$

and the normalized $e^\pm$ energy spectrum has the form

$$E(x) = 2x^2(3 - 2x). \tag{6}$$

Figure 4 shows the functions $E(x)/2$, $a(x)$, and their product. Note that a changes sign at low energy; however, very few positrons are

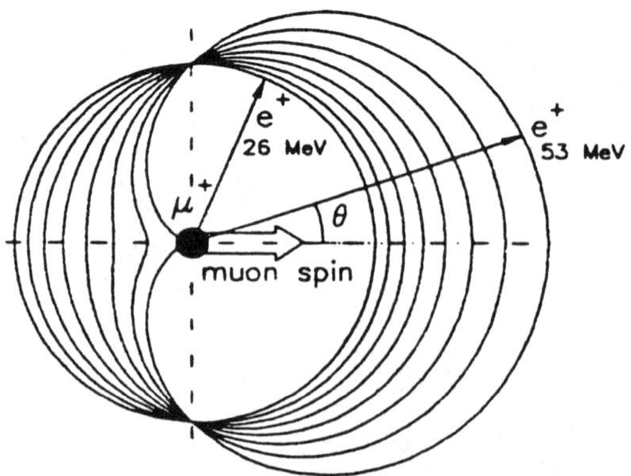

FIG. 3. Angular distribution of the e^+ from $\mu^+ \rightarrow e^+ + \nu_e + \bar{\nu}_\mu$ decay: The asymmetry (anisotropy) of the distribution is 100% for the highest $e^\pm$ energy $\epsilon_{\max} = 52.83$ MeV and zero (i.e., an isotropic distribution) for $\epsilon_e = \epsilon_{\max}/2$; for $\epsilon_e \rightarrow 0$ (not shown), the asymmetry is negative.

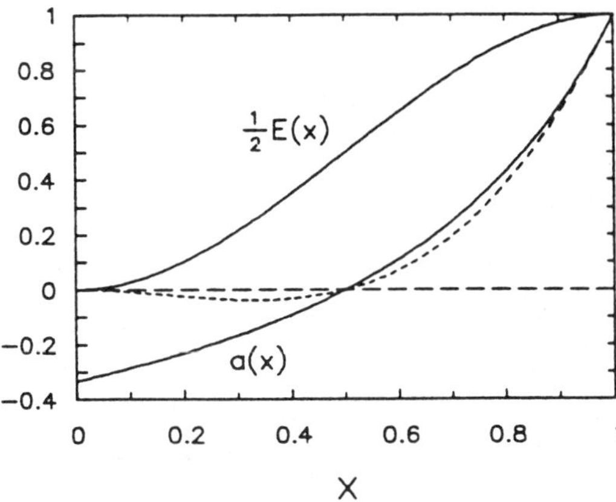

FIG. 4. Solid lines: energy spectrum $E(x)/2$ ($e^{\pm}$ energy $\epsilon_e = 52.83x$ MeV) of $e^{\pm}$ from normal $\mu^{\pm}$ decay and ϵ_e dependence of μ^+ decay asymmetry $a(x)$ (degree of correlation between e^+ momentum and μ^+ spin direction). (For $\mu^- \rightarrow e^- \bar{\nu}_e \nu_\mu$, a has the opposite sign) Dashed line: weighted $\mu^+ \rightarrow e^+ \nu_e \bar{\nu}_\mu$ asymmetry spectrum, the product of $E(x)/2$ and $a(x)$.

emitted with such low energies (see above), and those which are will usually not be detected (see below).

1.2.1 Asymmetry Calibration The theoretical average asymmetry,

$$\langle a \rangle = \int_0^1 a(x)E(x)dx = \pm\tfrac{1}{3}, \tag{7}$$

is never realized in practice (except by accident), for several reasons.

First, radiative corrections subtly distort the low-energy end of the spectrum (Sachs and Sirlin, 1975).

Second, the low-energy $e^{\pm}$ are easily stopped in even thin layers of material, such as the wrappings of scintillation counters: indeed, since this effect raises the average asymmetry, it is common practice to insert degraders between the muon stopping region and the $e^{\pm}$ detectors, when feasible, up to an optimum range (e.g., ≈ 3–4 cm of graphite) that maximizes the product A^2N, where A is the empirical asymmetry (defined below) and N is the $e^{\pm}$ rate reaching the detector. (One should generally use low-Z materials for degraders in order to minimize bremsstrahlung and pair production, which can greatly confuse the issue, unless one wishes to *use* such effects to bypass strong magnetic fields—see below.)

Third, a typical $e^{\pm}$ detector intercepts a rather large solid angle and, thus, averages $\cos\theta$ appreciably. The optimal detector geometry for general-purpose time-differential (TD)-μSR is a cube centered on the target where the muons stop, with each of the six faces a single detector. (Although, in principle, one could gain back some asymmetry by using position-sensitive detectors and weighting individual events according to the projections of the $e^{\pm}$ track along different axes, the complications of counting statistics—not to mention the expense of processing such information at very high rates!—are not usually justified by the marginal improvement.)

Finally, because the decay $e^{\pm}$ will follow a helical path in an applied magnetic field **H** (which is an essential part of many μSR experiments), the curling up of $e^{\pm}$ orbits causes the efficiency of their detection to be a function of both their energy and the applied field strength; the apparent direction of emission may also be affected. At p_e = 30 MeV/c (a typical $e^{\pm}$ momentum) the radius of curvature ρ_e of an orbit perpendicular to an applied magnetic field H = 1 T is exactly 10 cm, as given by the familiar formula

$$\rho\,(\text{cm}) = \frac{p\,(\text{MeV/c})}{0.3\,H\,(\text{kOe})}. \tag{8}$$

This effect becomes problematic at high fields ($H \gtrsim 5$ T $\Rightarrow \rho \lesssim 2$ cm), in which case μSR requires very small samples and detectors. Another possible approach to this problem is to generate bremsstrahlung photons intentionally from the decay $e^{\pm}$, using a lead converter near the target, and then reconvert the photons in front of the detector after they have traversed the high-field region.

Taken together, these systematic effects make the empirical asymmetry A in a given detector perplexingly dependent upon the thickness, geometry, and material of target and detectors, not to mention magnetic field. Attempts to correct analytically for $e^{\pm}$ energy, etc., are usually only reliable to within a few percent.

Fortunately, there are many materials (including most metals) in which the μ^+ suffers negligible depolarization; thus, the usual method for *calibrating the asymmetry* in μ^+SR is to measure first a dummy sample of dimensions identical to those of the real sample, but made from aluminum or silver. If great care is taken to arrange the two samples in exactly the same position, this will determine the *empirical maximum asymmetry* A_0 to within a few percent of itself; typical values are 0.2 to 0.3. However, materials of the same nominal thickness (in g/cm^2) will often exhibit different dE/dx and multiple scattering of the $e^{\pm}$, so that this calibration method cannot generally be trusted to better than ~1%.

In μ^-SR, as we shall see, such calibrations are much more difficult because of the loss of μ^- polarization in the initial cascade to the ground state of the muonic atom. To the author's knowledge, there is no satisfactory solution for the A_0 calibration problem in μ^-SR. The usual method is to calibrate on graphite, which gives a residual μ^- polarization of about (17 ± 2)% of the initial muon beam polarization (Kuno *et al.*, 1984).

1.3 The "Heavy Electron" versus the "Light Proton"

Asymmetry calibration is perhaps the least of the qualitative differences between μ^+SR and μ^-SR, all of which are consequences of the opposite charges of the μ^+ and μ^-. In elementary-particle physics, the muon is often described as a "heavy electron" in reference to its family resemblance to other leptons; it is the μ^- that is so designated. Indeed, the μ^- does just what one might expect of a heavy electron: It undergoes Coulomb capture into atomic orbitals analogous to those of electrons but 206.768 times closer to the nucleus and unrestricted by the Pauli exclusion principle; thus, the muon quickly (usually in ~10^{-14} s) cascades to the 1s ground state, emitting photons (known as muonic x rays) and/or Auger electrons (ejected from low-lying orbitals in the same atom by muonic x rays) on the way. Once there, the significant overlap between muon and nuclear wave functions (in heavy nuclei, the muon spends most of its time inside the mean nuclear radius) gives the weak interaction between the muon and the proton a chance to act, resulting in nuclear capture ($\mu^- + p \rightarrow n + \nu_\mu$), which appreciably reduces the μ^- lifetime (and the fraction of muons decaying to electrons for μ^-SR to detect).

By contrast, the positive muon avoids positively charged nuclei and will capture its own electron when it can, to form the hydrogen-like atom muonium (μ^+e^-, usually abbreviated Mu). In fact, the qualitative behavior of the μ^+ in matter resembles that of a light proton far more than that of its closer antilepton relative, the positron. The μ^+ mass is 0.1126 times that of the proton but 206.768 times that of the positron: its magnetic moment is 3.1832 times that of the proton but only 0.48363% of the positron's. The Mu atom is almost identical (except in mass) to the hydrogen atom, whereas the positronium atom (e^+e^-, or Ps) has no nucleus and half the binding energy of H. Positronium also annihilates very quickly, whereas weak electron capture in muonium ($\mu^+ + e^- \rightarrow \nu_\mu + \nu_e$) only shortens the μ^+ lifetime by ~1 part in 10^{10}.

As a result, μ^-SR differs qualitatively from μ^+SR.

1. *Lifetime(s):* The μ^+ lifetime is independent of its environment, while that of the μ^- depends strongly upon the Z of the nucleus to which it becomes attached: a sample containing several elements will have as many μ^- lifetimes, the ratios of whose probabilities cannot be reliably estimated from the Fermi-Teller "Z law" (Fermi and Teller, 1947) and, therefore, must be determined by fitting the time distribution of decay or capture products.
2. *Muon capture:* A μ^- that undergoes nuclear capture ($\mu^- + p \rightarrow n + \nu_\mu$) does not produce a decay electron for detection in μ^-SR; the μ^-SR event rate per incident muon is thus proportional to the μ^- lifetime, which in high-Z elements can be as little as 4% of the free muon lifetime. This rate loss can often be offset by increasing the μ^- beam intensity, except when the sample contains both heavy and light nuclei.
3. *Heterogeneous signal:* Whereas the μ^+ produces only positrons and undetected neutrinos, a μ^- entering some high-Z samples like ^{238}U is likely to cause emission of x rays and Auger electrons during the atomic cascade, fission fragments and neutrons (on average, several per muon) from either nuclear muon capture or nuclear internal conversion of muonic x rays, and the occasional decay electron. The μ^-SR signal is, therefore, noisy and heterogeneous. This does make it rather interesting from the points of view of atomic and nuclear physics, but such complica-

tions are so far unwelcome in historically typical μ^-SR applications.

4. *Lost polarization:* Spin-orbit coupling in the atomic cascade leaves the μ^- in its ground-state 1*s* orbital with a dramatically reduced spin polarization, typically $\lesssim$20%. Because statistical uncertainties shrink as $N^{-1/2}$, one must accumulate 25 times more events to achieve the same statistical accuracy in measuring a five times smaller signal. This seemingly irreducible disadvantage is the main limitation of μ^-SR applications and is the reason why μSR is often treated as synonymous with μ^+SR. However, since μ^-SR offers complementary and unique information, it will always be an essential part of the μSR repertoire. There is also a chance that new techniques, such as x-ray tagged μ^-SR, may help overcome this disadvantage.
5. *Hyperfine effects:* The hyperfine coupling between the muon and nuclear spins in nonzero-spin nuclei is enormous, often large enough to cause Auger emission of *K* and *L* electrons (Winston, 1963). The two hyperfine states F^+ and F^- of the 1*s* muonic atom (in which the muon spin is respectively parallel and antiparallel to the nuclear spin) therefore produce distinct μ^-SR signals, each with its own characteristic precession frequency, in such systems. This was long thought to limit useful μ^-SR to spinless elements, but quite a few nuclei with spin $I > 1/2$ have been found to exhibit potentially useful μ^- SR signals, albeit at reduced amplitudes (Brewer, 1984a, 1984b).

Taking into account the different lifetimes, cascade depolarization, and spin states of negative muons in different muonic atoms, we may generalize Eq. (1) to obtain the most general form for *electron-triggered* μ^-SR in a sample composed of different elements:

$$N(t) = B + \sum_i N_i^0 \exp(-t/\tau_i)[1 + A_i(t)], \qquad (9)$$

in which $N_i^0 = N_0 f_i(\tau_i/\tau_\mu)$, where N_0 is an overall normalization, f_i is the fraction of muons captured on the *i*th type of nucleus, and τ_i is the μ^- lifetime in that species of muonic atom and hyperfine state; $A_i(t) = A_0 P_i(t)$, where A_0 is a common maximal asymmetry factor and $P_i(t)$ is the polarization of muons in the *i*th type of muonic atoms. For TF-μ^-SR, this polarization will have the form $P_i(t) = P_i^0 G_i(t)\cos(w_i t + \phi_i)$, where P_i^0 is the polarization remaining in that atom's given hyperfine state after the cascade, $G_i(t)$ is the corresponding relaxation function (see below), ω_i is its characteristic precession frequency in the applied magnetic field, and ϕ_i is the initial phase.

In the interest of simplicity, further discussion of μ^-SR will be neglected in favor of μ^+SR, which occupies most of the attention of the μSR community.

2. μSR TECHNIQUES

Ideally (and often in reality), the μSR experimenter has control over the orientation of the detectors, the applied magnetic field, and (within some range) the muons' spin polarization. The beam momentum can be deflected as well, but this is rarely desirable. In order to designate different orientation choices consistently, the labeling conventions defined in Fig. 5 have been devised specifically for μ^+SR experiments using surface muons, which are originally polarized opposite to their momentum but whose spins can be rotated 90° by a Wien filter in the beamline (Brewer, 1981). (Such flexibility is not available for conventional or backward $\mu^\pm$ beams, which will be neglected here partly for that reason.) The standard detector array consists of six counters aligned with the positive and negative coordinate axes and labeled *F* (forward), *B* (backward), *U* (up), *D* (down), *L* (left), and *R* (right), according to a "beam's-eye view" naming convention. Note that the unrotated muon polarization points toward the *B* counter and, in the spin-rotated mode, toward the *U* counter; the latter depends, of course, on the orientation of the fields in the Wien filter.

2.1 Time-Differential μSR in Transverse Field

The simplest and most familiar time-differential (TD)-μSR technique is the transverse-field (TF) muon spin *rotation* experiment, in which an external magnetic field is applied perpendicular (transverse) to the muon polarization, causing the muon spins to precess (rotate) about the field.

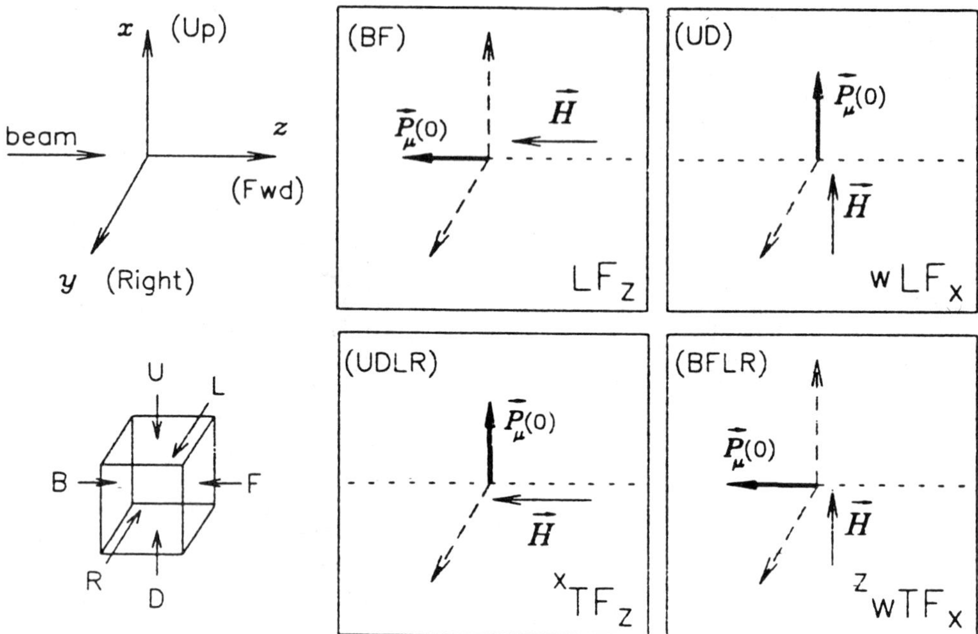

FIG. 5. Coordinate system and labeling conventions for surface-muon μSR experiments. Note that the superscript on the left indicates the direction of the incoming muon polarization, while the subscript on the right indicates the direction of the applied field (if any); for longitudinal field (LF) and by continuation for zero field (ZF), both will always be the same, but for TF, there are, in principle, two possible arrangements for each choice of super(sub)script (for instance, $^zwTF_y$ vs $^xwTF_y$). "Weak" (w) field means "not strong enough to deflect the muon beam appreciably."

2.1.1 The Basic Technique In transverse-field (TF)-μSR, a magnetic field **H** is applied perpendicular to the initial muon spin direction, causing Larmor precession of the muon polarization about **H**. This arrangement varies from the primitive version depicted in Fig. 1 to the 4-counter, spin-rotated ^xTF_z-μSR apparatus shown in Fig. 6 and the ^zTF_x-μSR arrangement in Fig. 7, each of which illustrates *quadrature* detection (two orthogonal pairs of opposing counters in the plane perpendicular to the applied field); but the time spectra from individual counters all have the qualitative appearance shown in Fig. 2.

2.1.1.1 Frequency. The most obvious parameter observable in a TF-μSR spectrum is the *muon precession frequency* $\omega_\mu = \gamma_\mu B$, which (thanks to our precise knowledge of γ_μ) is equivalent to the local magnetic field B at the muon. Since B may be affected by diamagnetism, paramagnetism, and contact hyperfine interactions with polarized electrons (Knight shifts) in the medium, B is generally different from the applied field H_0; this difference is often the main focus of the TF-μSR experiment. An example is shown in Fig. 8.

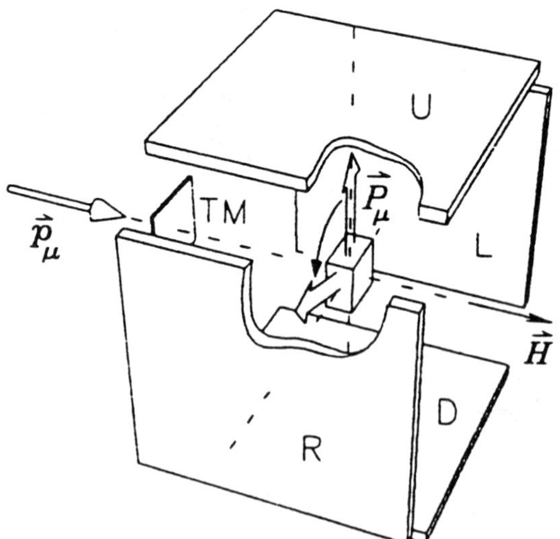

FIG. 6. Counter arrangements for ^xTF_z-μSR, in which the muon beam has been passed through a Wien filter to rotate the muon spins until they are perpendicular to their momenta. The momentum $\mathbf{p}_\mu$ is undeflected in the magnetic field $\mathbf{H} = H_z \| \mathbf{p}_\mu$, but the polarization $\mathbf{P}_\mu$ precesses in the x-y plane.

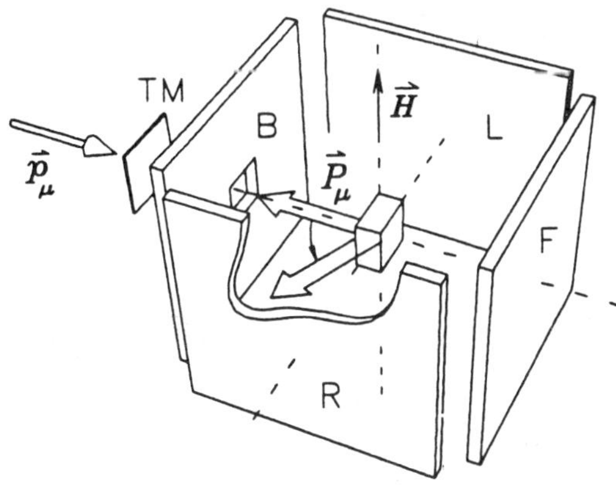

FIG. 7. Counter arrangements for ${}^{z}w\mathrm{TF}_x$-μ^+SR, in which the μ^+ beam still has its polarization antiparallel to its momentum. The field must therefore be perpendicular to both $\mathbf{p}_\mu$ and $\mathbf{P}_\mu$, which restricts use with surface muon beams to weak fields $H \lesssim 100$ Oe.

2.1.1.2 Asymmetry. The second obvious parameter characterizing muon precession in a transverse field is the amplitude, or *asymmetry*, of the precession signal. As mentioned earlier, the absolute calibration of this amplitude is tricky, but one can usually convert the initial amplitude into a *residual polarization* with an accuracy of a few percent. In metallic samples, the initial polarization is generally consistent with 100%, but in insulators or semiconductors, liquids, or gases, some fraction of the muons (ranging from 0 to 100%) either form muonium (Mu) or experience some other form of depolarization in the early ($\lesssim$1 ns) stages of thermalization in the sample.

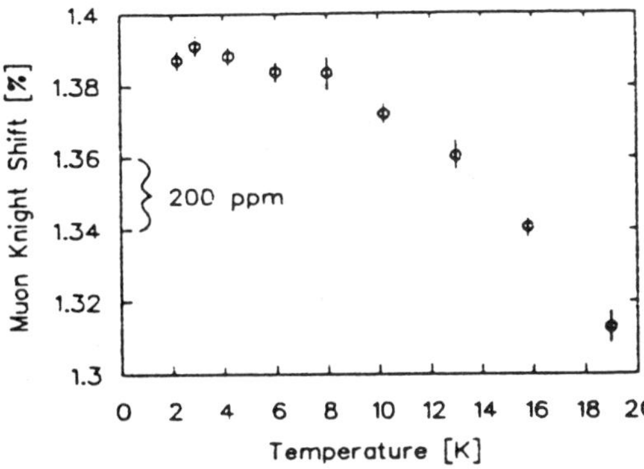

FIG. 8. Positive muon Knight shift $K_\mu \equiv (B_\mu - H_0)/H_0$ (where H_0 is the applied magnetic field and B_μ is the field at the muon) as a function of temperature in a single crystal of antimony with the *c* axis parallel to the applied field (1.5 T). The precision (approximately ±50 ppm) is typical of what can easily be achieved with routine methods in modest TF-μSR experiments; much higher precision is possible with refined techniques (see the later section on strobo-μSR).

The effect of Mu formation in *w*TF is to cause the muon polarization to precess in the opposite sense to that of muons in diamagnetic environments, roughly 103 times faster; this results in dramatic "dephasing" if the Mu atoms subsequently react at exponentially distributed times to enter diamagnetic states. If all this takes place within a few ns, all one can observe is the net effect on the subsequent diamagnetic μ^+ precession signal, namely, a reduction of asymmetry and a simultaneous shift of the apparent initial phase of precession. If the magnetic field is in the *z* direction and the initial μ^+ polarization is in the *x* direction, we may define the *complex* muon polarization $\tilde{P}(t) \equiv P_x(t) + iP_y(t)$, in terms of which the overall polarization of an ensemble of muons starting as Mu atoms and reacting at a rate Λ to form some diamagnetic species is given in the low-field limit (averaging over high-frequency hyperfine oscillations) by

$$\tilde{P}(t) \approx \tfrac{1}{2}\frac{\omega_{\mathrm{Mu}} + \omega_\mu}{\omega_{\mathrm{Mu}} + \omega_\mu - i\Lambda}\exp[-(\Lambda + i\omega_{\mathrm{Mu}}t)] + \tfrac{1}{2}\left[\frac{i\Lambda}{i\Lambda - (\omega_{\mathrm{Mu}} + \omega_\mu)} + \frac{\Lambda^2}{\Lambda^2 + \omega_0^2}\right]e^{i\omega_\mu t}, \tag{10}$$

where the diamagnetic precession frequency is ω_μ and the muonium precession frequency is $\omega_{\mathrm{Mu}} \approx 103\omega_\mu$ in the opposite sense. This sort of simple *residual polarization* picture also describes many other *delayed formation* scenarios.

2.1.1.3 Initial Phase of Precession. A more subtle aspect of the residual polarization picture is the shift of the apparent initial phase of the μ^+ precession due to (e.g.) the formation of short-lived Mu atoms precessing in the opposite sense. Measurement of such phase shifts has often proved valuable in sorting out "fast chemistry" effects (Brewer *et al.*, 1974).

2.1.1.4 "Relaxation." Following thermalization of translational degrees of freedom, which may take as little as 100 ps in solids, the muon precession signal may still decrease in amplitude because of dephasing

(T_2 effects) or true irreversible relaxation processes (T_1 effects). The latter are more often studied in longitudinal field (LF), where the distinction between T_1 and T_2 is not subject to so much semantic debate. (See Sec. 2.2.3.) Depolarization due to inhomogeneous magnetic fields often takes the form of a Gaussian relaxation function $\exp(-\frac{1}{2}\sigma^2 t^2)$, as shown in Fig. 9.

2.1.1.5 Muonium Precession. All the features just described for muon precession at the diamagnetic Larmor frequency $\omega_\mu = \gamma_\mu H$ are also often seen for *muonium precession* in ωTF-μSR experiments. The main differences are that $\omega_{Mu} = \gamma_{Mu} H$ is roughly 103 times larger than ω_μ in the same field H (because of the huge magnetic moment of the electron locked to the muon spin by the hyperfine interaction), that the sense of ωTF Mu precession is opposite to that of the free μ^+ (because of the opposite sign of the dominating e^- moment), that half the muon polarization appears lost in most experiments [because of the fast (4463 MHz for Mu in vacuum) hyperfine oscillations between the $|\Uparrow\downarrow\rangle$ and $|\Downarrow\uparrow\rangle$ states, where $\Updownarrow$ refers to the electron spin and $\updownarrow$ refers to that of the muon], and that ω_{Mu} splits into two frequencies for $H \gtrsim$ 20 Oe because the hyperfine interaction is finite. This phenomenon will be discussed in more detail below.

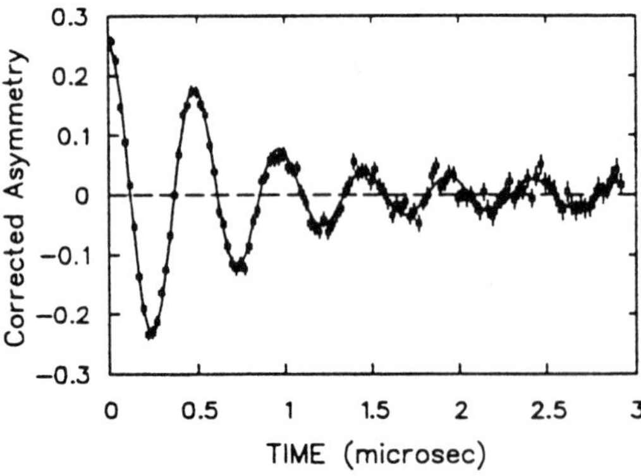

FIG. 9. Depolarization of muons in a sintered powder of superconducting $La_{1.85}Sr_{0.15}CuO_4$ at 10 K in an applied magnetic field of 150 Oe. The relaxation is due to local field inhomogeneities caused by flux exclusion and a vortex lattice in this type-II superconductor; it is adequately described by a Gaussian relaxation function in which $\sigma \propto \lambda^{-2}$, where λ is the London penetration depth. This has been much used for initial estimates of λ.

2.1.2 High Field Since the period of muon precession is 7.38 ns in a field of 1 T and is inversely proportional to H, it is not difficult to achieve a magnetic field strong enough to challenge the time resolution of most μSR spectrometers. Moreover, the radius of decay positron orbits shrinks with increasing field until experiments in $H \gtrsim 5$ T require small detectors within 1–2 cm of the sample. Nevertheless, the effort of miniaturization and state-of-the-art time resolution is often justified by the improved resolution for Knight shifts and/or access to intrinsically high-field phenomena. In high fields, a standard TD-μSR time spectrum consists of a large number of small time bins, each of which may have a rather low number of counts and a correspondingly large statistical uncertainty; this circumstance makes the usual laboratory-frame asymmetry plot rather uninformative to the eye and expensive to fit by χ^2 minimization, motivating new methods of data analysis.

2.1.2.1 Fourier Transforms. Often, especially when high frequencies are involved, one would rather see a frequency spectrum than a time spectrum. The art of converting the latter into the former is subject to continuing evolution and perpetual debate, but virtually all μSR facilities include the fast Fourier transform (FFT) in their standard data-analysis tools. A complete account of the merits and hazards of such treatments is beyond the scope of this article, but it is useful to point out several ubiquitous features that are ignored at one's peril.

First, as shown in Fig. 10, the fact that the time range of a μSR spectrum necessarily starts at $t = 0$ means that the imaginary (odd in t) part of the resultant frequency spectrum is dramatically broadened by the tacit presence of the Heaviside function; this has nothing to do with either the muon lifetime or the finite range of positive time. Thus, the real part of the amplitude (the proper objective of our FFT) may be quite narrow, but the square root of the power spectrum (the sum of the squares of the real and imaginary parts) will still be very broad. Various tricks are available for unmixing the real and imaginary parts.

Second, because of the finite muon lifetime, there are fewer statistics in the bins at late times, and consequently the "error bars" in the asymmetry spectrum grow exponen-

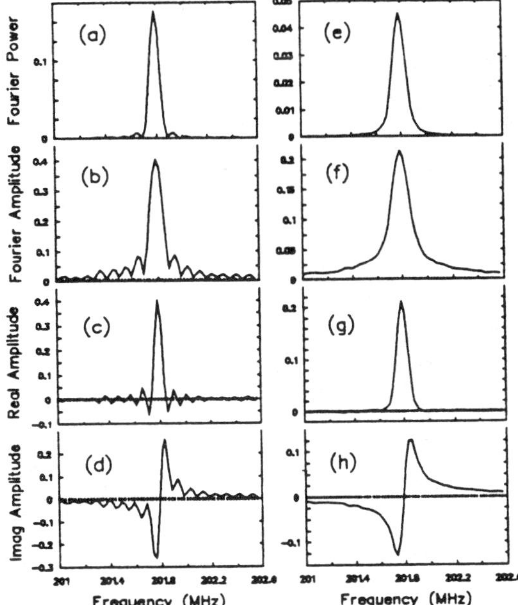

FIG. 10. Results of complex FFT of a quadrature μ^+SR time spectrum with a slow relaxation rate: (a)–(d) the power, the amplitude (square root of the power), the real and the imaginary parts of the frequency spectrum resulting from FFT of the raw complex time spectrum over 10.24 μs at a field of 1.489 T; (e)–(h) the same quantities for the same data *apodized* by a Gaussian envelope function with a 4-μs time constant.

tially with time as $\exp(+t/2\tau_\mu)$. (See Fig. 2.) This introduces noise in an annoyingly nonuniform way, which may, in principle, be corrected for using not-so-fast Fourier transform methods. These *noise effects* are always subject to suppression *without limit* (except for the finite patience of the experimenter) by simply taking more statistics: although the finite muon lifetime is responsible, it does not fix any absolute limits on precision—only practical ones.

Third, because experimenters' patience is indeed finite, most μSR time spectra only extend to 10–20 μs; thus, the FFT is cut off abruptly at some time limit, introducing the "ringing" effect evident on the left side of Fig. 10. This can be suppressed by *apodization*—multiplying the "raw" time spectrum by an *envelope function* that causes the product to go to zero gently before the abrupt end of the time range. As evident from Fig. 10, well-chosen apodization removes the ringing caused by FFT on a finite time interval but introduces a line-broadening effect that must be taken into account in the interpretation.

In general, there are no time-space ↔ frequency-space interconversions or interpretations that make everyone happy; hence, the earlier reference to such devices as "art." However, like other art forms, the FFT can produce some very pleasing results: for example, Fig. 11 shows the frequency (and, thus, the internal magnetic field) spectrum from μ^+SR in superconducting $YBa_2Cu_3O_{6.95}$.

2.1.2.2 Rotating Reference Frames. For TF-μSR in high magnetic field (HTF-μSR), another essential tool is the *rotating reference frame* (RRF) transformation, an example of which is shown in Fig. 12. The following description will omit the details of transforming discrete spectra and give only a simplified treatment in terms of an idealized continuous time spectrum. In fact, the RRF transformation works best on *orthogonal pairs* of spectra—as, for instance, when the magnetic field acts in the z direction and detectors are situated in the $\pm x$ and $\pm y$ directions defining the plane of precession of the muon polarization—where a *complex* asymmetry spectrum $\tilde{A}(t)$ can be defined as (e.g.) $\tilde{A}(t) \equiv A_x(t) + iA_y(t)$. In this case, the RRF transformation is simply

$$\tilde{A}_{\mathrm{RRF}}(t) = \tilde{A}(t)e^{-i\Omega t}, \qquad (11)$$

where Ω is the RRF frequency—chosen ar-

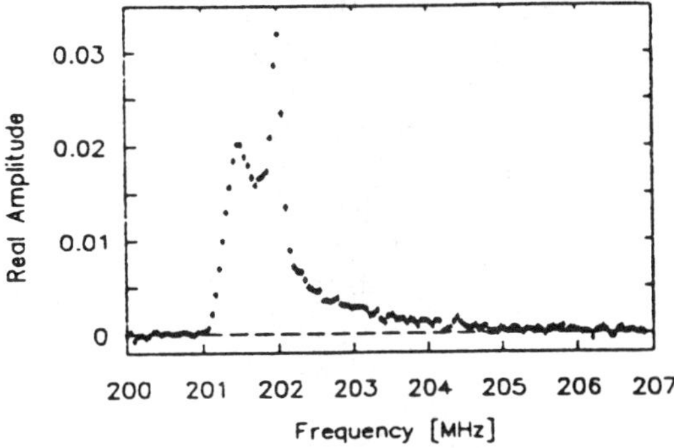

FIG. 11. Real part of frequency spectrum taken at a temperature of 6 K for positive muons in a single (3 × 3 × 0.2 mm^3) crystal of superconducting $YBa_2Cu_3O_{6.95}$ (a type-II superconductor with T_c = 92 K) cooled in a field of 1.489 T applied along the crystalline c axis. The sharp peak at 202 MHz is due to muons missing the superconductor and stopping in a normal region where the field distribution is not broadened by the vortex lattice.

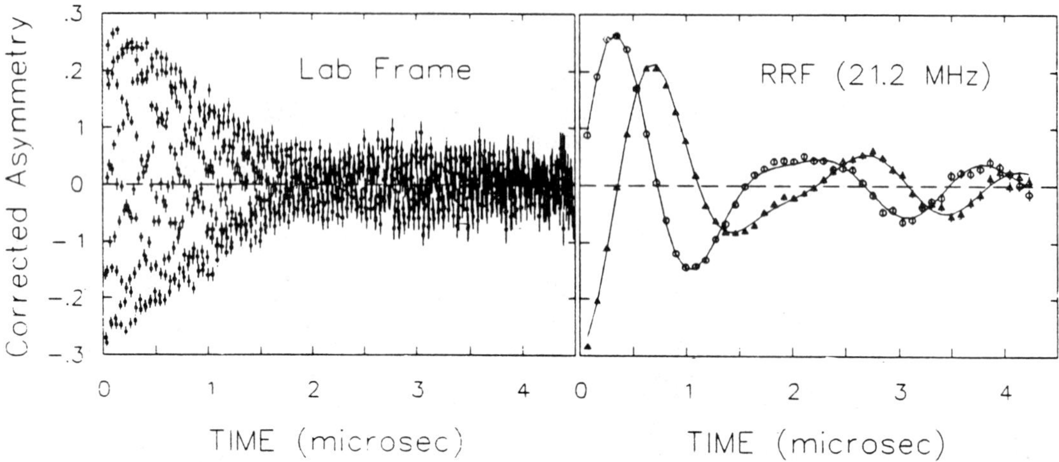

FIG. 12. The first 4.5 μs of a μ^+SR time spectrum taken at roughly 0.2 T in the laboratory frame where the data are recorded and in the rotating reference frame (RRF) to which it has been transformed numerically for display and fitting. The RRF frequency of 21.2 MHz is chosen to produce a slow precession signal for visual clarity. Because the original spectrum is real (no orthogonal detector arrays), the maximum amplitude in each of the complex RRF spectra (real part: circles; imaginary part: triangles) is actually a factor of 2 smaller than in the lab frame. The solid line in the RRF spectra is a fit by three frequencies; a single frequency (or even two) would result in a much poorer fit to the data.

bitrarily to produce a transformed $\tilde{A}_{\text{RRF}}(t)$ with the desired characteristics. For instance, if the muons precess at a high frequency ω_μ, one may select Ω slightly smaller than ω_μ to produce an $\tilde{A}_{\text{RRF}}(t)$ that varies relatively slowly with time; the resulting complex spectrum can be *packed* (n_p original time bins $\to$ one coarser time bin), in which case the statistical uncertainties of individual bins are reduced by a factor roughly equal to $\sqrt{n_p}$.

There are several advantages to such a transformation: first, and most obvious, is the reduction (by a factor of n_p) of the number of bins to be fitted in analysis programs, a straightforward improvement of efficiency; second, and probably more important, is the transformation of a large number of narrow time bins (with statistical uncertainties often larger than the signals under scrutiny) into a small number of well-defined bins representing the signal in a form that allows convenient display and visual inspection.

In the event that a single spectrum contains several signals at drastically different frequencies ω_i, separate RRF transformations at frequencies $\Omega_i \approx \omega_i$ can be used to isolate each signal for easy fitting.

It is also possible to perform a RRF transformation on a pure real spectrum (i.e., one without orthogonal detector axes) if care is taken to select Ω and n_p so as to "bin out" the spurious signal at the RRF frequency (Riseman and Brewer, 1990).

2.1.3 Paramagnetic States The use of Fourier transforms to reveal *line shapes* is actually a recent application in μSR; the oldest, and still most widespread, use of FFT in μSR is for extracting the frequency spectrum of muons in *paramagnetic* states such as muonium or radicals (molecules containing one or more unpaired electrons), in which the muon spin is strongly coupled to electron spins (or, in principle, orbital moments) by *hyperfine* interactions. The behavior of the simplest case, two spin-$\frac{1}{2}$ particles coupled by a scalar contact interaction in an applied magnetic field, is pictured in Fig. 13. Such couplings cause the μ^+e^- spin system (for example) to respond as a whole to applied magnetic fields, often producing rich and informative structure in the frequency spectrum (Patterson, 1988). A classic example is shown in Fig. 14 and a more recent case in Fig. 15.

2.2 Time-Differential μSR in Longitudinal and Zero Field

Referring again to Fig. 5, consider now the time evolution of the muon polarization in a magnetic field parallel to its initial direction

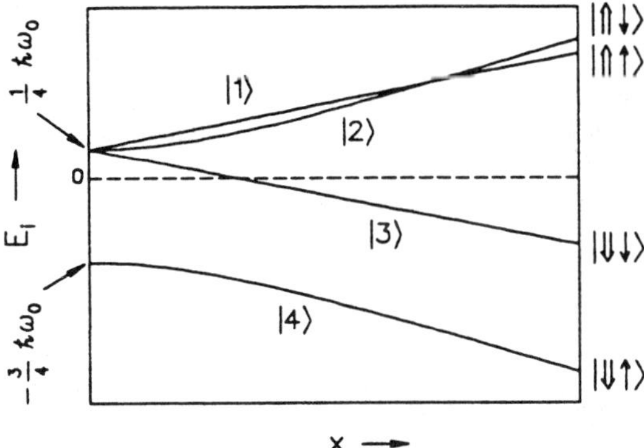

FIG. 13. Breit-Rabi diagram showing the energy levels of a system of two spin-$\frac{1}{2}$ particles of opposite sign and different magnetic moments—e.g. muonium—as functions of the reduced field $x \equiv H/H_0$, where H_0 (1585 Oe for Mu in vacuum) is a characteristic hyperfine field. For the purpose of illustration, unphysical values of moments and coupling constants have been used. The hyperfine frequency $\nu_0 \equiv \omega_0/2\pi$ has the value 4463 MHz for muonium in vacuum. In zero field, the three triplet (J = 1) eigenstates $|1\rangle$, $|2\rangle$, and $|3\rangle$ are degenerate, and the singlet (J = 0) ground state $|4\rangle$ is $\hbar\omega_0$ lower in energy. At high reduced field ($x \to \infty$), the eigenstates are $|1\rangle \to |\Uparrow\uparrow\rangle$, $|2\rangle \to |\Uparrow\downarrow\rangle$, $|3\rangle \to |\Downarrow\downarrow\rangle$, and $|4\rangle \to |\Downarrow\uparrow\rangle$ ($\Updownarrow$ refers to the electron spin, and $\updownarrow$ refers to that of the muon.) Note that state $|1\rangle$ is lower in energy than state $|2\rangle$ above $H_c = \omega_0(\gamma_e - \gamma_\mu)/2\gamma_e\gamma_\mu$ (16.386 T for Mu in vacuum).

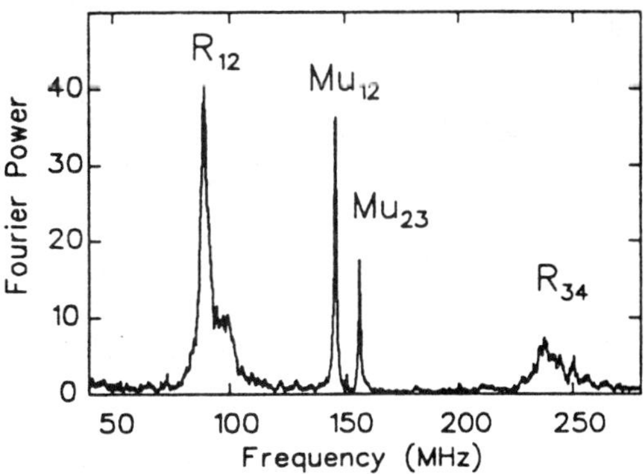

FIG. 15. Frequency power spectrum for positive muons in powdered buckminsterfullerite (crystalline C_{60}) at room temperature and 108 Oe (Kiefl *et al.*, 1992), showing simultaneously the signals from muons in the C_{60}Mu· radical (R_{12} and R_{34}) and the endohedral muonium (Mu@C_{60}) atom (Mu_{12} and Mu_{23}). In paramagnetic molecules with nuclear moments, such radical signals cannot be seen in low field because of nuclear hyperfine broadening.

(longitudinal field or LF). If the muon polarization initially has no components perpendicular to the local field, then none will develop, and only the polarization along the initial direction needs to be measured. Figure 16 illustrates the most common configuration, and Fig. 17 shows an alternative scheme sometimes used with spin-rotated beams.

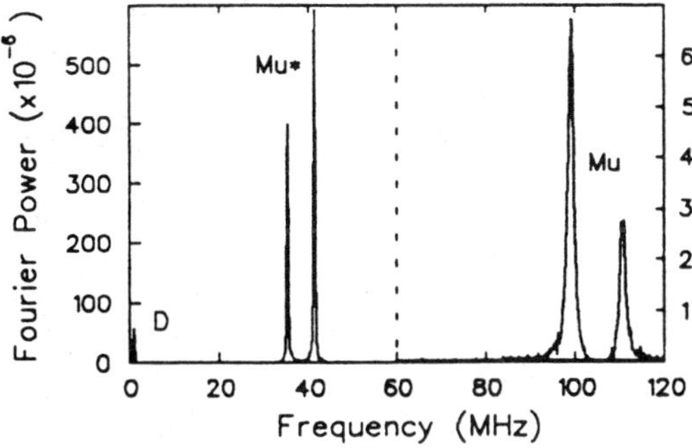

FIG. 14. Frequency power spectrum of positive muons in a pure silicon crystal at 10 K and 75 Oe transverse field showing the diamagnetic (D) signal, the rapidly diffusing muonium (Mu) signals, and the signals from so-called "anomalous" muonium (Mu*), which has been shown to be a muonium atom localized on a Si-Si bond center. Note vertical scale change at 60 MHz. After Brewer *et al.* (1973).

2.2.1 Two-Counter Asymmetry The perceptive reader will have noticed several figures with vertical axes labeled "Corrected Asymmetry." This is in reference to the fact that μSR time spectra taken in a longitudinal geometry cannot be converted to asymmetry spectra by use of Eq. (2), because neither N_0 nor B can be extracted numerically from the data without some model of the time dependence of the longitudinal polarization.

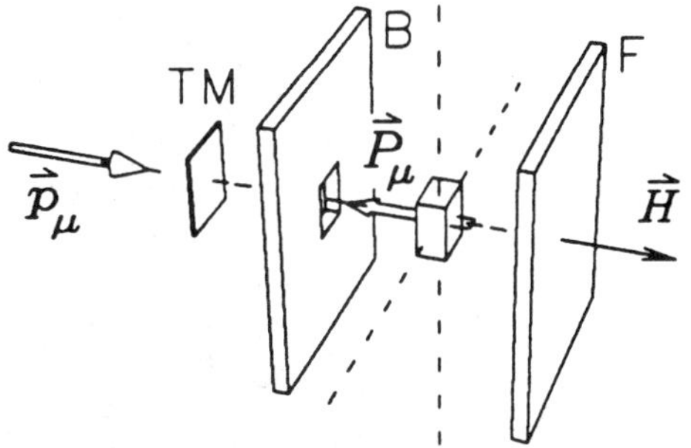

FIG. 16. Longitudinal-field μSR counter arrangement for LF_x-μ^+SR, in which the μ^+ beam is polarized antiparallel to its momentum. As $\mathbf{H} = H\hat{z} \parallel \mathbf{p}_\mu \parallel \mathbf{P}_\mu$, any magnetic field strength may be used.

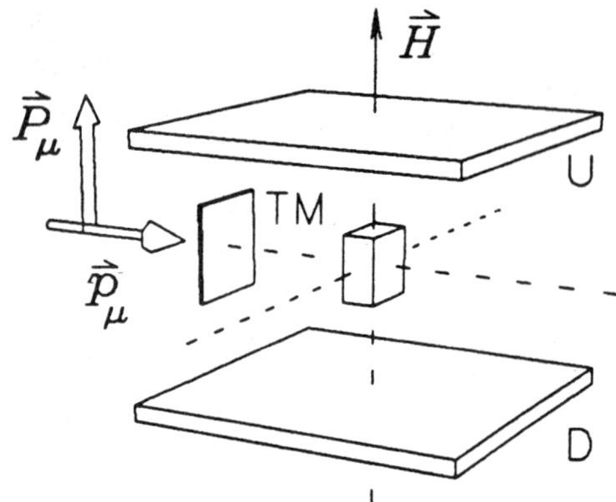

FIG. 17. Longitudinal-field μSR counter arrangement for LF$_x$-μSR, in which the muon beam arrives spin rotated by a Wien filter. Since the field is now perpendicular to $\mathbf{p}_\mu$, use with surface-muon beams is restricted to weak fields $H \lesssim 100$ Oe.

Instead, one combines the time spectra from two detectors on opposite sides of the sample, such as "U" and "D" in Fig. 17, in the following way. First, define the parameters $\epsilon_{U,D}$ = efficiency of U or D e detector, $B_{U,D}$ = background in said e detector (measured using "$t < 0$ bins"), $A_{U,D}$ = intrinsic asymmetry of e detector [count rate $\sim 1 \pm A_{U,D}$ for muons fully polarized along (opposite) detector symmetry axis], and $P_x(t)$ = muon polarization along axis. Thus,

$$N_{U,D}(t) = B_{U,D} + N_0\epsilon_{U,D}[1 \pm A_{U,D}P_x(t)],$$

where N_0 is a common normalization. The *experimental asymmetry* $a(t)$ is then obtained from either

$$a(t) \equiv \frac{[N_U(t) - B_U] - [N_D(t) - B_D]}{[N_U(t) - B_U] + [N_D(t) - B_D]} \quad \text{or}$$

$$a(t) = \frac{(1 - \alpha) + (1 + \alpha\beta)A_UP_x(t)}{(1 + \alpha) + (1 - \alpha\beta)A_UP_x(t)}, \tag{12}$$

where

$$\alpha \equiv \frac{\epsilon_D}{\epsilon_U} \quad \text{and} \quad \beta \equiv \frac{A_D}{A_U}. \tag{13}$$

Thus, $(1 - \alpha)/(1 + \alpha)$ is the "baseline" asymmetry for totally unpolarized muons.

2.2.1.1 The "Corrected" Asymmetry. In addition to the obvious "baseline shift," there is also a more subtle distortion in $a(t)$ for $\alpha\beta \neq 1$. A plot of $a(t)$ looks as if it has a nonlinear scale for the abscissa. To remove these distortions completely, one must somehow independently determine α and β (usually by fitting data taken in wTF with the same geometry) and then apply the correction

$$A_UP_x(t) = \frac{(\alpha - 1) + (\alpha + 1)a(t)}{(\alpha\beta + 1) + (\alpha\beta - 1)a(t)}. \tag{14}$$

These corrections apply equally to TF-μSR asymmetry spectra formed from opposing pairs of detectors; in fact, these are usually used to fit for α. However, β must be determined from simultaneous fits to the opposing raw spectra $N_{U,D}(t)$ in TF. It is not unusual to assume $\beta = 1$, although, in principle, one should always determine this empirical parameter as accurately as possible.

2.2.2 Zero Field By extension, the field may be zero (ZF-μSR), in which case all the same arguments hold as for LF-μSR, and one measures the time evolution of the muon polarization along its original direction. In μSR, this is just a routine extension of LF, but it bears emphasis since ZF is not so simple (though not impossible) in other magnetic-resonance techniques.

2.2.3 "Relaxation" One consequence of the ease with which one can reduce the field to zero in a LF-μSR experiment is that conventional notions of *longitudinal* (T_1) vs *transverse* (T_2) relaxation processes often become confused and subject to bitter semantic arguments. In NMR, longitudinal relaxation generally occurs in a strong LF, so that a change of polarization requires unambiguous transitions between Zeeman energy eigenstates of the probe spin. It is then easy to define the longitudinal relaxation rate T_1^{-1} in terms of such spin-lattice relaxation processes. Moreover, in strong transverse fields where the Zeeman energies are much greater than any local couplings (such as dipole-dipole interactions between the probe spin and nearby magnetic moments), it is easy to define a transverse relaxation rate T_2^{-1} in terms of the *dephasing* of probe spins precessing at slightly different frequencies as a result of small differences in the local field strength at different sites.

As noted by Kubo and Toyabe (1966), these distinctions are blurred and the terminology becomes less useful as the applied field becomes comparable to the local fields and eventually goes to zero. Basically, if the components of local fields transverse to the applied field are non-negligible in the vector sum forming the total field at the probe, then, even in this classical picture, the relaxation phenomena become quite complicated and are still being worked out today (Dalmas de Réotier *et al.*, 1992). A broad review of this subject is impossible here, but a few examples can help to illustrate the potential of ZF- and LF-μSR.

2.2.3.1 Nuclear Dipolar Relaxation. In the presence of strong electric-field gradients, nuclei with electric quadrupole moments (such as Cu) exert an effective classical dipolar field on the muon:

$$\mathbf{B}_{\text{dip}} = \hbar\gamma_n J_q[3(\hat{r}\cdot\hat{q})\hat{r} - \hat{q}]/r^3, \tag{15}$$

where γ_n is the nuclear gyromagnetic ratio, J_q is the component of nuclear spin along the electric-field gradient direction $\hat{q}$, and $\hat{r} = \mathbf{r}/r$ where $\mathbf{r}$ is the vector of length r from the muon to the nucleus. The muon usually has several near-neighbor nuclei generating a broad and nearly isotropic distribution of nuclear dipolar fields. Assuming a Gaussian distribution of internal fields with random orientation leads to a muon polarization function (Hayano *et al.*, 1979)

$$g_{zz}^{\text{KT}}(t) = \tfrac{1}{3} + \tfrac{2}{3}(1 - \Delta^2 t^2)\exp(-\tfrac{1}{2}\Delta^2 t^2), \tag{16}$$

which at early times ($t \ll \Delta^{-1}$) approaches a simple Gaussian form $G_{zz}(t) \sim \exp[-\Delta^2 t^2]$. The interpretation of Δ is defined by

$$\Delta^2/\gamma_\mu^2 = \tfrac{1}{2}(\langle B_x^2\rangle + \langle B_y^2\rangle); \tag{17}$$

thus, Δ is γ_μ times half the mean squared internal field in the plane perpendicular to the initial muon polarization (taken in this notation to be along the $\hat{z}$ direction).

2.2.3.2 Motional Narrowing. Figure 18 shows a famous example of nuclear dipolar relaxation of muons in copper metal for ZF and *w*LF. At the temperature of 45 K, the muons are almost perfectly static in the Cu lattice; at higher temperatures, they diffuse by thermally activated "hopping" between adjacent octahedral interstitial sites, causing a reduction of the relaxation rate and a change of its shape toward a slow exponential decay

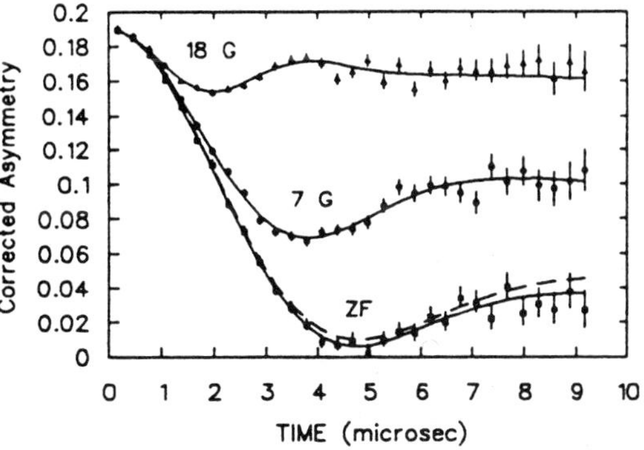

FIG. 18. Zero-field (ZF) and weak (7 and 18 Oe) longitudinal field (wLF)-μ^+SR in a single crystal of pure copper at 45 K with the muon polarization initially along the ⟨111⟩ axis of the crystal. At this temperature, the muons are almost static in the Cu lattice (hopping rate $\lesssim 0.1\ \mu s^{-1}$). The solid lines show fits by the exact spin Hamiltonian between the muon in an octahedral interstitial site and its six nearest-neighbor Cu spins; the dashed line shows a fit by a simple Gaussian Kubo–Toyabe function [Eq. (16)].

(often referred to as "motional narrowing," in reference to the analogous phenomena in NMR, where one observes a resonance line shape whose width is proportional to the relaxation rate). Considerable literature is devoted to the mathematics of "dynamicizing" static relaxation functions (Kehr *et al.*, 1978, Hayano *et al.*, 1979, Celio, 1987). At *lower* temperatures, the muons again begin to hop as quantum mechanical tunneling becomes important; this phenomenon (and others like it) is an important illustration of quantum diffusion in the presence of dissipation (Luke *et al.*, 1991), a major area of application of μSR, which provides a light interstitial probe ideally suited to testing theories of quantum dissipation (Kagan and Klinger, 1974; Kondo, 1986; Yamada *et al.*, 1986; Kagan and Prokof'ev, 1990, 1991, 1992).

2.2.3.3 Nuclear Dipolar Oscillations. In Fig. 18, we also see a slight difference between the simple Kubo–Toyabe function [Eq. (16)] and the exact quantum mechanical solution of the coupled equations of motion of the muon and its six nearest-neighbor Cu spins (Celio, 1986). This reflects the fact that the nuclear moments are not simply static dipoles producing a field at the muon site, but also precess in the dipolar field due to the muon. Such a classical picture quickly becomes useless in picturing the actual evolution of such spin systems, particularly when the nuclei are few in number and have no electric quadrupole interactions, as in the case of the $F\mu F^-$ ion formed when positive muons are implanted into any ionic fluoride compound such as LiF (shown in Fig. 19).

2.2.4 True Relaxation In a strong longitudinal field, the muon's *spin up* and *spin down* states are good eigenstates of the Zeeman Hamiltonian, and so the muon polarization will remain static ("locked" by the applied field H) unless some magnetic perturbation drives them in resonance at their Larmor frequency $\omega_\mu = \gamma_\mu H$. Such perturbations with finite spectral density at the muon's Larmor frequency can cause its spin to flip as in magnetic resonance and lead to true relaxation (involving irreversible transitions between energy levels) at a rate T_1^{-1}. For a simple spin-lattice relaxation process with a characteristic correlation time τ_c for fluctuations of local fields of strength δ/γ_μ (whether caused by fluctuations of the fields themselves or by hopping of the muon between sites with different fields), the longitudinal relaxation rate is given by

$$T_1^{-1} = 2\delta^2\tau_c/(1 + \omega_\mu^2\tau_c^2), \quad (18)$$

which leads to a "T_1 minimum" $T_1(\min) = \omega_\mu/\delta^2$ at $\omega_\mu\tau_c = 1$.

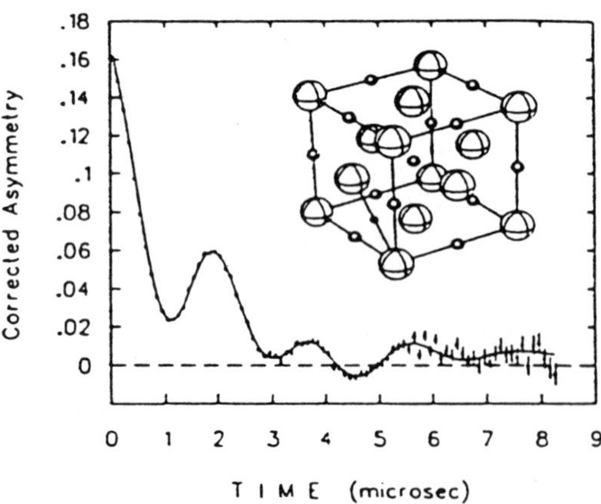

FIG. 19. Zero-field μ^+SR asymmetry spectrum in LiF at 89 K, showing the oscillatory behavior of the $F\mu F^-$ ion as a nearly isolated system of three spin-$\frac{1}{2}$ moments coupled by dipole-dipole interactions. Inset: The rock salt crystal structure showing the μ^+ site (black dot) between two F^- ions (large spheres), which are actually pulled in slightly toward the μ^+ (Brewer *et al.*, 1986).

2.2.4.1 Longitudinal-Field Muonium Relaxation. In very low field, the triplet component of the muonium atom spin system can be treated as a single spin-1 particle with approximately the magnetic moment of an electron. In this case, Eq. (18) applies as well for Mu as for the diamagnetic μ^+ (with $\omega_{\rm Mu} \approx 103\omega_\mu$ substituted for ω_μ), and we may see the behavior shown in Fig. 20 as the Mu hopping rate τ_c^{-1} changes with temperature.

In higher fields, the Mu spin system has four eigenstates, and transitions among levels become more complicated. Nevertheless, an analogous system of rate equations (Yen, 1988) can be used to extract the Mu hopping rate as a function of temperature, as shown in Fig. 21. This has allowed precise measurements of the quantum diffusion of muonium in insulators (Kadono, 1990), an important test ground for theories of quantum dissipation (see above).

2.3 Time-Integral μSR

In time-integral (I)-μ^+SR, one simply scales the total $e^\pm$ count rate in some direction with no regard for the time of arrival of the muons. As a result, there is no rate limitation due to "pileup": Any number of muons may be in the sample at once. As for TD-μSR, there are both TF and LF versions of I-μSR.

2.3.1 Transverse-Field I-μSRotation At cw cyclotron accelerators, the beam arrives at the production target in few-ns pulses separated by the cyclotron's rf period ($1/\nu_{\rm rf} \sim$

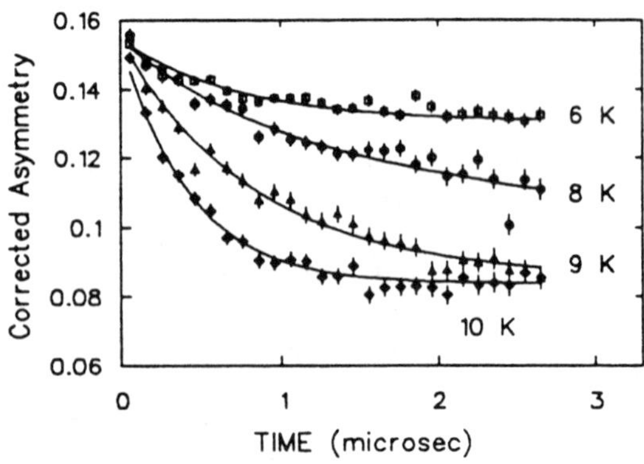

FIG. 20. Longitudinal relaxation of muonium atoms in solid nitrogen at 10 K (diamonds), 9 K (triangles), 8 K (circles), and 6 K (squares) for an applied LF of 8 Oe.

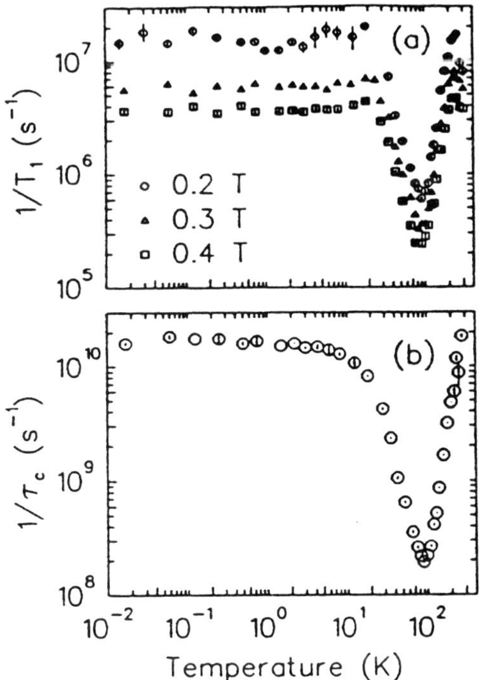

FIG. 21. (a) Longitudinal relaxation rate T_1^{-1} of muonium atoms in a pure GaAs crystal as a function of field and temperature. (b) The deduced Mu hopping rate τ_c^{-1} as a function of temperature. (Schneider *et al.*, 1992.)

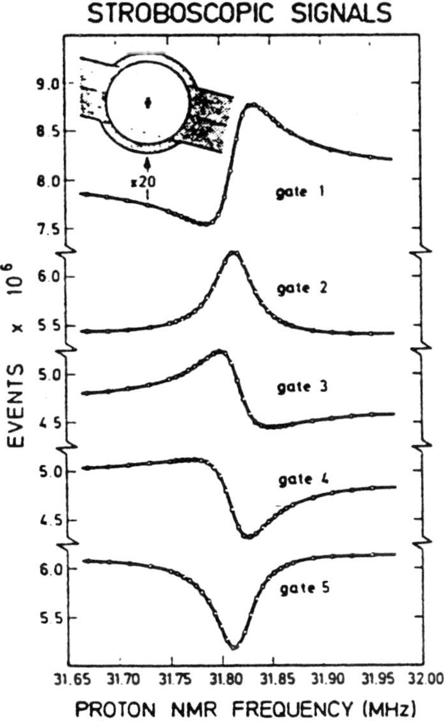

FIG. 22. Resonance signals in a strobo-μSR experiment.

20–50 ns), which must be stable to high precision. As a result, the muon beam also arrives with the same rf structure, though somewhat smeared out by the pion decay lifetime. If each muon stops in a magnetic field H of just the right strength to match an integral number n of muon-spin Larmor precession periods (n/ν_μ) to the rf period, then the polarization of muons arriving in subsequent rf "buckets" will be *in phase* with those arriving previously. Usually, n is limited to 1–3 because of the finite width of the muon buckets. The rf period is broken up into a small number of *timing gates*, during which muon decays in a given direction (determined by a counter) are accumulated in different scalers.

As illustrated in Fig. 22, the rate in a given gate as a function of magnetic field exhibits a resonance as the rf period matches an integral number of Larmor periods. In the absence of spin relaxation, the resonance line shape is Lorentzian with a width of $1/\tau_\mu$ due to the finite muon lifetime; as relaxation is included, the line shape becomes more complicated but may be fitted by a model function. All such resonance lines for the various gates are fitted simultaneously to determine the position and width of the line.

This stroboscopic μSR technique was developed to make a precise measurement of the muon's magnetic moment (Camani *et al.*, 1978) and has since been adapted to muon Knight-shift measurements (Gygax *et al.*, 1984), where its unlimited rate gives it an advantage over TD-μSR methods. "Strobo-μSR," as it is called, is best suited to situations where the muon precesses at a single well-defined frequency and where any relaxation is either uninteresting or fast enough ($\gtrsim 0.454\ \mu s^{-1}$) to be easily unfolded from the "natural linewidth." Splittings of less than this linewidth cannot be unambiguously resolved by strobo-μSR. However, single frequencies (and thus, e.g., muon Knight shifts) can be routinely measured with a precision of $\lesssim 5$ ppm by this method.

2.3.2 Longitudinal-Field IμSRelaxation

In ZF and LF, the I-μSR technique is much simpler, consisting of simply scaling decay counts in the forward (+) and backward (−) directions separately and measuring the resultant asymmetry as a function of independent variables like magnetic field or temperature. The relevant terms are defined as follows: R = rate of arrival of muons; $\epsilon_\pm$ = efficiency of e detector along (+) or opposite (−) μ polarization; $B_\pm$ = background rate in said e detector; $A_\pm$ = intrinsic asymmetry of e detector [count rate ~ $(1 \pm A_\pm)$ for muons fully polarized along (opposite) detector axis of symmetry]; $G_{zz}(t)$ = longitudinal relaxation function. The number of counts in time $T \gg T_\mu$ is

$$N_\pm = B_\pm T + RT\epsilon_\pm \pm RT\epsilon_\pm A_\pm \mathcal{L}(G_{zz}). \quad (19)$$

where

$$\mathcal{L}(G_{zz}) \equiv \int_0^\infty e^{(-t/\tau_\mu)} G_{zz}(t) \frac{dt}{\tau_\mu} \quad (20)$$

is the *Laplace Transform* of $G_{zz}(t)$.

The experimental asymmetry is

$$\mathcal{A} \equiv \frac{N_+ - N_-}{N_+ + N_-} \quad (21)$$

or

$$\mathcal{A} = \frac{b_- + (1-\alpha) + (1+\alpha\beta)A_+\mathcal{L}(G_{zz})}{b_+ + (1+\alpha) + (1-\alpha\beta)A_+\mathcal{L}(G_{zz})}, \quad (22)$$

where $b_\pm \equiv (B_+ \pm B_-)/R\epsilon_+$, $\alpha \equiv \epsilon_-/\epsilon_+$, $\beta \equiv A_-/A_+$. Useful approximations are $B_\pm \approx 0$, $A_\pm \approx A$, and $\epsilon_+ \approx \epsilon_-$, giving $\alpha \approx \beta \approx 1$, $b_\pm \approx 0$, and

$$\mathcal{A} \approx \frac{1-\alpha}{1+\alpha} + \frac{2}{1+\alpha} A_+\mathcal{L}(G_{zz}), \quad (23)$$

where $(1-\alpha)/(1+\alpha)$ is the baseline asymmetry for totally unpolarized muons.

2.3.3 Avoided Level-Crossing Resonance The LF I-μSR method discards all details of the time dependence of the muon polarization, preserving only the Laplace transform of $G_{zz}(t)$; thus, it may appear to be a retrogressive step. Indeed, at a *pulsed* μSR facility, where all the muons arrive at once (within some δt), there is little motive for ignoring the time dependence. However, most muon channels at meson factories are able to produce at least an order of magnitude more muons than can be accommodated in a conventional TD-μSR experiment because of the pileup ambiguity alluded to earlier. Thus, I-μSR trades off detailed information for higher sensitivity by accepting the full muon stop rate.

This is particularly useful when one is looking for *conditions of enhanced relaxation*, the most common being those values of the applied magnetic field for which the muon Zeeman splitting matches a transition between energy levels of the nuclear spins, leading to the possibility of a "flip-flop" of spins in which the muon and the nucleus simultaneously change levels with no change in total energy.

Such conditions are referred to as either "level-crossing resonance" (LCR) or "avoided level crossing" (ALC), depending mainly upon one's laboratory affiliation; the two acronyms refer to the same phenomenon, but one emphasizes the fact that levels never actually cross in quantum mechanics.

2.3.3.1 Nuclear Quadrupolar μALCR. The first such resonance studied in μSR was actually observed with TD-μSR methods, following a suggestion by Abragam (1984). In that experiment (Kreitzman *et al.*, 1986), the μ^+ Zeeman splitting at 81 Oe matches the (mostly) electric quadrupolar splitting between $m_I = \pm\frac{3}{2}$ and $m_I = \pm\frac{1}{2}$ levels of the spin-$\frac{3}{2}$ Cu nuclei, causing the resonant relaxation shown in Fig. 23.

This measurement and its later improvements (Luke *et al.*, 1991) allowed a precise determination of the electric-field gradient

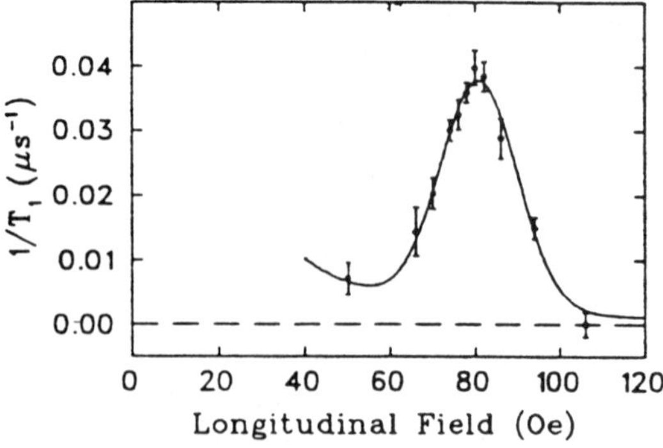

FIG. 23. Longitudinal relaxation rate of muons in a single crystal of copper with the applied field (and the muon polarization) along the ⟨111⟩ crystalline axis (Kreitzman *et al.*, 1986).

produced by the μ^+ at the Cu sites; similar experiments have since revealed nuclear quadrupolar μALCRs in solids containing ^{17}O, ^{14}N, and other nuclei. The main application is in determining the muon site in crystals.

2.3.3.2 "Muonated" Radicals. Shortly after the discovery of nuclear quadrupolar μALCR, it was realized that a much stronger resonance could occur through the hyperfine interactions of unpaired *electrons* with both the μ^+ and the nuclear spins. The quantum mechanics of this system is considerably more complicated, but the vague notion of level matching is still applicable: When the energy difference between two states in which both the muon and the nuclear spins have flipped approaches zero, resonant muon depolarization can occur.

One of the first applications of this principle to *paramagnetic* systems was in the case of the $C_6F_6Mu\cdot$ radical, a paramagnetic molecule formed by addition of a Mu atom to a double bond in the hexafluorobenzene ring (Kiefl *et al.*, 1986).

In these measurements, a time-integral method is always used, which is sometimes subject to "noise" as beam conditions shift without notice: to combat this problem and produce the smooth resonance patterns shown in Fig. 24, a *field-differential* technique is often employed: The experimental asymmetry $\mathscr{A}$ defined in Eq. (22) is measured for short time intervals (seconds) with a "toggle field" $\pm\delta H$ applied alternately along (+) or opposite (−) the main longitudinal field H_0, and the difference between these two asymmetries ($\mathscr{A}^+ - \mathscr{A}^-$) recorded as a function of H_0. This produces a signal analogous to a true differential resonance ($d\mathscr{A}/dH_0$), except that the finite difference ($\pm\delta H$) can actually be larger than the natural width of the resonance in some cases, producing a "bump" followed by a mirror-image "dip" $2\delta H$ later.

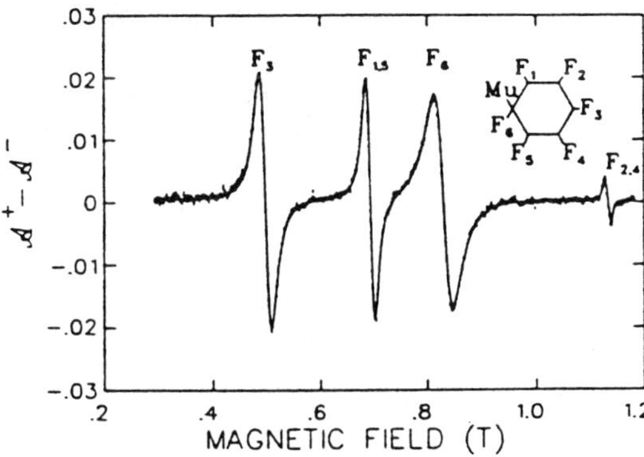

FIG. 24. Field-differential μALCR in the $C_6F_6Mu\cdot$ radical (Kiefl *et al.*, 1986)

Many such radicals have now been studied by this method, resulting in much detailed information about both the structure and the dynamics of these molecules, whose difference from the analogous radicals formed by H-atom addition is often insignificant. Several new species never observed in any other way have been discovered using μALCR spectroscopy.

2.3.3.3 Muonium in Semiconductors. Another paramagnetic system amenable to study by these methods is the muonium-like center in semiconductors, where the unpaired electron may have hyperfine interactions not only with the μ^+ but also with neighboring nuclei of the lattice. The resulting spectroscopy can be very rich, as illustrated in Fig. 25, and reveals the muonium site, as well as a great deal of information about its electronic wave function. This has been used to obtain most of what is known

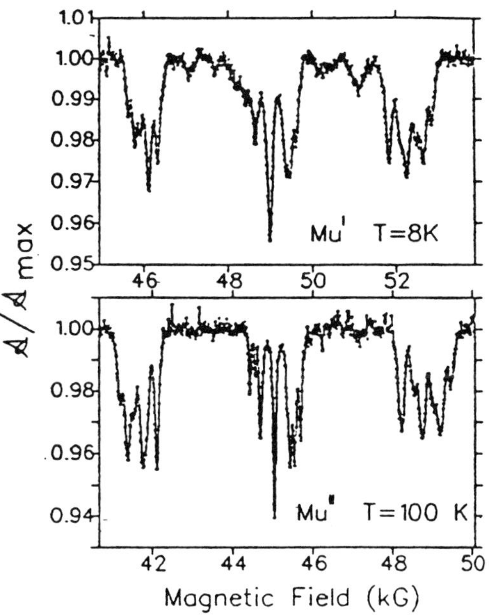

FIG. 25. Muonium-nuclear ALCR spectra in a single crystal of CuCl (Schneider *et al.*, 1990). The Mu^I (low temperature) and Mu^{II} (high temperature) muonium states have the same hyperfine interactions with neighboring nuclei but slightly different ω_0 values.

about isolated atomic hydrogen in Si, GaAs, GaP, and CuCl (Kiefl and Kreitzman, 1992).

2.4 Muon Spin Resonance

Traditional magnetic-resonance techniques involve intentional irradiation of the probe spin system with photons of energies equal to transitions between its eigenstates. None of the μSR techniques so far described have involved such irradiation, but true *resonance* techniques are also used in μSR, as outlined below. As in NMR, ESR, ENDOR, etc., the variety of irradiation schemes is huge and constantly growing; only a few simple generalizations are possible in limited space.

2.4.1 rf I-μSResonance The first and most familiar form of muon spin *resonance* uses an rf field in the frequency range of about 5–500 MHz to drive the polarization of muons in diamagnetic states (i.e., "bare" muons) in resonance at fields of 300 Oe to 4 T. The rf field may be generated in a simple coil at the lower frequencies or a cavity at the higher. In some cases, the rf frequency can be swept through resonance, but most often the main field H_0 is swept, as in NMR. A good recent survey of these techniques and applications may be found in Nishiyama (1992).

For the higher frequencies, the same apparatus may be used at lower fields to study paramagnetic species such as muonated radicals. An interesting variation of this is seen in Fig. 26.

The time scale for muon spin resonance is restricted by the muon lifetime; to be observed, a resonance effect must affect the muon polarization in $\lesssim$10 μs, implying an rf field $H_1 \gtrsim 2$ Oe, which is not difficult to achieve but represents fairly high rf power. Because of power supply limitations and Ohmic heating, such irradiations are less efficient at cw facilities than at pulsed-beam facilities (Nishiyama, 1992), where all the muons can be irradiated simultaneously with timed bursts of rf power; nevertheless, rf-μSR is quite feasible at cw facilities (Kreitzman, 1990) and has enjoyed a rapid growth of applications virtually everywhere.

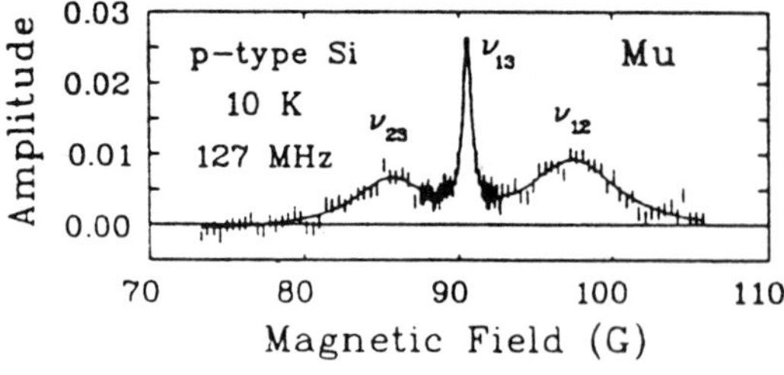

FIG. 26. Muonium resonance in mildly *p*-type Si at 10 K in a 127-MHz rf field, showing power-broadened resonances at the two muonium transition frequencies ν_{12} and ν_{23} (see Fig. 13), as well as a sharp two-photon resonance at the frequency ν_{13}.

One advantage of rf-μSR is that it can identify *final states* of muons in situations where stochastic processes render impossible the usual means of identification of these states by their distinctive time spectra. A typical example is the *delayed formation* of diamagnetic molecules following initial formation of Mu atoms: In TF-μSR, the early Mu precession quickly dephases the muon polarization so that no signal can be seen from the diamagnetic final state unless the reaction times are short compared to the Mu frequency; in strong LF, however, not only is the Mu polarization "held" by the applied field, but the population of the final state as a function of time can be determined by delaying the rf pulse (at the diamagnetic-resonance frequency) relative to the time of arrival of the muons.

Perhaps the most exciting feature of rf-μSR is its ability to reveal the time evolution of the muon polarization *during* the rf irradiation, a capability denied to conventional magnetic-resonance techniques, which detect the same sort of electromagnetic signal as they use to drive the probe spins. This is a consequence of the fact that μ^+SR detects high-energy positrons whose registration by the counters is unaffected by the rf fields.

2.4.2 Muon Spin Echoes Many modern NMR techniques involve *pulses* of rf power designed to rotate the probe's polarization through various angles in the rotating reference frame. This is more difficult for muons because of their short lifetime—to be useful in μSR, a $\pi/2$ pulse must be no more than $\lesssim$1 μs long, which means an rf magnetic field of $H_1 \gtrsim 75$ Oe. This is barely possible with today's rf techniques. It is particularly difficult at cw facilities, where the rf pulse must be *asynchronously triggered* by the arrival of individual muons, in order to occur at the same time (relative to "$t = 0$" when the muon arrives in the sample) for each. Nevertheless, such *spin echo* techniques (nicknamed μSE) were first successfully demonstrated at TRIUMF, a cw facility (Kreitzman *et al.*, 1988).

As mentioned earlier, μSR detection techniques allow observation of the time evolution of the muon polarization *during* the rf irradiation. This can be especially informative in the case of μSE pulse sequences. An example is shown in Fig. 27.

A bonus of rf-μSR at pulsed facilities is the chance to overcome the *time resolution* limitations of pulsed beams (muons arrive in bunches $\gtrsim$50 ns long) by stopping the muons in a strong LF and then using a high-power phase-locked $\pi/2$ pulse to rotate them all *in phase* into the plane perpendicular to $\mathbf{H}_0$, after which they will exhibit HTF-μSR precession with all the same observables as enjoyed at cw facilities.

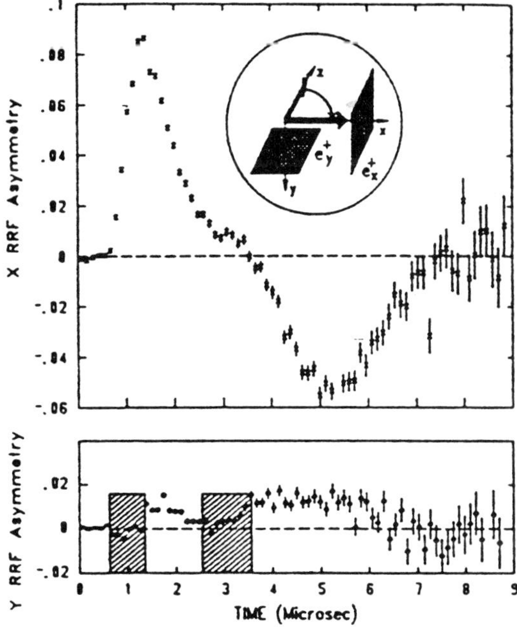

FIG. 27. Muon spin echo in the rotating reference frame during a $(\pi/2)_y$–τ–π_y pulse sequence (shaded regions) where the muons arrive initially polarized along $\mathbf{H}_0 = H_0\hat{z}$, and the detectors are in the x and y directions. The small signal observed along the y axis indicates that the rf was slightly off resonance. Inset: The geometry of the positron detectors and the polarization during the $\pi/2$ pulse (Kreitzman *et al.*, 1988).

3. μSR APPLICATIONS

Examples of selected experimental results having been shown above, this section will focus mainly on the "scientific demography" of μSR.

3.1 Muonium Chemistry

One of the main themes of μSR could be characterized as "The Chemistry of a Light Isotope of Hydrogen." The muonium atom and the H atom have the same size, and almost the same reduced mass and ionization potential, and both obey the Born-Oppenheimer approximation (the electron wave functions adiabatically adjust to the slow motions of the nuclei) that allows relatively simple calculations of chemical reaction processes and molecular structure.

3.1.1 Chemical Kinetics This strong similarity, combined with the unprecedented *isotopic* difference (a ratio of ~18 in mass between Mu and D) and the enhanced quantum behavior of the light Mu atom, makes muonium chemistry a very important testing ground for the few *ab initio* theories of chemical reaction kinetics (Fleming and Senba, 1992).

Once the *differences* between Mu and H chemistry are well understood (as is now the case for many types of reactions), measurements of Mu reactivity (which are often easy) can be used to predict reliably the reactivity of H under circumstances where H atoms cannot even be detected. This is now entirely feasible but has not yet attracted much attention from the "practical" side of the chemistry community.

3.1.2 Radicals Following chemical reaction, Mu is often incorporated into *radicals* (molecules with unpaired electrons) where a weakened hyperfine interaction persists between the electron and muon spins (and also between the electron and any other nuclei with magnetic moments). These molecules resemble their hydrogenic analogs (where the μ^+ is replaced by a proton) so closely that most of their reaction kinetics are nearly identical. This allows use of the "muonated" version to learn the chemical behavior of the "protonated" version even when information on the latter is unavailable by other methods. In some cases, μSR has confirmed the very *existence* of radicals that have never been observed by any other means.

3.1.3 Molecular Structure By using μALCR techniques, one can use μSR to measure not only the hyperfine coupling of the muon to the unpaired electron but also the couplings of that electron to the *other* nuclei; thus, the muon can be used to read out the molecular structure of regions of the molecule quite far from the muon's site, much as in a conventional ENDOR experiment (Percival *et al.*, 1987). The details of the *structure* of muonium- and hydrogen-substituted versions of radicals are sometimes influenced by the lighter mass of the muon: Bond lengths to Mu tend to be slightly longer, and bond angles are often changed by the larger zero-point motions of Mu. This, combined with the fact that the nuclear hyperfine couplings can often be measured more accurately by μSR than by ENDOR, makes μALCR a very important probe of molecular structure (Roduner, 1988).

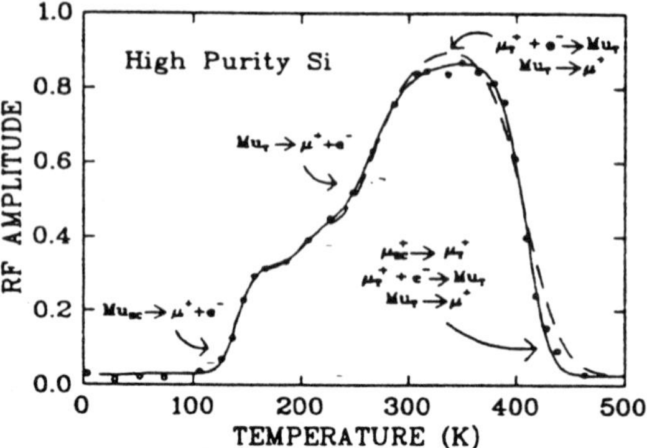

FIG. 28. The rf-μ^+SR diamagnetic-resonance amplitude in pure silicon, showing evidence for thermally activated transitions between various charge states and lattice sites of the μ^+: Mu_{BC} = anomalous muonium (also called Mu*) localized at a Si-Si bond-centered site: Mu_T = quasi-free Mu atom diffusing rapidly between tetrahedral sites; μ_{BC}^+ and μ_T^+ indicate freshly ionized muons at the corresponding sites.

3.2 Condensed-Matter Physics

The applications of μSR in condensed-matter physics also include many comparisons between the μ^+ and the proton, insofar as their behaviors in matter differ only by virtue of the muon's lighter mass; however, there are perhaps a greater variety of applications that might be titled, "The Muon as a Magnetic Probe." Both types will be surveyed briefly below.

3.2.1 Hydrogen in Semiconductors Closely related to the muonium chemistry applications discussed above are the uses of μ^+SR to learn the location, electronic structure, and dynamical behavior of Mu and/or H atoms in semiconductors, where H is known to be an ubiquitous impurity but where far more is known about isolated H atoms from μSR than from any method that observes H itself (Patterson, 1988, Kiefl *et al.*, 1988). Since H is known to affect the electronic properties of Si and GaAs and is now being used specifically for the purpose of *passivation* of electrically active impurities, this knowledge is vital to the semiconductor industry; consequently, μ^+SR spectroscopy in semiconductors is one of the most important applications of μSR in solids.

Figure 28 shows the temperature dependence of the probability of finding the muon in an ionized state (either μ^+ or Mu^-) at times comparable to the muon lifetime (long after any initial formation or reaction of paramagnetic states), obtained by rf-μSR at the diamagnetic frequency. Comparison of these results with those for highly doped *p*- or *n*-type Si reveals the interactions between H-like species and impurities (Hitti *et al.*, 1994).

3.2.2 Quantum Diffusion As mentioned earlier, the theory of *quantum dissipation* (quantum tunneling or bandlike propagation in the presence of stochastic or other interactions with the host medium) is currently a very important field of condensed-matter physics, having implications for the electronic and diffusive transport properties of all types of materials (Stamp and Zhang, 1991; Kagan and Prokof'ev, 1992). Because of its light mass, its affinity for electrons, and its repulsion from nuclei, the μ^+ (and its neutral version, Mu) is an apt probe of such behavior, the ideal "light interstitial" (Flynn and Stoneham, 1970; Kagan and Klinger, 1974; Petzinger, 1982). Study of the *quantum diffusion* of μ^+ and Mu in solids has therefore constituted a significant fraction of the μSR community's experimental effort over the last two decades (Kadono, 1992), and the results of such experiments have attracted considerable theoretical interest (Kagan and Prokof'ev, 1992).

3.2.2.1 Motion of the μ^+ in Metals. In impure or imperfect metallic crystals, the μ^+

tends to be trapped at defects or at least to be localized in their vicinity; the resultant temperature dependence of motional narrowing effects, used to measure μ^+ mobility, can be quite complicated (Möslang *et al.*, 1983, Petzinger, 1982). In pure, well-annealed crystals, however, one often observes the intrinsic interactions between the μ^+ and the lattice. The hopping of positive muons in various pure metals has been studied extensively by means of ZF- and wLF-μ^+SR measurements such as those shown in Fig. 18 (Luke *et al.*, 1991), as well as by TF-μ^+SR motional narrowing experiments. These studies have revealed a qualitatively consistent temperature dependence typified by the classic example shown in Fig. 29. At high temperatures (above about 80 K in the case of Cu), the muon exhibits semiclassical thermally activated hopping over the energy barriers between adjacent interstitial sites; at lower temperatures (below about 20 K for Cu), the μ^+ actually hops faster as T decreases, because of the increasing probability of quantum tunneling between sites, as the disorder due to lattice vibrations is reduced. The power law $\tau_c^{-1} \propto T^{-\alpha}$ is predicted by theory (Kondo, 1992) to have a much smaller exponent α for screened positive muons in metals than in insulators, as a result of the additional dissipation in electronic degrees of freedom (often referred to as "electron drag" effects). At still lower T, the intrinsic disorder of even pure high-quality crystals begins to inhibit tunneling once again. Similar results have been observed in aluminum using indirect methods involving trapping of the μ^+ at impurities.

3.2.2.2 Muonium Diffusion in Nonmetals. In ionic crystals, the μ^+ tends to form a hydrogen bond with the most negative species in the lattice, such as F^- or O^{--} ions. Muon diffusion is thus suppressed until quite high temperatures, typically ≈200 K (Kiefl *et al.*, 1990). If the muon forms muonium, however, the Mu atom is more or less decoupled from local electric fields and usually diffuses freely through the lattice; interestingly, Mu diffusion is quite similar in both covalent and ionic nonmetals. An example of Mu diffusion in the semiconductor GaAs was shown earlier in Fig. 21 (Schneider *et al.*, 1992); a startlingly similar result was obtained (Kiefl *et al.*, 1989) in an ionic insulator (KCl), as shown in Fig. 30.

Somewhat different Mu hopping behavior is observed in the van der Waals insulator s-N_2 formed by crystallization of solid nitrogen, as shown in Fig. 31. Such so-called *cryocrystals* have very low Debye temperatures and quite weak couplings between the neutral Mu atom and the host lattice; they are, therefore, interesting systems in which to test theoretical predictions of the temperature dependence of Mu quantum diffusion (Storchak *et al.*, 1994).

An interesting effect has been observed in *superconducting aluminum*, which can be driven normal by an applied magnetic field: At the same T, muon diffusion is considerably *faster* in the superconducting (SC) state than in the normal state, because the opening of a SC gap at the Fermi surface effectively quenches the dissipation due to screening electrons (Kondo, 1992). Thus, SC

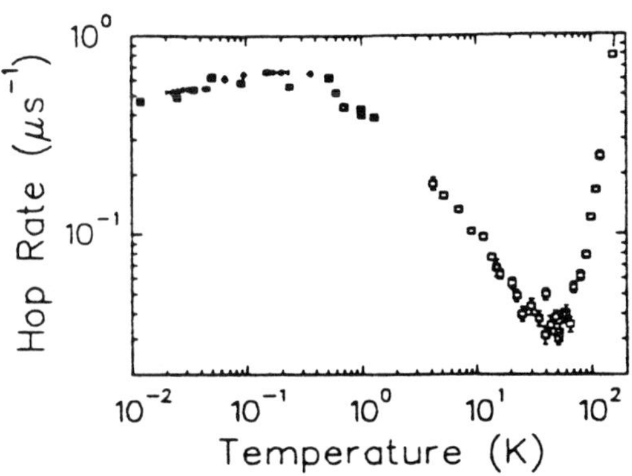

FIG. 29. Temperature dependence of the μ^+ hopping rate τ_c^{-1} in pure copper crystals (Luke *et al.*, 1991).

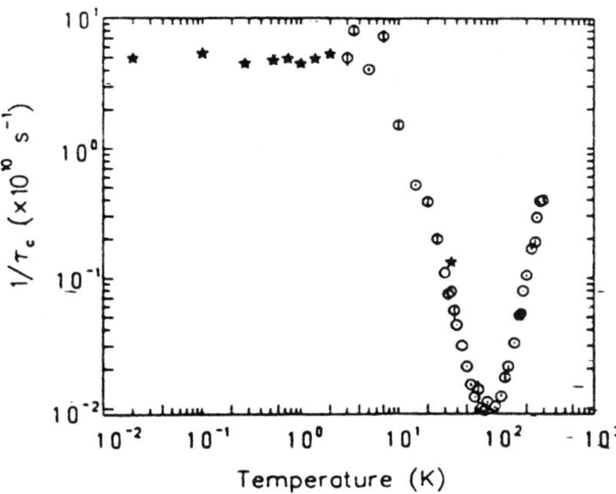

FIG. 30. Temperature dependence of Mu hopping rate in KCl (Kiefl *et al.*, 1989).

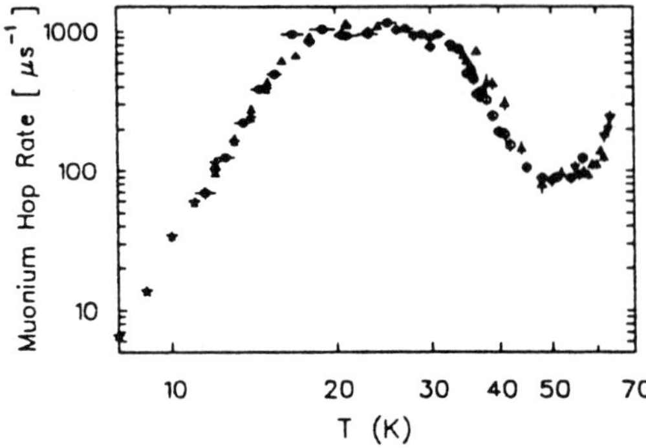

FIG. 31. Temperature dependence of the muonium hopping rate in solid nitrogen cryocrystals (Storchak *et al.*, 1994).

Al acts like an insulator with respect to muon quantum diffusion.

3.2.3 Magnetism Perhaps because it so easily generates rich and precise data, *magnetism* is currently the largest area of application of μSR.

In magnetically ordered materials, the muon (μ^+ or μ^-) generally samples the local magnetic field at some preferred site or sites and reveals that field through its characteristic Larmor precession frequency (Seeger and Schimmele, 1992). The local field $\mathbf{B}_\mu$ is composed of several contributions:

$$\mathbf{B}_\mu = \mathbf{H}_0 + \mathbf{B}_\mathrm{L} + \mathbf{B}_\mathrm{dip} + \mathbf{B}_\mathrm{hf}, \tag{24}$$

where $\mathbf{H}_0$ is the externally applied field (if any), $\mathbf{B}_\mathrm{L}$ is the *Lorentz field* produced by magnetic "charges" induced on the interior of a hypothetical spherical cavity around the muon site due to the average bulk magnetization of the medium, $\mathbf{B}_\mathrm{dip}$ is the *dipolar field* due to all the microscopic moments within the Lorentz cavity, and $\mathbf{B}_\mathrm{hf}$ is a *hyperfine field* transmitted to the muon by its Fermi contact interactions with conduction electrons. The last two components are usually the focus of experiments on magnetically ordered systems, where $\mathbf{B}_\mu$ is usually measured with $\mathbf{H}_0 = 0$, as well as measurements of the muon Knight shift $K_\mu \equiv (B_\mu - H_0)/H_0$ in paramagnets.

Interpretation of these fields requires a knowledge of the muon site, including zero-point motion and any tunneling between nearby sites, and can be quite difficult; nevertheless, in many cases, it is easier than trying to sort out the "core polarization" effects of various interacting atomic electrons that transmit the hyperfine field to the nucleus in (e.g.) NMR.

One attractive feature of μSR in magnetism studies is that measurements of internal magnetic fields can be performed on unaligned polycrystalline or powdered samples almost as easily as on perfect single crystals. In zero applied field, if all the local fields have the same *magnitude*, the only effect of random *directions* is to reduce the muon precession amplitude to $\frac{2}{3}$ of the value that would be observed if the fields were all perpendicular to the muon polarization. Thus, ZF-μSR, which requires a minimum of apparatus, is often adequate for very detailed investigations of the temperature dependence of magnetic order and dynamics. Moreover, using ZF-μSR, one can measure B_μ in crystals exhibiting antiferromagnetism or even more exotic forms of magnetic order as easily as in simple ferromagnets. A classic example is shown in Fig. 32 for the intermetallic compound MnSi, where the magnetic field due to itinerant-electron moments forms a long-wavelength spiral with a small component along the axis of the spiral—so-called *helimagnetic* order.

Disordered magnetic systems (such as *spin glasses*) can also be studied rather easily by μSR, leading to a strong μSR presence in the currently popular area of frustrated and low-dimensional magnetism. A classic example is shown in Fig. 33 and a more recent case in Fig. 34.

3.2.4 Superconductivity Because muon decay provides a nonelectromagnetic "sig-

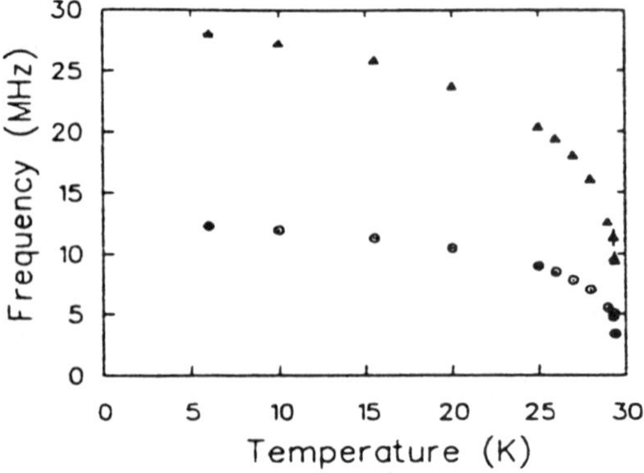

FIG. 32. Temperature dependence of the magnitudes of the internal fields at two different μ^+ sites in a single crystal of the itinerant-electron helimagnet MnSi.

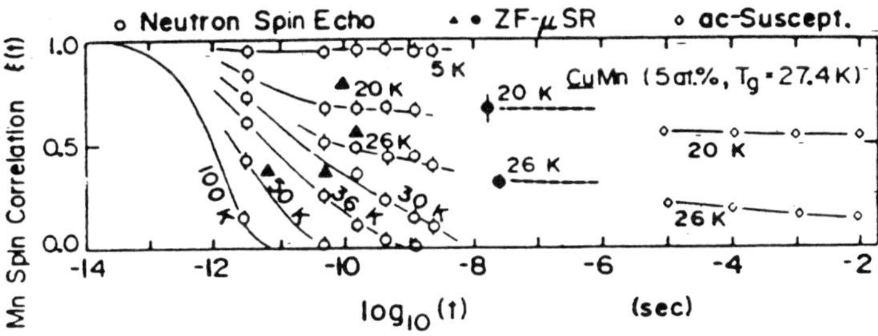

FIG. 33. Comparison of neutron spin-echo (NSE), ZF-μSR, and ac-susceptibility measurements on the spin glass CuMn (5 at. %), showing the time dependence of the static order parameter ξ (related to the fraction of Mn moments that have not yet moved appreciably) at several temperatures below the nominal glass transition at $T_g = 27.4$ K (Uemura *et al.*, 1984).

nal," μSR is able to probe magnetic fields inside superconductors without distortion. This gives μSR a unique window on the magnetic behavior of superconductors. Since 1987, the μSR community has exploited this advantage in the study of high-temperature superconductors (HTSC).

3.2.4.1 Phase Diagrams. All of the copper oxide superconductors show either antiferromagnetic (AFM) or spin-glass-like (SGL) order very close to (and sometimes overlapping) the superconducting (SC) phase, as first demonstrated for $YBa_2Cu_3O_x$ by μSR experiments in 1988. Since then, the temperature vs doping phase diagrams of most HTSC materials have been mapped out using μSR.

3.2.4.2 Magnetic Penetration Depth. The first application of μSR to HTSC was in measuring the magnetic penetration depth λ by fitting the high transverse field (HTF)-μSR time spectrum by a simple Gaussian relaxation function and thus estimating the second moment of the frequency distribution due to the vortex lattice of magnetic flux penetrating a type-II superconductor in the mixed state. Figure 35 shows the earliest results, obtained in January 1987 (Aeppli *et al.*, 1987). Since then, an enormous body of data has been accumulated by this method, which is still the most reliable way to estimating λ in sintered, random powders. However, as evident in Fig. 11, for well-characterized single crystals one finds a line shape much less symmetric than a Gaussian, as predicted by London and later Gor'kov models of the vortex lattice. The ability to make such measurements and interpret them correctly in terms of λ and ξ is a recent achievement of μSR, which is now poised to exploit these capabilities to legitimize (or, in some cases, debunk) the many published μSR results worldwide interpreting the Gaussian-fitted

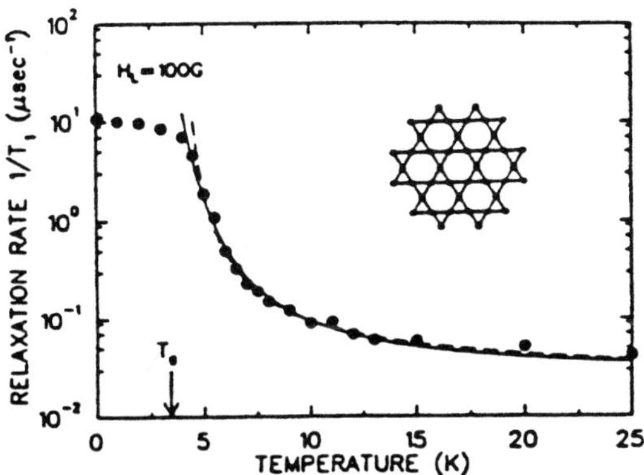

FIG. 34. Muon spin-relaxation rate T_1^{-1} due to fluctuating local fields in the antiferromagnetically frustrated two-dimensional kagomé lattice (inset) of $SrCr_8Ga_4O_{19}$ (Uemura *et al.*, 1994). The increase of T_1^{-1} with decreasing T is a manifestation of the gradual slowing down of Cr spin fluctuations; the saturation of this effect below T_g shows that the Cr moments never freeze even as $T \to 0$. This is in marked contrast with the situation in dilute spin glasses like CuMn, where all fluctuations cease as $T \to 0$.

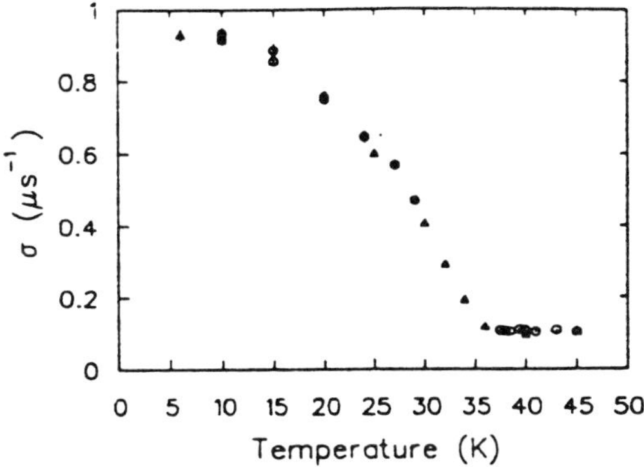

FIG. 35. First TF-μ^+SR measurement of the temperature dependence of the μ^+ dephasing rate σ (proportional to λ^{-2}, the inverse square of the magnetic penetration depth, which is, in turn, proportional to the superconducting carrier density n_s) in a high T_c cuprate superconductor—in this case, an unaligned sintered powder sample of $La_{1.85}Sr_{0.15}CuO_4$ in an applied field of 0.4 T (Aeppli *et al.*, 1987).

linewidths (for sintered powder or aligned HTSC samples) in terms of λ.

3.2.4.3 Universal Correlations between T_c and n_s. A collection of such results has been assembled by Uemura *et al.* (1989) to produce the plot shown in Fig. 36, commonly known as an "Uemura plot." The "universal" linear dependence of T_c on n_s in the *underdoped* (less than optimal n_s) regime gives way to a "turnover" and eventual linear *decrease* of T_c with increasing n_s in the *overdoped* regime (Niedermayer *et al.*, 1993).

***3.2.4.4* Non-*s*-Wave Pairing?** Whether it be crudely characterized in terms of a Gaussian relaxation rate σ or more accurately represented by the second moment of a line shape, the muon dephasing rate due to the vortex lattice in a type-II superconductor is proportional to the inverse square of the London penetration depth, λ^{-2}, which in turn is proportional to the superconducting carrier density n_s. Thus, a measurement of the temperature dependence of σ is (apart from a normalization) also a measurement of the excitation of normal carriers across the superconducting energy gap that characterizes the pairing mechanism. A conventional BCS "*s*-wave" superconductor has a prescribed T dependence of n_s, variations from which are taken as evidence for unconventional pairing mechanisms. While all this should, in the author's opinion, be taken with a grain of salt, still μSR has been an important contributor to the debate over the still-mysterious mechanism for HTSC. While the μSR data for the 18-K superconductor K_3C_{60} are consistent with *s*-wave BCS superconductivity, the μSR results on certain organic and heavy-fermion superconductors show a T dependence strongly suggestive of nodes in the gap (momentum-space directions in which the superconducting gap vanishes) (Uemura, 1992). A very similar T dependence has been observed for n_s in various samples of $YBa_2Cu_3O_{6.95}$—most notably in high-quality single crystals—as shown in Fig. 37.

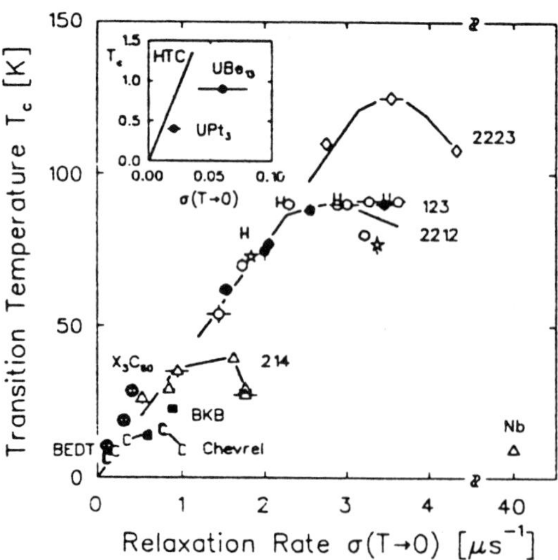

FIG. 36. Correlations between the superconducting carrier density n_s (as measured by the μ^+ dephasing rate σ) and the superconducting transition temperature T_c for a wide variety of "exotic" superconductors including many HTSC; conventional BCS superconductors have higher n_s with lower T_c.

4. μSR FACILITIES

No practical guide to μSR would be complete without some discussion of the *logistics* of μSR experiments and the *facilities* required for its pursuit.

4.1 Logistics of Accelerator Experiments

Because μSR requires muons, which must be produced by large, high-intensity accelerators, μSR beam time will probably always be a scarce commodity for which users will compete fiercely. Such competition has

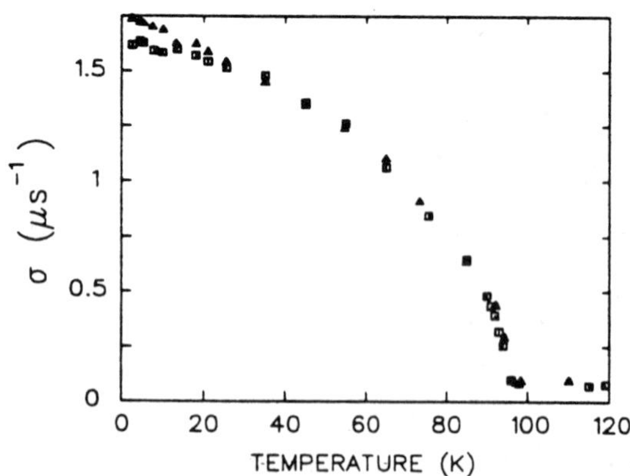

FIG. 37. Temperature dependence of the μ^+ dephasing rate σ (proportional to the superconducting carrier density n_s) in single crystals of $YBa_2Cu_3O_{6.95}$ in a TF-μ^+SR experiment at 0.5 T (triangles) and 1.5 T (squares). An "*s*-wave" BCS superconductor is expected to show very weak T dependence near $T = 0$; in contrast, these data indicate an almost linear decrease from $T = 0$.

a beneficial effect on the quality of the resulting research, but the scarcity of muons does regulate the growth of the field. It will probably always be the case that μSR experiments will be screened in advance by peer review committees and allocated beam time every few months for periods ranging from a few days to several weeks. While this sort of schedule may often hamper spontaneity, it is well matched to the cycle of experimental conception, design, execution, data analysis, and publication, as evidenced by the high output of most μSR facilities.

Furthermore, because a typical μSR experiment involves so many diverse technologies (from magnetic beam optics to charged-particle detectors to magnets and cryostats to fast electronics and extensive on-line computer software), it makes an excellent training ground for students, who, once having mastered the techniques of μSR, can apply them almost unaltered to a wide variety of scientific topics.

4.2 Standard μSR Hardware

Most accelerator laboratories with μSR programs also maintain some sort of μSR *user facility* providing standard μSR spectrometers, electronics, and data-acquisition computers; it is now rare for individual experimental groups to provide their own equipment, except for specialized cryostats, ovens, sample cells, or data-analysis software. Some of the standard components to be found at most μSR user facilities are listed below.

4.2.1 Detectors The workhorse of μSR is the plastic scintillation counter, in which a flash of light (generated as an ionizing particle passes through the scintillating plastic) is transmitted down clear plastic light guides by total internal reflection to a photomultiplier tube, which, in turn, emits an electrical pulse that is transmitted down a coaxial cable to a fast timing discriminator module in the counting room outside the experimental area (from which experimenters are excluded while the beam is on, because of the appreciable radiation levels in the muon beam itself). The result is a logical timing pulse whose arrival time corresponds within ~1 ns (and a fixed delay) to the time the particle passed through the detector. The *M* and *E* detectors in Fig. 1 would normally be plastic scintillation counters for the incoming muon and outgoing positron respectively, and would normally provide a resolution of 0.5–2 ns on the corresponding time interval. With careful design of the counters and special photomultiplier tubes, cables, and fast electronics, it is possible to achieve time resolutions of ~100 ps (Holzschuh, 1983; Keller *et al.*, 1987), but this is not yet common.

While scintillation counters are simple, reliable, and versatile and have excellent performance in the time domain, they have a number of frustrating features: First, even if the scintillator itself is small, the light guides are bulky and inflexible (flexible fiber optics can be used but so far only at the expense of timing degradation and/or loss of light-collection efficiency); this becomes increasingly inconvenient as increasing stopping luminosity reduces the scale of μSR samples and experiments. Second, most photomultiplier tubes are very sensitive to magnetic fields and must be magnetically shielded and/or removed to field-free regions by long light guides, which again reduce time resolution. Third, scintillation counters have only crude position sensitivity, which is generally of secondary importance to μSR experiments, but could be useful. For example, by distinguishing one incoming muon from another and matching the outgoing positrons to the muons from which they arise, one may overcome the pileup rate limitation normally imposed by allowing only one muon in the sample at a time; this technique has been demonstrated but never widely applied.

Potentially promising alternatives include modern proportional counters and solid-state barrier detectors, but neither of these have the required time resolution in their standard configurations; more development work is needed.

4.2.2 Electronics Raw counter pulses are fed to fast discriminators—usually constant-fraction discriminators (CFDs)—which generate uniform timing logic pulses if the raw pulse height is above a set threshold. The logic pulses are processed by a simple "fast electronics" circuit shown schematically in Fig. 38.

The incoming muon generates a pulse in the *M* counter, which ultimately sends a start pulse to a fast time digitizer ("clock"); later,

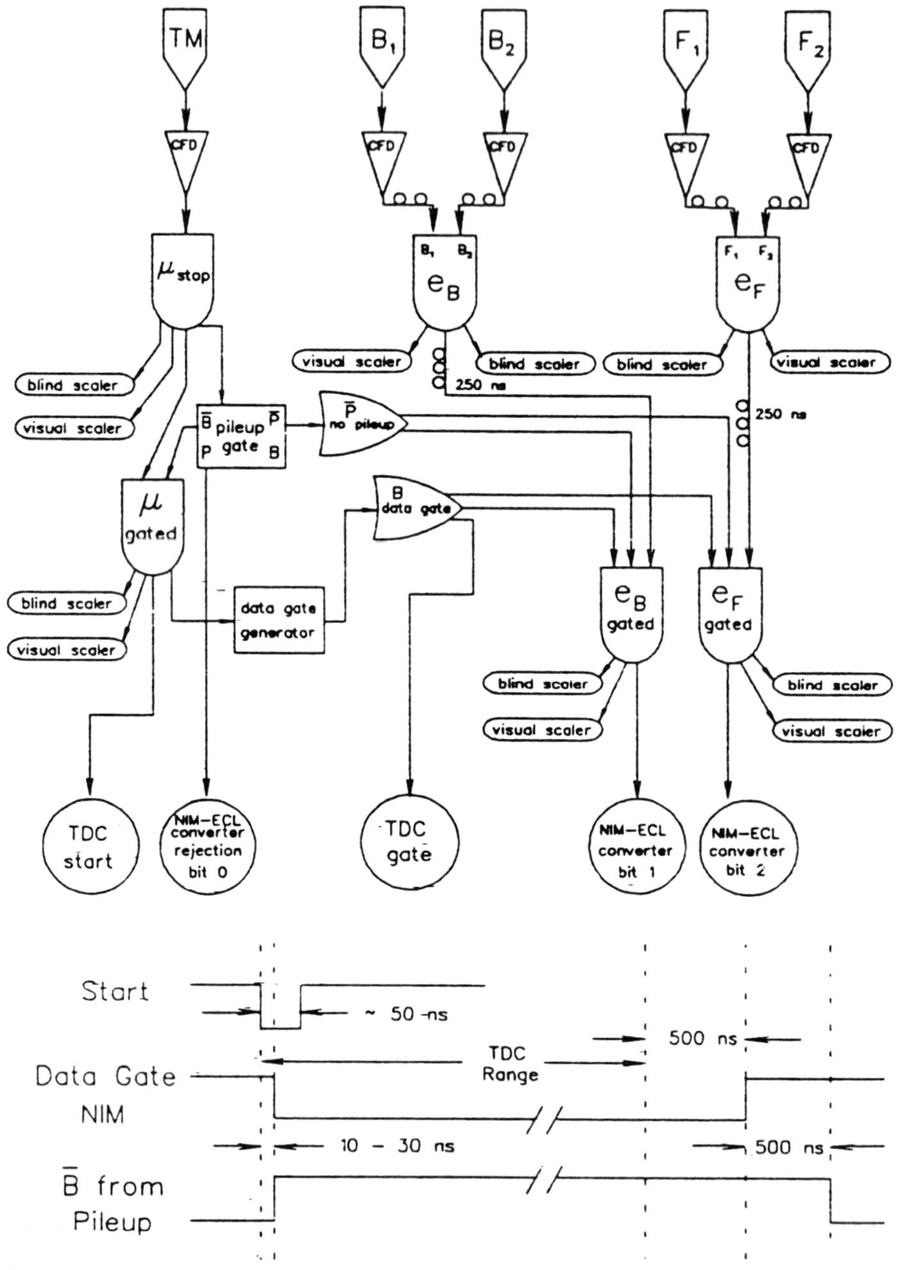

FIG. 38. Schematic electronics logic diagram for two "channels" of a typical time-differential (TD)-μSR experiment (see Fig. 1).

the muon decays, triggering a stop pulse in the E counter, which stops the clock; the time interval is digitized, and the corresponding bin in a time histogram is incremented. The additional logic modules ensure that events with "second muons" are rejected with uniform efficiency throughout the gate T, lest rate-dependent distortions spoil the resultant time spectra (histograms), which are the final output of the experiment.

Similar care must be taken to reject "second E" events in which a (possibly accidental) count in an E counter renders the identity of the stop pulse ambiguous. This function is handled internally in the LRS 4204 time-to-digital converter, which is now the standard μSR clock primarily because it provides this function along with excellent time resolution and the ability to directly increment time bins in a CAMAC Histogramming Memory, thus relieving the data-acquisition computer of all event-processing duties except periodic histogram and scaler readout.

There are numerous improvements and adaptations of this basic TD-μSR arrangement, of course, as well as several entirely different electronics setups for time-integral (I)-μSR techniques.

4.2.3 μSR Spectrometers The local facility almost always supplies a choice of several μSR spectrometers, whose function is to control the magnetic field at the sample (usually by means of three orthogonal sets of Helmholtz coils, one of which can generate a large main field) and to provide mounting brackets for counters, cryostats, or other paraphernalia. Often a standard set of counters is an integral part of the spectrometer, as well. There is such a huge variety of such spectrometers that it would be pointless to try to describe a "typical" version in any detail.

4.2.4 Target Vessels and Cryostats Since the muon's initial polarization is independent of the state of its environment, μSR samples may be gases, liquids, or solids at any temperature, pressure, or magnetic or electric field. This versatility is reflected in the variety of sample environments (target vessels) used in μSR experiments, ranging from large "gas cans" (for studying hydrogen gas at low pressure and high temperature) to miniaturized cryostats (for studying small crystals at low temperature and high magnetic field). The most common use of μSR is currently in low-temperature condensed-matter physics, so that most μSR spectrometers have general-purpose cryostats built in (but removable). However, in general, the user is responsible for the sample environment.

4.3 μSR Software

Because μSR data is always intimately connected with computers (indeed, one needs computer graphics just to see what one is measuring) and because there is such a premium on efficient use of beam time, which, in turn, demands that one know what one has learned from one run before setting the independent variables for the next (it is not unusual for a μSR experimenter to generate, while a spectrum is still accumulating, the figures that will be published later to describe it), one of the most important components of any μSR user facility is a suite of process-control, data-presentation, and data-analysis software to provide the user with all the mathematical tools described in this article, and more. Since implementation of such utilities is nontrivial, some effort is now being made to define international standards (data formats, process control, data analysis, *etc.*), so that the μSR community can cooperate on such software development and share the benefits thereof.

5. SPECULATIONS

The felicitous history of μSR has reached a point of crisis in the 1990s: Accelerator facilities rarely capture the fascination of the subatomic physicists who build them for more than about two decades, and the meson factories have reached that difficult age. While μSR facilities require only modest investments for qualitative improvements using existing accelerators, the big machines themselves need expensive upgrades to maintain world-class status as subatomic facilities. Thus, the future of μSR remains inextricably linked to that of subatomic physics—more specifically, to that of high-intensity, intermediate-energy, fixed-target hadron accelerators. This has always been the case and will remain so until someone invents a tabletop device capable of generating muon beams. In spite of the implied mutual liability, the author has always considered this unlikely symbiosis to be one of the greatest charms of μSR.

Works Cited

Abragam, A. (1984), *C.R. Acad. Sci. Ser. B* **299**, 95–99.

Aeppli, G., Ansaldo, E. J., Brewer, J. H., Cava, R. J., Kiefl, R. F., Kreitzman, S. R., Luke, G. M., Noakes, D. R. (1987), *Phys. Rev. B* **35**, 7129–7132.

Bowen, T. (1985), *Phys. Today* **38** (7), 22–34.

Brewer, J. H., Crowe, K. M., Gygax, F. N., Johnson, R. F., Patterson, B. D., Fleming, D. G., Schenck, A. (1973), *Phys. Rev. Lett.* **31**, 143–146.

Brewer, J. H., Crowe, K. M., Gygax, F. N., Johnson, R. F., Fleming, D. G., Schenck, A. (1974), *Phys. Rev. A* **9**, 495–507.

Brewer, J. H. (1981), *Hyperfine Int.* **8**, 831–834.

Brewer, J. H. (1984a), *Hyperfine Int.* **17–19**, 879–884.

Brewer, J. H. (1984b), *Hyperfine Int.* **17–19**, 873–878.

Brewer, J. H., Kreitzman, S. R., Noakes, D. R., Ansaldo, E. J., Harshman, D. R., Keitel, R. (1986), *Phys. Rev. B* **33**, 7813–7816.

Camani, M., Gygax, F. N., Klempt, E., Rüegg, W., Schenck, A., Schilling, H., Schulze, R., Wolfe, H. (1978), *Phys. Lett. B* **77**, 326–330.

Celio, M. (1986), *Phys. Rev. Lett.* **56**, 2720–2723.

Celio, M. (1987), *Helv. Phys. Acta* **60**, 600–610.

Combley, F. H., Farley, F. J. M., Picasso, E. (1981), *Phys. Rep.* **68**, 93–119.

Cronin, J. W. (1981), *Rev. Mod. Phys.* **53**, 373–383.

Dalmas de Réotier, P., Yaouanc, A., Meshkov, S. V. (1992), *Phys. Lett. A* **162**, 206–212.

Fermi, E., Teller, E. (1947), *Phys. Rev.* **72**, 399–408.

Fitch, V. L. (1981), *Rev. Mod. Phys.* **53**, 367–371.

Fleming, D. G., Senba, M. (1992), in: T. Yamazaki, K. Nakai, N. Nagamine (Eds.), *Perspectives of Meson Science*, Amsterdam: North-Holland, pp. 219–264.

Flynn, C. P., Stoneham, A. M. (1970), *Phys. Rev. B* **1**, 3966–3978.

Friedman, J. I., Telegdi, V. L. (1957), *Phys. Rev.* **105**, 1681–1682.

Garwin, R. L., Lederman, L. M., Weinrich, M. (1957), *Phys. Rev.* **105**, 1415–1417.

Gygax, F. N., Hintermann, A., Rüegg, W., Schenck, A., Studer, W., Van der Wal, A. J. (1984), *J. Less-Common Met.* **101**, 97–113.

Hayano, R. S., Uemura, Y. J., Imazato, J., Nishida, N., Yamazaki, T., Kubo, R. (1979), *Phys. Rev. B* **20**, 850–859.

Hitti, B., Kreitzman, S. R., Estle, T. L., Lichti, R. L., Chow, K. H., Schneider, J. W. (1994), *Hyperfine Int.*, **85–87**, in press.

Holzschuh, E. (1983), *Phys. Rev. B* **27**, 102–111.

Hughes, V. W. (1988), *Phys. Scr.* **T22**, 111–118.

Kadono, R. (1990), *Hyperfine Int.* **64**, 615–633.

Kadono, R. (1992), in: T. Yamazaki, K. Nakai, K. Nagamine (Eds.), *Perspectives of Meson Science*, Amsterdam: North-Holland, pp. 113–116.

Kagan, Yu., Klinger, M. I. (1974), *J. Phys. C* **7**, 2791–2807.

Kagan, Yu., Prokof'ev, N. V. (1990), *Phys. Lett. A* **150**, 320–326.

Kagan, Yu., Prokof'ev, N. V. (1991), *Phys. Lett. A* **159**, 289–294.

Kagan, Yu., Prokof'ev, N. V. (1992), in: A. J. Leggett, Yu. M. Kagan (Eds.), *Modern Problems in Condensed Matter Science*, Amsterdam: North-Holland.

Kehr, K. W., Honig, G., Richter, D. (1978), *Z. Phys. B* **32**, 49–58.

Keller, H., Kiefl, R. F., Baumeler, Hp., Kündig, W., Patterson, B. D., Imazato, K., Nishiyama, K., Nagamine, K., Yamazaki, T., Grynszpan, R. I. (1987), *Phys. Rev. B* **35**, 2008–2014.

Kiefl, R.F., Kreitzman, S. R. (1992), in: T. Yamazaki, K. Nakai, K. Nagamine (Eds.), *Perspectives of Meson Science*, Amsterdam: North-Holland, pp. 265–292.

Kiefl, R. F., Kreitzman, S. R., Celio, M., Keitel, R., Luke, G. M., Brewer, J. H., Noakes, D. R., Percival, P. W., Matsuzaki, T., Nishiyama, K. (1986), *Phys. Rev. B* **34**, 681–684.

Kiefl, R. F., Brewer, J. H., Kreitzman, S. R., Luke, G. M., Riseman, T. M., Estle, T. L., Celio, M., Ansaldo, E. J. (1988), in: G. Ferenczi (Ed.), *Proceedings of the 15th International Conference on Defects in Semiconductors, Budapest, Mater. Sci. Forum* **38–41**.

Kiefl, R. F., Kadono, R., Brewer, J. H., Luke, G. M., Yen, H. K., Celio, M., Ansaldo, E. J. (1989), *Phys. Rev. Lett.* **62**, 792–795.

Kiefl, R. F., *et al.* (1990), *Phys. Rev. Lett.* **64**, 2082–2085.

Kiefl, R. F., *et al.* (1992), *Phys. Rev. Lett.* **69**, 2005–2008.

Kondo, J. (1986), *Hyperfine Int.* **31**, 117–133.

Kondo, J. (1992), in: T. Yamazaki, K. Nakai, K. Nagamine (Eds.), *Perspectives of Meson Science*, Amsterdam: North-Holland, pp. 137–160.

Kreitzman, S. R. (1990), *Hyperfine Int.* **65**, 1055–1070.

Kreitzman, S. R., Brewer, J. H., Harshman, D. R., Keitel, R., Williams, D. Ll., Crowe, K. M., Ansaldo, E. J. (1986), *Phys. Rev. Lett.* **56**, 181–184.

Kreitzman, S. R., Williams, D. Ll., Kaplan, N., Kempton, J. R., Brewer, J. H. (1988), *Phys. Rev. Lett.* **61**, 2890–2893.

Kubo, R., Toyabe, T. (1966), in: R. Blinc (Ed.), *Magnetic Resonance and Relaxation*, Amsterdam: North-Holland, p. 810.

Kuno, Y., Imazato, K., Nishiyama, K., Nagamine, K., Yamazaki, T., Minamisono, T. (1984), *Phys. Lett. B* **148**, 270–284.

Lee, T. D., Yang, C. N. (1956), *Phys. Rev.* **104**, 254–258.

Lee, T. D., Yang, C. N. (1957), *Phys. Rev.* **105**, 1671–1675.

Luke, G. M., Brewer, J. H., Kreitzman, S. R., Noakes, D. R., Celio, M., Kadono, R., Ansaldo, M. (1991), *Phys. Rev. B.* **43**, 3284–3297.

Michel, L. (1950), *Proc. Phys. Soc. London, Sect. A* **63**, 514–531.

Möslang, A., Graf, H., Balzer, G., Recknagel, E., Weidinger, A., Wichert, Th., Grynszpan, R. I. (1983), *Phys. Rev. B* **27**, 2674–2681.

μSR Newsletter (1974), **1**, (15 January).

Niedermayer, Ch., Bernhard, C., Binninger, U., Glückler, H., Tallon, J. L., Ansaldo, E. J., Budnick, J. I. (1993), *Phys. Rev. Lett.* **71**, 1764–1767.

Nishiyama, K. (1992), in: T. Yamazaki, K. Nakai, K. Nagamine (Eds.), *Perspectives of Meson Science*, Amsterdam: North-Holland, pp. 199–218.

Patterson, B. D. (1988), *Rev. Mod. Phys.* **60**, 69–159.

Percival, P. W., Kiefl, R. F., Kreitzman, S. R., Garner, D. M., Cox, S. F. J., Luke, G. M., Brewer, J. H., Nishiyama, K., Venkateswaran, K. (1987), *Chem. Phys. Lett.* **133**, 465–470.

Petzinger, K. G. (1982), *Phys. Rev. B* **26**, 6530–6546.

Primakoff, H. (1975), in: V. W. Hughes, C. S. Wu (Eds.), *Muon Physics*, Vol. 2, *Weak Interactions*, New York: Academic, pp. 3–48.

Rasetti, F. (1944), *Phys. Rev.* **66**, 1–5.

Riseman, T. M., Brewer, J. H. (1990), *Hyperfine Int.* **63**, 249–252.

Roduner, E. (1988), *The Positive Muon as a Probe in Free Radical Chemistry*, Lecture Notes in Chemistry, Vol. 49, Berlin: Springer.

Sachs, A. M., Sirlin, A. (1975), in: V. W. Hughes, C. S. Wu (Eds.), *Muon Physics*, Vol. 2, *Weak Interactions*, New York: Academic, pp. 49–81.

Schneider, J. W., *et al.* (1990), *Phys. Rev. B* **41**, 7254–7257.

Schneider, J. W., *et al.* (1992), *Phys. Rev. Lett.* **68**, 3196–3199.

Seeger, A., Schimmele, L. (1992), in: T. Yamazaki, K. Nakai, K. Nagamine (Eds.), *Perspectives of Meson Science*, Amsterdam: North-Holland, pp. 293–382.

Stamp, P. C. E., Zhang, C. (1991), *Phys. Rev. Lett.* **66**, 1902–1905.

Stoker, D., Balke, B., Carr, J., Gidal, G., Jodido, A., Shinsky, K. A., Steiner, H. M., Strovink, M., Tripp, R. D., Gobbi, B., Oram, C. J. (1985), *Phys. Rev. Lett.* **54**, 1887–1890.

Storchak, V., Brewer, J. H., Hardy, W. N., Johnston, S., Kreitzman, S. R., Morris, G. D. (1994), *Phys. Rev. Lett.* **72**, 3056–3059.

Uemura, Y. J. (1992), in: T. Yamazaki, K. Nakai, K. Nagamine (Eds.), *Perspectives of Meson Science*, Amsterdam: North-Holland, pp. 87–112.

Uemura, Y. J., Harshman, D. R., Senba, M., Ansaldo, E. J., Murani, A. P. (1984), *Phys. Rev. B* **30**, 1606–1608.

Uemura, Y. J., *et al.* (1989), *Phys. Rev. Lett.* **62**, 2317–2320.

Uemura, Y. J., Keren, A., Le, L. P., Luke, G. M., Wu, W. D. (1994), *Hyperfine Int.*, **85–87**, in press.

Winston, R. (1963), *Phys. Rev.* **129**, 2766–2785.

Wu, C. S., Ambler, E., Hayward, R. W., Hoppes, D. D., Hudson, R. P. (1957), *Phys. Rev.* **105**, 1413–1415.

Yamada, K., Sakurai, A., Miyazima, S., Hwang, H. S. (1986), *Prog. Theor. Phys.* **75**, 1030–1043.

Yamanaka, T., Ishikawa, T., Ohtake, S., Iwasaki, M., Akiba, Y., Yamazaki, T., Hayano, R. S. Schnetzer, S. R., Taniguchi, T. (1984), *Hyperfine Int.* **17–19**, 901–906.

Yen, H. K. (1988). M.Sc. thesis, University of British Columbia.

Further Reading

Several excellent reviews of the *applications* of μSR have been written in recent years. The following are specially recommended:

Schenck, A. (1986), *Muon Spin Spectroscopy: Principles and Applications in Solid State Physics*, Bristol, UK: Adam Hilger.

Cox, S. F. J. (1987), *J. Phys, C* **20**, 3187–3319.

The Proceedings of all six International Conferences on μSR to date have been published in *Hyperfine Interactions* as special volumes **6** (1979), **8** (1981), **17–19** (1984), **31** (1986), **63–65** (1990), and **85–87** (1994). These (and any subsequent Proceedings) are the ideal sources for state-of-the-art developments in μSR.

MUONIC, MESONIC, AND BARYONIC ATOMS AND MOLECULES

F. J. HARTMANN, *Physik-Department, Technische Universität München, Garching, Germany*

INTRODUCTION

Muonic, mesonic, and baryonic atoms, usually also called exotic atoms, are atoms in which, besides electrons, an elementary particle of *negative* charge and sufficiently long lifetime is bound in an orbit around the nucleus. Table 1 gives information on the elementary particles for which exotic atoms have been observed. In *muonic atoms*, the exotic particle is a muon; in *mesonic* atoms, it is a meson (pion, kaon); and in *baryonic* atoms, it is a baryon (antiproton, sigma hyperon).

The existence of the respective exotic atoms usually was proven shortly after the discovery of the particle itself. muonic atoms, e.g., were postulated by Fermi *et al.* (1947) to explain observations by Conversi *et al.* (1947) that the disappearance rate of muons strongly depended on the element in which the muon was stopped. Characteristic x rays from pionic atoms were first observed in 1952 by Camac *et al.* (1952). Kaonic atoms were first seen at Argonne National Laboratory in 1965 (Burleson, 1965), and antiprotonic atoms were discovered in 1970 at CERN (Bamberger *et al.*, 1970), but only after a special antiproton beam line had been set up. Nowadays, muonic and

3-527-28133-9/94/$5.00 + .50

Table 1. Elementary particles that form exotic atoms (Particle Data Group, 1992). The symbols ν_e and ν_μ denote electron neutrino and muon neutrino, respectively. Antiparticles are marked with a bar.

Particle	Symbol	Mass (MeV/c^2)	Lifetime	Spin	Year of discovery	Main decay modes
Muon	μ^-	105.65839(3)	2.197 μs	1/2	1937	$\mu^- \to e^- + \bar{\nu}_e + \nu_\mu$
Pion	π^-	139.5679(7)	26.03 ns	0	1952	$\pi^- \to \mu^- + \bar{\nu}_\mu$
Kaon	K^-	493.646(9)	12.37 ns	0	1947	$K^- \to \mu^- + \bar{\nu}_\mu$ $\to \pi^- + \pi^0$
Antiproton	$\bar{p}$	938.2723(3)	Stable	1/2	1955	
Sigma hyperon	Σ^-	1197.43(6)	147.9 ps	1/2	1968	$\Sigma^- \to n + \pi^-$

pionic atoms may be easily obtained at so-called meson factories, proton accelerators with energies in the 600-MeV to 1-GeV region and a proton current of about 1 mA. Such factories are operated at Paul Scherrer Institut (PSI), Villigen, Switzerland, at the Los Alamos Meson Physics Facility, Los Alamos, New Mexico, U.S.A. (LAMPF), and at TRIUMF, Vancouver, British Columbia, Canada. Antiprotons to form antiprotonic atoms are produced abundantly in a dedicated antiproton facility [Low-Energy Antiproton Ring (LEAR)] at CERN, Geneva, Switzerland.

1. ENERGY LEVELS IN MUONIC, MESONIC, AND BARYONIC ATOMS

The mass of the elementary particle forming an exotic atom is at least 207 times as large as the electron mass; hence, the radii of the exotic-particle orbits in such an atom are much smaller and the binding energies much larger than for an electron in an ordinary atom. Already for rather large excitations, the exotic particle is inside the electron cloud, and the exotic-particle–nucleus system is hydrogenlike, with the electromagnetic interaction between the exotic particle and the nucleus screened only slightly by that part of the electron cloud that is between particle and nucleus. Therefore, all the effects known from hydrogenlike electronic atoms, like fine structure, hyperfine structure, etc., may show up in exotic atoms; they are, however, mostly of completely different importance. As in the hydrogen atom, the bound states in exotic atoms usually are characterized by the quantum numbers n (principal quantum number), l (orbital angular momentum quantum number), and j (total angular momentum quantum number). As a first guess, the binding energy of the exotic particle in an atom with atomic number Z may be calculated from the Bohr formula,

$$E_n = \tfrac{1}{2}\mu c^2 [\alpha Z/n]^2, \quad (1)$$

derived long before a quantum-mechanical treatment of the hydrogenlike atom was possible. The radius of the exotic particle's orbit comes out to be

$$r_n = a_0(m_e/\mu)(n^2/Z), \quad (2)$$

while the velocity is given by

$$v_n = v_0 Z/n. \quad (3)$$

Here μ, the reduced mass of the muon, meson, or baryon, reads

$$\mu = m_x M/(m_x + M), \quad (4)$$

with m_x and M the exotic-particle mass and the mass of the nucleus, respectively. The use of μ instead of m_x takes the movement of the exotic particle and nucleus around a common center of gravity into account. The electron mass is denoted by m_e, and α is the fine-structure constant ($\alpha = 1/137.036$); $a_0 = 52.92$ pm and $v_0 = \alpha c = 2.188 \times 10^6$ m/s are Bohr radius and velocity, respectively.

As can be easily seen from Eq. (3), the motion of the exotic particle in the lowest levels of heavy exotic atoms is relativistic. Hence, a relativistic wave equation has to be used to describe the interaction of the exotic particle and the nucleus. For particles with half-integer spin (like muons or antiprotons), this is the Dirac equation. The energy eigenvalues (total energy) $E(n,l,j)$ for a Coulomb ($1/r$) potential (valid for a pointlike nucleus) are given by

$$E(n,l,j) = \mu c^2 \left[1 + \left(\frac{Z\alpha}{n - |\kappa| + \sqrt{\kappa^2 - Z^2\alpha^2}} \right)^2 \right]^{-1/2}, \tag{5}$$

with $j = l \pm \frac{1}{2}$ and $|\kappa| = j + \frac{1}{2}$. Hence, for each l (except $l = 0$), there exist two states with different j and different binding energy; the l state has a fine structure.

For spin-0 particles (pions and kaons) the corresponding wave equation is the Klein–Gordon equation, sometimes also called the relativistic Schrödinger equation. Here the energy eigenvalues for a Coulomb potential are given by

$$E(n,l) = \mu c^2 \left[1 + \left(\frac{Z\alpha}{n - l - \frac{1}{2} + \sqrt{(l + \frac{1}{2})^2 - Z^2\alpha^2}} \right)^2 \right]^{-1/2}. \tag{6}$$

In the early 1960s, large discrepancies were found between the binding energies in heavy exotic atoms derived from the first experiments on the characteristic x rays from exotic atoms and the energy values calculated from Eqs. (5) and (6), respectively: In a more realistic picture, the finite size of the nucleus has to be taken into account, as the exotic particle is very near to the nucleus. The nuclear charge distribution is most widely assumed to be a Fermi distribution, with the radial charge density $\rho(r)$ given by

$$\rho(r) = \frac{N}{1 + \exp[4 \ln 3(r - a)/t]}.$$

In this equation, $N \approx (3Z/4\pi)[a^3 + (\pi t/4 \times \ln 3)^2 a]^{-1}$ is determined by the condition

$$4\pi \int_0^\infty \rho(r) r^2 \, dr = Z;$$

a is the half-density radius and t the skin thickness of the nucleus, i.e., the distance between the points where the charge distribution reaches 10 and 90%, respectively, of the maximum value.

There are other effects besides the nuclear size that influence the exotic-particle binding energy:

1. The vacuum polarization, i.e., the generation of virtual e^-e^+ (and $\mu^-\mu^+$) pairs in the Coulomb field of the nucleus, may be explained by quantum electrodynamics (QED) from the fact that even the vacuum has, under certain conditions, the properties of a dielectric medium. Vacuum polarization leads to an increase of the binding energy, which for elements in the lead region reaches about 70 keV.
2. The self-energy due to the interaction of the exotic particle with its own radiation field may also be explained by QED. This effect is less important than in electronic hydrogen.
3. The polarization of the nucleus is brought about by attraction of the protons in the nucleus by the negatively charged exotic particle.
4. The screening of the nuclear charge by that part of the electron cloud that is between exotic particle and nucleus diminishes the nuclear charge seen by the exotic particle and, hence, its binding energy.

As in electronic atoms, also in exotic atoms a nonspherical nuclear-charge distribution leads to an electric *hyperfine* splitting (hfs) of the levels. In most cases, only the interaction of the quadrupole moment of the nucleus with the exotic particle plays a role. The hfs reaches values in the keV region. For muonic Au, e.g., the quadrupole hfs splitting of the $2p_{3/2} \rightarrow 1s_{1/2}$ transition is nearly 20 keV.

Finally, again similar to the electronic atom, in an exotic atom with a spin-$\frac{1}{2}$ particle (like the muon) and a nucleus of nonzero spin, the interaction between nuclear and particle magnetic moment has to be taken into account. The resulting *magnetic* hfs of the levels reaches values up to a few keV, in muonic ^{209}Bi, e.g., roughly 15 keV for the magnetic hfs of the $2p_{1/2} \rightarrow 1s_{1/2}$ transition. Here the finite size of the nucleus (and of its magnetization distribution) leads to a reduction of the magnetic hfs compared to the values for a point nucleus (Bohr–Weisskopf effect).

For hadronic atoms, the strong interaction of the exotic particle with the nucleus results in a measurable shift of the lowest level populated before nuclear capture and also in a level broadening due to the very short lifetime of this level.

Typical level energies and corrections for some muonic atoms are given in Table 2; here the level shifts produced by all the electrodynamic effects just described are not hidden by strong-interaction shifts (cf. Sec. 4).

2. FORMATION OF EXOTIC ATOMS

When an energetic exotic particle enters a target (with a kinetic energy in the MeV region), it is slowed down by interaction with the target electrons. Down to velocities of the order of magnitude of the Bohr velocity (i.e., at energies of around 2.8 keV for muons or 25 keV for antiprotons, to give two examples), the slowing down is rather well described by the Bethe formula (Bethe, 1930). At lower energies, fewer and fewer atomic electrons contribute to stopping because they are too tightly bound to the nucleus to be knocked out of the atom by the exotic particle: So-called shell corrections have to be applied to the Bethe formula. Formation of exotic atoms takes place only after the negative elementary particle has been slowed down to energies around and below 100 eV. It is also commonly accepted that this *Coulomb capture* (called this to distinguish the process from nuclear capture mediated by strong interaction) takes place by the Auger effect: The exotic particle goes over from a free state to a bound orbit around the nucleus, and an electron is emitted that also takes the largest fraction of the energy released during the capture process. It is not so well known from which shell this Auger electron stems and which is the first bound orbit of the exotic particle in the atom. There are, however, indications that this first orbit has a rather large n. *Muonic* atoms with $n \geq 20$ have been detected via the emitted characteristic muonic x rays; furthermore, laser-stimulated de-excitation of a long-lived state with $n = 39$ and $l = 35$ in *antiprotonic* He, identified by the transition energy, established very recently that levels with very large n are populated by Coulomb capture in He.

The probability $p(l)$ for the population of states with angular momentum l by capture was for a long time assumed to follow the relation

$$p(l) \propto 2l + 1,$$

which means that each of the $2l + 1$ l-substates with different magnetic quantum number m ($m = -l, -l + 1, -l + 2, \ldots, l - 2, l - 1, l$) is populated with equal probability. A hand-waving argument for $p(l)$ to be roughly proportional to l is the following: If we assume the exotic particle to hit the atom

Table 2. Electromagnetic binding energies and energy corrections in muonic atoms (Barrett, 1977). The difference between binding energy and point-nucleus energy is mostly due to the finite-size effect. VP: vacuum-polarization correction; EN: corrections for electron screening and nuclear polarization; LS: correction for vacuum polarization by generation of virtual $\mu^-\mu^+$ pairs and for the self-energy of the muon.

Exotic atom	Level nl_j	Binding energy (keV)	Point nucleus energy (keV)	Corrections (eV)		
				VP	EN	LS
μ^7Li	$1s_{1/2}$	24.95	24.92	61	0	0
μ^{12}C	$1s_{1/2}$	100.37	100.38	398	6	−4
μ^{24}Mg	$1s_{1/2}$	397.59	403.97	2213	56	−46
μ^{56}Fe	$1s_{1/2}$	1731.57	1915.29	11 828	553	−420
μ^{56}Fe	$2p_{1/2}$	481.36	479.91	1882	19	3
μ^{56}Fe	$2p_{3/2}$	477.21	475.55	1823	19	−5
μ^{208}Pb	$1s_{1/2}$	10 596.51	20 992.36	67 154	5076	−2950
μ^{208}Pb	$2p_{1/2}$	4813.78	5385.41	32 436	1130	−331
μ^{208}Pb	$2p_{3/2}$	4628.41	4837.25	29 891	1116	−664
μ^{208}Pb	$4f_{5/2}$	1201.19	1197.37	3791	16	13
μ^{208}Pb	$4f_{7/2}$	1191.99	1188.30	3682	18	−9

with velocity v and impact parameter (i.e., distance from the center of the atom) b, it brings in an angular momentum $L = mbv$. As $L \approx \hbar l$ and the probability for the exotic particle to hit the atom at impact parameter b is proportional to the area of a thin ring of inner radius b, $p(l) \propto l$ follows. Deviations from the "statistical" distribution $p(l) \propto 2l + 1$ have been found for muonic atoms, with the distribution rather flat for the transition elements and also for elements in the rare-earth region.

Coulomb capture of exotic particles may be treated either by use of quantum mechanics or in a semiclassical approximation. The complex many-body problem of the exotic-particle interaction with the nucleus and all the atomic electrons has been too complicated up to now to be solved satisfactorily by quantum mechanics, except for the simplest atoms like hydrogen or helium. Semiclassical calculations, although certainly more superficial, have brought more understanding. They are the more adequate as the first bound orbits in exotic atoms have large n, and, hence, as a consequence of the correspondence principle, the results of quantum-mechanical and classical calculations should coincide.

In all semiclassical theories, the following assumptions are made:

1. The exotic particle moves along a classical trajectory in the electric field of the nucleus that is screened by the atomic electrons;
2. it loses energy by a frictional force proportional to its velocity $\mathbf{F} \propto -\mathbf{v}$, which is brought about by the Coulomb interaction of the particle with atomic (and valence) electrons;
3. the particle is assumed to be captured as soon as its total energy falls below zero during the passage through the atom.

The first model of this kind was developed by Fermi and Teller (1947). In their calculations, the probability for capture was simply assumed to be proportional to the rate of energy loss at very low energies in the electron gas of the respective atom. The per-atom probability $P(Z)$ for Coulomb capture came out in this model to be proportional to the atomic number Z, $P(Z) \propto Z$. In more recent calculations, a less pronounced Z dependence was found, with $P(Z)$ even oscillating with Z at periods correlated with those of the periodic table (see, e.g. Daniel, 1980, and Sec. 5.2).

3. DE-EXCITATION OF THE EXOTIC ATOM

The cascade of the exotic particle down the ladder of excited levels proceeds by a number of processes: emission of radiation, Auger effect, Stark mixing, and Coulomb and chemical de-excitation. For exotic hydrogen atoms, transfer to heavier atoms may interfere with the cascade. All processes are now described in more detail.

3.1 Radiative Transitions

In a radiative transition, the exotic atom de-excites, as is already indicated by the name, by emission of quantum radiation. The probability for an electric dipole (E1) transition—the normal transition in exotic as in normal atoms—from level (n_i, l_i) to level (n_f, l_f) is given by a formula derived from that valid for hydrogenlike electronic atoms (Bethe and Salpeter, 1977):

$$\Gamma_{\text{Rad}} = \frac{4}{3} \frac{\alpha c}{a_0} \left[\frac{m_e}{\mu}\right]^2 \left[\frac{\Delta E}{\alpha m_e c^2}\right]^3 \frac{\max(l_i, l_f)}{2l_i + 1} I^2_{i \to f}. \tag{7}$$

Here $I_{i \to f}$ is the matrix element

$$I_{i \to f} = \int_0^\infty R^*(n_i, l_i) r^3 R(n_f, l_f)\, dr, \tag{8}$$

with $R(n, l)$ the normalized wave functions for the hydrogenlike exotic atom and ΔE the transition energy. Only transitions with $|\Delta l| = |l_i - l_f| = 1$ are allowed. Because of the ΔE^3 factor, transitions with maximum possible change $\Delta n = n_i - n_f$ are favored, although $I_{i \to f}$ becomes smaller with increasing Δn. Radiative transitions constitute the main de-excitation mechanism for the late stage of the exotic-atom cascade. Besides E1 transitions, also E2 (electric quadrupole) transitions have been observed, but this only in the Lyman series, i.e., for transitions to the 1s level, in heavy atoms.

3.2 Auger Transitions

In an Auger transition, the exotic atom de-excites by emission of an atomic electron, either from the host atom itself (*internal* Auger effect) or from an atom the exotic atom collides with (*external* Auger effect). The expressions for the rate Γ_{Auger} for the internal Auger effect are rather complicated (for details, see Akylas and Vogel, 1978). A formula given by Ferrell (1960) connects Γ_{Auger} with the rate Γ_{Rad} for a radiative dipole transition between the same levels:

$$\frac{\Gamma_{\text{Auger}}}{\Gamma_{\text{Rad}}} = \frac{\sigma^{Z-1}(\Delta E)}{(Z-1)^2\sigma_{\text{T}}},$$

and may serve for a fast calculation of Γ_{Auger}. Here $\sigma^{Z-1}(\Delta E)$ is the cross section for the photoeffect at energy ΔE in the element with atomic number $Z - 1$, and σ_{T} denotes the Thomson cross section, i.e., the classical cross section for scattering of an electromagnetic wave on a free electron, $\sigma_{\text{T}} = 8\pi r_e^2/3 = 0.665 \times 10^{-28}\ \text{m}^2$ with $r_e = 2.818$ fm the classical electron radius.

Besides dipole transitions, monopole and quadrupole transitions play a role, however minor; changes in the angular momentum with $|\Delta l| \leq 2$ are possible. The Auger effect favors low-energy transitions, especially those with small Δn in the first stages of the exotic-atom cascade.

Figure 1 shows rates for radiative and Auger transitions between circular orbits for initial levels with $n \leq 6$. The rates are orders of magnitude larger than those for hydrogen-like electronic atoms.

The *external* Auger effect is most important for exotic hydrogen, as there the only electron has been ejected from the host atom during Coulomb capture; hence, the internal Auger effect can no longer take place.

3.3 Stark Mixing

Stark mixing is important for the cascade in exotic hydrogen and helium atoms because these systems are small and neutral or only singly charged and can closely approach other atoms in the target. Inside the electron cloud, they feel a strong electric field $\mathbf{E}$ from the nucleus that destroys the spherical symmetry of the exotic atom, as is well known from the Stark effect in electronic atoms. A dipole moment $\boldsymbol{\mu}$ is induced in the exotic atom by the electric field; by interaction of the field with this dipole moment, different angular-momentum states are mixed, as the angular momentum is no longer conserved; still conserved, however, is its projection in the direction of the electric field (characterized by the magnetic quantum number m).

The rate of Stark transitions from a state with quantum numbers n, l, and m to a state with quantum numbers n, l', and m' is proportional to the matrix element $M_{l,l'}$ for the interaction energy $\boldsymbol{\mu} \cdot \mathbf{E}$. A calculation of this

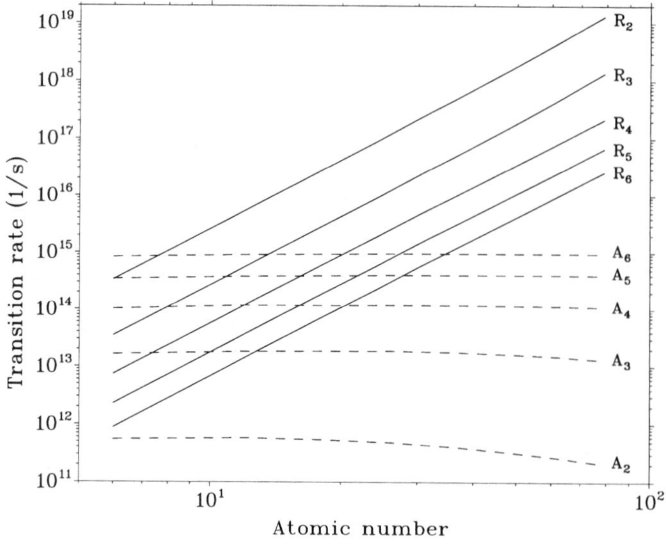

FIG. 1. Radiative (R_2 to R_6) and Auger (A_2 to A_6) rates in muonic atoms for the transitions $(n, l = n - 1) \rightarrow (n - 1, l = n - 2)$, $n = 2$ to 6, as a function of the atomic number.

matrix element with hydrogenlike wave functions leads to nonzero results only for $m = m'$ and $l = l' \pm 1$, with values given, e.g., by

$$M_{l,l+1} = \frac{3}{2} a_0 \frac{m_e}{\mu} e|\mathbf{E}|n \sqrt{\frac{(n^2 - l^2)(l^2 - m^2)}{4l^2 - 1}}.$$

To estimate the influence of the Stark mixing for exotic hydrogen atoms in hydrogen (certainly the most important case), one may approximate the electric field that the exotic atom feels during the collision with a hydrogen atom by a constant field that the atom experiences at the site of the Bohr orbit, $|\mathbf{E}| \approx e/4\pi\epsilon_0 a_0^2 = 5.14 \times 10^{11}$ V/m. For $\mu^- p$ and $\pi^- p$ atoms, e.g., the transition rate then becomes $\omega_{\text{Stark}} = M_{l,l'}/\hbar \approx 2 \times 10^{14} n\sqrt{n^2 - 1}\ \text{s}^{-1}$ if we take $l = 1$ and $m = 0$. The characteristic duration τ_{Stark} of a collision is roughly $\tau_{\text{Stark}} = a_0/v$, with a_0 standing for the dimension of the atom and v the exotic-atom velocity; thus, at a μp energy of 1 eV, $\tau_{\text{Stark}} \approx 5 \times 10^{-14}$ s and $\omega_{\text{Stark}}\tau_{\text{Stark}} \gg 1$ follows: There will be many Stark transitions between different l states with the same n during one collision (complete Stark mixing), and the angular-momentum states will be completely mixed already after one collision. The exotic-hydrogen atom will experience, however, such a large electric field only for central collisions with impact parameters smaller than a radius ρ_{lim} that is of the order of magnitude of the radius of the hydrogen atom.

Figure 2 compares radiative, Auger, and Stark transitions in pionic hydrogen for a gaseous target with 0.001 of the density of liquid hydrogen.

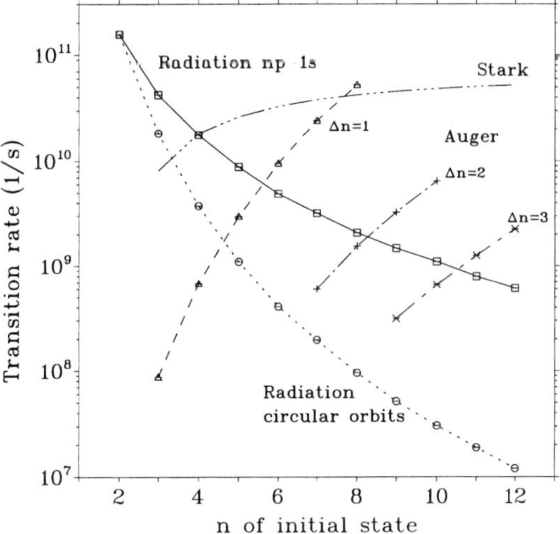

FIG. 2. Rates for radiative, Auger, and Stark transitions in pionic hydrogen for an atomic density that is one thousandth of the density of liquid hydrogen (Markushin, 1990).

3.4 Transfer Processes

Transfer is the reaction

$$X^- Z_0 + Z \rightarrow Z_0 + X^- Z,$$

with X^- an exotic particle. It plays a role mainly for exotic hydrogen atoms ($Z_0 = 1$), as here the *neutral* $X^- h$ system (where h stands for a proton, deuteron, or triton, respectively) can penetrate the electron shell of the atom Z (in the following, only this type of transfer will be discussed). We can simplify the problem if we assume the $X^- h$ atom to be in the ground ($1s$) state already and to have traversed the electron cloud of atom Z. The question then is the following: What is the probability for a transfer of X^- from the $1s$ orbit of the $X^- h$ atom to a bound orbit in $X^- Z$ while the nuclei h and Z come near and separate again? As has been shown, the probability for such a transfer becomes especially large if there are so-called crossings in a diagram which displays the energy of the systems ($X^- h(1s) + Z$) and ($h + X^- Z$), respectively, as a function of the distance R between h and the nucleus with charge Z. Such a diagram, also called correlation diagram and well known in molecular-orbital theory, is shown in Fig. 3 for the muonic proton-fluorine system. At the crossings, the energies of these two systems are equal.

The energy $\epsilon_{1s}^{(h)}$ for one limiting case, $X^- h$ in the ground state and the atom Z a distance R apart (point A in Fig. 3), is given approximately by

$$\epsilon_{1s}^{(h)} \approx -\frac{m_p}{1 + m_p}\left(\frac{1}{2} + \frac{9Z^2}{4R^4}\right),$$

with ϵ, R, and the proton mass m_p given in muonic Rydberg units, $R_{h_\mu} = 5.62$ keV, muonic Bohr units, $a_\mu = 256$ fm, and units of the muon mass, respectively. The second term in this equation is a (weak) van der Waals shift. In the second limiting case, X^- in an orbit with

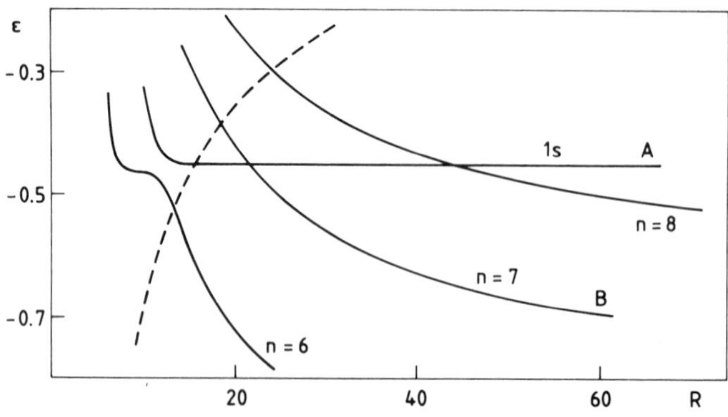

FIG. 3. Correlation diagram for the system $\mu + p +$ fluorine (Holzwarth and Pfeiffer, 1975). Radius R and total energy ϵ in muoatomic units (unit radius 256 fm, unit energy 5.6 keV). 1s: energy of the 1s level in the μp system; $n = 6$, $n = 7$; $n = 8$: energy of the levels with the respective n in the muonic fluorine atom.

principal quantum number n around nucleus Z and h some distance away (point B in Fig. 3, as an example), the energy ϵ_n^Z is given approximately by

$$\epsilon_n^Z \approx -\frac{m_Z}{1 + m_Z}\frac{Z^2}{2n^2} + \frac{Z - 1}{R},$$

with the same units and m_Z again given in muon-mass units. Here the second term corresponds to a Coulomb repulsion of the exotic atom (charge $Z - 1$) by the proton. Naturally, the two nuclei h and Z cannot come closer than a distance determined by the combined repulsion between h and Z and attraction between X^- and h and Z, respectively. It is shown in Fig. 3 as dashed line. Two crossings can be seen in the allowed R region, one between $\mu p(1s)$ and μF($n = 7$) at $r_c \approx 21$ muoatomic units (or 5.4 pm) and one between $\mu p(1s)$ and μF($n = 8$) at $r_c \approx 42$ muoatomic units (or 10.7 pm). A large rate is expected for the muon transfer from p to F, especially to levels with $n = 7$ and $n = 8$, and has been found by experiment.

The rate for transfer via crossing states, λ_{tr}, was recently recalculated (Sayasov, 1990) to be (for liquid-hydrogen density)

$$\lambda_{tr} = 2.6 \times 10^{10}\,(Z/\sqrt{M_p})\,F(\xi)\ \mathrm{s}^{-1},$$

with $\xi = (3\sqrt{2M_p}Z/8r_c)^{2/3}$, r_c the separation of the nuclei at the crossing point, and M_p the mass of the exotic hydrogen atom normalized to the μp mass. The term $F(\xi)$ is shown in Fig. 4 as a function of ξ, together with experimental $F(\xi)$ values derived from measured transfer rates. The model succeeds in explaining the large rate difference between muon transfer from p to Ne ($\lambda_{tr} = 5.6 \times 10^9$ s^{-1}) and that from d to Ne ($\lambda_{tr} = 1.4 \times 10^{11}$ s^{-1}), both at liquid-hydrogen density.

Transfer plays an especially prominent role in hydrogen *mixtures* where it is a step on the way from the exotic atom to exotic molecules (see Sec. 7). For energy reasons, transfer is only possible from the lighter to the heavier hydrogen isotope: Muons in the heavier muonic atom are, because of their larger reduced mass, more strongly bound.

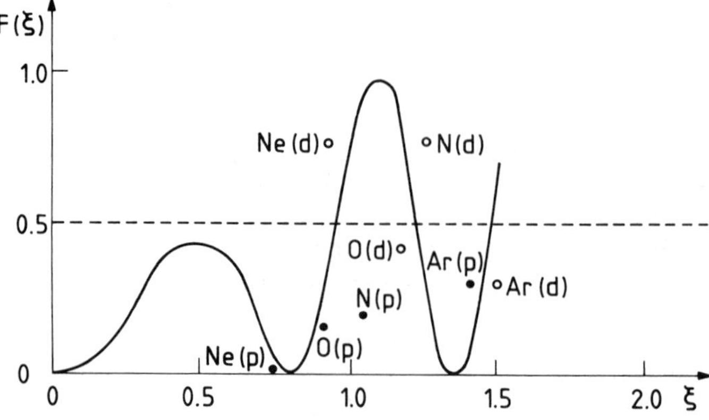

FIG. 4. Normalized transfer rate $F(\xi) = \lambda_{tr}/[2.6 \times 10^{10} Z M_p^{-1/2}]$ as a function of ξ (Gershtein, 1993). Solid line: calculations. Circles and dots: experimental results. For further explanations, see text.

3.5 Coulomb De-excitation

Coulomb de-excitation is a transition of the kind

$$(X^-p)_n + p \rightarrow (X^-p)_{n-k} + p'.$$

It is important only for the exotic hydrogen atom cascade because only the *neutral* exotic hydrogen atom may approach other atoms (mostly those with higher Z) close enough to make fast de-excitation possible. The mechanism is very similar to that for transfer, as also here the transition proceeds especially fast if crossings exist in the correlation diagram between the initial and the final levels of the exotic hydrogen atom.

3.6 Chemical De-excitation

At the early stage of the exotic-*hydrogen* atom cascade, chemical de-excitation, i.e., the de-excitation of the exotic atom by excitation of rotations and vibrations or by dissociation of hydrogen molecules, is possible.

3.7 The Exotic-Atom Cascade

The exotic-atom cascade shall now, as an example, be described in more detail for a muonic atom of medium Z, e.g., μ^-Ar. After capture into a highly excited orbit, the muon particle cascades down the ladder of excited states in the exotic atom. The only de-excitation mechanism in this stage is the Auger effect. By a series of successive Auger transitions, the exotic atom becomes highly ionized. Replacement of the missing electrons proceeds by electron capture into the multiply charged $\mu^-\mathrm{Ar}^{m+}$ ion in collisions with other Ar atoms in the gas. The collision rate depends strongly on the gas pressure and is, at low pressure at least, too small to keep up with the ionization of the exotic atom by the Auger effect. Thus, the μ^-Ar atom finally may become completely stripped of electrons. From this moment on, the cascade proceeds only by radiative transitions down to the $\mu^-\mathrm{Ar}(1s)$ level. It should be noted in parentheses that, for metals and even for insulators, replacement of electrons is fast enough to let the muonic atom be only weakly deprived of electrons during the cascade: The ionized exotic atom distorts the solid-state potential and strongly attracts electrons from the vicinity. As a consequence, radiative transitions become dominant in this case only when the muon has reached levels with $n \leq 10$ (depending on the atomic number of the host atom; see Fig. 1).

The exotic-atom cascade usually is finished within picoseconds. After reaching the ground state of the muonic atom, the muon either decays or reacts with a proton of the nucleus by weak interaction,

$$p + \mu^- \rightarrow n + \nu_\mu. \tag{9}$$

The muon accordingly disappears from the $1s$ level with a rate that is the sum of the muon-decay rate and the nuclear-capture rate. The probability for nuclear absorption, proportional to first order to the sum of the squared wave function at the sites of the protons in the nucleus,

$$w_{\mathrm{abs}} \propto \sum_{\text{all protons}} |\psi(r_{\mathrm{proton}})|^2,$$

should follow a Z^4 law because there are Z protons in the nucleus, and $|\psi(r_{\mathrm{proton}})|^2$ should be $\approx |\psi(0)|^2 \propto Z^3$ for a hydrogenlike atom with a point nucleus of charge Z. Saturation effects are expected for high Z because the large neutron excess in heavy elements limits the number of empty states in which the neutron formed by reaction (9) can be accommodated. The lifetime of the μ^- in the $\mu^-Z(1s)$ state decreases from 2.2 μs in μ^-H to 200 ns in μ^-Fe ($Z = 26$) and 75 ns in μ^-Pb ($Z = 82$).

Exotic He is a very special case. Here, as has been verified for pionic and kaonic He and especially for antiprotonic He (Yamazaki *et al.*, 1993), a small fraction ($\approx$3%) of the exotic He atoms are formed in long-lived states: The exotic atom can deexcite from circular orbits (i.e., those with $l = n - 1$) or nearly circular orbits only by the rather slow radiative transitions because Auger transitions with small angular-momentum transfer are forbidden for energy reasons, and Stark mixing is strongly suppressed for the $(\alpha X^- e^-)$ system, as a result of shielding by the remaining electron.

3.8 Exotic-Particle Polarization during the Cascade

One quantity with some relevance for applications [e.g., μ^- spin rotation (μ^-SR) experiments] has been left out of consideration up to now: the polarization of the exotic particle during slowing down, capture, and cascade to the 1*s* level in the exotic atom. Polarization is the correlation between the spin of a particle and a selected direction, e.g., the direction of flight; a particle is said to be completely polarized if its spin is parallel or antiparallel to this direction. Polarization clearly is most relevant for muons, as only they have a nonzero spin *and* survive the cascade down to the ground state of the exotic atom with very high probability.

Negative muons are usually produced in meson factories by the two-body decay of π^- (spin zero), $\pi^- \rightarrow \mu^- + \bar{\nu}_\mu$. In the pion rest frame, the μ^- spin is parallel to its momentum, as the $\bar{\nu}_\mu$ is completely polarized, the two decay particles fly in opposite directions, and the total spin after decay must be zero. In the laboratory frame, the polarization is no longer 100% because the center-of-mass velocity changes the direction of the muon but not its spin; polarization may be still rather high (around 70%), however. This means that the spin of the muons in the beam preferentially points into the direction of flight. During slowing down, the direction of the muon spin is not changed because the forces leading to the moderation do not act on the spin. The polarization after Coulomb capture has been shown (Mann and Rose, 1961) to be roughly $\pm\frac{1}{3}$ of the initial polarization for muon capture into the states with $j = l \pm \frac{1}{2}$ and for large l. Strong depolarization may occur during the muonic cascade, certainly not by dipole transitions (radiative or Auger), as they do not act on the muon spin, but by interaction between exotic-particle spin and angular momentum. As a result, the muons starting from the state $(n, l, j = l + \frac{1}{2})$ will preserve $\frac{1}{3}$ of their initial polarization if hyperfine interactions of the muon with the (possibly) nonzero spin of the nucleus do not lead to additional depolarization. Those muons starting from $(n, l, j = l - \frac{1}{2})$ will reach the 1*s* state completely depolarized, however.

Polarization of the muon becomes evident, through parity nonconservation in μ decay, by the asymmetry in the angular distribution $W(\theta)$ of the decay electrons with respect to the particle spin,

$$W(\theta) = 1 - a\cos(\theta). \qquad (10)$$

In this formula, a is an energy-dependent factor, which reaches 1 for maximum decay-electron energy (roughly 50 MeV).

4. HADRONIC ATOMS: THE STRONG INTERACTION

In all exotic atoms except those with muons, the exotic particle disappears by strong interaction with the nucleus—in all but the lightest atoms, before it has reached the 1*s* level. In order to describe the interaction, a *complex* strong-interaction potential has to be added to that of the electromagnetic interaction, complex because it has to describe the absorption of the hadron. It may be generated by superimposing the various hadron-nucleon potentials, taking the influence of the nuclear medium into account. If the hadron-nucleon interaction is not too strong, the potential can be assumed to be proportional to the product of an effective scattering length $\bar{a}$, which characterizes the strength of the strong interaction of the hadron with a single nucleon, and the nucleon density $\rho(\mathbf{r})$, the sum of proton and neutron densities, $\rho(\mathbf{r}) = \rho_p(\mathbf{r}) + \rho_n(\mathbf{r})$:

$$V_{\text{Strong}} = 4\pi/\mu[\bar{a}\rho(\mathbf{r})].$$

This *Ansatz* is usually used in antiprotonic atoms.

It has turned out that such a simple description is inadequate for the pion-nucleus potential. Rather elaborate potentials resulted, including not only local terms, i.e., those depending on $\mathbf{r}$, but also nonlocal terms, i.e., those containing the gradient of $\mathbf{r}$, $\nabla\mathbf{r}$. A widely used parametrization of the pion-nucleus potential is the Ericson-Ericson potential (Ericson and Ericson, 1966) given by

$$\begin{aligned} -(\mu/2\pi)V_{\text{Strong}}(\mathbf{r}) = {} & (1 + \mu/M) \\ & \times [b_0\rho(\mathbf{r}) + b_1\delta\rho(\mathbf{r})] \\ & + (1 + \mu/2M)B_0\rho^2(\mathbf{r}) \\ & - \nabla \frac{\alpha(\mathbf{r})}{[1 - \frac{4}{3}\pi\zeta\alpha(\mathbf{r})]}\nabla, \end{aligned}$$

with

$$\alpha(\mathbf{r}) = \frac{c_0\rho(\mathbf{r}) + c_1\delta\rho(\mathbf{r})}{1 + \mu/M} + \frac{C_0\rho^2(\mathbf{r})}{1 + \mu/2M}.$$

Here $\delta\rho(\mathbf{r}) = \rho_n(\mathbf{r}) - \rho_p(\mathbf{r})$, μ and M are the reduced hadron mass and the nucleon mass, respectively, b_0, b_1, c_0, and c_1 are adjustable complex parameters standing for the different possible interaction modes, and the terms with B_0 and C_0 represent pion absorption on two nucleons [for this reason, they are quadratic in $\rho(\mathbf{r})$]. The parameter ζ, finally, characterizes the strength of short-range correlations between the scattering nucleons in the term $1 - \frac{4}{3}\pi\zeta\alpha(\mathbf{r})$, which is the analog to the term found when calculating the susceptibility in dense polarizable media (Lorentz term). Even more elaborate potentials have been recently used (Konijn *et al.*, 1990).

As pointed out before, strong interaction results in an appreciable shift and width of the last level populated before absorption. But already the preceding x-ray transitions are attenuated by a nonzero probability for hadron absorption. Table 3 shows examples of shifts and widths of the lowest levels accessible in pionic and antiprotonic atoms.

Absorption of the pion by the nucleus (with an energy release ΔE approximately equal to the rest energy of the pion, $\Delta E \approx 139$ MeV) results in the evaporation of a number of light particles (neutrons, protons, deuterons, tritons). In antiproton annihilation, the energy release is more than an order of magnitude larger ($\Delta E \approx 1876$ MeV, twice the rest energy of the $\bar{p}$). Here part of the mean number of five pions (neutral and charged) generated during annihilation on *one* nucleon miss the nucleus, as the annihilation takes place at its outer periphery. Nevertheless, in the mean, three pions hit the nucleus, which, because of a strong resonance peak in the pion-nucleon cross section (the so-called Δ_{33} resonance), is black for these pions. The nucleus is heated up by the energy deposited by the pions, and cools down by the emission of fast light particles and later by neutron evaporation. For a medium-mass nucleus like Mo ($A \approx 100$), up to 30 nucleons are emitted. The energy spectra of protons, déuterons, tritons, and α particles emitted after $\bar{p}$ annihilation in ^{238}U are shown in Fig. 5.

Table 3. Typical strong-interaction shifts and widths in pionic and antiprotonic atoms (see Batty *et al.*, 1991, Heitlinger *et al.*, 1992, and Konijn *et al.*, 1990, for references). The shift is defined as the difference between measured binding energies and those calculated taking only the electromagnetic interaction into account. A negative energy shift corresponds to a repulsive strong-interaction potential.

		Strong interaction	
Exotic atom	Level (n, l)	Shift ϵ (keV)	Width Γ (keV)
π-^{3}He	$1s$	0.032(3)	0.028(7)
π-^{4}He	$1s$	−0.076(2)	0.045(21)
π-^{10}B	$1s$	−3.25(8)	1.58(16)
π-^{11}B	$1s$	−4.17(9)	1.77(16)
π-^{16}O	$1s$	−18.7(4)	7.9(3)
π-^{18}O	$1s$	−23.3(3)	6.3(4)
π-^{24}Mg	$1s$	−80.5(12)	24.3(16)
π-^{24}Mg	$2p$	0.125(4)	0.0725(18)
π-^{40}Ca	$2p$	1.77(8)	1.64(11)
π-^{93}Nb	$2p$	−11(3)	64(8)
π-^{93}Nb	$3d$	0.74(2)	0.402(16)
π-^{208}Pb	$3d$	22.7(24)	47(4)
π-^{208}Pb	$4f$	1.68(2)	0.98(5)
$\bar{p}$-^{1}H	$2p$	−0.730(20)	1.12(6)
$\bar{p}$-^{16}O	$3d$	−0.112(20)	0.495(45)
$\bar{p}$-^{17}O	$3d$	−0.140(46)	0.54(15)
$\bar{p}$-^{18}O	$3d$	−0.195(20)	0.64(4)
$\bar{p}$-^{23}Na	$3d$	−2.08(30)	2.9(11)

5. EXPERIMENTS ON EXOTIC ATOMS

Since the early days of exotic-atom research, investigators have mainly observed the properties of radiation and particles

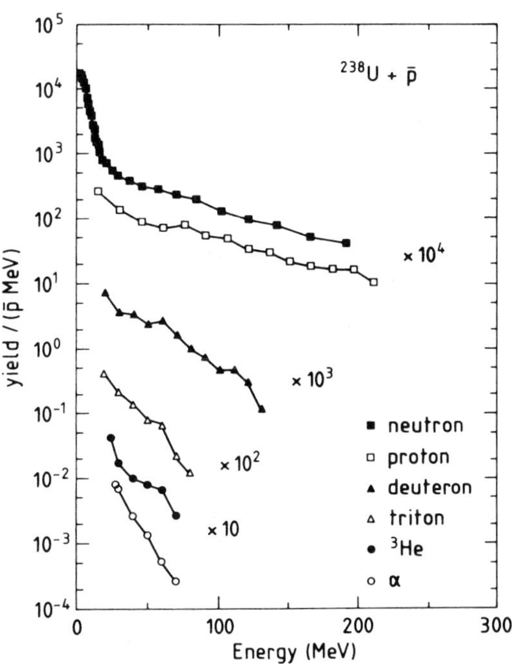

FIG. 5. Energy spectrum of annihilation products from antiproton annihilation in $\bar{p}$-^{238}U (Machner, 1993).

emitted during the atomic cascade and the decay or annihilation of the exotic particle:

1. energy, intensity, anisotropy, and polarization of x rays from exotic atoms (from now on, called exotic x rays),
2. energy and intensity of electrons emitted in the Auger effect,
3. emission-time delay, intensity, and anisotropy of electrons from μ decay in muonic atoms,
4. emission-time delay, energy, and multiplicity of annihilation products from hadronic atoms.

To collect such information, the following detectors have been employed:

1. scintillation counters with NaI(Tl), bismuth germanate (BGO), BaF, and CsI crystals for the detection of high-energy quantum radiation;
2. lithium-drifted Si and Ge detectors and hyperpure-Ge detectors mainly for the measurement of energy and intensity of medium- and low-energy quantum radiation;
3. crystal spectrometers, mainly for the precise determination of the energy of low-energy quantum radiation;
4. semiconductor spectrometers for the detection of electrons from exotic-particle Auger effect;
5. semiconductor and magnet spectrometers for the measurement of multiplicity and energy of charged particles generated by and following exotic-particle absorption by the nucleus;
6. scintillation counters with liquid organic scintillators for the detection of neutrons from exotic-particle annihilation and with plastic scintillators for the detection of electrons from muon decay.

Only in a very few cases has it been attempted to stimulate excitation or deexcitation of the exotic atom by laser irradiation because, on account of the restricted intensity of the exotic-particle beams, the experiments are tedious, and long measuring times are necessary. On the other hand, the unprecedented accuracy of the laser wavelength determination makes it possible to measure level-energy differences in exotic atoms with uncertainties orders of magnitude smaller than is possible with detectors; thus, such experiments are very valuable for the precise determination of the tiniest QED corrections in the level energies of exotic atoms. Examples of experiments aimed at the excitation or deexcitation of exotic atoms by laser irradiation are

1. the stimulation of the muonic ($2s \rightarrow 2p$) transition in muonic helium by laser irradiation,
2. laser-stimulated ($3d \rightarrow 3p$) transitions in muonic hydrogen,
3. the stimulated deexcitation of long-lived, high-lying states ($n \approx 38$) in antiprotonic helium, also by irradiation with a laser.

Representing the large number of experiments on muonic, mesonic, and baryonic atoms in the past decades, some examples will be described now.

5.1 Intensity of Exotic X Rays

A typical experimental arrangement for the detection of exotic x rays is shown in Fig. 6. A telescope with scintillation counters S_1, S_2, S_3, and S_4 is set up to detect the particles coming down the beam line. Coincidence of signals from S_1, S_2, and S_3 with no signal from S_4 indicates a stop of the particle in the target. With a moderator of adjustable thickness, the exotic particles are slowed down to stop in the target. The Ge detectors serve to measure the muonic x rays emitted within a very short ($\approx$50 ns) time window after the stop. A typical x-ray spectrum, in this case the Lyman series of muonic Fe, is shown in Fig. 7. Transitions from $l = 1$ levels with n up to $n = 20$ to the $1s$ level can be separated, but transitions from even higher levels can be seen.

To get an idea about the de-excitation of the exotic atom, the cascade of transitions is simulated with competition between radiative and Auger transitions, depletion of the electron shell by Auger transitions, and electron replacement taken into account. For hadronic atoms, nuclear absorption has to be taken into consideration too. One usually does *not* start the simulations at the highly excited states populated at atomic capture, because the transition rates are poorly known there. A good choice for the beginning of the simulation is a state with an orbit similar in size to the K-electron orbit because, from then

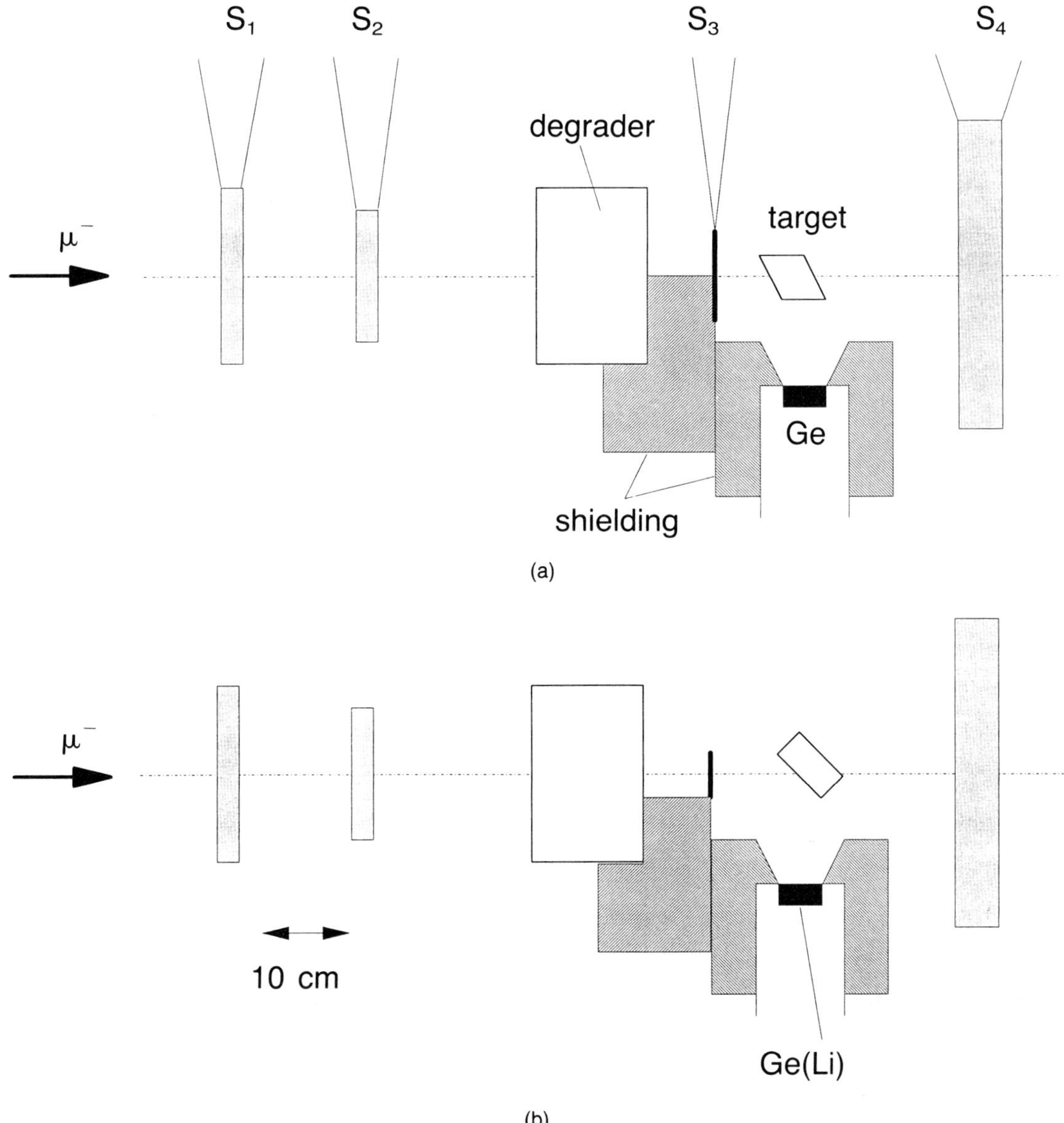

FIG. 6. Typical experimental arrangement for the measurement of muonic x-ray intensities. Upper half: side view: lower half: top view. S_1, S_2, S_3, S_4: scintillation counters; intrinsic Ge: hyperpure-Ge detector; Ge(Li): Li-drifted Ge detector.

on, the problem is hydrogenlike and the calculation of rates straightforward. In general, at the beginning of the simulation, angular-momentum states at only *one n* are assumed to be populated, with a probability distribution characterized by one or two adjustable parameters. The exotic x-ray intensities measured may then be compared with the results of the cascade calculations, and the parameters of the calculation, characterizing the initial distribution and depletion and replacement of the electrons, may be adjusted to get best agreement between simulated and measured x-ray intensities. More than 80 such intensities were recorded for muonic Fe, to give an example.

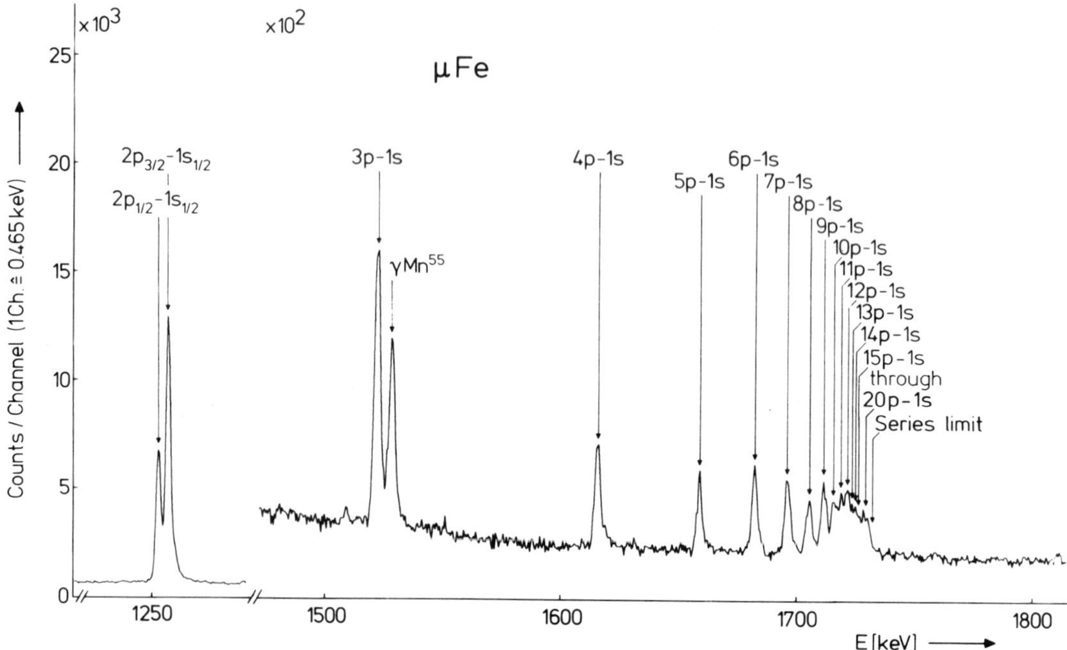

FIG. 7. Muonic Lyman series from muonic Fe (Hartmann *et al.*, 1976). Transitions with $n > 20$ are clearly visible.

5.2 Capture Ratios of Muons in Compounds

In muonic atoms, the sum of the Lyman-series x-ray intensities from a certain element is equal to the number of muons captured by the atoms of this element. This is because, except for atoms with $Z < 6$ or Z very large, each muon captured into the atom makes one and only one *radiative* transition to the $1s$ level of the muonic atom. In atoms with $Z < 6$, the muons reaching the $2s$ levels are trapped there because transitions to the $2p$ level are forbidden for energy reasons, and those to the $1s$ level are monopole transitions and, hence, strongly suppressed; for very heavy atoms, on the other hand, the transition energies in the final stage of the cascade become high enough to make de-excitation by neutron emission from the nucleus possible and probable.

Relative per-atom capture probabilities have been derived from Lyman-series x-ray intensities for 150 different compounds, alloys, and mixtures. From this large data set, it was possible to extract the per-atom capture probabilities for 65 elements (normalized to $Z = 8$). They are shown in Fig. 8. A pronounced periodicity of $P(Z)$ with Z is visible (cf. Sec. 2).

The chemical bond was shown to have an influence on the capture probability; this "chemical effect," however, changes $P(Z)$ by not more than 5%.

5.3 Precision Measurements of Transition Energies in Muonic and Pionic Atoms

Precision measurements of transition energies in exotic atoms serve to determine fundamental constants or quantities important for the understanding of properties of nuclei and of elementary particles.

One example is the measurement of the $3p \rightarrow 1s$ transition energy in pionic hydrogen with a crystal spectrometer. Such a measurement yields the strong-interaction shift ϵ_{1s} and width Γ_{1s} of the $1s$ level in pionic hydrogen and serves to deduce the very fundamental pion-nucleon strong-interaction parameters. A double-focusing crystal spectrometer with Si crystals cut in the (111) direction was used. Position-sensitive CCD (charged-coupled device) x-ray detectors served to detect the scattered x rays. The energy of the $3p \rightarrow 1s$ transition was measured

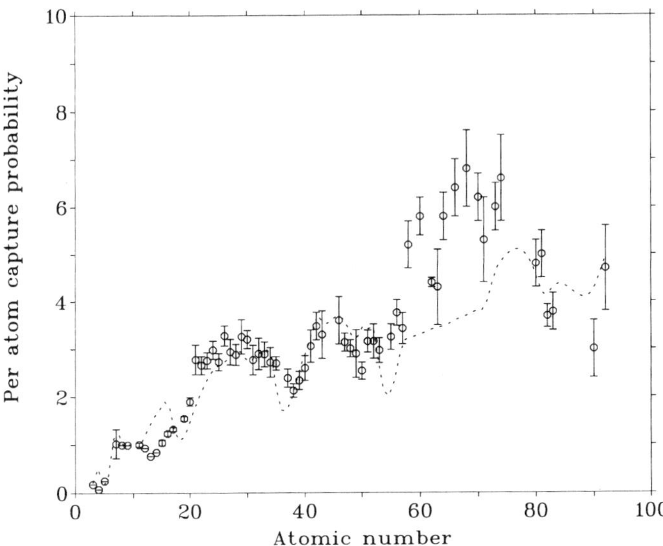

FIG. 8. Per-atom capture probability $P(Z)$ as a function of Z. Dashed line: result from a semiclassical calculation taking the atomic radii into account (Daniel, 1979).

with a relative error of only 100 ppm. Hence, the strong-interaction shift could be determined to better than 5%.

Similar measurements have been performed to determine the energy of the $4f \rightarrow 3d$ transition in pionic ^{24}Mg to derive the rest mass of the pion with high accuracy.

Numerous experiments with semiconductor detectors have been devised over the decades to measure muonic x-ray energies precisely, preferably from the Lyman series, in order to determine the finite-size energy corrections (cf. Sec. 1). From these corrections, nuclear-size parameters (radius, skin thickness, and deformation if applicable) could be deduced by comparing the energy eigenvalues of the Dirac equation for the potential of the extended nucleus with experiment. This method complements the method of high-energy electron scattering on nuclei.

5.4 μ^- Spin Rotation in Semiconductors

As outlined in Sec. 3.7, the spin polarization of the muon in the $1s$ level of a muonic atom is roughly $\frac{1}{6}$ of the initial polarization of the muon in the particle beam, but may strongly depend on hyperfine interactions between the muon and the electric and magnetic field at the site of the muonic atom. The muonic atom acts, so to say, as an extended nucleus with charge $Z - 1$ that feels the electronic structure of the surrounding medium. Thus μ spin rotation (μ^-SR) has become a tool in solid-state physics, although not nearly as extensively used as spin rotation with positive muons (μ^+SR) (see MUON SPIN ROTATION/RELAXATION/RESONANCE).

One such experiment performed only recently was μ^-SR in Si, pure and heavily doped with boron, respectively. An asymmetry of decay-electron emission as an indicator for the muon polarization (see Sec. 3.8) was visible at low temperatures and increased strongly with increasing boron-doping level. The result was interpreted as follows: The $2p \rightarrow 1s$ transition energy of 400 keV is large enough to let the μ-Si atom leave its place in the lattice through recoil, and, hence, a quasi-Al ion moves to an interstitial place in the lattice. Only at low temperatures does it stay at this place and is the muon polarization not immediately destroyed. Nevertheless, because of the interaction of the muon spin with an *unpaired* electron in a quasi-Al^{++} state, strong depolarization takes place. Strong boron doping apparently favors occupancy of a quasi-Al^{+} state with *no unpaired* electrons where the muon does not suffer strong depolarization.

6. APPLICATION OF EXOTIC ATOMS

6.1 Application of Muonic X Rays for Elemental Analysis

As rather high-energy x rays characteristic of the element that captured the muon are emitted when muons are stopped in a mixture or compound of elements and as the Coulomb-capture probability is rather well known for a large number of elements (see Sec. 5.2), the negative muon can be used as a probe for elemental analysis. The method is nondestructive, as only a tiny activation of the sample is to be expected; it is sensitive to all elements except (perhaps) the lightest ones; it may be applied (and has been applied) in surface analysis because muons with kinetic energies in the keV region, which can be produced in modern meson factories with a sufficient rate, are stopped in the first few layers of the sample; it certainly is not cheap, but rather simple, as the energy of the muonic x rays is in a region easily accessible to measurements with semiconductor detectors. The sensitivity of the method is about 10^{-4} atomic abundance. As an example, Fig. 9 shows the muonic x ray spectra from body and glaze of a bowl manufactured in Cairo in the 14th century. It is easily noted that Pb and Sn lines appear in the spectrum from the glaze, and the Ca intensity is increased, as well.

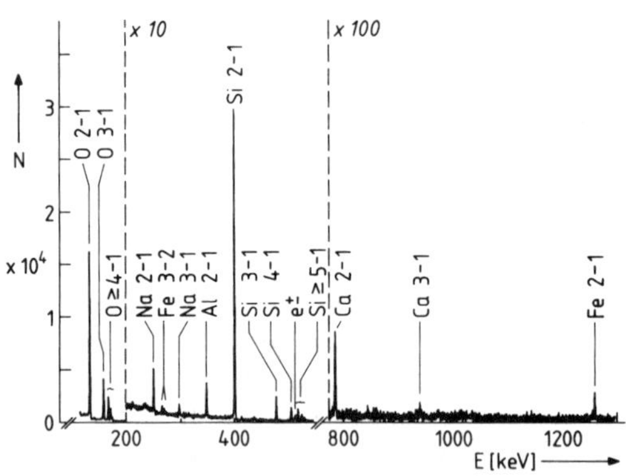

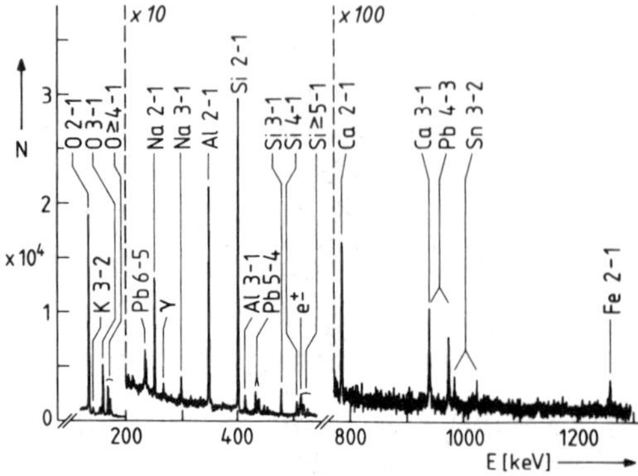

FIG. 9. Muonic x-ray spectra from body (top) and glaze (bottom), respectively, of an islamic bowl (Cairo, 14th century) (Köhler *et al.*, 1981).

6.2 Application of Muonic X Rays in Diagnostics

As pointed out more than 20 years ago, muons may also be used to determine the elemental composition of tissue and to replace neutron activation analysis because muon analysis is less destructive to the organism. Normal and pathological tissues are expected to differ, e.g., by their oxygen content. These expectations have been confirmed by experiments on the liver from a healthy dog and from a dog having suffered from liver cancer.

It seems to be feasible now to apply muons for the scanning of the human body. Both the path of the muon entering the body and the track of the decay electron may be determined with the help of multiwire chambers. The characteristic x rays from the atom that finally captures the muon may also be detected. Thus, even a complete scan of the elemental composition of the body might become feasible.

7. MUONIC MOLECULES

7.1 Definition

Muonic molecules are systems in which two nuclei are bound by the attracting force of a muon, rather similar to an electronic molecule where the binding force is exerted by electrons. In principle, exotic molecules could be formed with each kind of negatively charged exotic particle. The short lifetime and/or the strong interaction of the particle has made observation of exotic molecules other than those with muons impossible up to now. Muonic-molecule formation has been studied nearly exclusively for molecules formed by nuclei of the hydrogen isotopes (protium, deuterium, and tritium); besides these molecules, only those formed between hydrogen and He have found some interest. The large mass of the muon compared to the electron ($m_\mu \approx 207 m_e$) has two effects: The distances in these muonic molecules are about 207 times smaller, and the vibration frequency is 3000 times larger than in the corresponding electronic molecules. This in-

Table 4. Muonic molecules formed from hydrogen isotopes. The terms J and v denote the rotational and vibrational quantum numbers, respectively. The molecular formation rates quoted are normalized to liquid hydrogen density (4.25×10^{22} atoms/cm^3).

Exotic molecule	First bound level (J, v)	Binding energy (eV)	Molecular formation rate (μs^{-1})
$pp\mu$	(1,0)	107.312	2.2
$pd\mu$	(1,0)	97.587	5.9
$pt\mu$	(1,0)	99.190	6.5
$dd\mu$	(1,1)	1.967	...[a]
$dt\mu$	(1,1)	0.632	...[a]
$tt\mu$	(1,1)	45.236	3.0

[a]The molecule formation is resonant, and, hence, the formation rate is temperature dependent.

creases fusion probability by tens of orders of magnitude and leads to fast nuclear fusion. In the following, we will deal exclusively with muonic molecules with hydrogen nuclei.

7.2 Formation of Muonic Molecules

Muonic molecules are formed by collision reactions between muonic hydrogen atoms and electronic hydrogen molecules. Correspondingly, six different types of molecules may be formed between the three kinds of muonic hydrogen atoms and electronic hydrogen molecules. They are listed in Table 4. As transfer (cf. Sec. 3.4) from the lighter hydrogen isotope to the heavier one is much faster than molecule formation, the generation of muonic molecules is appreciable only for collisions of muonic hydrogen atoms with molecules formed by atoms of equal or smaller mass. Two formation mechanisms are possible:

1. Nonresonant formation. The formation energy of the muonic molecule is used to eject an electron from the partner molecule, e.g.,

$$d\mu + H_2 \rightarrow [(pd\mu)pe]^+ + e^-.$$

This process is comparatively slow and can take place for all kinds of muonic molecules between hydrogen isotopes (cf. Table 4).

2. Resonant formation. The energy released during the formation of the muonic molecule is used to excite rotations (quantum number K) and vibrations (quantum number v) in the muomolecular complex, e.g.,

$$t\mu + D_2 \rightarrow [(dt\mu)dee]^*_{vK}.$$

Here the $dt\mu$ molecular ion is one center of the complex, the second deuteron the other. Resonant formation becomes effective when the sum of the binding energy of the muonic molecule $|\epsilon_{11}|$ and of the kinetic energy ϵ_0 of the muonic atom matches the excitation energy ΔE_ν of the electronic molecule (cf. Fig. 10),

$$\epsilon_0 + |\epsilon_{11}| = \Delta E_\nu. \tag{11}$$

This may happen in two ways:

—During slowing down, the muonic hydrogen atom passes kinetic-energy regions where condition 11 is fulfilled.

—When the muonic atoms have been

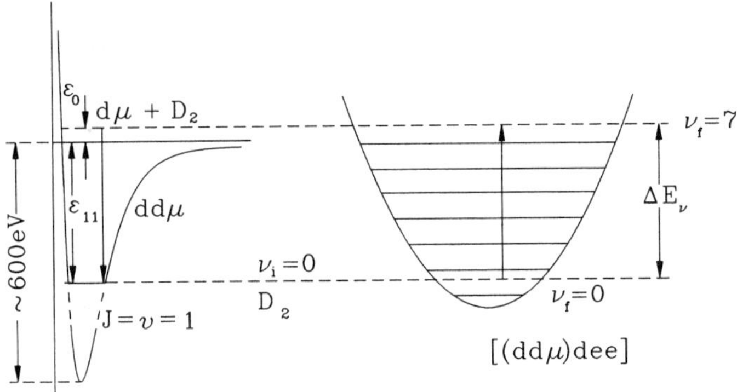

FIG. 10. The energy-level scheme for resonant $dd\mu$ molecule formation. The resonance condition is most easily met for the mesomolecular complex $[(dd\mu)dee]^*_{K,\nu}$ in the vibrational state with $\nu = 7$.

slowed down to thermal energies, which means that their kinetic-energy distribution is Maxwellian with a most probable energy $\epsilon_{max} = \frac{3}{2}kT$ (with T the absolute temperature), there are always muonic hydrogen atoms with energies for which Eq. (11) is satisfied. Their fraction, however, depends strongly on T and should reach a maximum at a temperature for which $\epsilon_{max} = \epsilon_0$.

For the most loosely bound levels in the $dd\mu$ and $dt\mu$ molecules, nonresonant formation is impossible, as the small energy gained in molecule formation (below 2 eV) is much smaller than the ionization energy of the host molecule (ionization potential ≈15 eV).

7.3 Energy Levels in Muonic Molecules

The calculation of binding energies in muonic molecules has been a successful application of three-body theory, and a high accuracy has been achieved during several decades of research. Typical results for the binding energy of the first state populated after muomolecule formation may be found in the third column of Table 4. Only small corrections, mainly for vacuum polarization and finite nuclear size (cf. Sec. 1), have to be added to the nonrelativistic binding energy. The respective values for the $dt\mu$ molecule, as an example, are nonrelativistic binding energy 660.2 meV, vacuum polarization and finite-size corrections 16.6 and 13.3 meV, respectively, total binding energy 631.8 meV.

The measurement of the muonic-molecule formation rate is a very sensitive tool to determine the binding energy of the (rotational quantum number $J = 1$, vibrational quantum number $v = 1$) level in the $dd\mu$ and $dt\mu$ molecules where resonant molecule formation takes place: As the energies of the rotational and vibrational states of the *electronic* host molecule are very well known, and the kinetic energy of the muonic atom *and* of the molecule it collides with are only some tens of meV in thermal equilibrium, the resonance condition poses a stringent condition on the binding energy of the muonic molecule. By changing the temperature of the mixture, one may realize small changes in ϵ_{max} and, hence, may scan slowly through the resonances. Thus, the level energy in muonic molecules may be determined to some tenths of a meV, a tiny fraction of the total binding energy.

7.4 Deexcitation of Muonic Molecules

As stated earlier (cf. Table 4) the muonic molecules are preferentially formed in excited rotational-vibrational states with $J = 1$. For muonic molecules with identical nuclei ($pp\mu$, $dd\mu$, $tt\mu$), electromagnetic E1 transitions to lower states are suppressed by the Pauli principle. In the $pd\mu$ and $pt\mu$ molecules, mostly formed in the ($J = 1, v = 0$) state, transitions by the Auger effect of the kind

$$[(pd\mu)_{J=1,v=0}\, pee] \rightarrow [(pd\mu)_{J=0,v=0}\, pe]^{+} + e^{-}$$

are fast (rates of the order of 10^{11} s^{-1}). In the $dt\mu$ molecule, finally, a competition takes place between de-excitation to the ground state ($J = 0$, $v = 0$) and the fusion reaction. Therefore, in 80% of the cases, fusion takes place from the ($J = 0$, $v = 1$) state, while in the other cases the molecule reaches the ground state.

7.5 Disappearance of Muonic Molecules: Muon-Induced Fusion

The muonic molecule may disappear by nuclear fusion or by muon decay. The rate λ_L^{Jv} for fusion from a state with orbital angular momentum L between the nucleons is given by

$$\lambda_L^{Jv} = K_L \rho_L^{Jv}.$$

In this expression, K_L is the nuclear-reaction constant, closely related to the cross section for nuclear fusion at zero kinetic energy between the nucleons, and ρ_L^{Jv} is the probability for the nuclei to be at zero distance. Only fusion reactions with $L = 0$ and $L = 1$ play a role. Fusion energies and rates for the different muon-induced fusion reactions are given in columns 3 and 4 of Table 5.

After muon-induced fusion, the muon is not necessarily lost for further fusions. Only in the reactions with gamma emission is the probability high (cf. column 5 of Table 5) that, immediately after fusion, the muon "sticks" to the fusion product ("initial sticking"), as the recoil kinetic energy of the generated nucleus is small. In all other cases (and especially for dt fusion), the probability of muon

Table 5. Fusion reaction(s) by strong interaction in muonic molecules.

Exotic molecule	Main fusion reactions	Energy gain (MeV)	Fusion rate (s^{-1})	Sticking probability
$pd\mu$	$pd\mu \to \mu^3\text{He} + \gamma$	5.5	2.6×10^5	0.99
$pt\mu$	$pt\mu \to \mu^4\text{He} + \gamma$	19.6	0.7×10^5	0.94
$dd\mu$	$dd\mu \to \mu + {}^3\text{He} + n$	3.3	4.3×10^8	0.12
	$\to t\mu + p$	4.0		
$dt\mu$	$dt\mu \to \mu + {}^4\text{He} + n$	17.6	1.2×10^{12}	0.0055
$tt\mu$	$tt\mu \to \mu + {}^4\text{He} + 2n$	11.3	1.5×10^7	0.14

"rejuvenation" is high. This is partly because the generated nucleus has too high a velocity to allow the muon to stick to it, but partly also because of the phenomenon that, in case the muon initially sticks to the nucleus, the muon is stripped with high probability from the muon-nucleus system during the deceleration ("reactivation"). The reactivation probability is the higher the larger the recoil energy of the μ-nucleus system generated by the fusion process.

Because the sticking probability is so low and muomolecule formation and muon-induced fusion are so fast, the cycle of muon-induced deuterium-tritium fusion is a very special case and is usually meant when one speaks about muon-catalyzed fusion, μCF.

7.6 The Cycle of Muon-Catalyzed *dt* Fusion

The cycle of muon-catalyzed *dt* fusion comprises muoatomic, muomolecular, and nuclear-fusion processes. It is shown in Fig. 11. Muons slowed down to an energy of around 10 eV in a mixture of deuterium and tritium form $D_2^+\mu^-$ and $T_2^+\mu^-$ complexes, which decay forming excited $d\mu$ or $t\mu$ atoms (kinetic energy $\approx$ 1 eV). After muon capture

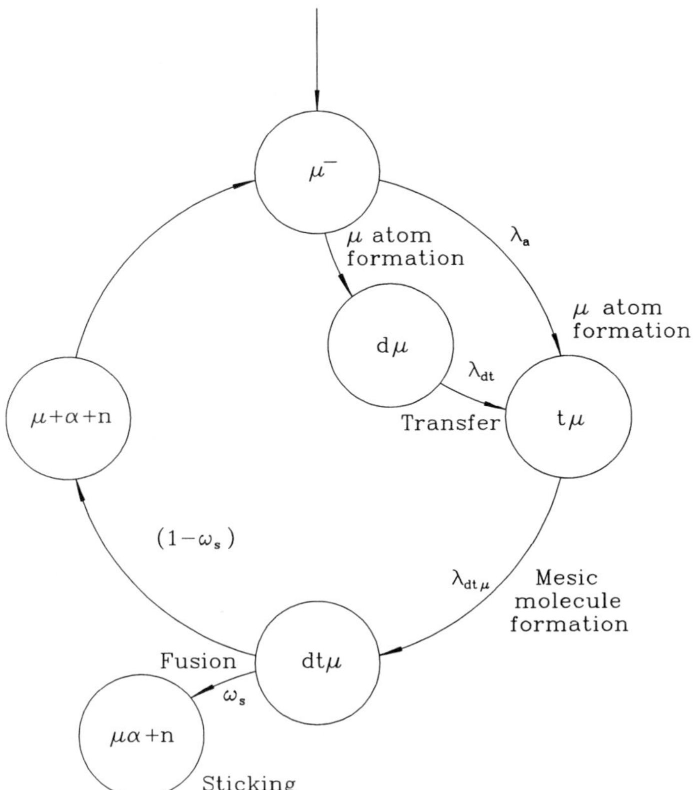

FIG. 11. The main muon catalysis cycle in a mixture of deuterium and tritium. λ_a, λ_{dt}, $\lambda_{dt\mu}$: rates for atom formation, transfer, and mesonic molecule formation, respectively; ω_s: probability for sticking.

to deuterium, fast transfer to tritium takes place (but, for energy reasons, not in the reverse direction; see Sec. 3.4). Calculations have revealed that transfer should be much faster if the muon is still in an excited state of the $d\mu$ atom. After transfer, the $t\mu$ system has appreciable kinetic energy as, e.g., the muon-binding energies in the $d\mu(1s)$ and $t\mu(1s)$ states differ by as much as 45 eV. The $t\mu$ system is slowed down by collisions with other atoms; the deceleration becomes rather slow at energies around 3 eV because there is a minimum in the scattering cross section (Ramsauer–Townsend effect). Finally, $dt\mu$ molecules are formed. Deexcitation and fusion compete in the muomolecule. After fusion, the muon sticks to the charged fusion product, an α particle, for only 0.7% of the cases. After this "initial" sticking, 30% of the trapped muons are reactivated during deceleration of the $\alpha\mu$ system. The rejuvenated muon can catalyze further fusions until it decays during the cycle or finally sticks to the fusion product where it is trapped until it decays. Up to 150 ± 20 fusions catalyzed by one muon were observed in liquid deuterium-tritium mixtures at tritium concentrations around 50%. Since, for each fusion event, 17.6 MeV of energy is produced, the energy release per muon is ≈2 GeV. Furthermore, for each fusion, a neutron of 14.1 MeV is produced and can be used to fission ^{238}U or to breed ^{239}Pu. This has posed the question of energy production by μCF.

7.7 The Prospects of Energy Production by Muon-Catalyzed *dt* Fusion

To judge today's perspective of energy production by muon-catalyzed fusion, one has to compare the energy necessary to produce one muon with the energy gain from about 150 catalyzed *dt* fusions. Muon production has to proceed via pion generation by nuclear reactions of GeV protons or deuterons with light nuclei (e.g., Be or C) and successive pion decay to convert pions into muons. Shooting deuterons of 4 GeV on carbon, for example, the mean energy necessary to create one π^- is about 6 GeV, and, at a conversion efficiency of 0.75, roughly 8 GeV have to be spent per produced muon. By increasing the accelerator efficiency, this amount can, perhaps, be decreased by another 25%, but still the energy necessary to produce one muon is three times larger than the energy gained by fusion processes catalyzed by this muon.

One way to make the energy balance in a μCF-based device positive is the muon-catalytic hybrid reactor (MCHR) (Petrov, 1985). The scheme is shown in Fig. 12. A beam of deuterons with ≈1 GeV/nucleon is directed to a target T, made from Be or Li. A fraction of 30% of the beam energy is used to create pions; the other 70% are used to fission ^{238}U or breed ^{239}Pu. The pions produced decay in a converter C into muons. The muons are guided to the so-called synthesizer S filled with a mixture of D_2 and T_2, where muon-catalyzed fusion takes place. Neutrons gen-

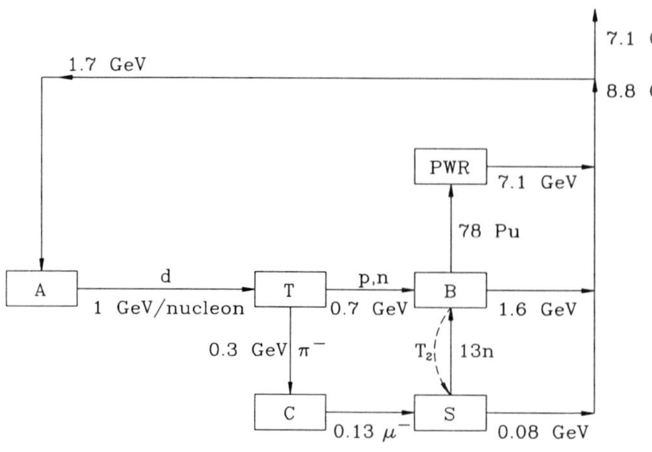

FIG. 12. Concept for a hybrid reactor based on muon-catalyzed *dt* fusion (Petrov, 1985). The energies given in the figure are energies needed or gained per nucleon. A: accelerator; T: target for pion production (Li or Be); C: converter $\pi^- \rightarrow \mu^-$, S: synthesizer (a high-density mixture of deuterium and tritium); B: blanket (a mixture of U and Li); PWR: nuclear power reactor.

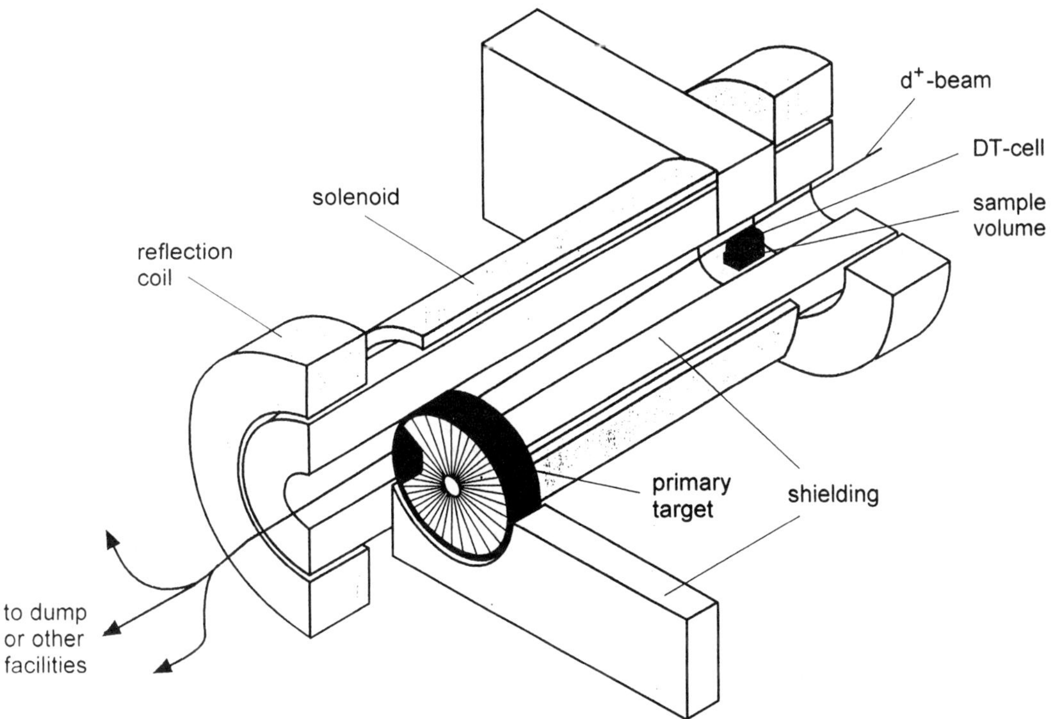

FIG. 13. Scheme for a high-flux source for 14.1 MeV neutrons based on muon-catalyzed fusion (Petitjean *et al.*, 1993).

erated in the synthesizer hit a blanket B (^{6}Li mixed with ^{238}U); tritium and plutonium are bred, which can be used to replace the tritium used in the fusion process and to feed thermal nuclear reactors that generate energy in a conventional way, respectively.

Doubts about the consequences of the MCHR for the environment and the very ambitious technology make it rather unlikely that this interesting scheme will become reality in the near future. Another system using muon-catalyzed fusion, which could be built using current technology, however, is a high-flux source for high-energy (14.1 MeV) neutrons necessary to test "first wall" and blanket materials for a future thermal-fusion reactor.

7.8 A High-Flux Neutron Source Based on Muon-Catalyzed *dt* Fusion

The conceptual design for an intense 14-MeV neutron source based on muon-catalyzed fusion (Petitjean *et al.*, 1993) is rather similar to the first stages of the MCHR (cf. Fig. 13). An intense deuteron beam interacts with a 30- to 50-cm-long carbon target. Muons from the backward decay of the generated pions are guided by a 5- to 10-T magnetic field to a target with a deuterium-tritium mixture of high density. With a deuteron beam of 12 mA at 1.5 GeV, about 10^{16} π^-/s can be generated, and up to 10^{15} μ^-/s may stop in the DT target. If we assume 100 fusion cycles to be catalyzed by one muon, a source strength of $\approx 10^{17}$ n/s results. The part of the beam not interacting with the carbon target (roughly 30–40%) can be used for other facilities (e.g., a spallation neutron source).

GLOSSARY

Antiproton: Antiparticle of the proton. Proton and antiproton differ in their charge. Annihilation of an antiproton with a proton sets free an energy of nearly 2 GeV, which goes mostly into pions.

Auger Effect: Transition of a particle (electron, exotic particle) from a free to a bound state or from one bound state to another one of lower energy with emission of an atomic electron.

Baryonic Atom: Exotic atom with a baryon (i.e., a strongly interacting elementary particle at least as heavy as the antiproton).

Cascade: The sequence of processes during the de-excitation of muonic, mesonic, and baryonic atoms. It comprises the de-excitation processes themselves (radiative, Auger, Stark transitions, etc.) and the refilling of the electron shell depleted by Auger transitions.

Coulomb Capture: Capture of a negatively charged elementary particle into an atom by virtue of the Coulomb interaction.

Coulomb Potential: Potential between two pointlike charges. The Coulomb potential decreases as $1/r$, where r is the distance between the charges.

Dirac Equation: Relativistic wave equation to describe particles with half-integer spin (like muons, antiprotons).

Fermi Distribution: Distribution function devised by Fermi. In exotic atoms used to describe the radial charge density in the nucleus.

Fine Structure: Splitting of energy levels in a muonic, mesonic, or baryonic atom with the same principal and angular-momentum quantum numbers. The fine-structure splitting is due to the interaction between spin and orbital angular momentum of the exotic particle and occurs only for particles with nonzero spin.

Hadronic Atom: Exotic atom with a strongly interacting elementary particle (pion, kaon, antiproton, Σ hyperon).

Hyperfine Structure: Splitting of energy levels in a muonic, mesonic, or baryonic atom with the same total angular momentum of the exotic particle; due to the interaction with the nuclear moments (e.g., the electric quadrupole moment or the magnetic dipole moment).

Kaon: Strongly interacting elementary particle with a mass of 966 electron masses and a life time *in vacuo* of $\approx$12 ns.

Klein–Gordon Equation: Relativistic wave equation to describe particles with integer spin (like pions or kaons).

Mesonic Atom: Exotic atom with a meson (pion, kaon).

Muon: Elementary particle. The muon is very similar to the electron, only its mass is $\approx$207 times as large as that of the electron, and its lifetime is only 2.2 μs.

Muon-Induced Fusion: Process of nuclear fusion of (hydrogen) nuclei bound in muonic molecules. Because of the small distance between the nuclei brought about by the (heavy) muon, the fusion process is very fast.

Muonic Molecules: Dinuclear molecules, usually with hydrogen isotopes, where a muon acts as the negative particle binding the hydrogen nuclei.

Nuclear Polarization: The polarization of the nucleus by the electric charge of the exotic particle in muonic, mesonic, and baryonic atoms. It plays a role only for the most tightly bound orbits of the exotic atom.

Pion: Elementary particle with a mass of $\approx$273 electron masses and a lifetime *in vacuo* of 26 ns. The particle interacts strongly with nucleons. Besides positive and negative pions, also neutral pions are known.

Ramsauer–Townsend Effect: Pronounced minimum in the cross section for the scattering of muonic hydrogen atoms by other hydrogen atoms.

Stark Mixing: Transitions between states with different orbital angular momentum (quantum number l)—but usually with the same principal quantum number—during encounters of an exotic atom with other atoms. In this case, the spherical symmetry of the exotic atom is destroyed, and l is no longer a good quantum number. Stark mixing plays a very important role in exotic hydrogen atoms.

Transfer: Transition of an exotic particle from an exotic hydrogen atom (μp, μd, or μt) to another (heavier) atom. Transfer plays a very important role in the muon-induced-fusion cycle.

Vacuum Polarization: The QED correction to the electromagnetic binding energy that is most important in muonic, mesonic, and baryonic atoms. It is due to the creation of virtual electron-positron (or $\mu^-\mu^+$) pairs in the electric field of the nucleus.

Works Cited

Akylas, V., Vogel, P. (1978), *Comp. Phys. Commun.* **15**, 291–302.

Bamberger, A., Lynen, U., Piekarz, H., Povh, B., Ritter, H. G., Backenstoss, G., Bunaciu, T., Egger, J., Hamilton, W. D., Koch, H. (1970), *Phys. Lett.* **33B**, 233–235.

Barrett, R. C. (1977), in: V. W. Hughes, C. S. Wu (Eds.), *Muon Physics*, Vol. 1, New York: Academic, pp. 309–322.

Batty, C. J., Friedman, E., Gal, A., Kalbermann, G. (1991), *Nucl. Phys. A*, **535**, 548–572.

Bethe, H. (1930), *Ann. Phys. (Leipzig)* **5**, 325–400.

Bethe, H. A., Salpeter, E. E. (1977), *Quantum Mechanics of One- and Two-Electron Atoms*, New York: Plenum.

Burleson, G. R., Cohen, D., Lamb, R. C., Michael, D. N., Schluter, R. A., White, T. O. (1965), *Phys. Rev. Lett.* **15**, 70–72.

Camac, M., McGuire, A. D., Platt, J. B., Schulte, H. J. (1952), *Phys. Rev.* **88**, 134; (1955), *ibid.* **99**, 897–905.

Conversi, M., Pancini, E., Piccioni, O. (1947), *Phys. Rev.* **71**, 209–210.

Daniel, H. (1979), *Z. Phys.* **A291**, 29–31.

Daniel, H. (1980), *Ann. Phys. (NY)* **129**, 303–319.

Ericson, M., Ericson, T. E. O. (1966), *Ann. Phys. (NY)* **36**, 323–362.

Fermi, E., Teller, E., Weisskopf, V. (1947), *Phys. Rev.* **71**, 314–315.

Fermi, E., Teller, E. (1947), *Phys. Rev.* **72**, 399–408.

Ferrell, R. A. (1960), *Phys. Rev. Lett.* **4**, 425–428.

Gershtein, S. S. (1993), in: L. Schaller, C. Petitjean (Eds.), *Muonic Atoms and Molecules*, Basel: Birkhäuser, pp. 169–186.

Hartmann, F. J., von Egidy, T., Bergmann, R., Kleber, M., Pfeiffer, H.-J., Springer, K., Daniel, H. (1976), *Phys. Rev. Lett.* **37**, 331–334.

Heitlinger, K., Bacher, K., Badertscher, A., Blüm, P., Eades, J., Egger, J., Elsener, K., Gotta, D., Morenzoni, E., Simons, L. M. (1992), *Z. Phys.* **A342**, 359–368.

Holzwarth, G., Pfeiffer, H.-J. (1975), *Z. Phys.* **A272**, 311–313.

Köhler, E., Bergmann, R., Daniel, H., Ehrhart, P., Hartmann, F. J. (1981), *Nucl. Instrum. Methods* **187**, 563–568.

Konijn, J., de Laat, C. T. A. M., Taal, A., Koch, J. H. (1990), *Nucl. Phys. A* **519**, 773–804.

Machner, H. (1993), Kernforschungsanlage Jülich Report No. KFA-IKP(1)-1993-4.

Mann, R. A., Rose, H. E. (1961), *Phys. Rev.* **121**, 293–301.

Markushin, V. M. (1990), in: L. M. Simons, D. Horvath, G. Torelli (Eds.), *Electromagnetic Cascade and Chemistry of Exotic Atoms*, New York/London: Plenum, pp. 73–96.

Particle Data Group (1992), *Phys. Rev. D* **45**, S1–S584.

Petitjean, C., Atchison, F., Heidenreich, G., Walter, H. K., Amelotti, F., Andreani, R., de Marco, F., Monti, S., Pillon, M., Vecchi, M., Markushin, V. E., Ponomarev, L. I., Niebuhr, C. (1993), Paul-Scherer-Institut Report No. PSI-PR-93-09.

Petrov, Yu. V. (1985), *Nature* **285**, 466–468.

Sayasov, Yu. S. (1990), *Helv. Phys. Acta* **63**, 547–548.

Yamazaki, T., Widmann, E., Hayano, R. S., Iwasaki, M., Nakamura, S. N., Shigaki, K., Hartmann, F. J., Daniel, H., von Egidy, T., Hofmann, P., Kim, Y.-S., Eades, J. (1993), *Nature* **361**, 238–240.

Further Reading

General:

C. J. Batty (1982), *Sov. J. Part. Nucl.*, **13**, 71–96.

Horvath, D., Lambrecht, R. M. (1984), *Exotic Atoms—A Bibliography 1939–1982*, Amsterdam: Elsevier.

Hughes, V. W., Wu, C. S. (Eds.) (1977), *Muon Physics*, Vols. I and III, New York: Academic.

Kim, Y. N. (1971), *Mesic Atoms and Nuclear Structure*, Amsterdam/London: North-Holland.

Schaller, L., Petitjean, C. (Eds.) (1993), *Muonic Atoms and Molecules*, Basel: Birkhäuser.

Simons, L. M., Horvath, D., Torelli, G. (Eds.) (1990), *Electromagnetic Cascade and Chemistry of Exotic Atoms*, New York/London: Plenum.

West, D. (1958), *Rep. Prog. Phys.* **21**, 271–311.

Yamazaki, T., Nakai, K., Nagamine, K. (Eds.) (1992), *Perspectives of Meson Science*, Amsterdam: North-Holland.

Section 1:

Engfer, R., Schneuwly, H., Vuilleumier, J. L., Walter, H. K., Zehnder, A. (1974), *At. Data Nucl. Data Tables* **14**, 509–597.

Section 2:

Leon, M., Seki, R. (1977), *Nucl. Phys.* **282**, 445–460.

Leon, M., Miller, J. H. (1977), *Nucl. Phys.* **282**, 461–473.

Vogel, P., Haff, P. K., Akylas, V., Winther, A. (1975), *Nucl. Phys.* **254**, 445–479.

Section 3:

Eisenberg, Y., Kessler, D. (1961), *Nuovo Cimento*, **19**, 1195–1210.

Evseev, V. S. (1977), in: V. W. Hughes, C. S. Wu (Eds.) (1977), *Muon Physics*, Vol. III, New York: Academic, pp. 235–298.

Section 5:

von Egidy, T., Jakubassa-Amundsen, D. H., Hartmann, F. J. (1984), *Phys. Rev. A* **29**, 455–461.

Section 6:

Sakamoto, K., Hosoi, Y., Nagamine, K. (1992), in: T. Yamazaki, K. Nakai, K. Nagamine (Eds.), *Perspectives of Meson Science*, Amsterdam: North-Holland, pp. 487–491.

Section 7:

Bogdanova, L. N. (1992), *Surv. High Energy Phys.* **6**, 177–221.

Breunlich, W., Kammel, P., Cohen, J. S., Leon, M. (1989), *Annu. Rev. Nucl. Part. Sci.* **39**, 311–356.

Ponomarev, L. J., Gershtein, S. S. (1977), in: V. W. Hughes, C. S. Wu (Eds.), *Muon Physics*, Vol. III, New York: Academic, pp. 141–233.

Ponomarev, L. I. (1990), *Contemp. Phys.* **31**, 219–245.

MUONIUM AND POSITRONIUM

TOSHIMITSU YAMAZAKI, *Institute for Nuclear Study, University of Tokyo, Tanashi, Tokyo 188, Japan*

YASUO ITO, *Research Center for Nuclear Science and Technology, University of Tokyo, Tokai, Ibaraki 319-11, Japan*

INTRODUCTION

Muonium (μ^+e^-, atomic symbol Mu) and positronium (e^+e^-, atomic symbol Ps) are the simplest atoms comprising only charged leptons: the electron (e^-,e^+) and the muon (μ^-,μ^+). They are hydrogenlike atoms with the proton in H replaced by either μ^+ or e^+. The proton has a finite size, which may affect the binding energies, while μ^+ and e^+ are believed to be point objects. Thus, Mu and Ps not only provide ideal means for testing quantum electrodynamics, but also serve as unique probes of matter. The basic properties of μ^+ and e^+ are compared to those of p^+ in Table 1, and those of Mu and Ps are given in Table 2 in comparison with the properties of H. The level diagrams of the three light isotopes of hydrogen, namely, Ps, Mu, and H, are presented in Fig. 1.

Both Mu and Ps are unstable atoms. Although studies of small hydrogen impurities in matter are very interesting and important, they are practically difficult. On the other hand, implanted muons form hydrogenlike impurities and probe their behavior and local electronic structures. The lifetime of Mu is limited by the free lifetime of μ^+ itself (2.2 μs). The characteristic features of muon polarization and decay asymmetry permit unique spectroscopic uses of muons and muonium for studying the local electronic structure, chemical bonding, chemical reactions, diffusion, trapping, magnetism, etc. (see MUON SPIN ROTATION/RELAXATION/RESONANCE).

The e^+ is a stable particle in vacuum, but

3-527-28133-9/94/$5.00 + .50

Table 1. Basic properties of positron, muon, and proton.

Property	e^+	μ^+	p^+
Rest mass (MeV)	0.511	$105.6596 = 206.7685\, m_e$ $= 0.1126\, m_p$	$938.2592 = 1836.4\, m_e$
Spin	$\frac{1}{2}$	$\frac{1}{2}$	$\frac{1}{2}$
Magnetic moment (erg/G)	$\mu_e = 9.2848 \times 10^{-21}$	$\mu_\mu = 4.490\,48 \times 10^{-23}$ $= 0.004\,836\, \mu_e$ $= 3.183\,34\, \mu_p$	$\mu_p = 1.410\,62 \times 10^{-23}$ $= 0.001\,52\, \mu_e$
g factor	2.002 319	2.002 332	2.675 20
Gyromagnetic ratio	2.802 47 MHz/G	13.5544 kHz/G	4.257 707 kHz/G
Lifetime	∞^a	2.1971 μs	∞
Decay	(Pair annihilation)	$e^+ \bar{\nu}_\mu \nu_e$	

[a]Stable in vacuum, but annihilates with e^- in matter. The lifetime of e^+ by pair annihilation depends on the electron density (see text).

Table 2. Basic properties of positronium, muonium, and hydrogen.

Property	Positronium	Muonium	Hydrogen
Mass	$2\, m_e$	$m_\mu + m_e = 207.8\, m_e$	$m_p + m_e = 1837.6\, m_e$
Reduced mass	$m_e/2$	$0.9952\, m_e$	$0.9994\, m_e$
Spin	0,1	0,1	0,1
Bohr radius (nm)	0.105	0.053 17	0.052 94
Ionization energy (eV)	6.77	13.549	13.595
de Broglie wavelength (300 K) (nm)	5.428	0.2979	0.1004
Thermal velocity (300 K) (m/s)	6.7×10^4	0.75×10^4	0.25×10^4
Hyperfine frequency (MHz)	203 400	4463.3	1420.4

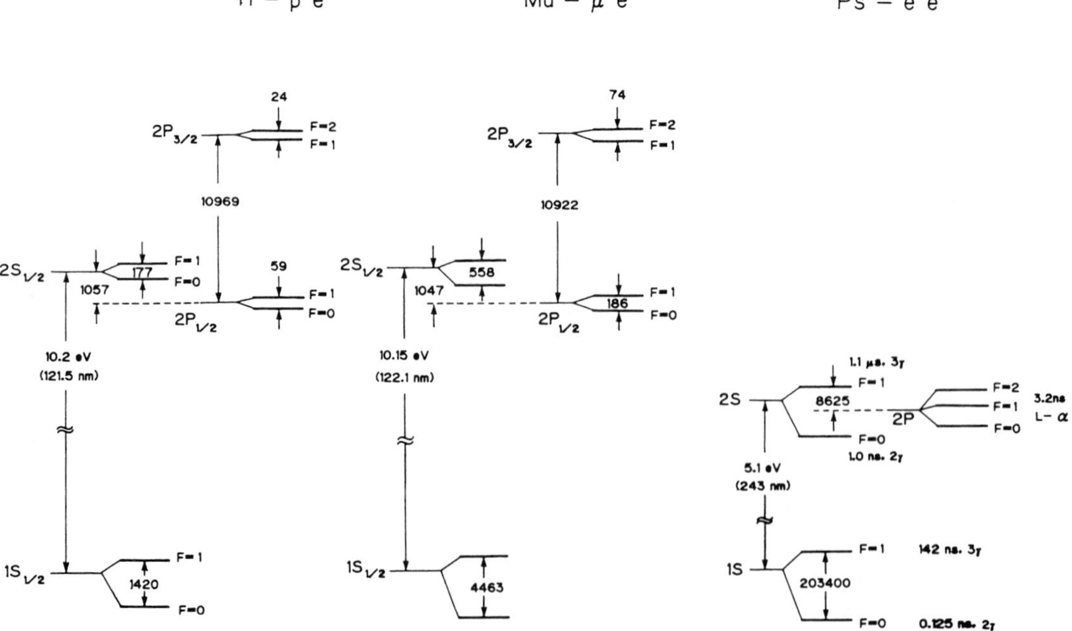

FIG. 1. Energy levels of muonium and positronium compared to those of the hydrogen atom. The energies for the fine and hyperfine structures are in MHz. For positronium, the lifetimes and annihilation modes, or the lifetime of Lyman-α emission, are given in parentheses.

annihilates with one of the electrons in a material substance. In vacuum, Ps has a mean lifetime of 1.25×10^{-10} s (*p*-Ps) and 1.42×10^{-7} s (*o*-Ps). The lifetime of e^+ in matter depends on the local electron density and is the source of microscopic information concerning the electronic states probed by the e^+ and Ps. Positron annihilation spectroscopy has been developed and is used in various fields of materials science (see POSITRON ANNIHILATION SPECTROSCOPY).

Although the existence of Mu and Ps was naturally expected since the discovery of μ^+ and e^+, it took quite some time until Mu and Ps were really discovered. Muonium was first identified by Hughes *et al.* (1960) in Ar gas from the characteristic precession frequency of 1.39 MHz/G in a small transverse field. Experimental evidence for Ps was first obtained by Shearer and Deutsch (1949) as a long-lived component in the slowing down of e^+ in nitrogen and other gases.

1. BASIC PROPERTIES OF FREE MUONIUM

Although the muonium mass is about nine times smaller than the hydrogen mass, its electronic properties (e.g., the Bohr radius and ionization energy) are very close to those of hydrogen, since the reduced mass of Mu is only 0.4% smaller (see Table 2).

1.1 Muonium Spin Dynamics

Muonium is a system of two Dirac particles with the spin Hamiltonian

$$\mathcal{H} = g_e\mu_e\mathbf{B}\cdot\mathbf{S} - g_\mu\mu_\mu\mathbf{B}\cdot\mathbf{I} + A\mathbf{I}\cdot\mathbf{S}, \qquad (1)$$

where g_e and g_μ are the electron and muon g factors, respectively, and μ_e and μ_μ are the corresponding Bohr magnetons. As in the hydrogen atom, the ground-state energy level is split into triplet (total angular momentum quantum number $F = 1$) and singlet ($F = 0$) states with a splitting equal to the hyperfine coupling constant (A), which in terms of the frequency corresponds to 4463 MHz (Fig. 1).

Muonium is formed when a μ^+ captures an electron. Under normal experimental conditions, the μ^+ is spin polarized. Since the electrons are unpolarized, muonium atoms with spin states $|\alpha\rangle_\mu|\alpha\rangle_e$ and $|\alpha\rangle_\mu|\beta\rangle_e$ are formed with a population of 50% in each state, where α and β denote up and down spin, respectively. The latter state (often mistakenly called "singlet Mu") oscillates between the two eigenstates $|F,m\rangle = |1,0\rangle$ and $|0,0\rangle$ (the true singlet state) at the hyperfine frequency of $\nu_0 =$ 4463 MHz. The former, accurately called "triplet Mu," is one of the eigenstates and is stable in a zero or longitudinal field. In general, Mu formed in matter may be different from free Mu because of possible coupling with neighboring atoms/molecules.

1.2 Muonium in Longitudinal Magnetic Field

When a magnetic field is applied in the direction of the μ^+ spin polarization, half of the initial muon polarization (i.e., in triplet Mu) does not evolve. At zero field, the observed μ^+ spin polarization is 50% of the initial μ^+ polarization, because μ^+ in singlet Mu evolves at the hyperfine frequency. The functions for the four spin states in an applied field B are

$$|1\rangle = |\alpha\alpha\rangle, \qquad (2)$$

$$|2\rangle = s|\alpha\beta\rangle + c|\beta\alpha\rangle, \qquad (3)$$

$$|3\rangle = |\beta\beta\rangle, \qquad (4)$$

and

$$|4\rangle = c|\alpha\beta\rangle - s|\beta\alpha\rangle, \qquad (5)$$

where

$$s = \frac{1}{\sqrt{2}}\left(1 - \frac{x}{\sqrt{1+x^2}}\right)^{1/2},$$
$$c = \frac{1}{\sqrt{2}}\left(1 + \frac{x}{\sqrt{1+x^2}}\right)^{1/2}. \qquad (6)$$

Here, $x = B/B_0$ is the magnetic field intensity measured in units of the internal magnetic field B_0, which is defined by

$$B_0 = \frac{h\nu_0}{g_e\mu_e + g_\mu\mu_\mu} = \frac{2\pi\nu_0}{\gamma_e + \gamma_\mu}. \qquad (7)$$

The value of B_0 for free Mu is 1585 G. The energy diagram, the so-called "Breit–Rabi" diagram, is shown in Fig. 2.

With increasing longitudinal field, the μ^+

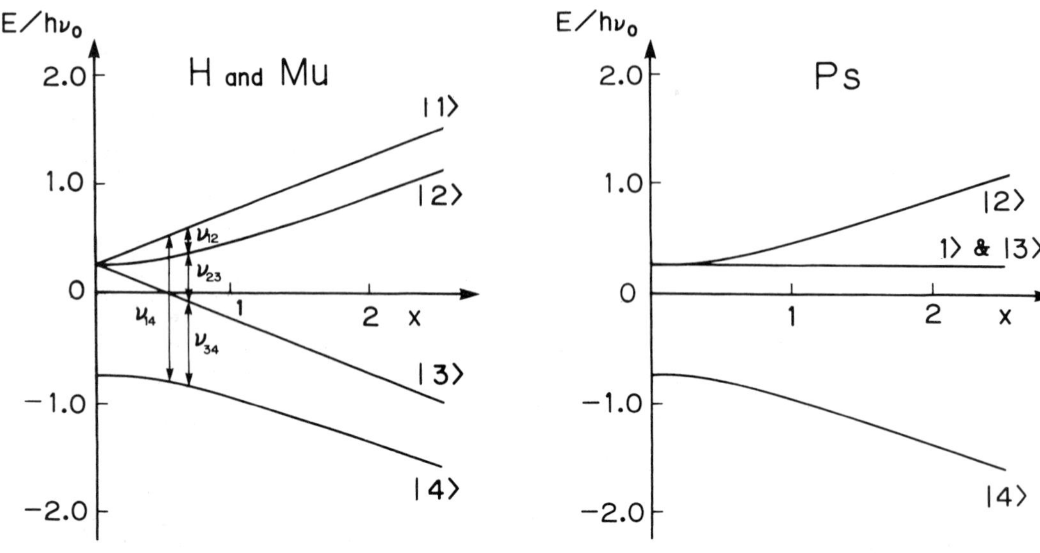

FIG. 2. Breit–Rabi diagram of H, Mu, and Ps. The energy levels of H, Mu, and Ps are described by common equations when the magnetic field is normalized to the internal magnetic field as $x = B/B_0$, and the energy normalized to the hyperfine coupling energy as $E/h\nu_0$. This results in very similar diagrams for H and Mu. For Ps, the levels $|1\rangle$ and $|3\rangle$ are degenerate at any magnetic field intensity as a result of a cancellation of the magnetic moments of positron and electron. B_0 = 505 G for H, 1585 G for Mu, and 36 245 G for Ps. ν_0 = 1420 MHz for H, 4463 MHz for Mu, and 203 400 MHz for Ps.

and e^- spins become decoupled as a result of the Paschen–Back effect, and the consequence is an increase in the observable μ^+ polarization,

$$P = \tfrac{1}{2}\left(1 + \frac{x^2}{1 + x^2}\right). \tag{8}$$

"Longitudinal field repolarization" refers to an experiment in which the μ^+ spin polarization is measured as a function of the longitudinal magnetic field. Here, not only the existence of Mu is tested, but also the *hf* coupling constant is roughly determined using Eq. (8).

1.3 Muonium in Low Transverse Field

In a low transverse magnetic field, the μ^+ spin in triplet Mu evolves (ν_{12} and ν_{23} in Fig. 2) according to the Larmor precession frequency,

$$\nu_M = (\gamma_M/2\pi)B, \tag{9}$$

where γ_M is the gyromagnetic ratio of Mu, namely,

$$\frac{\gamma_M}{2\pi} = \frac{1}{2}\frac{g_e\mu_e - g_\mu\mu_\mu}{h} = \frac{1}{2}\frac{\gamma_e - \gamma_\mu}{2\pi}$$
$$= -1.394 \text{ MHz/G}. \tag{10}$$

Since the magnetic moment of Mu is nearly equal to that of the electron, the Larmor precession of Mu is about 100 times faster than that of the μ^+:

$$\nu_\mu = (\gamma_\mu/2\pi)B, \tag{11}$$

where

$$\gamma_\mu/2\pi = g_\mu\mu_\mu/h$$
$$= 0.013\,5544 \text{ MHz/G}. \tag{12}$$

Thus, one can easily discriminate between μ^+ and Mu. At a moderate transverse field, ν_{12} and ν_{23} are not the same and the Mu precession contains a low-frequency beat with

$$\Delta\nu = \frac{\nu_{12} - \nu_{23}}{2} \sim -\frac{1}{4}\frac{\gamma_e^2 B^2}{2\pi\nu_0}. \tag{13}$$

From the observed beat, one can determine ν_0.

1.4 Muonium in High Transverse Field

At a sufficiently high field ($x^2 \gg 1$), the μ^+ spin evolution manifests two frequencies at

$$\nu_{12} = \frac{\nu_0}{2} - \frac{\nu_0}{2}\left[(x^2 + 1)^{1/2} - \frac{\gamma_e - \gamma_\mu}{\gamma_e + \gamma_\mu}x\right]$$
$$\simeq \frac{\nu_0}{2} - \nu_\mu \tag{14}$$

and

$$\nu_{34} = \frac{\nu_0}{2} + \frac{\nu_0}{2}\left[(x^2+1)^{1/2} - \frac{\gamma_e - \gamma_\mu}{\gamma_e + \gamma_\mu}x\right]$$
$$\simeq \frac{\nu_0}{2} + \nu_\mu. \quad (15)$$

From a pair of observed frequencies, ν_{12} and ν_{34}, one can precisely determine ν_0. A direct measurement of ν_0 is also possible by applying an rf field so as to cause transitions between the hyperfine doublet at zero field or between $|1\rangle$ and $|2\rangle$ ($|3\rangle$ and $|4\rangle$) at high field.

2. BEHAVIOR OF MUONIUM IN MATTER

Long-lived muonium has been identified by its characteristic precession frequency [Eq. (9)] in various substances: in gases [rare gases (except He and Ne), H_2, N_2, and vapors of molecular gases], in liquids [cryoliquids (including all liquified rare gases), water, alcohols, and some saturated organic liquids], in insulating solids (e.g., quartz, alkali halides, metal oxides, ice, and some organic solids), and in semiconductors (Si, Ge, GaAs, etc.). The muonium is formed when the ionization potential of host atoms/molecules is approximately less than the binding energy (13.5 eV) of Mu so that the electron transfer from a host atom to the μ^+ is energetically possible. The exceptional media are He and Ne, where the ionization energy (24.6 and 21.6 eV, respectively) is much larger than 13.5 eV. No muonium exists in metals, since the μ^+ charge is shielded by free electrons.

2.1 Muonium Chemistry

Like the hydrogen atom, Mu is a reactive species, and its reaction rate is measured from enhanced relaxation of the observed signal of Mu spin precession as a function of the amount of added reactants. This technique has been applied to studies of various reactions of Mu (substitution, abstraction, and spin-conversion reactions). The difference between the Mu chemical reactions, such as Mu + $F_2 \rightarrow$ MuF + F, and the analogous reactions of H atoms reveal remarkable isotope effects, which can be ascribed to quantum tunneling of Mu due to its light mass (Garner *et al.*, 1978).

When Mu reacts rapidly in a substance, it is chemically incorporated into a molecule. When it settles down in a diamagnetic species, such as MuOH, there is no hyperfine interaction for the μ^+, whose spin therefore precesses with approximately the frequency of a bare μ^+ [Eq. (11)]. Such Mu states are called "diamagnetic muons," and their observed polarization is denoted by "P_D." The fractions of diamagnetic muons and muonium formed in typical substances are given in Table 3.

Table 3. μ^+ and Mu formation in matter. P_D, polarization of diamagnetic muon; P_M, polarization of muonium; P_R, polarization of Mu-substituted free radicals; P_L, lost (or missing) polarization.

Substance		P_D	P_M	P_R	P_L
He	Gas	1.0	0	0	0
Ne	Gas	1.0	0	0	0
Ar	Gas	0.25	0.75	0	0
Kr	Gas	0	1.0	0	0
Xe	Gas	0	1.0	0	0
H_2	Gas, 3 atm	0.4	0.6	0	0
N_2	Gas, 1 atm	0.15	0.6	0	0
H_2O	Liq.	0.62	0.20	0	0.18
CH_3OH	Liq.	0.61	0.20	0	0.20
Hexane	Liq.	0.65	0.13	0	0.22
CCl_4	Liq.	1.0	0	0	0
C_6H_6	Liq.	0.15	0	0.65	0.20
Polyethylene		0.64	—	—	—
Quartz	Crystal	0.21	0.7	—	0.1
Graphite	Crystal	1.0	0	0	0
Diamond	Crystal, 4.2 K	0.1	0.2	0.12[a]	0.6

[a]The value of P_R for diamond refers to Mu*.

When Mu adds to unsaturated bonds, Mu-containing free radicals are produced. The μ^+ spin evolution in Mu-substituted free radicals is complicated because of the splitting of energy levels caused by superhyperfine coupling of the unpaired electron with the nuclear spins. At a sufficiently high field, however, the superhyperfine interactions are decoupled and the μ^+ spin evolves according only to its own hf coupling with the unpaired electron spin, showing the two frequencies given in Eqs. (14) and (15). The hyperfine frequency ν_0 is, however, considerably different from the free value. Such a state can be regarded as being a pseudo Mu with a larger μ^+–e^- distance. Different Mu-substituted free radicals can be identified from the hf constant ν_0 (Roduner *et al.*, 1978). Typical examples are shown in Fig. 3. Various Mu-containing free radicals are produced, and their structures have been studied. The fraction of Mu-substituted radicals, denoted by P_R, is presented also in Table 3.

The sum of P_D, P_M, and P_R is not necessarily equal to 1. The fraction of unobservable μ^+ spin polarization in a transverse field precession, $P_L = 1 - P_D - P_M - P_R$, is called "lost polarization" or "missing polarization" (see Table 3). Missing polarization may arise when the rate of the reaction that leads to diamagnetic muons or to Mu-substituted free radicals is not very high, i.e., on the order of the hyperfine frequency (4463 MHz) or less, and the μ^+ spin precession becomes out of phase. Missing polarization can be restored when a high longitudinal field is applied.

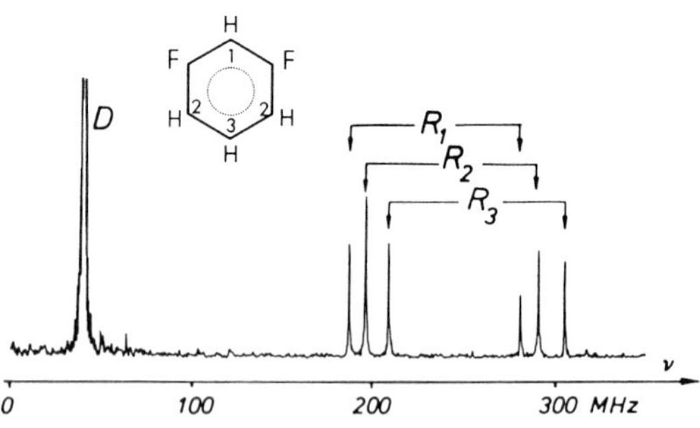

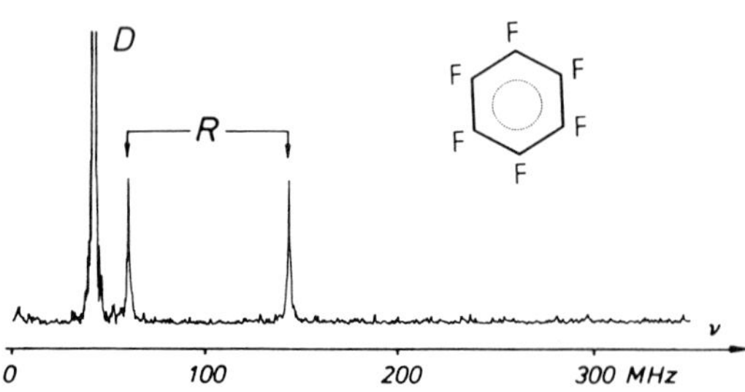

FIG. 3. Muon precession frequencies obtained with *m*-$C_6H_4F_2$ (top) and with C_6F_6 (bottom); from Roduner *et al.* (1982). *D* is the precession of muons in a diamagnetic environment. R_1–R_3 are the Mu-substituted radicals formed when Mu adds to the bond indicated by the number. For C_6F_6, only one kind of radical species is formed with Mu adding to any bonding.

Muonium states and Mu-substituted free radicals, when coupled with nearby nuclear moments, can be studied by the muon-level–crossing resonance (LCR) method, as first proposed by Abragam (1984), in which the μ^+ spin polarization is measured in a variable longitudinal magnetic field. At a particular magnetic field, where one of the Breit–Rabi levels crosses another one involving a nuclear spin flip, resonant transfer of polarization from μ^+ to the nuclear spin occurs. This results in a decrease in the μ^+ spin polarization. Each group of equivalent nuclear spins gives rise to a single resonance at a field B_R given by

$$B_R = \frac{1}{2}\left[\frac{A_\mu - A_n}{\gamma_\mu - \gamma_n} - \frac{A_\mu + A_n}{\gamma_e}\right], \quad (16)$$

where A_μ and A_n are the hyperfine coupling constants of the μ^+ and the nucleus, respectively, with the unpaired electron, and γ_μ, γ_e, and γ_n are the gyromagnetic ratios of the μ^+, e^-, and nucleus. The first successful application of this method was carried out by Kiefl *et al.*, (1986), who measured the modulated asymmetry $A^+ - A^-$ of integral counts of μ–e decay as a function of the longitudinal field. An example of the LCR spectrum is given in Fig. 4.

2.2 Muonium States in Solids

Muonium is formed in most insulators and semiconductors. Typical values of the hyperfine constant of Mu in solids are presented in Table 4. In most substances (quartz is exceptional), the hyperfine constant is smaller than the vacuum value; this is ascribed to the increased size of Mu in the crystal (Brewer *et al.*, 1973).

It is very important to study the location and local electronic structure of hydrogen impurities of an extremely small amount in

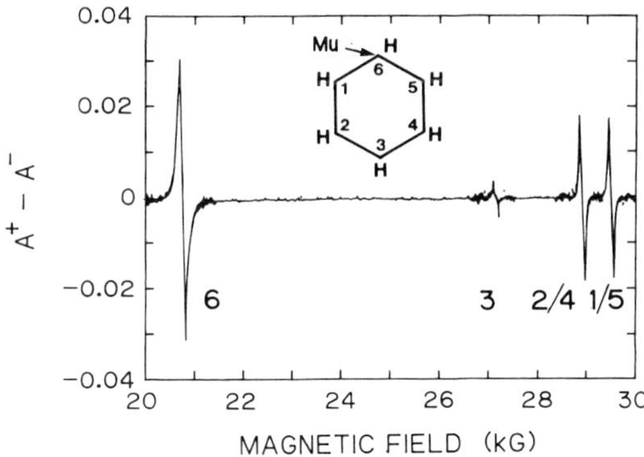

FIG. 4. Muon level-crossing spectrum of C_6H_6Mu in neat benzene at 300 K (Percival, 1987). The muon spin polarization is seen to be depolarized resonantly at four magnetic field intensities, from which the hyperfine frequencies are calculated by using Eq. (16) as 126.1 MHz for H(6), −25.1 MHz for H(1,5), 7.47 MHz for H(2,4), and 36.2 MHz for H(3), where H(*n*) indicates the hydrogen atom position.

semiconductors, and this can be done experimentally only by using the muonium probe. Comprehensive studies have been carried out, as reviewed by Patterson (1988). The μSR frequency spectrum at high transverse field for all the diamond-lattice semiconductors at low temperatures reveals the presence of not only a normal Mu but also an anomalous one, characterized by very low frequencies that depend on the crystal orientation with respect to the applied field. The hyperfine structure of the normal Mu is isotropic, while that of the anomalous one, which is denoted by Mu*, is anisotropic. The experimental data are well reproduced (Patterson *et al.*, 1978) by introducing an anisotropic hyperfine coupling term [generalization of Eq. (1)],

$$A_\perp(I_xS_x + I_yS_y) + A_\parallel I_zS_z. \qquad (17)$$

The hyperfine parameters $A_\perp$ and $A_\parallel$ obtained experimentally for typical solids are presented in Table 4.

The level-crossing resonance method was applied to Mu* in GaAs, and the hyperfine coupling constants of the nearest-neighbor ^{75}As and ^{69}Ga were determined (Kiefl *et al.*, 1987). From this information, the site of Mu* was unambiguously determined; while the normal Mu is in a tetrahedral interstitial site, the Mu* is located at or near the bond center with the unpaired electron centered on the two nuclei. Similar information was obtained also for Si using the hyperfine coupling with ^{29}Si. The dynamical behavior with characteristic temperature dependences of Mu and Mu* in semiconductors is an interesting problem.

The relaxation of the Mu signal in a solid is related to Mu diffusion and trapping characteristics. In quartz (Brewer, 1981), the small Mu relaxation rate ($\lambda = 0.2\ \mu s^{-1}$) below 80 K is due to an interaction of Mu with the nuclear dipole moment of ^{29}Si, whose natural abundance is 4.7%. Above 80 K, the relaxation rate increases rapidly with temperature until the precession signal becomes unobservable around 140 K. This relaxation is caused by Mu diffusion among sites that are equivalent but have different orientations of the principal axes of the hyperfine tensor. Above 150 K, the precession signal becomes observable and its relaxation rate decreases

Table 4. Typical Mu and Mu* states in solids. The hyperfine constants given here are at the lowest temperatures.

Substance	State	Hyperfine constant		
		A	$A_\parallel$	$A_\perp$
Quartz	Mu	4463		
Diamond	Mu	3721(21)		
	Mu*		167.983(57)	−392.586(55)
Si	Mu	2006.3(2)		
	Mu*		16.819(11)	92.59(5)
Ge	Mu	2359.5(2)		
	Mu*		27.269(13)	131.037(34)
GaP	Mu	2914(5)		
	Mu*		219.0(2)	79.48(7)
GaAs	Mu	2883.6(3)		
	Mu*		217.8(2)	87.74(6)
ZnS	Mu	3547.8(3)		

with temperature because of a "motional narrowing" effect. The temperature dependence of the Mu and Mu* relaxation rates has also been extensively studied for Si and Ge. The dynamic change of a Mu state to another Mu or μ^+ state can be studied by the rf resonance method under a longitudinal field (Nishiyama *et al.*, 1985).

2.3 Quantum Diffusion of Muonium

The diffusion of muons and muonium is expected to show anomalous quantum behavior because of the muon's unique mass, which is only $\frac{1}{9}$ of the hydrogen mass. Kagan and Klinger (1974) predicted a divergent behavior ($T^{-\alpha}$ dependence, α = 7–9) of the diffusion rate toward $T \rightarrow 0$. The diffusion of the μ^+ in metals shows such a unique behavior characteristic of quantum diffusion, namely, the diffusion rate (probed by nuclear dipolar relaxation) decreases with the decrease of temperature, but reaches a minimum at around 70 K and then increases according to a power law with $\alpha \sim 0.6$ (Kadono *et al.*, 1985). The small power was explained as the effect of conduction electrons by Kondo (1984) and by Yamada (1984). A longitudinal-field spin-relaxation experiment on Mu in alkali halides by Kiefl *et al.* (1989) revealed an even more dramatic effect with a sharper increase in the diffusion rate with $\alpha \sim 3$ toward lower temperature, as shown in Fig. 5.

The Mu was also observed to be diffusing in ice at temperatures as low as 8 K; the low-temperature diffusion is considered to be due to quantum tunneling (Leung *et al.*, 1987). Muonium radicals can traverse a chain of $(CH)_n$ bonds. In *cis*-polyacetylene, Mu radicals are stable, while in *trans*-polyacetylene, they traverse as a soliton wave (Nagamine *et al.*, 1984).

3. BASIC PROPERTIES OF POSITRONIUM

3.1 Structure and Annihilation of Positronium

Positronium comprises a positron and an electron and is the lightest hydrogenlike atom. Since its reduced mass ($\mu = m_e/2$) is approximately half that of the H atom, the fundamental atomic parameters can be compared to those of the H atom through a scaling factor of 2. Hence, its ionization potential is 6.8 eV, the Bohr radius is 0.105 nm, and so on (see Table 2). Since its center of mass is situated just between the positron and the electron, its Bohr radius is, itself, the diameter, and there is no nucleus on which the center of mass can be located; i.e., the Born–Oppenheimer approximation cannot be applied. Because of the extremely small mass and absence of a nucleus, the chemical properties of the Ps are substantially different from those of the Mu and H atoms.

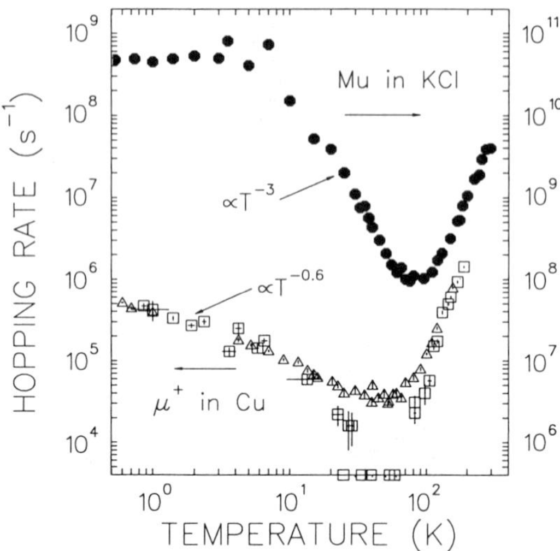

FIG. 5. Temperature dependence of the diffusion rate of Mu in KC*l* (Kiefl *et al.*, 1989) and μ^+ in Cu metal (Kadono *et al.*, 1985). Each shows a minimum at around 70 K, characteristic of quantum diffusion, though the power parameter in the low-temperature regime is different.

Ps is a system of two Dirac particles, and the spin Hamiltonian has four eigenstates as in the cases of H and Mu (Fig. 1). The singlet state ($|0,0\rangle$) is called parapositronium (*p*-Ps), and the triplet states ($|1,-1\rangle$, $|1,0\rangle$, and $|1,1\rangle$) are called orthopositronium (*o*-Ps). Following the selection rule of positron annihilation (see POSITRON ANNIHILATION SPECTROSCOPY), *p*-Ps annihilates with an intrinsic lifetime of

$$1/\lambda_S = 125 \text{ ps} \tag{18}$$

into two gamma rays, each having an energy of 511.0 keV. The *o*-Ps annihilates with an intrinsic lifetime of

$$1/\lambda_T = 142 \text{ ns} \tag{19}$$

into three gamma rays with a continuous energy spectrum ranging from 0 to 511.0 keV.

Various reactions can affect Ps annihilation, and the processes leading to a reduction in the *o*-Ps lifetime are collectively called "quenching." "Pickoff" annihilation is an important quenching process which takes place whenever an *o*-Ps is produced in matter, since the positron in the *o*-Ps can annihilate with one of the surrounding electrons having a singlet spin configuration to the e^+. The rate of pickoff annihilation is proportional to the overlap integral of the wave function of the outer electrons with the e^+. Although the pickoff annihilation mainly controls the lifetime of the *o*-Ps in condensed matter, it only slightly affects the lifetime of the *p*-Ps, since the *p*-Ps annihilation rate is already fast.

Other quenching processes are Ps compound formation, oxidation, and spin-conversion reactions. Examples are

$$o\text{-Ps} + \text{Cl} \rightarrow \text{PsCl} \quad \text{(compound formation)}, \tag{20}$$

$$o\text{-Ps} + \text{Fe}^{3+} \rightarrow e^+ + \text{Fe}^{2+} \quad \text{(oxidation)}, \tag{21}$$

and

$$o\text{-Ps} + \text{Fe}^{2+} \rightarrow p\text{-Ps} + \text{Fe}^{2+} \quad \text{(spin exchange)}. \tag{22}$$

The *o*-Ps lifetime can also be reduced in a magnetic field; this effect is called "magnetic quenching." Since e^+ and e^- have the same mass but opposite signs of the electric charge, the consequence of the Zeeman effect on the Ps involves a mixing of only the two $m = 0$ spin states, $|1,0\rangle$ and $|0,0\rangle$. At low fields, the field perturbed annihilation rates are approximated as

$$\lambda(|1,0\rangle) \sim \lambda_T + x^2\lambda_S/4 \tag{23}$$

and

$$\lambda(|0,0\rangle) \sim \lambda_S, \tag{24}$$

and at high fields, both reach the same rate,

$$\lambda(|1,0\rangle) = \lambda(|0,0\rangle) = (\lambda_S + \lambda_T)/2. \tag{25}$$

Here, $x = B/B_0$ is the field intensity measured in units of the internal field of the Ps, as in the case of Mu, Eq. (7),

$$B_0 = A/2g_e\mu_e. \tag{26}$$

3.2 Formation of Positronium

Positronium is formed when a positron picks up an electron from the substance. The simplest case is electron abstraction in a positron-molecule collision:

$$e^+ + M \rightarrow \text{Ps} + M^+. \tag{27}$$

The threshold (E_{th}) of this reaction is 6.8 eV below the ionization potential (I_M) of molecule M; $E_{\text{th}} = I_M - 6.8$ eV. Its cross section is of the same order as those of excitation and ionization. Ps formation through reaction (27) is considered to be efficient when the energy of the incident e^+ is in the Øre gap, i.e., between E_{th} and the first electronic excitation energy (E_{ex}), where Ps formation is the only possible electronic process. The idea that Ps formation takes place in the energy region of the Øre gap is called either the "Øre model" or the epithermal model (Goldanskii, 1968). This model is considered to be valid for rare gases, where the e^+ energy loss, being due solely to elastic collisions, is slow in the energy region of the Øre gap.

There is another model, called the spur model, where the Ps is considered to be formed as a result of the recombination of an e^+ with one of the excess electrons (e^-_{excess}) generated in the terminal positron spur (Mogensen, 1974),

$$e^+ + e^-_{\text{excess}} \rightarrow \text{Ps}. \tag{28}$$

Since excess electrons are involved in this mechanism, Ps formation should parallel the observations of radiation chemistry, such as electron and hole scavenging reactions as well as solvation effects. For the condensed phases of molecular substances, experimental results concerning Ps formation and radiation chemistry show strong parallelism in support of the spur model (Abbe *et al.*, 1981; Ito, 1988). In high-pressure gases, there is also some evidence for the spur reaction process (Jacobsen, 1986).

3.3 Methods for Studying Positronium

The common techniques of positron-annihilation spectroscopy, measurements of the lifetime, angular correlation, Doppler-broadened positron annihilation radiation (DBPA), and three-photon annihilation are also used in studying Ps (see POSITRON ANNIHILATION SPECTROSCOPY for the details of the techniques). Examples of the lifetime and angular correlation spectra in water are shown in Fig. 6. The lifetime spectrum is decomposed into the *p*-Ps, e^+, and *o*-Ps components in the order of increasing lifetime. In the angular correlation spectrum, e^+ annihilation and *o*-Ps pickoff annihilation produce broad components due to large momentum of the annihilating electron. Annihilation of *p*-Ps gives a narrow component, since its center-of-mass momentum is small. The energy of the DBPA carries momentum information that is equivalent to that obtained from an angular correlation measurement; however, because of insufficient energy resolution, DBPA is used only complementarily. The magnetic quenching effect provides an excellent means to confirm the existence of Ps. In special cases, the formation of an excited Ps is detected by measuring the Lyman-α photons (243 nm) of the Ps (Mills *et al.*, 1975).

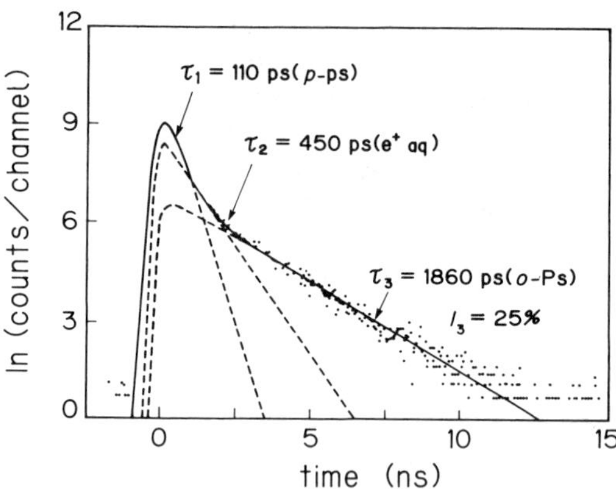

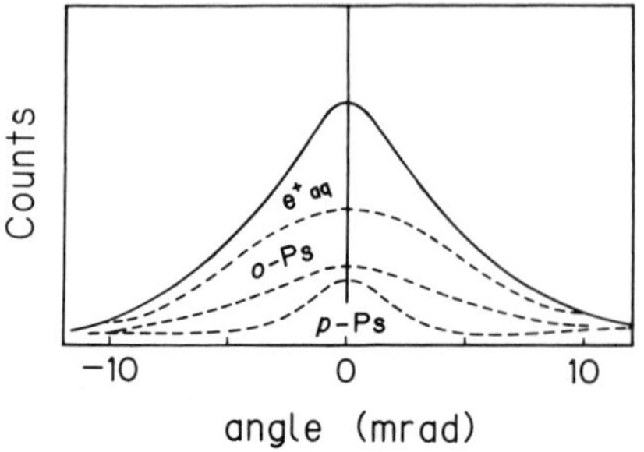

FIG. 6. Typical lifetime spectrum and angular correlation of positron-annihilation gamma rays in water. The observed spectra (solid lines) are decomposed into three components (broken lines) corresponding to different positron states: *p*-Ps, solvated positron e^+_{aa}, and *o*-Ps.

4. BEHAVIOR OF POSITRONIUM IN MATTER

4.1 Positronium in Gases

The positron lifetime spectrum in gases consists of three components: *p*-Ps, e^+, and *o*-Ps. The *p*-Ps annihilates with an intrinsic lifetime of 125 ps. The *o*-Ps annihilation rate is the sum of its intrinsic rate (λ_T = 1/142 ns^{-1}) and the pickoff annihilation rate, which is proportional to the gas density (n_M),

$$\lambda_t = \lambda_T + \pi r_e^2 \, c n_M {}^1Z_{\text{eff}}. \tag{29}$$

Here, ${}^1Z_{\text{eff}}$ is the effective number of electrons having a singlet spin configuration to e^+ per molecule, and $4\,{}^1Z_{\text{eff}}$ is to be compared with the total number of electrons (Z) in a molecule. In many cases, $4\,{}^1Z_{\text{eff}}$ is smaller than Z (see Table 5), since the e^+ in the Ps cannot penetrate into molecules because of a repulsion incurred by its partner electron. When $4\,{}^1Z_{\text{eff}} > Z$, as for O_2 and CCl_2F_2, it is an indication of a chemical reaction between the *o*-Ps and the molecules.

Positrons that have not formed a Ps further slow down and reach equilibrium annihilation. The shoulder in Fig. 7 corresponds to annihilation during this slowing-down process, and the width (τ_s) is the slowing-down time to the equilibrium state, which is inversely proportional to the gas density. This shoulder is observed mainly in monatomic gases; the slowing down is normally too fast to be observed in polyatomic gases. The equilibrium annihilation rate is proportional to the gas density as

$$\lambda = \pi r_e^2 \, c n_M Z_{\text{eff}}, \tag{30}$$

where Z_{eff} is the effective number of electrons per molecule. Normally, Z_{eff} is greater than the total number of electrons in a molecule, because the electron density at the e^+ is en-

Table 5. Annihilation parameters of e^+ and o-Ps in selected gases.

Gas	Molecular density (Amagat)	No. of electrons Z	$^1Z_{eff}$	Z_{eff}	$\tau_s n_M$ (Amagat · ns)
He	18–60	2	0.35	3.0	1600
Ne	7–39	10	0.23	6.0	1700
Ar	1–280	18	0.35	18–27	360
Kr	1–117	36	0.45	40–67	290
Xe	1–6	54	1.03	320	200
H_2	19–170	2	0.186	12–13	—
N_2	1–234	14	0.26	19–31	—
CO_2	1–48	22	0.48	50–120	—
O_2	7–215	16	14–86	19–26	—
NO_2	—	19	~10	—	—

hanced by e^+ interaction with the induced electric dipole moment. When Z_{eff} is far greater than Z, this is an indication of a strong interaction between the e^+ and the molecule.

The linearity in n_M of Eqs. (29) and (30) holds over a wide range of molecular densities. At high densities, both the o-Ps and e^+ lifetimes deviate from the linearity as a result of their localization. At high densities, a Ps can be localized in a hole that is made by the Ps's repulsive force. This is called a "Ps bubble" (Ferrell, 1957); the o-Ps lifetime becomes longer than that in the average gas density. The size of a Ps bubble in liquid He is about 1.5 nm, and the o-Ps lifetime is ≈100 ns. Positrons can be localized in a cluster or "droplet" of molecules formed by the positron's attractive force, in which case the e^+ lifetime becomes shorter.

A study of Ps-gas interactions (Griffith and Heyland, 1978) involves the difficulty of stopping high-energy positrons in a gaseous media; it has been customary to use high-pressure gases. By using silica aerogel, a low-density aggregate of ultrafine silica powder, as the media for producing Ps, it is possible to inject positrons in low-pressure gases at near to 1 atm (Kakimoto *et al.*, 1990). By combining this method of Ps formation with the magnetic quenching technique, the process of Ps thermalization at very low energies (from several tenths of eV down to thermal) was studied, and the momentum transfer cross section was obtained.

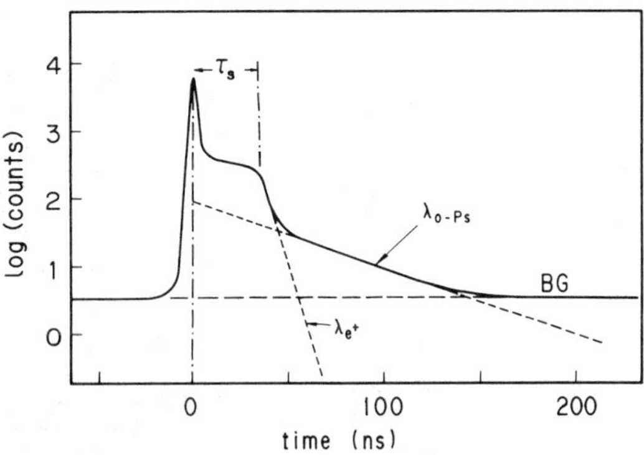

FIG. 7. Positron lifetime spectrum in He gas at 7.5 K (Canter *et al.*, 1975). The prompt component contains p-Ps annihilation and other fast annihilation in the source and wall of the chamber, and the longest-lifetime component corresponds to the annihilation of o-Ps. The intermediate component is due to annihilation of e^+, whose lifetime spectrum shows a nonexponential shoulder (time width τ_s) followed by an exponential decay corresponding to e^+ annihilation in equilibrium state.

4.2 Positronium in Insulators

Ps is formed in most insulating materials. When a p-Ps annihilates from its free state in crystals of insulating materials, the p-Ps exhibits a sharp, narrow component corresponding to its nearly zero momentum. Satellite peaks are also observed in single crystals of ice (Mogensen *et al.*, 1971; West *et al.*, 1981), quartz (Greenberger *et al.*, 1970; Berko *et al.*, 1977), and ionic crystals (Kasai *et al.*, 1988) at the projections of the reciprocal-lattice vectors, reflecting the symmetry of the crystal lattice. When p-Ps annihilates from a trapped state, the width of the peak becomes broader as a result of the zero-point momentum. The o-Ps is mostly identified from its long lifetime. The long lifetime, however, is not direct evidence for the existence of o-Ps, and for a critical identification, it is necessary to resort to the magnetic quenching effect.

There are at least three components in most

Table 6. Lifetimes of e^+ and *o*-Ps in selected substances.

Substance	e^+ lifetime (ns)	*o*-Ps lifetime (ns)	*o*-Ps intensity (%)
Al	0.166	—	No Ps formation
Si crystal	0.220	—	No Ps formation
Si monovacancy	0.270	—	No Ps formation
Si divacancy	0.320	—	No Ps formation
H_2O	0.4	1.85	28
CH_3OH	0.4	3.9	21
Hexane	0.4	3.9	42
Benzene	0.45	3.15	43

lifetime spectra of liquids and solids of insulators. They are *p*-Ps, e^+, and *o*-Ps in the order of increasing lifetime (see Fig. 6 and Table 6). When different phases are contained, as in polymers, there can be more positronium components. The longest lifetime component is normally due to *o*-Ps pick-off annihilation in the holes of insulators. Because of an exchange interaction between e^- in Ps and the electrons in the surrounding molecules, Ps is repelled to free volume regions. The *o*-Ps lifetimes in the holes in solids and in bubbles in liquids are well correlated with the radius (R) as

$$\frac{1}{\tau_3} = 2\left[1 - \frac{R}{R + \Delta R} + \frac{1}{2\pi}\sin\frac{2\pi R}{R + \Delta R}\right] \text{ns}^{-1}, \tag{31}$$

which describes the pickoff annihilation of *o*-Ps based on the overlap probability of the e^+ in *o*-Ps with one of the electrons in the inner surface of the hole with an effective thickness of ΔR (Tao, 1972; Eldrup *et al.*, 1981). In Fig. 8, the experimental and the theoretical [Eq. (31)] results for the correlation between the hole size and the *o*-Ps lifetime are shown. The *p*-Ps also enters the same holes, but mostly annihilates by itself. In the self-annihilation of *p*-Ps, the momentum associated with the zero-point oscillation in the holes is reflected in the linewidth of the angular correlation or the Doppler-broadened annihilation radiation. For a spherical potential well of radius R with infinite height, the width of the angular correlation curve (Γ (FWHM)) is given by

$$\Gamma R = 1.66 \text{ mrad nm}. \tag{32}$$

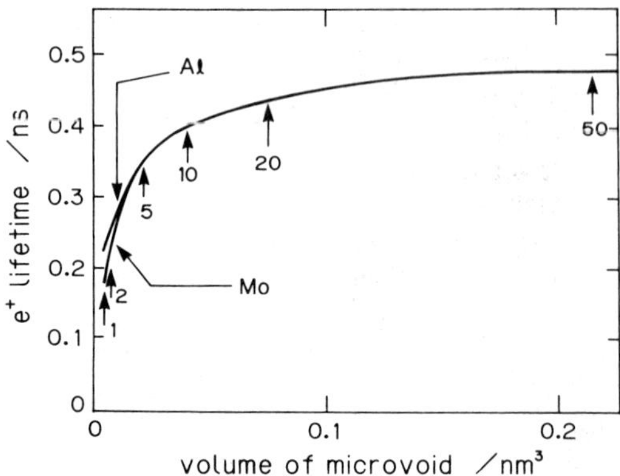

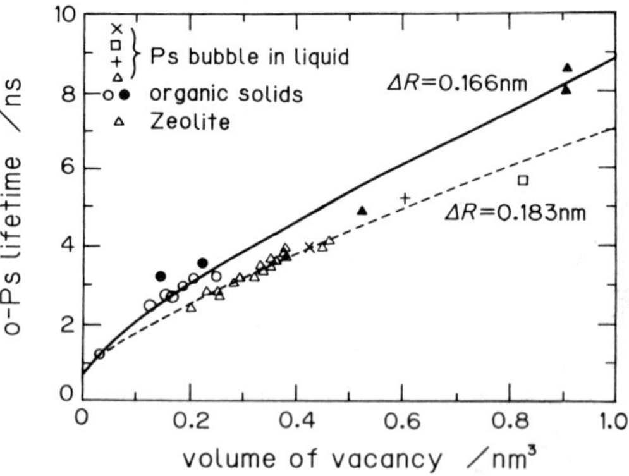

FIG. 8. Sensitivities of e^+ lifetimes toward microvoid size in metals (upper) and of *o*-Ps lifetimes toward vacancies or defects in insulators (lower). The upper figure is the theoretical expectation for Mo and Al (reproduced from Hautojarvi *et al.*, 1977), and the number in the figure describes the number of monovacancies contained in the microvoid. The lower figure contains experimental data points together with the results of fitting Eq. (31) (the solid and the broken lines) with the parameters indicated (Nakanishi *et al.*, 1988).

4.3 Positronium in a Metal Surface and in Voids

Positronium is not formed in the bulk of metals and alloys because of a shielding effect of the free electrons, and the positron exists as a Bloch state with the existence probability peaked at interstitial sites or, in the presence of vacancy-type defects, efficiently trapped in one of them. Because of the sensitivities of positron annihilation to the electron density and to the electron momentum, trapping results in an increased positron lifetime and a decreased momentum distribution (see POSITRON ANNIHILATION SPECTROSCOPY).

Positronium has been found in the voids of Al, Nb, and V (Jensen *et al.*, 1985; Hasegawa *et al.*, 1989), and the Ps states are known to be strongly affected by the internal and surface conditions of the voids where gas impurities are normally contained. Positronium is also formed when an e^+ is re-emitted from the surface of metals, by picking up a free electron (Mills, 1983). When an e^+ is injected at a low glancing angle to a metal surface, Ps is also produced by an electron pickup reaction (Gidley *et al.*, 1987). The Ps formation potential (E_{Ps}) at a surface is given by

$$E_{Ps} = \phi_+ + \phi_- - 6.8 \text{ eV}, \qquad (33)$$

where ϕ_+ and ϕ_- are the work functions of the positron and electron, respectively.

5. ADVANCED SOURCE OF MUONS AND POSITRONS

5.1 Ultralow-Energy Muons and Muonium Beams

When pions decay in flight, high-energy muons are produced. Muons with a smaller energy can be obtained when pions are stopped and made to decay near to the surface of the pion-producing target. The energy of these muons (so-called surface muons) is 4.1 MeV, which is normally sufficiently low for experiments in which a high μ^+ stopping rate in thin targets or in gases is required. For future applications, however, muons with a much lower energy are desired. An attempt to extract positive muons from the surface of materials, like extracting positrons of negative work function, has been made, and muons of about 10 eV were found to be emitted from a thin surface of rare gas solids (Harshman *et al.*, 1986, 1987). This phenomenon is interpreted as being due to the large band gap of the surface material; the ionization loss process in the final stage of muon stopping ceases, and still energetic (~10 eV) muons come out of the surface. The typical yield in solid Ar is around 10^{-5} per incident surface μ^+. Very recently, Morenzoni *et al.* (1994) observed that these μ^+'s retain the initial polarization.

Muonium can be produced as a very low-energy beam by an electron pickup reaction of low-energy μ^+,

$$\mu^+ + M \rightarrow \text{Mu} + M^*. \qquad (34)$$

Such a scheme was used for studies of the muonium Lamb shift (Oram *et al.*, 1984; Woodle *et al.*, 1990), though the intensity is very small.

A striking phenomenon has been found: when μ^+ is stopped in a hot W target (Mills, *et al.*, 1986) or in SiO_2 powders (Beer *et al.*, 1986), a fraction of the μ^+'s that have diffused to the surface is evaporated in the form of thermal-energy Mu. The yield of the thermal Mu is large, of the order of 10% of the incident surface muons. The thermal Mu emitted into a vacuum is important and useful, not only for fundamental studies (QED tests), but also as a possible source of a slow μ^+ beam, which may be formed after the ionization of thermal Mu by resonant multiphoton absorption (Chu *et al.*, 1988). A dedicated attempt to produce an ultralow-energy μ^+ beam is in progress (Nagamine and Ishida, 1992).

5.2 Advanced Techniques of Positronium Studies

Because of recent developments of slow positron beam techniques (see POSITRON ANNIHILATION SPECTROSCOPY), various new techniques for Ps studies have become possible. The cross sections of positron collision, including those of Ps formation, have already been measured for many molecules (Griffith and Heyland, 1978; Humberston, 1979). Attempts are being made to enhance the intensity and quality, i.e., monochromaticity and beam size, of a slow positron beam in order to extend the technique to measurements of the differential cross sections and positron reactions. It also becomes possible to obtain an energy-controllable Ps beam. When positron beams are passed through either dilute gases or very thin foils, a Ps beam can be generated via the forward-peaked nature of the cross sections of the charge-exchange reactions [Eq. (27)]. The energy of the Ps beam can be controlled by varying the energy of the incident positrons (Charlton and Laricchia, 1990). Such a Ps beam should not only augment our understanding of various fundamental interactions of Ps, but should also be used as a probe of new techniques, such as low-energy positronium diffraction (Roellig *et al.*, 1987).

The formation of cold positrons is also an important subject. A successful technique has been to use a kind of Penning trap, the upstream part of which comprises a concentration gradient of coolant N_2 gas; the downstream comprises a low potential and a high-vacuum trapping region. Positrons were cooled by exciting the vibrational energy levels of N_2 molecules down to room temperature and were stored for more than 100 s (Surko *et al.*, 1989). Cold positrons will become a powerful tool in Ps studies, which will be extended to the field of exotic-particle science.

6. TEST OF QUANTUM ELECTRODYNAMICS

Atomic hydrogen is the simplest nuclear atom, and its hyperfine splitting constant is one of the most accurately measured physical quantities. The Lamb shifts and the 2s → 1s energy have also been precisely measured. However, a theoretical treatment of the hydrogen atom cannot preclude uncertainties resulting from the finite size, mass, and polarizability of the proton; a comparison of the experimental data with theory to better than a few ppm has thus not been possible. As a result, special interest has been given to purely leptonic atoms, Mu and Ps, which are free of any nuclear effect. Compared to the hydrogen atom, however, Mu and Ps involve more experimental uncertainties because of their short lifetimes.

The QED test can be performed by precise measurements of any observable quantities—the hf splittings, the annihilation rates, the fine structure of the excited energy levels, and so on. See, for instance, Sapirstein and Yennie (1990) in the book of Kinoshita (1990). The presently known values of the hyperfine splitting, the $2S_{1/2} \rightarrow 2P_{1/2}$ classical Lamb shift, and the $n = 2 \rightarrow n = 1$ transition energy for H, Mu, and Ps, both experimental and theoretical, are summarized in Table 7. The comparison shows that in the case of the Mu and Ps, QED has been tested to higher orders. As the experimental precision improves, further meaningful information concerning QED corrections will be revealed.

6.1 Hyperfine Structure

The best experimental value of the ground-state hyperfine frequency [ν_{hf} (1s)] for muonium was obtained by an rf resonance observation of the hyperfine frequencies (ν_{12} and ν_{34}) at a high longitudinal field (Mariam *et al.*, 1982). Muonium was produced in 1.5-atm Kr gas. The experimental result has a precision of 0.04 ppm and shows excellent agreement with the theoretical value.

The most precise experiment for determining ν_{hf} for positronium was carried out by Ritter *et al.* (1984). The transition from *o*-Ps to *p*-Ps caused by an rf field can be signaled by the quenching of long-lived positronium. Although the precision is 3 ppm, it reveals some discrepancy with the quoted theoretical value. This difference may be ascribed to uncalculated higher order terms of QED.

6.2 1s–2s Energy Interval and Lamb Shift

The production of thermal Mu from SiO_2 powders into a vacuum was used to observe 3-photon ionization of Mu in a pulsed muon beam at KEK (Chu *et al.*, 1988) and later at ISIS of the Rutherford Appleton Laboratory (Maas *et al.*, 1994). A pulsed laser beam in the ultraviolet range (wavelength of 244 nm) was applied to the Mu emission region. The latest experimental value is shown in Table 7. It has a precision of 0.03 ppm. This value is in excellent agreement with the calculated value of 0.002 ppm precision, which includes various higher order corrections of QED.

The 1s–2s level splitting in positronium was determined by Chu *et al.*, (1984). A pulsed laser was used to induce 3-photon ionization of Ps. For this purpose, positrons were first

Table 7. Comparison of the hyperfine coupling constants (ν_{hf}), the Lamb shifts $S = E_{2S_{1/2}} - E_{2P_{1/2}}$, and the $2s(F = 1)$–$1s(F = 1)$ energies E_{2s-1s} of the three hydrogenic atoms, between experiment and theory. The theoretical values are taken from Sapirstein and Yennie (1990).

		H	Mu	Ps
$\nu_{hf}^{(1s)}$ (kHz)	Exp.	1420 405.751 7667(9)[a]	4463 302.88(16)[b]	203 389 100(740)[c]
	Theor.	1420 405(2)	4463 303.11(1.3)	203 404 500(930)
S (MHz)	Exp.	1057.845(9)[d]	1062(13)[e]	8621.6(28)[f]
	Theor.	1057.864(11)[g]	1047.49(1)	8625.14
E_{2s-1s} (MHz)	Exp.	2 466 061 413.8(1.5)[h]	2455 529 002(80)[i]	1233 607 218.9(11)[j]
	Theor.		2455 528 934(4)	1233 607 228.4(19)

[a]Hellwig *et al.* (1970); Essen *et al.* (1971)
[b]Mariam *et al.* (1982)
[c]Ritter *et al.* (1984)
[d]Lundeen and Pipkin (1981)
[e]Weighted average of Oram *et al.* (1984) and Woodle *et al.* (1990)
[f]Weighted average of Mills *et al.* (1975) and Hatamian *et al.* (1987)
[g]$\langle r_p^2\rangle^{1/2} = 0.833$ fm is assumed. The uncertainty of ± 0.030 fm in $\langle r_p^2\rangle^{1/2}$ causes a respective error of ± 0.009.
[h]Beausoleil *et al.* (1987)
[i]Maas *et al.* (1994)
[j]Chu *et al.* (1984)

stored and then injected pulsewise into a clean Al target surface, which desorbed free thermal Ps in a vacuum.

A low-energy muonium beam (described in Sec. 5.1) contains metastable 2s muonium. Such a Mu(2s) beam was used to obtain information concerning the "classical" Lamb shift, namely, $S = E_{2S_{1/2}} - E_{2P_{1/2}}$, by applying a quenching electric field along the flight region, where Lyman-α x rays were detected. A more accurate measurement was carried out by causing the $2s \rightarrow 2p$ transition by using resonating microwaves. The average experimental value from the two experiments (Oram *et al.*, 1984; Woodle *et al.*, 1990) is compared with the theoretical value in Table 7. The present experimental value of 0.13% accuracy does not allow a precision test of various corrections of QED.

The classical Lamb shift in positronium was measured by using a positronium beam containing Ps(2s) (Mills *et al.*, 1975; Hatamian *et al.*, 1987). The microwave resonance was probed by the emission of Lyman-α x rays. The experimental value with 300-ppm accuracy is in good agreement with theory.

6.3 Positronium Annihilation Rate

The theoretical annihilation rate of p-Ps (Harris and Brown, 1957) is

$$\Gamma(1^1S_0 \rightarrow 2\gamma) = \frac{\alpha^5 mc^2}{2\hbar}\left[1 - \alpha\left(\frac{5}{\pi} - \frac{\pi}{4} - \frac{2}{3}\alpha^2 \ln\alpha + \ldots\right)\right] = 7.986654(1) \times 10^9\ \text{s}^{-1}, \quad (35)$$

where the $\alpha^2 \ln \alpha$ term was added by Caswell and Lepage (1979). The experimental p-Ps annihilation rate obtained by a technique involving the Zeeman mixing of the two $m = 0$ states (Gidley *et al.*, 1982) is 7.994(11) ns^{-1}, in agreement with the theory.

The theoretical annihilation rate of o-Ps (Adkins, 1983; Caswell and Lapage, 1979) is

$$\Gamma(1^3S_1 \rightarrow 3\gamma) = \alpha^6 \frac{mc^2}{\hbar}\frac{2(\pi^2 - 9)}{9\pi}\left[1 - (3.273 \pm 0.001)\alpha + \frac{1}{3}\alpha^2 \ln\alpha + \ldots\right] = 7.03830(7) \times 10^6\ \text{s}^{-1}. \quad (36)$$

The o-Ps annihilation rate has been measured directly by producing o-Ps in low-pressure gases or at the surface of solid materials in a vacuum; the values have been interpolated to an ideal noninteracting condition in order to obtain the intrinsic rate. All of the results are larger than the theoretical value. The most accurate value to date is 7.0482(15) $\times 10^6$ s^{-1} (Nico *et al.*, 1992), which is 6 SD above the theoretical value. This discrepancy has not yet been explained.

Works Cited

Abbe, J. Ch., Duplatre, G., Maddock, A. G., Talamoni, J., Haesler, A. (1981), *J. Inorg. Nucl. Chem.* **43**, 2603–2610.

Abragam, A. (1984), *C.R. Acad. Sci. (Paris) Ser. II* **299**, 95.

Adkins, G. S. (1983), *Phys. Rev. A* **27**, 530–532.

Beer, G. A., Marshall, G. M., Manson, G. R., Olin, A., Gelbart, Z., Kendall, K. R., Bowen, T., Halverson, P. G., Pifer, A. E. (1986), *Phys. Rev. Lett.* **57**, 671–674.

Berko, S., Maghgooie, M., Mader, J. J. (1977), *Phys. Lett.* **63A**, 335–338.

Beausoleil, R. G., MacIntyre, D. H., Foot, C. J., Hidum, E. A., Couillaud, B., Hänsch, T. W. (1987), *Phys. Rev.* **A35**, 4878–4881.

Brewer, J. H., Crowe, K. M., Gygax, F. N., Johnson, R. F., Patterson, B. D., Fleming, D. G., Schenck, A. (1973), *Phys. Rev. Lett.* **31**, 143–149.

Brewer, J. H. (1981), *Hyperfine Int.* **8**, 375–380.

Canter, K. F., McNutt, J. D., Roellig, L. O. (1975), *Phys. Rev. A* **12**, 375–385.

Caswell, W. G., Lepage, G. P. (1979), *Phys. Rev. A* **20**, 36–43.

Charlton, M., Laricchia, G. (1990), *J. Phys. B* **23**, 1045–1078.

Chu, S., Mills, A. P., Jr., Hall, J. L. (1984), *Phys. Rev. Lett.* **52**, 1689–1692.

Chu, S., Mills, A. P., Jr., Yodh, A. G., Nagamine, K., Miyake, Y., Kuga, T. (1988), *Phys. Rev. Lett.* **60**, 101–104.

Eldrup, M., Lightbody, D., Sherwood, J. N. (1981), *Chem. Phys.* **63**, 51–58.

Essen, L., Donaldson, R. W., Bangham, M. J., Hope, E. G. (1971), *Nature* **229**, 110–111.

Ferrell, R. A. (1957), *Phys. Rev.* **108**, 167–168.

Garner, D. M., Fleming, D. G., Brewer, J. H. (1978), *Chem. Phys. Lett.* **55**, 163–167.

Gidley, D. W., Mayer, R., Frieze, W. E., Lynn, K. G. (1987), *Phys. Rev. Lett.* **58**, 595–601.

Gidley, D. W., Rich, A., Sweetman, E., West, D. (1982), *Phys. Rev. Lett.* **49**, 525–528.

Goldanskii, V. I. (1968), *At. En. Rev.* **6**, 3–148.

Greenberger, A., Mills, Jr., A. P., Thompson, A., Berko, S. (1970), *Phys. Lett.* **32A**, 72–73.

Griffith, T. C., Heyland, G. R. (1978), *Phys. Rep.* **C39**, 169–277.

Harris, I., Brown, L. M. (1957), *Phys. Rev.* **105**, 1656–1661.

Harshman, D. R., Warren, J. B., Beveridge, J. L., Kendall, K. R., Kiefl, R. F., Oram, C. J., Mills, A. P., Jr., Crane, W. S., Rupaal, A. S., Turner, J. H. (1986), *Phys. Rev. Lett.* **56**, 2850–2853.

Harshman, D. R., Mills, A. P., Jr., Beveridge, J. L., Kendall, K. R., Morris, G. D., Senba, M., Warren, J. B., Rupaal, A. S., Turner, J. H. (1987), *Phys. Rev. B* **36**, 8850–8853.

Hasegawa, M., Yoshinari, O., Matsui, H., Yamaguchi, S. (1989), *J. Phys. E: Cond. Matter* **1**, SA77–SA81.

Hatamian, S., Conti, R. S., Rich, A. (1987), *Phys. Rev. Lett.* **58**, 1833–1836.

Hautojarvi, P., Heinio, J., Manninen, M., Nieminen, R. (1977), *Philos. Mag.* **35**, 973–981.

Hellwig, H., Vessot, R. F. C., Levine, M. W., Zitzewitz, P. W., Allan, D. W., Glazer, D. J. (1970), *IEEE Trans. Instrum.* **IM-19**, 200.

Hughes, V. W., McColm, D. W., Ziock, K., Prepost, R. (1960), *Phys. Rev. Lett.* **5**, 63–65.

Humberston, J. W. (1979), *Adv. At. Mol. Phys.* **15**, 101–166.

Ito, Y. (1988), in: D. M. Schrader, Y. C. Jean (Eds.), *Positron and Positronium Chemistry*, Amsterdam, Oxford, New York, Tokyo: Elsevier, pp. 120–158.

Jacobsen, F. M. (1986), *Chem. Phys.* **109**, 455–464.

Jensen, K. O., Eldrup, M., Singh, B. N. (1985), *J. Phys. F: Met. Phys.* **15**, L287–L293.

Kadono, R., Imazato, J., Matsuzaki, T., Nishiyama, K., Nagamine, K., Yamazaki, T., Richter, D., Welter, J.-M. (1985), *Phys. Lett. A* **109**, 61–64.

Kagan, Yu., Klinger, M. I. (1974), *J. Phys.* **C7**, 2791–2807.

Kakimoto, M., Hyodo, T., Chang, T. B. (1990), *J. Phys.* **B23**, 589–597.

Kasai, J., Hyodo, T., Fujiwara, K. (1988), *J. Phys. Soc. Jpn.* **57**, 329–341.

Kiefl, R. F., Kreitzman, S. R., Celio, M., Keitel, R., Luke, G. M., Brewer, J. H., Noakes, D. R., Percival, P. W., Matsuzaki, T., Nishiyama, K. (1986), *Phys. Rev. A* **34**, 681–684.

Kiefl, R. F., Celio, M., Estle, T. L., Luke, G. M., Kreitzman, S. R., Brewer, J. H., Noakes, D. R., Ansaldo, E. J., Nishiyama, K. (1987), *Phys. Rev. Lett.* **58**, 1780–1783.

Kiefl, R., Kadono, R., Brewer, J. H., Luke, G. M., Yen, H. K., Celio, M., Ansaldo, E. J., (1989), *Phys. Rev. Lett.* **62**, 792–795.

Kinoshita, T. (Ed.) (1990), *Quantum Electrodynamics*, Singapore: World Scientific.

Kondo, J. (1984), *Physica B* **125**, 279–285; **B126**, 377–384.

Leung, S.-K., Brodovich, J.-C., Newman, K. E., Percival, P. W. (1987), *Chem. Phys.* **114**, 399–409.

Lundeen, S. R., Pipkin, F. M. (1981), *Phys. Rev. Lett.* **46**, 232–235.

Maas, F. E., Braun, B., Geerds, H., Jungmann, K., Matthias, B. E., zu Putlitz, G., Reinhard, I., Schwarz, W., Willmann, L., Zhang, L., Baird, P. E. G., Sandars, P. G. H., Woodman, G. S., Eaton, G. H., Matousek, P., Toner, W. T., Towrie, M., Barr, J. R. M., Ferguson, A. I., Persand, M. A., Riis, E.,

Berkeland, D., Boshier, M. G., Hughes, V. W., Woodle, K. A. (1994), *Phys. Lett.* **A 187**, 247–254.

Mariam, F. G., Beer, W., Bolton, P. R., Egan, P. O., Gardner, C. J., Hughes, V. W., Lu, D. C., Souder, P. A., Orth, H., Vetter, J., Moser, U., zu Putlitz, G. (1982), *Phys. Rev. Lett.* **49**, 993–996.

Mills, A. P., Jr., (1983), "Experimentation with Low-Energy Positron Beams," in: W. Brandt, A. Dupasquier (Eds.), *Positron Solid State Physics*, Amsterdam, New York, Oxford: North-Holland.

Mills, A. P., Jr., Berko, S., Canter, K. F. (1975), *Phys. Rev. Lett.* **34**, 1541–1549.

Mills, A. P., Jr., Imazato, J., Saitoh, S., Uedono, A., Kawashima, Y., Nagamine, K. (1986), *Phys. Rev. Lett.* **56**, 1463–1466.

Mogensen, O. E., Kvajic, G., Eldrup, M., Milosevic-Kvajic, M. (1971), *Phys. Rev. B* **4**, 71–73.

Mogensen, O. E. (1974), *J. Chem. Phys.* **60**, 998–1004.

Morenzoni, E., Kottmann, F., Maden, D., Matthias, B., Meyberg, M., Prokscha, Th., Wutzke, Th., Zimmermann, U. (1994), *Phys. Rev. Lett.* **72**, 2793–2796.

Nagamine, K., Ishida, K., Matsuzaki, T., Nishiyama, K., Kuno, Y., Yamazaki, T., Shirakawa, H. (1984), *Phys. Rev. Lett.* **53**, 1763–1766.

Nagamine, K., Ishida, K. (1992), in: T. Yamazaki, K., Nakai, K. Nagamine (Eds.), *Perspectives of Meson Science*, Amsterdam: North-Holland, Chap. 14.

Nakanishi, H., Jean, Y. C. (1988); in: D. M. Schrader, Y. C. Jean (Eds.), *Positron and Positronium Chemistry*, Amsterdam: Elsevier, Chap. 5.

Nico, J. S., Gidley, D. W., Skasley, M., Zitzewitz, P. W. (1992), *Mater. Sci. Forum* **105–110**, 401–409.

Nishiyama, K., Morozumi, Y., Suzuki, T., Nagamine, K. (1985), *Phys. Lett.* **A111**, 369–372.

Oram, C. J., Baily, J. M., Schmor, P. W., Fry, C. A., Kiefl, R. F., Warren, J. B., Marshall, G. M., Olin, A. (1984), *Phys. Rev. Lett.* **52**, 910–913.

Patterson, B. D. (1988), *Rev. Mod. Phys.* **60**, 69–159.

Patterson, B. D., Hintermann, A., Kündig, W., Meier, P. F., Walder, F., Graf, H., Recknagel, E., Weidinger, A., Wichert, T. (1978), *Phys. Rev. Lett.* **40**, 1347–1350.

Percival, P. W., Kiefl, R. F., Kreitzman, S. R., Garner, D. M., Cox, S. F. J., Luke, G. M., Brewer, J. H., Venkateswaran, K. (1987), *Chem. Phys. Lett.* **133**, 465–470.

Ritter, M., Egan, P. O., Hughes, V. W., Woodle, K. A. (1984), *Phys. Rev. A* **30**, 1331–1344.

Roduner, E., Brinksman, G. A., Louwrier, W. F. (1982), *Chem. Phys.* **73**, 117–130.

Roduner, E., Percival, P. W., Fleming, D. G., Hochmann, J., Fisher, H. (1978), *Chem. Phys. Lett.* **57**, 37–44.

Roellig, L. O., Weber, M., Berko, S., Brown, B. L., Canter, K. F., Lynn, K. G., Mills, A. P. Jr., Tang, S., Viescas, A. (1987), in: J. W. Humberston, E. A. Armour (Eds.), *Atomic Physics with Positrons*, NATO Advanced Study Institutes, Series B, No. 169, New York, London: Plenum Press, pp. 233–239.

Sapirstein, J., Yennie, D. R. (1990), in: T. Kinoshita (Ed.), *Quantum Electrodynamics*, Singapore: World Scientific, pp. 560–672.

Shearer, J. W., Deutsch, M. (1949), *Phys. Rev.* **76**, 462.

Surko, C. M., Leventhal, M., Passner, A. (1989), *Phys. Rev. Lett.* **62**, 901–904.

Tao, S. J. (1972), *J. Chem. Phys.* **56**, 5499–5510.

West, R. N., Mayers, J., Walters, P. A. (1981), *J. Phys. E: Cond. Matter* **14**, 477–488.

Woodle, K. A., Badertscher, A., Hughes, V. W., Lu, D. C., Ritter, M. W., Gladisch, M., Orth, H., zu Putlitz, G., Eckhause, M., Kane, J., Mariam, F. G. (1990), *Phys. Rev. A* **41**, 93–105.

Yamada, K. (1984), *Prog. Theor. Phys.* **72**, 195–201.

Further Reading

Ache, H. J. (Ed.) (1979), *Positronium and Muonium Chemistry*, Advances in Chemistry Series Vol. 175, Washington, D.C.: American Chemical Society.

Chappert, J., Grynszpan, R. I. (Eds.) (1984), *Muons and Pions in Materials Research*, Amsterdam, Oxford, New York, Tokyo: North-Holland.

Fleming, D. G., Senba, M. (1992), in: T. Yamazaki, K. Nakai, K. Nagamine (Eds.), *Perspectives of Meson Science*, Amsterdam: North-Holland, Chap. 8.

J. W. Humberston, E. A. G. Armour, (Eds.) (1987), *Atomic Physics with Positrons*, New York, London: Plenum Press.

Kadono, R. (1992), in: T. Yamazaki, K. Nakai, K. Nagamine (Eds.), *Perspectives of Meson Science*, Amsterdam: North-Holland, Chap. 4.

Kinoshita, T. (Ed.) (1990), *Quantum Electrodynamics*, Singapore: World Scientific.

Schultz, P. J., Lynn, K. G. (1988), *Rev. Mod. Phys.* **60**, 701–779.

Schenck, A. (1985), *Muon Spin Rotation Spectroscopy*, Bristol, Boston: Adam Hilger Ltd.

Schrader, D. M., Jean, Y. C. (Eds.) (1988), *Positron and Positronium Chemistry*, Amsterdam, Oxford, New York, Tokyo: Elsevier.

Walker, D. C. (1983), *Muon and Muonium Chemistry*, London, New York: Cambridge University Press.

MUSIC, ELECTRONIC

ROBERT MOOG, *Big Briar, Inc., Asheville, North Carolina, U.S.A.*

INTRODUCTION

Electronic music is both a body of technology and a medium of musical expression. Seen from the instrument designer's point of view, electronic music is a collection of electronic devices and systems for use by musicians to generate, process, and shape musical tones, and to organize the tones in time to produce musical works. From the musician's point of view, electronic music is a medium of artistic expression—a set of resources that enable one to work with sound in ways that are different from those of traditional acoustic musical instruments. Thus, our discussion of electronic music includes musical and cultural as well as technical developments.

The technology of electronic music has been under development since the beginning of the twentieth century, during which time the state of electronics has advanced through several important stages. The seminal development was the triode vacuum tube. A wide variety of oscillator and amplifier circuits were designed, and, starting in the 1920s, many novel electronic musical instruments were introduced as commercial products. Then, after World War II, high-quality tape recorders became available. In addition to conventional record-playback applications,

3-527-28133-9/94/$5.00 + .50

tape recorders enabled musicians to modify individual sounds, to splice sounds together with unprecedented precision, and to combine lines of sound into complete pieces of music. A few years later, transistors and then integrated circuits enabled equipment designers to increase the complexity and decrease the size and price of electronic equipment by orders of magnitude. And finally, digital computers, first as research installations and then more and more as desktop tools, have enabled musicians to use the power of digital data processing for musical ends.

For musicians, the advent of electronics bore cultural as well as technological consequences. As cultural leaders spoke glowingly of the benefits of electricity and electronics (Lenin proclaimed that "Communism is socialism with electrification"), musicians and inventors explored the new capabilities of electronics. Radio and the electric phonograph brought music of high technical quality into people's homes and fostered a common cultural datum. After World War II, experimentation with electronic music as a compositional medium accelerated as studios and research centers around the world explored the capabilities of tape recorders. At the same time, vernacular music utilized more and more electronic technology until, by the 1960s, rock and roll was virtually completely dependent on electronics. A furious spate of musical experimentation ensued throughout the world during the late 1960s and early 1970s, during which time awareness of the electronic music medium entered the popular culture. Then, during the 1980s and 1990s, the popular music industry adopted electronic music techniques, while the proliferation of personal computers and music software placed potent resources at the fingertips of professional musicians and amateurs alike. Most recently, the exponentially growing power of personal computers has made practical the integration of electronic music techniques with image and video processing, thus making possible the development and proliferation of desktop multimedia production systems.

Throughout the development of electronic music, several strands of commonality have characterized the medium, and have served to differentiate it from traditional acoustic music production. Two features of the medium have existed from the very beginning: The sounds of electronic music are produced by electronic oscillators or from electronically recorded data, and musical control over individual tones is generally specified in terms of general physical parameters such as frequency and amplitude. Other capabilities have become important more recently: data-storage capabilities of tape recorders, the ability allowed musicians by the computer media to compose music off line in a studio environment, and artificial intelligence software to establish interactive composing environments based on musical rules. Each of these features has its own set of advantages and limitations for musicians.

This article will first give some basic technical concepts and define terms related to music production. Then, in the remainder of the article, specific developments will be discussed. We will start with the first electronic musical instruments and generally follow the evolution of the medium. The evolutionary tree has two distinct branches: analog and digital. The analog branch extends to the 1970s and includes the tape recorder, "classical studio" signal generating and processing techniques, and voltage-controlled synthesizers. The digital branch begins in the 1950s and includes sound-production techniques, music composing systems, algorithmic composition, and real-time performance systems.

1. TECHNICAL PRINCIPLES OF MUSICAL TONE PRODUCTION

1.1 Pitch, Loudness, Brightness, and Timbre

Let us imagine that we hear a violinist playing a single note on the open A string as steadily as she can. We observe that the acoustic wave form thus produced repeats about 440 times per second, and has a reasonably steady amplitude. We also observe that the wave shape is complex and difficult to describe. These observations—of frequency, amplitude, and wave shape—are objective in nature.

Next, we imagine that we ask another musician to listen to the tone and to describe what he hears. He may say, "That is a violin tone. The pitch is A-440 and it is being played mezzo-forte." His observations are subjective. The variables that he uses to describe the tone—pitch, loudness, brightness, and timbre—are related but not identical to our objective variables of frequency, amplitude, and wave shape. The objective variable most closely related to pitch is average frequency, while the objective variable most closely related to loudness (dynamic level) is average amplitude.

What are the objective correlates of timbre, that property of a sound that enables a listener to identify the source from which it comes? Not too long ago, musical acousticians generally associated timbre with wave shape. Early textbooks on musical acoustics often showed oscillographs of a few cycles of a musical instrument's wave form, accompanied by text that linked the instrument's perceived timbre with specific features of the illustrated wave form. Or the spectral distribution (the wave form's Fourier transform) would be shown as a series of precise integral harmonics of the wave form's fundamental frequency, suggesting that the energy of each harmonic relative to the tone's total energy remained constant as the tone evolved.

It is now understood that wave shape *per se* (or harmonic content *per se*) is only weakly correlated with timbre. Suppose we electronically sample just one complete cycle of the violin tone, and then cause that one cycle to repeat continuously. If we play that signal for our musician-listener, he may say, "The pitch is A-440 and it is medium-loud. But it sounds like a buzzer, not a violin."

Our musician-listener is showing us that perception of timbre (tone quality) is determined not only by the steady-state shape of a sound wave form, but by how the objective variables, especially the wave form itself, change as the tone evolves. Every traditional musical instrument imparts characteristic fluctuations that we as listeners use to identify it. Put another way, if we regard a musical tone as a message quantum, then the steady-state properties of the wave form (or spectrum) are information carriers, and the temporal variations of these properties bear the musical information. Fig. 1 shows how the wave form of one particular violin tone changes as the tone evolves.

Rough relationships between musical variables and their objective correlates are shown in Table 1.

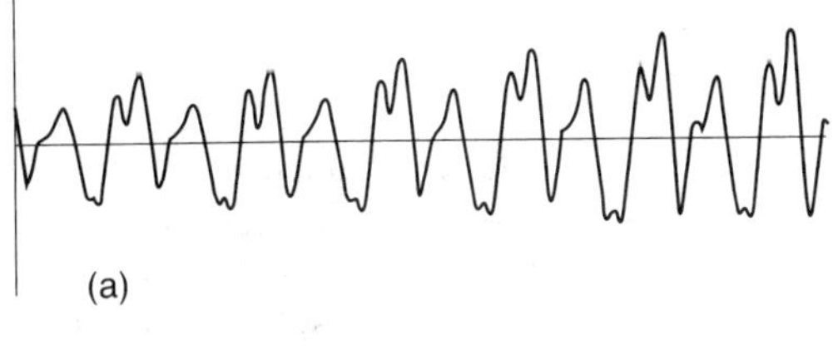

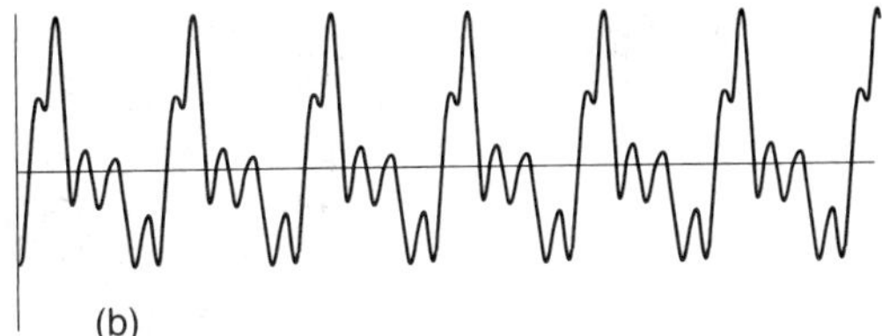

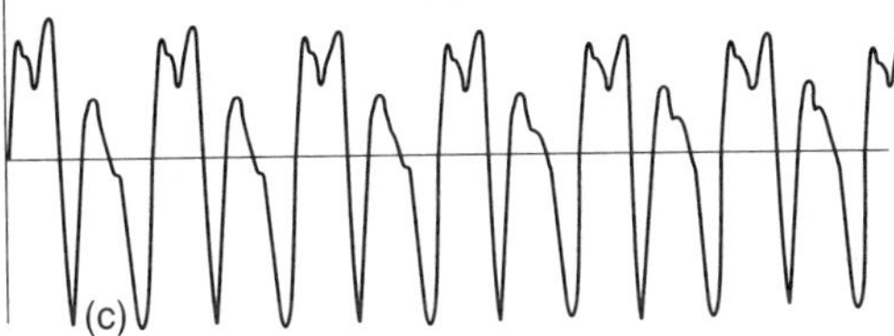

FIG. 1. Wave-form segments of a 660-Hz violin tone at three different times in its evolution: (a) 0.05 s after onset; (b) 0.50 s after onset; (c) 0.58 s after onset.

1.2 Envelopes

In general, musical tones are not steady, but evolve from onset to completion in characteristic ways. A violin tone, for instance, begins with a short burst of noise (complex,

Table 1.

Musical Variable	Objective Correlates
Pitch	Average frequency (strong correlation)
Loudness	Average amplitude (strong correlation)
	Average wave shape (some correlation)
Brightness	Average wave shape (strong correlation)
	Average amplitude (some correlation)
Timbre	Wave-shape fluctuations (strong correlation)
	Frequency fluctuations (some correlation)
	Amplitude fluctuations (some correlation)
	Average wave shape (weak correlation)

nonperiodic wave shape) as the bow scrapes the string, then evolves into a periodic wave form as the string begins to vibrate. As the tone evolves, the player imparts slowly varying changes to the tone's amplitude, frequency, and wave shape by changing the bow's position and velocity and the position of her finger on the string.

A graph of the temporal change of a tone's variable is called an envelope. Envelopes may describe how the overall amplitude of a tone evolves, how the amplitude of a component of the tone evolves, or how a single variable such as frequency fluctuates. Envelope features that are common to the tones of a given instrument largely define that instrument's timbre, while envelope features that vary from note to note carry the expressive nuance of the music. Four envelopes of a single cello tone are shown in Fig. 2.

Players of traditional acoustic instruments shape envelopes intuitively and continuously while they play. Electronic musicians, on the other hand, often specify envelope shapes explicitly and quantitatively, "off line," before the tones are actually produced.

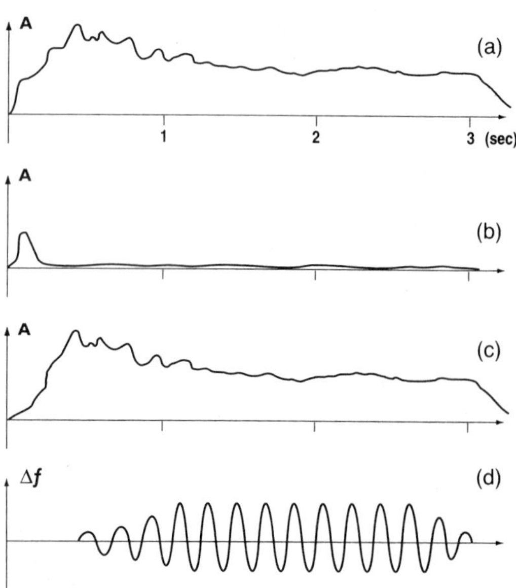

FIG. 2. Envelopes of a 3-s, 82-Hz cello tone: (a) overall average amplitude; (b) amplitude of noise (starting transient); (c) amplitude of periodic component; (d) frequency fluctuation (vibrato) of periodic component.

1.3 Modulation

In the parlance of electronic music, periodic or quasiperiodic fluctuation of a tone's variable is called modulation. Common types of modulation are vibrato, which is generated by varying frequency smoothly at a rate of about 6 Hz; trill, which is an abrupt frequency modulation; and tremolo, which is a smooth amplitude modulation. Modulations at subsonic rates create fluctuations that are slow enough for the ear to follow explicitly, whereas modulations at audio rates create sum and difference frequencies that are generally perceived as timbral changes. Frequency modulation will be discussed further below in Sec. 5.1.

It is often convenient to use envelopes to specify temporal changes of a modulation, just as it is useful to describe the evolution of a tone itself in terms of one or more envelopes. As an example, Fig. 3(a) shows the envelope of a synthetic vibrato (frequency modulation), the amplitude of which increases and

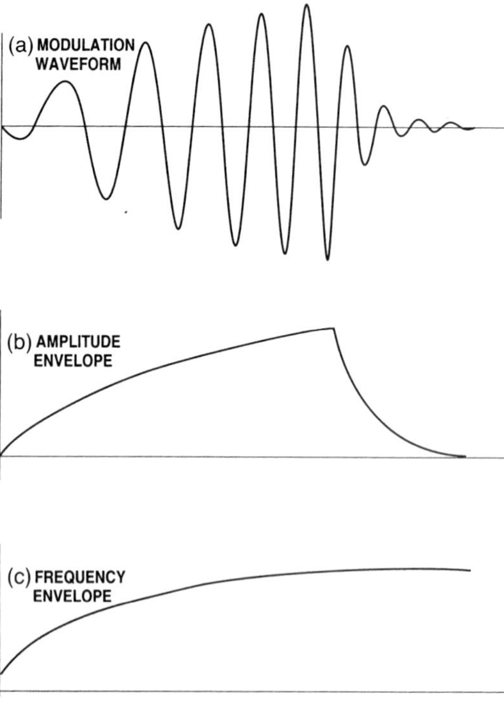

FIG. 3. Wave form of an electronically generated modulation function, with two envelopes that describe the temporal evolution of the modulation's amplitude and frequency: (a) modulation wave form; (b) amplitude envelope; (c) frequency fluctuation envelope.

then decreases, and the rate of which increases monotonically; Figs. 3(b) and 3(c) show the amplitude and frequency envelopes of the modulation. This example illustrates a hierarchy of envelopes, in which envelopes 3(b) and 3(c) specify temporal variations of function 3(a), which then defines changes in an audio signal.

1.4 Differences between Acoustic and Electronic Musical Instruments

Acoustic musical instruments differ inherently from electronic musical instruments, both in the tone-production mechanisms and in the means of musical control.

The tone-producing mechanisms of acoustic musical instruments are fundamentally different from those of electronic instruments. Acoustic musical instruments are distributed systems whose vibrations generally give rise to complex, constantly changing wave forms. Electronic oscillators, on the other hand, are generally nondistributed systems that produce simple, static wave forms.

With respect to musical control, acoustic instruments generally place the musician in intimate contact with the vibrating elements. This provides tactile feedback and enables the musician to exercise detailed control over the sound. On the other hand, electronic instruments are generally controlled through transducers that are physically separate from the tone producers, and provide little or no direct tactile feedback.

The history of the electronic music medium has been marked by ongoing technical developments to improve the musical richness of electronically produced tones, and to increase the amount of control that musicians can exert over the evolution of the tones. But it has also been marked by explorations by creative musicians who discovered new ways of producing music for which the capabilities of electronic instruments are admirably suited. The remainder of this article traces the remarkably multifaceted history of the electronic music medium.

2. EARLY ELECTRONIC MUSICAL INSTRUMENTS

Pitched musical instruments tend to fall into two classes: monophonic and polyphonic. Monophonic instruments produce only one tone at a time, but the pitch of that tone can be varied over a wide range. Polyphonic instruments can produce many tones at a time, but the pitch of each tone is fixed. The trumpet is an example of a monophonic instrument, while the piano is a polyphonic instrument. String instruments are midway between the two, in that they can produce a few tones at a time, and the pitch of each tone can be varied over a limited range.

2.1 Monophonic Instruments

In order to build a monophonic electronic instrument with a pitch range of several octaves (one octave = 2:1 frequency ratio), one needs an audio-frequency oscillator with a frequency range of at least 10:1, along with the means to change the frequency rapidly and precisely. The first instrument to meet these requirements was the Theremin, which was designed in 1920 by the Russian physicist Leon Theremin (Glinsky, 1992). Theremin used two nearly identical high-frequency (170 kHz) oscillators, plus a circuit that produced the difference (beat) frequency. By varying the frequency of one of the oscillators by only 1%, the beat frequency would change from zero to 1.7 kHz, about two octaves above A-440!

To enable a musician to control the beat frequency, Theremin connected an *LC* resonant circuit to one of the oscillators. The capacitive element of the circuit was a slim brass rod, called the pitch antenna. Bringing one's hand within a few inches of the pitch antenna would increase its capacitance by a fraction of a picofarad, lower the oscillator frequency by about 1%, and sweep the frequency of the beat tone through most of the musical pitch range!

To control the loudness of the tone, Theremin introduced a third high-frequency oscillator with its own "volume antenna" circuit, the output of which varied the gain of the instrument's audio amplifier.

Thus, Theremin not only provided an instrument with a wide pitch range but also introduced a completely novel means of control: motion of the player's hands in space, without touching the instrument at all! Professor Theremin is shown demonstrating his instrument in Fig. 4.

In 1929, Theremin licensed the Radio Corporation of America (RCA) to build and sell

FIG. 4. Leon Theremin demonstrating the Theremin (ca. 1927).

Theremins. RCA built several hundred instruments, using the same chassis that it had designed for its superheterodyne radio receivers. Fig. 5 is a simplified block diagram of the RCA circuitry. Fig. 6 is a photo of the main chassis of the RCA Theremin.

The Ondes Martenot is another early monophonic electronic instrument that uses a beat-frequency oscillator configuration. Developed by the French engineer Maurice Martenot, the Ondes Martenot employs a six-octave keyboard in conjunction with a flexible band to enable the "ondiste" to play either discrete or continuous pitch changes. Loudness and timbre control is achieved through a series of buttons and levers, which are played with the left hand (Laurendeau, 1990).

Although the Theremin and the Ondes Martenot both use a beat-frequency oscillator for tone production, their characteristic timbres are different. This is because of the difference in the control means. The theremin responds immediately to every motion of the player's hands, and therefore has a fluid, ethereal quality. The sound of the Ondes Martenot is more steady and less fluid because mechanical elements have to be moved in order to effect pitch changes.

The Trautonium, developed by Friedrich Trautwein, is another interesting scheme to provide a wide frequency range with novel pitch-control means. The Trautonium uses a gas-filled triode as a relaxation oscillator. The frequency-determining device is a long resistive element whose resistance the player varies by pressing on a conductive ribbon that is suspended over it, then sliding her finger along the ribbon. The oscillator wave form

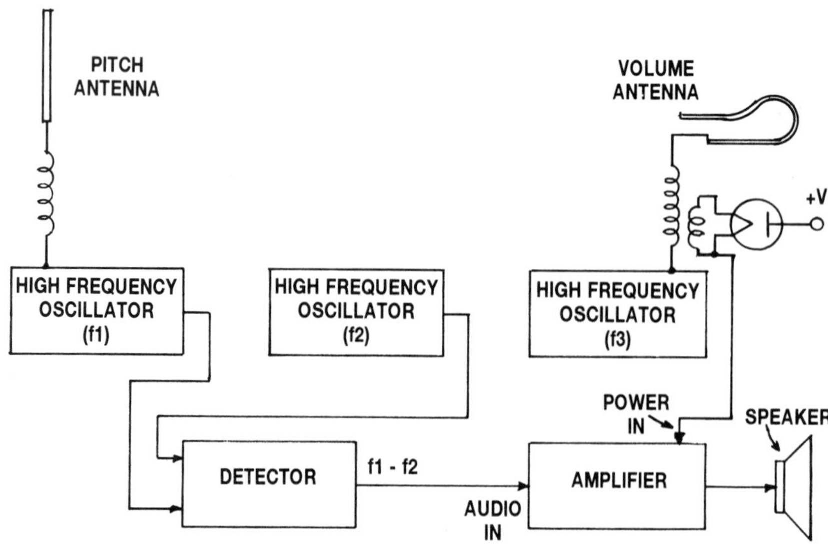

FIG. 5. Block diagram of the RCA Theremin (1929).

FIG. 6. Main chassis of the RCA Theremin. The volume antenna inductor is on the left, and the pitch antenna inductor is on the right.

has a sharp edge, which gives rise to the full range of harmonics. The harmonics may be filtered by one or more resonant circuits to simulate formants, the pronounced resonances in the spectra of vocal and certain acoustical musical instruments. The Trautonium was manufactured by Telefunken for a short period of time during the 1930s (Rhea, 1979b).

The Theremin, the Ondes Martenot, and the Trautonium were the three earliest electronic musical instruments to be manufactured commercially. With all three, the most distinguishing feature is the novel means with which the instrument's tone is controlled by the player. The originality and imagination displayed by the instruments' inventors appears to be a product of the optimism of the early days of electronic technology.

2.2 Polyphonic Instruments

Unlike the monophonic instruments described above, early polyphonic electronic instruments were close adaptations of their acoustic counterparts. In particular, the control interfaces of nearly all polyphonic electronic instruments were conventional organ-style or piano-style keyboards, and the instruments themselves were often called organs or pianos.

Many polyphonic instruments employed the motion of mechanical elements to generate periodic wave forms, and electronic amplification and loudspeakers to convert the wave forms into sound. The Hammond Organ, introduced commercially in 1935, used tiny rotating steel disks ("tonewheels") to generate a multiplicity of sine waves. Each key activated one switch for each of eight harmonics, while an array of drawbar switches above the keyboards determined the tones' brightness and timbre by setting the relative amplitudes of the harmonics. Vibrato was introduced with an *LC* delay line that was mechanically scanned at rates of up to 6 Hz. Overall volume was controlled with a foot-operated potentiometer. A simplified schematic is shown in Fig. 7, while an early model of the Hammond Organ is illustrated in Fig. 8.

Hammond tonewheel organs remained in production for nearly five decades, and have been widely used for popular and religious music performance (Rhea, 1977b). They have been notably ubiquitous on radio, perhaps because the instrument's audio bandwidth matches that of AM radio.

Continuous-tone polyphonic instruments using optical wave-form storage were developed during this period. In these instruments, transparent disks containing many concentric tracks were rotated at constant speed. Endless loops of tone wave forms were drawn in each of the tracks. Each disk was placed between a light source and an array of photodetectors. A mechanical shutter, opened by the depression of a key, activated a single track. Such instruments with rotating optical disks offer the advantages of ar-

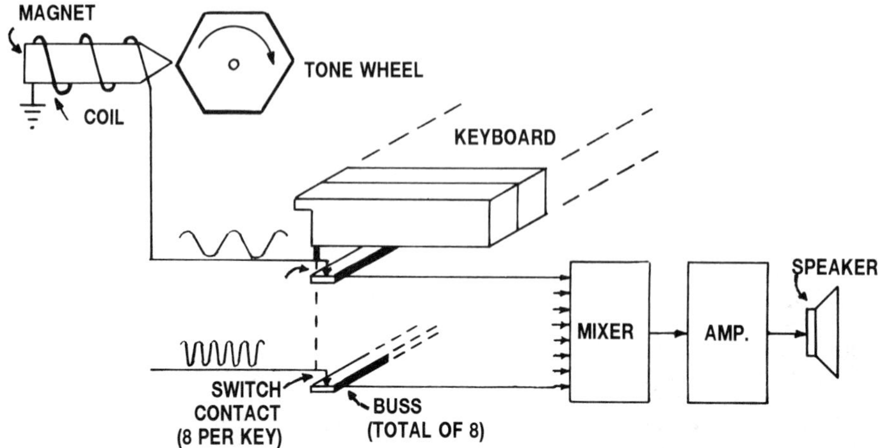

FIG. 7. Simplified diagram of the Hammond Organ.

FIG. 8. The Hammond Organ (1935).

bitrarily complex wave forms. One of the earliest examples of a "photoelectric organ" is the Cellulophone, constructed during the early 1930s in France by P. Toulon (Rhea, 1977a).

Many workers developed electronically amplified pianos in the 1930s. Design considerations for these efforts included not only transducer configuration but also transducer placement and hammer design. One of the most successful developers of amplified pianos was Benjamin Miessner, who introduced the Miessner Electronic Piano as a commercial product in 1932, and then licensed several piano manufacturers to build similar instruments. Miessner also developed an amplified pianolike instrument in which brass reeds instead of strings were struck with hammers. This design formed the basis of the Wurlitzer Electronic Piano, which soon became a staple of the popular music arsenal (Rhea, 1978a).

Because of the expense and relative instability of early vacuum tubes, the development of polyphonic instruments using electronic oscillators began somewhat later in the 1930s. In France, Edouard Coupleux and Joseph Givelet built several large scale all-electronic organs that were installed in several churches and radio stations (Rhea, 1978b). In 1939, the Hammond Novachord was introduced (Fig. 9). It was the first commercial polyphonic instrument to use electronic master oscillators for the top octave of notes, to which all lower octaves were synchronized. The Novachord included means for varying amplitude envelopes and formants, and achieved considerable popularity despite its notable complexity, weight, and price (Rhea, 1979a).

2.3 Early Encoded-Performance Instruments

Encoded-performance instruments are systems with which a musician may record explicit coded "performance instructions" off line, on a medium such as punched paper, marked paper, or rigid templates, and then realize a piece of music by having the machine scan the recording medium at a specified rate. The recording medium is directly analogous to a traditional composer's score, and the machine itself takes the place of one or more traditional performers.

The earliest successful electronic encoded-

FIG. 9. The Hammond Novachord (ca. 1939).

performance instrument was built and demonstrated by Coupleux and Givilet at the Paris exposition of 1929. The device included four *LC* oscillators and related wave-shaping circuitry, which were controlled by a paper roll and a player-piano-like pneumatic mechanism. The device was capable of continuous control of frequency (by moving iron cores in and out of inductors) and amplitude (by moving potentiometers), and discrete selection of several options (Rhea, 1979c).

In 1945, John Hanert developed an experimental encoded-performance instrument for the Hammond Instrument Company. The device, which became known as the Hanert Electrical Orchestra, employed a score reader mechanism that rolled down a long table on which cards containing the score were placed. The score consisted of conductive marks that the score reader detected as it rolled along. The score reader then generated keying and control signals that activated several racks of tone-production circuitry (Rhea, 1979d).

Whereas devices like those of Coupleux–Givilet and Hanert were primarily technological showpieces, other encoded-performance instruments were designed with specific compositional goals in mind. One such device is the Free Music Machine of Percy Grainger and Burnett Cross. Grainger envisioned a type of music devoid of beats, fixed pitches, and fixed harmonies. He and Cross designed an encoded-performance instrument consisting of four oscillators, each equipped with potentiometers for continuously changing frequency and amplitude. The potentiometers were attached to levers that followed contours cut onto rolls of heavy paper. As the paper scrolled, the levers followed the contours, thereby changing the frequencies and amplitudes (Rhea, 1979e).

The above-described instruments constitute only a tiny fraction of musical instruments and automata using electronic technology that were designed and built before the end of World War II. Most were highly experimental; a few were put into commercial production. Technically speaking, most were innovative and original. However, musically speaking, most (but not all) were designed for traditional modes of composition and performance. As World War II drew to a close, the advance of electronic technology accelerated, the world of experimental music blossomed, and the electronic music medium developed into a significant and innovative component of the musical mainstream.

3. THE IMPACT OF THE TAPE RECORDER

3.1 The Tape Recorder as a Composing Tool

Tape recorders became available to musicians a few years after the end of World War II. For the first time, musicians had a high-quality sound-recording medium that was convenient to use and manipulate. With a tape recording machine, razor blade, and some splicing tape, one could connect one sound to another and perform simple temporal transformations on individual sounds. Put another way, the tape recorder was a highly versatile, easy-to-use composing tool, offering a composer access to any sound whatsoever that could be picked up by a microphone or generated in an electronic instrument.

Musician experimenters quickly found that tape manipulation offered several very useful sound transformations. First, a length of tape on which a single sound is recorded can be cut into pieces to shorten or rearrange the sound. Or, the same piece of tape can be spliced into a loop to form a continuously repeating sound. Second, the tape can be turned around and run backward, thus reversing the evolution of the sound. And third, the speed of tape travel can be changed simply by changing the surface velocity of the capstan.

Just these three simple tape manipulations placed a wealth of new sound material at the disposal of experimental composers. To illustrate, suppose we have recorded a single piano tone that lasts 2 s. The tone has an initial "thump" attack transient, after which both its amplitude and harmonic content slowly decrease. We first cut off the attack, then select a 1-s segment of the remaining sound. We rerecord that segment, giving us two identical 1-s segments. We now splice the two segments together to form a loop, in such as way that the end of one segment is spliced onto the end of the second. Finally, we play back that loop at half speed, giving us a low-pitched sonorous drone that rises and falls in amplitude and brightness every 4 s. The drone does not sound like a piano tone at all, and in fact is a completely new musical sound.

3.2 Audio Signal-Processing Techniques

As experimental musicians learned the compositional capabilities of the tape recorder, they also explored the sound modification capabilities of other electronic instruments. They found that several basic signal processors that were originally developed for laboratory, recording studio, or military use were also capable of musically useful sound transformations. The more important of these signal-processing techniques will now be described.

Mixing is the linear addition of two or more audio signals. A mixer is basically an analog adder circuit, preceded by potentiometers (often called faders) at each of the inputs. A mixer enables one to combine two or more signals; by turning one fader up and another fader down, one can "crossfade" between two sounds.

Filtering is a signal-processing operation with a device, called a filter or equalizer, whose gain is a settable function of frequency. Filters are used to emphasize certain portions of the frequency spectrum, and attenuate other portions. Filters "color" sounds that are passed through them, in direct analogy to optical filters, which reshape the spectra of light passing through them. Common types of electronic music filters include low-pass, high-pass, and parametric filters, and bandpass filter banks. A low-pass filter permits frequencies up to the "cutoff frequency" to pass relatively unattenuated, and attenuates frequencies above cutoff. Similarly, a high-pass filter attenuates frequencies below its cutoff. A parametric filter has one or more resonances, each of the center frequencies and bandwidths of which can be varied. Bandpass filter banks (commonly called graphic equalizers) consist of an array of filters that pass contiguous bands of the frequency spectrum. The gains of the filters are individually variable, often by an array of slider potentiometers that provide a graphic indication of the overall frequency response of the bank.

Filters are capable of changing the timbre of a sound dramatically, especially in cases where a filter's operating points are changed as the sound evolves.

Ring modulation is the arithmetic multiplication of two signals. The output of a ring modulator contains the sum and difference frequencies of all frequency components of the two input signals, but not the input signals themselves. In a typical application, a sound to be processed is applied to input A while a sine wave of frequency f is applied to input B. The output signal contains two sets of partials. (Note: The term *partial* is used here to denote *any* component frequency of a musical sound, whereas the term *harmonic* generally refers to component frequencies that are integral multiples of the sound's fundamental frequency.) One set consists of the partials of input A shifted up in frequency by f, while the other set is the partials of input A shifted down by f. The term *partial* is used here instead of *harmonic* because the frequencies of harmonics are generally assumed to be integral multiples of a tone's fundamental frequency, or nearly so, whereas a partial may be of any frequency. Musical sounds with significant "out-of-tune" partial energy (i.e., containing strong partials whose frequencies are nonintegral multiples of the fundamental) are called *clangorous* sounds. They generally have a less "pure" quality and a more vague sense of pitch than tones arising from nearly periodic wave forms. Tympani and gongs are two examples of instruments that produce clangorous sounds. Ring modulators are particularly useful for producing clangorous sounds in which the frequency relationships among its partials can be quickly and precisely varied.

Frequency shifting (Klangumwandlung) is similar to ring modulation, but the output of a frequency shifter contains only one set of shifted partials. In general, *Klangumwandler*-produced tones are less clangorous than ring modulator tones.

Ring modulation and frequency shifting provide new sounds that retain the temporal variations of an existing sound, but have a different set of partials.

Reverberation and *echo* are two closely related methods of processing a signal by repeating it at different times. Reverberation simulates the multiplicity of closely spaced reflections that one hears in an enclosed space, while echo consists of fewer, more widely spaced repetitions of the original signal. *Tape echo* takes advantage of the fact that many tape recorders have a playback head that is separate from but spaced near to the recording head. If a sound to be processed is mixed with a certain amount of the playback head signal, and the mixture is fed back to the record head, then the recorded signal includes an attenuated replica (echo) of what was recorded a short time before. The strength and duration of the echoes can be controlled by varying the gain of the feedback loop, the spacing of the echoes can be controlled by varying the tape speed or the distance between the heads, and musically interesting evolutions of the echoes can be achieved by placing signal processors such as a filter or ring modulator in the feedback loop.

Electronic switching is the processing technique by which an output signal alternates periodically between two input signals. Electronic switching can proceed so rapidly that the qualities of the original input signals are completely obscured, and a new sonic texture emerges.

3.3 The Classical Tape Studio

During the late 1940s and early 1950s, experimental composers in Europe and the United States rapidly developed techniques and gathered equipment to compose complete pieces of music directly on tape. Their experimentations were both technical and aesthetic. As these explorers worked with their newly found equipment and honed their newly learned techniques, they discovered sonic relationships that went far beyond those of traditional instrumental music. Sounds that earlier had been considered musically irrelevant noises, when treated electronically, often acquired musical substance. At the same time, composition by tape recorder allowed music production to be done with precision and versatility.

In Europe, the first institutions to support the development of the electronic music medium were the radio stations. Tape composition studios were established at the French Radio in Paris in 1948, at the Northwest German Radio in Cologne in 1951, and at the Italian Radio in Milan in 1953. At the Paris studio, Pierre Schaeffer and his colleagues focused their efforts on composition with "natural" sounds—sounds that started out as recordings of acoustical events. They called their compositional approach "Musique Concrete" (Schaeffer, 1952). Werner Meyer-Eppler, Karlheinz Stockhausen, and Herbert Eimert took a diametrically opposite approach at the Cologne studio, preferring to create their music from the simplest of electronically generated waves, a compositional approach that they called "Elektronisches Musik" (Eimert, 1956).

Tape composition developed in the United States first in the music departments of universities. Perhaps the best-known American tape music pioneer is Vladimir Ussachevsky. He began his explorations in 1951 at Columbia University, and, with his colleagues, established the Columbia Princeton Electronic Music Center in 1961 in New York City (Moog, 1981).

Professional audio tape recorders were the basic pieces of composing equipment in all of these early studios. Signal-generating and -processing devices, generally off-the-shelf test gear, standard audio broadcasting and recording devices, and war-surplus communications equipment, were mounted in racks. Inputs and outputs were wired to patch panels for convenience of interconnection. The European studios were usually built to allow several people to work simultaneously. The composer would give instructions to one or more technicians, who would then actually manipulate the equipment. In contrast, the American studios tended to be more compact, so that one person could perform all operations (Fig. 10).

The above-described studio configuration, along with the techniques that were used, came to be known as the Classical Tape Composition Studio (Ussachevsky, 1958). As consumer-grade tape recorders became available and the prices of electronic equipment (especially war-surplus equipment and equipment in kit form) dropped, small schools and private musicians as well as large institutions set up tape music studios. Most of the work done in these studios was experimental in nature; by the early 1960s, the world of commercial music had only a few brief encounters with the world of experimental electronic music.

The classical tape studio offered musicians a wealth of new sounds, plus techniques for modifying and composing sounds that excited and inspired many composers. But a growing number of composers felt a need for more control over the evolution of individual sounds. The signal-generating and -processing instruments of the classical tape

FIG. 10. Vladimir Ussachevsky in a studio of the Columbia–Princeton Electronic Music Center (1965).

studio were settable by knobs or sliders. Rapid, precise change of parameters such as frequency of filter cutoff and gain was difficult if not impossible. Furthermore, few of these instruments were optimally suited for music production. For instance, most waveform generators were standard test oscillators with frequency ranges that were switched by factors of 10—a ratio that is appropriate for technical measurements but has virtually no musical significance. Starting in the early 1960s, these problems were addressed by equipment designers who were working with the newly available silicon junction transistors.

4. THE ADVENT OF VOLTAGE CONTROL

4.1 Technical Principles of Voltage Control

A voltage-controlled device is one in which one or more of its operating points can be varied by changing the magnitude of an externally applied voltage, called the control voltage. For instance, in a voltage-controlled oscillator (VCO), a control voltage changes the frequency, while in a voltage-controlled amplifier (VCA), a control voltage changes the gain.

The introduction of inexpensive, high-quality silicon junction transistors in the early 1960s facilitated the development of voltage-controlled devices with wide range and high accuracy. Silicon junction transistors are three-terminal amplifying devices with the following remarkable characteristics:

1. The range of collector currents over which the device operates accurately is well in excess of three decades.
2. Transconductance is directly proportional to the collector current over the entire operating range.
3. The collector current increases exponentially with respect to the base-to-emitter voltage over the entire operating range.
4. The device operates with high stability and low noise.

This combination of attributes has enabled equipment designers to build multiple–wave-form VCOs that cover the entire audio-frequency range as well as musically useful subsonic frequencies, voltage-controlled filters (VCFs) whose characteristic frequencies may be swept over the entire audio-frequency range, and VCAs with dynamic ranges that approach the limits of human hearing.

The exponential relationship between input voltage change and output current change has proven to be particularly useful in voltage-controlled electronic music devices. This is because physical variables such as frequency and amplitude are generally proportional to the exponentials of the corresponding subjective musical values. In particular, musical pitch intervals are frequency ratios. Stated another way, frequency is exponentially related to pitch. Similarly, audio signal amplitude is exponentially related to subjective loudness.

A VCO in which the frequency is exponentially related to the magnitude of its control voltage turns out to be ideal for electronic music applications. In a typical VCO, a control voltage increase of 1 V doubles the frequency, or raises the pitch by one octave. A 12-tones-to-the-octave equally tempered chromatic scale is generated simply by increasing the control voltage in $\frac{1}{12}$-V steps; pitch transposition is accomplished by adding multiples of $\frac{1}{12}$ V to the control voltage that produces the pitch pattern. Furthermore, timing relationships are just as simply manipulated when a VCO is operated in the subsonic frequency region. The tempo of a subsonic wave form can be doubled by a 1-V control voltage increase. The musician thinks of the control voltage as a measure of a musical value, and the exponential conversion within a voltage-controlled device automatically produces the correct operating point.

Another desirable property of voltage-controlled devices is that their operating points can be changed essentially instantly, and can therefore be used to impart rapid, precise variations to individual sounds. The task of designing new timbres with voltage-controlled equipment thus reduces to the specification of voltage variations to change the sound's variables as individual sounds evolve. These voltage variations take the form of envelopes, modulations, random variations, and arbitrarily shaped nuances (Moog, 1965).

Wave-form–generating devices such as voltage-controlled oscillators, signal-processing devices such as voltage-controlled filters and amplifiers, and generators of envelope, modulation, and other control signals may be assembled into a system for producing sounds of considerable complexity and musical interest. Control signals may be generated automatically in real time, prerecorded as a score of an encoded-performance instrument, or shaped continuously by a musician manipulating an appropriate set of transducers. The timbres thus produced may span the range of musical tones from pitched tones like those of orchestral instruments to sounds whose partials and modes of evolution are unlike any "natural" sounds. Systems such as these became known as Electronic Music Synthesizers, and beginning in the late 1960s, became commercially important resources for experimental and commercial musicians alike.

4.2 Electronic Music Synthesizers

The term "synthesizer" was coined before the development of transistorized voltage-controlled instruments. Applied to an electronic musical instrument, the term suggests that the sounds are produced, or synthesized, from separate elements.

One of the best-known precursors of voltage-controlled synthesizers is the RCA Electronic Sound Synthesizer, a large-scale experimental encoded-performance instrument, two of which were designed and built immediately after World War II at RCA Laboratories (Fig. 11). The RCA Synthesizer uses hundreds of vacuum tubes to produce four musical sounds simultaneously. The score is a paper roll, on which rows of holes are punched to specify the frequencies, envelopes, formants, and other variables of the four tones. The time resolution of the score is $\frac{1}{30}$ s. The score's drive mechanism is mechanically synchronized with a disc-recording lathe. The lathe partitions the disc surface into six separate bands, which may be recorded on one at a time and then played back simultaneously. Thus, a completed recording may have as many as 24 perfectly synchronized tones, which would then be mixed into a final two-track tape recording (tape recorders of more than two tracks had not yet been developed) (Babbitt, 1964).

The first voltage-controlled music production systems were designed not as stand-alone encoded-performance instruments, but as components of a classical tape studio. Each of the generating and processing instruments was built as an individual module. The composer would interconnect the modules with

FIG. 11. The RCA Electronic Sound Synthesizer (1955).

patch cords to produce a single sound or sequence of sounds, which would then be recorded on tape (Moog, 1967). However, musicians and equipment designers alike soon realized that, by adding programming and sequencing devices to produce control voltage changes for large-scale musical gestures, one could use voltage-controlled systems to produce complete musical works with little or no tape splicing. As the 1960s progressed, commercial and popular music producers as well as experimental composers came to use modular voltage-controlled synthesizers, and several companies began manufacturing them. Fig. 12 shows a representative commercial modular voltage-controlled synthesizer.

The late 1960s was a time of widespread musical experimentation. "Synthesizer music" was frequently heard on records and radio, in live performance, and as film and television backgrounds. Among these, the record album "Switched-on Bach," by W. Carlos and B. Folkman, stands out as a watershed event. Carlos and Folkman used a modular synthesizer, a small mixing board, and an eight-track tape recorder to realize a complete album of the music of J. S. Bach (Fig. 13). Using a conventional keyboard mechanism to generate $\frac{1}{12}$-V steps for the notes of the chromatic scale, Carlos and Folkman performed and recorded one line of music at a time onto a single track of the eight-track tape. When all tracks were satisfactorily recorded, they were mixed to a two-track master tape.

A student of Vladimir Ussachevsky, Carlos had the unique talents of understanding how to produce musically appealing timbres with electronic equipment, and how to use these timbres, which were distinctly different from those of acoustic instruments, to realize traditional music. Released at the end of 1968, "Switched-on Bach" captured the public's ear and soon became the largest-selling album of Bach's music (Berger, 1969).

"Switched-on Bach," and some other synthesizer records of the 1960s, inspired many popular musicians to perform live on synthesizers. Because of their complexity, modular synthesizers were ill-suited to the demands of the stage, and so equipment manufacturers began to produce instruments

FIG. 12. A modular voltage-controlled synthesizer (ca. 1970).

with simpler capabilities, and with fixed interconnections and fast-acting switches instead of patch cords. By the early 1970s, the considerable sonic resources and flexibility of modular, patch-cord-connected systems had evolved downward to an easy-to-use but relatively standard, fixed set of functions. Two or three VCOs for pitched sounds, plus a noise generator for pitchless or weakly pitched sounds, produced the starting audio wave forms. The VCO generally produced sawtooth and rectangular wave forms for high harmonic contect, and triangular wave forms when low harmonic content was called for. One VCF, usually a four-pole low-pass filter, shaped the spectral distribution of the combined wave forms, and one or two VCAs were used to shape the overall amplitude evolution. A low-frequency oscillator (LFO) produced subsonic sine control wave forms (for textures such as vibrato and tremolo) and square control wave forms (for textures such as trill). Envelope-defining control signals were produced by envelope generators (EGs); a common envelope shape had a four-segment attack-decay-sustain-release (ADSR) configuration (Fig. 14). The instrument was played by means of a keyboard that produced a frequency control voltage according to which key was depressed, and an envelope-start signal whenever any key was depressed. Most instruments had input ports for extra control signals from devices like foot pedals and joysticks (Fig. 15).

Despite the skepticism of the musical instrument retailing community, these "compact performance synthesizers" soon became extremely popular with performing musicians. With the circuit operating points immediately accessible with front panel knobs,

FIG. 13. The studio of W. Carlos, where "Switched-on Bach" was realized.

and most interconnections accessible with front panel switches, musicians quickly discovered the correlations between operating-point changes and timbral results. If, for instance, a rising envelope signal were used to raise the cutoff frequency of the low-pass filter over a time of a tenth of a second or so, then the resulting tone would have a horn-like attack. If, on the other hand, that same envelope signal were set up to descend at the beginning of the tone, thereby rapidly lowering the filter's cutoff frequency, then the resulting tone would have a plucked-string-like quality.

The first compact performance synthesizers produced one or two tones at a time and required the musician to set the front panel controls for a given timbre (Fig. 16). As the 1970s progressed, two technical advances contributed to the further evolution of commercial synthesizers. One of these is the microprocessor, while the other is the introduction of high-quality analog integrated circuits containing entire voltage-controlled signal-generating and -processing functions. Together, these advances enabled equipment designers to build "programmable polyphonic synthesizers." These instruments contained a small number of "voice circuits," each of which contained all the sound-generating and processing functions for one sound, plus a microprocessor-based scanning

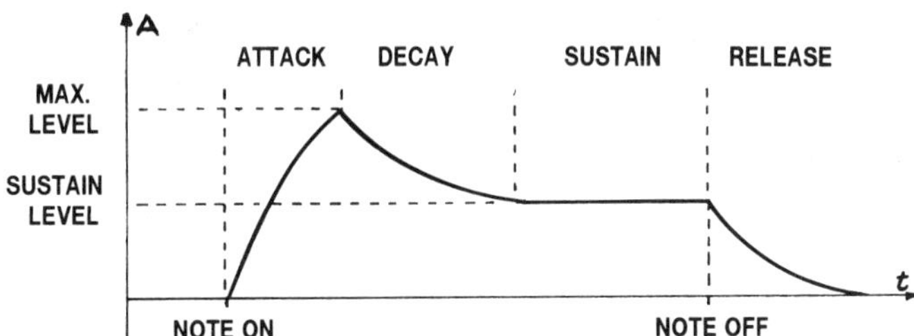

FIG. 14. An "ADSR" envelope.

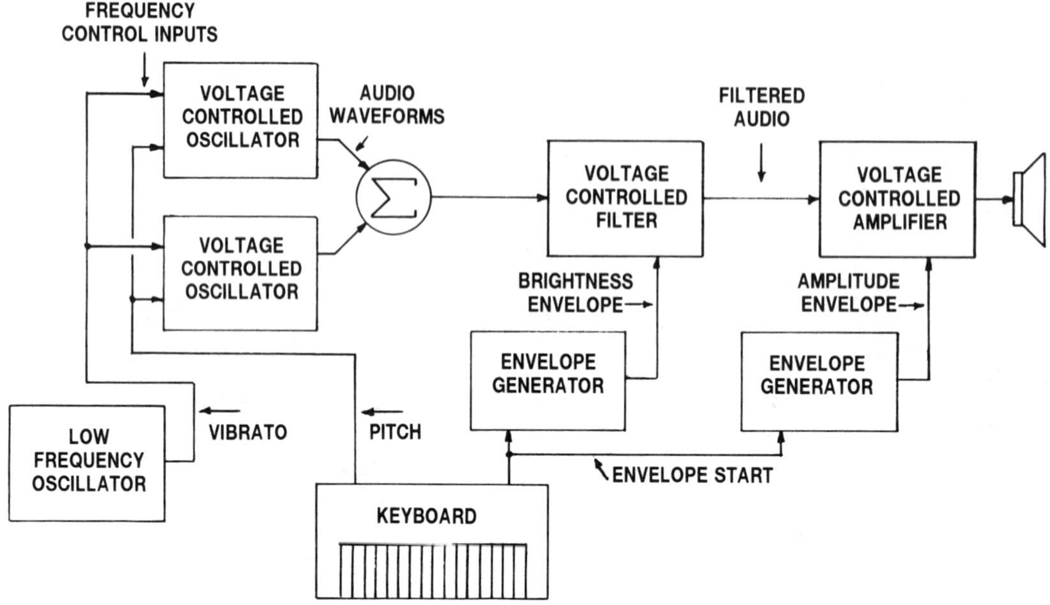

FIG. 15. Block diagram of a "standard synthesizer voice."

and data-storage system. The front panel controls were not connected directly to the sound-generating circuits. Instead, the microprocessor scanned the front panel controls, stored the settings for later editing or retrieval, and then routed the data to the voice circuits to set their operating points. The microprocessor also scanned the keyboard and, following nontrivial voice-assignment rules, associated a voice circuit with a newly depressed key.

Voltage-controlled devices were first de-

FIG. 16. A compact performance synthesizer (ca. 1972).

signed as instruments for the classical tape studio. Their users were experimental musicians who wanted maximum versatility, and were much less concerned with speed or ease of use. Over a period of no more than 15 years (1964–1979), they evolved into electronic-organ-like performance instruments. Instead of understanding the principles and practice of synthesis with voltage-controlled functions, players of latter-day voltage-controlled synthesizers are able to acquire complete banks of preprogrammed "synth sounds" on digital media. Beginning in the 1980s, many of the functions of voltage-controlled performance synthesizers have been ported over to all-digital instruments, which have become ubiquitous enough to be considered folk instruments.

Since the 1960s, experimental musicians seeking to develop their electronic resources further have gravitated to the digital computer. Advances in the realm of computer music have proceeded rapidly, with respect both to the performance/cost ratios of digital hardware and to the capabilities of computer software.

5. COMPUTER MUSIC

5.1 Early Application of Computers

Today's personal computers, together with a modest (by current standards) amount of sound-generation circuitry, are capable of producing complex, high-quality music in real time. Digital computers did not always offer this capability. Computers of the 1950s and early 1960s required hours just to compute the data from which a few seconds or minutes of music could be constructed. Then the data had to be processed by a digital-to-analog converter, an operation that taxed the then-available conversion hardware even to produce musical material of limited dynamic range and frequency response. For this reason, the earliest computer synthesis explorations centered on simple, sparse musical material.

The "Music *N*" (Music I–Music V) programs, developed during the 1960s at Bell Labs under the direction of Max Mathews, are representative of the music synthesis programs of this time. In writing the programs, Mathews addressed the problems of

1. how to minimize the vast amounts of data that are needed to generate musical sounds, and
2. what sort of a language could be employed to describe complex sequences of sounds.

Mathews's approach was to define the music in terms of *notes*, generate each note with one or more *unit generators*, and use stored wave forms as the starting material for the unit generators (Mathews, 1969).

Although Mathews's unit generators were subroutines of the Music *N* programs rather than hardware-based devices, they were similar in concept to voltage-controlled modules. Unit generators included such functions as oscillators (wave form generators with amplitude and frequency control inputs), multiple-input mixers (adders), and two-input multipliers. In the Music *N* programs, single-line instructions linked outputs to inputs, and numbers in decimal notation defined the wave forms, envelopes, and note durations. Most time-varying functions were defined in terms of straight line segments. Like a system of voltage-controlled-modules, a Music *N* "instrument" could produce tones whose complexity depended on how many unit generators were used. Unlike the analog counterpart, however, the complexity of a Music *N* instrument was limited not by hardware, but rather by constraints on available computing time. Figure 17 shows a very simple Music V instrument, along with the high-level instructions needed to set it up and have it produce two notes.

The program design principles articulated by Mathews became the basis for the development of many computer synthesis approaches. These range from simple (but computation intensive) additive synthesis schemes involving continuous control over the amplitudes of individual harmonics, to nonlinear schemes that produce complex musical tones with high efficiency. The latter is illustrated by John Chowning's explorations with frequency modulation (Chowning, 1973).

Electronic engineers are acquainted with frequency modulation (FM) as a communications technique. If the frequency of one sine wave (called the carrier) is varied in proportion to the amplitude of another sine wave

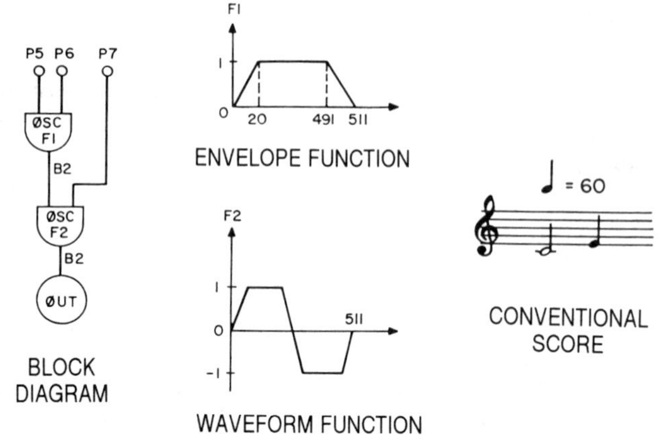


```
1  INS 0 1 ;
2  ØSC P5 P6 B2 F1 P30 ;
3  ØSC B2 P7 B2 F2 P29 ;
4  ØUT B2 B1 ;
5  END ;
6  GEN 0 1 1 0 0 .99 20 .99 491 0 511 ;
7  GEN 0 1 2 0 0 .99 50 .99 205 -.99 306 -.99 461 0 511 ;
8  NØT 0 1 2 1000 .0128 6.70 ;
9  NØT 2 1 1 1000 .0256 8.44 ;
10 TER 3 ;
```

COMPUTER SCORE: Lines 1-4 specify the instrument; line 6 specifies the waveform; line 7 specifies the envelope; lines 8-9 specify the notes.

FIG. 17. Example of a simple Music V program: (a) block diagram; (b) envelope function; (c) wave-form function; (d) conventional "score"; and (e) the computer score. In the computer score, lines 1 through 5 specify the block diagram, line 6 specifies the envelope function, line 7 specifies the wave-form function, and lines 8 and 9 "play" the two tones.

(called the modulator), then the resulting frequency-modulated wave contains a rich set of sum and difference frequencies. If both carrier and modulator are audio-frequency waves, then the sum and difference frequencies are also audible partials. By varying just the frequency ratio between the two sine waves, and the amplitude of the modulator, one can produce an extremely wide variety of complex, musically rich wave forms.

The results that Chowning and his colleagues obtained from their work with FM were subsequently licensed to the Yamaha Corporation, and have formed the technical basis for some of the most successful digital keyboard instruments to have been introduced. A representative model is the Yamaha DX7, whose sounds are widely heard in current popular music (Fig. 18). Figure 19 shows an illustration from the DX7 user's manual that explains, in terms that musicians have come to understand, how FM produces musical tones.

5.2 Sampling Technology

In synthesizing musical tones, one generally starts with simple wave forms and time-varying control functions, and then builds up complexity by combining components and processing operations. All of the synthesis methods described above work this way.

Another approach to achieving a desired timbre is to start with a sound that is already complex, and perform simple operations on it to change its character. The existing sound is usually a recording of a "natural" sound. This approach works well when the desired sound is close to some already existing sound. The technique of recording individual sounds, then playing them back at specified times and with specified modifications, is called *sampling*.

The advantages of sampling are obvious, especially when one considers the needs of popular-music performers. The first commercially viable sampling instruments used

FIG. 18. The Yamaha DX7 synthesizer (ca. 1983).

lengths of magnetic tape on which were recorded choral, string ensemble, and horn ensemble tones, all of which were difficult to synthesize. The tape segments were set in motion by pressing keys on a conventional keyboard; a single recording typically lasted 8 s or so, usually long enough to play one note in a piece of music. When the keys were released, the tape segments were quickly returned to their starting positions. These instruments bore the trade names Mellotron and Chamberlain, and were widely used throughout the 1960s and 1970s, despite their mechanical complexity and tendency for the tapes to break during performance.

Equipment designers understood that individual sounds could be stored as digital recordings in semiconductor memory and then read out in response to a signal such as a key depression. However, the high price of semiconductor memory made this approach impractical until the late 1970s, when the Fairlight Computer Music Instrument was introduced (Fig. 20). The Fairlight was a small general-purpose business computer that was equipped with a piano-style keyboard and a multiplicity of peripheral "voice" cards, each of which had enough RAM (random-access memory) for one sound, and a digital-to-analog converter. In addition to scanning the keyboard and providing appropriate commands to the voice cards to play their sounds, the Fairlight provided the means for recording individual sounds as digital files, graphically editing the wave forms of those sounds with a light pen, storing individual sound files on 8-in. floppy disks, and loading sounds from disk to the voice card memories. The Fairlight and other disk-based sampling instruments enjoyed great popularity among producers of commercial and popular music.

The Kurzweil K250 was the first sampling instrument to have a library of sound files stored in semiconductor ROM (read-only memory) instead of on disk (Fig. 21). The ROM storage allowed the player to switch sound files instantly, instead of having to change disks and wait for the new sounds to load.

Since the introduction of the K250, the popularity of ROM-based sample-playing instruments has grown rapidly. Commercially successful instruments range from digital pianos and "band-in-a-box" peripherals for personal computers, to sample-playing modules with a full range of high-quality orchestral timbres and a broad palette of signal-processing resources to modify the timbres of the samples.

What are the relative advantages of sample-playing instruments over instruments that synthesize sounds from simple components? Sample players produce realistic simulations

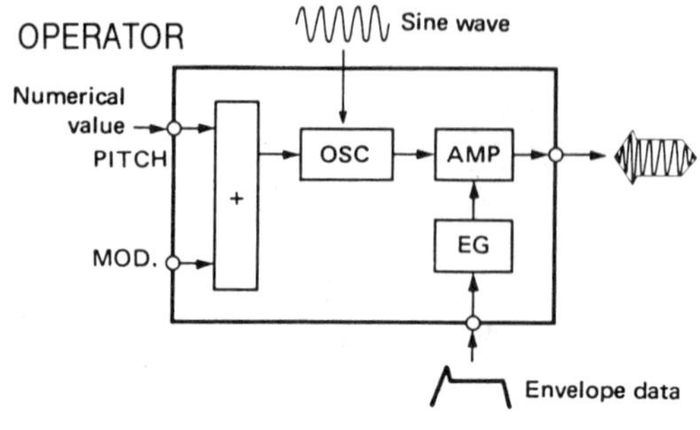

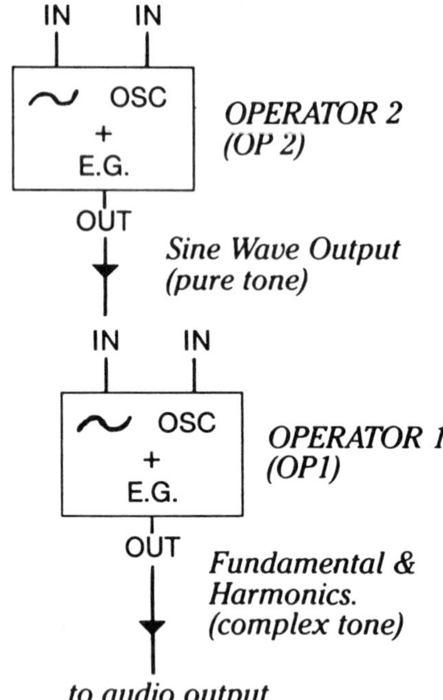

FIG. 19. An FM "operator," and a block diagram showing how two operators may be "patched" together to produce a complex tone.

of already familiar timbres. However, since their sounds are based on fixed recordings, the complex microstructure of each sound's evolution is difficult to change, and successive tones tend to sound alike. That is not a serious drawback in an instrument like a digital piano, because successive sounds of an acoustic piano also tend to sound alike. If an orchestral timbre such as a violin tone is played, however, successive tones should not sound exactly alike.

Synthesized timbres, on the other hand, are rarely as complex or "realistic" as sampled sounds. However, since they are generated "from scratch" in real time, it is generally possible for a musician to tailor the evolution of individual sounds, either by explicitly including small differences in the specifications of sounds as part of the composing process, or by using performance transducers to impart continuous changes in real time, as the tones are generated.

5.3 Contemporary Sound Synthesis Techniques

There are almost as many computer-based sound synthesis techniques in use as there are ways of defining time-varying periodic functions! Many of these are tailored to specific compositional goals, while others are intended to serve the needs of large groups of musicians. Four techniques, which are briefly described below, are representative examples of those that are currently in use.

Additive synthesis is synthesis of a periodic wave form by generating individual sinusoidal harmonics, imparting the appropriate slowly varying amplitude and frequency envelopes to them, and then combining them to achieve a desired sound (Snell, 1985). In theory, additive synthesis may be used to create any periodic function whatsoever. However, practical considerations tend to limit its usefulness. Since a complex low-pitched sound may contain hundreds of audible harmonics, additive synthesis tends to be extremely computation-intensive. Furthermore, additive synthesis cannot produce sounds with noise (random) components.

Nonetheless, several devices employing additive synthesis have been produced. Some enable the musicians to analyze a musical tone into its component harmonics, edit the envelopes of the harmonics, and then recombine the harmonics into a new sound that is closely related to the starting sound. This technique is called *resynthesis*, and is used by musicians who want to make small, carefully controlled timbral variations. The user interface of one such device, the Technos Axcel synthesizer, is a touch-sensitive surface on which a musician can draw envelope shapes directly with his finger.

Nonlinear distortion is a processing technique whereby a sound wave form is multiplied by a function that varies both with time and with the instantaneous amplitude of the

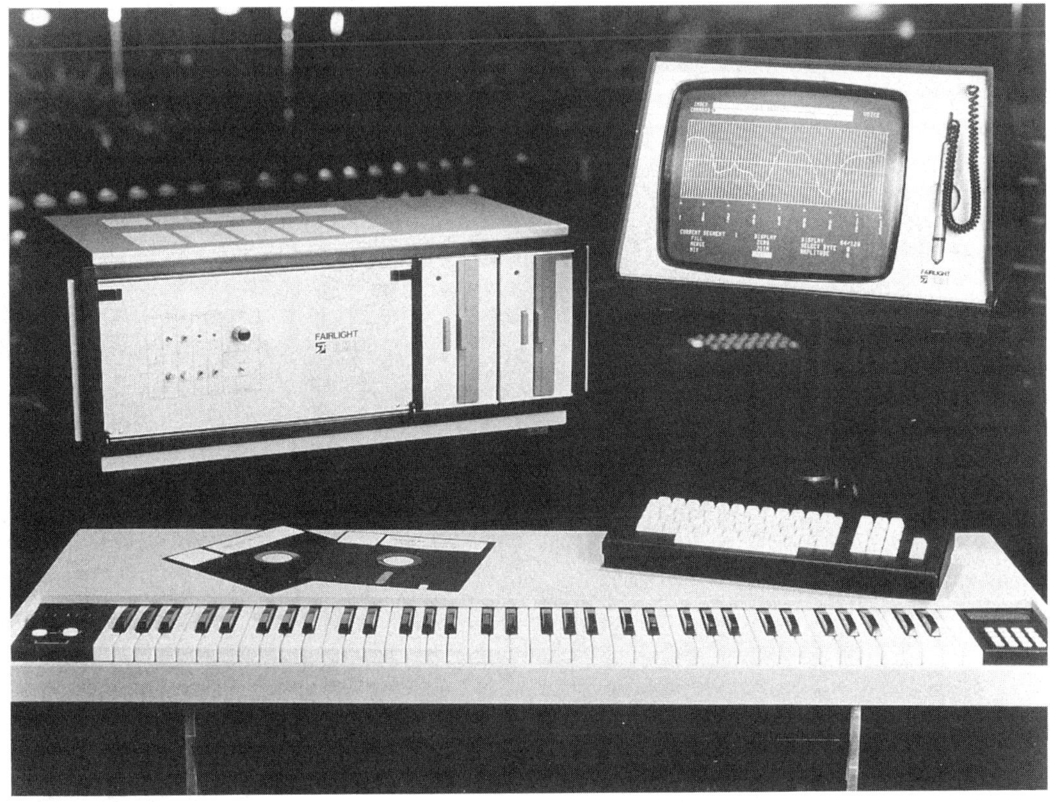

FIG. 20. The Fairlight "Computer Music Instrument" (ca. 1978).

starting wave form. This technique generally increases the harmonic content (Roads, 1985b). Useful distortion functions range in complexity from simple "clipping" of the starting waveform to complex, rapidly varying distortions that heighten the amplitudes of specific harmonics or groups of harmonics. Nonlinear distortion is especially useful for imparting timbral variation to sounds that are already defined, such as sounds produced by a sampling instrument.

Granular synthesis is the production of a wide variety of sonic textures by breaking continuous sounds up into very brief "granules." The composer specifies parameters such as the starting wave form, the average length and repetition time of the granules, the degree of randomness of length and repetition time, and the attack and decay times of the granules (Roads, 1985a). Music using granular synthesis tends to be pointillistic when the repetition time is long compared to the granule length, and shimmering and sonorous when the granules occur densely enough to overlap.

Models of distributed vibrating systems (either real or imagined) may be constructed with computer programs or dedicated digital circuitry. Properly excited and controlled, these models have the potential for producing complex wave forms of the type associated with acoustic musical instruments (Borin *et al.*, 1992). The Yamaha model VL1 Physical Modeling Synthesizer is an example of a commercial instrument employing such models (Fig. 22). *Waveguide modeling* is another modeling technique, in which waveguide boundaries, surface discontinuities, and exciting functions are specified, often as functions of time.

5.4 Linear Sound Processing Techniques

Just as many tape music composers preferred to compose by processing recorded "real-world" sounds, computer musicians

FIG. 21. The Kurzweil 250 (1985).

often prefer to start with recorded sounds, and then treat them with linear (nondistorting) processes before assembling them into musical structures. These processes are generally computation intensive, and rely heavily on the availability of fast multiplying devices and routines. Here are examples of linear processes that are currently in use.

Filtering is the process in which audio signals are passed through models of devices in

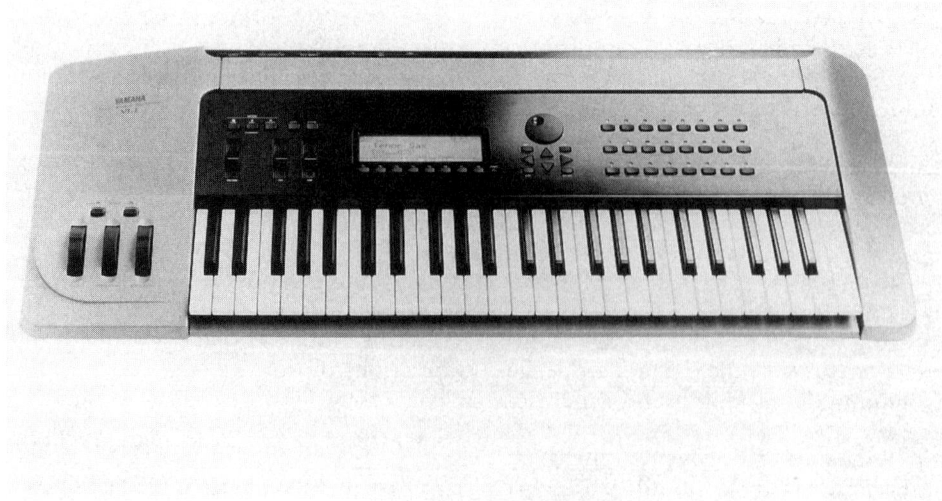

FIG. 22. The Yamaha VL1 synthesizer incorporates physical modeling of orchestral instruments (1994).

which the gain and phase shift are functions of frequency. Structures with multiple resonances (such as the bodies of string instruments) may be simulated, or filters with sharp cutoffs may be used to eliminate portions of the audio-frequency spectrum completely from the starting audio. All-pass filters are used as phase-shifting devices; when the phase-shifted output is added to the unshifted input signal, nulls occur in the composite frequency response, imparting the characteristic "comb filter" quality to the sound (Smith, 1989).

Linear predictive coding is a specialized filtering technique in which an input signal is analyzed to yield a time-based series of *n*-pole infinite impulse response (IIR) filters whose poles approximate the formants of the analyzed signal. The analysis is repeated as often as a hundred times a second, thereby providing a frequency response that follows the changing formant structure of the analyzed material. The frequency response is then used as a filter to process a second, relatively static signal (Makhoul, 1975). Linear predictive coding is especially valuable in analyzing vocal sounds and imparting vocal articulation and formant structure to simple wave forms with high harmonic content.

Reverberation is the creation and control of a multiplicity of closely spaced echos of the original audio material. Reverberation simulates the acoustic effect of the sound source that is enclosed in a reflective space. In digital reverberation systems (which consist of little more than a multiplicity of digital delay lines), the parameters of reverberation—spacing and decay rate of echoes, dependence of the decay rate on frequency, and elapsed time to the first echo—may be programmed to simulate listening spaces ranging in volume from that of a 55-gallon drum to Westminster Cathedral (Moorer, 1979). Used as a compositional resource, reverberation enables the composer to regard the perceived spatial environment as a manipulable compositional element. And, by making the delay times very long, say greater than 0.1 s or so, one can produce discrete echoes.

The perceived *spatial location and motion* of a sound can be controlled by a combination of simple processes. In addition to the standard bag of tricks employed by recording engineers to define left-to-right location, one can control the ratio of reverberated to unreverberated sound to influence the perception of distance between the listener and the sound source. One can also vary the sound's frequency and amplitude to simulate the Doppler effect of a sound source moving rapidly past the listener (Chowning, 1971).

5.5 Computer Systems as Composing Tools

At the present time, musicians are able to connect a wide variety of readily available electronic music devices into computer-controlled systems. This is made possible by MIDI (Musical Instrument Digital Interface), a communications protocol that has been adopted by the electronic musical instrument industry. MIDI employs a serial interface and simple two-conductor cables to interconnect system components. The MIDI codes enable information on when notes begin and end, how continuous variables are being changed, which "program" (set of parameter values that determine timbre) has been selected, and so forth, to be sent from one system device to the others (Rothstein, 1992).

The central component of a MIDI system is usually a personal computer that is equipped with MIDI-specific software. The most widely used class of MIDI programs are the so-called sequencer programs, which enable MIDI messages to be recorded, edited, and played back in loose analogy to a tape recorder. Sequencers have graphic interfaces that display the MIDI messages as musical scores, graphs, and alphanumeric lists. With a MIDI sequencer and a small number of MIDI-controlled sound-producing and -processing devices, one can realize a wide variety of music of moderate complexity.

MIDI was drawn up in response to the needs of musicians who are interested in popular, rather than experimental, music. For this reason, the orientation underlying the specification of MIDI codes tends to favor conventional pop music. Even so, many ingenious programs serving the needs of experimental musicians have been developed. Just a few examples are David Zicarelli's *MAX*, an unusually versatile screen-based patching and mapping system, Wayne Kirby's *Serious Composer*, an algorithmic composition program based on fractals, and Laurie Spiegel's *Music Mouse*, an improvisation environment in which one first specifies boundary conditions for a set of musical events, and then manipulates the events with a computer mouse.

5.6 Interactive Systems for Improvising

Improvising systems are algorithmic composition programs that are designed to be used in real time. They analyze inputs from acoustic or electronic tone sources, then apply composition rules to formulate a musical response. The composition rules are invariably complex, since the system must attempt to simulate the creative intuition of a human performer, anticipate the probabilities of a wide range of possible inputs, and continuously correct its output in a relatively graceful way. In a very real sense, the rules that guide an improvising system are the expression of a musical personality, and formulating the rules requires considerable musical insight.

A simple example of an improvising system is Roger Dannenberg's Blues Accompaniment program (Dannenberg and Mont-Reynaud, 1987). The program assumes that the live performer will play a blues line with the standard 12-measure blues chord progression in mind, and that he will play within a narrow tempo range. When the musician starts, the program seeks to determine the beat and location within the chord progression. Once it makes a decision, the system produces the chord pattern, while updating the likelihood of other chords being more or less "correct" with each successive beat.

Robert Rowe's program *Cypher* is another example of an interactive improvising system (Rowe, 1992). Rather than address the relatively rigid rules of a conventional blues improvisation, Cypher works on a looser set of composition rules. In Cypher, musical ideas are derived from the inputs provided by the improvising musicians, transformed algorithmically, and then played back to provide new musical ideas for the performers themselves to develop.

5.7 Performer Interfaces and Performance Techniques

The current state of the electronic music art is such that the means for creating sounds of high musical quality, and composing these sounds off line into musical works of moderate complexity, are widely available at relatively reasonable cost. However, real-time performance control over musical variables leaves something to be desired. The performance interfaces of most currently available electronic musical instruments are mechanically primitive organ-style or piano-style keyboards. Few instruments enable musicians to impart the sorts of complex, multivariable nuances that traditional acoustic musicians take for granted.

The development of alternative performance interfaces continues to proceed, however. In the commercial musical instrument world, several designs based on traditional performance interfaces are available. Among these are guitarlike, clarinet-like (Fig. 23), and drumlike devices whose outputs are MIDI messages rather than sound wave forms. Experimental musicians often favor more general, less conventional control interfaces such as touch-sensitive surfaces (Fig. 24).

Many experimental performers are working with "virtual reality" systems—multiple transducers, the outputs of which are used to create real-time computer models of a moving hand or body. The DataGlove is one such device. With a DataGlove and appropriate software, one can "play a virtual instrument" simply by moving one's fingers in space.

Work is also proceeding on adapting traditional keyboard design for use in electronic music systems. The complex tactile feedback provided by the keyboard of an acoustical piano has been electronically simulated by equipping each key of an experimental keyboard with a motion sensor and a linear motor, and providing the appropriate feedback from sensor to motor. In another development, an experimental keyboard has been equipped with sensors that continuously detect the vertical position of the key as well as two dimensions of the position of the performer's finger on the key surface, thereby providing continuous control over three variables of each sound (Fig. 25) (Moog and Rhea, 1990).

We conclude this section with a description of Max Mathews's Radio Baton. The Radio Baton is a conducting device. That is, it is intended to be used in a situation where a piece of music is already composed as a series of computer instructions, and is to be "performed" under the global control of a

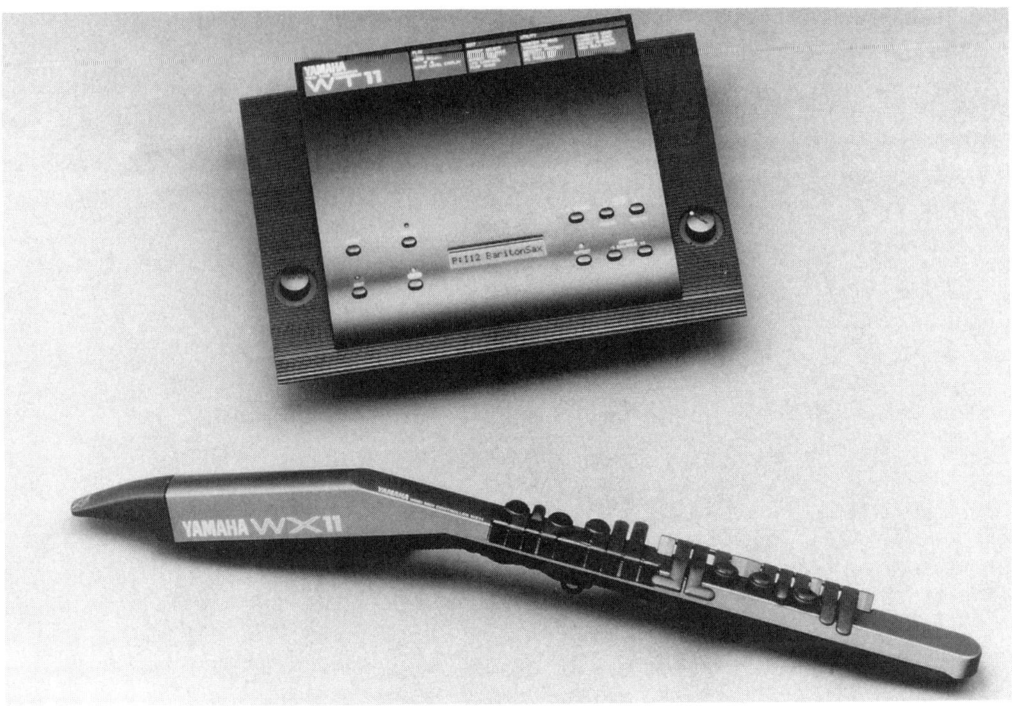

FIG. 23. The Yamaha WX11 wind controller. In addition to containing an array of switches for selecting pitches, the device contains a pressure transducer for continuously registering the player's breath pressure (1990).

"conductor." For instance, if the piece were of traditional tonal music, the score would specify the order of notes to be played, but not the overall tempo or dynamics. Mathews's Radio Baton consists of two drumsticklike batons, plus a surface over and on which the conductor moves the batons. As a transducer system, the Radio Baton uses radio-frequency fields to (continuously) determine the positions of the batons; from the conductor's point of view, the device enables him to use simple, intuitively appropriate gestures to control how the system "reads" the score (Fig. 26).

6. FUTURE DIRECTIONS FOR ELECTRONIC MUSIC TECHNOLOGY DEVELOPMENT

Despite its important role in contemporary experimental and art music, and its significant stake in the current musical instrument marketplace, electronic music technology has by no means reached its full maturity. The explosive development of digital electronics and software has opened up exciting new possibilities for musical sound generation, processing, performance, and composition. These possibilities have, in turn, underscored the need for better real-time human/machine interfaces, and encouraged researchers to explore further the psychoacoustic and cybernetic aspects of music production and listening. Some important opportunities for applying the principles of physics and engineering to the further development of electronic music technology follow.

6.1 Models for Sound Generation and Processing

The current trend in digital electronic music systems is toward increasing use of models of vibrating systems (as opposed to fixed wave forms) to generate and process musical tones. Models may be based on traditional acoustic musical instruments (for producing familiar timbres) or on imaginary vibrating devices (for exploring new timbres). The important requirements of such models

FIG. 24. A "touch-plate controller," which produces control signals proportional to two dimensions of the position of a player's finger on its surface, plus a third signal proportional to the area of the finger in contact with the surface (1984).

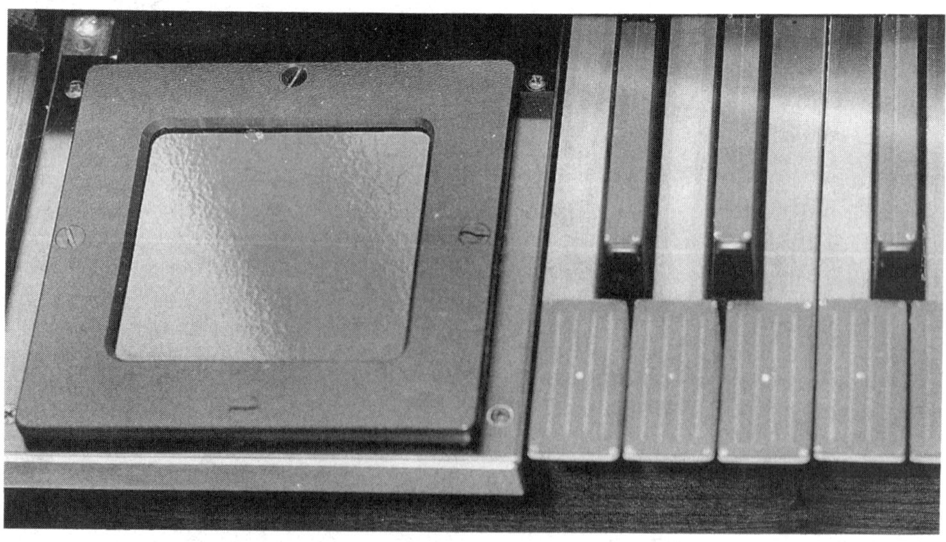

FIG. 25. A portion of a multiple touch sensitivity keyboard. The large touch plate on the left controls global parameters, while each of the keys on the right has its own small touch surface for shaping the parameters of a single sound event (1993).

FIG. 26. Max Mathews and his Radio Baton.

are that their output signals be musically interesting, and that they contain a multiplicity of continuously variable parameters that the musician can manipulate in order to impart musical gestures. The disciplines of musical acoustics, psychoacoustics, and vibration analysis are appropriate starting points for the development of models for musical tone generation.

6.2 Electroacoustic Tranducer Design

Electronic music always needs one or more electroacoustic transducers in order to be heard. Current loudspeaker and headphone technology has evolved to meet the requirements of two-channel (stereo) wideband audio signal reproduction. Electronic musical instruments and performance systems often require more application-specific electroacoustic transducers. One common example is the speakers used in guitar amplifiers, which are deliberately designed to be somewhat nonlinear and of narrow bandwidth. Another example is a system of a large number of speakers that are distributed throughout a listening space in order to create a multichannel acoustic field. Opportunities exist for collaborations among electronic musicians, speaker designers, and architectural acousticians.

6.3 Real-Time Control Interfaces

A good real-time control interface between a musician and his instrument

1. senses hand and other body motions accurately and rapidly, and
2. provides equally accurate and rapid tactile feedback.

Resolution and accuracy requirements are typically on the order of 1 mm for position detection, and on the order of 1 g for force detection. Mechanical bandwidth (response speed) requirements are on the order of hundreds of hertz. Sensors and tactile feedback mechanisms must be rugged, reliable, smooth-acting, aesthetically pleasing—and inexpensive. The studies of materials science, sensor design, actuator (motor) design, and ergonomics can be applied to the further development of user interfaces for electronic music production.

6.4 Digital Electronics, Computer Software, and Artificial Intelligence

For all of human history, musical instrument builders have looked to the most sophisticated technologies of their times. For this reason, virtually all of the new musical instruments and systems of our time are being built with digital electronics and software. As the distinction between hardware and software continues to blur, and ever-increasing percentages of the value of electronic music devices become directly traceable to the encoded intelligence of computer programs, workers in electronic music technology are gravitating toward the resources of software-based artificial intelligence. Already, many music producers, commercial as well as experimental, do much of their work at computer terminals. The instrument builder, traditionally an expert at fabricating and assembling physical materials, will tend more and more to build software-based music production environments. To do this, he will draw on the principles of physical modeling, psychoacoustics, and ergonomics.

Works Cited

Babbitt, M. (1964), "An Introduction to the R.C.A. Synthesizer," *J. Mus. Theory* **VIII**, 251–265.

Berger, I. (1969), "The 'Switched-on Bach' Story," *Saturday Rev.* **January 25**, 45–47.

Borin, G., De Poli, G., Sarti, A. (1992), "Algorithms and Structures for Synthesis Using Physical Models," *Comp. Mus. J.* **16** (4) 30–42.

Chowning, J. (1971). "The Simulation of Moving Sound Sources," *J. Audio Eng. Soc.* **19**, 2–6.

Chowning, J. (1973), "The Synthesis of Complex Audio Spectra by Means of Frequency Modulation," *J. Audio Eng. Soc.* **21**, 526–534.

Dannenberg, R., Mont-Reynaud, B. (1987), "Following an Improvisation in Real Time," *Proceedings of the 1987-International Computer Music Conference*, San Francisco: Computer Music Association, pp. 241–248.

Eimert, H. (1956), *Electronic Music*, Ottawa: National Research Council of Canada, Technical Translation TT-612.

Glinsky, A. V. (1992), *The Theremin in the Emergence of Electronic Music*, Ph.D. Thesis, New York University, New York.

Laurendeau, J. (1990), *Maurice Martenot, Luthier de l'Electronique*, Montreal: Louise Courteau Editrice Inc.

Makhoul, J. (1975), "Linear Prediction: A Tutorial Review," *Proc. IEEE* **63**, 561–580.

Mathews, M. (1969), *The Technology of Computer Music*, Cambridge, MA: MIT Press.

Moog, R. (1965), "Voltage-Controlled Electronic Music Modules," *J. Audio Eng. Soc.* **13**, 200–206.

Moog, R. (1967), "Electronic Music, Its Composition & Performance," *Electronics World* **February**, 42–46.

Moog, R. (1981), "The Columbia/Princeton Electronic Music Center," *Keyboard* **May**, 22–32.

Moog, R., Rhea, T. (1990), "Evolution of the Keyboard Interface," *Comp. Mus. J.* **14** (2), 52–60.

Moorer, J. (1979), "About This Reverberation Business," *Comp. Mus. J.* **3** (2), 13–28.

Rhea, T. (1977a) "The Cellulophone," *Keyboard* **September**, 56.

Rhea, T. (1977b), "The Hammond Organ," *Keyboard* **May**, 47, and **June**, 47.

Rhea, T. (1978a), "B. F. Miessner's 'Stringless Piano'," *Keyboard* **April**, 62.

Rhea, T. (1978b), "Early One-Tune-Per-Note Organs," *Keyboard* **December**, 69.

Rhea, T. (1979a), "The Novachord," *Keyboard* **January**, 62.

Rhea, T. (1979b), "The Trautonium," *Keyboard* **April**, 76.

Rhea, T. (1979c), "The First Synthesizer," *Keyboard* **August**, 70.

Rhea, T. (1979d), "The Hanert Synthesizer," *Keyboard* **September**, 78.

Rhea, T. (1979e), "The Grainger-Cross Free Music Machine," *Keyboard* **November**, 90.

Roads, C. (1985a), "Granular Synthesis of Sound," in: C. Roads, J. Strawn (Eds.), *Foundations of Computer Music*, Cambridge, MA: MIT Press, pp. 145–159.

Roads, C. (1985b), "A Tutorial on Nonlinear Distortion or Waveshaping Synthesis," in: C. Roads,

J. Strawn (Eds.), *Foundations of Computer Music*, Cambridge, MA: MIT Press, pp. 83–94.

Rothstein, J. (1992), *MIDI—A Comprehensive Introduction*, Madison, WI: A-R Editions, Inc.

Rowe, R. (1992), *Interactive Music Systems*, Cambridge, MA: MIT Press.

Schaeffer, P. (1952), *A la Recherche d'une Musique Concrete*, Paris: Editions du Seuil.

Smith, J. (1989), "Fundamentals of Digital Filter Theory," in: C. Roads (Ed.), *The Music Machine*, Cambridge, MA: MIT Press, pp. 509–519.

Snell, J. (1985), "Design of a Digital Oscillator That Will Generate up to 256 Low-Distortion Sine Waves in Real Time," in: C. Roads, J. Strawn (Eds.), *Foundations of Computer Music*, Cambridge, MA: MIT Press, pp. 289–325.

Ussachevsky, V. (1958), "The Process of Experimental Music," *J. Audio Eng. Soc.* **6**, 202.

Further Reading

Appleton, J., Perera, R. (Eds.) (1975), *The Development and Practice of Electronic Music*, Englewood Cliffs, NJ: Prentice-Hall, Inc.

Chamberlin, H. (1985), *Musical Applications of Microprocessors*, 2nd ed., Hasbrouck Heights, NJ: Hayden Book Co.

Davis, D. (1988), *Computer Applications in Music—A Bibliography*, Madison, WI: A-R Editions.

Davis, D. (1982), *Computer Applications in Music—A Bibliography, Supplement 1*, Madison, WI: A-R Editions.

Deutsch, D. (1982), *The Psychology of Music*, New York: Academic Press.

Mathews, M., Moore, F., Risset, J. (1974), "Computers and Future Music," *Science* **183**, 263–268.

Mathews, M., Pierce, J. (Eds.) (1989), *Current Directions in Computer Music Research*, Cambridge, MA: MIT Press.

Moog, R. (1977), "Electronic Music," *J. Aud. Eng. Soc.* **25**, 855–861.

Moog, R. (1986), "Digital Music Synthesis," *Byte* **June**, 155–168.

Moore, F. (1990), *Elements of Computer Music*, Englewood Cliffs, NJ: Prentice Hall, Inc.

Pellman, S. (1994), *An Introduction to the Creation of Electroacoustic Music*, Belmont, CA: Wadsworth.

Pierce, J. R. (1992), *The Science of Musical Sound*, New York: W. H. Freeman.

Prieberg, F. (1960), *Musica ex Machina*, Berlin: Verlag Ullstein.

Rhea, T. (1972), *The Evolution of Electronic Musical Instruments in the United States*, Ph.D. Thesis, George Peabody College for Teachers, Nashville.

Roads, C. (Ed.) (1990), *The Music Machine*, Cambridge, MA: MIT Press.

Roads, C. (Ed.) (1985), *Composers and the Computer*, Madison, WI: A-R Editions.

Rothstein, J. (1992), *MIDI—A Comprehensive Introduction*, Madison, WI: A-R Editions, Inc.

Strawn, J. (Ed.) (1985), *Digital Audio Signal Processing: An Anthology*, Madison, WI: A-R Editions, Inc.

Schwartz, E. (1973), *Electronic Music: A Listener's Guide*, New York: Praeger.

Strange, A. (1983), *Electronic Music: Systems, Techniques, and Controls*, 2nd ed., Dubuque, IA: William C. Brown.

Vail, M. (1993), *Vintage Synthesizers*, San Francisco: GPI Books.

Winckel, F. (1963), "The Psycho-Acoustical Analysis of Structure as Applied to Electronic Music," *J. Mus. Theory* **7** (2), 194–246.

MUSICAL INSTRUMENTS

THOMAS D. ROSSING, *Department of Physics, Northern Illinois University, Dekalb, Illinois, U.S.A.*

INTRODUCTION

Musical instruments have played an important role in nearly all cultures throughout history. First came the human voice (which we consider to be a musical instrument), then percussion instruments, and finally primitive string and wind instruments. All of these major families of instruments have developed along many different paths according to the culture that nurtured them. Together, they weave an interesting acoustical story.

The most common way to classify musical instruments is into the traditional families of string, wind, and percussion instruments, depending upon the nature of the primary vibrator. (Sometimes the alternative designations chordophones, aerophones, membranophones, and idiophones are used.) Of course, many instruments incorporate exci-

3-527-28133-9/94/$5.00 + .50

ters or resonators of a different nature from the primary vibrator: woodwind instruments make use of vibrating reeds, string instruments have rigid soundboards, xylophones and marimbas have air-filled resonators, etc.

An interesting acoustical characteristic of musical instruments is the positive or regenerative feedback that sustains oscillations in many instruments. In brass and woodwind instruments, the regenerative feedback is so strong that it creates regimes of oscillation. In bowed string instruments, it is also strong, but in plucked or hammered string instruments and in percussion instruments, it is hardly a factor at all. In instruments with little or no positive feedback, once the sound is initiated, the player has little means for changing the pitch or timbre. The human voice incorporates a modest amount of feedback.

The scientific study of musical instruments has attracted the attention of a number of famous scientists: Rayleigh, Helmholtz, Tyndall, Raman, Savart, and Saunders, to name a few. The study of musical instruments is an important part of the interdisciplinary field of musical acoustics. In addition to studying the production of sound by musical instruments, musical acousticians also study the acoustics of concert halls (see ACOUSTICS, ARCHITECTURAL), the recording and reproduction of sound (see SOUND REPRODUCTION), and the perception of musical sound (see ACOUSTICS, PSYCHOLOGICAL).

1. STRING INSTRUMENTS

1.1 Vibration of Strings

The equation for transverse motion of a completely flexible string can be written

$$\frac{\partial^2 y}{\partial t^2} = \frac{T}{\mu}\frac{\partial^2 y}{\partial x^2} = c^2\frac{\partial^2 y}{\partial x^2}, \tag{1}$$

where T is tension, μ is mass per unit length, and $c = \sqrt{T/\mu}$ is the speed of transverse waves. The general solution of Eq. (1) can be written in a form credited to d'Alembert (1717-1783):

$$y = f_1(ct - x) + f_2(ct + x). \tag{2}$$

The function $f_1(ct - x)$ represents a wave traveling to the right, and $f_2(ct + x)$ represents a wave traveling to the left. The nature of the functions f_1 and f_2 is arbitrary; they could be pulses, or they could be sinusoidal. In fact, these two independent functions can be chosen so that their sum represents any desired initial displacement $y(x,0)$ and velocity $\dot{y}(x,0)$.

A string of length L, fixed at its two ends, will have normal modes of vibration that can be written

$$y_n(x,t) = A_n \sin(2\pi f_n t + \phi_n)\sin(2\pi f_n x/c), \tag{3}$$

where $f_n = nc/2L$ is the frequency of the nth normal mode, A_n is its amplitude, and ϕ_n is its phase. The displacement y of the string can then be written as a sum of the normal modes:

$$y = \sum_n y_n. \tag{4}$$

The force of excitation determines the amplitude A_n and phase ϕ_n of each normal mode.

Plucking, bowing, or striking the same string would excite the same normal modes but with quite different values of the C_n's and ϕ_n's. Furthermore, the vibrational spectrum of a string depends not only on the manner of excitation but also on the point of excitation.

1.1.1 Plucked Strings When a string is plucked at its midpoint, the resulting vibration will be made up of the odd-numbered modes ($n = 1, 3, 5, 7, \ldots$) with amplitudes $C_n = 1, 1/9, 1/25, \ldots$ and alternating phases ($\phi_n = 0, 180°, 0, 180°, \ldots$). The mode amplitudes fall off as $1/n^2$ (12 dB/octave), and the string exerts a force on the end fixtures (e.g., the bridge of a guitar) that alternates in direction and whose waveform approximates a square wave (see Chap. 2 in Fletcher and Rossing, 1991).

Figure 1 shows the addition of modes to obtain the shape of a string plucked at one-fifth of its length. Note that the fifth mode is missing, also the tenth and fifteenth, as shown in the spectrum in Fig. 1. As a general rule, when a string is excited at a fraction $1/\beta$ of its length, the amplitude dips to zero for $n = \beta, 2\beta, 3\beta, \ldots$.

1.1.2 Struck Strings When a string is

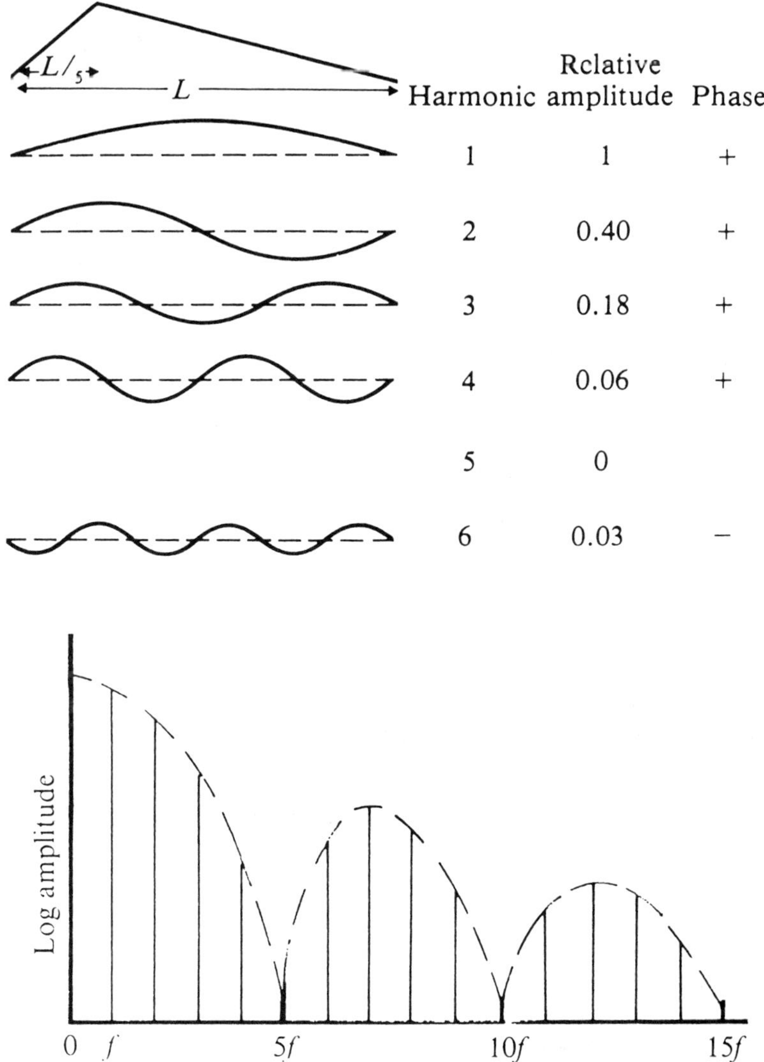

FIG. 1. Addition of modes to obtain the shape of a string plucked at one-fifth of its length. The mode spectrum at the right shows the relative amplitudes of the modes.

plucked, it is given an initial displacement $y(x,0)$, but the initial velocity is assumed to be zero. When a piano string is struck by a hammer, however, the opposite is true: it is given an initial velocity but the initial displacement is zero. Figure 2 shows the displacement and velocity for a long string struck by a hard narrow hammer having a velocity V.

A short time t after being struck, a portion of the string with length $2ct$ and mass $2\mu ct$ is set into motion. As this mass increases and becomes comparable to the hammer mass M, the hammer is slowed down and would eventually be stopped. With a string of finite length, however, reflected impulses return while the hammer still has appreciable velocity, and these reflected impulses react with the hammer, causing it to be thrown back from the string.

The interaction of a felt piano hammer with a piano string is considerably more complicated than suggested by the simple hard, narrow hammer approximation. In general, mode amplitudes in a struck string fall off approximately as $1/n$ (6 dB/octave), which is

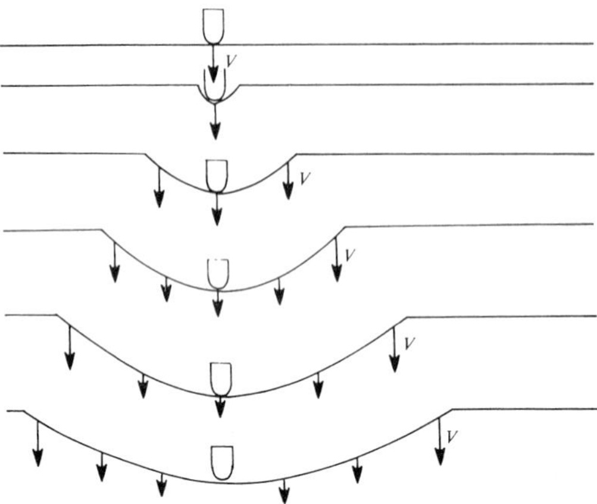

FIG. 2. Displacement and velocity of a long string at successive times after being struck by a hard narrow hammer having a velocity V. (From Fletcher and Rossing, 1991.)

considerably slower than those of a plucked string (which go nearly as $1/n^2$ or 12 dB/octave). If the string is struck at $1/\beta$ of its length, the amplitude drops to zero for $n = \beta, 2\beta, 3\beta, \ldots$, as in the plucked string case.

1.1.3 Bowed Strings When a bow is drawn across the string of a violin, the string appears to vibrate back and forth smoothly between two curved boundaries, much like a string vibrating in its fundamental mode. However, this appearance of simplicity is deceiving, as shown by Helmholtz (1877) over 100 years ago. The string more nearly forms two straight lines with a sharp bend at the point of intersection. This bend races around the curved path that we see, making one round trip for each period of the vibration.

The Helmholtz-type motion of a bowed string is shown in Fig. 3. As the bow moves ahead at constant speed, the bend races around a curved path. The graph on the right-hand side shows the position of the point of contact between bow and string at successive times; the letters correspond to the frames in the left half of the figure.

The action of the bow on the string is often described as a "stick and slip" action. The bow drags the string along until the bend arrives and triggers the slipping action [frame (a) in Fig. 3], which continues until the string is picked up once again by the bow [frame (c)]. From frames (c) to (i), the string moves at the speed of the bow. The bend moves along the string at the wave speed $c = \sqrt{T/\mu}$.

The envelope of the bend (dashed curve in Fig. 3) is composed of two parabolas, and it has an amplitude that is proportional, within limits, to the bow velocity. For a given bow velocity, it increases as the string is bowed closer to one end. Thus, a violinist has two ways to control loudness: bowing speed and bowing position. (Bow force has relatively little effect on amplitude, provided it stays within the range for which the stick and slip action works.)

The actual string motion may be a combination of several Helmholtz-type motions. The finer details of bowed string motion are elegantly treated in a book by Cremer (1981).

1.2 Plucked String Instruments

1.2.1 Guitars The modern guitar, which is a descendent of the 16th century Spanish vihuela, has 6 strings about 65 cm in length, tuned to E_2, A_2, D_3, G_3, B_3, and E_4 (f = 82, 110, 147, 196, 247, and 330 Hz). (Throughout this article, we will use the American Standard notation, in which A_4 has a frequency of 440 Hz. "Middle C" is denoted as C_4. The numbers change between B and C.) The top is usually cut from spruce or rosewood, planed to a thickness of about 2.5 mm, while the back is usually of hardwood with about the same thickness.

Acoustic guitars can be classified into four families of design: classical, flamenco, flat-top (or folk), and arch-top. Classical and flamenco guitars have nylon strings; flat-top and arch-top guitars have steel strings. Steel string guitars usually have a steel rod embedded inside their neck, and their soundboard has crossed bracing, whereas the braces on classical guitar soundboards are usually arranged in a fan-shaped pattern, as shown in Fig. 4.

A plucked guitar string exerts a series of impulses on the bridge, which is transmitted to the top plate. The top plate, in turn, transmits vibrational energy to the back plate and the enclosed air. At higher frequencies, sound is radiated mainly by the top plate, but at lower frequencies, sound radiation from the back plate and the sound hole are major contributors to the overall guitar sound. The mechanical frequency response of the guitar body to a large extent determines the quality of the radiated sound.

The mechanical frequency response and

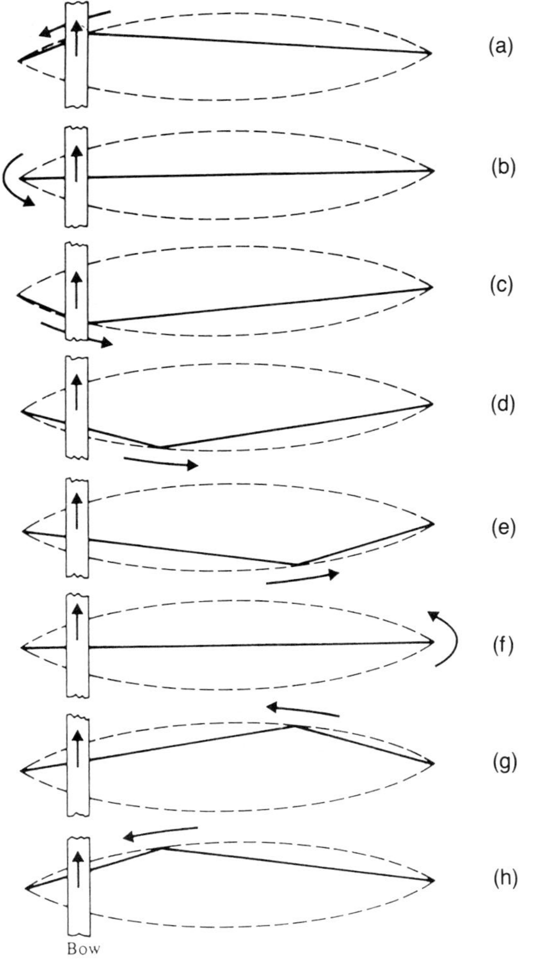

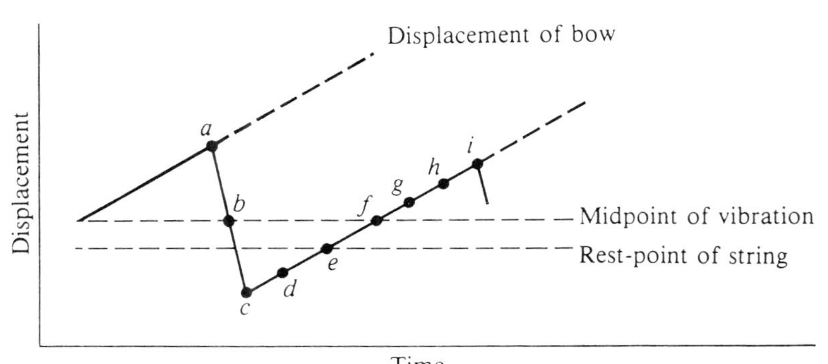

FIG. 3. Motion of a bowed string. Left: Time analysis of the motion, showing the shape of the string at eight successive times during the cycle. Right: Displacement of the bow (dashed line) and the string at the point of contact (solid line) at successive times. The midpoint of the string vibration is displaced slightly from the rest point of the unbowed string. The letters correspond to the letters in the time analysis at the left. (From Rossing, 1990.)

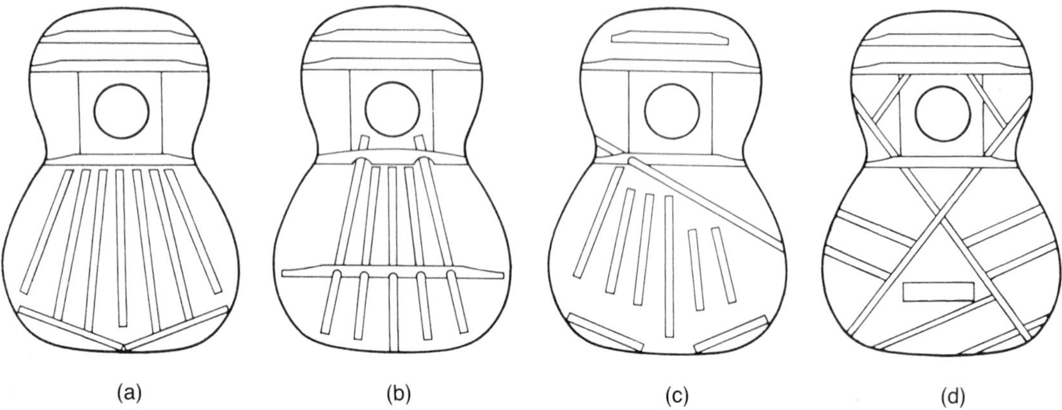

FIG. 4. Various designs for bracing a guitar soundboard: (a) traditional (Torres) fan bracing; (b) Bouchet (France); (c) Ramirez (Spain); (d) crossed bracing. (Rossing, 1990.)

radiated sound spectrum of a Martin D-28 folk guitar are shown in Fig. 5. The low-frequency sound spectrum is dominated by mechanical resonances at about 100, 190, and 205 Hz. These three resonances are due to coupled motions of the top plate, the back plate, and the enclosed air as shown in Fig. 6(a).

At 103 Hz, the top plate and back plate move in opposite directions, and so air is pumped in and out of the sound hole. At 204 Hz, on the other hand, the air flow through the sound hole is opposite in phase to the motion of the plates. This can be likened to the two modes of a mechanical system consisting of two masses connected by springs; in the mode of lower frequency, the masses move in the same direction, but in the mode of higher frequency, they move in opposite directions. At 193 Hz, the two plates move in the same direction.

The resonances at 376 and 436 Hz result from the "see saw" motion of the top and back plates around nodal lines running lengthwise. In the lower one, the top and back plates move in opposite directions so that air sloshes from one side to the other, while in the upper resonance, the two plates move together, resulting in less air motion.

The holographic interferograms in Fig. 7 show the motion of the top plate in a classical guitar at ten of its resonances. The lowest one at 268 Hz corresponds to the 376-Hz mode in the folk guitar, whose cross-bracing gives the top plate considerably more transverse stiffness.

1.2.2 Lutes Lutes of various types have

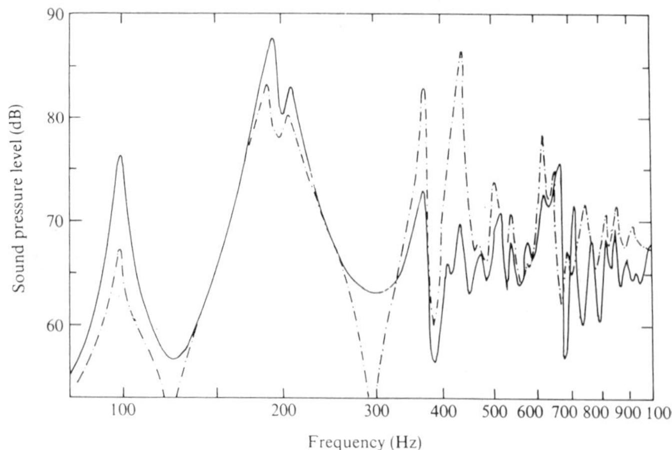

FIG. 5. Mechanical frequency response and sound spectrum 1 m in front of a Martin D-28 folk guitar driven by a sinusoidal force of 0.15 N applied to the treble side of the bridge. Solid curve, sound spectrum; dashed curves, acceleration level at the driving point. (Fletcher and Rossing, 1991.)

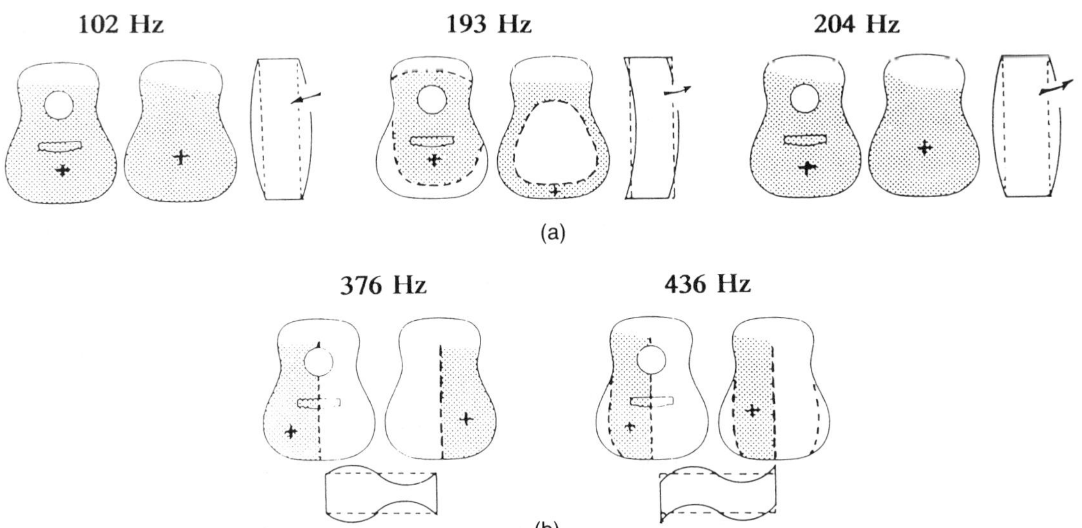

FIG. 6. Motion of the top plate, back plate, and enclosed air at five low-frequency resonances of a Martin D-28 folk guitar: (a) resonances in which the top and back plates each vibrate in a single phase; (b) resonances in which a nodal line divides the top and back plates into halves. (Fletcher and Rossing, 1991.)

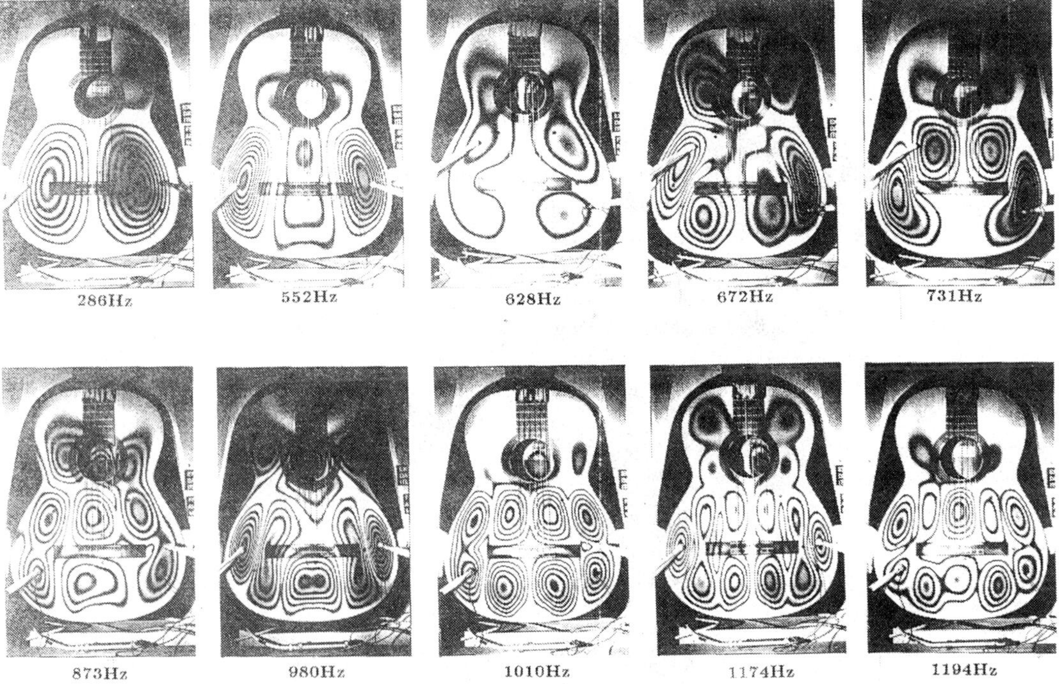

FIG. 7. Time-averaged holographic interferograms of top-plate modes of a classical guitar. The resonant frequencies of each mode are shown below the interferograms. (Richardson and Roberts, 1985.)

existed throughout the Near East and Far East for many centuries. The instrument that developed great popularity in Europe during the 16th and 17th centuries typically had 11 strings arranged in six courses (all but the uppermost consisting of two unison strings) that might be tuned A_2, D_3, G_3, B_3, E_4, and A_4, although the tuning was often changed to fit the music being played. The pear-shaped body of the lute is fabricated by gluing together a number of thin wooden ribs. The table or soundboard (of spruce, cedar, or cypress) is braced by transverse bars above, below, and at the soundhole.

Although the acoustics of the lute appears to be an interesting subject, only a few studies on their acoustical behavior have been reported. Firth (1977) measured the input admittance and sound radiation of a modern lute and also investigated the modes of vibration. Nodal patterns for several modes are shown in Fig. 8.

1.3 Bowed String Instruments

1.3.1 Violins The violin is the most studied of all musical instruments in the symphony orchestra. Outstanding contributions to our understanding of violin acoustics have been made by such distinguished scientists as Felix Savart (France, 1791–1841), Hermann von Helmholtz (Germany, 1821–1894), Lord Rayleigh (England, 1842–1919), C. V. Raman (India, 1888–1970), and Frederick Saunders (United States, 1875–1963). More recently, the work of Professor Saunders has been continued by members of the Catgut Acoustical Society, an organization that publishes an excellent journal describing current research on the acoustics of string instruments.

A violin has four strings of steel, gut, or nylon (wrapped with silver, aluminum, or steel), tuned to G_3, D_4, A_4, and E_5. The top plate is generally carved from Norway spruce (*Picea abies* or *Picea excelsis*) and the back, ribs, and neck from curly maple (*Acer platanoides*). The fingerboard is made of ebony and the bow stick of pernambuco (*Caesalpinia echinata*), a very dense wood. Running longitudinally under the top is a bass bar, and a sound post is positioned near the treble foot of the bridge.

Why are some violins, especially those crafted by the Italian masters, so highly treasured, while other instruments that appear to be virtually identical to them are considered inferior to them in quality? This is a question that has intrigued violin researchers for many years. One important fac-

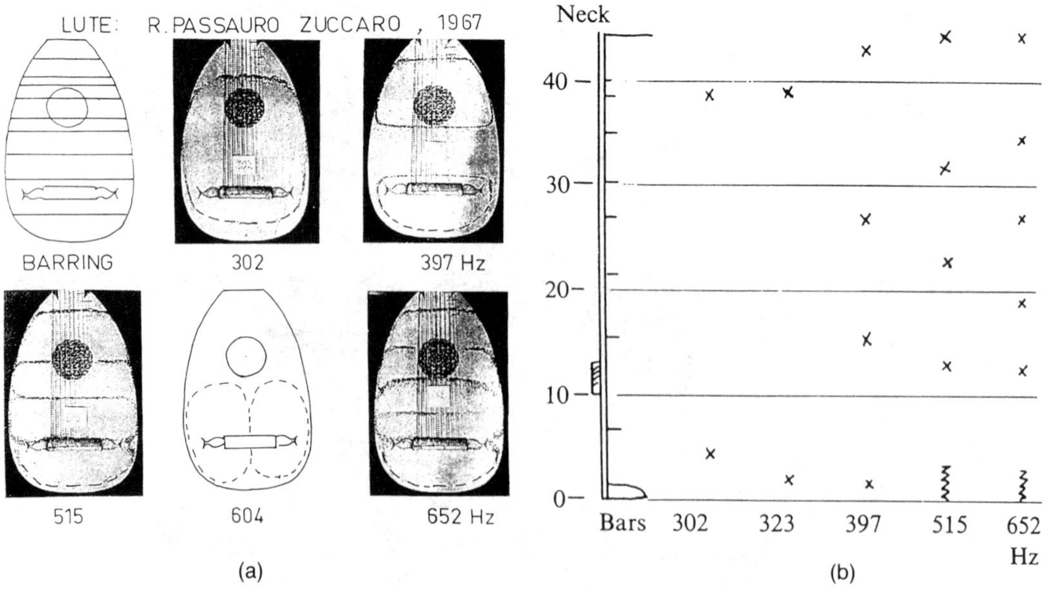

FIG. 8. (a) Barring pattern and nodal patterns in the top plate of the lute at five resonances and (b) locations of nodes compared to the bridge and the bars (Firth, 1977).

tor in the sound quality and playability of a violin is the vibrational behavior of its body. We often describe this in terms of normal modes of vibration, which consist of coupled motions of the top plate, the back plate, and the enclosed air. Smaller contributions are made by the ribs, neck, and fingerboard.

One way to determine the normal modes of vibration is to apply an oscillating force to the bridge and observe the motion of the various parts of the violin. Observing the motion may be done electrically, by means of accelerometers, or optically, by means of holographic interferometry.

Six low-frequency modes of vibration of a violin are shown in Fig. 9. The C_1 and C_2 body ("corpus") modes radiate little sound, although they contribute to the "feel" of the violin. The other four modes are indicated on the frequency response curve in Fig. 10, where the vertical coordinate is the velocity divided by the force (called mechanical admittance or mobility).

The A_0 ("air") mode and the T_1 ("top") mode both involve considerable motion of air in and out of the *f*-holes. Both modes radiate sound efficiently, and they dominate the low-frequency sound spectra of most violins. Above 1 kHz, the vibrational modes or resonances are bunched together, as can be seen in Fig. 10, and it is difficult to identify the individual modes.

One of the most critical steps in constructing a violin is carving the top and back plates. Since both plates have an arch, they are carved from blocks of wood thicker at the center. As the plates near their desired shapes, the violin-maker tests them by listening to "tap tones." To hear these, the plate is held in a particular way and tapped at certain spots. The trained ear of the violin maker can extract useful information by noting the pitch and the decay time of the tap tones.

The analysis of the vibrations of the plates has been refined considerably by the application of electronics and optics. One rewarding technique has been the observation of individual plate modes of vibration by Chladni patterns. The plate is usually mounted horizontally and excited by a loud-speaker. The two most important modes of vibration are those shown in Fig. 11. (The plates shown are for a viola, but violin and cello plates show similar patterns.) If the frequency of the lower mode (a) in the top plate matches that of the back plate, then tuning the upper modes (b)

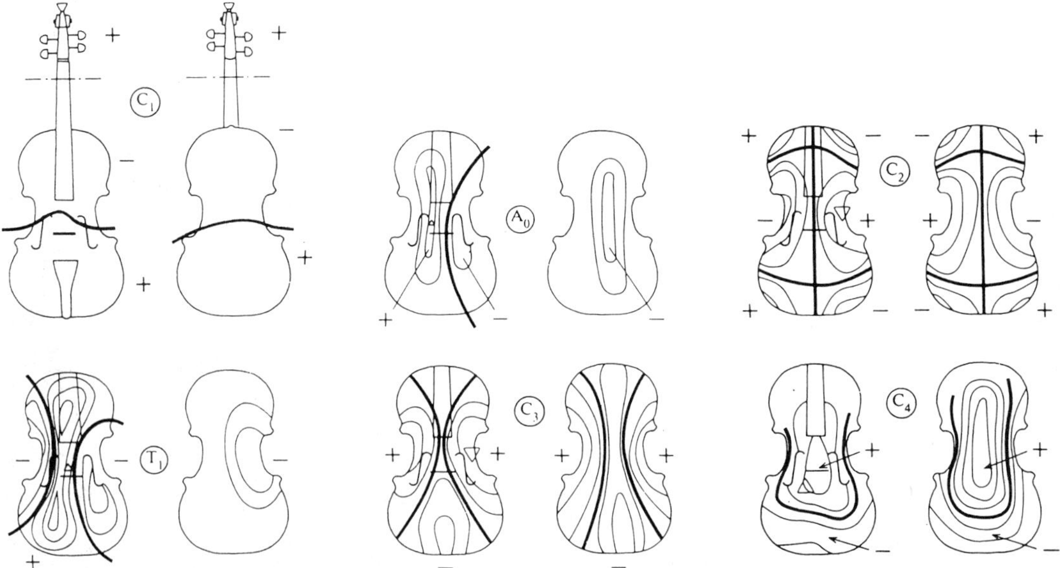

FIG. 9. Modal shapes for six modes in a violin. C_1 (185 Hz): one-dimensional bending; A_0 (275 Hz): air flows in and out of the *f*-holes; C_2 (405 Hz): two-dimensional flexure; T_1 (460 Hz): mainly motion of the top plate; C_3 (530 Hz), C_4 (700 Hz): two-dimensional flexure. Top plate and back plate are shown for each mode. The heavy lines are nodal lines; the direction of motion is indicated by + or −. The drive point is indicated by a small triangle. (After Alonso Moral and Jansson, 1982a.)

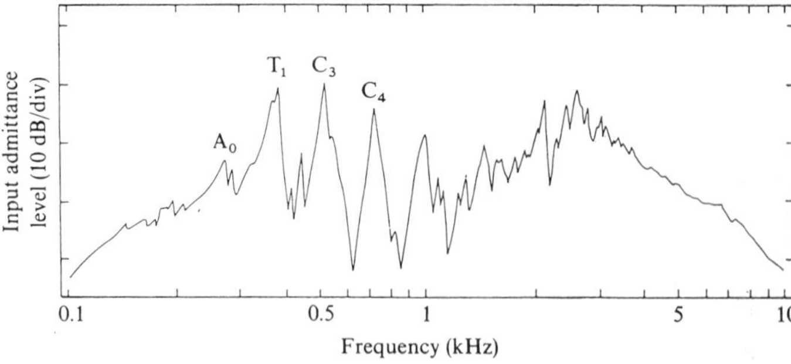

FIG. 10. Input admittance (driving-point mobility) of a Guarneri violin driven on the bass bar side (Alonso Moral and Jansson, 1982b).

to match each other in frequency or to be a semitone different will usually produce plates of high quality (Hutchins, 1977). The upper ("ring") mode (b) is often tuned an octave higher than the lower ("x") mode. Of course, the shapes of the nodal lines are equally as important as the frequencies of the modes.

The violin bridge transforms the motion of the vibrating strings into periodic driving forces applied by its two feet to the top plate of the instrument. Violin bridges typically have strong resonances around 3000 and 6000 Hz, as shown in Fig. 12. The lower resonance is due to a rocking-bending motion, whereas the upper one is due to a symmetrical up-and-down motion. These resonance frequencies can easily be changed by cutting away wood at the appropriate places, and thus, shaping the bridge is a convenient way to alter the frequency response of a violin.

In order to "darken" the sound of a violin, the player may attach a mute to the bridge. The additional mass of the mute shifts the bridge resonances to lower frequencies, changing the sound spectrum of the instrument. Typical violin mutes have masses of about 1.5 grams.

1.3.2 Other Instruments in the Violin Family The other string instruments in the violin family are the viola, cello (violoncello), and double bass (also known as contrabass, string bass, or bass viol). The general shape and construction of the viola and cello are similar to those of the violin. The strings, like those of the violin, are tuned in intervals of

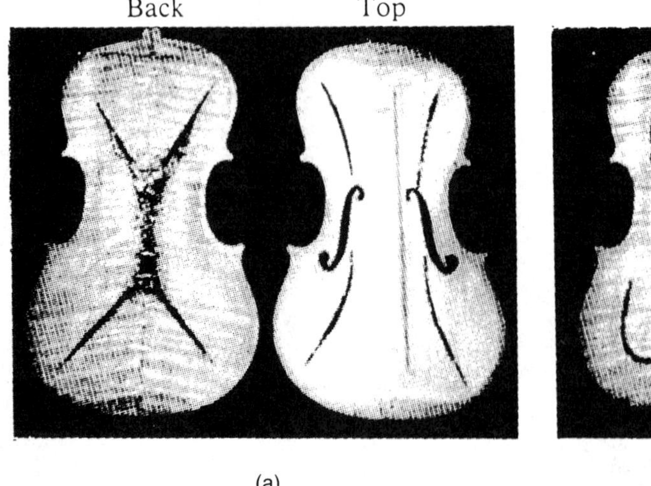

(a)

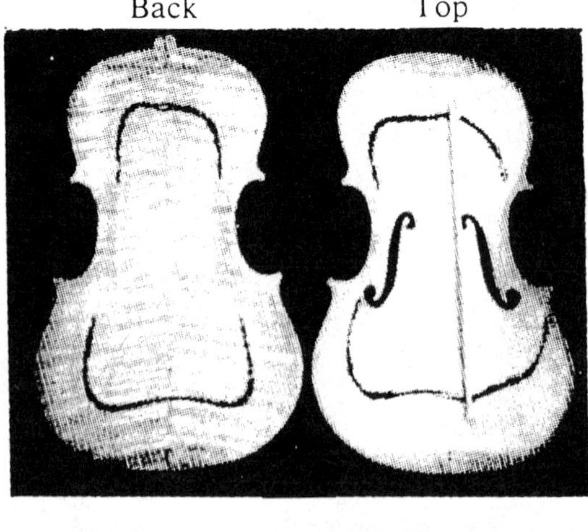

(b)

FIG. 11. Chladni patterns showing modes of vibration in the viola top and back plates (from Hutchins, 1977).

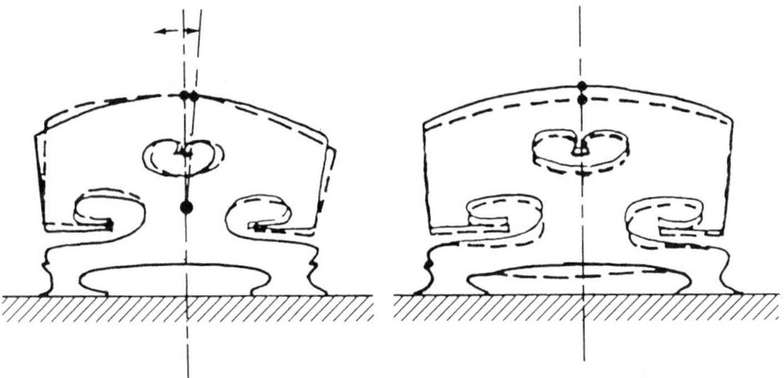

FIG. 12. First two vibrational modes (resonances) of a violin bridge.

fifths: C_3, G_3, D_4, A_4 in the viola and C_2, G_2, D_3, A_3 in the cello. Although the viola is tuned a fifth below the violin, its main air resonance is typically less than a third below that of the violin, and the main top plate resonance is a little over a third lower.

The lower resonances of a cello are related to its string frequencies in about the same way as those of the viola. However, the tall bridge of the cello results in a large driving force and a strong response near the main top plate (T_1) resonance. Also, the second air resonance is nearly an octave above the main air (A_0) resonance and thus reinforces this resonance by strengthening the second harmonic.

The double bass is tuned in fourths rather than fifths, which reduces the distances that the hand must travel along the long fingerboard. It has relatively narrow sloping shoulders and a flat back like members of the viol family.

In 1958, Carleen Hutchins and other members of the Catgut Acoustical Society set about to develop a new family of fiddles, with resonances scaled to those of a fine violin. Their research led to a family of eight new instruments. The family has two instruments pitched above the violin: the treble and the soprano. The alto, which is tuned like the viola, is either held under the chin (if one has long arms) or played vertically on a peg in the fashion of a cello. The tenor tunes between the viola and cello, and the baritone has cello tuning. Finally there are two basses, with their strings tuned in fourths. These new instruments have appeared in several concerts and have aroused considerable interest, both musically and scientifically.

1.3.3 Viols A viol differs from a violin primarily in that its back is flat and braced by three cross bars, the middle one supporting the sound post. The top plate has a shallower arch than that of the violin, and it has *c*-holes rather than *f*-holes. The ribs are considerably higher than those of the violin.

Viols were popular instruments during the 16th and 17th centuries, dropped from favor in the 18th century, and have enjoyed a considerable renaissance in the 20th century. The three most common members of the viol family are the treble, tenor, and bass. Each has six strings tuned in fourths except for a major third at the center.

1.4 Piano

The pianoforte, invented in 1709 by Bartolomeo Cristofori in Florence, was a direct descendent of the harpsichord. Many years of development led to the modern piano, which has become the most popular and versatile of all musical instruments. It has a range of more than seven octaves (A_0 to C_8) and a wide dynamic range. Pianos vary in size from small home uprights or spinets to large concert grands.

A typical concert grand piano has 243 strings, varying in length from about 2 m at the bass end to about 5 cm at the treble end. Included are eight single strings wrapped with one or two layers of wire, five pairs of strings (also wrapped), seven sets of three wrapped strings, and 68 sets of three unwrapped strings. Smaller pianos may have fewer strings, but they play the same number of notes: 88. The soundboard is nearly always made of spruce, about 1 cm thick, with its grain running the length of the piano. Ribs on the underside of the soundboard stiffen it in the cross-grain direction. The soundboard is the main source of radiated sound, as is the top plate of a violin or cello.

The upright piano developed about the middle of the 19th century. The nearly rectangular soundboard and the strings in upright pianos are vertical, and the hammers travel horizontally; thus, the action is different from that of the grand piano. In full-size upright pianos, which stand 130 to 150 cm ($4\frac{1}{2}$ to 5 feet) in height, the striking mechanism or action is located some distance above the keys and connected to them mechanically by thin rods called stickers. In studio uprights or console pianos, which stand about 100 to 130 cm ($3\frac{1}{4}$ to $4\frac{1}{4}$ feet) in height, the action is mounted directly over the keys without stickers. In small spinet pianos (less than 100 cm in height), the action is below the keys and drop stickers transmit key motion to the action.

Piano strings make use of high-strength steel wire. Efficiency of sound production calls for the highest string tension possible, while at the same time, minimizing inharmonicity calls for using the smallest string diameter (core diameter in a wrapped string) possible. This results in tensile stresses of around 1000 N/mm^2, which is about half the yield strength of steel wire. For steel with an elastic modulus of 2×10^{11} N/m^2, this results in an elongation of about $\frac{1}{2}$% when the string is under tension. Fortunately, when strings break, it is usually near the keyboard end so that the broken strings recoil away from the pianist.

A completely flexible string would vibrate in a series of modes that are exact harmonics of a fundamental. Actual strings have some stiffness, however, which provides a restoring force (in addition to the tension), slightly raising the frequency of all the modes. The additional restoring force is greater in the case of the higher modes because the string makes more bends. Thus the modes are spread apart in frequency and are no longer exact harmonics of a fundamental. In other words, a real string with stiffness is partly stringlike and partly barlike.

The inharmonicity of strings (that is, the amount by which the actual mode frequencies differ from a harmonic series) is found to vary with the square of the mode number. Thus, the second mode ("harmonic") is shifted four times as much as the fundamental.

Inharmonicity is especially noticeable in the case of the large bass strings, because the stiffness of a string increases sharply with its radius. Since wrapped strings are more flexible than are solid strings of the same diameter, the inharmonicity of the bass strings is reduced substantially by the use of wrapped rather than solid strings of the same weight. (The lower strings on a guitar and violin are wrapped rather than solid, for the same reason.)

An approximate expression for the inharmonicity $I_n = f_n/nf_1^0$ of the nth partial (spectral component) of a solid string can be written as

$$I_n = Bn^2, \tag{5}$$

where the inharmonicity constant B is given by

$$B = \frac{\pi^2\beta 2}{8} = \frac{\pi^3 a^4 E}{8L^2T}, \tag{6}$$

where a = radius, E = elastic modulus, L = length, and T = tension.

The inharmonicity in strings is the main reason why pianos are "stretch-tuned" (i.e., tuned so that octaves have frequency ratios slightly greater than 2). If they were not, the upper partials of a note would be slightly sharp with respect to the notes in the upper octaves to which they correspond, and undesirable beats would result when chords are played.

1.4.1 Piano Soundboards The soundboard is the main radiating member in the instrument, transforming some of the mechanical energy of the strings and bridges into acoustic energy. In addition, it opposes the vertical components of string tension that act on the bridges. This vertical force is in the range of 10–20 N per note, or a total of about 900–1800 N. The acoustical and structural functions are not totally independent of one another.

Nearly all piano soundboards are made by gluing strips of spruce together and then adding ribs that run at right angles to the grain of the spruce. The ribs are designed to add enough cross-grain stiffness to equal the natural stiffness of the grain along the wood, which is typically about 20 times greater than across the grain. In some pianos, the net cross-grain stiffness, including the contribution of the treble bridge, ends up being greater than the stiffness along the grain, a condition that is described as overcompensation. However, overcompensation is generally avoided in pianos of high quality. The unloaded soundboard is usually not a flat panel, but has a crown of 1–2 mm on the side that holds the bridges. When the strings are brought up to tension, the downward force of the bridge causes a dip in the crown, and in older pianos, the downward bridge force may have permanently distorted the soundboard.

Mode shapes of the soundboard of a 9-ft concert grand piano are shown in Fig. 13. At the lowest frequency, the entire soundboard moves in a single phase, but at the higher frequencies, the soundboard vibrates in several segments.

An idealized plate, clamped at its edges, has a radiation resistance that increases with increasing frequency and equals the characteristic impedance of air at high frequency. This assumes an optimum impedance match between the plate and its surroundings and no internal losses. Losses do occur, however, and these result in a decrease in sound radiation in the upper treble range. At the bass end, below the critical frequency (i.e., the frequency at which the speed of bending waves in the soundboard equals the speed of sound in air), the radiation efficiency drops dramatically. This results in a favored region for acoustic radiation between about 200 and 2000 Hz.

1.4.2 Sound Decay Since the strings are the principal reservoir for storing vibrational energy in a piano, the rate of sound decay is mainly determined by how rapidly energy is extracted from the strings. This depends upon the coupling between the strings, the bridge, and the soundboard, as well as the tuning of the unison strings.

Most piano notes show a compound decay curve: the initial decay rate is several times greater than the final decay rate. Initial decay rates may vary from about 4 dB/s at the bass end of the scale to 80 dB/s at the treble end. Decay times for a full 60-dB decay may vary from 0.2 to 50 s.

The compound decay curve gives rise to a very important feature of piano sound. The rapid transfer of energy to the soundboard initially results in a high rate of sound radiation and gives the listener the impression of loudness. At the same time, the slow final decay rate gives the impression of a sustained sound.

At least two different phenomena contribute to the compound decay: a change in the predominant direction of vibration of the string from perpendicular to parallel during decay and the coupling between the strings in a unison group. The mechanical impedance of the soundboard, which varies considerably with location and with frequency, is always much greater than the impedance of the strings. Typically, the impedance ratio is on the order of 200:1. Thus, vibrational energy is transferred rather slowly from the strings to the soundboard. The mismatch is even greater for vibrations parallel to the soundboard, since the soundboard stiffness is very large in this plane.

Figure 14 compares the decay rates for vibrations perpendicular to and parallel to the soundboard in the same string. Note the slower decay rate for parallel vibrations as a result of the greater stiffness of the soundboard in its own plane. The hammer excites mainly perpendicular vibrations in the string, but because of the spiral wrapping and the mechanical termination of the string at the bridge, the polarization changes in the course of time. This leads to a compound decay curve with a rapid initial decay (when perpendicular vibrations are dominant) and a slower final decay (due to poor coupling of the parallel vibrations).

When a hammer strikes a tricord (set of three unison strings), it sets all three strings into vibration with the same phase. Thus, they all exert vertical forces with the same phase on the bridge, and energy is transferred at a maximum rate. Because of very small differences in frequency, however, they soon get out of phase, and the resultant force on the bridge is diminished. Thus, the decay rate starts high and diminishes, a second reason for the compound decay curves. A skilled piano tuner can balance the fast initial sound decay and the slowly decaying aftersound in a piano by the tuning of the unisons. If the unison strings are closely tuned, the aftersound will develop more slowly since it will require a longer time for the unison strings to lose phase coherence.

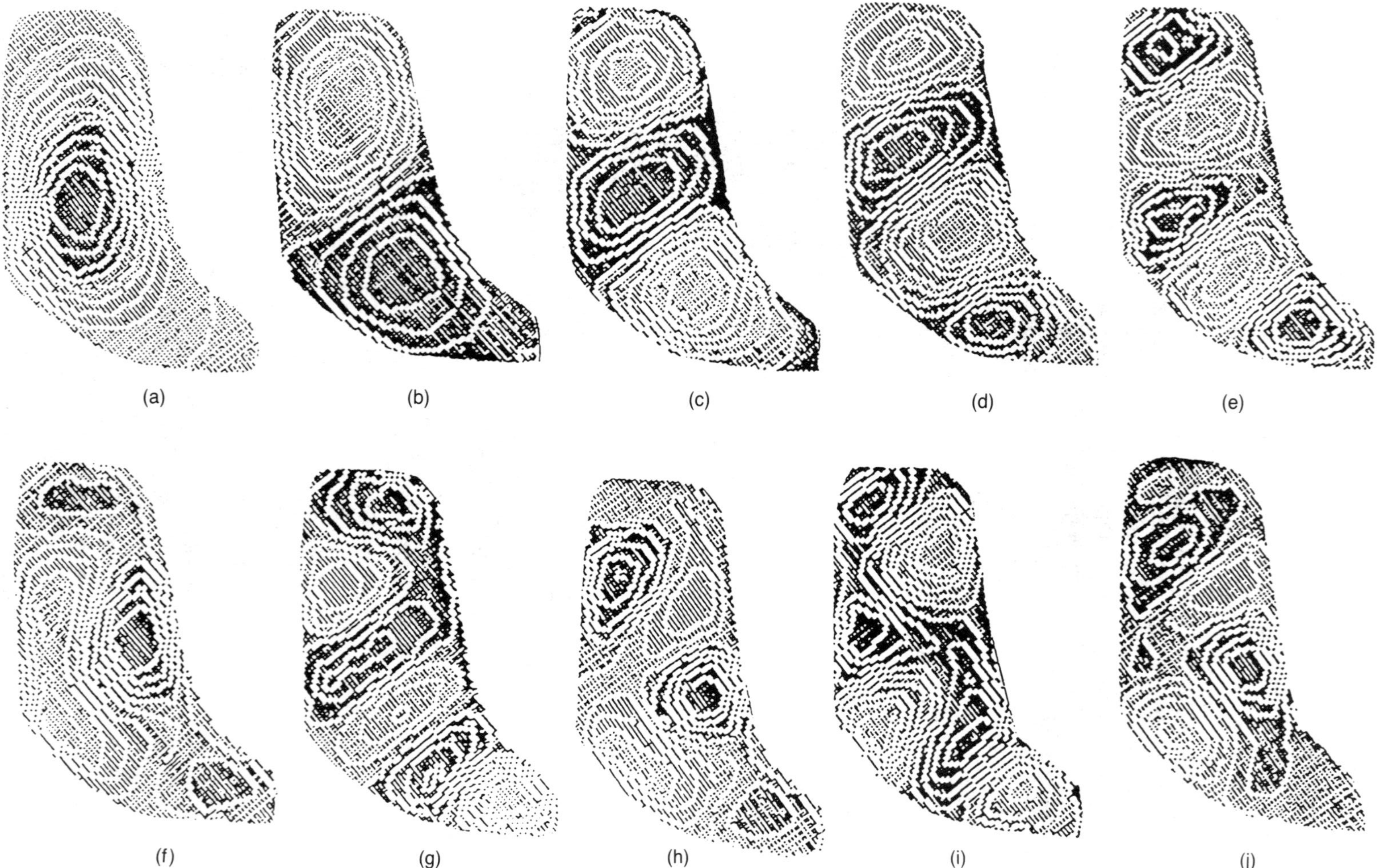

FIG. 13. Mode shapes on the soundboard of a 9-ft concert grand piano: (a) 52 Hz; (b) 63 Hz; (c) 91 Hz; (d) 106 Hz; (3) 141 Hz; (f) 152 Hz; (g) 165 Hz; (h) 179 Hz; (i) 184 Hz; (j) 188 Hz. (Kindel, 1989.)

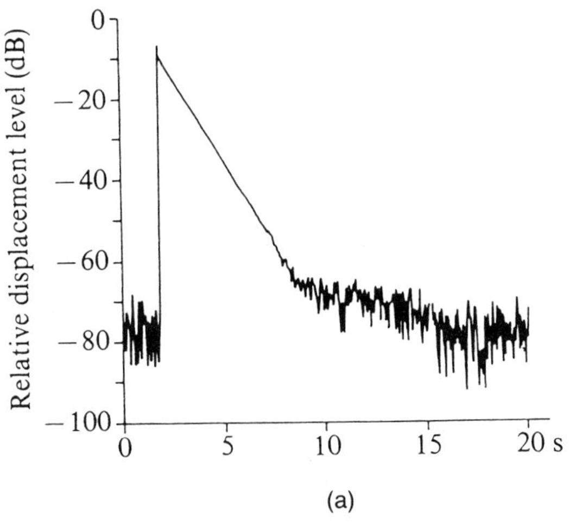

(a)

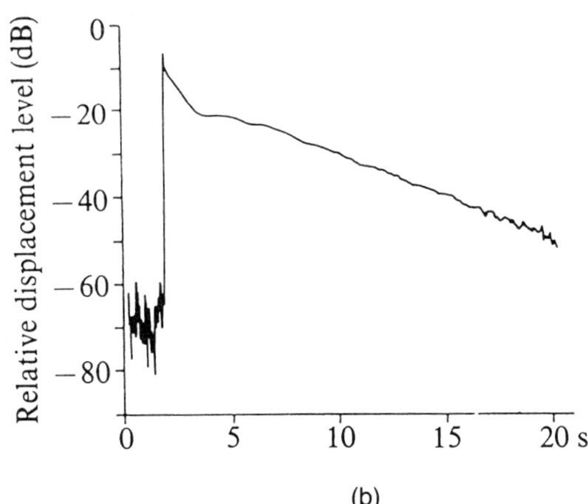

(b)

FIG. 14. Decay curves for vibrations in a single D: string (f = 311 Hz) of a grand piano: (a) perpendicular polarization and (b) parallel polarization (Weinreich, 1977.)

2. WIND INSTRUMENTS

2.1 Vibration of Air Columns

Longitudinal waves in an air column behave much the same way as the transverse waves on a string discussed in Section 1.1. The acoustic pressure p (the fluctuation of pressure above and below atmospheric pressure) can be written as

$$\frac{\partial^2 p}{\partial t^2} = \frac{K}{\rho}\frac{\partial^2 p}{\partial x^2} = c^2\frac{\partial^2 p}{\partial x^2}, \tag{7}$$

where K is the bulk modulus, ρ is the density, and c is the wave speed. For propagation of sound waves in air, the adiabatic bulk modulus is the appropriate one to use, and the wave velocity c (in m/s) can be written as

$$c = \sqrt{\gamma RT/M} = 20.1\sqrt{T}, \tag{8}$$

where T is absolute temperature. At 20 °C (T = 293 K), for example, c = 343 m/s.

A harmonic plane wave propagating in the x direction can be written as $p(x,t) = A \sin [k \times (ct - x)]$, where $k = 2\pi/\lambda$. In a pipe of length L with open ends (so that $p = 0$ at $x = 0$ and $x = L$), this leads to standing waves, and the pressure equation for the nth mode is similar in form to Eq. (3) for the transverse motion of a vibrating string:

$$p_n(x,t) = A_n \sin(2\pi f_n t + \phi_n) \sin(2\pi f_n x/c), \tag{9}$$

where $f_n = nc/2L$ is the frequency of the nth normal mode, A_n is its amplitude, and ϕ_n its phase.

In actual practice, the sound pressure does not drop to zero exactly at the end of a pipe but rather a small distance ΔL beyond the open end. The end correction ΔL for a cylindrical pipe of radius r without a flange is approximately 0.61 r (at low frequency when $r \ll \lambda$). Twice this amount should be added to the physical length of a pipe with two open ends (an "open" pipe) to obtain the acoustic length.

In a pipe with one open end and one closed end (a "closed" pipe), the mode frequencies are given by $f_n = nc/4L$ (n = 1, 3, 5 . . .). The lowest mode has half the frequency of the open pipe, and only odd-numbered harmonics are present. Note that at a closed end, where the acoustic pressure is maximum, the acoustic velocity and the displacement are minimum, while the opposite is true at an open end.

The modes or resonances of a conical pipe have essentially the same frequencies as an open pipe of the same length. This statement is true even if the cone is truncated, which may appear paradoxical at first glance. As a sound wave travels toward the small end of a cone, its pressure amplitude increases. Pressure distributions for the first three modes of a cone are shown in Fig. 15, along with the corresponding modes of an open pipe.

The cylindrical pipe in Fig. 15 is an example of an acoustic horn, a horn being de-

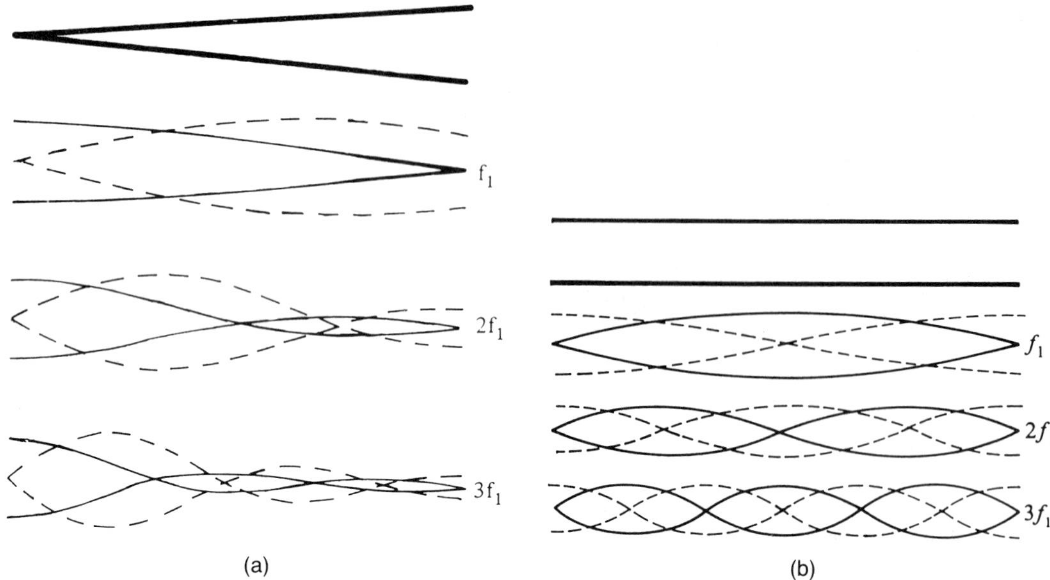

FIG. 15. Pressure (solid line) and velocity (dashed line) distributions for the first three modes of (a) a cone and (b) an open pipe. (From Rossing, 1990.)

fined as a conduit whose length is large compared with its lateral dimensions. Horns of various types are used in instrument bores, many of which can be approximated by Bessel horns whose radii are given by

$$r = bx^{-\epsilon} \tag{10}$$

If $\epsilon = 0$, the horn is cylindrical; if $\epsilon = -1$, it is conical. Most brass instruments have horn profiles that are nearly cylindrical for about the first half of their length, then expand with a flare that is well approximated by a Bessel horn (see Chap. 8 of Fletcher and Rossing, 1991).

2.2 Reed and Lip Vibrations

Wind instruments are made to sound either by blowing a jet of air across an opening (as in flutes and flue organ pipes) or by buzzing together the lips or a thin reed and its support (as in trumpets, clarinets, oboes, or reed organ pipes). In the latter case, the reed that controls the air flow (we can consider the brass player's lips as a type of reed) acts as an inward-striking valve that is blown closed [Fig. 16(a)] or as an outward-striking valve that is blown open [Fig. 16(b)]. A woodwind reed acts as a blown-closed valve, while the brass player's lips act as a blown-open valve.

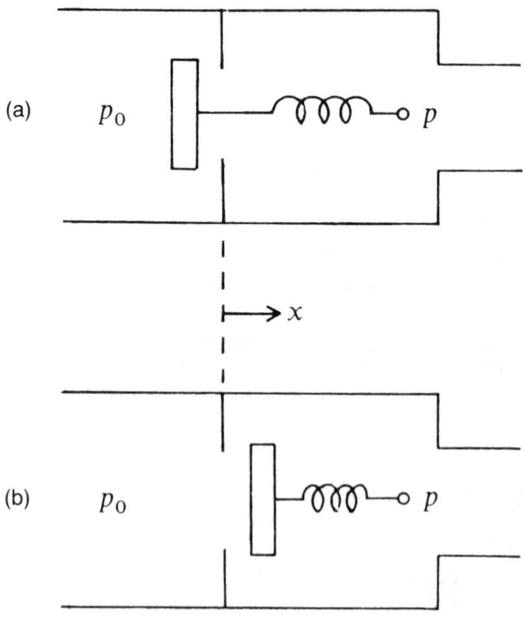

FIG. 16. The two basic types of a pressure-controlled valve. In each case, the valve cap is attached to a spring. In (a), the blowing pressure p_0 tends to blow the valve closed, while in (b), the blowing pressure tends to open the valve. The internal pressure p is the control variable and the coordinate x gives the position of the cap relative to its seat. (From Fletcher and Rossing, 1991.)

For the simple valves in Fig. 16, $x - x_0 = \gamma(p - p_0)$, where x is the position of the valve cap relative to its seat, and x_0 is its equilibrium position [$x_0 < 0$ in Fig. 16(a), $x_0 > 0$ in Fig. 16(b)]. The average volume flow U through the valve can be written as

$$U = |ax_0(p - p_0)^{1/2} + b(p - p_0)^{3/2}|, \qquad (11)$$

where a and b are positive constants.

For the blown-closed valve in Fig. 16(a), this leads to the solid curve in Fig. 17, which has a region of negative resistance extending from point A to point C. Over this range, regenerative feedback can occur over a wide frequency range up to the reed resonance, but a pipe resonator is generally required. For the blown-open valve in Fig. 16(b), x_0 is positive and Eq. (9) leads to the dashed curve in Fig. 17 which has no region of negative resistance. For such a valve, sound generation relies on phase shifts that occur near the resonance frequency of the valve (see Chap. 13 in Fletcher and Rossing, 1991).

Although in principle it is possible to generate a steady tone using only the mouthpiece of a wind instrument, in practice a stable tone requires positive feedback from the air column. Cooperation between the air column and the lips or reed is similar to the electrical feedback that sustains oscillations in an audio generator.

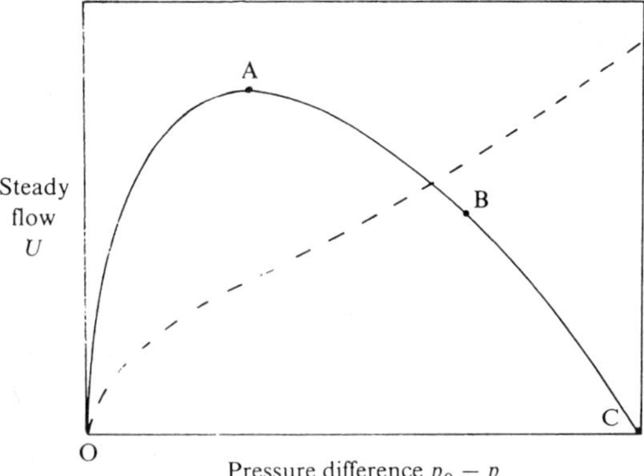

FIG. 17. The steady flow characteristic of a pressure-controlled valve of the type shown in Fig. 16(a) (solid line). The valve is blown completely closed at point C. The valve admittance is negative over the range AC. The broken line shows, to a different scale, the static flow behavior of a valve with a configuration as in Fig. 16(b). (From Fletcher and Rossing, 1991.)

2.3 Brass Instruments

The principal members of the brass family are the trumpet, French horn, trombone, and tuba. Their playing ranges are indicated in Fig. 18. Other important brass instruments are the cornet, fluegelhorn, bugle, and baritone horn.

Brass instruments typically have four sections: a mouthpiece, a tapered mouthpipe, a cylindrical section, and a bell, as shown in Fig. 19. The trumpet, French horn, and trombone have cylindrical sections of considerable length, while the fluegelhorn, baritone, and tuba, often characterized as instruments of conical bore, are tapered throughout much of their length.

2.3.1 The Trumpet The modern trumpet is a descendent of the valveless baroque or natural trumpet, which itself developed through many years of experimentation. The trumpet plays a very important role in symphony orchestras, military bands, jazz ensembles, and wind ensembles of all kinds. The most common trumpet is tuned in B♭, although trumpets tuned in other keys such as D, C, and F are also used. The B♭ trumpet bore is about 140 cm long and has a fundamental of B_2♭, although the lowest useful note is B_2♭.

One way to study the resonances of a pipe (or a wind instrument) is to measure its acoustic impedance as a function of frequency. (Acoustic impedance is defined as the sound pressure divided by the volume velocity.) The impedance of a B♭ trumpet is shown in Fig. 20. Note that the enhancement of impedance (pressure) peaks in the vicinity of the mouthpiece resonance (around 800 Hz). These large pressure peaks enhance the ability of the air column to control the rather massive lips of the player. Skillful design of the bore has brought the peak frequencies (except for the lowest one) very nearly into a harmonic relationship so that the air column resonances will respond to the partials of the desired tone.

Arrows at the top of Fig. 20 indicate the peaks (2, 4, 6, and 8) that correspond to the partials of B_3♭ ($f = 233$ Hz; this note is the written C_4 for an instrument tuned in B♭). When the note is blown very softly, the player's lips vibrate nearly sinusoidally, and only peak 2 is of importance. As the level of loudness increases, more harmonics are pro-

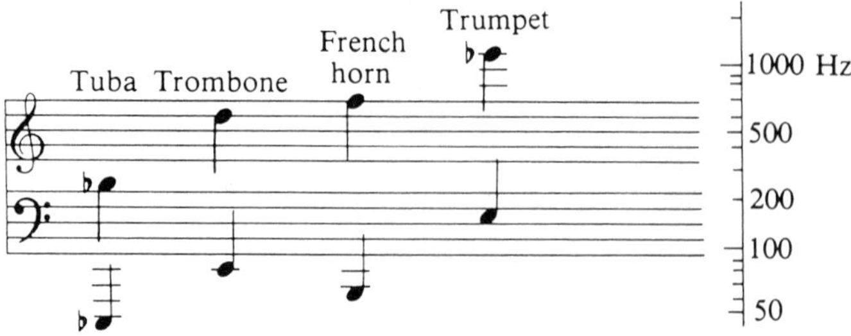

FIG. 18. Playing ranges of tuba, trombone, French horn, and trumpet.

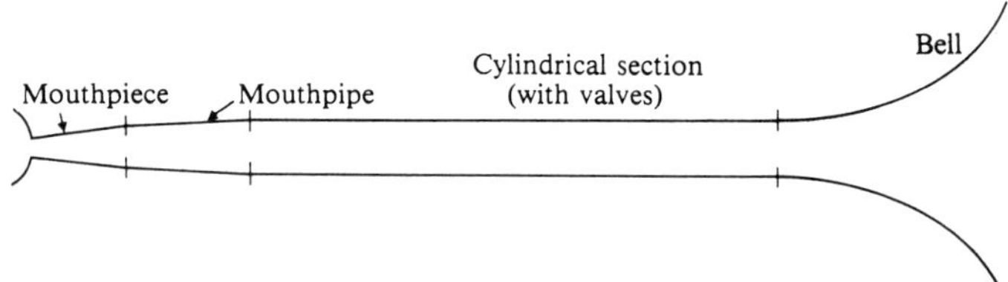

FIG. 19. A cross section of a brass instrument bore.

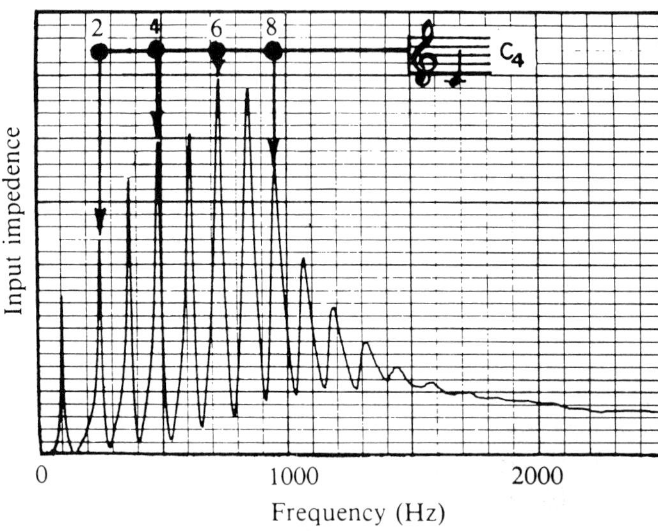

FIG. 20. Impedance curve of a B♭ trumpet, showing peaks that form a "regime of oscillation" for C_4 (from Benade, 1976).

duced, which excite peaks 4, 6, and 8. The tone takes on increasing stability as the lips become subject to control by pressure impulses corresponding to all four peaks.

The lowest peak in Fig. 20 lies below the fundamental frequency (about 117 Hz, corresponding to B_2♭). It is possible to play the "missing" fundamental note, sometimes called the pedal tone, by relying on feedback from other peaks having harmonically related frequencies. It is also possible to play at the frequency of the lowest peak, but that is not a useful note at all. Notes above F_5 (698 Hz) or so become increasingly difficult to play. The peaks that correspond to the harmonics are small, which indicates that the corresponding resonances are too weak to be of much help in controlling the lips and stabilizing the tone. Because peak 6 is very tall, F_5 is quite easy to play softly, indicating a strong resonance; little increase in stability is noted during crescendo, however, because peak 12, which lies near the second harmonic, is small. Above B_6♭ (written C_6), even the fundamental is weakly supported by the air column, and the lips must provide their own stability, which makes great demands on the player's ability to control his or her lips.

2.3.2 Valves and Slides On a brass instrument of fixed length (like the military bugle), it is possible to play eight or more notes, corresponding to the peaks in Fig. 20, and a few more may be added by skillful playing. To play the remaining notes of the scale, however, the acoustical length of an instrument must be varied. This is done by moving a slide (trombone) or by inserting lengths of additional tubing by means of valves (trumpet, French horn, baritone, tuba, etc.).

The playing positions of a trombone slide are shown in Fig. 21. Seven positions of the slide are needed to play all desired notes of the scale; note that the seven positions are not spaced equally. Going down a semitone decreases the frequency by about 6%, so the instrument should be lengthened by about this same amount. As the instrument gets longer, a 6% increase in length also becomes progressively longer.

In most brass instruments, however, a fundamental problem arises in designing the additional lengths of tubing for each valve. If the lengths of tubing are designed so that the first and second valves can lower the pitch by a whole tone and a semitone, respectively, then the combination will be inadequate to lower the pitch by three semitones, as illustrated by the following: To lower the pitch a whole tone, the first valve must add a length of tubing 12.2% of the total length L, or $0.122L$. To lower the pitch a semitone requires the addition of 5.9% of the total length, or $0.059L$. Thus the two valves together increase the length by 18.1%, or $0.181L$, but three semitones requires an increase of 18.9%.

Various compromises have been proposed to deal with this tuning problem. The valve slides for the first and second valves may be made slightly longer than the optimum length, thus accepting a slight flattening of the first two semitones in order to reduce the sharpening of the third one. Similarly, the third valve slide may be designed to give a shift slightly more than three semitones. Since the discrepancy is usually greatest when the third valve is used in combination with the others, most trumpets have a provision for moving the third valve slide with the left hand while the instrument is being played. Beyond that, the brass player must "lip" the various notes into tune, that is, make small corrections in frequency by changing the tension of the lips.

The valve problem becomes increasingly difficult in the lower brass instruments. A fourth valve is frequently added to the tuba; it lowers the pitch by a fourth, and thus substitutes for the troublesome combination of valves 1 and 3. Even a fifth or sixth valve is sometimes added, their function varying in the different tubas.

2.3.3 The French Horn The French horn in common use is a "double horn" composed of two horns tuned to F_2 and $B_2\flat$, which share the same mouthpiece and bell. The three main valves each have two sets of windways and valve slides, and a fourth valve is used to switch from one horn to the other. The B♭ horn is preferred for playing high notes, because it has stronger resonances at the higher frequencies, and they are spaced further apart.

Having a bore length of about 375 cm (30% longer than the trombone), the French horn has many resonances, and much of the time the horn is played at a pitch corresponding to one of the higher resonances. This means that many notes on a French horn lack the stability of notes normally played on a trumpet or trombone, where there are several prominent resonances corresponding to the overtones of the note played.

Placing the hand in the bell of the horn makes it easier to play the higher notes. By

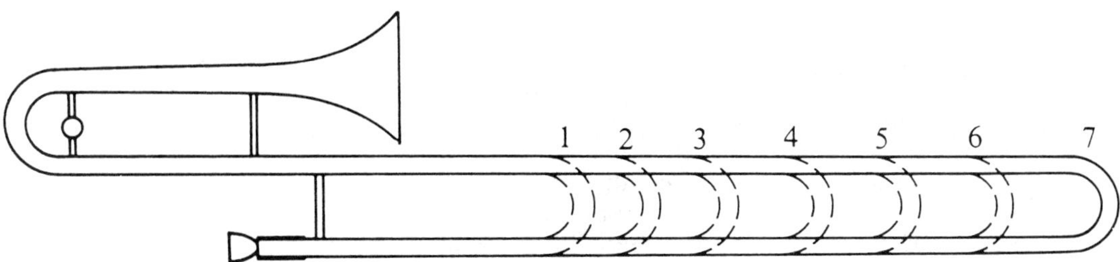

FIG. 21. Playing positions of a trombone slide.

inhibiting the radiation and increasing the reflection of the higher frequencies, it allows standing waves to build up and leads to more usable resonances at the higher frequencies. Placing the hand in the bell also lowers the pitch. If the hand is inserted as far as possible, a "stopped" tone is produced, which appears to be about a semitone higher than normal. Apparently all the resonances are lowered to the point where the next higher resonance is only a semitone above the desired note, which can be confirmed by holding a note as the hand is slowly inserted to the stopped position. If no attempt is made to hold the original pitch with the lips, the frequency will fall slowly. On the other hand, if the player tries to hold the pitch, the note jumps up to the next higher resonance at some point as the hand is inserted.

2.3.4 The Trombone The trombone, like the trumpet, has a bore that is predominantly cylindrical. Its tubing has about twice the length of a trumpet, however, so that its main resonances occur at frequencies that are approximately harmonics of a missing fundamental near $B_1\flat$ (58 Hz). The lowest note that is played in "first" position (i.e., with the slide fully retracted, see Fig. 21) is normally $B_2\flat$, although the $B_1\flat$ pedal note in a trombone is a more useful note than in a trumpet, and occasionally it is called for in musical scores.

The playing range of the ordinary tenor trombone extends up to D_5 (the tenth harmonic of the fundamental $B_1\flat$). The bass trombone has a slightly larger bore and a mouthpiece of larger volume than the tenor, making it easier to play the lower notes. Most bass trombones, and some tenor trombones as well, include a rotary valve and additional tubing to lower the fundamental to F_1. An excellent discussion of the acoustics of brass instruments appears in Chap. 9 of Campbell and Greated (1987).

2.4 Woodwind Instruments

The principal woodwinds in the symphony orchestra are the flute, clarinet, oboe, and bassoon; auxiliary instruments include the piccolo, English horn, bass clarinet, and contrabassoon. The saxophone is the principal woodwind in jazz bands and also plays an important role in marching and concert bands and, occasionally, in symphony orchestras. Old instruments, such as recorders and crumhorns, have recently experienced a revival in interest.

Most woodwinds were originally constructed from wood; hence their family name. Modern woodwinds may be either wood, metal, or plastic, although wood is still the preferred material for woodwinds other than flutes, piccolos, and saxophones. Like the brass instruments, woodwinds make use of feedback from an oscillating air column to control the flow of air input and maintain the oscillations. The flow-control valve may be either a vibrating reed or an oscillating stream of air. The feedback mechanisms in these two cases are quite different, as will be discussed in this section.

In woodwinds, the resonances of the air column are tuned by opening and closing tone holes with the fingers and with mechanical keys. Sound is radiated from the open tone holes, so that the radiation pattern becomes more complex than that of the brass instruments, which radiate virtually all their sound from the bell.

2.4.1 The Clarinet The clarinet is a single-reed instrument with a cylindrical bore about 15 mm in diameter. The cane reed is clamped by the ligature against a specially designed surface of the mouthpiece called the table. The opening between the tip of the mouthpiece and the reed is about 1 mm. When the clarinet is played, the lower lip pushes the reed in to about half this distance, and it vibrates around this position. For soft tones, the tip of the reed does not touch the mouthpiece; therefore, the flow of air is not interrupted. As the blowing pressure is increased, the amplitude of the reed increases, until for loud tones the tip of the reed touches the mouthpiece during about half of the cycle. Thus the flow of air, which is more or less sinusoidal for soft playing, takes on more harmonics as the blowing pressure increases.

A clarinet reed has a natural frequency of its own at around 2000–3000 Hz; oscillations at this frequency are prevented largely by the damping action of the lower lip pressed against the reed. If the clarinet is blown with the teeth against the reed rather than the lip, unpleasant squeaks and squeals appear.

The B♭ clarinet has a wide playing range of $3\frac{1}{2}$ octaves divided into three registers, as shown in Fig. 22. The low or *chalumeau* register extends from D_3 to E_4, and a second register, the *clarion*, plays a 12th above this, since the first overtone of a cylindrical closed tube is the third harmonic of the fundamental. Notes above $B_5\flat$ can be played in the *altissimo* register, which uses the third mode (fifth harmonic) of the air column. Often the "throat tones" (G_4 to $A_4\sharp$), which use open keys in the throat of the clarinet, are included in the chalumeau register.

The low tones of a clarinet are particularly rich in harmonics, although the second harmonic and other even-numbered harmonics are weak throughout the chalumeau register. The preponderance of odd-numbered harmonics gives the clarinet its "hollow" or "woody" quality. In the clarion register, the even-numbered harmonics are strong, resulting in quite a different tone quality from the lower register.

Orchestral clarinetists alternate between clarinets tuned in A and B♭, the choice usually depending on the key signature of the music. The bass clarinet, which plays an octave lower than the B♭ clarinet, is also used in orchestras and bands. Other clarinets are the E♭ soprano and E♭ alto. The C clarinet, used in the Classical period, is rarely seen today, but the basset horn in F (an alto clarinet with a narrow bore) is played in early-music ensembles.

When a pipe has more than one open hole, its acoustical behavior exhibits several interesting features. If the open holes are regularly spaced, they constitute a tone-hole lattice, not unlike a lattice of atoms in a crystal or a line of beads spaced equally on a string. The open tone-hole lattice acts as a filter that transmits waves of high frequency but reflects those of low frequency. The critical frequency above which sound waves can propagate through a lattice of tone holes is called the *cutoff frequency* of the lattice, which has been found to be an important factor in determining the timbre of a woodwind instrument.

The open-tone-hole cutoff frequency in most good clarinets is around 1500 Hz. A clarinet with a higher cutoff frequency tends to have a "bright" tone; an instrument with a lower cutoff frequency has a "dark" tone. In an experiment in which two matching clarinets were reworked to raise the cutoff frequency slightly in one and to lower it in the other, it was found that players of classical music preferred the one with the lower cutoff frequency, while jazz clarinetists chose the one with the higher cutoff and "bright" tone (Benade, 1976).

2.4.2 The Saxophone The saxophone is a conical-bore single-reed instrument invented in 1846 by Adolphe Sax. Although used occasionally in symphony orchestras, its main use has been in all types of bands and in ensembles that play jazz and popular music. The four main members of the saxophone family are the soprano in B♭, the alto in E♭, the tenor in B♭, and the baritone in E♭. Saxophones are characterized by a large bore diameter, a low input impedance, and a louder sound than the other woodwinds.

Since the saxophone has a conical bore, its spectrum has even-numbered harmonics as well as odd-numbered ones. Its cone angle (3° to 4°) is considerably greater than that of the oboe or bassoon, however, and so saxophone tone has fewer prominent harmonics than the double reed instruments. The larger bore of a saxophone allows correspondingly large tone holes that radiate efficiently, and thus it can be played loudly.

2.4.3 Double-Reed Instruments The family of orchestral double reeds includes the oboe, English horn, bassoon, and contrabassoon. Basically, they are conical tubes with

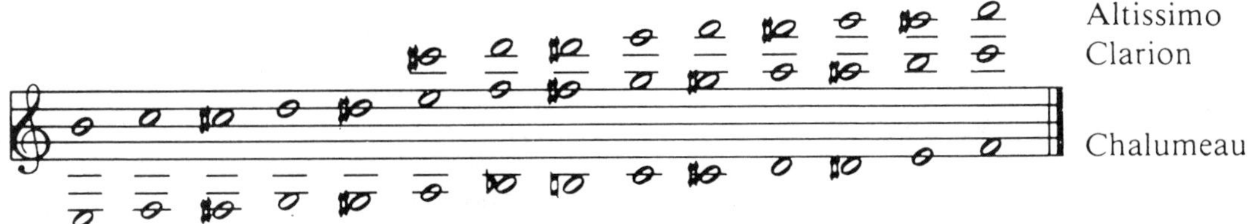

FIG. 22. Three registers of the clarinet. Shown are the notes fingered; notes sounded on a B♭ clarinet are one whole tone lower.

the tip of the cone cut off and a reed attached. All these instruments use a double reed consisting of two halves of cane beating against each other. Small mouthpieces with a single reed have been used in place of the double reed, but they have not become very popular because they change the quality of the instrument.

The oboe bore is a nearly straight cone about 60 cm in length with a cone angle of about 1.4°. Since the resonances of a cone are an octave apart, the registers of the oboe are also an octave apart, extending from D_4 to C_5 and D_5 to C_6, although additional keys and fingerings extend the playing range from $B_3\flat$ to G_6. Oboes are usually constructed from three pieces of wood: a top section, a bottom section, and a bell, with all three sections having tone holes.

Quality oboes have cutoff frequencies that are nearly constant throughout their playing range and, for different instruments, vary from about 1100 to 1500 Hz (Benade, 1976). A higher cutoff frequency results in a bright tone, and a lower cutoff frequency in a dark tone, just as in the case of clarinets. The spectra of oboe sounds show substantial amounts of the higher harmonics along with broad resonances or formants in the neighborhood of 1000 and 3000 Hz. These formants may be related to the mechanical properties of the reed.

The English horn or cor Anglais is an alto version of the oboe with a pear-shaped bell that gives a distinctive tone quality to certain notes near its resonance. Composers have taken note of this distinctive tone quality in writing solos for this instrument. The English horn is tuned in F, a fifth below the oboe.

The bassoon has a nearly conical bore with a cone angle slightly less than 1° and with a total length of about 254 cm. The tube makes several bends in order to be of a manageable size. The playing range of the bassoon extends from $B_1\flat$ to about C_5 (58–523 Hz), while the contrabassoon plays an octave lower. Open tone-hole cutoff frequencies for quality bassoons range from about 350 to 500 Hz (Benade, 1976). Like the oboe, the bassoon has many sharp resonances, and the tone is rich in harmonic overtones.

2.4.4 The Flute A flow-controlled air stream, such as that which occurs in a blown bottle or a flute, collaborates with an air column to oscillate at frequencies at which the air column has dips in its impedance curve (that is, where the pressure is at a minimum and, therefore, the flow-controlling jet deflection is at a maximum). This is opposite to the woodwind reed or brass player's lips, which are controlled by pressure peaks.

The flute is basically a cylindrical pipe, about 19 mm in diameter, except for the head joint, which tapers to about 17 mm at the mouth or embouchure hole. Its vibration modes thus form a series of frequencies that includes all the harmonics of the fundamental. The resonance frequencies of the pipe are influenced by the presence of the air stream, however. In particular, the frequency of the lowest resonance is raised when the blowing pressure increases, whereas the second resonance remains more or less unchanged. The range of the flute is normally B_3 to D_7.

The flute plays in three registers, and the player changes registers, without the aid of register keys, by adjusting the technique of blowing. The three parameters at the player's control are the blowing pressure, the length of the air jet, and the area of lip opening. The technique used by most flute players includes adjustments in all three of these (Fletcher, 1975).

The most efficient excitation of the fundamental takes place when the time for the air jet to travel across the embouchure hole is about half the period of an oscillation. If the jet travel time is shortened much below this, the fundamental will not sound. Thus, to sound the second register, the player moves the lips forward to decrease the jet length and/or increases the blowing pressure. At the same time the blowing pressure is increased, the size of the lip opening is decreased to maintain the loudness and tuning. Studies of a number of flute players by Fletcher (1975) show that to shift to a register an octave higher, a flute player will typically double the blowing pressure, reduce the jet length by 20%, and reduce the lip opening by about 30% (Doubling the blowing pressure increases the jet speed by 40%, since jet speed is approximately proportional to the square root of pressure.)

The angle of blowing depends on the shape of the player's lips, but typically ranges from 25° to 40° below the horizontal. Most players use a shallower angle for low notes and for loud playing.

The narrow end of the head joint is closed by a cork plug whose position can be adjusted by a screw. Changing the length of the small cavity between the cork and the embouchure hole can have a substantial effect on the tuning and "playability" of certain notes.

The piccolo is about one-half the length of the flute, and it sounds one octave higher. An alto flute, tuned a fourth lower, and a bass flute tuned an octave lower than the standard flute, are occasionally seen.

2.4.5 The Recorder The recorder, also known as the *Blockflöte,* flauto dolce, and English flute, is an early member of the flute family, now experiencing a revival. It has a reverse conical bore, tapering inward toward the foot, and a whistle-type mouthpiece with a fixed windway. The three most commonly played recorders are the descant (soprano) in C, the treble (alto) in F, and the tenor in C. A bass in F, a sopranino in F, and a great bass in C complete the family. Each instrument has a two-octave normal playing range.

The fixed windway makes the recorder easy to sound but somewhat difficult to play in tune at different dynamic levels. Without the flexibility to adjust the embouchure in the manner of the flautist, the recorder player changes to the upper register by half-opening the thumb hole and increasing the blowing pressure slightly. Wind pressure in an alto recorder varies from about 100 N/m^2 for the lowest note to about 500 N/m^2 for the highest.

The recorder has only eight tone holes. Therefore, in order to play a full chromatic scale, liberal use is made of cross-fingerings. Cross-fingerings leave one or more tone holes open and close other tone holes below these. In general, cross-fingered notes are not as stable as other notes, and the recorder player often changes the fingering slightly from instrument to instrument. Since the pitch produced by a given fingering may rise as much as a semitone (100 cents) from low to high blowing pressure, a skilled recorder player often changes fingerings with dynamic level.

2.5 Pipe Organs

The pipe organ has been called the "king of musical instruments." No other instrument can match it in size, range of tone, loudness, or complexity. The modern pipe organ includes a large variety of pipes, arranged into divisions or organs. The principal division of the organ is called the great organ, and it usually contains the most stops or voices. The pipes in the swell division are usually enclosed behind a set of shutters that can be opened or closed to change the loudness.

Organ pipes are organized into ranks of similar pipes. One rank of pipes will include one pipe for each note (61 in the divisions operated from a keyboard, 32 in the pedal division). Each stop on the organ usually corresponds to one rank of pipes, except in the case of mixture stops, which involve several ranks. On smaller organs, a rank of pipes may be included in more than one stop.

There are two basic types of organ pipes; flue (labial) pipes and reed (lingual) pipes. Flue pipes produce sound by means of a vibrating air jet, in a manner similar to the flute and the recorder. Reed pipes use a vibrating brass reed to modulate the air stream.

The essential parts of a metal flue pipe are shown in Fig. 23(a). The air jet passes through a flue or windway, a narrow opening between the languid and the lower lip. Sound is produced when the air jet encounters the upper lip and oscillates back and forth, sometimes blowing into the pipe, sometimes blowing out through the mouth of the pipe. The large flue pipes usually have ears on either side of the mouth to guide the air jet.

The stopped wooden flue pipe, shown in Fig. 23(b), uses a similar mechanism to produce sound. Since the the resonator is now a closed pipe, it need be only half as long as an open pipe in order to sound the same pitch. Wooden pipes usually have a square cross section and produce a sound with a flute-like quality.

There are three families of flue pipes: (1) diapasons, (2) flutes, and (3) strings. Pipes of the flute family usually have the least overtone content, while the bright-sounding strings have the most. String pipes are generally slender cylinders, whereas diapasons are open cylinders of somewhat greater diameter. Flute pipes come in several sizes and shapes, are constructed of either wood or metal, and may be open or closed. Closed pipes sound mainly the odd-numbered harmonics of the fundamental.

The reed pipe, shown in Fig. 23(c), has a

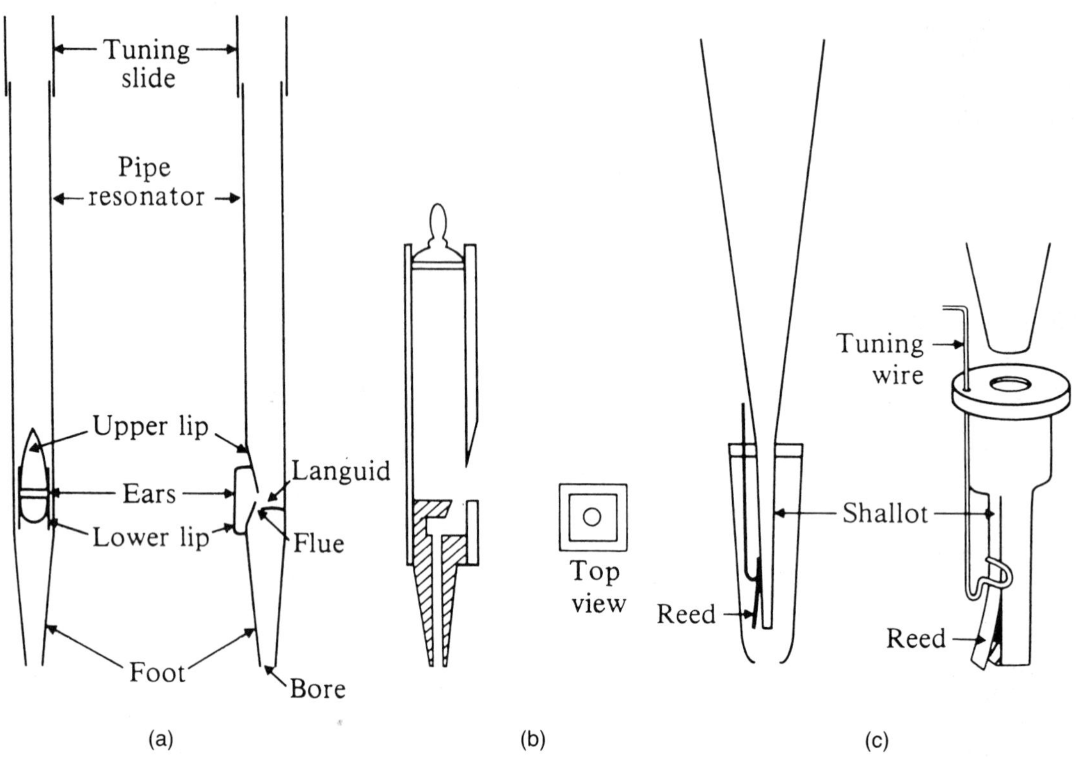

FIG. 23. Organ pipes: (a) open flue pipe of metal; (b) stopped wood flue pipe; (c) reed pipe, (From Rossing, 1990.)

vibrating reed or tongue, which modulates the flow of air passing through the shallot into the resonator. The reed is pressed against the open side of the shallot by a wire that can be adjusted up and down to tune the vibrating reed. Reed pipes are rich in harmonics. If the resonator is cylindrical, the odd-numbered harmonics will be favored, as in a clarinet. Conical resonators, however, will reinforce all harmonics, both odd and even.

Figure 24 shows several different pipes and the names given to the corresponding organ stops. Resonators of several different shapes are shown, including open and closed cylinders, cones, inverted cones, rectangular cylinders, and closed pipes with chimneys. (The chimney is tuned to boost one of the upper harmonics, around the fifth or sixth.)

Figure 25 compares the first four resonance frequencies of open and closed cylindrical, conical, and reverse conical pipes that can be used as organ pipe resonators. Note that the resonance frequencies of cone are essentially the same as those of an open cylinder, although the pressure amplitudes are quite different, as shown in Fig. 15 (Ayers *et al.*, 1985).

The *scale* of a rank of pipes refers to the ratio of diameter to length for the pipe of lowest pitch. Large-scale (large-diameter) pipes tend to have a dominant fundamental and fewer harmonics, whereas small-scale pipes have more harmonics. The reason for this is related to a frequency dependence of the end correction of a pipe. For an open cylindrical pipe, the end correction at the open end is approximately 0.6 times the radius, whereas at the mouth of a flue pipe, one adds approximately 2.7 times the radius to calculate the effective length (Strong and Plitnik, 1983). The end correction decreases with frequency, so the pipe effectively shortens for the higher partials; thus, the pipe resonances are slightly less than an octave apart. However, the spectrum of the sound source (the oscillating jet) has exact harmonics. Thus in a large-scale pipe, only the first few harmonics in the source are reinforced by the natural frequencies of the pipe. For pipes of small scale, such as the strings, the end correction

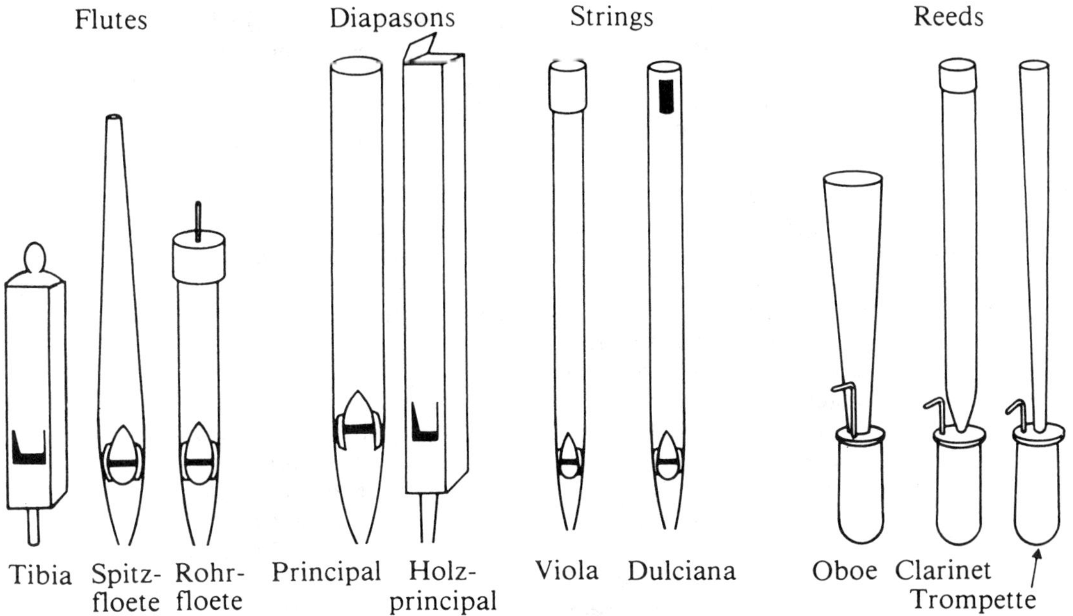

FIG. 24. Organ pipes of various families with the names of the corresponding stops (from Rossing, 1990).

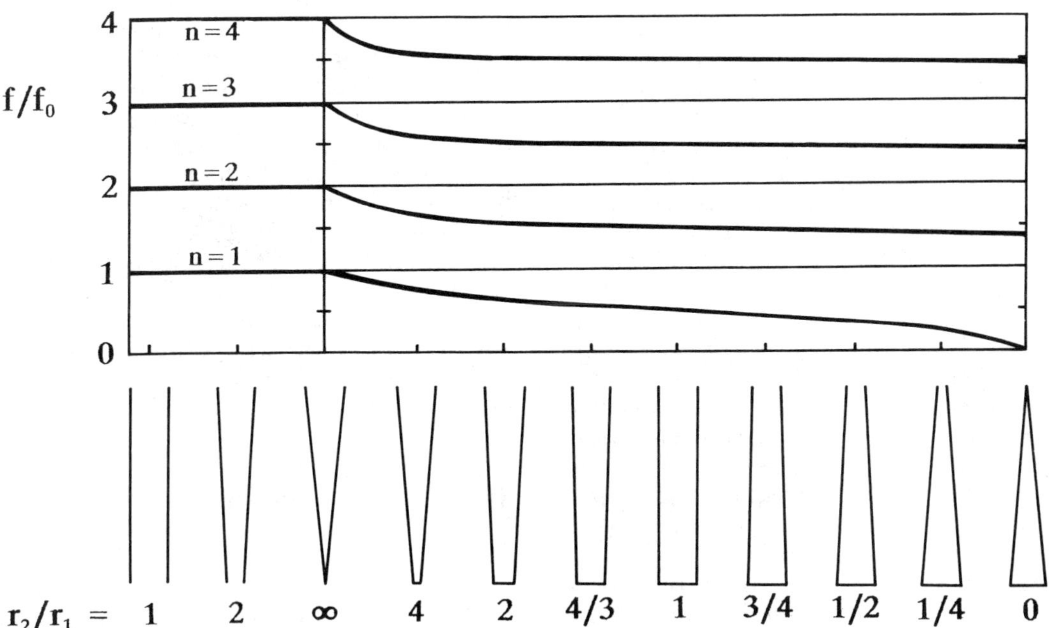

FIG. 25. First four resonance frequencies of open and closed cylindrical and conical pipes.

is much smaller to start with; thus, the pipe resonances more nearly match the harmonics of the oscillating jet.

Each rank of pipes is usually graduated in diameter according to a fixed relationship for that rank. For large-scale pipes, the diameter may reduce to one-half at the 17th note of the scale; for small-scale pipes, the diameter will reduce at a slower rate than this.

Normally, the mouth width, mouth height ("cut-up"), and width of the flue opening follow the same scale as the pipe diameter. The mouth width, for example, may vary from 0.6 times the pipe diameter (large-scale flute) to 0.8 times the diameter (string or diapason). Similarly, the cut-up may vary from 0.25 times the mouth width (diapason) to 0.4 times the width (flute); the flue width is typically about 0.03 times the mouth width.

2.5.1 Sound Generation in Flue Pipes When air is admitted at the foot of a flue organ pipe, it flows upward and forms a sheetlike jet as it emerges from the flue (see Fig. 23). The jet flows across the mouth of the pipe and strikes the upper lip, where it interacts both with the lip itself and with the air in the pipe resonator. The physics of this interaction is discussed in Chapter 16 of Fletcher and Rossing (1991).

Jets can be described as either *laminar* (arranged in layers or streamlines) or *turbulent* (characterized by eddies or vortices). Jets in organ pipes are almost completely turbulent, which makes it quite difficult to describe them mathematically. Somewhat paradoxically, however, a fully turbulent jet in an organ pipe is more stable than a laminar one, and organ builders go to some lengths to ensure fully turbulent jets by cutting fine nicks along the edge of the languid, for example.

A fully turbulent jet mixes gradually with the surrounding air, thus slowing down and broadening out in a rather simple way. The jet velocity, which is maximum at the center, falls off in a bell-shaped curve whose width increases linearly with distance from the flue. Conservation of flow momentum requires that the velocity decreases with the square root of the distance from the flue slit.

The air flow during the attack transient in an organ pipe, although quite complicated, can be studied by means of Schlieren photographs. Figure 26 shows the jet at intervals of 0.1 s as the jet first strikes the edge, forming vortices. Eventually, this settles down to an acoustically driven jet oscillation.

Assuming that standing waves already exist in the pipe resonator, the air motion at the mouth associated with these standing waves will be essentially at right angles to the jet. This sets up a wavelike disturbance in the jet that propagates along the jet at a little less than half the speed of the air at the center of the jet. The amplitude of this wavelike disturbance grows nearly at an exponential rate along the jet, typically doubling during each millimeter of travel. The growth rate is found to be greatest when the wavelength along the jet is about six times the width of the jet at that location. On the other hand, when the length of the wave is less than the width of the jet, it does not grow at all, so that only long wavelengths can propagate with a large amplitude along long jets.

The large acoustic standing waves in the pipe drive the tip of the jet alternately inside and outside the upper lip. If the puffs of air enter the pipe in step with the standing waves, energy will be added. The important relationship between the acoustical air motion in the pipe mouth and the time the air pulse arrives inside the upper lip is determined by the time required for a wave on the jet to travel from the flue slit to the upper lip. This is, in turn, determined by the mouth height (cut-up) and the blowing pressure.

One of the most critical of all the steps in organ building is voicing the pipes, which means making coarse and fine adjustments in the various parts of the pipe so that it "speaks" properly. Much of the voicing can be done in the organ-builder's shop, but the final voicing or finishing is done after the organ is installed, and it takes into account the acoustics of the room, as well as the acoustics of the organ. One of the objectives of voicing is to achieve a uniformity of loudness and timbre within each rank of pipes. Another is to adjust the initial transient or attack of each pipe.

3. PERCUSSION INSTRUMENTS

3.1 Vibrations of Bars, Membranes, and Plates

Percussion instruments generally use one or more of the following basic types of vibrators: strings, bars, membranes, plates, air

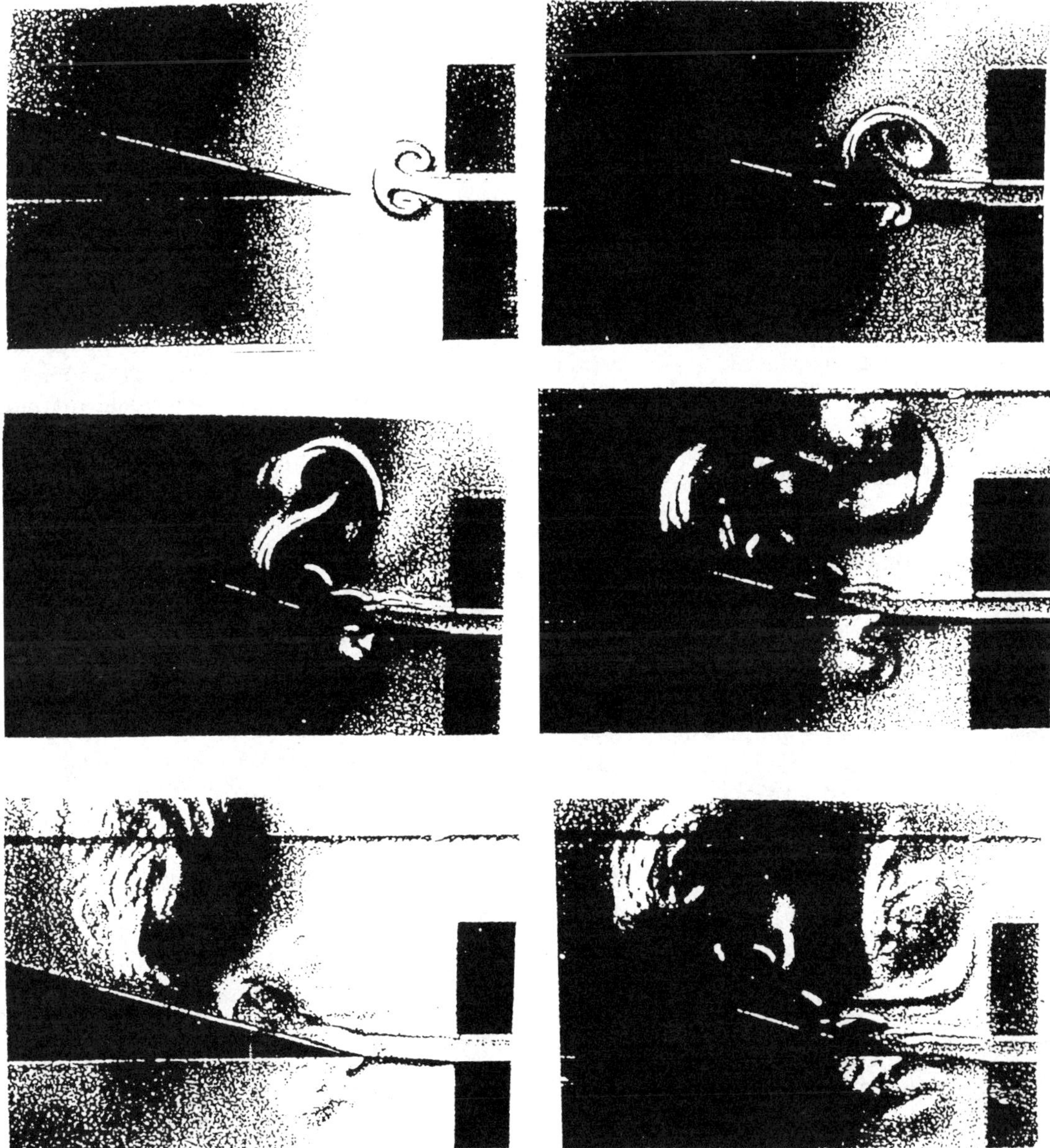

FIG. 26. Schlieren photographs showing air flow at intervals of 0.1 s during the attack when a jet strikes a sharp edge such as the labium of an organ pipe (photos furnished by A. Hirschberg).

columns, or air chambers. Strings and air columns tend to produce harmonic overtones; the others, in general, do not. Bars, membranes, and plates are three classes of vibrators whose modes of vibration are not related harmonically, and thus the overtones they sound will not be harmonics of the fundamental tone. The inharmonic overtones of these complex vibrators give percussion instruments their distinctive timbres.

3.1.1 Vibrations of Bars Although the compressional and torsional wave speeds in a bar show little or no dispersion (i.e., frequency dependence), this is definitely not the case with bending waves, whose speed is nearly proportional to $\sqrt{f}$. Bending waves involve both compressional and shear strains.

When a bar is bent, the outer side is stretched and the inner side is compressed; somewhere between is a neutral axis whose length remains unchanged. The simplest theory for such bending motion is the Euler-Bernoulli beam theory, which gives accurate results in thin bars and rods at low frequency. The Euler-Bernoulli equation of motion can be written

$$\frac{\partial^2 y}{\partial t^2} = \frac{-EK^2}{\rho} \frac{\partial^4 y}{\partial x^4}, \qquad (12)$$

where E is Young's modulus, K is the radius of gyration for the cross section (about the neutral axis), and ρ is density. This equation of motion differs from the wave equations for compressional and torsional waves in that it is a fourth-order equation; $f(vt \pm x)$ is not a solution. Harmonic solutions may be obtained by substituting $y(x,t) = Y(x)e^{j\omega t}$, giving a solution with four arbitrary constants to be determined from the end conditions,

$$Y(x) = A \cosh(\omega x/v) + B \sinh(\omega x/v) + C \cos(\omega x/v) + D \sin(\omega x/v). \qquad (13)$$

Harmonic waves propagate at a phase velocity that is proportional to the square root of frequency and also to the square root of the longitudinal wave speed c_L:

$$v = \sqrt{2\pi f c_L}. \qquad (14)$$

The dispersion relationship $\omega = c_L K k^2$ has the same form as that of de Broglie waves for a free particle. The group velocity is twice the phase velocity ($v_g = d\omega/dk = 2v$).

Although the relatively simple Euler-Bernoulli beam theory gives an accurate description of bending waves in a thin beam or rod at low frequency, it predicts too low a wave speed in a thick bar or a bar vibrating at high frequency. For one thing, it assumes that plane sections remain plane, which is equivalent to neglecting shear deformations. For another thing, it neglects rotary inertia. In Timoshenko beam theory, appropriate terms are added to the equation of motion to account for the effects of shear deformation and rotary inertia.

The modes of transverse vibration in a bar or rod depend upon the end conditions. Three different end conditions are commonly considered: free, simply supported (hinged), and clamped. For each of these, a pair of boundary conditions can be written. At a free end, there is no torque and no shearing force, so that the second and third derivatives are both zero; at a simply supported end, there is no displacement and no torque, so that y and its second derivative are zero; at a clamped end, y and its first derivative are zero.

There are six different combinations of these end conditions, each leading to a different set of vibrational modes. Three of the more common combinations are shown in Fig. 27.

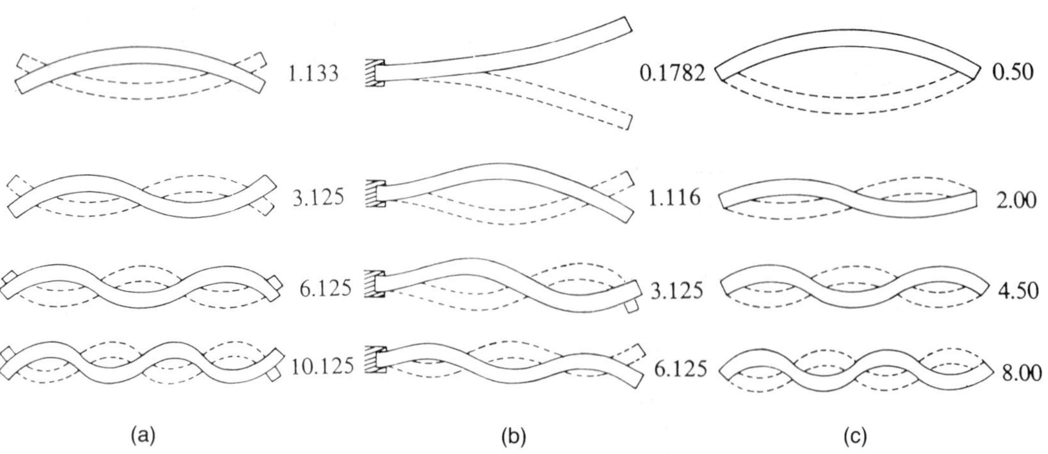

FIG. 27. Bending vibrations of (a) a bar with two free ends, (b) a bar with one clamped end and one free end, and (c) a bar with two supported (hinged) ends. The numbers are relative frequencies for a thin bar; to obtain actual frequencies, multiply by $(\pi K/L^2)/\sqrt{E/\rho}$.

In a bar with free ends, the vibrational frequencies are given by

$$f_n = \frac{\pi K}{8L^2}\sqrt{\frac{E}{\rho}}(2n+1)^2 = \frac{0.113h}{L^2}\sqrt{\frac{E}{\rho}}(2n+1)^2,\ n = 1, 2, 3 \ldots, \tag{15}$$

where h is the thickness of the bar (greater precision results from using 3.011 in place of $2n + 1 = 3.0$ when $n = 1$).

3.1.2 Vibrations of Membranes The wave equation for a thin membrane can be written as

$$\frac{\partial^2 z}{\partial t^2} = \frac{T}{\sigma}\nabla^2 z = c^2\nabla^2 z, \tag{16}$$

where T is the tension, σ is the mass per unit area, and c is the wave velocity. For a circular membrane, the solutions are generally written as Bessel functions of order m, and the zeros of these functions give the frequencies of the various modes of vibration. The first 14 modes of an ideal membrane are shown in Fig. 28. The (m, n) mode corresponds to the nth zero of the Bessel function of order m.

The normal-mode frequencies of real membranes may be quite different from those of an ideal membrane given in Fig. 28. The principal effects in the membrane acting to change the mode frequencies are air loading, bending stiffness, and stiffness to shear. In general, air loading lowers the modal frequencies, while the other two effects tend to raise them. In thin membranes, air loading is usually the dominant effect.

The effect on frequency of the air loading depends upon the comparative velocities for waves in the membrane and in air, and also upon whether the air is confined in any way. A confined volume of air (as in a kettledrum, for example) will raise the frequency of the axisymmetric modes, especially the (0,1) mode. When a membrane vibrates in open air, however, all the modal frequencies are lowered, the modes of lowest frequency being lowered the most. The confining effect of the kettle enhances this mass loading and resultant frequency lowering in the nonaxisymmetric modes such as (1,1) and (2,1).

3.1.3 Vibrations of Plates The equation of motion for bending or flexural waves in a plate is

$$\frac{\partial^2 z}{\partial t^2} + \frac{Eh^2}{12\rho(1-\nu^2)}\nabla^4 z = 0, \tag{17}$$

where ρ is density, ν is Poisson's ratio, E is Young's modulus, and h is the plate thickness. Bending waves in a plate are dispersive; that is, their velocity depends on frequency:

$$v(f) = \sqrt{1.8fhc_L}, \tag{18}$$

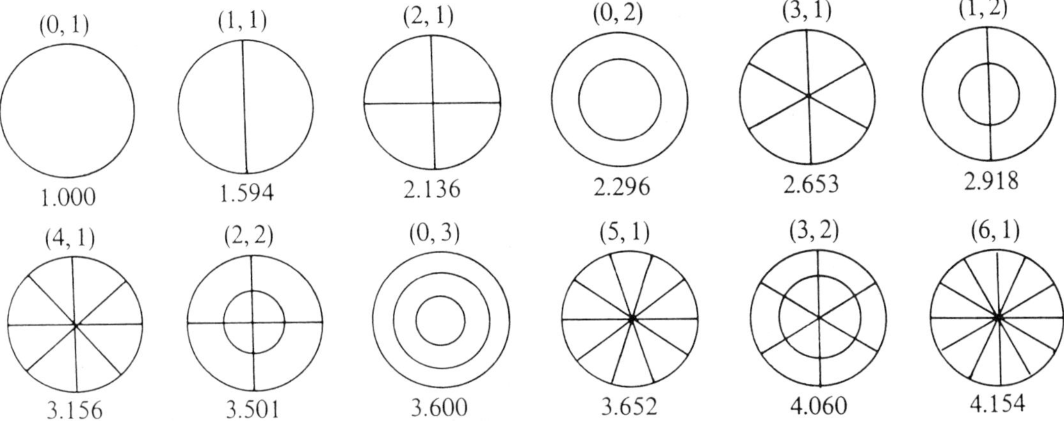

FIG. 28. First 14 modes of an ideal membrane. The mode designation (m,n) is given above each figure and the relative frequency below. To convert these to actual frequencies, multiply by $(2.405/2\pi a)\sqrt{T/\sigma}$, where a is the membrane radius.

where c_L is the longitudinal wave speed given by $c_L = \sqrt{E/\rho(1 - \nu^2)}$.

For a circular plate, ∇ is expressed in polar coordinates, and $Z(r,\phi)$ can be a solution of either $(\nabla^2 + k^2)Z = 0$ or $(\nabla^2 - k^2)Z = 0$. Solutions of the first equation contain the ordinary Bessel functions $J_m(kr)$; solutions to the second, the hyperbolic Bessel functions $I_m(kr) = j^{-m}J_m(jkr)$, where $j = \sqrt{-1}$. Thus the possible solutions are given by a linear combination of these Bessel functions times an angular function:

$$Z(r,\phi) = \cos(m\phi + \alpha)[AJ_m(kr) + BI_m(kr)]. \tag{19}$$

The hyperbolic Bessel functions are to the hyperbolic sines and cosines as the ordinary Bessel functions are to the ordinary sines and cosines.

If a plate is clamped at its edge $r = a$, then $Z = 0$ and $\partial Z/\partial r = 0$. The first of these conditions is satisfied if $AJ_m(ka) + BI_m(ka) = 0$, and the second if $AJ'_m(ka) + BI'_m(ka) = 0$.

A plate with a free edge is more difficult to handle mathematically. The boundary conditions used by Kirchhoff lead to a rather complicated expression for k_{mn}, which reduces to $(2n + m)\pi/2r$ for large ka. The modal frequencies f_{mn} are given in Table 1 for modes with m nodal diameters and n nodal circles.

Table 1. Vibration frequencies of a thin circular plate with free edge. Lowest mode frequency $f_{20} = 0.2413\ c_L h/a^2$.

m	n	f/f_{20}
2	0	1.00
0	1	1.73
3	0	2.33
1	1	3.91
4	0	4.11
5	0	6.30
2	1	6.71
0	2	7.34
3	1	10.07
1	2	11.40
4	1	13.92
2	2	15.97
5	1	18.24
3	2	21.19
4	2	27.18
5	2	33.31

3.2 Drums

Some drums, such as kettledrums, tabla, and boobams, convey a strong sense of pitch, and others do not; in the latter group are the bass drum, snare drum, tenor drum, tom-toms, bongos, congas, and many drums mainly of African and Oriental origin. As vibrating systems, drums can be divided into three categories: those consisting of a single membrane coupled to an enclosed air cavity (such as kettledrums), those consisting of a single membrane open to the air on both sides (tom-toms, congas) and those consisting of two membranes coupled by an enclosed air cavity (bass drums, snare drums). One can also categorize drums by their ethnic origin—for instance, Oriental, African, or Latin American—or according to the types of musical performance with which they are associated, such as symphonic, military, jazz, ethnic, dance, and marching bands.

3.2.1 Kettledrums Kettledrums, or timpani, are usually considered to be the most important drums in modern orchestras. Most modern timpani have a pedal-operated tensioning mechanism in addition to six or eight tensioning screws around the rim of the kettle. The pedal typically allows the player to vary the tension over a range of at least 3:1, which corresponds to a tuning range in excess of a musical sixth. At one time, most timpani heads were made of calfskin, but this material has gradually been replaced by Mylar (polyethylene terephthalate). Because of its homogeneity, Mylar is insensitive to humidity and is easier to tune than are heads made of natural materials. A thickness of 0.19 mm is considered standard for Mylar timpani heads. Timpani kettles are roughly hemispherical; copper is the preferred material, but fiberglass and other materials are also used.

Although the modes of vibration of an ideal membrane are not harmonic, a carefully tuned kettledrum is known to sound a strong principal note with two or more harmonic overtones. Lord Rayleigh recognized the principal note (at frequency f_1) as coming from the (1,1) mode and identified overtones about a perfect fifth ($f/f_1 = 1.50$), a major seventh (1.88), and an octave (2.00) above the principal tone (Rayleigh, 1894). He identified these overtones as originating from the (2,1), (3,1), and (1,2) modes, respectively, which in an ideal membrane should have frequencies of 1.34, 1.6, and 1.83 times the frequency of the (1,1) mode. Rayleigh's results are quite remarkable, considering the equipment available to him.

More recent measurements have indicated that the (1,1), (2,1), and (3,1) modes in a timpani have frequencies nearly in the ratios 1:1.5:2. The frequencies of the (4,1) and (5,1) modes are typically 2.44 and 2.90 times that of the fundamental (1,1) mode, within about half a semitone of the ratios 2.5 and 3, respectively. Thus the family of modes having 1–5 nodal diameters radiates prominent partial tones having frequency ratios of nearly 2:3:4:5:6, giving the timpani a strong sense of pitch.

How are the inharmonic modes of an ideal circular membrane shifted in frequency so that a series of prominent harmonic partials appears in the sound of a carefully tuned kettledrum? Four effects seem to contribute:

1. The membrane vibrates in a sea of air, and the mass of this air sloshing back and forth lowers the frequencies of the principal modes of vibration.
2. The air enclosed by the kettle has resonances of its own that interact with those modes of the membrane that have the right symmetry.
3. The bending stiffness of the membrane, like the stiffness of piano strings, raises the frequencies of the higher overtones.
4. Drumheads have a rather large stiffness to shear, and so they resist the type of distortion needed to deflect a membrane (as if it were being wrapped around a bowling ball) without wrinkling it.

Studies have shown that air loading, which lowers the low-frequency modes, is mainly responsible for establishing the harmonic relationship of kettledrum modes (Rossing, 1982). Other effects act merely to fine-tune the frequencies but may have a considerable effect on the rate of decay of the sound. The stiffness (or pressure) of the air enclosed in the kettle raises the frequencies of the axially symmetric modes, especially the (0,1) mode.

3.2.2 Snare Drums Snare drums are used in many types of music ensembles, including jazz bands, marching bands, and symphony orchestras. The orchestral snare drum is a two-headed instrument, about 35 cm in diameter and 13 to 20 cm deep. Strands of wire or gut stretch across the lower, or snare, head. When the upper, or batter, head is struck, the snare head vibrates against the snares. Alternatively, the snares can be moved away from the head to give a totally different sound.

In a two-headed drum, there is appreciable coupling between the two heads, especially at low frequency. This coupling may take place acoustically through the enclosed air or mechanically through the shell and leads to the formation of mode pairs. In the first two modes of vibration of the drum, the batter and snare heads move in the manner of the (0,1) membrane mode of an ideal membrane, as shown in Fig. 29. In the lower-frequency member of the pair, both heads move in the same direction, and in the higher-frequency mode, they move in opposite directions.

A simple two-mass model describes these first two modes reasonably well. The batter head is represented by a mass m_b and a spring

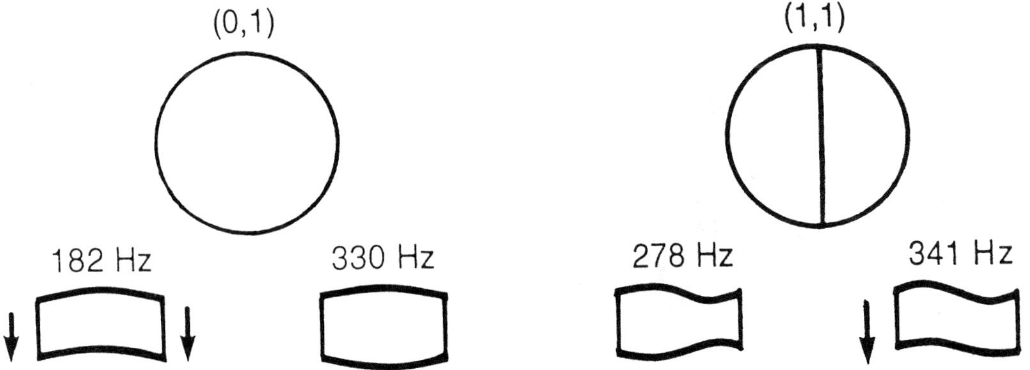

FIG. 29. Lowest four vibrational modes of a snare drum. The modal designations (*m*,*n*) refer to the membrane modes shown in Fig. 28. Mode frequencies shown are for a typical snare drum.

with stiffness K_b, the snare head by a mass m_s and a spring constant K_s; the enclosed air constitutes a third spring with constant K_c connecting the masses. This system is known to have two modes of vibration (Rossing, 1992), whose frequencies are given by

$$\omega^2 = \tfrac{1}{2}(\omega_b^2 + \omega_s^2 + \omega_{cb}^2 + \omega_{cs}^2) \pm \tfrac{1}{2}\sqrt{[(\omega_b^2 + \omega_{cb}^2) - (\omega_s^2 + \omega_{cs}^2)]^2 + 4\omega_{cb}^2\omega_{cs}^2} \quad (20)$$

where

$$\omega_b^2 = K_b/m_b, \quad \omega_s^2 = K_s/m_s,$$
$$\omega_{cb}^2 = K_c/m_b, \quad \omega_{cs}^2 = K_c/m_s. \quad (21)$$

3.2.3 Indian Drums Foremost among the drums of India are the tabla of northern India and the mridanga of southern India. The overtones of both of these drums are tuned harmonically by loading the drumhead with a paste of starch, gum, iron oxide, charcoal, or other materials.

The tabla has a rather thick head made from three layers of animal skin. (Calf, sheep, goat, and buffalo skins are used in different regions.) The shell is traditionally made of wood, but currently may be made of metal as well. The innermost and outermost layers of the head are annular, or ring shaped, and the layers are braided together at their outer edges and fastened to a leather hoop. Small straws or strings are placed around the edge between the outer and middle head. Tension is applied to the head by means of a long leather thong that weaves back and forth (normally 16 times) between the top and bottom of the drum. The tension in the thong can be changed by moving small wooden cylinders up or down. To tune the overtones of the tabla harmonically, one applies many thin layers of black paste to the center of the head, building up a circular patch. The paste consists of boiled rice and water with heavy particles such as iron oxide or manganese dust added to increase the density. After it is applied, each layer is allowed to dry and is then rubbed with a smooth stone until tiny cracks appear in the surface. The patch ends up with a slightly convex surface. Fine-tuning of the head is accomplished by upward or downward taps on the hoop with a small hammer.

The tabla we have described is usually played along with a larger drum, known by various names: banya, bayan, bhaya, dugga, or left-handed tabla. The head of this larger drum is also loaded (slightly off center), and the shell may be of clay, wood, or metal. To play the tabla, the drummer rests the edge of the palm on the widest unloaded portion of the membrane. This constraint causes the nodal patterns to be quite symmetrical. Releasing the palm pressure produces a sound of different quality.

The mridanga or mridangam is an ancient, two-headed drum that functions, in many respects, as a tabla and banya combined into one. The smaller head, like that of the tabla, is loaded with a patch of dried paste, while the larger head is normally loaded with a paste of wheat and water shortly before playing.

A succession of Indian scientists, beginning with C. V. Raman, have studied the acoustical properties of these drums. Raman and his colleagues recognized that the first four overtones of the tabla are harmonics of the fundamental mode. Later they identified these five harmonics as coming from nine normal modes of vibration, several of which have the same frequencies. The fundamental is from the (0,1) mode; the second harmonic is from the (1,1) mode; the (2,1) and (0,2) modes provide the third harmonic; the (3,1) and (1,2) modes similarly supply the fourth harmonic; and three modes, the (4,1), (0,3), and (2,2), contribute to the fifth harmonic.

Figure 30 shows the nodal patterns of the nine normal modes corresponding to the five harmonics. Also shown are some of the combination modes whose vibrational frequencies correspond to the five tuned harmonics.

3.3 Mallet Instruments

3.3.1 Marimba Marimbas and xylophones both make use of tuned bars of rosewood or fiberglass synthetic material. The *marimba* typically includes 3 to $4\frac{1}{2}$ octaves of tuned bars, graduated in width from about 4.5 to 6.4 cm ($1\frac{3}{4}$ to $2\frac{1}{2}$ in.). Beneath each bar is a tubular resonator tuned to the fundamental frequency of that bar. When the marimba is played with soft mallets, it produces a rich mellow tone. The playing range of a large concert marimba is A_2 to C_7 (f = 110 to 2093 Hz), although bass marimbas extend to C_2 (f = 65 Hz).

A deep arch is cut in the underside of marimba bars, particularly in the low register.

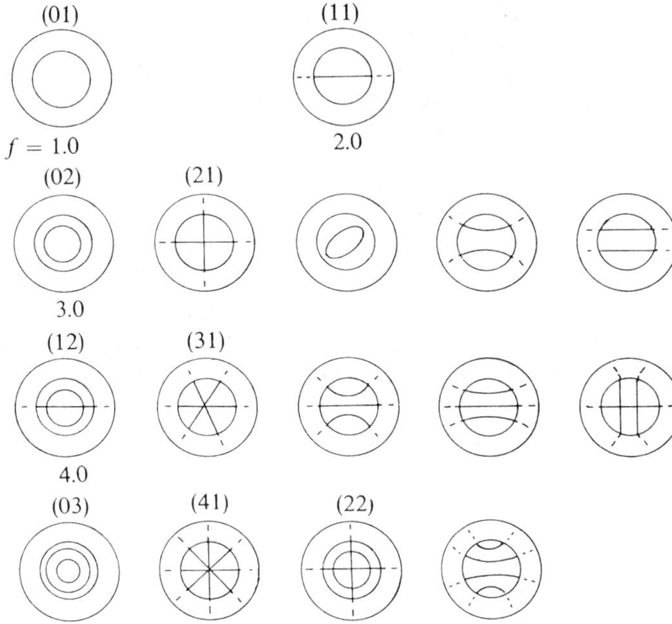

FIG. 30. Nodal patterns of the nine normal modes plus seven of the combination modes that correspond to the five harmonics of a tabla or mridanga head (from Fletcher and Rossing, 1991).

This arch serves two useful purposes: It reduces the length of bar required to reach the low pitches, and it allows tuning of the overtones (the first overtone is normally tuned two octaves above the fundamental). Figure 31 shows a scale drawing of a marimba bar, and it also indicates the positions of the nodes for each of the first seven modes of vibration.

Marimba resonators are cylindrical pipes tuned to the fundamental mode of the corresponding bars. A pipe with one closed end and one open end resonates when its acoustical length is one-fourth of a wavelength of the sound. The purpose of the tubular resonators is to emphasize the fundamental and also to increase the loudness, which is done at the expense of shortening the decay time of the sound.

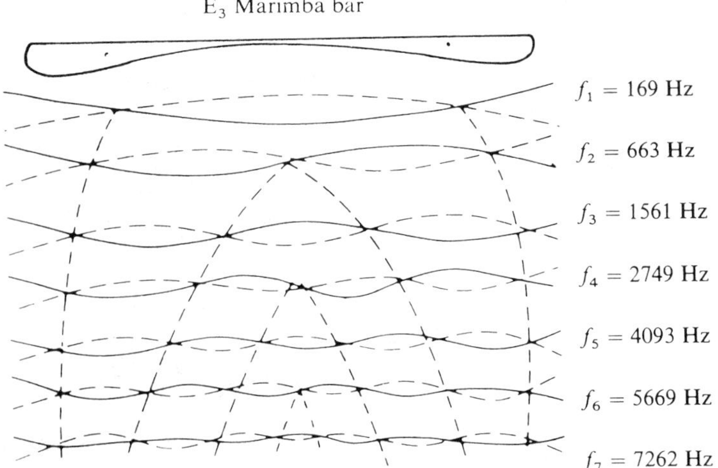

FIG. 31. Scale drawing of a marimba bar tuned to E_3 ($f = 265$ Hz). The dashed lines locate the nodes of the first seven modes.

3.3.2 Xylophone The term *xylophone* can refer to any percussion instrument made of wood, and most musical cultures feature such instruments. American xylophones typically cover a range of 3 to $3\frac{1}{2}$ octaves extending from F_4 or C_5 to C_8 (f = 349–4186 Hz) and, like the marimba, may have bars of synthetic material or rosewood. In some European countries, however, the term xylophone refers to marimbas as well as American xylophones. Modern xylophones are nearly always equipped with tubular resonators to increase the loudness of the tone.

Xylophone bars are also cut with an arch on the underside, but the arch is not as deep as that of the marimba, since the first overtone is tuned to a musical 12th above the fundamental (that is, three times the frequency of the fundamental). Since a pipe closed at one end can also resonate at three times its fundamental resonant frequency, a xylophone resonator reinforces the 12th as well as the fundamental. This overtone boost, plus the hard mallets used to play it, give the xylophone a much crisper, brighter sound than the marimba. We have found that careful overtone tuning is usually ignored in the upper register, as in the case of the marimba. (Rossing, 1976).

3.3.3 Vibes A very popular mallet percussion instrument is the *vibraphone* or *vibraharp*, as they are designated by different manufacturers. *Vibes*, as they are popularly called, usually consist of aluminum bars tuned over a three-octave range from F_3 to F_6 (f = 175–1397 Hz). The bars are deeply arched so that the first overtone has four times the frequency of the fundamental, as in the marimba. The aluminum bars tend to have a much longer decay time than the wood or synthetic bars of the marimba or xylophone, and so vibes are equipped with pedal-operated dampers.

The most distinctive feature of vibes, however, is the vibrato introduced by motor-driven discs at the top of the resonators, which alternately open and close the tubes. The vibrato produced by these rotating discs or pulsators consists of rather substantial fluctuation in amplitude ("intensity vibrato") and a barely detectable change in frequency ("pitch vibrato"). The speed of rotation of the discs may be adjusted to produce a slow vibe or a fast vibe. Often vibes are played without vibrato by switching off the motor. They are usually played with soft mallets or beaters, which produce a mellow tone, although some passages call for harder beaters.

Because vibraphone bars have a much longer decay time than do marimba and xylophone bars, the effect of the tubular resonators on decay time is more dramatic. At 200 Hz (A_3), for example, we have measured a decay time (60 dB) of 40 s without the resonator and 9 s with the tube full open. For A_5, we measured 24 s with the resonator closed and 8 s with it open.

3.3.4 Chimes Chimes, or tubular bells, are usually fabricated from lengths of brass tubing $1\frac{1}{4}$ to $1\frac{1}{2}$ in. in diameter. The upper end of each tube is partially or completely closed by a brass plug with a protruding rim. This rim forms a convenient and durable strike point.

An interesting acoustical property of chimes is that there is no mode of vibration with a frequency at, or even near, the pitch of the strike tone one hears. The frequencies excited when a chime is struck are very nearly those of a free bar described earlier. Modes 4, 5, and 6 appear to determine the strike tone. This can be understood by noting that these modes for a free bar have frequencies nearly in the ratio $9^2:11^2:13^2$, or 81 : 121 : 169, which are close enough to the ratio 2 : 3 : 4 for the ear to consider them nearly harmonic and to use them as a basis for establishing a pitch. The largest near-common factor in the numbers 81, 121, and 169 is 41, and so the subjective strike tone lies an octave below the frequency of mode 4.

3.4 Cymbals, Gongs, and Steel Pans

3.4.1 Cymbals Cymbals are among the oldest of musical instruments and have had both religious and military use in a number of cultures. The Turkish cymbals generally used in orchestras and bands are saucer shaped with a small dome in the center, in contrast to Chinese cymbals, which have a turned-up edge more like a tamtam. Orchestral cymbals are usually between 40 and 55 cm in diameter and are made of bronze. A leading manufacturer of cymbals, the Avedis Zildjian Company, claims that its secret process for treating cymbal alloys was discovered in 1623.

Many different types of cymbals are used in orchestras, marching bands, concert bands, and jazz bands. Orchestral cymbals are often designated as "French," "Viennese," and "Germanic" in order of increasing thickness. Jazz drummers use cymbals designated by such onomatopoetic terms as "crash," "ride," "swish," "splash," "ping," and "pang." Cymbals range from 20 to 75 cm (8 to 30 in.) in diameter.

Given the wide ranges in diameter and thickness, considerable variety of cymbal tone is available to the percussionist. Furthermore, a good cymbal can be made to produce many different tones by using a variety of sticks and striking it at several different places. A large cymbal struck gently near the rim produces a low sound, not unlike that of a small tamtam. The fullest sound is obtained by a glancing blow about one-third of the way in from the rim.

A cymbal may be excited in many different ways. It may be struck at various points with a wooden stick, a soft beater, or another cymbal. The onset of sound is quite dependent on the manner of excitation. The coupling between vibrational modes in a cymbal is strong, however, so that a large number of partials quickly appear in the spectrum, however it is excited.

By comparing sound spectra recorded immediately after striking and after varying time intervals later, we make the following observations about cymbal sound:

1. The sound level below about 700 Hz shows a rather rapid decrease during the first 200 ms, after which it decays slowly. This is apparently a result of conversion of energy into modes of higher frequency.
2. Several strong peaks in the 700–1000-Hz range build up between 10 and 20 ms, then decay.
3. Sound energy in the important 3- to 5-kHz range peaks about 50–100 ms after striking.
4. Sound in the range of 3–5 kHz, which gives the cymbal its "shimmer," is often the most prominent feature from about 1 to 4 s after striking.
5. The low frequencies again dominate the lingering sound, but at a much lower level, so that they are rather inconspicuous (Rossing and Shepherd, 1983).

3.4.2 Gongs and Tam-Tams Gongs play a very important role in Oriental music, but they enjoy considerable popularity in Western music as well. They are usually cast of bronze with a deep rim and a protruding dome. Gongs used in symphony orchestras usually range from 0.5 to 1 m (20 to 38 in.) in diameter, and are tuned to a definite pitch. When struck near the center with a massive soft mallet, the sound builds up relatively slowly and continues for a long time if the gong is not damped.

Massive gongs are central to every gamelan (ensemble) in Indonesia. Of considerable interest to the acoustician are the gongs used in Chinese opera orchestras, which glide downward or upward in pitch after being struck, because of their special shapes that lead to nonlinearities characterized by a hardening or softening spring constant (Rossing and Fletcher, 1983).

Tamtams are similar to gongs in appearance and are often confused with them. The main differences between the two are that tamtams do not have the dome of the gong, their rim is not as deep, and their metal is thinner. Tamtams sound a much less definite pitch than do gongs. In fact, the sound of a tamtam may be described as somewhere between the sounds of a gong and a cymbal. The sound of a large tamtam develops slowly, changing from a sound of low pitch at strike to a collection of high-frequency vibrations, which are described as "shimmer." These high-frequency modes fail to develop if the tamtam is not hit hard enough, indicating that the conversion of energy takes place through a nonlinear process. Figure 32 shows the shapes of a number of different gongs and tamtams.

3.4.3 Steel Pans An instrument of rather recent origin that has achieved great popularity is the Caribbean steel pan or steel drum. It has become the foremost musical instrument in its home country (Trinidad) and in other Caribbean countries; steel bands are becoming increasingly common in Europe and the United States as well. Steel pans are generally played in ensembles ranging in size from six or eight players to 100 or more. The instrument is occasionally used as a solo instrument accompanied by an orchestral ensemble.

Steel drums or pans are usually fabricated from 55-gal oil drums. The drums in a steel

FIG. 32. Typical shapes of gongs and tamtams. Note the deeper rims and tuning domes of the gongs.

band are known by various names such as single tenor, soprano, double tenor, guitar, cello, and bass. The single tenor or lead pan has from 26 to 32 different notes, but each bass drum has only 3 or 4; hence, the bass drummer typically plays on six drums in the manner of a timpanist.

The first step in making a steel drum is to hammer the end of the oil barrel to the shape of a shallow basin. Then a pattern of grooves is cut with a nail punch in order to define sections of the various notes. Next, each section is "ponged up" with a hammer, and after heating and hardening the drum, each section is tuned by the skillful use of the hammer.

Different pan makers use different patterns for their pans, and these patterns change from time to time. A steel band may typically employ the following instruments:

1. single tenor (also called lead, soprano, or ping-pong): one pan of $2\frac{1}{2}$ octaves (28 notes) ranging from D_4 to F_6;
2. double tenor: two pans of 15 notes each from $A_3\flat$ to $C_6\sharp$;
3. double second: two pans of 14 notes each from $F_3\sharp$ to A_5;
4. cello: three drums of seven notes each from B_2 to G_4; and
5. bass: six drums of three notes each from C_2 to F_3.

Other designs may include a guitar consisting of two drums of 10 notes each, a tenor bass consisting of four drums with five notes each, and a bass set of six drums with four notes each.

A skilled steel drum maker tunes at least one overtone of each note to a harmonic of the fundamental (usually the octave). Sometimes it is possible to tune another mode to the third or fourth harmonic by adjusting internal "tension" along a boundary.

The sound spectra of steel drums are surprisingly rich in harmonic overtones. These harmonic overtones appear to have three different physical origins:

1. radiation from higher modes of vibration tuned harmonically by the tuner;
2. radiation from nearby notes whose frequencies are harmonically related to the struck note; and
3. nonlinear motion of the note area vibrating at its fundamental frequencies.

3.5 Bells

Bells have been a part of nearly every culture in history. Bells existed in the Near East before 1000 B.C., and a number of Chinese bells from the time of the Shang dynasty (1600–1100 B.C.) are found in museums throughout the world. In 1978, a set of 65 tuned bells from the 5th century B.C. was discovered in the Chinese province of Hubei.

3.5.1 Carillon Bells Bells developed as Western musical instruments in the 17th century when bell founders discovered how to tune their partials harmonically. The founders in the Low Countries, especially the Hemony brothers (François and Pieter) and Jacob van Eyck, took the lead in tuning bells, and many of their fine bells are found in carillons today.

The carillon also developed in the Low Countries. Chiming bells by pulling ropes attached to the clappers had been practiced for some time before the idea of attaching these ropes to a keyboard or handclavier occurred to bell ringers in the 16th century. Many mechanical improvements during the 17th and 18th centuries, including the so-called breached wire system and the addition of foot pedals for playing the larger bells, led to development of the modern carillon. Today, the term carillon is reserved for an instrument of 23 (two octaves) or more tuned bells played from a clavier (smaller sets are called "chimes"). The largest carillon in existence is the 74-bell (6 octave) carillon at Riverside Church in New York whose largest bell is more than 18 000 kg (20 tons).

3.5.2 Vibrations of Carillon Bells and Church Bells Perhaps it is stretching the imagination a bit to think of a bell as being a plate, but the general principles of its vibrational behavior are similar. Although the mathematical description of the vibrations of a bell are understandably complex, the principal modes, at least, can be described by specifying the number of circular nodes and meridian nodes. The lowest mode of vibration (called the hum tone), for example, has four meridian nodes, so that alternate quarters of the bell essentially move inward and outward.

Figure 33 shows the principal vibrational modes for a carillon bell. The mode called the third is tuned a minor third above the strike tone, whereas the upper third is usually a major third above the octave. The strike tone is determined by the octave, the 12th, and the upper octave, whose frequencies have the ratios 2:3:4, just as in chimes. Unlike chimes, however, carillon bells have a mode called the prime or fundamental with a frequency at or near the strike tone. Careful studies have shown, however, that the pitch of the strike tone is determined by the three modes mentioned previously rather than by the prime.

Vibrational frequencies of groups O-X in a church bell with a D_5 strike note are shown in Fig. 34. Each group includes modes having the same number of nodal circles positioned at about the same height on the bell. Horizontal lines on the axis at the right indicate relative strengths of several partials in the bell sound (these vary considerably with location, of course). Arrows denote the three partials [octave or nominal, 5th (12th), and upper octave] that determine the strike note.

A new type of carillon bell has been developed at the Royal Eijsbouts Bellfoundry in The Netherlands. The new bell replaces the strong minor-third partial with a major-third partial, thus changing the tonal character of the bell sound from minor to major. This requires an entirely new bell profile. The new bell design evolved partly from the use of a technique for structural optimization using finite-element methods on a digital computer. This technique allows a designer to make changes in the profile of an existing structure and then to compute the resulting changes in the vibrational modes (Lehr, 1987).

3.5.3 Handbells Handbells also date back to at least several centuries B.C., although tuned handbells of the present-day type were developed in England in the 18th century. One early use of handbells was to provide tower bellringers with a convenient means to practice change ringing. In more recent years, handbell choirs have become popular in schools and churches—some 20 000 choirs are reported in the United States.

Although they are cast from the same bronze material and cover roughly the same range of pitch, the sounds of church bells, carillon bells, and handbells have distinctly different timbres. In a handbell, only two modes of vibration are tuned (although there are three harmonic partials in the sound), whereas in a church bell or carillon bell, at least five modes are tuned harmonically. A church bell or carillon bell is struck by a heavy metal clapper in order to radiate a sound that can be heard at a great distance, where the gentle sound of a handbell requires a relatively soft clapper.

In the so-called English tuning of handbells, followed by most handbell makers in England and the United States, the (3,0) mode is tuned to three times the frequency of the (2,0) mode. The fundamental (2,0) mode radiates a rather strong second-harmonic partial, however, so that the sound spectrum has prominent partials at the first three harmonics. Some Dutch founders aim at tuning the (3,0) mode in handbells to 2.4 times the frequency of the fundamental, giving their handbell sound a minor-third character somewhat like a church bell. Such bells are usually thicker and heavier than bells with the English-type tuning.

3.5.4 Ancient Chinese Two-Tone Bells Much interest has developed in the acoustical properties of ancient Chinese two-tone bells whose almond-shaped cross section consists of two segments of a circle joined at

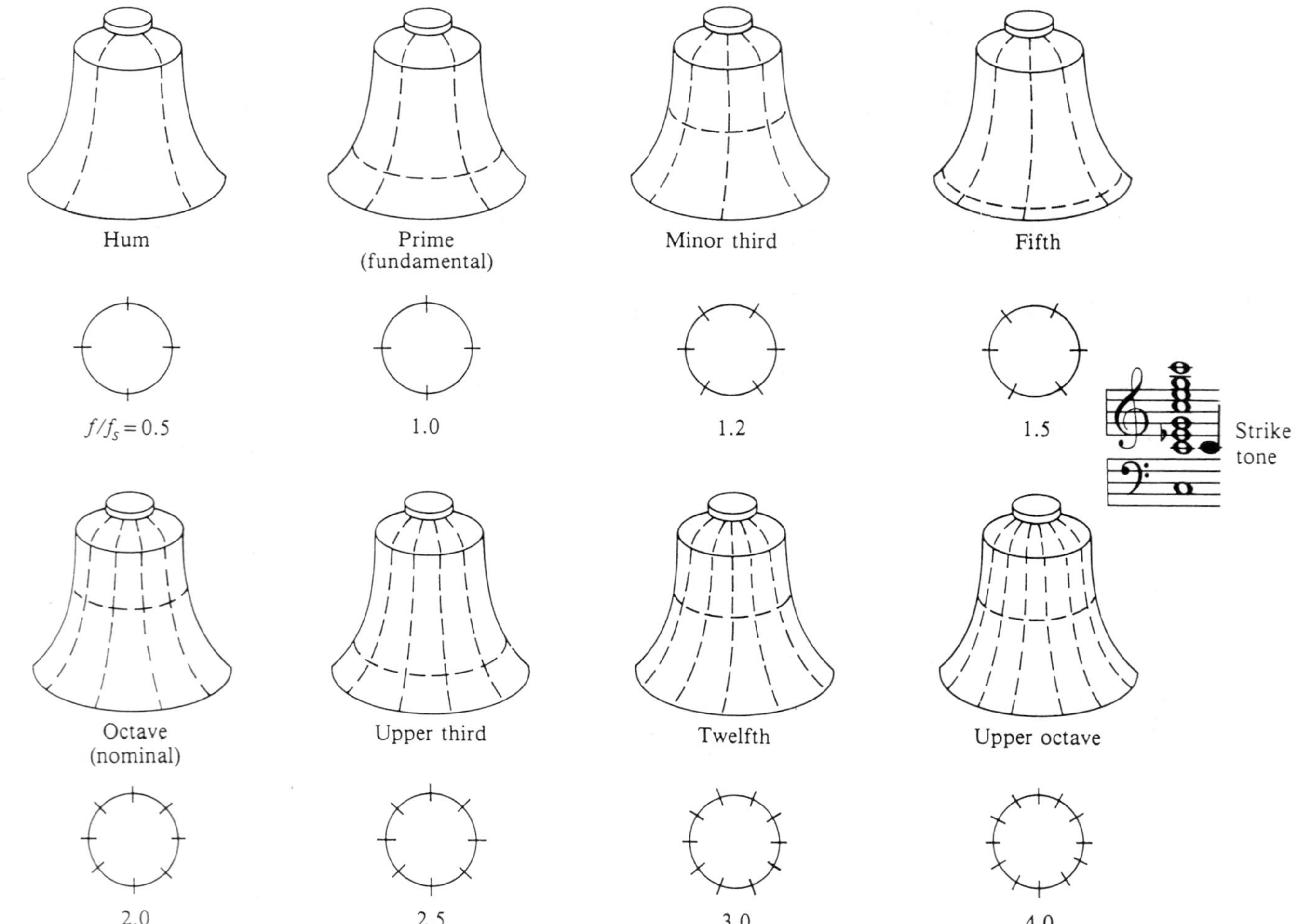

FIG. 33. The first eight vibrational modes of a carillon bell or tuned church bell. Dotted lines indicate the approximate locations of nodes. Frequencies relative to the strike tone are given. The notes that correspond to these in a C_4 bell are shown on a musical staff. In some bells, only five modes are tuned harmonically. (From Rossing, 1990.)

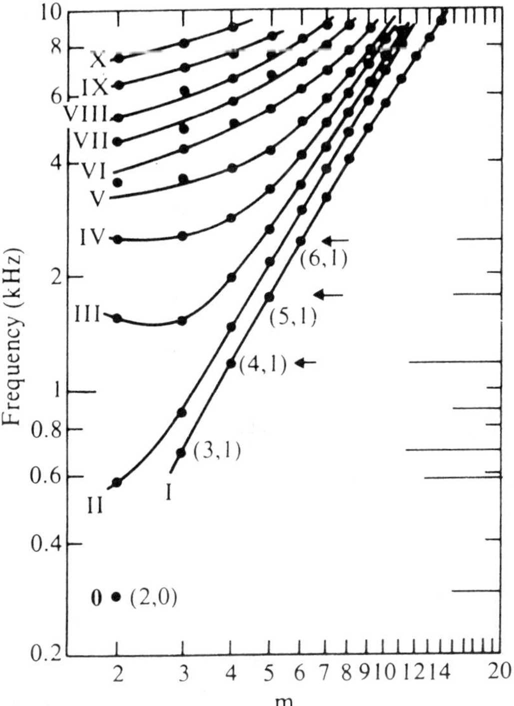

FIG. 34. Vibrational frequencies of groups O-IX in a D_5 church bell. Also shown on the right are the relative strengths at impact of several partials in the bell sound. Arrows denote the three partials in group I that determine the strike note. (From Rossing and Perrin, 1986).

the *xian* or spine. Because they have almond-shaped, rather than round, cross sections, the vibrational modes are split into frequency doublets, and thus, they sound tones with two different pitches when struck at two different points (sometimes referred to as the *sui-yin* and *gu-yin* strike points). This interest was stimulated by the discovery in 1977 of a rack of 65 bells in the tomb of the Marquis Yi of Zheng. These bells, carefully tuned and richly inscribed, dated from the 5th century B.C.

Holographic studies of two-tone bells have shown that the vibrational modes tend to occur in pairs, one with a node at the xian and one with an antinode at that location. The mode with a node at the spine generally has the higher frequency.

Some modes of vibration in a modern Chinese two-tone bell are illustrated in the holographic interferograms in Fig. 35. The interval between the two fundamental pitches, determined by the frequencies of the $(2,0)_a$ and $(2,0)_b$ modes, is a minor third ($f_{20b}/f_{20a} = 1.20$). This is the closest interval on the Western musical scale in about half the ancient bells that have been studied, the others varying from a minor second to a minor sixth. Modal frequencies are shown as a function of the number of complete nodal meridians m in Fig. 36.

4. RESEARCH OPPORTUNITIES

Support for research on musical instruments has always been difficult to obtain, and so many physicists who do research in this area earn their livelihood through teaching or research in other areas. Nevertheless, there is an active community of musical acousticians around the world, and international meetings offer opportunities to share research results. The Acoustical Society of America has a Technical Committee on Musical Acoustics that arranges technical sessions at most meetings of that society. The Catgut Acoustical Society stresses research on bowed string instruments, but its meetings include sessions dealing with other musical instruments as well.

Digital computers and digital instrumentation have led to great progress in musical acoustics research in recent years. Digital computers have themselves become musical instruments of great versatility (see MUSIC, ELECTRONIC). Not only can digital computers generate musical sounds, but they can be employed by composers to "write" music as well.

The analysis of musical instruments and the way in which they produce musical sound has been greatly aided by two techniques that rely on digital computers. Experimental modal testing or modal analysis makes use of a computer to determine the most plausible normal modes of vibration in a musical instrument based on its response to impact or steady-state excitation. Finite-element methods, using a digital computer, theoretically determine the modes of vibration and response function. Used together, these two methods constitute a very powerful technique for understanding sound production in existing musical instruments and the design of new instruments. One example of the latter is the development of carillon bells in which the dominant minor-third partial is replaced by a major third, thus producing an

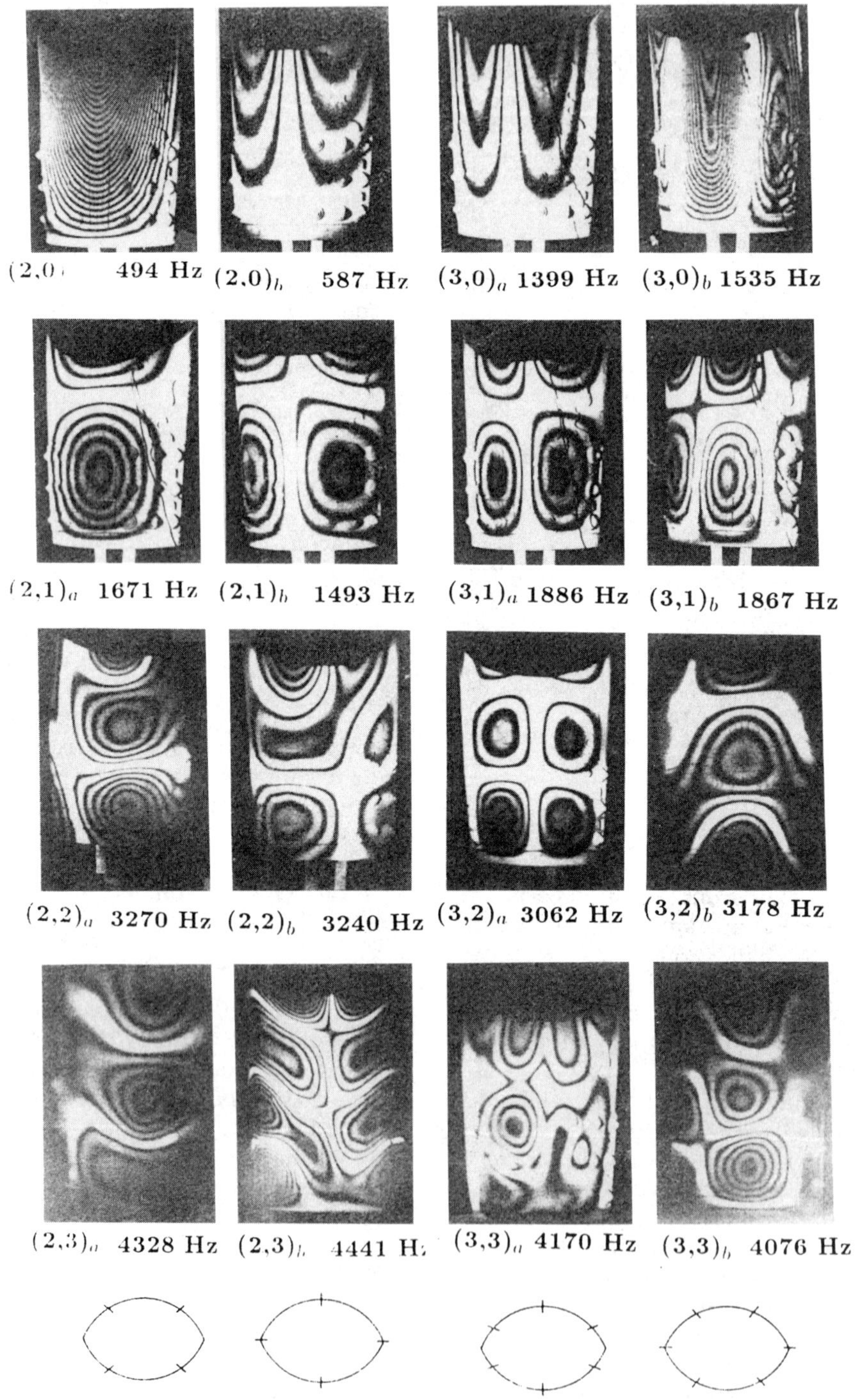

(2,0)$_a$ 494 Hz (2,0)$_b$ 587 Hz (3,0)$_a$ 1399 Hz (3,0)$_b$ 1535 Hz

(2,1)$_a$ 1671 Hz (2,1)$_b$ 1493 Hz (3,1)$_a$ 1886 Hz (3,1)$_b$ 1867 Hz

(2,2)$_a$ 3270 Hz (2,2)$_b$ 3240 Hz (3,2)$_a$ 3062 Hz (3,2)$_b$ 3178 Hz

(2,3)$_a$ 4328 Hz (2,3)$_b$ 4441 Hz (3,3)$_a$ 4170 Hz (3,3)$_b$ 4076 Hz

FIG. 35. Some modes of vibration in a modern Chinese two-tone bell (from Tsai, *et al.*, 1992).

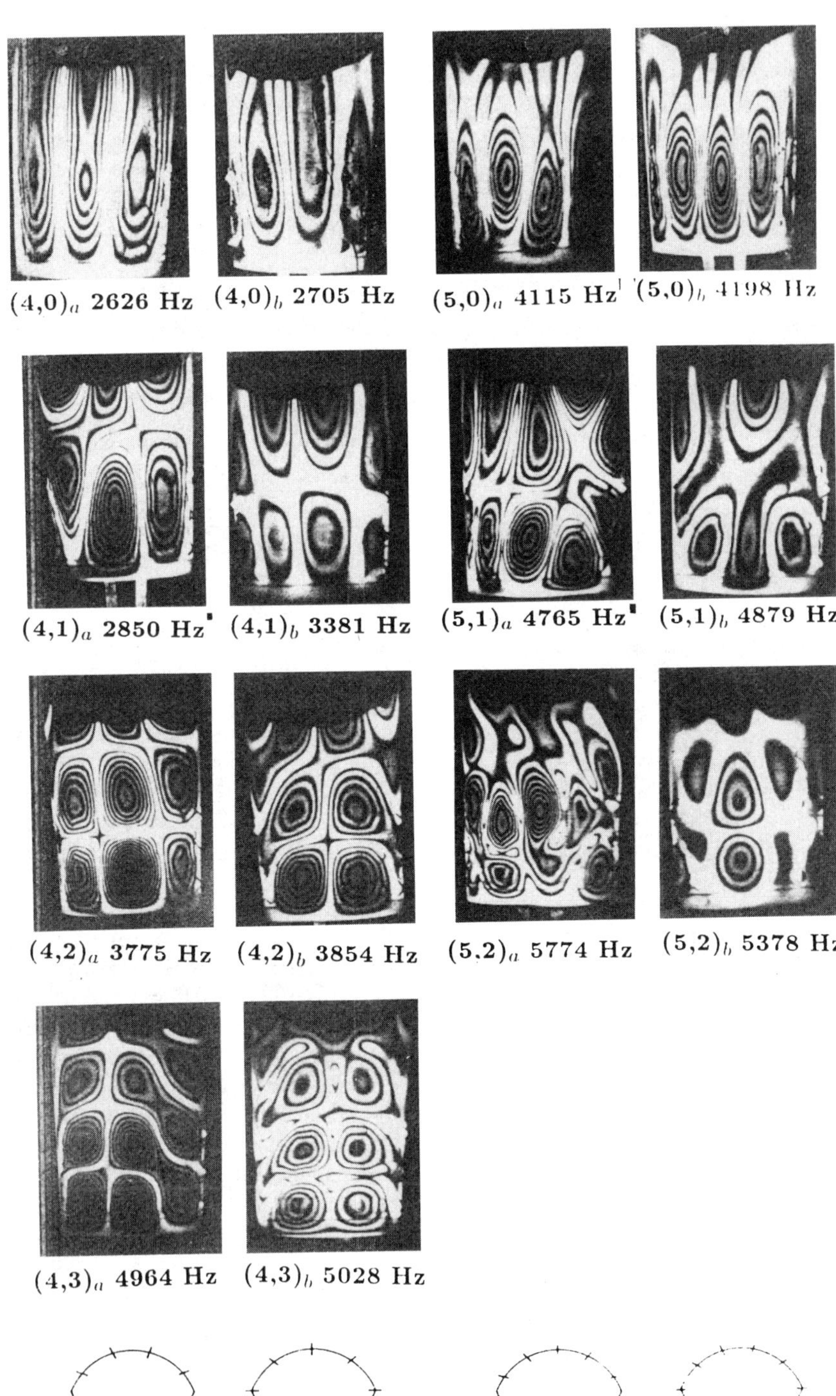

FIG. 35. Continued.

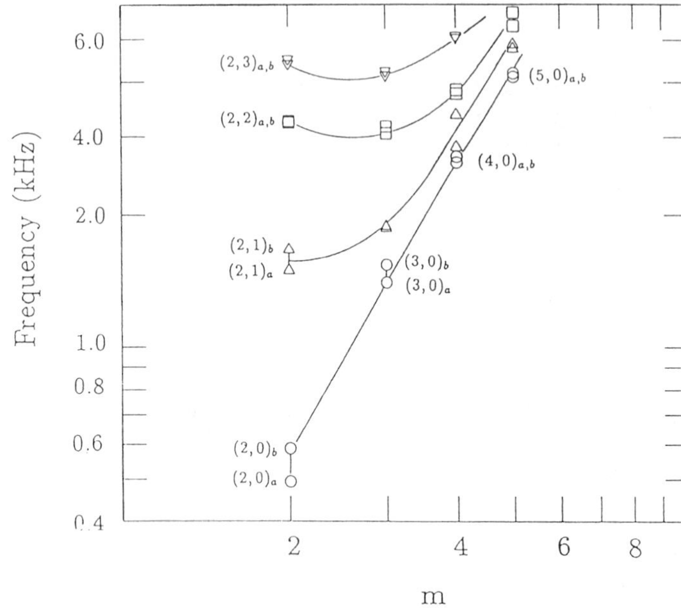

FIG. 36. Mode frequencies of a modern Chinese two-tone bell as functions of the number of nodal meridians *m* and nodal circles *n* (from Tsai, *et al.*, 1992).

entirely new musical character; this development was made possible by the use of finite-element methods on a digital computer.

Many musical instruments require a certain type of wood, and the growing scarcity of some wood is cause for alarm. The same is true of animal materials, such as gut for strings, hides for drumheads, and wool for felt hammers. This situation has led to extensive research on synthetic materials, but much more is needed.

Another concern of musicians is that high-quality handmade musical instruments are becoming unaffordable by many performers. More research is needed in order to raise the quality of affordable mass-produced instruments. Also, a better understanding of musical instruments will lead to the development of instructional aids that supplement or replace costly private instruction in learning to play musical instruments.

GLOSSARY

Admittance: Reciprocal of impedance. Mechanical admittance (mobility) is velocity divided by force. Acoustical admittance is flow velocity divided by pressure.

Bessel Horn: A family of horns whose radii r are given by $r = bx^{-\epsilon}$, which form the basis for the air columns in many wind instruments.

Carillon: A musical instrument consisting of 23 or more tuned bells whose clappers are mechanically coupled to a clavier or keyboard.

Cent: 1/100 of a semitone in the scale of equal temperament. An octave is equal to 1200 cents.

Chladni Patterns: A pattern created by sprinkling a fine powder on a vibrating plate. The powder tends to collect along the nodal lines, thus making them visible.

Cross Fingering or Fork Fingering: A woodwind fingering that leaves one or more tone holes open and closes other tone holes below this.

Harmonics: Tones or partials whose frequencies are integral multiples of a fundamental frequency.

Impedance: Reciprocal of admittance. Mechanical impedance is force divided by velocity. Acoustical impedance is pressure divided by flow velocity.

Mobility: Mechanical admittance. Velocity divided by force.

Partial Tones or Partials: Single-frequency components of a tone.

Radiation Resistance: Real part of the radiation impedance, which is a measure of the fluid loading on a vibrating surface.

Scale of Equal Temperament: A musical

scale in which each semitone increments the frequency by a factor $2^{1/12} = 1.05946 \simeq 1.06$. In this scale, each semitone interval is exactly 100 cents.

Sound Pressure Level: Logarithmic expression of sound pressure expressed in decibels (dB). The sound pressure level is $L_p = 20 \log (p/p_0)$, where p is sound pressure, and $p_0 = 2 \times 10^{-5}\ \mathrm{N/m^2} = 20\ \mu\mathrm{Pa}$.

Sound Spectrum: A "recipe" for a sound that expresses, in graphical form, the sound-pressure levels and the frequencies of the various partials or spectral components. In the spectrum of a musical sound, these partials are often (but not always) harmonics of a fundamental tone.

Tricord: A set of three piano strings tuned to the same frequency and struck by the same hammer.

Works Cited

Alonso Moral, J., Jansson, E. (1982a), *Acustica* **50**, 329–377.

Alonso Moral, J., Jansson, E. (1982b), Report STL-QPSP 203, Stockholm: Speech Trans. Lab., Royal Inst. Tech. (KTH), pp. 60–75.

Ayers, R. D., Eliason, L. J., Mahgerefteh, D. (1985), *Am. J. Phys.* **53**, 528–537.

Benade, A. H. (1976), *Fundamentals of Musical Acoustics*, New York: Oxford University Press. Reprint (1990), New York: Dover.

Campbell, M., Greated, C. (1987), *The Musician's Guide to Acoustics*, New York: Schirmer.

Cremer, L. (1981), *Physik der Giege*, Stuttgart: Hirtzal Verlag. English translation by J. S. Allen (1984), Cambridge, MA: MIT Press.

Firth, I. (1977), *Catgut Acoust. Soc. Newsletter* No. 27, 12.

Fletcher, N. H. (1975), *J. Acoust. Soc. Am.* **57**, 233–235.

Fletcher, N. H., Rossing, T. D. (1991), *Physics of Musical Instruments*, New York: Springer-Verlag.

Helmholtz, H. L. F. (1877), *On Sensations of Tone*, 4th ed.; translation by A. J. Ellis (1954), New York: Dover.

Hutchins, C. M. (1977), *Catgut Acoust. Soc. Newsletter* No. 28, 22–24.

Kindel, J. (1989), *Modal Analysis and Finite Element Analysis of a Piano Soundboard*, M. S. Thesis, Univ. of Cincinnati (unpublished).

Lehr, A. (1987), *The Designing of Swinging Bells in the Past and Present*, Asten, The Netherlands: Athanasius Kircher Foundation.

Richardson, B. E., Roberts, G. W. (1985), *Proc. SMAC 83*, vol. 2, Stockholm: Royal Swedish Academy of Music, p. 285–302.

Rossing, T. D. (1976), *Phys. Teach.* **14**, 546–556.

Rossing, T. D. (1982), *Sci. Am.* **247** (5), 172–178.

Rossing, T. D., Fletcher, N. J. (1983), *J. Acoust. Soc. Am.* **73**, 345–351.

Rossing, T. D., Perrin, R. (1986), *Appl. Acoust.* **20**, 41–70.

Rossing, T. D. (1990), *The Science of Sound*, Reading, MA: Addison-Wesley.

Rossing, T. D. (1992), *Phys. Today* **45** (3), 40–47.

Rossing, T. D., Shepherd, R. B. (1983), *Proceedings of the 11th International Congress on Acoustics*, vol. 6, Paris, pp. 329–333.

Strong, W. J., Plitnik, G. R., (1983), *Music, Speech, and High Fidelity*, 2nd ed., Provo, UT: Soundprint.

Tsai, J., Jiang, Z., Rossing, T. D. (1992), *Acoust. Australia* **20**, 17–19.

Weinreich, G. (1977), *J. Acoust. Soc. Am.* **62**, 1474–1484.

Further Reading

Rossing, T. D. (1987), "Resource Letter MA-2: Musical Acoustics," *Am. J. Phys.* **55**, 589–601. An annotated list of 414 references on musical acoustics.

Fletcher, N. H., Rossing, T. D. (1991), *Physics of Musical Instruments*, New York: Springer-Verlag.

Hutchins, C. M. (1975, 1976), *Musical Acoustics*, Parts 1 and 2, Stroudsburg, PA: Dowden, Hutchinson, and Ross. Reprints of 60 articles on bowed string instruments.

Elementary textbooks on musical acoustics:

Backus, J. (1977), *The Acoustical Foundations of Music*, 2nd ed., New York: Norton.

Benade, A. H. (1976), *Fundamentals of Musical Acoustics*, New York: Oxford; reprint (1990), New York: Dover.

Campbell, M., Greated, C. (1987), *The Musician's Guide to Music*, New York: Schirmer.

Hall, D. E. (1991), *Musical Acoustics*, 2nd ed., Pacific Grove, CA: Brooks/Cole.

Rossing, T. D. (1990), *The Science of Sound*, 2nd ed., Reading, MA: Addison-Wesley.

Strong, W. J., Plitnik, G. R. (1992), *Music, Speech, Audio*, Provo, UT: Soundprint.

NANOPHASE MATERIALS

Richard W. Siegel, *Materials Science Division, Argonne National Laboratory, Argonne, Illinois, U.S.A.*

INTRODUCTION

Nanophase materials, with their grain sizes or phase dimensions in the nanometer size regime, represent a subset of the broad class of nanostructured materials artificially synthesized with microstructures modulated in zero to three dimensions on length scales less than 100 nm that it has become possible to create over the past several years. The various types of nanostructured materials share three fundamental features: constituent atomic domains (grains, layers, or phases) spatially confined to less than 100 nm, significant atom fractions associated with interfacial environments, and interactions between the atomic domains. Nanostructured materials (see Fig. 1) thus include atom clusters and filaments—which can be considered as essentially zero-dimensionally modulated materials—, one-dimensionally modulated multilayers, two-dimensionally modulated ultrafine-grained overlayers, coatings, or buried layers, and their three-dimensionally modulated analogues, nanophase materials. Intermediate dimensionalities can also exist, and nanocomposite materials, containing multiple phases, can range from the most conventional case, in which a nanoscale phase (zero-dimensionally modulated) is embedded in a phase of conventional sizes, to the case in which all the constituent phases are of nanoscale dimensions (three-dimensionally modulated and nanophase or multilayered). While the man-made synthesis of these materials is rather recent, naturally formed nanostructured materials are found in the earliest meteorites and in a wide variety of biological systems.

Increasing interest in recent years has focused on a wide variety of nanostructured materials, since it is anticipated that their properties will be different from, and often superior to, those of conventional materials that have phase or grain structures on a coarser size scale (Kear *et al.*, 1989; Hadjipanayis and Siegel, 1994), and the initial investigations have indicated this to be the case. Moreover, it is possible to engineer these properties by controlling and varying the sizes of the constituent domains and the manner in which they are assembled (Siegel, 1993). This interest has been stimulated not only by the recent efforts and successes in synthesizing a variety of fascinating atom clusters, zero-dimensionality quantum-well structures, and one-dimensionally modulated multilayered materials with nanometer-scale modulations, but also by the rapidly developing potential for synthesizing three-dimensionally modulated, bulk nanophase materials via the assembly of clusters of atoms (Andres *et al.*, 1989) and a variety of nanocomposite materials (Roy *et al.*, 1988; Burggraaf *et al.*, 1989; Niihara, 1991). The possibilities to create new nanostructured materials with unique or improved properties are thus expected to have

3-527-28133-9/94/$5.00 + .50

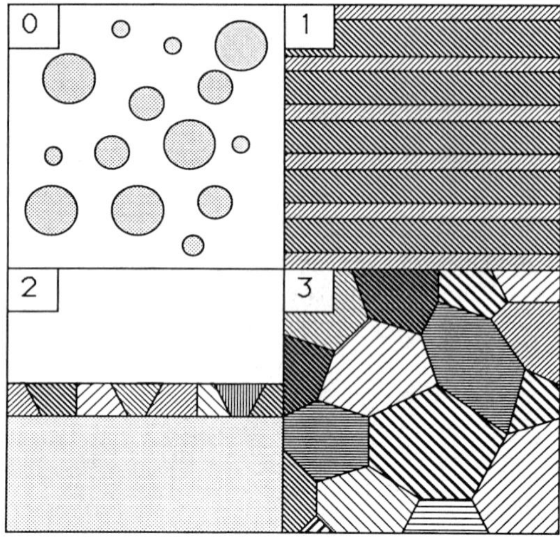

FIG. 1. Schematic of the four basic types of nanostructured materials, classified according to integral modulation dimensionality: zero, clusters of any aspect ratio from 1 to ∞; one, multilayers; two, ultrafine-grained overlayers (coatings) or buried layers; three, nanophase materials (Siegel, 1993).

a significant impact on our ability to engineer advanced materials with controlled chemical, electronic, magnetic, mechanical, and optical properties with attendant useful technological applications.

In the past few years, atom clusters with average diameters in the range of 5–50 nm of a variety of materials, including metals and ceramics, have been synthesized by evaporation and condensation in high-purity gases followed by consolidation *in situ* under ultrahigh-vacuum conditions to create nanophase materials. These new ultrafine-grained materials have properties that are often significantly different and considerably improved relative to those of their coarser-grained counterparts. The property changes result from their small grain sizes, the large percentage of their atoms in grain-boundary environments, and the interactions between their grains. Since their properties can be engineered during the synthesis and processing steps, cluster-assembled nanophase materials appear to have great technological potential beyond the current scientific interest in their grain-size–dependent properties. Recent research on nanophase materials is reviewed in this article and some future applications are considered.

1. BACKGROUND AND GENERAL PRINCIPLES

Evidence from the earliest meteorites that have been found and studied suggests that nanophase materials have been part of our universe from its beginnings after the Big Bang; it has been suggested that primordial materials with nanometer-scale phase structures condensed from our solar nebula (Blander and Katz, 1967; Blander and Abdel-Gawad, 1969; Grossman, 1972). Many biological materials, often constituents of a hierarchically structured system, are also found to be nanostructured (Aksay *et al.*, 1992) as they have evolved to optimize their performance. More recently, during the past two decades, the planned human synthesis of nanophase materials has begun. New methodologies for synthesizing ultrafine-grained nanophase materials are being developed to create new levels of property engineering through the sophisticated control of scale, morphology, and architecture.

Atom clusters in the nanometer size regime, containing hundreds to tens of thousands of atoms, can now be economically produced by either physical or chemical processes in sufficient numbers that they can be assembled into technologically useful nanophase materials. Such materials can take advantage of and incorporate a variety of size-related effects in condensed matter. These effects range from electronic quantum size effects caused by spatial confinement of delocalized valence electrons to altered cooperative (many-body) atom phenomena, such as lattice vibrations or melting, to the suppression of such lattice-defect mechanisms as dislocation generation and migration in confined grain sizes.

The desirable sizes of the constituent atomic ensembles are generally below 100 nm, since it is in this size range (and often below 10 nm) that various properties begin to change significantly as a result of a variety of confinement effects. A property will be altered when the entity or mechanism (or combination thereof) responsible for that property is confined within a space (defined by the size of the ensemble of atoms) smaller than some critical length associated with that entity or mechanism. Thus, for example, a metal that is conventionally ductile by virtue of the usual ease in creating and moving dislocations through its crystal lattice will become significantly harder when grain sizes are reduced to the point where dislocation sources are no longer able to operate at low levels of applied stress. Since the stress to operate a Frank–Read dislocation source, for example, is inversely proportional to the spacing between its pinning points, a critical length in this case is that for which the stress to operate this source becomes larger than the conventional yield stress for the given metal. Such confinement only appears to be different from that usually encountered in the technical literature, where, for example, the optical absorption properties of a quantum-well semiconductor device are blue shifted (to shorter wavelengths) because of the size of the well becoming comparable to and smaller than the effective size of the excitonic state responsible for this absorption. The specifics are quite different, but the underlying general principle of confinement is not.

Control of the size or sizes and morphologies of the phase domains or granular entities (e.g., clusters) being assembled is therefore of primary importance in any of the methods for the synthesis of nanostructured materials. Beyond this, chemical composition of the phases and cleanliness of the interfaces between phases or grains must be controlled as well. In bulk cluster-assembled nanophase materials, such control appears to be readily available. Before proceeding with a description of their synthesis and processing, it is useful to list some of the unique advantages of the assembly of nanophase materials under controlled atmospheres from gas-condensed clusters; they are as follows:

1. The <100-nm sizes and surface cleanliness of the atom clusters allow conventional restrictions of phase equilibria and kinetics to be overcome during material synthesis and processing; this results from the combination of short diffusion distances, high driving forces, and uncontaminated surfaces and interfaces available with cluster building blocks.
2. The large fraction of atoms residing in the grain boundaries and interfaces of these materials allows for the special interatomic arrangements in interfaces to constitute significant volume fractions of material; thus, novel material properties may result from these less than perfectly ordered atomic environments.
3. The reduced size scale and large surface-to-volume ratios of the individual nanophase grains can be predetermined and can significantly alter a variety of physical and chemical properties; the degree of porosity of the cluster assembly can also be controlled.
4. A wide range of materials can be produced in this manner, including metals and alloys, intermetallic compounds, ceramics, and semiconductors; and such materials can be synthesized to contain crystalline, quasi-crystalline, or amorphous structures.
5. The extensive possibilities for reacting, coating, and mixing *in situ* various types, sizes, and morphologies of clusters create significant future potential for the synthesis of a variety of new multicomponent composites with nanometer-sized microstructures and engineered properties that can be both multifunctional and hierarchical.

2. SYNTHESIS AND PROCESSING

2.1 Cluster Assembly

Many opportunities exist to create nanophase materials assembled from atom clusters of metals, ceramics, and semiconductors synthesized by means of a number of physical and chemical methods that appear to have very broad technological potential in the area of advanced materials. One of the traditional methods for synthesizing ultrafine powders or colloidal suspensions has been chemical precipitation. A wide variety of these methods, including sol-gel synthesis and the inverse micelle method, for example, have been successfully applied to the synthesis of nanometer-sized clusters with narrow size distributions. A number of high-temperature gas-reaction methods are also now becoming available for the synthesis of nanoscale clusters or larger powders with nanoscale substructures. Additional references to the variety of interesting methods available can be found elsewhere (Andres *et al.*, 1989; Aksay *et al.*, 1989; Rieke *et al.*, 1990; Averback *et al.*, 1991; Whitesides *et al.*, 1991; Heuer *et al.*, 1992; Komarneni *et al.*, 1993). Invariably, in each of the solution-chemical precipitation methods, the synthesized clusters are burdened by surface layers from the solutions in which they are formed. These surface layers are often intentionally introduced to cap the clusters at a given size or to keep them from agglomerating and forming large pores or to keep the clusters isolated from one another in order to control their interaction in the assembled material. If the clusters are to be assembled into monolithic bulk solids, where no such separation is desired, then the surface layers can become contaminants in the grain boundaries of the cluster-consolidated material, and processing and properties can be adversely affected. Such problems can be avoided in the gas-condensation method for synthesizing nanophase materials (Gleiter, 1989; Uyeda, 1991; Siegel, 1991), which will be highlighted in the present article.

The synthesis of ultrafine-grained materials by the *in situ* consolidation in vacuum of nanometer-size gas-condensed ultrafine particles or atom clusters first suggested by Gleiter (1981), was applied initially to metals (Birringer *et al.*, 1984, 1986). By consolidating clusters in this manner, materials with a large fraction of their atoms in grain boundaries could be formed. This method was subsequently applied to the synthesis of nanophase ceramics (Siegel and Hahn, 1987; Siegel *et al.*, 1988). The study of the gas condensation of ultrafine particles or atom clusters has, however, had a rather longer history stretching back to the formation and use of "smokes," such as carbon, or bismuth "blacks," for a variety of applications (Pfund, 1930). Scientific research into the controlled production of ultrafine particles by means of the gas-condensation method has been more recent (Kimoto *et al.*, 1963; Granqvist and Buhrman, 1976; Thölén, 1979), but still preceded cluster assembly and therefore provided an important basis for this work. In addition, of course, the previous knowledge of powder metallurgy and ceramics generated over an even longer period provided much needed background information for the work on nanophase materials to progress. The application of these ideas in recent years to the synthesis of a variety of nanophase metals and ceramics has built upon this broad scientific and technological base.

Early research on atom clusters formed via the gas-condensation method defined most of the important experimental parameters that control the sizes of the clusters formed in the conventional gas-condensation method that has been used to synthesize nanophase materials. The pioneering work of Uyeda and coworkers (Kimoto *et al.*, 1963) demonstrated that a wide range of metallic ultrafine particles (between 10 and 100 nm in diameter) could be condensed in a low-pressure Ar atmosphere and that the particle sizes could be controlled by varying the gas pressure in the range of about 1–30 Torr (0.13–4 kPa). Subsequent work by Thölén (1979) extended this method to a number of additional metals and to a study of the nucleation, growth, and coalescence of the clusters. However, the most detailed study to date of the conventional gas-condensation process for forming ultrafine metal particles or clusters via condensation in naturally convecting inert gases (He, Ar, or Xe) was carried out by Granqvist and Buhrman (1976). Some of their results will be discussed further below, but it was these early studies that elucidated the essential parameters (primarily type of gas, gas pressure, and evaporation rate) that control the formation of gas-condensed atom clusters that have made possible the synthesis of cluster-assembled nanophase materials in recent years.

The basic aspects of the generation of atom clusters via gas condensation can be described using the conceptual model shown in Fig. 2. A precursor material, either an element or compound, is evaporated in a gas maintained at a low pressure, usually well below one atmosphere. The evaporated atoms or molecules lose energy via collisions with the gas atoms or molecules and undergo a homogeneous condensation to form atom

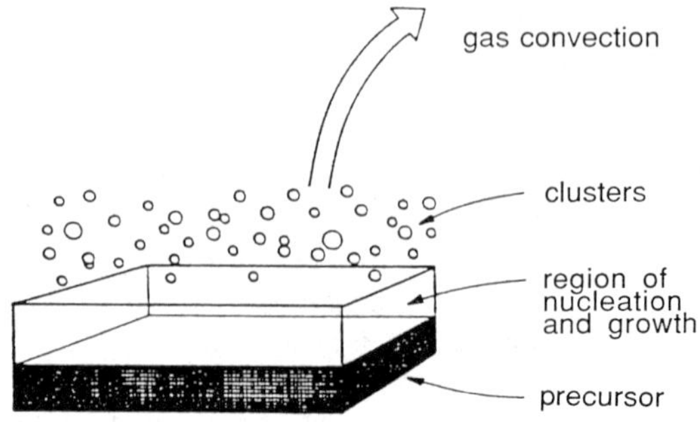

FIG. 2. Conceptual model for cluster formation via gas condensation (Granqvist and Buhrman, 1976).

clusters in the highly supersaturated vicinity of the precursor source. In order to maintain small cluster sizes, by minimizing further atom or molecule accretion and cluster-cluster coalescence, the clusters once nucleated must be removed rapidly from the region of high supersaturation. Since the clusters are already entrained in the condensing gas, this is readily accomplished by setting up conditions for moving this gas. Such gas motion has generally been driven by natural convection under the combined action of gravity and the temperature difference between the precursor source and a cooled thermophoretic cluster collection surface. However, a forced gas flow can also be used, with significant advantages in terms of both cluster-size control and process efficiency.

It should be emphasized that there are thus only three fundamental rates, which function relative to one another, that essentially control the formation of the atom clusters in the gas-condensation process. They are

1. the rate of supply of atoms to the region of supersaturation where condensation occurs,
2. the rate of energy removal from the hot atoms via the condensing medium, the gas, and
3. the rate of removal of the clusters once nucleated from the supersaturation region.

Other factors can also affect the clusters finally collected, particularly those that result in significant cluster-cluster coalescence, but these three rates represent the core of the process. Accordingly, the smallest cluster sizes for a given precursor are obtained for a low evaporation rate and condensation in a low pressure of a light inert gas, such as He. These conditions lead to a lower supersaturation of precursor atoms in the gas, slower removal of energy from the evaporated atoms (via the lighter gas atoms at lower pressure), and more rapid convective gas flow also as a result of the lower gas pressure. The rapid gas flow is significant, since it guarantees a shorter dwell time of the gas-entrained condensed clusters in the supersaturated region in which, if they remained, they would grow further. The effects of these controllable experimental parameters (gas type and pressure and evaporation rate) on the average cluster sizes produced under conditions of natural convective gas flow are shown in Figs. 3(a) and 3(b).

A typical apparatus (Siegel and Eastman, 1989) that uses such a process for the synthesis of nanophase materials via the *in situ* consolidation of gas-condensed clusters is shown schematically in Fig. 4. It consists of an ultrahigh-vacuum (UHV) system fitted with two resistively heated evaporation sources (which operate in the manner of Fig. 2), a cluster collection device (liquid-nitrogen–filled cold finger) and scraper assembly, and *in situ* compaction devices for consolidating the powders produced and collected in the chamber. Before making the powders, the UHV system is first evacuated by means of a turbomolecular pump to below 10^{-5} Pa and then back-filled with a controlled high-purity gas atmosphere at pressures of about a few hundred pascals. For producing metal powders this is usually an inert gas, such as He, but it can alternatively be a reactive gas or gas mixture if, for example, clusters of a ceramic compound are desired.

The clusters have rather narrow size distributions, as shown in Fig. 5, and are collected via thermophoresis on the surface of the cold finger in the form of very open, fractal structures. The clusters are weakly held on the collector surface by means of van der Waals type forces and are easily removed from this surface with a Teflon scraper. Upon removal, the clusters fall like snow from the surface and are funneled into a set of compaction devices (see Fig. 4) capable of consolidation pressures up to about 1–2 GPa, in which the nanophase samples are formed at room temperature, or at elevated temperatures if needed. The pellets formed in this conventional research apparatus are typically about 9 mm in diameter and 0.1 to 0.5 mm thick, depending upon the amount of material made (usually a few hundred milligrams) and the experiments to be performed. The sizes of these samples have been more a matter of laboratory convenience than any real limitation of the gas-condensation method itself. All of the fundamental rates involved in the cluster synthesis can be significantly increased above those in use in the laboratory, and scale-up leading to commercially viable production rates is already a reality. The scraping and consolidation are performed under UHV conditions after removal of the inert or reactive gases from the

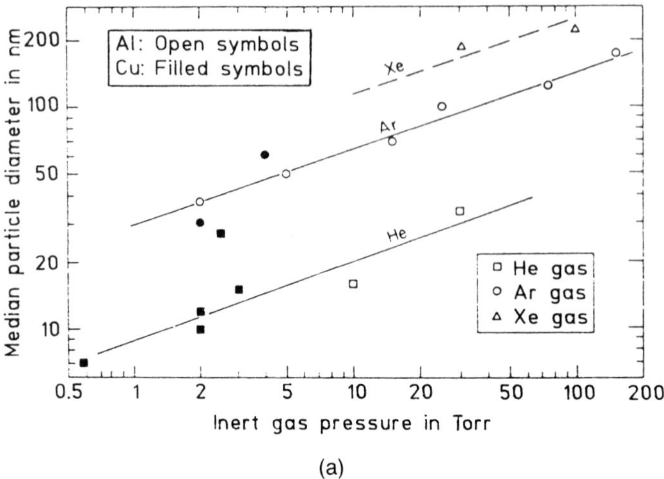

(a)

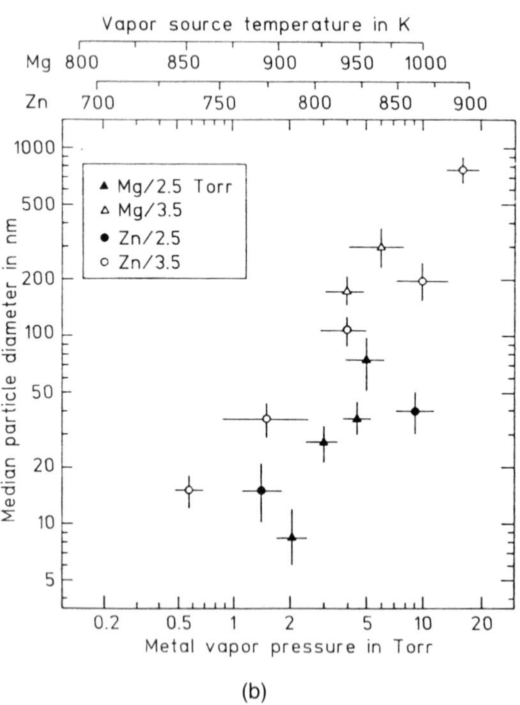

(b)

FIG. 3. (a) Median particle diameter versus pressure of He, Ar, or Xe gas for clusters of Al and Cu formed via gas condensation; the straight lines only serve to guide the eye. (b) Median particle diameter versus vapor pressure at the metal surface (or source temperature) for Mg and Zn evaporated and gas-condensed in two different Ar pressures (2.5 and 3.5 Torr) (Granqvist and Buhrman, 1976).

chamber, in order to maximize the cleanliness of the particle surfaces and the interfaces that are subsequently formed. Also, any possibility of trapping remnants of these gases in the nanophase compact is minimized by consolidation in vacuum. It should be noted in this regard that the total surface area of the nanophase powders produced in a given run is so great that, for a residual gas pressure of less than 10^{-5} Pa in a volume the size of the UHV chamber used, rather little gas contamination of the cluster surfaces would be expected.

Since the as-collected gas-condensed clusters are generally aggregated in rather open fractal arrays, their consolidation at pressures of 1–2 GPa is easily accomplished, even at room temperature. The difficulties in consolidating the hard equiaxed agglomerates of fine powders resulting from conventional wet-chemistry synthesis routes are mostly avoided. The sample densities resulting from cluster

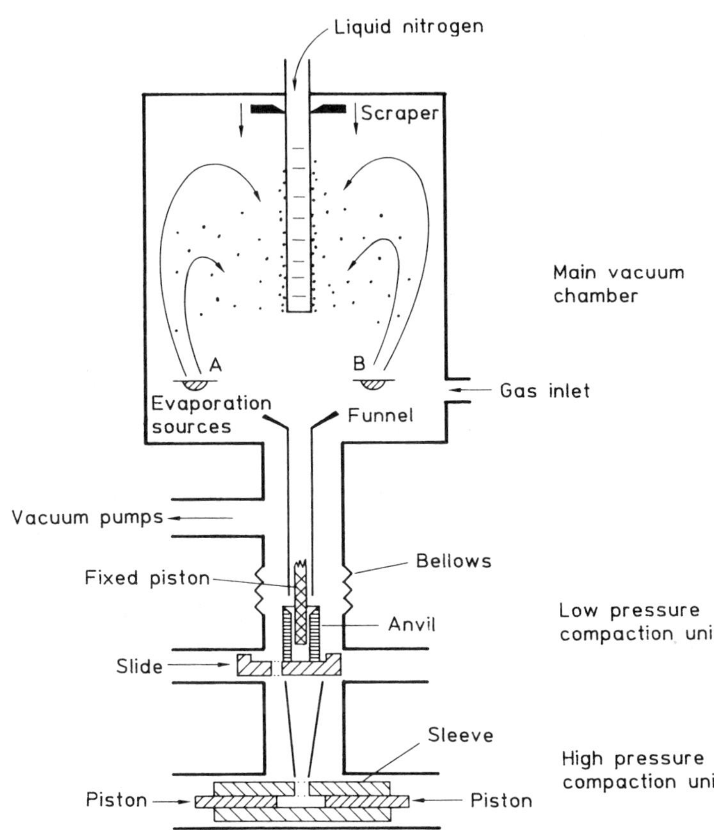

FIG. 4. Schematic drawing of a gas-condensation chamber for the synthesis of nanophase materials. Precursor material evaporated from sources A and/or B condenses in the gas and is transported via convection to the liquid-nitrogen–filled cold finger. The clusters are then scraped from the cold finger, collected via the funnel, and consolidated first in the low-pressure compaction unit and then in the high-pressure compaction unit, all in vacuum (Siegel and Eastman, 1989).

consolidation at room temperature have ranged up to about 97% of theoretical for nanophase metals and up to about 75–85% of theoretical for nanophase oxide ceramics. This as-consolidated, green-state porosity will be considered further in Sec. 3, but it probably represents (at least in part) a manifestation of powder agglomeration leading to voidlike flaws. Fortunately, these appear to be capable of being removed by means of cluster consolidation at elevated temperatures and pressures without significant attendant grain growth.

If an elemental precursor is evaporated in an inert gas atmosphere, then the atom clusters formed and collected are the same material, only in a reconstituted form. However, if clusters of a compound, such as a ceramic oxide, are desired, the process can become somewhat more complex. For example, in order to produce nanophase TiO_2 with a rutile structure and 12-nm average grain size, Ti metal clusters condensed in He were first collected on the cold finger and subsequently oxidized by the introduction of oxygen into the chamber (Siegel *et al.*, 1988). A similar method has been used to produce α-Al_2O_3 (Eastman *et al.*, 1989) with an 18-nm average grain size after oxidizing Al clusters in air at 1000 °C. If the vapor pressure of a compound is sufficiently large, as in the cases of MgO and ZnO, for example, it is possible to sublime the material directly from the oxide precursor in a He atmosphere containing, in addition, a partial pressure of O_2 to attempt to maintain oxygen stoichiometry during cluster synthesis. Such a method has been used (Eastman *et al.*, 1989) to produce nanophase MgO and ZnO with average grain sizes down to about 5 nm. Frequently, however, oxygen stoichiometry is not maintained.

In the case of nanophase TiO_2 cited above, the oxygen deficiency, while still present, is rather small and easily remedied as a result of the small grain sizes and short diffusion distances involved. Raman spectroscopy has been a useful tool in studying the oxidation state of nanophase TiO_2 because of the intense and well-studied Raman bands in both the anatase and rutile forms of this oxide and

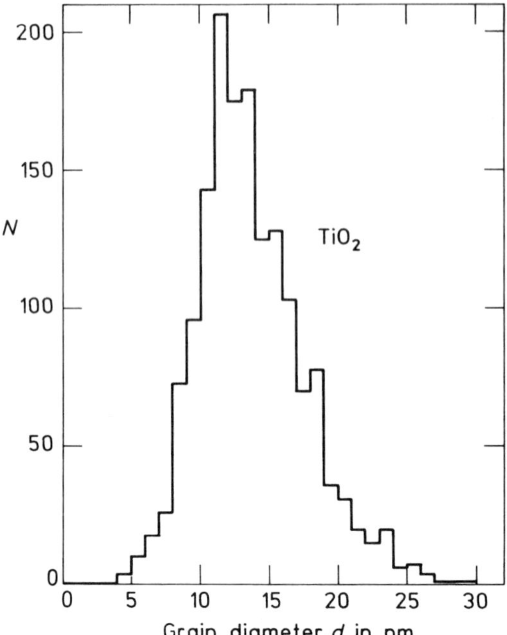

FIG. 5. Grain-size distribution for a nanophase TiO_2 (rutile) sample compacted to 1.4 GPa at room temperature, as determined by dark-field transmission electron microscopy (Siegel *et al.*, 1988).

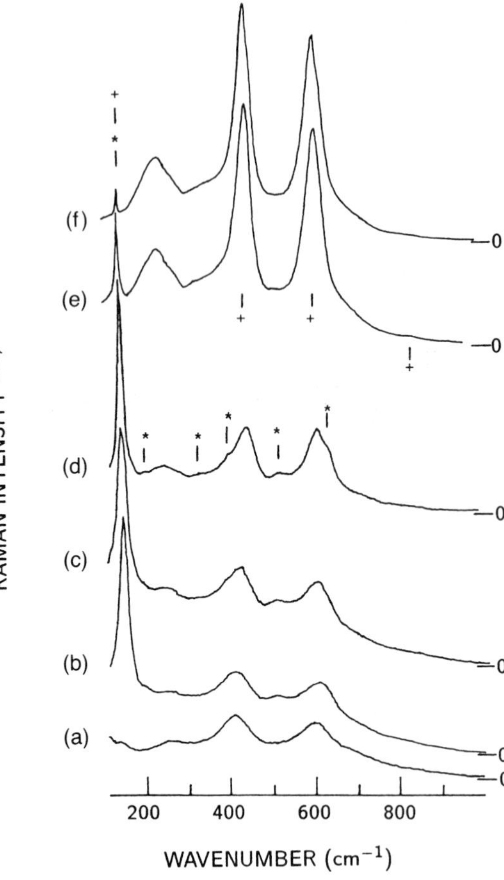

FIG. 6. (a), (b) Raman spectra of two areas of an as-compacted nanophase TiO_2 sample compared with spectra for the area used in (b) after successive annealing in air for 0.5 h at (c) 300 °C, (d) 500 °C, (e) 700 °C, and (f) 1000 °C (* indicates anatase $k = 0$ phonons; + indicates rutile $k = 0$ phonons) (Melendres *et al.*, 1989).

the observation that these bands were affected in nanophase samples (Melendres *et al.*, 1989). A series of Raman spectra from two as-consolidated titanium dioxide nanophase samples and from one of these samples annealed in air until fully oxidized to TiO_2 is shown in Fig. 6. The band broadening observed in the nanophase samples (and also band shifting in both the anatase and rutile phases) was confirmed (Parker and Siegel, 1990a) to be the result of an oxygen deficiency, which could be subsequently removed in these samples by annealing in air. A subsequent calibration of this deviation from stoichiometry (Parker and Siegel, 1990b), shown in Fig. 7, indicated that $TiO_{1.89}$ was the actual material produced in the apparatus of Fig. 4, but that it could be easily oxidized to fully stoichiometric TiO_2, if desired, without sacrificing its small grain size (12 nm). Also, if intermediate deviations from stoichiometry were sought, in order to select particular material properties sensitive to the presence of oxygen deficient defects, they could be readily achieved as well.

Most of the atom clusters assembled into nanophase materials to date have been generated from Joule-heated evaporation sources. However, such sources have limitations that need not be suffered, since a wide variety of other sources are also available. The primary limitations are source-precursor incompatibility, temperature range, uniformity and control, and dissimilar evaporation rates for different constituents in an alloy or compound precursor. Each of these limitations can be avoided by a host of alternative sources that have been developed over the years of ultrafine particle research, but that are only now beginning to enter the field of nanophase materials synthesis (Uyeda, 1991; Siegel 1991). Among other sources for bringing atom supersaturations into a condensing gas

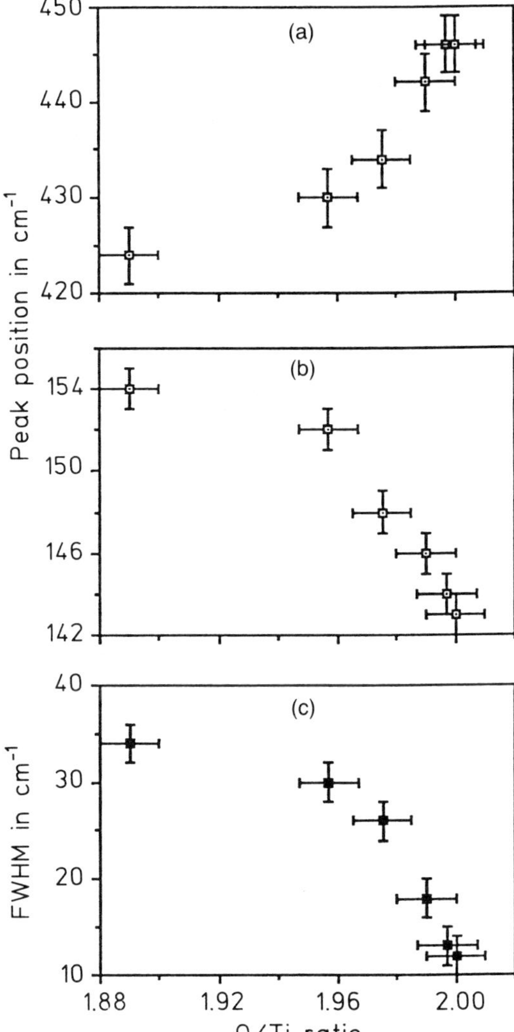

FIG. 7. Variation with O/Ti ratio of the peak position of (a) the rutile "447 cm^{-1}" vibrational mode and (b) the anatase "143 cm^{-1}" vibrational mode, as well as (c) this anatase mode's full width at half maximum (FWHM) (Parker and Siegel, 1990b).

medium that have been successfully used to produce clusters or ultrafine particles are Joule-heated ovens, sputtering, electron-beam heating, laser ablation, and plasma methods. This wide variety of evaporation methods will allow for greatly increased flexibility in the use of refractory or reactive precursors for clusters, and will be especially useful as one moves toward synthesizing technological quantities of more complex multicomponent or composite nanophase materials.

2.2 Mechanical Attrition

Before turning to the structure and properties of nanophase materials, it might be useful to mention the mechanical attrition method (Hellstern *et al.*, 1989; Koch *et al.*, 1989; Fecht *et al.*, 1989; Koch, 1993). Mechanical attrition produces its nanostructures, not by cluster assembly, but by means of the structural decomposition of coarser-grained structures induced by severe mechanical deformation. Nanometer-size grains nucleate within the localized shear bands of heavily deformed materials, converting a coarse-grained structure to nanophase. The heavy deformation is usually induced by means of high-energy ball milling, but can result as well from surface wear phenomena (Ganapathi and Rigney, 1990) and other methods for introducing high densities of deformation, such as the high-energy shear process (Valiev, 1993). A series of transmission electron micrographs and associated electron-diffraction patterns showing this grain refinement in AlRu as a function of ball-milling time is presented in Fig. 8. The individual grains resulting from mechanical attrition are never isolated clusters, and much synthesis and processing flexibility is lost. Nevertheless, ultrafine grain sizes can be readily achieved by this rather straightforward method, albeit with probable contamination from the sources of mechanical work. However, the method is capable of producing commercial quantities of material, and for those applications in which careful control of purity or grain morphology are not important, and for which the desired properties are indeed accessible in the mechanically attrited nanostructures, it can be a useful source of these materials.

The variation with grain size as a function of ball-milling time in Pd (Eckert *et al.*, 1992) is shown in Fig. 9(a). This type of saturation behavior is found to be typical for ball-milled metals and clearly shows that grain sizes down into the nanometer regime can be readily achieved, although it is more easily accomplished in relatively harder materials, as indicated by the relationship between minimum grain size and bulk modulus shown in Fig. 9(b). Mechanical attrition exercised at low temperatures (cryomilling) can effectively extend the range of applicability, as

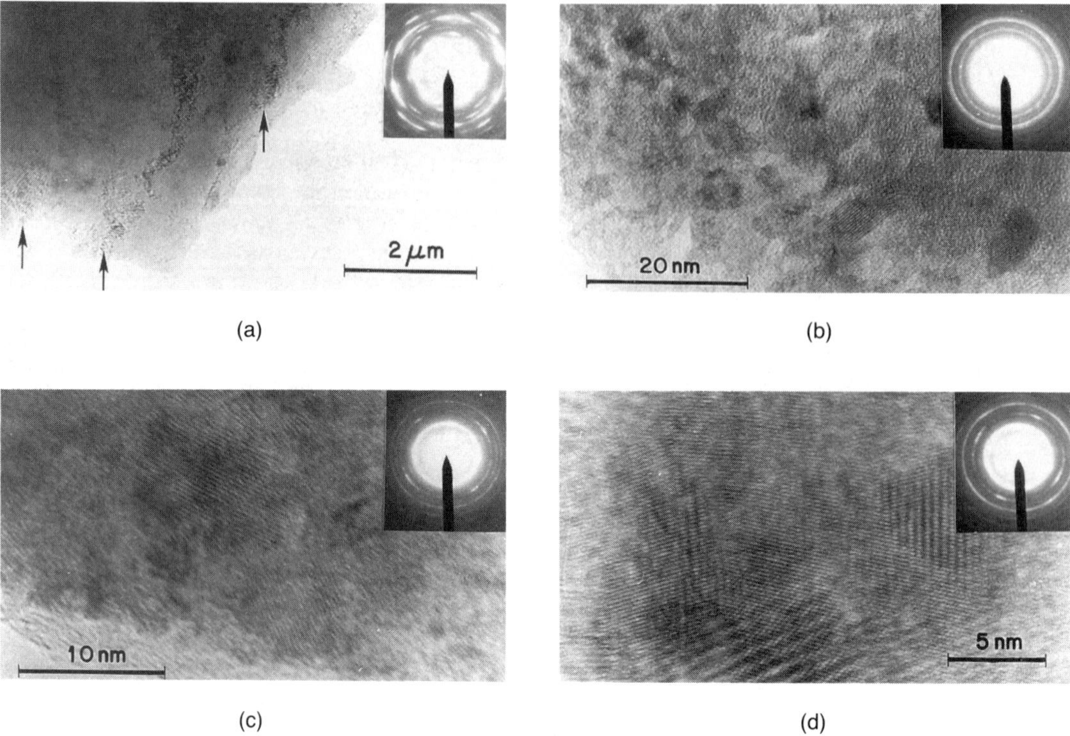

FIG. 8. Variation of the microstructure of AlRu with time of high-energy ball milling as seen by TEM and electron diffraction patterns shown in the insets (Fecht *et al.*, 1989): (a) After 10 min, arrows indicating highly deformed shear bands; high resolution images and diffraction patterns after (b) 10 min, (c) 2 h, and (d) 64 h.

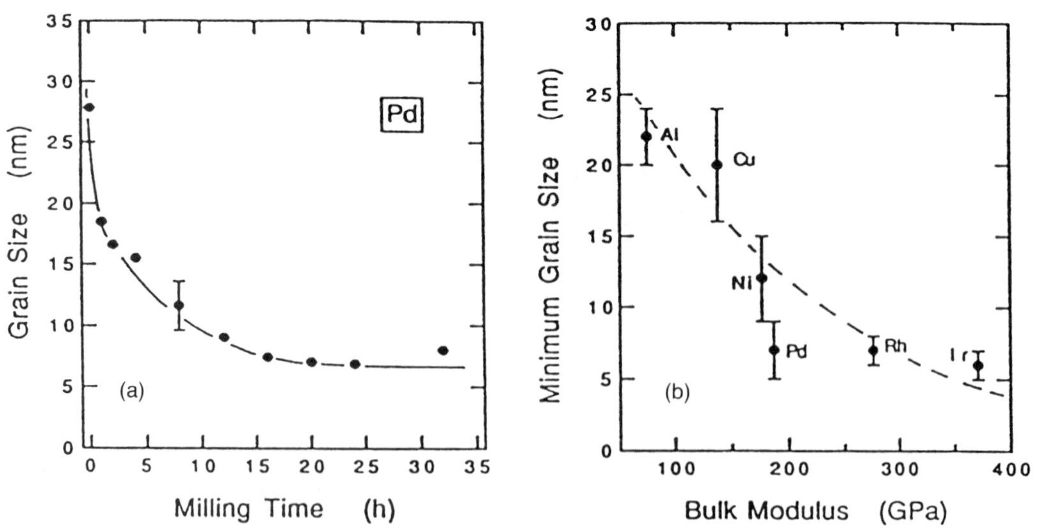

FIG. 9. (a) Average grain size, determined by x-ray line broadening, of Pd as a function of ball-milling time. (b) Minimum average grain size attained by ball milling of various fcc metals plotted against their bulk modulus values (Eckert *et al.*, 1992).

shown by Luton *et al.* (1989) in their work on dispersion-strengthened Al.

3. STRUCTURE

The structures of nanophase materials on a variety of length scales have an important bearing on their properties. They are dominated by ultrafine grain sizes and the large number of associated interfaces. In addition, however, other structural features, such as pores (and larger flaws), grain-boundary junctions, and other crystal-lattice defects that can depend upon the manner in which these materials are synthesized and processed, play a significant role. It has become increasingly clear that all of these structural aspects of nanophase materials must be considered in trying to understand fully the properties of these new materials.

3.1 Grains and Pores

Our present knowledge of the grain structures of nanophase materials, whether formed by cluster consolidation or by intense mechanical deformation, has resulted primarily from direct observations using transmission electron microscopy (TEM) (Siegel *et al.*, 1988; Thomas *et al.*, 1989, 1990; Wunderlich *et al.*, 1990). A typical high-resolution image of a nanophase Pd sample formed by the consolidation of gas-condensed clusters is shown in Fig. 10. TEM has shown that the grains in such nanophase compacts are essentially equiaxed, similar to the atom clusters from which they were formed, although departures from spherical structures are expected simply from the efficient packing of the clusters during consolidation. The grains also appear to retain the narrow (ca. ±25% FWHM) log-normal size distributions typical of the clusters formed in the gas-condensation method (see Fig. 5), since measurements of these distributions before or after cluster consolidation by dark-field electron microscopy yield similar results. Grain-size distributions in deformation-produced nanostructures can be somewhat broader.

The observations (Nieman *et al.*, 1991) that the densities of nanophase materials consol-

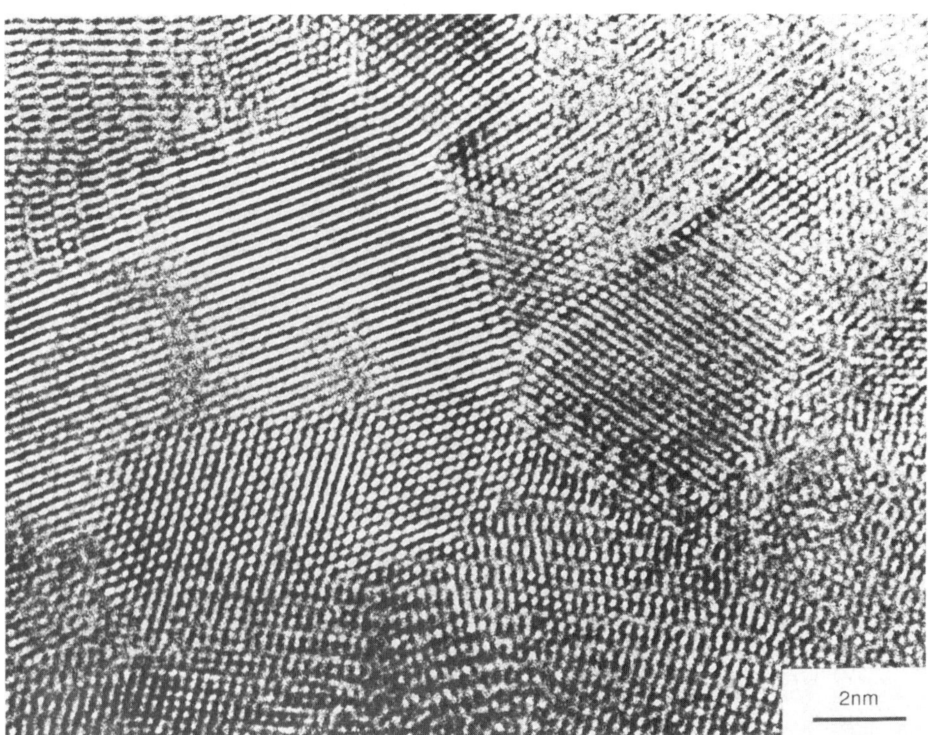

FIG. 10. High-resolution transmission electron micrograph of a typical area in nanophase palladium (Thomas *et al.*, 1990).

idated from equiaxed clusters extend well beyond the theoretical limit (78%) for the densest packing of undeformed identical spheres indicate that an extrusion-like deformation of the clusters must result during the consolidation process, filling (at least partially) the pores among the grains. Such observations indicate that cluster extrusion in forming nanophase grains may result from a combination of deformation and diffusion processes. These processes are also evident from scanning tunneling and atomic force microscopy observations (Sattler *et al.*, 1994) on nanophase Pd and Ag and from computer simulations of the consolidation of Pd clusters (Liu *et al.*, 1994).

Observations by both electron and x-ray scattering indicate, however, that no apparent preferred crystallographic orientations or texture of the grains results from their uniaxial consolidation and that the grains in the nanophase compact are essentially randomly oriented with respect to one another. This is interesting with respect to the fact that such deformation and annealing textures commonly observed in conventional grain-size materials are the result of dislocation glide processes, which seem to be suppressed in nanophase materials, as will be discussed in the following. Indeed, only very few dislocations are observed within the grains in these these ultrafine-grained materials, and these are normally in immobile or locked (i.e., sessile) configurations.

All of the nanophase materials consolidated to date from clusters at room temperature have invariably possessed some porosity. This porosity has ranged from about 15–25% for ceramics to less than 5% for metals. Significant porosity can also result from the deformation or crystallization synthesis of nanophase materials. Evidence for porosity has been obtained by positron annihilation spectroscopy (PAS) (Schaefer *et al.*, 1987, 1988; Siegel *et al.*, 1988), precise densitometry (Nieman *et al.*, 1991), and porosimetry (Hahn *et al.*, 1990a; Wagner *et al.*, 1991) measurements, and small-angle neutron scattering (SANS) (Wagner *et al.*, 1991; Sanders *et al.*, 1993).

PAS is primarily sensitive to small pores, ranging from single vacant lattice sites to larger voids, but PAS can probe these structures enclosed in the bulk of the material, even as a function of sintering as shown in Fig. 11 and discussed in Sec. 4.1. Porosimetry measurements using the BET (Brunauer–Emmett–Teller) N_2 adsorption method (Gregg and Sing, 1982), on the other hand, probe only pore structures open to the free surface of the sample, but BET can yield pore-size distributions (although with some questions regarding their validity at nanometer pore sizes), which are unavailable from PAS. Such a set of pore-size distributions in TiO_2 is shown in Fig. 12. Densitometry using an Archimedes method integrates over all structures in the sample, including grains, pores (open or closed), and density decrements at defect sites as well. SANS is quite sensitive to pores in the 1–100-nm size range, but deconvolution of the scattering data can be difficult when a broad spectrum of scattering centers is present, as is often the case in nanophase materials. However, even in such a case, it has been possible (Sanders *et al.*, 1993) to analyze SANS data from cluster-consolidated Pd in terms of a population of small (ca. 1-nm–diameter) voids presumed to be located at grain-boundary intersections (triple junctions) and grain-sized voids; however, the possible scattering from grain boundaries or larger flaws was not taken into account in the analysis.

A variety of such measurements to date have shown that the porosity in nanophase metals and ceramics is primarily smaller than or equal to the grain size of the material (although some larger porous flaws have been observed by optical and scanning electron microscopy). The porosity is frequently associated with the grain-boundary junctions (triple junctions) and especially, but not only, in ceramics, it is interconnected and intersects the specimen surfaces. Fortunately, consolidation at elevated temperatures should be able to remove this porosity uniformly without sacrificing the ultrafine grain sizes in these materials. Some evidence for this has already been obtained by means of the sinter forging of nanophase ceramics (Owen and Chokshi, 1993).

3.2 Dislocations

Dislocations are most often the type of lattice defect responsible for the permanent or plastic deformation in metals and nonmetal crystalline materials. As a dislocation moves through the crystal lattice under the influ-

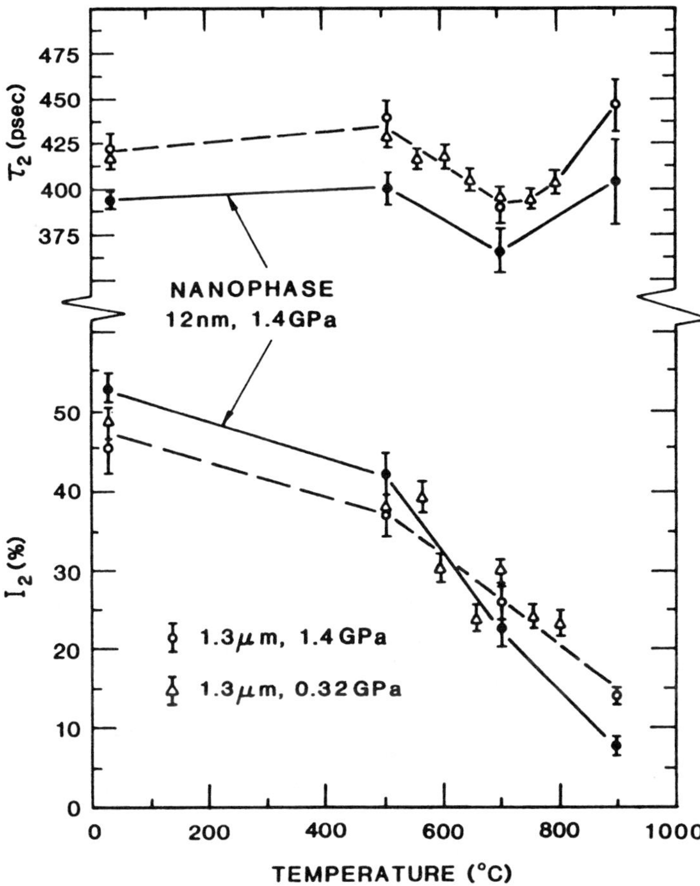

FIG. 11. Results of two-component lifetime fits to positron annihilation data from three TiO_2 samples as functions of sintering temperature; τ_2 corresponds to positrons trapped at voids (pores) and I_2 is the relative intensity of this component. A 12-nm grain-size nanophase sample (filled circles) compacted at 1.4 GPa is compared to 1.3-μm grain-size samples compacted at 1.4 GPa (open circles) and 0.32 GPa (triangles) from commercial powder. The PAS data were taken at room temperature; no sintering aids were used (Siegel *et al.*, 1988).

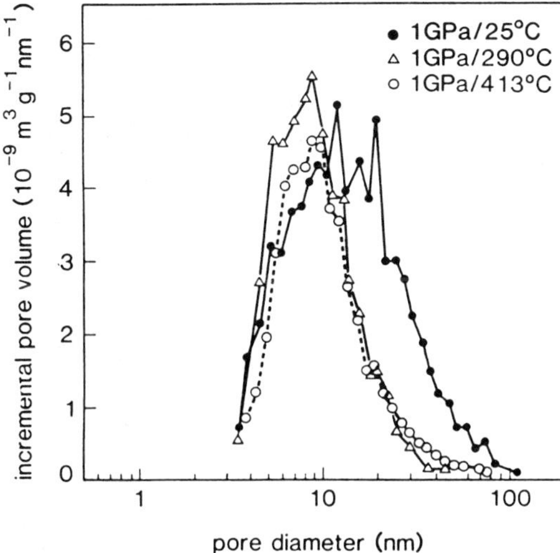

FIG. 12. Pore-size distributions in nanophase TiO_2 consolidated under 1 GPa pressure at room temperature, 290 °C, and 413 °C, as measured by the BET method (Wagner *et al.*, 1991).

ence of an externally applied stress, its motion imparts a permanent microscopic displacement that taken together with the motion of other dislocations can lead to a macroscopic shape change in the material. The presently available experimental evidence suggests that dislocations are seldom present in nanophase materials (Gao and Gleiter, 1987; Thomas *et al.*, 1990; Milligan *et al.*, 1993). When they are observed, it is primarily either in materials at the upper end of the grain size range (Morris and Morris, 1991) or in limited instances (Wunderlich *et al.*, 1990; Jain and Christman, 1994) in immobile or locked configurations. The reason for this substantial lack of dislocations is that image forces exist in finite atomic ensembles that tend to pull mobile dislocations out of the grains, especially when they are small, in analogy with the forces on a point electrical charge near a free surface of a conducting body. Since mobile dislocations are not initially available in sufficient numbers to effect plasticity in nanophase materials, new mobile dislocations must be created or other deformation mechanisms, such as grain-boundary sliding, must come into play.

Dislocations can be created or can multiply from a variety of sources. A simple but representative example is the Frank–Read dislocation source, shown in Fig. 13, in which a dislocation line, pinned between two pinning points that prevent its forward motion on a slip plane, can bow out between these pinning points to form a new dislocation, if the stress acting on the pinned dislocation is sufficient. The critical stress to operate such a Frank–Read source is inversely proportional to the distance between the pinning points and hence will also be limited by the grain size, which limits the maximum distance between such pinning points. This suggests that dislocation multiplication in nanophase metals will become increasingly difficult as the grain size decreases (see Sec. 4.2). The critical stress will eventually, at sufficiently small grain sizes, become larger than the yield stress (at which plastic deformation begins) in the conventional material and could even approach the theoretical yield strength of a perfect, dislocation-free single crystal.

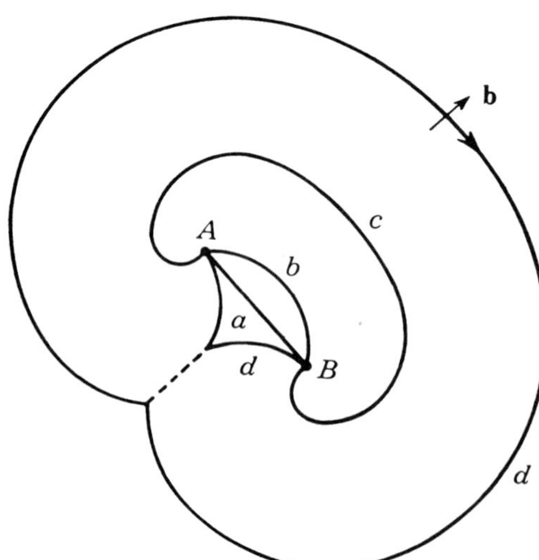

FIG. 13. Schematic representation of a Frank–Read dislocation source. A segment of dislocation loop **a** of length L and Burgers vector **b** is pinned between points A and B in a material of shear modulus G. When the stress applied to the dislocation line segment exceeds a critical stress, $\sigma_{crit} = kGb/L$, the dislocation bows out ($a \rightarrow b \rightarrow c \rightarrow d$) until it creates a new dislocation loop. The constant of proportionality, k, is 0.5 for an edge dislocation and 1.5 for a screw dislocation.

3.3 Grain Boundaries

As a result of their ultrafine grain sizes, nanophase materials have a significant fraction of their atoms in grain-boundary environments, where they occupy positions relaxed from their normal lattice sites. For high-angle grain boundaries (interfaces between grains with more than a few degrees of orientational mismatch) in conventional materials, these relaxations extend over about two atom planes on either side of the boundary, with the greatest relaxation existing in the first plane (Wolf and Lutsko, 1988; Wolf and Yip, 1992). Thus, for an average grain diameter range between 5 and 10 nm, where much of the research on nanophase materials has focused, grain-boundary atom percentages range between about 5 and 50% (see Fig. 14). Since such a large fraction of their atoms reside in grain boundaries, these interface structures can play a significant role in affecting the properties of nanophase materials.

Several early investigations of nanocrystalline metals, including x-ray diffraction (Zhu *et al.*, 1987), Mössbauer spectroscopy (Herr *et al.*, 1987), positron lifetime studies (Schaefer

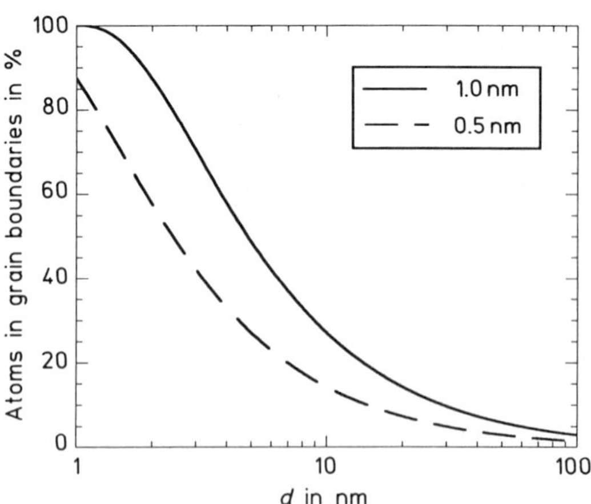

FIG. 14. Percentage of atoms in grain boundaries (including grain-boundary junctions) of a nanophase material as a function of grain diameter, assuming that the average grain-boundary thickness ranges from 0.5 to 1.0 nm (ca. 2 to 4 atomic planes wide) (Siegel, 1991).

et al., 1987, 1988), and extended x-ray absorption fine-structure (EXAFS) measurements (Haubold *et al.*, 1988, 1989), were interpreted in terms of grain-boundary atomic structures that may be random, rather than possessing either the short-range or long-range order normally found in the grain boundaries of conventional coarser-grained polycrystalline materials. This randomness was variously associated (Birringer and Gleiter, 1988) with either the local structure of individual boundaries (as seen by a local probe such as EXAFS or Mössbauer spectroscopy) or the structural coordination among boundaries (as might be seen by x-ray diffraction). However, direct observations by high-resolution electron microscopy (HREM) have indicated that their structures are rather similar to those of conventional high-angle grain boundaries. An extensive review of these results has appeared elsewhere (Siegel, 1992).

The direct imaging of grain boundaries with HREM can avoid the complications that may arise from porosity and other defects in the interpretation of data from less direct methods, such as x-ray scattering and Mössbauer spectroscopy. A HREM study (Thomas *et al.*, 1989, 1990) that included both experimental observations and complementary image simulations indicated no manifestations of grain-boundary structures with random displacements of the type or extent suggested by earlier x-ray studies on nanophase Fe, Pd, and Cu (Zhu *et al.*, 1987; Haubold *et al.*, 1988, 1989). Contrast features at the observed grain boundaries that might be associated with disorder did not appear wider than 0.4 nm, indicating that any significant structural disorder that may be present essentially extends no further than the planes immediately adjacent to the boundary plane. Such localized lattice relaxation features are typical of the conventional high-angle grain-boundary structures found in coarse-grained metals. HREM investigations of grain boundaries in nanophase Cu (Ganapathi and Rigney, 1990) and Fe alloys (Trudeau *et al.*, 1991) produced by surface wear and high-energy ball milling, respectively, appear to support this view. Recent nuclear magnetic resonance (NMR) studies of cluster-consolidated nanophase Ag (Suits *et al.*, 1993) have indicated that the grain boundaries in this material have an electronic structure consistent with that for conventional high-angle boundaries. However, there are indications (Tschöpe and Birringer, 1993; Valiev *et al.*, 1991) that metastable grain-boundary configurations do exist in some cases and that these configurations can be transformed by low-temperature annealing to more stable states. Whether this behavior is associated with the intrinsic grain-boundary structure itself or with extrinsic grain-boundary dislocation configurations and/or strains remaining from synthesis or processing remains to be clarified.

Indeed, as shown in Fig. 10 and in similar electron micrographs, the nanophase grain boundaries appear to be essentially low-energy configurations exhibiting flat facets interspersed with steps. Such structures could arise only if sufficient local atomic motion occurred during the cluster consolidation process to allow the system to reach at least a local energy minimum. These observations suggest at least two conclusions (Siegel and Thomas, 1992): first, that the atoms that constitute the grain-boundary volume in cluster-assembled nanophase materials have sufficient mobility during cluster consolidation to accommodate themselves into relatively low-energy grain-boundary configurations; and second, that the local driving forces for grain growth are relatively small, despite the large amount of energy stored in the many grain boundaries in these materials.

The foregoing discussion suggests that nanophase materials should be a valuable resource for studying the average properties of grain boundaries. The high number density of such defects in these materials enhances their influence on macroscopic properties, allowing these effects to be studied by a variety of experimental techniques. Indeed, a number of the effects observed in nanophase materials to date that have been deemed unusual may simply result from so many grain boundaries being available for study in a sample for the first time. For the careful study of grain-boundary properties to be successful in the future, however, specimen porosity will need to be removed, via consolidation at elevated temperature and/or pressure, so that its property contributions can be eliminated. By varying the grain size in a series of samples, the effects from interfaces and junctions in nanophase materials could be effectively separated in such future studies.

3.4 Stability and Strains

The experimental observations of narrow grain-size distributions, essentially equiaxed grain morphologies, and low-energy grain-boundary structures in nanophase materials suggest that the inherent resistance to grain growth observed for cluster-assembled nanophase materials (see Fig. 15) results primarily from a sort of frustration (Siegel, 1990). It appears that the narrow grain-size distributions normally observed in these nanophase materials coupled with their relatively flat grain-boundary configurations place these ultrafine-grained structures in a local minimum in energy from which they are not easily extricated. Such frustration would also likely be increased by the multiplicity of grain-boundary junctions in these materials. There are normally no really large grains to grow at the expense of small ones through an Ostwald ripening process, and the grain boundaries, being essentially flat, have no local curvature to tell them in which direction to migrate. Their stability thus appears to be analogous to that of a variety of closed-cell foam structures with narrow cell-size distributions, which are deeply metastable despite their large stored surface energy. Such a picture now appears to have some theoretical support (Rivier, 1992).

It should be pointed out that exceptions to this frustrated grain growth behavior would be expected if considerably broader grain-size distributions were accidentally present in a sample, which would allow a few larger grains to grow at the expense of smaller ones, or if significant grain-boundary contamination were present, allowing enhanced stabilization of the small grain sizes to further elevated temperatures. Occasional observations of each of these types of behavior have been made. One could, of course, intentionally stabilize against grain growth by appropriate doping by insoluble elements or composite formation in the grain boundaries. For cluster-assembled materials, such stabilization should be especially easy, since the grain boundaries are available as cluster surfaces prior to consolidation. The ability to retain the ultrafine grain sizes of nanophase materials is important when one considers the fact that it is their grain size and large number of grain boundaries that determine to a large extent their special properties.

Strains are a matter of fact in nanophase materials. Simply as a result of the large number of grain boundaries, and the concomitant short distances between them, the intrinsic strains associated with such interfaces (Wolf and Lutsko, 1988; Cammarata and Sieradski, 1989) are always present in these materials. Beyond these intrinsic strains, there may also be present extrinsic strains associated with the particular synthesis method. For example, intense plastic deformation synthesis of nanophase materials may lead to additional residual strains (Valiev, 1993) that can be subsequently relieved by low-temper-

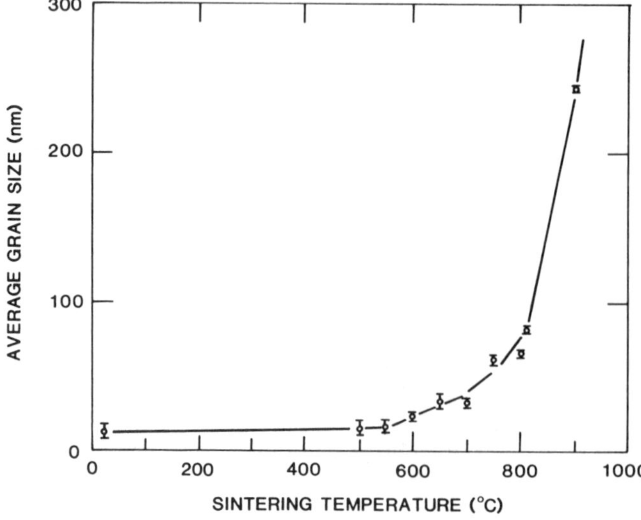

FIG. 15. Variation of average grain size with sintering temperature (0.5 h at each) for a nanophase TiO_2 (rutile) sample compacted to 1.4 GPa at room temperature, as determined by dark-field transmission electron microscopy (Siegel *et al.*, 1988).

ature annealing, leaving only those intrinsic strains due to the presence of the high-angle grain boundaries. Evidence for the strains in nanophase materials is now becoming available. X-ray line-broadening investigations of both cluster-consolidated (Nieman *et al.*, 1991) and ball-mill-attrited metallic nanophase samples (Eckert *et al.*, 1992) have indicated residual strains of about (0.1–1)%, which are consistent in magnitude with the strains expected from conventional high-angle grain boundaries (Wolf and Yip, 1992) in the grain size ranges investigated. Also, recent NMR measurements (Suits *et al.*, 1994) on cluster-consolidated nanophase Cu have yielded strain values in this range.

4. PROPERTIES

The unique properties of nanophase materials result from an interplay among their three fundamental features: constituent atomic domains (e.g., grains) spatially confined to less than 100 nm, significant atom fractions associated with interfacial environments (e.g., grain boundaries or free surfaces), and interactions between their domains. In some cases one of these features dominates, in other cases another. Research on a variety of chemical, mechanical, and other physical properties is beginning to yield a glimmer of understanding of just how this interplay manifests itself in the properties of these new materials.

4.1 Chemical Properties

Nanophase materials exhibit properties that are different and often considerably improved in comparison with those of conventional coarse-grained structures. Because of their small sizes and radii of curvature, coupled with their surface cleanliness, the constituent clusters of nanophase materials can react with one another rather aggressively, even at relatively low temperatures. For example, nanophase TiO_2 (rutile) exhibits significant improvements in both sinterability and resulting mechanical properties relative to conventionally synthesized coarser-grained rutile (Siegel *et al.*, 1988; Averback *et al.*, 1989; Hahn *et al.*, 1990a; Mayo *et al.*, 1990). Nanophase TiO_2 with a 12-nm initial mean grain diameter has been shown to sinter under ambient pressures at temperatures 400 °C to 600 °C lower than the conventional sintering temperature for coarse-grained rutile, without the need for any normally required compacting or sintering aid such as polyvinyl alcohol, as shown in Fig. 16. Furthermore, it has been demonstrated (Hahn *et al.*, 1990a) that sintering the same nanophase material under pressure (1 GPa) or with appropriate dopants such as Y can further reduce the sintering temperatures, while suppressing grain

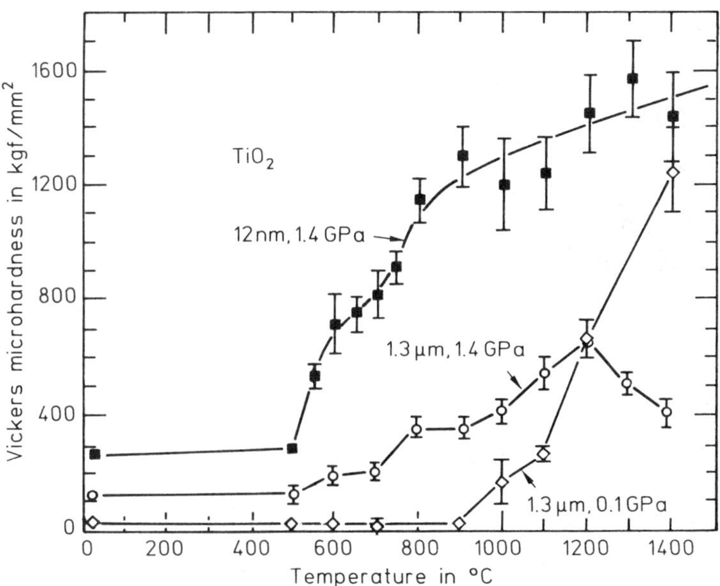

FIG. 16. Vickers microhardness of TiO_2 (rutile) measured at room temperature as a function of 0.5-h sintering at successively increased temperatures. Results for a nanophase sample (filled squares) with an initial average grain size of 12 nm consolidated at 1.4 GPa are compared with those for coarser-grained samples with 1.3-μm initial average grain size sintered with (diamonds) or without (circles) the aid of polyvinyl alcohol from commercial powder consolidated at 0.1 GPa and 1.4 GPa, respectively (Siegel *et al.*, 1988).

growth as well, thus allowing for the unique possibility of sintering nanophase ceramics to full density while retaining their ultrafine grain size. The resulting fracture characteristics (Li *et al.*, 1988b; Averback *et al.*, 1989; Mayo *et al.*, 1990; Höfler and Averback, 1990) developed for sintered nanophase TiO_2 are as good, or in some aspects improved, relative to those for conventional coarser-grained rutile.

Sintering behavior has also been followed by PAS and SANS measurements. It was already discussed in Sec. 3.1 how PAS can be a useful tool in the study of the ultrafine-scale porosity inherent in as-consolidated nanophase compacts; it can probe such porosity as a function of sintering temperature as well, to observe densification via the removal of voids. An example of PAS lifetime results (Siegel *et al.*, 1988) used to follow the sintering behavior of nanophase TiO_2 was already shown in Fig. 11. The intensity I_2 of the lifetime (τ_2) signal corresponding to positron annihilation from void-trapped states in the nanophase sample is seen to decrease rapidly during sintering above 500 °C as a result of the densification of this ultrafine-grained ceramic, even though rapid grain growth does not set in until above 800 °C (see Fig. 15). Furthermore, the variation of τ_2 with sintering indicates that there is a redistribution of void sizes accompanying this densification. Similar behavior is also observed for the coarser-grained samples investigated, but, as expected, the densification proceeds more slowly in these latter samples and the average pore sizes are larger according to the larger values of τ_2. The redistribution of pore sizes can also be monitored by means of BET measurements (if the pores are still open to the sample surfaces so that N_2 can enter) as demonstrated by Hahn *et al.* (1990a) and Wagner *et al.* (1991). Nanophase TiO_2 in its as-consolidated state and as a function of sintering in air was also followed by SANS (Epperson *et al.*, 1989). While SANS contrast can yield information regarding the nature of the intergrain nanophase boundaries, particularly their average density *vis-à-vis* the grain density (Jorra *et al.*, 1989; Epperson *et al.*, 1989, 1990), it can also yield information about the presence of voids (or porosity) and their removal during the sintering process. However, the clear separation of void and grain-boundary contributions to small-angle neutron scattering is still an open question, as discussed in Sec. 3.1, even though an attempt at using pressure-assisted sintering to remove the voids has recently been made (Wagner *et al.*, 1991).

Atomic diffusion in nanophase materials, which can have a significant bearing on their mechanical properties, such as creep and superplasticity, and other physical properties as well, has been found to be very rapid. Measurements of self-diffusion and impurity diffusion (Horváth *et al.*, 1987; Mütschele and Kirchheim, 1987; Horváth, 1989; Hahn *et al.*, 1989; Schumacher *et al.*, 1989; Averback *et al.*, 1989) in as-consolidated nanophase metals (Cu,Pd) and ceramics (TiO_2) indicate that atomic transport can be orders of magnitude faster in these materials than in coarser-grained polycrystalline samples, exhibiting surface-diffusion-like behavior. However, the very rapid diffusion in as-consolidated nanophase materials appears to be intrinsically coupled with the porous nature of the interfaces in these materials. It has been shown in at least one case (Hf in TiO_2) that the rapid surfacelike diffusivities can be suppressed back to conventional values by sintering samples to full density (Averback *et al.*, 1989). Nonetheless, there exist considerable possibilities for efficiently doping nanophase materials at relatively low temperatures by means of the rapid diffusion available along their ubiquitous grain-boundary networks and interconnected porosity, with only short diffusion paths remaining into their grain interiors, to synthesize materials with tailored chemical, mechanical, or other physical properties.

The chemical reactivity of nanophase materials, with their potentially high surface areas compared to conventional materials, can also be rather striking. Since clusters can be assembled by means of a variety of methods, for example, there can be an excellent degree of control over the total available surface area in the resulting self-supported ensembles. Thus, one can maximize porosity for obtaining very high surface areas, remove most of it via consolidation, but retain some to facilitate low-temperature doping or other processing, or fully densify the nanophase material. Also, composition control can be readily achieved, since rapid diffusion paths are plentiful and diffusion distances are short in the clusters. Measurements (Beck and Siegel,

1992) of the decomposition of H_2S over lightly consolidated nanophase TiO_2 at 500 °C demonstrate the enhanced chemical reactivity of nanophase materials rather well. Figure 17 shows the reaction rates for S removal from H_2S via decomposition for nanophase TiO_2 (rutile) compared with that for a number of other commercially available forms of TiO_2 having either the rutile or anatase structure. It can be seen that the nanophase sample was far more reactive than any of the other samples tested, both initially and after extended exposure to the H_2S. Indeed, only the nanophase titania demonstrated a nonzero reaction rate for S removal after 7 h. This greatly enhanced and sustainable activity was shown to result from a combination of unique features of the nanophase material, its high specific surface area combined with its rutile structure and its oxygen-deficient composition (see Sec. 2.1).

Another important aspect of materials with reduced spatial dimensions in the nanometer size regime is phase stability. As one reduces the sizes of constituent phases sufficiently, into the nanoscale (mesoscopic) regime where condensed matter behaves differently than either atomic or molecular species or bulk solids, equilibrium phase relations will change because of the expected changes in a variety of electronic and thermodynamic parameters of the confined atomic system. For example, the melting temperature of small clusters is suppressed well below that for the bulk solid (Buffat and Borel, 1976; Berry *et al.*, 1984; Han and Whetten, 1988; Goldstein *et al.*, 1992), the result of a higher effective pressure in the confined system, the Gibbs–Thomson effect. Related effects are also seen in cluster-consolidated materials in which high-pressure polymorphic structures have been stabilized (Li *et al.*, 1988a; Skandan *et al.*, 1992). Thus, the effects of reduced spatial dimensions on phase stability are important, since one cannot expect that the phase equilibria known for conventional materials systems will apply in general to those of significantly reduced dimensions.

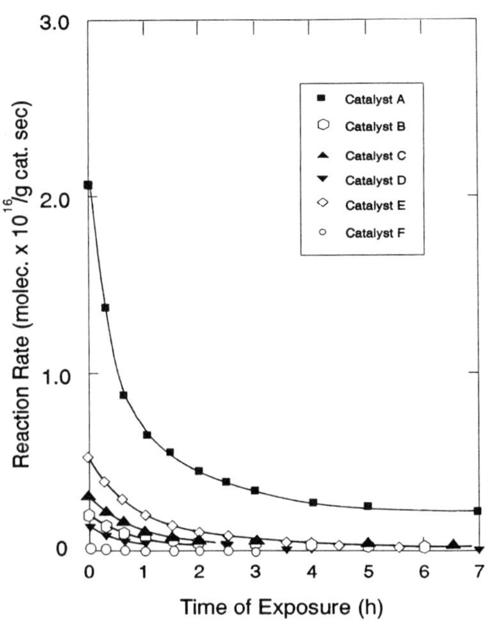

FIG. 17. Reaction rates for H_2S decomposition as a function of exposure time at 500 °C for nanophase TiO_2 compared with that from several commercially available TiO_2 materials and a reference (catalyst A, 76 m^2/g nanophase rutile; B, 61 m^2/g anatase; C, 2.4 m^2/g rutile; D, 30 m^2/g anatase; E, 20 m^2/g rutile; F, reference alumina) (Beck and Siegel, 1992).

4.2 Mechanical Properties

The mechanical properties of nanophase materials are generally rather different from those of their coarse-grained counterparts (Siegel and Fougere, 1994). This difference results primarily from spatial confinement and induced accommodation strains in nanophase metals and from small grain sizes and short diffusion distances in nanophase ceramics. Because of their ultrafine grain sizes, nanophase ceramics are easily formed, as has been clearly evident from the sample compaction process (Siegel and Hahn, 1987; Siegel *et al.*, 1988) and from early demonstrations via deformation (Karch *et al.*, 1987) as well. However, the degree to which nanophase ceramics are truly ductile is still only beginning to be understood.

Nanoindenter measurements on nanophase TiO_2 (Mayo *et al.*, 1990) and ZnO (Mayo *et al.*, 1992) have demonstrated that a dramatic increase of strain-rate sensitivity occurs with decreasing grain size, as shown in Fig. 18, that is remarkably similar for these two materials, even though they had as-consolidated densities of about 75% and 85%, respectively. Since this strong grain-size dependence was found for sets of samples in which the porosity was changing very little, it appears to be an intrinsic property of these

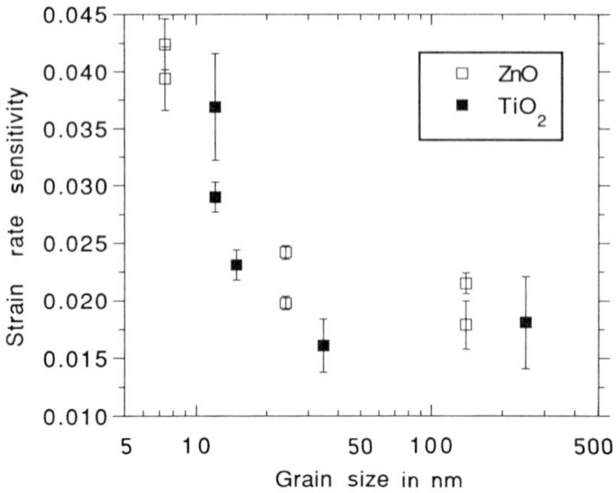

FIG. 18. Strain-rate sensitivity of nanophase TiO_2 (Mayo *et al.*, 1990) and ZnO (Mayo *et al.*, 1992) as a function of grain size. The strain-rate sensitivity was measured by a nanoindentation method, and the grain size was determined by dark-field TEM.

ultrafine-grained ceramics. The strain-rate sensitivity m is related to the stress σ applied to a sample and the resulting sample strain rate $d\epsilon/dt$ by the empirical expression $\sigma = K(d\epsilon/dt)^m$, where K is a constant. Values of m can vary from 0 to 1, the limits, respectively, for perfectly brittle and perfectly ductile behavior. The strain-rate sensitivity values at the smallest grain sizes yet investigated (12 nm in nanophase TiO_2 and 7 nm in ZnO) indicate not only ductile behavior of these nanophase ceramics at room temperature, but also a significant potential for increased ductility at even smaller grain sizes and at elevated temperatures. The maximum m values measured in these studies, about 0.04, are approximately one-quarter that for Pb at room temperature, for example. However, no superplasticity has yet been observed in nanophase materials at room temperature, which would be indicated by m values greater than about an order of magnitude higher than the maximum observed. Nevertheless, it already seems clear that in the future, at smaller grain sizes and/or at elevated temperatures, superplasticity of these materials will indeed be observed.

The possibilities for plastic forming of nanophase ceramics to near net shape appear to be well on their way to realization. Karch and Birringer (1990) have demonstrated that nanophase TiO_2 could be readily formed to a desired shape with excellent detail below 900 °C, and the fracture toughness was found to increase by a factor of 2 as well. The ability to deform fully dense nanophase TiO_2 extensively at elevated temperatures (ca. 800 °C) without cracking or fracture has also been demonstrated (Hahn *et al.*, 1990b; Guermazi *et al.*, 1991), as shown in Fig. 19. While these latter demonstrations have been accompanied by significant grain growth in the samples at the elevated temperatures employed, it can be expected that lower-temperature studies [below $(0.4–0.5)T_m$] in the future will also allow for near net-shape forming of nanophase ceramics with both their ultrafine grain sizes and their attendant properties retained.

The enhanced strain-rate sensitivity at room temperature found in the nanophase ceramics TiO_2 and ZnO (Mayo *et al.*, 1990, 1992) appears to result from increased grain-boundary sliding in this material, aided by the presence of porosity, ultrafine grain size, and probably rapid short-range diffusion as well. This behavior is therefore dominated by the presence of the numerous interfaces in these materials and the very short diffusion distances involved in effecting the necessary atomic healing of incipient cracks for grain-boundary sliding to progress at the strain rates utilized without fracturing the sample. Extrapolating from this apparently generic behavior, one can expect that grain-boundary sliding mechanisms, accompanied by short-range diffusion-assisted healing events, would be expected to dominate increasingly the deformation of a variety of nanophase materials. Enhanced forming and even superplasticity in a wide range of nanophase materials, including intermetallic compounds, ceramics, and semiconductors, might become a reality. Indeed, the ability to deform nanophase TiO_2 extensively at elevated temperatures near 800 °C without cracking or fracture, as indicated above, has already been clearly demonstrated, and similar results have been recently seen (Jain and Christman, 1994) in a mechanically attrited nanophase intermetallic compound (which is conventionally brittle), $Fe_{28}Al_2Cr$. Consequently, increased opportunities for the superplastic near-net-shape forming of many rather conventionally brittle and difficult-to-form materials could result.

The predominant mechanical property change resulting from reducing the grain sizes of nanophase metals, in contrast to the be-

FIG. 19. Nanophase TiO_2 sample before and after compression at 810 °C for 15 h (Hahn *et al.*, 1990b). The total true strains were as high as 0.6, which represents deformation to a final thickness of less than 2 mm from an initial length of 3.5 mm at about 0.5 of its melting temperature (1830 °C). The grain size increased from 40–50 nm to about 1 mm. The small rule divisions are millimeters.

havior found for nanophase ceramics, is the significant increase in their strength. While the microhardness of as-consolidated nanophase oxides is reduced relative to their fully dense counterparts (Fig. 16), as a result of significant porosity in addition to their ultrafine grain sizes, the case for nanophase metals is quite different. Figures 20 and 21 show microhardness and stress-strain results for nanophase Pd and Cu compared with similar results for their coarser-grained counterparts (Nieman *et al.*, 1989, 1990, 1991). In their as-consolidated state, nanophase Pd samples with 5–10-nm grain sizes have been observed to exhibit up to about a 500% increase in hardness over coarser-grained (ca. 100 μm) samples (Nieman *et al.*, 1989), with

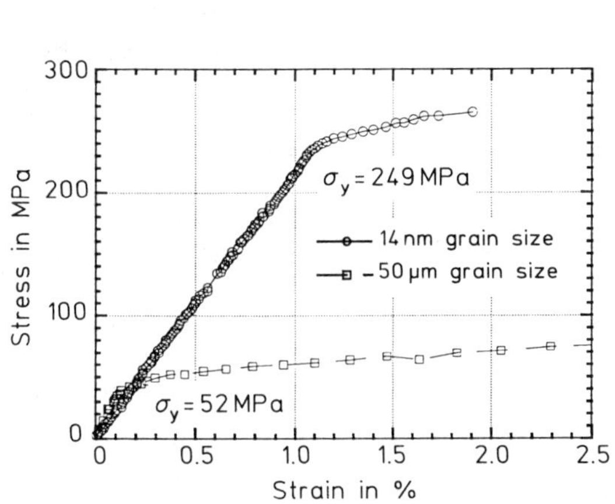

Stress in MPa
Strain in %
$\sigma_y = 249$ MPa
14 nm grain size
50 µm grain size
$\sigma_y = 52$ MPa

FIG. 20. Stress-strain curve for a nanophase (14 nm) Pd sample compared with that for a coarse-grained (50 μm) Pd sample (Nieman *et al.*, 1990). The strain rate $d\epsilon/dt = 2 \times 10^{-5}$ s^{-1}.

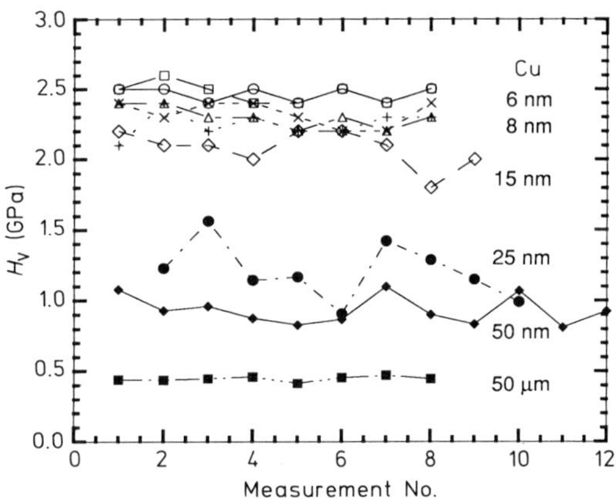

FIG. 21. Vickers microhardness (H_V) measurements at a number of positions across several nanophase Cu samples ranging in grain size from 6 to 50 nm, compared with similar measurements from an annealed conventional 50-μm grain-size Cu sample (Nieman *et al.*, 1991).

concomitant increases in yield stress σ_y, as shown in Fig. 20. Similar results have been observed in nanophase Cu as well, as shown in Fig. 21. The common strengthening behavior found in nanophase Pd and Cu indicates that this response is generic to nanophase metals, at least those with a fcc structure. The likelihood that such mechanical behavior is more broadly generic to nanophase metals is increased by the observations that nanophase metals and alloys produced via mechanical attrition also exhibit significantly enhanced strength. For example, Koch and coworkers (Jang and Koch, 1990; Koch and Cho, 1992) have found hardness increases of factors of 4 to 5 in nanophase Fe and a factor of about 1.2 in nanophase Nb_3Sn when the grain size drops from 100 to 6 nm.

In apparent contrast to these observations of enhanced strength with decreasing grain size, Chokshi *et al.* (1989) have reported a softening with decreasing grain size in the nanometer regime for cluster-assembled Cu and Pd samples, which was rationalized in terms of their expectation (Karch *et al.*, 1987; Birringer *et al.*, 1988) of room-temperature diffusional creep in these ultrafine-grained metals. A similar apparent softening was also reported (Chang *et al.*, 1991) for TiAl. The rapid atomic diffusion observed in nanophase materials (see Sec. 4.1), along with their nanometer grain sizes, has suggested that a large creep enhancement might result in nanophase materials, even at room-temperature. However, constant-stress creep measurements on nanophase Pd and Cu (Nieman *et al.*, 1990, 1991) show that the observed creep rates at room temperature are at least three orders of magnitude smaller than predicted on the basis of a Coble creep model, in which the creep rate varies as $D_b d^{-3}$, where D_b is the grain boundary diffusivity and d is the mean grain size. Such creep resistance will need to be explored further at elevated temperatures in these and other nanophase materials. However, it now appears that this apparent softening may only be a product of the manner in which the grain sizes were increased in these samples by annealing (Fougere *et al.*, 1992).

The increased strength observed in ultrafine-grained nanophase metals, although apparently analogous to conventional Hall–Petch strengthening (Hall, 1951; Petch, 1953; Siegel and Fougere, 1994) observed with decreasing grain size in coarser-grained metals, must result from fundamentally different mechanisms. The grain sizes in the nanophase metals considered here are smaller than the necessary critical bowing lengths for Frank–Read dislocation sources to operate at the stresses involved and smaller also than the normal spacings between dislocations in a pileup. It is therefore clear that an adequate description of the mechanisms responsible for the increased strength observed in nanophase metals will clearly need to accommodate to the ultrafine grain-size scale in these materials. As this scale is reduced, and conventional dislocation generation and migration become increasingly difficult, it is apparent that the energetic hierarchy of microscopic deformation mechanisms will become successively reached. Thus, easier mechanisms (such as dislocation generation from Frank–Read sources) will become frozen out at sufficiently small grain sizes and more energetically costly mechanisms will become necessary to effect deformation. Hence, grain confinement appears to be the dominant cause for the increased strength of nanophase metals. It can be generally expected that as nanophase grain sizes decrease and fall below the critical length scale for a given mechanism to operate, the associated property will be significantly changed. Clearly, much work in this area remains.

Nanocomposites consisting of metallic phases, ceramic and metallic phases (cermets), and ceramic phases in a variety of modulation dimensionalities also have considerably enhanced mechanical properties, including increased strength and fracture toughness. Prominent examples of these are ceramic nanocomposites in which nanoscale phases are combined with coarser-grained structures by means of mechanical milling (Niihara, 1991; Niihara *et al.*, 1993), and WC/Co nanophase cermets formed by a high-temperature thermochemical spray conversion method (Kear and McCandlish, 1993). This dramatically improved mechanical behavior, which should find use in a wide variety of wear and high-strength applications, results from a combination of the attributes of these nanocomposites including defect confinement, local and long-range strains, and increased complexity of fracture propagation.

4.3 Other Physical Properties

Rather limited research has been carried out so far on the other physical properties of cluster-assembled nanophase materials. However, there appear to be interesting prospects, based upon what little has been accomplished and on the expectations for confined systems of atoms in which the sizes of constituent domains fall below the critical length scales pertinent to a given property. A few examples of physical property changes that can occur in nanophase samples are presented here. Many more can be expected in the near future, as research begins to focus on these aspects of nanophase materials.

In magnetic multilayers, such as those formed by alternating layers of ferromagnetic Fe and Cr, the material can be nanostructured so that its electrical resistance is significantly decreased (by up to a factor of 2 depending upon the Cr-layer thickness) by the application of a magnetic field of 2 T (Baibach *et al.*, 1988). Such an effect, called giant magnetoresistance, occurs when the magnetic moments of the neighboring alternating layers (Fe/Cr) are arranged in an antiparallel fashion, so that application of the magnetic field overcomes the antiferromagnetic coupling and aligns the layers into a condition of parallel ferromagnetic ordering, strongly reducing the electron scattering in the system. It is important to control the nature of the interfaces between the layers, including both their structure and chemistry, in order to maximize this effect. Magnetoresistive materials are already being used in the magnetic recording industry as read heads, by virtue of their lower noise and improved signal-handling capabilities. It is now expected that nanophase materials will have a significant future impact in this area as new materials are developed with stable giant magnetoresistance at room temperature that can operate at low magnetic fields around 10^{-3} T. Recent discoveries that nanocomposites of magnetic cobalt clusters embedded in a nonmagnetic matrix of copper or silver (Berkowitz *et al.*, 1992; Xiao *et al.*, 1992) or magnetic platelike NiFe deposits embedded in silver (Hylton *et al.*, 1993) also exhibit giant magnetoresistance should hasten the advent of their useful application. The three-dimensional modulation and greater degrees of freedom in creating nanophase composite materials should allow for the engineering of these magnetic devices. Magnetic refrigeration applications of nanophase materials are also being explored (Shull *et al.*, 1993) with the anticipation that materials with usefully high efficiencies can be developed by nanostructuring.

The possibilities for other types of devices, such as chemical and physical sensors, resulting from nanophase materials are also apparent. Nanophase TiO_2 with a 12-nm grain size was doped at about the 1 at.% level with Pt diffused in from the surface. After annealing in air for 4 h at about 500 °C, the ac conductivity of the sample was measured as a function of temperature. The strongly nonlinear, and reversible, electrical response shown in Fig. 22, caused presumably by the Pt doping into the band gap of this wide–band-gap (3.2 eV) semiconductor, suggests that the rather easy impurity doping of nanophase electroceramics through rapid diffusion down their many grain boundaries and interconnected pores may lead to a wide range of interesting device applications. Also, the enhanced low-temperature sinterability of nanophase ceramics, without the need for any possibly contaminating additives, should help with device compatibility problems frequently encountered in such applications. However, much work remains to be done in this area.

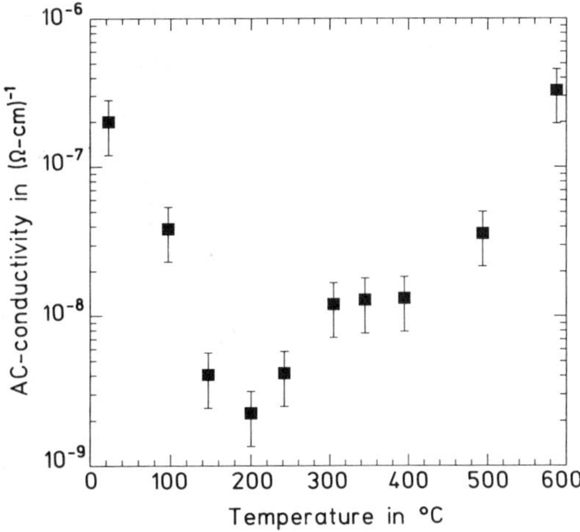

FIG. 22. The ac conductivity of Pt-doped nanophase TiO_2 as a function of temperature. The sample was preannealed in air at about 500 °C for 4 h prior to the conductivity measurements; the electrical response is reversible with temperature (Siegel, 1991).

The ability to control the porosity of nanophase ceramics will apparently lead to some rather useful optical applications. It has been demonstrated (Skandan *et al.*, 1991) that the oxidation step in the synthesis of cluster-consolidated nanophase Y_2O_3 can be controlled so that the as-consolidated material is effectively transparent. With such control, the resulting green-state porosity has a size distribution sufficiently small and narrow that light scattering is reduced and the material, although porous, is rather transparent as shown in Fig. 23. The ability thus to reduce the size of the porosity in such material to well below the wavelengths of visible or other radiation, coupled with the possibilities for efficient doping of these materials cited previously, should lead to a variety of interesting optical properties and related applications.

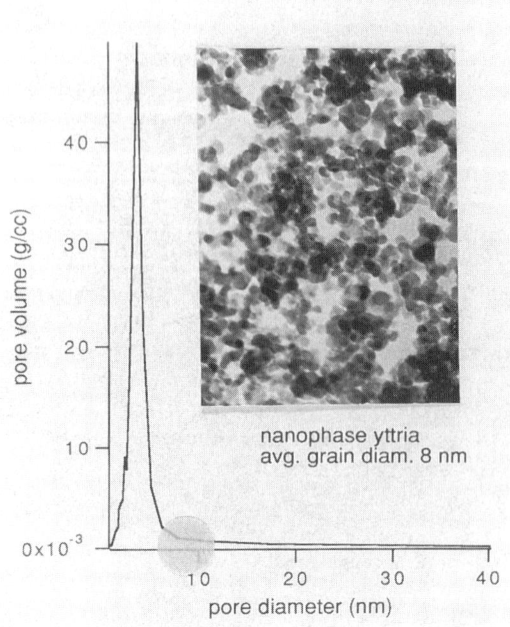

FIG. 23. Pore-size distribution in nanophase Y_2O_3 (the powder is shown in the inset TEM) consolidated at room temperature under 136 MPa pressure to a density of about 40% of the theoretical value. The transparent sample is shown on the axis at about its average grain diameter, 8 nm (courtesy of J.C. Parker, Nanophase Technologies Corporation).

5. FUTURE DIRECTIONS

It is very clear that in such a broad field as described in this article, we are just beginning to scratch the surface of the tremendous opportunities for synthesizing and applying nanophase materials. On the basis of the limited knowledge that has already been accumulated, the future appears to hold great promise for nanophase materials. The property changes that are found to occur when grain sizes are scaled down into the nanometer regime and interfaces take up significant volume fractions of the material can make significant differences in the range of use of a variety of materials. The cluster sizes utilized to date (down to about 5 nm) indicate that the high reactivities and short diffusion distances available in cluster-assembled materials can have profound effects upon the both the processing characteristics of these materials and their properties. These characteristics should be further enhanced as even smaller and more uniformly sized clusters become available in sufficiently large numbers to effect their assembly into usable and commercially viable materials.

The high diffusivities along their grain-boundary networks, with only a few atomic jumps separating grain interiors from grain boundaries, should make possible efficient impurity doping of these materials. Nanophase insulators and semiconductors, for example, could be easily doped with impurities at relatively low temperatures, allowing efficient introduction of impurity levels into their band gaps and control over their electrical and optical properties. Moreover, the ability to produce via cluster assembly fully dense ultrafine-grained nanophase ceramics with controlled or flaw-free microstructures that are readily formable and exhibit ductility should have a significant technological impact in a wide variety of applications. For example, near-net-shape forming of nanophase ceramic parts with complex and ultrafine detail would seem to be possible. Subsequent controlled grain growth could then be used to alter the grain-size–dependent properties of these ceramics.

Research on nanophase materials is now rapidly expanding, but much work still remains to be done. Further research on the synthesis of a broader range of nanophase materials, encompassing metals, alloys, ceramics, semiconductors, and composites, will have to be carried out in order to arrive at a fuller appreciation of just how broad an impact nanophase materials will have on materials technology. Investigations of their structure that will need to accompany such research will begin to elucidate the interplay between the effects of spatial confinement and large numbers of interfaces on the various chemical and physical properties of these new materials. A knowledge of the variation of such properties with the detailed structures and the synthesis and processing parameters of a variety of nanophase materials should eventually lead to an understanding of these new materials and consequently to the realization of their full impact on applied physics, materials science, and technology.

6. ACKNOWLEDGMENTS

This work was supported by the U.S. Department of Energy, BES-Materials Sciences, under Contract W-31-109-Eng-38. The author thanks his many collaborators without whose efforts and contributions this work would not have been possible.

Works Cited

Aksay, I. A., McVay, G. L., Ulrich, D. R. (Eds.) (1989), *Processing Science of Advanced Ceramics*, Materials Research Society Symposium Proceedings No. 155, Pittsburgh: MRS.

Aksay, I. A., Baer, E., Sarikaya, M., Tirrell, D. A. (Eds.) (1992), *Hierarchically Structured Materials*, Materials Research Society Symposium Proceedings No. 255, Pittsburgh: MRS.

Andres, R. P., Averback, R. S., Brown, W. L., Brus, L. E., Goddard, W. A., III, Kaldor, A., Louie, S. G., Moskovits, M., Peercy, P. S., Riley, S. J., Siegel, R. W., Spaepen, F., Wang, Y. (1989), *J. Mater. Res.* **4**, 704–736.

Averback, R. S., Hahn, H., Höfler, H. J., Logas, J. L., Chen, T. C. (1989), in: B. M. DeKoven, A. J. Gellman, R. Rosenberg (Eds.), *Interfaces between Polymers, Metals, and Ceramics*, Materials Research Society Symposium Proceedings No. 153, Pittsburgh: MRS, pp. 3–12.

Averback, R. S., Nelson, D. L., Bernholc, J. (Eds.) (1991), *Clusters and Cluster-Assembled Materials*, Materials Research Society Symposium Proceedings No. 206, Pittsburgh: MRS.

Baibach, M. N., Broto, J. M., Fert, A., Nguyen, F. D. V., Petroff, F., Etienne, P., Creuzet, G., Friederich, A., Chazelas, J. (1988), *Phys. Rev. Lett.* **61**, 2472–2475.

Beck, D. D., Siegel, R. W. (1992), *J. Mater. Res.* **7**, 2840–2845.

Berkowitz, A. E., Mitchell, J. R., Carey, M. J., Young, A. P., Zhang, S., Spada, F. E., Parker, F. T., Hutten, A., Thomas, G. (1992), *Phys. Rev. Lett.* **68**, 3745–3748.

Berry, R. S., Jellinek, J., Natanson, G. (1984), *Phys. Rev. A.* **30**, 919–931.

Birringer, R., Gleiter, H., Klein, H.-P., Marquardt, P. (1984), *Phys. Lett. A* **102**, 365–369.

Birringer, R., Herr, U., Gleiter, H. (1986), *Suppl. Trans. Jpn. Inst. Met.* **27**, 43–52.

Birringer, R., Gleiter, H. (1988), in: R. W. Cahn (Ed.), *Encyclopedia of Materials Science and Engineering*, Suppl. Vol. 1, Oxford: Pergamon, pp. 339–349.

Birringer, R., Hahn, H., Höfler, H., Karch, J., Gleiter, H. (1988), *Defect Diffusion Forum* **59**, 17–31.

Blander, M., Katz, J. L. (1967), *Geochim. Cosmochim. Acta.* **31**, 1025–1034.

Blander, M., Abdel-Gawad, M. (1969), *Geochim. Cosmochim. Acta* **33**, 701–716.

Buffat, Ph., Borel, J.-P. (1976), *Phys. Rev. A* **13**, 2287–2298.

Burggraaf, A. J., Keizer, K., van Hassel, B. A. (1989), *Solid State Ionics* **32/33**, 771–782.

Cammarata, R., Sieradzki, K. (1989), *Phys. Rev. Lett.* **62**, 2005–2008.

Chang, H., Höfler, H. J., Altstetter, C. J., Averback, R. S. (1991), *Scripta Metall. Mater.* **25**, 1161–1166.

Chokshi, A. H., Rosen, A., Karch, J. Gleiter, H. (1989), *Scripta Metall.* **23**, 1679–1683.

Eastman, J. A., Liao, Y. X., Narayanasamy, A., Siegel, R. W. (1989), in: I. A. Aksay, G. L. McVay, D. R. Ulrich (Eds.), *Processing Science of Advanced Ceramics*, Materials Research Society Symposium Proceedings No. 155, Pittsburgh: MRS, pp. 255–266.

Eckert, J., Holzer, J. C., Krill, C. E., III, Johnson, W. L. (1992), *J. Mater. Res.* **7**, 1751–1761; and in: W. A. T. Clark, C. L. Briant, U. Dahmen (Eds.), *Structure and Properties of Interfaces in Materials*, Materials Research Society Symposium Proceedings No. 238, Pittsburgh: MRS, pp. 745–750.

Epperson, J. E., Siegel, R. W., White, J. W., Klippert, T. E., Narayanasamy, A., Eastman, J. A., Trouw, F. (1989), in: L. E. McCandlish, D. E. Polk, R. W. Siegel, B. H. Kear (Eds.), *Multicomponent Ultrafine Microstructures*, Materials Research Society Symposium Proceedings No. 132, Pittsburgh: MRS, pp. 15–20.

Epperson, J. E., Siegel, R. W., White, J. W., Eastman, J. A., Liao, Y. X., Narayanasamy, A. (1990), in: S. M. Shapiro, S. C. Moss, J. D. Jorgensen (Eds.), *Neutron Scattering for Materials Science*, Materials Research Society Symposium Proceedings No. 166, Pittsburgh: MRS, pp. 87–92.

Fecht, H. J., Hellstern, E., Fu, Z., Johnson, W. L. (1989), *Adv. Powder Metall.* **1–2**, 111–122.

Fougere, G. E., Weertman, J. R., Kim, S., Siegel, R. W. (1992), *Scripta Metall. Mater.* **26**, 1879–1883.

Ganapathi, S. K., Rigney, D. A. (1990), *Scripta Metall. Mater.* **24**, 1675–1678.

Gao, P., Gleiter, H. (1987), *Acta Metall.* **35**, 1571–1575.

Gleiter, H. (1981), in: N. Hansen, A. Horsewell, T. Leffers, H. Lilholt (Eds.), *Deformation of Polycrystals: Mechanisms and Microstructures*, Roskilde: Risø National Laboratory, pp. 15–21.

Gleiter, H. (1989), *Prog. Mater. Sci.* **33**, 223–315.

Goldstein, A. N., Echer, C. M., Alivisatos, A. P. (1992), *Science* **256**, 1425–1427.

Granqvist, C. G., Buhrman, R. A. (1976), *J. Appl. Phys.* **47**, 2200–2219.

Gregg, S. J., Sing, K. S. W. (1982), *Adsorption Surface Area and Porosity*, New York: Academic Press.

Grossman, L. (1972), *Geochim. Cosmochim. Acta* **36**, 597–619.

Guermazi, M. Höfler, H. J., Hahn, H., Averback, R. S. (1991), *J. Am. Ceram. Soc.* **74**, 2672–2674.

Hadjipanayis, G. C., Siegel, R. W. (Eds.) (1994), *Nanophase Materials: Synthesis–Properties–Applications*, Dordrecht: Kluwer.

Hahn, H., Höfler, H., Averback, R. S. (1989), *Defect Diffusion Forum* **66–69**, 549–554.

Hahn, H., Logas, J., Averback, R. S. (1990a), *J. Mater. Res.* **5**, 609–614.

Hahn, H. Logas, J., Höfler, H. J., Kurath, P., Averback, R. S. (1990), in: M. J. Mayo, M. Kobayashi, J. Wadsworth (Eds.), *Superplasticity in Metals, Ceramics, and Intermetallics*, Materials Research Society Symposium Proceedings No. 196, pp. 71–76.

Hall, E. O. (1951), *Proc. Phys. Soc. London B* **64**, 747–753.

Han, M. Y., Whetten, R. L. (1988), *Phys. Rev. Lett.* **61**, 1190–1193.

Haubold, T., Birringer, R., Lengeler, B., Gleiter, H. (1988), *J. Less-Common Metals* **145**, 557–563.

Haubold, T., Birringer, R., Lengeler, B., Gleiter, H. (1989), *Phys. Lett. A* **135**, 461–466.

Hellstern, E., Fecht, H. J., Fu, Z., Johnson, W. L. (1989). *J. Appl. Phys.* **65**, 305–310.

Herr, U., Jing, J., Birringer, R., Gonser, U., Gleiter, H. (1987), *Appl. Phys. Lett.* **50**, 472–474.

Heuer, A. H., Fink, D. J., Laraia, V. J., Arias, J.-L., Calvert, P. D., Kendall, K., Messing, G. L., Blackwell, J., Rieke, P. C., Thompson, D. H., Wheeler, A. P., Veis, A., Caplan, A. L. (1992), *Science* **255**, 1098–1105.

Höfler, H. J., Averback, R. S. (1990), *Scripta Metall. Mater.* **24**, 2401–2406.

Horváth, J., Birringer, R., Gleiter, H. (1987), *Solid State Commun.* **62**, 319–322.

Horváth, J. (1989), *Defect Diffusion Forum* **66–69**, 207–228.

Hylton, T. L., Coffey, K. R., Parker, M. A., Howard, J. K. (1993), *Science* **261**, 1021–1024.

Jain, M., Christman, T. (1994), *Acta Metall. Mater.* **42**, 1901–1911.

Jang, J. S. C., Koch, C. C. (1990), *Scripta Metall. Mater.* **24**, 1599–1604.

Jorra, E., Franz, H., Peisl, J., Wallner, G., Petry, W., Birringer, R., Gleiter, H., Haubold, T. (1989), *Philos. Mag. B* **60**, 159–168.

Karch, J., Birringer, R., Gleiter, H. (1987), *Nature* **330**, 556–558.

Karch, J., Birringer, R. (1990), *Ceram. Int.* **16**, 291–294.

Kear, B. H., Cross, L. E., Keem, J. E., Siegel, R. W., Spaepen, F., Taylor, K. C., Thomas, E. L., Tu, K.-N. (1989), *Research Opportunities for Materials with Ultrafine Microstructures*, Vol. NMAB-454, Washington, DC: National Academy.

Kear, B. H., McCandlish, L. E. (1993), *Nanostruct. Mater.* **3**, 19–30.

Kimoto, K., Kamiya, Y., Nonoyama, M., Uyeda, R. (1963), *Jpn. J. Appl. Phys.* **2**, 702–713.

Koch, C. C., Jang, J. S. C., Gross, S. S. (1989), *J. Mater. Res.* **4**, 557–564.

Koch, C. C., Cho, Y. S. (1992), *Nanostruct. Mater.* **1**, 207–212.

Koch, C. C. (1993), *Nanostruct. Mater.* **2**, 109–129.

Komarneni, S., Parker, J. C., Thomas, G. J. (Eds.) (1993), *Nanophase and Nanocomposite Materials*, Materials Research Society Symposium Proceedings No. 286, Pittsburgh: MRS.

Li, Z., Hahn, H., Siegel, R. W. (1988a), *Mater. Lett.* **6**, 342–346.

Li, Z., Ramasamy, S., Hahn, H., Siegel, R. W. (1988b), *Mater. Lett.* **6**, 195–201.

Liu, C.-L., Adams, J. B., Siegel, R. W. (1994), *Nanostruct. Mater.* **4**, 265–274.

Luton, M. J., Janath, C. S., Disko, M. M., Matras, S., Vallone, J. (1989), in: L. E. McCandlish, D. E. Polk, R. W. Siegel, B. H. Kear (Eds.), *Multicomponent Ultrafine Microstructures*, Materials Research Society Symposium Proceedings No. 132, Pittsburgh: MRS, pp. 79–86.

Mayo, M. J., Siegel, R. W., Narayanasamy, A., Nix, W. D. (1990), *J. Mater. Res.* **5**, 1073–1082.

Mayo, M. J., Siegel, R. W., Liao, Y. X., Nix, W. D. (1992), *J. Mater. Res.* **7**, 973–979.

Melendres, C. A., Narayanasamy, A., Maroni, V. A., Siegel, R. W. (1989), *J. Mater. Res.* **4**, 1246–1250.

Milligan, W. W., Hackney, S. A., Ke, M., Aifantis, E. C. (1993), *Nanostruct. Mater.* **2**, 267–276.

Morris, D. G., Morris, M. A. (1991), *Acta Metall. Mater.* **39**, 1763–1770.

Mütschele, T., Kirchheim, R. (1987), *Scripta Metall.* **21**, 135–140; *ibid.*, 1101–1104.

Nieman, G. W., Weertman, J. R., Siegel, R. W. (1989), *Scripta Metall.* **23**, 2013–2018.

Nieman, G. W., Weertman, J. R., Siegel, R. W. (1990), *Scripta Metall. Mater.* **24**, 145–150.

Nieman, G. W., Weertman, J. R., Siegel, R. W. (1991), *J. Mater. Res.* **6**, 1012–1027.

Niihara, K. (1991), *J. Ceram. Soc. Jpn.* **99**, 945–952.

Niihara, K., Nakahira, A., Sekino, T. (1993), in: S. Komarneni, J. C. Parker, G. J. Thomas (Eds.), *Nanophase and Nanocomposite Materials*, Materials Research Society Symposium Proceedings No. 286, Pittsburgh: MRS, pp. 405–412.

Owen, D. M., Chokshi, A. H. (1993), *Nanostruct. Mater.* **2**, 181–187.

Parker, J. C., Siegel, R. W. (1990a), *J. Mater. Res.* **5**, 1246–1252.

Parker, J. C., Siegel, R. W. (1990b), *Appl. Phys. Lett.* **57**, 943–945.

Petch, N. J. (1953), *J. Iron Steel Inst.* **174**, 25–28.

Pfund, A. H. (1930), *Rev. Sci. Instrum.* **1**, 397–399.

Rieke, P. C., Calvert, P. D., Alper, M. (Eds.) (1990), *Materials Synthesis Utilizing Biological Processes*, Materials Research Society Symposium Proceedings No. 174, Pittsburgh: MRS.

Rivier, N. (1992), in: P. Jena, S. M. Khanna, B. K. Rao (Eds.), *Physics and Chemistry of Finite Systems: from Clusters to Crystals*, Dordrecht: Kluwer, pp. 189–198.

Roy, R., Komarneni, S., Yarbrough, W. (1988), in: J. MacKenzie, D. Ulrich (Eds.), *Ultrastructure Processing of Advanced Ceramics*, New York: Wiley Interscience, pp. 571–588.

Sanders, P. G., Weertman, J. R., Barker, J. G., Siegel, R. W. (1993), *Scripta Metall. Mater.* **29**, 91–96.

Sattler, K., Raina, G., Ge, M., Venkateswaran, N., Xhie, J., Liao, Y. X., Siegel, R. W. (1994), *J. Appl. Phys.* **76**, 546–551.

Schaefer, H. E., Würschum, R., Scheytt, M., Birringer, R., Gleiter, H. (1987), *Mater. Sci. Forum* **15–18**, 955–960.

Schaefer, H. E., Würschum, R., Birringer, R., Gleiter, H. (1988), *Phys. Rev. B* **38**, 9545–9554.

Schumacher, S., Birringer, R., Straub, R., Gleiter, H. (1989), *Acta Metall.* **37**, 2485–2488.

Shull, R. D., McMichael, R. D., Ritter, J. J., Bennett, L. H. (1993), in: S. Komarneni, J. C. Parker, G. J. Thomas (Eds.), *Nanophase and Nanocomposite Materials*, Materials Research Society Symposium Proceedings No. 286, Pittsburgh: MRS, pp. 449–453.

Siegel, R. W., Hahn, H. (1987), in: M Yussouff (Ed.), *Current Trends in the Physics of Materials*, Singapore: World Scientific, pp. 403–419.

Siegel, R. W., Ramasamy, S., Hahn, H., Li, Z., Lu, T., Gronsky, R. (1988), *J. Mater. Res.* **3**, 1367–1372.

Siegel, R. W., Eastman, J. A. (1989), in: L. E. McCandlish, D. E. Polk, R. W. Siegel, B. H. Kear (Eds.), *Multicomponent Ultrafine Microstructures*, Materials Research Society Symposium Proceedings No. 132, Pittsburgh: MRS, pp. 3–14.

Siegel, R. W. (1990), in: M. J. Mayo, M. Kobayashi, J. Wadsworth (Eds.), *Superplasticity in Metals, Ceramics, and Intermetallics*, Materials Research Society Symposium Proceedings No. 196, pp. 59–70.

Siegel, R. W. (1991), in: R. W. Cahn (Ed.), *Materials Science and Technology*, Vol. 15: *Processing of Metals and Alloys*, Weinheim: VCH, pp. 583–614.

Siegel, R. W. (1992), in: D. Wolf, S. Yip (Eds.), *Materials Interfaces: Atomic-Level Structure and Properties*, London: Chapman and Hall, pp. 431–460.

Siegel, R. W., Thomas, G. J. (1992), *Ultramicroscopy* **40**, 376–384.

Siegel, R. W. (1993), in: M. Nastasi, D. M. Parkin, H. Gleiter (Eds.), *Mechanical Properties and Deformation Behavior of Materials Having Ultra-Fine Microstructures*, Dordrecht: Kluwer, pp. 509–538.

Siegel, R. W., Fougere, G. E. (1994), in: G. J. Hadjipanayis, R. W. Siegel (Eds.), *Nanophase Materials: Synthesis–Properties–Applications*, Dordrecht: Kluwer, pp. 233–261.

Skandan, G., Hahn, H., Parker, J. C. (1991), *Scripta Metall. Mater.* **25**, 2389–2393.

Skandan, G., Foster, C. M., Frase, H., Ali, M. N., Parker, J. C., Hahn, H. (1992), *Nanostruct. Mater.* **1**, 313–322.

Suits, B. H., Siegel, R. W., Liao, Y. X. (1993), *Nanostruct. Mater.* **2**, 597–602.

Suits, B. H., Meng, M., Siegel, R. W., Liao, Y. X. (1994), *J. Mater. Res.* **9**, 336–342.

Thölén, A. R. (1979), *Acta Metall.* **27**, 1765–1778.

Thomas, G. J., Siegel, R. W., Eastman, J. A. (1989), in: B. M. DeKoven, A. J. Gellman, R. Rosenberg (Eds.), *Interfaces between Polymers, Metals, and Ceramics*, Materials Research Society Symposium Proceedings No. 153, Pittsburgh: MRS, pp. 13–20.

Thomas, G. J., Siegel, R. W., Eastman, J. A. (1990), *Scripta Metall. Mater.* **24**, 201–206.

Trudeau, M. L., Van Neste, A., Schulz, R. (1991), in: R. S. Averback, D. L. Nelson, J. Bernholc (Eds.), *Clusters and Cluster-Assembled Materials*, Materials Research Society Symposium Proceedings No. 206, Pittsburgh: MRS, pp. 487–492.

Tschöpe, A., Birringer, R. (1993), *Acta Metall. Mater.* **41**, 2791–2796.

Uyeda, R. (1991), *Prog. Mater. Sci.* **35**, 1–96.

Valiev, R. Z., Krasilnikov, N. A., Tsenev, N. K. (1991), *Mater. Sci. Eng. A* **137**, 35–40.

Valiev, R. Z. (1993), in: M. Nastasi, D. M. Parkin, H. Gleiter (Eds.), *Mechanical Properties and Deformation Behavior of Materials Having Ultra-Fine Microstructures*, Dordrecht: Kluwer, pp. 303–308.

Wagner, W., Averback, R. S., Hahn, H., Petry, W., Wiedenmann, A. (1991), *J. Mater. Res.* **6**, 2193–2198.

Whitesides, G. M., Mathias, J. P., Seto, C. T. (1991), *Science* **254**, 1312–1319.

Wolf, D., Lutsko, J. F. (1988), *Phys. Rev. Lett.* **60**, 1170–1173.

Wolf, D., Yip, S. (Eds.) (1992), *Materials Inter-*

faces: Atomic-Level Structure and Properties, London: Chapman and Hall.

Wunderlich, W., Ishida, Y., Maurer, R. (1990), *Scripta Metall. Mater.* **24**, 403–408.

Xiao, J. Q., Jiang, J. S., Chien, C. L. (1992), *Phys. Rev. Lett.* **68**, 3749–3752.

Zhu, X., Birringer, R., Herr, U., Gleiter, H. (1987), *Phys. Rev. B* **35**, 9085–9090.

Further Reading

Among the numerous references cited above, there are several comprehensive treatments of nanophase materials that can be usefully consulted. They are as follows:

Andres, R. P., Averback, R. S., Brown, W. L., Brus, L. E., Goddard, W. A., III, Kaldor, A., Louie, S. G., Moskovits, M., Peercy, P. S., Riley, S. J., Siegel, R. W., Spaepen, F., Wang, Y. (1989), *J. Mater. Res.* **4**, 704–736.

Gleiter, H. (1989), *Prog. Mater. Sci.* **33**, 223–315.

Hadjipanayis, G. C., Siegel, R. W. (Eds.) (1994), *Nanophase Materials: Synthesis-Properties-Applications*, Dordrecht: Kluwer.

Jena, P., Khanna, S. N., Rao, B. K. (Eds.) (1992), *Physics and Chemistry of Finite Systems: from Clusters to Crystals*, Dordrecht: Kluwer.

Kear, B. H., Cross, L. E., Keem, J. E., Siegel, R. W., Spaepen, F., Taylor, K. C., Thomas, E. L., Tu, K.-N. (1989), *Research Opportunities for Materials with Ultrafine Microstructures*, Vol. NMAB-454, Washington, DC: National Academy.

Nastasi, M., Parkin, D. M., Gleiter, H. (Eds.) (1993), *Mechanical Properties and Deformation Behavior of Materials Having Ultra-Fine Microstructures*, Dordrecht: Kluwer.

In addition, a series of symposia on the subject of nanophase and related materials has been sponsored by the Materials Research Society since 1988. The proceedings of these symposia are published in volumes 132 (1989), 206 (1991), 286 (1993), and 351 (1994) of the *MRS Symposium Proceedings* series. Current developments in the field can be followed in *Nanostructured Materials*, an Acta Metallurgica, Inc. international journal published by Pergamon Press since 1992. A series of International Conferences on Nanostructured Materials is also being held and the proceedings of the first (Cancun, 1992) and second (Stuttgart, 1994) of these are published in separate volumes (3 and 6, respectively) of *Nanostructured Materials*.

NATURAL GAS

ROBERT J. FINLEY, *Bureau of Economic Geology, The University of Texas at Austin, Austin, Texas, U.S.A.*

INTRODUCTION

In 1992, natural gas was the third leading primary source of energy in the world, following crude oil and coal, accounting for 71.2 quadrillion Btu (quads) (75.2×10^{12} MJ) of production, or 22.8% of a total of 311.8 quads (329.2×10^{12} MJ) (British Petroleum Com-

1-56081-070-X/94/$5.00 + .50

pany, 1993). One quad is approximately the energy equivalent of 28.3 × 10^9 m^3 of natural gas. The former Soviet Union was the major producer of natural gas in 1990, accounting for 39% of production, followed by the United States (24%) and Canada (5%). Although some natural gas is shipped internationally through pipelines and a small amount is shipped as liquefied natural gas (LNG), most natural gas is consumed in or near the country of origin. Thus, the three leading producers are also the leading consumers.

The prominence of natural gas will likely increase with increasing concern for the environmental consequences of fossil fuel use. Natural gas combustion results in minimal atmospheric emissions of sulfur oxides and particulates and far lower emissions of carbon monoxide, carbon dioxide, and nitrogen oxides than either coal or oil. In stationary (power plants, boilers, etc.) uses, natural gas emits roughly half as much carbon dioxide per unit of energy as coal and 30% less than oil.

Concerns about the natural gas resource base, especially in the United States, that existed in the late 1970s have largely dissipated. Such concerns led to restrictions on natural gas use in the United States that have now been lifted. Assessments of the resource base (proved reserves plus resources) in the United States (lower 48 states) have increased substantially from 1980 through 1992 to a level of 36.66 × 10^{12} m^3 (National Petroleum Council, 1992). Worldwide, proved reserves as of 1 January 1993 were 137.9 × 10^{12} m^3, of which 4.53 × 10^{12} m^3 were U.S. proved reserves (lower 48 states). With substantial reserves and even greater volumes of resources remaining to be developed, natural gas usage will increase as energy demand grows because of its low rate of combustion emissions, recognized availability, and improved technology for production, even from the geologically complex reservoirs that hold large natural gas resources.

1. DEFINITION OF NATURAL GAS

Natural gas is a naturally occurring mixture of hydrocarbon and nonhydrocarbon gases occurring in geologic formations beneath the earth's surface. The hydrocarbon gases are of very low molecular weight with the principal combustible component being methane, CH_4. A typical natural gas contains 70% to nearly 100% methane, 1% to 10% ethane, and smaller amounts of higher molecular weight hydrocarbons. Nonhydrocarbon components may include carbon dioxide, nitrogen, hydrogen sulfide, helium, water vapor, and hydrogen. Natural gas is referred to as sour if sufficient sulfur compounds are present to require removal for use of the gas. If referred to as wet, substantial and extractable quantities of C_2 to C_5 hydrocarbons, commonly referred to as natural gas liquids, are present. Most natural gas requires at least the removal of water vapor to bring it up to pipeline quality (American Gas Association, 1990; Kvenvolden, 1988).

1.1 Origin of Natural Gas

Natural gas forms thermogenically from the alteration and degradation of organic matter trapped within fine-grained sedimentary rocks. Rocks such as shales with less than 0.5% organic carbon content are poor source rocks, those with 1.0% to 2.0% are good, and those with over 4.0% organic carbon are very good. The type of original organic matter, temperature history of the source rocks, and geologic time during which source materials were buried determine the volume and chemical content of the resulting hydrocarbons forming natural gas. The same controls apply to the formation of oil. Natural gas may also form microbially through carbon dioxide reduction to produce methane; the resulting gas is quite dry with only very small amounts of hydrocarbons heavier than methane (Waples, 1985). A general progression of hydrocarbon generation with increasing time and temperature is from biogenic gas to oil, light oil, wet gas, and finally dry gas. Not all stages are present in the history of any one natural gas accumulation. The type of hydrocarbon generated is highly dependent on whether source materials are oil-prone or gas-prone. Most natural gas deposits in the earth contain thermogenic methane generated at the high-temperature end of this progression (Hunt, 1979). Another theory of the origin of natural gas holds that natural gas may be formed within the earth's crust exclusive of an organic source (Gold and Soter, 1980; Gold, 1987), but this theory is not widely held by petroleum geologists.

1.2 Migration and Trapping of Natural Gas

Once natural gas is generated, it may be expelled from the organic-rich source rock, such as a low-permeability shale, into a more permeable rock, such as a sandstone, and migrate in several stages until trapped by a low-permeability barrier. Such a barrier may be another shale or a chemical sediment such as salt or gypsum. The geometry of these barriers is quite varied, and trapping of migrated natural gas often depends on the structural configuration of the rock in addition to the rock type. The result is a natural gas reservoir from which natural gas can be technically and economically recovered. Migration is driven by the lower density and resultant buoyancy of natural gas relative to oil and formation waters. A reservoir rock, such as a sandstone, with permeability of >0.1 millidarcy from which gas can be readily extracted forms a *conventional* natural gas reservoir.

Some organic-rich source rocks such as coal and highly organic black shales retain recoverable gas that may be produced by tapping into the natural fracture systems within these source rocks. As gas (and, as in the case of coal, formation water) is withdrawn from the fracture system, reservoir pressure is reduced and additional gas desorbs from the rock matrix and can be produced. Such reservoirs require special treatments, such as the creation of a hydraulic fracture extending away from the wellbore or a cavity around the wellbore at the level of the coalbed, to produce gas economically. In some cases, sandstone reservoirs have such low permeability (<0.1 millidarcy) that they do not produce gas without special well treatments to dissolve minerals that have filled the reservoir pore space or to fracture the sandstone artificially near the well. Such low-permeability (tight) sandstones and potentially productive coals and shales are termed *unconventional* reservoirs. As conventional reservoirs are depleted, especially in maturely explored and developed regions such as onshore in the lower 48 states of the United States (compared to offshore production in the U.S. Gulf of Mexico), unconventional reservoirs are becoming increasingly important as natural gas sources.

1.3 Access to Natural Gas Resources

Natural gas resources will be exploited by continued exploration for undiscovered accumulations, by advanced recovery in known conventional reservoirs where the efficiency of recovery can be improved as more of a given field is drilled, and through development of unconventional reservoirs. The location of most unconventional resources is generally known for the United States, but only about 35% of the in-place resource may be exploited with current technology and well spacing. Economic extraction of this gas is highly dependent on continuing technological advancement. The same is true for advanced recovery of conventional resources where, in geologically complex reservoirs, only about 65% of the in-place gas may typically be reached. Additional gas supplies for U.S. and other major markets are available through pipeline imports and imported LNG.

Because transoceanic trade in LNG is limited by costs to liquefy, ship, and regasify, future gas supplies will be primarily dependent on exploration and production technologies disseminated on a worldwide basis to develop gas resources marketable through pipelines. By the year 2000, world LNG trade is expected to represent only about 4% of production, with supplies close to markets retaining a significant cost advantage (MacDougall and Linder, 1992).

1.4 Overview of World Gas Consumption, Production, and Reserves

World consumption of natural gas has grown steadily over the last two decades from about 1.08×10^{12} m^3 in 1970 to about 2.12×10^{12} m^3 in 1990. Total world consumption stood at 2.13×10^{12} m^3 in 1992. Between 1970 and 1990, North American consumption declined, while that of European countries not members of the Organization for Economic Cooperation and Development (OECD), notably the former Soviet Union, increased substantially. The former Soviet Union saw an increase in consumption of 92%, from 394×10^9 m^3 to 758×10^9 m^3 between 1980 and 1990 alone (British Petroleum Company, 1991, 1993).

Among the ten leading consuming nations, Germany, Japan, Italy, and France have substantial deficits between domestic production and consumption that must be made up by imports (Table 1). Among these ten nations, the former Soviet Union and the Netherlands exported natural gas to other parts of Europe, and Canada exported gas to the United States. Other exporting nations contributing to European supply are Algeria and Norway. Germany, the European nation with the largest consumption of imported natural gas, received 22.4×10^9 m^3 of imports from the former Soviet Union and 19.6×10^9 m^3 of imports from the Netherlands in 1990 (Energy Information Administration, 1993a). Gazprom, the Russian state-owned natural gas producer, is constructing new pipelines to increase its export capability to Western European markets. Gazprom aims to increase its total annual production to 739×10^9 m^3 in 1998 from 653×10^9 m^3 in 1992. In 1992, Gazprom delivered 113×10^9 m^3 to Western Europe and the same amount to the newly independent Baltic states (Wood, 1993).

Japan receives substantial volumes of LNG from Indonesia (24.4×10^9 m^3 in 1992) and from the Middle Eastern countries of United Arab Emirates, Iran, and Iraq (3.32×10^9 m^3 in 1992), with the balance of supply from other sources (Energy Information Administration, 1993a).

Consuming countries with significant reserves are led by the former Soviet Union (55.0×10^{12} m^3) and in particular Russia with 47.6×10^{12} m^3 of reserves as of the end of 1992. Saudi Arabia, Canada, and the Netherlands have significant reserve levels in relation to domestic consumption (Table 1). Iran has the most significant reserves of natural gas in the Middle East (19.8×10^{12} m^3, 1992) followed by Qatar (6.4×10^{12} m^3, 1992) and Abu Dhabi (5.3×10^{12} m^3, 1992). Other significant reserves are held by Nigeria (3.4×10^{12} m^3), Venezuela (3.6×10^{12} m^3), and Indonesia (1.8×10^{12} m^3), all as of the end of 1992 (British Petroleum Company, 1993). All of these countries already are, or plan to become, involved in international natural gas trade, either through pipeline exports or through LNG export facilities.

2. THE NATURAL GAS RESOURCE BASE

2.1 World Natural Gas Resources

Increased exploration and development of the resource base has increased world proved reserves of natural gas from 46×10^{12} m^3 in 1970 to 75×10^{12} m^3 in 1980 and 137.9×10^{12} m^3 in 1990 (British Petroleum Company, 1993). The former Soviet Union holds the largest proved reserves with 40% of the world total (Table 2). The former Soviet Union is also the world's leading natural gas producer, with Russia accounting for 79% of that production in 1991 (Anonymous, 1992a). Among Middle Eastern countries, Iran holds 19.8×10^{12} m^3 or about 14% of the world total in a region with 31.3% of the world total (Table 2). Among the North Sea producing countries, Norway and the Netherlands each

Table 1. Consumption (1991), production (1991), and reserves (1992) of natural gas for the 10 leading consuming countries (from British Petroleum Company, 1993; Energy Information Administration, 1993a).

Area	Consumption (10^9 m^3)	Production (10^9 m^3)	Reserves (10^{12} m^3)
Former Soviet Union	711.7	812.2	55.0
United States	541.4	504.3	4.7
Canada	70.0	112.5	2.7
Germany	105.2	25.8	0.3
United Kingdom	61.7	54.8	0.5
Japan	54.0	2.0	[a]
Italy	51.4	17.9	[b]
Netherlands	43.3	81.8	1.9
France	37.8	3.0	[b]
Saudi Arabia	33.0	33.2	5.2

[a]Less than 0.05.
[b]Less than 0.6 for OECD Europe other than Germany, Netherlands, Norway, and United Kingdom.

Table 2. World proved reserves of natural gas, 1 January 1993 (from Anonymous, 1992a; British Petroleum Company, 1993).

Area	10^{12} m^3	10^{12} ft^3	Share of total (%)	R/P ratio[a]
North America	7.4	262.8	5.4	12.0
USA	4.7	167.1	3.4	9.4
Canada	2.7	95.7	2.0	23.5
Latin America	7.3	259.5	5.4	75.8
OECD Europe	5.3	191.8	3.8	27.6
Former Soviet Union	55.0	1942.3	39.8	67.8
Non-OECD Europe	0.6	21.0	0.4	15.7
Middle East	43.1	1520.1	31.0	[b]
Africa	9.7	346.9	7.1	[b]
Asia and Australasia	9.5	341.0	6.9	52.5
World total	137.9	4885.4		64.8

[a]Reserve/production ratio at year end 1992.
[b]Over 100 years.

hold about 1.9 × 10^{12} m^3 of reserves, totaling 2.8% of world reserves. Norway has the potential to become a future exporter of LNG.

Assessments of world resources beyond proved reserves are not well defined outside of North America. The Canadian Energy Research Institute (CERI) calculated 221.9 × 10^{12} m^3 of potential reserves (proved undeveloped and undiscovered potential) outside of North America (MacDougall and Linder, 1992), yet this number seems conservative compared to the volume of proved reserves outside North America. Of the resources assessed by CERI, 37% are in the former Soviet Union and 33% are in the Middle East (Table 3). Major former Soviet Union fields are now in production in the Yamal Peninsula area of western Siberia. These fields contain about as large a volume in possible reserves as in already explored reserves (defined as proved plus probable), thereby substantially increasing reserves by additional development in these known fields alone, without an increment of undiscovered potential in the region. Yamal fields discovered to date contain approximately 16.6 × 10^{12} m^3 of gas and additional potential for undiscovered resources is foreseen (Anonymous, 1992b).

Table 3. World undeveloped and undiscovered natural gas resource potential outside North America (from MacDougall and Linder, 1992).

	Resources	
	10^{12} m^3	10^{12} ft^3
Latin America	14.9	528
Western Europe	9.6	338
Former Soviet Union	81.4	2877
Eastern Europe	0.6	21
Middle East	74.2	2619
Africa	15.1	534
Asia	11.7	415
Pacific Rim	14.4	508
Total	221.9	7840

2.2 Natural Gas Resources in Canada

Canada has significant natural gas reserves, and significant volumes of undiscovered resources lie almost entirely within the Western Canadian Sedimentary Basin (Alberta, northeast British Columbia, and southern Saskatchewan) and the northern Frontier areas (Mackenzie Delta/Beaufort Sea area and Arctic Islands). Of 2.7 × 10^{12} m^3 of proved reserves (1 January 1993), less than 2% are found in eastern Canada. Total resources assessed by the U.S. National Petroleum Council (NPC) amount to (17.6–21.0) × 10^{12} m^3 (Table 4), although the potential nonconventional resource may be more than double the volumes assessed (National Petroleum Council, 1992). Some opportunities for offshore resource development also exist in eastern Canada (offshore Newfoundland and Nova Scotia) and in western Canada (offshore British Columbia).

The U.S. represents a major market for Canadian natural gas production at the present time. In 1991, 47 × 10^9 m^3 was exported to the United States by pipeline, representing 45% of Canadian production. The re-

Table 4. Natural gas resources of Canada, 1 January 1991 (National Petroleum Council, 1992).

	Current technology		2010 technology[a]	
	10^{12} m^3	10^{12} ft^3	10^{12} m^3	10^{12} ft^3
Proved	2.0	72	2.0	72
Conventional				
Reserve appreciation	0.6	22	0.7	24
New fields	2.8	99	3.1	109
Frontier	8.3	293	9.0	317
Subtotal	11.7	414	12.8	450
Nonconventional				
Coalbed methane	2.3	80	3.7	129
Tight gas	1.6	55	2.5	89
Subtotal	3.9	135	6.2	218
Total resources	17.6	621	21.0	740

[a]Basis—Technically recoverable resources incorporating technology advancement through 2010.

maining Canadian production was split almost equally between Ontario and Québec and western Canadian markets.

2.3 Natural Gas Resources in Mexico

Mexico's natural gas production is predominantly associated with the nation's oil production. Current (1992) gas production is about 31.3 × 10^9 m^3 per year (British Petroleum Company, 1993), 87% of which comes from the onshore and offshore basins in the Southern Isthmus area. Proved reserves of Mexico are estimated at 2.0 × 10^{12} m^3 and total assessed resources at 7.1 × 10^{12} m^3 under assumptions of current technology (National Petroleum Council, 1992). The current reserves-to-production ratio is about 51:1.

Exports of natural gas from the United States to Mexico are presently growing as industrialization proceeds in the border area, thus increasing demand for electricity and for process heat. New cross-border interconnections are being added. United States exports to Mexico in 1991 were 1.71 × 10^9 m^3 and in the first half of 1992 were 1.25 × 10^9 m^3 (White, 1992a).

2.4 Natural Gas Resources in the United States

The most recent U.S. assessment of natural gas resources for the lower 48 states indicates 4.53 × 10^{12} m^3 of proved reserves (lower 48 states) and 32.13 × 10^{12} m^3 of resources (at levels of technology expected to be developed through 2010) as of 1 January 1991 (National Petroleum Council, 1992). The resources beyond proved reserves are about equally divided between conventional and unconventional reservoir types (Table 5). Larger volumes of in-place resources exist beyond those assessed within this recent evaluation but will only be recoverable with advanced technological development.

Resource estimates made in the period 1980–1992 have reflected progressively greater understanding of the resource base. In particular, this understanding has incorporated new knowledge of reserve growth, also termed reserve appreciation, of old fields and of new technology applicable to the production of all classes of natural gas resources. This is particularly true for the nonconventional re-

Table 5. U.S. lower 48 states natural gas resource base (National Petroleum Council, 1992).

	10^{12} m^3	10^{12} ft^3
Proved reserves	4.53	160
Conventional resources		
Reserve appreciation	5.75	203
New fields	11.69	413
Subtotal	17.44	616
Nonconventional resources		
Coalbed methane	2.77	98
Shales	1.61	57
Tight sandstones	9.88	349
Other	0.42	15
Subtotal	14.69	519
Total resources[a]	36.66	1295

[a]Technically recoverable resource base as of 1 January 1991 incorporating technology advancement through 2010.

sources such as those in low-permeability (tight) sandstones, coalbeds, and fractured shales. As the resource base has been more accurately assessed and as all categories of the resource base have been included in the assessment process, the total volume of assessed resources has grown. It is important to recognize that more than half, over 194 000 wells, of all the producing gas wells ever drilled in the United States [total 394 915 from 1918 to 1990 (DeGolyer and MacNaughton, 1991)] were completed between 1974 and 1990. As a result, substantially more is known in the early 1990s about the geology and engineering properties of natural gas reservoirs in the United States than just a decade ago. Much of this same knowledge will ultimately be applicable worldwide to advanced reservoir development.

2.4.1 United States Resource Estimates through Time Natural gas within any producing area, regional or national, consists of proved reserves in known fields, inferred reserves (the next most certain category of reserves) and undeveloped resources within these fields (the source of reserve growth), and undiscovered resources. Other than proved reserves, all volumes of future natural gas supply are estimates based on information derived from past and current experience in gas production and reservoir development. Even proved reserves are subject to periodic revision (Finley and Fisher, 1991). In 1988, the U.S. Department of Energy (DOE) evaluated the then-current resource assessments of the Potential Gas Committee, the U.S Geological Survey, and the Minerals Management Service to create a new, integrated assessment of U.S. natural gas resources (Finley *et al.*, 1988). An important element of the DOE-sponsored study was the first-time quantification of nonassociated gas reserve growth potential based on depositional complexity (reservoir heterogeneity) of many types of gas reservoirs. The 1988 DOE assessment of natural gas resources for the lower 48 states defined 29.98×10^{12} m^3 of reserves and recoverable resources with current technology. Proved reserves made up 4.50×10^{12} m^3 of the total. Of that total, 3.37×10^{12} m^3 represented nonassociated reserve growth resources related to depositional complexity within known reservoirs. The balance derives from such approaches as drilling deeper for new reservoirs or extending field boundaries. The total assessed reserve growth resource was 8.15×10^{12} m^3.

Many analysts regarded the DOE assessment as high in terms of the total volume of resources. Since that time, however, assessments by other organizations, public and private, have confirmed resource bases of over 28.3×10^{12} m^3. Enron Corp. estimated reserves plus recoverable resources of 33.69×10^{12} m^3 (with advanced technology) as of 1990, and the National Research Council (NRC) assessed 40.50×10^{12} m^3 of reserves plus resources as of year-end 1986 (National Research Council, 1990; Enron Corp., 1991). The Gas Research Institute (GRI), in their 1992 analysis of long-term trends in U.S. gas supply and prices, indicates 34.2×10^{12} m^3 of reserves plus resources as of 1 January 1991 (Holtberg *et al.*, 1992). Most recently, the NPC has assessed a supply base (reserves plus resources) of 36.66×10^{12} m^3 with advanced technology (Table 5). Including the NPC study, the average of the five assessments (DOE, Enron, NRC, GRI, and NPC) cited here is about 35×10^{12} m^3, a much more substantial resource than had been assumed prior to 1988. Note that somewhat different assumptions, such as those regarding the use of current or future technology, apply to each assessment. In comparison, the U.S. Geological Survey's assessment as of 1 January 1980 for the U.S. lower 48 states was only 23.33×10^{12} m^3 of reserves plus resources (Dolton *et al.*, 1981) and as of 1 January 1987 was 17.3×10^{12} m^3 (Mast *et al.*, 1989).

3. HISTORY OF NATURAL GAS RESOURCE DEVELOPMENT

3.1 United States Natural Gas Development

Total U.S. cumulative production of natural gas amounts to about 21.7×10^{12} m^3 as of 1 January 1991; only about 142×10^9 m^3 of that volume has been produced in Alaska and the balance in the lower 48 states (Potential Gas Committee, 1991). Annual dry gas production for the total United States reached 615×10^9 m^3 in 1973 and declined to 456×10^9 m^3 in 1986 before returning to about 503 $\times 10^9$ m^3 for 1991 and 1992 (Energy Information Administration, 1993b). The history of natural gas production in the United States has been one of interplay between government regulation from the wellhead to the local distribution company vs free market forces. Only as of 1 January 1993 have the last wellhead price controls been removed; however, since most gas was already deregulated, little effect on market prices was seen on that date. Transportation by pipelines is now predominantly a service provided for gas not owned by the pipelines, a transition that began in the mid 1980s and continues into the early 1990s.

The gas industry in the United States began in the mid to late 1800s with gas originally used for lighting and with much of that gas being manufactured gas from coal or wood. Manufactured gas continued to be important through World War II because there was no inexpensive means to transport natural gas long distances beyond the producing regions. By 1938, interstate movement of natural gas amounted to only 11.3×10^9 m^3 annually. That same year the first major piece of federal regulatory legislation was passed as The Natural Gas Act of 1938, designed to protect consumers from perceived monopolistic pricing practices of gas utilities and pipelines (National Petroleum Council, 1992).

A surplus of natural gas existed during the 1940s and 1950s as a by-product of oil production. The Railroad Commission of Texas took action in 1947–1948 to limit flaring of gas associated with oil production, thus reducing the waste of natural gas and promoting expanded natural gas use. Following World War II, long-distance interstate pipeline transportation became feasible beyond the six principal producing states of that time, namely Texas, Louisiana, California, Oklahoma, West Virginia, and Kansas. In 1945, interstate transportation volumes reached about 85×10^9 m^3. By 1960, volumes had increased to over 340×10^9 m^3 and by 1969 to 538×10^9 m^3. Much of this supply was destined for operation of a greatly expanded industrial base and for millions of new homes constructed during the post-war period.

One effect of federal regulation was a long period (1950–1970) of low gas prices and major demand growth. Exploration for and development of natural gas resources became less economically attractive during this period as a result of low wellhead prices and rising costs, and the result was a decline in total domestic gas production beginning in late 1973. Service curtailments developed, particularly during the severe winter of 1976–1977. Because federal market regulation did not apply within the states, a dual market developed consisting of higher-priced (more market sensitive) intrastate gas compared to lower-priced, regulated interstate gas. In response to supply and market issues, Congress passed the Natural Gas Policy Act (NGPA) in 1978, which established pricing categories for all gas produced and sold in the United States and helped spur resource development, particularly in some of the more difficult-to-produce resource categories such as gas from tight sandstones. The Fuel Use Act of 1978 was passed to prohibit use of natural gas in power plants and large industrial boilers. The perception was that natural gas was a limited resource and should be preserved for "higher" uses such as home heating.

As the United States entered the 1980s, price-induced conservation, long-term fuel switching, industrial restructuring away from gas-consuming heavy industries, and efficiency improvements reduced total consumption from around 563×10^9 m^3 in 1980 to about 477×10^9 m^3 in 1983 (Energy Information Administration, 1993b). Incentives under the NGPA supporting resource development combined with declining consumption to create a "gas bubble" of excess deliverability beginning in 1982–1983 (National Petroleum Council, 1992). This excess, and the orientation of the Federal Energy Regulatory Commission (FERC) toward a restructuring of the industry toward a freer and less regulated environment, led to gas-on-gas price competition and low spot-market natural gas prices that prevailed into the early 1990s. By the latter half of 1992, it began to appear that supply and increasing demand had come into a balance as the FERC continued to foster deregulation of the industry.

3.2 Canadian Natural Gas Development

Natural gas was discovered in Ontario and Alberta, Canada, in 1880 and 1884, respectively. In Alberta, some of the first pipelines were built to supply natural gas to the cities of Calgary and Edmonton between 1912 and 1923. Large-scale use of natural gas and widespread exploration of the Western Canadian Sedimentary Basin, mostly for oil, both began in 1947. Production from fields in British Columbia was connected to markets in the Vancouver area in the late 1940s and early 1950s, and the TransCanada pipeline was completed in 1958, linking western source areas to markets in the eastern provinces of Ontario and Quebec. A pipeline to the California market in the United States was completed in 1961 (National Petroleum Council, 1992).

Canadian natural gas production has increased steadily during the 1980s; 1984 production was 79×10^9 m^3, whereas 1991 production was 105×10^9 m^3. Of 1991 production, 47×10^9 m^3 was exported to the United States, more than doubling the export level of 22×10^9 m^3 of 1984 (Arthur Anderson & Co., 1987; National Petroleum Council, 1992). Canada's share of the U.S. market was 8.7% in 1991, increasing from 4.6% in 1986, as a result of expansion of cross-border pipeline capacity.

3.3 Mexican Natural Gas Development

Natural gas produced in Mexico has been primarily associated gas related to that nation's oil production from the Gulf of Campeche offshore and the Reforma area onshore between the cities of Veracruz and Villahermosa. Annual production in 1992 was about 31.3×10^9 m^3 (British Petroleum Company, 1993). Development of the gas resource was concurrent with the development of Mexico's oil resources during the 1970s and 1980s. The main trunk line system for natural gas distribution is a 1245-km-long pipeline system that extends from the Reforma area north to Monterrey. The pipeline system was completed in 1980 with additional lines extending farther north and also northwest and west to Mexico City and the Pacific coast.

In the early 1970s and from 1980 to 1984, the United States imported natural gas from Mexico; U.S. imports peaked at 3×10^9 m^3 in 1984. The U.S. exports to Mexico were never more than 0.11×10^9 m^3 annually between 1977 and 1988 but in 1989 jumped to 0.48×10^9 m^3 and in 1991 reached 1.7×10^9 m^3 annually. These supplies have been flowing into northern Mexico to meet demand from increased industrial development and also to replace supplies from southern Mexico that are now being used to displace fuel oil use in and around Mexico City. This displacement is an effort to improve air quality in that region. Industrial expansion possible under the North American Free Trade Agreement will likely lead to additional demand for U.S. gas exports to Mexico through the 1990s with that demand growing at an annual rate of 7% to 10% through 2000 (White, 1992a).

4. TECHNOLOGIES FOR NATURAL GAS DEVELOPMENT

4.1 Delineation of Reservoir Types

Technologies for natural gas development can be related to major categories of the natural gas resource base from which the gas is produced: associated reservoirs, nonassociated conventional reservoirs, and nonassociated unconventional reservoirs. Associated reservoirs are oil reservoirs that also produce natural gas. Such gas may be present in solution with oil or as a free gas cap or both. Associated gas may be produced and marketed or may be reinjected (cycled) to maintain reservoir pressure. The technologies related to associated gas resources are those of oil exploration and development of oil resources, many of which are identical for natural gas with the major exception of unconventional gas reservoirs.

Nonassociated conventional reservoirs have sufficient permeability so that the trapped gas will readily flow to the wellbore. They also typically have sufficient storage capacity in pores within the rock matrix so that they are not dependent on gas flow in natural fractures to provide a production pathway to the wellbore. The amounts of oil produced are minimal, although "wet" gas reservoirs contain economically recoverable quantities of natural gas liquids (NGLs) such as propane and butane. Currently (1992–1993) in the United States, readily producible reservoirs with a high content of NGLs are economically attractive because of the demand for NGLs as chemical feedstocks.

Nonconventional nonassociated reservoirs include low-permeability sandstones (and carbonates), coalbeds capable of gas production, and fractured shales. In the United States, reservoirs are considered low permeability, or tight, if their *in situ* matrix permeability is 0.1 millidarcy or less. Hydraulic fracturing to create a man-made fracture propped open with clean sand injected into the formation is the typical method used to produce natural gas from tight sandstones. This process may be helped or hindered by the presence of natural fractures. Coalbeds, shales, and tight sandstones have limited matrix porosity (storage capacity) but natural fractures ("cleat" in coal beds) act as flow pathways to the wellbore. Coalbeds produce gas that is adsorbed onto the coal surface after reservoir pressure is reduced, a process requiring varying amounts of initial water production. Fractured shales, some of which are quite silty, also produce gas from fractures and from adsorbed gas within the matrix of the rock. The amount of gas present in shale reservoirs is related to the organic content of the rock, and the shales act as the source, the reservoir, and the trap for the natural gas. The location of unconventional reservoirs is well known in many gas-producing regions of the United States and Canada, which contrasts with the need to search for conventional natural gas traps. Unconventional reservoirs are more sensitive to the levels of technology available for extraction from a known accumulation and whether or not such technology can be economically deployed at a given market price.

4.2 Technologies for Conventional Reservoir Development

4.2.1 Exploration Technologies Exploration for conventional natural gas today benefits from advanced geophysical technologies, new concepts of basin analysis and the sequences in which basin-filling rocks were deposited, satellite remote sensing, and a host of powerful computer-based information-handling and interpretation capabilities. Geophysical technologies include standard two-dimensional (2-D) seismic surveys but with improved resolution and the ability to estimate reservoir attributes and fluid content. Two-dimensional seismic surveying yields individual cross-section lines but without the ability to construct such sections at orientations other than that in which the line was acquired. Nor can 2-D data be used to construct images of continuous surfaces through a potentially gas-bearing rock volume to examine reservoir structure and quality.

4.2.1.1 Three-Dimensional Seismic Imagery Three-dimensional (3-D) data, acquired on closely spaced grid lines at right angles to each other, can be used to reconstruct seismic images at any orientation through the data volume and to image surfaces at any level within the volume. Typically, 3-D data are used for field development. Recently, however, 3-D seismic surveys are being deployed over large areas to define complex structures, especially in the U.S. Gulf of Mexico where migration of bedded salt has occurred. This offshore province yielded 26% of total U.S. natural gas production in 1991 and has, on average, produced about two-thirds gas and one-third oil in terms of its total supply mix. The expense of drilling wells in deep waters (200–1000 m) in the offshore U.S. Gulf of Mexico mandates that 3-D seismic surveying be utilized to reduce the risk of offshore exploration and development.

4.2.1.2 Basin Analysis and Sequence Stratigraphy Explorationists today are integrating seismic, well, and biostratigraphic data to understand the depositional framework and structural style of a basin. Sediment input and relative changes in the sea level are key controls on depositional sequences that define parts of a basin potentially prospective for the trapping of natural gas and oil. Such controls influence the distribution of source beds, the possible timing of hydrocarbon migration, and the relative position of reservoir rocks as they exist today. The process of evaluating all such controls and the resulting reservoir geometry is termed *sequence stratigraphy*.

4.2.2 Development Technologies Advanced technologies for development of natural gas reservoirs include innovations in reservoir geological characterization, geophysical detection, and drilling and production techniques. In mature producing trends in the United States and Canada, improved extraction from known reservoirs through advanced technologies has significant potential to sustain the yield of natural gas from known producing regions, thus extending the economic life of surface production infrastructure and the natural gas pipeline network.

4.2.2.1 Reservoir Characterization Technology Geologists and reservoir engineers are cooperating to an ever greater degree to define the detailed internal character of reservoirs, even where wells are separated by several hundred to 1000 m. Geological aspects of reservoir characterization involve predictions about the interwell area that can only be indirectly assessed from information collected at the wellbore or obtained seismically between wellbores. Nonuniform deposition of the rocks forming the reservoir unit, cementation of pore spaces between grains that reduces gas storage capacity, and structural changes after burial, such as faulting, affect flow of natural gas to the wellbore. Such variabilities in the reservoir rock, termed *reservoir heterogeneities*, have resulted in near-original reservoir pressures being encountered during drilling in reservoirs thought to have been largely depleted. Incomplete recovery of gas from a field because of flow boundaries within the reservoir can be overcome by additional selective drilling or recompletion of other, untapped reservoirs once the primary reservoir target has been depleted (see Sec. 4.4). Incremental gas recovery in mature fields in the United States is a major topic of current research that is expected to help sustain gas supplies well beyond current reserve estimates.

4.2.2.2 Geophysical Detection Technologies Advanced geophysical detection technologies applied to gas resource development are used to define the physical framework of the reservoir within a producing field. This framework includes the composition of the reservoir rock and distribution of pore space within the reservoir, and, under certain conditions, yields the detection of potential indicators of natural gas vs formation water within the reservoir. The latter is termed *bright spot technology*. Virtually all of this type of geophysical work is now done with 3-D seismic data (see above) as a result of the much greater detail available. Automated routines now available on workstations that have 2–10 (or more) gigabytes of memory allow rapid interpretation of a 3-D data set. Because of the importance of careful targeting of new wells in reservoir development, the proportion of 3-D seismic surveying to all seismic lines shot in the Gulf of Mexico and the North Sea has gone from 10% in 1980 to 60% in 1989 (Hansen *et al.*, 1989).

Future geophysical technologies are now being designed to avoid problems with surface conditions that absorb seismic energy. Cross-borehole tomographic techniques are emerging that use an energy source in one well and a receiver array in another well to image the interwell area. This approach is akin to computerized axial tomography (CAT) scanning, a diagnostic medical technique consisting of multiple x-ray scans through the body. The distance between wells with 1992 technology is limited to 200–350 m, somewhat less than typical gas well spacing. Higher-energy sources should extend application of this technique to wellbores 600 m or more apart, which will result in more frequent application in gas field development.

4.2.2.3 Horizontal Drilling Among advanced drilling and production technologies, horizontal drilling offers the opportunity to drain naturally fractured reservoirs more efficiently. It is also potentially effective to overcome depositional variations that compartmentalize reservoirs (reservoir heterogeneity) and prevent producible gas from flowing to the wellbore. Horizontal drilling has been a technical and economic success in the Austin Chalk trend, a carbonate oil reservoir with large volumes of associated gas, in south and central Texas, U.S.A. Experiments are now underway to horizontally drill fractured shale reservoirs, a type of unconventional gas reservoir, to increase production. Horizontal drilling of low-permeability sandstone reservoirs has been successful in the North Sea, offshore of the Netherlands, to increase both oil and gas production. There, multiple hydraulic fracture treatments have been carried out from the horizontal segment of the well, a technique that has widespread potential application to low-permeability natural gas reservoirs. Although horizontal drilling requires experience in each region where gas is produced, successful horizontal wells may have productivity 2.5–5 times that of vertical wells at 1.5–2.5 times the cost once that experience is gained.

4.3 Technologies for Unconventional Reservoir Development

Unconventional reservoirs of natural gas are ones whose location is usually well known as a result of previous exploration and development activities but that often have not been as extensively exploited as the conventional accumulations. The primary unconventional reservoirs include low-permeability (tight) sandstones, coalbed methane, and fractured shales, all of which are contributing to current world gas supplies. Other unconventional natural gas sources are known, such as gas entrained in geopressured brines, and gas hydrates, which are physical combinations of gas and water in which the gas molecules fit into a crystal structure resembling that of ice. Gas hydrates are extensive along some submerged continental margins and along the North Slope of Alaska. Very little is known about the economics of gas hydrate development. Extracting gas from geopressured brines suffers from the low energy content of the brines, typically 0.6–1.1 m^3 of gas per barrel (159 l) of brine, and the disposal problems associated with production of large volumes of saline fluids.

4.3.1 Advanced Extraction Technologies

Near-term commercially viable unconventional gas resources will be extracted through improved technology applied to tight sandstones, coalbeds, and fractured shales. The technologies are those of improved extraction from largely known accumulations and, therefore, are characterized as extraction techniques rather than either exploration or development technologies.

4.3.1.1 Hydraulic Fracturing The process of hydraulic fracturing involves pumping mixtures of fluids and a solid, such as loose quartz sand, into a natural gas reservoir with the objective of fracturing the reservoir rock. The resulting fracture in the rock, especially if held open by sand or other propping material carried into the reservoir, forms a direct, high-permeability pathway to the wellbore. Hydraulic fracturing is designed to overcome the low permeability of a rock formation that otherwise would not yield gas to a wellbore at economic rates sufficient to justify drilling and operating the well. A key issue in hydraulic fracturing is control of the geometry of the hydraulic fracture to maximize its extension in the reservoir, to limit its height to the target productive interval, and to know its orientation.

The geometry of the fracture is controlled by the stress state in the reservoir, which relates both to the regional basin-scale characteristics, such as the tension direction generally normal to the shoreline of the U.S. Gulf of Mexico, and to the physical character of individual beds (toughness relative to fracture propagation). If a hydraulic fracture penetrates upward rather than outward from a well, it achieves minimal contact with gas-saturated rock and production is inefficient. Excessive formation water may be brought into the well instead of natural gas, and more wells will need to be drilled to extract the gas resource from a given area.

Advanced techniques are being applied to model fracture propagation and to test formations in advance with very small-scale fracture treatments for analysis of fracture orientation and of pumping pressures required to induce fracturing. The most sophisticated techniques involve modeling fracture propagation through analysis of downhole pressures, flow rates, and fluid characteristics as the fluids and sand proppant are being pumped into the formation. Pumping rates, sand concentrations, and fluid properties are then varied to ensure that the maximum volume of treatment fluids penetrate the targeted reservoir interval.

4.3.1.2 Cavity Completions A new technique applied to coalbeds capable of yielding methane to wellbores is cavity completion. This technique involves pulsing a well by alternately injecting fluid and then allowing the fluid to flow back to the surface under rapid release of pressure. The target coalbeds immediately below the casing of the wellbore break down along their natural fractures, termed cleat, into particles of coal that are transported to the surface during the pulsing process. The result is a cavity that connects the wellbore to the coal cleat system. Methane is then produced from the well through desorption from the coal, typically after a period of dewatering that may be limited to days or that may take months to achieve economic gas production rates. Cavity completions have been successful in the San Juan Basin of northeast New Mexico, U.S.A., where extensive research into their development was carried out by GRI in 1990–1992.

4.4 Reserve Growth in Mature Fields

Not all improvements in natural gas production technologies in the early 1990s relate to field techniques for reservoir stimulation or well completion. The concept of reserve growth in mature fields targets a resource that has been assessed at between 6.68×10^{12} m^3 (National Petroleum Council, 1992) and 7.48 $\times$ 10^{12} m^3 (Energy Information Administration, 1990) but will probably be closer to 8.5 $\times 10^{12}$ m^3 in the U.S. lower 48 states once fully developed. The availability of this gas resource is dependent on understanding reservoir compartmentalization inherited from the original depositional system of the reservoir rock and its subsequent modification during burial. Thus, reserve growth is dependent on the concept of reservoir heterogeneity and an understanding of how the reservoir framework was deposited.

Because gas is a much more mobile fluid in the reservoir than oil, gas reservoirs of conventional permeability have been developed in the United States on relatively wide well spacings (259–130 hectares per well) compared to oil (16–8 hectares per well, or less). It was largely assumed that the mobility of natural gas would overcome reservoir heterogeneity at the production time scale. We now know that this is not the case in all reservoirs. Those with complex depositional geometries or those that have an areally or vertically variable reservoir quality cannot be uniformly drained by widely spaced wells. It has long been known that reservoirs broken up by faulting must be more extensively drilled, but recognition of the extensive gas reserve growth potential related to the former factors only emerged in the United States during the period 1988–1992. Certainly, the increasing maturity of the U.S. resource base, while seen as a problem from the viewpoint of exploration, now offers the required data base for reserve growth realization. Resultant reservoir development and redevelopment approaches will extend to applicable reservoirs beyond the United States as current world gas reserves mature and must be replaced.

5. GAS GATHERING, PROCESSING, AND DISTRIBUTION

Beyond the wellhead, gas gathering, processing, and distribution begins in the field where gas is produced. Water and associated oil may be separated from the gas stream in equipment serving one or several wells. A series of small-diameter pipelines then delivers gas to a natural gas processing plant. Natural gas liquids and impurities, such as carbon dioxide, hydrogen sulfide, and nitrogen, may be separated at a processing plant that serves one or more fields within a producing trend. The natural gas leaving such a plant would typically be of a quality suitable for consumption [approximately 1056 MJ (10^6 btu) per 28 m^3 (10^3 ft^3)].

In the United States, there are about 1.9 million km of pipelines that gather and move natural gas from the producing regions to distributors and end users. The U.S. natural gas demand has a high degree of seasonality with 60% of consumption occurring during the winter heating season (October through March). As a result, underground storage is needed to balance the relatively constant flow from natural gas reservoirs with seasonality, peak-month, and even peak-day usage that can vary with weather patterns. Peak-month delivery may be 4–8 times low-month delivery, and peak winter day usage may be 1.5–3 times greater than average winter day requirements. There are now about 370 underground storage fields in the United States with a working gas capacity of about 113 $\times 10^9$ m^3. These fields can provide (between 1.5 and 1.8) $\times 10^9$ m^3 per day on a peak day. The estimated 1991 peak-day U.S. capability to transport natural gas was 3.39 $\times 10^9$ m^3, whereas the estimated 1991 peak-day U.S. demand was 2.89 $\times 10^9$ m^3. The estimated annual delivery capability in 1991 was 680 $\times 10^9$ m^3, compared to actual 1991 consumption of 544 $\times 10^9$ m^3. Of the latter volume, 47 $\times 10^9$ m^3 was supplied through net pipeline imports from Canada (National Petroleum Council, 1992; Gas Research Institute, 1993).

The business of natural gas distribution changed markedly in the United States in the decade of the 1980s. Pipelines previously played the role of middleman between producers and the local gas utilities or local distribution companies (LDCs). Pipelines purchased the gas from producers, transported it as system gas owned by the pipeline, and sold it to the LDCs. Contractual relationships tended to be long term and involved relatively few parties, and the industry's interstate activities were heavily regulated. As the industry entered the 1990s, the majority of interstate gas transported by the pipelines was not owned by those pipelines; the pipeline company only provided the transportation service. Pipelines are now much less in the business of aggregating supplies. Independent gas marketers, individual producers, and

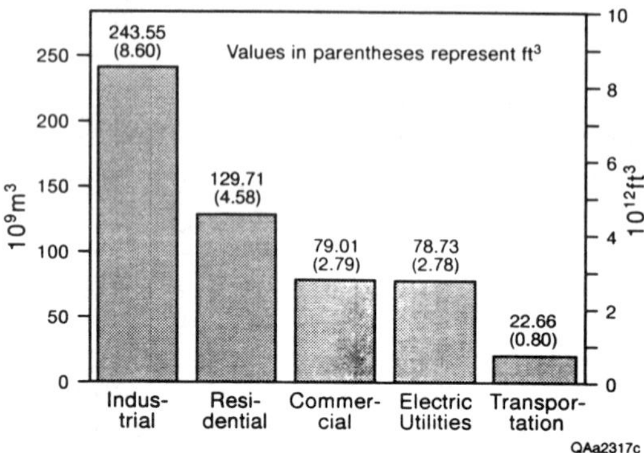

FIG. 1. Natural gas consumption in the United States by sector, 1991 (Energy Information Administration, 1992).

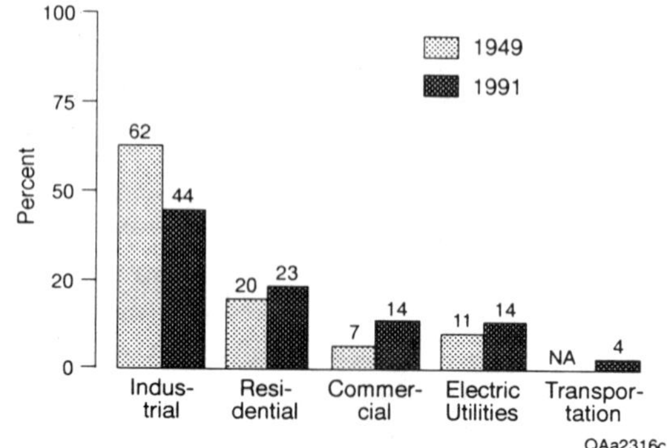

FIG. 2. Comparison of U.S. natural gas consumption by sector, 1949 and 1991 (Energy Information Administration, 1992).

groups of producers are selling gas, and the pipelines are under regulatory pressure to unbundle their services to distinguish, for example, costs of transportation and of storage. The LDCs and large end users are now in the business of building their own supply portfolios, arranging supply contracts, and balancing purchases with demand. Trends toward disaggregation of functions within the natural gas business are likely to continue particularly with respect to "open access" to space on pipelines for all gas shippers.

6. UTILIZATION AND APPLICATIONS

Natural gas can fill a variety of energy needs in fuel and power applications and as a raw material or feedstock for petrochemicals, fertilizers and other applications. Among the five natural gas–consuming sectors in the United States, industrial uses represented the greatest demand, followed by residential, commercial, electric utilities, and transportation (Fig. 1). Over the last 40 years, however, the volume of the industrial market has declined as a component of the total gas market, whereas the commercial sector has doubled its share of U.S. total gas use (Fig. 2). Total gas use peaked with top production in the early 1970s, then declined irregularly through 1986, but has seen a rebound up to 1991 to the consumption level of 1975 (Fig. 3) (Energy Information Administration, 1993a). These recent increases in U.S. gas use are in part driven by new end-use technologies for natural gas, removal of regulations on natural gas use in boiler and utility applications, and recognition of natural gas's benefits as an environmentally benign, domestically abundant fuel. It would be appropriate to consider current and future utilization of natural gas in each of the five major use sectors for the United States and other leading consuming nations (Tables 1 and 6).

6.1 Industrial Natural Gas Use

Industrial energy demand is the largest single category of natural gas use in the United States, Germany, and Canada (Fig. 1; Table

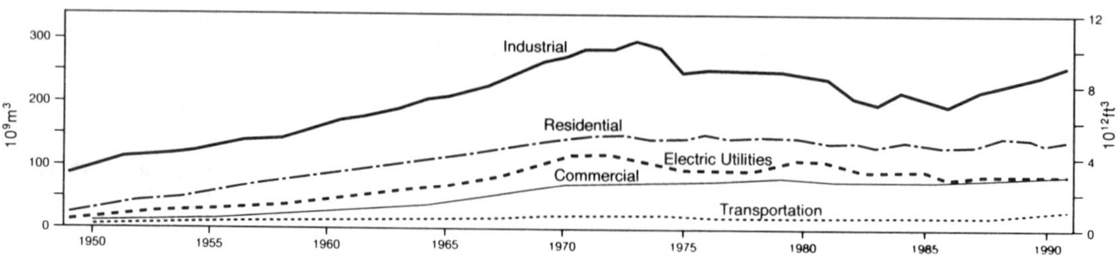

FIG. 3. History of U.S. natural gas consumption, 1949 through 1991 (Energy Information Administration, 1992).

6). To meet U.S. industrial demands, natural gas use for fuel and power grew between 5 and 10% annually between 1986 and 1991. This rate of growth is expected to slow from the mid-1990s to 2000 because natural gas prices are expected to increase at relatively greater rates than other fuels. Cogeneration of process steam and power represents 40% of the industrial power category of natural gas use and is the largest single category of use (Holtberg *et al.*, 1992).

Natural gas is used in the production of ammonia, hydrogen, and methanol, with hydrogen primarily being produced and used by oil refineries. Approximately 60% of U.S. natural gas use as feedstock is for ammonia production and about 25% is for hydrogen production. Methanol production accounts for about 17% of feedstock demand. Methanol is used in developing other chemicals as gasoline additives. Total industrial feedstock consumption of natural gas is projected to decline by about 15% by 2010, primarily as a result of reduced ammonia consumption for domestically produced fertilizers (Holtberg *et al.*, 1992).

6.2 Residential Natural Gas Use

Natural gas accounts for the largest portion, 45%, of total U.S. residential energy consumption, followed by electricity at 32%. Space heating accounted for 65% and water heating about 15% of 1990 residential gas consumption. Cooking, clothes drying, etc., accounted for the balance. The share of U.S. gas consumption used for space heating is expected to decline to 55% by 2010 as increased equipment efficiencies and improved thermal integrity of buildings more than offsets projected growth in the number of space-heating installations (Holtberg *et al.*, 1992).

The residential gas market is the major part of the "core" market for natural gas. Residential natural gas customers supplied by LDCs do not have alternative fuel capabilities and are thus captive with respect to their fuel choice. Such customers generally pay the highest prices for natural gas and benefit from the most highly developed parts of the LDCs distribution system. Because two-thirds of U.S. residential consumption is for space heating, year-to-year residential demand is highly dependent on winter weather severity, as it is worldwide for major countries using gas for residential consumption (Table 6).

6.3 Commercial Natural Gas Use

Natural gas consumed in the commercial sector (Table 6) is predominantly for space heating, accounting for 50% of gas use in this sector in the United States. The other half of U.S. commercial gas use is spread among water heating (5%), cogeneration (7%), and the balance for other applications such as cooking, process heat, etc. Gas consumption for space cooling in 1990 accounted for just over 1% of U.S. commercial gas consumption, but space cooling is a major potential growth area using technologies such as pack-

Table 6. Gas consumption by sector, 1990, in seven leading consuming nations, 10^9 m^3. Percentage indicates percent of total reported consumption accounted for by the five sectors indicated (from Energy Information Administration, 1993b; Japan Gas Association, 1993; United Nations, 1993).

	Former Soviet Union (71%)[a]	United States (93%)	Germany (88%)	Canada (94%)	United Kingdom (97%)	Netherlands (97%)	Japan (not available)
Industrial	49.88	198.75	22.89	37.02	18.28	9.23	4.02
Residential	9.29[b]	124.35	16.43	13.92	28.48	9.63	7.76
Commercial	14.80	74.28	7.77	10.82	8.33	7.78	2.56
Electric utility	247.56	78.93	12.01	2.21	1.10	9.96	43.30[c]
Transport[d]	0.37	18.69[e]	...[f]	0.07	...[f]	...[f]	...[f]

[a]Not included are petrochemical feedstocks, 14.8%, and losses in transport and distribution, 13.6%.
[b]1991 data; 1990 data not available.
[c]Estimated.
[d]Vehicles, except in U.S.
[e]Mostly pipeline compressor fuel (see text).
[f]Negligible volume.

aged cogeneration systems, heat pumps, and engine-driven chillers (Holtberg *et al.*, 1992). Use of such technologies would tend to balance the seasonal demand for natural gas, thus making more efficient use of the entire gas production and distribution infrastructure where climatic conditions lead to demand for cooling in commercial (and residential) facilities.

6.4 Electric Utility Use of Natural Gas

Among the leading consuming nations, the states of the former Soviet Union consume the most natural gas for electric generation, more than three times U.S. consumption (Table 6). Current (1991) U.S. natural gas usage of 78.93 $\times$ 10^9 m^3 per year for electricity generation is slightly greater than commercial sector consumption. Gas consumption in the U.S. electric utility sector is expected to grow as the result of the need to meet the Clean Air Act Amendments of 1990 and the need for new peaking and intermediate load service, particularly in the second half of the 1990s. Between 1995 and 2000, projected U.S. gas usage for electricity generation is expected to grow from 88 $\times$ 10^9 m^3 to 119 $\times$ 10^9 m^3, or at almost 6% per year. Most of this growth (31 $\times$ 10^9 m^3) is expected to be the result of cofiring of coal and natural gas (Holtberg *et al.*, 1992). Federal restrictions on use of natural gas for boiler fuel has held down gas use in this sector in the United States until the early 1990s.

6.5 Transportation Use of Natural Gas

This use sector includes both pipeline transportation and use in vehicles; vehicular use is the smallest use sector worldwide (Table 6). Three-quarters of the current natural gas used for transportation in the United States is consumed by natural gas pipeline compressors. Such use is expected to grow as part of the overall transportation use growth of 3% to 4% per year from 1990 through 2010. Use of compressed natural gas in U.S. vehicles will grow much more substantially at a rate of 26% per year from about 0.11 $\times$ 10^9 m^3 in 1990 to about 10.8 $\times$ 10^9 m^3 in 2010. Fleet vehicle use is the most likely source of this growth in view of public policy requirements for use of alternative fuels and given the limited number of public refueling stations. A projection of U.S. natural gas vehicle use sees 3.8 million out of 185 million vehicles using natural gas by 2010 (Holtberg *et al.*, 1992).

7. WORLD NATURAL GAS TRADE

Given the extensive reserves and resources of natural gas available in the world, international natural gas trade is predicted to expand from about 283 $\times$ 10^9 m^3 annually in 1990 to about 793 $\times$ 10^9 m^3 annually in 2015 (MacDougall and Linder, 1992). Pipeline delivery of gas accounted for 74% of such trade in 1990, and the balance was LNG. It is expected that increases in North American natural gas trade via pipeline will require that additional lines and interconnects be built. Current pipeline facilities are very near capacity, according to preliminary data for 1992 throughput volumes to the United States from Canada. With respect to U.S.–Mexico trade, more interconnects are planned to allow U.S. gas to flow into northern Mexico. Such trade has the potential to expand following the 1993 ratification of the North American Free Trade Agreement.

7.1 Liquefied Natural Gas Trade

Trade in LNG is dependent on facilities for liquefaction, shipping by LNG tanker, and regasification. In the Atlantic market, Algeria is currently the prime potential supplier. Facilities planned or under discussion in Venezuela, Nigeria, and Norway would add to the current 34 $\times$ 10^9 m^3 of Algerian liquefaction (design) capacity. The United States has four facilities capable of receiving LNG cargoes: Everett, Massachusetts; Cove Point, Maryland; Elba Island, Georgia; and Lake Charles, Louisiana. The Everett and Lake Charles facilities were operated in 1991 with an estimated combined volume for the year of 1.74 $\times$ 10^9 m^3 (White, 1992b).

Of major impact in the Atlantic market would be Nigeria's Bonney LNG project and Venezuela's Cristóbal Colón LNG project. The Bonney project is scheduled to begin deliveries in 1997 with the bulk of the shipments going to Italy, Spain, and France. The Cristóbal Colón project is designed to begin deliveries of over 5.7 $\times$ 10^9 m^3 per year to the United States and possibly Europe in 1997 or 1998. Venezuela has significant natural gas reserves (3.12×10^{12} m^3) but limited domestic demand, leading to the opportunity to export natural gas. Venezuelan LNG facilities will have the advantage of being closer to the United States than other LNG projects, which should decrease transportation costs. One assessment suggests, however, that the Venezuelan project will not be cost competitive with Algerian supplies until after 2010 (National Petroleum Council, 1992).

In the Pacific, Indonesia dominates the LNG market with a 50% share of liquefaction capacity. Malaysia, Brunei, Australia, Abu Dhabi, and the United States (Alaska) also export within the region, and together all countries have a total annual export capacity of 57×10^9 m^3. The U.S. LNG exports to Japan amounted to 1.5×10^9 m^3 in 1991 from a liquefaction plant located at Cook Inlet, Alaska. The Trans Alaska Gas System (TAGS) project is proposed to export initially about 10.1×10^9 m^3 per year of North Slope gas to Pacific nations, particularly Japan, South Korea, and Taiwan, beginning in 2000. The project will involve a 1300 km pipeline from North Slope oil fields to a terminal in Valdez, Alaska for tanker loading. Gas from these fields will reportedly not be available until 2000 (or even later) because of the need to keep North Slope oil reservoirs under pressure to maintain oil recovery efficiency.

8. SUMMARY AND FUTURE OUTLOOK

Natural gas, principally methane, is an abundant resource in the United States and worldwide. The United States (lower 48 states) has an assessed resource base as of 1 January 1991 of 36.66×10^{12} m^3 (including 4.53×10^{12} m^3 of proved reserves) produceable using advancing technology through 2010. Worldwide proved gas reserves as of 1 January 1993 were 137.9×10^{12} m^3. In the United States, natural gas use has increased 22% between 1986 and 1992, primarily in recognition of natural gas's suitability for industrial uses (low emissions, shorter lead times for gas-fired equipment), including cogeneration. Much of the future supply to continue this growth and to expand gas use in the future, such as an alternative fuel in the transportation sector, will be dependent on advanced recovery technology applied to a better geologically characterized resource base. Unconventional domestic resources such as low-permeability sandstones, coalbeds, and shales will provide increasing volumes of reserves, and old conventional reservoirs in mature fields will be reexplored and redeveloped using new concepts of reservoir geology. Advanced geophysics, horizontal drilling, improved reservoir characterization, improved hydraulic fracturing, and much greater capability to model and develop reservoirs will add to natural gas availability at moderate costs.

Natural gas use in the United States will remain strong in the industrial and commercial sectors but could be affected by continuing low to moderate world oil prices. Dual-fuel–capable facilities in the industrial and electric utility sectors are highly sensitive to delivered price on a Btu basis unless overridden by emissions requirements. End-user concern with natural gas supply has shifted from the volume of the resource base to issues of deliverability. Major U.S. companies have been shifting operations overseas, leading to concerns that the smaller companies and independent producers will be required to provide an increasing share of the domestic supply. The outlook for that supply remains good in the face of these changes as technology leads to improved per-well recoveries and technology-transfer activities by government, private research groups, universities, and trade organizations continue to expand. Such technologies are likely to become an increasingly important U.S. export commodity.

North American and world gas trade will increase as demand increases as a result of industrialization and concurrent concerns for the environment. Most such trade will be by pipeline, but new LNG projects are expected to become economic over the next two decades in several countries as demand grows and resources are developed beyond domestic needs. The United States itself may become an exporter of LNG from Alaska to the Pacific Rim unless pipeline facilities to the lower 48 states become economic and are constructed.

Whereas natural gas was once regarded as limited in availability, it is now known to be an abundant resource whose deliverability is dependent on efficient conversion of resources to proven reserves. In the United States, a balance between supply and demand, encouraged by reduced government regulation and improved efficiency in production and distribution, is emerging in 1992-1993. The U.S. oversupply that has persisted as the so-called "gas bubble" since 1982 has ended. Continued end-use market penetration by natural gas will likely be fostered by environmental benefits of natural gas use but remains fully dependent on the confidence in supply available to the end user.

GLOSSARY

Associated Gas: Gas produced concurrently with oil that was either in solution with oil in the reservoir or in contact with the crude oil as a gas cap within the reservoir.

Conventional Reservoirs: Typically, sandstones and carbonates with greater than 0.1 millidarcy permeability capable of readily flowing gas to a wellbore.

Conventional Resources: Resources included in this category are natural gas and natural gas liquids that exist in conventional reservoirs or in a fluid state amenable to extraction techniques employed in traditional development practices. They occur as discrete accumulations.

Field: A single pool or multiple pools of hydrocarbons grouped on, or related to, a single structural or stratigraphic feature that has trapped natural gas and from which the gas is being produced.

Inferred Reserves: That part of the identified economic resources, over and above measured and indicated reserves, that will be added through extensions and revisions and with the addition of new pay zones in discovered fields. Equivalent to probable resource.

Natural Gas: A naturally occurring mixture of hydrocarbon and nonhydrocarbon gases, with methane being the principal combustible component.

Nonassociated Gas: Free natural gas not in contact with oil, nor dissolved in crude oil, in the reservoir.

Reserve Growth: The addition of proved reserves from a developed reservoir, or series of reservoirs comprising a gas field, as a result of more intensive development (new drilling, well recompletions) within the established boundaries of the field. Undrained compartments, deeper previously undrilled reservoirs, and shallow bypassed reservoirs may all contribute new reserves.

Reserves: Part of the identified economic resource that is estimated from geologic evidence supported directly by engineering data. They are demonstrated with reasonable certainty to be recoverable in future years from known reservoirs under existing economic and operating conditions and are generally equivalent to "proved reserves."

Unconventional Reservoirs: Typically, low-permeability sandstones, shales, and coals with matrix permeability of less than 0.1 millidarcy requiring special treatment to flow gas to the wellbore. Flow in natural or hydraulically induced fractures is essential to economic production rates.

Undiscovered Resources: Resources estimated to exist, on the basis of broad geologic knowledge and theory, outside of known fields or known accumulations. Also included are resources from undiscovered pools within known fields to the extent that they occur as unrelated accumulations controlled by distinctly separate structural features or stratigraphic conditions.

Works Cited

American Gas Association (1990), *Changes in Natural Gas Recovery Technology & Their Implications,* a report of A.G.A. Planning and Analysis Group, September 1990, Arlington, Virginia, 34 pp.

Anonymous (1992a), "Worldwide Production Report," *Oil Gas J.* **90** (52), 39–43.

Anonymous (1992b), "Russia Pins Energy Hopes on Western Siberia Gas," *Oil Gas J.* **90** (36), 17–20.

Arthur Andersen & Company (1987), *Natural Gas Trends, 1987–88 Edition,* Chicago, IL: Cambridge Energy Research Associates, 154 pp.

British Petroleum Company (1991), *BP Statistical Review of World Energy: June 1991,* London: Ashdown Press Ltd.

British Petroleum Company (1993), *BP Statistical Review of World Energy: June 1993,* London: Leycol Printers Ltd.

DeGolyer and MacNaughton (1991), *Twentieth Century Petroleum Statistics: 1991,* 47th ed., Dallas: DeGolyer and MacNaughton.

Dolton, G. L., Carlson, K. H., Charpentier, R. R., Coury, A. B., Crovelli, R. A., Frezon, S. E., Khan, A. S., Lister, J. H., McMullin, R. H., Pike, R. S., Powers, R. B., Scott, E. W., Varnes, K. L. (1981), *Estimates of Undiscovered Recoverable Conventional Resources of Oil and Gas in the United States,* United States Department of the Interior, Geological Survey Circular 860, Alexandria, VA: U.S. Geological Survey.

Energy Information Administration (1990), *The Domestic Oil and Gas Recoverable Resource Base: Supporting Analysis for the National Energy Strategy,* Washington, DC: Office of Oil and Gas, U.S. Department of Energy, Service Report SR/NES/90-05, 56 pp.

Energy Information Administration (1992), *Geologic Distributions of U.S. Oil and Gas,* Washington, DC: Office of Oil and Gas, U.S. Department of Energy, DOE/EIA 0557, 137 pp.

Energy Information Administration (1993a), *Annual Energy Review 1992,* Washington, DC: Office of Energy Markets and End Use, U.S. Department of Energy, DOE/EIA-0384(92), 350 pp.

Energy Information Administration (1993b), *Monthly Energy Review, February 1993,* Washington, DC: Office of Energy Markets and End Use, U.S. Department of Energy, DOE/EIA-0035(93/02), 164 pp.

Enron Corp. (1991), *Enron Corp's Outlook for Natural Gas: Fueling the Future into the 21st Century,* Houston, Texas: Enron Corp.

Finley, R. J., Fisher, W. L. (1991), *Natural Gas Reserve Additions: Recent Trends and Future Potential in the Lower 48 States,* Society of Petroleum Engineers Hydrocarbon Economics and Evaluation Symposium, Dallas, Texas, April 11–12, 1991, SPE Paper No. 22037, pp. 209–216.

Finley, R. J., Fisher, W. L., Seni, S. J., Ruppel, S. C., White, W. G., Ayers, W. B., Jr., Dutton, S. P., Jackson, M. L. W., Banta, N. (1988), *An Assessment of the Natural Gas Resource Base of the United States,* Bureau of Economic Geology, Report of Investigations No. 179, Austin, Texas: The University of Texas at Austin.

Gas Research Institute (1993), *Policy Implications of the GRI Baseline Projection of U.S. Energy Supply and Demand to 2010,* Washington, DC: Gas Research Institute.

Gold, T. (1987), *Power from the Earth: Deep Earth Gas—Energy for the Future,* J. M. Dent and Sons Ltd., 208 pp.

Gold, T., Soter, S. (1980), "The Deep Earth-Gas Hypothesis," *Sci. Am.,* **242** (6), 154–161.

Hansen, T., Kingston, J., Kjellesvik, S., Lane, G., l'Anson, K., Naylor, R., Walker, C. (1989), "3-D Seismic Surveys," *Oilfield Rev.,* **1** (3), 54–61.

Holtberg, P. D., Woods, T. J., Lihn, M. L., Koklauner, A. B. (1992), *1992 Edition of the GRI Baseline Projection of U.S. Energy Supply and Demand to 2010,* Washington, DC: Gas Research Institute.

Hunt, J. M. (1979), *Petroleum Geochemistry and Geology,* San Francisco, CA: W. H. Freeman and Company.

Japan Gas Association, (1993), "Gas Industry in Japan," briefing paper, Tokyo, Japan.

Kvenvolden, K. A. (1988), in: L. B. Magoon (Ed.), *Petroleum Systems of the United States,* USGS Bulletin 1870, Washington, DC: U.S. Geological Survey, Department of the Interior, pp. 44–45.

MacDougall, M. W., Linder, P. T. (1992), *Long-Term Outlook for World Gas Trade: 1990–2015,* Calgary, Alberta: Canadian Energy Research Institute, Study No. 46, ISBN 0-920522-78-5, 271 pp.

Mast, R. F., Dolton, G. L., Crovelli, R. A., Root, D. H., Attanasi, E. D., Martin, P. E., Cooke, L. W., Carpenter, G. B., Pecora, W. C., Rose, M. B. (1989), *Estimates of Undiscovered Conventional Oil and Gas Resources in the United States—A Part of the Nation's Energy Endowment,* United States Department of the Interior, U.S. Geological Survey, Minerals Management Service, Washington, DC: U.S. Government Printing Office, 44 pp.

National Petroleum Council (1992), *The Potential for Natural Gas in the United States, Executive Summary,* Washington, DC: U.S. Department of Energy.

National Research Council (1990), *Fuels to Drive our Future,* Washington, DC: National Academy Press.

Potential Gas Committee (1991), *Potential Supply of Natural Gas in the United States (December 31, 1990),* Colorado School of Mines, Golden, Colorado: Potential Gas Agency.

United Nations (1993), *Annual Bulletin of Gas Statistics for Europe,* Geneva: Economic Commission for Europe.

Waples, D. W. (1985), *Geochemistry in Petroleum Exploration,* Geological Sciences Series, Boston, MA: International Human Resources Development Corporation.

White, B. (1992a), *The U.S.–Mexican Natural Gas Relationship,* Gas Energy Review, Vol. 20, No. 12, Arlington, Virginia: American Gas Association, pp. 18–27.

White, B. (1992b), *LNG Import/Export Projects—Current Status Report,* Gas Energy Review, Vol. 20, No. 10, Arlington, Virginia: American Gas Association, pp. 10–20.

Wood, P. (1993), *Russia and Ukraine Face Off Over Prices, Payments for Gas,* Natural Gas Week International, a supplement to Natural Gas Week, Washington, DC: The Oil Daily Company.

Further Reading

American Gas Association (1988), *The Future of Natural Gas in the United States,* Arlington, Virginia: Planning and Analysis Group, American Gas Association, April 1988. A general review of U.S. gas use and supply through 2000.

American Gas Association (1990), *Glossary for the Gas Industry,* prepared by Statistics and Load Forecast Methods Committee, Arlington, Virginia: Planning and Analysis Group, The American Gas Association, Fifth Edition. A glossary of natural gas industry terms that includes conversion factors and definitions of abbreviations.

American Petroleum Institute (1991), *1991-92 Guide to Petroleum Statistical Information,* 7th ed., Washington, DC: Central Abstracting & Information Services. A guide to public and private sources of statistical information related to natural gas and oil worldwide.

Barker, C. (1979), *Organic Geochemistry in Petroleum Exploration,* Tulsa, OK: American Association of Petroleum Geologists Department of Education Activities. An introduction to the fundamental ideas of organic geochemistry.

Energy Information Administration (1992), *Geologic Distributions of U.S. Oil and Gas: July 1992.* Washington, DC: U.S. Government Printing Office. The second report in an ongoing series that provides results of domestic oil and gas field-level data constructions, and analyses thereof, founded on EIA's unique collection of annual domestic proved reserves estimates.

Energy Modeling Forum (1988), *North American Natural Gas Markets, Volumes I–III,* Stanford University, Stanford, California, EMF Report 9, variously paginated. A summary of the research by an Energy Modeling Forum working group on the evolution of the North American natural gas markets between 1988 and 2010.

Enron Corp (1993), "The Outlook for Natural Gas," *The Enron Outlook,* March 1993, Houston, TX: Enron Corp. Enron Corporation's outlook for supply and demand for natural gas in North America.

Finley, R. J. (1993), "A Positive Assessment of the U.S. Natural Gas Resource Base," in: S. L. McDonald and M. Mohammadioun (Eds.), *The Role of Natural Gas in Environmental Policy,* Austin, TX: Bureau of Business Research, University of Texas at Austin, pp. 1–7. Describes U.S. natural gas supplies in the context of demand based on environmental policies.

Garrett, C. M., Jr., Hocott, C. R., Finley, R. J., Galloway, W. E., Siqueira, C., Davis, M. D., Murray, R. C. (1985), *Analysis of Negative Revisions to Natural Gas Reserves in Texas,* Bureau of Economic Geology, The University of Texas at Austin, Austin, Texas, Final Report prepared for the Gas Research Institute under Contract No. 5083-800-0908, GRI-85/0111. An assessment of why U.S. natural gas reserves were substantially revised downward.

Julius, D., Mashayekhi, A. (1990), *The Economics of Natural Gas: Pricing, Planning and Policy.* Oxford: Oxford University Press. A comprehensive survey of the major issues involved in the pricing of gas and in the planning of its development.

National Research Council (1991), *Undiscovered Oil and Gas Resources: An Evaluation of the Department of the Interior's 1989 Assessment Procedures.* Washington, DC: National Academy Press. A review, conducted by the National Research Council's Committee on Undiscovered Oil and Gas Resources, of the procedures of the Department of the Interior's most recent resource assessment, which was published in 1989.

U.S. Congress, Office of Technology Assessment (1985), *U.S. Natural Gas Availability: Gas Supply Through the Year 2000,* Washington, DC, OTA-E-245, February 1985. A thorough review of all categories of U.S. natural gas supply with excellent synopses of extraction technologies applicable to each supply category.

NETWORK THEORY, ELECTRICAL

EDWIN C. JONES, JR., *Electrical Engineering and Computer Engineering Department, Iowa State University, Ames, Iowa, U.S.A.*

3-527-28133-9/94/$5.00 + .50

INTRODUCTION

Network theory, also called circuit theory, is the study of the analysis of electric circuits or networks. In turn, these networks are interconnections of electrical devices such as resistors, capacitors, electronic devices, motors, sources of electrical energy, and many others. A few of these devices are described, and others are described elsewhere in this Encyclopedia. Such circuits are described mathematically by ordinary differential equations. The circuits may, however, be linear or nonlinear, time-invariant or time-varying.

The equations needed are given by the describing equations of the elements and two basic principles, known as Kirchhoff's laws. Kirchhoff's current law states that the sum of currents leaving a node is zero, while the voltage law states that the sum of the voltages around a closed path is zero.

Analysis methods proceed from the basic principles, and the choice of method depends on the complexity of the network, the needed information, and the preferences of the analyst. Some analysis methods depend on circuit reductions and equivalent circuits. Other methods require formal equation-writing methods, and include nodal analysis, loop analysis, and state-variable techniques.

Although classical differential equation theory can be used, most circuit analysts use operational methods, especially the Laplace transform. The Laplace transform is especially important for linear, time-invariant networks, and allows complete solutions that consider initial conditions as a part of the complete solution. Linear circuit equations are also solved frequently with computer techniques. An important program is SPICE, the acronym for *Simulation Program with Integrated Circuit Emphasis*. The Electronics Research Laboratory at the University of California at Berkeley developed SPICE in 1975, and made it available to the public. Many improvements and enhancements have been made, and the program is now the most widely used circuit-analysis program. The personal computer version of SPICE is known as PSPICE, available from the MicroSim Corporation. State-variable equations can be solved with Laplace transforms, but are typically solved by computer-based numerical methods when the circuits are nonlinear.

Virtually all of the electrical energy transmission and distribution in the world involves sinusoidal voltage sources. Because of the importance of the topic, special methods for analysis of networks operating in the sinusoidal steady state are needed. These require a special transform commonly called a phasor transform. This transform can be considered as a special case of the Laplace transform, or it can be developed independently. The latter approach is used here.

Electrical networks or circuits often are used to modify signals that carry information in electrical form. A very general type of network is the two-port, a circuit that typically has a pair of input terminals, a pair of output terminals, and the property that, at each pair, the current into one lead is equal to the current out of the other lead. It is often desirable to convert one two-port representation to another, and this process is described.

In recent years, circuits or networks described by difference equations have become quite important. They are the foundation for most digital signal-processing techniques, including image processing. Signals in such circuits are described by sequences in which the discrete quantity (independent variable) is an integral multiple of a time or space unit, and the signals are undefined between integers. Three operations are commonly used: addition, multiplication by a constant, and delay. Such circuits may be solved by techniques quite similar to the techniques used in ordinary differential equations. They may also be solved numerically. An operational method is the z transform. This transform is closely related to the Laplace transform, and it is defined and uses are developed.

1. BASIC CONCEPTS

Network theory, also known as circuit theory, refers to the body of knowledge that describes the behavior of interconnected electrical devices and components. The devices included may be resistors, capacitors, inductors, transformers, transistors, relays, motors, diodes—and there are many more, including sources of electrical energy. The terms network theory and circuit theory will be used interchangeably in this article. Circuits are used in the processing of energy and information in electrical form. Circuits of interest have the property that there is always at least one continuous path through the circuit; that is, it is possible to find a point at which one can start a traverse of the circuit, go through at least two elements, and return to the starting point.

Networks and circuits are described by ordinary differential equations. In general, the independent variable is time, and the circuits are sometimes known as continuous-time circuits. The use of ordinary differential equations results from a condition that is put on circuit theory, which is that the spatial effects of electrical quantities may be neglected. When spatial effects are significant, electromagnetic field theory must be used, a topic that requires the use of partial differential equations. The electrical properties of immediate interest are current and voltage.

1.1 Electric Current

In SI units, the fundamental electrical quantity is current. Current is considered to be the motion of charged particles. The smallest unit of charge is postulated as the equivalent of the charge on one electron, and there are two types of charge, called positive and negative charge. In SI units, the unit of charge is the coulomb (C). The charge on one electron is equal to 1.602×10^{-19} C. A charge flow rate of 1 C/s is defined as 1 ampere (A). By convention, current is considered to be the flow of positive charges. When it is necessary to consider the flow of negative charges, appropriate modifiers are used. A time-varying current will be written as $i(t)$.

1.2 Voltage

The motion of charged particles requires the expenditure of energy or results in the release of energy. The voltage between two points in space is defined as the work per unit charge (joules/coulomb) required to move a charge from one point to the second. Often one point is selected as a reference point, or a point at which the voltage is defined to be zero. When this is done, the voltage difference becomes the voltage at the second point. In general, the choice of reference point is arbitrary. In SI units, the unit of voltage is the volt (V). A time-varying voltage will be written as $v(t)$.

1.3 Power

The power delivered by an electrical source (sources are discussed in Sec. 3.6) to an electrical device is given by

$$p(t) = v(t)i(t), \tag{1}$$

where $p(t)$ is the power delivered to the device. See Fig. 1. The choice of sign is important—the choice made here is commonly called "the passive sign convention." When the power, as given by Eq. (1), is positive, then energy is being delivered to the device in question.

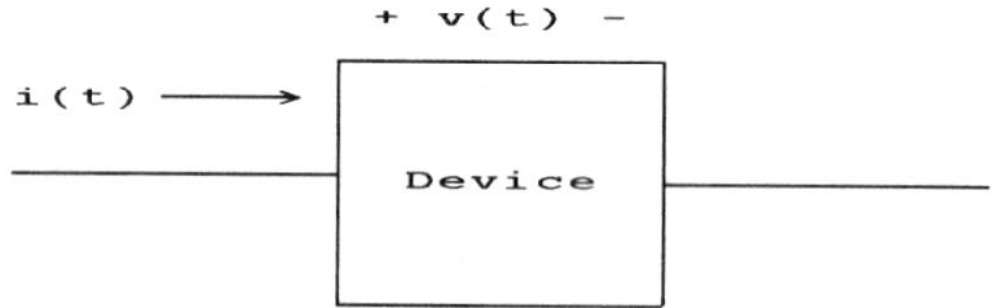

FIG. 1. Sign conventions for electrical devices.

1.4 Linear and Nonlinear Circuits

A linear circuit is a network or circuit that is completely described by a set or system of linear differential equations. In many circuits that are fundamentally nonlinear, the operating conditions are such that it is possible to use the mathematics of linear differential equations to analyze the circuit behavior. This technique is called "small-signal analysis." It depends on use of Taylor-series models, and the neglect of terms of order higher than one. It is necessary to explicitly state the limitations—an example might be that a certain current never goes negative.

A circuit that contains one or more nonlinear elements is fundamentally nonlinear. In most cases, such circuits must be analyzed numerically. While the principles of analysis of linear and nonlinear circuits are similar, the details differ substantially. In this article, the principal focus is on linear circuits.

1.5 Time-Varying Circuits

If one or more of the elements or devices in a circuit varies with time, then the circuit is said to be time-varying. The mathematical techniques used for analysis depend on the type of variation. In many cases the networks must be analyzed numerically. In this article, the principal focus is on circuits in which the component or element values are constant with respect to time, so that the networks are actually described by differential equations with constant coefficients.

1.6 Discrete-Time Circuits

The latter part of this article is devoted to networks or circuits that are described by a set or system of difference equations. The qualities of interest are known and defined at discrete points in time, but are not known or defined between those times. The times may or may not be uniformly spaced.

1.7 Superposition

If a circuit is linear, the principle of superposition applies. This principle states that, for a given system (network) having a general excitation $x(t)$ that yields a response $y(t)$, if the response to an excitation $x_1(t)$ is $y_1(t)$, and the response to an excitation $x_2(t)$ is $y_2(t)$, then any excitation

$$x(t) = a_1x_1(t) + a_2x_2(t)$$

yields the response

$$y(t) = a_1y_1(t) + a_2y_2(t) \quad (2)$$

for all values of the constants a_i.

2. CONTINUOUS-TIME CIRCUITS

In this section, some important terms in network analysis will be introduced. They include elements, nodes, loops, Kirchhoff's laws, and Tellegen's theorem.

2.1 Node

The junction of three or more wires from circuit elements in a circuit or network is called a *node*. (It should be noted that some writers used the term "node" to identify a junction of two or more elements, and reserve the term "essential node" for a junction of three or more elements.) One of the important laws of circuit analysis, Kirchhoff's current law, makes use of this definition of a node.

2.2 Kirchhoff's Current Law

Kirchhoff's current law, often abbreviated KCL, states that at any node in a circuit or network,

$$\Sigma i = 0. \quad (3)$$

The law states that the sum of currents leaving (or entering) a node is always zero. This principle results from Gauss's law or the principle of charge conservation. In circuit theory, charge conservation leads to the postulate that it is not possible for electrical charge to accumulate at a node.

2.3 Loop

In a circuit or network, there are one or more closed paths, or *loops*. Such paths start at a node, traverse two or more elements, and terminate at the starting point. The second law of circuit analysis, Kirchhoff's voltage law, makes use of this definition. If the circuit is planar, that is, if it may be drawn on a planar surface without any wires crossing, then the loop is frequently called a mesh.

2.4 Kirchhoff's Voltage Law

Kirchhoff's voltage law, often abbreviated KVL, states that around any loop in a circuit or network,

$$\Sigma v = 0. \quad (4)$$

The law states that the sum of voltages around a closed path or loop in a circuit is equal to zero, or that the voltage between any two points is independent of the path used between the two points.

2.5 Branch

A branch of a network or circuit is an element or combination of elements that provides a path from one node to another.

2.6 Tellegen's Theorem

This theorem (Tellegen, 1952) applies to any circuit, linear or nonlinear. It states: Let N be a network of b branches and n nodes. If KVL is satisfied by the branch voltages v_1, v_2, ..., v_b in every loop, and if KCL is satisfied by the branch currents i_1, i_2, ..., i_b at every node, then

$$\sum_{k=1}^{b} v_k i_k = 0. \quad (5)$$

When this theorem is applied, the sign convention of Fig. 1 must be observed.

3. CIRCUIT ELEMENTS

3.1 Resistors

Electrical components in which the voltage across the device is proportional to the current through the device are known as resistors. The proportionality constant is called the resistance, and the relation is known as Ohm's law,

$$v(t) = Ri(t). \quad (6)$$

In fundamental units, resistance has units of (kg m^2)/(ampere^2 second^3). The unit of resistance is called the "ohm," and the standard symbol is Ω. The reciprocal of resistance is called conductance, and the reciprocal unit is the siemens (S). Formerly, this unit was called "mho," ohm spelled backwards, but this term is now obsolete. The graphical symbol is shown in Fig. 2(a).

The power dissipated by a resistor is given by

$$p(t) = v(t)i(t) = i^2(t)R = v^2(t)/R. \tag{7}$$

3.2 Capacitors

A capacitor is a circuit element that stores energy in an electric field. This property is called capacitance, and the unit of capacitance is the farad (F). The two approved graphical symbols are shown in Fig. 2(b). The capacitor is described by the equations

$$i(t) = C\frac{dv}{dt}, \quad v(t) = \frac{1}{C}\int_0^t i(\tau)d\tau + v(0), \tag{8}$$

$$W = \tfrac{1}{2}Cv^2, \tag{9}$$

where W = energy stored in the capacitor in joules, τ = a dummy variable representing time in seconds, and C = capacitance in farads.

3.3 Inductors

An inductor is a circuit element that stores energy in a magnetic field. This property is called inductance, and the unit of inductance is the henry (H). The approved graphical symbol is shown in Fig. 2(c). The inductor is described by the equations

$$v(t) = L\frac{di}{dt}, \quad i(t) = \frac{1}{L}\int_0^t v(\tau)d\tau + i(0), \tag{10}$$

$$W = \tfrac{1}{2}Li^2, \tag{11}$$

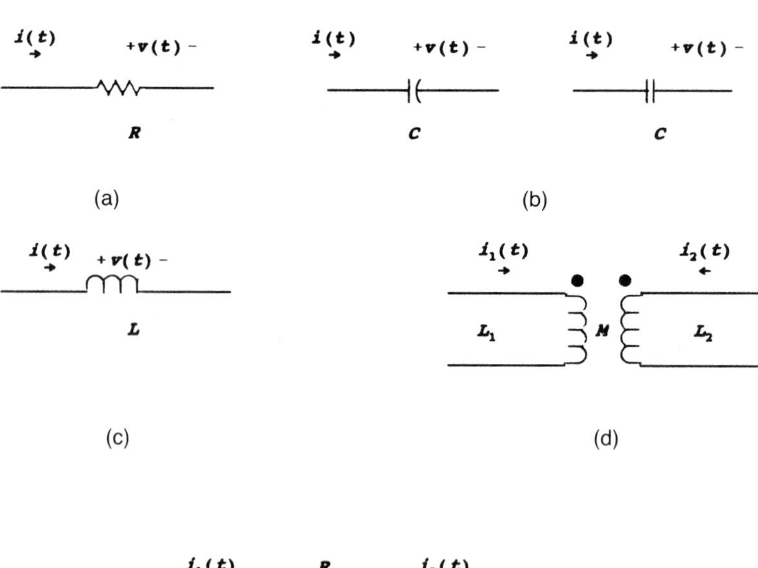

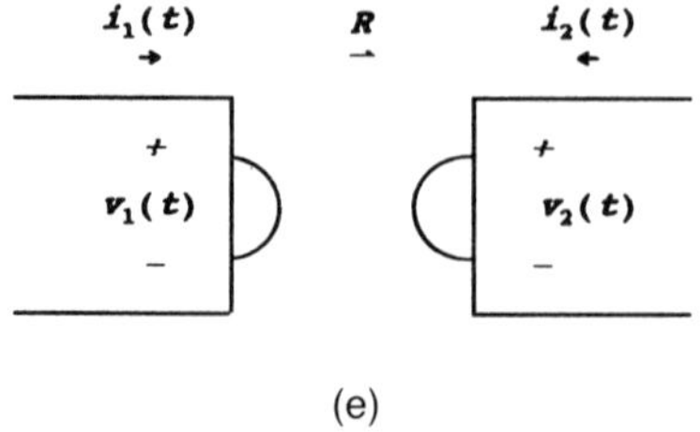

FIG. 2. Graphical symbols for circuit elements: (a) resistor, (b) capacitor, (c) inductor, (d) coupled coils, and (e) gyrator.

where W = energy stored in the inductor in joules, τ = a dummy variable representing time in seconds, and L = inductance in henrys.

3.4 Mutual Inductance

When two coils or inductors exist in close proximity, or are wound on the same coil form, a changing current in one coil will induce a voltage in the second. This effect forms the basis for transformers, a commonly employed electrical element. Figure 2(d) shows a symbolic representation of a pair of coupled coils. The dots represent the direction of the winding of the coils on the coil form. The convention is that currents entering (or leaving) dotted terminals develop additive fluxes. A pair of equations is needed to describe the coupled coil pair:

$$v_1(t) = L_1 \frac{di_1}{dt} + M \frac{di_2}{dt},$$
$$v_2(t) = M \frac{di_1}{dt} + L_2 \frac{di_2}{dt}, \qquad (12)$$

where the L's represent the self inductance of the coils, and M represents the mutual inductance.

Application of the principle of conservation of energy shows that $L_1L_2 - M^2 \geq 0$. A "coefficient of coupling"

$$k^2 = \frac{M^2}{L_1 L_2} \leq 1 \qquad (13)$$

may be defined as a figure of merit for coupled coils.

3.5 Gyrator

Tellegen (Tellegen, 1948) postulated the fundamental nonreciprocal electrical element in 1948, and named it the "gyrator." With modern technology excellent gyrators are built. The defining equations are

$$v_1 = Ri_2,$$
$$v_2 = -Ri_1, \qquad (14)$$

where R is known as the "gyration resistance." A commonly used symbol (there is no standard symbol) is shown in Fig. 2(e). Gyrators are used to control power in electronic circuits, especially microwave circuits. The concept is used by circuit theorists to develop circuit and network properties.

3.6 Sources

Electrical circuits must have an energy source. In network theory, the goal is to have mathematical models that describe the sources adequately, regardless of the types of sources represented. One convenient classification is to consider independent and dependent or controlled sources. Another classification, which applies to independent sources, is to consider constant (dc) sources, sinusoidal (ac) sources, and general time-varying sources.

3.6.1 Independent Voltage Sources The independent voltage source model postulates a device or system that supplies electrical energy at a voltage unaffected by the load or device that is connected to the source. Many devices are well described by combinations of an ideal source and one or more components such as a resistor. An ideal voltage source will supply whatever current is needed by the components connected to it.

A battery is modeled as a constant (with respect to time) voltage. Often these are called "dc" sources. While dc strictly means "direct current," the term is often used as a synonym for a constant source. Figure 3(a) shows the common graphical symbol for a constant voltage source.

Much electrical energy is distributed with sinusoidal voltage sources. Often these are called "ac" (for "alternating current") sources. In Sec., 8.2, a powerful method of circuit analysis with ac sources will be developed. It is known as "phasor analysis," and is based on the fact that a sinusoidal function is described by three quantities: amplitude V_m, frequency f, and phase α. A common representation of an ac voltage source is shown in Fig. 3(b).

In many applications, a more general model is needed. In Fig. 3(c) a general time-varying voltage source is indicated. The function for $v(t)$ may be known analytically, numerically, or by its statistical properties. Methods for analysis of circuits with such sources will be developed in later paragraphs.

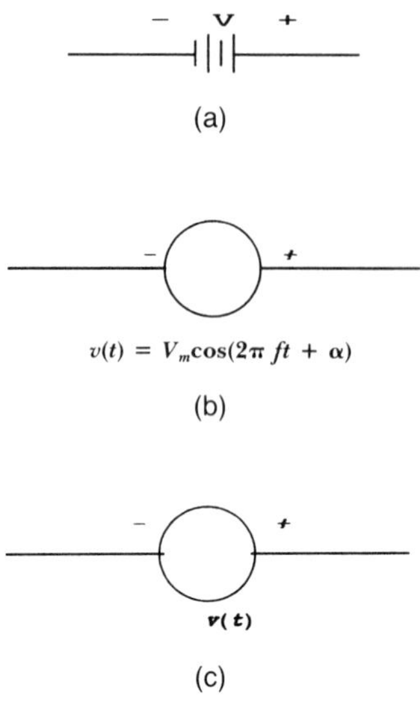

FIG. 3. Independent voltage sources: (a) constant (dc) voltage source, (b) sinusoidal (ac) voltage source, and (c) general time-varying voltage source.

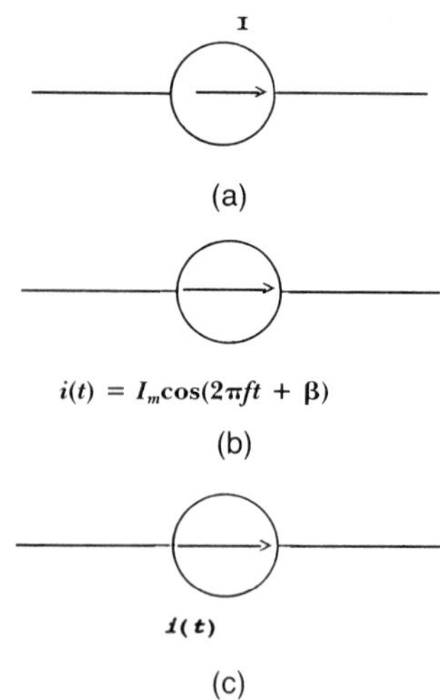

FIG. 4. Independent current sources: (a) constant (dc) current source, (b) sinusoidal (ac) current source, and (c) general time-varying current source.

3.6.2 Independent Current Sources The independent current source model postulates a device or system that supplies electrical energy at a current value unaffected by the load or device that is connected to the source. Many devices are well described by combinations of current sources and components such as resistors. An ideal current source will develop whatever voltage is needed by the components or load connected to its terminals.

A constant (dc) current source is indicated in Fig. 4(a). The current is constant with respect to time as well as connected load.

A sinusoidal (ac) current source is indicated in Fig. 4(b). As with ac voltage sources, the sinusoidal function is described by its amplitude, frequency, and phase. The phasor analysis techniques to be developed will be applicable to these functions.

A general time-varying current source is depicted in Fig. 4(c). As with voltage sources, the function $i(t)$ may be known analytically, numerically, or statistically.

3.6.3 Dependent or Controlled Sources Circuits that contain electronic devices such as amplifiers are frequently analyzed and designed using linear circuit models for the electronic devices. Such models contain a minimum of three terminals. In general there is an input terminal pair and an output terminal pair. One terminal may be common to both input and output pairs. Voltages can be measured between terminals, and there is current between terminals. If one such voltage or current is a known function of another current or voltage, then the second quantity is said to control the first, or the first is dependent on the second. Four such models can be identified. Examples of their use will be included later in examples.

3.6.3.1 Voltage-Controlled Voltage Source (VCVS). The model of Fig. 5(a) applies to the case in which the voltage across one terminal pair, v_{xz}, is proportional to the voltage across another pair, $v_{yz'}$. The proportionality factor is A, often but not always a constant. The device is, in general, not reciprocal.

3.6.3.2 Current-Controlled Voltage Source (CCVS). Figure 5(b) indicates that the voltage at one terminal pair, v_{xz}, is proportional to and controlled by the current $i_{yz'}$. The proportionality factor is shown as g, chosen because of a convention that will appear later in the section on two-ports. In general, the device being modeled is not reciprocal.

3.6.3.3 Current-Controlled Current Source (CCCS). When the current between one pair of terminals, i_{xz}, is proportional to the current between the terminals of another pair, $i_{yz'}$, the model is called a current-controlled current source, depicted in Fig. 5(c). The proportionality factor is β. In general, the device being modeled is not reciprocal.

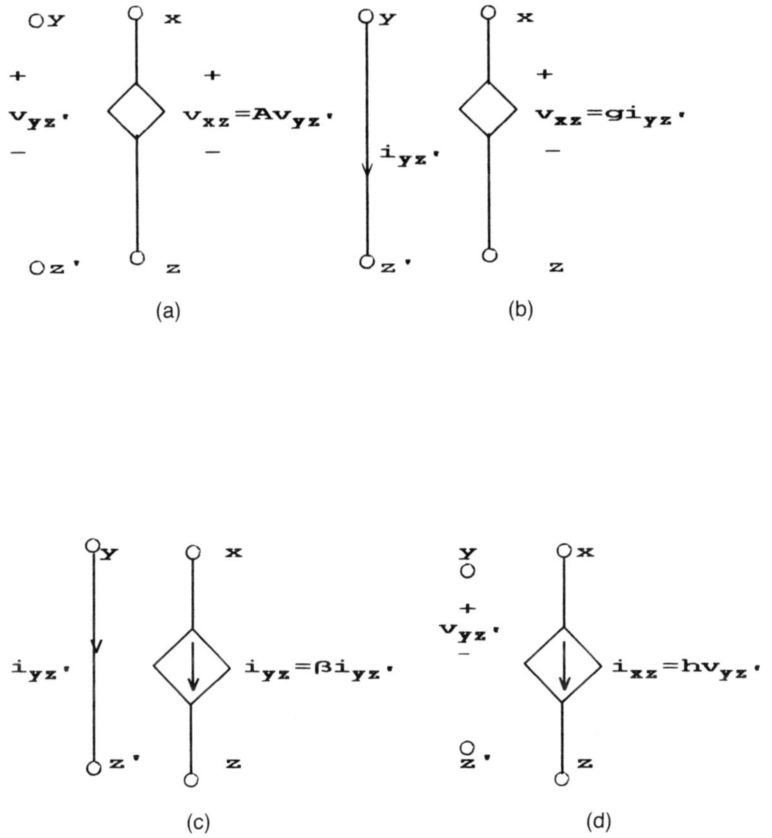

(a) (b)

(c) (d)

FIG. 5. Controlled sources: (a) voltage-controlled voltage source, (b) current-controlled voltage source, (c) current-controlled current source, and (d) voltage-controlled current source.

3.6.3.4 Voltage-Controlled Current Source (VCCS). When the current between the terminals of one pair, i_{xz}, is proportional to the voltage at another pair, $v_{yz'}$, the model is called a voltage-controlled current source. The proportionality factor is called h. The model is shown in Fig. 5(d). In general, the device being modeled is not reciprocal.

4. CIRCUIT-ANALYSIS TECHNIQUES

To solve an electric circuit or network means, at a minimum, to find the voltage across and the current through each element of the network. Many methods for doing this exist, and several important methods will be discussed in this section and in following sections. The first to be discussed are methods for replacing certain combinations of elements with simpler combinations. The reason for this is to reduce the complexity of the analysis and to give the analyst greater insight into the most important aspects of the overall circuit behavior. Equivalents for combinations of elements will be presented, followed by equivalents for sources and elements.

4.1 Series and Parallel Combinations

If two or more electrical circuit elements carry the same current, the elements are said to be in *series*. If two or more elements are connected across the same voltage, the elements are said to be in *parallel*. In neither case is being connected across equal or equivalent voltages, or carrying equal or equivalent currents, sufficient. Figure 6 shows equivalent network elements for series combinations of resistors, capacitors, and inductors. In the latter case, two cases must be distinguished when there is mutual inductance between the coils. Figure 7 shows equivalent network elements for parallel combinations of resistors, capacitors, and inductors. Though derivations are not included in this article, the equivalents may be developed by using the

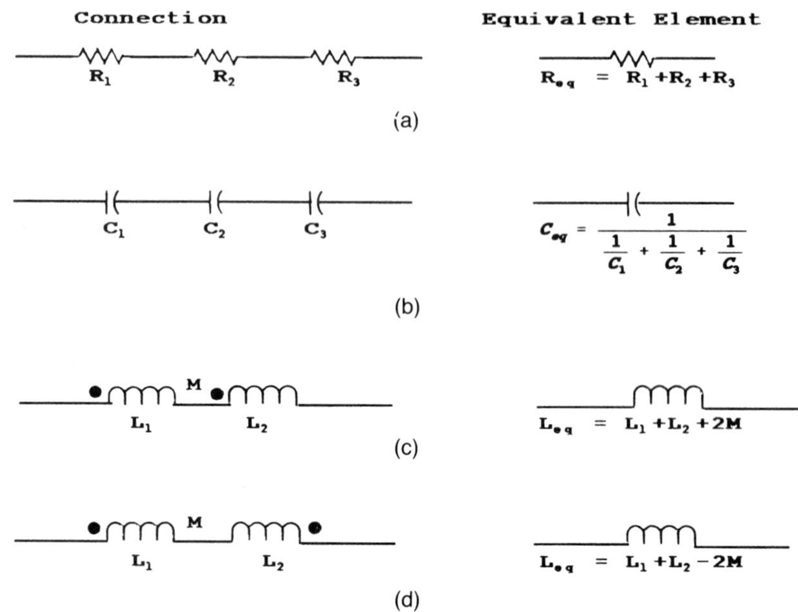

FIG. 6. Equivalent elements, series circuits: (a) resistances in series, (b) capacitors in series, (c) inductors in series, additive fluxes, and (d) inductors in series, subtractive fluxes.

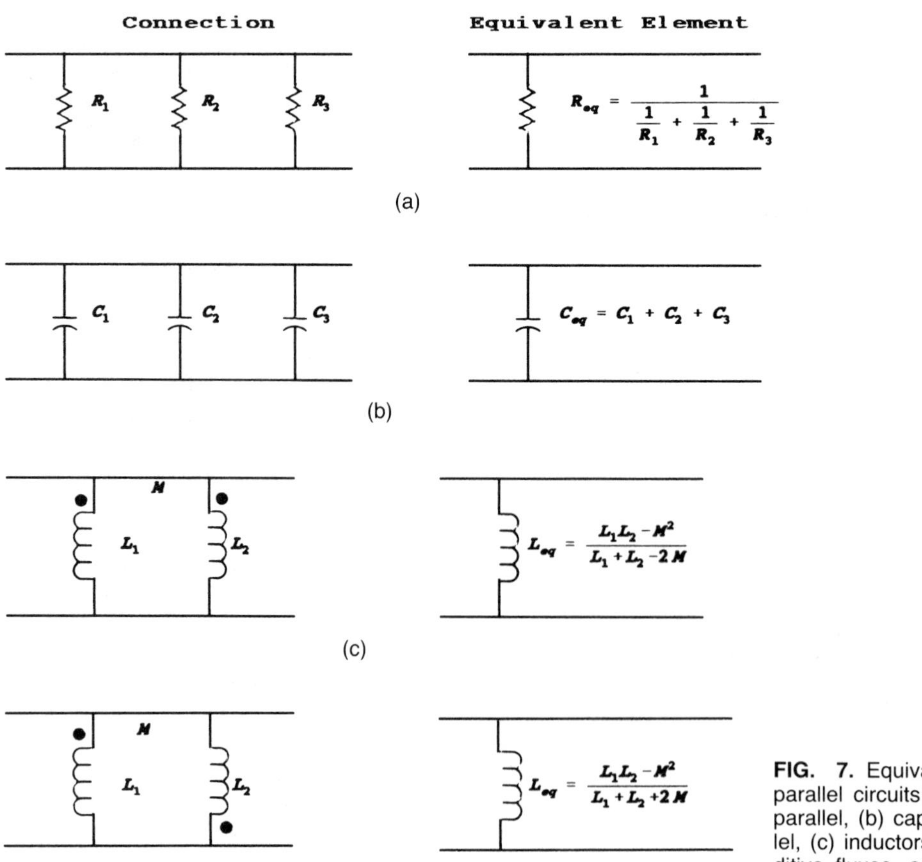

FIG. 7. Equivalent elements, parallel circuits: (a) resistors in parallel, (b) capacitors in parallel, (c) inductors in parallel, additive fluxes, and (d) inductors in parallel, subtractive fluxes.

defining equations for the elements in conjunction with Kirchhoff's laws.

4.2 Tee-Pi or Wye-Delta Conversions

Figure 8 shows two commonly used ways to connect three elements as parts of a circuit. The first, Fig. 8(a), is called a "tee" or "wye" network, the second, Fig. 8(b), a "pi" or "delta" network. When one or the other of these element combinations is apparent in a network, conversion to the other often enables the analyst to replace one combination of elements with an equivalent combination, which in turn leads to further simplifications, such as a possible series or parallel reduction. The equations are

$$R_a = (R_1R_2 + R_1R_3 + R_2R_3)/R_3,$$
$$R_b = (R_1R_2 + R_1R_3 + R_2R_3)/R_2,$$
$$R_c = (R_1R_2 + R_1R_3 + R_2R_3)/R_1; \quad (15)$$

$$R_1 = R_aR_b/(R_a + R_b + R_c),$$
$$R_2 = R_aR_c/(R_a + R_b + R_c),$$
$$R_3 = R_bR_c/(R_a + R_b + R_c). \quad (16)$$

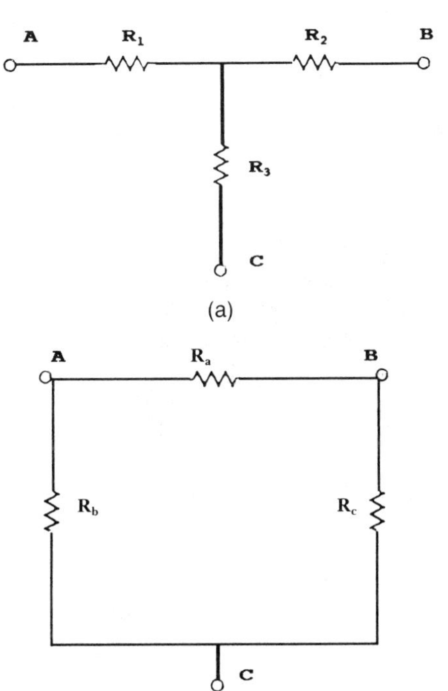

FIG. 8. Tee-pi or wye-delta conversion: (a) Tee- or wye-connected elements, and (b) pi- or delta-connected elements.

Though shown here for resistors, the equations may be generalized to impedances. More will be said about this point later in the article.

4.3 Thévenin Theorem

In many applications, an electrical network is used to control or modify an electrical signal delivered to some device. A common example is an amplifier-loudspeaker combination. An analyst might be interested in understanding how various loudspeakers perform with a given amplifier, but does not want to make repeated analyses of the amplifier, since the principal interest is in the behavior of the loudspeaker in its setting. This analyst needs a way to model the amplifier in a simplified but equivalent manner, and to use this model in conjunction with the electrical models of the various loudspeakers being considered. More specifically, since the loudspeaker normally has two terminals, then it would be convenient to represent the amplifier as a device with two terminals (a terminal pair) and other necessary components. A pair of theorems developed by Thévenin and Norton provide a means for doing this.

The Thévenin theorem states that, at a terminal pair, any linear network may be replaced by a series combination of a voltage source and a resistance (or impedance—to be discussed later); see Fig. 9(a). The voltage is equal to the voltage at the terminal pair when the external load is removed or open-circuited, and the resistance is equal to the resistance calculated or measured at the terminal pair with all of the independent sources de-energized.

De-energization means, in this context, that the source voltages and currents of *independent* sources are set to zero, but that the source resistances (impedances) are unchanged. *Controlled* or *dependent* sources are not de-energized or modified in any way.

4.4 Norton Theorem

This theorem, which is a companion to the Thévenin theorem, states that, at a terminal pair, any linear network may be replaced by the parallel combination of a current source and a resistance (or impedance); see Fig. 9(b). The current is equal to the current that flows through the short circuit when the external

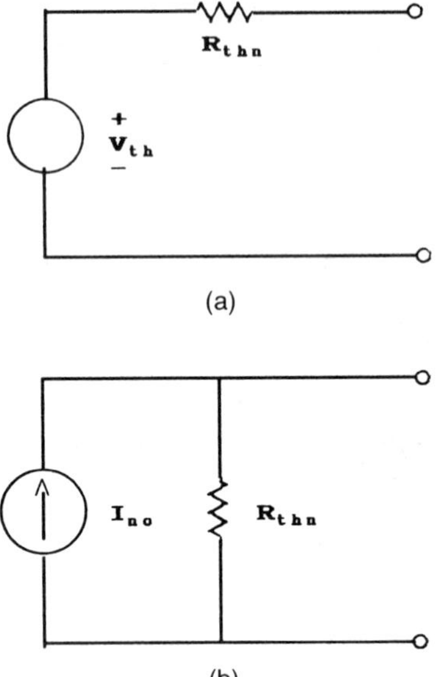

FIG. 9. (a) Thévenin and (b) Norton equivalent circuits.

load is short circuited. The resistance is equal to the resistance measured or calculated at the terminal pair with all independent sources de-energized. As before, de-energization means that independent source voltages and currents are set to zero but source resistances are unchanged. Dependent or controlled sources are not de-energized.

4.5 Thévenin–Norton Theorem Comparisons

The similarity of these two theorems suggests that they are closely related. Some authors combine them and speak of a Thévenin–Norton theorem. Since the definitions of the equivalent impedances are identical, these impedances themselves are identical. If the Thévenin equivalent for a circuit or network is known, then the Norton equivalent current is found from the relation

$$V_{\text{th}} = I_{\text{no}} R_{\text{thn}}. \tag{17}$$

4.6 Thévenin–Norton Theorem Example

Figure 10(a) shows a network with a controlled source. The Thévenin voltage is calculated, in this case, directly, without any modification of the circuit. The current through the 5-Ω resistor is zero because of the open circuit at terminals A and B. Thus, $V_{CB} = 15$ V. The resulting current through the dependent current source is (2 Ω^{-1}) (15 V) = 30 A. From this result, V_{AC} = (2 Ω)(30 A) = 60 V. Thus, $V_{AB} = 15 + 60 = 75$ V, and since this is the open-circuit voltage at the terminals, it is the Thévenin voltage.

In Fig. 10(b), the equivalent resistance is calculated. To begin, the independent source of 15 V is temporarily set equal to zero. Since the resistance of an ideal voltage source is zero, the voltage source is replaced by a short circuit. The Thévenin resistance can be calculated by applying, conceptually, a test source at the output terminal pair. The ratio of the test voltage to the resulting current is the needed resistance. Since the circuit is linear, any convenient value may be chosen for the source. Call the voltage source applied V_T, and the resulting current I_T. The current through the 5-Ω resistor is, by Kirchhoff's current law, also I_T. Thus, $V_{CB} = (5\ \Omega)I_T$. The current through the 2-Ω resistor is $I_T + (2\Omega^{-1})V_{CB} = 11I_T$. Therefore, $V_{AC} = (2\ \Omega)(11I_T) = (22\ \Omega)I_T$, and $V_T = V_{AB} = V_{CB} + V_{AC} = (27\ \Omega)I_T$. Finally, the ratio between this current and voltage is 27 Ω, which is the Thévenin–Norton resistance.

Figure 10(c) shows the circuit prepared for Norton current determination. The output terminal pair is short circuited, conceptually, and the current through this short circuit is calculated from the modified circuit. (It can also be measured when the current can be short circuited without damage.) The resulting current is labeled I_{no}. In this circuit, $V_{CB} = 15\text{ V} - (5\ \Omega)I_{\text{no}}$. Writing a Kirchhoff's voltage law equation around the closed path gives $(5\ \Omega)I_{\text{no}} + (2\ \Omega)[I_{\text{no}} - (2\ \Omega^{-1})V_{CB}] - 15\text{ V} = 0$. A simultaneous solution of these two equations gives $I_{\text{no}} = 25/9$ A, which is the required value.

Figure 10(d) shows the combination of these results into both the Norton and Thévenin equivalent circuits. The work may be checked by computing the ratio of the Thévenin voltage to the Norton current, which is 27 Ω, the same value that was computed in Fig. 10(d).

4.7 Ladder Networks

A ladder network is a special but frequently occurring circuit configuration. An example, with resistors only, is shown in Fig.

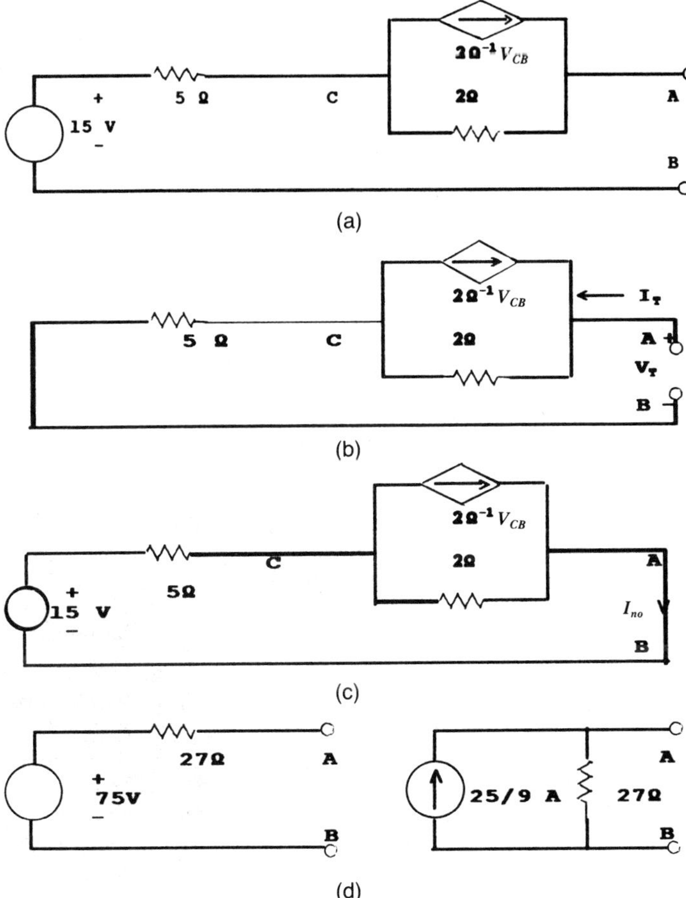

FIG. 10. Thévenin–Norton circuit determination. (a) Example circuit; (b) Thévenin–Norton resistance calculation; (c) illustrating Norton current calculation, and (d) Thévenin and Norton equivalent circuits.

11(a). The principal characteristics are that there is one terminal, labeled 2-2′, that is common to input and output, and that no node is "bridged." There would never be an element connecting, for example, terminal 1 to 4. Two special analysis methods are described for this configuration.

The first method starts with use of series and parallel combinations alternately to reduce the network to a single equivalent element and a source. To do this, the analyst starts at the element farthest from the source, and works toward the source. Currents through and voltages across each element are then found by reversing the process, starting at the source, and working toward the load.

The second method uses the principle of superposition. The analyst assumes a current (or voltage) at the output, computes the voltage (or current) at the output, and then uses KCL and KVL to work back toward the source. Eventually a source magnitude is determined, which is probably different from the actual source. However, because of superposition, the ratio of the actual current (or voltage) to the assumed current (or voltage) is equal to the ratio of the actual source to the source value resulting from the assumption.

Figure 11(b) shows an example of this solution method. In this example, voltages are stated with respect to the reference node. Assume a current of 1 A through the 9-Ω resistor, giving a voltage at terminal 2 of 9 V. The current through the 7-Ω resistor is also 1 A, so that the voltage at terminal 4 is 16 V. Then, the current through the 16-Ω resistor is 1 A, and the current through the 4-Ω resistor is 2 A. Next, the voltage at terminal 3 is found to be $16 + 2(4) = 24$ V. Thus, the current through the 6-Ω resistor is 4 A, leading to a current of 6 A through the 2-Ω resistor. Finally, the resulting voltage at terminal 1 would be 36 V. Since the actual voltage at terminal 1 is

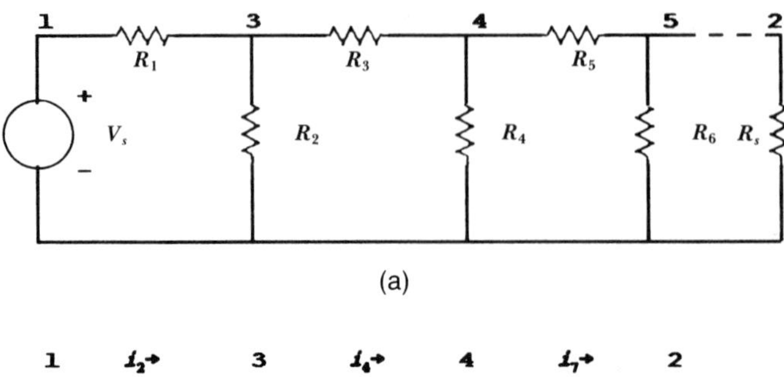

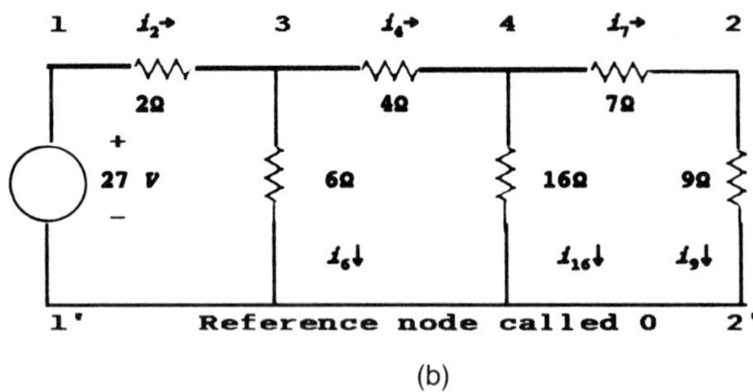

FIG. 11. Ladder networks. (a) General ladder network, (b) sample ladder network calculation.

27 V, the currents in all the resistors are scaled by $27/36 = \frac{3}{4}$. The solution for the actual current in the 9-Ω resistor is 0.75 A.

5. FORMAL CIRCUIT-ANALYSIS METHODS

5.1 Nodal and Loop Analysis

A combination of b branches or elements interconnected to form a circuit or network requires, for a complete analysis, the determination of b voltages and b currents, or $2b$ quantities. Thus, $2b$ equations are needed. Half of these are provided by the defining equations of the branches, e.g., Ohm's law. The remaining equations are provided by systematic application of Kirchhoff's laws. There are two methods, one based on each law. Both use the definitions of nodes and loops given earlier.

Suppose that a circuit having b elements has n (essential) nodes. Kirchhoff's current-law equations can be written at each, but one will be redundant, because it is the linear sum of the remaining $n - 1$ equations. Thus, in a circuit with n (essential) nodes, $n - 1$ independent KCL equations can be written. To complete the analysis, $b - (n - 1)$ KVL equations must be written. In general, this is possible, though in some cases, especially when the networks are nonplanar, care must be taken to ensure that the equations are independent. In practice, it is usually sufficient to use one set or the other. The choice depends on which technique yields the smaller number of equations, which technique yields the most useful and usable information, and the preference of the analyst.

5.2 Nodal Analysis

Figure 12 shows a typical node in a circuit that is isolated for study. The voltage at this node is determined with respect to some reference node in the circuit. The voltage at the

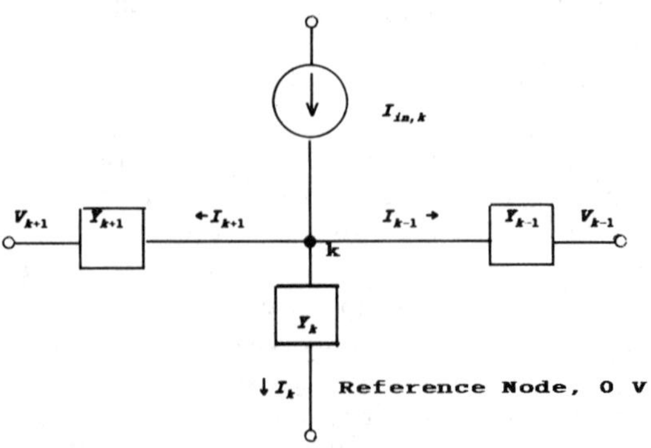

FIG. 12. Isolated node for KCL equations.

reference is arbitrarily called 0 V. One element is shown connected from the node in question to the reference node, though this is not necessary. Two other elements and a current source are shown.

Kirchhoff's current law written at this node yields

$$I_{k-1} + I_k + I_{k+1} = I_{in,k}. \tag{18}$$

A typical expression for the current through each element can be written

$$I_{k+i} = (V_k - V_{k+i})Y_{k+i}, \tag{19}$$

where the Y_k represent the branch relations, and are written in *admittance* form. Once all of the possible equations are written, simplified, and organized into matrix form, the result is

$$\begin{bmatrix} Y_{11} + Y_{12} + Y_{13} + \cdots & -Y_{12} & -Y_{13} & \cdot \\ -Y_{21} & Y_{21} + Y_{22} + Y_{23} + \cdots & -Y_{23} & \cdot \\ -Y_{31} & -Y_{32} & Y_{31} + Y_{32} + Y_{33} + \cdots & \cdot \\ \cdot & \cdot & \cdot & \cdot \\ \cdot & \cdot & \cdot & \cdots \end{bmatrix} \begin{bmatrix} V_1 \\ V_2 \\ V_3 \\ \cdot \\ \cdot \end{bmatrix} = \begin{bmatrix} I_1 \\ I_2 \\ I_3 \\ \cdot \\ \cdot \end{bmatrix}. \tag{20}$$

The square matrix is a complete description of the circuit. The column matrix (vector) of voltages represents the dependent variables, while the column matrix of currents represents the independent variables.

5.3 Nodal Analysis with Controlled Sources

When a branch contains a controlled source of any of the four types, it is usually convenient to use the Thévenin–Norton theorem to convert the controlled source to a voltage-controlled current source (unless it is already in this form). When this is done, the right-hand side of the previous equation will contain the dependent variables (voltages) as well as the independent current sources. The dependent-variable terms should then be transposed to the left-hand side of the matrix equation. A major result of this effort is that the square matrix is no longer symmetric.

5.4 Loop Analysis

Figure 13 shows a typical loop in a circuit that is isolated for study. A loop current is defined as indicated, and loop currents from neighboring loops are also indicated. As indicated, each element current is the difference between two loop currents. This formulation of the loop presumes that the network is planar, i.e., it can be drawn in a plane without wires crossing. If the network is not planar, then some element currents will be composed of three or more currents, and some signs in the final result may change. Kirchhoff's voltage law written around the loop states that

$$V_{k-1} + V_k + V_{k+1} + V_{k+2} = V_{in,k}. \tag{21}$$

An expression for each of the voltages is

$$V_{k+i} = (I_k - I_{k+i})Z_{k+i}, \tag{22}$$

where the Z_k describe the elements and are called *impedances*. When all of the possible equations are written and simplified, and the results are organized, a general result is

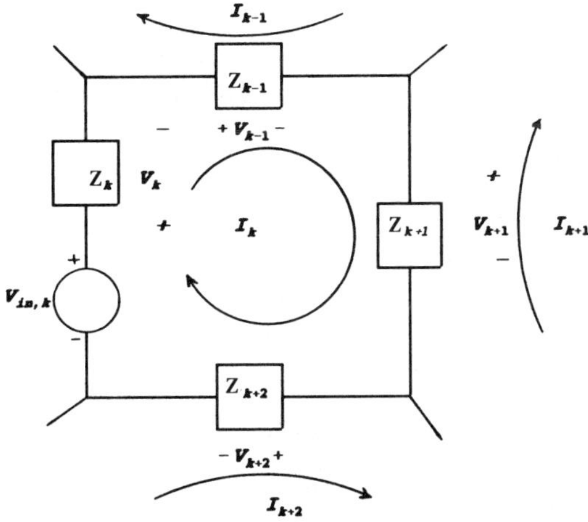

FIG. 13. Isolated loop for KVL equations.

$$\begin{bmatrix} Z_{11}+Z_{12}+Z_{13}+\cdots & -Z_{12} & -Z_{13} & \cdot \\ -Z_{21} & Z_{21}+Z_{22}+Z_{23}+\cdots & -Z_{23} & \cdot \\ -Z_{31} & -Z_{32} & Z_{31}+Z_{32}+Z_{33}+\cdots & \cdot \\ \cdot & \cdot & \cdot & \cdot \\ \cdot & \cdot & \cdot & \cdots \end{bmatrix} \begin{bmatrix} I_1 \\ I_2 \\ I_3 \\ \cdot \\ \cdot \end{bmatrix} = \begin{bmatrix} V_1 \\ V_2 \\ V_3 \\ \cdot \\ \cdot \end{bmatrix}. \quad (23)$$

The square matrix is a complete description of the network, while the column matrix or vector of currents shows the dependent variables, the loop currents, and the column matrix of voltages shows the independent variables or forcing functions in the various loops.

If the terms in square matrix are generalized admittances or impedances, as discussed in immediately following paragraphs on the solution of circuit equations by Laplace transforms, then solutions will be quotients of polynomials in the Laplace variable s.

5.5 Loop Analysis with Controlled Sources

When an element contains a controlled source, it is convenient to use the Thévenin–Norton theorem to convert all such sources to current-controlled voltage sources. When this is done and the equations are written, the right side of Eq. (23) will contain the dependent variables (loop currents) in addition to the independent voltage sources. These dependent-variable terms should then be transposed to the left side of the matrix equation. The resulting square matrix will no longer be symmetric, but it will describe the network completely.

5.6 Solution of Node-Voltage and Loop-Current Equations

If the sources are dc sources and the elements are resistive, the elements in the square matrices of Eqs. (20) and (23) are numerical terms, as are the vectors representing the dependent and independent variables. Calculators and computer programs that handle such equations are normally used to obtain the numerical solutions for the dependent variables, especially when there are three or more equations.

When the sources are sinusoidal (ac) and a steady-state solution is required, then all of the terms in the square matrices are complex numbers, and both the dependent and independent variables are phasors, also complex numbers. These concepts are discussed in Sec. 8.2. Computer programs and calculators for handling complex arithmetic facilitate the numerical solutions for the dependent variables.

6. LAPLACE TRANSFORMS IN CIRCUIT ANALYSIS

6.1 Introduction to Laplace Transforms

Since linear circuits or networks are described by linear differential equations, most often with constant coefficients, analysts can use any of the techniques available for solving such equations. A powerful technique is the Laplace transform. This technique, an important part of operational calculus, provides the analyst a powerful method for finding both steady-state and transient components of the solution simultaneously. The technique requires transformation of circuits and sources through application of certain theorems. Solution for unknown dependent variables is carried out algebraically, and inverse transforms are computed. It is necessary, however, to know initial conditions, the energy stored in capacitors and inductors.

6.2 Definition of a Laplace Transform

If a function of time $f(t)$ is known, and defined for $t \geq 0$, then the (single-sided) Laplace transform is given by

$$\mathcal{L}f(t) = F(s) = \int_{0^-}^{\infty} f(t)\epsilon^{-st}dt, \quad (24)$$

where s is a complex variable,* $s = \sigma + j\omega$,

*In these equations, j is used for the imaginary operator, $j = \sqrt{-1}$. Mathematicians and physicists typically use i for the imaginary operator, while electrical engineers use j. This choice is made to avoid confusion with the use of i, which is the standard symbol for current.

Table 1. Laplace transform theorems.

Operation	Theorem
Derivative	$\mathcal{L}\dfrac{df(t)}{dt} = sF(s) - f(0^-)$
nth-order derivative	$\mathcal{L}\dfrac{d^nf}{dt^n} = s^nF(s) - s^{n-1}f(0^-) - \cdots - f^{(n-1)}(0^-)$
Integral	$\mathcal{L}\displaystyle\int_{0^-}^{t} f(x)dx = \frac{F(s)}{s}$
Time shift	$\mathcal{L}[f(t-t_0)u(t-t_0)] = \epsilon^{-t_0s}F(s)$ where u is the unit step function
Frequency shift	$\mathcal{L}\epsilon^{-at}f(t) = F(s+a)$
Frequency scaling	$\mathcal{L}f(at) = \dfrac{1}{a}F\left(\dfrac{s}{a}\right),\ a > 0$
Initial value	$\lim_{t\to 0} f(t) = \lim_{s\to\infty} sF(s)$ provided the limit exists
Final value	$\lim_{t\to\infty} f(t) = \lim_{s\to 0} sF(s)$ provided the limit exists
Constant multiplier	$\mathcal{L}kf(t) = k\mathcal{L}F(s)$
Addition	$\mathcal{L}[a_1f_1(t) + a_2f_2(t)] = a_1F_1(s) + a_2F_2(s)$ a_1 and a_2 are constants
Convolution	$F(s) = F_1(s)\,F_2(s),\ f(t) = \displaystyle\int_0^t f_1(\tau)f_2(t-\tau)d\tau = \int_0^t f_2(\tau)f_1(t-\tau)d\tau$

chosen so that the integral will converge. In turn, σ is the real part of the variable, and ω is the imaginary part. The frequency of sinusoidal excitation functions, ω, is measured in rad/s.

6.3 Laplace Transform Theorems

If the function $f(t)$ has the Laplace transform $F(s)$, then the theorems of Table 1 apply. In these theorems, the term $f(0^-)$ represents the initial condition, or the value of f at $t = 0$.

6.4 Laplace Transform Pairs

Table 2 presents a listing of 12 common time functions and their Laplace transforms. These are sufficient for much of the analysis that is necessary. References include more extensive tables.

It should be noted that the time functions and their Laplace transforms form a biunique pair. In other words, a given time function has one and only one transform, and a given transform has one and only one inverse.

Table 2. Laplace transform pairs.

Name	Time function $f(t)$, $t > 0^-$	Laplace transform $F(s)$
Unit impulse	$\delta(t)$	1
Unit step	$u(t)$	$1/s$
Unit ramp	t	$1/s^2$
nth-order ramp	t^n	$n!/s^{n+1}$
Exponential	ϵ^{-at}	$1/(s+a)$
Damped ramp	$t\epsilon^{-at}$	$1/(s+a)^2$
Cosine	$\cos\omega t$	$s/(s^2+\omega^2)$
Sine	$\sin\omega t$	$\omega/(s^2+\omega^2)$
...	$(1/2\omega^3)(\sin\omega t - \omega t\cos\omega t)$	$1/(s^2+\omega^2)^2$
...	$(1/2\omega)t\sin\omega t$	$s/(s^2+\omega^2)^2$
Damped cosine	$\epsilon^{-at}\cos\omega t$	$(s+a)/[(s+a)^2+\omega^2]$
Damped sine	$\epsilon^{-at}\sin\omega t$	$\omega/[(s+a)^2+\omega^2]$

7. CIRCUIT ANALYSIS WITH ENERGY-STORAGE ELEMENTS

7.1 Initial Conditions

If a circuit has initial energy stored in it, that is, if any of the capacitors is charged when the circuit is energized or reconfigured, or if any of the inductors has a nonzero initial current when the circuit is energized or reconfigured, then these conditions must be determined before a complete analysis can be completed. In general, this will require knowledge of the circuit just before the circuit to be analyzed is connected. Normal circuit-analysis methods can be used to determine the initial conditions.

7.2 An Example with Nodal Analysis

Figure 14 shows a network with three (essential) nodes and three energy-storage elements. The earlier discussion shows that two independent KCL equations can be written, and are needed. The voltage source is a unit step function, and it is assumed that there is no initial energy stored in the network. When the defining equations for capacitors and inductors of Secs. 3.2 and 3.3 are combined with the theorems of Table 1, it is found that, for the inductor, $I(s) = V(s)Y(s) = V(s)(1/sL)$, and for the capacitors, $I(s) = V(s)sC$. For the resistors, $I(s) = V(s)\,(1/R)$. The terms from Table 1 that refer to initial conditions are zero.

The following treatment shows the writing of two Kirchhoff's current-law Laplace-domain equations for the network and a complete solution for $v_2(t)$. The two KCL equations are

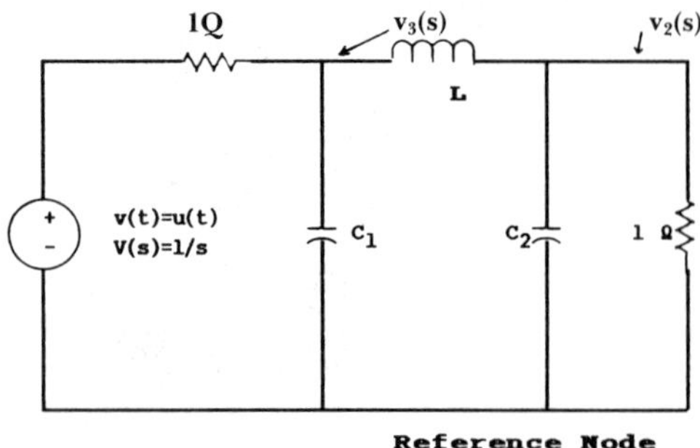

FIG. 14. Nodal analysis without initial conditions.

$$V_3\left(1 + sC_1 + \frac{1}{sL}\right) - V_2\left(\frac{1}{sL}\right) = \frac{1}{s},$$
$$-V_3\left(\frac{1}{sL}\right) + V_2\left(1 + sC_2 + \frac{1}{sL}\right) = 0. \tag{25}$$

The solution for $V_2(s)$ is

$$V_2(s) = \left(\frac{1}{LC_1C_2}\right) s^{-1}\left\{s^3 + s^2\left(\frac{1}{C_1} + \frac{1}{C_2}\right) + s\left[\frac{1}{C_1C_2} + \frac{1}{L}\left(\frac{1}{C_1} + \frac{1}{C_2}\right)\right] + \frac{1}{LC_1C_2}\right\}^{-1}. \tag{26}$$

The specific results will depend on the numerical values chosen for the two capacitors and the inductor. If $L = 1.345$ H, $C_1 = 0.680$ F, and $C_2 = 4.375$ F, Eq. 26 becomes

$$V_2(s) = \frac{0.25}{s(s^3 + 1.70s^2 + 1.60s + 0.50)}. \tag{27}$$

In general, the denominator polynomial must be factored, and computer techniques are employed in all but the simplest cases. When the denominator of Eq. (27) is factored, the result is

$$V_2(s) = \frac{0.25}{s(s + 0.50)(s^2 + 1.20s + 1.00)}. \tag{28}$$

Expansion into partial fractions provides a convenient method for finishing the solution. In this case, the result is

$$V_2(s) = \frac{0.50}{s} + \frac{-0.769}{s + 0.50} + \frac{0.135 - j0.139}{s + 0.60 + j0.80} + \frac{0.135 + j0.139}{s + 0.60 - j0.80},$$
$$V_2(s) = \frac{0.50}{s} - \frac{0.769}{s + 0.50} + \frac{0.270\,(s + 0.60) - 0.278(0.80)}{(s + 0.60)^2 + (0.80)^2}. \tag{29}$$

Finally, the inverse transform and solution is

$$v_2(t) = 0.50 - 0.77\,\epsilon^{-0.50t} + [0.27\cos(0.80t) - 0.28\sin(0.80t)]\,\epsilon^{-0.60t}. \tag{30}$$

7.3 An Example with Loop Analysis and Initial Conditions

Figure 15 shows a circuit with a voltage source, a load resistor, and three energy-storage elements. The excitation function is a unit step function. Assume that the initial energy stored in the capacitor is such that the initial voltage is +4 V, and that the initial energy in the $\frac{1}{6}$-H inductor is such that the initial current is 3 A. Table 1, in conjunction with the defining relations for an inductor, shows that, for the $\frac{5}{6}$-H inductor, $V(s) = I(s)(\frac{5}{6}s)$, while for the second inductor, $V(s) = \frac{1}{6}[sI(s) - 3]$. For the capacitor,

$$V(s) = \frac{I(s)}{(12/25)s} + \frac{4}{s},$$

The $4/s$ term represents the initial condition, and in Table 2, it represents the integral from $t = 0-$ to $t = 0+$. (Some authors define the integral from $t = 0+$ and show the initial condition explicitly.)

The equations become:

$$\frac{1}{s} = \frac{5s}{6} I_1 + \frac{1}{(12/25)s}(I_1 - I_2) + \frac{4}{s}, \tag{31}$$

$$0 = \frac{1}{(12/25)s}(I_2 - I_1) - \frac{4}{s} + \frac{1}{6} sI_2 - \frac{1}{6} 3 + I_2(1). \tag{32}$$

Solution for $I_2(S)$ gives, after some manipulation and simplification,

$$I_2(s) = \frac{3s^3 + 24s^2 + 7.5s + 15}{s^4 + 6s^3 + 15s^2 + 15s}. \tag{33}$$

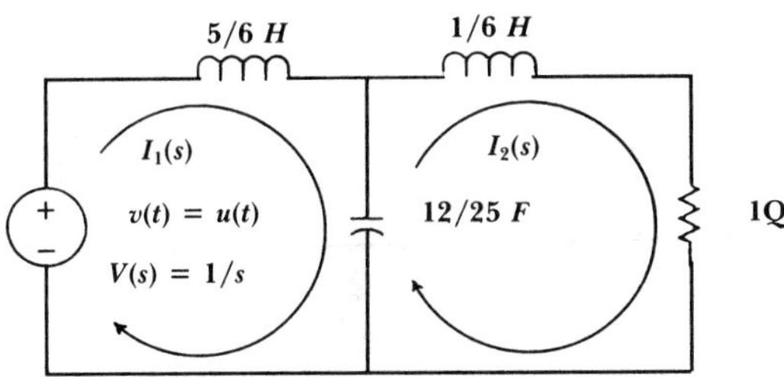

FIG. 15. Loop analysis with initial conditions.

Expansion in partial fractions gives

$$I_2 = \frac{1}{s} - \frac{11.64}{s + 2.32} + \frac{6.82 - j1.16}{s + 1.84 - j1.75} + \frac{6.82 + j1.16}{s + 1.84 + j1.75}, \tag{34}$$

$$I_2 = \frac{1}{s} - \frac{11.64}{s + 2.32} + \frac{13.64\,(s + 1.84) + 2.32(1.75)}{(s + 1.84)^2 + (1.75)^2}. \tag{35}$$

The inverse transform of Eq. (35) gives the time-domain solution that is required:

$$i_2(t) = 1 - 11.64\,\epsilon^{-2.32t} + \epsilon^{-1.84t}(13.64 \cos 1.75t + 2.32 \sin 1.75t). \tag{36}$$

It should be noted that this solution does satisfy the initial condition of 3 A in the $\frac{1}{6}$-H inductor and thus in the resistor.

7.4 State-Variable Methods

A technique for solving virtually all types of circuit problems is known as the state-variable method. It is valid for nonlinear and time-varying networks as well as for linear networks. It is readily adapted to digital computer solutions. The technique has become common both in control systems analysis and in network or circuit theory, and requires the analyst to put the circuit equations into a set of first-order differential equations. Invariably matrix analysis methods are used, as will be done here. The technique will be introduced in this article, and the references give much additional information.

Equations written in state-variable form are usually expressed as

$$\begin{aligned} \dot{\mathbf{x}}(t) &= \mathbf{A}\mathbf{x}(t) + \mathbf{B}\mathbf{u}(t) \\ \mathbf{y}(t) &= \mathbf{C}\mathbf{x}(t) + \mathbf{D}\mathbf{u}(t), \end{aligned} \tag{37}$$

where $\mathbf{x}$ is a matrix (vector) of dependent or response variables, $\mathbf{u}$ is a matrix of forcing or excitation functions, $\mathbf{A}$ is a matrix that describes the network, $\mathbf{B}$ is a matrix showing input connections to circuit, $\mathbf{C}$ is a matrix relating output variables to dependent variables, $\mathbf{D}$ shows direct input-output connections, and $\mathbf{y}$ is the matrix of output variables.

The key term in Eq. (37) is the matrix **A**, as it contains the information that relates currents and voltages in all network elements. If the network is linear and time-invariant, then all terms of the matrix **A** are constants. Otherwise these terms are nonlinear, time-varying, or both. The matrix **A** is square.

In this section, the formulation and solution of the equations for a commonly occurring case will be considered, and the technique will be presented by example. There are situations in which this method may not apply without modification. Since derivatives of dependent variables are needed, and since the basic element equations for inductors and capacitors require derivatives of capacitor voltages and inductor currents, the formulation of state-variable equations focuses on these quantities as an appropriate choice of state variables.

Figure 16(a) shows a network with a voltage source, a current source, two resistors, and two energy-storage elements. For this problem, the output signal is taken as the voltage across the resistor R. A good technique (Rohrer, 1970) for writing the equations is to draw an auxiliary circuit, as Fig. 16(b), in which the inductors are replaced by independent current sources, and the capacitors are replaced by independent voltage sources. Then, Kirchhoff's laws are written for the network. Generally, a mixture of the current and voltage laws is needed. For the circuit under consideration,

$$\begin{aligned} i_c(t) &= -Gv_c(t) + i_L(t) + i_{in2}(t), \\ v_L(t) &= -v_c(t) - Ri_L(t) + v_{in1}(t). \end{aligned} \tag{38}$$

After the element relations for the capacitor and inductor are substituted for the inductor voltage and capacitor current terms, and the equations are rewritten and combined into matrix form, the circuit description is

$$\begin{bmatrix} \dot{v}_C(t) \\ \dot{i}_L(t) \end{bmatrix} = \begin{bmatrix} -\dfrac{G}{C} & +\dfrac{1}{C} \\ -\dfrac{1}{L} & -\dfrac{R}{L} \end{bmatrix} \begin{bmatrix} v_C(t) \\ i_L(t) \end{bmatrix} + \begin{bmatrix} \dfrac{1}{C} & 0 \\ 0 & \dfrac{1}{L} \end{bmatrix} \begin{bmatrix} i_{in2}(t) \\ v_{in1}(t) \end{bmatrix},$$

$$v_0 = [0 \quad R] \begin{bmatrix} v_C \\ i_L \end{bmatrix} + 0. \tag{39}$$

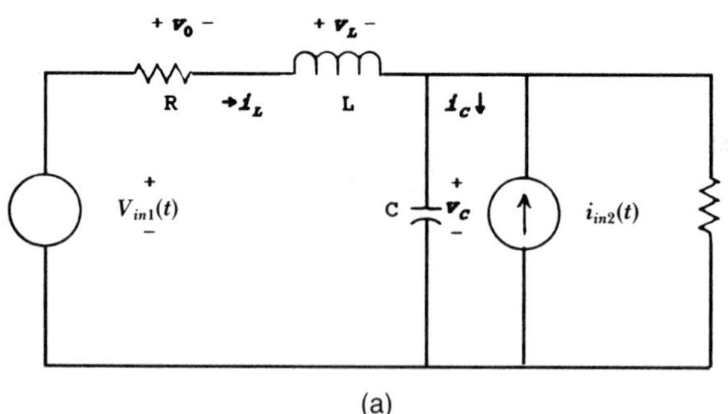

(a)

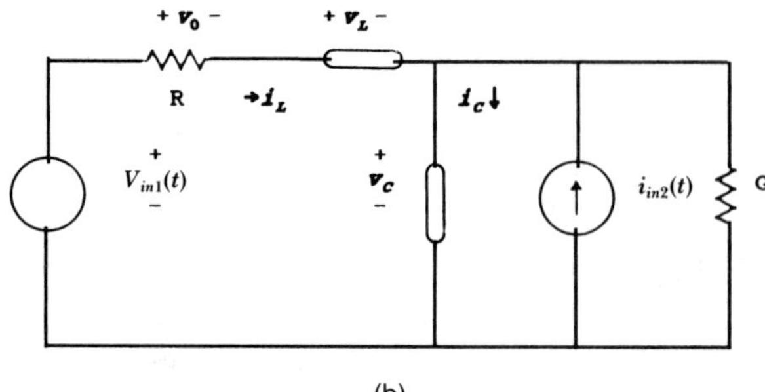

(b)

FIG. 16. Circuit to illustrate state-variable analysis techniques. (a) Original circuit, and (b) circuit with energy storage elements replaced by sources.

This example shows a technique that will work in most commonly encountered situations. If, however, a circuit or network contains a node at which the only branches connected are inductors and current sources, or if it contains a loop with only voltage sources and capacitors, then additional care is needed. Methods for dealing with this situation are discussed in many more advanced circuit theory textbooks, such as Rohrer (1970), Balabanian *et al.* (1969), or Desoer and Kuh (1969).

7.5 Solution of State-Variable Equations

Equations such as Eqs. (39) can be solved by a variety of methods. Laplace transform methods can be used. Extensions of classical differential equation solutions are possible, and will be illustrated here. Let $C = \frac{1}{3}$ F, $G = \frac{2}{3}$ S, $L = 1$ H, and $R = 6\ \Omega$. With these substitutions, Eqs. (39) become

$$\begin{bmatrix} \dot{v}_C(t) \\ \dot{i}_L(t) \end{bmatrix} = \begin{bmatrix} -2 & 3 \\ -1 & -6 \end{bmatrix} \begin{bmatrix} v_C(t) \\ i_L(t) \end{bmatrix} + \begin{bmatrix} 3 & 0 \\ 0 & 1 \end{bmatrix} \begin{bmatrix} i_{in2}(t) \\ v_{in1}(t) \end{bmatrix},$$
$$v_0 = [0 \quad 6] \begin{bmatrix} v_C \\ i_L \end{bmatrix} + 0; \tag{40}$$

that is,

$$\mathbf{A} = \begin{bmatrix} -2 & 3 \\ -1 & -6 \end{bmatrix}, \mathbf{B} = \begin{bmatrix} 3 & 0 \\ 0 & 1 \end{bmatrix},$$
$$\mathbf{C} = [0 \quad 6], \mathbf{D} = 0.$$

The zero-input response, which corresponds to the complementary function in conventional differential-equation solution techniques, is obtained from Eqs. (40) when the input functions are zero, or $u = 0$. This solution is

$$\mathbf{x}(t) = \epsilon^{\mathbf{A}(t)}\mathbf{x}(0), \quad t \geq 0;$$
$$\mathbf{y}(t) = \mathbf{C}\epsilon^{\mathbf{A}(t)}\mathbf{x}(t), \quad t \geq 0. \tag{41}$$

The exponential matrix in Eq. (41) is found from the eigenvalues of the matrix **A**, which in turn are found from its characteristic polynomial

$$\Delta(\lambda) = \det |\mathbf{A} - \lambda\mathbf{I}|. \tag{42}$$

Here **I** is the identity matrix, and from the values given, the characteristic polynomial is found to be

$$\Delta(\lambda) = \begin{vmatrix} -2-\lambda & 3 \\ -1 & -6-\lambda \end{vmatrix}$$
$$= \lambda^2 + 8\lambda + 15 = (\lambda + 3)(\lambda + 5). \tag{43}$$

These eigenvalues are distinct, and thus the exponential matrix function can be written as

$$\epsilon^{\mathbf{A}t} = \mathbf{f}(\lambda) = \alpha_0(t)\mathbf{I} + \alpha_1(t)\mathbf{A}, \tag{44}$$

where **I** is again the identity matrix. According to the Cayley–Hamilton (Balabanian *et al.*, 1969) theorem, any function of a square matrix can be satisfied as a scalar function by its eigenvalues. This fact is used to determine

$$f(\lambda) = \epsilon^{\lambda t} = \alpha_0(t) + \lambda\alpha_1(t), \tag{45}$$

where λ is an eigenvalue of **A**. When the eigenvalues -3 and -5, respectively, are substituted into Eq. (45), the results are

$$\epsilon^{-3t} = \alpha_0(t) - 3\alpha_1(t),$$
$$\epsilon^{-5t} = \alpha_0(t) - 5\alpha_1(t). \tag{46}$$

Thus,

$$\alpha_0(t) = (5\epsilon^{-3t} - 3\epsilon^{-5t})/2,$$
$$\alpha_1(t) = (\epsilon^{-3t} - \epsilon^{-5t})/2. \tag{47}$$

The exponential matrix of Eq. (44) is then found, and from this, the zero-input response is found as

$$\begin{bmatrix} v_C(t) \\ i_L(t) \end{bmatrix} = \begin{bmatrix} \dfrac{3\epsilon^{-3t} - \epsilon^{-5t}}{2} & \dfrac{3\epsilon^{-5t} - 3\epsilon^{-3t}}{2} \\ \dfrac{\epsilon^{-3t} - \epsilon^{-5t}}{2} & \dfrac{3\epsilon^{-5t} - \epsilon^{-3t}}{2} \end{bmatrix} \begin{bmatrix} v_C(0) \\ i_L(0) \end{bmatrix}, \tag{48}$$

where the term $[v_c(0) \ i_L(0)]^t$ is the initial state.

The preceding term corresponds to the familiar complementary function or homogeneous solution for differential equations. It is also necessary to compute the particular integral or particular solution. This is done with the generalized convolution integral, Table

1, which gives

$$\mathbf{x}(t) = \int_{t_0}^{t} \epsilon^{\mathbf{A}(t-\tau)}\mathbf{B}\mathbf{u}(\tau)d\tau. \tag{49}$$

The complete solution for $\mathbf{x}(t)$ is then the sum of Eqs. (41) and (49). The complete output expression $\mathbf{y}(t)$ is given by

$$\mathbf{y} = \mathbf{C}\int_{0}^{t} \epsilon^{\mathbf{A}(t-\tau)}\mathbf{B}\mathbf{u}(\tau)d\tau + \mathbf{D}\mathbf{u}(t). \tag{50}$$

If the current source is a step function of strength 2 A, and the voltage source is a step function of strength 4 V, then the excitation function becomes

$$\mathbf{u}(t) = \begin{bmatrix} 2 \\ 4 \end{bmatrix} u(t).$$

After some work, solution of Eq. (49) gives

$$\begin{bmatrix} v_C(t) \\ i_L(t) \end{bmatrix} = \begin{bmatrix} 3.2 - 5.0\,\epsilon^{-3t} + 1.8\,\epsilon^{-5t} \\ 2/15 - 1.8\,\epsilon^{-5t} + 5/3\,\epsilon^{-3t} \end{bmatrix} u(t) \tag{51}$$

as the zero-state response, and the complete solution for the network is the sum of Eqs. (48) and (51).

8. SINUSOIDAL STEADY-STATE ANALYSIS

8.1 Description of the Technique

An important forcing function to consider is a sinusoidal signal. One reason is the fact that, throughout the world, virtually all electric energy is distributed by voltages and currents that are sinusoidal. A second reason is that designers find that analysis of electrical systems with sinusoidal forcing functions yields good insight to the understanding of how the system will process other signals. The statement is true for high-power electrical distribution systems and for relatively low-power communications and related systems. The phrase *sinusoidal steady state* refers to the behavior of the circuit after the transient terms (terms arising from the complementary function) have become negligible.

When the circuit or network is linear, then the steady-state response to a sinusoidal forcing function is another sinusoidal signal at the same frequency. The relative amplitudes and phases may vary, but the frequency does not. This observation is crucial to development of the techniques. In essence, the process is one of finding an equivalent circuit in the frequency domain, analyzing it algebraically rather than through the use of differential equations, and interpreting the final results as needed. Since the frequency of the response signal is identical with that of the input signal, the analyst need only find the amplitudes and phases of the response signals, whether currents or voltages.

There are two approaches to the topic. One is to develop the topic from basic principles. The second is to consider sinusoidal steady state as a special case of Laplace analysis in which the complex variable s is replaced by its imaginary component, $s = j\omega$. The results are interpreted in sinusoidal terms. The first approach will be the principal approach used in this article. Further, $\omega = 2\pi f$, where f is the frequency in hertz (Hz).

The process is called *phasor analysis*. It makes use of the imaginary operator and Euler's relation:

$$j = \sqrt{-1},$$
$$\epsilon^{jx} = \cos x + j\sin x. \tag{52}$$

Euler's relation may be solved for the trigonometric terms, giving

$$\cos x = (\epsilon^{jx} + \epsilon^{-jx})/2,$$
$$\sin x = (\epsilon^{jx} - \epsilon^{-jx})/j2. \tag{53}$$

A form of Euler's relation that often appears in phasor analysis is

$$\epsilon^{j(\omega t+\alpha)} = \cos(\omega t + \alpha) + j\sin(\omega t + \alpha). \tag{54}$$

8.2 Definition of a Phasor

The phasor representation or phasor transform of a sinusoidal function is, by definition, a complex number that contains both the amplitude and phase information of the original. Specifically, when

$$v(t) = V_m\cos(\omega t + \alpha) = V_m\sin(\omega t + \alpha + \pi/2), \tag{55}$$

the phasor representation of $v(t)$ is

$$V = V_m \epsilon^{j\alpha}. \quad (56)$$

It should be noted here that electric power engineers usually use a scaling factor for phasors. The alternative definition is given as

$$V_{\text{eff}} = (V_m/\sqrt{2})\,\epsilon^{j\alpha}, \quad (57)$$

and the justification will be presented in the section on power calculations. The definition may be used in reverse to convert a phasor to a time function, provided the frequency is known.

8.2.1 Phasor Algebra Phasors are complex numbers, and it is frequently necessary to perform extensive arithmetic computations with phasors. Phasor algebra follows all of the rules of complex-number arithmetic and algebra, or of two-dimensional vector algebra. Modern calculators and mathematical computer programs have greatly facilitated the arithmetic calculations that are required, as have circuit-analysis programs.

8.2.2 Differentiation and Integration Suppose a phasor is represented by

$$P_1 = P_m \epsilon^{j\omega t} \epsilon^{j\theta}. \quad (58)$$

where the frequency term is included in order to study differentiation and integration. Differentiation with respect to time becomes

$$\frac{dP_1}{dt} = j\omega P_m \epsilon^{j\omega t} \epsilon^{j\theta} = \omega P_m \epsilon^{j\omega t} \epsilon^{j(\theta+\pi/2)}, \quad (59)$$

while integration becomes

$$\int P_1 dt = \frac{1}{j\omega} P_m \epsilon^{j\omega t} \epsilon^{j\theta} = \frac{1}{\omega} P_m \epsilon^{j\omega t} \epsilon^{j(\theta-\pi/2)}. \quad (60)$$

8.3 Susceptance and Reactance

Table 3 shows the ratios of voltage phasors to current phasors, or their reciprocals. The former are called *reactances*, while the latter are called *susceptances*. When the element being considered is a capacitor, the adjective capacitive is used. When the element is an inductor, the adjective used is inductive.

Table 3. Capacitor and inductor equivalents in phasor domain.

Phasor ratio	Capacitor equivalents	Inductor equivalents
$\dfrac{V(j\omega)}{I(j\omega)}$ (Reactance)	$\dfrac{1}{j\omega C}$	$j\omega L$
$\dfrac{I(j\omega)}{V(j\omega)}$ (Susceptance)	$j\omega C$	$\dfrac{1}{j\omega L}$

8.4 Admittance and Impedance

Analysis of circuits operating in the sinusoidal steady state is typically begun by replacing each capacitor and each inductor by its susceptance or reactance. These terms may be single numbers or may be functions of frequency (in rad/s). Resistors are unaffected by this transformation. In turn, these terms may be combined following the series/parallel rules of Figs. 6 and 7, or the tee-pi conversions of Fig. 8. The result is a complex number that may or may not be a function of frequency. If the term is the ratio of a voltage phasor to a current phasor, it is called *impedance*. If the term is the ratio of a current phasor to a current phasor, it is called *admittance*.

8.5 Network Theorems and Solution Techniques

The circuits being studied are linear. The Thévenin and Norton theorems apply. Kirchhoff's laws still apply. The process of writing either loop-current or node-voltage equations still applies. These features make the method attractive and useful. The results of the analyses will again be phasor currents and voltages, of the same frequency as the excitation function. Sometimes these are reconverted to time-domain functions, but often they are left as phasor quantities and interpreted directly. This is possible because, in a linear circuit, the frequency components in the response functions are identical with those of the source. New signal frequencies can be generated only by nonlinear networks.

Figure 17 shows an analysis of a circuit that has three energy-storage elements and two nodes. The current source has a frequency of 1.25 rad/s. Specifically, Fig. 17(a) shows the basic circuit, while Fig. 17(b) shows the circuit prepared for phasor analysis at the spec-

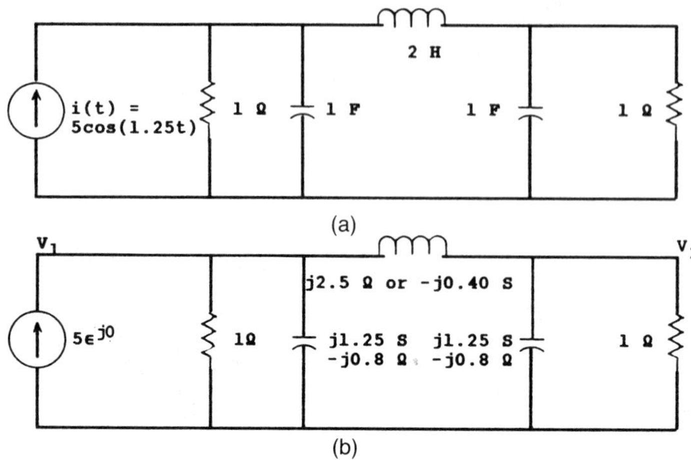

FIG. 17. (a) Two-node circuit for phasor analysis. (b) Phasor-domain equivalent circuit.

ified frequency. The following steps show the writing of Kirchhoff's current-law equations at the two nodes in matrix form, the solution for the two phasor voltages, and the reconversion to the time domain for the final answer:

$$\begin{bmatrix} 5 \\ 0 \end{bmatrix} = \begin{bmatrix} (1.00 + j1.25 - j0.40) & -(-j0.40) \\ -(-j0.40) & (1.00 + j1.25 - j0.40) \end{bmatrix} \begin{bmatrix} V_1 \\ V_2 \end{bmatrix}, \tag{61}$$

$$\begin{bmatrix} V_1 \\ V_2 \end{bmatrix} = \begin{bmatrix} 3.055 - j2.155 \\ -1.103 - j0.284 \end{bmatrix} = \begin{bmatrix} 3.378\ \epsilon^{-j0.614} \\ 1.139\ \epsilon^{-j2.890} \end{bmatrix}, \tag{62}$$

$$\begin{bmatrix} v_1(t) \\ v_2(t) \end{bmatrix} = \begin{bmatrix} 3.738\cos(1.25t - 0.614) \\ 1.139\cos(1.25t - 2.890) \end{bmatrix}. \tag{63}$$

8.6 Power and Energy Considerations

When a sinusoidal voltage $v(t) = V_m \cos(\omega t + \alpha)$ is impressed across a resistor of resistance R, then there is a current $i(t) = I_m \cos(\omega t + \alpha)$ through the resistor, where $I_m = V_m/R$. The instantaneous power is given by Eq. (1), and, over an integral number of cycles, the average power is given by

$$P_{\text{avg}} = \frac{V_m I_m}{2} = \frac{V_m^2}{2R} = \frac{I_m^2}{2} R. \tag{64}$$

This amount of power is equal to the power that would be delivered by a constant voltage of magnitude $V_{\text{eff}} = V_m/\sqrt{2}$ or a constant current of magnitude $I_{\text{eff}} = \sqrt{2} I_m$. This leads to the definition of *effective values* of ac voltages and currents, which are equal to the peak values divided by $\sqrt{2}$.

The concept of effective values is in quite common use. For example, a typical light bulb available in North America might be rated at 60 W, 120 V. The 120-V specification is an effective-value voltage specification, while the average power consumed by the bulb is 60 W.

When Eqs. (56) and (57) were introduced, it was noted that the latter gives an alternative definition of a phasor. Equation (57) uses effective values in its definition, as it facilitates power and energy calculations without the necessity of retaining the scaling factor. There is not a basic difference, however, but it is necessary that the analyst be aware of which definition is being used, and to be consistent.

A more general definition for effective values is known as the *root mean square* (rms) value of the function, which applies to any signal, not simply sinusoidal signals. The term describes the defining equation, which is

$$F_{\text{rms}} = \left[\frac{1}{T} \int_{t_0}^{t_0+T} [f(t)]^2 dt \right]^{1/2}. \tag{65}$$

8.6.1 Power Factor When circuits or systems operate in the sinusoidal steady state, the current and voltage are generally not in phase; in other words, there is a phase difference between them. One effect of this fact is that the average power delivered to such systems is less than the product of effective values of current and voltage, though that term is significant. Specifically, if

$$v(t) = V_m \cos(\omega t + \alpha) \tag{66}$$

and

$$i(t) = I_m \cos(\omega t + \beta), \tag{67}$$

a computation of the average power delivered over an integral number of cycles gives

$$\begin{aligned} P_{\text{avg}} &= \frac{V_m I_m}{2} \cos(\alpha - \beta) \\ &= V_{\text{eff}} I_{\text{eff}} \cos(\alpha - \beta). \end{aligned} \tag{68}$$

The angle $\alpha - \beta$, which is the phase difference between the voltage and current, is called the *power factor angle*, and its cosine is called the *power factor*. It represents the ratio of the average power to the product of current and voltage for a device. For example, a motor rated at 120 V, 500 W, and 6.5 A would have a power factor of $500/(120 \times 6.5) = 0.641$.

8.6.2 Reactive Volt-Amperes When the voltage across and the current through a device are given by Eqs. (66) and (67), respectively, a computation of the power delivered to the device as a function of time shows

$$\begin{aligned} p(t) = \frac{V_m I_m}{2} [&\cos(\alpha - \beta) \\ &+ \cos(2\omega t + \alpha + \beta)]. \end{aligned} \tag{69}$$

There is a constant term and a double-frequency term that represents the energy that is interchanged between the electric and magnetic fields of the source and device. This new term is called *reactive volt-amperes* (vars). This quantity must be considered in the analysis and design of electrical systems, especially large power systems.

8.6.3 Power and Vars It may be shown that

$$\begin{aligned} \text{vars} &= \frac{V_m I_m}{2} \sin(\alpha - \beta) \\ &= V_{\text{eff}} I_{\text{eff}} \sin(\alpha - \beta). \end{aligned} \tag{70}$$

A combination of Eqs. (68) and (70) shows that, if the phasor voltage across and the phasor current through a device are given, respectively and in effective values, by

$$\begin{aligned} V_1 &= V_{\text{eff}} \epsilon^{(j\alpha)}, \\ I_2 &= I_{\text{eff}} \epsilon^{(j\beta)}, \end{aligned} \tag{71}$$

then the equation

$$\begin{aligned} VA &= V_{\text{eff}} I_{\text{eff}} [\cos(\alpha - \beta) + j \sin(\alpha - \beta)] \\ &= V_{\text{eff}} I_{\text{eff}}^*, \end{aligned} \tag{72}$$

where the asterisk denotes the complex conjugate, may be used to find both average power and vars. The real part of the expression is the average power, while the imaginary part is the vars.

8.7 Three-Phase Systems

Virtually all bulk electric energy distribution is accomplished through the use of three-phase systems. There are several important reasons for this. The first is that a three-phase system distributes energy with less losses than a single-phase system would. The second is that it uses materials more efficiently. The third is that it can be shown that a (balanced) three-phase system distributes electric power at a constant rate, i.e., without a double-frequency term such as appears in Eq. (69). As a consequence, three-phase generators develop and deliver constant power, and three-phase motors develop constant torque. The double-frequency torque component in a single-phase motor does not exist in a three-phase motor. Throughout this section, effective values of voltages and currents will be used.

8.7.1 Balanced Three-Phase Systems A balanced three-phase source consists of three voltage (or current) sources that are sinusoidal, equal in magnitude, and separated in phase by 120° or $2\pi/3$ rad. In time-domain and phasor format, a typical set of voltages is given by Table 4. (At a frequency of 60 Hz,

Table 4. Phasor representations of three-phase voltages.

Time format	Phasor format
$v_{ab} = V_{\text{eff}} \cos(377t) = v_{ao}$	$V_{\text{eff}} \epsilon^{j0}$
$v_{bc} = V_{\text{eff}} \cos(377t - 2\pi/3) = v_{bo}$	$V_{\text{eff}} \epsilon^{(-j2\pi/3)}$
$v_{ca} = V_{\text{eff}} \cos(377t + 2\pi/3) = v_{co}$	$V_{\text{eff}} \epsilon^{(j2\pi/3)}$

the frequency in rad/s is $2\pi \times 60 \cong 377$ rad/s.) These three sources may be connected in either of two ways to form a balanced three-phase source. One is called a delta (or mesh or pi) connection, the second a wye (or star or tee). Both are shown in Fig. 18. Three wires or lines will be needed to connect the sources to a load. These are labeled A, B, and C in Fig. 18. The individual source voltages are known as *phase voltages*, and the voltages between any pair of lines are known as *line voltages*. For the wye connection, a fourth line, labeled "O," is possible. Sometimes it is utilized in practice, but not always. For the delta system, it can be shown that, in phasor notation only,

$$\begin{aligned} V_{AB} &= V_{ab} = V_{\text{eff}}\epsilon^{(j0)}, \\ V_{BC} &= V_{bc} = V_{\text{eff}}\epsilon^{(-j2\pi/3)}, \\ V_{CA} &= V_{ca} = V_{\text{eff}}\epsilon^{(j2\pi/3)}, \end{aligned} \tag{73}$$

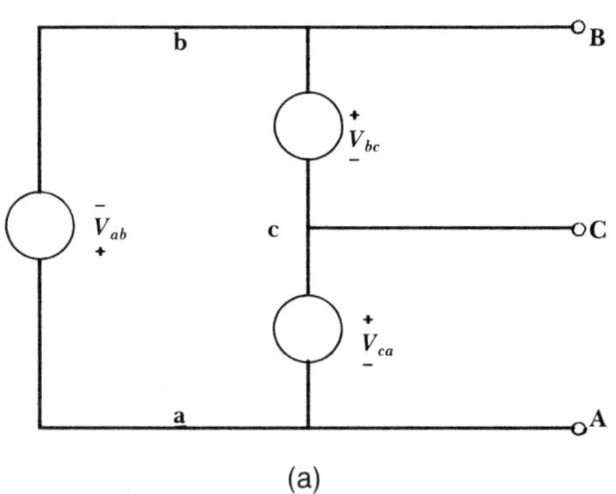

(a)

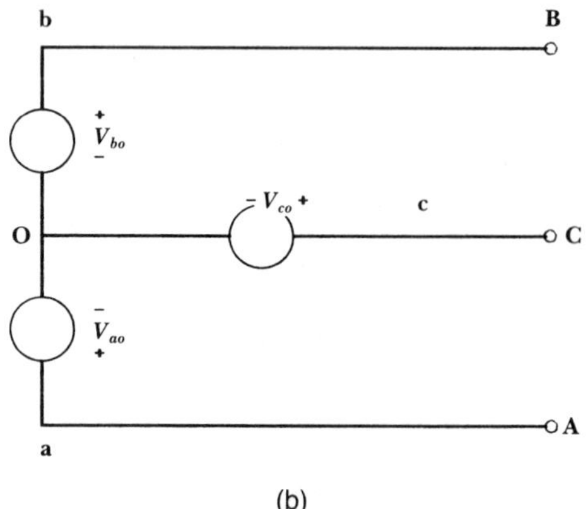

(b)

FIG. 18. (a) Delta-connected three-phase source. (b) Wye-connected three-phase source.

whereas in the wye system,

$$\begin{aligned} V_{AB} &= V_{ao} - V_{bo} = \sqrt{3}\, V_{\text{eff}}\epsilon^{(j\pi/6)}, \\ V_{BC} &= V_{bo} - V_{co} = \sqrt{3}\, V_{\text{eff}}\epsilon^{(-j\pi/2)}, \\ V_{CA} &= V_{co} - V_{ao} = \sqrt{3}\, V_{\text{eff}}\epsilon^{(j5\pi/6)}. \end{aligned} \tag{74}$$

These results show that in a balanced wye system, the line voltages are equal to the phase voltages multiplied by $\sqrt{3}$, and are shifted in phase by 30°. In a balanced delta system, the line and phase voltages are equal. Similarly, it can be shown that, in a balanced delta system, the line currents are equal to the phase currents multiplied by $\sqrt{3}$, and they are shifted in phase by 30° with respect to the phase currents. In a balanced wye system, the line and phase currents are equal.

8.7.2 Balanced Loads A balanced three-phase load consists of three equal impedances connected either in delta or wye. Equations (15) and (16) can be used to convert from one to the other. Because of the symmetry of a balanced source and a balanced load, analysis reduces to analysis of a single phase, with appropriate shifts in phase for the other two phases.

8.7.3 Unbalanced Loads and Symmetrical Components A three-phase system that has nonsymmetrical loads, or sources that differ in magnitude, or sources that differ in phase by other than 120°, is said to be unbalanced. Circuits of this type may be analyzed by conventional nodal or loop analysis, or by any other conventional technique. There is a technique for solution of unbalanced networks that makes use of symmetrical components. In essence, this technique requires the analyst to formulate unbalanced systems as the sum of three balanced systems, and to use the techniques of balanced network analysis on the equivalent networks. The technique is described in several books (Anderson, 1973; Stevenson, 1982).

9. TWO-PORT ANALYSIS

9.1 Definition of a Port

A common use of electrical circuits is to connect a source of electrical energy or information to a load, often in such a way as to modify the signal in a prescribed way. In its basic form, such a circuit has one pair of input terminals and one pair of output terminals. Each of the four wires will have a current in or out of the circuit. Between any pair of terminals there will be a voltage.

If the structure of the circuit is such that the current into one terminal is equal to the current out of a second terminal, then that terminal pair is called a *port*. By Kirchhoff's current law, the second pair of terminals will also be a port. The circuit of Fig. 19 has two such ports, and is called a two-port. The input-output circuit variables are completely characterized, at the terminals, by the two currents and two voltages indicated in Fig. 19.

A particularly important type of two-port is called a filter. In the article ELECTRONIC CIRCUITS filters are discussed at length. In this section, a variety of relations among the terminal voltages and currents will be discussed. Six such sets are found to be useful. They are known as the open-circuit impedance parameters, short-circuit admittance parameters, hybrid parameters (two types), and transmission-line parameters (two types). For resistive circuits, all sets of parameters will be composed of constants. For circuits operating in the sinusoidal steady state, the parameters will be complex numbers for single-frequency operation, or functions of ω when the frequency is a variable. If Laplace analysis is being done, the parameters will be functions of the Laplace variable s. In the following, functional notation will be suppressed.

9.2 Open-Circuit Impedance Parameters

If the two currents are considered to be the independent variables and the two voltages are the dependent variables, then one can write

$$\begin{bmatrix} V_1 \\ V_2 \end{bmatrix} = \begin{bmatrix} z_{11} & z_{12} \\ z_{21} & z_{22} \end{bmatrix} \begin{bmatrix} I_1 \\ I_2 \end{bmatrix}, \quad \text{or} \quad [V] = [z][I]. \tag{75}$$

The four numbers or functions in the square matrix characterize the network. They may be computed using any conventional method of circuit analysis, and often are directly computed by computer-based software. They may also be measured. For example, the term z_{21} can be computed or measured as the ratio V_2/I_1 when I_2 is set equal to zero (for computations), or made zero by an appropriate open circuit when measurements are being taken.

9.3 Short-Circuit Admittance Parameters

If the two voltages are in independent variables, and the two currents are the dependent variables, then one can write

$$\begin{bmatrix} I_1 \\ I_2 \end{bmatrix} = \begin{bmatrix} y_{11} & y_{12} \\ y_{21} & y_{22} \end{bmatrix} \begin{bmatrix} V_1 \\ V_2 \end{bmatrix}, \quad \text{or} \quad [I] = [y][V]. \tag{76}$$

Measurements and computations follow principles similar to those of open-circuit impedance parameters.

9.4 Hybrid Parameters

Voltages and currents may be mixed in their roles as independent and dependent variables in two ways, as indicated:

$$\begin{bmatrix} V_1 \\ I_2 \end{bmatrix} = \begin{bmatrix} h_{11} & h_{12} \\ h_{21} & h_{22} \end{bmatrix} \begin{bmatrix} I_1 \\ V_2 \end{bmatrix} \quad (h \text{ parameters}), \tag{77}$$

$$\begin{bmatrix} I_1 \\ V_2 \end{bmatrix} = \begin{bmatrix} g_{11} & g_{12} \\ g_{21} & g_{22} \end{bmatrix} \begin{bmatrix} V_1 \\ I_2 \end{bmatrix} \quad (g \text{ parameters}). \tag{78}$$

9.5 Transmission-Line Parameters

The voltage and current at the input port may be used as independent variables with the output quantities as dependent variables, or the roles may be reversed.

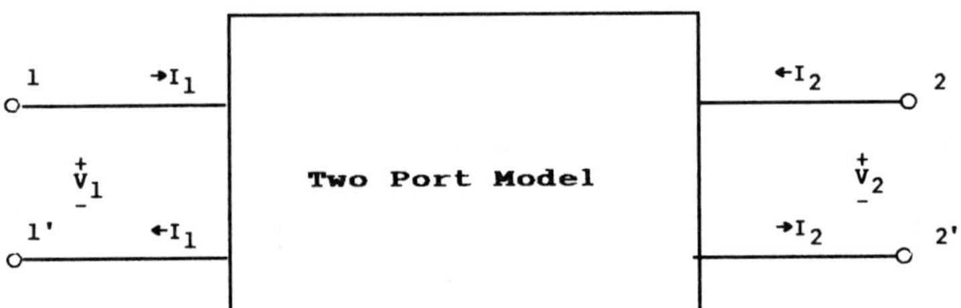

FIG. 19. Prototype to define ports of a network.

$$\begin{bmatrix} V_1 \\ I_1 \end{bmatrix} = \begin{bmatrix} A & B \\ C & D \end{bmatrix} \begin{bmatrix} V_2 \\ I_2 \end{bmatrix}$$
(chain or transmission-line parameters), (79)

$$\begin{bmatrix} V_2 \\ I_2 \end{bmatrix} = \begin{bmatrix} \alpha & \beta \\ \gamma & \delta \end{bmatrix} \begin{bmatrix} V_1 \\ I_1 \end{bmatrix}$$
(inverse transmission-line parameters). (80)

9.6 Two-Port Parameter Conversions

Any set of two-port parameters may be converted to any other set. For example, Eq. (75) may be solved for the two currents, and the result compared with Eq. (76). This comparison shows that

$$\begin{bmatrix} y_{11} & y_{12} \\ y_{21} & y_{22} \end{bmatrix} = \begin{bmatrix} \frac{z_{22}}{|z|} & \frac{-z_{12}}{|z|} \\ \frac{-z_{21}}{|z|} & \frac{z_{11}}{|z|} \end{bmatrix}, \tag{81}$$

where $|z| = z_{11}z_{22} - z_{12}z_{21}$. Complete tables are given in Huelsman (1963) and in Fink and Beaty (1993), Article 2.

9.7 Equivalent Circuits for Two-Ports

Equivalent circuits may be drawn for any set of two-port parameters. The process is shown in Fig. 20 for the $[g]$ parameters, Eq. (78), but the technique is quite similar for the other combinations. Figure 20(a) shows the equivalent circuit.

9.8 Two-Port Analysis

From the equivalent circuit for a two-port, circuit analysis is possible. Figure 20(b) shows a two-port, with hybrid-g parameters, a current source, and a load resistor. As shown here, the current source has a very high output resistance or impedance. If it were present and of such a size as to be significant in circuit operation, it would be added to g_{11}. The output voltage is $V_2 = -I_2R_L$. Analysis of the equivalent circuit using Kirchhoff's laws and some simplification shows that

$$\frac{V_{\text{load}}}{I_{\text{source}}} = \frac{V_2}{I_1} = \frac{R_L g_{21}}{(g_{11}g_{22} - g_{12}g_{21}) + g_{11}R_L}. \tag{82}$$

Similar analyses can be performed for any configuration that may arise.

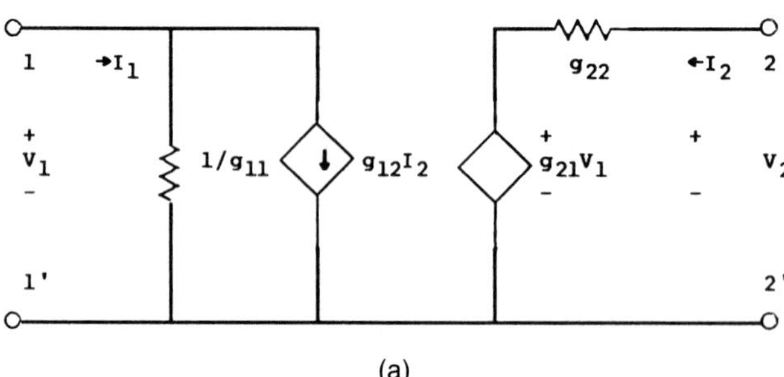

(a)

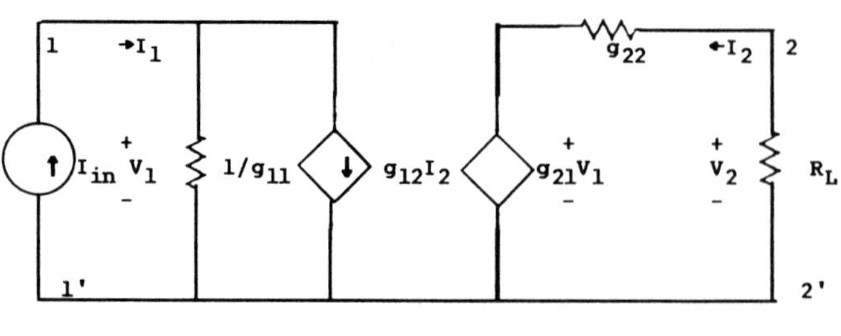

(b)

FIG. 20. Two-port analysis techniques. (a) Equivalent circuit for hybrid-*g* parameters, and (b) two-port with source and load.

10. DISCRETE CIRCUITS

10.1 Introduction

Discrete circuits are circuits are circuits or networks used to process discrete data. In turn, discrete data are data defined at specific instants, but undefined at other times. Discrete data may also be defined in spatial terms, but such topics are not considered in this article. As a nontechnical example, the number of people attending each of a sequence of concerts in a theater may be known, but no data are given—and it is possible that there is no interest—in the number of spectators in the theater at times when there is no concert. Another, often used, example is the set of equations that leads to the radar equations (Cadzow and Van Landingham, 1985). The data arise, in one scenario, when a radar transmitter on one airplane sends pulses that are reflected back from a second airplane. Specifically, the time for return reflections from each pulse is measured. No data are collected or available between pulses. The equations themselves will be developed in Sec. 10.7.

Discrete data are available in two ways. It is important to emphasize that the concept of discrete data applies to the independent variable—time or a spatial measurement. If the dependent variables are expressed in or converted to digital formats, the process is called *digitization* or *quantization*. Continuous-time data may be converted to discrete-time data formats through some sort of sampling process. An important principle that applies here is Shannon's sampling theorem, discussed in Sec. 10.10. This is a necessary step when the data are to be processed by a digital computer, often the mode of implementation of a discrete circuit. The second is from transducers that give discrete data directly.

Discrete circuits are implemented with electronic circuits or with digital computers. In this discussion, however, the interest is in the terminal behavior of the circuits, not in the detailed operation of the circuits. This section considers linear, time-invariant circuits only.

10.2 Discrete Signal Representation

In this discussion, it will be assumed that the discrete data are taken at integral multiples of some time unit that is usually called the *sampling interval*. The integer k will denote the time, kT, where T is the sampling interval. This regular spacing of the data is not necessary, but it is convenient.

10.2.1 Analytic Representation Discrete signals may be represented in analytic form. Several examples are shown in Table 5.

10.2.2 Sequence Representation Discrete signals may be represented as a sequence of numbers, as

$$f(k) = \{5, 4, \underset{\uparrow}{-3}, -2, 8, 9, \ldots\} \qquad (83)$$

The vertical arrow below one element of the sequence denotes the value at $k = 0$. This is an example of a sequence with nonzero values for negative k.

10.2.3 Numerical or Graphical Representation Discrete data may be represented in numerical form, and stored numerically. The sample space may be multidimensional, for example time and space, and then several indices are needed. The data may also be represented or shown graphically, with the index k being the independent variable. Figure 21 shows a possible representation. It is essential that the form of the graph not suggest any sort of line or connection between ordinates as k varies.

10.3 Discrete-Signal Operations

There are three basic operations in linear, time-invariant discrete-circuit analysis and design. The first of these, illustrated in Fig.

Table 5. Analytic representation of discrete functions.

Name	Analytic representation
Exponential	$f(kT) = f(k) = A\epsilon^{-kT}$
Step	$u(kT) = u(k) = 1$, all $k \geq 0$
Ramp	$f(kT) = f(k) = kT,\ k \geq 0$
Impulse	$f(kT) = f(k) = \begin{Bmatrix} 1, k = 0 \\ 0, k \neq 0 \end{Bmatrix}$
Complex periodic exponential (N defines the period)	$f(kT) = f(k) = A\epsilon^{j(2\pi/N)kT}$
Sinusoidal	$f(kT) = f(k) = A\cos\left(\dfrac{2\pi kT}{N} + \alpha\right)$

Note: k takes on integral values only.

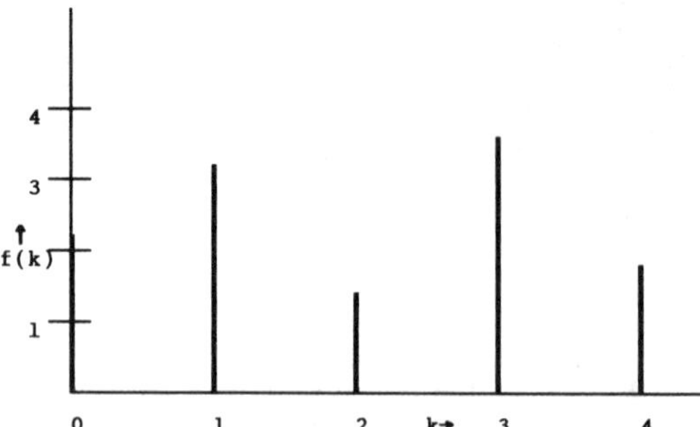

FIG. 21. Graphical representation of discrete data.

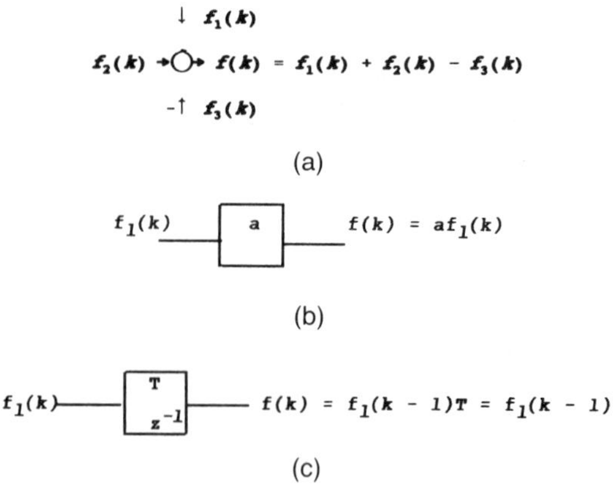

22(a), is addition. Addition includes subtraction if signs are appropriate. The second, illustrated in Fig. 22(b), is multiplication by a constant. Sometimes, this is called gain or amplification. (If the gain were a function of the index k, then the circuit would be time-varying. If two signals were multiplied together, the circuit would be nonlinear.) The third operation, illustrated in Fig. 22(c), is delay. A general term that includes both gain and delay is transmittance. The output of a delay block is identical with the input, but it is delayed in time, normally by one sampling interval, though this is not essential. Delay, like energy-storage elements in ordinary circuits, provides memory. The notation z^{-1} is a commonly used symbol for delay, because of its association with the z transform, to be introduced in Sec. 10.9.

10.4 Signal-Flow and Block Diagrams

Discrete circuits are commonly shown by block diagrams or by signal-flow diagrams. The two ideas are quite similar, and are used in similar ways. The block diagram uses geometric figures to describe the three basic operations, and directional lines to show how signals move through the system. A typical block diagram is shown in Fig. 23(a). Figure 23(b) shows the same system described by a signal-flow graph, which uses directional lines to show signal propagation through a system, and coefficients on or near the lines.

10.4.1 Signal-Flow Graph Definitions It is convenient to define a number of terms associated with signal-flow graphs.

Node: A point in a signal-flow graph at which two or more elements connect.

↓ $f_1(k)$

$f_2(k)$ →○→ $f(k) = f_1(k) + f_2(k) - f_3(k)$

−↑ $f_3(k)$

(a)

$f_1(k)$ [a] $f(k) = af_1(k)$

(b)

$f_1(k)$ [T z^{-1}] $f(k) = f_1(k - 1)T = f_1(k - 1)$

(c)

FIG. 22. Discrete-circuit operations: (a) summing, (b) multiplication or gain, and (c) delay.

Branch: An element that shows signal propagation from one node to another. A branch is directional.

Input node: A node that has only outgoing branches.

Output node: A node that has only incoming branches.

Path: A set of branches traversed in the same direction. In this definition, an element may be included more than once.

Forward path: A path that starts at an input node, ends on an output node, and does not include any branch or node more than one time.

Loop: A path that starts and terminates on the same node, and does not traverse any node or branch more than once.

Forward path gain: The product of branch gains or transmittances is called the path gain. If the path is a forward path, then the path gain is called forward path gain.

Loop gain: The product of branch gains or transmittances around a loop.

10.4.2 Signal-Flow Graph Reductions At times it is useful to simplify or to expand a signal-flow graph either to facilitate analysis or to enhance system understanding. Figure 24 illustrates three equivalents that are commonly used. They may be extended to a wide variety of situations.

10.4.3 Mason's Rule or Formula A powerful rule was developed by Mason (1953, 1956) for use in the analysis of signal-flow graphs. Although presented here in connection with discrete circuits, it is equally ap-

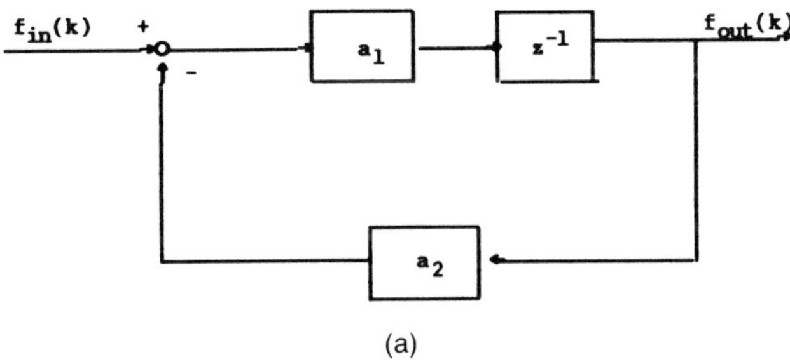

(a)

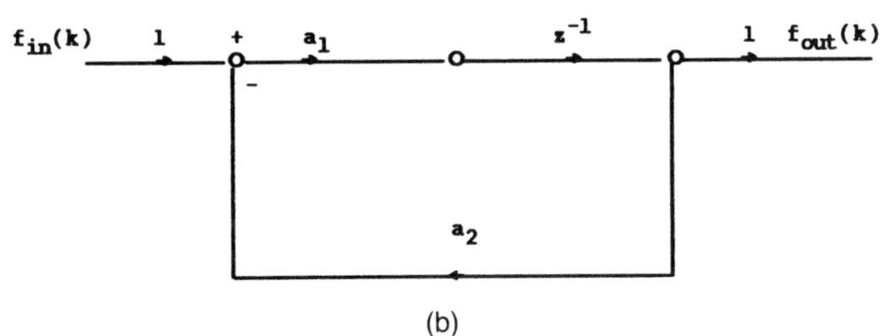

(b)

FIG. 23. Equivalent graphical representations of a single discrete circuit: (a) block diagram, and (b) signal-flow diagram.

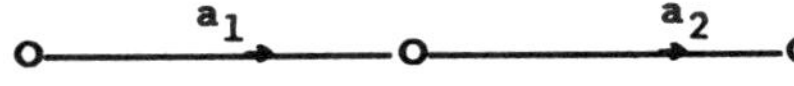

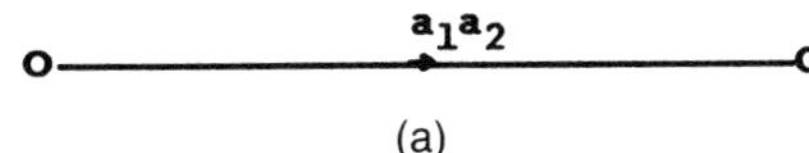

(a)

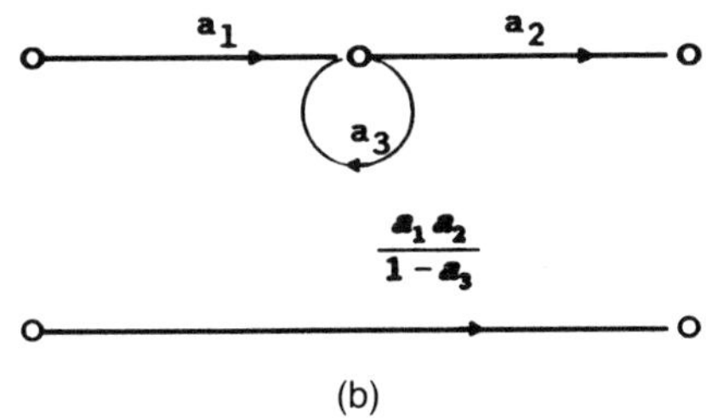

(b)

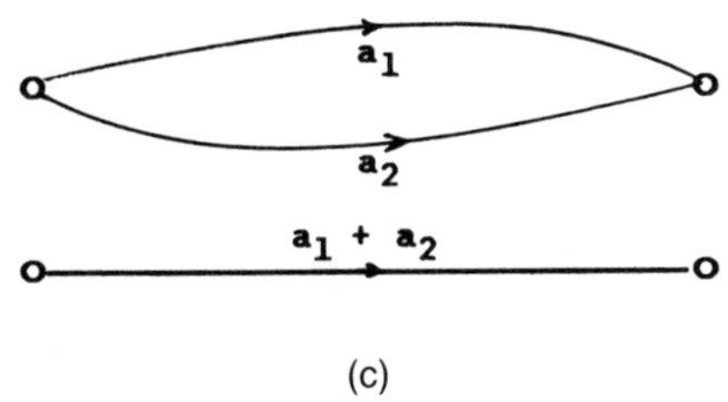

(c)

FIG. 24. Signal-flow graph reduction techniques: (a) paths in cascade, (b) removal of self loop, and (c) parallel paths.

plicable to continuous-time circuits. The rule is general, easily applied, and readily applicable to digital computer analysis. The rule states that the overall or general transmittance from an input node to an output node is given by

$$M = \frac{y_{\text{out}}}{x_{\text{in}}} = \frac{1}{\Delta}\sum_k M_k\Delta_k, \tag{84}$$

where $\Delta = 1 -$ (sum of all individual loop transmissions) + (sum of the products of the transmittances of loops that do not touch, taken two at a time) − (sum of the products of the transmittances of loops that do not touch, taken three at a time) + $\cdots$, M_k = path transmittance of the kth forward path, and Δ_k = value of that part of Δ that arises from terms that do not touch the kth forward path.

As an example of the use of Mason's rule is shown in Fig. 25. The diagram shown has seven nodes and ten transmittance paths. The input node is labeled x, the output y. There are three loops, one of which (this type of loop is called a "self-loop") does not "touch" the other two. There are two transmittance paths, one of which touches all three loops, the other of which does not touch two of the loops. Thus,

$$\Delta = 1 - a_8 - a_4a_9 - a_4a_5a_{10} + a_8a_4a_9 + a_8a_4a_5a_{10}, \tag{85}$$

$$M_1\Delta_1 = (a_1a_3a_4a_5a_6a_7)(1), \tag{86}$$

$$M_2\Delta_2 = (a_1a_2)(1 - a_4a_9 - a_4a_5a_{10}), \tag{87}$$

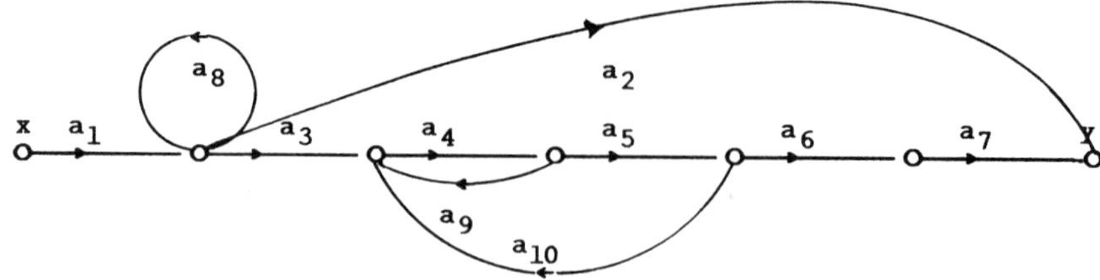

FIG. 25. Signal-flow graph for analysis.

so that

$$M = \frac{y_{out}}{x_{in}} = \frac{a_1a_3a_4a_5a_6a_7 + a_1a_2(1 - a_4a_9 - a_4a_5a_{10})}{1 - a_8 - a_4a_9 - a_4a_5a_{10} + a_8a_4a_9 + a_8a_4a_5a_{10}}. \quad (88)$$

10.5 Discrete Equations

A general type of expression that describes a finite discrete circuit is

$$\begin{aligned} y(k) = {} & \alpha_n x(k) + \alpha_{n-1}x(k-1) + \alpha_{n-2}x(k-2) \\ & + \cdots + \alpha_0 x(k-n) - \beta_{n-1}y(k-1) \\ & - \beta_{n-2}y(k-2) - \cdots - \beta_0 y(k-n). \end{aligned} \quad (89)$$

This is a generic nth-order *difference equation* with constant coefficients. It relates the output or dependent variable $y(k)$ to the input or independent variable $x(k)$. In general, $x(k)$ is known in some way—analytically, numerically, or stochastically. To solve the equation means then to find a $y(k)$ that satisfies Eq. (89) for all values of k.

Several ways for solving equations like Eq. (89) exist. Occasionally, solutions can be developed by algebraic methods. It is possible to build a theory of difference equations that is analogous in nearly every respect to the theories of differential equations. Concepts like complementary function and particular solution apply. Similarly, a powerful transform method has been developed. It is known as the z transform, and is discussed in Sec. 10.7. Numerical techniques are also commonly employed.

10.6 A Compound-Interest Example

The calculations for compound interest are a good illustration of a discrete equation and its solution. Suppose a person deposits an amount of money PR regularly at the beginning of an interest period, and that all money on deposit earns interest at a rate R over this period. If the annual interest rate is r, compounded N times annually, then $R = r/N$. Call the total amount on deposit y. The following set of equations illustrates the basic compounding process.

$$\begin{aligned} y(1) &= PR, \\ y(2) &= (1 + R)y(1) + PR = (1 + R)PR + PR, \\ y(3) &= (1 + R)y(2) + PR \\ &= (1 + R)^2PR + (1 + R)PR + PR, \\ &\vdots \\ y(n) &= [(1 + R)^{n-1} + (1 + R)^{n-2} \\ &\quad + \cdots + (1 + R) + 1]PR, \end{aligned} \quad (90)$$

where $y(n)$ is the amount on deposit at the end of the nth compounding period immediately after a deposit has been made. There is an algebraic identity that states

$$\sum_{k=0}^{n} \alpha^k = \frac{1 - \alpha^{n+1}}{1 - \alpha}.$$

If this identity is used with the preceding equation, with $\alpha = 1 + R$, then, after some manipulation, an expression for the amount on deposit is

$$\begin{aligned} y(n) &= \left[\frac{(1 + R)^n - 1}{R}\right]PR \\ &= \left[\frac{(1 + r/N)^n - 1}{r/N}\right]PR. \end{aligned} \quad (91)$$

Suppose a person deposits \$100 into a savings account at the beginning of a month, and makes 24 such deposits. Further, consider a 9% interest rate, with monthly compounding, so that $N = 12$. Equation (91) shows that, immediately after making the 24th deposit, the saver will have \$2618.85 in the account.

10.7 The α-β Radar Equations

A frequently quoted example of discrete equations is a set of equations obtained for the purpose of analyzing data from radar pulses sent from one airplane to another (Cadzow and Van Landingham, 1985). One purpose of the system might be to avoid collisions. In a basic representation, the transmitter on one airplane sends short pulses of electromagnetic energy. These are reflected by the second airplane, and received by the first. Figure 26(a) typifies the sent and received pulses. The received pulses are contaminated with electrical noise, and thus are not rectangular. The time of pulse travel to and from the "target" ΔT must be measured in the presence of noise, and these estimates are the only data obtained.

Define the following terms:

$u(k) = (c/2)\Delta T$, the measurement of distance to the target, where c is the velocity of light.

$y_v(k)$ = an estimate of the velocity of the target.

$y_p(k-1)$ = a prediction of the distance to the target, based on earlier data.

$y(k)$ = present value of the range of the target.

The basic ideas are to

1. use the preceding value of position, the latest value of velocity, and T to predict where the target will be at the kth pulse;
2. use the predicted range to the target and the new measurement of range to determine a new value for the range; the two pieces of data are weighted by factors $1 - \alpha$ and α respectively;
3. use the preceding velocity and an estimate of the velocity change based on the new target range and the predicted range, weighted by a factor β, to compute a new value of velocity.

The equations are

$$\begin{aligned} y_p(k) &= y(k-1) + Ty_v(k-1), \\ y(k) &= y_p(k) + \alpha[u(k) - y_p(k)], \\ y_v(k) &= y_v(k-1) + (\beta/T)[u(k) - y_p(k)]. \end{aligned} \tag{92}$$

The determination of the factors α and β has been studied a great deal as radar systems have developed and improved. Figure 26(b) shows a flow graph of these equations.

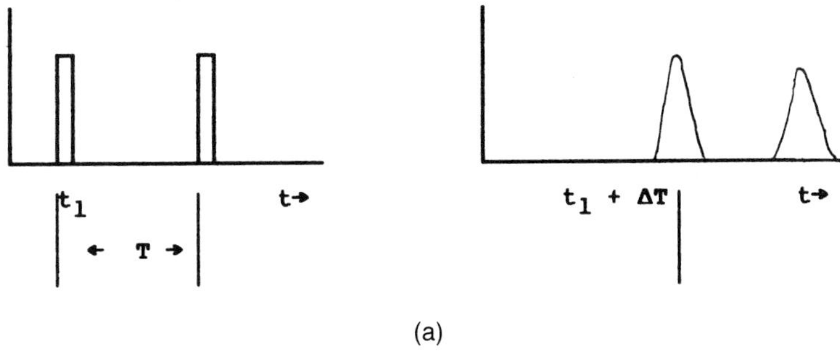

(a)

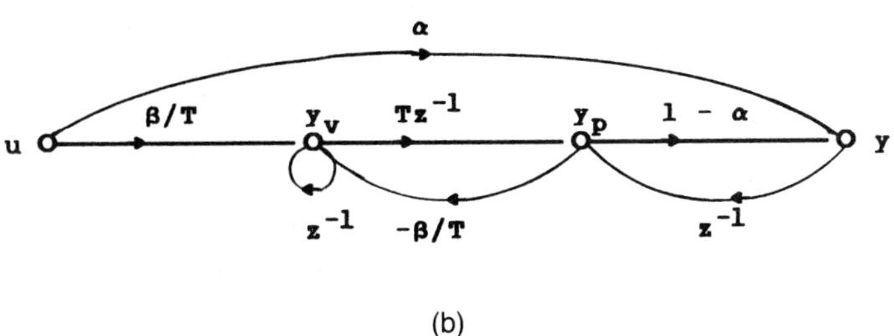

(b)

FIG. 26. (a) The transmitted and received pulses in radar. (b) The signal-flow diagram for deducing the velocity of the target.

10.8 Numerical Solutions of Difference Equations

Numerical solutions are carried out in a sequential fashion, and are readily adapted to digital computer solutions. The technique is illustrated by an example. Let a sample of Eq. (89) be given by

$$y(k) = 2x(k) - x(k-1) - 1.5x(k-2) - 0.5y(k-1), \tag{93}$$

where

$$x(k) = \begin{cases} 2, & k \geq 0 \text{ even}, \\ 1, & k \geq 0 \text{ odd}, \end{cases}$$

and the initial condition is $y(k-1) = 0$. Further, $x(k-1) = x(k-2) = 0$. The solution becomes

$$y(0) = 2x(0) = 4$$

$$y(1) = 2x(1) - x(0) - 0.5y(0) = 2 - 2 - 2 = -2$$

$$y(2) = 2x(2) - x(1) - 1.5x(0) - 0.5y(1) = 4 - 1 - 3 + 1 = 1$$

$$y(3) = 2x(3) - x(2) - 1.5x(1) - 0.5y(2) = 2 - 2 - 1.5 - 0.5 = -2$$

$$y(4) = 2x(4) - x(3) - 1.5x(2) - 0.5y(3) = 4 - 1 - 3 + 1 = 1.$$

10.9 The *z* Transform

A powerful transform method has been developed for analysis and design of discrete data systems. It has much in common with the Laplace transform, and it allows the use of algebraic methods in many aspects of the circuit analysis and design. In this section, the topic will be defined and introduced, theorems will be stated, tables will be given, and sample problems will be worked.

10.9.1 Definition of the *z* Transform Suppose a discrete function sequence $\{e(k)\}$, $k \geq 0$, is being considered. The z transform of $\{e(k)\}$, denoted by $E(z)$, is defined as a power series in z^{-k} whose coefficients are equal to the values of the sequence. Thus,

$$E(z) = Z[\{e(k)\}] = \sum_{k=0}^{\infty} e(k)z^{-k}. \tag{94}$$

In general, z is a complex number.

10.9.2 Examples of *z* Transforms Let the sequence be given as $e(k) = \{1, -\frac{1}{2}, \frac{1}{4}, -\frac{1}{8}, \frac{1}{16}, \ldots\}$. From Eq. (94),

$$E(z) = 1 - \tfrac{1}{2}z^{-1} + \tfrac{1}{4}z^{-2} - \tfrac{1}{8}z^{-3} + \cdots \tag{95}$$

If the identity

$$\frac{1}{1-x} = 1 + x + x^2 + x^3 + \cdots, \quad |x| \leq 1 \tag{96}$$

is used, then

$$E(z) = \frac{1}{1 - (-\frac{1}{2})z^{-1}} = \frac{z}{z + \frac{1}{2}}. \tag{97}$$

As a second example, suppose that the sequence is

$$e(k) = \epsilon^{-akT}. \tag{98}$$

Then,

$$E(z) = 1 + \epsilon^{-aT}z^{-1} + \epsilon^{-2aT}z^{-2} + \cdots \tag{99}$$

Application of Eq. (96) to this sequence gives

$$E(z) = \frac{z}{z - \epsilon^{-aT}}. \tag{100}$$

10.9.3 *z*-Transform Theorems Table 6 presents, without proof, a variety of theorems found to be useful with z transforms. Proofs of these theorems follow from the definitions. In this table, it is assumed that $E(z)$ is the z transform of the sequence $e(k)$.

10.9.4 *z*-Transform Tables Table 7 presents, without proof, a limited set of z transforms for a commonly used set of sequence functions. Additional tables may be found in some of the references.

Table 6. z-transform theorems.

Operation	Theorem
Addition and subtraction	$Z[\{e_1(k)\} \pm \{e_2(k)\}] = E_1(z) \pm E_2(z)$
Multiplication by constant	$Z[a\{e(k)\}] = aE(z)$
Real translation, part 1	$Z[\{e(k-n)\}] = z^{-n}E(z)$
Real translation, part 2	$Z[\{e(k+n)\}] = z^n[E(z) - \sum_{k=0}^{n-1} e(k)z^{-k}]$
Complex translation	$Z[\{\epsilon^{ak}e(k)\}] = E(z\epsilon^{-a})$
Initial value	$e(0) = \lim_{z\to 0}E(z)$ (provided limit exists)
Final value	$\lim_{k\to\infty}e(k) = \lim_{z\to 1}(z-1)E(z)$ (provided limit exists)

10.10 Formulation and Solution of z-Transform Equations

Example 1. Equation (93), which has been solved numerically, will be used to illustrate a z-transform solution. It is repeated here:

$$y(k) = 2x(k) - x(k-1) - 1.5x(k-2) - 0.5y(k-1) \tag{101}$$

where

$$x(k) = \begin{cases} 2, & k \text{ even}, \\ 1, & k \text{ odd}. \end{cases}$$

Denote the z transform of $y(k)$ by $Y(z)$, and the z transform of $x(k)$ by $X(z)$. If the theorems from Table 6 are used, and a transform of the equation is taken, with initial conditions (values for $k < 0$) that are zero, then the transformed equation becomes

$$Y(z) = 2X(z) - z^{-1}X(z) - 1.5z^{-2}X(z) - 0.5z^{-1}Y(z). \tag{102}$$

The input sequence $x(k)$ can be written as the sum of two sequences, $\{\frac{3}{2}, \frac{3}{2}, \frac{3}{2}, \ldots\}$ and $\{\frac{1}{2}, -\frac{1}{2}, \frac{1}{2}, -\frac{1}{2}, \frac{1}{2}, \ldots\}$, or in another form, as $\frac{3}{2}\{1, 1, 1, \ldots\}$ and $\frac{1}{2}\{1, -1, 1, -1, 1, \ldots\}$. The latter two sequences have, from Table 7, z transforms of

Table 7. A short table of z transforms

Name	Function	z transform
Unit impulse	$\delta(k) = 1, k = 0$; $=0, k \neq 0$	1
Unit step	$u(k) = 1, k \geq 0$	$\frac{z}{z-1}$
Alternating step	$u(k) = (-1)^k, k \geq 0$	$\frac{z}{z+1}$
Unit ramp	$k, k \geq 0$	$\frac{Tz}{(z-1)^2}$
nth-order ramp	$k^n, k \geq 0$	$\lim_{a\to 0}(-1)^n \frac{\partial^n}{\partial a^n}\left[\frac{z}{z-\epsilon^{-aT}}\right]$
Exponential	ϵ^{-akT}	$\frac{z}{z-\epsilon^{-aT}}$
Damped ramp	$t\epsilon^{-akT}$	$\frac{Tz\epsilon^{-aT}}{(z-\epsilon^{-aT})^2}$
Cosine	$\cos(akT)$	$\frac{z(z-\cos(aT))}{z^2 - 2z\cos(aT) + 1}$
Sine	$\sin(akT)$	$\frac{z\sin(aT)}{z^2 - 2z\cos(aT) + 1}$
Damped cosine	$\epsilon^{-akT}\cos(bkT)$	$\frac{z(z-\epsilon^{-aT}\cos(bT))}{z^2 - 2\epsilon^{-aT}\cos(bT)z + \epsilon^{-2aT}}$
Damped sine	$\epsilon^{-akT}\sin(bkT)$	$\frac{z\epsilon^{-aT}\sin(bT)}{z^2 - 2\epsilon^{-aT}\cos(bT) + \epsilon^{-2aT}}$

$$\tfrac{3}{2}\frac{z}{z-1} \quad \text{and} \quad \tfrac{1}{2}\frac{z}{z+1}, \tag{103}$$

respectively. When the sum of these two terms is substituted into Eq. (102) and the resulting equation is solved for $Y(z)$, the result is

$$Y(z) = \tfrac{3}{2}\frac{2z^2 - z - 1.5}{(z-1)(z+0.5)} + \tfrac{1}{2}\frac{2z^2 - z - 1.5}{(z+1)(z+0.5)}. \tag{104}$$

Because the degree of the numerator is equal to the degree of the denominator in both parts of Eq. (104), a direct partial-fraction representation is not possible. Examination of Table 7 shows that invariably a factor z appears in the numerator in the transform. Thus, a technique that virtually always works is to expand $Y(z)/z$ in partial fractions, and then to multiply through by z at the end. When this is done, the result is

$$Y(z) = 4.5 - \tfrac{1}{2}\frac{z}{z-1} - \frac{z}{z+0.5} - 1.5 + \frac{z}{z+0.5} + \frac{1.5\,z}{z+1}. \tag{105}$$

As it happens, some cancellations are possible. When these are done and the information in Table 7 is used to take the inverse transform, the result is

$$y(k) = 3\delta(k) - \tfrac{1}{2}u(k) + 1.5(-1)^k. \tag{106}$$

Evaluation of Eq. (106) gives the same numerical results that were given earlier in the numerical solution.

Example 2. This example will show some computational details, especially some that arise in the computation of the inverse z transform. When roots are complex, then the residues in the inverse z transform expansion are complex numbers, which may be unwieldy to deal with. Euler's relation [Eqs. (52) and (53)] may be used to arrange the results in real-number terms only. Suppose that the transfer function for a discrete system, as a specific example of Eq. (88) and Fig. 25, is given by

$$\frac{Y(z)}{X(z)} = \frac{z^3 + 2z^2 + 5z - 5}{z^3 - 1.70z^2 + 1.59z - 0.623} \tag{107}$$

and that the forcing function $X(z)$ is a step function, so that

$$X(z) = \frac{z}{z-1}. \tag{108}$$

When $Y(z)$ is computed by first factoring the denominator, finding $Y(z)/z$, and then multiplying through by z, the result is

$$Y(z) = \frac{11.24\,z}{z-1} + \frac{0.868\,z}{z-0.70} + \frac{(-5.552 + j\,0.995)z}{z - 0.50 - j\,0.80} + \frac{(-5.552 - j\,0.995)z}{z - 0.50 + j\,0.80}. \tag{109}$$

While the first two terms need no special attention, the latter are difficult to interpret without modification. Consider the auxiliary equation

$$Y(z) = \frac{a+jb}{z-\alpha-j\beta} + \frac{a-jb}{z-\alpha+j\beta} \tag{110}$$

whose inverse transform can be written as

$$y(k) = (a+jb)\,\gamma^k\epsilon^{jk\theta} + (a-jb)\,\gamma^k\epsilon^{-jk\theta}, \tag{111}$$

where

$$\gamma = \sqrt{\alpha^2 + \beta^2}, \quad \theta = \arctan(\beta/\alpha). \tag{112}$$

When the two real terms from Eq. (111) are collected and the two imaginary terms are collected, and in addition, the numerator and denominator of the real terms are multiplied by 2, while for the imaginary terms, $j2$ is used, then

$$y(k) = 2a\gamma^k\frac{\epsilon^{jk\theta} + \epsilon^{-jk\theta}}{2} - 2b\gamma^k\frac{\epsilon^{jk\theta} - \epsilon^{-jk\theta}}{j2}. \tag{113}$$

With this auxiliary result, the inverse transform from Eq. (109) can be written

$$y(k) = 11.24u(k) + 0.87(0.70)^k - 11.10\,(0.94)^k\cos(1.01k) + 1.99\,(0.94)^k\sin(1.01k). \tag{114}$$

Figure 27 shows a graph of this function for

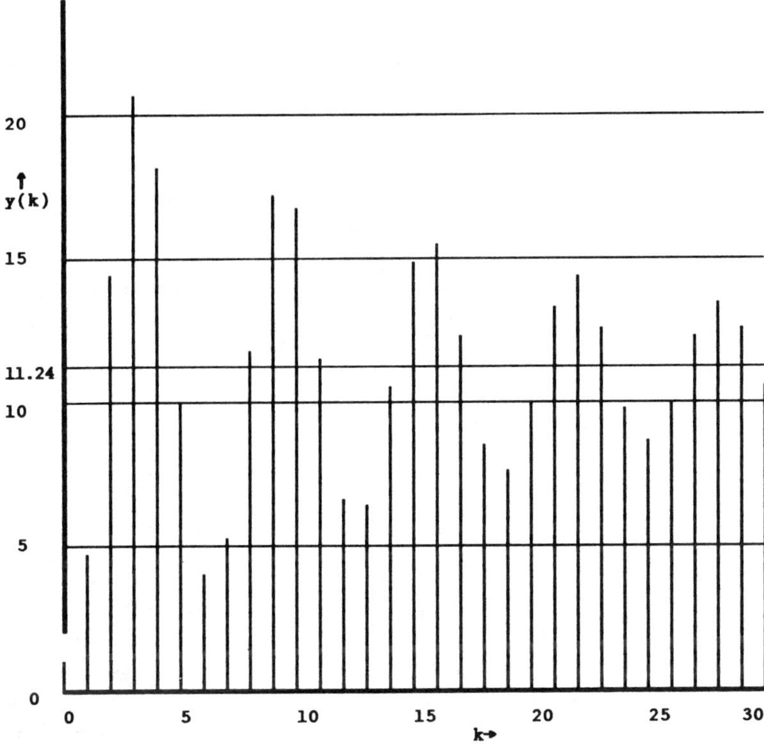

FIG. 27. Graph of Eq. (114).

values of k up to 30. It is seen that this function approaches its final value of 11.24 slowly, primarily because of the large coefficients and the fact that the magnitude of the complex roots is close to unity.

10.11 Shannon's Sampling Theorem

An important question that arises in the design of systems that require sampling of a continuous-time system signal in order to obtain a discrete representation of that signal is the question of how often the signal must be sampled in time in order to obtain a satisfactory representation of that signal. Shannon answered this question in a classic paper (Shannon, 1949) that shows the minimum sampling frequency must be at least twice the highest frequency component of the signal. For example, if a signal has frequency components from very low frequency to, say, 20 kHz, then the minimum sampling frequency must be 40 kHz, or the maximum sampling interval can be no longer than 25 μs. This rate is commonly called the Nyquist rate, because of earlier work done on the problem by Nyquist (1928).

11. SUGGESTIONS FOR FURTHER STUDY

11.1 Nonlinear and Time-Varying Networks

It was indicated at the beginning that this article would focus on linear, time-invariant networks. The study of nonlinear networks is interesting and challenging. Time-varying networks present a different set of challenges, and the combination is still rich with areas of research. Some of the items listed as suggestions for further reading include excellent discussions of these concerns.

11.2 Computer-Based Circuit Analysis

The advent of the digital computer has given circuit analysts and designers powerful tools to work with. These tools are refined regularly and rapidly, so that the specific details change rapidly. Software is available for desktop computers and for centralized computing facilities. The reader interested in this subject is advised to check books that have been published quite recently, and to also study software manuals carefully. Some suggestions are included.

11.3 State-Variable Methods for Continuous-Time Circuits

The techniques presented here for state-variable analysis work well in most cases, but, as indicated, loops with nothing but capacitors and voltage sources, or nodes with nothing but current sources and inductors, require special treatment.

11.4 Topology

Circuits or networks have been studied extensively using topological methods. These methods yield great insight into analysis and design methods, ensure that the equations written are independent, and make possible the sophisticated circuit analysis and design programs that are available. Some titles are suggested.

11.5 Stability

Active networks, in their basic form, are networks or circuits that contain controlled sources. Because of the energy such sources can provide, it is possible that such networks can become unstable. In general, this is not a desirable behavior. Methods exist for predicting such instability, and there are design methods for eliminating it.

11.6 State-Variable Methods for Discrete-Time Circuits

It is possible to develop a state-variable technique for discrete circuits. The study is particularly useful in the design and analysis of digital control systems and in signal-processing techniques. The technique has a lot in common with what was presented for continuous-time circuits, but significant differences in details must be considered.

11.7 The Modified *z* Transform

The *z*-transform techniques presented do not allow for study between sampling intervals. The modified *z* transform (Phillips and Nagle, 1990) allows such study when needed.

GLOSSARY

Admittance: The ratio of the transform (phasor or Laplace) of a current to the transform of a voltage.

Block Diagram: A graphical representation of the interconnection of the elements of a circuit or system. See **Signal-Flow Graph**.

Branch: A circuit element connecting two nodes in a circuit.

Capacitor: An electric circuit element having the property that the current through the device is proportional to the time derivative of the voltage across the device.

Circuit: An interconnected collection of electrical devices such as resistors, capacitors, electronic devices, and electrical energy sources. Also called **networks**.

Continuous-Time Circuit: A circuit or network that is described by a set of continuous-time integro-differential equations.

Current: The flow of electrically charged particles.

Dependent or Controlled Source: A source of electrical energy or signal whose current or voltage is controlled by or proportional to another current or voltage in the circuit.

Discrete-Time Circuit: A circuit or network that is described by a set of discrete-time difference equations.

Effective value: Also known as the root mean square (rms) value of the function:

$$F_{\mathrm{rms}} = \left[\frac{1}{T}\int_{t_0}^{t_0+T} (f(t))^2\, dt\right]^{1/2}.$$

Gyrator: The fundamental nonreciprocal electrical element. The defining equations are

$$v_1 = R\, i_2,$$
$$v_2 = -R\, i_1,$$

where R is known as the "gyration resistance."

Impedance: The ratio of the transform (phasor or Laplace) of a voltage to the transform of a current.

Independent Source: A source of electrical energy or signal that is controlled by an agent outside the circuit.

Inductor: An electric circuit element having the property that the voltage across the device is proportional to the time derivative of the current through the device.

j: The imaginary operator, defined by $j^2 = -1$. Also denoted by i.

Kirchhoff's Current Law (KCL): The statement that the sum of currents leaving (or entering) a node is zero.

Kirchhoff's Voltage Law (KVL): The statement that the sum of voltages around a loop is zero.

Laplace Transform: If a function of time $f(t)$ is known, and defined for $t \geq 0$, then the (single-sided) Laplace transform is given by

$$\mathscr{L}f(t) = F(s) = \int_0^\infty f(t)\,\epsilon^{-st}\,dt,$$

where s is a complex variable, $s = \sigma + j\omega$, chosen so that the integral will converge. In turn, σ is the real part of the variable, and ω is the imaginary part.

Linear Circuit: A circuit or network described by a set of linear integro-differential equations.

Loop: A path through a circuit that begins and ends at the same node.

Mason's Rule: A technique for formulating input-output relations from flow graphs or block diagrams.

Mesh: A path in a planar circuit, i.e., one that can be drawn on a planar surface without wires crossing.

Node: In a circuit or network, the junction of three or more wires from circuit elements. Sometimes called "essential node."

Network: A term that is synonymous with circuit.

Norton Theorem: A theorem stating that, at a terminal pair, any linear network may be replaced by the parallel combination of a current source and a resistance or impedance. The current is equal to the current that flows through the short circuit when the external load is short circuited. The resistance is equal to the resistance measured or calculated at the terminal pair with all independent sources de-energized.

Path: A traverse through two or more branches in a circuit.

Phasor: A representation of a sinusoidal function that is a complex number that contains both the amplitude and phase information of the original. When

$$v(t) = V_m \cos(\omega t + \alpha) = V_m \sin(\omega t + \alpha + \pi/2),$$

the phasor representation of $v(t)$ is

$$V = V_m \epsilon^{j\alpha}.$$

The definition may be used in reverse to convert a phasor to a time function, provided the frequency is known. Sometimes called a phasor transform.

Port: A circuit or network with one or more terminal pairs having the property that the current into one wire is equal to the current out of a second wire. These two wires form a terminal pair that is called a port.

Resistor: An electric circuit element having the property that the voltage across the device is proportional to the current through the device.

Shannon's Sampling Theorem: A theorem that states that, in order to preserve the original content, the sampling frequency used must be at least twice the highest frequency component of the signal being sampled. Sometimes called the Nyquist rate.

Signal-Flow Graph: A graphical representation of the interconnection of the elements of a network or system. See block diagram.

Superposition: A principle stating that, for a given system (network) having a general excitation $x(t)$ that yields a response $y(t)$, if the response to an excitation $x_1(t)$ is $y_1(t)$, and the response to an excitation $x_2(t)$ is $y_2(t)$, then any excitation

$$x(t) = a_1 x_1(t) + a_2 x_2(t)$$

yields the response

$$y(t) = a_1 y_1(t) + a_2 y_2(t)$$

for all values of the constants a_i.

Tellegen's Theorem: This theorem states: Let N be a network of b branches and n nodes. If KVL is satisfied by the branch voltages v_1, v_2, ..., v_b in every loop, and if KCL is satisfied by the branch currents i_1, i_2, ..., i_b at every node, then

$$\sum_{k=1}^{b} v_k i_k = 0.$$

Thévenin Theorem: A theorem stating that, at a terminal pair, any linear network may be replaced by a series combination of a voltage source and a resistance or impedance. The voltage is equal to the voltage at the terminal pair when the external load is removed or open-circuited, and the resistance is equal to the resistance calculated or measured at the terminal pair with all of the independent sources de-energized.

Two-Port Parameters: A set of equations that describes a circuit having two ports by its terminal behavior.

Voltage: The energy, per unit charge, required to move a charge from one point in space to another.

***z* Transform:** A power series in z^{-k} whose coefficients are equal to the values of a sequence. Thus,

$$E(z) = Z[\{e(k)\}] = \sum_{k=0}^{\infty} e(k)\, z^{-k}.$$

In general, z is a complex number.

Works Cited

Anderson, P. M. (1973), *Analysis of Faulted Power Systems*, Ames: Iowa State University Press.

Balabanian, N., Bickart, T. A., Seshu, S. (1969), *Electrical Network Theory*, New York: John Wiley and Sons.

Cadzow, J. A., Van Landingham, H. F. (1985), *Signals, Systems, and Transforms*, Englewood Cliffs, N.J.: Prentice-Hall Book Company, Inc.

Desoer, C. A., Kuh, E. S. (1969), *Basic Circuit Theory*, New York: McGraw-Hill Book Company.

Fink, D., Beaty, W. (1993), *Standard Handbook for Electrical Engineers*, 13th ed., New York: McGraw-Hill Book Company.

Huelsman, L. P. (1963), *Circuits, Matrices, and Linear Vector Spaces*, New York: McGraw-Hill Book Company.

Mason, S. J. (1953), "Feedback Theory—Some Properties of Signal Flow Graphs," *Proc. Inst. Radio Eng.* 41:9, pp. 1144–1156.

Mason, S. J. (1956), "Feedback Theory—Further Properties of Signal Flow Graphs," *Proc. Inst. Radio Eng.* **44** (7), 920–926.

Nyquist, H. (1928), "Certain Topics in Telegraph Transmission Theory," *Trans. Am. Inst. Electrical Eng.* **47** (4), 617–644.

Phillips, C. L., Nagle, H. T. (1990), *Digital Control Systems: Analysis and Design*, 2nd ed., Englewood Cliffs, N.J.: Prentice-Hall Book Company.

Rohrer, R. A. (1970), *Circuit Theory: An Introduction to the State Variable Approach*, New York: McGraw-Hill Book Company.

Shannon, C. E. (1949), "Communication in the Presence of Noise," *Proc. Inst. Radio Eng.* **37** (1), 10–21.

Stevenson, W. D., Jr. (1982), *Elements of Power System Analysis*, 4th ed., New York: McGraw-Hill Book Company.

Tellegen, B. D. H. (1948), "The Gyrator, a New Electric Network Element,"*Philips Res. Rep.* **3**, 81–101.

Tellegen, B. D. H. (1952), "A General Network Theorem, with Applications," *Philips Res. Rep.* **7**, 259–269. [A related article is Tellegen, B. D. H. (1953), "A General Network Theorem, with Applications," *Proc. Inst. Radio Eng., Australia* **14**, 265–270.]

Further Reading

Chan, Shu-Park (1969), *Introductory Topological Analysis of Electrical Networks*, New York: Holt, Rinehart and Winston.

Chan, Shu-Park, Chan, Shu-Yun, Chan, Shu-Gar (1972), *Analysis of Linear Networks and Systems*, Reading, MA: Addison-Wesley Publishing Company.

Chua, L. O. (1969), *Introduction to Nonlinear Network Theory*, New York: McGraw-Hill Book Company.

Chua, L. O., Desoer, C. A., Kuh, E. (1987), *Linear and Nonlinear Circuits*, New York: McGraw-Hill Book Company.

Conant, R. (1993), *Engineering Circuit Analysis with PSPICE and PROBE*, New York: McGraw-Hill Book Company.

Franklin, G. F., Powell, J. D., Workman, M. L. (1990), *Digital Control of Dynamic Systems*, 2nd ed., Reading, MA: Addison-Wesley Publishing Co.

Gardner, M. F., Barnes, J. L. (1942), *Transients in Linear Systems*, New York: John Wiley and Sons, Inc.

Hayt, W. H., Kemmerly, J. E. (1993), *Engineering Circuit Analysis*, 5th ed., New York: McGraw-Hill Book Company.

Keown, J. L. (1991), *PSPICE and Circuit Analysis*, New York: Merrill Publishing Company.

Mayhan, R. J. (1984), *Discrete-Time and Continuous-Time Linear Systems*, Reading, MA: Addison-Wesley Publishing Co.

McGillem, C. D., Cooper, G. R. (1991), *Continuous and Discrete Signal and System Analysis*, 3rd ed., Philadelphia: Saunders College Publishing (a division of Holt, Rinehart, and Winston).

Monssen, F. (1993), *PSPICE with Circuit Analysis*, New York: Merrill, Macmillan Publishing Company.

Nilsson, J. W. (1993), *Electric Circuits*, 4th ed., Reading, MA: Addison-Wesley Publishing Co.

Rashid, M. H. (1990), *SPICE for Circuits and Electronics Using PSPICE*, Englewood Cliffs: Prentice-Hall Book Company.

Seshu, S., Reed, M. B. (1961), *Linear Graphs and Electrical Networks*, Reading, MA: Addison-Wesley Publishing Company.

Spence, R. (1970), *Linear Active Networks*, New York: Wiley-Interscience, Inc.

Strum, R. D., Kirk, D. E. (1988), *First Principles of Discrete Systems and Digital Signal Processing*, Reading, MA: Addison-Wesley Publishing Company.

Strum, R. D., Kirk, D. E. (1994), *Contemporary Linear System—Using MATLAB*, Boston: PWS Publishing Company.

Thorpe, T. W. (1992), *Computerized Circuit Analysis with SPICE*, New York: John Wiley and Sons.

Tuinenga, P. W. (1988), *A Guide to Circuit Simulation and Analysis Using PSPICE*, Englewood Cliffs: Prentice-Hall Book Company.

Van Valkenburg, M. E. (Ed.) (1974). *Circuit Theory: Foundations and Classical Contributions*, Stroudsburg, PA: Dowden, Hutchinson, and Ross Publishing.

Van Valkenburg, M. E. (1982), *Linear Circuits*, Englewood Cliffs: Prentice Hall Book Company.

NETWORKS, COMPUTER

EHUD GAVRON, *Aces Research, Inc., Tucson, Arizona, U.S.A.*

INTRODUCTION

The explosion in both the popularity and the use of computers has created a situation where it is no longer simply enough to process data on one computer. The need frequently arises to share data and resources between many computers. This is the primary purpose of utilizing a computer network.

As network use becomes more widespread, however, its use is also becoming more difficult. The advantages of using computer networks are such that the added difficulty is viewed merely as a hurdle to overcome. With companies attempting to minimize costs it becomes apparent that sharing expensive peripherals such as printers and disk drives between many users is a goal that is highly desirable. Making a large number of devices available to a large number of users increases the utilization of resources while only marginally increasing costs. Additionally, a computer network can eliminate dangerous dependences on a single resource and allow for more robust infrastructure. No longer is the failure of a single computer likely to halt work in a department, because there are several other machines or servers that can perform the same task. Networks have allowed the computing world to step away from the peer-to-peer computing model, where all computers are equal, into client-server relationships, where servers offer their resources to systems that desire to use those services (clients). Client-server computing and its tremendous popularity are easily linked to the advances in available network technologies. Networks have advanced from their infancy in the 1970s into large and often complex constructs. Understanding these networks, their uses, and their limitations will become an increasingly critical skill in business and society.

1. A BRIEF HISTORY OF COMPUTER NETWORKS

The first real computer network was designed and built in the 1950s at the Massachusetts Institute of Technology's (M.I.T) Lincoln Laboratory. The Semi-Automatic

3-527-28133-9/94/$5.00 + .50

Ground Environment (SAGE) system consisted of 23 computer networks, each of which connected roughly 100 radar sites to a single computer. Data were sent through voice-grade phone lines at a rate of approximately 1300 bits per second (bps).

In 1964 the first large commercial computer network was built by IBM for American Airlines. The SABRE system heavily built upon work done for the Department of Defense (DoD) on the SAGE system, and indeed involved many of the same individuals who had worked on SAGE.

1.1 ARPANET

In 1971 the Advanced Research Projects Agency (ARPA) (now known as DARPA) heavily promoted the creation of the ARPA Network (ARPANET) to connect military installations and universities with widely varying environments. The technique used to connect these institutions was called "packet switching" (see Sec. 4.3). Work on ARPANET was often groundbreaking. It was the test-bed for analytical and performance models for networks. Additionally, many techniques such as flow control, fault-tolerant performance, and layered protocols received much of their initial testing on the ARPANET. In 1983 ARPANET was divided into two separate networks: the Defense Data Network (DDN) and a new ARPANET. Today ARPANET is almost completely phased out as hosts have been moved to direct Internet connectivity.

1.2 Internet

The Internet is a worldwide internetwork of hosts communicating via the TCP/IP protocol suite. Internetworks are multiple networks connected such that data can pass between the various networks. The Internet is the only current global-scale internetwork. TCP/IP is a developing suite of communication protocols that was formed as a direct result of the DARPA research. Its strengths are in fast and simple scalability, although as local-area networks have expanded into the global Internet, a potential shortage of unique worldwide addresses is possible in the near future. This same global Internet is widely being touted as the Information Superhighway of the future. The Internet of today owes much of its structure and existence to the ARPANET of old, both in structure and in design.

1.3 Ethernet

Ethernet appeared from Digital Equipment Corporation, Intel, and Xerox (DIX) in 1974 and remains today the dominant Local-Area Network (LAN) technology. Its original goal was to devise a way to send information between many office machines. Its basic protocol was a refinement of earlier work with wireless data transmission from Hawaii to the mainland. Originally specifying a one megabit per second network (1Mb/s), the DIX consortium later succeeded in revising the standard by an order of magnitude, making Ethernet a 10-Mb/s LAN medium. The development of other local-area network protocols (token-ring, Apple's localtalk, Banyan's vines) occurred at about the same time, but these other protocols have never gained the popularity and use that Ethernet has.

2. TYPES OF COMPUTER NETWORKS

Computer networks come in many shapes and sizes. The different characteristics of networks results in several different categories of networks. The first characteristic often used to separate networks is the size and complexity of the network. This segmentation, in general, leads to three different classifications of networks: Local-Area Networks (LANs), Metropolitan-Area Networks (MANs), and Wide-Area Networks (WANs). Additionally, there are internetworks that bridge the gap between individual networks.

2.1 Local-Area Networks

LANs are by far the most common type of telecommunication network. As its name implies, a LAN is typically a network that spans a small geographic region. Devices, or nodes, in a LAN are typically never further apart than a few kilometers. LANs are generally owned by a single organization and contain no more than a few hundred nodes. The cabling between the nodes is typically installed and owned by the organization that manages the LAN. The data transmission speeds are typically in the 10-Mb/s range. However, as fiber-optic links become more prevalent, these speeds are constantly increasing.

2.2 Metropolitan-Area Networks

MANs are the next largest category of networks. A MAN typically covers a larger geographic region than a LAN. However, there is no set distance or size at which a LAN and a MAN are distinct; thus the line between them is often blurred. One characteristic that does often distinguish the two is that the cable for a MAN is often owned by local public data carriers (i.e., local phone companies) and leased by the organization. Because of the higher costs of leasing cable, data transmission rates are typically between 56 kb/s and 1.544 Mb/s (referred to by telephone companies as T-1 circuits). These are also available in Europe under the slightly different configurations of 64 kb/s and 2.048 Mb/s (known as E-1 circuits).

2.3 Wide-Area Networks

WANs are the largest of the discrete networks. WANs often cross state and even country boundaries to create a single network for large organizations. The costs associated with such networks are high as, in the United States, the cable must invariably be leased from interexchange carriers (IXCs), also known as "long-distance carriers." These carriers (e.g., AT&T) provide these high-bandwidth lines at a relatively high cost. The speeds of such connections generally range from 56 kb/s to 45 Mb/s (T-3 circuits). Several intracontinental fiber-optic connections are currently available, promising significantly more bandwidth for future use. These larger data pipes (150 Mb/s to 1.5 Gb/s) have necessitated the design of protocols capable of rapidly processing data at such rates. These are discussed in more detail under ATM and SONET.

2.4 Internetworks

The concept of connecting various networks together provides a framework for integrating the advantages of multiple networks with simple scalability. Such integrated networks are known as internetworks. Internetworks are designed for circumstances when applications or needs transcend a single network and must be able to span several networks. These networks connect several small networks to create larger networks. Examples of this might be the connection of several MANs to create a WAN. Internetworks are usually only used by large organizations that can afford not only the cabling costs associated with the connections, but also the cost in personnel needed to administer the smaller networks and the larger networks simultaneously.

3. GEOMETRIES OF COMPUTER NETWORKS

Computer networks may also be characterized by their geometry in addition to their size. There are many different possible topologies. Different geometries may often be used for different sized networks, and for networks with different functions. In general there are four different types of computer network geometries: bus, ring, star, and mesh. These geometries are illustrated in Fig. 1.

Based on the same technology found inside the modern computer, connecting all "cards" to a chassis or "motherboard," a bus geometry is characterized by a single com-

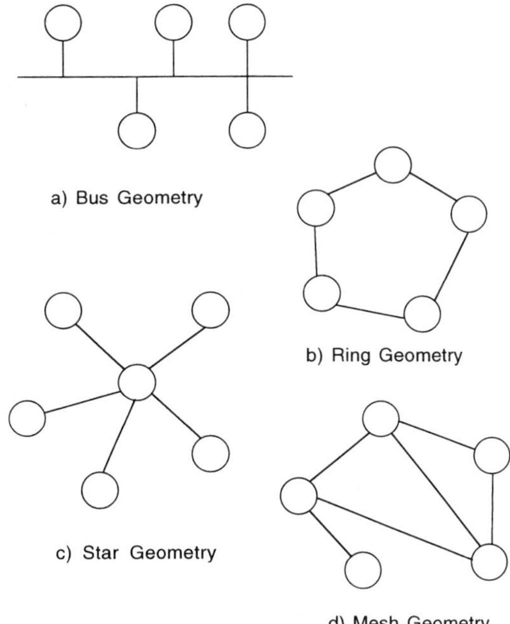

FIG. 1. Computer network geometries.

mon cable, or bus, in which nodes are connected via short cables. This is the most common configuration for LANs as it is an architecture of lesser complexity which performs its functions better under low network loads.

The ring (or loop) topology consists of all the nodes being connected in series along a closed path. Normally this path involves data transmission in a single direction, either clockwise or counterclockwise. However, there are technologies, primarily Fiber Distributed Data Interface (FDDI), that utilize a double ring where data flows on one ring clockwise and on the other ring counterclockwise. The advantage of using dual rings is that not only does it increase the amount of data that can be sent and received in a given time, but it is also much more fault tolerant. A FDDI ring by design includes some fault tolerance, in that segmentation of the ring will cause FDDI frames to reverse direction at the break point. Thus a single break is not sufficient to eliminate more than the general throughput of the ring. Multiple failures, however, will isolate the ring into broken segments. A dual-ring configuration provides redundancy via a completely separate fiber-optic path for data to follow. A failure on one ring is compensated by the existence of a second complete ring. The advantage to using rings at all is that they are normally implemented in such a manner as to allow them to do very well at high network loads.

A star topology, based on the model of a single computer with many outlying connections to terminals, has a single central node that is connected via point-to-point links to the other nodes. This single node acts as a central switching point for all of the other nodes. The disadvantage of this configuration is that the central node is a single point of failure; should it fail, the entire network will be useless. The star's advantage is that it does an excellent job of keeping very busy nodes from bogging down slower nodes with their network traffic.

Finally, the mesh (or general) geometry is a general configuration that allows for multiple routes for data. This is the topology that is most commonly seen for WANs. In particular, the ARPANET was developed using this configuration and the current NSFNET is still of this configuration type. Its main advantage is fault tolerance; the failure of a single node is normally acceptable because there are multiple routes to each node. However, in general, it is more expensive in terms of both cabling costs and network equipment costs because of this redundancy.

4. SWITCHING TECHNIQUES FOR COMPUTER NETWORKS

In computer networks, there are three primary switching techniques. These techniques allow sharing of communication facilities by multiple users. These techniques are circuit switching, message switching, and packet switching. A form of packet switching utilizing fixed-length packets of information is known as cell-switching, and is the basis for such protocols as ISDN (Integrated Services Digital Network) and ATM (Asynchronous Transfer Mode). The emphasis is therefore on covering the details of packet switching as it is currently the dominant technique. Figure 2 shows the different times and techniques involved with each switching technology.

4.1 Circuit Switching

Circuit switching is what is used by the average individual most because it is used for ordinary telephone calls. The basic construct is the simplest of the three techniques. Use is first initiated by a connections phase, during which a circuit is established between a source and destination. The second phase is the actual message phase, where data are transmitted back and forth. Finally, there is the disconnection phase, where the circuit is reset. An example of circuit switching can be shown in a telephone call. The connection phase occurs when the phone number is dialed and while waiting for the recipient to answer. The message phase is the actual conversation, and the disconnection phase is when both of the parties hang up. The disadvantage of circuit switching technology is that in computer time the connection phase is relatively long. This automatically burdens the end system with a high overhead before a message is even sent. However, as the connection phase times become shorter, circuit switching may become more prevalent.

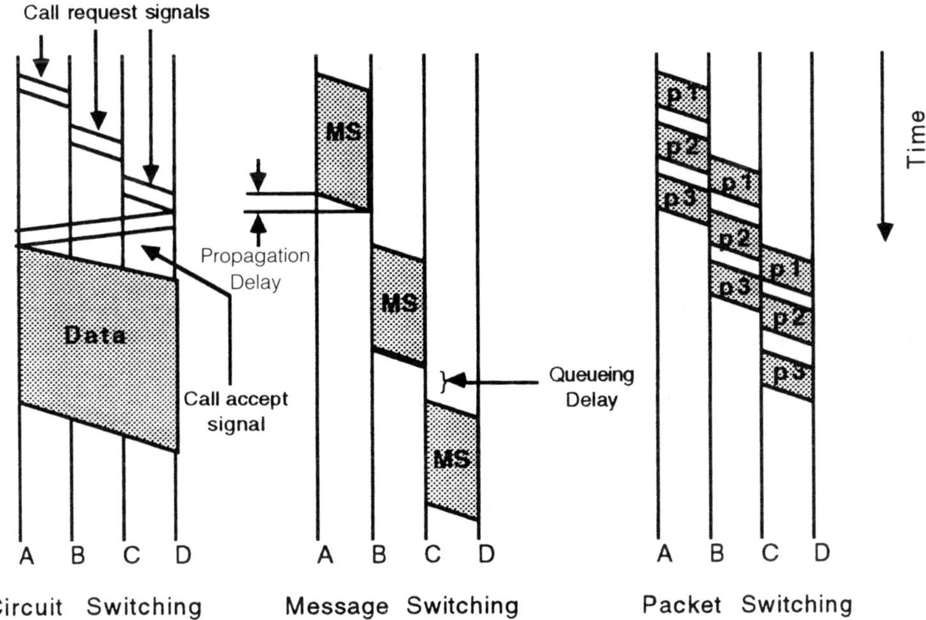

FIG. 2. Switching techniques.

4.2 Message Switching

Message switching shares a single connection between multiple users. In this scheme, there is no connection or termination phase because a message is simply sent from link to link until it reaches its destination. It is sometimes also referred to as store-and-forward switching, since messages are stored at intermediate nodes on their way to their destinations. The basic concept can be easily illustrated. If a message needs to get from point A to point C, and it passes through point B, it can be sent to point B first where it will then be forwarded to point C. The message would first go from A to B. It would then sit at B until a connection to C became available. In the interim someone else can also send a message from A to B. Finally, when the B to C link becomes available the message would be sent, and hopefully shortly thereafter any other messages would also continue on to point C. The advantage to this system is that it allows multiple users to use a single channel with little added delay over circuit switching. However, the time gained in not having a connection phase can often be lost in queuing delay as messages sit at intermediate nodes.

4.3 Packet Switching

Finally, there is packet switching. Packet switching is equivalent to message switching for short messages. Any message that is larger than some specified length is broken into smaller pieces, known as packets. These packets, each with a certain amount of extra data known as a header, are then transmitted through the network individually. The main advantage of packet switching is that it allows for an effect known as pipelining. Pipelining occurs when the gaps that are normally associated with message switching on a network get filled by parts of the packets, thus leaving little idle time on any of the segments. This effect is visible in Fig. 2.

There are two basic approaches to packet switching: virtual circuit and datagram. Virtual circuits are again analogous to setting up a telephone call. One host determines a route to another host and then all subsequent packets follow that same route. The advantage to virtual-circuit packet switching is that it allows for smaller headers on the packets, as they simply must know a circuit number. The disadvantage is that if some node in the circuit should go down, all packets will be lost. Datagram routing is more closely analogous to sending a letter. There is no time required to set up a route, and different packets may take different routes to get to their location. As with mailing a letter, datagram routing is next-hop routing. In next-hop routing, each host only needs to know the next host responsible for routing the datagram to its eventual destination. Just as the letter is given to the carrier who gives it to the post office, etc., next-hop routing makes possible large internetworks with minimal routing path information storage. The advantage of datagram packet switching is that it is more robust; if a node on the way to a destination should go down, the packets will simply use a different route if it is available. The disadvantages are that each packet must have more information about the destination, and the order of packet delivery could vary. This requires more robust network programs on the receiving end node.

A variant of packet switching is called "cell-switching" or "cell-relay," and it uses fixed-length data packets known as "cells." The chief advantage to the use of the fixed-length cells as compared to variable-length packets is that equipment can be optimized to buffer, copy, and route data by quickly scanning multiple headers in multiple cells.

5. ATM AND SONET

In the early 1990s, the introduction of globally available fiber-optic technology had prompted many of the IXCs to create nationwide (and in some cases international) fiber-optic networks. These networks typically used FDDI at its initial rate of 100 Mb/s. As technology advanced, SONET (Synchronous Optical NETwork) was introduced to provide the much-needed higher rates of speed. SONET rates are measured in scales from OC-1 to OC-12, with the number being a multiple of 50 Mb/s. Thus OC-3, which is typically discussed as appropriate for replacing FDDI, is roughly 150 Mb/s.

With the availability of faster networking, the technology to do high-speed cell routing became greatly needed. The telephone company ISDN offering was heavily modified and changed to become ATM. ATM is expected to be the backbone networking standard into the next century as it provides access to bandwidth-intensive video workstations, sound transmissions, and data networking.

6. COMPUTER NETWORK MEDIA

In the 1980s, one of the hot topics of discussion was the choice of media for use in LANs, MANs, and WANs. In the early 1980s, the only real choice in network media was coaxial cable. However, in recent years, the shift has been away from coaxial cable for the LAN applications and toward twisted-pair copper wire, because of its lower cost. On the MAN and WAN side the trend has definitely accelerated toward optical fiber.

Coaxial cable is so named because of its construction. It consists of a center conducting cable, surrounded by an insulator, which is in turn surrounded by another conductor, which is finally covered with an outer insulator. This construction gives coaxial cable very good immunity to electromagnetic interference (EMI) that often affects other cables. It is capable of being used for cable runs anywhere from a few meters to a few kilometers. However, its cost effectively prevents it from achieving widespread use in both MAN and WAN applications.

Twisted-pair cable has rapidly become the cable of choice for LAN installation. Twisted pair is constructed with two copper wires that are insulated, and then twisted around each other. This twisting helps to eliminate the effect of outside electromagnetic interference on the cable. As the number of twists per inch increases, so does the effectiveness of the noise reductions. Unfortunately, this increase in the number of twists increases the cost of the wire. The technology has evolved to the point where twisted pair can be used for cable runs up to 1 km and data rates of up to 100 Mb/s, with two current schemes competing for the opportunity to provide this standard. CDDI (Copper Distributed Data Interface) and 100-Mb/s Ethernet comprise these two camps.

Optical fiber is the medium of choice for MANs and WANs. Optical fiber is essentially a long transparent core of glass or plastic compounds, surrounded by another layer of the material with different optical characteristics than the core. The advantages of optical fiber are numerous. First, it is immune to electromagnetic interference. Second, there is little signal loss to the medium itself. As a result, cable runs can be longer (up to roughly 90 km) and tend to have fewer errors. Finally, the data rates for optical fiber can exceed 1 Gb/s! The main disadvantage of optical fiber used to be the cost. However, by the late 1980s fiber cost began to drop rapidly as better production methods and high volumes lowered manufacturing costs. A relevant disadvantage of optical fiber is that the devices that convert the optical signals to electrical signals and vice versa are still relatively expensive.

Connecting remote locations has become easier with the use of radio networking. Broadband radio networking utilizes large-bandwidth allocations of frequency space to transmit the full bandwidth of the network over radio waves. Spread spectrum radio networking utilizes multiple smaller-bandwidth channels for similar effect. A large contributor to the growth of the latter has been the Federal Communication Commission (FCC) deregulation of low-power spread-spectrum communication.

7. DIGITAL SIGNAL-ENCODING TECHNIQUES

Digital signal encoding is used to modify the binary data that computers and networks communicate with in order to improve communication performance. There are three main criteria used to determine which encoding techniques are to be used, with a goal of maximum signal with minimal noise. These are spectrum shaping, clock extraction, and error-detection capability. Spectrum shaping is designed to minimize the impact of low-frequency noise on a signal. The ability for the receiver to determine the clocking frequency of the sender with no additional data is defined as clock extraction. Finally, error detection is the characteristic that allows the receiver to determine whether or not the data on the line are actually valid. The four encoding schemes covered are polar nonreturn to zero (NRZ), unipolar return to zero (RZ), bipolar RZ, and Manchester. These four techniques are illustrated in Fig. 3.

Polar NRZ is the most common encoding technique between computers and remote terminals. A full-width positive pulse indicates a one, and a full-width negative pulse indicates a zero. The weakness of polar NRZ is that it does poorly on computer network media that do not use low frequencies (i.e., coaxial cable and twisted pair). Additionally, error detection for polar NRZ is poor and clock synchronization is not inherent to the technique.

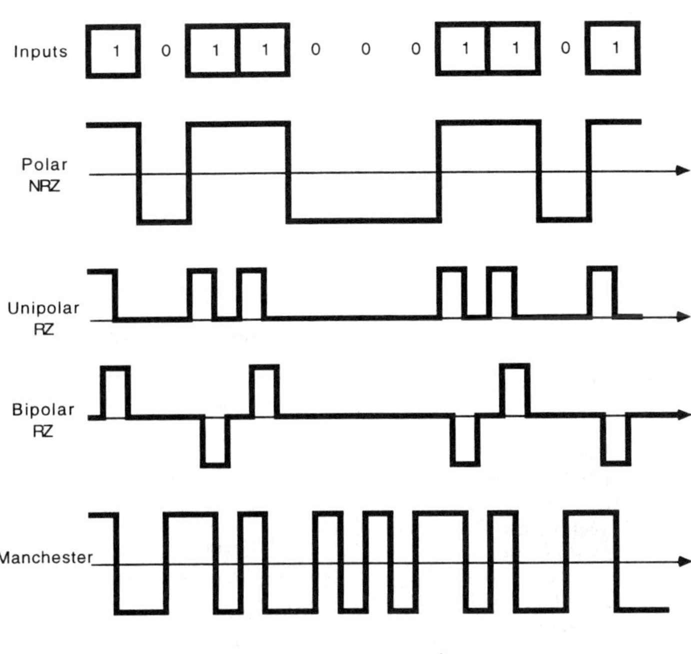

FIG. 3. Digital signal-encoding techniques.

Unipolar RZ encoding uses a half-width positive pulse to represent a one and no pulse to represent a zero. Its clock extraction is poor, although it does have moderately good error detection. However, it shares the disadvantage of polar NRZ and is not particularly useful for networks that are even as small as LANs.

Bipolar RZ uses alternating positive and negative half-width pulses to indicate ones, and no pulse to indicate zeros. A noteworthy feature of this technique is that it helps eliminate the effects of dc voltages on the lines. Clock extraction is not inherent to the encoding, but error detection is good, although not as good as for Manchester encoding. This technique is often used for WANs that utilize cable as opposed to satellites.

Manchester (or biphase) encoding uses a positive and negative half-width pulse to represent a one, and a negative and positive half-width pulse to represent a zero. This is in fact the dominant encoding technique for LANs. Clock extraction for Manchester is simple and error detection is excellent.

8. NETWORK PROTOCOL MODELS

Commonly agreed upon conventions for communications are typically referred to as protocols. A protocol is a set of rules stating how information is going to be exchanged. In most situations in life, fairly informal protocols may be used. However, for computer networks, a very formal protocol set must be used in order to ensure that all computers involved can communicate effectively. In computer networks, the protocol models that are typically used are referred to as layered protocols. These protocols consist of several layers, each with its own function and definition, that combine to make up the entire protocol model. By far the most common network protocol model is the International Standards Organization (ISO) Open Systems Interconnection (OSI) architecture. This is typically referred to as the ISO/OSI model, or more informally as the Seven-Layer Model. However, a second networking protocol developed for the United States Department of Defense (DoD) is the most widely implemented model. This model is a direct result of the ARPANET project from the 1970s and 1980s. Figure 4 shows both of these models.

The ISO/OSI model is broken into seven layers: physical, data link, network, transport, session, presentation, and application. Each of these layers has a separate function, and can communicate with the layers both above and below it bidirectionally. A brief description of the seven layers follows.

1. The physical layer provides media-level (e.g., electrical, light-oriented, radio waves, etc.) functional and procedural character-

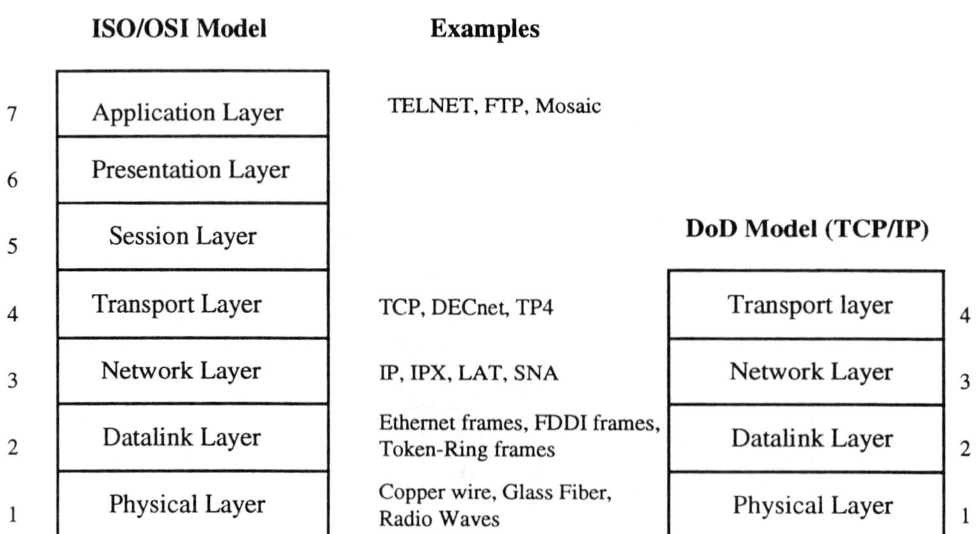

FIG. 4. Network protocol models.

istics involved in transparently passing the bits of computer data across the network.

2. The data-link layer provides functional means of transferring data between network components and may do some error correction. It also regulates flow control, and media access control.
3. The network layer provides switching and routing functions to different network hosts and masks data transfer details from higher layers.
4. The transport layer provides data transfer to upper layers and ensures reliability and cost-effective data transfer. This layer does so without presenting lower-layer information to the higher layers. It is therefore said to provide a seemingly transparent connection between two communication endpoints.
5. The session layer provides a mechanism for structuring dialogs between different application processes.
6. The presentation layer serves to make differences in data representation between applications transparent to the application.
7. The application layer is concerned only with the semantics of the applications. Applications all reside in this layer.

The DoD model is much more flexible and allows applications to skip layers that would otherwise be required in the ISO/OSI model. Where the ISO model emphasizes the creation of a structured model, the DoD model demonstrates true functional internetworking. This may be partly because the ISO model was constructed by theoreticians with network ideals, while the DoD model was designed by practitioners documenting an existing model. This DoD advantage makes it especially suited for use on internetworked WANs and MANs. With the emphasis on internetworking, the DoD model is a connectionless model designed to handle short interactive data transfers. The DoD model consists of four layers, which are defined as follows:

Layer I (network access layer) contains protocols that control the flow of data on the network from host to host.

Layer II (Internet layer) provides the functions necessary to allow data to traverse between networks from host to host. Additionally, this layer may handle error correction, message segmentation, and assembly and message identification.

Layer III (host-host layer) contains functions needed to process data between processes on different network hosts. It may include setting up logical connections, error control, and flow control.

Layer IV (process/application layer) contains protocols specific to a given process or application. This may include connecting to remote hosts, transferring files, and exchanging messages.

The DoD model is used heavily by advocates of client-server computing because of its inherent efficiencies. In the early days of computing, all processing was done on one machine in a batch environment. The same machine read the input, processed the data, and generated output. Over time, multiple machines were used in such a fashion, with one doing input, one doing processing, and yet another doing output.

In the last few years, the norm has changed to utilize machines of differing power, where the large, powerful machines make their processing power available to smaller, less powerful machines. In such an event, the large machine is said to be "providing a service" to the small machine by making its (the large) CPU horsepower available.

This usage has expanded now; if the hardware offers cycles then it is a "compute server." If the software offers other software programs the ability to process data for them, it is an "application server." Programs and hardware that use these services are called "clients."

For example, if there are 30+ IBM PC clones, and 100+ Macintoshes, each of these have local disks, but it would be unreasonable to buy everybody large disks. Instead, one can have one machine dedicated to providing everyone disk space; it would be purchased with large disk storage facilities. These it would provide to anyone who wished to use them over the network. Thus it is a "file server," and any PC or Macintosh that uses it is called a "client."

In general, there are two philosophies of structuring information processing. Peer-processing, where all computers are created equal and all are equally capable of performing a job, is the first such philosophy or model. This is often found in large computer centers filled with minicomputers or mainframes. The other is client-server, where a minority of large systems provide services (disk, CPU, memory, data processing, results processing, application data, etc.) to smaller systems. This latter philosophy works well with cost-conscious computing establishments. As the global Internet has grown to include clients and servers of all kinds, the concept of a "global backplane" has emerged, where the physical location of the clients and servers is not at all important, but rather their connection to the global network, and being able to access available resources, are. In the "global backplane" topology, all computers are connected together, and a suitably authorized client may access any available server.

9. NETWORK SECURITY

As computer networks become more prevalent and higher volumes of data flow on them, the security of both the network and the data becomes a greater concern. In the infancy of computer networks, dependency on computers was minimal, and as a result a loss of service was not crippling. However, this has changed and today's organizations have yet another mission-critical component to keep a watchful eye over.

Threats to computer network security are numerous. The primary concerns are as follows: the physical integrity of the network, the confidentiality of the data, denial of services, and recovery plans. These issues are briefly discussed, and methods of minimizing the threat are provided. It should be noted that this discussion is in no way complete or all-encompassing, as this is the subject of numerous books.

9.1 Physical Integrity

The physical integrity of the network is probably the issue that should be first addressed when looking at network security issues. Even after eliminating all other threats, no matter how sophisticated the attack, if the equipment is left without adequate protection, the network can be shut down with nothing more complicated than a pair of wire cutters or the stroke of a key. The key to physical security is to locate network-critical items in restricted areas. If an organization has a single server, then placing it in the hallway where anyone could turn it off would not be as wise a decision as placing it in a locked room. The same holds true for the cables between computers. The normal practice is to run the cables underground where they are difficult to get to, or through the actual structure of a building. Unfortunately, with the increasing use of the global internetworks by anyone with the financial resources, control of all access to the cables or even interconnected machines is difficult to achieve. In this case, the best practice is to isolate the network in a manner that still allows the realization of most of the benefits without the liabilities. This is normally done by placing a "firewall" between the organizational network and the networks under external control. A firewall is a computer that is designed to stop external access and prevents outsiders from being able to glean information that they could use to attack the organizational network.

9.2 Data Confidentiality

Data confidentiality is perhaps the issue that people concern themselves with most. The thought of proprietary data getting into the hands of competitors or external agencies is normally a fear that will motivate decision makers to ensure that their network is secure. However, most attacks do not attempt to get proprietary data, but instead attempt to get access control mechanisms such as passwords and user names that will allow someone access to the network at a later time. In part, this vulnerability is a direct result of the networking protocols themselves. The protocols normally transmit the data with no sort of basic encryption used. This makes the "sniffing" or viewing of the data a relatively trivial task for someone who is knowledgeable. The extremely paranoid network users (primarily governments) actually place their cables in gas-pressurized lines so that they can detect if someone attempts to tap into the cable in order to view their data. However, this is impractical for most nongovernmental organizations. The best practical method of preventing these types of attacks is to make sure that the networks are physically secure, and to remove some level of trust from the users and replace it with wary education. Most security attacks are the direct result of an internal user either giving their password out to someone, or doing the actual damage themselves.

9.3 Denial of Service

Denial of service attacks are relatively rare, but need to be discussed, if only briefly. A denial of service attack does not attempt to access the data, but instead it attempts to prevent effective use of the equipment from occurring. An example might be to try to fill up the disk drives so that data storage can no longer be done. Another example might be to produce CPU-intensive applications in an attempt to prevent real processing from being done. Again, the best method of preventing these attacks is to make sure that the network is physically secure, and that users are educated against providing access control mechanisms to unauthorized parties. An additional method of gaining access is the physical one, where denial attacks may come from programs brought in by individuals using the network. These programs may contain hidden programs within them, called "Trojan horses," that can destroy important data as well as infiltrate other systems. Programs that spread in this manner are also referred to as "viruses."

9.4 Recovery Plans

The final portion of network security is a backup and recovery plan. No matter how good the security is, it is not possible to plan for all eventualities. In this case, a well thought-out method of backing up data and storing them so that they can be recovered later is invaluable. A very good practice is to back up the data (normally to digital tapes) and then to remove them from the physical proximity of the normal computing facilities or to put them in fireproof/waterproof safes. Off-site storage is by far the safest practice, but is sometimes expensive.

10. NETWORK ALTERNATIVES

While a majority of the users of computer networks seek to connect their networks to larger networks that can provide those services they seek, alternative providers have also proliferated. Such companies as Compuserve, Delphi, America On-Line, and others have packaged network-oriented services for those not directly networked. Using a standard telephone modem (MOdulator, DEModulator), a device that converts digital data into audio, customers of these providers can access the provider database, which is itself linked into the global information base. Some companies also offer connection via local telephone numbers that connect via X.25 protocol networks to the provider's base system.

11. CONCLUSIONS

Computer networks are becoming more prevalent and necessary elements of not only business, but society. They are available in homes, businesses, schools, and governmental organizations. Networks today are already very large, and look only to get larger. Fortunately, the basic concepts of LANs, MANs, and WANs are sufficiently broad, and the concepts are sufficiently advanced, to allow computer networking technology to continue to evolve without being limited by older, legacy technologies.

GLOSSARY

ARPA: Advanced Research Projects Agency.

ARPANET: ARPA-funded research network.

ATM: Asynchronous transfer mode.

CDDI: Copper distributed data interface.

DARPA: Defense Advanced Research Projects Agency.

DIX: Digital, Intel, and Xerox.

DDN: Defense Data Network.

FDDI: Fiber distributed data interface.

IP: Internet protocol.

ISDN: Integrated services digital network.

IXC: Interexchange carriers.

LAN: Local-area network.

MAN: Metropolitan-area network.

NSFNET: National Science Foundation-funded research network.

OSI: Open Systems Interconnect.

SONET: Synchronous optical network.

TCP: Transport control protocol.

WAN: Wide-area network.

Further Reading

Abrams, M. D., Podell, H. J. (1986), *Computers and Network Security*, Washington, DC: IEEE Computer Society Press.

Comer, D. E. (1991), *Internetworking with TCP/IP*, Englewood Cliffs, NJ: Prentice-Hall.

Halsall, Fred (1992), *Data Communications, Computer Networks and Open Systems*, Reading, MA: Addison-Wesley.

Kessler, G. C., Train, D. A. (1991), *Metropolitan Area Networks; Concepts, Standards and Service*, Reading, MA: Addison-Wesley.

Nemzow, Martin, A. W. (1992), *The Ethernet Management Guide: Keeping the Link*, Reading, MA: Addison-Wesley.

Spragins, J. D., Hammond, J. L., Pawlikowski, K. (1991), *Telecommunications Protocols and Design*, Reading, MA: Addison-Wesley.

Stallings, W. (1993), *Networking Standards: A Guide to OSI, ISDN, LAN, and MAN Standards*, Reading, MA: Addison-Wesley.

Stevens, R. W. (1990), *Unix Network Programming*, Englewood Cliffs, NJ: Prentice-Hall.

NEURAL NETWORKS

TOMMASO TOFFOLI, *Laboratory for Computer Science, Massachusetts Institute of Technology, Cambridge, Massachusetts, U.S.A.*

INTRODUCTION

Neural networks are digital or analog circuits consisting of a large number of simple elements, and designed in such a way as significantly to exploit aspects of collective behavior—rather than rely on the precise behavior of each individual element. Historically, interest in neural networks arose from the realization that conventional digital computers, in spite of their enormous raw processing power, compare poorly in many tasks with the nervous system of animals. By mimicking structural aspects of the latter, would one more easily reproduce the strong points of its behavior? Thus, neural networks are ostensibly an alternative type of computing hardware, loosely patterned (both in the nature of the circuit elements and in the way they are interconnected) after the animal nervous system.

Today, however, it is becoming clear that—rather than just another type of computing medium—neural networks represent a different *conceptual approach* to computation, depending in an essential way on the use of statistical concepts. In this sense, the theory of neural networks plays in information processing a role analogous to that of statistical mechanics in physics.

From a descriptive viewpoint, we shall be concerned with what neural networks *are* ("structure") and what they *do* ("behavior"). However, we shall never leave far behind questions of a more critical nature: What makes it possible for a certain kind of network to perform certain kinds of tasks? Intuitively, *how* does it work, and *why* should

3-527-28133-9/94/$5.00 + .50

it work at all? This theme is well developed in Hertz *et al.* (1991), to whom we are greatly indebted here.

A typical application for neural networks is to help in making decisions based on a large number of input data having comparable *a priori* importance—for instance, reconstructing the characters on a license plate (a few bits of information) from the millions of pixels of a noisy and somewhat blurred and distorted camera image.

In general, the neural-network approach seems best suited to computational problems of large width and moderate depth—"democratic" rather than hierarchical algorithms. Keep in mind that word recognition out of connected speech—which is a hard task for conventional computers—is performed by our brain with a latency of just a fraction of a second, and thus cannot involve more than a few levels of neurons.

Neural-network design and analysis typically assume a regime of high hardware redundancy. It then becomes both possible and desirable to program a network for a given task by indirect methods (training by example, successive approximations, simulated annealing, etc.). Indeed, the metaphor of a network "learning" its task instead of being "programmed" for it is one of the most appealing—and elusive—aspects of this discipline. By empirical means, it is not hard to come up with a neural-network design that works for a certain toy problem. It is much harder to prove the correctness of the design, and rationally determine its potential and limitations. The importance of theoretical work in this context cannot be overstated.

In spite of the nominal emphasis traditionally being placed on neural networks as a distinguished kind of *hardware*, it must be clear that ultimately one is speaking of a distinguished class of *algorithms*; as a matter of fact, many neural-network applications are routinely and satisfactorily run on ordinary digital computers.

1. NEURONS AS CIRCUIT ELEMENTS

1.1 Stylized Neurons

The human brain consists of about 10^{11} neurons of various types; each neuron typically connects, via an axon that eventually branches out into strands and substrands, to many thousand neurons (Fig. 1). The firing of a neuron is mostly an all-or-nothing business; this discrete character is retained as the pulse travels down an axon. However, upon arrival at a destination neuron the pulse is handled by a synaptic interface characterized by an analog parameter (typically, an excitation or inhibition weight) whose value may be to some extent history-dependent. The complete physiological picture is rather complex.

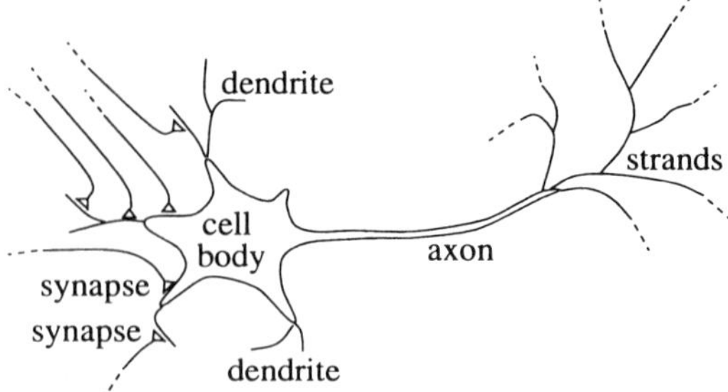

FIG. 1. Schematic structure of a typical brain neuron.

Though neuronlike circuit elements are not strictly essential to neural networks in the modern sense of the term (see Secs. 6.4 and 1.7), historically they have played a dominant role. A drastically simplified model of a neuron, proposed by McCulloch and Pitts (1943), is shown in Fig. 2. The neuron can be in one of two states, +1 and −1, which may be thought of as "on" and "off," or "true" and "false," this state appears at the neuron's output. The inputs may come from other neurons or from external stimuli. State updating may be synchronous (all neurons are

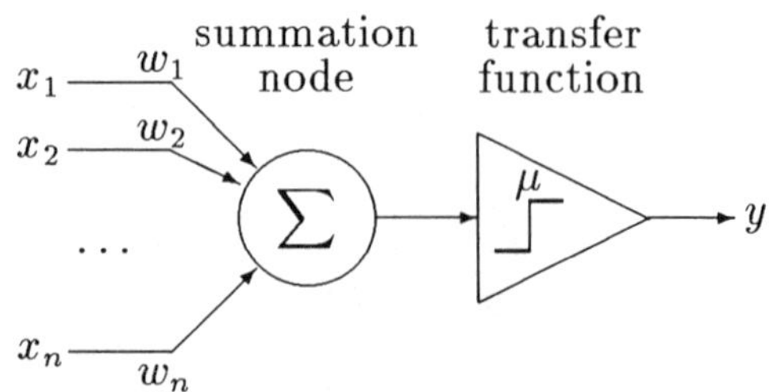

FIG. 2. McCulloch–Pitts neuron. The summation node constructs the weighted sum (with coefficients w_1, w_2, . . .) of the inputs. Depending on whether or not this sum exceeds a threshold μ, an output of +1 or −1 is returned by the transfer function.

updated simultaneously at times $t = 0, 1, 2, \ldots$) or asynchronous (each neuron is updated at random times with a given probability per unit time). The new state of the neuron is determined by the inputs as follows. Input x_j is multiplied by a weight w_j, representing the strength of the corresponding synaptic connection (positive weights correspond to excitatory synapses, negative weights to inhibitory ones). The contributions from all inputs are added and compared with a threshold μ; the neuron turns on if the threshold is exceeded, and off otherwise.

The overall response of the neuron is thus of the form

$$y = g\left(\sum_{j=1}^{n} w_j x_j\right), \tag{1}$$

where g—the *transfer function*—is given by

$$g(x) = \mathrm{sgn}(x - \mu). \tag{2}$$

[By $\mathrm{sgn}(x)$ one denotes the signum function, which returns $+1$ or -1 depending on whether $x \geq 0$ or $x < 0$.]

It is often convenient to replace the step function (2) by a more general transfer function. A continuous function such as the *sigmoid* (Fig. 3), defined by

$$g(x) = \tanh(\beta x) \equiv \frac{e^{\beta x} - e^{-\beta x}}{e^{\beta x} + e^{-\beta x}} \tag{3}$$

(where β is an adjustable parameter), allows one to write formally the partial derivatives of the neuron's output with respect to its inputs (cf. Sec. 5.3): the signum function is recovered in the limit as $\beta \to \infty$.

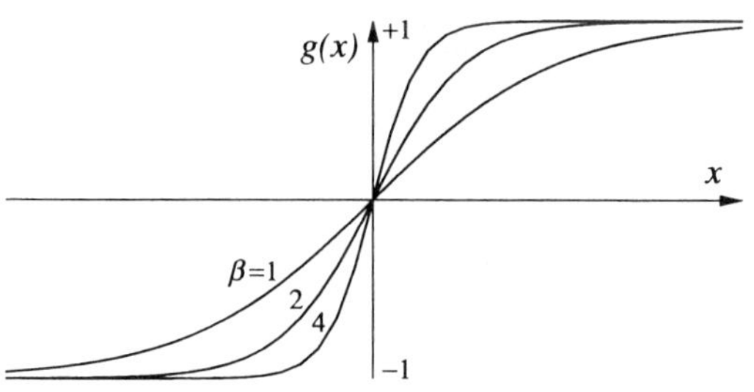

FIG. 3. Sigmoid transfer functions are often used in analog networks. The parameter β gives the *small-signal gain*. The *saturation levels*, ± 1, are independent of β.

1.2 Why Neurons?

McCulloch–Pitts neurons are computation universal circuit elements (Minsky, 1967); in fact, they can simulate conventional logic gates such as AND, OR, and NOT. (For instance, with a two-input neuron, weights of 1 for each input, and a threshold of $\frac{1}{2}$ the neuron will fire only when both inputs are turned on, yielding an AND gate.) But why, then, not use gates to begin with?

The neuron is optimized for a different circuit architecture—one that demands elements capable of responding in a nontrivial way to a *large* number of inputs. A generic Boolean function of n inputs can be thought of as a lookup table with one independent entry for each combination of input values; its complexity (intuitively, its material cost) grows exponentially with n. To have a complexity that grows only linearly with n—and thus a general design that is materially realizable even when n is large—one must drastically restrict the class of functions computable by the circuit element. This is achieved by the two-stage design of Fig. 2: a summation node followed by a transfer function. The first stage deals with all n inputs, but only in an additive way; while the second stage, which has only one argument, contributes the required nonlinearity.

2. HISTORICAL NOTES

Neural networks started out, with McCulloch and Pitts' neurons, as an exercise in mathematical biology. The first networks to use many-input neurons systematically were the perceptrons (Rosenblatt, 1962), in which neurons are arranged in regular layers with no feedback from a layer to previous ones (Sec. 4.2). The very book that celebrated perceptrons (Minsky and Papert, 1969) also brought about their untimely death by pointing at a serious limitation of one-layer perceptrons while casting doubts about the existence of systematic learning algorithms for multilayer ones.

After that, interest in neural networks remained sparse for 20 years, with occasional contributions from physiologists (Marr, 1969) and physicists (Caianiello, 1961). The 1980s saw a sweeping revival with new ideas from statistical mechanics and dynamical systems, such as energy function and stable attractors (Hopfield, 1982); new programming techniques, such as the back-propagation learning algorithm (Rumelhart *et al.*, 1986); and, of course, the availability of computing machinery of ever increasing performance.

Today, neural networks are used routinely in many specialized applications, chiefly in low-level image and speech processing, and sensor–actuator integration in motor control; they are also widely used for a variety of noncritical tasks where adequate training by example can be imparted rapidly and economically by nonspecialists: data presorting, screening of applications, poll analysis, quality control, etc. (see Sec. 6.6). However, it still is not clear whether they will ever represent a major component of everyday computation. On the theoretical side, much of the initiative and of the conceptual machinery for fresh developments have been coming from the statistical mechanics community. On the architectural side, computer designs based on elements that are simpler, more numerous, and more heavily interconnected than in traditional architectures are preached by many with almost religious fervor; neural networks fit in very well with the "connectionist gospel" (Hillis, 1985). However, the tenets of this gospel are constantly being revised under the pressure of technological expediency and, ultimately, of fundamental physical constraints: massively parallel architectures tend to favor uniform and local interconnections and limited fanout of signals. Perhaps a final assessment of the potential of neural networks will only come when we know how the brain actually manages to do what it does.

In the meantime, neural networks have matured enough to provide substantial conceptual and practical contributions to the study of the brain itself. This is the domain of *computational neuroscience,* with which we shall not deal here (but see Further Reading).

3. TUTORIAL

As an introductory tutorial, we shall present the Hopfield model (Hopfield, 1982) in its use as an associative memory.

3.1 Associative Memories

In a conventional random-access memory, p patterns $x^1, x^2, \ldots, x^p$ (entries) are stored in such a way that, when presented with the address μ (*key*), the memory returns the entry x^μ. The association between entries and keys is arbitrary; i.e., there are no semantic connections between a key and the corresponding entry. In *associative* (or "contents-addressable") memories retrieval is keyed by a pattern x rather than by an explicit address μ. The memory's task is to return the entry $\bar{x}$ that, according to given criteria, is the "closest match" to key x. It may be reasonable to have the associative memory return "no match" in certain ambiguous situations—for example, if the key is equidistant from the two closest entries ("is \$100 closer to \$101 or to \$99?"), or if it is so far from all of the entries that a match would be meaningless ("is a turnip closer to a book or to a bicycle?").

For sake of illustration, we will take patterns to be bitmap images, each consisting of N pixels (values $+1$ and -1) indexed by $j = 1, 2, \ldots, N$. The "closeness" of the match between two patterns x and x^μ will be measured in terms of their *Hamming distance,* i.e., the number of bit positions on which the two patterns disagree.

3.2 The Hopfield Model

The desired Hopfield model will be constructed with one neuron per pixel position; thus, the collection x_j of neuron states is a pattern. Every neuron is formally connected to all other neurons with a synaptic weight w_{ij} from neuron j to neuron i (of course, lack of physical interconnection can be represented by $w_{ij} = 0$). There are no external inputs. Since information may flow from one neuron back to itself, directly or through other neurons, this is a network with feedback; indeed, since there are no external inputs, it is an *autonomous* network (Sec. 4.4). A threshold at 0 will be used as a transfer function. Thus, the dynamics of the network is given by

$$x_i := \operatorname{sgn}\left(\sum_j w_{ij} x_j\right). \tag{4}$$

Updating, represented by the assignment symbol ":=", is done asynchronously in this case.

Started with an arbitrary initial pattern x and driven by the above dynamics, the network will describe a trajectory through the space of all possible patterns. Our goal is to program the network's weights so that, for any reasonable pattern x used as a key (cf. Sec. 3.1), the trajectory starting from it will eventually settle on the matching entry $\bar{x}$. Intuitively we want to construct around each of the entries a *basin of attraction* (Fig. 4) such that any trajectory starting within a basin will come to rest on the entry itself (the basin's *attractor*).

Analogy with plausible neurological mechanisms (Hebb, 1949) suggests the following weight assignment ("Hebb rule"):

$$w_{ij} = \frac{1}{N} \sum_{\mu=1}^{p} x_i^{\mu} x_j^{\mu}. \tag{5}$$

It turns out that this assignment substantially achieves the goal, *provided* that the entries are sufficiently distant from one another.

Note that increasing the number N of pixels makes more room for distance between the entries, and thus tends to increase the storage capacity of the associative memory. It has been shown that, for a given number N of pixels, recall errors are rare for $p \ll N$ but grow catastrophically as soon as p approaches $0.138N$. If no errors are tolerated, the storage capacity is proportional to $N/\log N$. The cost of the network (measured in terms of the number of synapses, N^2) grows as the square of its storage capacity.

For unreasonable keys (see above), instead of "no match" this network may return a spurious entry (i.e., a pattern not in the set $x^1, x^2, \ldots x^p$). This issue will be further examined in Sec. 5.3.

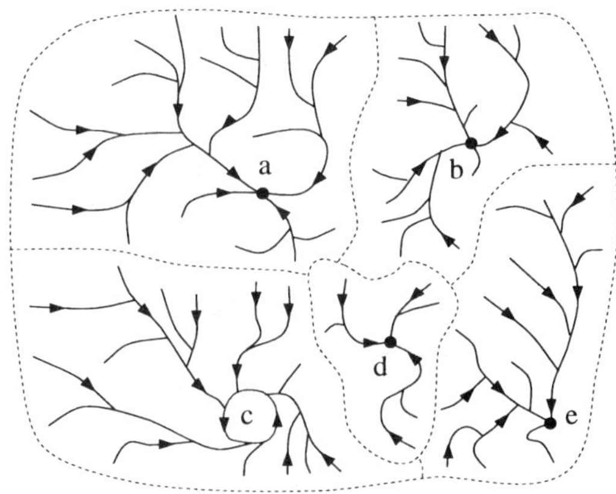

FIG. 4. Basins of attraction. Attractor c is a cycle rather than a point.

3.3 Discussion

Let us review the types of activities that were involved in this application, as they are representative of the whole neural network approach.

We started with a well defined *task*—to realize an associative memory for patterns consisting of arrays of N bits.

We then constructed a *structure* hopefully suitable for the task. This was an autonomous, fully interconnected network of McCulloch–Pitts neurons, having the same geometric layout as the pattern themselves.

Next, we *programmed* the network with the desired collection of patterns, using the Hebb rule as a programming algorithm. One could compute and enter the weights [Eq. (5)] by hand. What is remarkable about this algorithm, however, is that it could be easily carried out in parallel by the network itself in a special *learning* mode, as follows. On entering this mode, all weights are cleared. The first entry pattern is then "shown" to the network, i.e., the neuron states are set to the values of the corresponding pixels. Within each neuron, each synapse will compare its input signal with the neuron state; if they are equal, the synapse weight increments by $1/N$; otherwise, it decrements by the same amount. This teaching procedure is repeated for every entry pattern. By the time we are done, the network will have "learned" the correct weights.

At this point, the network can be *used* in its normal operating mode. A "key" pattern is written in the initial state of the network; the network is allowed to evolve according to its internal dynamics until it attains equilibrium; the matching "entry" pattern is read off the final state.

Finally, we *analyzed* the performance of the network in the given task (error probability, storage capacity, cost). We discovered that, as we attempt to cram in more and more patterns, at a certain point the network undergoes a phase transition and suddenly loses its functionality as an associative memory.

4. STRUCTURES

Below, we give a brief classification of neural networks by structure.

4.1 Generalities

In terms of interconnection, a computing network is usually drawn as a directed graph whose nodes represent the computing elements and whose arcs represent the signals flowing in and out of these elements—including external input and output signals (Fig. 5). The names of signal variables, such as *a* and *b* in the figure, are placed next to the corresponding arcs. A fanout node is one that produces at its outputs multiple copies of its single input. A directed path from a node to itself is a *cycle*; in computational terms, a cycle represents feedback.

In the neural-network literature, somewhat different conventions are used in drawing these graphs. In general, a computing element may have many distinct outputs. In a neural network, each neuron has a single output—which reflects the neuron's state (Sec. 1.1). However, copies of this output are typically sent to many destinations. This *fanout* of signals is usually represented as taking place within a neuron; thus, in graphs such as Fig. 6, it is understood that all the arcs leaving a node carry the same signal. It then becomes more practical to place signal labels not on the arcs that carry them but on the nodes that originate them. Since also external input signals usually fan out to different destinations, the source of each external input is indicated by an auxiliary node (a black dot in Fig. 6)—which, however, is only a graphical device and is not counted among the network nodes. Moving signal names from arcs to nodes leaves room for placing on each arc, as a label, the weight of the synapse to which the arc is attached.

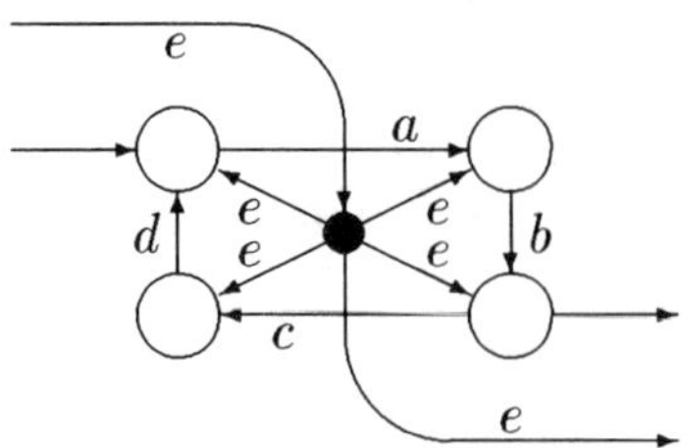

FIG. 5. Nodes and arcs in a directed graph. Here, arcs *a*, *b*, *c*, *d* make up a feedback cycle; the black dot represents a fanout node.

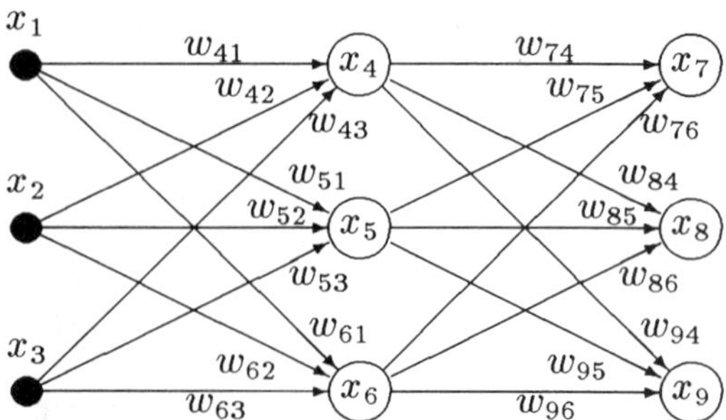

FIG. 6. Two-layer perceptron.

4.2 Feedforward Networks

A network is called *feedforward* if it has no cycles. Such a network acts as a memoryless transducer: information enters the network through its inputs, flows in one direction from node to node, and leaves the network through its outputs. A typical example is the *perceptron* (Fig. 6), where the nodes are arranged in regular layers, each corresponding to a different processing stage.

In principle, a two-layer perception can be programmed to compute an arbitrary Boolean function of its inputs, with the first layer being dedicated to ANDing appropriate subsets of the network inputs or their complements, and the second to ORing the intermediate results. However, this approach corresponds to brute-force lookup. Depending on the nature of the computing task, extra levels may make it possible to achieve the same results in a more structured way and with fewer nodes and wires.

For example, in the *cognitron* (Fukushima, 1975) successive layers are meant to deal with a hierarchy of features in image processing. Within each level, the cognitron architecture supports the idea of *locality*—which nodes are "near" to which other nodes: on each level, a neuron connects only to a small number of neurons of the previous level that are near to one another. At the bottom level, locality is interpreted geometrically; on higher levels, it may refer to distance in some parameter space (slope, contrast, etc.).

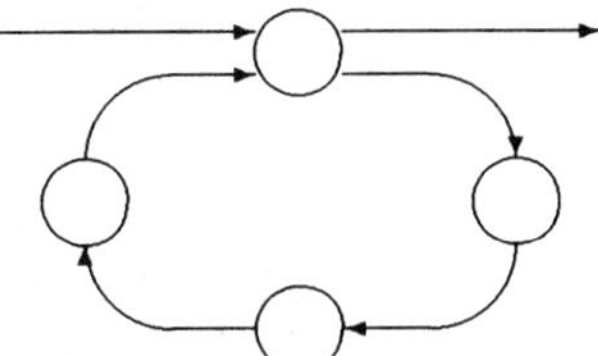

FIG. 7. In a convolution network, some information is recirculated through feedback loops.

4.3 Convolution Networks

Much as feedforward networks, convolution networks have external inputs and outputs, with information generally traveling in the direction from the former to the latter. However, some information is recirculated through feedback loops (Fig. 7), allowing more recently acquired information to interact with earlier information. Convolution networks may be viewed as transducers with memory.

For instance, if the input stream is connected speech and the task is speech segmentation, a convolution network may be used to implement a "sliding window" a few seconds wide. Through taps on this window, the segmentation algorithm can evaluate each input sample in the context of a few earlier and later samples.

4.4 Autonomous Networks

An autonomous network has no external inputs or outputs (Fig. 8); it can be thought of as an isolated dynamical system governed by time-independent dynamics.

For computational purposes, this isolation is broken at time t_0, when the current state is flushed out and replaced by an arbitrary initial state, and then at time t_1, when the system is halted and its final state examined. In this context, an autonomous network is equivalent to a feedforward network consisting of an indefinite number of identical layers (Fig. 9). Each iteration of the original network corresponds, in the new network, to going through one more processing layer.

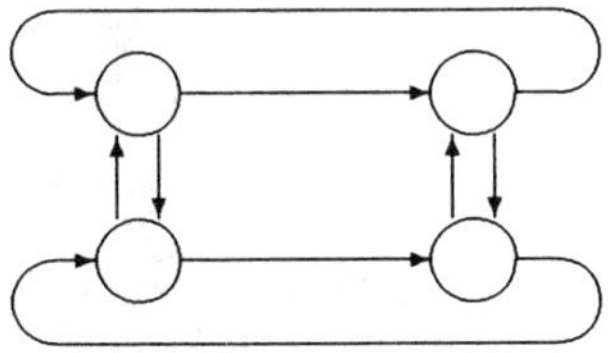

FIG. 8. Autonomous network.

Depending on a delicate balance between reversibility and irreversibility, linearity and nonlinearity, excitability and refractoriness, short- and long-range interconnections, etc., certain autonomous networks may display remarkable invariant features of a statistical character (equilibrium states, attractors, limit cycles, phase transitions, etc.). These features, which are implicitly contained in the dynamical laws, are explicitly "developed" by letting the system evolve; the result may depend in a highly nontrivial way on the initial conditions. When understood and controlled, this correspondence between initial state and developed features provides a rich computational repertoire (cf. Sec. 3).

4.5 Boltzmann Machines

A "canonical" type of autonomous network is represented by Boltzmann machines (Hinton and Sejnowski, 1986). These are networks with symmetric connections and stochastic dynamics, with probability P for a neuron x_i to fire given by an expression of the form

$$P[x_i := 1] = f\left(\beta \sum_j w_{ij} x_j\right), \tag{6}$$

where β is an adjustable parameter (see discussion in Sec. 5.3). Boltzmann machines (Fig. 10) are more general than Hopfield networks in that they may include among the x_i, beside the variables in whose terms a problem is stated, additional "slack" variables that rep-

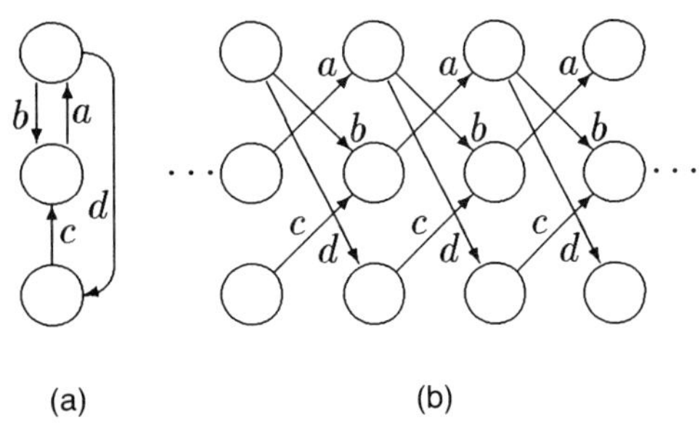

FIG. 9. By "unrolling in time" the autonomous network in (a) one obtains the equivalent feedforward network in (b).

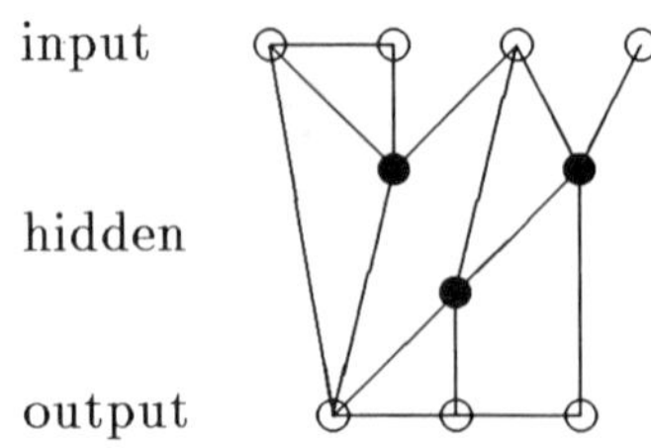

FIG. 10. Boltzmann machine. The hidden units represent slack variables—additional neurons whose state in not initialized or read out by the user. No direction arrows are drawn on the links, since the connections must be symmetrical. The distinction between hidden and visible nodes, and, for the latter, between input and output nodes, is not a structural one, but just reflects the intended use of the network.

resent extra degrees of freedom. The nodes carrying these variables are called *hidden units*—as contrasted to the other nodes, or *visible units*.

Because of the symmetry of connections, Boltzmann machines have a well-defined energy function E (see Sec. 5.2). The corresponding equilibrium distribution is the Boltzmann–Gibbs distribution

$$P(x) = Z^{-1}e^{-\beta E(x)}. \tag{7}$$

The function that expresses the normalization coefficient Z, namely

$$Z = \sum_{\text{all } x} e^{-\beta E(x)}, \tag{8}$$

is called the *partition function*, and plays a fundamental role in statistical mechanics (Baierlein, 1971). From distribution (7) one can in principle derive all the statistical properties of the equilibrium state.

The presence of hidden units, which during training are not forced to particular input values and are not requested to converge to particular output values, gives the network more freedom to adjust visible-unit values so as to obey all the given constraints. Without such flexibility, certain problems have no solutions (they are overspecified), and the network cannot be expected to learn to solve them.

4.6 Physical Models

Often, the activity of a neural network can be motivated or illustrated by a physical model—mechanical, electrical, optical, chemical. Sometimes, the physical process itself can be used as a practical way to carry out the desired computation.

Let us consider an array of raw data, $\bar{x}_i$, which are to be smoothed to values $\bar{x}_i$ (Sec. 6.3) by some form of low-pass filtering. In the mechanical model of Fig. 11, the x's are represented by the vertical positions of a number of unit masses, and $\bar{x}$'s by the positions of an equal number of anchor points. The mass at x_i is connected to the corresponding anchor point at $\bar{x}_i$ by means of a spring of stiffness α, and to its neighbors at $x_{i\pm1}$ by springs of stiffness β. Higher-order filtering can be achieved by adding connections to more distant neighbors such as $x_{i\pm2}$.

The cost function to be minimized is the energy stored in the springs, given by

$$E = \tfrac{1}{2}\alpha \sum_i (x_i - \bar{x}_i)^2 + \tfrac{1}{2}\beta \sum_i (x_i - x_{i-1})^2 \tag{9}$$

(this is the usual sum of the squares of the errors). A larger value for β gives greater smoothness; a larger value for α, greater fidelity to the original data. The configuration of least energy, corresponding to the solution of the problem, is easily found by physical means. In fact, as soon as one introduces some damping in the system [viz., a dissipative term in Eq. (9)], this configuration becomes an attractor of the dynamics (cf. Sec. 3.2): from any initial configuration for the x_i, the network will converge toward it. The fastest convergence rate is achieved with critical damping.

The same solution can be arrived at by means of the electrical model of Fig. 12. The $\bar{x}$ and x data are represented by voltages. The current through each resistor is proportional to the voltage difference between the two ends of the resistor, while the dissipation through the same resistor is proportional to the square of the current. At equilibrium, the voltages

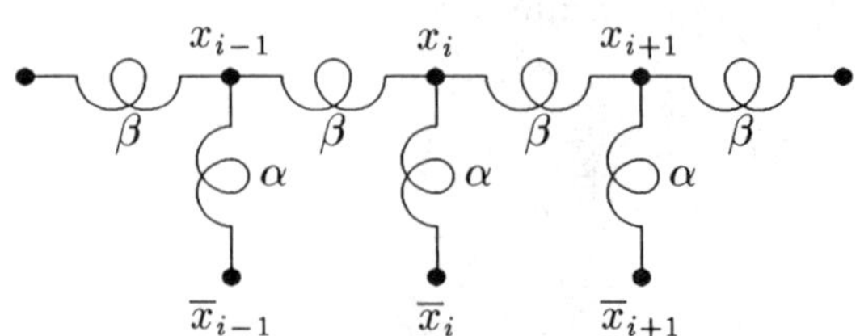

FIG. 11. A mechanical network for "smoothing" a one-dimensional array of raw data. The $\bar{x}_i$ are the raw data, the x_i, the conditioned ones.

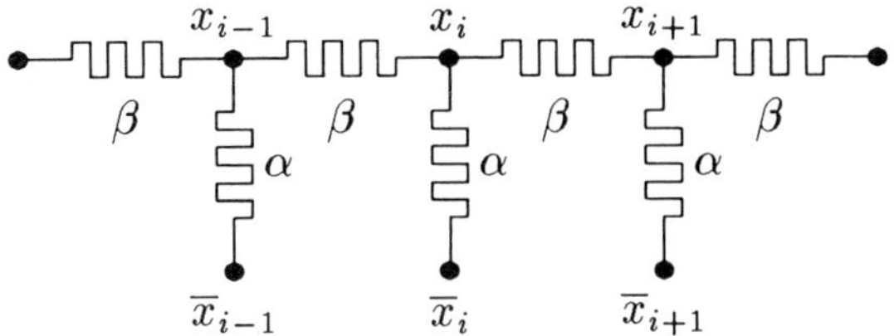

FIG. 12. An electrical model for data smoothing.

are such as to minimize the total dissipation in the network, which is given by a formula identical to Eq. (9).

A typical visual scene contains smooth areas bounded by sharp edges; the latter would be degraded by a smoothing-only algorithm. In order to enhance the edges, a cost function such as Eq. (9) could be augmented with appropriate "reward" and "penalty" terms. A more sophisticated approach is to let overstressed springs "snap" (or, in the electrical model, overheated wires "burn"); on a second pass through the optimization algorithm, the two severed ends would be free to converge to different values. See Koch *et al.* (1986) for an extensive discussion of methods of this kind in early vision processing.

4.7 Cellular Automata

In contrast to the "engineering" flavor (springs, resistors, etc.) of the above models, cellular automata (Burks, 1970) provide a computing-network metaphor for more universal aspects of physics—such as uniformity, locality, spacetime structure, inertia, microscopic reversibility, etc. (Toffoli and Margolus, 1987).

For our purposes, cellular automata are autonomous networks (Sec. 4.4) obeying the following constraints (Fig. 13). Nodes are arranged in a uniform array. To each node, or *cell*, there is associated a state variable ranging over a finite set—typically just a few bits' worth of data. Time advances in discrete steps, and the dynamics are given by an explicit rule—say, a lookup table—through which at every step each cell determines its current state from the current state of its neighbors. Gone are two more typically "neural" features of neural networks, i.e., arbitrary interconnection, and provisions for an arbitrary number of inputs at each node (cf. Sec. 1.2). Communication between distant nodes is still possible, but through a sequence of action-by-contact interactions rather than by action-at-a-distance.

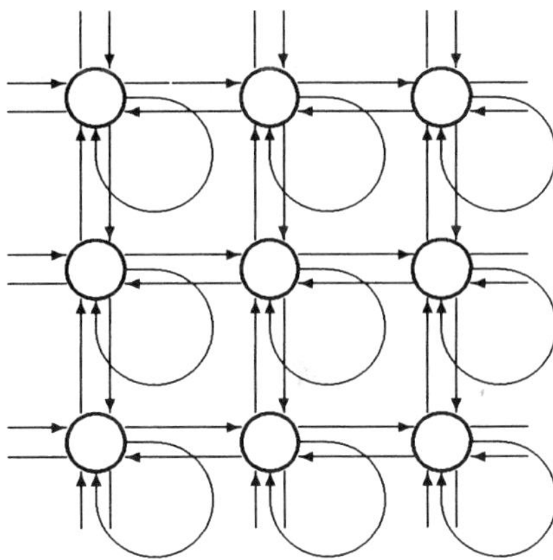

FIG. 13. A two-dimensional cellular automaton with first neighbor interconnection. Note the feedback loop from each node to itself.

In spite of important differences in motivation and emphasis, the cellular automata approach has much to share with that of neural networks; they are both concerned with the synthesis of complex macroscopic behavior out of the collective action of a large number of simple primitives (Wolfram, 1986).

Within the connectionist tradition, the cellular-automation scheme of computation is unique in the way it lends itself to extremely efficient concrete implementations. This feature provides a strong incentive to recast many neural-network techniques into a cellular-automaton format. Of course, cellular automata are the ideal substrate for many computing activities having a strong spatial component (e.g., image manipulation).

5. PROGRAMMING

An ordinary computer is programmed by writing an explicit sequence of instructions: the programmer should be able to describe and justify what each line of code "does." The typical neural network is programmed for its task by assigning the synaptic weights. Note, however, that a synaptic weight reflects the projection on a one-dimensional scale of parameters that inhabit a multidimensional space, and in turn will affect the values of a large number of computation variables. The programming of such weights is typically done by indirect means, usually by iterating a training procedure that is applied in parallel to the entire network. For example, we have seen in Sec. 3.3 that an associative memory *à la* Hopfield may be programmed by "showing" to it, in learning mode, the desired target patterns; each pattern leaves an "imprint" of itself in the form of an additive contribution to all the network weights.

A simple analogy for this situation is found in the Fourier transform $\mathcal{F}:\{f \to g\}$. Information about each point of the original function f is thinly and uniformly spread over the entire range of the new functions g. Intuitively, every point of g contains an infinitesimal amount of information about every point of f, and vice versa.

The most general indirect method of programming is, of course, trial and error: the weights are adjusted until the desired input/output behavior (or a minimum in an appropriate error function) is attained. Given the exponential size of the network parameter space, such an approach is hopeless unless one can rely on some sense of direction. Usable convergence criteria vary enormously depending on the nature of the network, the nature of the application, and the sophistication of the available theoretical tools. Except in very special cases (e.g., linear programming), in order to achieve a *global* minimum of the error function one needs strategies to avoid entrapment in *local* minima. These two concerns, namely local optimization by continuous means and global exploration by judiciously timed jumps, are illustrated below by, respectively, the back-propagation algorithms and the simulated annealing algorithm.

5.1 Back-Propagation

In the two-layer perceptron of Fig. 6, let the indices i, j, and k run respectively over the output nodes (second layer of neurons), the "hidden" nodes (first layer), and the network inputs; thus, for instance, w_{ij} will denote the synaptic weight of the connection from hidden neuron j to output neuron i.

Let x^μ denote an input pattern ($\mu = 1, \ldots, p$), y^μ the corresponding output pattern for a given set of weights, and $\bar{y}^\mu$ the desired output pattern for that input. The overall error between actual and desired response will be measured by the following error function (sum of squares):

$$E = \tfrac{1}{2} \sum_{ij} (y_i^\mu - \bar{y}_i^\mu)^2. \tag{10}$$

With a neuron of the form in Eq. (1), the variable term in Eq. (10) is

$$y_i^\mu = g\left(\sum_j w_{ij} g\left(\sum_k w_{jk} x_k^\mu\right)\right). \tag{11}$$

Under the assumption that g is a differentiable transfer function such as the sigmoid (Fig. 3), E is a continuous, differentiable function of every weight, and can be minimized by letting the w's slide in the direction of the gradient ("gradient descent"). Similar considerations apply to the case of more than two layers.

Because of the nesting of weights from different layers in Eq. (11), it is convenient to compute the errors and consequently adjust the weights layer by layer, starting with the output layer and proceeding toward the input. This *back-propagation algorithm* is local (weight adjustments for each neuron can be performed on the basis of the information available at the neuron's input and output), and is not computationally very demanding (if n is the number of synapses, it requires calculating order n derivatives, rather than order n^2 as in the minimization of E by simultaneously adjusting all the weights).

Many variants of the back-propagation algorithm have been studied, the main objectives being faster convergence and avoidance of local minima. Overall, theoretical and empirical results confirm the viability of this general strategy.

5.2 Energy Function

The following discussion is complemented by the entries on annealing, dissipation, energy, energy function, entropy, and temperature in the Glossary.

The Hamming distance between two patterns (Sec. 3.1) is defined independently of what patterns we are interested in retrieving. Can we define a real-valued function on individual patterns such that the resulting ordering of patterns would provide a "sense of direction" in solving a specific problem? In physics, energy plays such a role. For instance, in a conservative system, state b can be reached from state a only if $E(b) = E(a)$; in a dissipative system, only if $E(b) \leq E(a)$. Moreover, in a dissipative system, the states for which E has a local minimum are the attractors of the dynamics.

The idea of energy function in neural networks was pioneered by Hopfield (1982). In the example of Sec. 3, with weights w_{ij} given by the Hebb rule as in Eq. (5), let us define the *energy* of a pattern x by

$$E(x) = -\tfrac{1}{2} \sum_{ij} w_{ij} x_i x_j. \tag{12}$$

Then

1. the stored patterns $x^1, \ldots, x^p$ correspond to local minima for E, and
2. the dynamics in Eq. (4) are suitable for finding these local minima, since on each trajectory the energy E decreases or remains constant.

A sufficient condition for a quantity of the form of Eq. (12) to be an energy function is that the weights be symmetrical, i.e., $w_{ij} = w_{ji}$ (which is the case in the Hopfield model). See Sec. 4.5 for further details.

The importance of the energy function is that it gives a descriptive rather than a prescriptive characterization of the solutions (cf. variational principles vs. dynamical laws in physics). The problem is stated in terms of an energy landscape, and one is free to develop different strategies to explore this landscape in a search for solutions. Even one's idea of what should constitute an "acceptable solution" may be influenced by the nature of the landscape.

5.3 Simulated Annealing

Simulated annealing (Kirkpatrick *et al.*, 1983) is a broad strategy for searching for deep minima in an abstract energy-function landscape. It mimics the controlled, gradual cooling of a sample of material aimed at relieving internal stresses. If the sample is cooled too rapidly ("quenching"), it may remain frozen in a high-energy metastable state very close to the initial state; if it is cooled too slowly, it may tumble all the way down to the ground state (the absolute minimum of energy), thus erasing all memory of its initial state. Different annealing schedules allow one to "dial in," as it were, terminal states having different degrees of stability.

In physics, temperature measures the intensity of the randomizing influences that affect a system's nominal dynamics. As remarked above, in the deterministic neuron dynamics of Eq. (4) the value of a pixel is complemented if and only if the new configuration has less energy—no matter how small the energy difference. To represent temperature in the network, we introduce a stochastic dynamics derived from Eq. (4) as follows.

The output from the summation node in neuron i,

$$h_i = \sum_j w_{ij} x_j, \tag{13}$$

represents the "tendency" (on a scale from $-\infty$ to $+\infty$) for the neuron to fire. We rescale this tendency to the interval $(-1,+1)$ by sending h_i through a sigmoid of the form of Eq. (3), and we interpret the result as the expectation value of the neuron state. The corresponding probability P for neuron i to fire is given by

$$P[x_i := 1] = 1/(1 + e^{-2\beta h_i}). \tag{14}$$

For $\beta = \infty$, Eq. (14) yields back Eq. (4): there is no thermal agitation. For $\beta = 0$, the neuron will go with equal probabilities to states $+1$ or -1: thermal agitation completely swamps any internal tendencies. In fact, β can be identified with the temperature parameter—and its inverse, $T = 1/\beta$, with the absolute temperature—of a statistical mechanical system (Katz, 1967).

In simulated annealing, one starts the search for a minimum of the energy function with the system at a high temperature, and gradually reduces it. The details of this thermal schedule influence not only the speed with which a solution is reached, but also the nature of the solution.

Suppose there is a path of monotonically decreasing energy from point P to point Q; even if P and Q are near to one another, this path may be long and circuitous because of energy barriers that block more direct paths. (Given the high dimensionality of the typical problem space, the number of paths from P to Q may be astronomical.)

If the temperature is high, the system can easily go across high (but thin) barriers, thus availing itself of much shorter paths between P and Q. Intuitively, at temperature T the system sees an energy landscape smoothed out (cf. Sec. 4.6) on a scale $\Delta E \approx T$, and may be made to follow an energy-function gradient averaged out on this scale; finer details are virtually ignored. Once the approximate bottom of a valley is thus rapidly reached, one may reduce the temperature, thus exposing ripples on a finer scale and allowing further descent.

On the other hand, if the initial temperature is too high, the system may escape from the original basin of attraction and land in a different basin. Though desirable in certain circumstances, such wholesale erasure of initial-state information would clearly defeat the purpose in the associative memory task discussed above.

5.4 Learning

We have seen that neural networks are typically programmed by indirect means (i.e., without direct external access to the synaptic weights), for which "teaching" is an appropriate metaphor. Actually, the term usually employed for this process is "learning"—which emphasizes the network's rather than the programmer's viewpoint. Intuitively, the programmer influences the network's "genotype" by applying pressure on its "phenotype." Unlike biological evolution, besides mechanisms of a Darwinian kind—or *genetic algorithms* (Holland, 1975)—like simulated annealing (Sec. 5.3), it is sometimes possible to set up more efficient, Lamarckian mechanisms, like back-propagation (Sec. 5.1).

In supervised learning, the network's role is to internalize appropriately information presented by the teacher during a training session: explicit data to be memorized (cf. Sec. 3), examples of the desired behavior, or scores evaluating the current behavior of the network and aimed at steering it toward the desired one (cf. Sec. 6.4). Once the training phase is over, the values of the network's parameters are "set"; the network is then ready for the operational phase, where the network's state variables come into play.

Some neural networks may be designed to operate in an unsupervised learning mode. Typically, the network will be expected to identify and extract significant features of the input stream without the complicity of a teacher, and possibly to apply this knowledge to subsequent segments of the input stream (cf. Sec. 6.2). Here, of course, the distinction between network parameters and state variables depends on the point of view and on the time scale.

5.5 Realization Capabilities

Without loss of generality, the target task for a neural network can be thought of as a function f to be computed, and the programmed network as an approximate realization of this function, obtained by composition of given primitives. Thus, the study of what can be computed by a neural network is basically a chapter of functional analysis.

It is true that the approximation error ϵ_{approx}—i.e., the distance between the target function and the function computed by the network when optimally programmed—can be made as small as desired by increasing the number n of nodes; however, as n increases, the number N of trials required to train the network must increase as well.

Typically, one does not explicitly know the target function. Moreover, though the actual network function is in principle knowable, a detailed knowledge of it becomes more and more expensive as the number of nodes increases. What one can reasonably demand is an estimate of this function, subject to a certain estimation error ϵ_{estim}. Thus, a reliable programming methodology must have access to a way to assess both the approximation error and the estimation error.

Such assessments have been made by Barron (1991) for networks with a sigmoidal transfer function (Fig. 3), under specific smoothness assumptions for f, yielding

$$\begin{aligned}\epsilon_{\text{approx}} &\approx 1/n,\\ \epsilon_{\text{estim}} &\approx (nd/N)\log N\end{aligned} \tag{15}$$

(where d is the number of arguments of f). The approximation error is inversely proportional to the number n of nodes; however, for the two errors to remain comparable, one needs

$$N/\log N \approx n^2,$$

i.e., the number N of trials must increase somewhat faster than the square of the number of nodes. This sets a practical limit on the number of nodes.

The above bounds provide qualitative indications of the trade-offs that one may expect in more general situations.

6. TASKS

Below, we briefly review typical tasks for neural networks. The same kind of abstract task (e.g., optimization) can be found in different applications (e.g., vision, control, scheduling). Vice versa, many applications entail, at different stages, different kinds of tasks; for example, automated vehicle driving may require first recognition of the landscape and then appropriate control responses.

6.1 Recognition, Classification, Generalization

Recognition and, more generally, classification are among the tasks that are most commonly pressed upon neural networks. Usually, the classes one is interested in are not predefined by formal criteria, but happen to be distinguished from one another by the values of a large number of variables—even though some overlap may occur in some of the variables. (For example, an overwhelming number of factors may lead one to classify a kiwi as a "bird," even though the lack of wings, by itself, would tend to disqualify it.)

In this context, writing an explicit computer program to do the classification would be an immense and at the same time ill-defined task. Too much cleverness in devising compact decision tables might lead to untenable criteria—classification dead ends—as new data accumulate. In a sense, specifying a neural network (such as a perceptron or a Hopfield model) for this task is a way of giving an implicit definition of the classes themselves, based on the training data's statistics and "internal evidence" rather then externally imposed criteria.

When the basins of attraction (Fig. 4) are large, so that they encompass patterns that must be very different from the "type" attractor, classification turns into generalization: is an airplane more like a car or like a bird? Again, a specific neural network in this context must be viewed as a creative way of *proposing* generalization criteria where none are extant, rather than *implementing* given ones. In special situations, this generalization mechanism may lead to such consistent, objective results that it qualifies as an inference tool (cf. Sec. 6.4).

6.2 Discovery

Most unsupervised learning is aimed at identifying information-rich descriptors for a collection of objects or a distribution of samples. For example, in police files, "hair color" conveys much information in France but little in Iceland or China. The prototypical task in this area is *principal-component analysis*: in a distribution over a vector space of dimension N, is there a subspace of dimension $M < N$ that accounts for a large fraction of the data variance?

A simple approach to this problem, involving an M-node, one-layer, linear feedforward network, was proposed by Sanger (1989). The node's response is

$$y_i = \sum_j w_{ij} x_j, \tag{16}$$

and the incremental learning rule is

$$\Delta w_{ij} = \eta y_i \left(x_j - \sum_{k=1}^{i} y_k w_{kj} \right) \tag{17}$$

(where η is a learning rate parameter). By iterating this rule, the M vectors that make up the $M \times M$ matrix w_{ij} converge to a set of orthonormal vectors that represent the first M principal components of the distribution.

The principal application of this kind of analysis is adaptive data compression.

6.3 Optimization

In classification, one seeks a pattern (from a collections of "types") from which the given pattern has minimal distance. In optimization, one seeks the (unknown) pattern that minimizes the value of an assigned *cost function*. Optimization has long been the chief concern of operations research (*q.v.*), and has spawned a prodigious amount of literature.

In this field, one can expect novel contributions (whether conceptual or practical) from neural networks only in special circumstances: a cost function that depends on a large number of variables, modest coupling between variables, willingness to accept as a solution a "reasonably low" value of the cost function (rather than an absolute minimum), or stress on practical implementability and real-time performance.

The benchmark for combinatorial optimization methods is the traveling salesman problem, which provides nontrivial constraints. Neural-network approaches to this problem have been proposed by Hopfield and Tank (1986) and many others. The emphasis is, for the moment, conceptual, since the results are not generally competitive with more conventional algorithms. Useful results based on optimization algorithms are routinely achieved in more down-to-earth tasks, such as are found in early visual processing (cf. Sec. 4.6).

6.4 Inference

What comes after 1, 2, 3, 4? After 2, 4, 6, 8? After 1, 4, 9, 16? For each of these sequences, though many continuations are defensible, one stands out so compellingly that we view it as the "right" answer. Can one make neural networks reason in this fashion, and infer from a limited number of examples the entire set or sequence from which the examples "must" have been drawn?

A good paradigm for this kind of task is provided by an experiment discussed by Carnevali and Patarnello (1987), where the objective is to teach a network to perform binary addition by showing to it a number of examples. For definiteness, consider a box with two eight-bit inputs, x_1 and x_2, and one eight-bit output, y. The box contains a few hundred NAND gates (here is an example of a "neural network" consisting of Boolean gates rather than McCulloch–Pitts neurons), which are to be connected as a feedforward network (Sec. 4.2) from the inputs x_1, x_2 to the output y. For any given interconnection pattern, the box will compute a certain function f of the arguments x_1, x_2. The domain of f consists of $K = 2^8 \times 2^8$ points, and f itself can be identified with a lookup table of K eight-bit entries.

Our goal is to arrive at an interconnection pattern for which the box will compute the function

$$y = \bar{f}(x_1,x_2) \equiv x_1 + x_2 \,(\text{mod } 2^8). \tag{18}$$

[For simplicity, we ask that addition in Eq. (18) be performed modulo 2^8—i.e., ignoring the carry from the most significant digit—so as to have an eight-bit result.]

In the experiment, the network interconnection pattern, initially random, is gradually steered toward the goal by a simulated annealing procedure (Sec. 5.3). However, our only guide in this steering will be provided by the value of f on a small number k of "test points" (to be specific, let $k = \sqrt{K}$), where we strive to make f coincide with $\bar{f}$. That is, as an energy function we shall use the Hamming distance between f and $\bar{f}$, but computed only on the k test entries rather than on all the K entries of the table. The details of the thermal schedule are not essential here. The training will terminate when the network yields the correct results (i.e., $y = x_1 + x_2$) on the k test points.

Once the training is over, almost invariably the network will be found to compute $\bar{f}$ correctly not only on the k test points, but on all the K input points: the network has successfully inferred what function the samples came from!

Is such degree of success an exception or the rule? We defer discussion of this remarkable result to Sec. 7.3.

6.5 Control

Consider a robotic arm consisting of n consecutive segments, as in Fig. 14. The near end of the first segment is hinged on a base plate, while the far end of the last segment carries a tool. The tool's position in space x

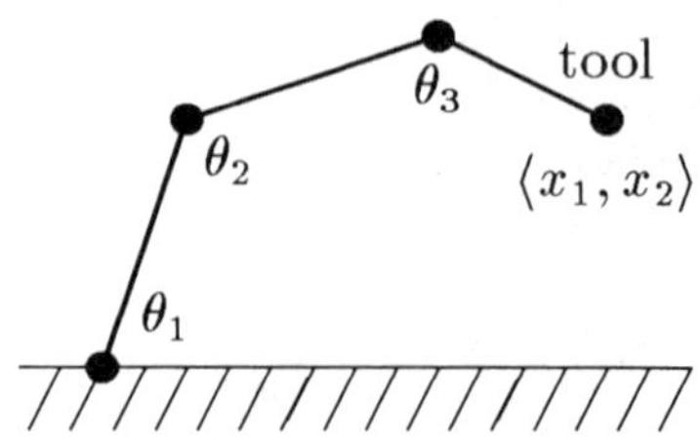

FIG. 14. Robotic arm with three joints, in two dimensions.

$= \langle x_1, \ldots, x_m \rangle$ is a function $x = f(\theta)$ of the set of joint angles $\theta = \langle \theta_1, \ldots, \theta_n \rangle$ (where θ_1 is the angle between the base plate and the first segment, θ_2 that between the first segment and the second, and so forth). The number of joints is usually somewhat larger than the number of degrees of freedom of the tool end, so that the same tool position may be achieved by different settings of the joint angles.

While the driving software has direct control of the joint angles θ, the desideratum is the tool position x. A typical control problem, which we shall briefly discuss here for sake of illustration, is the *inverse kinematics* problem (Craig, 1989); namely, for any desired x, to find a θ for which $x = f(\theta)$.

The function f itself is explicitly given in closed form and is easy to compute (the θ_i are handled one at a time in sequence). However, inverting f (a nonlinear, many-to-one function of many variables) is an awkward task, for which semiempirical methods based on neural networks are well suited.

Learning feedback may be provided by a camera that compares the actual position x' achieved by the tool with the target position x. In the specialized learning approach discussed by Psaltis *et al.* (1988), one makes do without a camera by taking advantage of the fact that f is a known, continuous function of x. In Fig. 15, g represents a feedforward neural network which, given a target position x, strives to output an appropriate value for θ. Learning feedback for g is provided by another feedforward network, f, which encodes the known function $x = f(\theta)$ and does not need any learning itself (its parameters are fixed). If the value of θ produced by g is such that $x = f(\theta)$, then $x' = x$; if not, the error $\epsilon = x' - x$ is back-propagated first through the stages of f, so as to suggest a new value for the intermediate target θ, and then through g itself, so as to suggest new settings for g's synapses.

Kinematics, is, of course, but a first step.

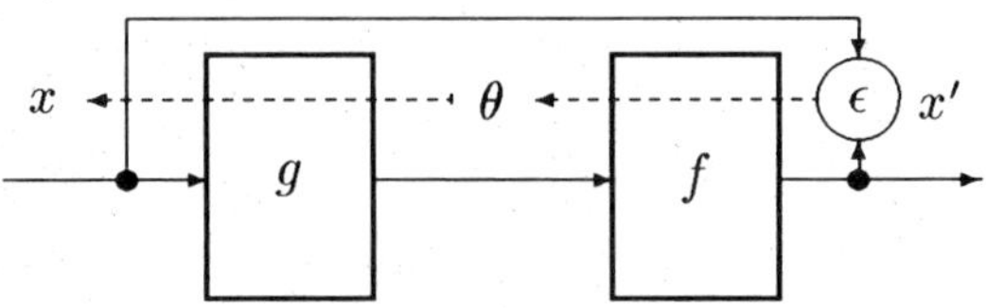

FIG. 15. Feedback loop arrangement in specialized learning (after Kröse *et al.*, 1993).

More demanding control tasks to which neural networks are being applied include dynamics, trajectory planning, and integration of vision with mechanical control.

6.6 Routine Applications

In numerical terms, perhaps the most common application of neural networks is coarse, routine, noncritical classification of objects or events, where training (typically, programming "by example" using some variant of the back-propagation method) can be carried out in a casual and inexpensive way by nonspecialists. For example, which tax returns should be earmarked for a second look by the auditor? The selection criteria are easily updated simply by adding to the training set any returns that seem for some reason remarkable. If the network errs by selecting an unsuspicious return, a few minutes of auditor time will be wasted; if it errs by overlooking a "fishy" return, no liabilities will be incurred.

Inexpensive hardware and software are available that can handle a variety of nondemanding recognition, classification, generalization, and inference tasks. The usefulness of these tools is, of course, highly dependent on the quality of the documentation and the skill of the user.

Besides professionally approached applications, there is room for an enormous variety of applications where some "roll your own" preprocessing by neural networks, no matter how rudimentary and how naively programmed, may have (at least for some of the parties involved) an economical advantage over no preprocessing at all: real-estate searches, mailing lists, insurance application screening, preventive health screening, quality control, inventory control.

7. NEURAL "ALCHEMY"

Many a hardware or architecture feature was at one time or another touted as "the magic element" that somehow would give neural networks properties that transcended those of conventional computers. Here we shall try to separate the real relevance of these features from their mythological aspects.

7.1 Analog versus Digital

Analog circuitry vied seriously with digital circuitry in the early days of scientific computation. Today, analog circuitry is still competitive in certain specialized real-time tasks, such as TV signal processing.

Analog elements are capable of impressive nonlinear feats: for example, a matched transistor pair can be used to compute accurately logarithms or exponentials with a dynamic range of 10^8. However, neural networks are not geared to exploiting such sophisticated "calculus" building blocks; all the nonlinearity they need can be provided by low-accuracy "squashing" functions (Figs. 2 and 3). In neural networks, most of the action is in a linear analog operation, i.e., the weighted sum of n arguments (cf. Sec. 1.2); this can be implemented by a simple passive device, namely, a network of n resistors tied together into a summing junction (Fig. 16). This device only need be followed by an inverting amplifier/limiter to yield a complete neuron (Fig. 16) having computing capabilities equivalent to those of a McCulloch–Pitt neuron. (The adjustable resistors of Fig. 16 only provide positive weights; the effect of a negative weight is achieved by driving an input with an inverted signal. Note that the inverting amplifier already provides an inverted output signal; for a noninverted signal, one would use one more inverting stage.)

Note that, with fixed input resistors of equal value and appropriate threshold bias, this circuit becomes an n-input NOR gate. Indeed, RTL (Resistor Transistor Logic) digital gates are realized in this fashion.

The threshold provided by the single transistor stage in Fig. 16 is not very sharp; indeed, one can think of it as a sigmoid with a maximum slope of approximately ten (cf. Fig. 3). While two or three cascaded limiting stages would give an excellent approximation to a digital output, the single-stage circuit can be used as an analog-input, analog-output neuron.

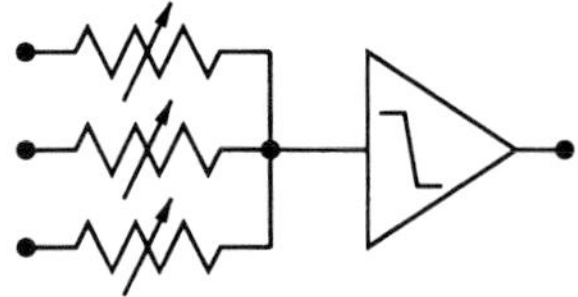

FIG. 16. A neuron consisting of a voltage summing junction (the adjustable resistors set the synapse weights), and an inverting, limiting amplifier.

If, with the same amount of machinery as a digital circuit, a neural network can handle real-valued variables, does it not have, in a sense, "infinitely more" computing power? Can something like this account for the brain's performance?

There is no doubt that, given comparable resources, a verbatim physical implementation of an analog computation may in certain cases outperform by a few orders of magnitude a digital simulation (e.g., by floating-point circuitry) of the same computation. This point has been well argued by Mead (1989). What should be clear, however, is that the advantage—both in terms of capacity and speed—is in the form of a *constant* factor. In fact, because of thermal noise and fabrication tolerances, the nominally continuous range of an analog variable is actually equivalent to a modest number of distinguishable states. Moreover, when one changes one of the inputs in Fig. 16, the new voltage at the summing node is approached exponentially with a time constant τ_{analog}: to have k significant digits, one must wait for a time $\approx k\tau_{\text{analog}}$. If the same input data were encoded as binary strings and processed by a serial digital adder with a clock period τ_{digital}, one would get k digits in a time $\approx k\tau_{\text{digital}}$.

Thus, the analog aspects of neural networks may be good heuristics and may be retained to advantage in a physical implementation, but they do not hold the key computational capabilities that transcend those of digital networks.

7.2 Random Interconnection

Formally, a neuron can directly connect with as many other neurons—and as distant—as desired. Random interconnection is often regarded as a distinguishing trait of neural networks, and one that could give them special computational capabilities. However, even though random interconnection may be a useful metaphor at the conceptual design stage, eventually the material cost of interconnection must be reckoned with.

With no constraints on interconnection, the number of distinct nodes that can be reached from a given node in n steps grows exponentially with n. In a three-dimensional world, if we assume that a "step" represents a fixed unit of cost (such as transmission time or dissipated power), this number can only grow polynomially, i.e., as n^3. The cost of embedding an exponential network in a polynomial space soon becomes a dominant aspect of design: our brain devotes perhaps nine-tenths of its resources to interconnection.

In light of the above considerations, the strict discipline of the cellular-automaton network format (Sec. 4.7) no longer appears a capriciously imposed handicap: the contrast between "random interconnection" and "local and uniform interconnection" is one of heuristic approach, not of computational capabilities.

7.3 Programming by Example versus Hard Coding

A special appeal of neural networks is the new programming style they bring in (cf. Sec. 5.4). By giving a number of concrete examples, one might be able to tell a network *what* to do without having to explain in detail *how* to do it.

To evaluate the reasonableness of these claims in a particular context, one must ask a number of questions. How precisely is the desired task identified by a given set of examples? Does the network have, to begin with, adequate resources to carry out the task it is supposed to learn? Is the training procedure capable of reaching its goal, and of doing so in an acceptable time?

These questions are hard to answer without making unduly restrictive hypotheses. Typical of the state of the art is a thorough analysis by Baum and Haussler (1989). This is concerned with the task of generalizing, from training examples, a pass/fail criterion to be applied to items occurring with a certain probability distribution; its conclusions may be used as an indication of what one can reasonably expect in a wider range of tasks.

Here, we shall paraphrase a study by Carnevali and Patarnello (1987), chosen for its direct intuitive appeal, which throws light on the potential and the intrinsic limitations of programming by example.

With reference to the model of Sec. 6.4, consider the space S of all possible interconnection patterns (which here we shall call *structures*) achievable with n gates; the size N of this space increases exponentially with n. For a given n, certain functions will be realized by many different structures; others by few; while others may have no realization at all. Let S_f be the set of structures that realize a function f, and N_f its size. Intuitively, if we are wandering at random through S, the chances of finding a structure that implements f are proportional to N_f/N. Indeed, we can use $\Delta S = \log(N/N_f)$—the amount of information that must be supplied by the training procedure to home in on the subset S_f—as a measure of the complexity of the problem f. Remarkably, numerical evidence suggests that ΔS is to a good approximation independent of n.

Given an appropriate energy function E for f, simulated annealing will lead one toward structures of lower and lower energy, i.e., structures that compute f to greater and greater accuracy. Indeed, if S_f is not empty, one will eventually reach a structure for which $E = 0$: the network will have "learned" f, essentially by rote memorization.

To attempt inference, we provide an energy function E' that only reflects the network's error on a fraction k/K of f's domain. Annealing will then home in on a superset S'_f of S_f, of size N'_f. The probability that a structure for which $E' = 0$ will also be one for which $E = 0$ is N_f/N'_f. Thus, inference will be successful only on those problems for which the number of solutions increases slowly as we relax the specifications.

In conclusion, both the cost of learning and the probability of successful inference depend on the underlying combinatorics—how many different structures realize a given function—and thus widely vary from problem to problem. Learning a problem by rote has a cost ΔS that reflects the relative difficulty (with respect to all possible problems) of putting together a solution with the given resources; once that price is paid, learning is all but guaranteed. On the other hand, the chances of correct inference depend on the rate of change of a problem's difficulty as its specifications are relaxed; this rate is presumably very hard to estimate *a priori*.

8. COMPLEX-SYSTEM DYNAMICS AND EMERGENT BEHAVIOR

Historically, much neural network research has been conducted with ostensibly utilitarian goals; a neural network was supposed to be "for something." Complex-system dynamics may be viewed as the study of neural networks for their own sake. What are the modes by which complex macroscopic behavior may emerge from the collective action of microscopic components? What kinds of behavior are more likely to arise in this way?

8.1 Collective Behavior

Here we shall use the term *assembly* for any system consisting of a large number of simple, identical nodes interconnected in a simple, regular way. In other words, an assembly is a network whose structural "texture" is the same—or at least has the same statistics—everywhere. Macroscopic properties of an assembly are, by definition, those obtained by coarse-grain averaging—i.e., by counting the occurrences of individual node states over volume elements that encompass a large number of nodes but are still much smaller than the entire assembly. Finally, by an *architecture* we shall mean the set of all assemblies that have a given basic structure but differ in the values of certain adjustable parameters. For instance, the set of all two-dimensional cellular automata having two states per cell and first-neighbor interconnection is an architecture whose variable "parameter" is the cell rule itself (see Sec. 4.7).

In Sec. 7.3, we concluded that a predisposition for a network to learn certain tasks must be balanced—in what is in effect a constant-sum game—by an aversion to learning other tasks. Analogous trade-offs must be expected in emergent phenomena. Only if the behavior we desire belongs to the "natural" macroscopic repertoire of a given architecture do we have a good chance to elicit it by simply adjusting a few parameters. (Otherwise, to obtain that behavior we will have to program the network explicitly at a microscopic level—which, of course, is what is done in traditional digital-circuit design.)

In this regard, cellular automata provide instructive and well-understood examples (cf. Fig. 17). It must be noted that, for a given cellular automata architecture, almost all rules (see Glossary) lead to trivial macroscopic behavior: i.e., from most initial states the system immediately attains equilibrium in the form of a maximally random state. For nontrivial behavior to emerge, the rule must obey constraints that commute with coarse-grain averaging. Typical constraints of this kind are conservation laws and symmetries. For example, if a microscopic rule is translation-invariant, so must be all macroscopic properties.

For two-state cellular automata, let us consider the small subset of rules that conserve the number of 1's (thought of as "particles" in a "vacuum" of 0's). Particle conservation clearly withstands the ravages of coarse-grained averaging. It turns out that from almost all these rules there emerges a characteristic macroscopic behavior; namely, the evolution of particle density obeys the diffusion equation [Fig. 17(a)]. Further, within this subset of particle-conserving rules, let us

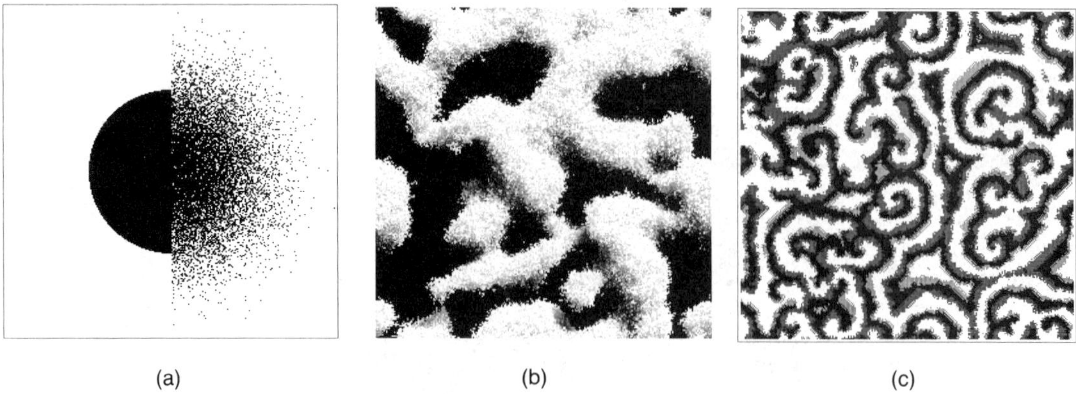

FIG. 17. Samples of emergent behavior in cellular automata. (a) Diffusion; this is a composite figure, the left half from the initial state (a solid disc), the right half from a later state. (b) Phase separation, in three dimensions, as a consequence of a "voting majority" rule. (c) Traveling waves in an excitable medium.

look at the small fraction that depart from diffusive behavior. These are rules that obey additional constraints. For example, those that conserve momentum (the x-component of momentum is the number of particles traveling in the positive x direction minus that traveling in the opposite direction) macroscopically lead to the Navier–Stokes equation, i.e., hydrodynamic behavior (Frisch *et al.*, 1986)—except, of course, for a small subset that obey further constraints; and so forth.

In Sec. 7.3, the structures that computed binary addition correctly 10% of the time were very few (with respect to the set of all possible structures), but, of these, almost all did it correctly a full 100% of the time: intuitively, on this task it takes some rare perversity to manage to go part of the way without going all the way. Analogously, in cellular automata one has to look hard for rules that conserve particles; but when one is found, it is a rare conspiracy that will make the rule not yield the diffusion equation.

In brief, while certain traits of the rule are essential to obtaining a desired type of macroscopic behavior, others are irrelevant: a whole class of microscopically distinct rules are, for macroscopic purposes, equivalent. The chances of finding a rule of this class are proportional to the size of the class relative to the larger volume of rule space through which one is conducting the search. Fixing at least some of the essential traits drastically decreases this volume and thus increases the chances of success.

8.2 Neural-Network Programming and Statistical Mechanics

In physics, thermodynamical states are equivalence classes (established according to certain natural criteria) of microscopic states. The dynamical aspects of thermodynamics are but the expression of counting arguments: basically, a system will progress from state A to state B if

1. the size of class B is much larger than that of class A, and
2. there are no constraints to prevent the microscopic state from reaching class B from class A.

Similar considerations apply to the programming of neural networks. The role of microscopic states is played by microscopic (i.e., completely specified) rules (such as the lookup table for a cellular automaton, the list of synaptic weight for a network of neurons, or the circuit diagram for a network of Boolean gates); the role of thermodynamical state is played by the network's macroscopic behavior. During training, a network will progress from behavior A to behavior B under conditions analogous to **1** and **2** above.

Thus, much of the enormous wealth of concepts and techniques of statistical mechanics is directly applicable to neural networks, even though the latter may have nothing to do with physics. Though ordinarily found only in physics textbooks, entropy, energy, and temperature are, at the bottom, combinatorial concepts that belong to everybody. If it was physicists that happened to discover, name, and use them, it is because their discipline confronted them with massive combinatorics problems before anybody else.

9. CONCLUSIONS AND PERSPECTIVES

Much of the expertise developed in dealing with neural networks as tools actually has to do with a better understanding of the tasks to which they were addressed. Much as statistical mechanics rapidly outgrew its thermodynamical origins, this understanding transcends the original breeding ground of "network of neurons" and is transferable to a wider range of problems and tools. Specifically, neural networks retain their relevance as a high-level programming environment even when the target hardware is a conventional digital computer.

Neural networks are not an improved "brand" of hardware or programming style, to be used—on the same tasks—as a drop-in replacement for more traditional brands. They address different tasks, and thus are meant to play a complementary rather than a competitive role. However, the range of tasks for which neural networks are best suited is still poorly known.

The "market share" of neural networks is likely to be very responsive to the nature and the availability of different kinds of "raw computational materials." Just to illustrate a possible scenario, nanoscale device physics may soon offer "quantum dot" arrays (Reed, 1993)—near–atomic-scale devices fabricatable in astronomical numbers but having very primitive computing capabilities and connected only on a local scale. Initially, quantum dots are expected to operate in a quasi-analog regime, and to be affected by substantial noise; extracting useful computation from such devices would be a challenge well suited for neural networks. On the other hand, as soon as quantum dots attained a digital, low-noise regime, the door would be open for competition from more traditional, gate-level design approaches.

GLOSSARY

Additive: In the present context, synonymous with **Linear**.

Almost All: Strictly, this means "all, except a subset of measure zero." More loosely, when the set s under consideration is seen as a generic element of a sequence $\{s_i\}$ of sets (e.g., increasingly large neural networks of a given kind), it is convenient to say that a property is true for "almost all" elements of s if the fraction of s_i on which the property is true goes to 1 as $i \rightarrow \infty$.

Analog: An analog signal is one designed to convey a choice out of a continuous range (e.g., any voltage between 0 and 5 V). The actual number of distinguishable choices is of course limited by noise and other tolerances. See **Digital**.

Annealing: A procedure by which a system is subjected to a randomizing procedure (e.g., shaking, heating) of controlled intensity in order to remove internal stresses and leave the system in a more stable state.

Attractor: In a dynamical system, a point or a cycle that is asymptotically stable: if the system is on an attractor, after a small perturbation it will tend to return toward it. See **Basin of Attraction**.

Autonomous: A system is autonomous if its evolution is not dependent on the state of variables external to the system itself. Intuitively, "isolated" or "self-contained."

Basin of Attraction: The set of all initial states whose trajectories converge toward a given attractor. See **Attractor**.

Computation Universal: A set of circuit elements is computation universal if, for any computable function, out of these elements one can synthesize a computing network capable of computing that function. The concept of "computable function" is independently defined and reasonably well agreed upon.

Deterministic: A system is deterministic if its state at any future time is uniquely determined by its current state. See **Stochastic**.

Digital: A digital signal is one designed to convey one of two choices, or, more generally, one of a finite, fixed (and typically small) number of choices. See **Analog**.

Dissipation, Dissipative: A mechanism consisting of macroscopic parts that interact according to specified laws is, on a finer scale, made up of an astronomical number of microscopic parts. Perfect decoupling between the nominal, macroscopic "modes of operation" and the microscopic modes is in practice impossible to achieve. Energy injected into the macroscopic modes eventually redistributes itself as thoroughly as possible through the system, without regard for nominal boundaries between macroscopic and macroscopic modes; thus, there is a tendency for energy to flow from macroscopic to microscopic modes until equilibrium is attained. This tendency is called dissipation. An analogous situation arises in any system consisting of a large number of weakly coupled parts. See **Energy Function**.

Emergent: A quantity, property, or behavior that, though not defined in terms of microscopic dynamics, is well characterized at the level of macroscopic statistics. For example, "pressure"—the average collision rate—is meaningless for an individual molecule but well defined for a gas near equilibrium.

Energy: A physical quantity whose main property is that it is conserved in a wide range of circumstances even in systems that are not perfectly decoupled from the rest of the world. This property greatly constrains—and in certain contexts completely determines—the range of admissible dynamical or statistical behavior. See **Entropy, Temperature, Energy Function**.

Energy Function: The amount of energy stored in the mechanical modes of operation, as a function of the system's mechanical state (see **Dissipation**). Knowledge of the energy function allows one to predict the path that the mechanical state will follow on its way to equilibrium (cf. **Energy**).

Entropy: Entropy quantifies by how much a partial description of a system's state fails to be a complete description; basically, it counts (on a logarithmic scale) how many possibilities are consistent with the given description. In a physical system, this number of possibilities depends, among other things, on the ways one can apportion energy, momentum, material particles, etc. between different parts of the system (see **Temperature, Energy**), and leads to important statistical predictions derivable from purely combinatorial arguments. Similar arguments can be used to study the statistics of more abstract systems.

Exponential: A function $f(x)$ is exponential if it grows approximately as a^x, for some fixed a; it is polynomial if it grows approximately as x^b, for some fixed b. For large enough x, an exponential function grows faster than any polynomial function.

Feedback: Feedback occurs whenever the current state of a system variable depends to some extent on the state of the same variable at a previous time. A feedback loop is a path, in the system's structure, through which this dependence make itself felt.

Hamming Distance: Given two sets of binary digits in a one-to-one correspondence (e.g., two strings of the same length, two arrays of the same format), their Hamming distance is simply the number of positions in which they do not agree.

Hidden Unit: A neural-network node not used as an input or an output by the algorithm under consideration. See **Visible Unit, Slack Variable**.

Inversion: Inverting a function f means finding all values x such that $f(x) = y$.

Linear: A system is linear if

1. for any two states a, b the "sum" (in some specified sense) $a + b$ of these states is also a state, and
2. the evolution law f is such that $f(a + b) = f(a) + f(b)$.

Monotonic: A function $y = f(x)$ is monotonic if the increment Δy corresponding to a given increment Δx is always of the same sign (either ≥ 0 or ≤ 0) throughout the range of x.

Network: In the present context, the set of interconnections or pathways through which system variables influence one another.

Polynomial: See **Exponential**.

Reversibility: A system is reversible if its state at any previous time is uniquely determined by its current state. See **Deterministic**.

Slack Variable: Strictly, a variable such that the given problem has solutions for a whole range of values of the variable itself. More loosely, a variable added in the hope that, with this extra degree of freedom, solutions will be easier to find. Adding **hidden units** to a network increases the number of ways a solution compatible with the given inputs can be constructed.

Stochastic: If the evolution in time is not uniquely determined by the current state (see **Deterministic**), but nonetheless definite probabilities can be assigned to the different possibilities, the system is stochastic. Intuitively, "nondeterministic but statistically well-characterized."

Symmetry: A system is symmetric with respect to a given group of transformations if its behavior is unchanged under those transformations. For example, addition is symmetric with respect to permutations of its input terms, while subtraction is not.

Temperature: Measures by how much the amount of disorder in a system can be increased as a new amount of energy becomes available to be "reshuffled" between its parts (see **Energy, Entropy**). Analogs of temperature arise whenever a fixed amount of something has to be distributed between many parts.

Transfer Function: Intuitively, the analog equivalent of a conversion table such as a tax table. The level, range, or scale of a variable x constructed within part of a system may not be suitable for input to another part; a transfer function $y = f(x)$ performs the appropriate conditioning (e.g., level-shifting, range-compression, shaping).

Visible Unit: A neural-network node used as an input or an output by the algorithm under consideration. See **Hidden Unit**.

Works Cited

Baierlein, R. (1971), *Atoms and Information Theory—An Introduction to Statistical Mechanics*, San Francisco: W. H. Freeman.

Barron, A. R. (1991), "Approximation and Estimation Bounds for Artificial Neural Networks," in: L. C. Valiant, M. K. Warmuth (Eds.), *Proceedings of 4th Annual Workshop on Computation Learning Theory (COLT '91)*, San Mateo, CA: Morgan-Kaufman.

Baum, Eric B., Haussler, David (1989), "What Size Net Gives a Valid Generalization?" *Neural Comput.* **1**, 151–160.

Burks, A. (Ed.) (1970), *Essays on Cellular Automata*, Urbana, IL: University of Illinois Press.

Cajaniello, E. R. (1961), "Outline of a Theory of Thought and Thinking Machines," *J. Theor. Biol.* **1**, 204–235.

Carnevali, P., Patarnello, S. (1987), "Exhaustive Thermodynamic Analysis of Boolean Learning Networks," *Europhys. Lett.* **4**, 1199–1204.

Craig, J. J. (1989), *Introduction to Robotics*, Reading, MA: Addison-Wesley.

Frisch, U., Hasslacher, B., Pomeau, Y. (1986), "Lattice-Gas Automata for the Navier-Stokes Equation," *Phys. Rev. Lett.* **56**, 1505–1508.

Fukushima, K. (1975), "Cognitron: A Self-Organizing Multilayered Neural Network," *Biol. Cybernetics* **20**, 121–136.

Hebb, D. O. (1949), *The Organization of Behavior*, New York: Wiley.

Hertz, J., Krogh, A., Palmer, R. G. (1991), *Introduction to the Theory of Neural Computation*, Redwood City, CA: Addison-Wesley.

Hillis, D. (1985), *The Connection Machine*, Cambridge, MA: MIT Press.

Hinton, G. E., Sejnowski, T. J. (1986), "Learning and Relearning in Boltzmann Machines," in: D. E. Rumelhart, J. L. McClelland (Eds.), *Parallel Distributed Processing: Explorations in the Microstructure of Cognition*, Vol. 1, Cambridge, MA: MIT Press, Chap. 7.

Holland, J. (1975), *Adaptation in Natural and Artificial Systems*, Ann Arbor, MI: University of Michigan Press (republished by MIT Press, 1992).

Hopfield, J. J. (1982), "Neural Networks and Physical Systems with Emergent Collective Computational Abilities," *Proc. Nat. Acad. Sci. USA* **79**, 2554–2558.

Hopfield, J. J., Tank, D. W. (1986), "Computing with Neural Circuits: a Model," *Science* **233**, 625–633.

Katz, A. (1967), *Principles of Statistical Mechanics*, San Francisco: W. H. Freeman.

Kirkpatrick, S., Gelatt, C. D., Vecchi, M. P. (1983), "Optimization by Simulated Annealing," *Science* **220**, 671–680.

Koch, C., Marroquin, J., Yuille, A. (1986), "Analog 'Neuronal' Networks in Early Vision," *Proc. Nat. Acad. Sci. USA* **83**, 4263–4267.

Kröse, B. J. A., van der Smagt P. P. (1993), *An Introduction to Neural Networks*, Amsterdam: The University of Amsterdam.

Marr, D. (1969), "A Theory of Cerebellar Cortex," *J. Physiol.* **202**, 437–470.

McCulloch, W. S., Pitts, W. (1943), "A Logical Calculus of Ideas Immanent in Nervous Activity," *Bull. Math. Biophys.* **5**, 115–133.

Mead, C. (1989), *Analog VLSI and Neural Systems*, Reading, MA: Addison-Wesley.

Minsky, M. (1967), *Computation: Finite and Infinite Machines*, Englewood Cliffs, NJ: Prentice-Hall.

Minsky, M. L., Papert, S. A. (1969), *Perceptrons—An Introduction to Computational Geometry*, Cambridge, MA: MIT Press (republished with additions in 1988).

Psaltis, D., Sideris, A., Yamamura, A. A. (1988), "A Multilayer Neural Network Controller," *IEEE Control Systems* **8** (4), 17–21.

Reed, M. (1993), "Quantum Dots," *Scientific American* **268** (1), 118–123.

Rosenblatt, F. (1962), *Principles of Neurodynamics*, Portage, MI: Spartan.

Rumelhart, D. E., Hinton, G. E., Williams, R. J. (1986), "Learning Representations by Back-Propagating Errors," *Nature* **323**, 533–536.

Sanger, T. D. (1989), "Optimal Unsupervised Learning in a Single-Layer Linear Feedforward Neural Network," *Neural Networks* **2**, 459–473.

Toffoli, T., Margolus, N. (1987), *Cellular Automata Machines—A New Environment for Modeling*, Cambridge, MA: MIT Press.

Weisbuch, G. (1991), *Complex Systems Dynamics*, Redwood City, CA: Addison-Wesley.

Wolfram, S. (ed.) (1986), *Theory and Applications of Cellular Automata*, Singapore: World Scientific.

Further Reading

The book by Hertz *et al.* (1991) is one of the best all-around introductions to the theory of neural networks. More descriptive works are Kröse and van der Smagt (1993), for neural networks, and Weisbuch (1991), for complex systems. Articles on neural networks are regularly to be found in a number of journals, including *Complex Systems, Neural Computation, Physical Review Letters, Physical Review A, Neural Networks, Europhysics Letters, Journal of Physics, Applied Optics, Cognitive Science, Biological Cybernetics*, and *Neural Networks*, and occasionally in journals such as *Science* and *Nature*. The bimonthly *Neural Computation* deals with the physiological aspects of neural computation. Another rich source of articles is the proceedings of several IEEE conferences on neural networks—specifically, *Advances in Neural Information Processing Systems*, published annually (since 1988) by Morgan-Kaufman.

Katz (1967) is a sophisticated introduction to statistical mechanics from the information-theory viewpoint; Baierlein (1971), a more leisurely one. A direct road from information theory to inference theory—with thermodynamics as a corollary rather than as a starting point—has been traced by E. T. Jaynes in a series of papers. We recommend "Where Do We Stand on Maximum Entropy?" reprinted in his collection, Jaynes, E. T. (1983), *Papers on Probability, Statistics, and Statistical Mechanics*, Dordrecht: Reidel, pp. 210–315, as well as "How Does the Brain Do Plausible Reasoning?" in: G. J. Erickson and C. R. Smith (Eds.) (1988), *Maximum-Entropy and Bayesian Methods in Science and Engineering*, Vol. 1, Boston: Kluwer Academic, pp. 1–24.

NEUROBIOPHYSICS

ANDREW A. MARINO,* *Department of Orthopaedic Surgery and Department of Cellular Biology and Anatomy, Louisiana State University Medical Center, Shreveport, Louisiana, U.S.A.*

ILKO G. ILIEV, *Department of Orthopaedic Surgery, Louisiana State University Medical Center, Shreveport, Louisiana, U.S.A.*

INTRODUCTION

Neurobiophysics is the application of basic physical principles to the operation of the nervous system. The methods of neurobiophysics are identical to those of other branches of quantitative science:

1. observations of phenomena under controlled conditions and subsequent replication of the phenomena;
2. elaboration of models to which physical laws can be applied and unambiguously shown to explain pertinent quantitative aspects of the observations.

A basic assumption in neurobiophysics is that all neuronal activity is susceptible of an explanation based on the application of known physical laws. In this view, the morphological complexity of the neuron and the structural complexity of neuronal interconnections are practical barriers to an understanding of the nervous system, but it is not expected that as-yet-undiscovered laws will be needed to explain nervous-system activity. Because electricity is the currency of the nervous system, most models of neural activity are electrical in nature.

In 1902, Julius Bernstein hypothesized that cells were ionic solutions surrounded by thin membranes having permeability properties that resulted in the establishment of an electrical potential across the membrane. Further, during nervous-system activity, the permeability of the membrane changed in such a way as to lower the membrane potential. These ideas were subsequently developed by many investigators, culminating in the fluid-mosaic model of the membrane (Singer and Nicolson, 1972) and in the work of Alan Hodgkin and Andrew Huxley on the biophysical basis of the time-dependent permeability changes in nerve axons (Hodgkin and Huxley, 1952). The quantitative description of synaptic transmission at the neuromuscular junction, and Wilfrid Rall's development of methods for describing the flow of electrical activity in nerve dendrites (Rall, 1960), were further important contributions to the present-day theoretical framework of neurobiophysics. Beginning around 1975, the experimental technique *patch-clamping* was developed; it permitted direct observation of the kinetics of the *ion channel*, the membrane-level effector in the nervous system.

*Name and address for all communications.

1-56081-070-X/94/$5.00 + .50

Much activity is presently devoted to characterizing the broad range of ionic channels present in the nervous system, finding both new channels and previously unrecognized functions of known channels, and establishing the mechanisms by which ion channels are regulated and synchronized to produce the specific phenomena observed in the nervous system.

1. STRUCTURE AND FUNCTION OF THE NERVOUS SYSTEM

The complexity of the nervous system varies with the degree of development of the organism; in the human being, it is a structure composed of more than 10^{10} cells that carries out sensory and regulatory functions, facilitates a variety of behaviors, and subserves what we recognize as memory, emotion, will, and intellect. The overall organization of the vertebrate nervous system is depicted in Fig. 1.

The *neuron* is the electrically active cell of the nervous system. There are many types of inputs to neurons, including deformation, light, and temperature, but probably the most important neuronal input signal consists of a flow of ions that enters the cell via a *synapse*, a specialized junction with a neighboring neuron. Two general types of synapses are recognized, depending on the source of the ions (Fig. 2). In an electrical synapse, the presynaptic and postsynaptic cells are linked by conducting channels that permit ionic flow between them. Electrical synapses (also called *gap junctions*) are rare in mammals, but are found between neurons in lower vertebrates and in invertebrates. The chemical synapse is a more complicated junction, and is the characteristic linkage between neurons in the mammalian nervous system. In chemical synapses, a 1–2-μm diameter region of the plasma membranes of the cells is separated by a 20–30-nm gap that must be traversed by a *neurotransmitter*, a chemical agent synthesized and secreted by the presynaptic cell, to effect a communication between the cells. The neurotransmitter diffuses across the gap and binds to receptors on the postsynaptic cell, resulting in the opening or closing of membrane channels and thereby altering ion flow between the interstitial fluid and the neuron. In some cases, the neurotransmitter receptor and the pore through which the ions pass are each part of a unitary transmembrane protein. In other cases, however, the receptor and the pore are distinct proteins, and the events at each site are coupled by an intracellular second messenger (the neurotransmitter is the first messenger). Since direct gating of ion channels involves only a change in the conformation of a single macromolecule, it can occur on the order of milliseconds. In contrast, channels activated by second messengers are slow (seconds to minutes) because they involve a series of sequential reactions. In both cases, the net result of the ion flow is to produce a change in the membrane potential of the neuron in the vicinity of the point of entry of the ions into the cell. If Cl^- enters the cell, the resting membrane potential (typically about −70 mV) usually becomes more negative, resulting in a hyperpolarization; entry of cations—Na^+, for example—produces a depolarization. For reasons that will be discussed, hyperpolarization of the neuronal membrane inhibits

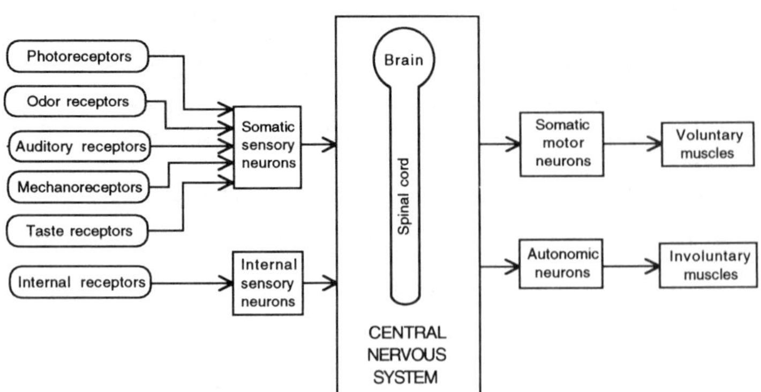

FIG. 1 General organization of the vertebrate nervous system. The input to the central nervous system (CNS) consists of propagating electrical signals called action potentials that convey information to the brain from the sensory organs or from receptors located in internal tissues. The CNS output consists of action potentials that propagate along the motor neurons from the brain to the muscles. Additionally, the CNS communicates with the endocrinological and immune systems via chemical signals. (Adapted from Alberts *et al.*, 1989.)

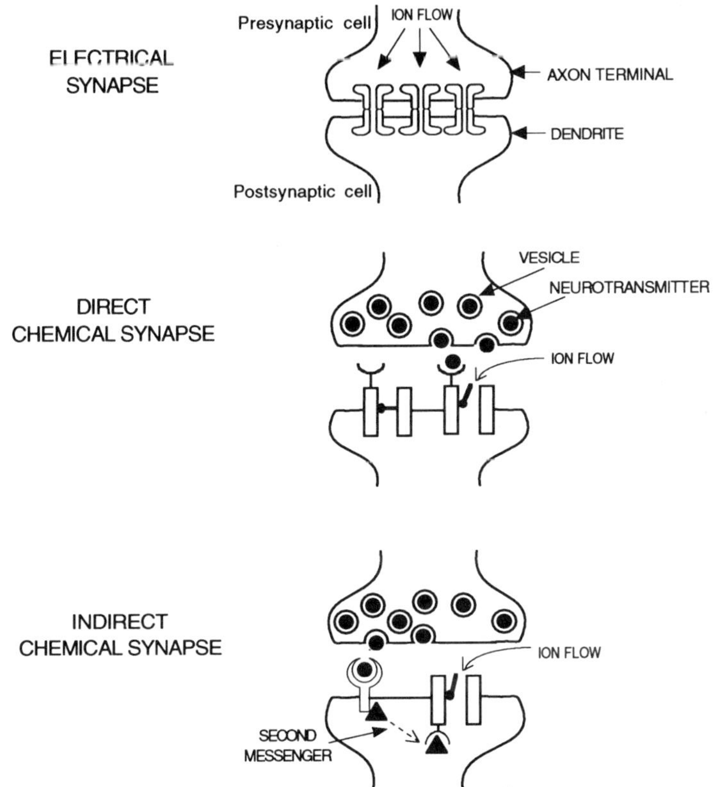

FIG. 2. Types of synapses in the nervous system. A synapse is a functional junction between two neurons, and it is their defining characteristic. The presynaptic and postsynaptic neurons are defined by the direction of information transfer between them. The signal to a postsynaptic neuron consists of a flow of ions that enters the cell in the region of the dendrites or cell body. In the chemical synapse, the cells are separated by a narrow gap across which a neurotransmitter diffuses; it binds to receptors on the postsynaptic cell, thereby triggering the opening of otherwise closed membrane channels that permit the flow of ions into the postsynaptic cell from the interstitial fluid. The receptor, membrane pore, and gate may consist of a unitary protein structure (*direct chemical synapse*), or the receptor and pore portions of the channel may be physically separated, linked by various intermediary substances known as second (or higher-order) messengers (*indirect chemical synapse*). The conductance of individual channels at electrical synapses is relatively large (100–200 pS) compared with those at chemical synapses. Typically, directly gated channels mediate neuronal activity, whereas indirectly gated channels modulate the excitability of neurons.

neuronal activity, whereas depolarization produces the opposite effect.

The signal transmitted by the neuron has several different forms, depending on the morphological region of the cell (Fig. 3). A typical neuron simultaneously receives numerous excitatory and inhibitory input signals at synapses on the dendrites or cell body, but it responds only to the instantaneous sum of the individual changes in membrane potential. The components of the summation signal propagate passively from their point of initiation to the cell's trigger zone, usually the axon hillock; consequently, the contribution of a particular input signal to the summation signal is inversely proportional to the distance between the location of the input and the axon hillock. If the net change in the membrane potential at the axon hillock induced by the summed inputs is a depolarization that exceeds a threshold value, an *action potential* is generated, which propagates along the cell axon.

The action potential does not undergo amplitude diminution such as occurs during propagation of the postsynaptic potential to the axon hillock. Action potentials are all-or-none phenomena in the sense that their occurrence depends upon whether the cell threshold is exceeded. The electrical characteristics of individual propagating action potentials are determined by the physical properties of the neuronal axon, not by the characteristics of the stimulus; thus, within particular neurons, the action potentials are essentially identical in amplitude, pulse width, and propagation velocity. Quantitative information regarding the stimulus intensity is coded by the repetition rate of the action potential: An increase in the magnitude of the stimulus results in an increase in the repetition rate. Some axons are wrapped with an

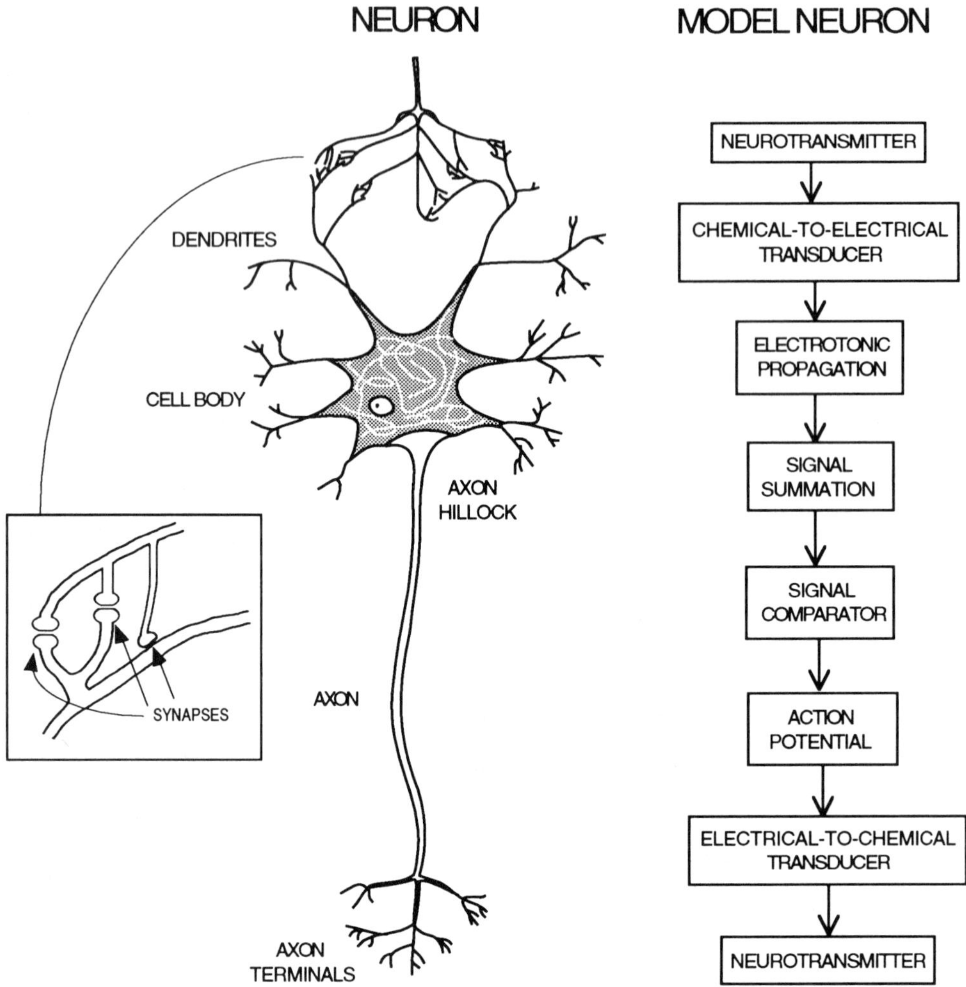

FIG. 3. Information transfer in the neuron. The initial signal consists of vesicles containing a neurotransmitter (NT) that are received by the neuron at synapses in its dendritic region. A neuron may have many dendrites, but only one axon.

insulating material called *myelin* that is interrupted at intervals by the *nodes of Ranvier*; electrical conduction in such neurons is the result of a combination of the processes underlying propagation of the summation signals and the action potential, and it proceeds more quickly than would be the case in the absence of myelin (Deutsch and Micheli-Tzanakou, 1987). Arrival of the action potential in the terminal portion of the axon triggers a series of biochemical events leading to secretion of the neurotransmitters that function as the input signal for the next neuron. Thus, in a typical mammalian neuron, neurotransmitters constitute the signal conveyed to and emitted by the neuron, but intraneuronal transmission is essentially electrical in nature.

The biophysical mechanisms underlying the complementary transduction events involving the interconversion of chemical and electrical signals are not well understood, but adequate descriptions of the phenomena that must be explained are now available. Cable theory provides a good description of the passive propagation of the postsynaptic potentials and the subsequent formation of the summation signal. The physical substrate of the signal comparator (Fig. 3) is the density and functional characteristics of the ion channels located at the axon hillock. The best-understood aspect of neuronal signaling is the formation and propagation of the action potential; the explanation of this phenomenon on the basis of ionic permeability changes was a significant development in neurobiophysics.

2. ION CHANNELS

2.1 Overview

The electrical activity of neurons arises from the flow of ions through transmembrane proteins called channels. There are probably several hundred different channel proteins in the nervous system, but an individual neuron contains only some of them. A gated channel is a functional unit containing a pore through which ions may pass, a gate, and a sensor capable of opening or closing the gate in response to a signal; there are two classes of channels, depending on the nature of the signal to which they are responsive. *Voltage-gated* channels have an ion conductance that depends on the membrane potential, whereas the conductance of *ligand-gated* channels is dependent, directly or indirectly (Fig. 2), upon the binding of a neurotransmitter (the most important class of ligands in the nervous system) to the sensor portion of the channel (which, in the case of chemical signals, is called a *receptor*). Most gated channels exhibit only one form of gating behavior. Nongated channels are essentially membrane pores lacking gates and sensors.

The *selectivity* of a channel refers to the ion species that will pass through the channel pore. Voltage-gated channels are denoted by the ion that passes through most readily; the main types are Na^+, K^+, Ca^{2+}, Cl^-, and all small cations. Ligand-gated channels are labeled by a ligand that is effective in opening the channel. Thus, a Na^+ channel has Na^+ as the main permeant ion, and a nicotinic acetylcholine channel is a transmembrane protein having a receptor capable of binding the neurotransmitter acetylcholine, resulting in the passage of ions (Na^+ and K^+) through the channel pore. The response of a neuron to its environment is mainly determined by the gating and selection characteristics of its ion channels, and by the density and distribution of each channel type in the neuronal membrane. Neurons contain

1. nongated channels that serve to establish the membrane potential;
2. ligand-gated channels that subserve reception of the input signal;
3. voltage-gated Na^+ and K^+ channels that function in a synchronized fashion to permit propagation of an action potential; and
4. voltage-gated Ca^{2+} channels that participate in the transduction of the action potential into the chemical signal that constitutes the neuron's output.

In addition, there is an important class of channels that are sensitive to various intracellular chemical signals (Hall, 1992). Elucidation of the mechanisms underlying channel permeability (a measure of the selectivity of an ion pathway), channel conductance (a measure of the interaction between the ion and the channel structure), and channel gating are the central problems of neurobiophysics.

2.2 Methods of Study

The patch-clamp technique permits measurements of currents from individual ion channels in biological membranes (see BIOPHYSICS). This is achieved using a heat-polished micropipette having an opening of about 0.5–1 μm, and filled with a solution whose composition is compatible with the cell cytoplasm and the particular purposes of the measurement. The micropipette is gently placed against the cell membrane under direct visualization, and application of a slight negative pressure through the pipette results in an intimate contact between the membrane and the micropipette tip (Fig. 4). The electrical resistance of the seal is about 10–100 GΩ, indicating that the membrane is firmly attached to the pipette, thereby ensuring that the measured current actually flows between the cell interior and the micropipette, and is not shunted to the bath solution. Several measurement configurations—relationships between the micropipette and the membrane under measurement—are employed for the study of membrane kinetics. In the *cell-attached* (also called *on-cell*) configuration, the micropipette is sealed onto an intact cell. Several mechanical manipulations can be performed at this stage. The membrane patch may be ruptured by a brief pressure pulse applied through the micropi-

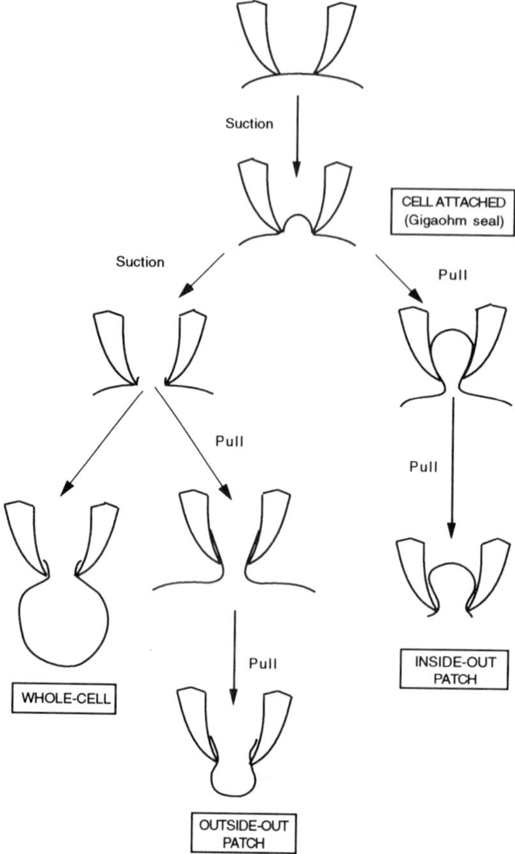

FIG. 4. Procedures leading to the various patch-clamp configurations (adapted from Neher, 1992).

pette, thereby establishing an electrical connection between the interior of the micropipette and the cell cytoplasm; the resulting *whole-cell* configuration is electrically equivalent to a conventional microelectrode penetration. Withdrawal of the pipette from the cell after the gigohm seal has been formed usually leads to one of two other configurations, depending on the mechanics of the process; the seal can be formed in such a way that the cytoplasmic surface of the membrane comes to face the bath solution or the micropipette solution—respectively termed *inside-out* and *outside-out* patches (Fig. 4).

Each of the four patch-clamp configurations has properties that are useful for studying different kinds of problems. Recording from cell-attached patches provides a minimal perturbation of the cell under study, and allows single-channel currents from nearly all known types of ion channels to be resolved. The inside-out and outside-out patches permit high resolution of channel currents and provide the opportunity to control the ion milieu on both sides of the membrane. The whole-cell configuration (and a recently developed modification using the antibiotic nystatin) allows conventional voltage-clamp and current-clamp measurements to be performed on small cells.

In principle, any ion channel can be isolated, purified, chemically modified, and reinserted into artificial lipid bilayers. The result is a well-characterized membrane system in which the biochemical properties likely to be important in regulation and transport can be monitored, controlled, and studied with patch-clamp techniques. In another molecular approach, mRNA encoding a channel protein is injected into the oocyte of the frog *Xenopus*, which then synthesizes the protein and inserts it into the cell membrane, where it becomes accessible for study using patch-clamp techniques. The relation between structure and function can be systematically studied by mutating the mRNA and observing the resulting effect on the encoded protein after its insertion in the membrane. This strategy can be used to identify the portions of the proteins that are directly involved in voltage-dependent and ligand-dependent interactions. Recombinant DNA techniques permit creation of chimeric genes that code for channels made up of subunits from differing species; such preparations permit study of the conservation of the determinants of ion transport.

The various methods described above can be combined to permit

1. analysis of the kinetic pattern of specific channels during normal and pharmacologically induced activity;
2. study of the molecular structure of ion channels;
3. studies of the mechanisms of the regulation of the ion channel activity and identification of the cellular components involved in the regulation; and
4. identification and description of the roles of the channels in cellular processes.

2.3 Gating and Structure

The electrophysiological and molecular paradigms have not yet been fully carried out for any channel, but intensive work is under way, and good progress has been achieved for several channels in the voltage-gated and in the ligand-gated families (Hille, 1992).

The Na^+ channels respond rapidly to depolarization; their basic role in the nervous system is to generate the initial portion of the action potential, and neither their pharmacology nor their structure varies significantly from tissue to tissue. The Na^+ channel consists of three subunits (Fig. 5); the α subunit has four homologous domains, each of which contains 6–8 hydrophobic amino-acid sequences, and an occludable pore selective for the passage of Na^+. It appears from site-directed mutagenesis studies that the locus of the voltage sensitivity is contained in an amino-acid sequence known as S4, one of which is contained in each of the four-domains. When amino acids in S4 are replaced by either neutral or negatively charged residues, the reduction in positive charge decreases the relative sensitivity for activation, which is direct evidence that the positive charge in S4 forms part of the voltage sensor for channel activation.

The K^+ family of voltage-gated transmembrane proteins is the largest and most diverse of the voltage-gated channels. The K^+ channels subserve a variety of functions, including return of the membrane potential to a pre-existing level, formation of trains of action potentials, and the occurrence of rhythmic activity. There are at least three major types of voltage-gated K^+ channels in the nervous system and perhaps ten times that number of K^+-channel subtypes. The main types are

1. the *delayed rectifier*, which is the axonal K^+ channel that opens with depolarization and is largely responsible for repolarizing the axon membrane following an action potential;
2. the K_A^+ channel, an axonal channel that opens rapidly upon depolarization and then quickly closes; and
3. the inward-rectifying K^+ channel, which opens only with hyperpolarization.

Another important class of K^+ channels is sensitive to levels of intracellular Ca^{2+}; the channels appear to be voltage-gated, with the Ca^{2+} acting to shift the voltage dependence (Hall, 1992; Hille, 1992). The K_A^+ channel has been well studied and is believed to be a prototype for the other K^+ channels; there is only about a 10% variation in the total amino-acid sequence among the various K^+ channels. The molecular weight of the protein encoded by K_A^+ DNA is about 25% of that of the Na^+ channel protein, and consists of six transmembrane segments; consequently, by analogy with the known structure of the Na^+ channel, each K^+ channel is thought to be a tetramer (Fig. 5).

Among the more remarkable features of some voltage-gated K^+ channels is the combination of high conductance and high ionic selectivity. The conductance of some K^+ channels appears to exceed the value associated with movement in aqueous solutions.

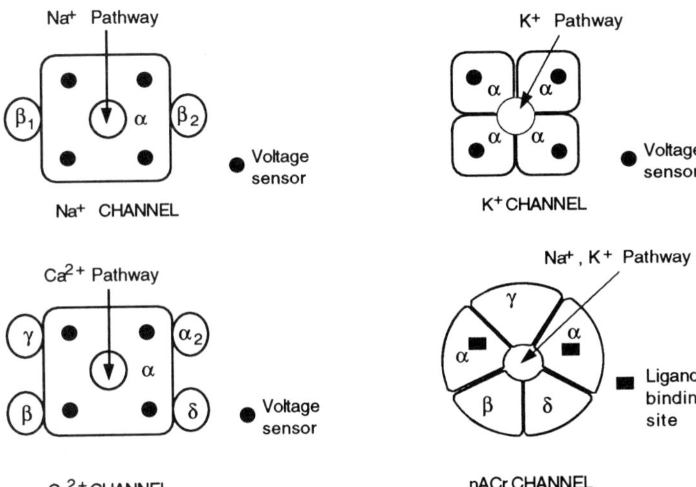

FIG. 5. Subunit composition of typical voltage-gated and ligand-gated ion channels. The channels are depicted as they appear when viewed from above the plasma membrane. (Na^+, K^+, and Ca^{2+} channels adapted from Catterall, 1988; nACr channel adapted from Alberts *et al.*, 1989.)

It is difficult to understand a mechanism that could be faster than the free-diffusion process, and yet could facilitate high ion selectivity. One possibility is that the channel can accommodate more than one ion at a time. If so, the presence of the first ion might reduce the electrostatic forces on the second ion, thereby facilitating its passage through the pore. It is not possible to calculate the magnitude of such an effect because K^+ channel structure is not known in sufficient detail.

Axons from mollusks, arthropods, annelids, and vertebrates have been found to contain essentially identical Na^+ and K^+ channels participating in conduction of the action potential. Thus, the evolution of these channels was essentially complete at the time of the common ancestor of these phyla, approximately 500 million years ago (Hille, 1992).

There is a wide range of voltage-gated Ca^{2+} channels within the nervous system; they all open with depolarization and appear to have a common subunit composition (Fig. 5), but they differ in voltage dependence, ionic selectivity, and pharmacology. Neurotransmitter secretion at nerve terminals is a well-studied Ca^{2+}-dependent process. Voltage-gated Ca^{2+} channels in the presynaptic terminal open following the depolarization produced by the arrival of the action potential, thereby triggering some of the membrane-bound vesicles containing neurotransmitter to fuse with the membrane surface, releasing neurotransmitter into the synapse. The Ca^{2+} channels in the axon terminal provide the only mechanism whereby the action potential can be transduced into a form capable of carrying information across a chemical synapse.

What mechanism might explain voltage-dependent activation of ion channels? A voltage sensor is the part of the transmembrane protein that undergoes a conformational transformation as a consequence of a change in membrane potential. From measurements of the time dependence of the response of channel proteins to changes in membrane potential, it can be inferred that such a conformational change is equivalent to the movement of 2–6 positive charges from the inner to the outer surface of the membrane (the *gating current*). Flow of the gating current is completed before ions flow through the channel. The S4 segments consist largely of positively charged amino acids (usually arginine), followed by two hydrophobic amino acids. In a proposed model (Catterall, 1988), the S4 segment is viewed as consisting of a helix in which each peptide bond is regularly hydrogen-bonded to nearby peptide bonds in the helix (α helix), with the positive charge of the arginine residue forming a spiral pattern around the core of the helix. Prior to activation, each positive charge is paired with a negatively charged amino acid residue located in another transmembrane segment. The ion pairing is in equilibrium with the resting membrane electric field, but, upon depolarization, the stabilizing field is reduced, resulting in rotation of the S4 segment about its axis in an outward-going direction. This helical motion corresponds to a 60° rotation and a 5-Å outward translation, and is equivalent to the movement of one positive charge a distance of 5 Å. When this process occurs in each of the four S4 segments (transfer of four gating charges or more, depending on the number of positively charged residues), the transmembrane pore is created. Upon return to the initial resting membrane potential, the sets of ion pairs formed as a consequence of depolarization are broken, and the original ion pairing is reestablished.

The helix-screw gating model is speculative and is only one of a number of possibilities for gating. Whatever the actual mechanism, it is important to note that, in general, ion channels are voltage dependent for two reasons:

1. The voltage provides the driving force for each ion;
2. the probability that a channel is open depends on the voltage.

Ligand-gated ion channels are specialized for converting neurotransmitters into graded electrical signals; the similarity in structure of ligand-gated channels suggests that they are a superfamily of proteins. The channels open transiently following binding of a neurotransmitter, thereby producing a postsynaptic potential as a consequence of the ion flux; ligand-gated channels are usually voltage insensitive. Each ligand-gated channel (whether gated directly or indirectly; see Fig. 2) has one or more binding sites for a particular neurotransmitter or second messenger, and a characteristic ion selectivity. Neurons contain four general classes of ligand-gated ion channels, distinguished on the basis of the chemical class of neurotransmitter to which they are responsive (Table 1).

The most studied ligand-gated channel is the nicotinic acetylcholine receptor (nACr) channel (Fig. 5), which is found at the neuromuscular junction and at other locations in the nervous system; the nACr channel serves as a prototype for the less studied ligand-gated channels with regard to function, structure, and underlying biophysical mechanisms. When two acetylcholine molecules bind to the nACr, a conformational change is induced that opens an aqueous pore for an average of about a millisecond. Thereafter, the acetylcholine molecules disassociate from the receptor and are hydrolyzed by acetylcholinesterase. Each channel is probably about 9 nm in diameter, and protrudes from the membrane surfaces about 6 nm into the extracellular space and about 2 nm into the cytosol; when activated, the pore itself is about 2–3 nm in diameter. The nACr channel excludes anions, possibly because of the negatively charged amino acids at its mouth. The channel is formed from five subunits having the stoichiometry shown in Fig. 5; the α subunits bind acetylcholine with high affinity, one molecule of which must bind to each α subunit for the channel to open efficiently. Site-directed mutagenesis of the cDNAs of the α subunits has led to the identification of the binding sites for acetylcholine near two cysteine residues on the extracellular portion of the subunit. The four subunits are encoded by different but homologous genes; each subunit appears to consist of four membrane-

Table 1. The four major classes of neurotransmitters. Acetylcholine is the only member of its class; representative neurotransmitters in the other three classes are listed. The localization, gating, subunit composition, and amino-acid sequence are known to varying degrees for many of the neurotransmitters in the first three classes. Much less is known regarding the larger class of peptide neurotransmitters. In many cases, one or more peptide neurotransmitters are colocalized at synapses with a classical neurotransmitter (as indicated). Although many substances can act as neurotransmitters by activating specific receptors on the cell surface, there are fewer second-messenger pathways (Fig. 2).

Transmitter class	Neurotransmitter	Gating	Permeant ion
Acetylcholine	Acetylcholine (A)	Direct (nicotinic)	Na^+/K^+
		Indirect (muscarinic)	K^+, Ca^{2+}
Amino acids	GABA (B)	Direct	Cl^-
		Indirect	K^+, Ca^{2+}
	Glycine	Direct	Cl^-
	Glutamate	Direct	Na^+, K^+, Ca^{2+}
Monoamines	Epinephrine (C)	Indirect	K^+, Ca^{2+}
	Norepinephrine (D)	Indirect	K^+, Ca^{2+}
	Dopamine (E)	Indirect	K^+, Ca^{2+}
	Serotonin (F)	Indirect	K^+, Ca^{2+}
Peptides	ACTH		
	Angiotensin		
	β-endorphin		
	Bombesin		
	Camosine		
	Cholecystokinin, +E		
	Endorphins		
	Dynorphin		
	Luteinizing-hormone releasing hormone		
	Enkephalins, +A, +D, +F		
	Motilin		
	Neuromedins		
	Neuropeptide Y, +D		
	Neurotensin, +C, +E		
	Oxytocin		
	Prostaglandin		
	Somatostatin, +B		
	Substance K		
	Substance P, +A, +F		
	Thyroid-hormone releasing hormone, +F		
	Vasoactive intestinal peptide, +A		
	Vasopressin		

spanning regions arranged in such a way that specific regions of each of the subunits face each other to create the membrane-spanning pore.

3. BIOPHYSICS OF NEURONS

3.1 Membrane Potential

The membrane potential of the neuron plays an important role in its bioelectric processes; instantaneous deviations from the steady-state membrane potential can be viewed as the basic physical change in the neuron underlying information transfer. The mechanism responsible for the establishment of the steady-state membrane potential in the neuron is the same as that in other cells (Fig. 6). The Na^+-K^+ pump is a transmembrane energy-consuming enzyme that moves three Na^+ out of the cell and two K^+ into the cell for each molecule of ATP that is converted to ADP. Nongated ion channels in the cell membrane permit transmembrane flow of ions, principally Na^+ and K^+, down their concentration gradients. Since the cell membrane is more permeable to K^+ than to Na^+, a membrane potential is generated oriented to retard K^+ diffusion and enhance Na^+ diffusion, resulting in equal diffusion rates and thereby preserving electroneutrality. Under steady-state conditions (no net current) with no applied voltage V_m, the membrane potential is called the *resting membrane potential*

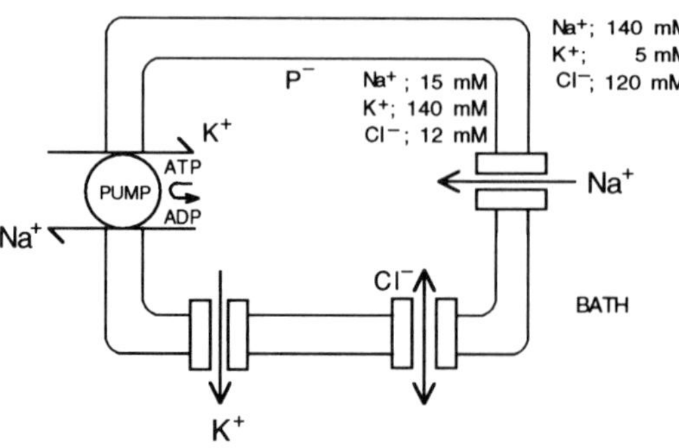

FIG. 6. The pump–leak model for the origin of the cell resting membrane potential. The interaction among the pump and leak channels results in a negative membrane potential (cell interior with respect to bath solution). P^-, negatively charged impermeant proteins. Typical intracellular and extracellular concentrations of several important ions are shown.

and is given by the Goldman–Hodgkin–Katz (GHK) equation (Hille, 1992):

$$E_m = \frac{RT}{F} \ln \frac{P_{Na}[Na]_o + P_K[K]_o + P_{Cl}[Cl]_i}{P_{Na}[Na]_i + P_K[K]_i + P_{Cl}[Cl]_o}, \tag{1}$$

where R is the universal gas constant, T is the absolute temperature, F is Faraday's constant, the internal and external concentrations of Na^+ are represented by $[Na]_i$ and $[Na]_o$, respectively, and the membrane permeability of Na is P_{Na}. The K^+ and Cl^- concentrations and permeabilities are represented in a similar fashion. If the permeability of any two of the ions is zero, the GHK equation reduces to the Nernst equation for the third ion, and the resulting potential is the *reversal potential* for that ion.

The GHK equation describes the resting membrane potential of the neuron in terms of the various concentrations and permeabilities. An equivalent-circuit model of the neuronal membrane permits consideration of other conditions, such as the response of a cell to an applied voltage or to injection of current from a microelectrode or from a natural cellular event (Kuffler and Nicholls, 1976). In this case, the membrane potential departs from that predicted by the GHK equation, and the equivalent circuit can be used to analyze the resulting time-dependent and steady-state current changes that occur: Assume, for example, that a step voltage is applied to the membrane. Application of Ohm's law to the membrane equivalent circuit (Fig. 7) yields $I_m = g_m(V_m - E_m)$, where V_m is the total potential across the membrane, g_m is the membrane conductance, and I_m is the current flowing across the membrane, and where it is assumed that enough time has passed following a voltage step for the system to have reached steady state (zero capacitive current); $V_m - E_m$ is known as the *driving force*. The permeability mechanisms can be investigated by studying the relationship between membrane current and voltage. If $g = g(V)$, the current-voltage relationship of a membrane will be nonlinear; if multiple conductance pathways exist in the membrane, complex relationships may be observed. Under conditions of zero net membrane current, the membrane potential is that derived from the GHK equation.

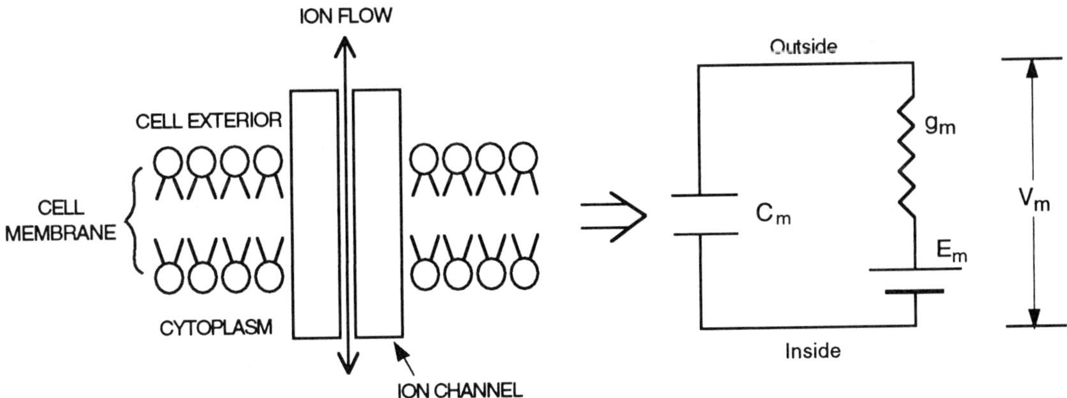

FIG. 7. A membrane equivalent circuit. Each conducting pathway contributes to the membrane potential: a lumped presentation of the permeability pathways is depicted.

3.2 Chemical-to-Electrical Transduction

Signal transduction at the postsynaptic membrane is mediated by transmembrane ion flow occurring in response to the interaction of a neurotransmitter (NT) secreted by the presynaptic cell and a receptor in the membrane of the postsynaptic cell. Neurotransmitter binding to the receptor results in either formation of a conductance pathway in the receptor, or initiation of an intracellular second messenger, which, in turn, leads to transmembrane ion flow (Fig. 2). The ion flow produces a transient change in the resting membrane potential at its point of entry into the neuron.

The nACr channel is a representative neurotransmitter-gated ion channel. It shows little selectivity among cations, and, consequently, the relative contributions to the channel current are determined by the cationic driving forces. Here K^+ is near Nernst equilibrium for a typical neuron, but the Na^+ concentration gradient and the membrane voltage both act to drive Na^+ into the cell; thus, nACr current is carried mostly by Na^+.

In a membrane region containing both nACr channels and nongated Na^+, K^+, and Cl^- channels, the membrane potential at no net current is

$$V_m = \frac{g_K}{g_T} E_K + \frac{g_{Na}}{g_T} E_{Na} + \frac{g_{Cl}}{g_T} E_{Cl} + \frac{g_{nACr}}{g_T} E_{nACr}, \tag{2}$$

where g_T is the sum of the individual channel conductances. Under resting condition, the membrane potential is about -70 mV, and $g_{nACr} = 0$. Following activation of the acetylcholine receptor, g_T increases significantly because of the large contribution of g_{nACr}; since E_{nACr} (the reversal potential of the nACr channel) is near 0 (Kandel *et al.*, 1991), the net result of the acetylcholine-induced conductance is the production of an electrical depolarization in the membrane containing the nACr channels. The total transmembrane current at the synapse is the sum of the ion flow through several hundred thousand such transmitter-gated channels, each of which has an identical conductance, but an open time that is governed by stochastic processes that render progressively longer open times correspondingly less likely.

The biophysical principles governing synapses at the neuromuscular junction also apply to synapses in the central nervous system (CNS). However, signal transduction at central synapses is complicated by several factors.

1. A typical CNS neuron receives many simultaneous excitatory and inhibitory inputs (synapses involving vertebrate skeletal muscle are always excitatory).
2. Many different neurotransmitters may be involved in signal transduction by one neuron, and a given neurotransmitter may have more than one kind of receptor at the cell membrane.
3. The actual role of the neuron in the signaling pathway is determined by the sum of its excitatory and inhibitory inputs, and not merely by the occurrence of the inputs

(at the neuromuscular junction, each synaptic potential produces an action potential).

4. The indirect mechanism (Fig. 2) for the effect of neurotransmitters on membrane potential can result in (a) channels that open or close at the resting potential; (b) transient changes in membrane voltage that last much longer than those caused by directly gated channels; or (c) second (and higher-order) messengers that cause effects in addition to those on channel conductance (alterations in receptors for other neurotransmitters, and in gene expression, for example).

3.3 Signal Summation

The transient voltage changes that occur at synapses as a consequence of the action of neurotransmitters have no individual significance with regard to the information actually transferred by the postsynaptic neuron: Physiological significance resides in the relative versions of the transients that propagate to the axon hillock. The essential features of the propagation process in the dendritic portion of the neuron can be modeled in terms of the response of a lossy insulated wire embedded in a conducting medium (Fig. 8) (Deutsch and Micheli-Tzanakou, 1987; Shepherd, 1990). The dendrite is conceptually divided into a series of isopotential segments represented by the membrane capacitance per unit length, C, in parallel with the transmembrane resistance R_{sh} (expressed in units of ohms × distance), with each adjacent pair of segments connected by a resistor R determined by the axonal resistance, where $R = 4\rho/\pi d^2$ (units of ohms/distance) and ρ is the resistivity of the cytoplasm. A voltage transient occurs at a synapse on the distal portion of the dendrite as a result of an ion flux, and a portion of the resulting current charges the capacitance of the membrane adjacent to the synapse, and thereby increases or decreases the membrane potential, depending on the charge of the permeant ion. The remaining current splits and either charges the capacitance of the second segment, passes through the membrane, and completes the circuit back to its source, or continues on to the next segment. This process continues until all the current has leaked out and returned to its source via the extracellular fluid, which is assumed to have negligible resistance.

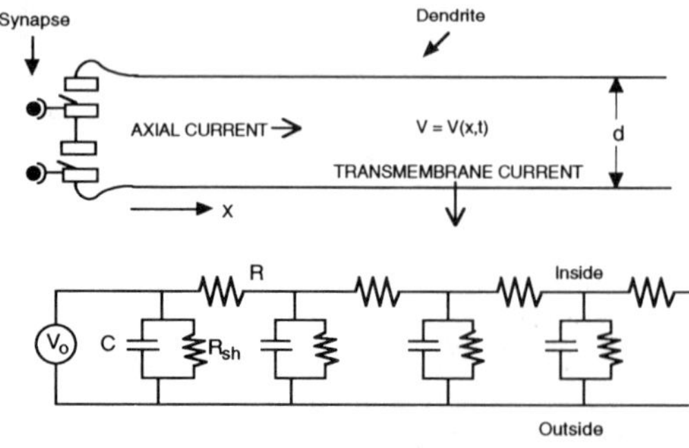

FIG. 8. Electrotonic spread of voltage in a dendrite and the corresponding equivalent circuit. At $t = 0$, a transient change V_0 in the membrane potential is induced by ion flow through neurotransmitter-gated channels in the dendrite. V_0 can be modeled as a voltage spike having a duration that is negligible with respect to typical propagation times. Thereafter, dendritic membrane potential is given by $V(x,t)$.

From Ohm's and Kirchoff's laws, the equation governing the spread of the potential $V(x, t)$, measured from the resting potential, is

$$\lambda^2 \frac{\partial^2 V}{\partial x^2} = \tau_m \frac{\partial V}{\partial t} + V, \qquad (3)$$

where $\lambda = \sqrt{R_{sh}/R}$ is the length constant, and $\tau_m = R_{sh}C$ is the time constant. In terms of C_m and R_m, which are the capacitance and resistance per unit area, respectively, we have $\lambda = \sqrt{R_m d/4\rho}$ and $\tau_m = \sqrt{R_m C_m}$; C_m is usually assumed to be about 1 μF/cm^2. Both λ and τ_m depend on the type of neuron; for the hippocampal pyramidal cell, typical values are τ_m = 15–70 msec, and $\lambda = (0.5–1.5)l_o$, where l_o is the length from the tip of the dendrite to the cell body.

The general solution for $V(x,t)$ has been given (Rall, 1960), but the behavior of the potential can be inferred from a consideration of the simplest special cases. The steady-state solution for an infinitely long dendrite (this model is applicable to slow synaptic potentials and to background depolarizations, as in cells in the retina) is $V = V_0 e^{-x/\lambda}$. Since $\rho \approx 100\ \Omega$ cm in all neurons, the spatial characteristics of the potential are determined by the membrane resistance and the dendrite diameter.

The spread of the potential in finite dendritic systems depends on the nature and extent of the branching that occurs. If the diameters of the dendrites at a branching point are such that (Rall, 1960)

$$d_o^{3/2} = \sum_i d_i^{3/2}, \tag{4}$$

where d_o is the diameter of the parent dendrite and d_i are the diameters of each of the daughter branches, then the branches are electrically equivalent to the stem, and the potential spreads through the entire system as it would in an infinite dendrite. This *equivalent cylinder* model, developed by Rall, is useful in modeling the passive spread of potential in dendritic systems (*electrotonus*); in conjunction with measurements of transient responses, it can be used to estimate λ of the dendritic tree.

The membrane time constant is an important determinant of the time course of the postsynaptic potential, but a precise description depends upon many factors, including the structural model chosen for consideration. The essential feature of all such models is the prediction that the postsynaptic potential diminishes in peak amplitude and increases in pulse width as it propagates from its point of origin. Thus, synapses near the cell body facilitate relatively large and rapid responses, whereas distant inputs lead to weaker and slower changes in the membrane potential at the cell body. Neurons have evolved mechanisms by which they can receive synapses at their distal dendrites but, nevertheless, transmit large and rapid postsynaptic potentials to the region of the axon hillock (Fig. 3), which is the site of generation of the action potential. These mechanisms include

1. the presence of a high specific membrane resistance;
2. production of a particularly large postsynaptic potential; and
3. the presence, in dendrites, of active membrane processes such as voltage-gated ion channels similar to those that participate in production of the action potential.

When two neighboring synapses on a cell are activated simultaneously, the conductance changes interact nonlinearly, thereby precluding a general analysis of the response as a superposition of the effects associated with the individual postsynaptic potentials; postsynaptic potentials produced at widely separated synapses may behave linearly because spatial separation lessens mutual interaction. Thus, the net effect of simultaneously activated synapses is greater if the synapses are located in different regions of the dendritic tree.

Dendritic spines are narrow projections from the dendrites; they are found on many types of neurons (see Fig. 9, for example), and can be the locus of synaptic inputs. The structural and electrophysiological characteristics of spines are not well understood; they may exhibit properties and functions not occurring elsewhere in the dendritic system. For example, the dendritic spine might provide a high-resistance path for a postsynaptic potential into the dendritic tree; this would have the effect of electrically isolating its postsynaptic potential from those induced at synapses on the dendrite itself. Consequently, the dendrite-synapse postsynaptic potential and the spine-synapse postsynaptic potential would add in a more nearly linear fashion than if both potentials occurred side by side directly on the dendrite.

Although the basic response of each portion of the dendritic membrane to a transient voltage change consists of the reasonably well understood phenomena of a propagating and diminishing voltage transient, the overall response of the neuron is difficult to characterize because of its highly complex morphology. A typical spinal neuron from a monkey is shown in Fig. 9; such a cell may contain 20 000–30 000 synapses, any combination of which may simultaneously transmit either inhibitory or excitatory postsynaptic potentials. Since the input and output of each region of the dendritic tree are processed in parallel, as in a network, the computation carried out in the dendritic tree is determined by dendritic morphology and the specific spatial and temporal relations of the synaptic sites to each other and to the axon hillock. Specialized computer programs have been developed to accommodate the many degrees of freedom needed to apply the cable equation to realistic models of actual neurons (McKenna *et al.*, 1992).

3.4 Action Potential

The biophysical process underlying development of the action potential was elucidated by Hodgkin and Huxley in a classic se-

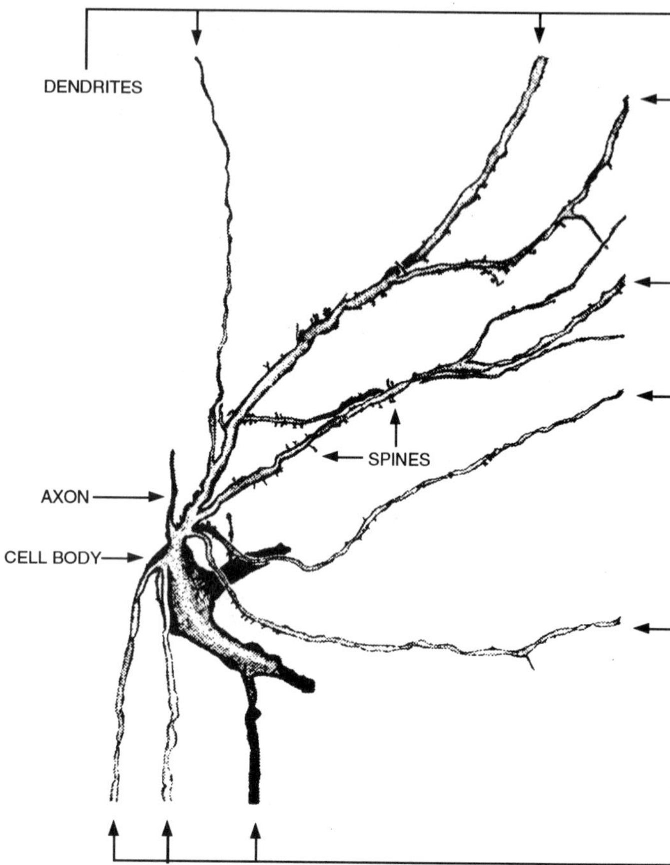

FIG. 9. Drawing of a sensory neuron from the dorsal horn of the spinal cord of a Macaque monkey. Only a portion of the dendritic tree and the axon is shown. The cell morphology was determined from serial reconstructions, using the Golgi technique; the shading indicates relative depth. A single such neuron can be affected by simultaneous excitatory and inhibitory stimuli from more than 20 000 axons. (Courtesy of John A. Beal.)

ries of studies on the giant axon of the squid, employing the *voltage-clamp* technique (Fig. 10). Since voltage dependence of membrane conductance proved to be the basic explanation for the action potential, the voltage clamp was particularly useful because it permitted direct control over the fundamental variable.

The current across the membrane was described in terms of a capacitive and three ionic components (Noble, 1966) (Fig. 11):

$$I_m = C\frac{dV_m}{dt} + g_{Na}(V_m - E_{Na}) + g_K(V_m - E_K) + g_L(V_M - E_L), \qquad (5)$$

where g_L is a leakage conductance of undetermined ionic basis; the electrical excitability of the membrane is contained in the time- and voltage-dependent conductances, g_K and g_{Na}. By varying ion concentrations in the bath solution and in the axon and the magnitude of the displacement of the membrane potential, and by using radioactive K^+, Hodgkin and Huxley showed that the early inward current in the voltage-clamped squid axon was due to Na^+ entering the axon, and the later-appearing outward current was due to K^+ leaving the axon (Fig. 12).

To provide a basis for reconstructing the action potential, Hodgkin and Huxley measured the variations of g_{Na} and g_K with time for various values of the membrane potential, using an ion-substitution method (pharmacological methods are now used for dissecting the Na^+ and K^+ currents). The mathematical description of the observations was based on the maximum possible conductance for both Na^+ and K^+. The K^+ conductance was expressed as a constant, $\bar{g}_K$ (the maximum conductance value), multiplied by a coefficient whose magnitude varied between 0 and 1, and which contained all of the voltage- and time-dependent characteristics of the K^+ conductance. The data showed that depolarization of the membrane

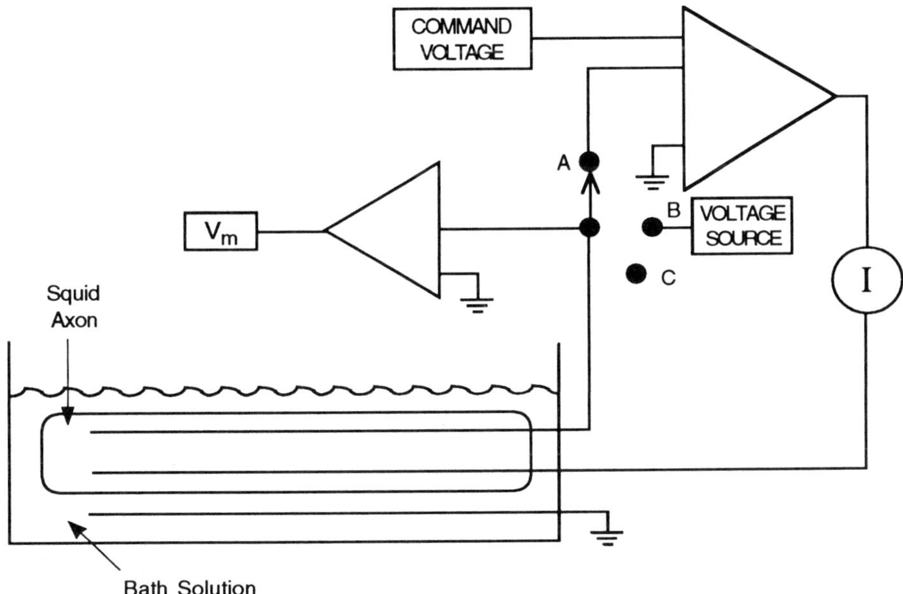

FIG. 10. The voltage-clamp technique and measurement of the action potential in the squid axon. For voltage-clamp measurements (switch position A), a voltage-recording electrode and a current-delivering electrode are placed intracellularly. The output of the feedback amplifier is determined by the difference between the command voltage and the membrane potential. For nonzero differences, the amplifier drives a current through the membrane in such a direction as to reduce the difference to zero. When a voltage step is applied, the membrane capacitance becomes charged in a time on the order of microseconds; thereafter, the capacitive current does not contribute to the membrane current. The circuit permits an abrupt displacement of the membrane potential that can be maintained indefinitely at the new value as the membrane current is measured. The design of the electrodes prevents the flow of longitudinal currents. In switch position B, the voltage clamp is removed, and a brief (100 μs) voltage pulse is applied between the voltage-measuring and bath-solution electrodes; if the switch is then moved to the open-circuit position C, the axon develops an action potential or returns to baseline, depending on whether the depolarizing pulse reached the threshold level.

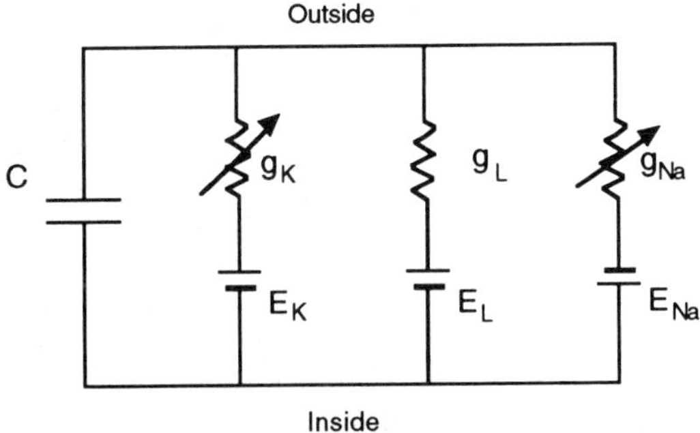

FIG. 11. Pathways for current flow in the Hodgkin–Huxley model of the axon. In voltage-clamp studies, capacitive current flows only during the time the potential is changing; thereafter, the membrane current consists solely of an ion flow through the three ion-conductive branches represented by Na^+, K^+, and leak conductances. The batteries depict the Nernst equilibrium potentials; time- and voltage-dependent conductances are employed to explain the response of the membrane to changes in voltage.

changed K^+ conductance such that both the amount and rate of change increased with depolarization. One interpretation was that the K^+ conductance was determined by a first-order process whose rate constants depended on the membrane potential. If α and β are two possible conformational states of a molecule, and n is the fraction in the active state (α), then

$$\frac{dn(V, t)}{dt} = \alpha_n(V)[1 - n(V, t)] - \beta_n(V)n(V, t), \tag{6}$$

where α_n is the rate constant for conversion from β to α and β_n is the rate constant for the reverse process. Hodgkin and Huxley modeled the observed data as $g_K \propto n^x$ and found that the best fit occurred for $x = 4$. Consequently, they chose $g_K = n^4\bar{g}_K$.

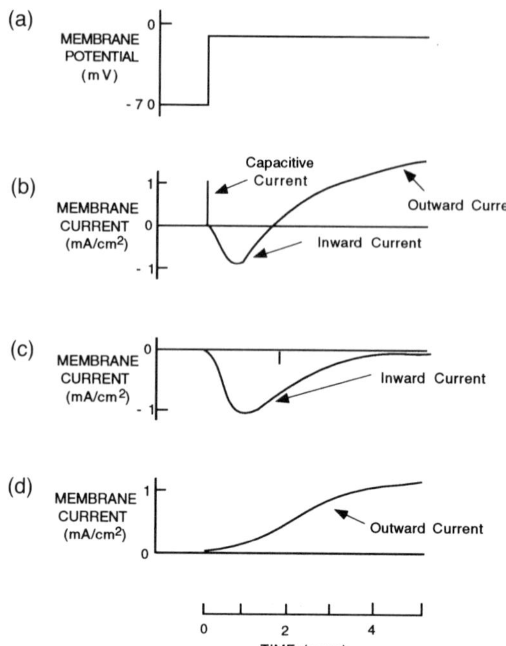

FIG. 12. Current flow across the squid axon under voltage-clamp conditions. During a depolarization of the membrane potential (a), an early inward and later outward current is observed following the capacitive transient. The ionic current flow (b) can be resolved into (c) an inward Na^+ current and (d) an outward K^+ current. (Adapted from Kuffler and Nicholls, 1976.)

Except for the rate of increase, the initial behavior of the Na^+ conductance following depolarization was similar to that of K^+. Consequently, g_{Na} could also be modeled in terms of a first-order reaction:

$$\frac{dm(V,t)}{dt} = \alpha_m(V)[1 - m(V,t)] - \beta_m(V)m(V,t). \tag{7}$$

In contrast with the behavior of g_K, the increased value of g_{Na} was not maintained, even though the membrane was clamped at the depolarized value (Fig. 12). This effect was accounted for by assuming a voltage-dependent reaction in the opposite direction to the m and n processes,

$$\frac{dh(V,t)}{dt} = \alpha_h(V)[1 - h(V,t)] - \beta_h(V)h(V,t), \tag{8}$$

where α_h decreased and β_h increased on depolarization of the membrane, in contrast to the changes in the rate constants of the other two processes. The Na^+ conductance was assumed to be determined by both m and h: $g_{Na} = \bar{g}_{Na}m^3h$. The processes that determine the K^+ and Na^+ conductances in the chosen model are depicted schematically in Fig. 13.

In this manner, the voltage-clamp data were converted into analytical expressions for g_{Na} and g_K as functions of membrane potential and time. Then the change in membrane potential that would occur with time following a brief superthreshold voltage pulse was calculated, using a piecewise-linear approximation at successive 0.01-msec intervals (Fig. 14); the calculation was shown to account for the time course of the action potential, thereby establishing that voltage-dependent permeabilities and ionic gradients were sufficient to explain electrical excitability.

The voltage-clamp data consisted of curves that described the voltage- and time-dependent membrane conductances of Na^+ and K^+. The measurements were made under highly controlled conditions, in the absence of longitudinal currents (hence, no propagation), with the membrane potential held at one or another specific value, at various Na^+ and K^+ concentrations. In a remarkable effort, employing some *ad hoc* but plausible assumptions, Hodgkin and Huxley synthesized the data obtained under the disparate experimental conditions, and succeeded in reconstructing the action potential. The Hodgkin–Huxley model, together with the cable equation, is sufficient to explain propagation of the action potential, the existence of the voltage threshold, and the velocity of propagation. Despite some limitations, the model is the generally accepted explanation of the origin and characteristics of the action potential in mammalian peripheral nerve, and in the central neurons of invertebrates and vertebrates; the mathematical variables employed in the model appear to have molecular structural counterparts. The complex pattern of electrical activity found in most neurons arises, ultimately, from the Hodgkin–Huxley mechanism, neuronal structural complexity, and ion-channel diversity.

3.5 Electrical-to-Chemical Transduction

When the action potential arrives at the axon terminal, the depolarization opens voltage-gated Ca^{2+} channels, which are concen-

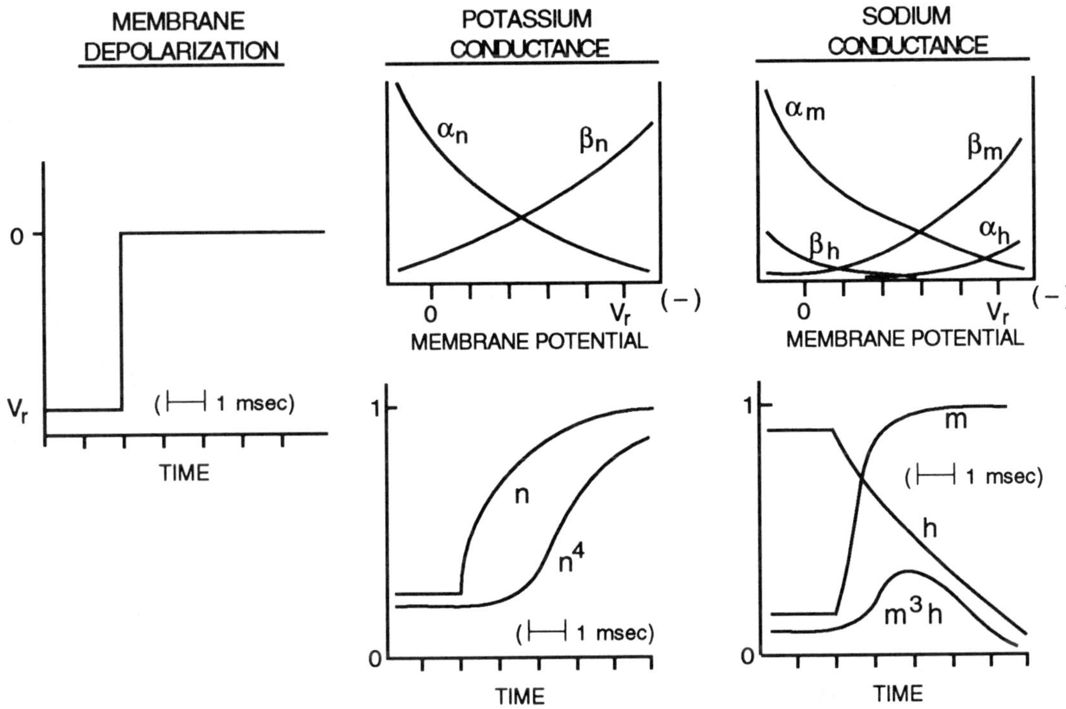

FIG. 13. Schematic representation of the processes assumed by Hodgkin and Huxley to determine the K^+ and Na^+ conductances. The K^+ conductance is determined by n, the K^+ activation variable (the fraction of activated molecules), which is itself determined by the rate constants for the forward and reverse processes; α_n is the rate constant for conversion from the inactive to the active state, and β_n is the rate constant for the reverse process; both constants were assumed to follow first-order kinetics (see CHEMICAL KINETICS). At the resting potential, β_n is greater than α_n; consequently, the fraction of molecules in the active state is small. Depolarization increases α_n and decreases β_n; consequently, for any specific depolarization step, n rises exponentially to a larger value. Hodgkin and Huxley employed n^4 because it provided a better fit to the experimentally observed conductance changes. Modeling of the Na^+ conductance changes requires both an activation and an inactivation variable, and both processes are assumed to follow first-order kinetics. V_r, membrane resting potential. (Adapted from Noble, 1966.)

trated in the terminal region. Neither the influx of Na^+ nor the efflux of K^+ is necessary for transmitter release. A higher frequency of action potentials results in a longer time during which the membrane is depolarized, thereby permitting a greater Ca^{2+} influx; thus, neurotransmitter release is a graded response.

The neurotransmitter is stored in specialized organelles called synaptic vesicles. Under resting conditions, a baseline level of neurotransmitter release occurs spontaneously. The entry of Ca^{2+} increases the amount of released neurotransmitter by increasing the probability that an individual intracellular vesicle will be released. The effect of the arrival of each action potential at the axon terminal can be viewed as a Bernoulli trial with regard to the fate of individual vesicles. If p is the probability that a given vesicle will be released, then the probability that x vesicles of a total population of n will be released is

$$p(x) = \frac{n!}{(n-x)!x!} p^x q^{n-x}, \tag{9}$$

where $q = 1 - p$. Such a relationship has been established at the neuromuscular junction (del Castillo and Katz, 1954) and is thought to be a model applicable to the central nervous system.

When the vesicles fuse with the cell membrane, the resulting increase in membrane area results in an increased membrane capacitance; that increase, and the decrease that occurs as the excess membrane is retrieved, can be measured (Fernandez *et al.*, 1984). The

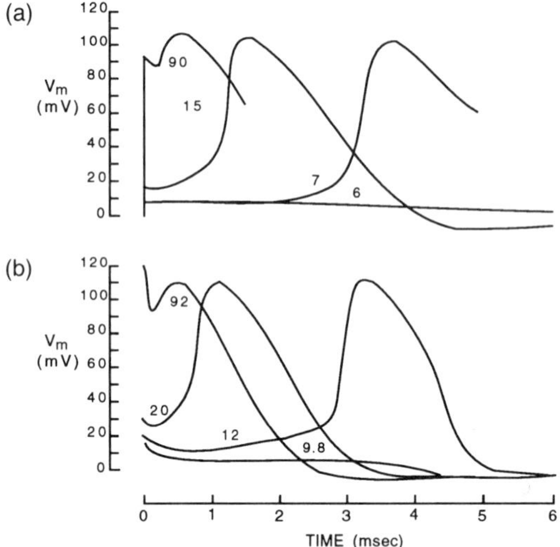

FIG. 14. Comparison of observed and calculated action potentials in squid axon. (a) Action potentials seen by Hodgkin and Huxley in the squid action preparation. The data were obtained by applying a brief depolarizing potential of the magnitude indicated on the various curves, and then measuring the subsequent time course of the membrane potential (cf. Fig. 10). (b) The predicted time course of the action potential based on the assumed model for current flow (Fig. 11) and the analytical expressions for g_K and g_{Na} derived from modeling the conductance data under voltage-clamp conditions. Small depolarizations of the membrane have a marginal effect on g_{Na}; for large depolarizations, g_{Na} increases to the point that the inward Na^+ current overcomes the outward K^+ current, thereby permitting the regenerative Na^+ response to occur. V_m, membrane potential measured from the resting potential. (Adapted from Kuffler and Nicholls, 1976.)

mechanism by which the vesicles fuse with the membrane is unknown, but it may involve formation of an electrical synapse between the vesicle and the cell membrane (Breckenridge and Almers, 1987).

4. NEURONAL SYSTEMS

4.1 Overview

Neurons receive inputs from receptors that detect stimuli external to the nervous system (Fig. 1). Following transduction into the language of the nervous system (Fig. 3), the information is processed by groups of neurons in local neuronal networks whose activity may be synchronized with that occurring in other regions of the nervous system. Synchronization is accomplished by neurons that are hardwired between local networks, and possibly by *volume transmission*, mediated by electrical and chemical messengers. The ultimate result of the sensory inputs and subsequent regional interactions is the behavior manifested by the organism. Not surprisingly, the extent of knowledge of the biophysical basis of the activity of neuronal systems is inversely related to the complexity of the level of the activity under consideration: There is convincing evidence that receptor function is mediated by changes in conductance of ion channels, but the physical basis of higher functions, such as behavior, memory, and consciousness, is poorly characterized.

4.2 Sensory Systems

Information enters the nervous system via *receptor* cells, which are specialized neurons or cells in intimate contact with neurons. At these sites, objectively characterizable chemical, mechanical, or energetic stimuli are transduced into electrical activity that controls the autonomic regulatory systems in the body, or that serves as the basis of conscious perception of the environment. The main types of receptor cells in mammals and the particular sense to which each corresponds are listed in Table 2. Other life forms possess sensory capabilities not shown to occur in human beings. Examples include the ability of some fish to detect weak electric fields, and the ability of some birds and bacteria to detect weak magnetic fields.

The mechanism underlying transduction by sensory receptors is the change in conductance of membrane-bound ionic channels (Fig. 15). The membrane receptor protein is indirectly coupled to a channel protein via a second-messenger system in the cases of chemical and light stimuli, but mechanical forces act directly on the channel to produce a deformation that alters conductance. In all three cases, the stimulus transduction ultimately produces a change in channel conductance that gives rise to a change in membrane potential called the *receptor potential* (also called a *generator potential*). The receptor potential is smoothly graded in proportion to the strength of the stimulus, and propagates electrotonically to the site of impulse generation. Thus, it is the receptor that determines the specificity of the sensory re-

Table 2. Main types of mammalian sensory modalities. Modified from Shepherd (1988).

Sense	Stimulus	Receptor location	Receptor cell
		Chemical	
Arterial oxygen	O_2 tension	Carotid body	Cells and nerve terminals
Toxins (vomiting)	Molecular	Medulla	Chemoreceptors
Glucose	Glucose	Hypothalamus	Glucoreceptors
*p*H (cerebrospinal fluid)	Ions	Medulla	Ventricle cells
Taste	Ions and molecules	Tongue and pharynx	Taste-bud cells
Smell	Molecules	Nose	Olfactory receptors
Pain	Various	Skin and various organs	Nerve terminals
		Mechanical	
Osmotic pressure	Osmotic pressure	Hypothalamus	Osmoreceptors
Pressure	Mechanical	Skin and deep tissue	Nerve terminals
Vascular pressure	Mechanical	Blood vessels	Nerve terminals
Muscle stretch	Mechanical	Muscle spindle	Nerve terminals
Muscle tension	Mechanical	Tendon organs	Nerve terminals
Joint position	Mechanical	Joint capsule and ligaments	Nerve terminals
Balance (linear and angular acceleration)	Mechanical	Vestibular organ	Hair cells
Hearing	Mechanical	Cochlea	Hair cells
		Energetic	
Vision	Photons	Retina	Photoreceptors
Temperature	Temperature	Skin, hypothalamus	Nerve terminals and central neurons

sponse, because the change in channel conductance, whether induced by the natural stimulus for the channel or by an agonist, results in the same electrical consequences.

4.3 Neural Networks

Receptor cells have no functional significance within nervous systems unless they are part of a network. Network complexity varies across the phylogenetic scale from a simple two-cell reflex response to the highly complex neural networks that mediate human consciousness and behavior. Various mathematical models of neuronal networks have been proposed to explain the learning abilities of the human brain (see NEURAL NETWORKS). In most such networks, the neuron is modeled as a summing node that weights each input and transforms the sum in a predetermined manner to yield an output activity level (Fig. 16). A model network typically consists of three or more layers of model neurons interconnected in such a way that the outputs of the model neurons in one layer are the inputs to the next layer. By following a systematic iterative procedure (the *back-propagation algorithm*), the weights and transfer function of each model neuron can be adjusted so that the input pattern is reproduced at the network output. Thus, the artificial network can mimic the learning behavior of biological neurons.

Artificial neural networks have considerable practical utility, but it is doubtful that they simulate the actual learning procedures used by the brain. Artificial networks operate via backward flow of information in the network, in distinction to the forward flow of information in biological neurons. The time required for an artificial network to learn increases rapidly as the size of the network increases, and there seems to be no biological parallel to this property of artificial neural networks. Finally, artificial networks require specification of both the input and output before learning can be accomplished. Again, this is not a property of biological networks, because no agency provides specific instructions regarding the particular pattern of receptor or action potentials that must occur before learning occurs.

Actual neuronal networks are spatially localized functional units; in addition to input

(a) CHEMICAL RECEPTOR

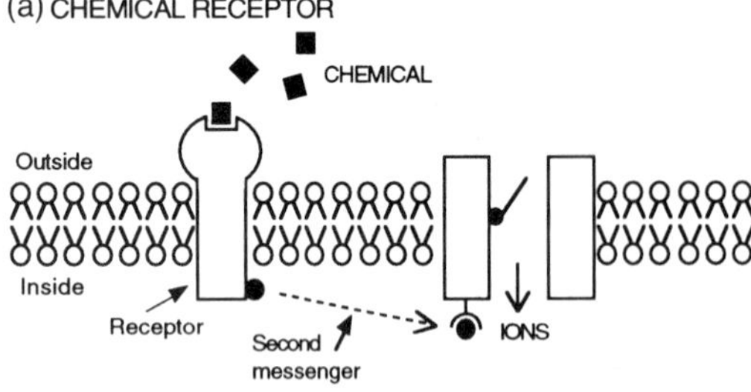

(b) LIGHT RECEPTOR

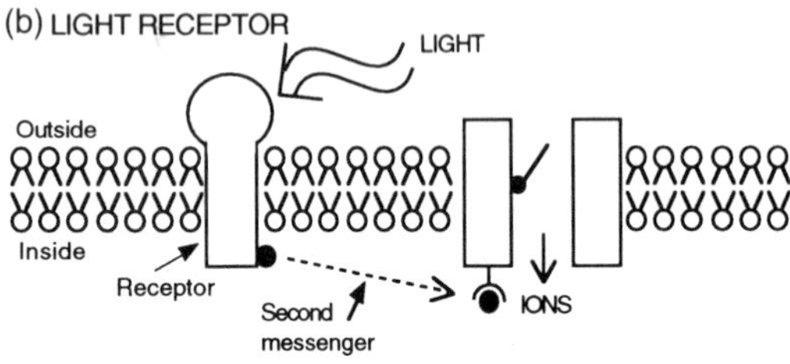

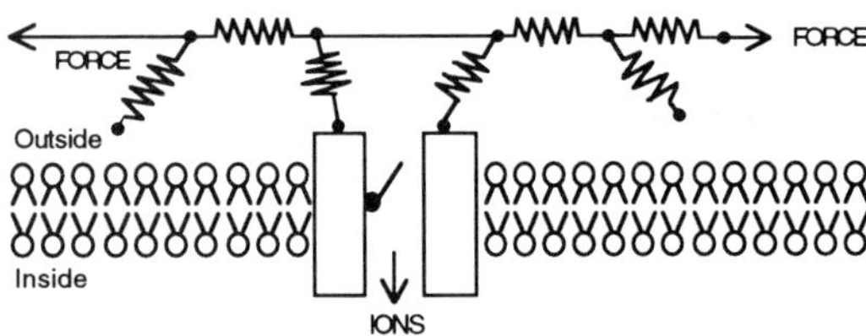

FIG. 15. Signal transduction in the three main classes of sensory receptors. (a), (b) In detection of chemicals or light, channel conductance is altered by intracellular second (or higher-order) messengers released following molecular events triggered by the interaction of the stimulus and the receptor. (c) In contrast, force transduction is directly coupled to channel conductance via a mechanically induced conformational change of the channel. (Adapted from Shepherd, 1988.)

and output neurons, such units usually contain local neurons that are involved exclusively in information processing within the network. The operations within the local networks are mediated by a combination of basic excitatory and inhibitory synaptic patterns (Fig. 17). Synaptic divergence is a morphological pattern that amplifies neuronal activity by distributing it to many cells. Neurotransmitter release does not necessarily occur at each synapse within a particular morphological unit; in the neuromuscular junction, for example, only about 10% of the synapses are activated by entry of the action potential into the presynaptic axon terminal (about 90% are *silent synapses*) (Shepherd, 1990). Synaptic convergence [Figs. 17(b) and 17(c)] is the pattern that makes possible the spatial and temporal integrative function of the dendritic tree. Presynaptic inhibition is a neuronal connectivity pattern in which a specific cell may be simultaneously presynaptic to one cell and postsynaptic to a second cell [Fig. 17(d)]; this arrangement permits a neuron to modify the activity of a cell without actually synapsing with it. The combination of the dendritic architecture and the types of possible synapses provides for a wide range of computational possibilities.

A local network that is characteristic of the cerebral cortex is depicted in Fig. 18. Various combinations of the elements depicted have been shown to result in rhythmic activity, directional sense, spatial contrast, and many other behavioral and physiological manifestations; presumably, they also underlie more complicated behaviors. Even though the network (Fig. 18) is highly simplified, it nevertheless exemplifies several important properties of biological neuronal networks:

1. A variety of specific biophysical processes that interact nonlinearly are responsible for the function of the local network;
2. performance of the network is degraded if one or more of the elements fail but catastrophic failure does not occur.

Networks of biological neurons differ fundamentally from the models developed thus far; analysis of real neuronal networks is a

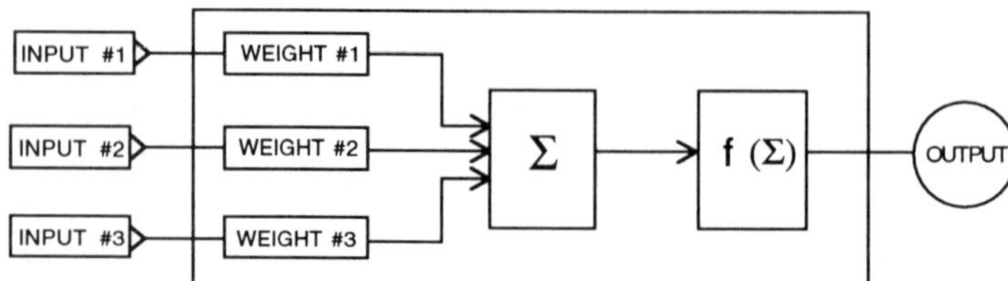

FIG. 16. Typical model neuron in an artificial neural network. The output is the transformed sum of the weighted inputs. A threshold is a common transfer function; in such a case, the weighted sum becomes the output only if it exceeds a specified level.

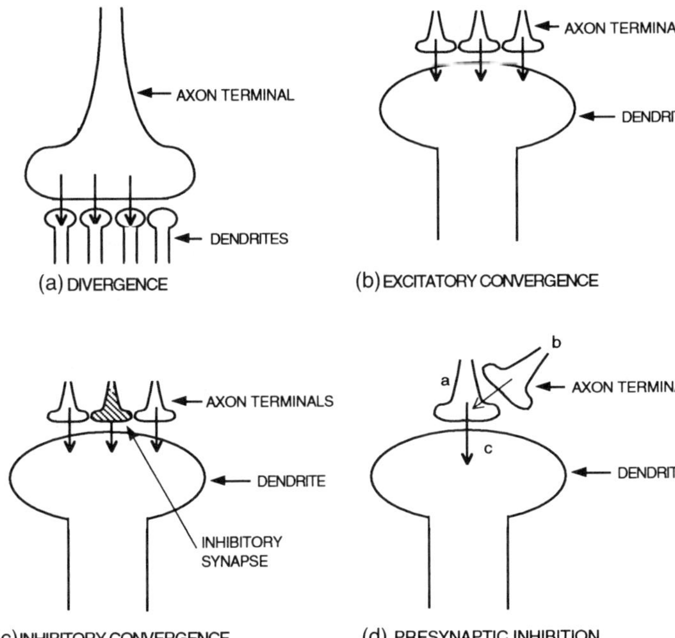

FIG. 17. Basic form of neuronal connectivity. (a) A divergent pattern produces a gain in activity unless all but one of the synapses are silent. (b), (c) The convergent patterns mediate neuronal integration. (d) In presynaptic inhibition, the effect of cell *a* on cell *c* is modified by cell *b*, which does not directly affect cell *c*. (Adapted from Shepherd, 1988.)

prerequisite to formation of more realistic models (Shepherd, 1990).

4.4 Volume Transmission

In the hard-wired model of the central nervous system, information transfer is viewed as being mediated by synapses between adjacent cells (Fig. 2), and neuronal complexity is ascribed to the complexity of the interconnections (Fig. 9) and their connectivity patterns (Fig. 17). Neurotransmitters (Table 1), in addition to their recognized role of diffusion across the synaptic cleft, may have other physical consequences. Some molecules may be released in a region not containing clefts, or may escape from the cleft region and diffuse to other parts of the central nervous system through the cerebrospinal fluid in the extracellular space of the brain (which occupies about 20% of the brain's volume). In this manner, a neurotransmitter could reach many different and distant targets, thereby constituting a parallel information transfer

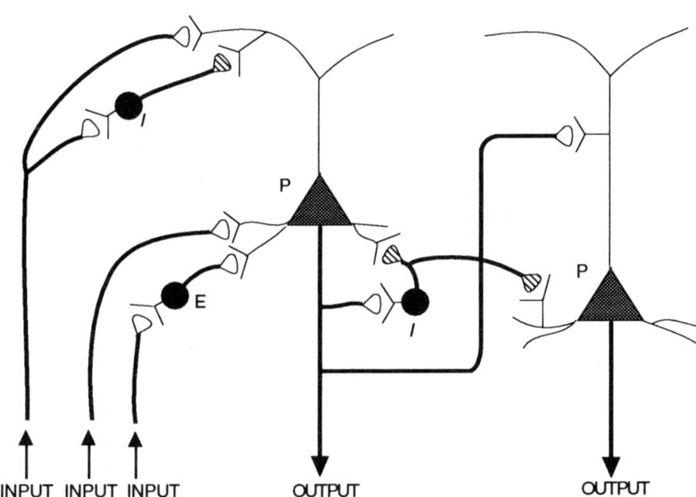

FIG. 18. Basic circuit organization of the cerebral cortex. Pyramidal neurons *P* receive input, generate outputs, and interact with one another. Local neurons may be inhibitory (*I*) or excitatory (*E*). Because of the cell density, dendritic architecture, and synaptic organization, there exist a vast number of computational possibilities. (Adapted from Shepherd, 1988.)

system within the central nervous system based on *volume diffusion*.

Neuroactive substances such as neurotransmitters can migrate through brain tissue from their point of secretion to distant binding sites, and theoretical models for the extracellular diffusion have been proposed to describe the concentration profiles as functions of time and position (Fuxe and Agnati, 1991). For example, neuropeptide Y is a neurotransmitter that is released in certain regions of the thalamus and hypothalamus; the receptors for neuropeptide Y are located several mm from the points of release. Since neuropeptides diffuse through the brain at about 1 mm/h, the spatial separation of secretion and binding of neuropeptide Y suggests that it may mediate slow-channel information transfer within the brain.

As discussed previously, axial flow of ions in the neuronal cytoplasm and propagation of the action potential require a corresponding ionic flow in the cerebrospinal fluid by virtue of Kirchoff's law. The potentials produced by these local currents propagate electronically, thereby altering the electrical environment of distant neurons by *volume conduction*, resulting in the *electroencephalogram* (EEG). This passive electrical consequence of the activity of individual neurons carrying out their hard-wired role in information transfer could, in principle, convey information to more distant neurons. As measured on the scalp, the EEG is a nonstationary, time-dependent voltage that can be observed between any two points; the spontaneous activity of the EEG exhibits characteristic changes during sensory processing (Fig. 19) and sleep, and in some clinical conditions, including brain tumors, epilepsy, and infection. Some of the processes thought to underlie these characteristic patterns include a predominance of electrical activity in a group of neurons functioning in a synchronized fashion, standing waves such as the normal modes of a vibrating membrane, and various metabolic alterations. For the special case of n point-current sources in a homogeneous conductor, the potential is

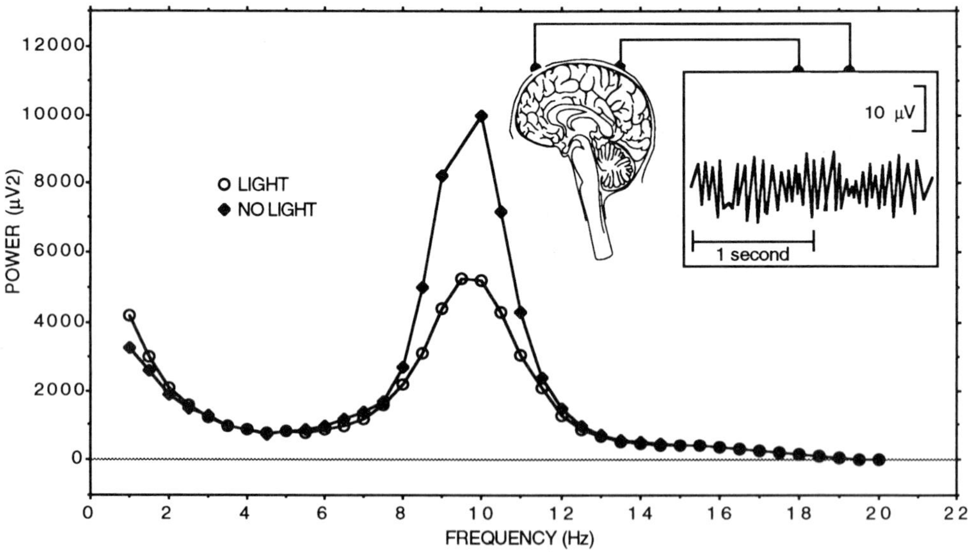

FIG. 19. Change in Fourier representation of the EEG during sensory processing. The EEG is a summation signal consisting of contributions from myriad action potentials and synaptic potentials occurring throughout the brain, particularly in the cerebellum. The signal is nonstationary in the sense that the amplitudes of its component frequencies vary with time, even under conditions of sensory deprivation. For analysis, an interval of EEG may be Fourier transformed, resulting in a representation of the EEG in terms of its component frequencies. The top curve depicts the results of Fourier transformations of 2-s epochs from an isolated subject (no sensory clues); the lower curve was obtained while the subject was presented with a weak light stimulus (in both cases, the results are averaged over 50 epochs). The data show that, as a direct consequence of detection of the light, an alteration occurred in the pattern of neuronal currents, resulting in a diminution in the amplitude of the EEG components in the frequency range 8–12 Hz.

$$V(\mathbf{r}, t) = \frac{1}{4\pi\sigma} \sum_{i=1}^{n} \frac{I_i(t)}{R_i}, \tag{10}$$

where $I_i(t)$ is the total current flowing from the ith point source into a medium of conductivity σ, and R_i is the distance of the ith source from the field point $\mathbf{r}$. Thus, in principle, ionic flow in each neuron makes a contribution to the electrical environment of the other approximately 10^{10} neurons in the human central nervous system. It seems plausible to expect that these interactions are physiologically significant, and there is evidence in the case of cells from some fishes that volume conduction has such consequences (Faber, 1991).

4.5 Learning and Memory

Memory (accessible storage of information) is a prerequisite for learning; the hippocampus is generally thought to be the most significant locus of memory in the vertebrate (Levitan and Kaczmarek, 1991). Memory involves an acquisition stage in which information is somehow encoded in neural networks in the hippocampus and stored as a consequence of a change in synaptic pattern (Shepherd, 1990). If the change is permanent, the information is essentially imprinted on the organism, resulting in *long-term* memory; if the *plasticity* of the synaptic patterns is more labile, then the result is *short-term* memory (Hall, 1992). Plasticity is hypothesized to be the basis of cognitive neuroscience (Kandel, *et al.*, 1991).

Associative learning involves the connection between two stimuli. The response of Pavlov's dog is an example of the *classical conditioning* form of associative behavior; another form involves *operant conditioning*, which is a paradigm in which an animal learns a task (such as pressing a lever), and is then studied to determine details regarding the nature of the learned behavior. The two forms of *nonassociative learning* are habituation and sensitization. *Habituation* is the decrease in the magnitude and duration of a response while the intensity of the stimulus remains constant over time. *Sensitization* involves the occurrence of an enhanced response to a stimulus. Historically, it was argued that associative learning was a systemic response distinct from specific physiological and biophysical processes occurring at the cellular and molecular levels and that, consequently, there was no necessary conceptual link between behavior and the underlying physical processes (Skinner, 1966). Subsequently, associative learning came to be perceived as a specific cellular process, but a diffuse property of the brain, not present in any specific cell or group of cells (Lashley, 1950). The modern view is that associative learning is a result of the activity of groups of cells assembled in neuronal networks, and that it occurs because of the formation of a time-dependent pattern of synaptic connections within particular circuits (*synaptic plasticity*) (Shepherd, 1988). The general hypothesis is that strengthening of synaptic connections, by either increased channel density or chemical modification of channel proteins, is the physical basis underlying learning and memory (Shepherd, 1988; Shepherd, 1990).

Mechanisms important in mediating some forms of nonassociative learning in some invertebrates have been identified. From studies of the gill-withdrawal reflex in *Aplysia*, for example, it appears that alterations in Ca^{2+} channel currents mediate both habituation and sensitization via an effect on neurotransmitter release. Because of the complexity of the mammalian nervous system, experimental efforts have concentrated largely on animals with relatively simple nervous systems and on *in vitro* preparations. In such systems, it is sometimes possible to locate specific neuronal pathways involved in particular behaviors, and to identify specific changes in synaptic connections or membrane conductances that are associated with learning or memory. But large gaps in knowledge remain, particularly with regard to criteria governing permissible extrapolation of results from lower to higher organisms. Presently, it remains unclear how the identified mechanisms could, even in principle, be synthesized into a deterministic explanation of behavior of the higher life forms.

GLOSSARY

Action Potential: A propagating electrical signal that may be initiated in a neuron when the membrane potential becomes depolarized beyond a threshold value.

Associative Learning: Learning involving the perception of a connection by the subject between two stimuli.

Back-Propagation Algorithm: An iterative mathematical procedure for training an artificial neural network.

Cell-Attached Configuration: A patch-clamp technique in which the micropipette is sealed onto an intact cell (also called on-cell configuration).

Classical Conditioning: A form of associative learning in which the subject is not required to act on its environment.

Delayed Rectifier: An axonal K^+ channel that opens with depolarization and is largely responsible for repolarizing the axon membrane following an action potential.

Dendritic Spines: Narrow projections from dendrites of many types of neurons; they can be the locus of synaptic inputs.

Driving Force: The total membrane potential minus the resting membrane potential.

Electroencephalogram (EEG): A nonstationary electrical signal consisting of the sum of the action potentials and synaptic potentials occurring throughout the brain.

Electrotonus: The passive spread of electrical potential in the neuron.

Equivalent Cylinder Model: A model of the dendritic tree, developed by Rall, used to model the passive spread of potential.

Gap Junctions: Electrical synapses.

Gating Current: The movement of electrical charge through a transmembrane protein that subserves the protein conformational change in voltage-gated channels that occurs during formation of a transmembrane ion pore.

Generator Potential: A change in membrane potential in a receptor cell that determines the specificity of a sensory response. Also known as a receptor potential.

Habituation: A form of nonassociative learning in which a constant stimulus produces a decrease in response over time.

Inside-Out Patch: A patch-clamp configuration in which, upon withdrawal of the micropipette, the cytoplasmic surface of the membrane comes to face the bath solution.

Ion Channel: The transmembrane protein that is the membrane-level effector in the nervous system.

Ligand-Gated Channels: Ion channels whose conductance depends directly or indirectly on the binding of a neurotransmitter to the channel.

Long-Term Memory: Information storage in the nervous system that causes a permanent change in synaptic patterns.

Memory: Accessible storage of information in the brain.

Myelin: An insulating material covering some axons.

Neuron: The electrically active cell of the nervous system.

Neurotransmitter: A chemical agent synthesized and secreted by a presynaptic cell which effects a communication with another cell.

Nodes of Ranvier: Periodic gaps in the myelin sheath of axons.

Nonassociative Learning: Learning not involving the perception of a connection by the subject between two stimuli; the principal types are habituation and sensitization.

On-Cell Configuration: A patch-clamp technique in which the micropipette is sealed onto an intact cell (also called cell-attached configuration).

Operant Conditioning: A form of learning behavior in which the subject is required to act on its environment.

Outside-Out Patch: A patch-clamp configuration in which, upon withdrawal of the micropipette, the cytoplasmic surface of the membrane comes to face the micropipette solution.

Patch-Clamping: An experimental technique for directly observing ion channels.

Plasticity: The tendency of synapses and neural circuits to change as a result of activity.

Receptor: In the case of chemical signals, the sensor portion of ion channels.

Receptor Cell: A cell that directly transduces a chemical, mechanical, or energetic stimulus to produce the receptor potential.

Receptor Potential: A change in membrane potential in a receptor cell that determines the specificity of a sensory response. Also known as a generator potential.

Resting Membrane Potential: The membrane potential under steady-state conditions; i.e., no net current, and no applied voltage.

Reversal Potential: For a channel with one permeant ion, the equilibrium potential determined from the Nernst equation.

Selectivity: The permeant ion species of a channel pore.

Sensitization: A form of nonassociative learning in which a response to a stimulus is enhanced over time.

Short-Term Memory: Information storage resulting in a temporary change in synaptic patterns.

Silent Synapse: Synapses that are not activated by entry of the action potential into the presynaptic axon terminal.

Synapse: A specialized junction between neighboring neurons.

Synaptic Plasticity: A time-dependent pattern of synaptic connections within particular circuits; used to explain learning.

Voltage-Clamp Technique: A technique for studying transmembrane ion kinetics in which the membrane potential is held at a predetermined value.

Voltage-Gated Channels: Ion channels whose conductance depends on the membrane potential.

Volume Conduction: A form of volume transmission consisting of the electrotonic propagation of postsynaptic and action potentials, resulting in an alteration of the electrical environment of distant neurons.

Volume Diffusion: A form of volume transmission in which neurotransmitters diffuse through the cerebrospinal fluid, thereby reaching distant receptors.

Volume Transmission: Information transfer within the nervous system by means other than a synapse.

Whole-Cell Configuration: A patch-clamp technique in which the cell membrane patch may be ruptured with the micropipette, thereby establishing a connection between the micropipette and the cell interior.

Works Cited

Alberts, B., Bray, D., Lewis, J., Raff, M., Roberts, K., Watson, J. D. (1989), *Molecular Biology of the Cell*, 2nd ed., New York: Garland Publishing.

Breckenridge, L. J., Almers, W. (1987), *Nature* **328**, 814–817.

Catterall, W. A. (1988), *Science* **242**, 50–61.

del Castillo, J., Katz, B. (1954), *J. Physiol. (London)* **124**, 553–559.

Deutsch, S., Micheli-Tzanakou, E. (1987), *Neuroelectric Systems*, New York: New York University Press.

Faber, D. S., Korn, H., Lin, J. W. (1991), *Brain Behav. Evol.* **37**, 286–297.

Fernandez, J. M., Neher, E., Gomperts, B. D. (1984), *Nature* **312**, 453–455.

Fuxe, K., Agnati, L. F. (Eds). (1991), *Volume Transmission in the Brain: Novel Mechanisms for Neural Transmission*, New York: Raven Press.

Hall, Z. W. (1992), *Molecular Neurobiology*, Sunderland, MA: Sinauer Assoc., Inc.

Hille, B. (1992), *Ionic Channels of Excitable Membranes*, 2nd ed., Sunderland, MA: Sinauer Assoc., Inc.

Hodgkin, A. L., Huxley, A. F. (1952), *J. Physiol. (London)* **117**, 500–544.

Kandel, E. R., Schwartz, J. H., Jessell, T. M. (1991), *Principles of Neuroscience*, 3rd ed., New York: Elsevier.

Kuffler, S. W., Nicholls, J. G. (1976), *From Neuron to Brain*, Sunderland, MA: Sinauer Assoc., Inc.

Lashley, K. S. (1950), *Symp. Soc. Exp. Biol.* **4**, 454–482.

Levitan, I. B., Kaczmarek, L. K. (1991), *The Neuron: Cell and Molecular Biology*, New York: Oxford University Press.

McKenna, T., Davis, J., Zornetzer, S. F. (Eds.) (1992), *Single Neuron Computation*, Boston: Academic Press.

Neher, E. (1992), *Science* **256**, 498–502.

Noble, D. (1966), *Physiol. Rev.* **46**, 1–50.

Rall, W. (1960), *Exp. Neurol.* **2**, 503–532.

Shepherd, G. M. (1988), *Neurobiology*, 2nd ed., New York: Oxford Press.

Shepherd, G. M. (Ed.) (1990), *The Synaptic Organization of the Brain*, 3rd ed., New York: Oxford Press.

Singer, S. J., Nicolson, G. L. (1972), *Science* **175**, 720–731.

Skinner, B. F. (1966), *Science* **153**, 1205–1213.

Further Reading

Appenzeller, O. (1990), *The Autonomic Nervous System*, 4th ed., Amsterdam: Elsevier.

Cole, K. S. (1968), *Membranes, Ions and Impulses: A Chapter of Classical Biophysics*, Berkeley: University of California Press.

Deutsch, S., Micheli-Tzanakou, E. (1987), *Neuroelectric Systems*, New York: New York University Press.

Fuxe, K., Agnati, L. F. (Eds). (1991), *Volume Transmission in the Brain: Novel Mechanisms for Neural Transmission*, New York: Raven Press.

Hille, B. (1992), *Ionic Channels of Excitable Membranes*, 2nd ed., Sunderland, MA: Sinauer Assoc., Inc.

Kandel, E. R., Schwartz, J. H., Jessell, T. M. (1991), *Principles of Neuroscience*, 3rd ed., New York: Elsevier.

Levitan, I. B., Kaczmarek, L. K. (1991), *The Neuron: Cell and Molecular Biology*, New York: Oxford University Press.

McKenna, T., Davis, J., Zornetzer, S. F. (Eds.) (1992), *Single Neuron Computation*, Boston: Academic Press.

Nicholls, J. G., Martin, A. R., Wallace, B. G. (1992), *From Neuron to Brain*, 3rd ed., Sunderland, MA: Sinauer Associates, Inc.

Niedermeyer, E., Lopes da Silva, F. (1987), *Electroencephalography: Basic Principles, Clinical Applications and Related Fields*, 2nd ed., Baltimore: Urban and Schwarzenberg.

Nunez, P. L. (1981), *Electric Fields in the Brain: The Neurophysics of EEG*, New York: Oxford University Press.

Peters, A., Palay, S. L., Webster, H. (1991), *The Fine Structure of the Nervous System*, 3rd ed., New York: Oxford University Press.

Schmitt, F. O., Worden, F. G. (Eds.) (1979), *The Neurosciences: Fourth Study Program*, Cambridge, MA: MIT Press.

Shepherd, G. M., (Ed.) (1990), *The Synaptic Organization of the Brain*, 3rd ed., New York: Oxford Press.

Siegel, G., Agranoff, B., Albers, R. W., Molinoff, P. (Eds.) (1989), *Basic Neurobiochemistry*, 4th ed., New York: Raven Press.

NEUTRAL ATOMIC AND MOLECULAR COLLISION PROCESSES

M. Kimura, *Argonne National Laboratory, Argonne, Illinois, U.S.A., and Department of Physics, Rice University, Houston, Texas, U.S.A.*

L. K. Johnson, *The Aerospace Corporation, Los Angeles, California, U.S.A.*

INTRODUCTION

Historically, the interactions between neutral atoms and molecules have been of the greatest interest to chemists and engineers seeking to determine reaction rates as functions of various parameters such as temperature and pressure. An applied chemical process is typically investigated by modeling the complex of possible reactions, using experimentally determined rates to parametrize the system of equations constituting the model. Our understanding of a process of applied interest (combustion of gasoline in air, for example) thus relies directly on the measured rate constants. Other complex processes that are not principally chemical also rely on a database of rate constants or collision cross sections; good examples would be the behavior of fusion-confinement plasmas, molecular-beam-induced epitaxy of semiconductors, or the characteristics of the earth's atmosphere from ground level to space. Atomic and molecular experimental studies, along with

3-527-28133-9/94/$5.00 + .50

quantum solid-state and surface physics (which are all similar in technique), frequently are the principal connections between our best physical understanding of the microscopic world and our most sophisticated and profound technological challenges. Coupled with recent experimental advancements such as laser technologies, the understanding of these processes is increasingly of fundamental importance for manipulating materials at atomic and molecular levels. We are now in an era of nanometer technology.

1. OVERVIEW

Processes that may be possible upon collisions of neutral particles can be classified into one group involving species with quantum states at the ground (lowest) levels and another group involving species with quantum states at excited levels. Species in the two groups are formed as follows:

1. Ground state:

$$A + BC(v,J) \rightarrow A + BC^{+} + 1e^{-}, \quad \text{ionization;} \quad (\text{p1a})$$
$$\rightarrow A^{+} + BC + 1e^{-}, \quad \text{stripping;} \quad (\text{p1b})$$
$$\rightarrow A + BC^{*},\ A^{*} + BC, \quad \text{electronic excitation;} \quad (\text{p2a})$$
$$\rightarrow A^{-} + BC^{+},\ A^{+} + BC^{-}, \quad \text{electron transfer;} \quad (\text{p2b})$$
$$\rightarrow A + BC^{*}(v',J'), \quad \text{rovibrational excitation;} \quad (\text{p3})$$
$$\rightarrow AB + C,\ AC + B,\ ABC, \quad \text{chemical reaction.} \quad (\text{p4})$$

2. Excited state: all processes [(p1) to (p4)], plus

$$A^{*} + BC(v,J) \rightarrow A + BC^{*}, \quad \text{excitation transfer;} \quad (\text{p5})$$
$$\rightarrow A + BC(v',J'), \quad \text{quenching;} \quad (\text{p6})$$
$$\rightarrow A + BC^{+} + 1e^{-}, \quad \text{Penning ionization;} \quad (\text{p7})$$
$$\rightarrow ABC^{+} + 1e^{-}, \quad \text{associative ionization.} \quad (\text{p8})$$

The species indicated by A, B, and C may each represent a simple atom or a complex molecule. The asterisk represents a species in an excited state [an electronic state or, for a molecule, a rotationally and vibrationally (rovibrationally) excited state that is designated by a set of vibrational and rotational quantum numbers (v,J)]. Processes (p1a) and (p1b) correspond to ionization events, with an electron being released from a target (ionization) or from a projectile (stripping). Similarly, electronic excitation (p2a) is possible for a target or a projectile or both. Upon collision, electron(s) may be captured either by a projectile to form $A^{-} + BC^{+}$ or by a target to form $A^{+} + BC^{-}$. In chemical reaction (p4), reaction mechanisms are classified as either complex mechanisms, where reactants pass through a transition-state complex as an intermediate state, or direct mechanisms, where no complex is formed. The criterion defining these two types of mechanisms is whether the lifetime of a reaction complex is longer (in a complex mechanism) or shorter (in a direct mechanism) than its rotational time ($\sim 10^{-12}$ s). Chemical reaction (p4) (reactive scattering) frequently accompanies rovibrational excitation in products. The quenching process (p6) includes both electronic deexcitation to a lower level and rovibrational deexcitation. Penning ionization (p7) occurs when the excitation energy of A^{*} exceeds the ionization potential of the BC target. When the final products form a molecular ion, the process is an associative ionization (p8). An electronically excited molecule normally decays by dissociation or, for highly excited cases, by ionization.

Figure 1 is a schematic diagram showing

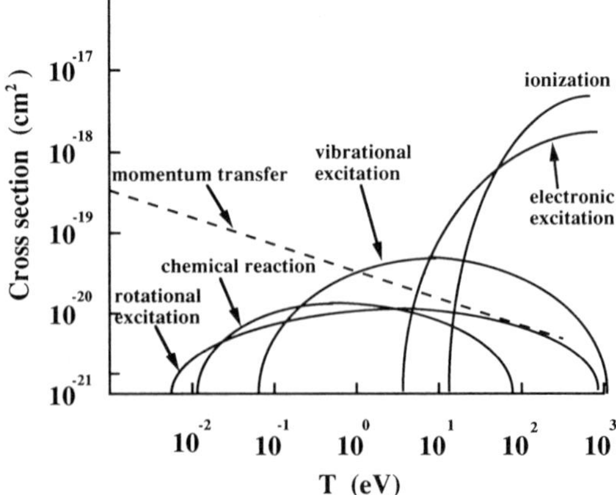

FIG. 1. Schematic diagram of probable processes and the magnitudes of their cross sections for a hypothetical collision species as a function of collision energy (T).

the probable processes and the approximate magnitudes of the corresponding cross sections for a hypothetical collision species as a function of collision energy. The actual magnitudes of the cross sections vary from system to system. When a polar molecule or a highly polarizable atom is involved in the collision, the inelastic cross sections are normally large, with magnitudes as large as about 10^{-15} cm^2 for chemical reaction and 10^{-17} cm^2 for vibrational excitation. When a nonpolar molecule or a slightly polarizable atom is a collision partner, the magnitudes of these cross sections are reduced significantly. Upon collision, electronic excitation and ionization (stripping) may be the dominant processes above a few tens of electron volts and up to the high-energy (>10 keV) region. If one member of the pair is a molecular species and an additional internal degree of freedom (e.g., a heavy nucleus) is involved, then vibrational and rotational (rovibrational) excitation processes become important in the low-energy ($\lesssim$10 eV) region. In the same energy region ($\lesssim$10 eV), chemical reaction (reactive scattering) may compete, if a colliding pair approaches close enough at sufficiently low energy to exchange a particle. Elastic (momentum transfer) processes are always associated with any collision and have their own interesting features, which are closely connected with various types of resonances. For neutral pairs, the interaction is weaker than for ionic pairs, with a long-range attractive van der Waals interaction with a dominant asymptotic term of R^{-6} when the two particles are separated by a distance R. When one of the two particles is an ion, the dominant asymptotic interaction is due to attractive dipole polarization and is of the form R^{-4}. At small R, strong overlap of electron charge clouds causes the interaction to become repulsive. At intermediate R, a balance between attractive and repulsive interactions sometimes causes a small minimum (bonding) at the equilibrium distance, forming a stable molecule.

As an actual example, Fig. 2 shows cross sections of processes resulting from collisions of H atoms with H_2 molecules. Because the H_2 molecule is nonpolar, the magnitudes of the cross sections are relatively small. Processes for ionization, dissociative ionization, collisional dissociation, and electronic exci-

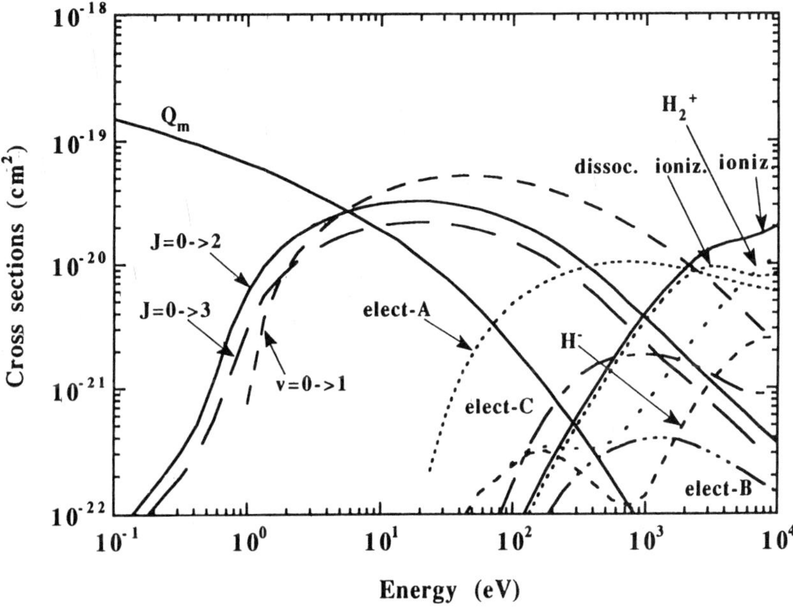

FIG. 2. Representative examples of various inelastic processes resulting from collisions of H atoms with H_2 molecules (adapted from Phelps, 1990). Notations in the figure are as follows: Q_m, momentum transfer; $J = m \rightarrow n$, rotational transition from m level to n level; $v = m \rightarrow n$, vibrational transition from m level to n level; elect-A, electronic excitation of the H_2 to the $A\Sigma$ state; elect-B, electronic excitation to $B\Sigma$; elect-C, electronic excitation to $C\Sigma$; H^-, H^- formation.

tation (elect-A, elect-B, and elect-C) are dominant above a few tens of electron volts. The H^- formation resulting from charge transfer is also a large contributor in this energy domain. Various vibrational excitation processes are dominant in the energy region between a few tens of electron volts and a few tenths of an electron volt; below this energy region, various rotational excitations dominate.

Theoretically, in the high-energy domain (>1 keV), the collision takes place impulsively, and the projectile leaves the interaction region very quickly. Hence, the interaction time is short (i.e., weak interaction), and the couplings among the various final states are considered to be weak. Thus, perturbation theory (such as the Born series), which normally accounts for only the initial and final states, can be applied to describe the collision dynamics reasonably well. As the energy decreases to a few kiloelectronvolts and further to electron volts, the interaction times become longer, causing a greater disturbance to the target. Therefore, various collision processes may occur simultaneously in a single collision. Because these processes couple strongly with each other in a complex manner as a function of collision energy and time, the processes all must be treated equally in any theoretical model. The model based on this physical picture is termed a close-coupling method. In this energy domain, the relative velocity of heavy particles (nuclei) is so much smaller than that of orbital electrons that the colliding particles are considered to form a collision complex or a quasi-molecule during the collision. Therefore, a complete understanding of collision dynamics in this energy domain requires an accurate knowledge of the adiabatic potential curves (surfaces in three dimensions) of the colliding system, obtained by a large-scale quantum-chemical procedure.

Experimentally, the study of neutral atomic and molecular collisions is a relatively mature field, although recent advances in experimental technique have brought renewed attention to previously impossible measurements. The emphasis in recent experimental work has been on well-prepared and well-analyzed quantum states of collision partners and products, as well as on examination of collisions under single-collision conditions. Early collision work examined the characteristics of atomic and molecular mixtures over a range of temperatures. These techniques usually are not sufficiently specific about which among processes (p1) to (p4) dominates. The evolution of the charged-particle beam, available over a wide range of energies, and the fact that an ion beam can be neutralized via process (p2b) provide experimentalists with good control over the initial dynamics of the neutral collision partners. Advances in beams produced by supersonic jets, pulsed or continuous, provide good control over the initial dynamics and states at low interaction energies, where cooling in the expansion of the jet flow conveniently restricts the number of populated rovibrational molecular states. In addition, techniques to detect relatively small (1–1000) numbers of atoms and molecules allow investigators to isolate collisions sufficiently to distinguish among the host of possible collision paths. Finally, the desire to prepare a specific quantum state or to use spectroscopic sensitivity to isolate product states (or both) has led to the widespread adoption of tunable lasers in experimental neutral-collision physics.

2. THEORETICAL FRAMEWORK

2.1 Inelastic Processes

2.1.1 High-Energy Collisions (>10 keV) When the velocity (v_{rel}) of the incident atom is larger than the orbital velocities (v_{el}) of the electrons in the target atom or molecule, electron transfer (p2b) rapidly becomes unimportant compared with ionization (p1a) and electronic excitation (p2a). This situation simplifies the theory substantially. Cross sections for this case can be evaluated on the basis of a Born-type expansion or its distorted-wave generalization. When the particle is sufficiently fast (but still nonrelativistic) and it interacts through the potential V, the general solution $\Psi(\xi)$ (where ξ represents all coordinates collectively) of the Schrödinger equation can be written formally by using the Green's function $G(k,\xi,\xi')$, as (atomic units are used throughout)

$$\Psi(\xi) = \Phi(\xi) + \int G(k,\xi,\xi')V(\xi',\xi')\Psi(\xi')d\xi', \qquad (1)$$

where $\Phi(\xi)$ is a solution of the homogeneous equation of total angular momentum J,

$$\left[\frac{d^2}{d\xi^2} + k^2 - \frac{J(J+1)}{\xi}\right]\Phi(\xi) = 0. \tag{2}$$

Starting from the incident plane wave $\Phi_{k_i}(\xi)$ with wavelength k_i as a zero-order approximation, we may write $\Psi(\xi)$ in the Born series (which is a perturbation expansion) in powers of the interaction potential,

$$\Psi_k(\xi) = \Phi_{k_i}(\xi) + \int G(k,\xi,\xi')V(\xi')\Phi_{k_i}(\xi')d\xi' + \int G(k,\xi,\xi')V(\xi')G(k,\xi',\xi'')V(\xi'') \times \Phi_{k_i}(\xi'')d\xi' d\xi'' + \ldots . \tag{3}$$

Correspondingly, with Eq. (3), the scattering amplitude can be written in the Born series as

$$f = -\frac{1}{4\pi}\langle \Psi_{k_f}|V + VGV + VGVGV + \ldots|\Psi_{k_i}\rangle. \tag{4}$$

By substituting the zero-order plane-wave approximation for Ψ_{k_i} and Ψ_{k_f} and retaining only the first term of V, we obtain the first Born scattering amplitude.

The cross section for a transition from state i to state f, calculated in the lowest order in the interaction potential V between the projectile and target (the first Born approximation), is specifically

$$\sigma_{fi} = \frac{\mu^2}{2\pi k_i^2}\int_{K_{\min}}^{K_{\max}} KdK \int_0^{2\pi} d\phi |T_{fi}^B(K)|^2, \tag{5}$$

with

$$T_{fi}^B(\mathbf{K}) = \int d^3R\, e^{i\mathbf{K}\cdot\mathbf{R}} \int d^3r \varphi_f(r)V(r)\varphi_i(r). \tag{6}$$

Here, μ is the reduced mass of the colliding system, r represents the collective position of the electrons centered at the target, R is the internuclear separation, ϕ is the azimuthal angle, $\hbar\mathbf{k}_i$ and $\hbar\mathbf{k}_f$ are the momentum of the projectile before and after the collision, $\hbar\mathbf{K} = \hbar(\mathbf{k}_i - \mathbf{k}_f)$ is the momentum transfer, $\varphi_i(r)$ and $\varphi_f(r)$ are electronic wave functions of the target for the initial and final atomic states, and $V(r)$ is the interaction potential between two particles.

As the energy decreases, the first Born approximation [Eqs. (5) and (6)] is considered to be less accurate. For such a case, agreement can be achieved by retaining higher-order terms in the perturbation series in Eq. (4); however, rigorous evaluation of higher-order terms is extremely difficult, and only some approximate attempts have been made to date.

2.1.2 Intermediate-Energy Collisions (10 eV–10 keV) When $v_{\rm rel}$ approaches or becomes less than $v_{\rm el}$, the interaction time becomes longer, causing stronger disturbance to the target. The condition $v_{\rm el} \simeq v_{\rm rel}$ is also favorable for a jump by a target electron to the projectile (electron capture). Therefore, transition probabilities for many inelastic events become comparable in magnitude, as exemplified in Fig. 2. Furthermore, these processes couple strongly with one another. This situation requires the theory to treat of all the involved processes equally. The close-coupling method has been developed specifically for this purpose. In this energy region, a semiclassical treatment, in which electronic states are solved quantum mechanically while nuclear motions are treated classically, is still considered to be valid. The assumption is that the nuclei move along a classical path specified by the internuclear separation vector $\mathbf{R}(t) = \mathbf{v}_{\rm rel}t + \mathbf{b}$, where $\mathbf{v}_{\rm rel}$ is the relative velocity and $\mathbf{b}$ is the impact parameter, and that the electronic wave function $\Psi(\mathbf{r},t)$ satisfies a time-dependent Schrödinger equation of electronic Hamiltonian $H_{\rm el}$,

$$i\frac{\partial \Psi(\mathbf{r},t)}{\partial t} = H_{\rm el}\Psi(\mathbf{r},t). \tag{7}$$

The solution of Eq. (7) can be obtained by expanding the total scattering wave function $\Psi(\mathbf{r},t)$ in terms of a known basis set of electronic wave functions $\varphi_i(\mathbf{R}(t),\mathbf{r})$ in the form

$$\Psi(\mathbf{r},t) = \sum_i a_i(t)\varphi_i(\mathbf{R}(t),\mathbf{r}) \times \exp\left(-i\int^t E_i(\mathbf{R}(t'))dt'\right). \tag{8}$$

By substituting Eq. (8) into Eq. (7) and projecting onto any basis function $\varphi_j(\mathbf{R},\mathbf{r})$, one obtains the set of first-order coupled equations for time-dependent coefficients $a_i(t)$,

$$i \sum_j S_{ji} \frac{da_j(t)}{dt} = \sum_{i \neq j} V_{ij} a_i, \tag{9}$$

where S is the overlap matrix $S_{ji} = \langle \varphi_j | \varphi_i \rangle$, and V_{ij} describes the coupling matrix

$$V_{ji} = \left\langle \varphi_j \middle| H_{\mathrm{el}} - \frac{\partial}{\partial t} \middle| \varphi_i \right\rangle.$$

The coupled Eqs. (9) are normally solved numerically, subject to the initial condition $a_i(t \to -\infty) = \delta_{ij}$, to extract the scattering amplitude $a_k(t \to +\infty)$ of a transition to state k. The square of the scattering amplitude gives a transition probability, and the cross section is defined as the integration of the probability over impact parameter b,

$$\sigma_k(E) = 2\pi \int b db |a_k(E,b)|^2. \tag{10}$$

To describe the collision dynamics correctly and hence obtain accurate transition probabilities, selecting a best-suited basis set $\varphi_i(\mathbf{R}(t),\mathbf{r})$ for the expansion in Eq. (8) is important. In the kiloelectronvolt regime, the collision time is sufficiently short that the colliding particles are considered to retain their atomic character during the collision (a distant collision). In this circumstance, using an atomic orbital (AO) for $\varphi_i(\mathbf{R}(t),\mathbf{r})$ as an expansion basis set is adequate and correct; however, as energy decreases into the electronvolt region, the electronic wave function has time to adjust to the changing potential field of two nuclei, and the colliding particles are regarded correctly as forming a quasi-molecule during the collision (a close collision). In this case, a Born–Oppenheimer molecular orbital (MO) is a suitable choice for the expansion basis set to describe the collision system correctly. The theoretical treatment based on this MO is summarized in Sec. 2.1.3.

When the AO expansion is used and only initial and final states are retained (two-state approximation), the coupled Eq. (9) is known to recover the first Born approximation of Eqs. (5) and (6) in the impact-parameter representation.

2.1.3 Low-Energy Collisions (≤10 eV) When $v_{\mathrm{rel}} \ll v_{\mathrm{el}}$ the electronic wave functions have time to adjust partially to the changing nuclear potential field (i.e., to near-adiabatic collision conditions). The incompleteness of this adjustment causes mixing of the molecular states and leads to electronic transitions. For an extremely slow collision, the electrons completely adjust to the slowly changing potential field; the electronic state behaves adiabatically, with its energy staying always on a single Born–Oppenheimer energy surface, which corresponds to an eigenstate of the molecular electronic Hamiltonian. In this case, no electronic transition takes place; however, even if no electronic transition occurs in the collision, rovibrational excitation and chemical reactions are possible because of their smaller energy defects (electronic excitation, 1–10 eV; vibrational excitation, 0.01–1 eV; and rotational excitation, <0.01 eV). In this energy region, two difficult tasks must be accomplished to calculate the transition probabilities accurately.

First, the semiclassical approximation discussed in Sec. 2.1.2 is no longer valid because the de Broglie wavelength of the relative motion of the heavy particles (nuclei) is no longer small compared with atomic dimensions; in principle, a fully quantum mechanical treatment for both electronic and nuclear motions is required. Many electronic and rovibrational motions couple strongly, making it extremely difficult to solve rigorously the multistate quantum-mechanical close-coupled equations.

Second, the details of collision dynamics are very sensitive to every aspect of computation and are particularly sensitive to the quality of the adiabatic potential (the energy surface) of the collision system. Thus, highly precise calculation of molecular electronic states as functions of the distance and geometry of the colliding particles is essential for accurate determination of collision dynamics and cross sections. To illustrate the complexity of solving the coupled equations, a formal form of the coupled equations (a simplified version) is given as

$$\left[\frac{1}{2\mu} \frac{d^2}{dR^2} - E - \frac{J(J+1)}{2\mu R^2} \right] \Psi_\alpha(R) = \sum_{\alpha'} \langle \alpha | V | \alpha' \rangle \Psi_{\alpha'}(R). \tag{11}$$

In Eq. (11), α designates a set of all quantum numbers at a given state specifying the electronic and rovibrational states of the colliding system. The summation of the coupling matrix element on the right-hand side includes, in principle, all combinations of the other quantum states coupled. The number of states needing to be considered easily exceeds a few hundred when one tries to calculate rigorously, for example, rovibrational excitation cross sections at collision energies of a few electron volts. Therefore, to reduce the number of coupled equations to a numerically tractable level, various types of approximations are used at various levels of computation.

Very powerful quantum-chemistry structure codes developed by various groups enable us to determine potential curves (surfaces) for simpler systems like atom–diatomic-molecular collision systems; however, even with these powerful computer programs, obtaining adiabatic potentials with reasonable accuracy is still exceedingly difficult and time consuming, particularly when excited states must be determined.

Additional difficulty arises when electronic continuum (ionized) states couple strongly with other degrees of motion. A good example of this is Penning ionization (p7), in which the initial excited state is embedded in the final continuum state, and ionization is always possible. High-quality computation of excited molecular states is still far more difficult than that for the ground state because of the strong mixing of many states lying nearby. Rigorous treatment of continuum states is also extremely difficult, if not impossible. A procedure normally proposed is a discretized sampling of continuum energies, followed by treatment of continuum states as if they were discrete; however, this procedure has not been applied extensively to studies of the continuum problem arising from low-energy collision systems.

2.2 Elastic Processes

Low-energy elastic-scattering cross sections (total or differential cross sections) resulting from neutral-particle collisions often show various types of structures, due to interferences or resonances, as functions of energy or scattering angle. Careful analysis of these structures provides detailed information on the interaction potential between colliding particles. In addition, the momentum-transfer cross section, the cross section that accounts for large-angle elastic scattering, is an essential physical quantity for describing diffusion and mobility of a particle through a gas or condensed matter and hence is practically important.

In the classical two-body collision, the classical deflection function X is defined as

$$X(E,b) = \pi - 2b \int_{R_0}^{\infty} \frac{dR}{[1 - V(R)/E - b^2/R^2]R^2}, \tag{12}$$

where b is the impact parameter, E is the collision energy at the center of mass of the system, and $V(R)$ represents the interaction potential. The lower limit of the integrand, R_0, is the classical turning point. $X(E,b)$ can easily be evaluated if the interaction potential $V(R)$ is known. When $V(R)$ is repulsive or attractive, $X > 0$ or $X < 0$, respectively.

The classical differential scattering cross section can be expressed as a function of scattering angle θ,

$$\sigma(\theta) = \frac{2\pi b db}{2\pi \sin\theta d\theta}. \tag{13}$$

By combining Eqs. (12) and (13), this cross section is re-expressed as

$$\sigma(\theta) = \frac{b}{\sin\theta \, |dX/db|}. \tag{14}$$

Several important features in the differential cross section are summarized subsequently.

2.2.1 Glory Scattering Because the particle feels a balance between the attractive nature and the repulsive nature of the potential $V(R)$, at a certain impact parameter (or scattering angle θ_g), $\sin\theta_g = 0$. For such a circumstance, from Eq. (14), the classical differential cross section diverges and gives an infinite value, called the glory scattering; however, this argument is purely classical, and if the quantum-mechanical treatment is used, the value of the differential cross section remains finite with a maximum, because many partial waves contribute and cause interference.

2.2.2 Rainbow Scattering Because the potential has attractive and repulsive natures, the classical deflection function possesses either a maximum or a minimum. Therefore, at a certain impact parameter, the condition $dX/db = 0$ is met, leading to the divergence of the differential cross section Eq. (14). This peak is called the rainbow scattering. Again, this argument is purely classical; and if the quantum-mechanical treatment is used, the differential cross section remains finite, with merely a shallow dip at the rainbow angle because of contributions of many partial waves.

2.2.3 Orbiting Resonance In the classical picture, when the projectile approaches the target with a certain impact parameter, the projectile may circle the target several times before departing from the interaction region. This phenomenon, the orbiting resonance, is particularly seen when the incident energy equals the magnitude of the effective potential (the sum of the interaction potential and the centrifugal potential). Quantum mechanically, the phenomenon corresponds to resonances because the projectile stays for a longer period of time near the target.

3. EXPERIMENTAL FRAMEWORK

Collision processes are typically examined by performing scattering experiments, which conventionally consist of several common experimental elements, regardless of interaction energy. The collision partners must be prepared in well-defined initial states before the collision occurs. One would like to know the initial momentum of each collision partner as well as the complete quantum state information. After the partners are allowed to interact, some product of the collision process must be detected. A complete measurement of a scattering event includes the final states and the momentum of each exiting particle. In addition, for an absolute measurement, the quantities of initial and final particles must be measured independently during the experiment. In practice, such definition is almost never achieved. For example, absolute results can be obtained by scaling relative data to absolute measurements available in the literature. However, the states and momenta of the collision system are rarely fully measured; consequently, assumptions must be made for comparison with theory or to use experimental results in a predictive model for applied work. Nearly every experiment has substantial shortcomings, and conscientious authors candidly note such limitations.

The desire to restrict the momenta of particles in the collision system under study has led to several common experimental techniques. Atoms and molecules participating in the collision under study are initially widely separated (centimeters to meters); they come together to interact (10^{-9} m) and subsequently separate again. Thus, these experiments take place, in some sense, in the gas phase. Because no other collisions are allowed except the interaction under study, experiments generally require the long mean free paths available in high-vacuum apparatus. Initial particle trajectories are typically restricted to one dominant direction in the apparatus, forming them into a *molecular beam*; such a beam is amenable to restriction of particle speeds in the beam. Thus, a collision partner can be prepared with a well-defined momentum before the collision. Techniques for selecting formation and velocity vary with energy and can influence the initial quantum state of the colliding particles, as discussed subsequently.

A collision event can take place via the interaction of two beams or by placing a cell containing the other collision partner in the path of the beam. In the latter case, the nomenclature *projectile* and *target* is appropriate. In nearly every case, the laboratory reference frame is not the center-of-mass (c.m.) frame, and a transformation to the c.m. frame is required to compare experimental results with theory.

The interaction between particles takes place under carefully controlled conditions. In particular, the process under study must be distinguishable from other competing processes. The initial particle densities or fluxes are kept low so that multiple collisions do not dominate. More will be said about this subsequently.

After the particles separate, some product of the collision must be collected. This collection can be somewhat artificially separated into analysis and detection. The products typically pass through the analyzer, which restricts their properties, and then arrive at the products detector. Analyzers can restrict the quantum state or the momentum of products (or both). A large variety of analyzers exists. Some of the more common analyzers are discussed subsequently. The detector converts the arrival of the analyzed product into an electrical signal, which is conventionally stored as a count in a computer data file. Sometimes the product of interest is a photon, and so optical filters or spectrometers constitute the analyzer, and detectors like photomultipliers or charge-coupled devices are used. More frequently, the product of interest is an atom, molecule, ion, or electron. In some cases, an ion or electron current can conveniently be collected directly from an electrode inside the apparatus. In other cases, the flux is so small that amplifying detectors based on secondary electron ejection and cascade (such as a Channeltron or microchannel plate) must be used. Products that do not carry an electrical charge can be analyzed and detected in several ways. High-velocity neutral particles may transfer enough energy to a detector to generate a secondary electron, which is amplified and detected in the usual way. Neutral products can be ionized for analysis and detection by using electron impact or multiphoton techniques, but the effect of the ionization process (such as dissociative ionization) must be taken into account. Finally, the neutral product can be excited with a probe photon into an upper state that emits upon transition to a lower state, where the emission is spectroscopically detected. This technique generally requires the high photon flux of a laser and is consequently called laser-induced fluorescence. The technique must be used carefully, because competing mechanisms also depopulate the upper state and because the technique relies on published oscillator strengths.

The experimentalist often struggles to produce enough signal to overcome instrumental noise. This problem can be overcome by improving the efficiency of detection or the efficiency of the analyzer (or both) or by increasing the number of collisions occurring. However, the number of collisions cannot be raised indiscriminately, because unwanted multiple collisions will affect the results.

3.1 High to Superthermal Energies (10 eV–1 MeV)

Five orders of magnitude are considered as "high energy" because the experimental techniques for beam generation and detection in this range are similar. Atomic and molecular physics and nuclear physics overlap at the high end of this energy region. Our discussion is limited to electronic interactions between neutral atoms and molecules.

3.1.1 Charge-Transfer Beams Neutral beams with energies above 10 eV are typically generated by passing an ion beam of the desired energy through a cell containing gas. Charge-transfer events at high impact parameters produce fast neutral particles whose trajectories are virtually unchanged. The neutral beam coming from the target cell passes through a downstream aperture to select these coaxial neutral particles. Charge transfer occurs most readily with an identical species; producing neutral rare-gas beams is therefore relatively straightforward. Charge transfer is also favored when the ionization potentials of the target and projectile are similar. For this reason, krypton (ionization potential = 14.0 eV) is used as a target for protons to produce H atom beams.

3.1.2 Sputtered Beams A fast neutral beam can be produced by bombarding solid targets with an intense ion beam of kiloelectronvolt energy. More than 90% of the sputtered particles that leave the target are neutral, falling into a broad energy spectrum ranging from 0.1 to 100 eV. Therefore, a velocity selector is essential for obtaining a monoenergetic beam.

3.1.3 Seeded Supersonic Beams A high-energy beam of neutral atoms and molecules can be obtained by supersonic expansion of a gas. In this method, a small fraction of the desired heavy gas is mixed with light inert carrier gas and expanded through a nozzle of 0.1-mm diameter that is heated to about 1000 °C. The final beam, which has an energy range up to a few tens of electron volts, is produced by skimming and collimation. This technique is particularly useful when it is applied to fragile molecules that would easily be broken up in a high-temperature or violent environment.

3.1.4 Other Methods Another technique for the production of neutral beams from ions involves photodetachment from a negative-ion state. The negative-ion beam intersects an intense pulsed or continuous laser; barring multiphoton effects, the resulting neutral beam is entirely in the ground electronic state. The cross section for photodetachment is low enough, however, that intense laser fields (intracavity Ar^+, for example) are required.

Methods of ion-beam production will not be discussed here, except to note that the characteristics of ion sources (such as electron energy, pressure, size of the ion extraction region, and impurities or mixtures of source gases) do affect the qualities (particularly the energy spread and quantum state) and flux of the extracted beam (see PARTICLE SOURCES, ION). For example, ionized oxygen extracted from a conventional plasma source at kiloelectronvolt energies will be in a mixture of ion states, including the $O^+(^4S)$ ground and the $O^+(^2D)$ metastable excited states. Various measures have been developed for preparing ion beams in a known quantum state or with a low energy spread; the applicability of these techniques to neutral studies is limited to cases where the preparation technique does not adversely affect the neutral beam qualities (such as intensity or divergence).

3.2 Low to Thermal Energies (<10 eV)

Most thermal-energy-beam sources employ a flow system in which the beam material flows as a vapor from an oven into a chamber of lower pressure. In the chamber, collimation and differential pumping are used to produce a final well-defined beam for experimentation (in the collision chamber). Two different regimes characterize the beam flow, depending on the size of the mean free path of atoms or molecules in the source relative to the slit width. For flow in which collisions among atoms or molecules are rare, or molecular effusion, molecular motions are mutually independent; for flow in which atoms or molecules undergo many collisions, or hydrodynamic flow, energy transfer among various internal motions and translational motions occurs frequently.

A molecular-effusion source has an advantage for predicting the beam properties easily and accurately because it operates at thermal equilibrium. The beam leaves a chamber, which contains the beam material as a vapor, via a narrow slit. The gas pressure in the source can be controlled to produce molecular effusion by making the width of the slit smaller than the mean free path of the molecules. The beam thus produced has the Maxwellian velocity distribution, which can control the energy resolution of the beam. For some purposes, some type of velocity selection may be needed.

The hydrodynamic source has an advantage of higher beam intensities than those in molecular-effusion sources. This beam, produced by replacing the thermal equilibrium gas in the molecular-effusion source with a supersonic jet of gas, leaves a chamber of high pressure through a nozzle. Therefore, a large jet of hydrodynamic gas flow can be obtained.

3.3 Beams of Excited Species

The production of beams of excited (electronic or rovibrational) species is also important for applications. The lifetimes of these species vary significantly, from 10^{-1} s for metastable electronic states from which transitions are optically forbidden to less than 10^{-5} s for electronic states that can make optically allowed transitions. Usually, the lifetimes for rovibrational excited states lie between these extremes; however, in general, the techniques are simply a combination of the beam production method already described with a method to produce excited states.

Depending on whether the decay of the desired excited state is optically allowed or forbidden, the specific technique and device to produce the species vary. Thermal dissociation, electron impact, optical excitation, and chemical enhancement are the most commonly used techniques.

3.3.1 Thermal Dissociation Thermal dissociation is one of the simplest techniques for producing atomic beams of excited species with translational kinetic energies of kT in an oven or arc source. The device simply consists of a tungsten tube into which the desired gas is introduced, then heated for dissociation. With this technique, reasonably high-intensity beams of atomic H gas can be obtained.

3.3.2 Electron Impact Electron impact is particularly useful for producing, through electron exchange, optically forbidden metastable states that have a longer lifetime of 10^{-3} s. Negative ions formed temporarily are also known to decay to specific rovibrational states of a molecule; hence, negative-ion resonance can be used selectively for this purpose. As alternative techniques with similar electron scattering properties, radio frequency and microwave discharge are commonly used.

3.3.3 Optical Excitation Optical excitation is essentially the same as electron impact, but a photon is used as the exciting source. This method is suited for production of optically allowed excited states. However, its usefulness depends on absorption intensity.

3.3.4 Chemical Enhancement By controlling, with the help of a laser, a chemical process in which the specific reaction cross section is known, chemical enhancement is useful for production of beams with state-selected quantum states (vibrational and rotational excited states) of molecules.

All methods for beam production discussed previously are also useful as cluster beam sources.

Detection systems for beams of excited species are essentially a combination of those described previously.

4. OBSERVABLES RELEVANT TO APPLICATIONS

Various physical observables that can be determined rigorously, through experimental and theoretical work, are useful for applications. Some of the most fundamental quantities are summarized here.

4.1 Differential and Total Cross Sections and Rate Constants

The differential cross section, usually determined in an experiment, is defined as the number of particles scattered at the scattering center in direction (θ,ϕ), per unit solid angle, per unit incident flux. The total intensity of scattering, the total cross section, defined by integrating the differential cross section over all solid angles, has a dimension of area or $(\text{length})^2$. Since the incident particles normally have a distribution of velocity, $f(v)$, the rate constant $k(v)$ may be defined as

$$k(v) = \int_0^\infty dv\, v f(v)\, \sigma(v). \tag{15}$$

4.2 Stopping Power

When energetic particles traverse a material, they lose their kinetic energies in a series of collisions with constituent atoms and molecules in the material. The average rate of energy loss in a collision that causes a transition from i to j is an important measure characterizing energetic particles and is defined as

$$-\left.\frac{dE}{dt}\right|_{i\to j} = v_A \Delta E \sigma_{i\to j} n_B \equiv v_A n_B S_{ij}. \tag{16}$$

Here, v_A is the velocity of the incident particle A, n_B is the density of B atoms or molecules in the material, and ΔE is the average energy loss of the incident particle A for the process $i \to j$ with cross section $\sigma_{i\to j}$. The quantity S_{ij}, sometimes referred to as the stopping cross section, has units of (area × energy). By summing over all possible processes, one can obtain the energy loss per unit of path length (dx) or the stopping power:

$$-\frac{dE}{dx} = n_B \sum_j \Delta E_j \sigma_{i\to j}. \tag{17}$$

The stopping power is directly measurable experimentally.

4.3 Diffusion Constant and Mobility

As a slow neutral particle or ion moves through a gas or condensed matter under the influence of a static, uniform electric field or as a result of density or temperature gradients that act as thermodynamic driving forces, the particle or ion gains energy from the field and gains or loses energy through collisions, finally attaining a steady-state condition. In a steady state, the particle flux is proportional to the applied field $\mathscr{E}$ (external field or density/temperature gradient) with a multiplication constant K. The constant K is called the mobility (or diffusion constant) of the atoms and is usually expressed in units of cm^2/$\mathcal{E}$-s.

5. APPLICATION AND RELEVANCE

Knowledge of the spectroscopy, dynamics, and production of specific species in collision processes involving neutral atoms and molecules is important because of its intrinsic scientific interest; its significance in other fields of science; and its application in technological and engineering areas. Here we summarize representative examples of scientific and technological applications of this knowledge.

5.1 Scientific Applications

5.1.1 Dynamics: The Determination of Interaction Potential Measurement of the differential cross section as a function of scattering angle at a given collision energy provides information on the nature of the interaction between two particles. For simpler, but difficult, systems like H + H and H + He, a rigorous quantum-mechanical calculation for elastic and inelastic processes can easily be performed. The result thus obtained can be compared readily with the measurement as a test of the precision of the interaction potential used. Repeating the calculation with refined potential parameters allows one to achieve better agreement with the measurement and, hence, better understanding of the electronic structure of the system and its collision dynamics. For example, an experimental differential cross section for the small-angle elastic collision of He atoms with He atoms at 0.5 keV is shown in Fig. 3(a), along with the results of quantum-mechanical calculations using different types of interaction potentials. Accurate determination of the interaction potential of the He-He system is crucial in calculations of bulk properties of liquid and solid He and transport properties of He vapor in conjunction with superfluidity and Bose–Einstein condensation. The potentials used in the calculation are shown in Fig. 3(c). The experimental data agree well with calculations using three types of potentials in the scattering angle range below 0.1°, suggesting that all three potentials at large R (the long-range attractive part) are reasonable; however, at angles larger than 0.1°, the experi-

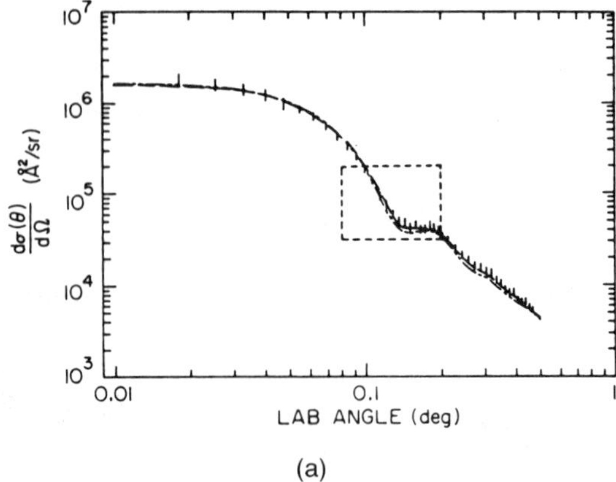

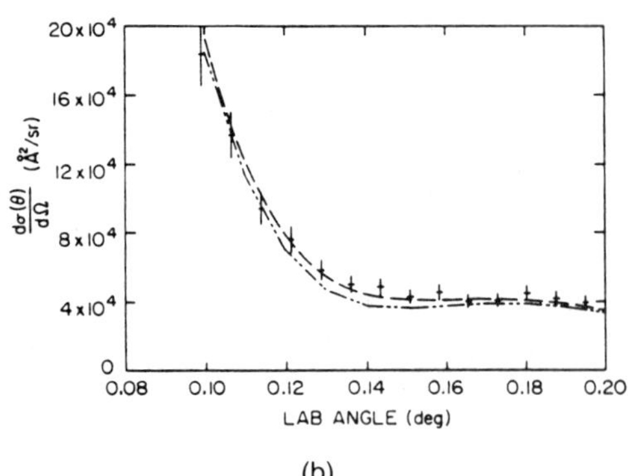

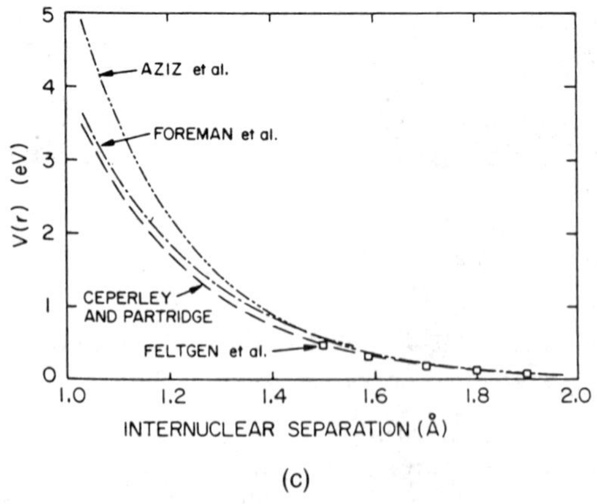

FIG. 3. (a) Measurements and theoretical calculations of differential elastic cross sections in collisions of He atoms with He atoms (from Nitz *et al.*, 1987). (b) Enlarged portion at $0.08 < \theta < 0.2$ of the differential elastic cross section shown in (a). (c) Proposed potential curves of the He-He ground state (from Nitz *et al.*, 1987).

mental data compare more favorably to the results of the calculation using the least steep portion of the potential at $R < 1.8$ Å.

5.1.2 Spectroscopy: The Determination of Physical Constants and Identification Probing molecular beams, coupled with photons that considerably increase spectral resolution over a wide energy range, are a central feature of modern spectroscopy. The interactions between atoms or molecules and a radiation field can be analyzed to reveal a variety of physical properties that characterize atomic or molecular structures and electronic states with higher precision, permitting higher efficiency and optimization in the design of experimental and engineering apparatus. Magnetic resonance, electron resonance, infrared laser, microwave, infrared, ultraviolet, and photofragment spectroscopies have particularly benefitted.

5.2 Technological Applications

5.2.1 Quantum Electronics: The Gas Laser Taking the He-Cd^+ ion laser as an example, we describe here the basic principle of the laser and its relationship with collisions of neutral atoms. The Cd^+ ions excited during collisions with He atoms decay to lower states, emitting photons that are known to be an efficient laser source. The fundamental goal for this laser is to maintain a large fraction of excited Cd^{+*} states so that collision-induced photon emission occurs continuously (inversion of population). To this end, two processes are known to be efficient. One is based on charge transfer by the ion (p2b),

$$He^+ + Cd(5s^2\,{}^1S_0) \rightarrow He + Cd^{+*}(6f\,{}^2F, 6g\,{}^2G). \tag{18}$$

The excited Cd^{+*} ion decays to a lower level by emitting a photon ranging in wavelength from 6360 to 7237 Å. The second process is based on Penning ionization (p7):

$$He^*(2^3S) + Cd(5s^2\,{}^1S_0) \rightarrow He(1^1S) + Cd^{+*}(4d^9 5s^2\,{}^2D) + 1e^-. \tag{19}$$

Here, the excited $Cd^{+*}(4d^9 5s^2)$ ions emit laser photons of 3250 and 4416 Å. As this example shows, knowledge of the various inelastic processes in the atomic collision is essential for developing an efficient laser.

5.2.2 Plasma and Fusion Research Neutral-particle beams are frequently used for plasma diagnosis. Injection of neutral particles into a plasma or fusion reactor and study of the density distribution, charge distribution, and emitted photons or the scattering angles of the neutral-particle flux exiting from the plasma can provide information on the plasma's temperature and density and various parameters that characterize the spatial and time distribution of the plasma. In addition, careful analysis of this information may lead to an understanding of the distribution of an impurity and its mean charge distribution in the plasma environment.

5.2.3 Surface Sciences Neutral-atom scattering from surfaces has become a vitally important technique for the study of surfaces. In particular, He scattering has several advantages because it is a highly surface-sensitive, nondestructive probe. Unlike what happens in electron or heavy-ion scattering, low-energy (millielectronvolt) He atoms colliding with a surface do not penetrate deep into the bulk. Thus, any information taken away by the scattered He atom can be related to the nature of the outermost surface layer of the bulk solid, whether the information concerns structure or dynamics. Furthermore, analysis of the spectrum is easier because the He beam is inert and nonreactive and has no internal degrees of freedom except large electronic excitation energies. Hence, the surface structure or dynamics of virtually any material, in the form of a static state, under the influence of an external field, or during surface processing, can be studied without altering the material.

Structural and dynamic phenomena are directly accessible to He scattering. By measuring elastic scattering, we can reconstruct the map of the electronic charge density of the atoms on the surface. In inelastic He scattering, phonons, surface vibrations, and even adsorbate vibrations leave their signatures in the scattered-He spectrum. As a representative example, a He scattering intensity spectrum from a Pt(111) surface with and without a low density of randomly adsorbed CO molecules is illustrated in Fig. 4(a). This experiment provides information on the density and location of adsorbed CO molecules. This type of experiment also gives a useful

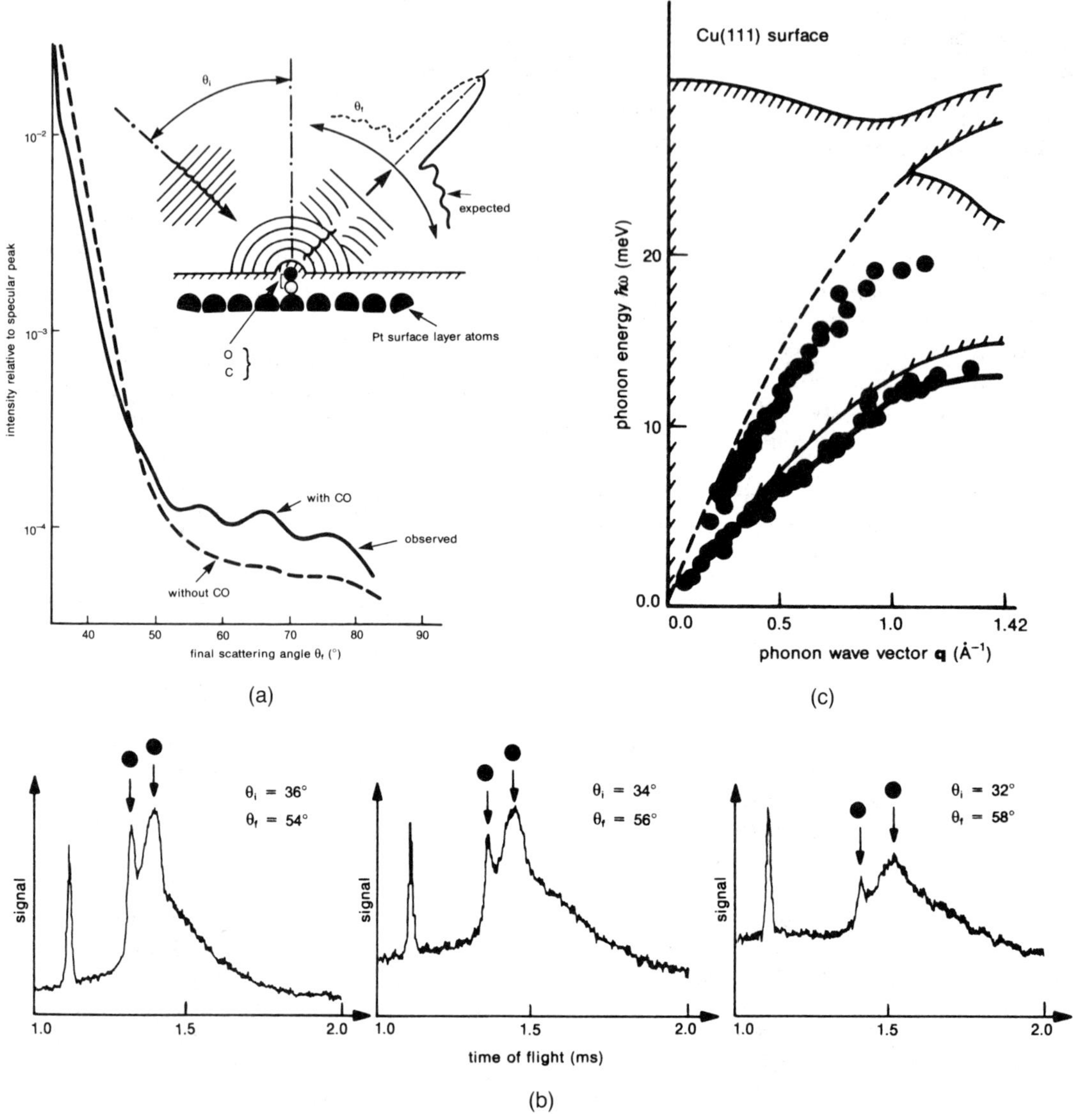

FIG. 4. (a) Helium scattering intensity spectrum from Pt(111) surface (from Lahee and Toennies, 1993). (b) Time-of-flight spectra for He atoms scattered from Cu(111) surface (from Lahee and Toennies). (c) Surface phonon dispersion relation curves (from Lahee and Toennies).

method for analysis of the structures and densities of defects. Figure 4(b) displays a series of time-of-flight energy-loss spectra for He atoms scattered from a Cu(111) surface. Each peak shows the signal of energy loss of the He atom. The relative shape of the peaks represents the signature of single-phonon creation. From these energy-loss spectra, the dispersion curves can be constructed, as shown in Fig. 4(c). The dispersion curves contain much useful information on the forces binding atoms at the surface.

5.2.4 Energy *Combustion.* Combustion is essentially a gas-phase reaction. Detailed kinetic knowledge is required to understand the reaction process, and this knowledge is an important ingredient in models to optimize fuel efficiency, for example. Knowledge needed includes the identities and concentrations of combustion species and unwanted products (pollutants) in both ground and excited states, the reaction dynamics, and the reaction cross section or rate coefficients.

Magnetohydrodynamic electrical power. The

essential idea in magnetohydrodynamics (MHD) is to extract an electrical current from a hot gas without a mechanical device. A hot gas, produced, for example, by burning a fossil fuel, is exhausted, after ionization, through a magnetic field. The interactions of the magnetic field and the flow velocity create an electric field transverse to the flow direction, producing an electric current across the field that can be extracted for commercial use. One of the major advantages of MHD electrical power is its increased efficiency over conventional mechanical generators. Penning ionization (p7) is known to play an essential role in the physics of this process.

6. CONCLUSION

We have provided the information necessary for understanding a variety of phenomena arising from the application of neutral atomic and molecular processes in experimental and theoretical instrumentation. Recent advances in technologies, such as the production of high-power, high-precision lasers and high-intensity beams, have significantly improved our basic understanding of chemical processes. We are now in an era of femtochemistry, in which an experiment can directly monitor bond breaking and bond formation during a chemical reaction in units of femtoseconds (10^{-15} to 10^{-12} s). We can witness every movement of the atoms and molecules in the chemical process at any instant. These refinements in experimental information on chemical processes demand more refined and precise theoretical treatments to match the experimental advances. With the recent development of the scanning tunneling microscope, nanometer technology or angstrom technology has emerged and is becoming the standard in high-tech industries such as the semiconductor industries. Combining femtochemistry with nanometer technology is the ultimate goal in the application of chemical processes at atomic and molecular levels within the foreseeable future.

ACKNOWLEDGMENT

This work was supported by the U.S. Department of Energy, Office of Energy Research, Office of Health and Environmental Research, under Contract No. W-31-109-Eng-38.

GLOSSARY

Born Approximation: One of the perturbation theories that is frequently used for solving quantum-mechanical scattering problems. Interaction potentials between a projectile and target are often chosen as perturbations to cause a transition in the system. When the interaction is of the lowest order, it is termed the first Born approximation.

Born–Oppenheimer Approximation: An approximate method to solve the Schrödinger equation for a molecular system on the basis that large differences in masses of electrons and nuclei leads to the separation of the electronic wave function and the nuclear wave function.

Charge-Transfer Beam: A neutral-particle beam produced by effectively using the charge transfer of an ion beam through a neutral gas.

Close-Coupling Method: A theoretical approach for treating many processes simultaneously and equally. Usually, the coupled equations are solved numerically to extract scattering amplitudes or the scattering S-matrix.

Momentum-Transfer Cross Section: Cross section for a collision with momentum transfer but no change in the internal state of a target. Momentum transfer occurs because momentum is conserved in the system. The momentum-transfer cross section σ_{MT} for scattered angle θ is defined as

$$\sigma_{\mathrm{MT}} = 2\pi \int d\theta \sin\theta(1 - \cos\theta) \frac{d\sigma_{\mathrm{el}}}{d\Omega}, \tag{20}$$

where $d\sigma_{\mathrm{el}}/d\Omega$ is the differential elastic cross section per unit of solid angle.

Penning Ionization: In collisions between an excited species and a ground-state species, ionization by interaction may be possible if the excitation energy of the excited species exceeds the ionization energy of the target. This process is called Penning ionization. A good example is $\mathrm{He}^*(2^2P) + \mathrm{Ar} \rightarrow \mathrm{He} + \mathrm{Ar}^+ + 1e$, where the excitation energy of $\mathrm{He}(2^2P)$ is 21 eV, and the ionization potential of Ar is 16 eV.

Perturbation Theory: In quantum mechanics, the total Hamiltonian is divided into two parts. The Schrödinger equation corresponding to one part can be solved exactly for the energy and the wave function (the unperturbed wave function). The other part corresponds to a correction that is approximately calculated by using the unperturbed wave function. The results from the unperturbed and perturbed portions are added to give an approximate total energy of the system.

Seeded Supersonic Beam: A neutral-particle beam obtained by using supersonic expansion of a gas.

Sputtered Beam: A fast neutral-particle beam produced by bombarding solid targets with energetic ions.

Stopping Power: An experimentally measurable quantity that describes the rate of average energy loss of a particle passing through a material.

Works Cited

Lahee, A., Toennies, P. (1993), *Phys. World* **6** (4), 61–66.

Nitz, D. E., Gao, R. S., Johnson, L. K., Smith, K. A., Stebbings, R. F. (1987), *Phys. Rev. A* **35**, 4541–4547.

Phelps, A. V. (1990), *J. Phys. Chem. Ref. Data* **19**, 653–675.

Further Reading

Fluendy, M. A. D., Lawley, K. P. (1973), *Chemical Applications of Molecular Beam Scattering*, London: Chapman and Hall.

Kimura, M., Lane, N. F. (1989), in: Bates, D., Bederson, B. (Eds.), *Advances in Atomic, Molecular and Optical Physics*, Vol. 26, New York: Academic Press, p. 79.

Lahee, A., Toennies, P. (1993), *Physics World* **6** (4), 61–66.

Massey, H. S. W., McDaniel, E. W., Bederson, B. (Eds.), (1982), *Applied Atomic Collision Physics*, Vols. 1–5, New York: Academic Press.

Nitz, D. E., Gao, R. S., Johnson, L. K., Smith, K. A., Stebbings, R. F. (1987), *Phys. Rev. A* **35**, 4541–4547.

Phelps, A. V. (1990), *J. Phys. Chem. Ref. Data* **19**, 653–675.

Prigogine, I., Rice, S. A. (Eds.), (1975, 1992), *Advances in Chemical Physics*, Vols. 30 and 83, New York: John Wiley & Sons.

Scoles, G. (Ed.), (1988), *Atomic and Molecular Beam Methods*, Vols. 1 and 2, Oxford, UK: Oxford University Press.

NEUTRON DIFFRACTION

THOMAS VOGT, *Physics Department, Brookhaven National Laboratory, Upton, New York, U.S.A.*

INTRODUCTION

Neutrons constitute a very important probe for condensed matter research. This is a consequence of their unique properties. The neutron is an uncharged particle. To detect neutrons one must therefore rely on nuclear reactions producing charged particles. The nuclear reaction on which the most common detector for thermal neutrons, the ^{3}He gas detector, is based is

$$n + {}^3\text{He} \rightarrow {}^3\text{H} + {}^1\text{H} + 0.764\ \text{MeV}. \quad (1)$$

The proton (^{1}H) and triton (^{3}H) move away in opposite directions from the neutron absorption site and collide with other gas molecules, ionizing them (see DETECTORS, PARTICLE, TRACKING). The neutron has a mass of 1.0087 u. As for any elementary particle, the wave-particle formalism using de Broglie's equation $\lambda = h/mv$ allows the energy to be expressed as

$$E(\text{meV}) = 0.08617\, T\,(\text{K}) = 5.227\, v^2\,(\text{km s}^{-1}) = 81.81/\lambda^2\,(\text{Å}). \quad (2)$$

Its mass allows the neutron to be slowed down after its creation in a fission or spallation process. This so-called moderation process occurs mostly via inelastic neutron-proton collisions. The neutrons adapt to the temperature of the hydrogen-containing moderator. In a research reactor, this moderation process leads to a distribution of neutron wavelengths between 1 and 3 Å (for 300 K, λ = 1.78 Å) and energies between 80 and 10 meV. These values correspond to typical interatomic distances and energy values of thermal excitations (phonons, magnons) in condensed matter. Inelastic neutron scattering first made the phonon "visible" and provided various crucial experiments exploring the dynamical nature of condensed matter (see NEUTRON SCATTERING).

Here we shall focus on structural investigations using neutron diffraction. Neutrons do not primarily interact with the electrons of matter but instead with the nucleus. The nuclear forces have a very short range, on the order of a few femtometers (1 fm = 10^{-15} m).

Thermal neutrons "see" very little when passing through matter, since the size of the scattering centers is about 10^5 times smaller than the distances between them. An important consequence of this is that the nucleus can be effectively considered as a point scatterer and the neutron is therefore scattered isotropically. The differential cross section describing the interaction of neutrons with the nucleus is a constant, $\delta\sigma/\delta\Omega = b^2$. In contrast to this, x rays are not scattered isotropically since the size of the "electron cloud" is comparable to the size of the x-ray wavelengths typically used in a diffraction experiment. This is why one observes a decrease

3-527-28133-9/94/$5.00 + .50

of the interaction between x rays and electrons at high scattering angles, and it is difficult to measure intensities at very high scattering angles (see X-RAY DIFFRACTION).

The constant b is the so-called scattering length and is a property of the nucleus with a given atomic number Z and weight A as well as the spin state relative to that of the neutron. The interaction of neutrons with the nucleus allows negative scattering lengths corresponding to a phase change of π during the scattering process.

Two main fields of applications in scattering experiments arise from the above:

1. The low-Z ("lighter") elements may scatter more strongly than the "heavier" ones. Neighboring elements may vary quite a bit in their scattering lengths. To distinguish, for instance, between Mn and Fe, which have one electron difference, using x rays is virtually impossible without measuring at an absorption edge. With neutrons, however, there is a big difference in their coherent scattering length ($b_{Mn} = -3.73$ fm, $b_{Fe} = 9.45$ fm) as shown in Fig. 1. Neutron diffraction is an indispensable tool to determine cation distributions in ceramics and minerals containing elements with little difference in Z.
2. Isotopes of the same atom (same Z but different atomic weight A) may also vary dramatically in their scattering power and might even have scattering lengths with opposite signs, as in the case of hydrogen (-3.74 fm) and deuterium (6.674 fm) or the isotopes ^{62}Ni (-8.7 fm) and ^{61}Ni (7.60 fm) or ^{48}Ti (-6.08 fm) and ^{50}Ti (6.18 fm). Experiments exploiting this by changing the scattering length of a particular element via isotopic substitution are called contrast variation experiments and are extensively used in structural investigations of a wide variety of matter, including liquids, metals, polymers, and glasses. In biological applications, the variation of the H_2O/D_2O ratio changes the "visibility" of water in a diffraction experiment and might give important clues to its location, for instance in proteins.

When low-energy ("cold") neutrons scatter at small angles from biological samples in solution, the measured intensity depends on the "contrast" between the molecules and the solvent; that is, the difference between their scattering-length densities. If one is dealing with a multicomponent solution, as one often does in biology, one can match the scattering-length density of the solvent by adjusting the ratio $H_2O:D_2O$ and thus eliminate the scattering contribution of that par-

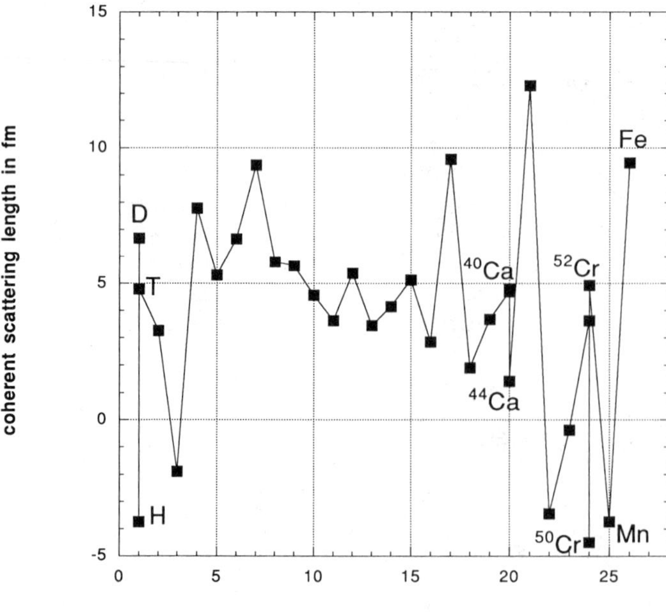

FIG. 1. The variation of the coherent scattering length in femtometers for the elements $Z = 1$ (H, D, T) to $Z = 26$ (Fe). For the cases of $Z = 1, 20, 24$ the coherent scattering lengths of the major isotopes are shown. For further details see text.

ticular component. This effect can also be used when studying metal or polymer alloys, liquids, and glasses by varying the isotopic ratio of a particular element that has isotopes with opposite scattering lengths (e.g., Ni, Ti). See Jacrot (1976) for more details.

Since the interaction of neutrons with the nuclei of matter occurs within a much smaller area than the one between x rays and electrons, and since there is no Coulomb barrier to overcome because the neutron is uncharged, the penetration depth of neutrons is much bigger than that of x rays. Neutrons are a true bulk probe. This has the advantage that sophisticated sample environments such as pressure cells, furnaces, cryostats, and magnets can be more easily penetrated and, provided good collimation, their scattering will not mask the actual measurement. *In situ* experiments are generally more cumbersome to do with x rays.

On the other hand, the typical flux of a monochromated neutron beam is in the order of 10^6 to 10^8 neutrons cm^{-2} s^{-1}. This is comparable to the flux of molecules in a vacuum and explains why in neutron scattering experiments—especially when measuring weak signals as in inelastic scattering—typical sample volumes of about 1 cm^3 have to be used to achieve reasonable counting rates.

One property that really put neutrons in the spotlight of attention about 50 years ago was the fact that the neutron has a magnetic moment of $1.9132\mu_N$, which interacts via a dipole-dipole interaction with the magnetic moments of unpaired electrons. The first solid-state magnetic structure investigation using neutrons was performed by Shull and Smart (1949). They showed that at low temperature the MnO diffraction pattern reveals extra Bragg peaks ("magnetic" Bragg peaks) because the magnetic moments of the manganese are aligned along the edges of the unit cell pointing in opposite directions. This leads to a magnetic cell twice as big as the chemical cell. This was the experimental proof of antiferromagnetism as predicted by Néel. It is safe to say that the scientific contributions that neutron scattering provided in the field of magnetism are of paramount importance and will continue to be. Neutron diffraction is for the moment the only method to determine magnetic structures of bulk samples with atomic resolution.

The magnitudes of magnetic and nuclear scattering are comparable. However, because of the spatial dimension of the unpaired electron density, the scattering process is no longer isotropic, and scattering-angle–dependent magnetic form factors describing the strength of the scattering as a function of the scattering angle exist similar to the ones observed in x-ray scattering. This form factor is the Fourier transform of the unpaired electron density. For more details, see MAGNETIC ORDERING IN SOLIDS.

The neutron has a spin angular momentum of $\frac{1}{2}$. One is able to polarize, that is, separate, neutrons into beams with spins of $\pm\frac{1}{2} \times h/2\pi$. One can polarize neutrons using Bragg reflections if the nuclear and magnetic contributions to the scattering amplitude compensate exactly for one of the two spin states. This is achieved, for instance, when using the (111) reflection of Heusler alloys such as Cu_2MnAl and CoFe.

Since only the perpendicular component of the magnetic moment with respect to the scattering vector contributes to magnetic scattering, nuclear and magnetic scattering contributions can be separated (see NEUTRON SCATTERING).

1. SCATTERING THEORY

Independent of the specific type of interaction a probe has with matter (electrostatic for electrons, electromagnetic for x rays, and nuclear for neutrons), a similar formalism is used for scattering phenomena, upon which we will rely in the following (see Fig. 2).

There are two principal types of scattering: elastic scattering, where the probing particle is deflected without energy loss or gain during the scattering process, and inelastic scattering, where the probing particle loses or gains energy. The energy gain or loss is called the energy transfer (for more details see NEUTRON SCATTERING). In both cases, the particle is scattered by the scattering angle 2θ. Scattering processes are often depicted using scattering triangles where a scattering vector $\mathbf{Q}$ is defined by the vector relationship $\mathbf{Q} = \mathbf{k} - \mathbf{k}'$ as shown in Fig. 2. Here $\mathbf{k}$ and $\mathbf{k}'$ are the vectors representing the incident and diffracted spherical wave fronts of the neutron. As can easily be demonstrated by simple trigonometric arguments, in the elastic case $\mathbf{Q}$ equals $(4\pi \sin\theta)/\lambda$. The scattering vec-

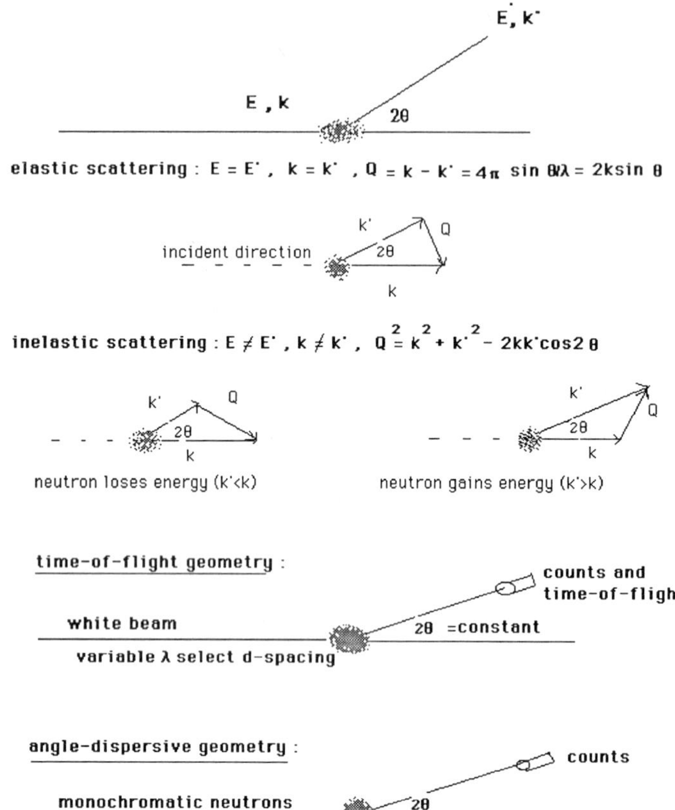

FIG. 2. Basic scattering terms used in the text. For more details see text.

tor multiplied by $h/2\pi$ is referred to as the momentum transfer.

If the probing particles interact with the scattering centers in such a way that the scattered waves have a phase relationship with each other, one speaks of coherent scattering and interference is possible. Diffraction or Bragg scattering is the simplest form of coherent scattering. If the probing particles interact in an independent, not phase-related manner with the different scattering centers, one refers to this as incoherent scattering. Here no interference is possible. The intensities from the different scattering centers simply add up. In the case of neutrons this is very important, since one can obtain information about a specific scattering species at different positions and times (see NEUTRON SCATTERING). In diffraction experiments the background can be dominated by incoherent scattering, often due to the presence of hydrogen in the material, and when performing, e.g., powder diffraction experiments, where the Bragg reflections are much weaker than those of single crystals, the samples containing hydrogen will often have to be deuterated to improve the signal-to-noise ratio.

2. NEUTRON DIFFRACTION

The basis of all diffraction is Bragg's law (see CRYSTALLOGRAPHY):

$$\lambda = 2d \sin\theta. \tag{3}$$

There are two principal ways of doing diffraction experiments:

1. If one measures with a monochromatic beam obtained by the use of an appropriate single-crystal monochromator (see MONOCHROMATORS), thus leaving the selected narrow wavelength band $\Delta\lambda/\lambda$ constant and varying the scattering angle θ,

one is using a so called angle-dispersive or single-wavelength setup.

2. If, on the other hand, one leaves the scattering angle θ constant using a beam with a broad distribution between λ_{min} and λ_{max}, one is measuring in an energy-dispersive setup. When using neutrons this is also referred to as the time-of-flight technique.

The angle-dispersive setup is mostly used at steady-state research reactors and x-ray synchrotrons.

Differentiating Bragg's law one obtains

$$\Delta\lambda/\lambda = \Delta d/d + \Delta\theta \cot\theta. \quad (4)$$

For both energy and time-of-flight setup, the resolution is given by

$$\Delta d/d = \Delta\theta \cot\theta. \quad (5)$$

To obtain a good resolution (small $\Delta d/d$) a high scattering angle has to be chosen. Neglecting collimation, the intensity depends on the wavelength band reflected off the monochromator:

$$\Delta\lambda/\lambda = \Delta\theta_M \cot\theta_M. \quad (6)$$

The quantity $\Delta\theta_M$ is the mosaic spread of the monochromator crystal reflecting at a monochromator scattering angle $2\theta_M$, referred to as the takeoff angle (for more see MONOCHROMATORS).

The advantage of the time-of-flight setup is that one can choose the scattering angle so that little or no parasitic scattering from the sample environment masks the pattern of the sample. This is very important, for instance, in high-pressure work: in the angle-dispersive technique one has to use a very small beam to reduce the background scattering from the pressure cell. In the time-of-flight technique one can position the detector at a scattering angle of 90°, where one does not detect any Bragg reflections originating from the pressure cell. This does, however, lower the resolution and limit the **Q** range. Most of the high-pressure work using neutrons these days is done using neutrons created in spallation sources.

In the time-of-flight technique, the time it takes the neutrons to travel from the source to the detector ("time of flight") allows the separation of the Bragg peaks corresponding to their different d spacings. If L is the flight path, ΔL its uncertainty (since the neutrons are generated in a finite moderator slice and detected in a finite detector), t the time of flight, Δt its uncertainty, and θ the scattering angle with its uncertainty $\Delta\theta$, then the resolution of a time-of-flight powder diffractometer is given by

$$\Delta \mathbf{Q}/\mathbf{Q} = [(\Delta t/t)^2 + (\Delta L/L)^2 + (\cot\theta\Delta\theta)^2]^{1/2}, \quad (7)$$

and the flight time t can be related to the flight path L, the d spacing d_{hkl}, and the wavelength λ as follows:

$$t = 505.555\, L\,[\mathrm{m}]\, d_{hkl}\lambda\,[\text{Å}] \sin\theta_0. \quad (8)$$

This explains why a long flight path L and a high scattering angle (near 180°, the so-called backscattering geometry) are advantageous. The flight path on the high-resolution powder diffractometer at the ISIS neutron spallation source in the United Kingdom, for instance, is 96 m. If one were to use only an array of detectors covering a small angular range close to 180°, the largest d spacing that could be observed using a neutron pulse with a wavelength distribution $\lambda_1 < \lambda < \lambda_2$ is $\lambda_2/2$, whereas at $2\theta = 90°$ a larger d spacing of $\lambda_2/2^{1/2}$ can be detected. Since the observation of large d spacings is crucial for the determination and refinement of magnetic structures as well as for the indexing of unknown phases during the course of *ab initio* structure determinations, constant-wavelength diffractometers offer an advantage for this kind of investigations. On the other hand, constant-wavelength instruments are not able to provide intensities in such a wide range of low d spacings. This is important for the accurate determination of atomic displacement parameters, strain tensors, and particle size distributions, provided all wavelength-dependent corrections are taken into account appropriately.

2.1 Medium-Range Order in Glasses

Silicate glasses were assumed to be built of silicate frameworks consisting of SiO_4 tetrahedra, in between which the cations were randomly distributed as network modifiers. Neutrons are of tremendous use when investigating structural features of glasses, liquids, and dense gases since one can obtain intensities from high scattering angles and separate the various partial structure factors (see CRYSTALLOGRAPHY) of a multicomponent system by measurements using different isotopic compositions.

Neutron scattering experiments using three specimens of a calcium silicate glass with different isotopic concentrations of calcium revealed that nonrandom correlations beyond the nearest-neighbor distances exist among the cations. (See Fig. 1 for the different scattering lengths of the Ca isotopes.) Gaskell *et al.* (1991) provided the first experimental evidence of medium-range order in the cation distribution of this calcium silicate glass and sparked a series of other investigations. By measuring the differences between the scattered intensities of the glass containing the natural isotopic mixture of calcium and one with equal amounts of natural Ca and ^{44}Ca and the difference between a glass containing a mixture enriched with ^{44}Ca and pure ^{44}Ca one removes all terms contributing to the total structure factor other than the Ca-centered correlations such as Ca-Si, Ca-O, and Ca-Ca. Then, by taking the difference of these differences, all terms except the Ca-Ca correlations cancel. Even though this double-difference signal is very weak and subject to systematic and random errors, the authors clearly observed a second neighbor distance near 6.4 Å. This is consistent with a calcium ordering scheme in which each calcium is located at the center of an octahedron of oxygens. They share edges and create sheetlike regions within the silicate matrix. This arrangement resembles the connectivity in the mineral wollastonite. If all the sheets were connected smoothly beyond the medium-range order, this crystalline structure would be realized. This work has led to a substantial change of our view on glasses: the cations, which have been previously regarded as network discontinuities, are actually ordering elements, and the long-range disorder in glasses coexists with short- and medium-range order. Glasses and their corresponding crystalline structures may have more in common than previously thought.

2.2 Neutron Powder Diffraction

The recent discovery of high-temperature superconducting ceramics and the unraveling of their structural details demonstrated the indisputable utility of neutron powder diffraction. Structure solution and refinement are no longer the exclusive domains of single-crystal diffraction. Powder diffraction is the only tool to obtain structural information of many samples like catalytically active zeolites, and low-temperature and high-temperature phases where single crystals are not available. At the same time, information about the mesoscopic properties (e.g., texture, particle size, stacking faults) of the sample can be obtained. Modern powder methods have developed from a qualitative method previously used for phase identification to a quantitative method to detect phases and determine their volume fraction and atomic and mesoscopic structure.

2.2.1 *Ab Initio* Structure Determination

This area has created a lot of excitement in recent years, since historically it was the exclusive domain of single-crystal diffraction. A powder diffraction experiment provides two types of information: the position of the Bragg peaks, which is determined by the unit cell size and symmetry, and the intensity, which depends on the position of the atoms within the unit cell. Once a structural model is established (for further details see CRYSTALLOGRAPHY) this model will have to be adjusted by refining its crystallographic parameters (e.g., positions of the atoms within the unit cell) with respect to the measured quantities.

The availability of highly collimated monochromatic neutron and synchrotron powder diffractometers made it possible to obtain good enough data to tackle the task of structure determination from powder diffraction data. The first goal in that process is always the determination of the unit cell. Here, many times electron diffraction can provide help. After determining the unit cell and narrowing down the possible space groups, the deconvolution of the individual integrated intensities is the next step. This is referred to as pattern matching or whole pattern fitting and does not require a structural model. Given a space group, the diffraction pattern is refined using the individual intensities of the Bragg reflections, the cell constants, which determine their position, and the so-called machine parameters of the diffractometer describing the resolution (*UVW*, peak-shape parameters, etc.) as variables. The obtained intensities can then be used to obtain a structural model via direct methods or Patterson procedures (see CRYSTALLOGRAPHY for more details). The actual refinement is then usually done with the Rietveld technique once a structural model has been established. If one has a reasonable model, a Rietveld refinement can be used to explore the fit of various models to the data and reject or approve them. Zachariasen (1949) solved the monoclinic structure of α- and β-UF_5 this way already in 1949. This trial-and-error approach is still used quite often. An example is given by Cockcoft and Fitch (1988). When solving the low-temperature structure of the molecule SF_6, they used the intramolecular distances and angles as chemical restraints in the least-squares refinements to ensure that the SF_6 octahedra did not distort beyond unphysical limits while exploring various intermolecular configurations.

The use of these subsidiary conditions by introducing prior chemical knowledge (e.g., distance and/or angle restraints) forces the minimized quantity of the least-squares refinement to satisfy the diffraction profile *and* the imposed model. This can be used, as in the above-mentioned case, to reduce the number of parameters or to obtain accurate structural parameters when refining very complex structures such as the one mentioned in Fig. 3.

One of the first structure determinations

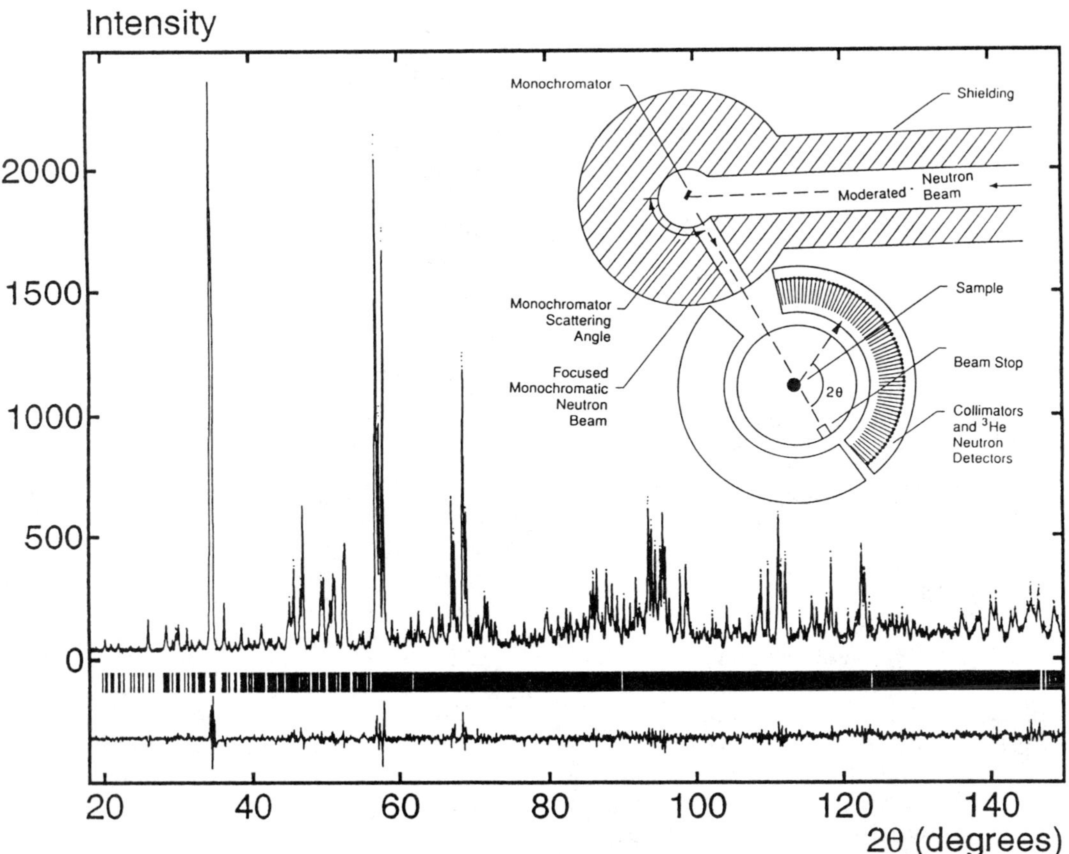

FIG. 3. A typical high-resolution neutron powder diffraction pattern recorded with the High-Resolution Neutron Powder Diffractometer at the High-Flux Beam Reactor at Brookhaven National Laboratory using a wavelength of 1.88 Å. A sketch of the basic scattering geometry of the instrument is shown. The material examined is the high-temperature form of Bi_2MoO_6, a prototypical catalyst. It has a monoclinic unit cell with a = 17.2623(1) Å, b = 22.4297(2) Å, c = 5.584 86(5) Å, and β = 90.4975(6)°. The structure was refined using the Rietveld technique (see text for details) using 2559 reflections whose positions in 2θ are indicated by the tick marks below the pattern and refining 146 parameters. The differences between the refined model and the data are given in the lower part of the plot.

in modern days that focused the general attention to solving structures from neutron powder diffraction data was done by Attfield in 1986. One of the more complicated structures solved was $Ga_2(HPO_3)_3 \cdot 4H_2O$, a novel framework structure with 29 atoms in the asymmetric unit cell and 117 structural parameters (Morris *et al.*, 1993). When solving structures using only powder diffraction the complementarity of neutron and x ray synchroton radiation is often very helpful. For a recent review see Cheetham *et al.* (1992).

One of the main reasons why "classical" direct methods to solve structures from powder data alone are not always successful is their need for atomic resolution and an extensive sampling of reciprocal space. The most crucial step in extracting information out of a complex powder pattern is the way heavily overlapped reflections are dealt with. Instead of ignoring them or assigning equal intensities to each particular reflection, maximum-entropy methods attempt to deal with them so that they are as noncommittal as possible with respect to the other data. For an example see Tremayne *et al.* (1992). The basic theory of the use of maximum entropy in crystallography and in powder diffraction in particular has been laid out by Bricogne (1991).

Besides solving structures, maximum-entropy methods are also used to reconstruct Patterson and Fourier densities. Papoular *et al.* (1991) established the superior quality of reconstructed densities when using maximum-entropy methods compared to standard Fourier methods. The low noise and smaller truncation effects allow the display of the disordered density.

2.2.2 The Rietveld Method The Rietveld technique makes use of the fact that the peak shape of Bragg reflections can be described analytically and the variation of their widths (FWHM, full width at half-maximum) can be expressed as a convolution of the optical characteristics of the diffractometer (takeoff angle θ_M, collimation, etc.) with sample-related effects (strain, particle size broadening, etc.). This allows the least-squares refinement of crystal structure parameters by fitting the diffraction profile without explicitly extracting structure factors or integrated intensities. Assuming a Gaussian shape for each Bragg reflection and referring to the original notation of Rietveld (1969), the contribution of a reflection k to a measured profile point y_{ik} at the position $2\theta_i$ is

$$y_{ik} = aj_k L_k S_k^2 \times 2[\ln 2/H_k \pi^{1/2}]^{1/2} \times \exp[-4\ln 2\{(2\theta_i - 2\theta_k)/H_k\}^2], \quad (9)$$

with a being the scale factor relating the integrated intensities to the modulus of the structure factor, j_k the multiplicity of the reflection, L_k the Lorentz factor, S_k the sum of nuclear and magnetic contributions, $2\theta_k$ the calculated position of the Bragg reflection, and H_k the FWHM, which in the simplest case can be expressed as a function of 2θ according to Cagliotti *et al.* (1958) in the following way:

$$H_k = (U\tan^2\theta_k + V\tan\theta_k + W)^{1/2}. \quad (10)$$

Rietveld refinement nowadays is the method with which neutron and x-ray powder diffraction data are analyzed. David *et al.* (1993) demonstrated the structural insight and details that modern high-resolution neutron powder diffraction can reveal when determining the structure of buckminsterfullerene (C_{60}) as a function of temperature. A very careful analysis of the low-temperature phase showed that the C_{60} molecules reorient between two energetically very close orientations. In one of the orientations, the pentagons face carbon-carbon double bonds in a neighboring molecule; in the other, energetically less favorable orientation the hexagons face the carbon-carbon double bonds. These results underline the fact that modern neutron powder diffraction can yield very accurate results, comparable in some cases to those obtained by single-crystal diffraction.

2.2.3 The Analysis of Mesoscopic Structures Using High-Resolution Powder Diffraction With the development of high-resolution neutron powder diffractometers in the range of $\Delta d/d \sim 5 \times 10^{-4}$, it was quickly realized that the resolution was so good that the mesoscopic structure of the sample became accessible and sometimes gave problems that had to be dealt with to be able to determine the atomic structure correctly.

Under mesoscopic structures we summarize structural inhomogeneities in materials beyond interatomic distances on length scales of tens to hundreds of angstroms, e.g., twin and domain structures, antiphase boundaries, or stacking faults. Since these effects manifest themselves on length scales beyond atomic resolution they reveal their presence in powder pattern mainly as deviations from the "ideal" symmetrical peak shape and resolution. The modeling of the peak profile parameters becomes more complicated than indicated in the simple expression (9) by Rietveld, since the isotropic peak shape, which is independent of the lattice direction, has to be replaced by an anisotropic one. The physical origin of these profile shape variations with the lattice direction can be very diverse and include variations of interatomic distances due to internal stresses, variations in stoichiometry, stacking faults, dislocations, and many more. These variations are then convoluted with the optical properties of the diffractometer (e.g., collimation, takeoff angle). This is why the knowledge of the peak shape and resolution function (variation of the full width at half-maximum with the scattering angle 2θ) of a given high-resolution powder diffractometer is of crucial importance for these studies.

The two most commonly encountered mesoscopic structures in high-resolution powder diffraction are strain and particle size effects, which lead to a broadening of reflections. If one compresses a polycristalline material in a particular direction, then the d spacings are varied depending on the direction and strength of the stress with respect to the individual crystallites. If one measures the d spacings in different orientations of the unit cell to the stress, one observes variations. These strains can be converted into stresses using the elastic constants of the probed material. They are typically in the order of 10^{-4} in $\Delta d/d$, the resolution now accessible with high-resolution neutron powder diffractometers. The knowledge of the stress distribution in matter is not only important for applied science (engineering parts such as railway tracks, welds, etc.) and earth sciences (e.g., stress distribution in rocks), but may also play a major role in investigations of phase transitions, lattice defects, and compositional variation in materials.

Differentiating Bragg's law one obtains

$$\Delta d/d = -\Delta 2\theta(2\,\tan\theta) \tag{11}$$

or

$$\Delta 2\theta = -2\epsilon\,\tan\theta \quad \text{with} \quad \epsilon = \Delta d/d.$$

This shows that strain will manifest itself in the high-angle reflections.

One of the basic assumptions of the kinematical scattering theory applied in general to analyze diffraction data is that of an infinite lattice that allows the reciprocal lattice points to be treated as δ-functions. If the particle size diminishes, the coherently diffracting domains have a finite size, smearing out all the reciprocal lattice points to the same spread. The size broadening is described by the Scherrer relation:

$$\Delta 2\theta = \text{const}\ \lambda/\cos\theta\, T, \tag{12}$$

T being the thickness of the coherently diffracting domain. Both effects, particle size and strain broadening, can be present in the same material, and both will manifest themselves at high angles, where they can be separated as a result of their different variation with 2θ (size broadening increases with $1/\cos\theta$, strain broadening increases with $\tan\theta$). Equation (12) shows that particle size broadening is wavelength dependent. Time-of-flight neutron data have to be corrected for this, if the quantities that are to be determined like accurate atomic displacement parameters are not to be corrupted by systematic errors. One of the basic assumptions of the Rietveld method was the simple angular variation of the FWHM of a reflection profile with 2θ usually expressed using Cagliotti's formula (Eq. 10). With the continuing efforts of improving reactor- and spallation-source–based high-resolution neutron powder diffractometers, this simple angular variation will have to be replaced by more complex ones.

The investigation of mesoscopic features in materials will play a major role in the future. Their importance, especially in superconducting ceramics, has been recognized. The following examples illustrate this further.

1. Medarde *et al.* (1992) were able to model the anisotropic broadening of all reflections *hkl* with $h > 0$ in $La_{1.5}Sr_{0.5}NiO_{3.6}$ and understand its origin in an oxygen vacancy ordering introducing microstrain in the NiO_2 basal plane.
2. Roessli *et al.* (1993) found linewidth and shape anomalies *only* in the magnetic reflections of the low-temperature magnetic phase of $HoBa_2Cu_3O_7$ at 7 mK. These asymmetric line shapes could be ex-

plained as magnetic stacking faults along the *c* axis of the structure resulting from finite magnetic correlations that are in the order of 29 Å. The same type of magnetic mesoscopic structure was found at 10 K in samples of $PrBa_2Cu_3O_6$ by Guillaume *et al.* (1993).

The last two examples again underline the unique information one can obtain from neutron diffraction data with regard to magnetic structures.

2.3 Single-Crystal Neutron Diffraction

Classical areas of single-crystal neutron diffraction are

1. the location of "light" or neighboring elements;
2. the determination of magnetic structures;
3. the high-resolution structural investigations to detect subtle changes of crystal structures near structural phase transitions by analyzing atomic displacement parameters or to obtain accurate atomic positions for electron density studies.

In materials science and structural chemistry, a lot of problems concerned with the first two points can nowadays be tackled using powder diffraction. However, when one attempts to solve magnetic structures or refine biological structures, single-crystal diffraction will always be the method of choice. Protein crystallography using neutrons has not shown its full power yet, since it is difficult to grow single crystals of sufficient size and the fluxes of monochromatic beams available at the moment are often not high enough to tackle interesting problems. Even if oversimplified, the physicist's description of a protein as a molecule made up to roughly 50% by hydrogen serves to make the point concerning the inherent potential of neutron diffraction in the field of biological structures. The availability of new sources providing high-flux neutron beams together with advances in instrumentation—and here especially in two-dimensional position-sensitive detectors—will provide a way out of the current dilemma.

The high-resolution single-crystal experiments profit from the accuracy of high-angle intensities. The Fourier transform of the atomic displacement parameters (see CRYSTALLOGRAPHY) represents the probability of locating an atom in real space. This so-called probability density function reveals the dynamic and static disorder of atoms. At high temperatures or close to phase transitions the underlying harmonic approximation of the potential function may no longer be valid. By using higher expansion terms of the atomic displacement parameters these disorder phenomena can be quantified. The fact that the neutron scattering length is constant in **Q** space results in very accurate high-angle structure factors needed to determine reliable atomic displacement parameters. For more details see Kuhs (1992).

2.3.1 Two-Dimensional Position-Sensitive Detectors Single-crystal neutron diffraction is changing rapidly as a result of improvements in instrumentation and particularly of the availability of two-dimensional position-sensitive detectors. The classical method of measuring a diffraction pattern is to measure the intensity step by step using a single counter. This is very time-consuming and not very efficient since in general the number of reflections "available" in a given solid angle is in the dozens to hundreds in the case of proteins. The gain by measuring simultaneous reflection is enormous. Multidetectors can be one-dimensional when measuring simultaneously in one direction. They are used in powder diffraction where a whole powder pattern can be recorded "at once" without stepping the counter. For single-crystal diffraction one needs two directions along which one would like to measure simultaneously.

The use of two-dimensional position-sensitive detectors (2D-PSD) in single-crystal neutron diffraction has been shown to be very suitable in single-crystal time-of-flight measurements at spallation sources (Schultz *et al.*, 1984) and in reactor-based macromolecular crystallography (Schoenborn *et al.*, 1984), since many Bragg reflections are excited simultaneously and thus the time of an experiment can be reduced considerably. In protein crystallography this can determine if an experiment can be done at all.

Recently Lehmann *et al.* (1989) and McIntyre (1992) made the case for using a two-dimensional detector when analyzing single crystals with smaller unit cells. Besides the obvious advantages of a much faster measurement time, better separation from undesired parasitic scattering due to sample environments such as furnaces and pressure cells, and a more accurate integration of weak Bragg intensities because of the much better determination of the background, they demonstrated that when studying crystals with poor quality or twinning, the 2D-PSD offers enormous advantages. McIntyre and Renault (1989) were able to separate overlapping reflections of a twinned sample of $YBa_2Cu_3O_{7-\delta}$ and integrate the four domain orientations individually (see Fig. 4). They revealed that the difference in intensity between certain pairs of individual overlapping reflections was sensitive to the difference in the occupancy of oxygen sites in the plane between pairs of closest barium atoms.

Another neutron scattering technique that has benefited from the use of 2D-PSDs is small-angle neutron scattering (SANS). With SANS one probes mesoscopic features like aggregates of atoms or structural inhomogeneities in matter (voids in solids, precipitates, etc.). One uses long-wavelength neutrons, typically beyond 10 Å, and detects the elastic coherent scattering to obtain information about size and shape of the studied aggregates. The installation of cold sources (e.g., liquid-hydrogen moderator) to obtain higher fluxes of cold neutrons and the development of 2D-PSDs has allowed exciting new experiments in this field. Recently the detection of flux lines in high-temperature superconducting single crystals using SANS was a major step forward in our understanding of these materials (see NEUTRON SCATTERING). Neutron scattering is the only method by which one can measure *directly* the spatial variation of a magnetic field in the bulk of a superconductor.

2.4 Neutron Reflectometry

Scattering from a surface or interface without changing the energy of the probing particle and at the same glancing angle of incident and reflected beam is referred to as elastic, specular scattering (see Fig. 5). If the wave vector $\mathbf{Q}$ is such that $2\pi/Q$ is much bigger than the interatomic distances in the medium, it can be described as a continuum.

In vacuum, the neutron has no potential energy and the kinetic energy is $(h/2\pi)k_0^2/2m$. The potential energy of a neutron in the continuum limit is $(h/2\pi)bN/m$, N being the number of scattering nuclei and b the scattering amplitude. In a medium, the total energy is the sum of potential and kinetic energy. In analogy to classical optics, the medium is characterized by a refractive index:

$$n = [(1/4\pi)bN/K_0^2]^{1/2}. \tag{13}$$

One can see that the variation of the refractive index is a measure of the density variation $\rho(z) = bN$ perpendicular to the surface. By measuring the reflectivity as a function of $\mathbf{Q}$ one can determine the thickness of a thin film and its scattering density ρ since in the Born approximation the reflectivity is given as

$$|R|^2 = [(4\pi)^2/\mathbf{Q}^4]2\rho^2\,(1 - \cos Qd). \tag{14}$$

Even though it is not possible to resolve the atomic structure of the film if $\mathbf{Q}/2\pi$ is much bigger than the interatomic distances, changes in the scattering density $\Delta\rho$ with angstrom resolution can still be observed. One can measure reflectivities at $\mathbf{Q}/2\pi$ values in the region of interatomic distances and then can determine the atomic structure. This is then a normal diffraction experiment.

Neutron reflectometry is a rapidly expanding field and has allowed, for instance, a better understanding of block copolymers—polymers made out of two different types of monomers linked to each other. They are used as, e.g., surfactants, adhesives, and compatibilizing agents in polymer blends. The different polymer blocks tend to separate, leading to either a lamellar structure of alternating layers or a rod structure with one block being embedded in a matrix of the other block. The lamellar structures are ideal study objects for neutron reflectometry, since varying the contrast through selective deuteration allows one or the other block to be highlighted.

If the material is saturated magnetically, the scattering length b incorporates the mean magnetic and nuclear scattering lengths of the material. So the product $(b_n \pm b_m)N$ is a measure of the sum of the coherent nuclear and magnetic scattering-length densities. The sign depends on the spin component of the neutron and is determined by whether the magnetic moment is aligned parallel (spin up) or antiparallel (spin down) to an applied ex-

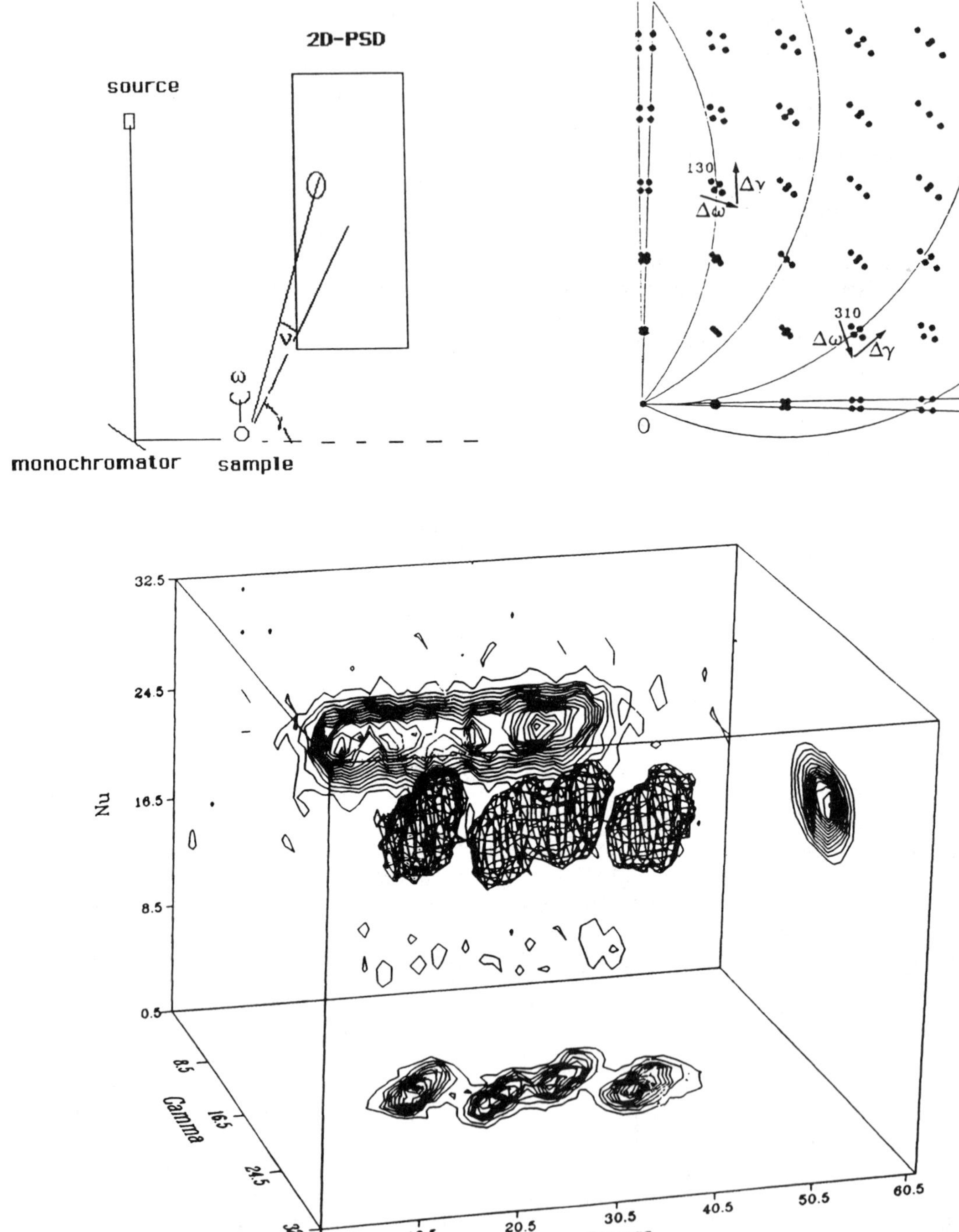

FIG. 4. The basic scattering geometry of a single-crystal diffractometer using a two-dimensional position-sensitive detector. The 420 and 240 Bragg reflections of a twinned $YBa_2Cu_3O_{7-\delta}$ crystal obtained when turning around ω are shown. MacIntyre and Renault (1989) were able to separate nearly a complete set of untwinned single-crystal intensities using 2D-PSD data. The ω projections show that a single detector would record reflections that are not resolved when performing the same scan.

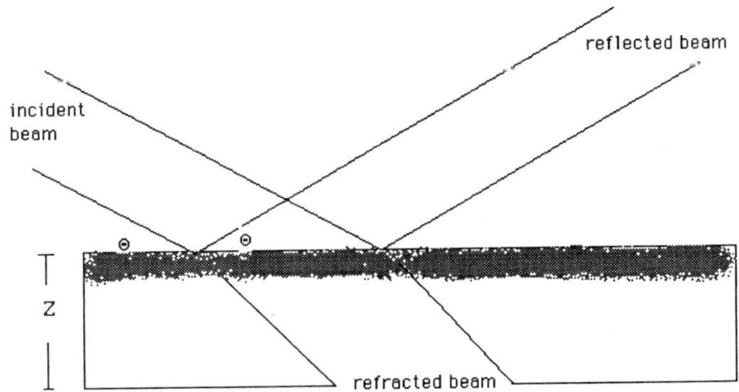

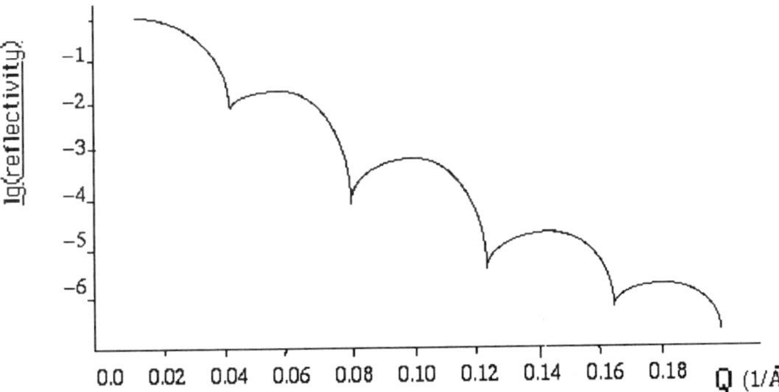

FIG. 5. The principal arrangement of specular reflection experiments is shown. The glancing angles θ of incident and reflected beams are the same. The density profile along z is probed in such an experiment. Below, the variation of the neutron reflectivity of a thin Ni film on a silicon substrate is sketched. The thickness of the film can be obtained directly from the period of oscillation of the reflectivity, since the reflectivity $|R|^2 = (16\pi^2/\mathbf{Q}^4)2\rho^2(1 - \cos Qd)$, where d is the film thickness and ρ the scattering density along z.

ternal field. Thus the critical angle for the two spin states is different and allows their separation.

It is straightforward to calculate a specular reflectivity for a given model, but to solve the inverse problem of determining a chemical and/or magnetic density profile from measured reflectivity curves is much more demanding and not always successful. For details see Felcher and Russell (1991).

If one is interested in the structure of an interface, and particularly in its roughness, it is sometimes not possible to distinguish between different models by measuring only the specular reflectivity, since one averages over the surface perpendicular to **Q**. By measuring in a nonspecular setting one might be able to distinguish between different interface models by the shape and amplitude of the diffuse scattering, resulting from a different in-plane roughness.

By applying Snell's law and setting the refracted angle in the medium equal to zero, one obtains a critical value of the wave-vector transfer $Q_c = (16\pi\rho)^{1/2}$ below which total reflection will occur. Below Q_c the kinetic energy of the neutron is smaller than the potential barrier of the medium. This is the principle of neutron guides now widely used to transport neutrons by internal reflections to experiments situated up to 100 m away from the core or target. By curving them one can reduce the background consisting of γ rays and fast neutrons. They are highly polished, nickel-plated glass tubes with rectangular cross sections. With natural nickel $\theta_c \sim 0.1°$ at $\lambda = 1$ Å. The value of θ_c can be increased by 20% by the use of ^{58}Ni. A further increase can be obtained by using multilayer materials (e.g., NiTi with $\theta_c = 3\theta_{c\text{Ni}}$). Multilayers are composites made of repetitive blocks of discrete material characterized by a lattice spacing d separated by another block of crystalline or amorphous material. By uniformly varying the layer distance, the Bragg reflection can be separated in such a way that its tail runs into the edge of the total reflection at θ_C. This leads to a substantial gain of reflectivity in comparison to normal guides and can be used to make magnetic as well as nonmagnetic mirrors. The challenge for the future of neutron scattering will be whether the tremendous potential of polarized neutron scattering using high-flux spin-polarized beams can be realized. This requires substantial instrumental developments as well as a new generation of high-flux reactors and spallation sources.

Works Cited

Attfield, J. P., Sleight, A. W., Cheetham, A. K. (1986), *Nature* **322**, 620–622.

Bricogne, G. (1991), *Acta Crystallogr.* **A47**, 803–829.

Cagliotii, G., Paoletti, A., Ricci, F. P. (1958), *Nucl. Instrum. Methods* **3**, 223.

Cheetham, A. K. (1992), *Angew. Chem., Int. Ed. Engl.* **32**, 1557–1570.

Cockcroft, J. K., Fitch, A. F. (1988), *Z. Kristallogr.* **184**, 123–145.

David, W. I. F., Ibberson, R. M., Matsuo, T. (1993), *Proc. R. Soc. London A* **442**, 129–146.

Felcher, G. P., Russell, T. P. (1991), *Physica B* **173**, 1–210.

Gaskell, P. H., Eckersley, M. C., Barnes, A. C., Chieux, P. (1991), *Nature* **350**, 675–677.

Guillaume, M., Fischer, P., Roessli, B., Podlesnyak, A., Schefer, J., Furrer, A. (1993), *Solid State Commun.* **88** (1), 57–61.

Jacrot, B. (1976), *Rep. Prog. Phys.* **39**, 911.

Kuhs, W. F. (1992), *Acta Crystallogr.* **A48**, 80–98.

Lehmann, M. S., Kuhs, W. F., McIntyre, G. J., Wilkinson, C., Allibon, J. R. (1989), *J. Appl. Crystallogr.* **22**, 652–568.

McIntyre, G. (1992), *Neutron News*, **3** (2), 15–19.

McIntyre, G., Renault, A. (1989), *Physica* **B156/157**, 880–883.

Medarde, M., Rodriguez-Carvajal, J., Obradors, X., Sayagues, M. J., Vallet, M., Gonzalez-Calbet, J. (1992), *Physica B* **180/181**, 402.

Morris, R. E., Harrison, W. T. A., Nicol, J. M., Wilkinson, A. P., Cheetham, A. K. (1993), *Nature* **359**, 519–522.

Papoular, R. J., Prandl, W., Schiebel, P. (1991), in: C. R. Smith *et al.* (Eds.), *Maximum Entropy and Bayesian Methods*, Norwell, MA: Kluwer Academic Publishers, pp. 359–376.

Rietveld, H. (1969), *J. Appl. Crystallogr.* **2**, 65.

Roessli, B., Fischer, P., Staub, U., Zolliker, M., Furrer, A. (1993), *Europhys. Lett.* **23** (7), 511–515.

Schoenborn, B. P. (1984), in: B. P. Schoenborn (Ed.), *Neutrons in Biology*, New York: Plenum, pp. 261–279.

Schultz, A. J., Srinivasan, K., Teller, R. G., Williams, J. M., Lukehart, C. M. (1984), *J. Am. Chem. Soc.* **106**, 999–1003.

Shull, C. G., Smart, J. S. (1949), *Phys. Rev* **76**, 1256.

Tremayne, M., Lightfott, P., Metha, M. A., Bruce, P. G., Harris, K. D. M., Shankland, K., Gilmore, C. J., Bricogne, G. (1992), *J. Solid State Chem.* **100**, 191–196.

Zachariasen, W. H. (1949), *Acta Crystallogr.* **2**, 296.

Further Reading

Bacon, G. E. (1975), Neutron Diffraction, Oxford: Clarendon Press.

Lovesey, S. W. (1984), *Theory of Neutron Scattering from Condensed Matter*, Vols. 1 and 2, Oxford: Clarendon Press.

Price, D. L., Skold, K. (1986), *Introduction to Neutron Scattering in Methods of Experimental Physics*, Vol. 23, Part A, Orlando, FL: Academic, p. 1.

Rossat-Mignod, J. M. (1987), "Magnetic Structures," in: *Methods of Experimental Physics*, Vol. 23, Part C, Orlando, FL: Academic, p. 69.

White, J. W., Windsor, C. G. (1984), *Rep. Prog. Phys.* **47**, 707.

Windsor, C. G. (1981), *Pulsed Neutron Scattering*, London; Taylor and Francis.

NEUTRON SCATTERING

Harold G. Smith, *Oak Ridge National Laboratory, Oak Ridge, Tennessee, U.S.A.*

INTRODUCTION

Neutron-scattering techniques are now used in many branches of science—e.g., physics, chemistry, metallurgy, biology, and geology and their numerous sub-branches (see NEUTRON DIFFRACTION). Some of the unique attributes of thermal neutrons that make them so useful are described below. One is their neutrality, which permits deep penetration into materials. Only a few of the isotopes of the elements have highly absorbing properties, such as ^{3}He, ^{6}Li, ^{10}B, ^{113}Cd, and ^{157}Gd; but then, these properties can be useful in thermal-neutron detectors and/or shielding. A neutron of wave vector **k** scattered by a nucleus can be represented by a spherical wave of the form

$$\Psi = -be^{ikr}/r, \quad (1)$$

where b is the scattering length (or amplitude) of the nucleus and is a complex quantity; however, except for a few high neutron absorbers, b may be treated as a real quantity and may be either positive or negative. The scattering cross section is defined as the ratio of the scattered neutron flux to the flux of the incident beam and is denoted by

$$\sigma = 4\pi b^2. \quad (2)$$

For an assembly of atoms, as in a crystal, the scattering cross section is a sum over all the atoms, taking care of the interference conditions between the atoms. A complete description of the various scattering processes—spin and isotopic, coherent and incoherent—are clearly described by Bacon (1975).

The absorption and scattering cross sections vary widely from element to element and isotope to isotope. The isotopes ^{4}He, ^{7}Li, ^{11}B, ^{114}Cd, and ^{160}Gd are transparent or nearly transparent to thermal neutrons. Many of the other stable isotopes can be exploited for their unique scattering lengths. The almost random variation of the scattering lengths among

3-527-28133-9/94/$5.00 + .50

the elements and isotopes is due to the varying spin states of the nuclei and the spin-$\frac{1}{2}$ state of the neutron. While most of the lengths are positive, some are negative, such as those for ^{1}H, ^{7}Li, and ^{62}Ni. The most useful combination of isotopes exploiting the positive and negative scattering lengths are hydrogen and deuterium, where $b = -3.74 \times 10^{-15}$ m for the former and $b = +6.67 \times 10^{-15}$ m for the latter (often referred to as units in fermis, where 1 fm = 10^{-15} m). These properties are extremely useful in studying polymers and biological compounds.

The energies and momenta of the thermalized neutron beams can be tailored for their optimum interactions in the materials to be studied—relatively high thermal energies for vibrational and magnetic excitations, to very low energies for small-angle scattering studies and extremely low-energy dynamical motion, as in rotational tunneling and diffusive motion in solids.

Another unique property of the neutron is its magnetic moment, −1.9135 nuclear magnetons, which interacts strongly with the magnetic moments of the atoms and magnetic molecules. The large amount of knowledge of the magnetic properties of materials—their magnetic structures and their magnetic interactions—would not have been possible without the extensive neutron-scattering investigations of the wide variety of magnetic materials. Examples are presented in a later section.

The small-angle neutron scattering (SANS) facilities at both reactor and pulsed accelerator sources are the most utilized of all the instruments, with hundreds of users (chemists, physicists, metallurgists, and biologists) performing thousands of experiments each year, many of the users coming from industry. As will be described more fully below, polymer scientists rely heavily on the SANS facilities, and many are industrial users, who find that SANS complements nicely their other measuring tools, such as x-ray and light scattering and nuclear magnetic resonance (NMR). In contrast to experiments that are measured in days and weeks, such as phonon and magnon scattering, polymer experiments are often measured in terms of hours and days, so that many users can be accommodated in a relatively short length of time.

1. THERMALIZED NEUTRONS

Most neutron-scattering measurements utilize thermal neutrons—those fast neutrons (MeV) from either reactor sources or pulsed accelerator sources that have been moderated or "thermalized" to near room-temperature energies (~25 meV). Most research reactors require a moderator temperature of the order of 40 °C, which produces a broad thermal-neutron spectrum with wavelengths from about 0.05 to 1.0 nm with a peak in the spectrum at about 0.15 nm, or about 25 meV. These energies are comparable to the energies of most crystal excitations—phonons, magnons, and crystal field levels—and the wavelengths are of the order of atomic dimensions. As indicated in the article NEUTRON DIFFRACTION, these wavelengths are ideal for elastic diffraction studies, but the thermal energies and momenta are ideal for inelastic-scattering studies because the energy and momenta changes are large and are easily measurable.

In more recent years, interest has been in neutrons of much lower energy (cold neutrons) than the thermal reactor spectrum and in energies above the reactor spectrum (hot neutrons). The former are obtained by inserting a cold source (liquid hydrogen) in the moderator and the latter by inserting a hot source (hot graphite block) in the moderator. Figure 1 shows the estimated neutron flux distribution for a 5×10^{15} $n/cm^2\ s^{-1}$ D_2O-moderated reactor with a hot source and a cold source (Moon, 1985). Depending on the type of guides, mirrors, and monochromators, cold-source fluxes would be expected to be between curves A and B.

In the case of pulsed neutron sources from accelerators, the neutrons are produced as spallation products of high-energy protons (or electrons) striking a heavy-metal target, such as tungsten or uranium. As in the case of reactor fission neutrons, they must be moderated from MeV to meV and eV energies. Various moderators are selected for desired energy changes. Because of the nature of accelerator sources, there are more very hot neutrons (>1 eV) than are available from even a high-flux reactor with a hot source, such as at the Institut Laue-Langevin. This permits inelastic-scattering experiments of higher energy transfer that would be difficult if not impossible at the reactor sources.

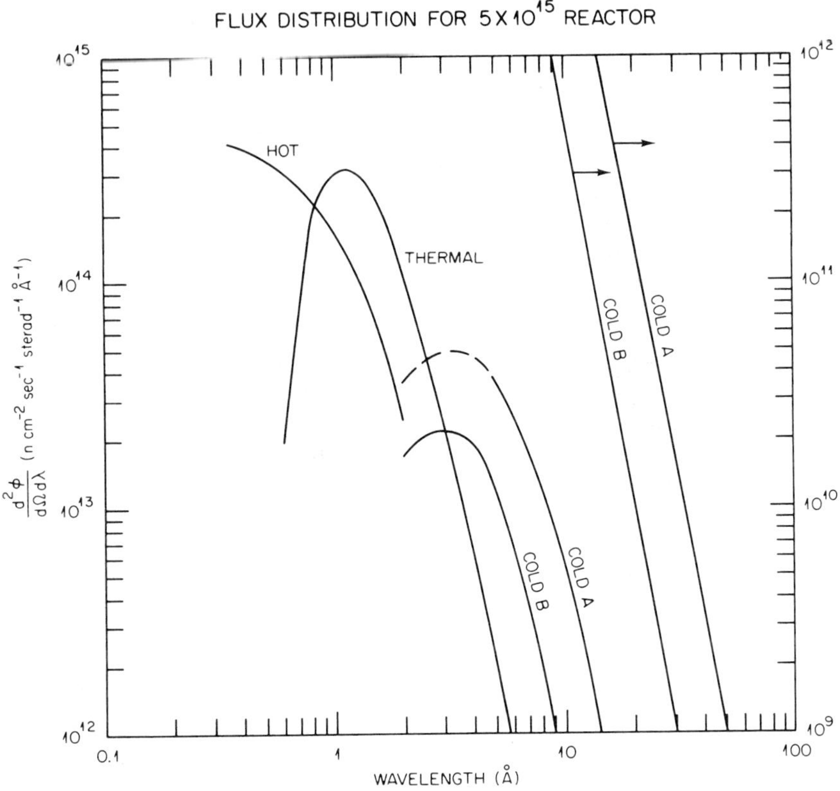

FIG. 1. Estimated flux distributions for a 5×10^{15} $n/(\mathrm{cm}^2\ \mathrm{s})$ D_2O-moderated reactor with hot and cold sources. (Reproduced with permission of R. M. Moon.)

2. INELASTIC NEUTRON SCATTERING

The inelastic scattering of neutrons is a very powerful technique for studying the various excitations in crystalline solids—vibrational, rotational, and magnetic. It also offers a means of studying the vibrational properties of liquids and amorphous solids, although not nearly to the extent that is possible in single crystals. Studies of diffusive motions, particularly of hydrogen atoms, are also possible, as is the rotational tunneling motion of molecular entities. The energy of the incoming neutrons can be chosen to be comparable to the energies of the excitations to be measured so that the neutron energy changes are large and easily measurable. Similarly, the wave vectors of the neutrons are comparable to the wave vectors of the excitations, and the excitation spectrum can be measured throughout the Brillouin zone. This is in contrast to the infrared and Raman techniques, which involve small wave vectors near the Brillouin-zone center.

The most detailed information about the interatomic interactions between neighboring and more distant neighbors is obtained with single crystals, where the relatively simple relations of conservation of energy and crystal momentum for the one-phonon or one-magnon scattering process can be readily applied.

2.1 Phonons

The dynamical properties of crystalline solids and their relation to the interatomic forces in the crystals are best studied by measuring the phonon dispersion relations in high-symmetry directions. A general introduction to lattice dynamics based on the Born-von Kármán formalism can be found in the classic treatise by Born and Huang (1954). The equations of motion can be compactly expressed in matrix notation as

$$\omega^2 \mathbf{u} = \mathbf{D}\mathbf{u}, \tag{3}$$

where

$$D_{ab}(dd') = \sum_{l'} \phi_{ab}(ld,l'd') \times \exp[i\mathbf{q} \cdot \mathbf{r}(l')]/\sqrt{M_d}\sqrt{M_{d'}}, \quad (4)$$

and the ϕ_{ab} represent the force constants between atom d of mass M_d in unit cell l and atom d' of mass M_d' in unit cell l'. The wave vector is denoted by $\mathbf{q}$ and the atomic displacements by $\mathbf{r}(l')$, and the sum is over l'. Applying group-theory principles to the symmetrical properties of the crystal lattice greatly reduces the number of force constants that need to be determined.

The conditions for conservation of crystal momentum and energy for the one-phonon coherent scattering process in single crystals require that

$$\mathbf{Q} = \mathbf{k} - \mathbf{k}' = \boldsymbol{\tau} + \mathbf{q}, \quad (5)$$

$$\Delta E = h\nu(\mathbf{q}) = \hbar^2(k^2 - k'^2)/2m, \quad (6)$$

where $\mathbf{Q}$ is the crystal momentum vector, $\mathbf{k}$ is the wave vector of the incoming neutron, $\mathbf{k}'$ is the wave vector of the scattered neutron, $\boldsymbol{\tau}$ is the reciprocal lattice vector, and $\mathbf{q}$ is the wave vector of the measured excitation. The frequency of the excitation is $\nu(\mathbf{q})$, $\hbar$ is Planck's constant h divided by 2π, and m is the mass of the neutron. The dispersion relation for phonons is given by $\nu_j(\mathbf{q})$, where j denotes the polarization vector of the phonons for a given direction in the crystal.

2.1.1 Triple-Axis Spectrometer The triple-axis neutron spectrometer pioneered by B. N. Brockhouse is particularly well suited for measuring and identifying the various branches of the phonon dispersion curves. There are, in general, $3n$ branches, where n is the number of atoms in the primitive unit cell. However, in high-symmetry directions, the number of branches is often reduced by degeneracies, and the experimental measuring time is considerably reduced. A schematic diagram of a triple-axis spectrometer at the Oak Ridge High Flux Isotope Reactor is shown in Fig. 2. The monochromator selects a narrow band of neutron energies from the "white" beam from the reactor. The neutrons are scattered by the lattice vibrations in the specimen crystal, and the energy change is measured by the analyzer crystal.

A simulated phonon scan is illustrated in Fig. 3, where a transverse phonon in the [100] direction with wave vector $\mathbf{q}$ is depicted; the angles correspond to those of Fig. 2. A neutron of incident wave vector $\mathbf{k}$ is scattered by the lattice vibrations (phonons) of the sample crystal at an angle Φ with an energy gain or loss and final wave vector $\mathbf{k}'$, as shown in Fig. 3(a) (energy loss in this case). The wave vector $\mathbf{k}$ (hence the incident energy) is varied by moving the spectrometer angle $2\Theta_M$ through a series of steps while the wave vector $\mathbf{k}'$ is held fixed; its value is determined by the Bragg diffraction conditions of the analyzer crystal. The detector angle $2\Theta_A$ is fixed to accept the diffracted neutrons from the analyzer crystal. The $H0L$ reciprocal lattice plane is shown in Fig. 3(a) along with the crystal-momentum $\mathbf{Q}$-vector diagram of Eq. (3). For convenience and simplicity, the vector $\mathbf{Q}$ and phonon wave vector $\mathbf{q}$ are held constant during the scan. As the locus of $\mathbf{k}$ moves in steps from A to B, the detector counts increase dramatically over the background count [Fig. 3(b)], the peak indicating the energy for that particular phonon group, 3.2 THz.

The direction of the phonon wave vector $\mathbf{q}$ in this case is along the [100] reciprocal lattice direction. For the symmetry shown, the phonon eigenvectors are either purely longitudinal (atomic displacements along the direction of propagation) or purely transverse (atomic displacements normal to the direction of propagation). As discussed below in Sec. 2.1.2, the phonon mode simulated here is a transverse mode. To measure the longitudinal mode, the crystal would need to be oriented with the $\mathbf{q}$ vector along the [001] direction. A series of such scans maps out the longitudinal and transverse branches of the dispersion curves in different directions of the crystal [Fig. 3(c)].

2.1.2 Lithium Fluoride Lithium fluoride is a good example of a simple ionic crystal that illustrates some of the dynamical properties that can be determined by inelastic neutron-scattering methods (Dolling *et al.*, 1968). Figure 4 shows the phonon dispersion curves in the three high-symmetry directions in the crystal. Although the crystal is face-centered cubic with eight atoms in the unit cell, there are only two atoms in the primitive unit cell; therefore, there are six phonon branches in each direction in the crystal, three

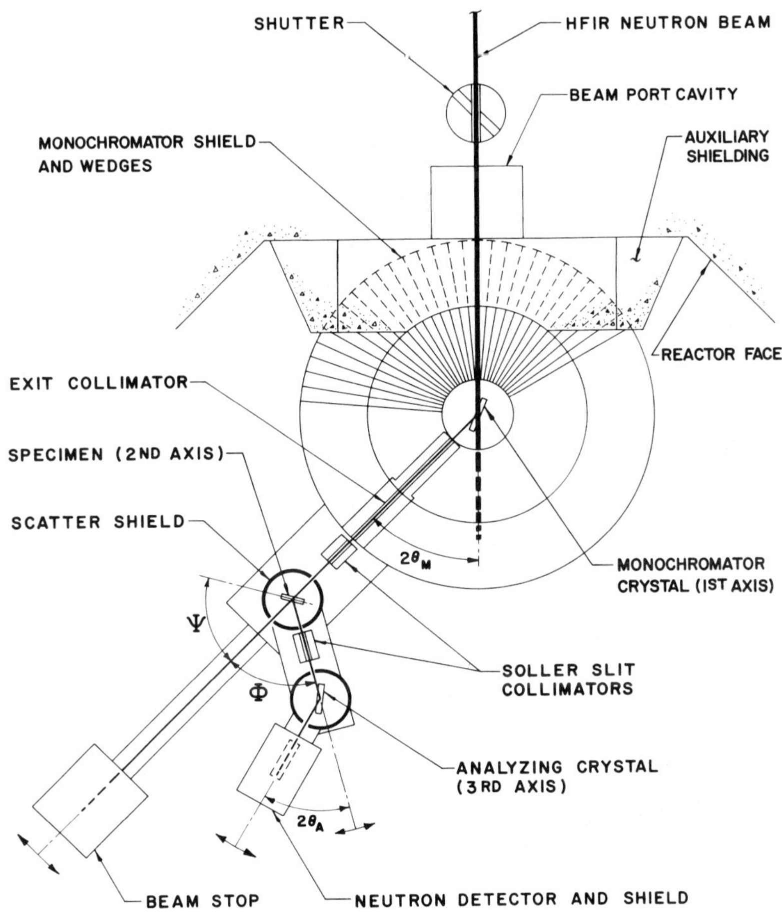

FIG. 2. Schematic diagram of a triple-axis spectrometer at the High Flux Isotope Reactor. (Reproduced with permission of H. G. Smith.)

acoustic and three optic branches. The branches are labeled by standard group-theoretical notation. The transverse optic Δ_5 and Σ_3 branches are each doubly degenerate, as required by the crystal symmetry.

The slopes of the acoustic branches at small q are the group velocities of the phonons and are simply related to the elastic constants of the crystal, C_{11}, C_{44}, and C_{12}. The transverse-optic (TO) mode at $q = 0$ is the infrared-active mode (or *Reststrahlen* frequency). The highest-frequency mode is the longitudinal optic (LO) mode. The large difference between the LO and TO modes is due to the large macroscopic electric field in the crystal and is related to the static and high-frequency dielectric constants, e_0 and e_∞, of the crystal. This is the Lyddane–Sachs–Teller relation, where

$$e_0/e_\infty = (\nu_{\rm LO}^2/\nu_{\rm TO}^2)_{q=0}. \tag{7}$$

For more complex crystals or crystals of lower symmetry, the relations are not as simple. For a metallic crystal with the NaCl type of structure, the conduction electrons screen the electric field and there is no splitting of the optic modes; i.e., the modes are triply degenerate. The dispersion curves in Fig. 4 have been successfully fitted with a simple shell model involving the charges and polarizabilities of the ions. A rigid-ion model was not sufficient to fit the data.

There are no selection rules as such in the neutron–phonon scattering process as there are in Raman scattering and infrared ab-

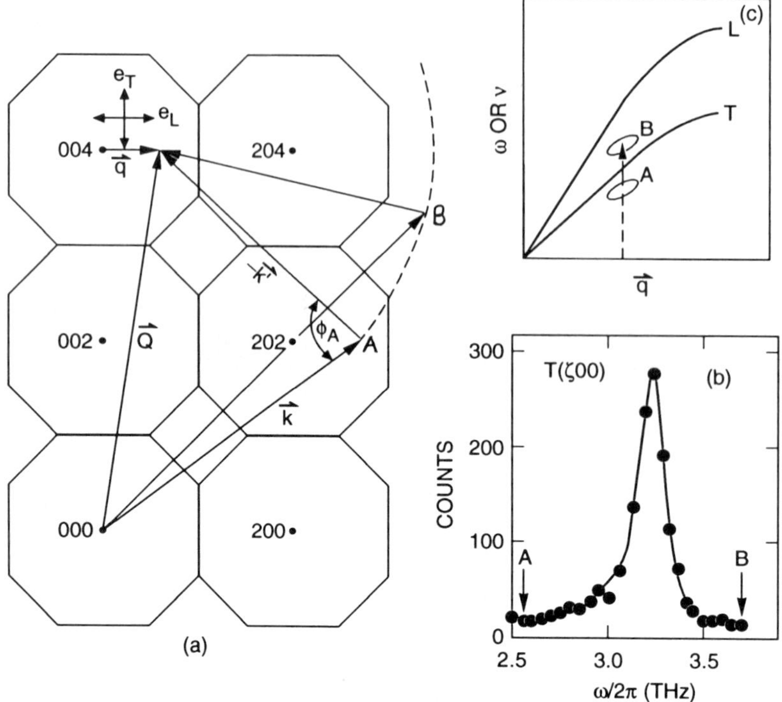

FIG. 3. A simulated phonon scan, where a transverse phonon in the [100] direction with wave vector **q** is depicted. (a) The *H0L* reciprocal plane is shown along with the crystal momentum **Q**-vector diagram of Eq. (3). (b) As the locus of **k** moves in steps from A to B, the detector counts increase dramatically over the background count, the peak indicating the energy for that particular phonon group (3.2 THz). (c) A series of similar phonon scans maps out the longitudinal and transverse branches of the dispersion curves in different directions of the crystal. (Reproduced with permission of R. M. Nicklow.)

sorption. Nevertheless, there are terms in the one-phonon scattering cross section that allow the identification and separation of the longitudinal and transverse modes. This is due to the term $|\mathbf{Q} \cdot \mathbf{e}|^2$, where $\mathbf{Q}$ is the wave vector of the crystal momentum (see Fig. 3) and $\mathbf{e}$ is the phonon eigenvector. In the example shown, the value of $|\mathbf{Q} \cdot \mathbf{e}|$ for the longitudinal mode is essentially zero and only the transverse mode in the plane will be measured. The intensity of the other transverse mode (out of the plane) will also be zero. In general, for more complex crystals than the alkali halides, the intensities of the phonons must also be measured in order to identify many of the phonon branches. This is particularly true for the A-15 compounds, which exhibit up to 24 branches in the lower-symmetry directions. The calculated intensities are very model dependent, and so it is very important to have an approximate model early on in the investigation.

2.1.3 The Electron–Phonon Interaction

It was some years after Bardeen *et al.* (1957) proposed that the mechanism for superconductivity was mediated through the electron–phonon interaction that McMillan (1968) proposed a more explicit expression relating the temperature of the superconducting transition temperature T_c, to the electronic and vibrational properties of the lattice through an electron–phonon coupling parameter λ:

$$T_c = \frac{\Theta \exp[-1.04(1 + \lambda)]}{1.45\,[1 - \mu^* (1.062\,\lambda)]}, \tag{8}$$

where Θ is the characteristic Debye temperature, μ^* is the Coulomb pseudopotential, and λ is defined by

$$\lambda = 2 \int \alpha^2(\omega) F(\omega) \omega^{-1}\, d\omega. \tag{9}$$

$\alpha^2(\omega)F(\omega)$ is the electron–phonon spectral

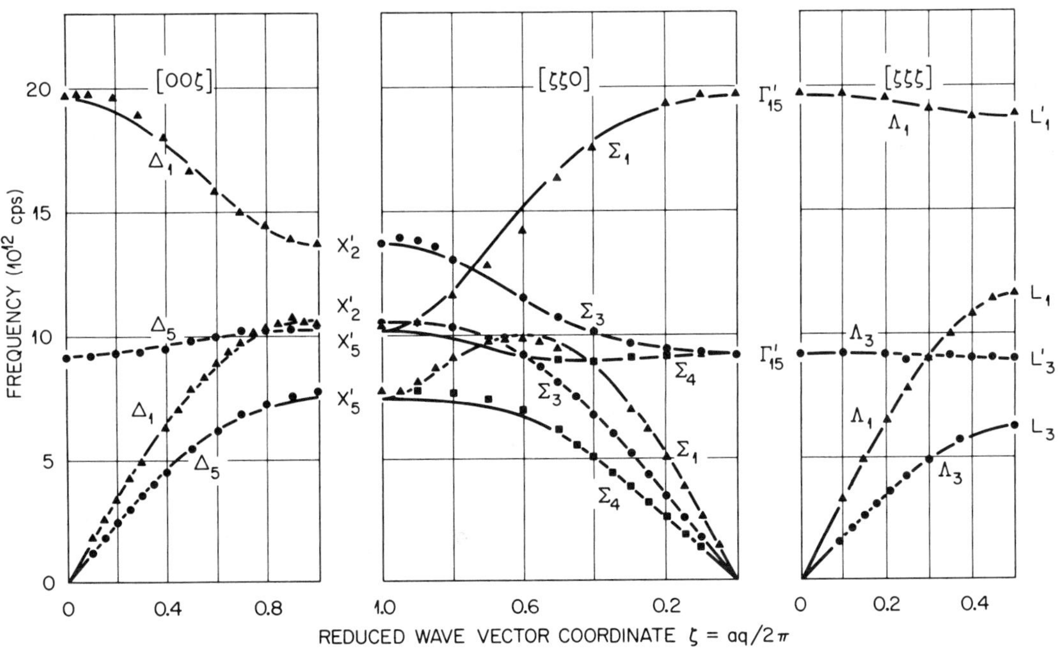

FIG. 4. Dispersion curves for lithium fluoride at 298 K. (Reproduced with permission of H. G. Smith.)

function and $F(\omega)$ is the phonon density of states. Both functions are obtainable by experimental means, the former by electron tunneling experiments on superconductors and the latter from inelastic neutron-scattering measurements on polycrystalline materials.

It was subsequently observed that the phonon dispersion curves of single-crystal superconductors showed anomalous dips in certain branches compared to similar, but nonsuperconducting, metals, alloys, and compounds. An example is shown in Fig. 5, where the longitudinal acoustic branches of superconducting TaC are compared with the same branches of nonsuperconducting HfC. The differences have been shown theoretically to be due to the differences in the band structures of the materials. In general, the higher the T_c, the more pronounced are the anomalies.

Perhaps even more dramatic is the information contained in the phonon linewidths. It has been shown that λ, and hence T_c, can be directly and simply related to the electron–phonon contribution to the phonon linewidths and phonon frequencies. Even for what are now considered low-T_c materials, such as Nb, the observed linewidths (see Fig. 6) are large and measurable and agree well with the theoretical calculations, which preceded the experiments (Butler *et al.*, 1977). A more sensitive measurement, but not as well formulated theoretically, is the change in linewidth for certain phonon modes in Nb_3Sn as the temperature is varied over the transition-temperature region (Fig. 7). Interestingly, the phonon linewidth in Nb_3Sn is suppressed if the phonon energy is less than the superconducting gap energy (Axe and Shirane, 1973).

There has been much speculation on whether the electron–phonon interaction is the dominant mechanism for superconductivity in the more recently discovered very high-T_c superconductors (T = 100 K) of the $BaY_2Cu_3O_{7(1-x)}$ type, or whether the high T_c is due to a more exotic mechanism. The crystal structures of these compounds are more complex than most of the low-T_c materials and, therefore, have many more phonon branches, which are very difficult to measure and identify, even as larger crystals have become available. Nevertheless, numerous phonon groups and branches have been measured in the above compound, and a few others, with some success. It has also been possible to measure the phonon density of

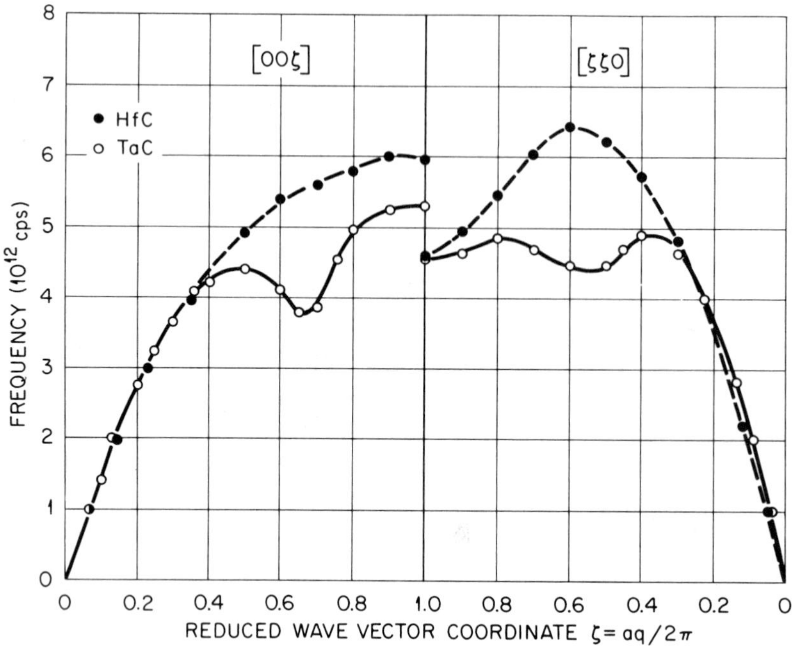

FIG. 5. Longitudinal acoustic modes in tantalum and hafnium carbides, illustrating the phonon anomalies in superconducting TaC vs the smooth phonon curves in nonsuperconducting HfC. (Reproduced with permission of H. G. Smith.)

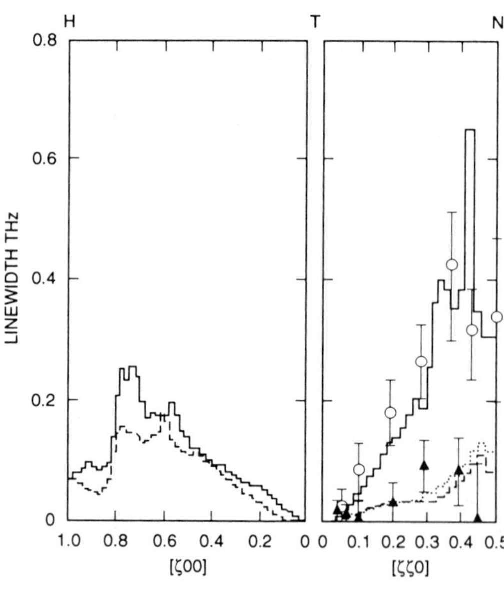

FIG. 6. Phonon linewidths in Nb. Histograms indicate calculated linewidths. Circles (triangles) are experimental longitudinal (transverse) mode linewidths. (Reproduced with permission of H. G. Smith.)

states (PDOS) of some high-T_c materials. For example, there clearly are differences between the PDOS of superconducting $BaY_2Cu_3O_{6.7}$ and the PDOS of nonsuperconducting $BaY_2Cu_3O_6$, particularly at the high-energy end of the spectrum, where the vibrations involve the Cu—O breathing modes. Nevertheless, it is premature to conclude whether the *e*-ph interaction is the mechanism or merely an effect of an unproven mechanism.

In contrast to the situation before the discovery of the high-T_c Cu-oxide superconductors, where superconductivity and magnetism were incompatible, many of the new superconductors appear to have unusual but interesting magnetic properties existing in the superconducting state, which suggests a mechanism or mechanisms entirely different from the electron–phonon interaction of BCS origin. The existence of magnetic excitations in a single crystal of $YBa_2Cu_3O_7$ has been elegantly demonstrated by Mook *et al.* (1993), employing polarized neutron beams on a triple-axis polarized-beam spectrometer and showing that the excitations in question were indeed of magnetic origin.

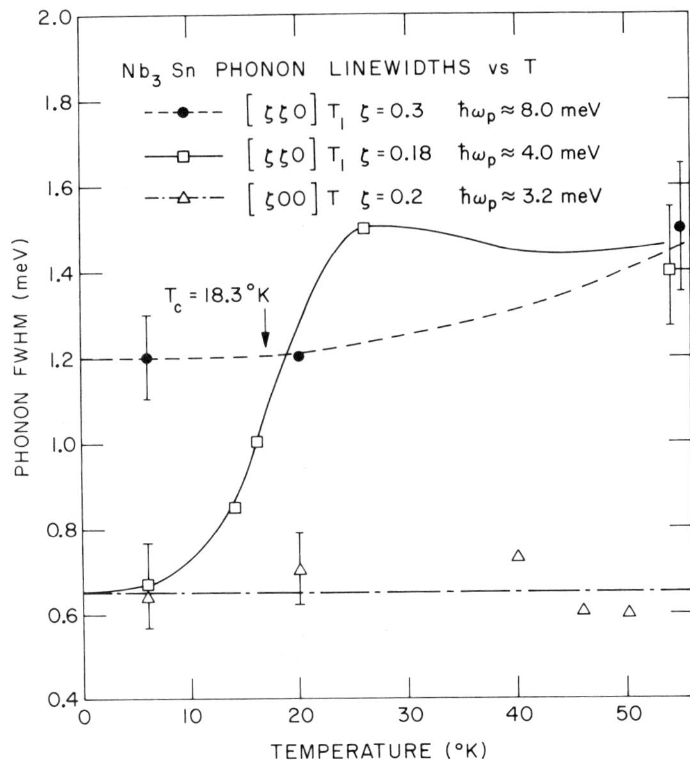

FIG. 7. Transverse acoustic linewidths in Nb_3Sn near the superconducting transition temperature. (Reproduced with permission of J. Axe.)

2.1.4 Phase Transitions Inelastic neutron scattering is also a valuable tool for studying phase transitions in solids—crystal structure changes and/or magnetic structure changes—and whether the transitions are first order, second order, or of intermediate order. Precise measurements of critical scattering near the transition temperature have provided valuable information regarding the dimensionality of some transitions. The complex nature of many martensitic phase transformations has been elucidated by studying their phonon dispersion curves, which often show anomalous features reminiscent of the phonon anomalies in some of the superconducting materials described above. Of interest are whether the transitions are first order, or *nearly* first order, and if there exist intermediate structures between the initial and final structures. An interesting example is the Ni-Al alloy system, which shows unusual phonon and structural behavior over a very narrow composition range.

Shapiro (1990) and co-workers have studied in great detail the temperature-dependent behavior of the Σ_4 branch of the alloy $Ni_{62.5}Al_{27.5}$, which shows a pronounced phonon softening and subsequent central peak growth (i.e., $\Delta E = 0$) at a wave vector between $\frac{1}{6}$ and $\frac{1}{7}$ of the distance to the zone boundary (Fig. 8). A long-range, seven-layer modulated structure develops below T_c. Not all martensitic phase transitions exhibit this type of precursor behavior. The alkali metals Li and Na undergo phase transitions to complex close-packed structures with a high degree of stacking-fault disorder, yet show no precursor effects above T_c, in either their phonon spectra, diffuse scattering, elastic constants, or ultrasonic attenuation. Pure first-order phase transitions do not require any phonon-softening phenomena. However, some elements, such as α-U at low temperatures, do show precursor effects, charge-density waves, discommensurations, and both second-order and first-order phase transitions. These studies and others are extremely valuable in efforts to understand the mechanisms of phase transitions. They could not have been possible without the neutron-scattering techniques described above.

2.1.5 Time-of-Flight Spectrometer Another technique for measuring inelastic neu-

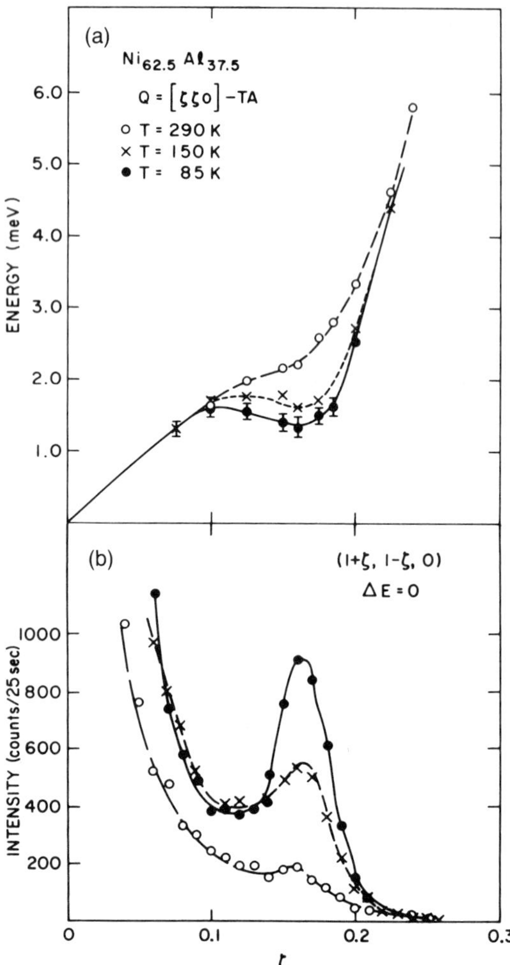

FIG. 8. (a) The temperature dependence of the [110]-⟨110⟩ phonon branch for $Ni_{62.5}Al_{37.5}$. (b) The temperature and q dependence of the elastic scattering associated with the phonon branch. (Reproduced with permission of S. M. Shapiro.)

tron scattering, and one that is particularly suited for pulsed neutron sources, is to measure the interval between the time that a neutron is incident upon a sample and the time that the scattered neutron travels a given distance and is detected at a detector (usually a whole bank of detectors at several scattering angles). Those neutrons that give up energy to the sample, i.e., create phonons or magnons, arrive at the detectors at a later time than in purely elastic scattering, and the change in energy on scattering is readily determined. This is a very efficient process for studying the dynamics of liquids and polycrystalline materials, but there are some difficulties when it is applied to one-phonon or one-magnon processes in single crystals. It is not so straightforward to identify properly the numerous phonon or magnon groups with their eigenvectors and their respective dispersion curves. However, new hybrid spectrometers that utilize multiple analyzer crystals, each associated with a given detector, can be positioned to record phonon groups in desired high-symmetry directions in the single crystal. A schematic diagram of the PRISMA spectrometer at the ISIS pulsed neutron source at the Rutherford Appleton Laboratory in England is shown in Fig. 9. A beam of white neutrons from the neutron moderator is incident on the single-crystal sample and is scattered through angles ϕ_1 to ϕ_{16}. The crystal analyzers and detectors (up to 16 may be used) are prearranged to observe phonons (and/or magnons) in a particular direction in the crystal. Each detector has a fixed final wave vector $\mathbf{k}_f$. The wave vectors $\mathbf{k}_i$ of the incident neutrons can then be obtained knowing the total time of flight (TOF) from the moderator to the detector. The energies and momenta of the excitations can then be plotted as dispersion curves along certain directions (usually of high symmetry) in the crystal. The PRISMA instrument is described in detail by Steigenberger *et al.* (1991), and a more complete description of the use of indirect geometry time-of-flight spectrometers, in general, is given by Robinson and Pynn (1988).

An impressive example of the simultaneous measurement of multiple phonons is shown in Fig. 10 for $CaCO_3$ (Dove *et al.*, 1992). Three room-temperature scans utilizing 11 detectors produced 33 spectra, which were identified and plotted as phonon dispersion curves in the [1 0 −4] direction through the (300) Bragg peak. The dashed line in the figure shows the energy-vs-momentum relationship obtained by one of the detectors in one of the scans. Its spectra at room temperature (closed circles) and 773 K (triangles) are shown in Fig. 11.

2.2 Magnetic Scattering

In a magnetic system that exhibits long-range magnetic order, be it a ferromagnet, ferrimagnet, antiferromagnet, or spiral structure, there are small fluctuations of the instantaneous magnetic moment directions

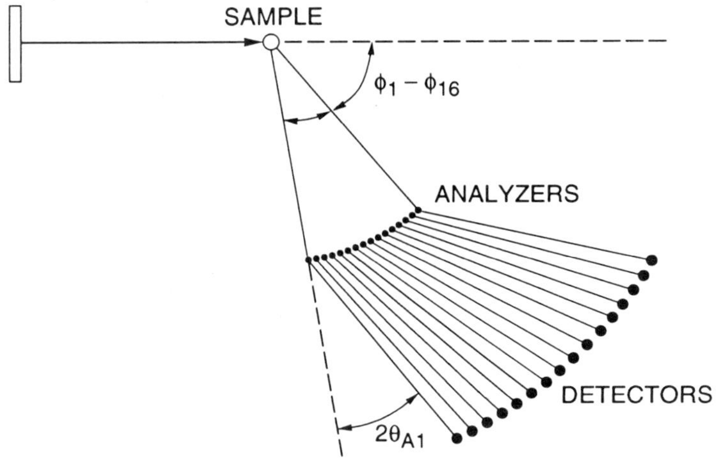

FIG. 9. Schematic diagram of the multianalyzer PRISMA spectrometer at ISIS. A beam of white neutrons from the accelerator's neutron moderator is scattered by the lattice vibrations of the sample crystal onto the analyzer crystals and then diffracted into the detectors. (Reproduced with permission of U. Steigenberger.)

about the average moment direction, similar to the small vibrations of atoms about their average equilibrium positions. These collective magnetic excitations are called spin waves or magnons and exhibit magnon dispersion curves with energies comparable to phonon dispersion curves. Since the neutron has a magnetic moment, the magnetic spectra can be easily measured by the same standard inelastic-scattering techniques used for measuring the phonon spectra in crystals. Single crystals are required to map out the magnon dispersion curves in the different directions in the crystal. The magnon energies can range from a few meV in, say, the simple antiferromagnetic MnF_2 to several hundred meV in a strong magnet, such as iron. A good introduction and thorough review of magnetic excitations in general has been given by Stirling and McEwen (1987).

2.2.1 Magnons in Dysprosium Dysprosium is a magnetic system of moderate complexity that illustrates (Nicklow *et al.*, 1971) the behavior of both a ferromagnetic phase (below 87 K) and a spiral magnetic phase (179–87 K) in the same crystal as a function of temperature. Dysprosium has the hexagonal close-packed structure, and the magnetic moment directions are in the hexagonal planes; but the moment directions vary from layer to layer in a spiral arrangement, and

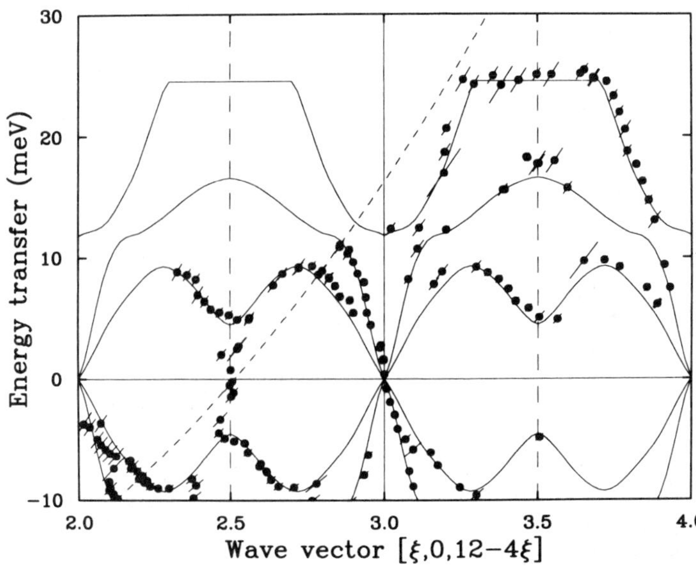

FIG. 10. The phonon dispersion relations for $CaCO_3$ measured on PRISMA along the [1 0 −4] direction through the (300) Bragg peak at room temperature. The dashed line in the figure shows the phonon energy-vs-momentum relationship obtained by one of the detectors in one of the scans. (Reproduced with permission of M. Hagen.)

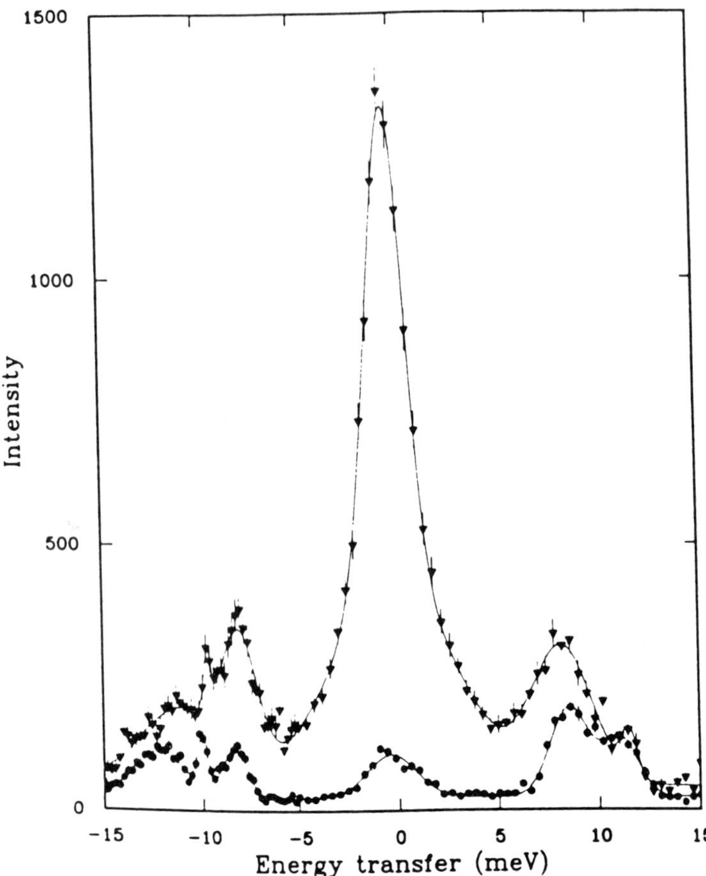

FIG. 11. The spectra measured along the path shown by the dashed line in Fig. 10 at room temperature (closed circles) and at 773 K (triangles). (Reproduced with permission of M. Hagen.)

the magnetic periodicity is incommensurate with the crystal lattice. At the spiral-to-ferromagnetic transition, the magnetic moments in the basal plane all lock into the same crystallographic direction and the periodicity is the same as the atomic periodicity, namely, every two layers in the c direction.

In this low-temperature ferromagnetic phase, in addition to the strong exchange interactions, there are also terms in the magnon energy expression that include the magnetoelastic strains and crystal-field anisotropies, axial and planar. These additional terms produce a pronounced energy gap at $q = 0$, as shown in Fig. 12(a) for the ferromagnetic magnon dispersion curves in the c direction at 78 K. If one accepts the value for the (macro) magnetoelastic strain constant, the interplanar exchange constants $J^c(q)$ can be determined as functions of layer distance by a Fourier analysis of the c-direction magnon dispersion curve. The results are shown in Fig. 12(b). In the spiral phase, the magnetoelastic interaction vanishes and the planar anisotropy averages to zero, and so there is no longer a gap at $q = 0$. The magnon dispersion curves have been analyzed in terms of the interplanar force constants for seven layers, and the results are also shown in Fig. 12(a) for comparison with those obtained in the ferromagnetic phase. The maximum in the ferromagnetic curve at $\zeta = 0.16$ in Fig. 12(b) indicates that, even in the ferromagnetic phase, the exchange interaction favors a spiral structure.

In the spiral phase, there is little indication of effects due to magnon–phonon interactions, but in the ferromagnetic phase, the magnon energies increase significantly at low temperatures and the magnon and phonon branches are split where they would otherwise cross. This is a measure of the strong magnetoelastic interactions in dysprosium and can be studied in more detail by using polarized neutrons and polarization analysis.

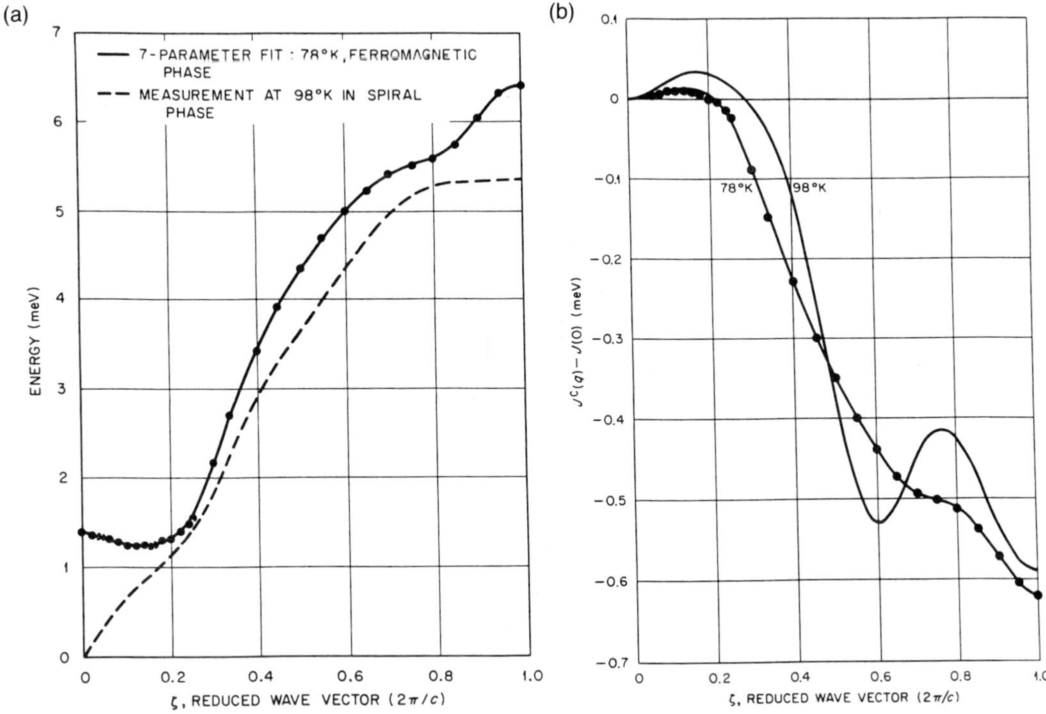

FIG. 12. (a) Magnon dispersion relation for dysprosium in the *c* direction. (b) $J^c(q) - J(0)$ vs reduced wave vector ζ derived from curves measured in (a). (Reproduced with permission of R. M. Nicklow.)

2.2.2 Crystal-Field Excitations and Intermultiplet Transitions In some magnetic systems, where the exchange interactions between neighbors are very weak, collective spin waves as such do not exist. The observed magnetic excitations are neutron-induced transitions between levels of the ground-state multiplets, where the degeneracy has been removed by the crystalline electric field. Many rare-earth and actinide compounds fall into this category. The transition energies are usually low, between 1 and perhaps 20–30 meV, but they can be much larger when the crystalline electric fields are very strong. They are nearly dispersionless (with some exceptions), and large polycrystalline samples can be studied. The splitting and ordering of the levels can usually be deduced from the crystal symmetry, the measured intensities, and their temperature dependence. They have been widely studied at both reactor and pulsed sources.

There are also transitions between the ground state and higher multiplet states of different total angular momentum that have recently become accessible by neutron-scattering techniques. Previously, these intermultiplet transitions were studied by optical means because the transitions were of the order of an eV and higher. With the advent of high-intensity pulsed neutron sources with copious beams of very hot neutrons (>1 eV), several metals and compounds have been studied on the high–energy-transfer (HET) spectrometer at the ISIS pulsed spallation neutron source of the Rutherford Appleton Laboratory in England.

There are several advantages to using neutrons for these studies, but perhaps the greatest advantage is the measurement of the excitations as a function of wave vector, for the cross sections are often a strong function of the wave vector. This serves to identify the transition levels properly. A good example of this technique is shown in Fig. 13, where the intermultiplet transitions in praseodymium metal have been observed by Taylor *et al.* (1988). The four peaks have been assigned to the $^3H_4 \rightarrow {}^3H_5$ (dipolar) and $^3H_4 \rightarrow {}^3F_{2,3,4}$ (nondipolar) transitions. The latter transitions have somewhat lower energies than the free-ion values in the vapor state and in the

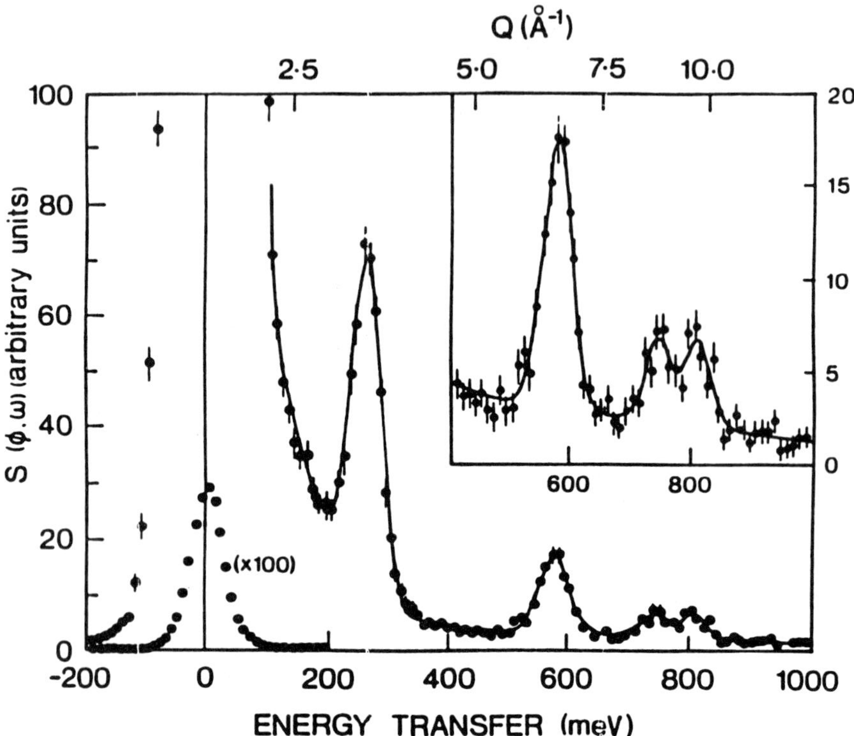

FIG. 13. Neutron-scattering cross section of praseodymium at 17 K, measured with an incoming energy of 1300 meV. Inset shows the results above 400 meV on an expanded scale. (Reproduced with permission of A. D. Taylor.)

Pr-doped $LaCl_3$ insulator, and this is thought to be due to the screening of the Coulomb potential by the conduction electrons in the metal. Intermultiplet transitions with energies up to nearly 2 eV have been observed in other rare-earth metals and compounds. These and similar experiments in the future will provide much valuable information regarding spin-orbit coupling in the rare-earth and actinide compounds and serve as a means of studying their magnetic wave functions.

2.3 Tunneling Motion and Diffusive Motion

2.3.1 Tunneling Motion in Molecular Crystals Most lattice vibrational frequencies occur between 10^{11} and 10^{13} s^{-1}, although internal molecular vibrations, particularly those involving light atoms, can be another order of magnitude higher. However, there are some molecular groups, such as CH_4, CH_3, NH_3, and NH_4^+ groups, that may undergo free or nearly free rotational motion at high temperatures and undergo librational motions at low temperatures. In some circumstances, depending on the potential the atoms or ions feel, the groups may reorient by tunneling from one site to another. These tunneling frequencies are extremely low (10^9–10^{10} s^{-1}) and to measure them by inelastic neutron scattering requires very low-energy neutrons, of the order of 1 meV, obtainable only from a cold source at a reactor or accelerator.

An example of particular fundamental interest is the tunneling spectra in the quantum solid CH_4 (Asmussen *et al.*, 1992). At temperatures near 5 K, the methane groups of one crystallographic set are locked firmly in the lattice, but those of a nonequivalent set are able to reorient by tunneling motion to equivalent sites with frequencies of about 10^9 s^{-1} (i.e., energies of several μeV). The tunneling spectra for CH_4 at 5 K are shown in Fig. 14(a). The pronounced variation of the

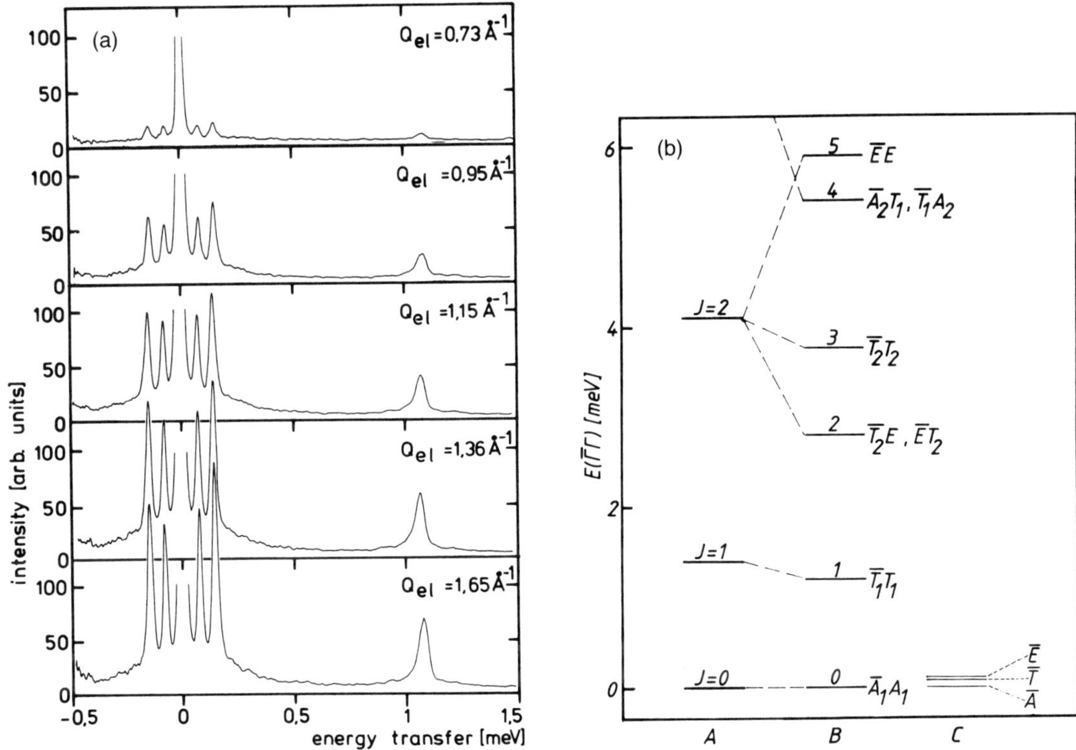

FIG. 14. (a) Tunneling spectra of CH_4-II at $T = 5$ K at several values of momentum transfer Q_{el} collected on the IRIS spectrometer at ISIS. The lower energy peaks at ±75 and ±145 μeV represent the tunneling transitions $T \rightarrow E$ and $A \rightarrow T$. The + (−) signs represent energy gain (loss) of the incident neutrons. The peak at zero energy is incoherent *elastic* scattering from the nuclear spin and position disorder. The higher-energy excitation at 1.07 meV is due to the 0 → 1 transition of the disordered methane molecules. (b) Rotational energy-level diagram of the completely free CH_4 rotor (A), the CH_4 molecules in solid methane at the disorder sites (B), and the tunneling molecules in CH_4-II (C). (Reproduced with permission of B. Asmussen.)

spectral intensity with the momentum transfer Q_{el} is shown for several values of Q_{el}. The peak positions and separations are directly related to their energy-level diagram and are shown in Fig. 14(b). Experiments such as these are essential for a theoretical understanding of tunneling in quantum solids. Similar striking results have been obtained for other molecular groups, such as CH_3, NH_3, and NH_4^+.

2.3.2 Diffusive Motion of Hydrogen in Metals The diffusive motion of hydrogen in metal hydrides has been extensively studied by many techniques. Most experimental measurements give the macroscopic diffusion constants at essentially infinite wave lengths, whereas neutron-scattering measurements give both the macroscopic diffusion constants and the microscopic diffusion constants as functions of wave vector in the Brillouin zone. Their signature is a broad quasielastic diffuse peak of Lorentzian line shape beneath a sharp Gaussian peak of the purely elastic neutron scattering. The anisotropy of the motion and whether the diffusive motion is continuous or occurs in discrete jumps can be ascertained. The diffusion of the isotopes of H, D, and T in palladium metal is a prime example of the type of information available from the unique neutron-scattering measurements. This information is valuable for many practical applications of the diffusion of hydrogen in metal-hydride systems, e.g., in metal-hydride batteries. Figure 15 illustrates some of the results for $PdH_{0.03}$ for different directions of Q, which indicate that the hydrogen jumps are between octahedral sites rather than the tetrahedral sites (Rowe *et al.*, 1972).

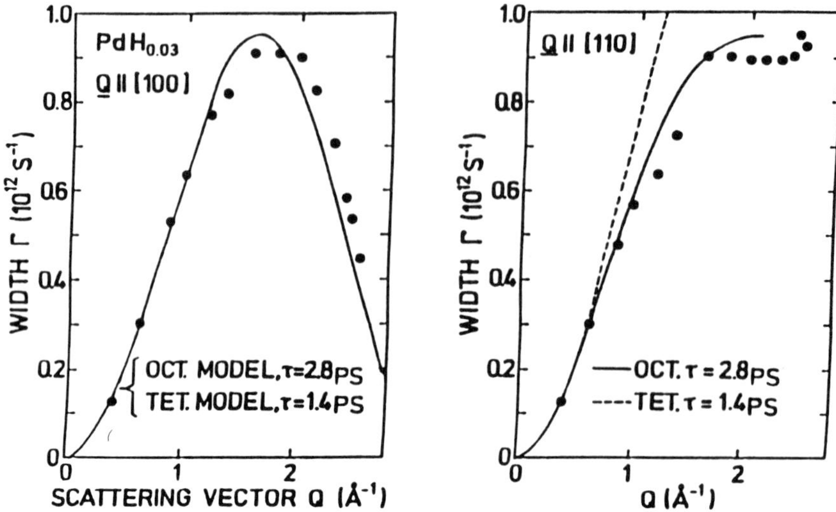

FIG. 15. Widths of quasielastic spectra vs Q for α-$PdH_{0.03}$. Solid and dashed lines are from theory with jumps between octahedral and tetrahedral sites, respectively. (Reproduced with permission of J. J. Rush.)

3. DIFFUSE SCATTERING OF NEUTRONS

Even in an ideal perfect crystal, the background between Bragg peaks is not expected to be zero. There is always thermal diffuse scattering (TDS) from the fundamental vibrations of the crystal lattice that are responsible for the Debye–Waller factor in the Bragg peak intensities; however, this scattering can be minimized at very low temperatures. It is proportional to $|Q|^2$, and so it increases at high scattering angles. Most elements and their isotopes have a spin incoherent scattering cross section and their interactions with the spin $\frac{1}{2}$ of the neutron produces a uniformly isotropic background. (Each spin state of a nucleus has a distinct scattering length, but except at ultralow temperatures, there are no correlations between their positions, and this gives rise to an incoherent scattering background. See Bacon, 1975, for a more extensive description.) There are a few isotopes with zero nuclear spin (even-even nuclei) that, therefore, produce no spin incoherent scattering background. ^1H and ^{51}V are isotopes that have very large spin incoherent scattering cross sections compared to their coherent scattering cross sections, and they produce large unwanted backgrounds (there are some exceptions). H compounds are often deuterated to minimize the effects of poor signal-to-noise ratio in their diffraction patterns, particularly in polycrystalline studies (see NEUTRON DIFFRACTION). Although the incoherent scattering in ^{51}V is large, its coherent scattering cross section is almost zero and, therefore, the intensities of the Bragg peaks are nearly zero, making it a good sample container for furnaces and cryostats that is nearly free from contaminating peaks. However, most elements are composed of isotopes with different cross sections, and they must be properly averaged to give their coherent and isotopic incoherence scattering cross sections (see Bacon, 1975, for a general but basic discussion). With some random substitutional alloys, such as Ti (negative scatterer) and Zr (positive scatterer), a null matrix can be prepared, where the average scattering length is zero and there are no observable Bragg peaks. The null-matrix alloy is sometimes used as a sample container in place of vanadium. Nevertheless, like vanadium and its large spin incoherent background, the null $Ti_{0.60}$–$Zr_{0.40}$ alloy has a large isotopic incoherent background as a result.

The above types of background scattering are generally unwanted, but only the temperature diffuse scattering can be mostly eliminated, and that by employing an analyzing crystal on a triple-axis spectrometer set for elastic scattering or a TOF spectrometer to separate the elastic from the inelastic background. Several other types of diffuse scattering that can be observed are described below. They give interesting and useful information about deviations from perfect crystals, or even ideally imperfect crystals.

3.1 Temperature Diffuse Scattering and Huang Scattering

Although the temperature diffuse scattering from the lattice vibrations occurs throughout reciprocal space, it is most intense underneath and near the Bragg peaks. This is because the intensities are proportional to $1/\nu^2$ and the frequencies of the acoustic modes approach zero as the phonon wave vectors approach zero. There are interesting systems with low-lying phonon branches or soft modes condensing near a phase transition, where the diffuse scattering is far from the Bragg peaks and can be readily studied. The resolution conditions can be greatly relaxed in order to obtain as much scattering intensity as possible. An example is the Ni-Al alloy mentioned in Sec. 2.1.4. Other well-known examples are the condensation of the R-point phonon in $SrTiO_3$ and the X-point phonon in $La(Sr)Cu_2O_4$. The condensation can also occur at an incommensurate point, as in K_2SeO_4.

A type of scattering that appears to be very similar to TDS is Huang scattering, which is caused by small static distortions from defects near lattice sites. The defects may be, for example, vacancies, interstitials, or substitutional size effects in disordered alloys. With neutron elastic-scattering techniques (triple-axis or TOF), they can be separated from the TDS, which cannot always be eliminated by going to very low temperatures. Information regarding the static forces surrounding the defects is obtained.

3.2 Disorder Scattering

Just as the long-range order (LRO) in crystals is studied by carefully measuring the intensities of the Bragg reflections, much valuable information about the short-range order (SRO) can be obtained by carefully measuring the diffuse scattering between the Bragg reflections. As noted above, there may be several different scattering processes contributing to the total diffuse intensity, and it is not always a trivial matter to separate the various contributions.

A very active area that is widely studied by both x-ray and neutron scattering methods is SRO in metallic alloys and compounds. The two methods complement each other, and it is desirable to have both types of studies made on the same material if possible. It is also an area of intense theoretical activity, where, by applying first-principles total-energy calculations, crystal structures and phase diagrams can be predicted. However, at this stage of development, experimental confirmation is essential to differentiate among the various approaches and approximations.

The experimental diffuse intensities must be accurately measured and standard, but nontrivial, corrections made. The data can then be analyzed in terms of the Cowley–Warren SRO (Cowley, 1950). In some cases, distortion scattering due to size effects needs to be considered. Neutrons are particularly suitable for studying the nonstoichiometric transition-metal carbides and nitrides, for the scattering lengths of carbon (6.65 fm) and nitrogen (9.36 fm) are of the same order or larger than the metal values. This permits the C-C or N-N displacements to be determined much more accurately than can be obtained with x rays or electrons. The degree of SRO and lattice distortion varies with the type of heat treatment given to the samples. Figure 16 shows the diffuse scattering pattern due to SRO and static displacements in $NbC_{0.76}$, and the sharp Bragg reflections in $VC_{0.83}$ due to the LRO of the carbon vacancies in the crystal. In the latter crystal, the intensities are due mainly to the carbon atoms, since the scattering length of V is almost zero (−0.39 fm). The two-dimensional diffraction patterns shown in Fig. 16 were recorded in only a few seconds time by a neutron-sensitive TV-image–intensifier system (Davidson and Smith, 1984), but these patterns are only a qualitative record to guide further studies. To analyze the diffuse scattering properly in terms of SRO and static displacements, conventional spectrometers and detectors must still be used to obtain data with sufficient precision to separate out the various contributions to the scattering. This also requires the use of analyzer crystals or TOF methods to eliminate the inelastic scattering due to the lattice vibrations. The reader is referred to

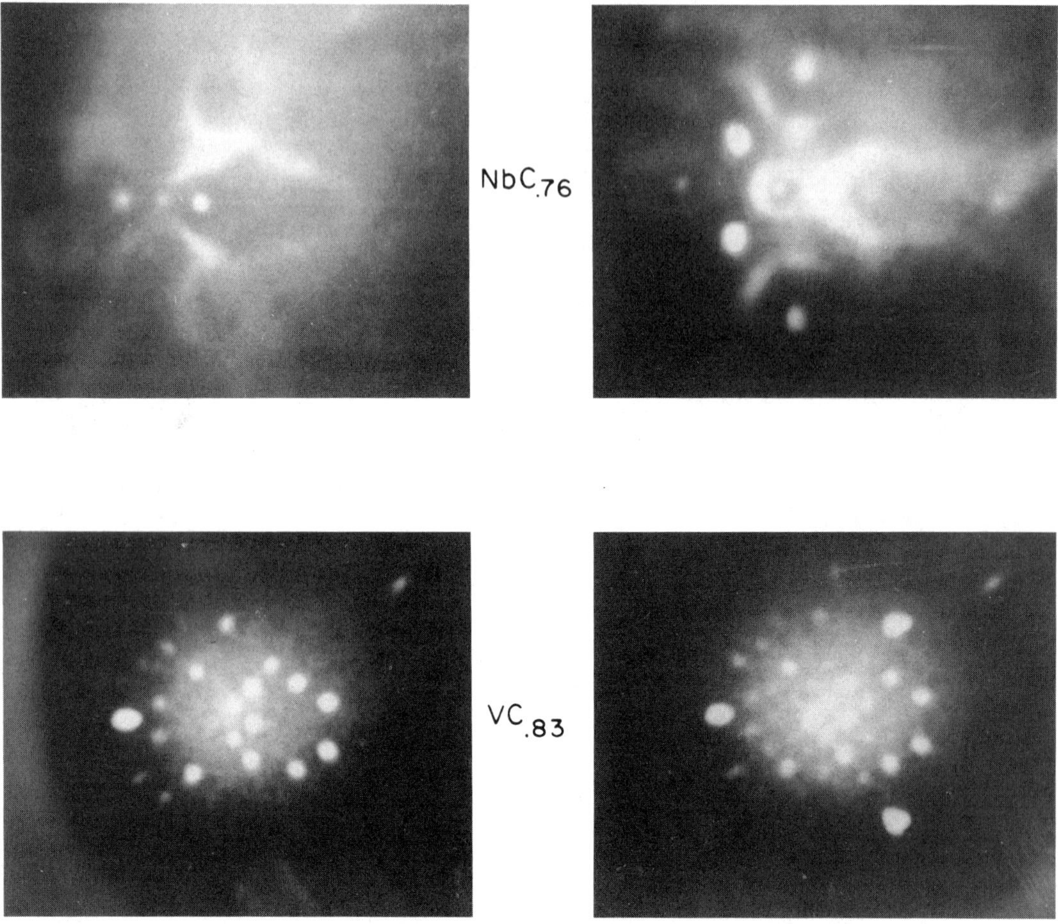

FIG. 16. The diffuse scattering pattern in $NbC_{0.76}$ due to SRO and static displacements, and the sharp Bragg reflections in $VC_{0.83}$ due to the LRO of the carbon vacancies in the crystal. (Reproduced with permission of H. G. Smith.)

the study by Priem *et al.* (1989), in which they determined the SRO parameters, static displacements, and effective interatomic pair energies of several Ti and Nb carbides and nitrides by neutron-scattering techniques.

Diffuse scattering studies of other materials, such as superionic conductors at elevated temperatures, reveal defects related to clustering and diffusion paths of the ions through the lattice.

3.3 Magnetic Diffuse Scattering

Halpern and Johnson (1939), in one of their classic papers on the magnetic scattering of neutrons, derived the differential scattering cross section $d\sigma$ for a random distribution of the moment directions of paramagnetic ions:

$$d\sigma = \tfrac{2}{3}\, S(S + 1)\, (e^2/mc^2)^2 \gamma^2 f^2, \qquad (10)$$

where S is the spin quantum number of the ion; e, m, and c are the electron charge, electron mass, and velocity of light, respectively, γ is the neutron magnetic moment in nuclear magnetions; and f is the form factor of the magnetic electrons of the paramagnetic ions. This form factor is similar to that in x-ray scattering except that it arises from the outer electrons and falls off more rapidly with scattering angle than the total x-ray form factors.

The correctness of the above equation was beautifully demonstrated by Shull *et al.* (1951) in their studies of the diffuse background scattering in polycrystalline MnF_2, after they corrected for TDS, nuclear spin, and isotopic incoherent scattering and any multiple scattering that was present. It was the first measurement of a magnetic form factor and, hence, the wave functions of d electrons. For ions with orbital moments, Eq. (10) must be modified to include the orbital contributions. More precise form factors, however, can be determined by measuring the magnetic Bragg peaks with polarized neutrons (NEUTRON DIFFRACTION).

Even above the Curie and Néel temperatures (ferromagnetic and antiferromagnetic systems, respectively), there is often short-range magnetic order that produces diffuse peaks in the vicinity of the LRO peaks. In magnetic systems with only one-or two-dimensional magnetic order, the magnetic scattering consists of diffuse spots and streaks, but here too, the magnetic scattering can be distinguished from any nuclear disorder scattering by using polarized neutrons.

An important area of magnetic studies has been the measurement of the magnetic diffuse scattering very close to the magnetic phase transition temperature (within a fraction of a degree). The experiments have been used to test the scaling laws and critical exponents of various theories of phase transitions.

4. SMALL-ANGLE NEUTRON SCATTERING

By small angles is usually meant scattering angles (2θ) between zero and a few degrees from the incident beam direction. For crystalline samples, using Bragg's law $\lambda = 2d \sin \theta$, these angles correspond to d spacings of the order of 60 nm for a wavelength of 1.0 nm. (For one degree, $d = \lambda/2\theta \sim 60\lambda$). These wavelengths are appropriate for studying diffraction by large viruses, DNA, and collagen molecules, for example. The scattering cross section is usually plotted against wavevector units, $Q = 4\pi\theta/\lambda$, instead of scattering angle 2θ.

For noncrystalline materials, similar scattering angles and wavelengths represent length scales of the particles or molecules that can be probed. For an isotropic medium, the near-forward scattering is zero, but for nonisotropic media, such as mixtures, defects, voids, etc., the near-forward scattering is not zero but is a function of the fluctuations of the scattering length density of the sample. The fluctuations can be quite large in some systems, such as voids in irradiated materials, precipitates in metal alloys, and large molecules in solution, for example, and they can be studied by x-ray and/or neutron small-angle scattering. However, in polymers, the contrast, or fluctuation in the scattering-length density, is not so great, and there is less information in the scattering cross section. The power in neutron small-angle scattering is the ability to vary the scattering-length density by employing isotopic substitution for some of the atoms. Nowhere is this more useful than in the study of the small-angle scattering of polymers by long-wavelength neutrons. This subject is expanded upon in much greater detail in the review article by Wignall (1993).

The limitations of the weak neutron beams compared to those of x-ray rotating anodes and synchrotron sources have been partially overcome by using large samples ($\sim 10^{-4}$ m^2) and, by necessity, larger detectors and longer sample-to-detector distances—up to 20 m separation. A schematic diagram of a SANS instrument is shown in Fig. 17. The instrument at the High Flux Isotope Reactor uses a double monochromator consisting of two banks of pyrolytic graphite crystals at scattering angles of 90°. The wavelength in this case is fixed at 0.475 nm, and it has been used successfully for numerous experiments. Nevertheless, it is highly desirable to have variable wavelengths up to 2 nm in order to increase the resolution in Q and to access scattering by particles and molecules of larger size. Cold sources of liquid hydrogen or methane, and even D_2O ice cooled to 25 K, are used to moderate the neutron energies to the wavelengths desired (see Fig. 1). The neutron beams are extracted from the reactor cold source by long guide tubes, where the long-wavelength neutrons are totally reflected by the mirrored surfaces of the guide. The neutrons are further monochromatized by a mechanical velocity selector, which produces a more intense beam than a crystal monochromator but is usually of lower resolution.

4.1 Polymers

With the building of the SANS instruments in Europe in the 1970s and the realization of the power of substituting positive-

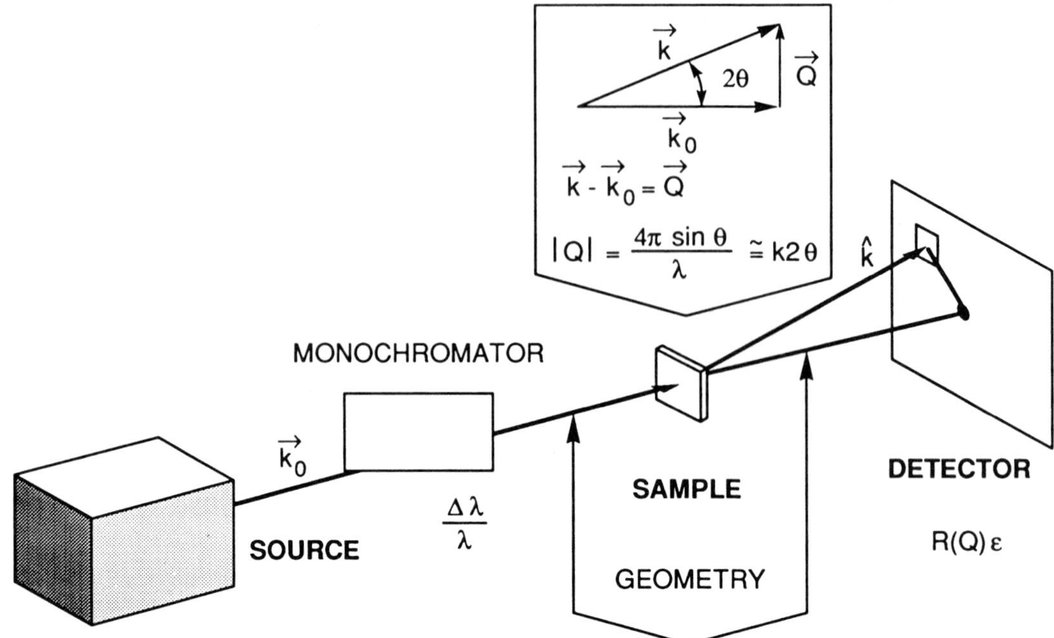

FIG. 17. Schematic diagram of a small-angle scattering spectrometer. The scattering angle 2θ is typically less than a few degrees. (Reproduced with permission of G. Wignall.)

scattering deuterium for negative-scattering hydrogen in polymer chains or segments of chains, the field grew very rapidly, and now every major neutron research facility, including pulsed neutron sources, has state-of-the-art SANS facilities, which are often oversubscribed by the scientific community, industrial as well as academic.

As mentioned in the Introduction, deuterium has a scattering length of +6.67 fm compared with −3.74 fm for hydrogen. For polymers containing hydrogen (and most do), the scattering-length density can be varied dramatically (contrast variation) with striking results in the scattering cross section. Figure 18 illustrates the pronounced effect that partial deuteration of polyethylene (2% PED in a protonated polymer matrix) has on the differential scattering cross section as compared with the undeuterated polymer (PEH). Careful analysis of the intensity vs Q reveals the size of the polymer chains and their conformational properties.

Of significant interest to the polymer industry is an understanding of the behavior of block copolymers. A simple system, called a di-block copolymer, consists of one block of type-A polymer joined end to end with a block of type-B polymer. Such systems can exist in different phases and undergo order-disorder phase transitions (ODT). By deuterating one of the blocks, sufficient contrast is obtained for the SANS measurements. An example is shown in Fig. 19, where not only does the intensity increase as the temperature is lowered toward the ODT in 1,2-1,4 polybutadiene, but also the peak position shifts to lower Q values, indicating that the polymer units (coils) are stretching, contrary to theoretical expectations for polymers in the weak segregation limit (Bates and Hartney, 1985). These and similar experiments on more complex copolymers have been the basis for new theoretical developments in this important area of polymer science. Similar contrast variation techniques are widely used in the study of biological compounds (see NEUTRON DIFFRACTION), micellular structures, and ferrofluids.

4.2 Defects in Metals

As with the study of polymers by chemists, physical metallurgists have found the study of alloys by SANS extremely useful (see Kostorz, 1988). Alloys, particularly commer-

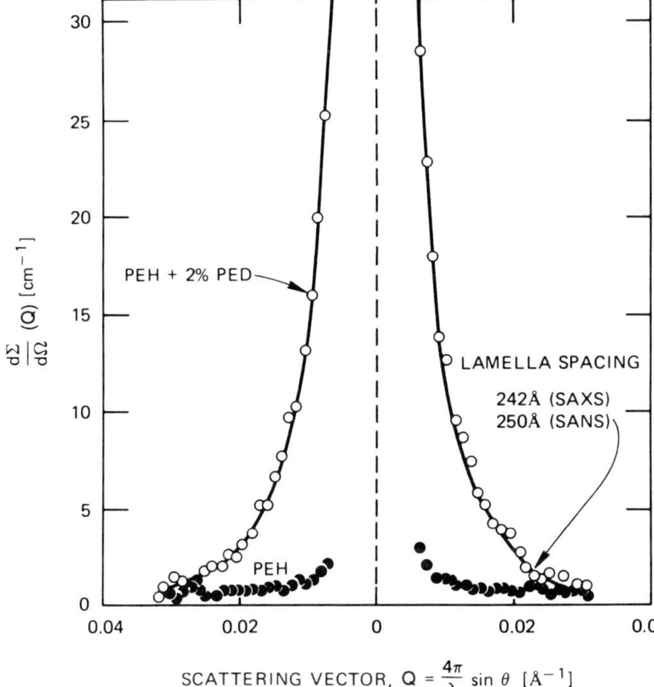

FIG. 18. Small-angle scattering cross section of partially deuterated polyethylene (2% PED in a protonated polymer matrix) as compared with the PEH. (Reproduced with permission of G. Wignall.)

cial alloys, are often far from perfect crystalline specimens. In fact, imperfections are often purposely introduced in order to improve some of the alloys' physical properties, such as their strengths. Many imperfections, introduced or natural, are of the order of 10 nm to 1000 nm in size and are, therefore, very amenable to small-angle scattering techniques if the contrasts are sufficient. For the important transition-metal alloys, the neutron scattering lengths vary from small (2.78 fm for Co) to large (10.8 fm for Ni), and some are even negative (−3.37 fm for Ti) (see NEUTRON DIFFRACTION), and there is usually sufficient contrast for the experiments to be successful. The penetrating ability of the neutrons is very important, for sizable samples can be investigated and the bulk properties are sure

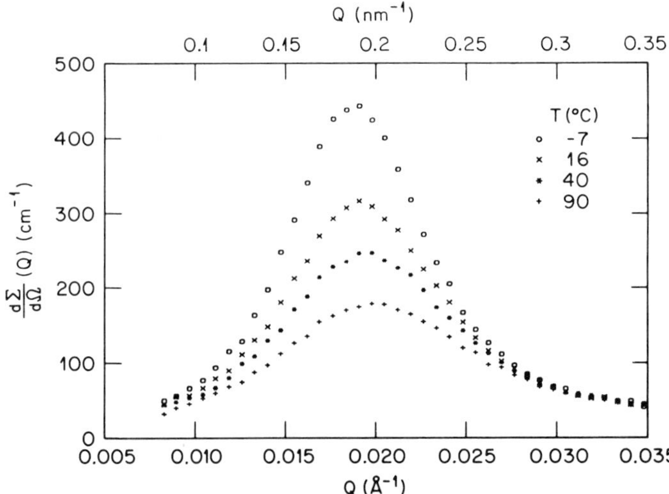

FIG. 19. Scattering cross section vs Q for 1,2-1,4 polybutadiene block copolymer as a function of temperature. (Reproduced with permission of F. S. Bates.)

to be probed. The effects of sample environmental chambers, such as furnaces and cryostats, usually present no problems, in contrast to small-angle scattering with x-rays.

SANS techniques have played an important role in studying the irradiation effects on nuclear reactor and fusion reactor materials. The investigation of the swelling in reactor-based steels caused by voids produced by the radiation has led to improved alloys to be used in future reactor systems. Grain-boundary cavitation has also been observed in fatigued Cu and Ni, and the size of the cavities or voids has been shown to be strongly influenced by temperature. Large-scale inhomogeneities have been observed in several amorphous alloys by SANS techniques.

SANS studies of the kinetics of alloys as a function of heat treatment, aging, quenching, and concentration have yielded valuable information regarding their metallurgical properties. Spinodal decomposition has also been widely studied by these methods.

Of course, magnetic defects as well as magnetic domains and domain walls can be uniquely studied with neutrons, and with the development of magnetic supermirrors, intense beams of long-wavelength polarized neutrons are becoming available.

4.3 Flux-Line Lattices in Superconductors

In type-I superconductors, an applied magnetic field is completely excluded from its interior (the Meissner effect) until a critical magnetic field is reached and the superconductivity is quenched (suppressed). Abrikosov (1957), in his theory of type-II superconductors, predicted the existence of the "mixed" state, where magnetic flux carried by quantized flux lines or vortices penetrates the bulk material, causing parts of the superconducting material to become normal. This occurs for applied fields (B) such that $H_{c1} < B < H_{c2}$, where H_{c1} and H_{c2} are the lower and upper critical fields, respectively. The interaction between vortices is repulsive; hence, vortices form a lattice that maintains the maximum distance between them (of the order of 50–200 nm). Under suitable treatment (alloying, work hardening, etc.), the magnetic flux-line regions can be pinned and the two-dimensional flux-line lattice is stable even after the applied field is removed. It was first suggested by de Gennes and Matricon (1964) that neutron beams with their magnetic moment can be diffracted by these flux-line lattices and should be readily observed at very small angles on a neutron spectrometer. This was immediately confirmed by Cribier *et al.* (1964) on a single crystal of niobium in an applied field in the superconducting state. Much information, both fundamental and applied, can be obtained from studying these flux-line lattices, or vortex lattices as they are sometimes called.

A recent study of the flux-line lattice in a twinned single crystal of a high-T_c superconductor, $YBa_2Cu_3O_7$, by Yethiraj *et al.* (1993) revealed interesting and unexpected behavior. Plate 1 shows the Bragg peaks of the flux-line lattice with the applied magnetic field parallel to the c axis and 30° from the c axis. The patterns in the figure are actually difference patterns between the patterns with and without flux lines in the sample, for there was appreciable small-angle scattering from the sample itself.

In an ideal untwinned superconducting crystal, a trigonal fluxoid lattice would be expected and this would produce a hexagonal (i.e., sixfold) diffraction pattern. The orthorhombic crystal was twinned along the {110} planes, and it is thought that strong pinning centers along the twinning interfaces are responsible for the fourfold symmetry observed in the left side of Plate 1. There are weak flux-line lattice peaks about 45° from the main peaks. The distorted hexagonal pattern observed in the right-hand side suggests that other pinning mechanisms are present. The rapid falloff in Bragg intensities as the temperature is increased cannot be explained by either a BCS model or a two-fluid model and is not well understood at this time (see Sec. 2.1.3).

5. SUMMARY

A number of topics in the neutron scattering field that are not associated with Bragg diffraction *per se* have been presented. After a brief introduction on the general properties and advantages of neutron scattering, the neutron thermalization process is mentioned. The inelastic scattering processes that are most unique to neutron techniques are discussed in Sec. 2. This section involves phonon-scattering experiments, with both steady-state reactor sources and pulsed spallation sources. Magnetic excitations of spin waves, crystal-field levels, and very high-energy intermultiplet transitions are also included. Section 3 is concerned with elastic diffuse scattering between Bragg peaks; and a few topics were chosen that complement the x-ray scattering activities in this field. Magnetic diffuse scattering was also briefly discussed. In Sec. 4, the emphasis is on small-angle neutron scattering from polymers with the contrast variation techniques available with the opposite-sign scattering lengths of hydrogen and deuterium. SANS studies of defects in metals also complement this very active area in x-ray scattering. Finally, the application of SANS, with the neutron's unique magnetic moment, to studying magnetic flux-line lattices in high-T_c superconductors is discussed.

GLOSSARY

Acoustic Modes: Atomic vibrations where the phonon energies approach zero as their wave vectors approach zero (gives the velocity of sound).

Analyzer: A crystal used to determine the energy of neutrons scattered by a sample.

BCS Theory: Bardon–Cooper–Schrieffer theory for the electron–phonon interaction responsible for the mechanism for superconductivity.

Coherent Scattering: Scattering of neutrons by an assemblage of atoms, where phase relationships exist between the atoms.

Cold Neutrons: Neutrons moderated to very low energies by a cold source, such as liquid hydrogen.

Constant *Q*: A method of measuring phonons or magnons by varying the energy while keeping the wave vector constant.

Contrast Variation: Use of selected isotopes to enhance signals over background; mostly used in polymers and biological compounds with H and D substitutions.

Crystal-Field Excitations: Single-particle magnetic excitations.

Dispersion Relation: The relationship between the energies of phonons or magnons and their wave vectors throughout the Brillouin zone.

Electron–Phonon Interaction: The attractive interaction responsible for the formation of Cooper pairs in the BCS mechanism for superconductivity.

Flux-Line Lattice: Lattice formed by concentrated quantized magnetic lines of force in a type-II superconductor.

Force Constants: A measure of the interactions between neighboring atoms.

Huang Scattering: Scattering by atoms in a crystal that have small static displacements from their equilibrium positions.

Hot Neutrons: Thermal neutrons from a reactor that have been increased in energy by moderation in a hot source, or undermoderated neutrons from a spallation neutron source.

Incoherent Scattering: Scattering of neutrons by atoms that have no phase relationship to other atoms.

Isotopic Incoherent Scattering: Scattering produced by isotopes of an element with different scattering lengths for each isotope (there is no ordering of the isotopes).

Lattice Dynamics: Study of the interatomic forces in crystals from their lattice vibration spectra.

Longitudinal Mode: Mode in which the atoms vibrate in the direction of propagation of the phonon wave vector.

Magnons: Collective magnetic excitations.

Monochromator: A crystal used to select neutrons of a narrow energy band from the white spectrum.

Neutron Absorption: Some nuclei absorb neutrons, thereby removing them from the scattering process.

Optic Modes: Vibrations of atoms with out-of-phase relationships.

Phase Transition: For solids, a change of crystal structure as a function of temperature and/or pressure.

Phonons: Collective lattice vibrations.

Reactor Neutrons: Neutrons produced by the fission processes in a reactor.

Rigid-Ion Model: A lattice-dynamical model of ions that ignores the polarizabilities of the ions and their influence on the frequencies of the lattice vibrations.

SANS: Small-angle neutron scattering.

Scattering Length: A measure of the strength of the interaction between a neutron and a nucleus.

Shell Model: A lattice-dynamical model of ions in a crystal that allows for the deformation of the outer valence electrons.

Short-Range Order: Coherent relations between atoms extending only a few nanometers and giving rise to diffuse peaks, in contrast to sharp peaks associated with LRO that extends for hundreds or thousands of nanometers.

Spallation Neutrons: Neutrons produced in an accelerator by protons (or electrons) striking a heavy-metal target.

Spin Waves: Magnons.

Spin Incoherent Scattering: Except for even–even nuclei with zero spin, each spin state of the nucleus has a different scattering length which bears no coherence with other similar nuclei, except for those rare cases where the nuclei order at ultralow temperatures.

Spiral (Helical) Magnetic Structure: A magnetic structure in which the magnetic moment direction spirals about a crystallographic direction with a period that is usually incommensurate with the underlying atomic lattice.

TAS: Triple-axis spectrometer.

TDS: Temperature diffuse scattering.

Thermal Neutrons: Neutrons moderated to thermal energies.

TOF: Time-of-flight.

Transverse Mode: Mode in which the atomic vibrations are at right angles to the direction of propagation of the phonon wave vector.

Works Cited

Abrikosov, A. A. (1957), *Soviet Phys.—JETP* **5**, 1174–1182.

Asmussen, B., Prager, M., Press, W., Blank, H., Carlyle, C. J. (1992), *J. Chem. Phys.* **97**, 1332–1342.

Axe, J., Shirane, G. (1973), *Phys. Rev. Lett.* **30**, 214–216.

Bacon, G. E. (1975), *Neutron Diffraction*, Oxford: Clarendon Press.

Bardeen, J., Cooper, L. N., Schrieffer, J. R. (1957), *Phys. Rev.* **108**, 1175–1204.

Bates, F. S., Hartney, M. A. (1985), *Macromolecules* **18**, 2478–2486.

Born, M., Huang, K. (1954), *Dynamical Theory of Crystal Lattices*, London: Oxford University Press.

Butler, W. H., Smith, H. G., Wakabayashi, N. (1977), *Phys. Rev. Lett.* **39**, 1004–1007.

Cowley, J. M. (1950), *J. Appl. Phys.* **21**, 24–30.

Cribier, D., Jacrot, B., Rao, L. M., Farnoux, B. (1964), *Phys. Lett.* **9**, 106–107.

Davidson, J. B., Smith, H. G. (1984), in: O. K. Harling, Lincoln Clark, Jr., P. von der Hardt (Eds.), *Use and Development of Low and Medium Flux Research Reactors*, Supplement to Vol. 44, Munich: Atomkernenergie-Kerntechnik, pp. 767–775.

de Gennes, P. G., Matricon, I. (1964), *Rev. Mod. Phys.* **36**, 45–48.

Dolling, G., Smith, H. G., Nicklow, R. M., Vijayaraghavan, P. R., Wilkinson, M. K. (1968), *Phys. Rev.* **168**, 970–979.

Dove, M. T., Hagen, M., Harris, M. J., Steigenberger, U., Powell, B. M. (1992), *J. Phys. Cond. Mater.* **4**, 2761–2774.

Halpern, O., Johnson, M. H. (1939), *Phys. Rev.* **55**, 898–923.

Korstorz, G. (1988) in: M. M. Elcombe, T. J. Hicks (Eds.), *Neutron Scattering, Advances and Applications*, Materials Science Forum Vols. 27, 28. Aedermannsdorf, Switzerland: Trans Tech Publications Ltd., pp. 325–344.

McMillan, W. L. (1968), *Phys. Rev.* **167**, 331–344.

Mook, H. M., Yethiraj, M., Aeppli, G., Mason, T. E., Armstrong, T. (1993), *Phys. Rev. Lett.* **70**, 3490–3493.

Moon, R. M. (1985), in: G. H. Lander, V. J. Emery (Eds.), "Scientific Opportunities with Advanced Facilities for Neutron Scattering, Summary of a Workship held at Shelter Island, New York, USA, October 23–26, 1984." *Nucl. Instrum. Methods* **B12**, 556–560.

Nicklow, R. M., Wakabayashi, N., Wilkinson, M. K., Reed, R. E. (1971), *Phys. Rev. Lett.* **26**, 140–143.

Priem, T., Beuneu, B., de Novion, C. H., Chevrier, J., Livet, F., Finel, A., Lefevbre, S. (1989), *Physica B* **156, 157**, 47–49.

Robinson, R. A., Pynn, R. (1988), *Nucl. Instrum. Methods* **A272**, 758–762.

Rowe, J. M., Rush, J. J., de Graaf L. A., Ferguson, C. A. (1972), *Phys. Rev. Lett.* **29**, 1250–1253.

Smith, H. G. (1972), in: D. H. Douglas (Ed.), *Superconductivity in* d- *and* f-*band Metals*, AIP Conference Proceedings No. 4, Woodbury, NY: American Institute of Physics, pp. 321–338.

Shapiro, S. M. (1990), in: B. C. Muddle (Ed.), *Martensitic Transformations*, Materials Science Forum Vols. 56–58, Aedermannsdorf, Switzerland: Trans Tech Publications, pp. 33–34.

Steigenberger, U., Hagen, M., Caciuffo, R., Petrillo, C., Cilloco, F., Sachetti, F. (1991), *Nucl. Instrum. Methods* **B53**, 87–96.

Stirling, W. G., McEwen, K. A. (1987), in: K. Skold, D. L. Price (Eds.) *Methods of Experimental Physics*, Vol. 23C. Orlando: Academic Press Inc., Chap. 20.

Shull, C. G., Strauser, W. A., Wollan, E. O. (1951), *Phys. Rev.* **83**, 333–345.

Taylor, A. D., Osborn, R., McEwen, K. A., Stirling, W. G., Bowden, Z. A., Williams, W. G., Balcar, E., Lovesey, S. W. (1988), *Phys. Rev. Lett.* **61**, 1309–1312.

Wignall, G. D. (1993), *Physical Properties of Polymers*, Washington, D.C.: American Chemical Society, Chap. 7.

Yethiraj, M., Mook, H. A., Wignall, G. D., Cubitt, R., Forgan, E. M., Paul, D. M., Armstrong, T. (1993), *Phys. Rev. Lett.* **70**, 857–860.

Further Reading

Bacon, G. E. (1975), *Neutron Diffraction*, Oxford: Clarendon Press.

Bilz, H., Kress, W. (1979), *Phonon Dispersion Relations in Insulators*, Berlin, New York: Springer-Verlag.

Golub, R., Richardson, D., Lamoreaux, S. K. (1991), *Ultra-Cold Neutrons*, Bristol: Adam Hilger.

Lovesey, S. W., Springer, T. (Eds.) (1977), *Dynamics of Solids and Liquids by Neutron Scattering*, Berlin, Heidelberg, New York: Springer-Verlag.

Nicklow, R. M. (1979), "Phonons and Defects," in: G. Kostorz (Ed.), *Neutron Scattering*, Treatise on Materials Science and Technology, Vol. 15, New York: Academic Press, Inc., pp. 191–226.

Price, D. L., Skold, K. (Eds.) (1987), *Methods of Experimental Physics*, Vols. 23 A, B & C. *Neutron Scattering*, Orlando, San Diego: Academic Press, Inc.

NEUTRON SPECTROMETERS

See SPECTROMETERS, NEUTRON

NMR

See MAGNETIC RESONANCE, NUCLEAR

NOISE, ACOUSTICAL

See SONIC NOISE

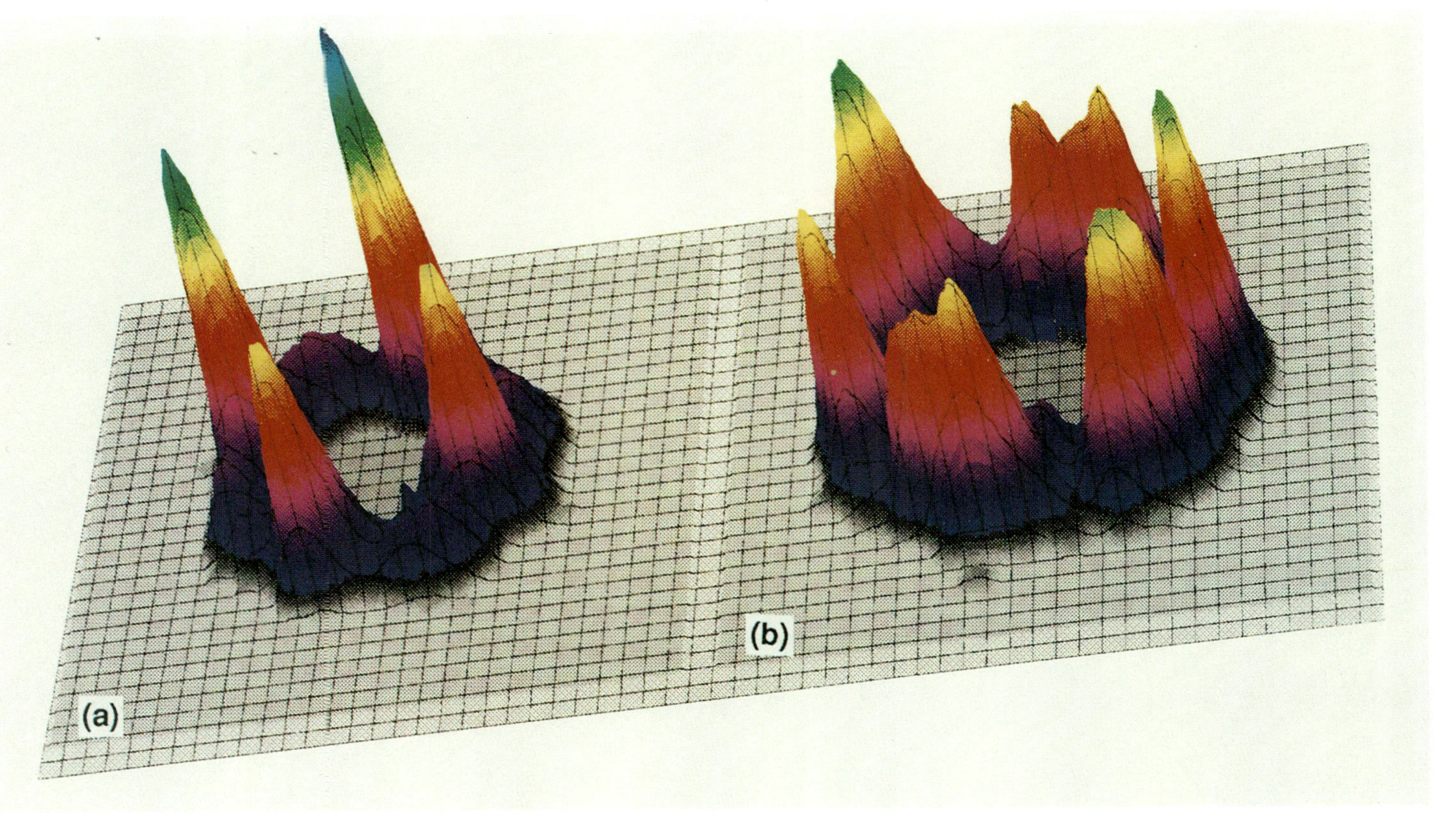

PLATE 1. Flux-line lattice in $YBa_2Cu_3O_7$ with an applied field of 8 kOe. (a) Applied field parallel to c axis. (b) Applied field 30° from c axis. (Reproduced with permission of M. Yethiraj.)

NONHOMOGENEOUS FLOWS

H. G. E. Hentschel, *Emory University, Atlanta, Georgia, U.S.A.*

INTRODUCTION

Any flow involving more than a single phase—liquid, solid, or gas—may be classified as nonhomogeneous. If the interface between the phases is macroscopic, we are interested in the dynamics of this evolving interface or, more generally, in surface flows—flows in porous media, flow on two-dimensional surfaces, or flow on one-dimensional fibers. Such nonhomogeneous flows are dominated by surface tension and wetting effects. The phenomenon of deterministic mixing by chaotic advection can also be considered a surface evolution phenomenon.

On the other hand, if the interfacial structure between the phases occurs on a microscopic scale, we have nonhomogeneous materials with rheological behavior very different from that encountered in the case of a simple Newtonian fluid. The material can then be considered as a dispersion of several differ-

3-527-28133-9/94/$5.00 + .50

ent pure phases, such as granular materials (solid in gas), foams (gas bubbles in liquids), slurries (solid in liquid), or emulsions (droplets of one immiscible fluid in another). Orientational ordering at the macroscopic scale can also result in non-Newtonian anisotropic flows, as is the case for nematic liquid crystals or electrorheological fluids.

A special, but technologically very important, case is that of fluid flow in porous media. The explicitly nonuniform boundary conditions in porous media, due to the tortuous connectivity of its pores and throats, result in a wealth of nonhomogeneous flows and interfacial evolution.

In Sec. 1, we consider flows involving macroscopic surfaces. These flows may involve either two-fluid hydrodynamic instabilities, such as the Taylor and Saffman–Taylor instabilities, or interfacial flow phenomena on solid substrates, such as droplet motion on surfaces and fibers. At the interface between liquids and gases, both surface gravity and capillary waves can occur. Surface evolution due to deterministic mixing and diffusion will also be described.

In Sec. 2, we consider the rheological behavior and flow exhibited by a variety of multicomponent systems, including foams, slurries, and granular materials. The flows exhibited by complex molecular materials, such as polymer solutions, and orientationally ordered fluids, such as nematic liquid crystals and electrorheological fluids, are also briefly discussed.

Finally, in Sec. 3, we consider a variety of flows observed in porous media. Topics such as the interfaces observed during drainage of wetting fluids by nonwetting fluids, imbibition of wetting fluids by porous media, and the anomalous transport of passive scalars in porous media will be examined.

Historically, the field of nonhomogeneous flows has been an active area of research since at least the middle of the 19th century. Lamb in his *Hydrodynamics*, originally published in 1879, considered such topics as cavitation, the motion of solids through liquids, and tidal and surface waves. G. I. Taylor contributed greatly to our understanding of surface instabilities and dispersion in porous media in the first half of this century. More recently, de Gennes has emphasized the importance of scaling arguments in the analysis of nonhomogeneous flows. In particular, scaling arguments have been used to extract the relevant time and length scales, as well as the dimensionless parameters describing various nonhomogeneous flow regimes, and, in so doing, have helped to emphasize the relative importance of various physical mechanisms in creating the particular flow. In addition, topics such as flow in porous media and multiphase flows have been greatly enriched by new concepts in statistical mechanics, such as fractals, self-affine surfaces, percolation, and self-organized criticality. Technologically, nonhomogeneous flows have been central to an astounding variety of industrial processes.

In the oil industry, secondary enhanced oil recovery is a major task. When oil wells are first dug, perhaps only 10% of the oil in an oil-rich sediment will flow back under its own pressure. To release the remaining oil deposits, water is often pumped into the porous sediments to displace the more viscous residual oil. This immediately leads to a classic instability in two-phase fluid flow, the Saffman–Taylor instability, where displacing a more viscous fluid (oil) through a porous medium using a less viscous one (water) leads to a fingering interface, which, from a technological viewpoint, reduces the areal sweep efficiency for oil compared to that for a flat interface, resulting in oil being left in the sediment. Typically, even after secondary oil recovery, only one-third of the oil has been recovered.

All processes in which mixing is used involve nonhomogeneous flows. These include materials blending in the polymer, chemical, and food industries. In addition, mixing flows affect boundary-layer drag in pipes, nozzles, and jets and in naval technology, such as in flow near ships' hulls. The technological importance extends well beyond these areas, however, as any process involving combustion will also be concerned with turbulent mixing, as will such fields as bioengineering and physiology, where knowledge of the flow of blood in fluid vessels is required. Indeed, when geological, oceanographic, and astronomic mixing processes are included, such as mixing in the earth's mantle, atmospheric turbulence, and mixing in the cores of stars, 40 orders of magnitude in Reynolds number may be involved.

Wetting of surfaces and fibers constitutes another class of flows—this time on dimensionally reduced domains—that have important technological consequences. Thus, technologies such as lubrication, processing of fibers, droplet creation, and imbibition by porous media are all controlled by the surface tensions and long-range forces at phase interfaces.

Multiphase flows, such as foams, slurries, and granular flows, form the basis of many industries.

Foams, which are liquid–gas mixtures, are used in fire fighting, as well as to transport pulverized coal in pipelines. In the oil industry, as foam has a low mobility in porous media, this property can be used for blocking and diverting oil flows in desired directions, as well as to transport sand in the technology of enhanced oil recovery. Formations are fractured to enhance oil flowback, and these fractures are kept open using proppants such as sand that are carried to the site of action without settling by foams that then degrade, allowing easy cleanout. Foams are also used in the paper industry to disperse pigments and other paper coatings, while, in the polymer industry, liquid foam flow is an important intermediary in the formation of synthetic cellular solids.

Liquid–solid mixtures in the form of slurries and suspensions also have many industrial applications. They appear in the food and paint industries in the form of pulps, inks, paints, coatings, and emulsions, while, in biotechnology, such problems as coagulation of blood, water purification, and the flows of enzymes all depend on the rheology of suspensions.

Granular flows are two-phase solid–gas mixtures, and such flows are intrinsic to the mining and building industries, among others. Tasks such as funneling flows in hoppers, material crushing for buildings and its resulting flows under stresses, and segregation of granular materials of many different sizes all depend on the very unusual rheologies displayed by granular flows.

Flows in porous media varying in scale from geological deposits to membranes are vital to technologies as varied as the water industry and biotechnology. In the field of hydrogeology, the demand for pure water supplies means that it is vital to have knowledge of the manner in which effluents and other environmental pollutants are carried and dispersed by water in porous media. The related field of filtration is required for water purification. For biotechnology, chromatography and electrophoresis are of ever-increasing importance for the separation of large polymeric molecules, and the manner in which such polymers move in porous media in the presence of electric fields will determine sample purity, while semipermeable flows through membranes are especially important for dialysis.

1. SURFACE FLOWS

By surface flows, we mean all flows that involve in one way or another an evolving macroscopic interface between two phases. We shall consider three types of nonhomogeneous flows under this heading: the evolution of interfaces in liquid–liquid immiscible flows; waves at liquid–gas interfaces; and capillary flow of liquids on solid substrates.

In immiscible two-fluid flows, the behavior of this interface is all important, and in those cases where the interface is unstable to small perturbations in curvature, such as is the case when a less viscous fluid is used to drive a more viscous one, an initially flat surface can develop into a complex asymptotic pattern. The final structure is strongly dependent on boundary conditions, and both asymptotic fingering and dendritic branching patterns in the same universality class as two-dimensional diffusion-limited aggregation have been observed.

At liquid–gas interfaces, wave motion is often observed. Wave motion is due to the interaction between an inertial term tending to keep a system moving at a constant velocity when disturbed, and a restoring force tending to bring the system back to equilibrium. The inertial term, however, causes the system to overshoot the equilibrium point, setting up an oscillatory motion. At liquid–gas interfaces, the restoring forces include gravity and surface tension.

Another class of flows involving surfaces comprises those where the flows actually occur on a solid substrate, such as on two-dimensional planes or one-dimensional fibers . In this case, the flow is strongly influenced by surface tension and the wetting properties of the liquid.

1.1 Dynamics of Liquid–Liquid Interfaces

Here we examine the time evolution of interfaces between two immiscible liquids, each liquid segregated into a single simply connected domain of macroscopic scale.

1.1.1 The Saffman–Taylor and Taylor Instabilities How does such an interface between two immiscible fluids develop with time under an applied pressure? The important criterion for surface evolution is stability. If an initially flat interface is stable, then it will remain flat with time, and any small perturbations due to noise will decay. Under certain flow conditions, however, the flat interface can become unstable, and the interface develops with time into a complex curved surface. Two important causes of such instabilities are the Taylor and Saffman–Taylor instabilities.

Consider the experimental situation when one immiscible fluid displaces another in a porous medium. Taylor (1950) showed that the interface can become unstable when a fluid of higher density ρ_2 is forced, usually by gravity, into another fluid of lower density ρ_1 (for example, the interface between a heavy oil lying on top of less dense water is unstable). Saffman and Taylor (1958) pointed out that a related instability occurs when a fluid of lower viscosity η_2 is driven with a velocity U into a fluid of greater viscosity η_1. The flat surface is then replaced by a highly irregular contour with fingerlike projections [see Fig. 1(a)]. Such convoluted interfaces also occur in enhanced oil recovery when water is pumped in at a point [see Fig. 1(b)] into sediments to release the residual oil. The importance of this fingering in enhanced oil recovery is apparent: The fraction of oil recovered, or areal sweep efficiency, is reduced compared to that expected for a flat displacing interface.

The full set of equations describing the Taylor and Saffman–Taylor instabilities are the following:

1. Darcy's law in a gravitational field, which relates the flow rate $\mathbf{v}$ of a liquid of viscosity η to the pressure gradient ∇p in a porous medium of permeability κ (equivalent to Ohm's law for the electric current in a resistor),

$$\nabla p_\alpha = -(\eta_\alpha/\kappa_\alpha)\mathbf{v}_\alpha - g\rho_\alpha \cos\theta\mathbf{i}. \tag{1}$$

(a)

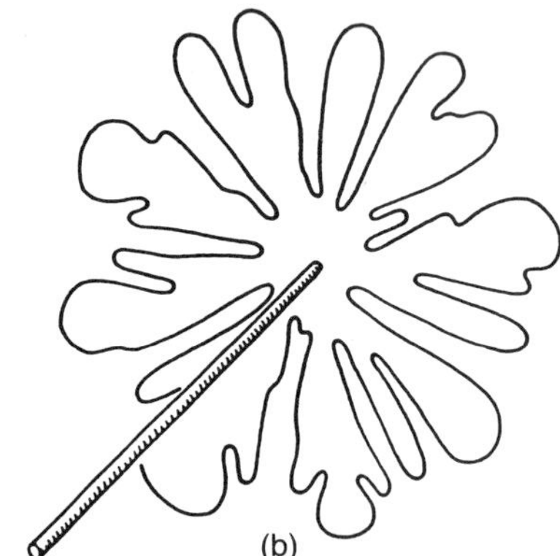

(b)

FIG. 1. The developing interface of immiscible two-fluid flow: (a) dyed water displacing glycerine in a porous medium; (b) air injected in at a point in a Hele–Shaw cell filled with glycerine.

Here the subscripts $\alpha = 1,2$ represent respectively the displaced and displacing fluids; also, the sediment is assumed to be so aligned that an angle θ exists between the Earth's gravitational field g and the sediment, which is taken to lie in the x-y plane (see Fig. 2).

2. The incompressibility conditions

$$\nabla \cdot \mathbf{v}_\alpha = 0. \tag{2}$$

3. The pressure difference across any point $\mathbf{s}$ of the two-dimensional curved interface due to the effective surface tension σ in the porous medium,

$$p_2(\mathbf{s}) - p_1(\mathbf{s}) = \sigma[R_1^{-1}(\mathbf{s}) + R_2^{-1}(\mathbf{s})], \tag{3}$$

where $R_1(\mathbf{s})$ and $R_2(\mathbf{s})$ are the two principal radii of curvature of the interface at $\mathbf{s}$.

The origin of surface tension lies at the

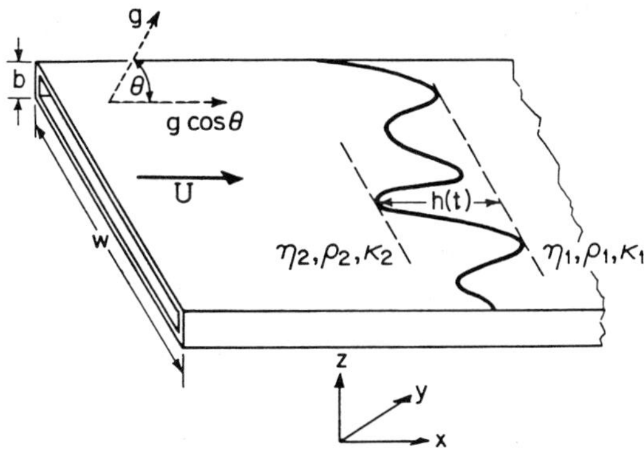

FIG. 2. Schematic diagram of the Saffman–Taylor and Taylor instabilities in a Hele–Shaw cell.

molecular level in the observation that the interaction energy per molecule deviates from its bulk value near an interface, and, in general, this deviation is in the positive direction because of the broken attractive bonds that exist near an interface. Consequently, the energy per unit area of a surface $\sigma = (\partial G/\partial A)_{T,P}$ is positive, or, equivalently, a normal force of size $df = \sigma dl$ exists on any line dl drawn on the surface, tending to it tear apart. This second definition gives rise to the term "surface tension." For curved surfaces with principal radii of curvature R_1 and R_2, the pressure drop Δp required across the surface to keep such a conformation in being is then given by Eq. (3). That this statement is likely to be true can be most easily appreciated by considering the case of a surface defined by the function $z(x, y)$. Consider for the moment the case where the surface is only curved in the x direction $z(x)$. Draw two lines on the surface of length δy parallel to the y axis a distance δx apart, and consider the total force δf_z in the z direction on this rectangle of area $\delta y \delta x$ due to surface tension. By definition,

$$\delta f_z = \sigma \delta y \left[\left(\frac{\partial z}{\partial x} \right)_x - \left(\frac{\partial z}{\partial x} \right)_{x+\delta x} \right] \approx -\sigma \delta y \delta x \frac{\partial^2 z}{\partial x^2}.$$

As the pressure is force/area, we see that the surface tension has induced a pressure $p_{\text{tension}} = df_z/\delta y \delta x = -\sigma\, \partial^2 z/\partial x^2$ across the interface for stability. But on a slowly varying surface, the curvature in the x direction is given by $R_x^{-1}(\mathbf{s}) \approx -\partial^2 z/\partial x^2$. With a more rigorous argument, Eq. (3) can be derived exactly.

The initial instability of a flat surface if surface tension is neglected can be studied using linear-stability analysis. When such a procedure is carried out, the condition for unstable growth is

$$[(\eta_1/\kappa_1) - (\eta_2/\kappa_2)]U + g \cos \theta(\rho_2 - \rho_1) > 0, \tag{4}$$

which includes the Taylor instability and Saffman–Taylor instability conditions as special cases. Experimentally, the gravity-dominated Taylor instability has been observed in oils of similar viscosity but different densities, while the Saffman–Taylor instability has been studied in detail in Hele–Shaw cells.

1.1.2 Flow in Hele–Shaw Cells The Taylor and Saffman–Taylor instabilities are very relevant to the oil industry, concerned as it is with the extraction of oil from geological sediments using water pumped in at high pressure. To study such flows in the laboratory, scaled-down versions of the resulting flow must be created. As such oil-bearing sediments are often effectively two dimensional in character—they have a depth b much less than their width W, or length L (typically, $b \sim 10$ m, $W \sim L \sim 1000$ m)—motion in oil sediments is often analyzed by considering two-fluid flow between glass plates, a device known as a Hele–Shaw cell (typically, $b \sim 0.1$ cm, $W \sim 10$ cm, $L \sim 100$ cm). This use of Hele–Shaw cells is possible because Darcy's law $\mathbf{v} = -(\kappa/\eta)\nabla p$ also applies to these cells with the permeability κ replaced by $b^2/12$ (typically, for porous media, 10^{-7} cm$^2 < \kappa < 10^{-4}$ cm^2, depending on whether the sand is fine or coarse packed, while $b^2/12 \sim 10^{-3}$ cm^2 for a Hele–Shaw cell).

In Saffman–Taylor experiments in Hele–Shaw cells, the less viscous fluid is forced in with a velocity U, and, as expected, an initial instability appears (see Fig. 3). Asymptotically, one finger becomes dominant (see Fig. 4) and grows at the expense of the rest. This finger has a well-defined shape with width λW. Experimentally, the parameter λ has been found to vary from $\lambda \to 1$ as $U \to 0$, to, apparently, $\lambda \to \frac{1}{2}$ as $U \to \infty$, though at high velocities spatiotemporal fluctuations appear.

A great deal of theoretical effort (see, for example, Bensimon *et al.*, 1986, and references therein) has gone into trying to understand the shape and stability of this asymptotic finger for two reasons. First, flow in

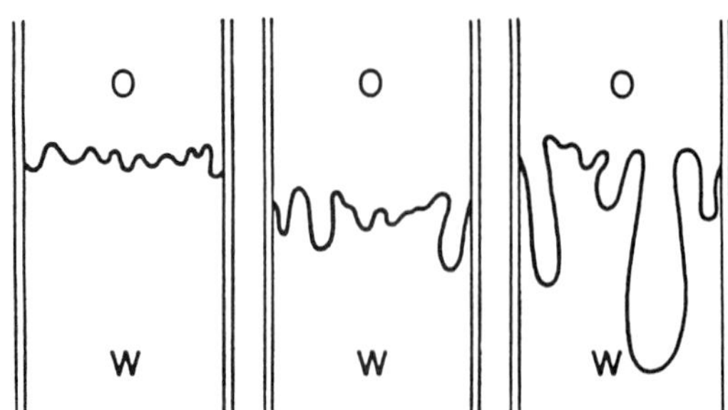

FIG. 3. The growth of an unstable flat interface in the Saffman–Taylor instability (W, water; O, oil); redrawn from Maher (1985).

Hele–Shaw cells represents perhaps the simplest pattern-forming flow that can be seen experimentally and that may yet be theoretically tractable because of its quasi-two-dimensional nature. Second, this finger is closely related to a large class of important problems involving pattern formation in interfaces, including dendrite formation and directional solidification.

Originally, Saffman and Taylor tried to find a relationship between λ and U for the asymptotic finger, while neglecting surface tension, by formulating the flow as a free-boundary problem in the plane, which they solved using conformal mapping. By convention, the glass plates in a Hele–Shaw cell are taken to be separated in the z direction, and the flow is taken to be in the x direction; then the shape of the finger found by Saffman and Taylor can be written

$$\frac{x - Ut}{W} = \frac{1 - \lambda}{\pi} \ln \cos (\pi y/\lambda W). \tag{5}$$

Unfortunately, there are two separate problems with the solution given by Eq. (5): It is not unique, as, for each U, a continuous family of solutions parametrized by λ exist; and all these solutions are linearly unstable to small perturbations.

Both problems can be remedied by keeping surface tension (Chuoke *et al.*, 1959). Obviously, in the experimental situation, surface tension does exist, and a unique finger is selected for each U. The effect of surface tension is to stabilize the smallest-wavelength fluctuations that have the largest curvatures. A linear-stability analysis of the flat interface confirms physical intuition. Adding a small perturbation to the flat interface $h(y, t) = A_k(t) \cos ky$ results in growth of the k wavevector mode as

$$\frac{dA_k}{dt} = kA_k\left[U - \frac{\sigma b^2}{12(\eta_1 - \eta_2)} k^2\right]. \tag{6}$$

As the minimum value for a wave vector is $k = 2\pi/W$, for very small forcing velocities $U < (\pi^2/3)(b/W)^2\sigma/(\eta_1 - \eta_2)$, the interface is stable. Notice that the stability appears to depend on only one dimensionless parameter $d = (\pi^2/3)(b/W)^2[\sigma/(\eta_1 - \eta_2)U]$, rather than two separate dimensionless parameters—the capillary number $Ca = (\eta_1 - \eta_2)U/\sigma$ [which measures the ratio of viscous $(\eta_1 - \eta_2)U/r$ to capillary forces σ/r for an immiscible two-fluid interface of curvature $1/r$ flowing with velocity U] and the aspect ratio W/b, which is dependent solely on the Hele–Shaw cell geometry. Thus, the condition for stability of a flat interface can be rewritten $d > d_c = 1$. Also, by keeping the surface tension σ, a unique scaling relationship between λ and U can be found (Pitts, 1980) if one equates the change in pressure due to the flow with the change in pressure across the interface at the tip and sides of the finger due to variations of transverse curvature. Again, this relationship depends solely on the parameter d,

$$\lambda = \lambda(d), \tag{7}$$

and obeys the asymptotic conditions $\lambda \to 1$ as $d \to \infty$ and $\lambda \to \frac{1}{2}$ as $d \to 0$.

But what about the stability of this asymptotic finger? Using the flat-interface results as a first estimate, one might expect that, for $d \to \infty$, the finger would be stable, but, for $d < d_{\text{finger},c} \approx 1$, the interface would become unstable. Certainly, experimentally, for high flow rates U, the asymptotic finger

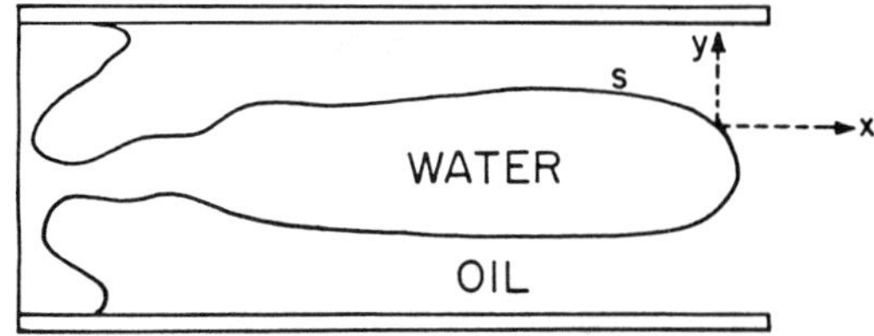

FIG. 4. The asymptotic finger seen in two-fluid flow confined to a Hele–Shaw cell.

appears to become unstable, and time-dependent chaotic fluctuations as well as finger bifurcations set in that distort the shape (Maher, 1985; Tabeling and Libchaber, 1986). Recent theoretical work (DeGregoria and Schwartz, 1986; Bensimon, 1986), however, would suggest that as $d \rightarrow 0$, the asymptotic finger remains stable to infinitesimal perturbations, though unstable to finite perturbations. The size of the finite perturbations that can destroy the finger becomes, however, asymptotically small: $\delta h_{\text{noise}} \sim \exp(-c\ d^{-1/2})$ as $d \rightarrow 0$. Thus, as any experimental setup will always have some noise, there will always appear to exist a critical value for d dependent on the noise such that, for $d < d_c(\text{noise}) \sim 1/(\ln \delta h_{\text{noise}})^2$, the finger will show spatiotemporal fluctuations.

1.1.3 Laplace's Equation and Diffusion-Limited Fingering Asymptotic fingers are generated in a Hele–Shaw cell in the presence of surface tension. But what happens when the viscosity ratio is much greater than one, $\eta_1/\eta_2 > 1$, and there is no surface tension, so that $d = 0$? It is observed experimentally that bifurcation in fingering occurs on all scales, and the flow takes on a dendritic structure reminiscent of diffusion-limited aggregation.

Diffusion-limited aggregation was originally introduced by Witten and Sander (1981) as a model for aggregation of soot particles. Imagine that carbon atoms performing a Brownian walk travel until they touch the soot aggregate and that, at this point, they stick irreversibly. What will be the shape of the resulting aggregate? Probably, before the first simulations were performed, one would have guessed that a compact cluster with a rough surface would be seen. In fact, a fractal dendritic object occurs [see Fig. 5(b)]. The reason for such growth is screening. The particle performing a random walk sticks on first contact and consequently cannot enter deep into the resulting fjordlike voids to create a compact object. Diffusion-limited aggregation might have remained a model for a certain class of aggregation processes, except that it was shown by Niemeyer *et al.* (1984) to be isomorphic to a solution of Laplace's equation (because of the random walk taken by the aggregating particles) with moving boundary conditions and, thus, a basic universal paradigm controlling the structure of such apparently diverse phenomena as electrochemical deposition (Brady and Ball, 1984; Matshushita *et al.*, 1984), dielectric breakdown (Niemeyer *et al.*, 1984), and even aspects of biological growth, such as retinal vasculature (Family *et al.*, 1989). Only the nature of the field obeying Laplace's equation changes: for example, the concentration field in the case of aggregation or the electrical potential in the case of dielectric breakdown.

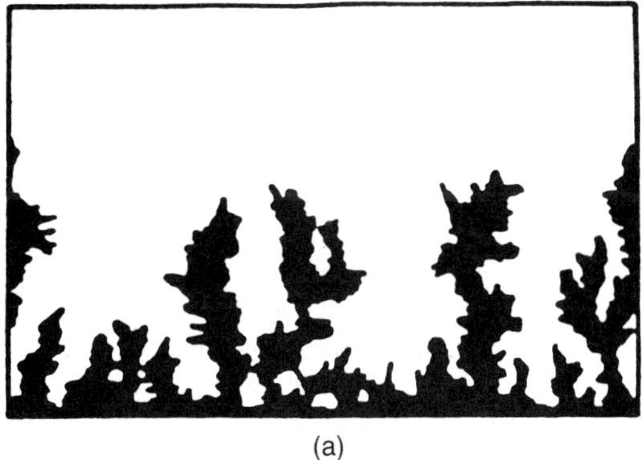

(a)

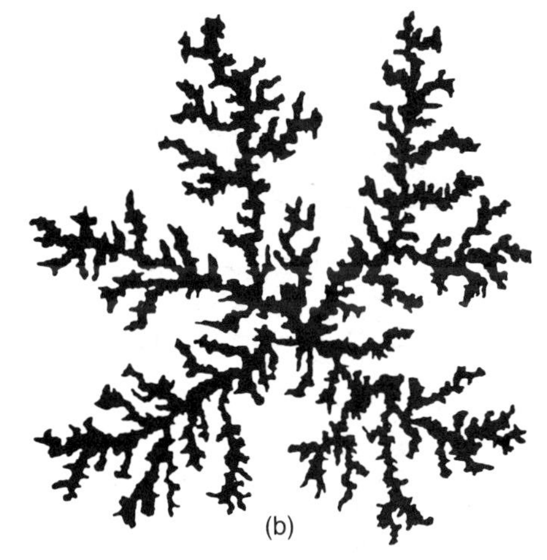

(b)

FIG. 5. Diffusion-limited aggregate grown (a) on a flat interface, (b) from a point seed. Compare with the two-fluid flow experiments in Fig. 1.

Two-fluid flow in porous media was also shown by Paterson (1984) to obey Laplace's equation in a certain limit, with the field this time being the pressure field (compare Fig. 5, which shows computer simulations of diffusion-limited aggregation with various boundary conditions, with Fig. 1). The evolving interface in immiscible two-fluid flow can therefore be shown to be in the same universality class as diffusion-limited aggregation. To see this, consider Eqs. (1–3).

First, neglect g and η_2, in which case Eq. (1) reduces to $\nabla p_2 = 0$ for the displacing phase. This has the solution $p_2 = 0$, and specifically, at the interface, $p_2(\mathbf{s}) = 0$. Now neglect surface tension σ; then $p_1(\mathbf{s}) = 0$ for the displaced fluid interface also. Using Eq. (2), it can now be seen that in the more viscous phase, the pressure field $p_1(\mathbf{x})$ and interfacial velocity $\mathbf{v}_1(\mathbf{s})$ obey identical equations to diffusion-limited aggregation (DLA):

$$\nabla^2 p_1 = 0, \tag{8}$$

with $p_1(\mathbf{s}) = 0$ and p_1(other boundaries) constant, while the boundary moves as $\mathbf{v}_1(\mathbf{s}) = -(\kappa_1/\eta_1)\nabla p_1(\mathbf{s})$. And, indeed, the argument appears to be correct. Experiments on two-fluid flow (Nittmann *et al.*, 1985) with $\eta_1 \gg \eta_2$ and negligible surface tension σ, by using water ($\eta_2 \approx 10^{-2}$ P) to displace polysaccharide solutions ($\eta_1 \approx 10^2$ P) in a Hele–Shaw cell, yield a fractal interface with DLA exponents.

For finite values of the capillary number $Ca = (\eta_1 - \eta_2)U/\sigma$ and the viscosity ratio $M = \eta_2/\eta_1$, a more complex crossover behavior is seen. At short times $t \ll t_c(Ca, M)$, the flow appears to be a dendritic fractal in the same universality class as DLA with an inner length scale $l(Ca, M)$ controlled by capillarity and viscosity ratio. At long times, however, $t \gg t_c(Ca, M)$, crossover to compact flow is observed, though the interface may have a rough self-affine form (see Sec. 3).

1.2 Dynamics of Liquid–Gas Interfaces

Interfaces between gas and liquids also exist, and the most commonly observed dynamics for this case are waves.

1.2.1 Surface Gravity Waves The large-scale surface evolution to be observed on seas and lakes is in the form of surface gravity waves. This wave motion is caused by the interaction of an inertial term, the liquid density ρ_0, tending to keep the liquid moving at a constant velocity, while gravity $\mathbf{g}$, taken to act in the z direction, is the anisotropic resorting force leading to oscillations. Assume that the equilibrium liquid–gas interface (typically water–air) lies in the $z = 0$ plane and consider a small perturbation $h(\mathbf{r}, t)$. This perturbation is assumed small enough that linear theory is applicable, and, specifically, the interface is single valued at each point (a condition violated by the large-scale wave motions seen in stormy seas).

In the water phase $z < h(\mathbf{r}, t)$, the velocity field $\mathbf{v}(\mathbf{x}, t)$ may be taken to obey

$$\rho_0 \frac{\partial \mathbf{v}}{\partial t} = -\nabla p + \rho_0 \mathbf{g}, \tag{9}$$

where $p(\mathbf{x}, t)$ is the pressure field in the water and dissipation has been neglected. The additional constraint of incompressibility gives $\nabla \cdot \mathbf{v} = 0$. Looking for irrotational potential flow solutions $\mathbf{v} = \nabla\phi$ implies that the velocity potential obeys Laplace's equation

$$\nabla^2 \phi = 0, \tag{10}$$

while the pressure field is given by

$$p = p_a - \rho_0 \frac{\partial \phi}{\partial t} - \rho_0 g z. \tag{11}$$

Here, the constant of integration p_a is the air pressure at the interface.

Eq. (11) can be used to yield the boundary condition for the velocity potential at the interface,

$$\left.\frac{\partial \phi}{\partial t}\right|_{z=h} = -gh, \tag{12}$$

while, if the water is of depth d, then the normal component of the velocity is zero at this depth $v_z(z = -d) = 0$, or, in terms of the velocity potential,

$$\left.\frac{\partial \phi}{\partial z}\right|_{z=-d} = 0. \tag{13}$$

As the boundary condition at the interface Eq. (12) is expressed in terms of the surface conformation h, the equation of motion for the interface itself, given by $dh/dt = v_z(z = h)$, is needed to complete the free-boundary problem involved. In terms of the velocity potential, the interface therefore obeys

$$\frac{\partial h}{\partial t} + \nabla\phi \cdot \nabla h = \left.\frac{\partial \phi}{\partial z}\right|_{z=h}, \tag{14}$$

and, if we combine Eqs. (12) and (14), the explicit surface profile can be deleted in favor of the velocity potential ϕ alone. In the linear approximation, this yields the interface boundary condition

$$\left.\frac{\partial^2\phi}{\partial t^2}\right|_{z=0} = -\left.\frac{g\partial\phi}{\partial z}\right|_{z=0}. \tag{15}$$

Laplace's equation, together with the boundary conditions at the interface [Eq. (15)] and on the sea bottom [Eq. (13)], yields traveling-wave solutions $\phi(\mathbf{r}, t) = \Phi(z)\exp(ikx - i\omega t)$ provided

$$\omega^2 = gk \tanh kd. \tag{16}$$

Thus, in contrast to sound waves in bulk fluids, which are nondispersive and longitudinal in nature, surface gravity waves can be strongly dispersive, in addition to which the resulting velocity field during propagation has both longitudinal and transverse components.

Two limiting cases can be derived from Eq. (16). For not too long-wavelength surface waves on deep water, $kd \to \infty$, and the dispersion relation reduces to $\omega^2 = gk$. Thus, the phase velocity of of deep-water waves increases with wavelength, $c_k = \omega/k = (g/k)^{1/2}$, as does the group velocity, which is exactly half the phase velocity: $u_g(k) = d\omega/dk = c_k/2$. This increase of velocity with wavelength can be seen when a stone is thrown into a pool: After a time t, the localized initial disturbance spreads into a sinusoidal pattern in which the long-wavelength modes have traveled further, $u_g(k)t$, than the shorter wavelengths, which remain near the origin of the disturbance.

The other limit consists of long-wavelength waves in shallow channels. In this case, $kd \to 0$, and the dispersion relation reduces to $\omega = (gd)^{1/2}k$. Such channel waves are nondispersive with a speed of propagation $c = (gd)^{1/2}$, which, for typical channel depths, is a much slower mode of propagation than sound. In addition, for such long-wavelength modes, and in contrast to sound, the wave motion is essentially transverse. The dependence of speed on depth also gives an indication of the way waves break on a shoreline. The closer to the shoreline, the slower the speed of wave propagation, and this will result in a shocklike phenomenon of waves breaking on the ocean front as the faster, deep-water waves catch up with slower waves near the shoreline.

1.2.2 Capillary Waves Including surface tension results in a pressure drop $\Delta p = \sigma(1/R_1 + 1/R_2)$ across the liquid–gas interface. Clearly, the influence of surface tension becomes ever more significant at larger curvatures, and, therefore, small-wavelength oscillations must finally be dominated by surface tension. Such waves, known as capillary waves or surface ripples, can be seen when the surface of a glass of water is gently disturbed.

In the linear approximation for waves traveling in the x direction, the only nonzero surface curvature lies in the x direction and is given by $\Delta p = \sigma(1/R_1 + 1/R_2) \approx -\sigma\,\partial^2h/\partial x^2$. Therefore, the excess pressure in the water above the air pressure is given by

$$p_e(z) = \rho_0 gh - \sigma\frac{\partial^2 h}{\partial x^2}. \tag{17}$$

In consequence, for an oscillation of wave number k, the dispersion behavior for surface gravity waves may be generalized to include the effects of surface tension by replacing g by $(g + \sigma k^2/\rho_0)$:

$$\omega^2 = (g + \sigma k^2/\rho_0)\,k \tanh kd. \tag{18}$$

It is apparent from Eq. (18) that crossover to surface-tension–dominated ripples occurs at $k_c \approx (\rho_0 g/\sigma)^{1/2}$. For water, this corresponds to a wavelength of about 1.7 cm.

For deep-water waves, the dispersion relation [Eq. (18)] reduces to $\omega^2 = (g + \sigma k^2/\rho_0)k$, and, therefore, the phase velocity can be written

$$c_k = \left[\frac{g+\sigma k^2/\rho_0}{k}\right]^{1/2}. \tag{19}$$

Note that Eq. (19) implies that the speed of the wave motion has a minimum value at the crossover to capillary-dominated behavior, $k_c \approx (\rho_0\, g/\sigma)^{1/2}$. For water, this corresponds to a speed of 23 cm/s. At even smaller wavelengths, surface tension dominates surface fluctuations entirely, in which case the dispersion curve for pure capillary waves finally appears as

$$\omega^2 = \sigma k^3/\rho_0. \qquad (20)$$

Despite the rich variations that surface waves have been shown to exhibit, the whole discussion here has been within the linear approximation, and both dissipation and nonlinear wave phenomena have been neglected. The complexity of real sea-wave motion attests to the rich store of new phenomena to be expected if such nonlinear behavior is included [see, for example, Lighthill (1978)].

1.3 Flow on Surfaces

In flows on solid substrates, once again surface tension plays a dominant role in controlling both form and flow. Surface tension will sometimes tend to minimize surface areas, accounting for such observations as the fact that droplets are spherical, or that wetted fibers tend to stick together, while in other cases it encourages wetting of the complete substrate.

1.3.1 Wetting, and Capillary Flow Consider a column of liquid in a capillary tube of radius r, and specifically, its curvature at the top of this column where the liquid and vapor phases meet. It could either be concave, in which the case the liquid is wetting the solid substrate forming the capillary tube (water in glass, for example), or convex, in which case the liquid is nonwetting (mercury in glass, for example). For a nonwetting fluid to enter a capillary tube and displace the air, an external pressure must be applied to overcome capillary forces resisting entry of the liquid, while for a wetting fluid, the capillary forces will encourage entry even against an external gravitational field. The same phenomena of drainage and imbibition occur during flow in porous media (see Sec. 3).

To answer the question as to whether a liquid will or will not wet a solid surface, the angle between the tangent to the liquid–vapor interface where it meets the substrate and the substrate itself is required. This angle is known as the static wetting angle θ [see Fig. 6(a)]. For mechanical equilibrium to exist for a droplet on a solid substrate, the sum of all the forces due to the surface tension for each interface, $\sigma_{LV} = \sigma$, σ_{LS}, and σ_{SV} (L = liquid, V = vapor, S = substrate), must vanish at the contact point or

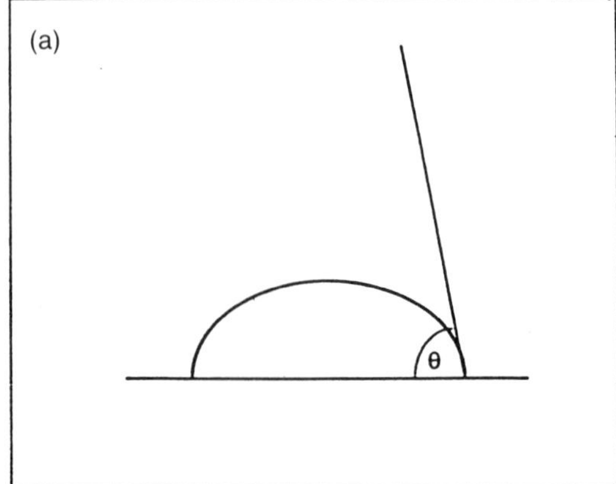

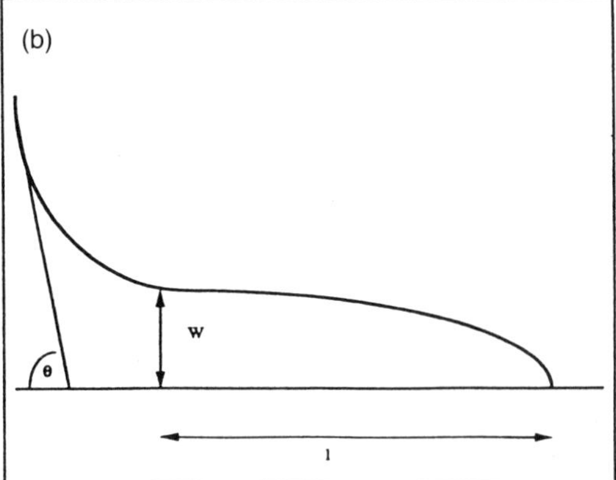

FIG. 6. Schematic diagram showing (a) a droplet in mechanical equilibrium, (b) a spreading droplet on a flat substrate.

$$\sigma_{SV} - \sigma_{LS} - \sigma \cos\theta = 0. \qquad (21)$$

This is Young's equation for the wetting angle θ of a stable nonwetting liquid–vapor interface lying on a substrate. If we use a similar argument for the dynamic wetting angle θ_d during the motion of a liquid up a capillary tube, then there exists a pressure $p_{\text{tension}} = 2(\sigma_{SV} - \sigma_{LS} - \sigma\cos\theta_d)/r$ due to capillarity. Thus, the pressure drop across the column of height h is $\Delta p = [2(\sigma_{SV} - \sigma_{LS} - \sigma\cos\theta_d)/r - \rho g h]$, and, if Poiseuille flow $v = r^2\,\Delta p/8\eta h$ is occurring in the capillary tube, then

$$\frac{dh}{dt} = \frac{r^2}{8\eta h}\left[\frac{2(\sigma_{SV} - \sigma_{LS} - \sigma\cos\theta_d)}{r} - \rho g h\right]. \qquad (22)$$

Wetting fluids may therefore be expected to rise a distance $h_{\max} = 2(\sigma_{SV} - \sigma_{LS} - \sigma \cos \theta_d)/r\rho g$ in capillary tubes, though this will occur with an ever-decreasing velocity as $h_{\max}$ is approached, and to reach this height requires theoretically an infinite amount of time.

1.3.2 Droplet Flow If we define a spreading coefficient $S = \sigma_{SV} - \sigma_{LS} - \sigma$, then Eq. (21) implies that, for $S > 0$, a stable droplet is impossible, as Young's equation cannot be obeyed for any static wetting angle, and the droplet will flow over the solid surface, forming a thin film of microscopic dimensions. The dynamics of this wetting process are rather complex, for even the question of the existence of a dynamic wetting angle θ_d has to be treated with some care. A cursory definition of a macroscopic dynamic contact angle $\theta_{d,\text{macro}}$ as the angle between the nominal contact line and the substrate surface would suggest that this contact angle has a singularity as the spreading velocity U tends to zero, for the static contact angles defined in the limits of receding ($U \to 0^-$) and advancing ($U \to 0^+$) contact lines are often observed to be different. A microscopic examination of the region near the spreading droplet in the vicinity of the apparent contact point at which the liquid, vapor, and substrate phases meet, however, would reveal that the nominal macroscopic contact line is actually preceded by a mesoscopically thin precursor film [see Fig. 6(b)] due to a disjoining pressure Π that reflects the existence of attractive long-range van der Waals forces between the liquid and substrate phases (Joanny and de Gennes, 1984; de Gennes, 1985). Thus, the apparent singularity is a macroscopic artifact, and a separate microscopic dynamic contact angle $\theta_{d,\text{micro}}$ can be defined by the angle between the spreading surface and the tangent between the macroscopic droplet and the tip of this precursor film. The thickness W of this precursor film, as opposed to its spreading length l, is controlled by a balancing of the van der Waals forces, which will tend to make this film microscopically thin, indeed of atomic dimensions, with the short-range capillary forces that will tend to discourage the existence of such a film at all.

Consider a droplet spreading on a solid surface whose normal lies in the z direction, and suppose the droplet is spreading in the x direction. The dynamic droplet shape $h(x, t)$ during spreading can be related to the spreading velocity $U(t)$ [see, for example, Joanny (1990)]. If the velocity field in the droplet has a parabolic profile in the z direction as in Poiseuille flow, then the average velocity $U(x, t)$ is given by

$$U(x, t) = \left(\frac{h^2}{3\eta}\right)\frac{dp}{dx}, \tag{23}$$

where the pressure $p = \sigma d^2h/dx^2 + \rho gh + \Pi(h)$. Solving the equations of motion for sessile drops, that is, small drops where the hydrostatic pressure can be neglected, yields

$$h(x, t) = [9\eta U(t)/\sigma]^{1/3}\, x \log^{1/3}(kx), \tag{24}$$

where k is a constant of integration. Thus, apart from logarithmic correction, the drop profile near the interfacial contact point is wedgelike, with a dynamic contact angle

$$\theta(t)_{d,\text{macro}} = [9\eta U(t)/\sigma]^{1/3}. \tag{25}$$

Equation (25) is Tanner's law, which states that a universal relation exists between the contact-line velocity and the dynamic contact angle.

Equation (25) can also be used to find the scaling behavior of droplet radius with time, $R(t)$. For a nonvolatile liquid, the total volume V is conserved during spreading, but for a wedge-shaped drop, $V \sim R^3(t)\theta_{d,\text{macro}}(t)$, and, as $U(t) \sim dR/dt$, the droplet radius increases as

$$R(t) \sim V^{3/10}[\sigma t/\eta]^{1/10}. \tag{26}$$

Similar calculations can carried out for pendant-shaped drops dominated by gravity, yielding $R(t) \sim V^{3/8}[\rho g t/\eta]^{1/8}$.

It should be emphasized that the results above imply a very idealized form of wetting dynamics. First, to study mathematically the dynamics of wetting, it is not enough to assume Poiseuille flow for the droplet. Clearly, stick boundary conditions on the substrate surface are incompatible with a moving contact line, but what are the correct boundary conditions? Possibilities would include a frictional force proportional to the surface velocity v introduced at the substrate surface, resulting in a stress tensor $\sigma_{x,z} = \gamma v$; another possibility would be that the main body of fluid in the droplet before the nominal contact line does indeed obey stick boundary conditions, but slip boundary conditions are applied to the precursor film, whose length diverges as $l = vt$, and, consequently, the molecules in the precursor film are free to roll at the substrate surface.

Second, volatility and contamination have been neglected; these can lead to temperature and surface-tension gradients, resulting in additional forces (the Marangoni effect). In fact, the alteration in dynamics can be dramatic: For example, if surfactant solutions are spread on moist surfaces, large surface-tension gradients will appear, leading to a fingering of the contact line (Marmur and Lelah, 1981); this spreading is an order of magnitude faster than surface-tension- or gravity-dominated wetting. Also, if the substrate surface is not smooth, then the wetting dynamics may resemble more closely flow in porous media than the creeping flow described here. Finally, liquids have been assumed to be Newtonian; for non-Newtonian rheologies, the shape of a spreading droplet may vary considerably from the caplike droplets seen in simple wetting. For example, shear-thinning liquids typically develop a protrusion that extends beyond the cap, forming a macroscopic foot (Brochard and de Gennes, 1984).

So far, the wetting of two-dimensional surfaces has been examined. But, technologically, the wetting of fibers is just as important. One of the most interesting instabilities concerns the wetting–nonwetting transition on fibers. Will a liquid prefer to form an annulus of thickness l enveloping the fiber, which can be taken to have the form of a cylinder of radius r and length L, or to form a series of droplets instead, as is often seen on spider's webs after rain? The total capillary energy per unit length of fiber E/L above that for the dry fiber may be written

$$E/L = 2\pi r(\sigma_{LS} - \sigma_{SV}) + 2\pi(r + l)\sigma, \tag{27}$$

and, therefore, when $E/L < 0$, or the spreading coefficient $S > \sigma l/r$, the fiber will be wetted. Thus, it is more difficult to wet a fiber than a surface made from the same material.

1.4 Surface Evolution and Mixing

Surfaces also evolve in a complex manner when mixtures are stirred. The process occurs in two stages. First, deterministic stirring can lead to a stretching and folding of material surfaces reminiscent of the manner in which a baker creates millefeuille pastry. Such stretching and folding lie at the heart of chaotic mixing. Then, when the folded layers of the liquids give rise to large enough gradients in their respective concentrations, diffusion takes over from deterministic mixing as the most powerful mechanism equilibrating the separate liquid phases, resulting ultimately in a homogeneous mixture.

1.4.1 Deterministic Mixing Consider the evolution of immiscible interfaces by external deterministic mixing. It would certainly not surprise anyone that with random stirring, efficient mixing of liquids can be made to occur, but what may surprise one is that periodic stirring can be very efficient in mixing. Under periodic stirring, a stretching and folding of the material surfaces can occur in which the interfacial region grows with time, forming a highly interwoven convoluted structure having fractal properties, though often, ordered flow regions also coexist with the chaotic mixing regions and appear to be separated from them by impenetrable barriers. Clearly, from a technological viewpoint, such barriers are undesirable if enhanced mixing is required.

Such mixing behavior has not only been observed in the laboratory but also on a large scale in atmospheric and oceanographic flows. For example, vortices and jets have been observed to act as trapping barriers to mixing and may be the cause of the persistence of the Great Red Spot on Jupiter. In addition, vortexlike flows in the Atlantic have been observed to trap for years water and animal life from the Mediterranean where these eddies originated.

Typically [see Ottino (1990) for a comprehensive review], during a mixing process, the material can be considered to have no back reaction on the flow field, so that the material behaves as a passive scalar, and the resulting concentration of one component $c(\mathbf{r}, t)$ is given by

$$\frac{\partial c}{\partial t} + \mathbf{v} \cdot \nabla c = D\nabla^2 c, \tag{28}$$

where the velocity $\mathbf{v}(\mathbf{r}, t)$ is a known deterministic function of space and time. In addition, the flow may be taken to be incompressible, and, therefore, $\nabla \cdot \mathbf{v} = 0$. The term $D\nabla^2 c$ represents diffusion due to thermal fluctuations, which are the dominant contribution to mixing at small scales (see Sec. 1.4.2).

How can this completely deterministic stirring flow field $\mathbf{v}(\mathbf{r}, t)$ chaotically advect the material surfaces and induce the observed complexity and inhomogeneities that characterize such stirred fluids (see Fig. 7)? The answer appears if, for the moment, we neglect diffusion and consider the motion of any marked particle in this flow field. It will obey the deterministic dynamical system

$$\frac{d\mathbf{x}}{dt} = \mathbf{v}(\mathbf{x}, t), \tag{29}$$

which, as $\nabla \cdot \mathbf{v} = 0$, is the equation of motion for a conserved flow and therefore reminiscent of N-dimensional nonintegrable Hamiltonian systems obeying

$$\frac{\partial p_\alpha}{\partial t} = -\frac{\partial H}{\partial q_\alpha}, \quad \frac{\partial q_\alpha}{\partial t} = \frac{\partial H}{\partial p_\alpha}, \tag{30}$$

where p_α, q_α $(1 \le \alpha \le N)$ are the canonical momentum and position variables describing the dynamics. Such mechanical systems also conserve volume in phase space by Liouville's theorem. Conserved flows obeying Eq. (29) can have chaotic trajectories for the autonomous case $\mathbf{v}(\mathbf{x}, t) = \mathbf{v}(\mathbf{x})$ in three or more dimensions, and, even in two dimensions, chaotic trajectories are possible for time-periodic flows $\mathbf{v}(x, y, t + \tau) = \mathbf{v}(x, y, t)$. Indeed, two-dimensional periodic flows can be made isomorphic to periodically forced one-dimensional Hamiltonian systems with the real space of the fluid motion being replaced by the phase space $(x \rightarrow p,\ y \rightarrow q)$. Thus, by studying the dynamics of Hamiltonian systems, we can understand aspects of deterministic mixing.

Flows in integrable N-dimensional Hamiltonian systems have N constants of motion and are therefore confined to N-dimensional submanifolds of the complete $2N$-dimensional phase space with the topology of a torus. Thus, using the analogy introduced above, for integrable stirring, one would expect material trajectories to be confined to submanifolds of the stirred domain. This is not what is observed experimentally. Real stirring is equivalent to a nonintegrable Hamiltonian system, and the question arises of what dynamical systems theory can tell us about this case. Perhaps the most important property of

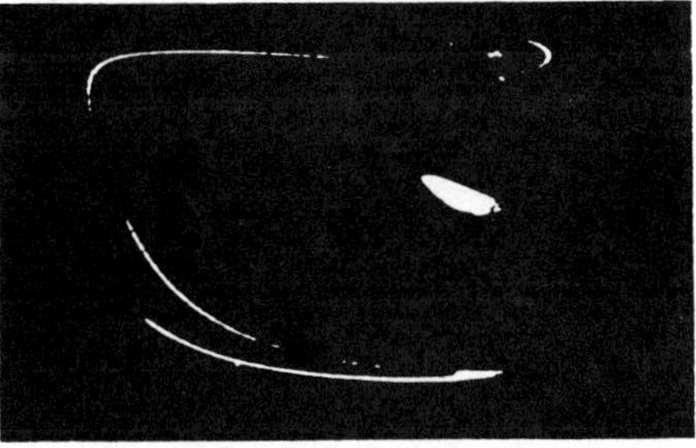

FIG. 7. Stretching and folding of two passive blobs under periodic mixing; note the appearance of regular regions within the chaotic flow; redrawn from Ottino (1990). Reproduced with permission from the *Annual Review of Fluid Mechanics*, Volume 22, ©1990, by Annual Reviews Inc.

nonintegrable Hamiltonian flows is given by the KAM (Kolmogorov–Arnol'd–Moser) theorem, which describes what happens to integrable quasiperiodic flows on tori as nonintegrable terms are added to the Hamiltonian. The motion on an N-torus consists of N in general incommensurate periodic motions and is therefore quasiperiodic. Nonintegrable terms destroy all tori eventually, and commensurate-frequency tori immediately (though they are of measure zero compared with the incommensurate case); but what is important is that most tori with incommensurate frequencies are at first only distorted, the most stable tori, which last longest under nonintegrable perturbations, being those with the most incommensurate frequencies. Once a torus is destroyed, a trajectory is free to wander in an apparently random manner, resulting in great sensitivity of the trajectories to their initial conditions. Thus, typical regions of a deterministic chaotic flow are separated by surviving tori through which no trajectory can penetrate, accounting for the observed spatial structure in chaotic advection.

As smaller and smaller length-scale structure is created during this stretching and folding process, however, there must come a point where this analogy with Hamiltonian flows breaks down, for diffusion will destroy all tori at small enough scales (the dissipative term dominates the inertial transport term, $D\nabla^2 c \gg \mathbf{v}\cdot\nabla c$, at small enough scales), and the final stage of any mixing process will be dominated by diffusion.

1.4.2 Mixing by Diffusion Diffusion is the transport mechanism by which thermal fluctuations equilibrate density inhomogeneities in a multiphase medium. Consider a drop of ink placed in a quiescent glass of water; with time, the fluids will mix totally, and, on a macroscopic level, the densities of both water and ink will appear constant. To find the equations of motion for such diffusion, we need to consider the equations of continuity for multicomponent miscible fluids.

Miscible fluids can be described by their local mass densities $\rho_i(\mathbf{r}, t)$ (n components in the general case). The current of species i is given by the sum of convective and diffusive contributions:

$$\mathbf{J}_i = \rho_i\mathbf{v} + \mathbf{J}_{i,\text{diff}}, \tag{31}$$

where the assumption is being made in Eq. (31) that the local velocity of each species is identical, $\mathbf{v}_i = \mathbf{v}$.

The diffusive flux can be related to the thermodynamic force by

$$\mathbf{J}_{i,\text{diff}} = -\lambda\nabla\mu_i = -\lambda\frac{\partial\mu_i}{\partial\rho_i}\nabla\rho_i = -D_i\nabla\rho_i. \tag{32}$$

In Eq. (32), μ_i is the chemical potential of species i, and the mass diffusion constant for species i is therefore given by $D_i = \lambda\partial\mu_i/\partial\rho_i$. With these constitutive relations, the equations of continuity for a multicomponent miscible fluid become

$$\frac{\partial\rho_i}{\partial t} + \nabla\cdot(\rho_i\mathbf{v}) = D_i\nabla^2\rho_i. \tag{33}$$

The total density of the fluid is $\Sigma_i\rho_i = \rho$, and this total fluid density obeys $\partial\rho/\partial t + \nabla\cdot(\rho\mathbf{v}) = 0$, as $\Sigma_i\,\mathbf{J}_{i,\text{diff}} = 0$. This last result is a consequence of the fact that the densities of only $n - 1$ components in an n-component mixture can be independent. It is also possible to define the specific density of each species as $\rho_i^* = \rho_i/\rho$, in which case $\Sigma_i\,\rho_i^* = 1$. The specific density obeys the related equation of continuity

$$\frac{\partial\rho_i^*}{\partial t} + \mathbf{v}\cdot\nabla\rho_i^* = D_i\nabla^2\rho_i^*, \tag{34}$$

accounting for Eq. (28) describing mixing.

To estimate the power of diffusion in mixing surfaces over a distance L, note that the time T_{diff} taken for an inhomogeneity to spread a distance L by diffusion scales as $T_{\text{diff}} \sim L^2/D$. As the deterministic mixing time is $T_{\text{det}} \sim L/U$, where $U \sim \max|v(\mathbf{x}, t)|$, diffusion must win at small scales $L \ll D/U$.

2. MULTIPHASE AND ANISOTROPIC FLOWS

Fluids having Newtonian rheologies (see RHEOLOGY) are isotropic and have a stress proportional to rate of strain, and, indeed, such assumptions describe remarkably accurately the flow properties of simple molecular fluids. Here, we consider the flows of that large class of technologically important nonhomogeneous materials that break one or both of these assumptions.

An important class of flows comprises those in materials that contain dispersions of one pure phase in another, separated by interfaces of microscopic or mesoscopic size. Such multiphase flows can on a macroscopic scale be treated as continuous but usually have non-Newtonian rheologies. This would include foams, which consist of a gaseous phase separated by liquid interfaces; emulsions, which consist of droplets of two immiscible liquid phases; slurries, which are high concentrations of solid particles dispersed in a liquid phase; and granular flows, which on a mesoscopic scale are solid particles separated by a gas phase. Coarse-graining of the mesoscopic dynamics results in a new effective rheology whose properties are often far from Newtonian, showing strong shear-thinning or -thickening behavior; or, if Newtonian, then having transport coefficients that are strongly dependent on the structure at this mesoscopic scale.

Macroscopic anisotropic flows can also occur in molecular fluids and mixtures. Such fluids are orientationally ordered on a macroscopic scale and include liquid crystals and electrorheological fluids.

2.1 Liquid–Solid Phase Flows

Numerous technologically important materials can be thought of as consisting of solid particles immersed in a liquid phase. This applies not only to suspensions and slurries, where the suspended particles are of macroscopic or mesoscopic size, but also to polymer solutions (see POLYMER DYNAMICS), which have a somewhat analogous structure in the semidilute regime, with their radius of gyration replacing the hard-sphere radius.

2.1.1 Slurry Flows A concentrated mixture of solid particles immersed in a liquid forming a suspension can be argued to lead to even more complex flow fields than are to be expected for flows in porous media (see Sec. 3). The reason for this is that, while porous media present a quenched random topography (that is, a fixed-in-time though random environment) in which the flow must occur, the suspended-particle geometry in which liquid flow occurs in a slurry is controlled by a complex feedback process between flow and geometry. In principle, for a fixed multiparticle geometry, the complete Stokes flow must be solved; then this flow in turn will result in hydrodynamic forces on the suspended particles, leading to a complex suspension dynamics [see, for example, Batchelor (1967)].

The resulting macroscopic rheology of concentrated suspensions can be studied as a function of the volume fraction ϕ of solid particles suspended in the liquid phase, provided one is interested in shear flows that are varying over much larger length scales than the interparticle distance. In the semidilute suspension limit, the effective viscosity can be expanded in a power series in this volume fraction:

$$\eta = \eta_0[1 + 5\phi/2 + O(\phi^2)]. \tag{35}$$

The first term in this expansion, derived originally by Einstein (1906), can be found by calculating the effect on the flow field of a single spherical particle and then summing these contributions from all particles, as though independent, into a new effective stress tensor. To first order the effect of the suspension on rheology is to increase the fluid viscosity in the universal manner given by Eq. (35); that it should increase is clear, for the effect of the suspended particles can only increase dissipation during flow. The term quadratic in the volume fraction is the first contribution that requires for its calculation that the suspended particles are no longer assumed to be independent; and such calculations lead to additional direct interaction and Brownian forces on the particles, as well as a changed hydrodynamic stress.

The coefficients of the higher-order terms in the volume fraction are not universal constants but depend on the Peclet number $Pe = \gamma a^2/D$, where γ is the shear rate, a is the particle radius, and D is the particle diffusion constant. The Peclet number here measures the ratio of two time scales—a diffusive time scale $t_D \sim a^2/D$ to a shear time scale $t_{\text{shear}} \sim 1/\gamma$—for particle motion. Because of this dependence of the effective viscosity on Peclet number, the flow rheology cannot be Newtonian and, in general, can be expected to show shear-thinning or shear-thickening behavior. Indeed, measurements of η/η_0 vs the Peclet number typically show a dramatic shear-thinning behavior for suspensions at Pe

$\sim$ 10, with separate plateaus in the effective viscosity for much lower and much higher Peclet numbers (see Fig. 8).

For concentrated solutions, we move into the slurry flow regime [see, for example, de Kruif (1990)]. Such flows are highly shear thinning, but the viscosity begins to diverge,

$$(\eta/\eta_0) \sim [1 - \phi/\phi_m]^{-2}, \tag{36}$$

as the volume fraction ϕ approaches the value $\phi_m \approx 0.63$, which is the limiting value for the volume fraction of random close-packed spheres. Indeed, even before this limit has been reached, phase transitions to ordered solid phases for $\phi > 0.5$ have been observed, and at high shear rates, the suspension can become anisotropic, as light- and neutron-scattering studies confirm.

2.1.2 Flows in Polymer Solutions and Melts Consider a semidilute polymer solution formed from the polymerization of a number concentration c of monomers into polymer molecules consisting of N contiguous monomers. It is known (Flory, 1969) that in a good solvent, such linear polymer chains take up the conformation of a self-avoiding random walk of radius of gyration $R_g \sim N^{3/5}$. This observation allows one to estimate the effect of the polymer on the solvent viscosity. By treating the polymers as spheres of radius $\sim R_g$ and using the results for solid suspensions [Eq. (35)] to calculate their rheology in the semidilute limit (in the concentrated regime, this scaling analogy will break down as individual polymer molecules begin to interpenetrate), one would expect

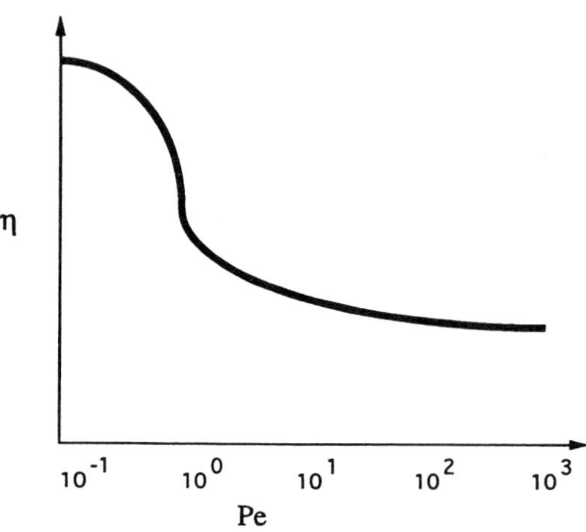

FIG. 8. Schematic diagram showing the dependence of the effective suspension viscosity on Peclet number.

$$\delta\eta/\eta \sim \phi \sim cR_g^3/N \sim cN^{4/5}. \tag{37}$$

Such scaling arguments, which were first introduced by Flory (1971) and de Gennes (1984) into polymer physics, are very powerful and have been verified for numerous static and dynamic properties of polymers.

As a further example of dynamical scaling behavior, consider the self-diffusion D. For a solid particle of radius R, Stokes calculated that $D = kT/6\pi\eta R$. Therefore, for a polymer molecule, one may estimate $D \sim kT/6\pi\eta R_g$ (de Gennes, 1977), implying a scaling of the diffusion coefficient with polymer size of $D \sim N^{-3/5}$, which again has been confirmed by light-scattering experiments (de Gennes, 1984). Using this diffusion coefficient, one may introduce a time scale $\tau_{\text{Zimm}} = R_g^2/D \sim 6\pi\eta R_g^3/kT$ for polymer molecule self-diffusion across a distance of the polymer size. Thus, only shear flows greater than the inverse time scale $\gamma \gg 1/\tau_{\text{Zimm}}$ can be expected to influence polymer dynamics significantly.

Semidilute polymer solutions are also shear thinning and are often represented by a viscosity which has a constitutive form of the Bird–Carreau type,

$$\frac{\eta}{\eta_0} \approx \left[1 + \frac{dv/dz}{\gamma_c}\right]^n, \tag{38}$$

with $n < 0$ for shear-thinning behavior. At high shear rates ($\gamma \gg 1/\tau_{\text{Zimm}} \sim kT/6\pi\eta R_g^3$) and in turbulent flows, however, a strongly shear-thickening behavior may be expected, as dynamic coil–stretch transitions of the polymer configuration can occur, resulting in an enhanced viscosity.

In contrast with dilute or semidilute polymer solutions is the flow behavior that occurs in polymer melts. Here the individual polymer chains interpenetrate much like a bowl of spaghetti. From the standpoint of rheology, a strong viscoelastic behavior is observed—polymer melts behave like highly viscous fluids when perturbed slowly, but any high-frequency perturbation will result in rubberlike behavior. The crossover time scale $\tau \sim \tau_0 N^a$ for such viscoelastic behavior can be of the order of seconds, and the exponent is approximately $a \approx 3$ or slightly larger. A simple argument leading to a value of $a = 3$ is a consequence of de Gennes's (1984) reptation model for polymer dynamics in melts. The idea is that each polymer molecule moves by Brownian motion in a tube created by the local configuration of the other polymers. The viscoelastic time scale can be estimated as the time for a typical molecule to travel its own length in such a tube,

$$\tau \sim L^2/D_{\text{tube}} \sim N^2/D_{\text{tube}} \sim N^3. \tag{39}$$

In deriving Eq. (39), the scaling behavior of $D_{\text{tube}} \sim N^{-1}$ was used, and this scaling behavior can be estimated from the frictional force f_{friction} that the tube exerts on the polymer. By the fluctuation-dissipation theorem, $D_{\text{tube}} \sim kT/f_{\text{friction}}$. As the frictional force can be expected to be proportional to the length of the polymer chain, we indeed find that $D_{\text{tube}} \sim N^{-1}$, and, therefore, $\tau \sim N^3$, as advertised.

2.2 Solid–Gas Phase Flows

2.2.1 Flows in Granular Material Granular materials consist of a high volume fraction of solid particles separated by a gaseous phase. In contrast to slurry flow, nearly all of the momentum is carried by these solid particles. The viscoelastic properties of granular materials can be expected to depend strongly on the local applied stresses. As stress is increased, local bonds between grains are broken, and this failure leads to quasistatic flow of moving particle clusters. At very high stresses, a "melting" transition occurs, leading to a rapid flow regime. Indeed, in many respects, granular flows have analogies to molecular dynamics. Apart from the "phase transitions" mentioned above, in the free-flow regime, the velocity of individual grains can be considered as a superposition of a local mean velocity and a random "thermal" motion

$$\mathbf{v}_i = \langle \mathbf{v} \rangle(\mathbf{r}) + \delta \mathbf{v}_i, \tag{40}$$

where a "granular temperature" can be defined by $kT = \rho a^3 \langle (\delta \mathbf{v}_i)^2 \rangle$, in analogy with normal thermal fluctuations (Ogawa, 1978). Here, ρ is the grain material density, and a is the grain radius. A fundamental difference from molecular systems, however, is that all granular collisions are inelastic, and, therefore, energy is dissipated during intergrain collisions and transformed into heat. Thus, the granular temperature $T(\mathbf{r})$ is, in contrast to simple fluids, a local property of the flow considered.

The experimentally observed behavior of the stress tensor for granular flows and slurries is

$$\sigma_{ij} = \rho a^2 f_{ij}(\phi) \gamma^2, \tag{41}$$

where γ is the shear rate (Campbell, 1990).

The dependence on shear of both the stress tensor and the granular temperature can be derived from scaling arguments based on the observation that, whereas for molecular gases, $|\langle \mathbf{v} \rangle| \ll |\delta \mathbf{v}_i|$, in granular flows, $|\langle \mathbf{v} \rangle| \sim |\delta \mathbf{v}_i|$. This observation allows us to estimate the stress tensor as $\sigma \sim \rho \langle \mathbf{v} \rangle^2 \sim \rho a^2 \gamma^2$, in agreement with Eq. (41), and the dependence on shear of the granular temperature as

$$kT = \rho a^3 \langle (\delta \mathbf{v}_i)^2 \rangle \sim \rho a^3 \langle \mathbf{v} \rangle^2 \sim \rho a^5 \gamma^2.$$

This last scaling relation can be used to estimate the effective granular viscosity for simple shear flows as

$$\eta \sim \rho a^2 \gamma \sim [\rho kT/a]^{1/2}, \tag{42}$$

and, thus, granular flows are shear thickening, with a viscosity proportional to shear or to the square root of the granular temperature. Similar arguments for the diffusion of "heat" in the presence of temperature gradients in granular flows yield, for the thermal conductivity, $\kappa \sim T^{1/2}$. These estimates are similar to those derived in the kinetic theory of gases for the transport coefficients, with the substitution of a granular temperature for the real temperature.

The dependence of the stress tensor and, consequently, the effective viscosity, on the solid volume fraction $f_{ij}(\phi)$ is also complex. As most of the momentum is carried by the solid, one might estimate $f_{ij}(\phi) \sim \phi$, and, indeed, there is a range of ϕ around $\phi \approx 0.5$ where this estimate appears to hold (see Fig. 9). But, for both limits, $\phi \to 0$ and $\phi \to 1$, $f_{ij}(\phi)$ diverges. As for the case of slurries [see Eq. (36)], at high volume fraction, no shear flow is possible, and this accounts for the divergence at the high end. At low volume frac-

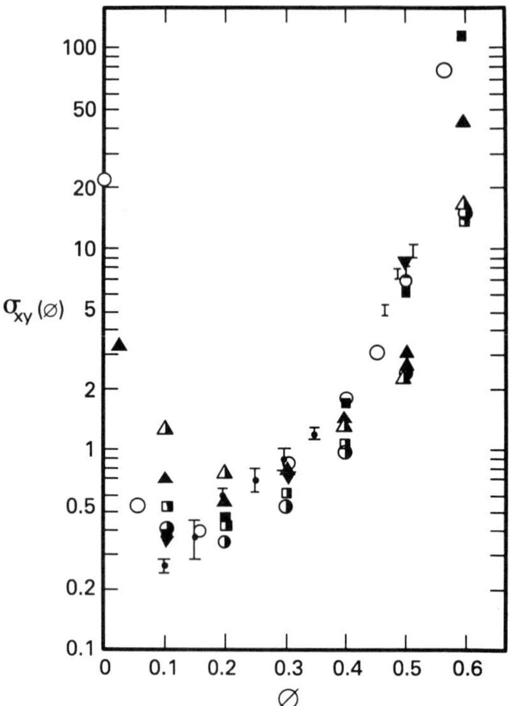

FIG. 9. Dependence of the stress tensor on the solid fraction ϕ in granular flows from various experimental studies and computer simulations; note the divergence for $\phi \to 0$ and $\phi \to 1$; redrawn from Campbell (1990). Reproduced with permission from the *Annual Review of Fluid Mechanics*, Volume 22, ©1990, by Annual Reviews Inc.

tion, the effective granular temperature diverges, as dissipation tends to zero at a much faster rate $\sim\phi^2$ (because of the lack of inelastic collisions) than energy input through gravity and the granular boundaries $\sim\phi$, resulting in an effective shear-thickening flow.

When granular materials are shaken, a fluidization transition occurs, followed by convective motions by the granular material reminiscent of the Rayleigh–Bénard convection. Consider shaking a container containing a granular material. Typically, a type of viscoelastic behavior is observed. At low frequencies, the granular material acts as a solid, at high frequencies, a fluidization transition occurs, and, at higher frequencies still, convective cells appear. If the motions of individual grains are followed in computer simulations, then at low frequencies, smooth trajectories are seen, while at higher frequencies, the granular temperature increases, and the individual grains perform Brownian motion.

An important property of polydisperse granular flows is segregation. In granular flows consisting of varying sized particles, the small grains will tend to migrate to the bottom of a container when shaken, while the large grains will be found at the top. This phenomenon, which is the opposite of diffusive mixing, can be understood in terms of the effect of gravity in the presence of hard-core repulsive interactions. The only way a large grain could possibly sink in the container would be for the random shaking to create a vacancy large enough to accommodate the grain volume. Clearly, the larger the volume, the more improbable that this event occurs—as small grains will immediately tend to fill any small vacancy due to gravity—resulting in the observed segregation.

2.2.2 Granular Topography and Self-Organized Criticality So far, we have been considering the rheology of granular flows. Granular systems and powders do not, however, only support fluidized motion; they also can behave like solids and support finite shear stresses without flow. Specifically, they can support surface structures with hill-like topographies without volume flow. If the angles of these hills are denoted by θ, then it is possible to define two angles: θ_r, the angle of repose, and θ_s, the maximum angle of stability, so that for $\theta < \theta_r$, any hill is stationary, and for $\theta > \theta_s$, the angle is unstable, leading to avalanches resulting in a new angle of tilt $\theta < \theta_s$; for $\theta_r < \theta < \theta_s$, the hills are marginally stable. The dynamics of avalanche-like behavior in this marginally stable region have recently become an active area of research with the introduction of the concept of self-organized criticality.

Self-organized criticality is the process whereby a nonequilibrium system self-organizes itself into a marginally stable macroscopic state having no intrinsic length scale and showing self-similar critical fluctuations on all length scales. Such self-organization is argued to be robust, similar to the attractor of a dynamical system with a large basin of attraction, so that for a large set of initial conditions, the system will always end up in the self-organized critical state. Typical examples of phenomena said to exhibit such self-organized criticality include earthquakes, the structure of macroscopic terrains such as mountains and sand piles, and diffusion-limited aggregation.

In the case of sand piles, the idea is that on continually adding grains to the sand surface, columns of grains will form until a gravity-induced instability results in their collapse. The resulting occurrence of an avalanche will self-organize the surface into a new marginally stable form with a self-similar distribution of hill sizes. A variety of models for avalanche flow have been simulated. The original discrete sandpile model of Bak *et al.* (1987) for self-organized criticality consisted of adding grains randomly to a lattice to form columns of height h_i at each site i. To simulate the instability due to gravity of tall columns on a two-dimensional lattice, the rule was introduced that if $h_i > h_{\text{crit}}$, then the column would lose four grains, $h_i \rightarrow h_i - 4$, while, in parallel, the nearest-neighbor columns would each gain one grain, $h_i \rightarrow h_i + 1$, conserving the granular number in the absence of external addition. This rule does indeed lead to avalanche-like behavior when a series of neighboring columns all reach marginal stability together. In addition, the fluctuations of loading times, that is, the times between avalanches, do occur with a $1/f$-like power spectrum associated with processes involving no intrinsic time scale, while the resulting topography, as well as the size distribution of avalanches, has no intrinsic length scale associated with it. Other models followed, such as replacing a critical column height h_{crit} by a critical slope $|h_i - h_j| = S_{\text{crit}}$ for avalanches to occur (Kadanoff *et al.*, 1989); they all exhibited similar types of scaling behavior, showing that self-organized criticality is a robust and ubiquitous phenomenon.

In concert with simulations of models, theoretical analyses of the current fluctuations and avalanches in a flowing sandpile can be made using driven Langevin equations incorporating the symmetries and conservation laws of the discrete models (see Sec. 3.4.3). For example, the sand forming a hill has a macroscopic flow direction, and we consider small undulations $h(\mathbf{r}, t)$ transverse to the macroscopic slope. Hwa and Kardar (1989) showed that the resulting equation for these spatiotemporal fluctuations is

$$\frac{\partial h}{\partial t} = \nu_{\parallel}\partial_{\parallel}^2 h + \nu_{\perp}\nabla_{\perp}^2 h - \frac{\lambda}{2}\,\partial_{\parallel}h^2 + \eta(\mathbf{r}, t), \tag{43}$$

which is anisotropic because of the existence of a macroscopic flow direction; and the noise due to the added sand grains is taken to be white, $\langle \eta(\mathbf{r}, t)\eta(\mathbf{r}', t')\rangle = 2\,D\delta(\mathbf{r} - \mathbf{r}')\delta(t - t')$.

Experiments on natural sandpiles present important differences from the predictions of self-organized criticality because dissipation effects are not correctly incorporated in the simple sandpile models described above, but qualitative similarities remain, including the fact that characteristic quantities such as angles of tilt and the loading times have no characteristic angle or time scales, but rather are given by distributions.

2.3 Liquid–Gas and Liquid–Liquid Phase Flows

2.3.1 Bubble and Emulsion Rheology

Closely related in structure and rheology to suspensions of solid particles in liquids are emulsions consisting of drops of one immiscible liquid in another, and suspensions of gas bubbles in a liquid medium.

If Stokes's equations are solved for the case of droplets of viscosity η_{drop} dispersed in a continuous phase of viscosity η_0, and the droplets are assumed to be perfectly spherical with no deformation, then Taylor (1932) first showed that, to first order in the volume fraction, Einstein's expression for the effective viscosity given by Eq. (35) is generalized to

$$\eta = \eta_0\left[1 + \frac{5\lambda+2}{2(\lambda+1)}\,\phi + O(\phi^2)\right], \tag{44}$$

where $\lambda = \eta_{\text{drop}}/\eta_0$. An examination of Eq. (44) shows that, as $\lambda \rightarrow \infty$, Einstein's result for solid particles is recovered, while for bubbles, the limit $\lambda \rightarrow 0$ needs to be taken with the result

$$\eta_{\text{bubbles}} = \eta_0[1 + \phi]. \tag{45}$$

Equations (44) and (45) are only valid to lowest order in the volume fraction, and for spherical droplets. For small departures from sphericity, the effective viscosity can also be expected to depend on the capillary number $Ca = \eta a\lambda/\sigma$; and, as for solid suspensions, the higher-order terms will depend on the Peclet number $Pe = \lambda a^2/D$ in the case of simple shear flows. For more complex flow patterns, droplet interaction with and deformation in the flow can be expected to complicate the rheology.

2.3.2 Cavitation Cavitation is the tendency of liquids to form bubbles spontaneously when subjected locally to negative pressures, as they cannot withstand the tension involved. Indeed, a liquid, when subject to pressures $p(\mathbf{r})$ below the vapor pressure p_v of the gaseous phase, will respond by creating bubbles to relieve this tension. Typically, such bubbles will nucleate around microbubbles of the vapor phase already dissolved in the liquid, or other impurities, and the conditions for cavitation to occur are typically created by the motion of solid bodies through the liquid phase.

To see that this is the case, consider a constant flow field with velocity $\mathbf{v}$ in a gravitational field $\mathbf{g}$. The energy density e at any point $\mathbf{r}$ in the flow is related to the pressure field $p_0(\mathbf{r})$ by Bernoulli's equation $e = \rho v^2/2 + p_0(\mathbf{r}) + \rho\mathbf{g}\cdot\mathbf{r}$. Now place a streamlined obstacle in the flow field. In the presence of this obstacle, the velocity field changes to $a(\mathbf{r})\mathbf{v}$. Consequently, the energy density of the flow may now be written $e' = \rho\alpha(\mathbf{r})^2v^2/2 + p(\mathbf{r}) + \rho\mathbf{g}\cdot\mathbf{r}$, where $p(\mathbf{r})$ is the pressure field in the liquid in the presence of the obstacle. For steady potential flow, however, the energy density $e' = e$ remains unchanged, and therefore

$$\frac{p(\mathbf{r}) - p_v}{\rho v^2/2} = \frac{p_0(\mathbf{r}) - p_v}{\rho v^2/2} - \frac{\alpha(\mathbf{r})^2 - 1}{2}. \tag{46}$$

As cavitation is likely to occur whenever the local pressure is below the vapor pressure in the gas phase, a local dimensionless cavitation number $K(\mathbf{r}) = [p_0(\mathbf{r}) - p_v]/(\rho v^2/2)$ can be defined, which, whenever it falls below some critical value K_c that depends on the shape and size of the obstacle involved, will make it likely that cavitation occurs at that point.

Typical conditions under which cavitation would be observed would be near the blades of a rotating propeller. As the rotation speed of the propeller increases, the cavitation number decreases, and greater regions of the flow lie below K_c.

2.3.3 Foam Structure and Stability As the volume fraction of gas droplets in a liquid phase, or *foam quality*, tends to one, a transition in structure takes place from a dispersed phase of bubbles into a cell-like structure consisting of regions of the gaseous phase separated by domain walls of the liquid phase. Usually, the liquid phase will contain surfactants, that is, surface-active agents such as soap, consisting of molecules that tend to gather at the liquid–gas interface and reduce surface tension. Such surface-active additives are usually necessary in order to create the large interfacial areas that are characteristic of foams.

Two-dimensional foam patterns are best described as a distribution of cell-like domains, most of which are pentagons or hexagons, though small numbers of other polygons are also seen (see Fig. 10). The number of foam cells N_{cell}, number of edges N_{edge}, and number of vertices N_{vert} are not independent but related topologically by Euler's formula

$$N_{\text{vert}} + N_{\text{cell}} = N_{\text{edge}} + 1. \tag{47}$$

For three-dimensional foams, the gas bubbles consist of polyhedra with cell-like faces, and, again, a topological relationship can be derived, relating the number of polyhedra N_{poly} to the number of vertices, edges, and cells:

$$N_{\text{vert}} + N_{\text{cell}} = N_{\text{edge}} + N_{\text{poly}} + 1. \tag{48}$$

For two-dimensional foams, three cells meet at each vertex, forming angles of 120° (this structure can be understood on the basis of surface tension, which will tend to reduce surface area—thus, if four cells meet at a vertex, then surface energy will be reduced if this configuration is replaced by two vertices, each connected to three neighboring cells). This topology can be used to calculate the average number of faces n in a typical cell. For a large foam pattern where boundary effects can be neglected, $N_{\text{vert}} \approx nN_{\text{cell}}/3$, while $N_{\text{edge}} \approx nN_{\text{cell}}/2$. Substituting these estimates into Eq. (47) yields $n = 6$, and, therefore, the typical cells in foam are hexagons.

This extraction of structure from topology can be extended to the stability of individual cells. As was first pointed out by von Neumann (1952), to lowest-order approximation, whether an individual cell grows or decays depends only on the number of edges n in that cell. In fact, the cell area a_n of an n-sided cell grows or shrinks as a result of gaseous dif-

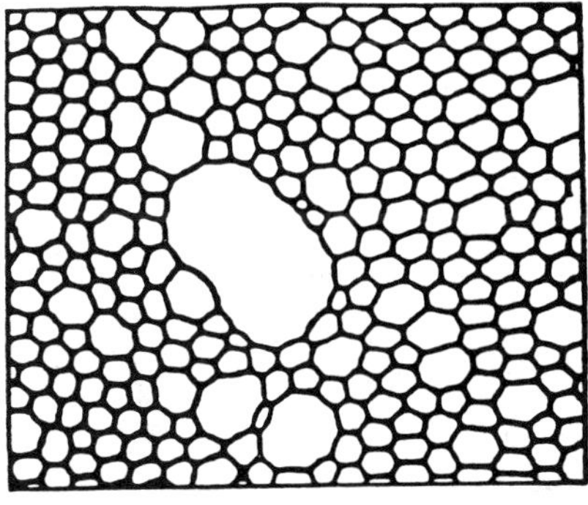

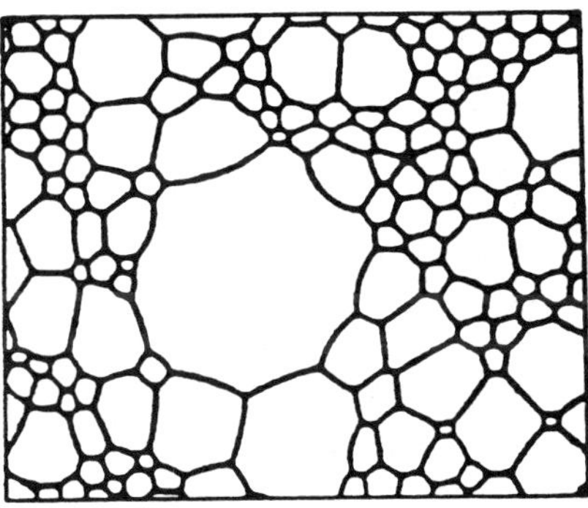

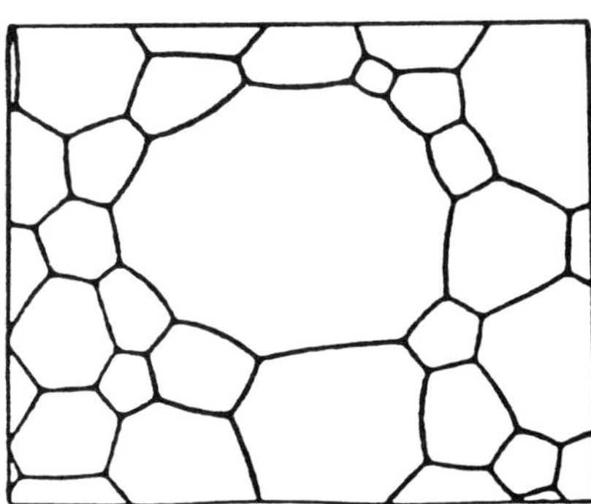

FIG. 10. Coarsening of a two-dimensional foam pattern; note also the topology of the foam cells, edges, and vertices; redrawn from Glazier (1989).

fusion through the cell membrane as

$$\frac{da_n}{dt} = \kappa(n - 6). \tag{49}$$

The derivation of Eq. (49) can be understood by treating the cell as a regular polygon consisting of n edges, each of length l and radius of curvature R. Then one would expect the rate of change of area due to gaseous diffusion to be proportional to both the total surface length of the polygon nl and the pressure drop $\Delta p \sim \sigma/R$. Thus, $da_n/dt \sim nl\Delta p \sim nl\sigma/R$, but for an n-sided polygon, from geometric arguments, $R = 3nl/(6 - n)$, yielding Eq. (49).

2.3.4 Foam Rheology and Texture Coarsening Because foam is not in a state of equilibrium, foam rheology is time and history dependent. Nevertheless, certain general properties about foam rheology can be made [see, for example, Kraynik (1988)].

Foams have strong viscoelastic properties and can behave either as elastic media or as liquids, depending on the stress applied. For small applied stresses below a material-dependent yield stress, $\sigma_{\alpha\beta} < \sigma_{\text{yield}}$, foam will deform elastically and therefore have an infinite viscosity. Above this yield stress, $\sigma_{\alpha\beta} > \sigma_{\text{yield}}$, however, foam behaves as a shear-thinning liquid with an effective viscosity that can approximately be parametrized as

$$\eta = \sigma_{\text{yield}}/\gamma + \eta_\infty. \tag{50}$$

The transport parameters in Eq. (50) are strongly dependent on foam quality and, therefore, time dependent as coarsening takes place in the foam in question.

The boundary conditions for foam flows also tend to differ from those applied to typical Newtonian fluids. Instead of stick boundary conditions as in Poiseuille flow, foams tend to slip at walls with a slip velocity $v_{\text{slip}} = \Psi_w\sigma_w$, where σ_w is the wall shear stress and Ψ_w is a slip coefficient. In addition, for small applied stresses below the yield stress, $\sigma_{\text{slipyield}} < \sigma < \sigma_{\text{yield}}$, though no relative motion of the cell or polyhedral domains will occur, plug flow is still possible. For even smaller stresses $\sigma < \sigma_{\text{slipyield}}$, even plug flow is impossible.

Foams cannot be treated as systems with a static structure and a consequent static rheology, as they are constantly coarsening with time. Coarsening is the property of foams whereby large cells tend to grow at the expense of small ones, which can also burst, leading to fewer but larger cells with time, a well-known property of foam, as anyone who has studied the head of a glass of beer knows.

In real foams, this has many causes: Free-energy minimization will tend to reduce surface area, and this will encourage large bubbles to grow at the expense of small ones; many-faceted cells will be favored at the expense of few-faceted ones, as Eq. (49) suggests; liquid films between cells will rupture and bubbles coalesce, again encouraging coarsening. One main cause of such rupturing is the thinning of bubble walls due to drainage in the presence of gravity. Gravity will also tend to cause an overall spatial variation in foam texture and, consequently, transport properties.

von Neumann's equation [Eq. (49)] can be used to estimate this coarsening process in a foam of total area A. Assume that polygons with greater than six sides grow, while three-, four-, and five-sided polygons are continually bursting, so that their numbers are decreasing in time. The manner in which the average number of cells $N(t)$ in this piece of foam diminishes in time as a result of this coarsening process and the concomitant increase in their average polygonal area $a(t)$ can be estimated from scaling arguments. First, conservation of area implies that $N(t)a(t) = A$. Second, assuming that the loss of cells comes solely from the disappearance of three-, four-, and five-sided polygons at a rate set by von Neumann's law, one would expect

$$\frac{dN}{dt} \sim \sum_{n=3,4,5} a_n^{-1}(t)N_n(t)\frac{da_n(t)}{dt} \sim -\kappa \sum_{n=3,4,5} a_n^{-1}(t)N_n(t)(n-6),$$

where $N_n(t)$ is the number of n-sided polygons at time t. To estimate the size of this sum, assume that in the scaling state, $a_n(t) \sim a(t)$ and $N_n(t) \sim N(t)$, in which case $dN/dt \sim -N(t)^2$. Thus, the number of cells diminishes with time as $N(t) \sim t^{-1}$, while their average area grows linearly with time, $a(t) \sim t$.

The effect of surfactant on texture coarsening can also be important. The relationship between the concentration of surfactants c in the bulk of the liquid, the concentration $\Gamma(c)$ at the liquid–gas interface, and the reduction in surface tension is $d\sigma/d(\ln c) = -kT\Gamma(c)$. For small concentrations of surfactant, $\Gamma(c) \sim c$, and, therefore, the reduction in surface tension is linear with volume surfactant concentration; but this relationship must break down, as only a finite concentration of surfactant molecules can occupy the interface, and, therefore, there exists an asymptotic limit to surface-tension reduction. Thus, during drainage of fluid from films in the presence of gravity, surface-tension gradients will appear, which will result in forces (the Marangoni effect) that can change the drainage process significantly.

The same considerations apply to bubble collapse due to gaseous diffusion through the cell wall under the pressure $\Delta p(t) = 2\sigma/R(t)$. Solving for the dependence of radius on time, one finds $R_0^2 - R(t)^2 \sim \sigma t$, where R_0 is the initial bubble radius, and, therefore, the bubble disappears in a finite time. In the presence of surfactant, however, as the bubble decreases in radius and therefore in surface area, the surfactant concentration at the surface increases. Therefore, the surface tension decreases, resulting in a bubble that lasts a much longer time than the simple estimate $\tau \sim R_0/\sigma$ would suggest. A good review of foam structure and dynamics can be found in Glazier (1989).

2.4 Anisotropic Flows

Though clearly all the flows we have considered thus far involve interfaces between phases at either a macroscopic or a mesoscopic scale, in the main they remain macroscopically isotropic. That is, their flow properties do not have any macroscopic directionality with flow. There are classes of liquids, however, that do have a specific macroscopic orientation associated with their structure, for example, nematic liquid crystals (see LIQUID CRYSTALS, STRUCTURE OF) and electrorheological fluids (see ELECTRORHEOLOGY), and, in consequence, their flow properties are profoundly changed from simple Newtonian flow.

2.4.1 Nematic Liquid-Crystal Rheology Liquid crystals are liquids consisting of long rigid organic molecules that, because of short-range repulsion, tend to align, like logs in a river, forming a phase intermediate between a simple liquid and a crystal. If the ordering is in both the position and orientation of the molecules, then the phase is called a smectic, while if the positions of the molecules are random but orientational ordering survives, then a nematic phase exists. Such a phase can be described macroscopically by an order parameter $\mathbf{n}(\mathbf{r})$, called the director, which gives the local average orientation of the molecules at point $\mathbf{r}$. The existence of such a director will affect flow in several ways, as the liquid possesses macroscopic orientational order.

First, the stress tensor can no longer be expected to be symmetric as is the case for simple fluids, because macroscopic torques

$$\Gamma_\alpha(\mathbf{n}, \nabla\mathbf{n}) = \epsilon_{\alpha\beta\chi}\sigma_{\beta\chi} \tag{51}$$

can be applied to to nematic liquid crystals whose magnitude can be calculated from the distortion energy density of the nematics [see, for example, de Gennes (1974)]. These torques will distort the constant orientation of the nematic phase; and if in the undistorted state the director is taken to be a constant vector in the z direction, $\mathbf{n}(\mathbf{r}) = n_z\mathbf{k}$, then the basic distortions in orientation displayed by the nematic phase are splays $\partial n_x/\partial x$ or $\partial n_y/\partial y$, twists $\partial n_y/\partial x$ or $\partial n_x/\partial y$, and bends $\partial n_x/\partial z$ or $\partial n_y/\partial z$ in the director.

Second, the dissipative contributions to the stress tensor $\sigma'_{\alpha\beta}$ in the equation of motion for the momentum density

$$\rho dv_\alpha/dt = -\partial_\alpha p + \partial_\beta\sigma'_{\alpha\beta} \tag{52}$$

no longer depend only on the symmetric rate-of-strain tensor $S_{\alpha\beta} = \partial_\alpha v_\beta + \partial_\beta v_\alpha$ (assuming the nematic to be incompressible), but will also depend on the rate of rotation of the director

$$\mathbf{N} = d\mathbf{n}/dt - \mathbf{\Omega} \times \mathbf{n} \tag{53}$$

additional to that ($\mathbf{\Omega} = \frac{1}{2}\nabla \times \mathbf{v}$) induced by vorticity alone.

To write down the constitutive relation for $\sigma_{\alpha\beta}$, one symmetry can, however, be invoked. Because the flow for nematics will depend on the orientation of the rodlike molecules, but will remain unaffected by a flip through 180° (consider again the analogy to logs in a river; the tops and bottoms of the log are identical in contrast to dipole moments), it follows that the stress tensor must be invariant under the flip $\mathbf{n}(\mathbf{r}, t) \rightarrow -\mathbf{n}(\mathbf{r}, t)$ and the associated flips in the rate of rotation $\mathbf{N}(\mathbf{r}, t) \rightarrow -\mathbf{N}(\mathbf{r}, t)$. Therefore, the Leslie stress tensor for nematics can be written

$$\sigma'_{\alpha\beta} = \alpha_4 S_{\alpha\beta} + \alpha_1 n_\gamma S_{\gamma\delta} n_\delta n_\alpha n_\beta + \alpha_5 n_\alpha n_\gamma S_{\gamma\beta} + \alpha_6 n_\gamma S_{\gamma\alpha} n_\beta + \alpha_2 n_\alpha N_\beta + \alpha_3 N_\alpha n_\beta, \tag{54}$$

where the α's are transport coefficients of which only $\alpha_4 = \eta$, the shear viscosity, survives for a simple fluid. Notice that the Leslie stress tensor obeys the flip symmetry and is linear in the rate of strain $S_{\alpha\beta} = (\partial_\alpha v_\beta + \partial_\beta v_\alpha$. For a nematic, therefore, effective shear viscosities can be defined depending on the direction of the rate of shear relative to the director, and, in consequence, despite the linearity in rate of shear of the stress tensor, nematics can be expected to display a non-Newtonian orientational dependence in their flow characteristics.

2.4.2 Dynamics of Electrorheological Fluids In contrast with nematic liquid crystals, electrorheological fluids consist of polarizable particles dispersed in nonpolar solvents, such as silica gel in oil. They also exhibit orientational characteristics, forming in the presence of strong electric fields columns of particles that aggregate together at long times (see ELECTRORHEOLOGY).

The dynamics of column formation and the observed coarsening in electric fields have been studied extensively recently (Halsey, 1992). The orientational ordering appears to involve two time scales: the time $\tau_1(E)$ after switching on the field for columns of the polarizable particles to form; then a longer time scale $\tau_2(E)$, on which the columns appear to drift together as a result of thermal fluctuations, ending finally in one giant aggregate. Scaling arguments can be made to estimate these time scales. If the polarizable particles are taken to have a radius r_p and polarizability α, then the drift velocity $\mathbf{v}$ of the particles in an electric field can be found by balancing the attractive dipole-dipole force $f \sim d^2/r_p^4$, causing chain formation (the induced dipole moment per particle $\mathbf{d} = \alpha\mathbf{E} \sim \beta r_p^3\mathbf{E}$), to the Stokes friction $f \sim \eta v r_p$ leading to $v \sim d^2/\eta r_p^5 \sim E^2/\eta$. Thus, the time for column formation in a container of length scale L can be expected to scale as $\tau_1(E) \sim L/v \sim L\eta/E^2$. The sharp decrease in column formation time with increasing applied field has been observed experimentally (Ginder and Elie, 1992). From scaling, we may also expect that $\tau_2(E)/\tau_1(E) = f(\lambda)$, where the dimensionless constant $\lambda = (\beta E)^2 r_p^3/kT$ measures the ratio of dipole-dipole interactions d^2/r_p^3 to thermal fluctuations kT.

In consequence of this chain formation, in large electric fields the viscosity has been observed to increase by a factor of $\eta(E \text{ large})/\eta \sim 10^5$. The actual response of electrorheological fluids in applied electric fields E to stress is viscoelastic, with a yield stress that scales as $\sigma_{\text{yield}} \sim CE^2$. Under small stresses $\sigma \ll \sigma_{\text{yield}}$, the fluid behaves as an elastic solid, while if large stresses $\sigma \gg \sigma_{\text{yield}}$ are applied, the electrorheological fluid will flow, though with a markedly increased effective viscosity $\eta(E)$.

3. FLOW IN POROUS MEDIA

Studies of both miscible and immiscible flows in porous media represent a rich field of present-day research dependent on many of the concepts introduced previously.

In the case of immiscible flows, the structures of the rough interfaces occurring during drainage and imbibition experiments often turn out to be fractal or self-affine (see FRACTAL GEOMETRY). This interfacial structure can be expected to be dependent on many variables, including whether the flow is wetting or nonwetting, because of the large liquid–solid interface involved; whether the strength of capillary forces is significant, because in porous media of low permeability κ and consequent small pore sizes $a \sim \kappa^{1/2}$, the pressure drops due to surface tension σ in pores, $\Delta p_{\text{tension}} \sim \sigma/a \sim \sigma/\kappa^{1/2}$, can be large compared to the viscous pressure drop, $\Delta p_{\text{viscous}} \sim U\eta/a \sim U\eta/\kappa^{1/2}$; the magnitude of the viscosity ratios $M = \eta_1/\eta_2$, because of the existence of the Saffman–Taylor instability; and whether gravity interacts on the evolving interface, because of the Taylor instability—gravity may be expected to have a significant effect when the hydrostatic pressure drops $\Delta p_{\text{gravity}} \sim \Delta\rho g\kappa^{1/2}$ that occur across pores become large in comparison with capillary and viscous pressure drops.

In the case of miscible flows, such questions as the nature of the anomalous diffusion, which controls dispersion of passive scalars, and how these transport coefficients scale with flow velocity in porous media have been studied.

3.1 Microscopic Equations of Motion

A porous medium can be thought of as consisting of a tortuous, multiply connected domain of pores connected by throats. Such porous media are created naturally by grains of sand or other material, typically random in shape and coming in a distribution of sizes (though in laboratory simulations, often monodisperse glass beads are used instead), that are packed into a random configuration. Typically, these configurations do not vary over time, and, hence, the porous medium forms a quenched (that is, time independent) random topography through which the fluid must flow. The terms pores and throats are not well-defined concepts, but represent the idea that the topology of a porous medium can usually be considered to consist of fairly large interstitial holes connected by long thin channels. The Reynolds numbers for typical flows in porous media are usually very small, $Ua/\nu \ll 1$, and, for such creeping flows, neglect of inertial terms is valid. Therefore, in principle, for the case of multiphase flow, the full set of equations that need to be solved are Stokes flow in a gravitational field $\mathbf{g}$,

$$\frac{\partial \mathbf{v}_\alpha}{\partial t} = -\frac{1}{\rho_\alpha}\boldsymbol{\nabla} p + \nu_\alpha \nabla^2 \mathbf{v}_\alpha + \mathbf{g}, \tag{55}$$

supplemented by the incompressibility conditions $\boldsymbol{\nabla}\cdot\mathbf{v}_\alpha = 0$. Here, the subscripts $\alpha = 1,2$ represent respectively the displaced and displacing fluids [oil (1) being displaced by water (2), for example] in the case of two-phase flow.

The boundary conditions are all important for flows in these contorted domains with large liquid–liquid and liquid–solid interfaces. At any point $\mathbf{s}$ of the curved liquid–liquid interface between the immiscible fluids, the normal component of the velocity across the interface must be continuous, while the pressure difference due to surface tension σ is discontinuous, leading to the boundary conditions

$$\mathbf{v}_{1,n}(\mathbf{s}) = \mathbf{v}_{2,n}(\mathbf{s}) = 0,$$
$$p_2(\mathbf{s}) - p_1(\mathbf{s}) = \sigma[R_1^{-1}(\mathbf{s}) + R_2^{-1}(\mathbf{s})], \tag{56}$$

where $R_1(\mathbf{s})$ and $R_2(\mathbf{s})$ are the two principal radii of curvature of the interface at $\mathbf{s}$. In contrast, at the liquid–solid interface, stick boundary conditions can be assumed:

$$\mathbf{v}_1\,(\text{substrate}) = \mathbf{v}_2\,(\text{substrate}) = 0. \tag{57}$$

The complexity of this flow lies in the stick

boundary conditions at the liquid–substrate interface and the free boundary conditions at the liquid–liquid interface. The shape of the domain is typically so tortuous that analytical approaches to studying such flows starting from the microscopic equations Eqs. (55)–(57) are hopeless, and the major theoretical tools developed to deal with this class of flows have been the derivation of macroscopic phenomenological equations to account for gross flow behavior. More recently, this approach has been supplemented by computer simulations on simplified models, such as invasion percolation and gradient percolation, coupled with scaling arguments to account for both interfacial roughness and transport.

3.2 Macroscopic Transport in Porous Media

Structure and transport in porous media have been described macroscopically in terms of a few parameters, such as porosity ϕ, permeability κ, effective surface tension σ_{eff}, and saturation S, which represent simple macroscopic statistical averages of the very complicated nonhomogeneous flow taking place in the porous medium [see, for example, Adler and Brenner (1988)]. Phenomenologically, Darcy's law,

$$\mathbf{v} = -(\kappa/\eta)\nabla p, \tag{58}$$

has been used to describe the flow dynamics. It represents a equation similar to Ohm's law, describing the linear relationship between flux and applied pressure for single-phase flow in porous media. Such equations of motion and the macroscopic transport coefficients involved are averages on length scales L much larger than the pore size, $L \gg a$, and, consequently, can only be expected to be valid on macroscopic flow scales. The porous medium is also usually supposed to be homogeneous and isotropic on macroscopic scales. If any of these conditions are not met, then the phenomenology must be altered to take the new conditions into account. For example, if the porous medium is heterogeneous on all scales considered, then all averages will depend on sample size, while if the medium is macroscopically anisotropic, then tensor equations can be expected to describe the flow, and Darcy's law will be generalized to $v_i = -(\kappa_{ij}/\eta)\, \partial p/\partial x_j$.

The porosity ϕ is the volume fraction not occupied by the solid matrix, that is, the volume fraction of the throats and pores making up the flow domain in a fully connected porous medium. Typical values for the porosity might vary from 5% to 20% in sandstones or limestone.

The permeability to single-phase flow κ can be found from Darcy's law by measuring the flux of liquid as a function of the pressure drop. Its dimensions are length squared, $[\kappa] = L^2$, and its magnitude can be estimated for many porous media as $\kappa \sim a^2$, where a is a typical pore size. It must be remembered, however, that permeability depends not only on porosity but also on connectivity, and this estimate may break down for poorly connected media—a porous medium can have a high porosity, yet if no connected path exists from one end of the porous medium to the other, its permeability is zero. For typical porous media forming geological reservoirs, the pore sizes can vary from 10^{-6} to 10^{-2} cm, and, therefore, the permeability will vary to a much greater degree than porosity, covering up to eight orders of magnitude in the range $10^{-12}\ \text{cm}^2 < \kappa < 10^{-4}\ \text{cm}^2$, depending on pore size and connectivity [see, for example, Pittman (1983)].

The pressure differences across interfaces when measured on a macroscopic scale represent a weighted average of the pressure differences locally at each throat and pore, which will depend on the instantaneous position of the interface and the wettability of the underlying substrate. Consequently, the effective surface tension σ_{eff} measured macroscopically in a porous medium can be very different from the microscopic value σ between two pure phases. For example, in the case of oil and water, the ratios $\sigma_{\text{eff}}/\sigma \approx 5.5$ for oil wetted media and $\sigma_{\text{eff}}/\sigma \sim 306$ for water-wetted media have been found.

In the case of immiscible two-phase flow, a permeability for both the displaced and displacing species can be defined from the pair of Darcy's-law equations $\mathbf{v}_1 = -(\kappa_1/\eta_1)\nabla p$ and $\mathbf{v}_2 = -(\kappa_2/\eta_2)\nabla p$; but, in this case, the permeability of each species will be a function of the saturation S, the volume fraction of each species in the interstitial flow domain. Calling the saturation of the displaced fluid S_1 and that of displacing fluid S_2, then, naturally, $S_1 + S_2 = 1$. The two-phase permeabilities κ_1 and κ_2 are much more complex than those for single-phase flow and depend not only on saturation, but on the surface tension between the phases σ, on the viscosity ratio M, and also on the history of the flow, as strong evidence for hysteresis is often observed during drainage and imbibition experiments.

3.3 Drainage and Imbibition

Two of the most important classes of flow in porous media are drainage and imbibition. In a drainage experiment, a porous medium is first saturated with a wetting liquid ($S_1 = 1, S_2 = 0$), and then a nonwetting fluid is used to displace the wetting fluid. For this to occur, an external pressue p must be applied. For small pressures $p < p_{c2}$, small amounts of the wetting fluid are displaced, to be replaced by small nodules of nonwetting fluid, but there is no macroscopic flow of the displacing liquid, and the permeability of the displacing phase is zero, $\kappa_2 = 0$ (see Fig. 11). When a pressure is applied sufficient to overcome the capillary forces, $p > p_{c2}$, then a macroscopic flow of the nonwetting phase is suddenly seen, and $\kappa_2 > 0$. Thus, a critical pressure p_{c2} exists for drainage to occur, and one can estimate this critical pressure, which is due to surface tension, as $p_{c2} \sim \sigma/a$. One can also define X_1 and X_2, the volume fractions of interstitial region occupied by the displaced and displacing phases, which are parts of domains spanning the entire sample and through which macroscopic flow can therefore occur; in contrast, the saturations S_1 and S_2 will not distinguish between spanning clusters and small isolated nodules of fluid. Clearly, $X_2(p) = 0$ for $p < p_{c2}$, and $X_2(p) > 0$ for $p > p_{c2}$. A percolation transition has occurred at p_{c2}.

Percolation transitions are geometric phase transitions that involve the clusters formed from sites (or bonds) on a lattice that are occupied randomly with a probability P [see, for example, Stauffer (1985)]. For very small P, only isolated clusters of occupied sites exist. But, as $P \to P_c$, the size of the typical clusters diverges as $\xi \sim (P_c - P)^{-\nu}$, and an incipient infinite cluster that spans the lattice appears (see Fig. 12). That such a cluster must appear is clear because when $P = 0$, no such cluster exists, $X(0) = 0$, and when $P = 1$, every site belongs to this cluster, $X(1) = 1$. A geometric phase transition is involved because the behavior of $X(P)$ is not analytic; rather $X(P) = 0$ for $P < P_c$, and $X(P) \sim (P - P_c)^\beta$ for $P > P_c$. One of the most important developments in modern physics has been the realization that certain properties of percolation, as well as other phase transitions, are universal (see UNIVERSALITY) and, therefore, independent of the microstructure of the lattice—for example, critical exponents, such as β and ν—while

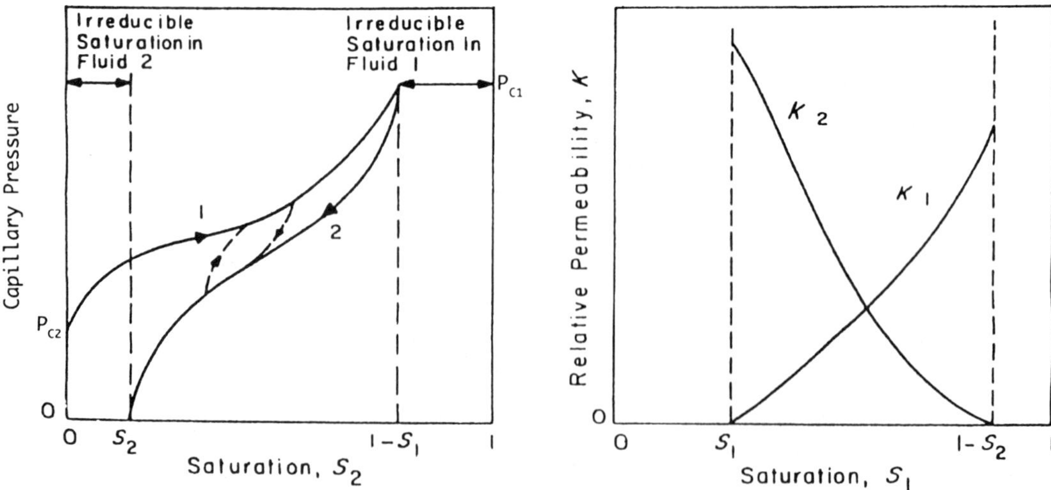

FIG. 11. Plots of the capillary pressure and relative permeabilities as function of saturation during drainage and imbibition; redrawn from Adler and Brenner (1988). Reproduced, with permission, from the *Annual Review of Fluid Mechanics*, Volume 20, ©1988, Annual Reviews Inc.

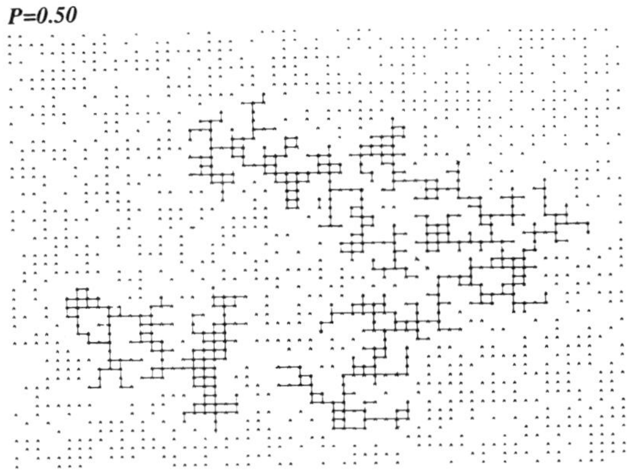

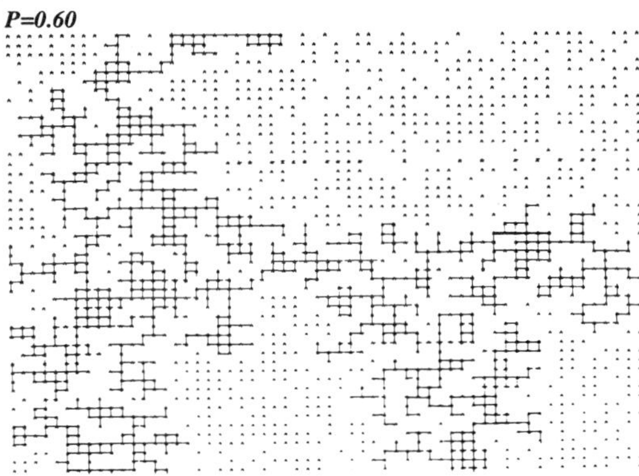

FIG. 12. The percolation transition on a square two-dimensional lattice. Here $P_c = 0.5928 \ldots$, and the largest clusters for (a) $P = 0.5$ and (b) $P = 0.6$ are shown; redrawn with permission from Stauffer (1985).

others are strongly lattice dependent, such as P_c.

Returning now to flow in porous media, the universality does not reside in the pressure p_{c2}—indeed, the estimate $p_{c2} \sim \sigma/a$ is dependent on both pore size and surface tension—but rather in the exponents t and β, which control respectively the manner in which the permeability $\kappa_2(p) \sim (p - p_{c2})^t$ and the probability $X_2(p) \sim (p - p_{c2})^\beta$ that a piece of the displacing fluid belongs to a spanning cluster increase for pressures above the critical pressure $p > p_{c2}$. The exponents t and β, among many others, are the same as for percolation and depend only on dimension. For example, the exact value for the percolating fraction exponent is $\beta = 5/36$ in two dimensions, while in three dimensions, it is $\beta \approx 0.417$; similarly, the permeability exponent t, which is identical to the conductivity exponent in random resistor networks, is $t \approx 1.3$ in two dimensions, while in three dimensions, it is $t \approx 2.0$. One's intuitions might suggest that $\kappa(p) \sim X(p)$ and, therefore, $t = \beta$, but only a fraction of the pores and throats constituting the spanning cluster actually have pressure drops across them and, therefore, carry flowing fluid. Thus, t depends not only on the spanning cluster geometry but also on the fluid flow on that cluster, though, as these scaling relations are valid as $p \rightarrow p_{c2}$, we can expect $t > \beta$.

As the pressure of the nonwetting fluid is increased well beyond p_{c2}, more and more of the wetting fluid is displaced from the porous medium, and both $\kappa_1 > 0$ and $\kappa_2 > 0$. At some pressure $p_{c1} > p_{c2}$, a second percolation transition occurs when the displaced wetting fluid itself no longer spans the porous medium [$X_1(p) = 0$ for $p > p_{c1}$, and $X_1(p) > 0$ for $p < p_{c1}$], but rather forms a set of disconnected clusters. At this point, $\kappa_1 = 0$, and an irreducible fraction of the wetting fluid $S_{1,\text{irred}}$ is left behind in the porous medium.

If the external pressure is now reduced again, the wetting fluid will begin to replace the nonwetting fluid. This reverse process is called imbibition. More generally, an imbibition experiment would begin with a porous medium filled with a nonwetting fluid, which is replaced by a wetting fluid—for example, water soaking into a piece of paper. No external pressure needs to be applied here because of the existence of a positive spreading coefficient $S = \sigma_{SV} - \sigma_{LS} - \sigma$.

It might appear that a cyclic reversible process of drainage followed by imbibition could be set up. This is not the case; the process is hysteretic because, though an experiment with a porous medium soaked with a wetting fluid could be set up, after one drainage and imbibition cycle an irreducible amount of nonwetting fluid $S_{2,\text{irred}}$ would remain in the medium at the point when the capillary pressure is zero. After one cycle, at least an irreducible minimum of each phase must exist in the porous medium, $S_{1,\text{irred}} < S_1 < 1 - S_{2,\text{irred}}$ and $S_{2,\text{irred}} < S_2 < 1 - S_{1,\text{irred}}$, and in this region, each phase spans the porous medium. The irreducible minimum of each phase is the saturation at the percolation transition; thus, at $S_1 \approx S_{1,\text{irred}}$ where the wetting permeability tends to zero as $\kappa_1 \sim (p_{c1} - p)^t$, and $S_2 \approx S_{2,\text{irred}}$, where the nonwetting permeability tends to zero as $\kappa_2 \sim (p - p_{c2})^t$.

3.4 Models for Fluid Flow in Porous Media

The discussion on drainage and imbibition above combines empirical observations with percolation theory. But the theoretical isomorphism drawn between conductivity on a percolating lattice and two-fluid flow in a porous medium is tentative: For example, in contrast to exact percolation, there is no reason in two-fluid flow below the percolation transition why locally large pressure gradients cannot exist capable of mobilizing individual blobs of fluid that can then flow or fuse with other clusters. In addition, real two-fluid flow has a large number of variables that can be expected to affect the interfacial structure and flow but that do not appear in normal percolation theory, including surface tension σ, gravity g, viscosity ratios M, flow rate U, the nature of the porous medium, and wetting properties. All these variables, which will give rise to a large phase diagram, need to be investigated. What is more, the saturation of each species in the porous medium and the permeability only give partial information on structure. More recently, fractal and self-affine properties of interfaces in two-phase flow have been the subject of detailed investigations.

To simulate aspects of this phase diagram and to study interfacial structure, new models and approaches have been introduced. Invasion percolation (de Gennes and Guyon, 1978; Wilkinson and Willemsen, 1983) has been used to study low–capillary-number [$Ca = (\eta_1 - \eta_2)\, U/\sigma \ll 1$] creeping flows dominated by surface tension. In the opposite limit of high capillary numbers, surface tension can be neglected, and diffusion-limited fingering is observed. In the presence of gravitational fields, the relevant dimensionless control variable is the Bond number $B = \Delta\rho g \kappa/\sigma$, and the model of gradient percolation (Sapoval *et al.*, 1985; Gouyet *et al.*, 1990) can be used to study its influence on flow.

Hand in hand with new models has gone the development of new theoretical approaches to understanding flow in porous media. For example, the influence of the porous medium on two-fluid flow has been analyzed theoretically in terms of effective Hamiltonians that were shown to lie in the same universality class as random-bond and random-field Ising models (Brochard and de Gennes, 1983), while the realization that surfaces in imbibition experiments are often self-affine can be accounted for in terms of the self-affine solutions of a class of forced partial differential equations (Edwards and Wilkinson, 1982; Kardar *et al.*, 1986).

3.4.1 Invasion Percolation How do drainage interfaces depend on capillary number Ca and the viscosity ratio M of the fluids involved? To answer this and related questions, models for porous media and for the flows in such quenched random topographies are introduced and simulated.

To simulate fluid flow in porous media, lattice models are often used to represent the pores and throats. The pores can be represented by sites on this lattice, while the throats are represented by bonds. The quenched random nature of the porous medium is represented by assigning a set of random variables r_{ij} to the bonds $\langle ij \rangle$ between sites i and j, representing the variable radius of the throat considered.

To model drainage of a wetting fluid by a nonwetting fluid in the limit of low capillary numbers involves creeping flows, and the contribution of viscosity to the pressure drop across a throat can be neglected. Therefore, the process of drainage in this limit is entirely due to the capillary forces $p_{ij} = 2\sigma/r_{ij}$ across each throat. Assume that an external pressure p is applied and that flow can occur in any throat provided $p > p_{ij}$. If all throats obeying $p > p_{ij}$ were filled with the nonwetting fluid, the model would be isomorphic to the pure percolation problem discussed in Sec. 3.3. But, in drainage, the external pressure is applied along one edge of the porous medium, and the displacement of wetting fluid by the nonwetting invader begins at this edge. To model this flow, the invasion percolation model was introduced (de Gennes and Guyon, 1978; Wilkinson and Willemsen, 1983) describing the process of filling the active ($p > p_{ij}$) sites. An invasion edge is chosen (hence the name invasion percolation) where the external pressure is presumed to be applied, and, sequentially, the most active bond available at that point in time, that is, the bond with largest radius r_{ij}, is filled with nonwetting fluid. In consequence, several differences exist from normal percolation.

First, invasion percolation is a dynamic process. Second, throats can only be occupied by the nonwetting fluid if there exists a contiguous path of active sites from the initial invasion edge of the nonwetting fluid. Therefore, for $p < p_c$, there exist no disconnected clusters of occupied sites as in true percolation, but, rather, a finite invasion of the porous medium by the nonwetting fluid occurs. Third, trapping of the wetting fluid by the nonwetting phase can occur because of the incompressibility of the fluids involved. Consider an active site that is a candidate for the replacement of the wetting phase by the nonwetting phase. If a contiguous path exists to the opposite face of the porous medium from which the drainage of the wetting phase is occurring, invasion is possible; but if the candidate active throat filled with wetting fluid is surrounded by the invading nonwetting phase, then invasion of the throat is impossible, as the ejected wetting fluid would have nowhere to go. Simulations (Chandler *et al.*, 1982) appear to yield exponents close to those one would predict on the basis of pure percolation arguments, and so do real experiments involving displacement of glycerol by air (Lenormand and Zarcone, 1985). For example, the saturation $S_2(W)$ on a finite-size lattice of width W scales at breakthrough on a two-dimensional lattice as $S_2(W) \sim W^{-0.17}$. Finite-size scaling arguments would predict $S_2(W) \sim W^{-\beta/\nu}$, where $\beta/\nu = 15/144 \approx 0.11$ for normal percolation.

Invasion percolation should be valid at $Ca = 0$. For finite flow rates, the pressure drops due to viscosity must be included in simulations (Chen and Wilkinson, 1985; Lenormand *et al.*, 1988). If Poiseuille flow is assumed in each throat, then the flux J_{ij} of fluid across each bond $\langle ij \rangle$ if pressures p_i and p_j exist at sites i and j is

$$J_{ij} = \pi r_{ij}^4(p_i - p_j - p_{ij})/8l\eta \tag{59}$$

when the pressure drop is greater than the capillary pressure, $p_i - p_j > p_{ij}$, and the flux is zero, $J_{ij} = 0$, otherwise. In Eq. (58), l is the length of the throat, and η is the average viscosity. Incompressibility demands that $\Sigma_j J_{ij} = 0$, where the sum is over nearest-neighbor sites j of each site i. This set of equations has been simulated by Lenormand *et al.* (1988) as a function of the capillarity Ca and viscosity ratio $M = \eta_2/\eta_1$. Three qualitative "phases" of the flow can be seen in both numerical simulations and real experiments (see Fig. 13): a stable flat interface between the immiscible fluids, viscous fingering, and capillary fingering.

In the limit of very small capillary numbers, capillary fingering is observed, described by invasion percolation. As the capillary number increases and, therefore, surface-tension effects are less dominant, a stable flat interface appears, provided a more viscous fluid is used to displace a less viscous one, $M > 1$, as the Saffman–Taylor stability analysis would suggest. What is perhaps most surprising about these simulations is that crossover does not occur when $Ca \approx 1$, but rather when $Ca \lessapprox 10^{-2}$. In the opposite limit, $M \ll 1$ and $Ca \gg 10^{-2}$, viscous fingers appear, as diffusion-limited arguments would suggest.

3.4.2 Gradient Percolation The influence of gravity on the stability of two-fluid flow has been shown by Taylor (1950) to be substantial. What are the consequences for drainage in a gravitational field? Without a gravitational field, breakthrough of the nonwetting phase always occurs whatever the system size under the application of a finite pressure. Against-gravity breakthrough will never occur in an infinite system. Rather, a macroscopic invasion front of length z_m can be achieved, depending on the applied pressure, and the interface of the front is rough—indeed, fractal. Experimentally (Clement *et al.*, 1985; Gouyet *et al.*, 1990), the structure of the invasion front can be studied by, for example, forcing nonwetting Wood's metal, a liquid alloy, through packs of crushed glass, then freezing the metal and examining slices at various heights z (assume that the invasion front begins at $z = 0$, and that the slices are taken for $z > 0$, in a gravitational field pointing downward, $\mathbf{g} = -g\mathbf{k}$).

The saturations of both the displaced $[S_1(z)]$ and displacing $[S_2(z)]$ phases are functions of the height z, and the same applies to the probability of belonging to a spanning cluster for both the displaced $[X_1(z)]$ and displacing $[X_2(z)]$ phases. For small z, the crushed glass is almost completely saturated by the Woods metal, and, therefore, $S_2(z) \approx 1$, with only a few unconnected blobs of the displaced medium, $S_1(z) \approx 0$, in this case, air [similarly, $X_1(z) = 0$, while $X_2(z) > 0$]. At some critical height z_{c1}, a percolation transition occurs in the displaced phase, and, for $z_{c1} < z < z_{c2}$, both the wetting and nonwetting

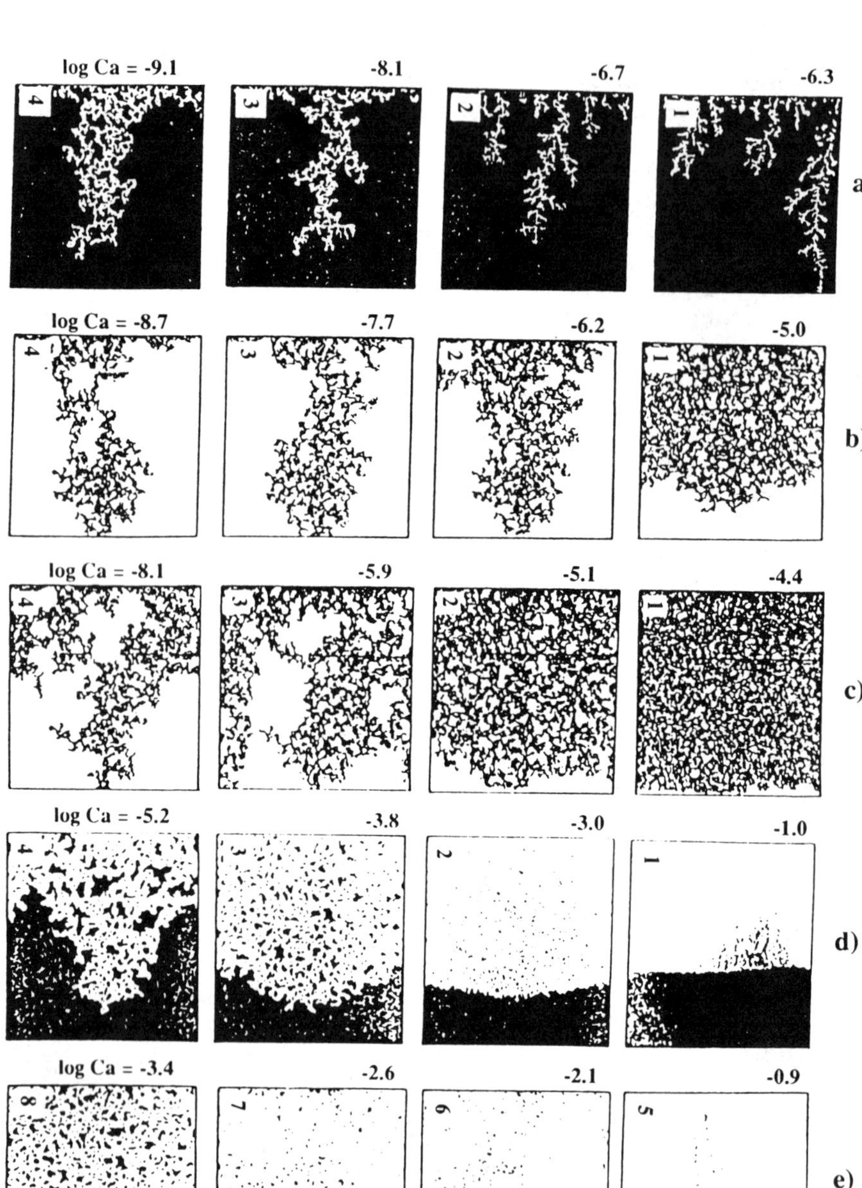

FIG. 13. The two-phase interface at various capillary numbers and viscosity ratios M. (a) The transition from viscous fingering (1) to capillary fingering (4) for an unstable Saffman–Taylor interface ($\log M = -4.7$); (b),(c) the interfaces are stable according to the Saffman–Taylor criterion ($\log M = 0.7$ and $\log M = 1.9$), but the low capillary numbers yield capillary fingers modeled by invasion percolation; (d),(e) the interfaces are even more stable according to the Saffman–Taylor criterion ($\log M = 2.0$ and $\log M = 2.9$), and the interfaces vary between apparently flat on the large scale (though perhaps having a self-affine structure at smaller scales) and incipient capillary fingering; redrawn with permission from A. Coniglio in *Hydrodynamics of Dispersel Media*, Eds. J. P. Hulin, A. M. Cazabat, E. Guyon, F. Carmona Amsterdam: North-Holland (1990).

phases are connected [and, therefore, both $X_1(z) > 0, X_2(z) > 0$]. But the displacing phase becomes ever more sparse as the displacing phase percolation transition at $z = z_{c2}$ approaches; and for $z > z_{c2}$, only the displaced phase is connected [in this region, $X_1(z) > 0$, while $X_2(z) = 0$], with the displacing phase forming a critical tail in the form of a rough interface with a fractal dimension $D \approx 2.5$ and a width $W(B)$ that depends on the Bond number $B = \Delta\rho\gamma\kappa/\sigma$, a measure of the ratio of hydrostatic to surface-tension forces on the fluid in a pore.

The creation and properties of this nonhomogeneous flow structure can be understood in terms of invasion percolation in a gravitational field or gradient percolation (Sapoval *et al.*, 1985; Gouyet *et al.*, 1990). The terminology gradient percolation comes from the influence of the gravitational field on the probability $P(z)$ for an active bond to exist at a height z. In normal invasion percolation, we have $P(z) = P$, a constant dependent only on the constant applied pressure. In a gravitational field, however, the effective hydrostatic pressure at a height z is $\Pi(z) = \rho g(z_m - z)$, where the constant z_m depends only on the applied pressure. Therefore, if $P^*(\Pi)d\Pi$ is the probability for the existence of capillary pressures in the porous medium between Π and $\Pi + d\Pi$, then the way in which the active bond probability $P(z)$ depends on z is

$$P(z) = \int_0^{\rho g(z_m - z)} P^*(\Pi)d\Pi. \qquad (60)$$

If we assume that the throat sizes are equally distributed, then $P(z)$ decreases with a constant gradient $|dP/dz|$ as $P(z) = 1 - |dP/dz|z$.

Using this gradient dependence for $P(z)$ and assuming the validity of the usual percolation exponents, several estimates can now be made for the properties of the invasion front during drainage in a gravitational field. The two percolation transitions can be seen to occur at heights $z_{c1} = (1 - P_{c1})/|dP/dz|$ and $z_{c2} = (1 - P_{c2})/|dP/dz|$, while the width of the critical tail can be estimated as $W(B) \sim \xi \sim [P(z) - P_{c2}]^{-\nu}$. To find the Bond number dependence of the width, one estimates from scaling that $P(z) - P_{c2} \approx |dP/dz|(z - z_{c2}) \sim |dP/dz|W(B)$, and, therefore, one finds self-consistently that $W(B) \sim |dP/dz|^{-\nu/(1+\nu)}$. Finally, as the gradient is proportional to the ratio of the hydrostatic pressure $\Delta\rho g\kappa^{1/2}$ to the capillary pressure $\sigma/\kappa^{1/2}$ and, therefore, to the Bond number, $|dP/dz| \sim B = \Delta\rho g\kappa/\sigma$, one finds a singular form for the width of the interface (Wilkinson, 1984), which should diverge as the gravitational field tends to zero as

$$W(B) \sim \kappa^{1/2} B^{-\nu/(1+\nu)}. \qquad (61)$$

The observed fractal dimension of the rough interface can also be predicted from percolation theory as the fractal dimension of percolation clusters $D = d - \beta/\nu$. In three dimensions, $D \approx 2.55$, in close agreement with experimental observation.

3.4.3 Macroscopic Langevin Equations

When the process of imbibition rather than drainage is considered, then the interface on a very large scale appears to be flat. Microscopically, this makes good sense. If the interface of a nonwetting fluid is considered, it will have great difficulty in flattening neighboring fingers, as this may involve flow through small pores, which is discouraged; for a wetting fluid, however, capillary forces will encourage such flow and, in so doing, relax incipient fingers (Lenormand and Zarcone, 1985). But, on closer examination, at smaller scales, the rough surfaces are not really flat at all, but to a good approximation can be described as single-valued fluctuations in height $h(\mathbf{r}, t)$ on a flat substrate; that is, overlaps are unimportant. Even more importantly, they appear to be self-affine. That is, they have a type of self-similarity under changes of scale, as do fractals, but, in contrast with fractals, they are not isotropic: If surfaces in windows of length L and length λL are compared, then the height fluctuations do not become statistically self-similar if an equal change of scale in the direction of the fluctuations $h \rightarrow \lambda h$ is made, as would be the case for fractals, but, rather, the change of scale $h \rightarrow \lambda^\alpha h$ must be made, where α, known as the Hurst exponent, is a property of the self-affine interface.

The growth and form of such interfaces have been the subject of extensive investigations in recent years and have resulted in the recognition (Family and Vicsek, 1985) that the width fluctuations $W(L, t) = \langle|h(\mathbf{r} + \mathbf{L}, t) - h(\mathbf{r}, t)|\rangle$ of interfaces grown on substrates of dimension L^{d-1} all obey a special form of self-affine dynamic scaling: Initially, the fluctuations increase with time,

$$W(L, t) \sim t^{\beta} \quad \text{if } t \ll t_L, \tag{62a}$$

and then saturate to a length-scale–dependent final value,

$$W(L, t) \sim L^{\alpha} \quad \text{if } t \gg t_L. \tag{62b}$$

The crossover time scale $t_L \sim L^z$ is also length-scale dependent, with an exponent $z = \alpha/\beta$. This last identity can be derived from the existence of a scaling form $W(L, t) = t^{\beta}f(t/L^z)$.

In addition to scaling, a great degree of universality was discovered in the actual values of the critical exponents α, β, and z seen in such apparently unrelated phenomena as the interface of two-phase flow during imbibition, surfaces grown by molecular beam epitaxy, and the structure of surfaces in growing bacterial colonies. This universality was confirmed by computer simulations of various microscopic models for surface roughening, which concluded that universality classes exist for apparently unrelated models creating self-affine interfaces, with the values within a given class dependent only on certain conservation laws and symmetries and on the spatial dimensionality d.

To account for this universality, the underlying kinematics of rough surface growth were incorporated into Langevin equations. These equations incorporate the notion that, under driving with a random force $\eta(\mathbf{r}, t)$, the height fluctuations $h(\mathbf{r}, t)$ in interfacial growth grow with a velocity λ normal to the interface and relax at a rate D because of surface diffusion or surface tension (surface tension would be the relaxation mechanism in imbibition, while an isomorphic equation with the surface tension replaced by surface diffusion describes molecular beam epitaxial growth of surfaces). Typical of such Langevin equations is the Kardar–Parisi–Zhang (KPZ) equation (1986) for the height fluctuations $h(\mathbf{r}, t)$ in an interface growing with a velocity λ normal to the interface:

$$\frac{\partial h}{\partial t} = D\nabla^2 h + \frac{\lambda}{2}(\nabla h)^2 + \eta(\mathbf{r}, t), \tag{63}$$

where $\langle\eta(\mathbf{r}, t)\eta(\mathbf{r}', t')\rangle = \Gamma\delta(\mathbf{r} - \mathbf{r}')\delta(t - t')$. The KPZ equation is one of a zoo of such equations, which differ depending on the symmetries and conservation laws that need to be incorporated. The Hwa–Kardar equation (43) describing height fluctuations in flowing sandpiles is another example. The exponents α and β are therefore not independent, and for the KPZ equation, the scaling relation $\alpha + \alpha/\beta = 2$ can be derived.

The actual values of the exponents require further analytical or numerical work, but, in $d = 2$, the dynamic renormalization group (Medina *et al.*, 1989) gives the exact result $\alpha = \frac{1}{2}$ and $\beta = \frac{1}{3}$. There are no exact results in $d > 2$, but, on the basis of numerical evidence alone, Kim and Kosterlitz (1989) have suggested $\alpha = 2/(d + 2)$ and $\beta = 1/(d + 1)$.

How well do such equations agree with experimental data on imbibition? The results are mixed. Typical values for the exponents in two-dimensional experiments on imbibition (Horvath *et al.*, 1991) involving water displacing air in paper gives $\alpha \approx 0.75$, $\beta \approx 0.6$. Therefore, scaling laws such as $\alpha + \alpha/\beta = 2$ are obeyed to great accuracy, but the individual exponents are very far from theoretical predictions.

There could be many causes for this variance. The noise may not be δ correlated in time, or the porous medium itself can be considered to contribute a quenched random noise $\zeta(\mathbf{r}, h)$ obeying $\langle\zeta(\mathbf{r},h)\zeta(\mathbf{r}',h')\rangle = G\delta(\mathbf{r} - \mathbf{r}',)\ \delta(h - h')$ to the growth process. This noise could cause local pinning of the interface, leading when depinning occurs to sudden bursts of growth, or even the existence of a global pinning-depinning transition. This is an active area of present research, and a definite reason for the variance between theory and experiment remains, at present, unknown.

3.5 Hydrodynamic Dispersion in Porous Media

The discussion of flow of wetting fluids in porous media has focused so far on the structure of the self-affine interface during imbibition. Such fluids can, however, also carry dispersions of solute particles. Examples would include sediments or pollutants in groundwater. The manner in which such suspensions are dispersed by flow is also a matter of technological importance. Here, we consider the dispersion of such solutes dissolved in a liquid flowing in a porous medium and the magnitude of the diffusion coefficients involved.

3.5.1 Dispersion of Conserved Solute Particles Consider a pulse of passive scalar particles, that is, particles whose motion is controlled by the velocity field in the porous medium but which do not react back on the flow, released inside the fluid flowing in a porous medium. They will disperse under the combined action of convection and molecular diffusion with time. This dispersion will be anisotropic because of the existence of a macroscopic flow U, say in the x direction. In consequence, the convection-diffusion equation for the particle density $c(\mathbf{r}, t)$ can be written

$$\frac{\partial c}{\partial t} + U\frac{\partial c}{\partial x} = D_{\parallel}\frac{\partial^2 c}{\partial x^2} + D_{\perp}\nabla_{\perp}^2 c, \qquad (64)$$

where, from scaling, one would expect the longitudinal and transverse effective diffusion constants to be of the same order of magnitude, $D_{\perp} \sim D_{\parallel}$. How do these effective diffusion constants $D_{\perp} = D_{\perp}(D, U, \kappa)$ and $D_{\parallel} = D_{\parallel}((D, U, \kappa)$ depend on the molecular diffusion constant D, the macroscopic flow velocity U, and the permeability κ? A great deal of information can be derived from scaling arguments.

First, as the flow velocity tends to zero, $U \to 0$, the dispersion must be isotropic, and, as there is no convection, molecular diffusion is the only source of dispersion:

$$D_{\parallel} \approx D_{\perp} \approx D. \qquad (65)$$

In the opposite limit $U \to \infty$, the dispersion is convection dominated, and, from scaling, one expects

$$D_{\perp} \sim D_{\parallel} \sim U\kappa^{1/2}. \qquad (66)$$

This result can be understood as follows: The streamlines in porous media are extremely complex—streamlines in different parts of the same pore or throat will separate into different pores and throats; velocity in large pores and throats is slow, while in small pores and throats, it is fast; the complex connectivity of the porous medium allows streamlines to separate and then rejoin later. The consequence is that mechanical mixing will dominate, and the contribution of molecular diffusion to dispersion can be neglected. To estimate the effective diffusion constant in the high-flow limit, one can argue that the relaxation time scale $\tau \sim a/U$ is the time necessary for a particle to cross a pore, and, therefore, $D_{\perp} \sim D_{\parallel} \sim \langle x^2\rangle/\tau \sim a^2/(a/U) \sim U\kappa^{1/2}$.

This scaling argument depends on the existence of a single flow rate U in the porous medium for its validity. Experimentally, logarithmic corrections are often observed to scaling:

$$D_{\perp} \sim D_{\parallel} \sim U\kappa^{1/2}\log(U\kappa^{1/2}/D), \qquad (67)$$

and these logarithmic deviations from scaling can be understood on the basis of the existence of a distribution of flow velocities in the pores. There does not exist a single transit time $\tau \sim a/U$, but, rather, a distribution with moments given by $\langle t^n\rangle \sim \int(a/v)^n P(v)dv$, where $P(v)dv$ is the probability of a velocity in the pore between v and $v + dv$. If $P(v) \sim v$ as $v \to 0$, then logarithmic corrections to the high–flow-rate effective dispersion may be expected.

At intermediate flow rates, an alternative scaling form is also possible. Taylor (1953) considered the problem of dispersion in a pipe and pointed out that because Poiseuille flow in a pipe has a parabolic profile in velocity (maximum at the center of the pipe, zero at the edges because of stick boundary conditions), molecular diffusion across streamlines will enhance dispersion above that for pure molecular diffusion alone. A similar argument should be valid for porous media: In a time $\tau \sim a^2/D$ that the particle diffuses across the pore as a result of molecular diffusion, the actual variance in position between particles near the edge of the pore and the pore center is $\langle x^2\rangle \sim (U\tau)^2 \sim U^2a^4/D^2$. Therefore, dispersion in this regime is controlled by an effective Taylor diffusivity of size

$$D_{\perp} \sim D_{\parallel} \sim \langle x^2\rangle/\tau \sim a^2U^2/D \sim U^2\kappa/D. \qquad (68)$$

In conclusion, a scaling form for the effective longitudinal and transverse diffusivities may be written:

$$D_{\parallel}(D, U, \kappa) = U\kappa^{1/2}f_{\parallel}(Pe), \qquad (69)$$
$$D_{\perp}(D, U, \kappa) = U\kappa^{1/2}f_{\perp}(Pe),$$

where $Pe = U\kappa^{1/2}/D$ is the dimensionless Peclet number, which is the ratio of the convective time scale $\tau_{\text{convective}} \sim a/U$ to the diffusive time scale $\tau_{\text{molecular}} \sim a^2/D$ in the porous medium.

Thus far, the dispersion in porous media has been assumed to be Gaussian, albeit with an effective anisotropic diffusion constant depending on the flow. Actual experiments, such as injecting glycerol with colored dyes into two-dimensional porous media, suggest that while the average concentration of tracer particle does indeed obey a convective-diffusive equation, fluctuations exist about this average. Feder *et al.* (1990) studied the equi-concentration surfaces of the dispersed dye and found that the surfaces are not smooth lines, but self-affine, with $\alpha \approx 0.58$ in two dimensions.

3.5.2 Grain Trapping in Porous Media

The material forming the solute in Sec. 3.5.1 was implicitly assumed to be of microscopic dimensions, certainly much smaller than any pore size. But what happens to grains of material that start to approach the pore size or throat radius as they flow through the porous medium? Clearly, the possibility exists that such material will be trapped or pinned at a site. Perhaps later this same grain will be released as a result of Brownian noise, only to be retrapped somewhere else at a later time. The total effect of such trapping is to slow down greatly the average velocity of such grains from the average flow velocity U.

de Gennes (1979) gave an estimate for this slowing down. Suppose the grains have a radius R, while the pores have a distribution of sizes with a probability $P^*(a)da$ that their radius lies between a and $a + da$. Then the probability of being trapped in a pore is $P_{\text{trap}} = \int_0^R P^*(a)da$. Therefore, the average distance traveled between traps is $\langle x\rangle \sim \kappa^{1/2}/P_{\text{trap}}$, while the typical time between such trappings is $\tau = \langle x\rangle/U \sim \kappa^{1/2}/UP_{\text{trap}}$. Using the high–flow-rate form for the effective diffusion constants $D_\perp \sim D_\parallel \sim U\kappa^{1/2}$, one can estimate the dispersion in these distances as $\langle y^2\rangle \sim \langle x^2\rangle - \langle x\rangle^2 \sim \kappa/P_{\text{trap}}$.

Once it is trapped, the next question is how long such a grain is likely to remain trapped before it is released as a result of thermal fluctuations. This can be estimated from the Fokker–Planck equation $\partial P/\partial t + U\partial P/\partial x = D\partial^2 P/\partial x^2$ for the probability $P(x,t)$ that a grain travels a distance x in a time t against the flow U, where the diffusion constant $D \sim kT/6\pi\eta R$. The equation has an equilibrium solution $P(x) \sim \exp(-Ux/D)$, and, therefore, the probability of moving a pore distance $a \sim \kappa^{1/2}$ is $P(a) \sim \exp(-U\kappa^{1/2}/D)$. If the escape attempt rate is D/κ due to Brownian motion, then the release time can be estimated as $\tau_{\text{release}} \sim \kappa/D \exp(U\kappa^{1/2}/D)$, and, therefore, the average grain velocity is reduced from its nominal value U due to trapping to

$$U_{\text{grain}} \sim U\tau/\tau_{\text{release}} \sim (D/\kappa^{1/2}P_{\text{trap}}) \exp(-U\kappa^{1/2}/D). \tag{70}$$

The swift decrease in grain mobility with increasing Peclet number is very evident in expression (70).

GLOSSARY

Attractor: A term from dynamical systems theory to describe the region of phase space that a trajectory approaches asymptotically. To be an attractor, this region must be stable under small perturbations of the trajectory. The attractor itself could be a point or a limit cycle, or have some more complex topology, such as a fractal set.

Basin of Attraction: The basin of attraction for a given attractor is that region of phase space for which all trajectories beginning in this basin will asymptotically reach the attractor.

Brownian Motion: The noise due to thermal fluctuations. At the microscopic level, this noise manifests itself as random forces acting on the individual molecules with a variance proportional to the temperature, causing the molecules to perform a random walk. At the macroscopic level, the noise is manifested as diffusion.

Chaos: The state of a deterministic dynamical system whose time evolution shows extremely sensitive dependence on initial conditions—a consequence of which is that detailed prediction of the space and time development of the flow is impossible, and statistical methods have to be used to extract average properties of both spatial and temporal fluctuations.

Conformal Mapping: A mapping using an analytic function in the complex plane. Its importance lies in the fact that conformal mapping techniques can be used to solve Laplace's equation in the presence of complex boundary conditions by transforming the boundary into a simpler shape.

Constitutive Relation: The equations of motion derived for hydrodynamic densities in the form of the equations of continuity are all expressed in terms of stress tensors $\sigma_{\alpha\beta}$ and fluxes and, therefore, do not form a closed set of equations. To achieve this eminently desirable result, explicit expressions for these stresses and fluxes, called constitutive relations, are required. Typically, fluxes are assumed to be proportional to thermodynamic forces (Darcy's law gives velocity fields proportional to pressure gradients, Ohm's law gives electric currents proportional to electric-potential gradients, for example). For Newtonian fluids, the phenomenological assumption made is that the stress $\sigma_{\alpha\beta}$ is proportional to the rate-of-strain tensor $\partial_\alpha v_\beta + \partial_\beta v_\alpha$; the constant of proportionality is the viscosity.

Dynamical Systems: Systems described by deterministic simple differential equations, or mappings. Recent developments in the study of the time evolution of nonlinear dynamical systems have led to the realization that even very simple deterministic equations can lead to very complex trajectories that show extreme sensitivity to initial conditions (see **Chaos**).

Fokker–Planck Equation: Stochastic processes, that is, dynamical systems in the presence of noise, will lead to an ensemble of dynamical behaviors, one set of trajectories for each realization of the noise, and, consequently, unique trajectories have to be replaced by a probability distribution for dynamical variables with time. The equations of motion for these probability distributions are known as Fokker–Planck equations.

Fractals: A word coined by Mandelbrot to describe objects with a box-counting dimension D larger than their topological dimension. If a fractal of size L is covered with boxes of size l, and the number of such boxes required to cover the object scales as $N(l) \sim (L/l)^D$, then its fractal dimension is D. In general, such objects are exactly or stochastically scale invariant and often created by some recursive process. This means that there is no length scale associated with the object in question, and it appears self-similar under magnification.

Hamiltonian Dynamics: Dynamical laws describing a special subset of dynamical systems that conserve energy and phase-space volume. They come originally from a generalized form of Newton's law of motion and describe the motion of mechanical systems without friction. Because of phase-space conservation, there cannot be any attractors in a Hamiltonian flow. In general, a system with N degrees of freedom will be described by N coordinates q_α and N conjugate momenta p_α. A function known as a Hamiltonian $H(p_\alpha,q_\alpha)$ can be built from these coordinates and momenta, such that the dynamical system obeys $\partial p_\alpha/\partial t = -\partial H/\partial q_\alpha, \partial q_\alpha/\partial t = \partial H/\partial p_\alpha$.

Hysteresis: When a physical system does not show reversible behavior coming back to its original state as a control parameter is varied in a cyclic manner, the system is said to show hysteresis. Originally, the term was applied to the magnetic behavior of ferromagnets in external magnetic fields.

Langevin Equations: Equations of motion for dynamical systems to which random fluctuations and also dissipative frictional forces have been added. The fluctuating and dissipative terms are not usually independent, but intimately related via the fluctuation-dissipation theorem. Langevin equations are a different language in which to describe the same stochastic phenomena that Fokker–Planck equations do.

Linear Stability Analysis: The analysis of the manner in which infinitesimal perturbations applied to a system grow or decay. As small perturbations always exist, even a single growing mode implies instability. Linear stability analysis can only show that a system is stable to infinitesimal fluctuations; it says nothing about finite perturbations.

Phase Diagram: The behavior of physical systems changes as control variables such as temperature, magnetic field, temperature gradients, flow rates, etc. are altered. Typically, however, a given phase or spatiotemporal pattern changes smoothly and is stable for a region of values of these control parameters, followed by abrupt bifurcations to new phases or flow patterns. Phase diagrams are schematic representations of these various stable regimes and the lines of marginal stability between them.

Phase Space: An abstract space whose coordinates are all the variables needed to describe the temporal behavior of a dynamical system. For example, for a Hamiltonian system with N degrees of freedom, its dimensionality is $2N$. The instantaneous state of the system is then given by a single point in this phase space, and the motion of this point gives the trajectory of the system with time.

Phase Transition: The macroscopic phase of a system can be described by its symmetry and by the existence of a set of macroscopic variables that describe the degree of order (translational, rotational, magnetic, etc.) present in the system. As control parameters change, bifurcations between phases can occur, called phase transitions. These can either be discontinuous, resulting in the appearance of a nonzero new order parameter (a first-order phase transition), or continuous, in which case the new order parameter grows from zero as the system is brought through the transition (a second-order phase transition).

Pinning: In porous media, quenched random impurities can lock the interface at a given position and only release it when surface-tension forces become very large. In certain cases, this local pinning can become so strong and occur at so many points that the interface becomes globally pinned.

Polydisperse: Often used in the context of polymers and granular materials to describe polymer chains or grain sizes that are not all of the same magnitude but follow some distribution.

Power Spectrum: Fluctuating temporal signals can be thought of as consisting of a linear combination of modes fluctuating at different frequencies. The power spectrum $S(f)$ gives the power of fluctuations at each frequency f. Therefore, a single periodic mode would have a power spectrum consisting of a single δ function at that frequency, while the opposite extreme would be represented by white noise with a constant power spectrum.

Quenched Randomness: Any randomness that does not change with time, typically spatial randomness. The opposite is annealed randomness, which, over the time scale of the experiment, samples the complete distribution of random ensembles, as is the case for Brownian noise.

Renormalization Group: The group formed by the set of operators used to integrate out all features below a given length scale in, and therefore to coarse-grain, a physical system. From their influence with length scale on the parameters describing the new coarse-grained system, critical exponents can be found.

Scaling Arguments: A theoretical approach to understanding the spatial or temporal properties of a physical process by combining dimensional analysis with the relevant length or times scales involved. Often, exponents can be derived uniquely by such arguments, which depend for their validity on deep intuition about the physical processes involved.

Self-Affine Surfaces: A self-affine surface is self-similar under magnification, but, in contrast with fractals, this magnification must be anisotropic. Typically, a surface of height $h(\mathbf{x})$ is self-affine if $|h(\mathbf{X} + \mathbf{L}) - h(\mathbf{X})| \sim L^{\alpha}$, and, therefore, a change in magnification by a factor λ in the plane must be supplemented by a factor λ^{α} in the height direction for the surface to remain self-similar. Surfaces which are stochastically self-affine often characterize interfacial growth.

Stress Tensor: Normally written $\sigma_{\alpha\beta}$, it is the force per unit area in direction α applied by the rest of the fluid on a material particle surface whose normal is in direction β (thus, the normal stress is the negative of the pressure exerted by the fluid particle $\sigma_{\alpha\alpha} = -p$, while $\sigma_{\alpha\beta}$ with $\alpha \neq \beta$ is the tangential stress applied to the material particle surface). For a simple Newtonian fluid, the stress tensor must be symmetric, $\sigma_{\beta\alpha} = \sigma_{\alpha\beta}$.

Tensors: Mathematical objects that transform under groups of symmetry operations in the same manner as products of vectors.

Universality: In phase transitions and critical phenomena, the idea that certain quantities—critical exponents, amplitude ratios—are independent of the detailed microscopic interactions present, but depend only on very general properties, such as symmetry of the interactions, and dimensionality.

Works Cited

Adler, P. M., Brenner, H. (1988), *Annu. Rev. Fluid Mech.* **20**, 35–59.

Bak, P., Tang, C., Weisenfeld, K. (1987), *Phys. Rev. Lett.* **59**, 381–384.

Batchelor, G. K. (1967), *An Introduction to Fluid Dynamics*, Cambridge, UK: Cambridge University Press.

Bensimon, D. (1986), *Phys. Rev. A* **33**, 1302–1308.

Bensimon, D., Kadanoff, L. P., Liang, S., Schraiman, B. I., Tang, C. (1986), *Rev. Mod. Phys.* **58**, 977–999.

Brady, R. M., Ball, R. C. (1984), *Nature* **309**, 225–229.

Brochard, F., de Gennes, P. G. (1983), *J. Phys. Lett. (Paris)* **44**, L785–L791.

Brochard, F., de Gennes, P. G. (1984), *J. Phys. Lett. (Paris)* **45**, L597–L602.

Campbell, C. S. (1990), *Annu. Rev. Fluid Mech.* **22**, 57–92.

Chandler, R., Koplik, J., Lenormand, R., Willemsen, J. F. (1982), *J. Fluid Mech.* **119**, 249.

Chen, J. D., Wilkinson, D. (1985), *Phys. Rev. Lett.* **55**, 1892–1895.

Chuoke, R. L., van Meurs, P., van der Poel, C. (1959), *Pet. Trans. AIME* **216**, 188.

Clement, E., Baudet, C., Hulin, J. P. (1985), *J. Phys. Lett. (Paris)* **46**, L1163–L1171.

DeGregoria, A. J., Schwartz, L. W. (1986), *J. Fluid Mech.* **164**, 383.

Edwards, S. F., Wilkinson, D. R. (1982), *Proc. Roy. Soc. London, Ser. A* **381**, 17–31.

Einstein, A. (1906), *Ann. Phys. (Leipzig)* **19**, 289.

Family, F., Vicsek, T. (1985), *J. Phys. A* **18**, L75–L81.

Family, F., Masters, B. R., Platt, D. E. (1989), *Physica D* **38**, 98–103.

Feder, J., Maloy, K. J., Jossang, T. (1990), "Experiments on Diphasic Flows: Front Dynamics in Model Porous Media," in: J. P. Hulin, A. M. Cazabat, E. Guyon, F. Carmona (Eds.), *Hydrodynamics of Dispersed Media*, Amsterdam: North-Holland, pp. 231–248.

Flory, P. J. (1969), *Statistics of Chain Molecules*, New York: Interscience Publishers.

Flory, P. J. (1971), *Principles of Polymer Chemistry*, Ithaca, NY: Cornell University Press.

de Gennes, P. G. (1974), *The Physics of Liquid Crystals*, Oxford, UK: Oxford University Press.

de Gennes, P. G. (1977). *Makromolecules* **9**, 587.

de Gennes, P. G., Guyon, E. (1978), *J. Mech.* **17**, 403.

de Gennes, P. G. (1979), *C.R. Acad. Sci. (Paris)* **289**, 329–331.

de Gennes, P. G. (1984), *Scaling Concepts in Polymer Physics*, Ithaca, NY: Cornell University Press.

de Gennes, P. G. (1985), *Rev. Mod. Phys.* **57**, 827–863.

Ginder, J. M., Elie, L. D. (1992), in R. Tao (Ed.), *Proc. Conf. Electrorheological Fluids*, Singapore: World Scientific, pp. 23–36.

Glazier, J. A. (1989). *Dynamics of Cellular Patterns*, Ph.D. Thesis, The University of Chicago.

Gouyet, J. F., Rosso, M., Clement, E., Baudet, C., Hulin, J. P. (1990), "Invasion Percolation in Model Porous Media under Gravity," in: J. P. Hulin, A. M. Cazabat, E. Guyon, F. Carmona (Eds.), *Hydrodynamics of Dispersed Media*, Amsterdam: North-Holland, pp. 179–192.

Halsey, T. C. (1992), *Science* **258**, 761–766.

Horvath, V. K., Family, F., Vicsek, T. (1991), *J. Phys. A* **24**, L25–L29.

Hwa, T., Kardar, M. (1989), *Phys. Rev. Lett.* **62**, 1813–1816.

Joanny, J. F., de Gennes, P. G. (1984), *C. R. Acad. Sci. (Paris)* **299**, 279.

Joanny, J. F. (1990), "Spreading Kinetics," in: J. P. Hulin, A. M. Cazabat, E. Guyon, F. Carmona (Eds.), *Hydrodynamics of Dispersed Media*, Amsterdam: North-Holland, pp. 17–27.

Kadanoff, L. P., Nagel, S. R., Wu, L., Zhou, S. (1989), *Phys. Rev. A* **39**, 6524–6537.

Kardar, M., Parisi, G., Zhang, Y. (1986), *Phys. Rev. Lett.* **56**, 889–892.

Kim, J. M., Kosterlitz, J. M. (1989), *Phys. Rev. Lett.* **62**, 2289–2292.

Kolmogorov, A. N. (1954), *Dok. Akad. Nauk SSSR* [*Sov. Phys. Dokl.*] **98**, 527–532.

Kraynik, A. M. (1988), *Annu. Rev. Fluid. Mech.* **20**, 325–357.

de Kruif, C. G. (1990), "The Rheology of Colloidal Dispersions in Relation to Relation to their Microstructure," in: J. P. Hulin, A. M. Hulin, A. M. Cazabat, E. Guyon, F. Carmona (Eds.), *Hydrodynamics, of Dispersed Media*, Amsterdam: North-Holland, pp. 79–101.

Lamb, H. (1879), *Hydrodynamics*; republished (1945), New York: Dover.

Lenormand, R., Touboul, E., Zarcone, C. (1988), *J. Fluid Mech.* **189**, 165.

Lenormand, R., Zarcone, C. (1985), *Phys. Rev. Lett.* **54**, 2226–2229.

Lighthill, J. (1978), *Waves in Fluids*, Cambridge, UK: Cambridge University Press.

Maher, J. (1985), *Phys. Rev. Lett.* **54**, 1498–1501.

Marmur, A., Lelah, M. D. (1981), *Chem. Eng. Commun.* 13, 133.

Matshushita, M., Sano, M., Hayakawa, Y., Honjo, H., Sawada, Y. (1984), *Phys. Rev. Lett.* **53**, 286–289.

Medina, E., Hwa, T., Kardar, M., Zhang, Y. (1989), *Phys. Rev. A* **39**, 3053–3075.

Moser, J. (1973), *Stable and Random Motions in Dynamical Systems*, Princeton, NJ: Princeton University Press.

Niemeyer, L., Pietronero, L., Wiesmann, H. J. (1984), *Phys. Rev. Lett.* **52**, 1033–1036.

Nittmann, J., Daccord, G., Stanley, H. E. (1985), *Nature* **314**, 141–144.

Ogawa, S. (1978), in: *Proc. U.S.-Japan Seminar Contin.-Mech. and Stat. App. Mech. Granular Mater.*, Tokyo: Gukujutsu Bunken Fukyukai, pp. 208–217.

Ottino, J. M. (1990), *Annu. Rev. Fluid. Mech.* **22**, 207–253.

Paterson, L. (1984), *Phys. Rev. Lett.* **52**, 1621–1624.

Pittman, E. D. (1983), in: D. L. Johnson, P. N. Sen (Eds.), *Physics and Chemistry of Porous Media*, New York: AIP, pp. 1–19.

Pitts, E. (1980), *J. Fluid Mech.* **97**, 53.

Saffman, P. G., Taylor, G. I. (1958), *Proc. Roy. Soc. London, Ser. A* **245**, 312–329.

Sapoval, B., Rosso, M., Gouyet, J. F. (1985), *J. Phys. Lett. (Paris)* **46**, L149–L156.

Stauffer, D. (1985), *Introduction to Percolation Theory*, London: Taylor & Francis.

Tabeling, P., Libchaber, A. (1986), *Phys. Rev. A* **33**, 794–796.

Taylor, G. I. (1932), *Proc. R. Soc. London, Ser. A* **138**, 41–48.

Taylor, G. I. (1950), *Proc. R. Soc. London, Ser. A* **201**, 192–196.

Taylor, G. I. (1953), *Proc. R. Soc. London, Ser. A* **219**, 186–203.

von Neumann, J. (1952), "Discussion," in: *Metal Interfaces*, Cleveland: American Soc. Metals, pp. 108–110.

Wilkinson, D., Willemsen, J. F. (1983), *J. Phys. A* **16**, 3365–3376.

Wilkinson, D. (1984), *Phys. Rev. A* **30**, 520–531.

Witten, T. A., Jr., Sander, L. M. (1981), *Phys. Rev. Lett.* **47**, 1400–1403.

Further Reading

The Works Cited represent but a minute fraction of the available literature on nonhomogeneous flows and were chosen with three purposes in mind: as a historical paper representing the beginning of a new approach or area in nonhomogeneous flows; because the paper contains a specific piece of information deemed important for the particular areas mentioned in this article; or because the work, usually a book, represents a significant source of information and can be used either as a starting point for further reading or as a database of references.

In this last instance, however, it should be kept mind that research in nonhomogeneous flows is a very rapidly developing subject, and the latest references will only be found in the latest volumes of such journals as *Physical Review A* and *E*, *Physical Review Letters*, and *Journal of Fluid Mechanics*. Review articles on various aspects of nonhomogeneous flows may be found in *Reviews of Modern Physics*, *Annual Reviews of Fluid Mechanics*, and Conference proceedings.

Several books and articles are particularly appropriate to special areas of this article. Batchelor (1967) is a good general reference on fluid mechanics. Pelce, P. (Ed.) (1990), *Dynamics of Curved Fronts*, Perspectives in Physics, Boston: Academic Press, is a good collection of classic papers in interfacial evolution. Lighthill (1978) is probably the best all-round modern text on waves in fluids. Ottino (1990) is a good reference on mixing, and its foundation in Hamiltonian mechanics can be found in Arnol'd, V. I. (1978), *Mathematical Methods of Classical Mechanics*, Berlin: Springer-Verlag, and Moser (1973). de Gennes is a useful reference for many areas mentioned in this article, including liquid crystals (1974), and the dynamics of wetting (1985). Finally, as an example of the application of scaling methods to modern statistical mechanics with which fluid mechanics is becoming ever more tightly bound, it is impossible to find a superior text to de Gennes (1984), where polymer configurations and flows are considered. The classic text on flow through porous media is Bear, J. (1972), *Dynamics of Fluids in Porous Media*, New York: American Elsevier.

NONLINEAR ACOUSTICS

See ACOUSTICS, NONLINEAR

NONLINEAR DYNAMICS

See CHAOTIC PHENOMENA

NONLINEAR OPTICS

See OPTICS, NONLINEAR

NUCLEAR MAGNETIC RESONANCE

See MAGNETIC RESONANCE, NUCLEAR

NUCLEAR MAGNETIC RESONANCE, DETERMINATION OF STRUCTURES BY

See STRUCTURE DETERMINATION BY NUCLEAR MAGNETIC RESONANCE

NONLINEAR SYSTEMS

HENRY D. I. ABARBANEL,* *Department of Physics and Marine Physical Laboratory, Scripps Institution of Oceanography, University of California–San Diego, La Jolla, California, U.S.A.*

INTRODUCTION

The study of the nonlinear behavior of physical systems dates back in some sense to Newton (1687), where the planetary equations of motion under $1/r^2$ gravitational forces were the first example of nonlinear interaction among the degrees of freedom

$$\mathbf{x}_j(t), \frac{d\mathbf{x}_j(t)}{dt}, \tag{1}$$

for the $j = 1, 2, \ldots, N$ planetary bodies. The remarkable fact that the case of two bodies ($N = 2$) with $1/r^2$ forces is exactly integrable gave rise to a worldview that for over two centuries limited inquiry into the fascinating aspects of the nonlinear properties of these equations when $N \geq 3$. The difficulty of the problem was known to many of the great nineteenth-century mechanicists and was chronicled by Whittaker (1937). The first recognition of the insolubility of the problem on Newtonian terms was perhaps that of Poincaré (1892). His work stands clearly as the beginning of the study of the effects of nonlinearity in a qualitative fashion, and the slow recognition that we are unlikely to repeat the

*Institute for Nonlinear Science.

3-527-28133-9/94/$5.00 + .50

wonderful accident of Newton's discovery of integrable dynamics and must instead turn to statistical and qualitative versions of our description of physical systems.

Since the work of Lorenz (1963), it has become clear that dissipative systems are equally as rich as the pure, friction-free equations of Newton and Hamilton. Since dissipation appears in all applications in physics, the properties of these systems are worthy of study on their own.

The physical world is replete with nonlinear systems, and, indeed, the view that only nonlinear systems are of interest to physicists is easy to defend. This article cannot cover all of physics, of course, and so the scope of the article is limited to examples from a few areas of physics and to the simplest examples of nonlinear behavior.

Our plan is to describe in physical terms the origin of the critical unexpected phenomenon appearing in nonlinear physical systems: *nonperiodic evolution of the orbits* in any system with three or more degrees of freedom. This is absolutely the most striking distinction between nonlinear behavior and the familiar linear behavior, which is composed of superpositions of sinusoidal modes or resonances of the source of observations. The origin of this nonperiodic behavior is *instability* in the physical processes arising because there are alternative, more efficient ways to transfer energy through the system than linear conduction or diffusion. The way in which these instabilities are tamed and used by physics is the most amazing part of the tale, and we shall attempt to convey that while presenting some of the tools that have been developed to analyze signals composed of bounded and folded instabilities.

This motion has been called *chaos*, which would seem to imply it is no different from "noise" and unpredictable. These motions are completely deterministic, however, and are slightly predictable. Indeed, chaos is evolution of a physical system that is intermediate between the completely irregular signals we characterize as noise and the completely regular signals we see in superpositions of sine waves. Chaos has a broad Fourier power spectrum and is nonperiodic, just as noise would be, but it is structured in the space of its own state variables, and noise is not (Abarbanel *et al.*, 1993). This structure in state space and the limited predictability of chaos are the keys to utilizing chaotic physical systems in applications.

1. NONLINEAR SYSTEMS IN PHYSICAL PROBLEMS

1.1 Thermal Convection

A paradigm for nonlinear systems has emerged in the Lorenz model, which is based on the partial differential equations for convection in the lower atmosphere when it is considered as a fluid heated from below. One starts with the Navier–Stokes equations for fluid velocity, coupled with the heat equation for the temperature, and makes a three-mode approximation to the motion. The rescaled mode amplitudes satisfy

$$\begin{aligned} \frac{dx(t)}{dt} &= \sigma[y(t) - x(t)], \\ \frac{dy(t)}{dt} &= rx(t) - x(t)y(t) - y(t), \\ \frac{dz(t)}{dt} &= x(t)z(t) - bz(t), \end{aligned} \tag{2}$$

where

$$\sigma = \nu/\kappa$$

is the Prandtl number,

$$r = g\alpha\Delta T h^3/\kappa\nu \tag{3}$$

is a reduced Rayleigh number, and b is a horizontal length scale.

The solutions to this equation are stable and decay to the point $(x, y, z) = (0, 0, 0)$ when $r < 1$. They are stable and decay to one of the points $(\pm\sqrt{b(r-1)}, \pm\sqrt{b(r-1)}, r-1)$ when $1 < r < \sigma(\sigma + b + 3)/(\sigma - b - 1)$; and when r exceeds this value, the orbits of the system become *nonperiodic*. The Fourier power spectrum of an orbit becomes broad and continuous; there are no peaks indicating strong periodic behavior, and the spectrum falls off rapidly with Fourier frequency. For very large r, the orbit again becomes periodic.

The solution to the Lorenz equations with $\sigma = 10$, $r = 45.92$, and $b = 6$ has been widely studied as an example of chaotic motions. In

Fig. 1, we display the time trace of $x(t)$ resulting from the Lorenz equations integrated with sampling time $\tau_s = 0.01$ in dimensionless units. The data displayed are $x(n) = x(t_0 + n\tau_s)$. In Fig. 2, we have the Fourier spectrum of this time series showing its broadband, continuous nature. In Fig. 3 is the phase-space portrait for the Lorenz orbits in $(x(t), y(t), z(t))$ space. The irregularity of the orbits for $x(t)$ results from the projection of the geometric figure here onto the $x(t)$ axis.

The Lorenz equations also arise in several other settings. In the case of a liquid confined to a torus, with the lower part of the torus heated to $T_0 + \Delta T$ and the upper half to T_0, the Lorenz equations with $b = 1$ are a nearly exact representation of the motion. For a laser near the lasing threshold, one again finds the Lorenz equations with the variables representing light intensity, population inversion, and medium polarization. The widespread appearance of the equations and their simple form and rich solution structure have given impetus to their detailed study over the past three decades.

1.2 Nonlinear Circuits

The appearance of chaotic motions that are nonperiodic has been known in nonlinear circuits for over 50 years (VanderPol, 1922). The dynamical origin of the ensuing "noise" was not recognized for some time. Indeed, the main effort for decades in work with electronics and circuits has been to banish this "noise" and emphasize the linearity of circuit response. Since all circuit elements are nonlinear at some level, the opportunity to take advantage of the chaotic behavior of circuits in applications has only emerged with the knowledge we now have of how to characterize the chaos and how to model it without detailed circuit equations, especially for the critical nonlinear elements. An example of the observed signal from a circuit with a hysteretic nonlinearity is shown in Fig. 4. It is broadband spectrally. Using a technique to be introduced later, its portrait in phase space is shown in Fig. 5. Here we see the state-space structure, even though the time trace is complicated and the Fourier spectrum is broad.

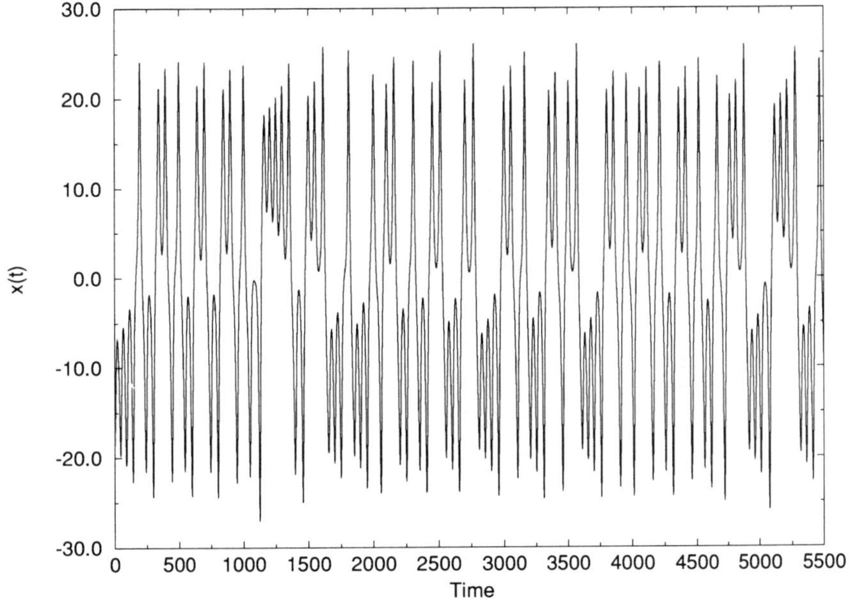

FIG. 1. The time series for one of the components of the Lorenz equations (2). The equations were solved with a Runge–Kutta integrator using a time step of 0.01. $\sigma = 16$, $b = 4$, $r = 45.92$. The irregularity in time is one of the characteristics of chaos.

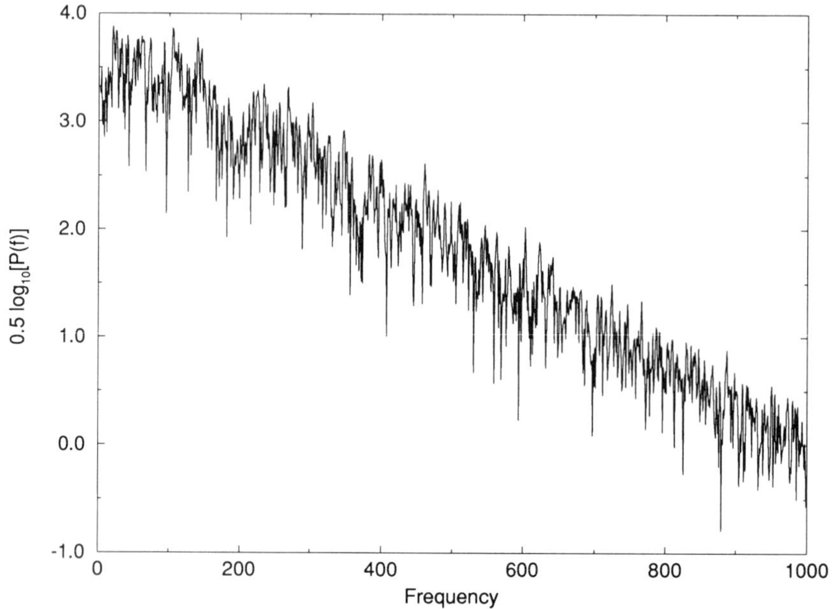

FIG. 2. The Fourier power spectrum of the time series in Fig. 1. The power spectrum is broad and continuous. No sharp peaks indicating periodicity are seen.

The possibility of applications flowing from chaotic operation of such circuits is based on the geometry seen in this figure. "Noise" would be isotropic and unstructured in phase space—here we see structure that can be exploited.

1.3 Planetary Motions

The motions of the bodies in our solar system are, in detail, far from the simple periodic motions characteristic of Keplerian orbits. The planetary motions are well described without dissipation, so that the basic equations are those of Hamiltonian mechanics. The motions become chaotic through sequences of instabilities associated with the resonances in mechanical systems. In the 1950s and 1960s, through the work of Kolmogorov, Arnol'd, and Moser (Arnol'd and Avez, 1968), it became clear that associated with these resonances are orbits that find their way throughout phase space rather than being

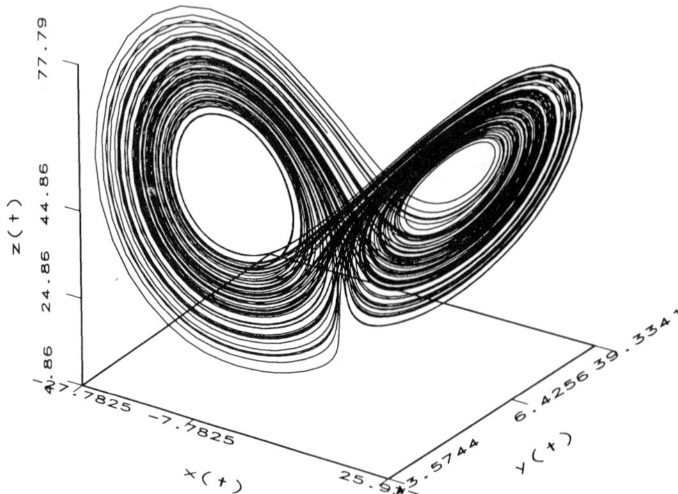

FIG. 3. The Lorenz attractor as seen in three dimensions using as coordinates the solutions ($x(t)$, $y(t)$, $z(t)$) of the Lorenz equations.

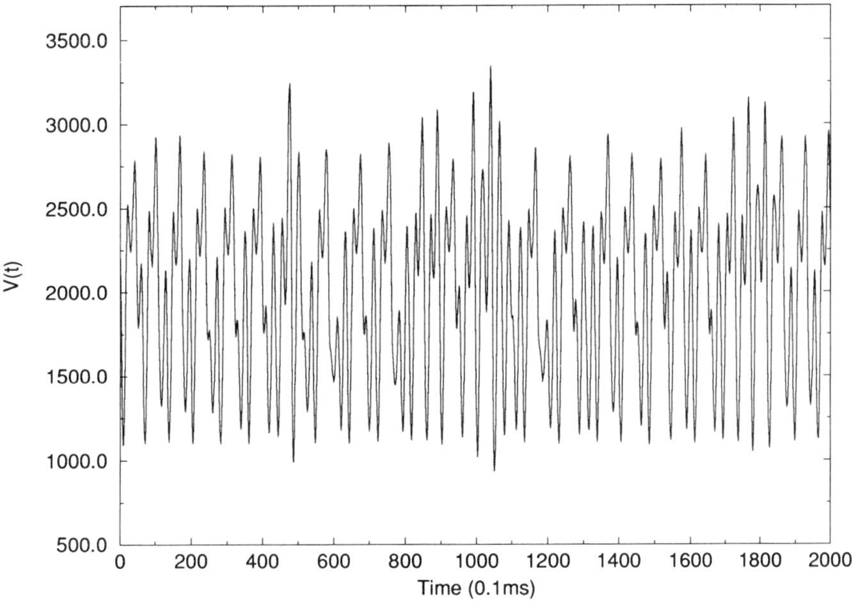

FIG. 4. The time series of a voltage from a nonlinear hysteretic circuit. The sampling time $\tau_s = 0.1$ ms in this experiment. The irregularity of the time series is characteristic of chaotic behavior.

confined to small subportions of phase space, as they would if the motion were regular and integrable. Indeed, in almost every problem of planetary motions in our solar system (Wisdom, 1987), there should be some regions of phase space where chaotic motions are seen. This leads to substantial deviations from regular behavior, and it points out the underlying reason why efforts to solve the three-body problem or to apply classical perturbation theory to many-body dynamics often led to disappointing results or outright failure. Basically, the motions that were described by that perturbation theory were so far from the assumed orbit, to which small corrections were to be found, that power-series expansions had no chance of succeeding. Pioncaré showed that the standard power-series approximations would at best be asymptotic and thus eventually fail, regardless of

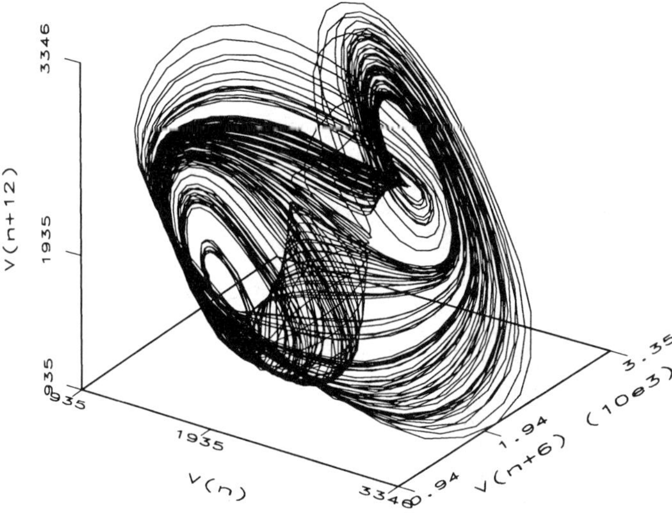

FIG. 5. Using time-delayed values of the voltage in Fig. 4, **(**$V(t)$, $V(t + T)$, $V(t + 2T)$**)**, as coordinates we display the attractor associated with the nonlinear hysteretic circuit. $T = 6\tau_s$ and is chosen by the use of average mutual information. See Fig. 8.

how accurate they might be in the short run. Wisdom (1987) also points out that the time scale for chaotic phenomena to exhibit themselves can often be much longer than tens of millennia and might easily be missed if one were not aware either in simulations or other calculations of their possibility.

1.4 Other Chaotic Systems

Chaos appears in many other physical systems: in plasma physics, in the motion of particles in accelerators, in the description of earthquake motions, and so forth. By restricting ourselves to the examples above, we are trying to avoid a simple listing of almost all of physics in order to allow us to illustrate how one would think about specific issues. Chaos also appears in biological systems and chemical systems, many of which are of substantial interest. We shall have little to say about these systems, as the details of their structure would take us too far afield. It is important to note that the hallmark of the presence of chaotic behavior is that, when a physical system is out of equilibrium and is transporting energy or momentum through it from some source, such as heat in the case of convection, to some sink, then it now appears that as the ratio of energy transported to energy dissipated increases, chaos is the mode preferred to transport these quantities most effectively. Transport through many harmonic modes is unstable to chaos, which is nonperiodic motion.

2. HAMILTONIAN SYSTEMS AND DISSIPATIVE SYSTEMS

Physical systems divide rather naturally into those that are closed and are described by the conservative dynamics of Hamilton and open systems where energy flows in at some scales and flows out at others. The Lorenz system, which has its origin in viscous fluid flows, is of the second kind, while planetary motions and motions associated with large-scale atmospheric and oceanic dynamics are often nearly Hamiltonian. The key distinction from the point of view of analyzing the nonlinear dynamics is that, as they evolve, conservative systems preserve volume in phase or state space, while dissipative systems do not. This apparently small distinction gives rise to a world of interesting phenomena unanticipated by Newtonian mechanics.

2.1 Phase-Space Volume

Conservative systems can be described by Hamilton's equations. In most physical problems, we can find canonical coordinates $\mathbf{q}(t) = (q_1(t), \ldots, q_N(t))$ and $\mathbf{p}(t)$ that, given a scalar function $H(\mathbf{q}, \mathbf{p})$, the Hamiltonian, evolve as

$$\frac{d\mathbf{q}(t)}{dt} = \frac{\partial H(\mathbf{q}, \mathbf{p})}{\partial \mathbf{p}}, \quad \frac{d\mathbf{p}(t)}{dt} = -\frac{\partial H(\mathbf{q}, \mathbf{p})}{\partial \mathbf{q}}. \tag{4}$$

In these canonical coordinates, the volume of phase space is preserved. This follows from the general result that, if a dynamical system with variables $\mathbf{x}(t) = (x_1(t), \ldots, x_M(t))$ satisfies

$$\frac{d\mathbf{x}(t)}{dt} = \mathbf{F}(\mathbf{x}(t)), \tag{5}$$

then a volume $V(t)$ evolves as

$$\frac{dV(t)}{dt} = \int_{V(0)} d^M x \nabla_{\mathbf{x}} \cdot \mathbf{F}(\mathbf{x}(t)). \tag{6}$$

For Hamiltonian mechanics,

$$\mathbf{F}(\mathbf{q}, \mathbf{p}) = \left[\frac{\partial H}{\partial \mathbf{p}}, -\frac{\partial H}{\partial \mathbf{q}}\right] \tag{7}$$

and

$$\nabla_{\mathbf{x}} \cdot \mathbf{F}(\mathbf{x}) = 0. \tag{8}$$

For the Lorenz system given before, $\nabla_{\mathbf{x}} \cdot \mathbf{F}(x, y, z) = -(\sigma + b + 1) < 0$, and so volumes in phase space for that dynamics obey $V(t) = V(t = 0) \exp[-(\sigma + b + 1)t]$, which goes to zero rapidly as time unfolds. Volumes in dissipative system shrink to zero with time, so that any collection of initial conditions will evolve to a set with dimension less than that of the state space. In a Hamiltonian system, the volume is preserved in canonical coordinates, so that any set of initial conditions in dimension $2N$ will remain of dimension $2N$, however twisted the individual orbits become.

This would be a fact of no special importance, except that Hamiltonian systems are rare in physics. Indeed, they are not generic, since even the slightest dissipation of energy from the spatial and temporal scales describing some phenomenon leads to the decrease in phase-space volume associated with those scales. In viscous fluid flow, macroscopic motions dissipate their energy into microscopic motions, resulting in no motion on macroscopic scales unless energy from outside the system is available. In the Lorenz system, energy is input from the temperature difference and lost to viscosity and thermal dissipation. The balance between these effects has remarkable consequences.

When phase-space volumes shrink, the points on any orbit evolve to a set with dimension less than that of the overall phase space in which the equations lie. For example, suppose the phase space is one dimensional, with a dynamical variable $x(t)$ satisfying

$$\frac{dx(t)}{dt} = -\nu x(t), \quad \nu > 0. \tag{9}$$

Then the solution $x(t) = x(0)e_{-\nu t}$ rapidly goes to the zero-dimensional limit $x = 0$, regardless of $x(0)$. Here $\nabla_{\mathbf{x}} \cdot \mathbf{F}(\mathbf{x}) = -\nu$. The dimension of the solution after a long time is less than the dimension of the space in which the solution wanders.

2.2 Attractors

The shrinking of phase-space volume leads to the notion of an *attractor*, which is the set of points in the phase space to which solutions go after transients have died out. The part of phase space from which all points go to the attractor is called the *basin of attraction*. In the simple example $\dot{x}(t) = -\nu x(t)$, the attractor is the origin, and the basin of attraction is the whole real line. A system can have many attractors with complex boundaries between their basins of attraction. The attractors can be as simple as a point, which has dimension zero, or as interesting as others we now discuss. A Hamiltonian system has no attractor, but phase-space points wander all over the surface of constant $H(\mathbf{q}, \mathbf{p})$, since $H(\mathbf{q}(t), \mathbf{p}(t)) = 0$. In fact, Gibbsian statistical mechanics starts with the hypothesis that this surface is uniformly populated and with this assumption introduces the microcanonical ensemble. Dissipative systems do not follow this notion.

Physically, a dissipative system, when driven weakly, will move toward a state of no motion. So the attractor would just be a point, which has dimension zero, regardless of the dimension of the overall space. When the driving or energy input is increased, instabilities arise, and the geometry of the motion changes and becomes richer. Among the first things to happen is that a periodic orbit may emerge from the instability.

2.3 Regular and Periodic Behavior

Evolution to a point attractor can become unstable when the driving forces are increased. This is seen in the simple example of motion in a plane with polar coordinates (r, θ):

$$\frac{dr(t)}{dt} = ar(t) - r^3(t),$$
$$\frac{d\theta(t)}{dt} = 1. \tag{10}$$

For $a < 0$, all points in the plane move to the origin, which is the attractor of dimension zero. For $a > 0$, all points in the plane move to and then around the circle of radius $\sqrt{a}$. The stable motion for long times, namely the attractor, is periodic with frequency unity, and the attractor has dimension one. In each case, the basin of attraction is the whole plane. This change in the structure and topology of time asymptotic orbits is called a *bifurcation*.

Seen in terms of the cartesian coordinates $x = r \cos \theta$, $y = r \sin \theta$, the fixed point $(0, 0)$ for the $\dot{x}$, $\dot{y}$ dynamics is stable for $a < 0$. Near $(0, 0)$, small deviations satisfy

$$\frac{dx(t)}{dt} = ax(t) - y(t),$$
$$\frac{dy(t)}{dt} = ay(t) + x(t), \tag{11}$$

which has solutions proportional to $\exp[\lambda t]$ with $\lambda = a + i$. So two complex conjugate eigenvalues of the linearized problem cross the imaginary axis in the λ plane when the fixed point becomes unstable. This Hopf–Andronov bifurcation or instability is generic for sets of differential equations describing physical phenomena. When we place the convection problem described in Sec. 1.1 in a rotating frame of reference, the first instability encountered as r increases can be a periodic orbit, which is said to bifurcate from the fixed point describing pure conduction.

The possibility of transition from a fixed long-term behavior to periodic orbits has been understood for a long time. In fact, a popular theory about turbulence held that it was composed of a sequence of Hopf–Andronov instabilities resulting in more and more complex time traces associated with many, many new frequencies in a fluid. The attractor would progress from a circle to a torus to tori in higher dimensions. It would always be regular, and its dimension would be integer. Experimentally, this is now known not to be correct (Gollub and Benson, 1980).

2.4 Strange Attractors and Nonperiodic Behavior

It was realized somewhat over 20 years ago (Ruelle and Takens, 1971) that when one has a dynamical evolution which shrinks phase-space volumes, the attractor need not be a regular geometrical figure with an integer dimension. These arguments, connected with observations of Smale (1967), suggested that, in fact, regular motions were unlikely, namely were not generic, for realistic dynamical evolution and in physical systems should rarely be observed. The catchy name of *"strange" attractors* was given to the geometric figures that would appear in these circumstances.

Several features of strange attractor motion soon became apparent. First, the dimension of the attractor was typically not an integer. Second, the motion on the attractor was nonperiodic, as observed by Lorenz in 1963. Third, the attractors could be compact in phase space, even while all motions were unstable. Fourth, the orbits on the attractor were "sensitive to initial conditions," meaning that two orbits very nearby would separate rapidly, exponentially fast actually, while moving about on this compact object.

Amazingly, this set of attributes, so divorced from Newtonian expectations, has been verified over and over again in experimental observations. Some of the evidence will be touched on as we proceed.

2.5 Unstable Periodic Orbits

One may wonder what happened to the periodic orbits that were emerging from the instabilities as the forcing parameters of a system were increased. These orbits remain in the phase space, but they are unstable. This means that any orbit from an arbitrary initial condition will wander near an unstable periodic orbit and then wander away to another and then to another forever. Since the Fourier spectrum of orbits on strange attractors is broad and may have no sharp peaks, unstable periodic orbits of all frequencies can be contained within the motion on a strange attractor. The collection of all of these unstable periodic orbits may be used to describe the attractor and may be viewed as some kind of nonlinear basis set for all motions. In any case, it suggests that many applications that are based on *stable* periodic motions could well be realized even in the presence of chaos using the periodic but unstable motions that abound in them. Below, we shall see that the presence of these unstable periodic orbits allows one to "control" chaos while exploiting the benefits of the broad regions of phase space visited by orbits on a strange attractor.

In Plate 1, we display the unstable periodic orbits up to period nine that lie within the strange attractor generated by the map of the plane to itself $(x(n), y(n)) \rightarrow (x(n+1), y(n+1))$,

$$\begin{aligned} x(n+1) &= 1.4 + 0.3y(n) - x(n)^2, \\ y(n+1) &= x(n), \end{aligned} \tag{12}$$

which is due to Hénon (1976). The plethora of unstable orbits is rather clear from this Plate. The unstable periodic orbits are detected within the strange attractor by placing small circles about each of the points generated by iterating the map and asking at how many steps, if at all, the orbit returns to the small circle. If it does return as the circle is made small, this locates the unstable periodic orbit.

3. MAPS AND FLOWS

All nonlinear systems depend on time in a continuous fashion. This corresponds to our daily experience, and we describe the movement of a system in time with a multivariate state space vector $\mathbf{x}(t) = (x_1(t), x_2(t), \ldots, x_d(t))$. When we actually make observations, we see the state of the system only at discrete times that are multiples of some sampling time τ_s characteristic of our measurement instruments. This means we see only $\mathbf{x}(n) = \mathbf{x}(t_0 + n\tau_s)$, that is, discrete samples of the state of the system. The dynamics associated with this changes from differential equations to *maps* or nonlinear systems in discrete time. So the continuous time dynamics, known as a *flow*,

$$\frac{d\mathbf{x}(t)}{dt} = \mathbf{F}(\mathbf{x}(t)), \qquad (13)$$

becomes

$$\mathbf{x}(n+1) = \mathbf{f}(\mathbf{x}(n)). \qquad (14)$$

If τ_s is short, then $\mathbf{f}(\mathbf{x}) \approx \mathbf{x} + \tau_s\mathbf{F}(\mathbf{x})$.

Maps also arise when we view the system as it passes through some designated subspace of the whole phase space. For example, if we have a nonlinear system in three dimensions $\mathbf{x}(t) = (x(t), y(t), z(t))$ and view the state every time $z(t_n) = 0$, then we produce a two-dimensional map $(x(t_n), y(t_n)) \to (x(t_{n+1}), y(t_{n+1}))$ that replaces the dynamics. The qualitative features of the discrete map are often similar enough to that of the original flow that they may be used as diagnostics for the flow itself.

3.1 Stability of Maps or Flows

Maps and flows typically depend on parameters describing physical properties of the system, such as viscosities, conductivities, length scales, etc., and aspects of external forcings, such as temperature, external magnetic or electric field strengths, and so forth. As we vary these parameters, the stability of solutions to the continuous or discrete dynamics will change. For example, if we have a time-independent solution $\mathbf{x}_0$ of some discrete dynamics $\mathbf{x}_0 = \mathbf{f}(x_0, \mu)$, where μ is a set of parameters, then we can examine the evolution of this system in the neighborhood of $\mathbf{x}_0$ by setting $\mathbf{x}(n) = \mathbf{x}_0 + \delta\mathbf{x}(n)$, where, to linear order in $\delta\mathbf{x}(n)$, we have in vector notation

$$\delta\mathbf{x}(n+1) = \mathbf{Df}(\mathbf{x}_0)\cdot\delta\mathbf{x}(n), \qquad (15)$$

and

$$Df_{ab}(\mathbf{x}) = \frac{\partial f_a(\mathbf{x})}{\partial x_b}, \qquad (16)$$

is the Jacobian matrix for this dynamics. The linear stability of the solution $\mathbf{x}(n) = \mathbf{x}_0$ rests on the eigenvalues of $\mathbf{Df}(\mathbf{x}_0)$. If they all lie within the unit circle, then

$$\delta\mathbf{x}(n+L+1) = \underbrace{\mathbf{Df}(\mathbf{x}_0)\cdots\mathbf{Df}(\mathbf{x}_0)}_{L \text{ times}}\cdot\delta\mathbf{x}(n) \qquad (17)$$

decays rapidly to zero, since $\delta\mathbf{x}(n + L + 1)$ behaves as $(\text{eigenvalue})^L\delta\mathbf{x}(n)$. If any eigenvalues lie outside the unit circle, the solution $\mathbf{x}_0$ is unstable. These eigenvalues move with the parameter settings μ, and, as they move out of the unit circle, the fixed point $\mathbf{x}_0$ loses stability, and a new solution dictated by the nonlinear properties of $\mathbf{f}(x)$ emerges. This kind of change in the solution to a dynamical evolution, discrete or continuous, is called a *bifurcation* and usually results in a change in the geometric structure of the solution.

The physical reason associated with this change in solution type is connected with the passage of energy through the nonlinear system. As the system is forced more and more vigorously, the optimum method for passage of this energy through the system changes. One can see this quite clearly in the free-energy arguments given by Chandrasekhar (1961) connected with the transition from heat conduction to fluid convection in a fluid heated from below. The dimensionless ratios of energy input to energy loss or dissipation characterize the allowed states of the nonlinear system, and as they are varied, the system will undergo bifurcations from one form of solution to another.

The various kinds of allowed bifurcations are well studied (Drazin, 1993). Many have been seen in physical systems. The simplest bifurcations are associated with changes in a single parameter, such as a Reynolds number or a Rayleigh number. The patterns of changes in solutions are well classified for this case. When more than one parameter is altered at a time, the class of allowed solutions is much richer (Guckenheimer and Holmes, 1986).

3.2 One-Dimensional Maps: Universality

A class of bifurcations that has been studied in detail is that of smooth one-dimensional maps $x(n + 1) = f(x(n))$, where $f(x)$ is differentiable. The paradigm for these maps is the logistic map $f(x) = rx(1 - x)$, which undergoes a sequence of bifurcations that results in periodic solutions with twice the period of the previous solution. This period-doubling bifurcation sequence (Collet and Eckmann, 1980) has *universality* properties that often appear in physical experiments. The fixed point at $x_0 = 1 - 1/r$ remains stable until $r = 3$; then two solutions (period doubling) emerge. At $r \approx 3.449$, this becomes unstable to four solutions, then eight, etc. If the lth solution appears at r_l, then

$$r_l \rightarrow K - \text{const.} \times \delta^{-l}, \tag{18}$$

where $K \approx 3.5700$, and $\delta = 4.6692 \cdots$ is a new constant appearing for all $f(x)$ with smooth maximum. The other constants vary with the map. Another universal constant is associated with the rescaling properties of the solutions for the logistic map. The universal behavior translates into expected strengths for the Fourier modes of the period-doubled solutions, and these are confirmed by experiments where the dynamics appears to lie nearly on a one-dimensional subspace of the whole state space.

The universality properties of one-dimensional maps stand out as a special feature of the analysis of nonlinear systems. In most realistic systems, a sequence of period doubling may occur in some part of parameter space, but many other sequences of bifurcation or instability are also present. If one sees a period-doubling sequence in observations, then we expect to see both universality properties of period doubling:

1. the exponential decrease in the value of the bifurcation parameter between bifurcations; and
2. a specific order in which various periodic solutions occur (Devaney, 1989).

These are then useful for further analysis.

The sequence of bifurcations for the logistic map $x \rightarrow rx(1 - x)$ is shown in Fig. 6. The final 200 points after 450 iterations of the map

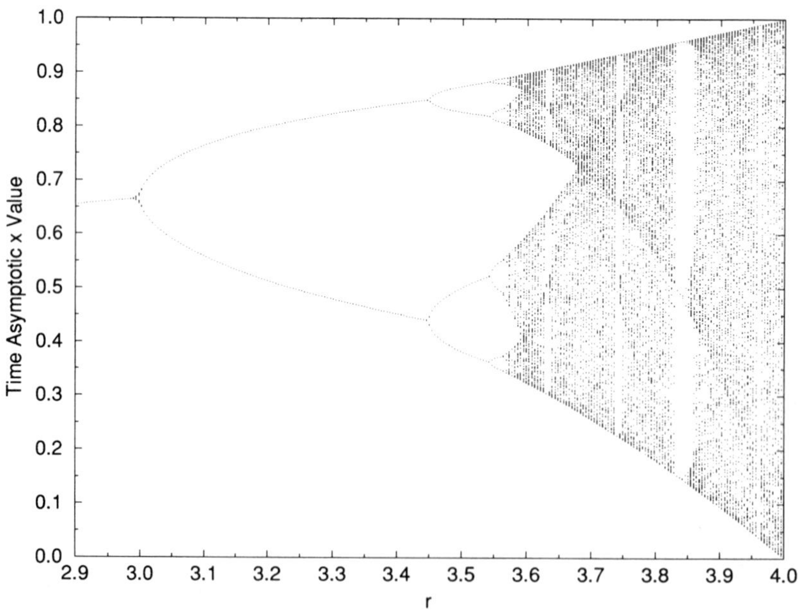

FIG. 6. Bifurcation diagram for the logistic map $x \rightarrow rx(1 - x)$ for $2.9 \leq r \leq 4$. The final 200 points from iterating the map for 450 steps are displayed at each value of r. $x(0) = 0.54$ in each case. The apparent "gap" near $r \approx 3.85$ is a region of period-three orbits, each part of which then undergoes a sequence of period-doubling bifurcations. Indeed, periodic orbits and chaotic regions are thoroughly mixed in this diagram. With only 450 points at each value of r, the full structure is not revealed.

starting from $x_0 = 0.54$ are displayed. The structure in the region $r \geq 3$ is clear.

4. DIMENSIONS OF STRANGE ATTRACTORS

A solution of a nonlinear system describes a sequence of points $\mathbf{x}(n)$ in a d-dimensional state space. One of the clearest properties of a set of points is its dimension. A surprise about chaotic dynamics of nonlinear systems is that the set of points it visits may have a dimension d_A that is not integer. This sounds peculiar or strange at first, thus the name strange attractor. Recall that the orbits are nonperiodic and cannot close, yet they must lie on a subspace of the whole state space because the dynamics dissipates energy, and so the orbit must visit an infinite number of distinct points in the state space. Under such circumstances, the dimension of that infinite set of points may be unexpected.

The idea of sets of points with noninteger dimension is really quite old (Cantor, 1883), but until experiments in physics revealed fractional dimensional sets, called *fractals*, in observations (Brandstater and Swinney, 1987), the idea remained a mathematical curiosity. Its appearance in experiments makes it a compelling notion for physics, and the idea of dimension of the long-time orbit of a dynamical system turns out to be quite useful for classifying the source of signals.

4.1 Box Counting

The crudest notion of a dimension of a set is associated with counting the number of boxes of linear side r required to capture the whole set when r becomes small. This number for any regular set (a point with $d = 0$, a line segment with $d = 1$, an area with $d = 2$, etc.) behaves as

$$N(r) \approx \text{const.}/r^d, \tag{19}$$

and, taking logarithms to eliminate the constant, we are led to expect the dimension of a set of points to be

$$d_A = \lim_{r\,\text{small}} \frac{\log N(r)}{\log(1/r)}. \tag{20}$$

It is not hard to find sets of an infinite number of points having dimension d_A that is not an integer. Take the set formed by successively removing from a unit line segment the middle $\frac{1}{5}$ of what remains. At the first stage, we remove the piece from $\frac{2}{5}$ to $\frac{3}{5}$. Then we remove the pieces from $(\frac{2}{5})^2$ to $\frac{6}{25}$ and the middle fifth of the piece from $\frac{3}{5}$ to 1. Continuing in this manner, we see that, at the mth stage, we have $N(r) = 2^m$ and $r = (\frac{2}{5})^m$; this leads to

$$d_A = \frac{\log 2}{\log(\frac{5}{2})} \approx 0.7564\cdots. \tag{21}$$

This *Cantor set* (Cantor, 1883) is quite typical of the structures one encounters in the results of numerical calculations of nonlinear evolution rules and from experiment.

For example, if one takes data from the Lorenz attractor for $(x(t), y(t), z(t))$ and evaluates the number of boxes required to cover the data points as the size of the boxes shrinks, then $d_A \approx 2.06$ emerges. This tells us that the geometric structure in Fig. 3 is nearly two dimensional, but the excursions of the orbit out of the plane sparsely fill the third dimension, and a cross section of the orbit would reveal a Cantor set of dimension about 0.06. The importance of the fractal dimension d_A lies in its being one of the characteristic invariants of the strange attractor. The orbit of the system as it moves around the strange attractor is enormously sensitive to small perturbations. No two orbits are identical. A quantity such as d_A is a statistical property of the density of points on the attractor $\rho(x)$ that characterizes its geometry. To explore this further, we turn to more general concepts of dimension.

4.2 Generalized Dimensions; D_q

The density of points $\rho(\mathbf{x})$ on an attractor is inhomogeneous in a typical case arising in physics. We can characterize this inhomogeneity in the distribution with appropriately chosen moments. If the orbit visits the $\mathbf{x}(n) = \mathbf{x}(t_0 + n\tau_s)$, then, absent any error in measurement or computation, the density is a sequence of delta functions at the location of the orbit points. This means that a natural density on the attractor for M data points will be

$$\rho(\mathbf{x}) = \frac{1}{M}\sum_{k=1}^{M} \delta^d(\mathbf{x} - \mathbf{x}(k)). \tag{22}$$

This is too singular to deal with, but we expect that noise or any errors will coarse-grain our ability to measure these points, and, thus, we can associate with the small measurement volume V_j near a point $\mathbf{x}(j)$ the normalized frequency of visit to that volume

$$p(\mathbf{x}(j), r) = \int_{V_j} d^d x \rho(\mathbf{x})$$
$$= \frac{1}{M}\sum_{k=1}^{M} \theta(r - |\mathbf{x}(k) - \mathbf{x}(j)|), \tag{23}$$

where the step function $\theta(w) = 0$, $w < 0$, and $\theta(w) = 1$, $w > 0$. The length r is a typical size of the volume V_j. This probability or frequency is no longer singular. As M becomes large and r becomes small, it captures the essential idea we have of number of points in a small volume on the attractor. The moments of this quantity are useful for defining generalized dimensions. We consider (Grassberger, 1983)

$$C_q(r) = \sum_{j=1}^{M} p(\mathbf{x}(j), r)^q. \tag{24}$$

For $q = 0$, this gives just the number of boxes needed to cover the set of orbit points and scales with r as $r^{-D_0} = r^{-d_A}$. This defines the fractal dimension d_A, which we have identified with the "box-counting dimension" D_0. For $q \neq 0$, we expect this to scale as

$$C_q(r) \approx r^{(q-1)D_q}, \quad r \text{ small}, \tag{25}$$

defining the generalized dimension D_q.

In this discussion, we encounter quite a few "dimensions." The integer dimension of the phase space of a system we have been calling d. The fractal dimension of an attractor that we observe in the phase space we have called d_A and identified with the power D_0, also fractional, entering in the small-r behavior of $C_0(r)$. The generalized dimension D_q is also fractional, as a rule.

Since $p(\mathbf{x}(j), r) \le 1$, $p(\mathbf{x}(j), r)^{q_1} \ge p(\mathbf{x}(j), r)^{q_2}$ if $q_2 \ge q_1$, implying $D_{q_1} \ge D_{q_2}$. This means we have a whole spectrum of dimensions, which are all equal in the rare case when the attractor is uniformly populated. The various D_q can be calculated using the "correlation integrals" $C_q(r)$, and one finds that the D_q vary smoothly in the neighborhood of $q = 0$. Here $q = 1$ is a special case worth noting. Since

$$D_q = \lim_{r\,\text{small}} \frac{1}{q-1}\frac{\log C_q(r)}{\log r}, \tag{26}$$

we need to take the limit to find

$$C_1(r) = \sum_{j=1}^{M} p(\mathbf{x}(j), r)\log[p(\mathbf{x}(j), r)],$$

and

$$D_1 = \lim_{r\,\text{small}} \frac{\log C_1(r)}{\log r}. \tag{27}$$

This defines the information dimension.

In a typical experiment involving fluid flow between rotating cylinders, it is possible to evaluate the $C_q(r)$ using an "embedding" method to be described below. Values for D_1 and D_2 have been extracted from the observations. They vary with the relative rotation rates of the cylinders and increase with this rate, as one might expect from their interpretation as a sense of the number of degrees of freedom excited in the fluid. The observed values (Brandstater and Swinney, 1987) range smoothly between nearly 2 and about 4.5 as the speed of the rotating cylinders is increased by about a factor 2. Because of the accuracy of real data from these experiments, the difference between D_1 and D_2 is not distinguishable.

All of these dimensions are of importance because they serve to classify the attractor and thus are useful in experiment for identifying the source of a chaotic signal, so that one can recognize it in any future measurement. The point here is that any given orbit of the system is unstable to small perturbations of the orbit. Such perturbations, as we shall see in the next section, grow exponentially rapidly, so that the original orbit and any orbit near it separate in phase space while both are moving about on the attractor. They move quite differently on the attractor, so that they rapidly become uncorrelated. This means that every measurement of an orbit will yield a different answer in detail, and orbits from experiment and from models cannot be compared. One must compare items not sensitive to initial conditions or perturbations. The D_q are among such quantities.

4.3 Statistics on Strange Attractors

Since any given orbit $\mathbf{x}(t)$ or $\mathbf{x}(n)$ is extremely sensitive to perturbations or changes in initial conditions, we cannot compare the computed orbits from a model to any that are observed. We must have some basis on which to test models and to compare one experiment with another, and this basis is given by the statistical properties of strange attractors (Abarbanel *et al.*, 1993; Eckmann and Ruelle, 1985).

The statistics are based on the distribution of points $\rho(\mathbf{x})$ we saw above. Any function on the state space $g(\mathbf{x})$ allows us to define a quantity, its mean over the attractor weighted with the density of points, that is unchanged under changes in the initial condition. The mean of $g(\mathbf{x})$ is

$$\begin{aligned}\bar{g} &= \int d^d x g(\mathbf{x})\rho(\mathbf{x}) \\ &= \frac{1}{M}\sum_{k=1}^{M} g(\mathbf{x}(k)). \end{aligned} \tag{28}$$

If we now estimate this using $g(\mathbf{f}(\mathbf{x}))$, where we assume $\mathbf{x}(k+1) = \mathbf{f}(\mathbf{x}(k))$, that is, we start one step further along on the orbit, we find that $\bar{g}$ is unaltered to order $1/M$. So, as M becomes large, the mean taken with $\rho(\mathbf{x})$ as the probability density is an invariant of the dynamics. For the correlation functions $C_q(r)$ that define the generalized fractal dimensions D_q, we must take

$$\begin{aligned} g(\mathbf{x}) &= \frac{1}{M}\sum_{n=1}^{M} \theta(r - |\mathbf{x} - \mathbf{x}(n)|), \\ C_q(r) &= \int d^d x \rho(\mathbf{x}) \frac{g(\mathbf{x})^{(q-1)}}{q-1}. \end{aligned} \tag{29}$$

This means that the D_q are invariants of the motion and can be used to classify the source of an observation. This is very useful, as we can now compare the D_q, or, actually, the entire function $C_q(r)$, as seen in observations with other observations to establish if we are seeing a known system, or with computations on a model as part of the verification process for the model itself.

5. LYAPUNOV EXPONENTS

The D_q and associated moments of other physically interesting quantities are properties of the static, time-evolved distribution of points on the attractor, as represented by $\rho(\mathbf{x})$. There are other invariants associated with the dynamical evolution on the attractor, and they give a sense of the instability associated with the formation of the attractor, as well as of the ability to make predictive models of the nonlinear system.

The quantities that govern the stability of an observed orbit $\mathbf{x}(n)$ are called *Lyapunov exponents*. They are defined by looking at the linearized evolution of perturbations to a solution to $\mathbf{x}(n+1) = \mathbf{f}(\mathbf{x}(n))$, which we have seen obeys

$$\delta\mathbf{x}(n+1) = \mathbf{Df}(\mathbf{x}(n)) \cdot \delta\mathbf{x}(n), \tag{30}$$

so that

$$\begin{aligned}\delta\mathbf{x}(N+L+1) &= \mathbf{Df}(\mathbf{x}(L+n)) \\ &\quad \cdot \mathbf{Df}(\mathbf{x}(N+L-1)) \\ &\quad \cdots \mathbf{Df}(\mathbf{x}(n)) \cdot \delta\mathbf{x}(n) \\ &= \mathbf{Df}^L(\mathbf{x}(n)) \cdot \delta\mathbf{x}(n). \end{aligned} \tag{31}$$

The eigenvalues of the L times composed Jacobian matrices $\mathbf{Df}(\mathbf{x})$, which we have called $\mathbf{Df}^L(\mathbf{x})$, determine the stability of the observed orbit under small perturbations.

5.1 Oseledec's Multiplicative Ergodic Theorem

In 1968, Oseledec (Oseledec, 1968) proved the remarkable result that, if we consider the orthogonal matrix (T denotes transpose)

$$\mathbf{OSL}(\mathbf{x}, L) = \{\mathbf{Df}^L(\mathbf{x})^T \cdot \mathbf{Df}^L(\mathbf{x})\}^{1/2L}, \tag{32}$$

its eigenvalues as $L \to \infty$ are independent of $\mathbf{x}$ for $\mathbf{x}$ within the basin of attraction of any attractor. These eigenvalues, written as $\exp[\lambda_1]$, $\exp[\lambda_2]$, . . ., $\exp[\lambda_d]$, define the Lyapunov exponents λ_a, which we order as $\lambda_1 \geq \lambda_2 \geq \cdots$. The independence of $\mathbf{x}$ on the λ_a is known as the multiplicative ergodic theorem. The remainder of the theorem proves that the eigendirections associated with the $\exp[\lambda_a]$ exist. It is very important that associated with the linearization of the nonlinear dynamics is a set of quantities that are independent of starting location within a basin of attraction. This means that each of the λ_a serves as a characteristic of the attractor itself and is not associated with any particular orbit. Further, it is easy to show that, under any smooth change of coordinates $\mathbf{x} \rightarrow H(\mathbf{x})$ where derivatives of $H(\mathbf{x})$ exist in phase space near the attractor, the values of the λ_a are unchanged. To be more precise, when one makes a smooth change of variables, the limit $L \rightarrow \infty$ acquires a piece that is of order $1/L$ associated with the change of coordinates. This vanishes as $L \rightarrow \infty$, of course, but changes the interpretation of local properties on the attractor.

If the orbit satisfies a differential equation, that is, $\mathbf{x}(t)$ comes from a flow, then at least one of the λ_a is identically zero. This is seen by considering a perturbation $\delta\mathbf{x}(n)$ that is precisely along the direction of $\mathbf{f}(\mathbf{x}(t)) = \dot{\mathbf{x}}(t)$. If the system producing the signal is Hamiltonian, then the symmetry of Hamiltonian dynamics guarantees that the sum of the exponents is zero (nonshrinking volumes), and the exponents come in pairs (λ, $-\lambda$). This tells us that, when we have observed data from which we evaluate Lyapunov exponents, we can determine without knowing the equations of motion if the system is a flow or a map and whether it is Hamiltonian or not.

5.2 Positive Exponents and Strange Attractors

If one or more of the λ_a is positive, the perturbation is unstable. It does not grow without bound, however, but is folded back into the region of the attractor by the dissipation associated with the dynamics or, in the case of a Hamiltonian system, moves out of a region of resonance and wanders through phase space. What is critical is that, since the error $\delta\mathbf{x}(n)$ grows exponentially rapidly and soon reaches the size of the attractor itself, predictability is lost on a time scale of order $1/\lambda_1$.

Line segments in the d-dimensional phase space grow as $\exp[\lambda_1]$; areas grow as $\exp[\lambda_1 + \lambda_2]$; three-dimensional volumes as $\exp[\lambda_1 + \lambda_2 + \lambda_3]$; d-dimensional volumes, as $\exp[\lambda_1 + \lambda_2 + \cdots + \lambda_d]$. Since we know that volumes in the whole space must shrink, we must have $\lambda_1 + \lambda_2 + \cdots \lambda_d < 0$. This means that, somewhere between line segments and d-dimensional volumes, there is a sense of a dimension associated with a subvolume of the whole space where the subvolume neither grows nor shrinks. This allows one to define a *Lyapunov dimension* D_L (Kaplan and Yorke, 1979), which should be essentially the dimension of the attractor. The Lyapunov dimension is defined by considering that integer K where $\Sigma_{a=1}^{K} \lambda_a > 0$ but $\Sigma_{a=1}^{K+1} \lambda_a < 0$; then

$$D_L = K + \sum_{a=1}^{K} \lambda_a/|\lambda_{K+1}|. \tag{33}$$

The term D_L is conjectured to be equal to D_1, the information dimension.

5.3 Global and Local Exponents

The Lyapunov exponents λ_a tell us what happens to a perturbation a long time after it is made. More to the point in considering prediction, we need to know what happens a finite time after a perturbation is made, since we can only predict for a short time in any case. The answer to this question is given by the local Lyapunov exponents $\lambda_a(\mathbf{x}, L)$, which are the logarithms of the eigenvalues of the matrix $\mathbf{OSL}(\mathbf{x}, L)$ given above. The $\lambda_a(\mathbf{x}, L)$ satisfy $\lambda_a(\mathbf{x}, L) \rightarrow \lambda_a$ for large L, but the variations of the $\lambda_a(\mathbf{x}, L)$ as we visit parts of the attractor are quite large for small L. The variance of the values of $\lambda_a(\mathbf{x}, L)$ about the mean falls to zero as $1/L^p$ where $\frac{1}{2} \leq p \leq 1$. Local exponents can be evaluated from observed data, as we will discuss below.

6. INFORMATION THEORY AND STRANGE ATTRACTORS

The presence of positive Lyapunov exponents but finite-size attractors distinguishes nonlinear systems from linear systems in a striking way. Linear systems have all $\lambda_a \leq 0$, or they make no sense. Associated with this is the notion of a chaotic nonlinear system as a generator of information in the direct sense of Shannon (1948). The idea is that if we have a resolution in any observation given by a size R in phase space, then two state points $\mathbf{x}_1$ and $\mathbf{x}_2$ within the ball of size R cannot be distinguished. A time T later, the distance between these two points has grown to $|\mathbf{x}_1 - \mathbf{x}_2| \exp[\lambda_1 T]$, and when this is larger than $2R$, we can distinguish the two points that were hidden by our instrument resolution at the initial moment. The instability associated with $\lambda_1 > 0$ has revealed information not measurable at the start.

This suggests that the theory of information may well play a role in the description of nonlinear systems that is not realized in linear analyses. This led Fraser and Swinney (1986) to suggest that when determining when two events are independent in a nonlinear system, one should use the notion of *mutual information* (Gallager, 1968). We associate with the state-space sequence $\mathbf{x}(j)$ a probability distribution $P_x(\mathbf{x})$. Similarly, with another sequence $\mathbf{y}(n)$, we associate $P_Y(\mathbf{y})$. If we measure $\mathbf{x}(j)$, how much do we learn about $\mathbf{y}(n)$? The answer (in bits) is

$$I_{XY}(\mathbf{x}(j), \mathbf{y}(n)) = \log_2 \left[\frac{P_{XY}(\mathbf{x}(j), \mathbf{y}(n))}{P_X(\mathbf{x}(j)) P_Y(\mathbf{y}(n))} \right], \tag{34}$$

where $P_{XY}(\mathbf{x}, \mathbf{y})$ is the joint distribution of both kinds of events. If the measurements $\mathbf{x}$ and $\mathbf{y}$ are independent, then $P_{XY}(\mathbf{x}, \mathbf{y}) = P_X(\mathbf{x}) P_Y(\mathbf{y})$, and $I_{XY}(\mathbf{x}, \mathbf{y}) = 0$. This is to be seen as a statistic on the attractor where $\mathbf{x}$ and $\mathbf{y}$ are drawn from points on the attractor. As pointed out earlier, the average of this over the attractor with the natural density leads to an invariant. This yields the *average mutual information*

$$I_{XY} = \sum_{j,n} I_{XY}(\mathbf{x}(j), \mathbf{y}(n)) P_{XY}(\mathbf{x}(j), \mathbf{y}(n)), \tag{35}$$

which is characteristic of the systems X and Y and not of any particular orbits $\mathbf{x}(j)$ or $\mathbf{y}(n)$.

In analyzing chaotic systems, we may ask what "correlation" two events $\mathbf{x}(j)$ and $\mathbf{y}(n)$ have with each other. The average mutual information statistic allows us to probe the nonlinear connection between these two events, and there is some evidence that this probe can be much more sensitive than the familiar linear cross correlation $\Sigma_n \mathbf{x}(n + m)\mathbf{y}^T(n)$.

If the measurements $\mathbf{x}(j)$ and $\mathbf{y}(n)$ are the same, then the average mutual information is

$$h_X = -\sum_n P_X(\mathbf{x}(n)) \log P_X(\mathbf{x}(n)), \tag{36}$$

which looks like the Boltzmann entropy for this system. It is called the Kolmogorov–Sinai entropy, and a result due to Pesin (Eckmann and Ruelle, 1985) connects it with the positive Lyapunov exponents of the source:

$$h_X = \sum_{\lambda_a > 0} \lambda_a. \tag{37}$$

This corresponds to our heuristic explanation of the role played by information in describing nonlinear systems.

7. ANALYSIS OF EXPERIMENTAL DATA

The developments discussed above are useful as insights into how one thinks "nonlinearly." When confronting experiments on physical systems and when choosing how to model or control or predict in physical systems, we must understand how to extract from these observations characteristics such as fractal dimensions and Lyapunov exponents.

The challenge is more severe since in most observations, we measure one, or sometimes a few, variables of the physical dynamics. In a fluid dynamical experiment, we may measure the fluid velocity or temperature or density at some points, perhaps over a long period of time. Unfortunately, we do not measure the fluid velocity everywhere in the fluid or even usually measure all components of the velocity. Rarely does one also measure pressure and density and velocity and temperature, except at a few locations within the fluid. In a nonlinear circuit, we typically observe one or a few voltages in the circuit and rarely measure all independent voltages and currents describing the circuit.

7.1 Time-Delay Embedding

Nonetheless, there is a striking result due to Ruelle and elaborated on by Takens (1981) and Mañé (1981) that accurate measurement of a single variable, call it $x(n)$, will suffice for reconstructing all of the statistical information we seek. The idea is that knowledge of all the time derivatives of the observed variable would capture the full operating phase space of the system. We could try to create these derivatives from the $x(n)$ via

$$\dot{x}(n) = \frac{x(t_0+(n+1)\tau_s)-x(t_0+n\tau_s)}{\tau_s}; \quad (38)$$

for finite τ_s, this would be a linear filter of our data and one that requires very accurate knowledge of the data $x(n)$ because the filter involves differences. Accurate evaluations of small differences between large numbers require very accurate measurements. Ruelle observed that for purposes of providing coordinates for the phase space of the system, we do not need the derivatives, but the time-delayed variables $x(t_0 + (n + 1)\tau_s)$ would do just fine; indeed, that is the only new information in the filter we call $\dot{\mathbf{x}}(n)$. This suggests building d-dimensional spaces from vectors composed of time delays of the observations:

$$\begin{aligned}\mathbf{y}(n) &= \{x(t_0 + n\tau_s), x(t_0 + (n + T)\tau_s), \\ &\quad \ldots, x(t_0 + [n + (d - 1)]T\tau_s)\} \\ &= [x(n), x(n + T), \ldots, x(n + (d - 1)T)]. \end{aligned} \quad (39)$$

The Takens–Mañé results tell us on the basis only of geometry that if we choose $d > 2d_A$, where d is integral while d_A need not be, then we have sufficient coordinates to represent faithfully the space of the source of the $x(n)$. The essential idea behind the result is that we are observing the strange attractor of the nonlinear system as projected on the observation axis $x(n)$. To "unfold" this projection, we must provide enough coordinates to allow the orbit to evolve without intersecting itself. In a space of dimension d, subspaces of dimension d_1 and d_2 intersect generically in a space of dimension $d_1 + d_2 - d$. Self intersections of a space with itself, $d_1 = d_2 = d_A$, are avoided if the dimension of the intersections is less than zero; so $d > 2d_A$. This is a sufficient condition and is independent of the value of the time lag T, itself a multiple of τ_s.

To use this result, we must find rules that allow us to choose a good time delay T and a dimension d, called the embedding dimension, using properties of the data.

7.2 Average Mutual Information

Since the unfolding result holds for arbitrary T, we need another principle to choose it. The choice of T too small would be uninteresting, since then the coordinates $x(n)$ and $x(n + T)$ of our data vector $\mathbf{y}(n)$ are essentially the same; the dynamics would not have had enough time to act to distinguish the two states at those times. If the lag T is too large, however, then the instabilities in the system should have destroyed the connection between $x(n)$ and $x(n + T)$, and so T too long is not helpful. We require a way to choose an intermediate time lag T.

For this, we turn to average mutual information, as that tells us the nonlinear "correlation" between $x(n)$ and $x(n + T)$. We form the distribution of $x(n)$, $P(x(n))$, which is also $P(x(n + T))$, if there are enough data. Then we form the joint distribution $P(x(n), x(n + T))$, and the average mutual information

$$\begin{aligned} I(T) = \sum_{x(n+T)} & P(x(n), x(n + T)) \\ & \times \log_2 \left[\frac{P(x(n), x(n+T))}{P(x(n))P(x(n+T))}\right]. \end{aligned} \quad (40)$$

When this quantity, which satisfies $I(T) \geq 0$, has its first minimum, we have a compromise time delay that is quite useful for forming data vectors $\mathbf{y}(n)$ that have rather independent components $x(n + kT)$, $k = 1, 2, \ldots, d - 1$.

For data from the hysteretic circuit discussed earlier, we have the average mutual information curve shown in Fig. 7. Look again at Fig. 5 to see the strange attractor in phase space made out of vectors $(x(n), x(n + T_0), x(n + 2T_0))$ where $T_0 = 6$ is chosen from the average mutual information curve.

7.3 Global Embedding Spaces

The choice of the embedding dimension d required by the data is not as prescriptive as the choice of time lag T. In fact, the route to take is guided by the Takens result about unfolding the attractor by adding dimensions to the data vector. As one adds dimensions, we undo apparent intersections of the orbit with itself due to projection of a higher dimensional object into a lower dimension. If we keep track of these intersections and establish when they disappear, we should be

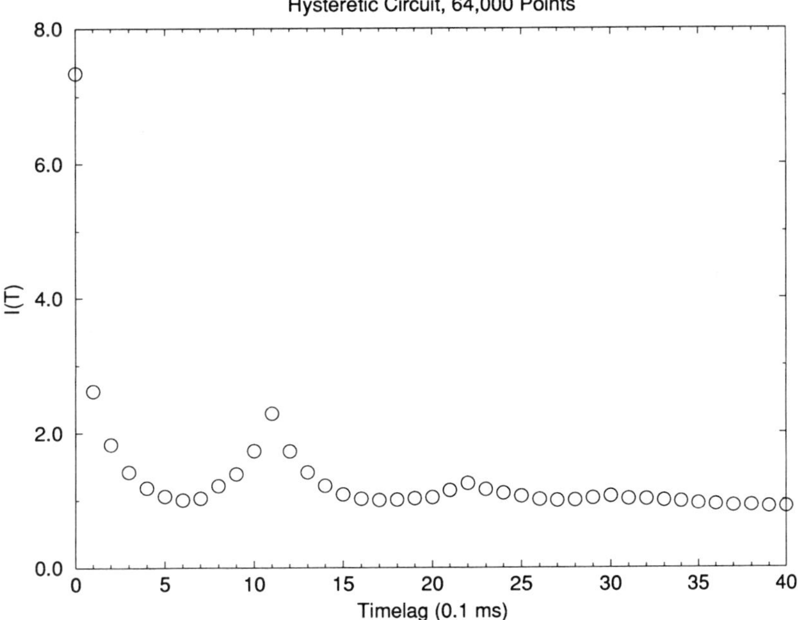

FIG. 7. Average mutual information $I(T)$ for data from the nonlinear hysteretic circuit as shown in Fig. 4. The first minimum of $I(T)$ is at $T = 6$, and this value is used in creating time-delayed coordinates for the display of the attractor in Fig. 5.

able to establish an embedding dimension for the data vectors that is necessary according to the data. The criterion $d > 2d_A$ is sufficient and generic.

It is important from a practical point of view to establish the necessary dimension, since all additional components of the data vector will be occupied by contamination or noise, which respects no finite dimension. To alleviate the effects of this, we must work in as low a dimension as is permitted by the data. It is also true that the amount of calculation one needs to do to evaluate fractal dimensions or Lyapunov exponents rises rapidly with the embedding dimension, so that having the smallest d one can utilize is quite important.

7.3.1 True Vector Fields; False Nearest Neighbors The key to establishing the necessary dimension for a set of data is the idea that orbit points should be near each other only by virtue of the dynamics, not because of a projection from another part of the attractor because we work in too small a space. Dynamically nearby orbit points will not move away from each other as we increase the dimension, but points nearby because of projection will move apart and stay apart. This suggests the method of *false nearest neighbors*, which systematically adds components to the time-delay data vector $\mathbf{y}(n)$, asking at each dimension what percentage of neighbors at the previous dimension are no longer neighbors. When this number drops to zero, the necessary dimension for unfolding has been reached. The key to this procedure is a systematic search for the neighbors of any data point, and this can be done quite efficiently (Kennel *et al.*, 1992). In Fig. 8, we have the percentage of false nearest neighbors for the data from the hysteretic circuit discussed before. Clearly, the number of false neighbors vanishes at $d = 3$; the use of $T = 6$ comes from mutual information.

The other geometric test for embedding dimension rests on the observation that the direction of the vector field $\mathbf{F}(\mathbf{x}(t))$ is unique in small regions of phase space. Thus, if one can carve up the phase space surrounding an attractor and determine when the local vector field is unique in each small sector, then the minimum embedding dimension will have been uncovered (Kaplan and Glass, 1992). This procedure works well in low dimensions and in the absence of any contamination of the

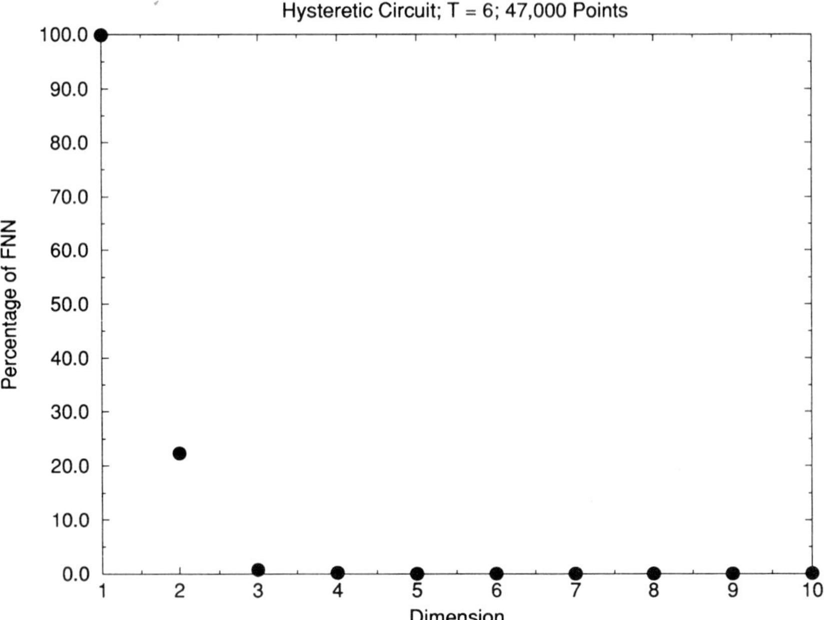

FIG. 8. The percentage of false nearest neighbors for the data from the nonlinear hysteretic circuit seen in Fig. 4. Using the time delay determined by average mutual information shown in Fig. 8, we see that false neighbors go to zero at $d = 3$. This tells us that the attractor can be unfolded in three dimensions, and the display of Fig. 5 is based on this information.

data. In high dimensions, the number of regions one must explore becomes quite large. Further, the accuracy with which one can determine by finite differencing the direction of $\dot{\mathbf{x}}(t)$ becomes quite dependent on the number of data, as distances on an attractor are approximately proportional to

$$(1/N)^{1/D_1} \tag{41}$$

for N data points.

7.3.2 Local Dimensions Even when one has determined the global embedding dimension required to unfold the attractor, the number of degrees of freedom required to describe the dynamics has been given an upper bound. As a contrived example, one can image a flow in two dimensions which evolves on a twisted surface or a Möbius strip. The global embedding dimension is $d = 3$, but the dynamics is two dimensional. If one wishes to evaluate Lyapunov exponents for these data, we had better know that only two of them are dynamical and the third purely geometric.

To discover the local dimension, one needs to work in the global space that separates all points properly and then move locally onto the attractor to establish what dimension captures the evolution of points from neighborhood to neighborhood. Adding this dynamical aspect to the determination of a dimension works rather well and is rather robust to external noise (Abarbanel and Kennel, 1993).

7.4 "Noise" versus Chaos

The whole discussion of dimension raises the very interesting issue of what constitutes "noise" and how we are to distinguish it from chaos. It should be clear that noise in a classical setting, compared to quantum mechanical spontaneous emission, for example, may well be seen as very high-dimensional dynamics. High is rather much in the eye of the beholder or in the ability of the computer. If the dimension of a signal is determined to be larger than $d = 25$, let us say, then, for practical purposes, it may as well be considered infinite dimensional and treated as a stochastic source of contamination of other signals; that is, we treat it just as a traditional view of noise would dictate. Somewhere between a low dimension, $d = 5$, say, and this higher dimension, we lose our ability in a practical sense to analyze and use properties of the structure in state space that are an intrinsic part of the solutions to nonlinear systems. This is an artificial barrier, which we may overcome dimension by dimension as computational capability increases. One should recognize that there is no fundamental difference between noise and chaos, as the former is simply the latter in high dimensions, and, as time goes by, the boundary between our treatment of the two is likely to become less and less clear. Basically, one can say that, when it is possible in a practical sense to utilize the state-space structure of a signal, one should do it. When it appears not to be possible, statistical methods of a traditional sort are all we can turn to.

7.5 Dimensions from Data

To determine fractal dimensions from observations of data, the route is now clear. From the observed scalar data $x(n)$, we need to find an appropriate time delay $T\tau_s$ and an appropriate embedding dimension d with which to make data vectors $\mathbf{y}(n) = [x(n), x(n + T), \ldots, x(n + (d - 1)T)]$. With these data, we need to form the number density in any sphere of radius r about each phase space point $\mathbf{x}$. This is

$$n(\mathbf{x}, r) = \frac{1}{M} \sum_{n=1}^{M} \theta(r - |\mathbf{y}(n) - \mathbf{x}|), \tag{42}$$

as we noted before. The moments of this with the natural density $\rho(\mathbf{x})$ give the correlation functions

$$C_q(r) = \frac{1}{q-1} \int d^d x n(\mathbf{x}, r)^{(q-1)} \rho(\mathbf{x}), \tag{43}$$

and this scales for small r as $r^{(q-1)D_q}$. Computing the $C_q(r)$ has become a kind of cottage industry. The most popular is the evaluation of D_2, for which an enormous body of literature exists (Abarbanel *et al.*, 1993).

The accurate computation of $C_q(r)$ and the extraction of the D_q by looking at plots of log $C_q(r)$ vs log r involve many numerical issues, not the least of which has to do with the fact that this plot is often not even linear. However, when one is quite fortunate, one can extract from these plots acceptable values of the fractal dimensions and use these values for classifying the attractor. Many have focused on these numbers as a sign of the low dimensionality of the source producing the observations, and while that is true, the utility of the D_q as classifiers is much more significant. If it is only low dimensionality one wishes to identify, the geometrical methods above (Sec. 7.3.1) are less computationally demanding and more robust against signal contamination.

The collection of D_q's serves to identify the attractor with which one is dealing. What constitutes a "complete" collection of these invariants is an unanswered question at this time. The utility of a collection of different quantities with which to classify an attractor is clear, and it is quite likely that, absent a clear answer on the completeness of some set of descriptors, one will proceed in a practical sense to use all that can be evaluated reliably.

7.6 Lyapunov Exponents from Data

The route to the determination of Lyapunov exponents from data is also reasonably clear. One must construct data vectors $\mathbf{y}(n)$ as described and then use these to form numerical estimates of the Jacobians $\mathbf{Df}(\mathbf{x})$ at various points in the phase space. The general formulation of this requires creating local maps in the state space that evolve whole phase-spatial neighborhoods into other neighborhoods. In this way, temporal data can be converted into data in phase space for the numerical evaluation of the space derivatives required in the Jacobians. If the local map $\mathbf{f}(x)$ is represented in terms of some basis set of known functions $\phi_k(\mathbf{x})$,

$$\mathbf{f}(x) = \sum_{k=1}^{K} \mathbf{a}_k \phi_k(\mathbf{x}), \tag{44}$$

then the coefficients $\mathbf{a}_k$ can be determined by looking at each data point $\mathbf{y}(n)$ and its neighbors and the temporally subsequent data point $\mathbf{y}(n + 1)$ and its neighbors. Then a least-squares fit to the evolution of the set of neighbors near $\mathbf{y}(n)$ into the set of neighbors near $\mathbf{y}(n + 1)$ gives the $\mathbf{a}_k$. The local Jacobian follows as

$$Df_{jm}(\mathbf{y}(n)) = \sum_{k=1}^{K} (a_k)_j \left. \frac{\partial \phi_k(\mathbf{x})}{\partial x_m} \right|_{\mathbf{x}=\mathbf{y}(n)}. \tag{45}$$

The Oseledec matrix can then be evaluated as a product of local Jacobians and with some care (Abarbanel *et al.*, 1993) can be diagonalized to determine both the local and global Lyapunov exponents.

7.7 Using Invariants of the Dynamics

Of what use can these invariants be?

7.7.1 Classification and Identification First, as we have suggested, these quantities such as fractal dimensions and Lyapunov exponents are averages over the attractor of physically motivated statistics. They are independent of initial conditions, as the particular orbits of the system are not. Since one may evaluate these quantities from observed data and since they are independent of the coordinate system in which one evaluates them, they naturally have a role in classifying the system that produced the signal. If one is concerned with knowing whether the signal that she sees this week is from the same source as the signal seen last week, then these are the identifiers which she must use for those purposes. From the point of view of system design for this or that specific application, these are the parameters toward which one can direct the design procedures. Suppose one wants a source of signals that has less or more predictability; then attaching the capability for evaluating Lyapunov exponents to the design tools will enable this aspect of the design to be achieved.

At the very least, whether one wishes to manipulate these quantities for design purposes or catalog them for identification uses, they provide the correct taxonomy for recognition of the source and the basis for listing its physical properties. In this fashion, they replace the familiar resonant frequencies of a system, which is a concept tuned to linear dynamics, while the material we have been discussing is focused on nonlinear systems.

7.7.2 Models and Observations The critical role played by the invariants arises when one is trying to make models of the physics of the source of observations. The models which one can make basically fall into two categories:

1. Black-box-like models, which provide local or global "fits" to the data in the multidimensional reconstructed phase space of the data vectors $\mathbf{y}(n)$. Such models consist of parametric rules for reproducing the observed $\mathbf{y}(n) \rightarrow \mathbf{y}(n+1)$, and the parameters are chosen by some kind of minimization procedure,most efficiently a least squares estimation. For local modeling, we create rules that fit the way the neighborhood of points in phase space around the point $\mathbf{y}(n)$ evolves into the neighborhood of points around $\mathbf{y}(n+1)$. Some basis set of functions is chosen for this fit, but if the neighborhoods are small enough, polynomials will do fine. If not, more sophistication is called for (Abarbanel *et al.*, 1993).

   Global models perform the same tasks of a least-squares fit to parameters but use all the data on the attractor at once. The benefit of the global models is their compact form, which is used to "explain" all of the data, but one must have a sensible choice of model to begin with. The number of models is clearly infinite, and incorrect choices may be hard to distinguish from one another. Local models are less elegant and use more computations, but can basically be used to fit the evolution of nearly anything in phase space. Local models put together can certainly reproduce any dynamics, but at the cost of a large lookup table, which essentially translates the neighborhood-to-neighborhood behavior into a numerical rule. Enormously accurate predictive models can be achieved with high enough order local models.

   Since each form of model works by evaluating its parameters by a least-squares fit, there is no guarantee that the properties such as fractal dimensions or Lyapunov exponents will be properly reproduced, since they represent other aspects of the dynamics. This allows one to use the invariants as constraints on model making and may usefully limit the variation in parameters that one need explore. Indeed, it happens that unconstrained modeling may result in very bad values for Lyapunov exponents, while constrained models, differing only slightly from the very best least-squares fit, do substantially better.

2. Physically based models. From the point of view of doing physics with chaotic nonlinear systems, the most important aspect of the invariants is to allow one to compare observations with models. Suppose one has observations of a plasma under some circumstances where the signals are chaotic and has a proposed model for the dynamics based on approximations to the electrodynamics and possibly thermonuclear reactions of the constituents of the system. Then using the invariants extracted from the observations and comparing these to the invariants evaluated directly from the model may allow one to exclude the model or change parameters in the model to correspond to the experiment or verify one's physical sense of the dynamics. In some sense, the usual activities of model making and verification on the part of physicists are moved to this arena and away from the Fourier-based tradition that is appropriate for linear dynamics.

7.8 "Control" of Chaos

It is not always a welcome discovery to learn that the physical system one has created is operating in a chaotic mode. If this is not the desired operating condition, one can utilize methods to "control" that chaotic motion. The basic idea is to locate some of the infinite number of unstable periodic orbits that comprise a strange attractor, and, in a locally linear analysis about these periodic orbits, identify the stable and unstable directions associated with perturbations near the orbits. Then one can devise small changes in the parameters of the system that drive observed evolution onto the stable direction of one of these periodic orbits (Ott *et al.*, 1990). Repeated application of the required small parameter changes drives the system onto the stable direction of the unstable periodic orbit that was chosen. Basically, one is turning an autonomous problem where a parameter is held constant into a nonautonomous problem where an external force is altering the parameter according to the system location in phase space. Thus, we create a new system in which the previously unstable periodic orbit has become stable.

This method has proven successful in stabilizing the unstable periodic orbits of chemical reactions, cardiac cells, infrared lasers, magnetoelastic ribbons, and nonlinear circuits, among other systems. It suggests that one might actually wish to drive a system of interest into a chaotic regime where it visits an enormously larger piece of state space than in regular motion and then stabilize the evolution to one or more chosen periodic orbits at will. This kind of flexibility will no doubt be quite useful in applications of these methods.

GLOSSARY

Attractor: The set of points in phase space visited by the solution to an evolution equation long after transients have died out. An attractor can have an integer dimension—a regular attractor—or a fractional dimension—a *strange* attractor.

Average Mutual Information: The amount (in bits if base-two logarithms are used) one learns about one observation from another observation on the average over all measurements. This is a statistic introduced in communications theory by Shannon (1948), which describes chaotic nonlinear systems as they are information sources.

Basin of Attraction: The set of points in phase space that, taken as initial conditions for an evolution rule, lead to the same attractor. The boundaries between basins of attraction can be fractional dimensional.

Bifurcation: A change in the state or topology of the solution to an evolution equation when a parameter is varied. A bifurcation is associated with changes in stability of solutions to dynamical nonlinear equations.

Cantor Set: An uncountably infinite set of points in R^d (a) which contains no subinterval of R^d, (b) which is closed, and (c) each point of which is an accumulation point. The dimension of such a set is typically fractional. We have called this dimension d_A, since, in physical applications, the Cantor set is an attractor of a dynamical system.

Chaos: Deterministic evolution of a nonlinear system that is between regular behavior and stochastic behavior or " noise." This motion of nonlinear systems is slightly predictable and nonperiodic, and specific orbits change exponentially rapidly in response to changes in initial conditions or orbit perturbations.

Embedding Dimension: The dimension of phase space required to unfold the attractor of a nonlinear system from the observation of scalar signals from the source. This dimension is integer.

Flows and Maps: Flows are evolution rules in continuous time. Maps are evolution rules in discrete time. Maps can come from flows by finite sampling associated with real experiments or from Poincaré sections, which view a continuous orbit stroboscopically.

Fractal Dimensions: The dimensions of an infinite set of points associated with the way the density of points scales with small volumes surrounding the points. Fractional values for the dimensions are typical for strange attractors of nonlinear systems.

Hamiltonian Systems: Laws of evolution corresponding to closed systems or systems without dissipation. The equations of motion come from a scalar "Hamiltonian" and, when expressed in canonical coordinates, the Hamiltonian flow preserves volumes in phase space.

Lyapunov Exponents: The rate at which nearby orbits diverge from each other after small perturbations when the evolution of a nonlinear system is chaotic. There are M exponents for evolution of an M-dimensional flow or map. For a dissipative system, the sum of these exponents is negative. For a flow, one exponent must be zero.

Phase Space or State Space: Nonlinear systems are described by multidimensional vectors labeled by continuous or discrete time. The space in which these vectors lie is called state space or phase space. The dimension of phase space or state space is integer.

Works Cited

Abarbanel, H. D. I., Brown, R., Sidorowich, J. J. ("Sid"), Tsimring, Lev Sh. (1993), "The Analysis of Observed Chaotic Data in Physical Systems," *Rev. Mod. Phys.* **65**, 1331–1392.

Abarbanel, H. D. I., Kennel, M. B. (1993), "Local False Nearest Neighbors and Dynamical Dimensions from Observed Chaotic Data," *Phys. Rev. E* **47**, 3057–3068.

Arnol'd, V. I., Avez, A. (1968), *Ergodic Problems of Classical Mechanics*, New York: Benjamin.

Brandstater, A., Swinney, H. L. (1987), "Strange Attractors in Weakly Turbulent Couette-Taylor Flow," *Phys. Rev. A* **35**, 2207.

Cantor, G. (1883), "Über Unendliche, Lineare Punktmannichfaltigkeiten," *Math. Ann.* **21**, 545–591; see also Hausdorff, F. (1919), "Dimension und Äusseres Mass," *Math. Ann.* **79**, 157–179.

Chandrasekhar, S. (1961), *Hydrodynamic and Hydromagnetic Stability*, Cambridge, UK: Cambridge University Press.

Collet, P., Eckmann, J.-P. (1980), *Iterated Maps on the Interval as Dynamical Systems*, Boston: Birkhauser.

Devaney, R. L. (1989), *An Introduction to Chaotic Dynamical Systems*, 2nd ed., Redwood City, CA: Addison-Wesley.

Drazin, P. G. (1993), *Nonlinear Systems*, Cambridge, U.K.: Cambridge University Press.

Eckmann, J.-P., Ruelle, D. (1985), "Ergodic Theory of Chaos and Strange Attractors," *Rev. Mod. Phys.* **57**, 617.

Fraser, A. M. (1989a), "Information and Entropy in Strange Attractors," *IEEE Trans. Info. Theory* **35**, 245.

Fraser, A. M. (1989b), *Physica D* **34**, 391.

Fraser, A. M., Swinney, H. L. (1986), *Phys. Rev. A* **33A**, 1134.

Gallager, R. G. (1968), *Information Theory and Reliable Communication*, New York: John Wiley and Sons.

Gollub, J. P., Benson, S. V. (1980), "Many Routes to Turbulent Convection," *J. Fluid Mech.* **100**, 49–470.

Grassberger, P. (1983), "Generalized Dimensions of Strange Attractors," *Phys. Lett. A* **97**, 227; see also Hentschel, H. G. E., Proccacia, I. (1983), "The Infinite Number of Generalized Dimensions of Fractals and Strange Attractors," *Physica D* **8**, 435; see also Renyi, A. (1970), *Probability Theory*, Amsterdam: North Holland.

Guckenheimer, J., Holmes, P. J. (1986), *Nonlinear Oscillations, Dynamical Systems, and Bifurcations of Vector Fields*, 2nd ed., New York: Springer-Verlag.

Hénon, M. (1976), "A Two Dimensional Mapping with a Strange Attractor," *Commun. Math. Phys.* **50**, 69.

Kaplan, D. T., Glass, L. (1992), "Direct Test for Determinism in a Time Series," *Phys. Rev. Lett.* **68**, 427.

Kaplan, J. L., Yorke, J. A. (1979), "Chaotic Behavior in Multidimensional Difference Equations," *Lecture Notes in Mathematics* **730**, 228.

Kennel, M. B., Brown, R., Abarbanel, H. D. I. (1992), "Determining Minimum Embedding Dimension using a Geometrical Construction," *Phys. Rev. A* **45**, 3403–3411.

Lorenz, E. N. (1963), "Deterministic, Nonperiodic Flow," *J. Atmos. Sci.* **20**, 130–141.

Mañé, R. (1981), in: D. Rand, L. S. Young (Eds.), *Dynamical Systems and Turbulence, Warwick, 1980*, Lecture Notes in Mathematics Vol. 898, Berlin: Springer, p. 230.

Newton, I. (1687), *Philosophia Naturalis Prin-*

cipia Mathematica, Imprimatur S. Pepys, Reg. soc. prases. Julii 5. 1686. Londini, Jussu Sociaetatis Regia ac Typis Josephi Streater. Prosat apud plures Bibliopolas. Anno, 1687.

Oseledec, V. I. (1968), "A Multiplicative Ergodic Theorem. Lyapunov Characteristic Numbers for Dynamical Systems," *Trudy Mosk. Mat. Obsc.* **19**, 197; *Moscow Math. Soc.* **19**, 197.

Ott, E., Grebogi, C. Yorke, J. A. (1990), "Controlling Chaos," *Phys. Rev. Lett.* **64**, 1196–1199.

Poincaré, H. (1892), *Les Méthodes Nouvelles de Méchanique Celeste*, Paris: Gauthier-Villars.

Ruelle, D., Takens, F. (1971), "On the Nature of Turbulence," *Commun. Math. Phys.* **20**, 167; see also Newhouse, S., Ruelle, D., Takens, F. (1978), "Occurrence of Strange Axiom A Attractors Near Quasiperiodic Flows on T^m (m = 3 or more)," *Commun. Math. Phys.* **64**, 35.

Shannon, C. E. (1948), "A Mathematical Theory of Communication," *Bell Syst. Tech. J.* **27**, 379–423 [Part I], 623–656 [Part II].

Smale, S. (1967), "Differentiable Dynamical Systems," *Bull. Am. Math. Soc.* **73**, 747.

Takens, F. (1981), in: D. Rand, L. S. Young (Eds.), *Dynamical Systems and Turbulence, Warwick, 1980*, Lecture Notes in Mathematics Vol. 898, Berlin: Springer, p. 366.

VanderPol, B. (1992), "On Oscillation Hysteresis in a Simple Triode Generator," *Philos. Mag.* **43**, 700–719.

Whittaker, E. T. (1937), *A Treatise on the Analytical Dynamics of Particles and Rigid Bodies*, 4th ed., London, New York: Cambridge University Press.

Wisdom, J. (1987), "Chaotic Dynamics in the Solar System," *Icarus* **72**, 241–275.

Further Reading

Abarbanel, H. D. I., Rabinovich, M. I., Sushchik, M. M. (1993), *Introduction to Nonlinear Dynamics for Physicists*, Singapore: World Scientific.

Arrowsmith, D. K., Place, C. M. (1990), *An Introduction to Dynamical Systems*, Cambridge, U.K.: Cambridge University Press.

Baker, G. L., Gollub, J. P. (1990), *Chaotic Dynamics*, Cambridge, U.K.: Cambridge University Press.

Lichtenberg, A. J., Lieberman, M. A. (1983), *Regular and Stochastic Dynamics*, Berlin: Springer-Verlag.

Moon, F. C. (1992), *Chaotic and Fractal Dynamics*, New York: John Wiley and Sons.

Ott, E. (1993), *Chaos in Dynamical Systems*, Cambridge, U.K.: Cambridge University Press.

Ottino, J. M. (1989), *The Kinematics of Mixing: Stretching, Chaos, and Transport*, Cambridge, U.K.: Cambridge University Press.

Shinbrot, T., Grebogi, C., Ott, E., Yorke, J. A. (1993), *Nature* **363**, 411–417.

Thompson, J. M. T., Stewart, H. B. (1986), *Nonlinear Dynamics and Chaos*, New York: John Wiley and Sons.

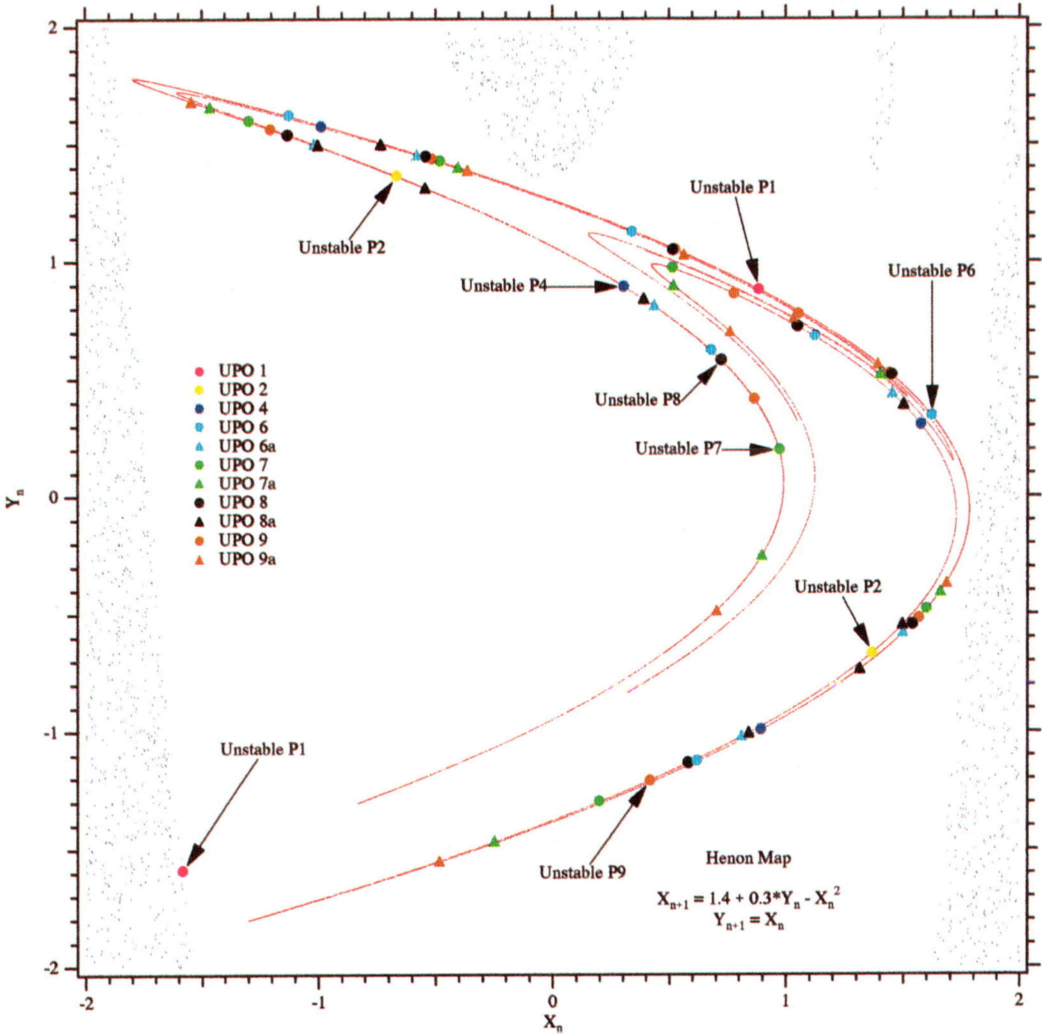

PLATE 1. The unstable periodic orbits in the Hénon map of the plane to itself for periods up to period nine. The orbits are labeled as P*N* for period *N*. The designation UPO stands for Unstable Periodic Orbit.

NUCLEAR ENERGY, FISSION

WALTER Y. KATO, *Brookhaven National Laboratory, Upton, New York, U.S.A.*

INTRODUCTION

2 December 1992 marked the 50th anniversary of the initiation of the first self-sustaining nuclear chain reaction of a manmade nuclear reactor in the world. That reactor, called Chicago Pile No. 1 (CP-1), was constructed in secrecy by the Manhattan Project at the beginning of World War II underneath the west stands of the University of Chicago squash court by a team of physicists under the leadership of Enrico Fermi (Smyth, 1945; Rhodes, 1986). CP-1 was composed of graphite blocks of varying lengths and pressed natural uranium oxide and uranium metal cylinders. It was designed and constructed to prove that a self-sustaining fission chain reaction could exist and to confirm theoretical physics calculations. The success of the self-sustaining chain reaction reactor in 1942 led to the design and construction of a large, graphite-moderated, natural-uranium–fueled, water-cooled plutonium production reactor at Hanford, Washington.

The plutonium production reactor used the nuclear reaction

$$^{238}U + \text{neutron} \rightarrow {}^{239}U \xrightarrow{\beta^-} {}^{239}Pu$$

for the production of plutonium for nuclear weapons. The chemical separation of plutonium from irradiated natural uranium appeared to be a much easier process at that time than the isotopic separation of the ^{235}U isotope from natural uranium. The Manhattan Project also initiated in 1942 the design and construction of a large ^{235}U isotope separation plant based upon a diffusion process at Oak Ridge, Tennessee. This led in 1945 to the production of sufficient quantitites of highly enriched uranium containing ^{235}U for the construction of the first atomic weapon.

The Soviet Union also started on the construction of a graphite-moderated natural-uranium reactor under the leadership of I.V. Kurchatov in Moscow in 1943. This resulted in the Φ-1 reactor achieving a self-sustaining chain reaction on 25 December 1946 at what is now known as the I.V. Kurchatov Institute of Atomic Energy in Moscow. Following attainment of the self-sustaining chain reaction in the Φ-1 reactor, the Soviet scientists and engineers designed and constructed graphite-moderated, water-cooled, natural-uranium–fueled reactors for the production of plutonium in the Urals region of the Soviet Union by 1948.

Although the first nuclear reactors were constructed for plutonium production for nuclear weapons, it was recognized early that the energy released from ^{235}U fission could be used for the production of electrical power, motive power for ships, and district heating. With the end of World War II, large quantities of separated ^{235}U isotopes, which could be used to fuel nuclear power reactors, became available from the separation plants at Oak Ridge.

Nuclear reactors to produce self-sustaining chain reactions utilize the following

3-527-28133-9/94/$5.00 + .50

nuclear reactions:

$^{235}U + n \rightarrow$ fission products
+ kinetic energy + $2.42n$

or

$^{239}Pu + n \rightarrow$ fission products
+ kinetic energy + $2.87n$.

The first nuclear reactor that produced electricity was Argonne National Laboratory's Experimental Breeder Reactor No. 1 (EBR-1) (Lichtenberger *et al.*, 1955), which was a 1.2 MW(t) [200 kW(e)] sodium-potassium–cooled fast reactor located at the National Reactor Testing Station (NRTS), which is now known as the Idaho National Engineering Laboratory (INEL), located about 60 miles west of Idaho Falls, Idaho. EBR-1 first produced electricity on 20 December 1951.

In the middle 1940s, the U.S. Navy under the leadership of Adm. Hyman Rickover became interested in developing nuclear power for powering submarines. This led to the establishment of the Knolls Atomic Power Laboratory (KAPL) near Schnectady, New York, and the Bettis Atomic Power Laboratory (BAPL) near Pittsburgh, Pennsylvania, for the development of a nuclear power system for submarines. The General Electric Company (GE) operated KAPL and the Westinghouse Electric Corporation operated BAPL for the U.S. Atomic Energy Commission. BAPL developed the concept of a pressurized, water-cooled and -moderated nuclear reactor, while KAPL developed the concept of a liquid-metal–cooled reactor for motive power for the submarines. The BAPL water-moderated and -cooled reactor became the forerunner of the commercial pressurized-water reactors (PWR) that were later manufactured by Westinghouse, Combustion Engineering, and Babcock & Wilcox corporations.

Argonne National Laboratory (ANL) originated the concept of the boiling-water reactors (BWR). The concept of the BWR was demonstrated by the BORAX [15 MW(t)] Reactor series of experiments (Dietrich *et al.*, 1955) at the NRTS begining around 1952, which led to the construction of the Experimental Boiling Water Reactor (EBWR) (Harrer *et al.*, 1955), a 20 MW(t) [5 MW(e)] prototype reactor at the ANL site near Chicago in 1961. The EBWR and the Vallecitos BWR (VBWR), later designed and constructed by GE in Vallecitos, California, were the forerunners of the large commercial BWRs.

Although fission reactors essentially produce heat energy, it is important to note the significant differences between fission reactors and chemical-combustion, fossil-fueled plants for electricity generation. Fission reactors produce heat through the fission process as a result of a nuclear interaction between a neutron and the fissile isotope nucleus. The fissioning of ^{235}U results in the energy release of 1 MW-d (24 000 kW h) per gram of ^{235}U. On the other hand, a chemical combustion (oxidation) process of methane gas, for example, is a result of the oxygen molecule combining chemically with carbon molecules, resulting in the heat energy release of about 1.6 kW h per gram of methane (CH_4) gas. This has the following consequences:

1. Fission reactors, in addition to producing heat energy, produce neutron and gamma radiation that must be contained by shielding.
2. Nuclear reactors produce gaseous and particulate fission products, which are radioactive and decay with the emission of beta, gamma, or alpha rays on varying time scales. The decay rate of a radioactive material represents the activity of the material. The time it takes for the activity to decay by a factor of 2 is known as the half-life of the radioactive material. Radioactive materials must also be contained by shielding. They must be handled carefully and prevented from being released to the environment for long periods of time.
3. Nuclear reactors have the ability to produce greater than design-basis heat energy; therefore, they require well-designed and reliable control mechanisms.
4. Finally, nuclear reactors, because of their varying half-life fission products, produce significant amounts of heat energy called decay heat even though the fission process has been terminated. This is in contrast to chemical-combustion plants, which produce no heat once the combustion process is stopped. The amount of decay heat produced after termination of the fission process is illustrated in Table 1. The existence of decay heat results in the requirement that nuclear fuel elements must be cooled for months or years after

Table 1. Decay heat from fission reactors.

Time after reactor shutdown	Approximate percentage of full power production
1 sec	6
1 h	1
28 h	0.5
12 d	0.3
4 m	0.13

the termination of the fission process in order to avoid melting of the fuel and releasing of the fission products.

The ability of the ^{235}U isotope nucleus to capture a neutron and either be fissioned or become the ^{236}U isotope and emit a gamma ray depends on the probability of interaction between the neutron and target nuclei, which is measured by the *interaction cross section*. Various types of interactions are possible depending on the kinetic energy of the neutron. Three principal reactions for the ^{235}U isotope and neutrons are scattering, capture, and fission at low neutron energies. Neutrons have a kinetic velocity or energy measured in m/s or eV. As an example, neutrons at equilibrium at room temperature have a velocity of 2200 m/s or an energy of 0.025 eV. Cross sections, which are conventionally designated by the symbol σ, are measured in units of cm^2 as a function of the kinetic energy of the neutron. A cross section of 10^{-24} cm^2 is called a cross section of 1 barn. The capture (σ_c) and fission (σ_f) cross sections for ^{235}U at 0.025 eV are 98 and 582 barns, respectively. Isotopes such as ^{233}U and ^{239}Pu, also, readily fission at low neutron energies since they have fission (σ_f) cross sections of 527 and 742 barns, respectively, at 0.025 eV, whereas ^{238}U fissions with neutrons having energies greater than about 0.68 MeV.

The fissioning of the nucleus of a fissionable isotope such as ^{235}U produces on the average two fragments called fission products plus on the average 2.44 neutrons. Fission products are isotopes having atomic mass about half of the uranium atomic mass and are usually unstable, which means that they are radioactive and decay by emitting a beta particles, gamma rays, or possibly alpha particles. The neutrons emitted at fission have an average energy of 1.98 MeV, and these neutrons could cause additional fissions if another ^{235}U isotope is in proximity. This could create a self-sustaining chain reaction. Unfortunately, natural uranium is a mixture of the two isotopes ^{235}U (0.72%) and ^{238}U (99.28%), which have fission cross sections of about 4.2 and ~0.6 barns at 0.025 eV and 2 MeV, respectively, and capture cross sections of about 2.7 and 0.4 barns at 0.025 eV and 2 MeV, respectively. The significance of these values is that it is not possible to have a self-sustaining chain reaction with natural uranium alone because the capture cross section at 2 MeV is not significantly less than the fission cross section, which means that most of the neutrons would be captured or escape out of the system without causing additional fissions.

In order to produce a self sustaining chain reaction, it is necessary either to separate the fissionable ^{235}U isotope from natural uranium and use uranium enriched in ^{235}U or to decrease the energy of the fission neutrons by neutron moderation. The original nuclear reactor CP-1 used natural uranium as fuel and pure graphite as moderator. The graphite, having a low capture cross section and low atomic mass, would slow the neutrons from 2 MeV to 0.025 eV without being captured by the graphite or escaping out of the reactor system. Neutrons, when slowed down, would be captured by the ^{235}U nucleus and cause fission. The trick in making a self-sustaining chain reaction is to minimize leakage and captures by the moderator or structural materials while maximizing fissions by the ^{235}U nucleus. It is necessary to use high-purity graphite because industrial graphite contains small amounts of such materials as boron and vanadium, which have high neutron absorption cross sections. Other materials, in addition to high-purity graphite, that could act as moderators are light water, heavy water (D_2O), and beryllium because of their low atomic mass numbers. Unfortunately, hydrogen has a fairly high capture cross section, and therefore it is not suitable as a moderator with natural uranium. Heavy water can be used as a moderator with natural uranium, but it requires considerable effort to separate heavy water from ordinary water. If natural uranium could have its ^{235}U content increased to between 2.5 and 3.2%, then you could have a self-sustaining chain reaction using light water as moderator.

Today, natural uranium consists principally of the two isotopes ^{238}U (99.2%) and ^{235}U (0.72%) with half-lives of 4.47×10^9 yr and

7.04×10^8 yr, respectively. Natural uranium had a higher content of ^{235}U (i.e., 1.6%) a billon (10^9) years ago because of the difference in the ^{238}U and ^{235}U half-lives. Two billion years ago the ^{235}U content in natural uranium was closer to 3%. Under proper conditions, the natural uranium of two billion years ago could have produced a self-sustaining chain reaction with light water as moderator.

In May 1972, H. Bouzigues, a staff member of the French nuclear fuel processing plant at Pierrelatte, discovered that some of the natural uranium coming from the Oklo region of the Republic of Gabon in West Africa had a ^{235}U content of 0.7171% instead of 0.7202% found in normal natural uranium (Cowan, 1976). Other samples contained only 0.44% of U^{235}. A serious investigation into this discrepancy by the French Commissariat à l'Energie Atomique (CEA) resulted in the discovery that nature had constructed the world's first nuclear fission reactor possibly 1.7 to 1.9 billion years ago in a rich deposit of uranium ore in the Oklo region. That a self-sustaining fission reactor had existed was deduced from the other stable isotopes that were located near the region where the uranium depleted in ^{235}U was found. A study of the absolute amount of such rare earth elements as neodymium, lanthanum, and gadolinium and metals such as zirconium, ruthenium, and rhodium, and of their isotopic composition, led to the firm conclusion that a natural uranum fission reactor had existed there. The concentration of the uranium ore, the availability of water possibly in the form of water of crystalization in sedimentary ore or saturated water in the ground, and the lack of neutron absorbers, such as boron or lithium compounds, resulted in a self-sustaining chain reaction, which released an estimated 15 000 MW y of energy. Estimates of the duration of the fission reactors range from 150 000 to 1.5×10^6 years, $(1.7 \text{ to } 1.9) \times 10^9$ years ago. In the Oklo region, the French have found six reactor zones that were located in regions with exceptionally rich ores. In all six areas, the ^{235}U was found to be significantly depleted, leading to the conclusion that reactors had operated at more than one site. It is, thus, concluded that CP-1, the graphite natural uranium reactor, was the first manmade fission reactor but not the world's first fission reactor.

1. REACTOR TYPES

One approach to catagorizing reactors is by their function or use. Three principal categories are research, power generation, and material production. In the research category can be included zero-power or critical experiments, training reactors, and reactors that are operated for the production of neutrons. Examples of research reactors are the High Flux Beam Reactor (HFBR), as shown in Fig. 1 (Hendrie, 1965), at Brookhaven National Laboratory (BNL), the High Flux Isotope Reactor (HIFR) (Swartout *et al.*, 1965) at Oak Ridge National Laboratory (ORNL); and the Advanced Test Reactor (ATR) at the Idaho National Engineering Laboratory (INEL) (IAEA, 1989). These reactors have beam ports through which beams of neutrons can be removed for neutron scattering and activation experiments. They can also be used for radioisotope production. Research reactors are operated at power levels from a few watts or kilowatts to tens of megawatts. The heat that is produced by the fission process in these reactors is essentially waste heat and is disposed to the environment. Included in the research category are zero-power reactors or critical experiments, which are flexible-geometry, low-power reactors and operate at a few milliwatts to about 100 W for short periods of time to verify reactor physics calculations. CP-1, the first manmade reactor, was essentially a critical experiment, since it was operated at very low power levels and was used to verify reactor physics calculations. The objective of critical experiments is to measure such parameters as critical mass, neutron flux and power distribution, neutron energy spectrum, etc., while varying the fuel composition or geometry. Since handling of fuel is usually carried out by hand, it is necessary to minimize the total amount of fission energy produced by the reactor for any given run to minimize radiation levels from the fuel. An example of an operating critical experiment is the Zero Power Plutonium Reactor (ZPPR) (Davey, 1972), as shown in Fig. 2, operated by ANL at INEL since 1967.

Power generation reactors are those reactors that have been designed for the generation of electricity or process heat. These reactors operate at power levels of a few hundred to thousands of megawatts. Since the thermal efficiency of power reactors is about $\frac{1}{3}$, it requires about a 3000-MW power plant

FIG. 1. Experimental area of BNL High Flux Beam Reactor (HFBR)

for the generation of 1000 MW of electrical power. There are a number of different types of power reactors, namely, light-water moderated and cooled reactors (LWRs), heavy-water moderated and cooled reactors (HWRs), heavy-water moderated light-water cooled reactors, graphite-moderated light-water cooled reactors, graphite-moderated gas-cooled reactors (HTGRs), and unmoderated sodium-cooled fast reactors (LMRs). Examples of these commercial power generation reactors will be given later.

A third category comprises the material production reactors. These reactors are designed for the production of large quantities of plutonium, tritium, or possibly ^{233}U isotopes for weapons use. These reactors are either graphite-moderated light-water cooled reactors, such as the production reactors at Hanford, Washington, or heavy-water moderated and cooled reactors, such as at the Savannah River Plant in South Carolina, and are operated at a few hundred to a few thousand megawatts.

2. NUCLEAR PHYSICS

An understanding of some nuclear physics is essential to the understanding of this brief exposition of nuclear reactor theory and operation. The reader is referred to such texts as Glasstone and Sesonske (1987), Lamarsh (1977), or Stamm'ler and Abbate (1983) for more detailed discussions on nuclear physics and nuclear reactor theory.

Neutrons and protons make up the the principal particles in atomic nuclei. Fission or breakup of heavy atomic nuclei, such as lead, tungsten, uranium, and heavier nuclei can occur by bombardment of these elements by particles such as neutrons, protons, alphas, or high-energy gamma rays. The higher the energy of the incident particle, the easier it is for fission to occur. Charged particles because of the Coulomb barrier must be accelerated to high energies to cause fission. Only neutrons, however, can cause fission at energies below 2 MeV and only in uranium or higher atomic mass nuclei. Neutrons can be

FIG. 2. Zero Power Plutonium Reactor (ZPPR) at ANL.

produced through such nuclear interactions as the (α, n) reaction using naturally occurring radioactive elements such as radium or polonium, which emit alpha particles during their decay, and elements such as beryllium. The incidence of alpha particles on beryllium nuclei results in the production of neutrons. Radium-beryllium neutron sources produce neutrons whose kinetic energy is about 4.8 MeV. With the availability of various isotopes of plutonium, plutonium-beryllium neutron sources have become available.

Although ^{235}U can fission at most neutron energies, it has a higher probability of causing fission when the incident neutrons have energies of about 0.025 eV. ^{238}U, on the other hand, will only fission with neutrons greater than 0.68 MeV. The probability of fission occurring in ^{238}U at energies greater than 0.68 MeV is considerably less than the fission of ^{235}U at 0.025 eV. The reason that the compound nucleus ^{236}U, which is formed by the absorption of a neutron by the ^{235}U nucleus, fissions is that the binding energy for the ^{236}U nucleus is less than the energy of the combined ^{235}U nucleus and neutron. The same is true for such isotopes as ^{233}U, ^{239}Pu, and ^{241}Pu, which are also fissile isotopes.

The breakup of the nuclei in the ^{235}U fission process produces on the average 2.44 neutrons and two nuclei, each having about half of the atomic mass of the original nuclei. The neutrons that are emitted at fission have a distribution of kinetic energy, as shown in Fig. 3, with an average value of 1.98 MeV. The technical problem in designing a self-sustaining reactor system is how to slow the neutrons to about 0.025 eV without the neutrons leaving the system or being absorbed by some other non-fissioning nuclei.

In addition to neutrons causing fission in ^{235}U or other isotopes, there are other reactions such as capture, elastic and inelastic scattering, and alpha, proton, or two-neutron production that can occur. Illustrations of absorption cross section values for various

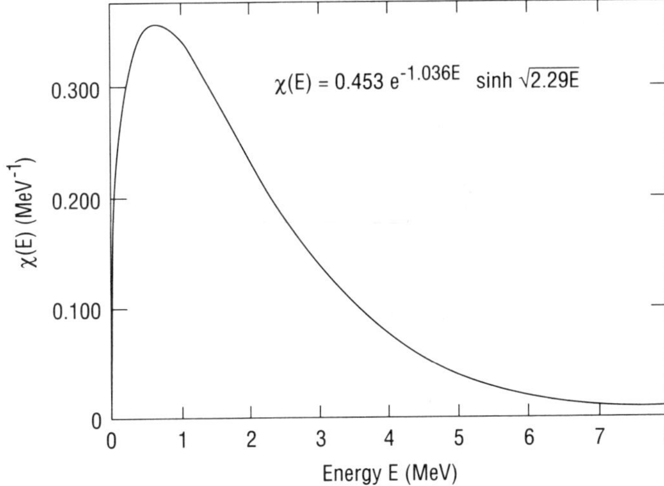

FIG. 3. Prompt neutron spectrum from fission.

isotopes at incident neutron energies of 0.025 eV are given in Table 2. The interaction of neutrons with different isotopic nuclei varies with the nuclei as well as with the energy of the incident neutrons.

Illustrations of the total neutron cross sections for ^{238}U and fission cross sections for ^{235}U nuclei are shown in Figs. 4 and 5, respectively. Typical neutron parameters for ^{235}U and ^{239}Pu are given in Table 3. Although ^{235}U fission has been used as an illustration in this discussion, isotopes such as ^{233}U and ^{239}Pu will also fission with neutrons at thermal energies. These isotopes do not exist in nature in any significant quantitites since they have shorter half-lives than ^{238}U and ^{235}U and have decayed since the beginning of the universe. Both ^{233}U and ^{239}Pu have been made in large quantities through the use of reactors.

Table 2. Thermal absorption cross sections of typical isotopes at $E = 0.025$ eV.

Isotope	Cross section (b)
^{1}H	332×10^{-3}
^{2}H	0.53×10^{-3}
^{6}Li	940
^{7}Li	37×10^{-3}
^{10}B	3837
^{11}B	5.5×10^{-3}
^{12}C	3.4×10^{-3}
^{54}Fe	2.25
^{197}Au	98.8
^{232}Th	7.4
^{235}U	680
^{238}U	2.7
^{239}Pu	1011

It can be seen from these illustrations that there are two components to the cross sections: a smoothly varying 1/v component and a resonance structure component. Single-level and multilevel resonance formulations based on the Breit–Wigner resonance formula have been developed to fit the resonance structure. The single-level Breit–Wigner resonance formula for neutron capture is

$$\sigma_c = \frac{\lambda_r^2 g}{4\pi} \frac{\Gamma_n \Gamma_\gamma}{(E - E_r)^2 + \Gamma^2/4},$$

where σ_c is the capture cross section of the resonance at energy E_r, λ_r is the wavelength of the neutron with energy E_r, g is a statistical factor, Γ_n and Γ_γ are the neutron width and radiation width, respectively, and Γ is the total width of the resonance at one-half its height.

Measurements, evaluations, and compilations of neutron cross sections for different reactions as functions of incident neutron energy for most isotopes have been carried out. In the United States, the National Nuclear Data Center (NNDC) at BNL is responsible for the compliation of neutron cross sections and resonance parameters for all isotopes. The NNDC has developed a CSISRS (Cross Section Information Storage and Retrieval System), which is an automated compilation of unevaluated experimental data. Originally, NNDC developed the Evaluated Nuclear Data Files (ENDF) A and B. ENDF/A contains both complete and incomplete data sets for many

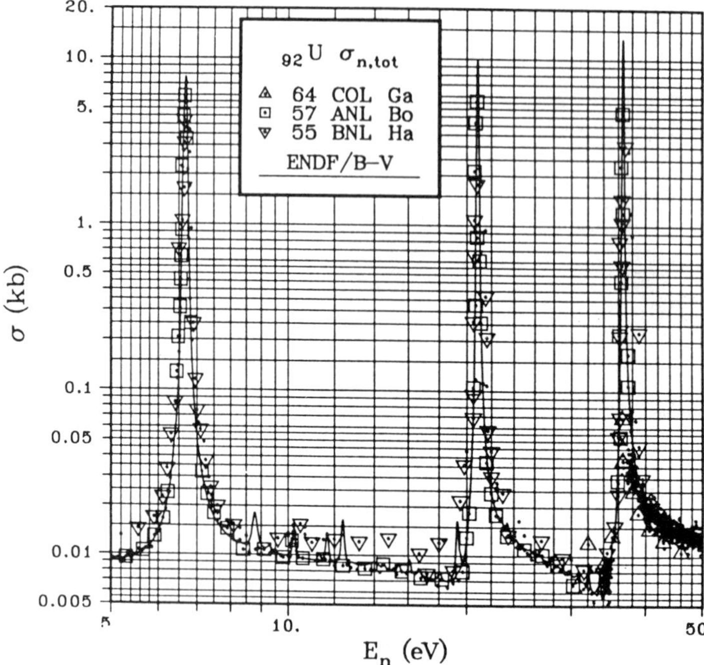

FIG. 4. Total cross section for ^{238}U

isotopes. ENDF/B, on the other hand, contains a unique complete cross section set needed for reactor design. Today, ENDF/B is the principal repository for evaluated neutron-induced–reaction data in the energy range from 10^{-5} eV to 20 MeV. The evaluation of the data for ENDF/B is carried out by the Cross Section Evaluation Working Group (CSEWG), a cooperative effort by technical experts from national laboratories,

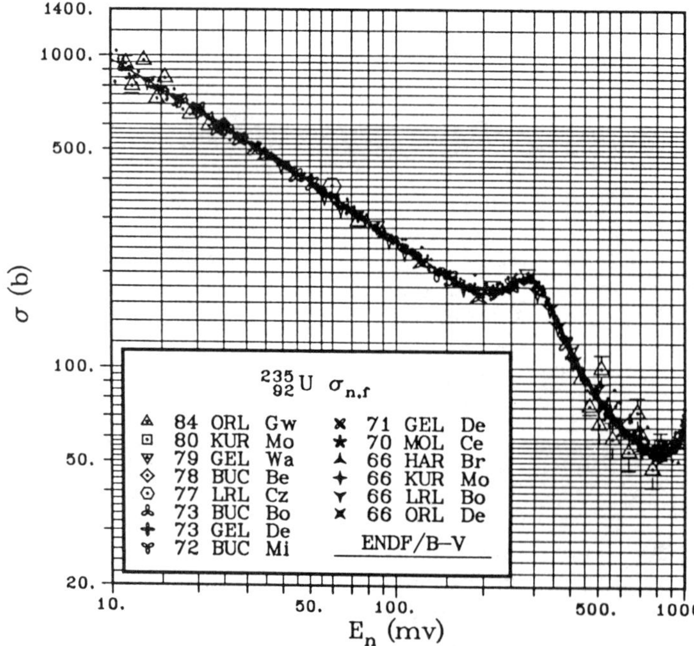

FIG. 5. Fission cross section for ^{235}U.

Table 3. Neutron parameters for ^{235}U and ^{239}Pu.

	Thermal $E = 0.025$ eV		Fast E = 2 MeV	
	^{235}U	^{239}Pu	^{235}U	^{239}Pu
Fission cross section (b)	582	742	2.7	2.2
Capture cross section (b)	98	269	1	0.71
No. of neutrons/fission	2.418	2.871	2.44	2.88
No. of neutrons/capture	2.068	2.108	2.21	2.60
Fraction of delayed n's	0.0065	0.0021	0.0064	0.00206

industry, and academia. The ENDF/B file and other data sets, in addition to being accessible on-line via computer terminals have been published by the NNDC (Mughabghab *et al.*, 1981; McLane *et al.*, 1988). The ENDF/B neutron cross section files are used extensively in reactor physics calculations to calculate fundamental parameters, such as critical mass and power distributions, for reactors. Measured cross sections and evaluations are routinely exchanged with nuclear data compilation centers in Europe, the Former Soviet Union, Japan, and China. Periodically, conferences are held among technical experts in nuclear data to exchange data, methods for evaluation, and low-energy nuclear theory.

When a low-energy neutron interacts with a ^{235}U nucleus and causes fission, on the average, 2.44 neutrons are emitted instantaneously and a small number of neutrons are emitted a fraction of a second or seconds later from the decay of unstable fission-product nuclei. These delayed neutrons enable us to control the chain reactions in fission reactors more easily. The total energy released during fission is on the average about 200 MeV, which includes the contributions from gamma rays, kinetic energy of the fission products, and neutron kinetic energy. It is this energy that is utilized for the generation of heat, which is subsequently converted to electrical power.

3. REACTOR DESIGN

Since 2.44 neutrons are produced on the average whenever the ^{235}U nuclei fissions, it is possible to have a continuing chain of fission events if at least one of the neutrons produced at fission causes fission in another ^{235}U nucleus. The job of a reactor designer is to maximize the probability of neutrons born in fission to survive and cause fission in a ^{235}U or other fissile nucleus by

a. minimizing the probability of a neutron leaving the system;
b. minimizing a neutron from being captured by nonfissile nuclei;
c. maximizing a fission event occurring by increasing the number of ^{235}U or other fissile nuclei per unit volume; and
d. slowing down the average velocity of the neutrons from that at 1.98 MeV to 0.025 eV.

The designer does a trade-off study of these four approaches in designing a reactor. To minimize the probability of a neutron escaping the system, the designer can surround the reactor core, which contains the fissile and moderator materials, with a reflector consisting of material that has a large scattering cross section and a low absorption cross section. Materials such as water, graphite, beryllium, iron, or stainless steel have been used as reflectors. To minimize the probability of a neutron being captured by nonfissile nuclei, it is necessary for the designer to use low-absorption materials that have high strength at operating temperatures, such as aluminum or zirconium for structural and fuel cladding materials, to minimize the amount of nonfissile material utilized. To increase the number of fissile materials per unit volume, the designer would have to use as high a density as possible of fissile materials, generally metals or oxides, with uranium enriched in the ^{235}U isotope. Increasing uranium enrichment increases the cost of the material as well as fabrication costs because of production and materials accountability issues. Finally, the designer selects the operating neutron energy spectrum of the reactor, which determines the type of coolant and moderator used. For a thermal energy spectrum reactor, whose average neutron energy is around 0.025 eV, the designer has the choice of moderating materials such as water, heavy water, and graphite. If water or heavy water is chosen, these fluids can also act as coolants or heat removal fluids at the same time. If beryllium or graphite is chosen as moderator, gases such as helium or liquids such as water could be selected as the coolant.

Some of the essential parameters that a reactor designer must determine by analysis so that the reactor can be started up, maintained at steady-state operation over its lifetime, and shut down as required are listed here:

1. Criticality: A reactor is said to attain criticality when the chain reaction in the system is self-sustaining.
2. Critical mass: The amount of fuel required to maintain a steady-state or self-sustaining chain reaction.
3. Multiplication: The ratio of the number of neutrons in one generation over the number in the previous generation is defined as the multiplication of the system and is usually designated by k. When $k = 1$, the reactor is said to be critical. The ratio of the change in multiplication, due to effects such as fuel or control-rod movement, material density changes resulting from temperature change, or fissile material depletion or buildup, to the multiplication itself ($\Delta k/k$) is called the reactivity of the system and is a measure of the supercriticality or subcriticality of the system.
4. Power distribution: The spatial distribution of heat energy generation throughout the reactor core and reflector as a function of fuel burnup.
5. Control worth: The amount of neutron absorber materials, such as boron, cadmium, etc., in solid or in liquid solution required in the core to control the reactor. The absorber materials are often called poisons since they tend to terminate the chain reactions. Absorber material, such as soluble burnable poisons, is introduced into the coolant at the initial stages of a new core to balance the later buildup of fission product poisons.
6. Temperature effects on operation: The effect on reactivity due to changes in the system due to temperature effects is called temperature coefficient of reactivity. Temperature effects on the system can be due to thermal expansion of fuel and structural materials, density changes in materials such as fuel and coolant due to temperature, and changes in the width of neutron resonance cross sections for the fertile and fissile material resulting from the Doppler effect.
7. Fuel burnup: The destruction of the fissile isotopes due to fission is called burnup.
8. Fission product production, which introduces absorption into the reactor core.
9. Kinetic parameters, such as neutron lifetime and effective delayed neutron fraction, for prediction of the transient behavior of the reactor under normal and abnormal conditions.
10. Effects of radiation damage on reactor core materials.
11. Radiation levels from direct core radiation, and activation of coolants and coolant impurities.

To give the reader an idea of how a reactor is designed, a brief description of the thinking process that a designer undertakes in reactor design is presented. Assuming that only purified natural uranium metal is available to the designer, he would have to select a neutron-moderating material that would have the smallest capture cross section and at the same time would have the largest slowing-down potential per collision. The average logarithmic fractional loss in energy due to an elastic collision of the neutron and a nuclei is given by

$$\xi \equiv \ln \frac{E_1}{E_2} = 1 + \frac{\alpha}{1-\alpha} \ln \alpha,$$

where

$$\alpha \equiv \left(\frac{A-1}{A+1}\right)^2.$$

Here, A is the atomic mass of the scattering nuclei, E_1 is the energy of the incident neutron, and E_2 is the energy of the neutron after scattering. Some of the scattering properties for the lighter elements are given in Table 4. It can be seen that the designer has choices among heavy water (D_2O), graphite, and beryllium for a nuclear reactor using natural uranium. Of these, purified graphite is the most plentiful, least expensive, and easiest to manufacture. D_2O is very plentiful; however,

Table 4. Scattering properties of various materials.

Element	Mass no.	Average change in lethargy	Average number of collisions to 1 eV from 2 MeV
H	1	1.0	15
D	2	0.726	20
Be	9	0.207	70
C	12	0.158	92
O	16	0.120	121
U	238	0.0083	1700

it would require an isotope-separation plant to separate the D_2O from H_2O since D_2O constitutes only 1.5% of ordinary water in nature.

Now that the designer has selected graphite as the moderator and natural uranium metal as the fuel, he must choose the shape and dimensions of the fuel. Depending upon the purpose of the reactor (i.e., research, electrical power generation, or material production), the designer must decide upon the total amount of power to be generated by the reactor and its power density. He must also decide the type of cladding that the uranium metal fuel will require. The cladding is to prevent the oxidation of the uranium as well as to contain the fission products and avoid possible chemical interactions of the uranium with the coolant. The choice of a cladding material is determined by the desired operating temperature and the neutronic and structural properties of the proposed cladding material. The cladding material must be able to maintain its strength at operating temperatures as well as be able to withstand radiation damage due to neutron interactions with the cladding material during the lifetime of the fuel in the reactor. The fuel cladding material and the structural material must be chemically compatible with the proposed coolant. In order to minimize parasitic neutron losses to the cladding material, it is desirable to use materials such as aluminum or zircaloy, a zirconium–aluminum alloy. High-strength materials such as stainless steel would not be acceptible for natural uranium reactors because of high neutron capture cross section unless the designer decided to use higher enriched uranium as fuel. Even zirconium would have to be highly purified in order to remove the small amount of hafnium that normally occurs with zirconium, since hafnium has very large absorption resonances.

Assume that the designer has selected air as the coolant and aluminum as the cladding material. He must decide upon the shape and dimensions of the fuel and the cladding thickness. Uranium metal must be cast and machined in an inert atmosphere, since uranium metal is pyrophoric and will oxidize very rapidly in air. The designer must also conduct many reactor physics analyses to determine the shape and amount of fuel that must be embedded in the graphite in order to make a self-sustaining chain reaction of fission events.

Earlier, a chain reaction was described as one initiated by a neutron from a neutron source or cosmic-ray event causing a chain reaction by producing a fission event in a ^{235}U nucleus, which in turn produced neutrons that again caused fission in another ^{235}U nucleus. The occurrence of many parallel chain reactions without dying out gives rise to a self-sustaining nuclear reactor.

The self-sustaining chain reactions may be described quantitatively by the multiplication factor, k. If k is greater than 1, the number of fissions is increasing from one generation to another, which would mean that the total number of fissions in the system or reactor is increasing, therefore, the total power (energy being generated) of the reactor is increasing. If k is less than 1, then the power of the reactor is decreasing and the chain reactions will be terminated and the reactor is said to be shut down. If k is exactly equal to 1, the reactor is said to be stable and at a constant power level. The operator of a reactor must, therefore, bring together enough fuel in the form of ^{235}U or other fissile material and moderator material and remove any absorbing materials that might be used as control rods in order to achieve a $k = 1$ critical system. In order to be able to control the reactor, the designer will have designed into the reactor control rods, which consist of removable neutron absorber material such as boron, or cadmium. The operator attains criticality ($k = 1$) by the removal of the control rods after the reactor has been loaded with fuel. After attaining criticality, the operator can reach an operating power level by removing additional control rods, making the reactor supercritical, or $k > 1$. As the total power generation of the reactor is increased, the operator must now introduce coolant into the reactor to remove the heat generated. The operator balances the heat-removal capability of the system to the heat generation of the reactor. When the heat removal from the reactor is balanced with the heat being generated, the operator then inserts control rods so that $k = 1$ and the reactor operates at constant power. To shut down the reactor, the operator inserts the control rods so that $k < 1$ and the chain reaction is terminated.

The multiplication factor and other reactor parameters, such as power distribution, neutron flux distribution, etc., of a system can be analytically calculated by using the Boltzmann neutron transport theory to describe the geometry and material composition of the system (Lamarsh, 1977; Bell and Glasstone, 1970). The simplest approach would be to use a one-dimensional, one–energy-group diffusion-theory approximation to the Boltzmann transport equations to calculate the k of a system. If fission neutrons were born at thermal energy instead of with a distribution with an average energy of 1.98 MeV, it would be possible to describe a reactor analytically with a one–energy-group diffusion theory, where the rate of change with time of number of neutrons in a system is equal to the production of neutrons minus the leakage and the absorption of neutrons in the system.

Unfortunately, since fission neutrons are born with a continuous energy distribution up to about 10 MeV with an average energy of 1.98 MeV, it is necessary to solve multiple–energy-group transport or multiple–energy-group diffusion-theory equations to determine the reactor-physics parameters. In addition to being multiple-group, these equations may have to be multidimensional in order to describe a complex geometry system. Such systems of multidimensional, multiple-group partial differential equations were initially solved using finite-difference methods using fast, large-memory, mainframe computers. Today, diffusion-theory equations are usually solved using various nodal methods, while transport-theory equations still use finite-difference methods. Examples of reactor physics computer codes that numerically solve the diffusion- or transport-theory equations are given in Table 5 (National Energy Software Center, 1991). As work stations have become more powerful with larger memories and with higher computational speeds, many reactor physics computer codes have been adapted for use on work stations.

Table 5. Examples of reactor analysis computer codes.[a]

Title	Description
	Cross section generation
NJOY	Nuclear data processing code system for producing pointwise and multigroup neutron and photon cross sections from ENDF/B data.
	Static design studies
TWODANT	One and two-dimensional, multigroup, discrete-ordinates transport theory code.
1DB	One-dimensional, multigroup diffusion theory code.
3DB	Three-dimensional, multigroup diffusion theory code.
	Space-time–dependent kinetics
DIF3D	Three-dimensional, multigroup diffusion-theory, time-dependent code.
	Monte Carlo
MCNP	Monte Carlo neutron and photon transport code.

[a] For additional reactor codes, see Radiation Shielding Information Center (1993) or Energy Science and Technology Software Center (1993).

In the diffusion-theory approximation, the neutron is considered as a function of space and energy only. Because there is an angular dependence of elastically and inelastically scattered neutrons, it is necessary to take into consideration this angular dependence for accurate analysis. Simplifying assumptions in the usual diffusion equations introduce inaccuracies near interfaces of the reactor core with vacuum or at core-reflector, core-control rod, or core-blanket interfaces. Interfaces of the core and strongly neutron-absorbing media and fuel and coolant cells are also difficult to model accurately. In such cases, differential transport (Boltzmann equation) theory models, which include higher order angular elastic scattering effects, are used.

In order to determine the k of a system, it is first necessary to develop averaged group cross sections for each energy group that is planned to be used for the diffusion- or transport-theory equations. The averaged group cross sections are developed by averaging the microscopic cross sections in the ENDF/B file over an energy spectrum in each energy group. Computer codes such as NJOY (RISC Software Directory, 1993) may be used to generate the group cross sections. Of course, increasing the number of energy groups increases the complexity of the computation and the computer time required for the solution of the problem. Following the generation of the group cross sections, reactor-physics computer codes, as shown in Table 5, are used to calculate the reactor-physics parameters needed in the design, such as the multiplication, power distribution, etc.

Neutron movement from production, through migration, to absorption or leakage from the system can be considered a stochastic process or a random-walk process with a probability distribution for each type of interaction, such as angular scattering, energy loss or gain, or fission neutron energy, in the system. A process of following the transport of neutrons stochastically in the reactor system has been developed and is called the Monte Carlo method. This method essentially follows the life of a single neutron until it is absorbed or leaks from the system. Neutrons that are produced as a result of fission from this neutron are also followed as a chain of events or reactions. It is necessary to follow the history of many hundreds of thousands of initiating neutrons in order to obtain a statistically meaningful value for the multiplication constant or other physics parameter. Computer codes such as MCNP (Briesmeister, 1986) have been developed using the Monte Carlo method to model complex-geometry reactor systems. In order to model reactor systems accurately, it is necessary that a sufficient number of neutron interaction histories be calculated. The Monte Carlo method is a large consumer of computing time on the fastest mainframe computer, so that obtaining a million histories to achieve a convergence of k to $\pm 0.1\%$ is very time-consuming and expensive. With the advent of parallel computers and fast and relatively inexpensive work stations, using the Monte Carlo method to solve the neutron-transport problem has become routinely possible. It is anticipated that, in the near future, a fast parallel computer having 1000 or more parallel computing units could produce 10^7 to 10^9 histories in a short time. The reactor designer will then be able to carry out more accurate design analysis of complex reactor systems less expensively.

The use of diffusion or transport equations, discussed above to describe a reactor, results in a description of the steady-state condition of the reactor. In the steady-state condition, the neutron flux is independent of time and only a function of position in the reactor. The total power generated in the reactor is determined by integrating over space and energy the neutron flux, the number of fissile atoms, and their fission cross section. The reactor can be critical at any power level from a few microwatts or milliwatts to many thousands of megawatts provided that the heat generated is removed from the reactor to avoid melting of the fuel. The power level of the reactor is controlled by inserting or removing neutron absorbers from the reactor. Decreasing the neutron absorbers in the reactor will increase the k_{eff} so that, when the multiplication becomes greater than 1, the neutron flux will start to increase exponentially. The fact that some of the neutrons produced at fission are delayed and are not emitted instantaneously allows the reactor power to be increased slowly under control. It would be difficult without delayed neutrons to control the reactor, because the reactor power would increase very quickly exponentially with a short reactor period when the multiplication became greater than one. The reactor period is defined as the time required for the power to increase by a factor e. The rate of increase of the neutron density is a function of the prompt-neutron lifetime of the system and the degree to which k_{eff} is greater than 1.0. The prompt-neutron lifetime, l_p, is the average time elapsed between the time when the neutron is born at fission and the time it is absorbed or lost out of the reactor. It can be considered as having two components: the time it takes for the neutron to be slowed down to thermal energy from fission energy and the time it takes for the neutron to diffuse and become absorbed or lost out of the reactor. The slowing-down time is much smaller than the diffusion time, so that the prompt-neutron lifetime may be equated to the diffusion time. The prompt-neutron lifetime for thermal reactors can vary from a few tenths of a millisecond to over a hundred milliseconds, depending on the moderator. The neutron lifetime for fast-spectrum reactors is in the range of a few tenths of microseconds to a few microseconds.

In a thermal reactor, where most of the fissions occurring in the reactor are from ^{235}U, the fraction of delayed neutrons is approximately 0.0065. When the k_{eff} is >1.0 but <1.0065, the reactor is said to be delayed critical and the reactor power will increase with a period of seconds to minutes. An illustration of how rapidly the reactor power would increase is shown in Table 6, where the reactor period is given as a function of neutron lifetime and the reactivity of the system. When k_{eff} is >1.0065, the reactor is said to be prompt critical and the reactor power will increase exponentially with a very short period, as can be seen in the last line of Table 6. The rate of increase of the reactor power above prompt critical is dependent on the degree of prompt criticality and the neutron lifetime of the system. The reactor that becomes prompt critical will increase its power level by many orders of magnitude in a very short time until other mechanisms, such as temperature effects or destruction of the core configuration, terminate the reactor power excursion.

The k_{eff} of the reactor at any given time is determined by the geometry and the materials in the reactor, as well as the temperatures of the fuel, moderator, and coolant of the reactor. The temperature changes the density of the materials as well as their position in the reactor and, hence, the k_{eff} of the system. An increase in temperature also changes the effective width and height of the resonance cross sections in such materials as ^{238}U and ^{235}U through the Doppler effect. This effect increases the number of resonance level neutrons absorbed in these materials. An increase in ^{238}U absorptions would tend to decrease the reactivity, while increasing the number of fissions in ^{235}U would tend to increase reactivity. This change in reactivity due to the broadening of the resonance levels by a change in fuel temperature is called the Doppler temperature coefficient. The reactivity effect due to the creation of voids in the coolant due to vaporization of the coolant is called the void coefficient of reactivity. In calculating changes in power level due to changes in reactivity, it is necessary to take into consideration the reactivity coefficients due to such effects as Doppler, void formation, expansion effects, and density changes.

The reactor designer must choose the appropriate geometries and materials for the reactor core so that it will have an overall negative power reactivity coefficient, so that when the power increases the reactivity of the reactor decreases. Usually, for thermal reactors, where the ^{235}U enrichment is only a few percent, the ^{238}U Doppler effect will provide a negative Doppler fuel-temperature coefficient. The coolant or moderator void coefficient could be positive or negative depending on the degree of moderation. In the western world, LWRs are designed with a negative void coefficient so that the reactor's reactivity will decrease with an increase in power due to decreasing moderator density and increasing fuel temperature. On the other hand, the Soviet-designed RBMK, a graphite-moderated, light-water–cooled reactor, had a positive void coefficient of reactivity, which led to the catastrophic Chernobyl-4 accident.

The study of the behavior of the time variation of the neutron flux when the multiplication is greater or less than unity is called reactor kinetics. To calculate accurately the neutron flux when the reactor becomes superdelayed critical or subdelayed critical, it is necessary to solve a time-dependent diffusion or transport set of equations. For simple geometry systems, it can be assumed that the spatial and energy distribution of the neutron flux does not change with time. This significantly eases the effort to calculate the neutron flux change with time. It is, however, necessary to take into consideration the fact that the rate of emission of delayed neu-

Table 6. Reactor period (s) for various values of reactivity for ^{235}U system.

Reactivity	Lifetime = 10^{-3} s	Lifetime = 10^{-4} s	Lifetime = 10^{-5} s
0.001	60	60	60
0.003	10	10	10
0.004	3.5	2.5	2.5
0.007	0.8	0.2	0.14
0.010	0.3	0.04	0.0003

trons falls into six different time-decay groups, each of which is characterized by a definite exponential decay rate. The average energy of the delayed neutrons emitted is also substantially lower than the average prompt neutron energy at fission. The variation of the neutron flux with time is calculated by solving the point reactor kinetics equation (Glasstone and Sesonske, 1987), which incorporates the effects of the six groups of delayed neutrons.

In complex-geometry reactor systems where there is asymmetry in reactivity changes due to nonuniform control rod movements or temperature effects, it may be necessary to use one-dimensional or multidimensional kinetics equations in order to analyze power transients accurately. Large neutron flux asymmetries due to nonuniformities in neutron absorber distribution or spectral changes due to temperature effects require the use of multidimensional and multiple–energy-group time-dependent solutions of the transport equations for transient analysis. The transient excursion of the Chernobyl-4 reactor is a good example of a transient that requires the solution of such multidimensional, time-dependent equations.

4. POWER REACTORS

The development of ^{235}U enrichment production plants during World War II made possible the development of large nuclear power plants for electricity generation in the 1950s and 1960s because of the availability of large quantities of enriched uranium. Although highly enriched uranium or 93% ^{235}U was used primarily for weapons and naval reactors, enriched uranium also became available for the development of research, test, and demonstration power plants. Eventually in the United States, the utility and reactor vendor industry settled down on the choice of either boiling-water or pressurized light-water reactors using 2–5% ^{235}U-enriched uranium oxide fuel. The light-water reactors have become the predominant type of reactors used for electric power generation in the world. In the late 1960s through the 1980s, GE was the BWR manufacturer in the United States, while Westinghouse, Combustion Engineering Co., and Babcock and Wilcox were the PWR manufacturers. Since the early 1980s, only Westinghouse, GE, and ABB (ASEA Brown Boveri) Combustion Engineering have remained in the reactor vendor business. Combustion Engineering was taken over by ABB, which formed ABB-CE.

In 1991, there were approximately 111 nuclear power plants licensed to operate in the United States with a total operating capacity of 101 GW(e), which resulted in about 22% of the total electricity being generated from nuclear plants. There are nuclear power plants operating in 32 countries worldwide, which supply more than 17% of the world's electricity. In 1992, there were at least 411 reactors with a net electrical power capacity of over 320 GW(e). In 1991, France generated 72.7% of its electricity from nuclear plants, as compared with Belgium with 59.3%, and Sweden with 51.6%. A list of nuclear capacity, in countries worldwide, is shown in Table 7.

As of 1992, there were 237 PWRs [207 GW(e)] and 88 BWRs [72 GW(e)] out of a total of 411 nuclear power plants in worldwide operation. In both categories, ordinary water is used as moderator and coolant. The fuel is uranium dioxide enriched in ^{235}U content to between 2.5 and 5%. In the BWR case, the coolant pressure and temperature are such that water is in a boiling state as it exits from the core region, and steam is directly passed into a turbine for electric power generation. A schematic diagram of a BWR is shown in Fig. 6. A detailed schematic of the reactor portion is shown in Fig. 7, followed by a schematic arrangement of the steam supply system in Fig. 8.

In the case of the PWR, high coolant pressures are maintained so that water does not reach the boiling temperature and the heated water is fed directly into a steam generator. The secondary side of the steam generator produces steam, which is then fed into a turbine for electricity generation. A schematic diagram of a PWR is shown in Fig. 9. A schematic diagram of the reactor coolant system for a four-loop PWR is shown in Fig. 10, while the details of the core of a PWR are shown in Fig. 11. A typical PWR fuel assembly is shown in Fig. 12, while a steam-generator cross section is shown in Fig 13.

The Soviet Union developed light water reactors designated VVER, which are similar to the western pressurized-water reactors.

Table 7. World commercial reactors in 1991.

Country	Number of operating reactors	Capacity [GW(e)]	Electricity generation % nucl., 1990
S. Africa	2	1.8	5.6
Argentina	2	0.69	19.8
Belgium	7	5.5	60.1
Brazil	1	0.62	1.0
Bulgaria	6	3.7	35.7
Canada	20	14.0	14.8
China	1	0.3	—
Czech and Slovak Republics	8	3.2	28.4
Finland	4	2.3	35.0
France	55	56.5	74.5
Germany	21	22.5	33.1
Hungary	4	1.7	51.4
India	9	1.8	2.2
Japan	43	33.2	27.1
S. Korea	9	7.22	49.1
Lithuania	2	2.7	55
Mexico	1	0.65	2.6
Netherlands	2	0.5	4.9
Pakistan	1	0.13	1.1
Russia	28	17.8	11
Slovenia	1	0.62	25
Spain	9	7.1	35.9
Sweden	12	10	45.9
Switzerland	6	4.88	42.6
Taiwan	6	5.14	34.6
Ukraine	14	12.1	25
United Kingdom	35	11.9	19.7
United States	108	98.2	20.6

Significant differences between the early VVERs and the western PWRs include the lack of a containment, the lack of multiple-loss-of coolant emergency cooling systems, and horizontal steam generators. Containment buildings are constructed around the newer VVERs.

Three other types of reactors have been built in western countries for electric power generation. These include the CANDU, a heavy-water–moderated, heavy-water–cooled tubular reactor designed and built by Atomic Energy of Canada, Ltd. (AECL), and a high-temperature gas-cooled reactor (HTGR) built by General Atomics Corporation of San Diego, California, which is shown schematically in Fig 14. Liquid-metal–cooled reactors (LMRs) (Waltar, 1981) have been built by ANL (EBR-I and EBR-II), GE (SEFOR), and Westinghouse (FFTF) (Peckinpaugh, 1982, 1986). A schematic view of a liquid-metal–cooled reactor (LMR) is shown in Fig 15.

The liquid-metal–cooled reactors designed and constructed by ANL, under the sponsorship of the Atomic Energy Commission, are based upon a metallic fuel element with a sodium-potassium (NaK) alloy as coolant for the EBR-I and sodium as coolant for the EBR-II. Both EBR-I (Lichtenberger *et al.*, 1955) and EBR-II (Lentz *et al.*, 1986) were built at the National Reactor Testing Station (NRTS) in Idaho. The EBR-I had a small compact core and produced 1.2 MW of heat energy. The NaK alloy conducted heat away from the reactor core to a steam generator, where the heat energy produced steam. A small turbine produced about 200 kW of electricity in EBR-I. EBR-I operated with neutrons whose spectrum had an average energy of about 100 keV as opposed to thermal reactors whose average neutron energy is 0.025 eV. In addition to demonstrating the possibility of electricity generation, EBR-I showed that it was possible to produce more plutonium than the amount of ^{235}U consumed. It was a converter of ^{238}U atoms into plutonium, ^{239}Pu, with a conversion ratio of a little greater than 1.0. The conversion ratio, or the breeding ratio, is defined as the ratio of the average of number of fissile atoms pro-

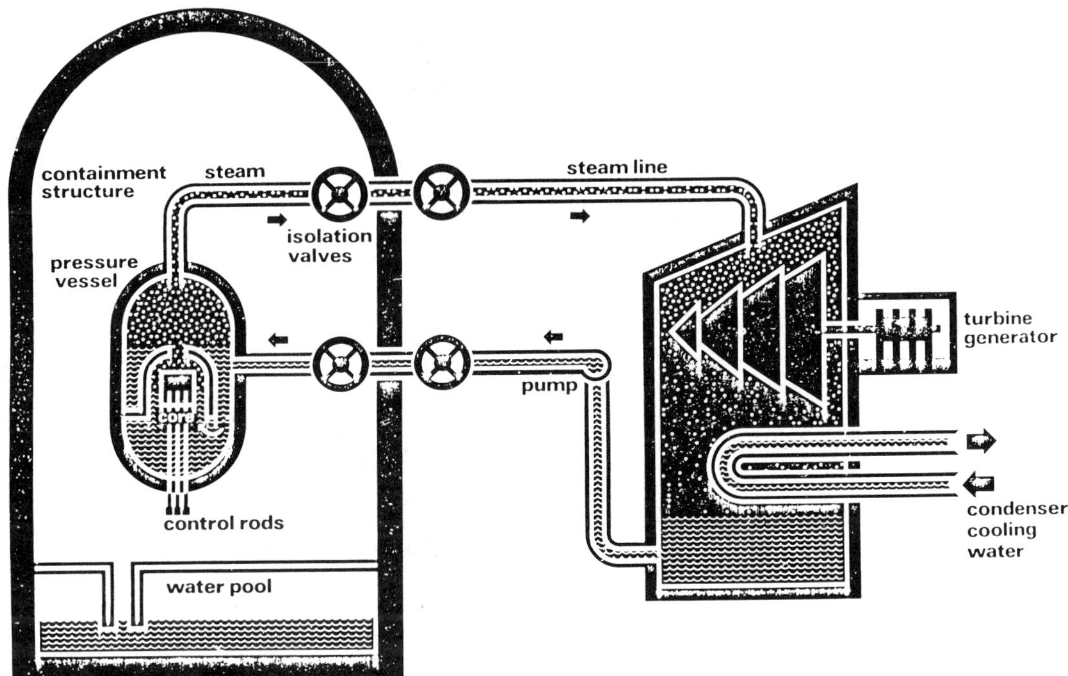

FIG. 6. Schematic diagram of BWR.

duced per fissile atom consumed in the reactor. When the new fissile atoms produced are different from the fissile atoms consumed, the reactor is said to be a converter, while if the new fissile atoms are the same as the fissile atoms consumed, the reactor is said to be a breeder reactor. Since ^{235}U constitutes only 0.72% of natural uranium, a breeder reactor could utilize the remaining ^{238}U to produce fissile atoms. During the early development of the nuclear industry, it had been forecast that there would be many 1000 MW(e) nuclear power reactors in operation. It was anticipated that these thermal reactors would consume the world's available ^{235}U in about half a century, indicating the necessity for near-term plutonium breeder reactors in order to maintain the nuclear option. Following the accident at the Three Mile Island–2 (TMI-2) nuclear power plant in April 1979, there was a considerable slowdown in the construction of power plants, because it was necessary for the existing plants and plants under construction to meet additional regulatory requirements. This gave rise to significant construction cost increases as well as operation and maintenance costs. As of 1993, there had been no new orders for nuclear power plants in the United States since 1978 because of a number of factors including no urgent need for additional generation capacity, the uncertainy in financing large capital projects, the uncertainty in regulations, and increased public opposition.

The FSU, in addition to the VVER series, which is similar to the western PWRs, built the RBMK type of power generation plant, which is a variation of their plutonium production reactor. It is a graphite-moderated, light-water–cooled, boiling-water, channel-type reactor. The Chernobyl-4 reactor, which exploded catastrophically in April 1986, is a typical RBMK-type reactor, which generated about 1000 MW(e). Instead of the reactor being contained in a thick pressure vessel, fuel elements were located in 1661 vertical high-pressure pipes or tubes. The reactor is designed so that water reaches saturation temperature about $\frac{1}{3}$ of the way up the channel with nucleate boiling in the remaining $\frac{2}{3}$ of the channel. The fuel was 2% enriched uranium oxide clad with zirconium containing 1% niobium. It was designed so that it could have on-line refueling, thus increasing its op-

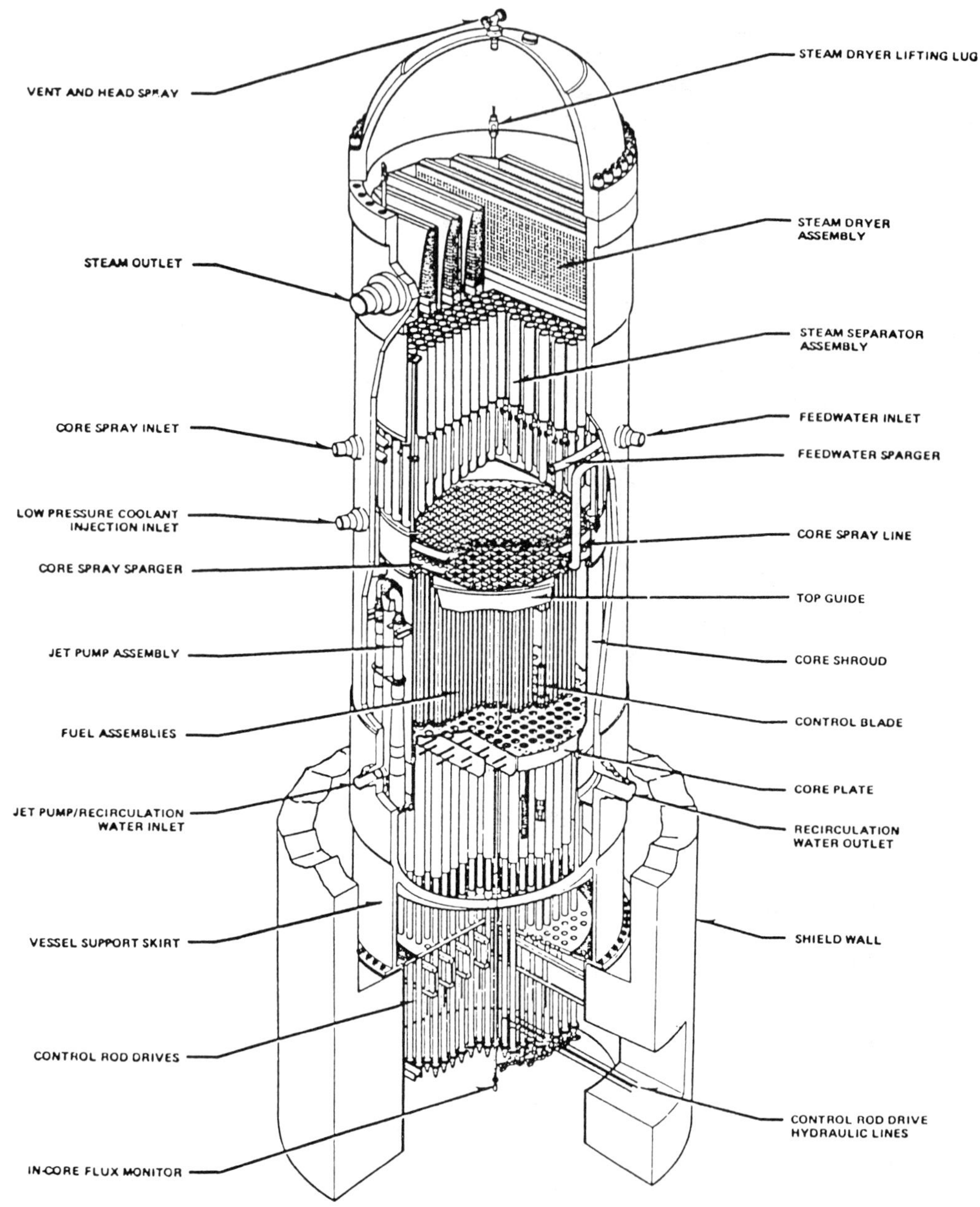

FIG. 7. BWR reactor assembly.

erating capacity. Its inlet temperature was 270° C and average outlet temperature was 284° C, with an operating pressure of 986 psig. Although the reactor itself was contained in a steel compartment, the entire reactor did not have a containment building in the Western sense. The RBMK was also designed with a positive void coefficient of reactivity, as well as having a control-rod design such that, when the rod was being inserted into the core from its full out position, it would add reactivity instead of decreasing reactivity. Following the accident at Chernobyl, the RBMK reactors were modified so that the insertion of the

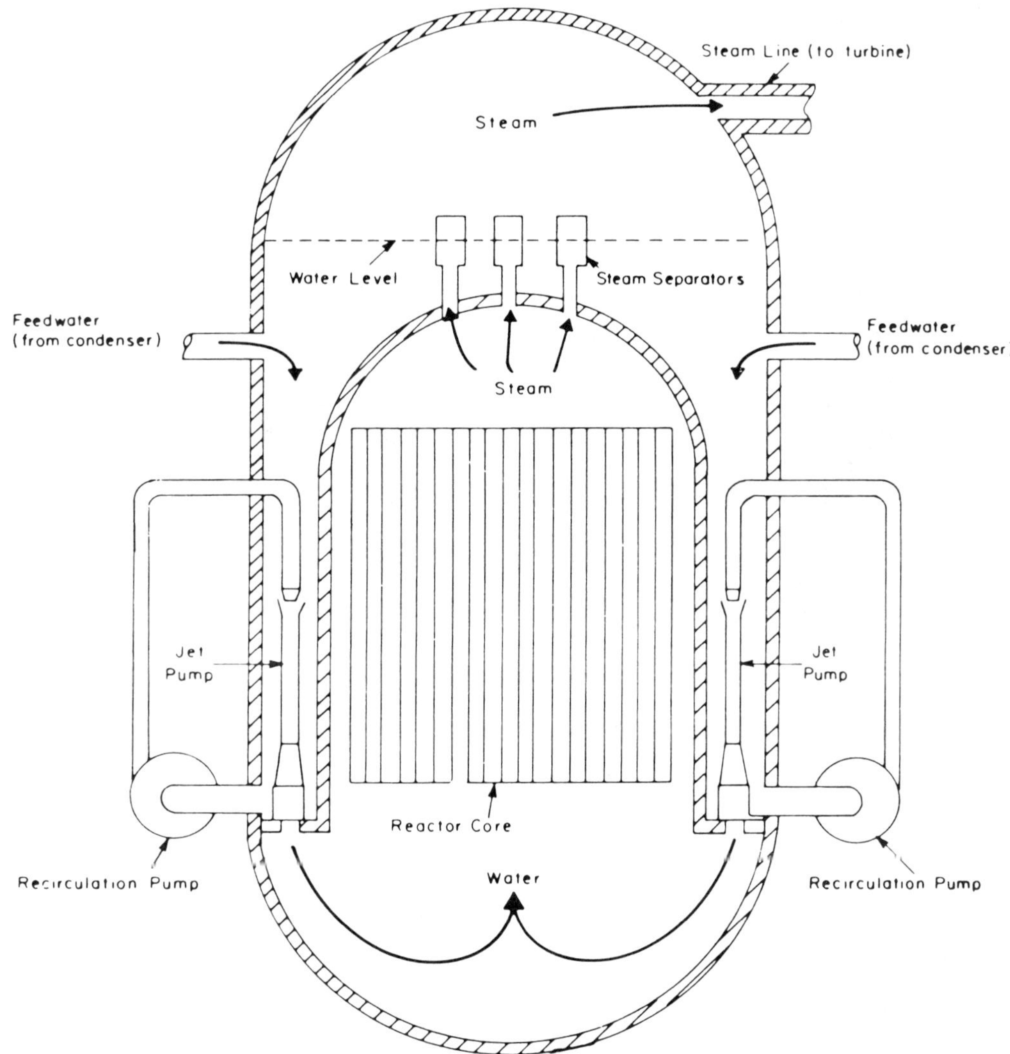

FIG. 8. Schematic diagram of BWR nuclear steam supply system.

control rods from the full out position does not initially add reactivity.

5. REACTOR SAFETY

Fissile isotope fuel in commercial nuclear power plants is consumed generating electricity in the range of 30 000 to 50 000 MW-d per metric ton of uranium. The energy produced by the fuel in MW-d per metric ton is a measure of the fuel consumed or burned. Since about a third of the fuel load is replaced per year after a year's operation, there is considerable buildup of fission products in the fuel at any given time. It is essential that these radioactive fission products be prevented from being released to the environment. This is accomplished by maintaining a defense-in-depth concept by, first, using an oxide fuel which has a high melting point. Second, the fuel has zircalloy cladding, which has high heat-transfer characteristics and high mechanical strength. Third, various engineered safeguards are built into the design of the plant to maintain core cooling under most accident conditions. Fourth, the reactor core is placed inside a high-strength pressure ves-

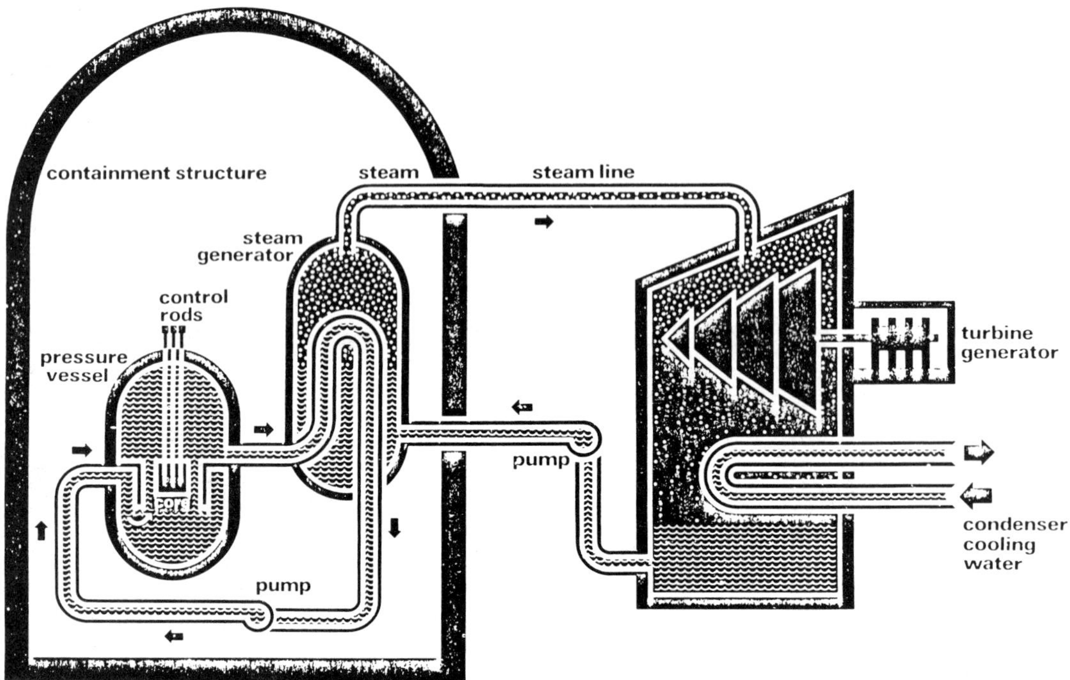

FIG. 9. Schematic diagram of PWR System.

sel and primary circuit. Fifth, the entire reactor and primary circuit are placed inside a steel and/or concrete containment building. Sixth, instrumentation and controls are built into the system so that automatic shutdown of the reactor and automatic operation of the engineered safeguards will take place without operator intervention. Highly trained operators, maintenance personnel, and supervisors complete the safeguards for the safe operation of a nuclear power plant. In addition, the Nuclear Regulatory Commission (NRC) has resident site inspectors at each nuclear power plant to audit the operations of the plant continuously.

It is of interest to consider the engineered safeguards that have been built into the power plant to maintain adequate cooling of the reactor core at all times. Maintaining core cooling of the reactor for a significant period of time after shutdown is important because of the need to remove the decay heat associated with fission product decay.

The engineered safeguards consist of instrumentation and control circuits, redundant heat-removal systems, emergency core-cooling systems, redundant emergency power sources, containment cooling systems, and strong steel and concrete containments. The instrumentation and control system is probably the most important engineered safeguard, since it is through this system that the reactor can be automatically shut down and the operators can detect abnormal situations and take corrective action. Since a reactor can generate more power than its design cooling capability, it is necessary to have instrumentation that continuously monitors the rate of power increase as well as controlling its operating power level. Usually, neutron and fission detectors are used to monitor the rate of change of power as well as the power levels. Neutron detectors are usually ion chambers filled with boron trifluoride gas and operated in an ion-current or pulse mode. Neutrons interacting with ^{10}B produce alpha particles that cause ionization in the ion chamber. Fission detectors are ion chambers filled with argon gas and internal plates or cylinders coated with fissile material. Neutrons causing fission in the ion chamber result in ionization of the gas in the chamber by the fission products. Again, the ionization current in the chamber, which is proportional to the fission

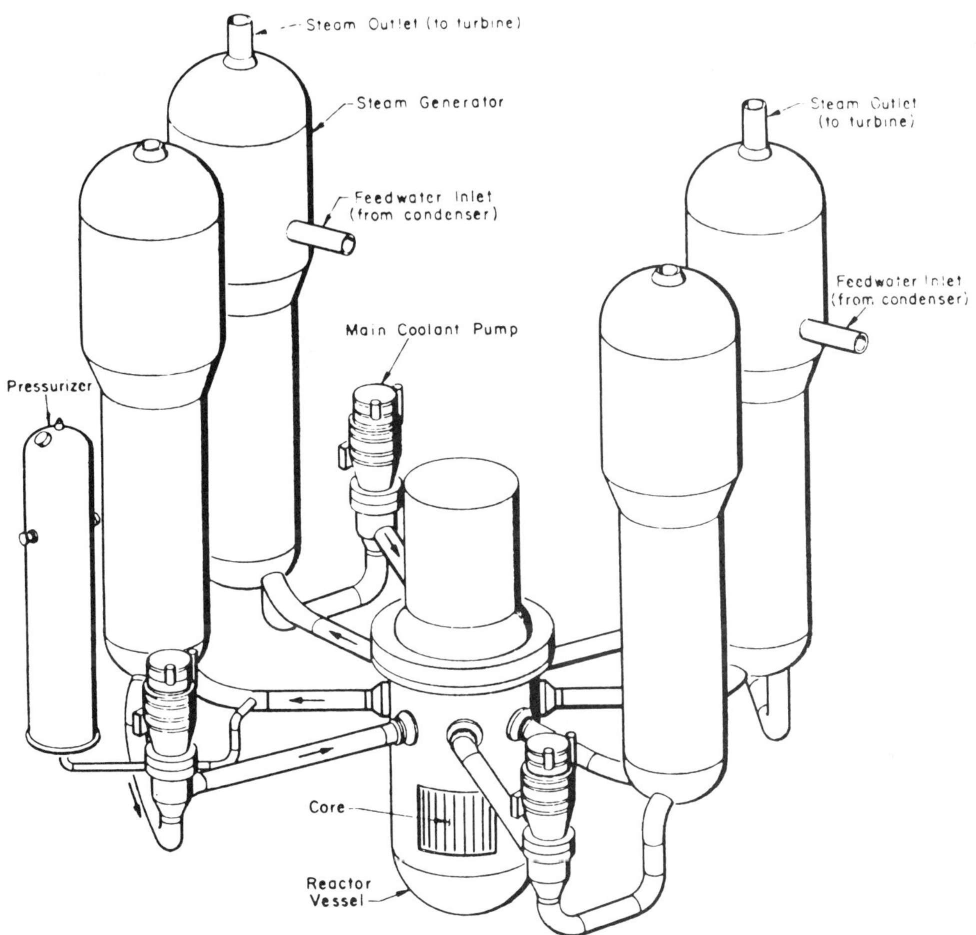

FIG. 10. Schematic diagram of PWR reactor-coolant system.

rate, is measured. In addition, instrumentation is available to monitor the distribution of power being generated in the core. Reactors have automatic power trip or "scram" circuits that will automatically cause insertion of the control and safety rods when the power level exceeds preset limits, or when the rate of increase of power exceeds preset limits. There are also temperature gauges that monitor coolant temperatures in various parts of the primary circuit and can also cause reactor scram when the values exceed preset limits. Radiation detectors throughout the containment building and turbine building monitor these areas for possible inadvertent release of radioactivity through leaks in piping coupled with defective fuel cladding or cladding rupture, which would allow fission products to escape from the fuel.

Since it is essential to maintain coolability of the reactor core by avoiding loss of circulation capability or loss of coolant, a number of redundant systems to maintain coolability have been installed in LWRs. The operation of valves and pumps requires electrical power; therefore, off-site power is brought into the plant by two redundant approaches. In addition, a battery system and emergency diesel generators are provided at the power station for operation in the event of loss of off-site power.

In the case of PWRs, which have primary and secondary coolant systems for heat removal, two or more auxiliary feed-water pumps and corresponding feed-water circuits and supply are provided to maintain adequate secondary water supply to the steam generators. During certain modes of normal

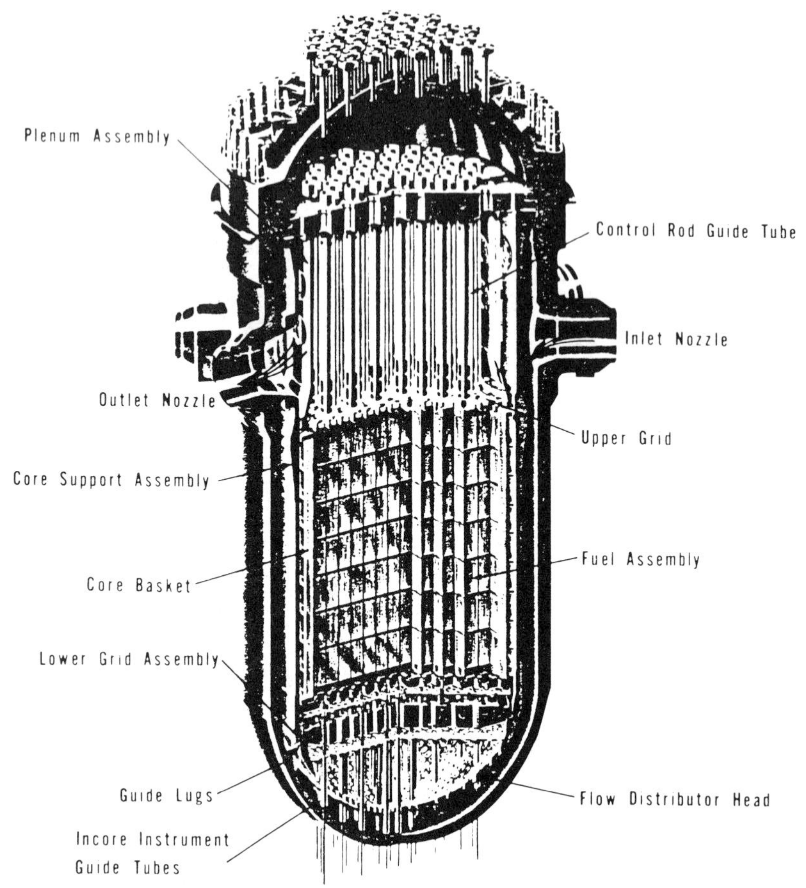

FIG. 11. PWR reactor core assembly.

operation and in emergency situations, steam can be released to the atmosphere as a heat sink. Two separate emergency core-cooling systems (ECCS) are also provided that can introduce water into the primary circuit. One system is composed of several large-volume accumulators containing borated water, which will automatically supply water to the primary system when its pressure drops below preselected values. A second ECCS system is available, where a pump supplies borated water from a large-volume supply tank. The ECCS was designed to handle large pipe-break loss-of-coolant accidents (LOCA). Since the Three Mile Island Unit 2 (TMI-2) accident in April 1979, where the core was uncovered because of a stuck-open pressurizer valve resulting in a loss of primary coolant water, consideration is also given to the prevention of small-break LOCAs (SBLOCAs). In addition to these mechanisms for heat removal, there is a smaller capacity residual-heat–removal system that is available to the primary system for removing heat from the core after it has been shut down. Since large pumps require their own cooling water for operation, there are redundant service water systems for maintaining the operation of pumps. These varied and redundant safety systems have been introduced to be able to maintain core cooling under all circumstances.

Before a nuclear power station can be constructed and operated, the NRC requires for review and approval technical information about a proposed nuclear power plant, such as detailed descriptions of the site, design of the power plant, operation, maintenance, and management procedures; probabilistic safety assessment; accident analysis, including deterministic normal and abnormal transients;

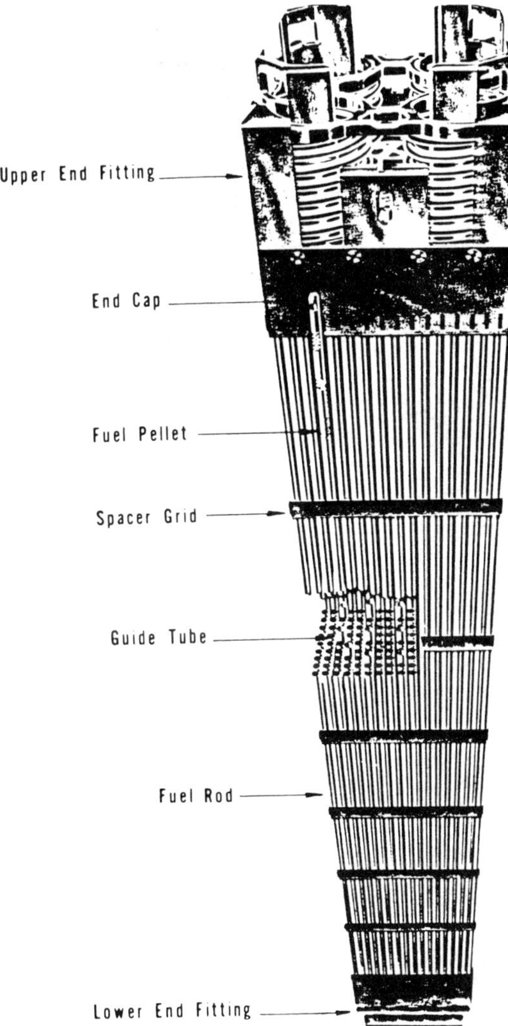

FIG. 12. PWR fuel assembly.

and beyond-design-basis accident analysis. To carry out reactor accident analysis, it is necessary to have an understanding of reactor physics (i.e., neutronics) of the reactor as well as the thermal hydraulics or heat-transfer mechanisms of the plant. In some instances, coupled neutronics and thermal hydraulics analysis is essential for the analysis of transients during operation.

A great deal of LWR safety research under the sponsorship of the U.S. Atomic Energy Commission (USAEC), the U.S. Nuclear Regulatory Commission (USNRC), industry, and the Electric Power Research Institute (EPRI) has been carried out since the late 1960s in the development and verification of complex thermal hydraulics accident analysis computer codes. The RELAP (Ransom *et al.*, 1985, 1987) and TRAC (Schnurr *et al.*, 1992; Borkowski *et al.*, 1992) family of computer codes have been developed by the Idaho National Engineering Laboratory (INEL) and Los Alamos National Laboratory (LANL), respectively, under the sponsorship of the USNRC for the thermal hydraulics analysis of LOCAs and SBLOCAs. RELAP-5, for example, is a complex computer code that solves six thermal hydraulic equations and considers energy, mass, and momentum for water in the liquid and vapor phases for the modeling of various components of the primary and secondary circuits of a PWR. A version of RELAP-5 also exists that can model BWRs. The reactor core is usually modeled on a three-dimensional basis, whereas the remainder of the components are modeled on a one-dimensional basis. Such complex computer codes as RELAP-5 to simulate SBLOCAs have required, until recently, the use of large mainframe computers such as the IBM 3090 or Cray high-speed computers. With the introduction of high-speed and large-memory work stations, such codes as RELAP-5 have been adapted to these smaller computers. A significant amount of experimental work, such as the LOFT (Loss of Flow Test) series of experiments at INEL as well as separate effects experiments, has been carried out to verify RELAP-5 analyses. In addition, international cooperative programs under the leadership of the USNRC for RELAP-5 verification have been carried out by research groups in the European Community, Japan, Korea, and Taiwan.

The TMI-2 accident showed the necessity of improving our understanding of severe accidents, since a large fraction of the core became molten and flowed to the bottom of the pressure vessel. Thermal hydraulic computer codes such as RELAP-5 are adequate for analyzing a transient to the onset of core uncovery but breaks down when there is massive degradation of the fuel. In order to simulate the degradation of the fuel mathematically, computer codes such as MELCOR (Summers *et al.*, 1991) have been developed by Sandia National Laboratory (SNL) and SCDAP/RELAP5 MOD2 (Allison *et al.*, 1989) by the INEL under the sponsorship of the USNRC. MELCOR is an attempt to follow the severe-

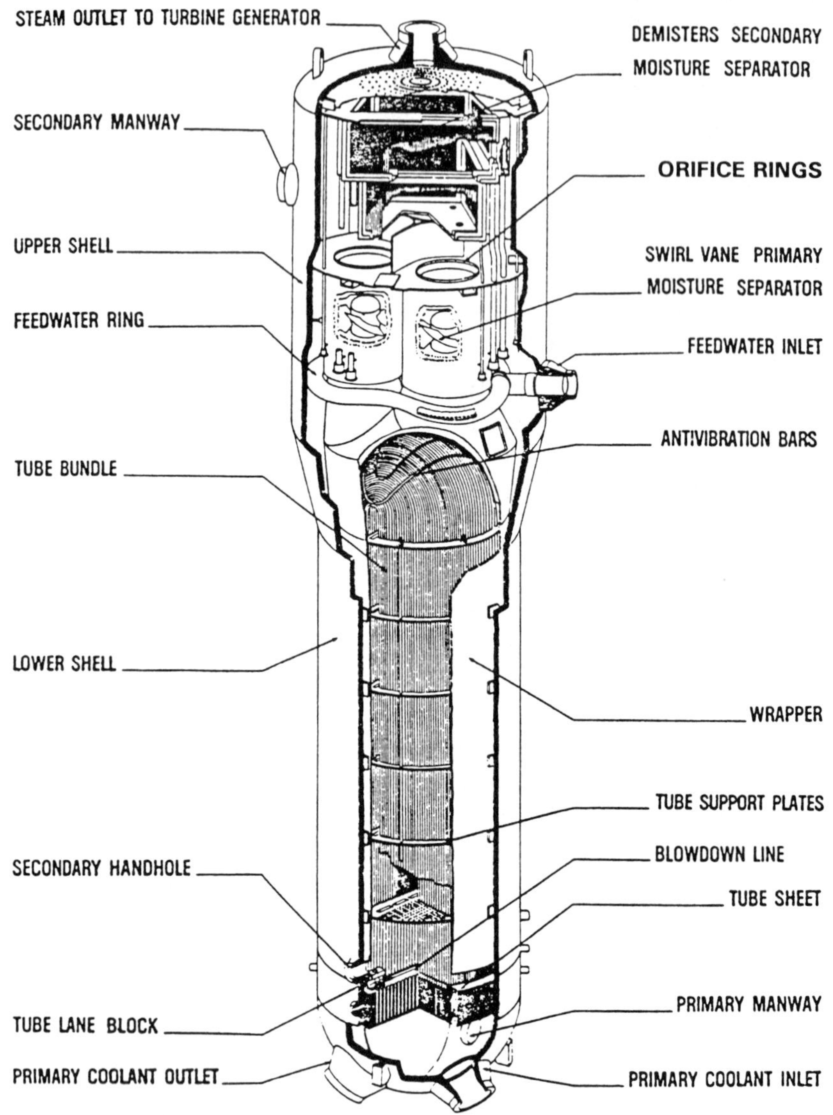

FIG. 13. PWR steam generator.

accident phenomenon from the overheating of the fuel rods to the molten core slumping to the bottom of the pressure vessel, interacting with the steel pressure vessel, and finally flowing out of the pressure vessel onto the reactor cavity concrete pad beneath the reactor. Core–concrete interaction is then followed by a buildup of pressure and rupture of the containment due to combustion of hydrogen produced from the oxidation of the zirconium of the cladding combined with other phenomena such as direct containment-atmosphere heating by molten-core particles. The rupture of the containment with the subsequent release of fission products to the environment and its consequences are simulated by other computer codes such as the Melcor Accident Consequence Code System (MACCS) (Chanin *et al.*, 1990). Severe-accident analysis is an attempt to provide a quantitative basis to define emergency preparedness and evacuation zones surrounding a nuclear power plant, as well as to determine the design safety margins in the con-

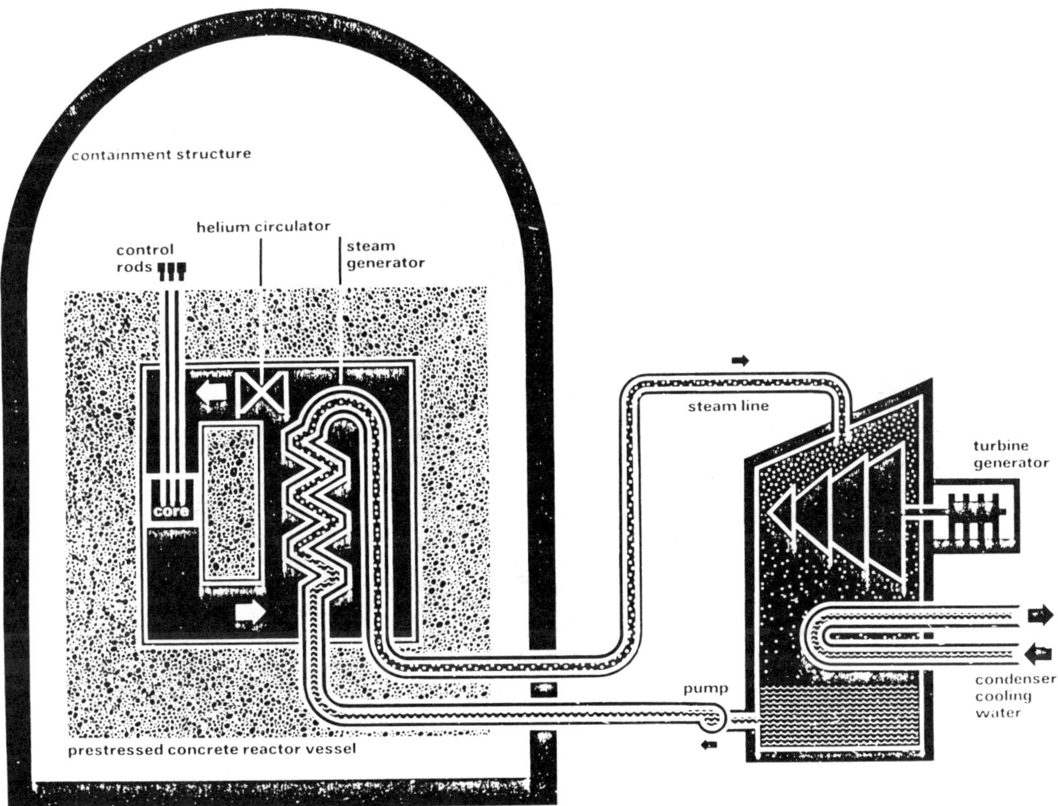

FIG. 14. Schematic diagram of HTGR.

tainment structures and other barriers to fission product release.

Since the mid-1970s probabilistic risk analysis (PRA) (Fullwood and Hall, 1987), more recently called probabilistic safety assessment (PSA), has become a popular way to analyze complete power plants to identify design deficiencies. The approach is to develop fault trees for all of the systems in a plant and to try to quantify overall failure probabilities of systems by providing failure probabilities of components based upon experience. The probability of a core becoming molten, or core melt probability, is calculated. The rupture of containment and resulting release of fission products to the environment can also be calculated on a probabilistic basis. Finally, with use of wind pattern frequencies and other weather statistics and population densities surrounding the plant site, consequences such as deaths and early cancers can be calculated on a probabilistic basis. The first complete analysis of power plants using PRAs was carried out in the Reactor Safety Study, WASH-1400 (NUREG 75/014) in 1973–1975. A more recent study on the assessment of severe accident risks for a set of commercial LWRs in the United States by the USNRC resulted in the report "Severe Accident Risks: An Assessment for Five U.S. Nuclear Power Plants," published by the USNRC as NUREG 1150 in October 1990. Although these studies indicated that the probability of a core melt was in the range of 10^{-5} to 10^{-6} per reactor year with accompanying release of fission products, there has been a significant decrease of public confidence in commercial nuclear power as a result of the accidents at TMI-2 in April 1979 (Rogovin *et al.*, 1980) in the United States and Chernobyl Unit 4 (U.S. Nuclear Regulatory Commission, 1987; IAEA, 1992) in the Soviet Union in April 1987.

In the case of TMI-2 core, uncovery took

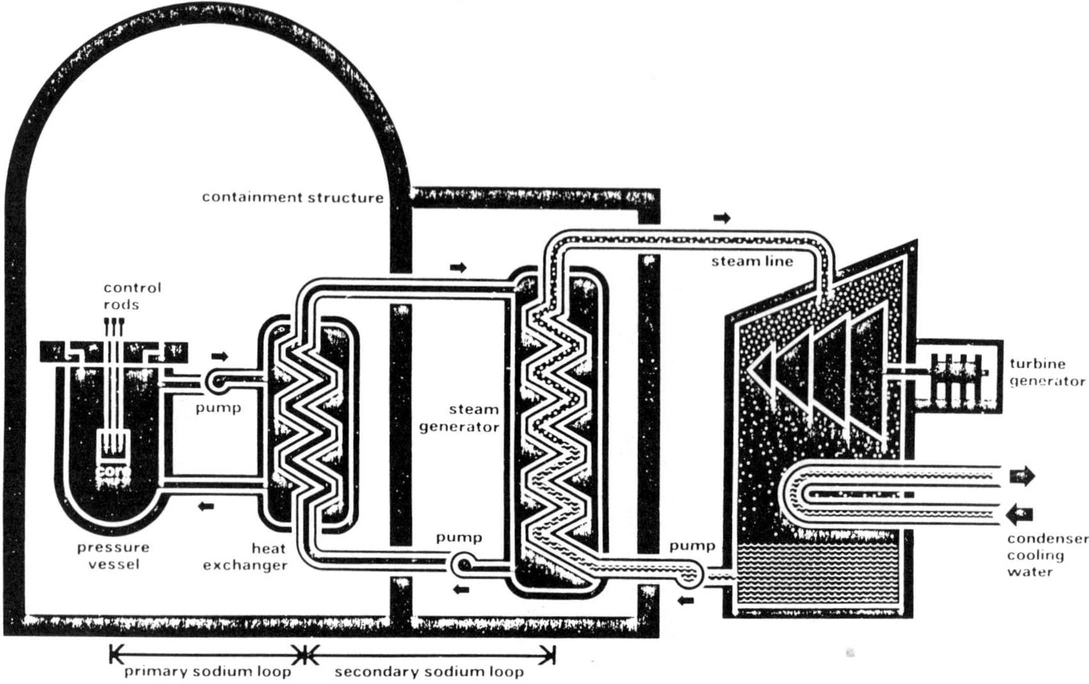

FIG. 15. Schematic diagram of LMFBR.

place as a result of a stuck-open motor-operated pressurizer relief valve coupled with operator errors. As a consequence, a large fraction of the core became molten and settled in the bottom of the pressure vessel. Fortunately, a sufficient amount of core cooling water was poured into the pressure vessel so that the vessel remained intact. Although the core was completely destroyed, most of the fission products were contained in the primary circuit. Some gaseous fission products such as the noble gases and radioiodine were released to the containment building, with only a small quantity of radioactivity being released to the atmosphere and surrounding environment. The report of Rogovin *et al.* (1980) estimates that about 2.4×10^6 Ci of short-lived noble gases and 15 Ci of ^{131}I in gaseous form were released to the environment. Because of the uncertainty in the status of the reactor in the early stages of the accident and media confusion, there was consideration given to large-scale evacuation of the population surrounding the TMI site. However, this was not carried out because the primary circuit and the containment remained intact and most of the fission products were contained. Although the reactor plant was a financial loss and the cleanup required large expenditure of funds, no lives were lost. The probability of any latent cancers has been estimated to be very small, such that it is not possible to distinguish cancers resulting from the accident from cancers due to other sources. A consequence of the TMI-2 accident has been the requirement of an emergency preparedness zone, as well as emergency procedures for evacuation and public notification, around nuclear power plant sites.

The Chernobyl-4 accident, on the other hand, released large quantities of radioactivity to the environment, required significant evacuation of the population around the site, and has taken considerable land out of use because of radioactive contamination of the soil. The Chernobyl nuclear power station is located about 60 miles north of the city of Kiev in Ukraine, one of the former Soviet Union republics. There were four RBMK-type nuclear power plants in operation at the station with an additional two under construction at the time of the accident in April 1986. The Chernobyl-4 accident was a result of design deficiencies and gross operator error. The lack of an adequate containment compounded the consequences because of the release of such long-lived radioactive isotopes as cesium-137 and strontium-90 and short-lived iodine-131 on a worldwide basis in the northern hemisphere. The Chernobyl-4, as described earlier, was a graphite-moderated, light-water–cooled, pressure-tube reactor. Design deficiencies, such as the reactor having a positive coolant void coefficient, control rods adding reactivity when being inserted from their full out position, and reactor instability at low power, coupled with operator error, such as bypassing scram circuits, operating with too many control rods in their full out position, and conducting an experiment without fully understanding the physics and thermal hydraulics of the system, caused the catastrophic accident. Decreasing the flow in the pressure tubes by deactivating the coolant pumps during a special experiment caused water boiling in the core, which resulted in increased reactivity. Power, pressure, and temperature increased. Coupled with the addition of reactivity from control rod insertion, this caused the rupture of pressure tubes, adding further voiding which in turn added more reactivity. The addition of large amounts of reactivity made the reactor superprompt critical, which essentially blew the reactor apart, resulting in molten fuel, fission products, and hot graphite being expelled out of the building and destroying the building structure itself. It has been estimated that about 50×10^6 Ci of noble gases and another 50×10^6 Ci of other radionuclides, which constituted about (3–4)% of the total fission-product inventory, were released to the environment (U.S. Nuclear Regulatory Commission, 1987). The initial wind pattern at the time distributed the fission products and fine vaporized fuel toward the west and northwest. Significant quantities of radioactivity were detected a few days after the accident in Finland and Sweden. Later, winds dropped the fission products in Eastern and Western Europe.

The magnitude of the potential property loss and health effects from the release of large quantitites of fission products and other radioactive reactor components make it mandatory that designers, constructors, operators, and regulators of nuclear power plants take utmost care in carrying out their respective responsibilities, with emphasis on reactor safety, to avoid future release of significant quantities of radioactivity.

6. ADVANCED WATER REACTORS

Since 1985, EPRI has led a utility, government, and industry advanced light-water reactor program. This program includes the simplification of overall design, including the design of passive emergency core-cooling, decay-heat–removal, and containment-cooling systems. The objective of the program is to seek greater safety margins, greater ease of construction, improved reliability and availability, improved maintainability, lower costs, and greater ease of operation over existing large LWRs (Stahlkopf *et al.*, 1988). The U.S. Department of Energy (DOE) since 1986 has had contracts with Westinghouse and GE for the detailed design, development, and certification for the APWR, AP600, and ABWR and the SBWR, respectively. The Westinghouse APWR, GE ABWR, and ABB Combustion Engineering's System 80 PWR are evolutionary LWRs improving upon the current version of LWRs. The GE SBWR and Westinghouse AP600 are mid-sized LWRs with passive safety features.

In addition to the LWRs being developed in the United States, other reactors such as the Modular High Temperature Gas-Cooled Reactor (MHTGR) by General Atomic; the advanced liquid-metal–cooled fast-spectrum reactor (PRISM) by GE; the light-water reactor (PIUS) by ABB Atom; and the heavy-water–moderated and –cooled reactor (CANDU-3) by Atomic Energy of Canada, Ltd. (AECL) are some of the advanced reactors that are being designed. A compilation of the characteristics of some of these reactors is provided in Table 8. Institutes of the Commonwealth of Independent States (CIS), states of the former USSR, which are forming private and independent organizations, are also working on advanced pressurized light-water reactors such as the VVER 91 and 92 (Gagarinski *et al.*, 1992). A French and German consortium, NPI, is also working on advanced PWRs.

An ABWR (Anon., 1992) of 1350 MW(e) has been developed jointly by GE, Hitachi, and Toshiba. The Tokyo Electric Power Company initiated the construction of two ABWRs in

Table 8. Design characteristics of new reactors (Nuclear News, 1992).

Design characteristics	System 80+	ABWR	SBWR	APWR	AP600	MHTGR	CANDU-3	PIUS	ALMR
Reactor type	Pressure l. water	Boiling l. water	Boiling l. water	Pressure l. water	Pressure l. water	Gas-cooled	Pressurized D_2O	Pressurized H_2O	Liq. metal fast breeder
Lead designer	ABB-CE	GE	GE	Westhse.	Westhse.	General Atomics	AECL	ABB ATOM	GE/ANL
MW(e) (net)	1300	1300	640	1050	600	538(4)	450	640	1440(3)
MW(t)	3817	3926	2000	3150	1940	1400(4)	1440	2000	4245(3)
Coolant	L. water	L. water	L. water	L. water	L. water	Helium	D_2O	H_2O	Liq. sodium
Moderator	L. water	L. water	L. water	L. water	L. water	Graphite	D_2O	H_2O	None
Fuel material	UO_2/PuO_2	UO_2	UO_2	UO_2	UO_2	UCO	Nat. UO_2	UO_2	U25%/Pu10%/Zr metal alloy
Cladding material	Zircal-4	Zircal-2	Zircal-2	Zircal-4	Zircaloy	Refractory coated particles	Zircaloy-4	Zircaloy-4	HT-9 ferritic alloy
Fuel geometry	16 × 16	8 × 8 or 9 × 9	8 × 8 or 9 × 9	17 × 17	17 × 17	Hexagonal graphite blocks	37-element fuel bundle	18 × 18	217 pin in hexagonal bundle
No. fuel assemblies	241	872	732	193	145	660	232 fuel channels	213	66 driver fuel
Pin diameter (mm)	9.7	12.3	12.3	9.5	9.5	13	13.1	9.5	7.2
Equivalent core diameter (mm)	3650	3708	4880	3370	2922	1650 3500	4912	3760	1570
Fuel length (mm)	3810	3708	2743	3658	3658	7925	5944	2500	1350
Average power density (kW/L)	95.5	50.6	41.0	96.2	78.8	5.9	12.78	72.3	180
Coolant inlet temperature (°C)	292	216	216	287	276	258	268	260	338
Coolant outlet temperature (°C)	324	288	288	325	312	687	310	290	485
System pressure (MPa)	15.41	7.07	7.07	15.51	15.41	6.38	9.90	8.99	Unpressurized
No. of loops	2	N.A.	N.A.	3	2				

1989 at its Kashiwazaki–Kariwa Nuclear Power Station. Some features that improve construction costs and maintainability and provide additional margins of safety include internal recirculation pumps, which eliminate external recirculation pumps and piping; no large coolant nozzles below the top of the core; new pressure suppression containment, which permits rapid vessel depressurization; improved core and fuel design that maximizes performance and fuel economy; and a triple-redundant ECCS core protection system.

The APWR (Anon., 1992) that has been designed by Westinghouse, Mitsubishi Heavy Industries of Japan, and five Japanese PWR owner utilities is a four-loop, 1350-MW(e) power plant. Its most significant difference from the exisiting PWRs is that it has a 15–20% lower power density, which provides a greater margin of safety. In addition, a larger pressure vessel provides a larger core inventory of water above the core, which enhances safety and decreases the required capacity of the ECC system. A larger diameter containment building provides greater ease in maintenance. Economic goals are to reduce the capital costs by 15% and fuel-cycle costs by 20% over existing PWRs. The reduction in the fuel-cycle cost is achieved by the reduction of the power density, which leads to a reduction of the initial enrichment required to achieve the same energy output. A further reduction of the initial enrichment is accomplished by shifting the spectrum of the core during early operation of the core by displacing the moderator with a mechanical water displacer. This results in increased production of plutonium by slightly hardening the spectrum for increased resonance neutron absorption. This plutonium is then burned later in the cycle when the displacer is removed, producing a softer spectrum and increased core reactivity.

The AP600 (Anon., 1992; Vijuk and Bruschi, 1988; Bruschi and Anderson, 1991) is a two-loop, 600-MW(e) PWR system that includes a simplified reactor coolant system and a balance of plant, passive safety, and plant arrangement systems with a modular construction approach. Some of the key design features include a reactor coolant system that uses sealed canned motors mounted directly on the steam generators, a core design resulting in a low core power density (80 kW/L), a reactor core surrounded by a stainless steel and water reflector, and passive safety features that operate following plant transients or accidents. The passive safety system provides for residual heat removal, reactor coolant makeup for inventory control, safety injection, and containment spray cooling. A passive residual-heat–removal heat exchanger (PRHR-HX) has been provided to remove the decay heat by natural circulation from the core in the event that the normal or startup feedwater systems (i.e., steam-generator heat-removal systems) are not functioning. In the AP600, if there is steam released to the atmosphere inside the steel containment, natural circulation is used to transfer the heat to the steel shell. The AP600 has been designed to have significantly fewer valves and large pumps, less piping, and less control cabling and ducting in order to minimize construction costs.

General Electric has been working on the design of the Simplified Boiling Water Reactor (SBWR) (McCandless and Redding, 1989) since 1982 with the objectives of providing power generation costs cheaper than coal, simplified safety systems, shorter construction schedules, and a 600-MW electrical rating. In the SBWR design, the power density was decreased to 42 kW/L, and natural circulation was chosen for providing coolant flow through the core. A gravity-driven core-cooling system provides a simple approach to emergency core cooling, which eliminates the need for pumps and emergency diesels. It has a gravity driven cooling pool that can keep the core covered. The SBWR pressure vessel is designed so that there are no large pipes attached to the pressure vessel near or below the core elevation, which ensures core coverage for all design basis events. In addition to the passive cooling pool, the SBWR uses a pressure suppression pool, isolation condenser, and passive containment cooling system for its passive emergency response.

The PIUS (Process Inherent Ultimate Safety) reactor concept by ABB Atom of Sweden (Hannerz, 1988) is a 640-MW(e) PWR-type reactor that is located near the bottom of a deep reactor pool. The pool is compartmentalized so that the reactor core is near the bottom of a chimneylike structure submerged in a large pool. The pool contains water with a high boron concentration in a prestressed concrete vessel. The reactivity is controlled by the boron concentration in the coolant water and by temperature so that there are no control rods in the reactor. The moderator temperature coefficient is strongly negative throughout the operating cycle. During normal operation, the power is controlled by using the negative moderator temperature coefficient without adjusting the coolant boron concentration. The PIUS is an innovative concept since fuel integrity is maintained by an inherent passive coolant system, and there should be no core uncovery or fuel overheating in any credible accident because of the large volume of available coolant water.

7. ECONOMIC CONSIDERATIONS

The generation of electricity through the use of nuclear energy must be competititve with the cost of power generation from the burning of fossil fuels such as coal and oil. The cost of electricity generation may be considered to be composed of capital carrying charges, operation and maintenance costs, and fuel costs. The capital carrying charges include the depreciation and amortization of the costs of building and financing the plant. The capital costs of a plant may also change depending on the amount of additional investments required during its operating life. In the case of nuclear plants, the capital carrying charges are the predominant costs for the generation of electricity. The large capital carrying costs include the high cost of construction, as well as the interest charges accumulated during the length of time required to construct the plant, and decommissioning costs.

The Ahearne Committee Report on Nuclear Power (National Research Council, 1992; Ahearne, 1993) presents the Energy Information Administration's 1988 figures for the comparison of nuclear- and coal-generated electricity. Table 9 lists the highest, lowest, and average total generation costs for nuclear and coal in 1988.

The reasons that existing nuclear power plants have higher capital carrying charges than coal plants are that the nuclear power plant equipment and building have been more expensive to construct, they have taken longer to construct and hence accumulate more interest charges, and finally they have decommissioning costs that must be taken into consideration. In the late 1960s and early 1970s, at the beginning of the commercial nuclear power industry, it was believed that operation and maintenance (O&M) costs would be fairly low. After the TMI-2 accident in April 1979, there was an increase in regulatory requirements, which significantly increased the O&M costs.

Table 9. Average electricity generation costs for 1988 (cents/kWh) (National Research Council, 1992).

	Nuclear	Coal	Highest/Lowest Nuclear	Highest/Lowest Coal
Capital carrying	3.4	1.1	9.4/0.4	5.4/0.3
Operation & maintenance	1.5	0.4	1.2/0.8	0.7/0.5
Fuel	0.8	1.7	0.7/0.5	2.5/1.4
Total	5.6	3.1	11.3/1.6	8.5/2.2

The Ahearne committee reports that plants that went into commercial operation during the period 1978–1989 required an average construction time of 11.1 years in the United States as compared with 5.9 years in France and 4.7 years in Japan. Since a lengthy construction period with the attendant interest charges has a significant impact on the capital costs, there is a strong incentive to minimize the construction time in the United States.

As there have not been any orders for new nuclear power plants since 1979, it is difficult to estimate the cost of electricity generation from new nuclear power plants because of the uncertainty of the capital costs for the advanced light-water reactors (ALWR) that the reactor vendors would probably supply to the utilities. The ALWRs are being designed with simplification as well as ease in construction, operation, and maintenance in mind. In addition, with the NRC tending toward a single construction and operating license approval process, which would decrease the construction time, the capital cost should be more predictable for the new plants.

In 1992, there were a couple of nuclear power plants that the utility owners decided to shut down because of the costs to replace steam generators, such as in the case of Portland General Electric's Trojan power plant whose alternate source of electricity from hydropower is cheaper. The Yankee Rowe plant in Massachusetts was shut down because of the uncertainties in radiation embrittlement of the pressure vessel, which caused concern in the NRC. As the nuclear power plants begin to approach the end of their 40-year operating license period, the owners must consider the cost effectiveness of pursuing license extension. Such an extension might require additional capital investment, which, together with higher O&M costs, may make nuclear electricity less competitive compared with electricity generated from other sources. Nuclear electricity is highly competitive in countries such as Japan and France, which have no indigenous fossil fuels. Limitations on the production of greenhouse gases, such as CO_2, and acid rain restrictions on fossil fuel combustion in both electricity generation and transportation may result in the increased use of nuclear power plants for electricity generation in the future in the United States.

8. RADIOACTIVE WASTE MANAGEMENT

Although the production of electricity using fission energy has a number of advantages, one of the disadvantages is that the resultant radioactive substances created must be prevented from being dispersed in the environment for many years. In addition to the high-level waste (i.e., fission products and transuranium isotopes) that is produced during reactor operation, there are other low-level wastes generated during the fuel cycle that must be managed (see NUCLEAR WASTE MANAGEMENT).

During the fuel production process from the mining and milling operation through the conversion to uranium hexafloride, the gaseous-diffusion, centrifuge, or atomic-vapor–laser enrichment processes, and the fuel fabrication process, there are mill tailings, sludge from chemical processes, scrap, and fabrication residue containing uranium and its daughter products that must be considered as low-level waste. During reactor operation, resins from the cleanup of the primary cooling water, filters from air cleaning, sludge and waste from evaporator concentrates, and airborne releases of radioactive noble gases and tritium are additional sources of low-level radioactive waste. If the spent fuel from commercial nuclear power plants is reprocessed, as it is in France, the United Kingdom, Japan, and the Former Soviet Union, and as is the fuel from military reactors in the United States, there is high-level waste consisting of the fission products and transuranic elements that must be disposed of. In addition, there will be cladding hulls with residual fuel, transuranic solids from the chemical processing, and gaseous radioactive wastes such as the noble gases, tritium, and iodine. Finally, in the decontamination and decommissioning of the nuclear plant, there will be large components, such as the pressure vessel and the reactor internals including the control rods, internal shield, etc., which will have become radioactive from interactions with neutrons. In all of the components of the primary circuit, there will be activated residue from small impurities in the water, which is difficult to remove. These components in total must be considered for radioactive material disposal. The concrete shielding and other support structures surrounding the pressure vessel will also have become activated to some small degree so that they may have to be considered as low-level waste.

The United States national policy in 1994 is to not reprocess spent fuel from commercial nuclear power plants but to develop a final storage facility for the spent fuel to discourage the use of plutonium so as to encourage nonproliferation. Although it had been planned by the Department of Energy (DOE) to characterize three possible sites for permanent storage of spent fuel, the U.S. Congress in 1987 enacted legislation restricting site characterization to the Yucca Mountain site in the State of Nevada. Because of public opposition from the State of Nevada and its citizens, initiation of the characterization studies was delayed until late 1991 by the refusal of the State to issue various permits required for the study. The spent fuel contains long-lived fission products and transuranic elements, which must be contained for tens to hundreds of thousands of years. The proposed concept is to encapsulate the spent fuel and then store the fuel in underground storage facilities at Yucca Mountain. The controversy involves the question of whether the Yucca Mountain area will remain stable against earthquakes or volcanic action for a long period so that the disposal facility will be free from ground water.

In countries such as France and Japan, where fuel reprocessing is carried out, the fission products are separated from the transuranic elements. The fission products are then vitrified (i.e., embedded in glass) and then encapsulated for final disposal in underground salt mines. The transuranic elements such as plutonium are separated for use as fuel in liquid-metal–cooled breeder reactors or are mixed with enriched uranium oxide for light-water reactor fuel [mixed oxide (MOX) fuel]. A high degree of separation of the transuranic elements from the spent fuel in fuel reprocessing would ease the requirements for the high-level waste respository, since most of the fission products have significantly shorter half-lives than the alpha-active transuranic elements.

The DOE, in 1993, carried out alternative studies for the disposal of the transuranic elements. Concepts such as developing reactors specifically for burning or transmuting the transuranic elements, or using high-current, high-energy proton accelerators for transmuting the transuranic elements to fission products with shorter half-lives, are being considered. The principle in transmutation is to design a system with as hard a neutron energy spectrum as possible so that the capture-to-fission ratio for the transuranic elements is as low as possible. This minimizes the number of higher isotopes being produced and maximizes the fission of the fissionable transuranic elements. In addition, it is desirable, from a safety point of view, to have a system which has as high a power density as possible to minimize the total amount of transuranic inventory in the system. The concept of using accelerators for transmutation is based on the ability to construct a linear proton accelerator that produces high-energy protons in the range of 1.6 GeV with currents of 25 mA or greater. The high-energy, high-current protons are targeted on a heavy element such as lead or tungsten, which produces copious amounts of high energy neutrons. These neutrons then cause the fission of the transuranic elements in the system. The long–half-life transuranic elements are, thus, transmuted by fission into shorter–half-life fission products. Since the accelerator system can be operated with a multiplication of $k < 1$, or less than critical, the concern for supercriticality accidents would be diminished.

Very hard-spectrum reactors using liquid-metal cooling could be designed for the transmuting of transuranic elements. Such reactors could also be designed to produce electrical power.

9. NUCLEAR REGULATION

9.1 USNRC

The federal government in the United States has the responsibility for regulating the production and utilization of atomic energy in accordance with the Atomic Energy Act of 1946 and the Atomic Energy Act of 1954. The Energy Reorganization Act of October 1974 (Public Law 93–438) established the Nuclear Regulatory Commission (NRC) and separated the regulatory functions from the promotional functions of the Atomic Energy Commission (AEC). The Energy Reorganization Act, at the same time, created the Energy Research and Development Administration (ERDA), which took over the other functions of the AEC including its promotional developmental functions. The ERDA eventually became the Department of Energy by the Act of 1977. The NRC is composed of five members appointed by the President of the United States. Its principal functions are the licensing and regulation, including safeguards, of the construction and operation of nuclear reactors and the facilities for processing, handling, and transporting nuclear materials, including special nuclear materials, and the transport and storage of radioactive waste. It also has the responsibility for conducting research necessary for the performance of its licensing and regulatory functions. In carrying out its regulatory functions, the Commission reviews designs and construction quality assurance procedures, as well as conducting inspections of nuclear facilities and enforcing its regulations. The NRC maintains site inspectors at each of the commercial nuclear power plants in the United States.

The Commission carries out its responsibilities by promulgating rules and regulations, which become a part of the Code of Federal Regulations (CFR), a manual for the executive departments and the agencies of the federal government. Title 10, Chapter 1 of the CFR has been reserved for the NRC regulations. This chapter is divided into various parts, as shown in Table 10. The NRC also issues Regulatory Guides which

1. describe methods which the regulatory staff finds acceptable in implementing regulations;
2. discuss techniques used by staff to evaluate specific problems; and
3. provide guidance to license applicants.

The Commission has the three statutory offices of Nuclear Reactor Regulation, Nuclear Regulatory Research, and Nuclear Material Safety and Safeguards to carry out its principal duties. It has an Executive Director of Operations who coordinates the activities of the three principal offices, as well as five regional offices, as of 1993, which provide oversight of the nuclear facilities in their respective geographical areas. It also has the Advisory Committee on Reactor Safeguards (ACRS), the Advisory Committee on Nuclear Waste, and Atomic Safety and Licensing Boards to provide the Commission with advice.

The Commission had, up to 1990, required the submission of a preliminary safety analysis report (PSAR) for review and approval prior to the initiation of construction of a nuclear facility and a final safety analysis report (FSAR) before initiation of operation of the facility. The content requirements for SARs are given in NRC Regulatory Guide 1.70 and include such items as detailed description of the site, plant, reactor, engineered safety features, radioactive waste management, radiation protection, conduct of operations, safety assessments including accident analyses, and technical specifications.

Table 10. Code of Federal Regulations, Title 10, Chapter 1.

Part 1:	Statement of Organization and General Information
Part 2:	Rules of Practice
Part 20:	Radiation Protection Standards
Part 50:	Licensing of Production and Utilization Facilities
Part 51:	Licensing and Regulatory Policy and Procedures for Environmental Protection
Part 52:	Early Site Permits; Standard Design Certifications; and Combined Licenses for Nuclear Power Plants
Part 55:	Operators' Licenses
Part 70:	Special Nuclear Materials
Part 100:	Reactor Siting Criteria

The objective of the safety assessment of nuclear power plants is to provide an analysis and evaluation of the design and performance of the structures, systems, and components of the facility to assess the risk to public health resulting from operation of the plant. The analysis must include margins of safety during normal and transient conditions anticipated during the life of the facility and adequacy of the facility to prevent accidents and mitigate the consequences of accidents.

The range of accidents which should be considered are

1. events of moderate frequency with no abnormal radioactive releases from the facility,
2. events of small probability with potential for small radioactive release from the facility,
3. potentially severe accidents of extremely low probability to establish performance requirements of engineered safety features and to evaluate the acceptability of the site.

For new nuclear power plants, the Commission now also requires the submission of a Probabilistic Safety Assessment (PSA) [formerly called Probabilistic Risk Assessment (PRA)] based upon the detailed design of the plant. It should also be recognized that the Environmental Protection Agency (EPA) requires an Environmental Impact Statement for new large projects.

Following a detailed review of the SAR by the Commission staff, a review by the ACRS is conducted. Before a license is granted for the start of construction, the Commission usually conducts a public hearing at which any opposition views are heard.

In 1990, the NRC promulgated 10CFR52, "Early Site Permits; Standard Design Certifications; and Combined Licenses for Nuclear Power Plants," which significantly changed the rules and regulations governing the licensing of nuclear power plants. The NRC now issues a Combined Operating License (COL), which essentially combines the former construction permit and the operating license. Reactor vendors now submit an application for design certification of a specific nuclear reactor design. In parallel, a utility may submit an application for an early site permit. Following completion of the design certification process and an early site permit, an application for the COL may be considered. The technical information that is required of applicants for a COL is essentially the same as for the previous construction permits and operating licenses except for information technically relevant to a specific site. In addition, such analysis as a design-specific probabilistic risk assessment and proposed tests, inspections, and acceptance criteria are also required. The NRC staff and ACRS conduct a thorough review of the COL application, which is followed by a mandatory public hearing before a COL is issued. Construction of the nuclear power plant may be started after the issuance of the COL. Following completion of the construction, the NRC must issue a finding that the plant as constructed meets all acceptance criteria before operation is permited.

9.2 IAEA

The International Atomic Energy Agency (IAEA) (Scheinman, 1987) was originally established to prevent the proliferation of nuclear weapons and to encourage the use of nuclear energy for peaceful uses. Since the Chernobyl accident in 1986, there has been increased emphasis on helping member states with reactor safety. Before Chernobyl, the IAEA organized international conferences and workshops on nuclear safety, as well as providing operational safety review teams (OSART) for the evaluation of the safety of nuclear power plants at the request of member states. In the late 1970s, the IAEA developed an Incident Reporting System (IRS) for the compilation and distribution of unusual events occuring at nuclear power plants. The IAEA IRS contains unusual-event reports compiled by the Agency as well as those compiled by the Nuclear Energy Agency (NEA) of the Organization of Economic Cooperation and Development (OECD). Also included in the complilation of the unusual events is the analysis of unusual events.

Following the Chernobyl-4 accident in 1986, an International Nuclear Safety Advisory Group (INSAG) was established by the Director General of the IAEA with the intention of strengthening the IAEA's contribution to ensuring the safety of nuclear power plants. The INSAG's responsibilities include "serving as a forum for the exchange of information on nuclear safety issues of international significance and formulating where possible common safety concepts." One of its first products has been the issuance of the "Basic Safety Principles for Nuclear Power Plants" (IAEA, 1988), which outlines specific safety principles that should be followed for the design, construction, and operation of nuclear power plants.

The IAEA at the request of member states will conduct peer review of safety activities in those states. Peer review of three categories of activities has been carried out. These are regulatory activities, engineering matters, and operational safety. The IAEA has organized independent regulatory review teams and developed nuclear safety standards under the heading of regulatory activity peer reviews. They have conducted peer reviews of PSAs, safety analysis, siting, and design concepts under the category of engineering safety review service. In the operational safety category, they have conducted reviews of the operating process, event assessment, and feedback of experience.

Another area where the IAEA has been very active is in the organization of Assessment of Safety Significant Events Teams, which are sent to plants that have had unusual events. They have had over 50 separate missions to over 21 countries.

In the early 1990s, experts from the IAEA and NEA [the Nuclear Energy Agency of the Organization for Economic Cooperation and Development (OECD)] developed the International Nuclear Event Scale (INES), shown in Table 11, which would categorize unusual events so that it would be easier for the public to understand the severity of unusual events or incidents. The INES has categories

Table 11. International nuclear event scale (INES).

0	Below scale (deviation, no safety significance)
1	Anomaly
2	Incident
3	Serious incident
4	Accident without significant off-site risk
5	Accident with off-site risk
6	Serious accident (significant off-site impact)
7	Major accident (widespread health and environmental effects)

from 0 to 7, where 7 represents a major accident that has widespread health and environmental effects. The TMI-2 accident would be a category 5, whereas the Chernobyl-4 accident would be classified as a 7. Although some member states such as Japan, France, and the United States have developed their own nuclear event scale, the IAEA INES is becoming accepted on a worldwide basis.

10. CONCLUSION

This has been a brief review of the development of nuclear fission reactors over the past 50 years. Although nuclear energy was born of military necessity, commercial nuclear electricity with over 411 large nuclear power plants in operation worldwide in 1993 has become a mature, major new source of energy for the world. In 1953, 10 years after the construction of the first nuclear reactor, President Eisenhower of the United States initiated the Atoms for Peace Program, which has since spread the knowledge of nuclear fission technology worldwide. It is up to mankind to determine whether nuclear fission will be used as a source of clean energy for the improvement of man's living standards throughout the world or for destructive purposes through its use for nuclear weapons. If used carefully and properly, nuclear fission, as described in this article, is an environmentally benign technology which can provide clean heat and electrical energy to help fulfill the increasing energy demands by the world's current population of about 5×10^9 people and for a projected $(8–9) \times 10^9$ people in the years 2020–2030 and beyond. With the depletion of fossil fuels such as wood, oil, and coal and the concern over the increasing greenhouse gases in the atmosphere from combustion of fossil fuels, it is urgent that people universally accept the use of nuclear fission for peaceful purposes. Issues such as prevention of nuclear weapons proliferation, radioactive waste management, safety, and economics must be addressed to the satisfaction of the concerned communities before there will be universal acceptance of this very important and useful technology.

Works Cited

Ahearne, J. F. (1993), "The Future of Nuclear Power," *Amer. Sci.* **81**, 24–35.

Allison, C. M., Johnson, E. C., Berna, G. A., Gheng, T. C., Hagrman, D. L. (1989), *SCDAP/RELAP5/MOD2 Code Manual*, Vols. 1 and 2, NTIS No. NUREG/CR-5273-V1/HDM, Springfield, VA: National Technical Information Service.

Anon. (1992), "The New Reactors," *Nucl. News* **35** (12), 66–90.

Bell, G. I., Glasstone, S. (1970), *Nuclear Reactor Theory*, New York: Van Nostrand Reinhold.

Borkowski, J. A., Wade, N. L., Giles, M. M., Rouhani, S. Z., Shumway, R. W. (1992), *TRAC-BF1/MOD1: An Advanced Best Estimate Computer Program for BWR Accident Analysis*, Vols. 1 and 2, NTIS No. NUREG/CR-4356-V1/HDM, Springfield, VA: National Technical Information Service.

Briesmeister, J. F. (Ed.) (1986), *MCNP—A General Monte Carlo Code for Neutron and Photo Transport. Version 3A, Rev. 2*, NTIS No. DE87000708/HDM, Springfield, VA: National Technical Information Service.

Bruschi, H., Anderson, T. (1991), "Turning the Key," *Nucl. Eng. Int.* **36** (448), 15–22.

Chanin, D. I., Spring, J. L., Ritchie, L. T., Jow, H. N. (1990), *Melcor Accident Consequence Code Systam (MACCS), Users Guide*, Vol. 1, NTIS No. NUREG/CR-4691-V1/HDM, Springfield, VA: National Technical Information Service.

Cowan, G. A. (1976), "A Natural Fission Reactor," *Sci. Amer.* **235** (1), 36–47.

Davey, W. G. (1972), "The Demonstration Reactor Benchmark Program," in: *Proceedings of the Topical Meeting on New Developments in Reactor Physics and Shielding*, Vol. 2, Rept. No. CONF-720901, Washington, DC: U.S. Atomic Energy Commission, p. 789.

Dietrich, J. R., Lichtenberger, H. V., Zinn, W. H. (1955), "Design and Operating Experience of a Prototype Boiling Water Power Reactor," in: R. A. Charpie *et al.* (Eds.), *Proceedings of the International Conference on the Peaceful Uses of Atomic Energy*, Vol. 3, New York: United Nations, p. 56.

Energy Science and Technology Center (1993), *Software Listing*, Report No. ESTSC-2/Rev. 1, Oak Ridge, TN: Oak Ridge National Laboratory.

Fullwood, R. R., Hall, R. E. (1987), *Probabilistic Risk Assessment in the Nuclear Power Industry*, Oxford, United Kingdom: Pergamon.

Gagarinski, A. Y., Ignatiev, V. V., Novikov, V M., Subbotin, S. A. (1992), "Advanced Light Water Reactors: Russian Approaches," *IAEA Bull.* **2** (2), 37–40.

Hannerz, K. (1988), "Making Progress on PIUS Design and Verification," *Nucl. Eng. Int.* **33** (412), 29–31.

Harrer, J. M., Jameson, A. S., West, J. M. (1955), "The Engineering Design of a Prototype Boiling Water Power Plant," in: R. A. Charpie *et al.* (Eds.), *Proceedings of International Conference on the Peaceful Uses of Atomic Energy*, Vol. 3, New York: United Nations, p. 250.

Hendrie, J. M. (1965), "The Brookhaven High Flux Beam Reactor (HFBR)," in: A. de Calmès *et al.* (Eds.), *Proceedings of the International Confer-*

ence on the Peaceful Uses of Atomic Energy, Vol. 7, New York: United Nations, p. 373.

IAEA (1988), *Basic Safety Principles for Nuclear Power Plants*, Safety Series No. 75-INSAG-3, Vienna: International Atomic Energy Agency.

IAEA (1989), *Directory of Nuclear Research Reactors*, Vienna: International Atomic Energy Agency.

IAEA (1992), *The Chernobyl Accident: Updating of INSAG-1*, Report No. INSAG-7, Vienna: International Atomic Energy Agency.

Lamarsh, J. R. (1977), *Introduction to Nuclear Engineering*, Reading, MA: Addison-Wesley.

Lentz, G. L., Buschman, H. W., Smith, R. N. (1986), "EBR-II: Twenty Years of Operating Experience," in: *Proceedings of the Symposium on Fast Breeder Reactors: Experience and Trends*, Vol. 1, Vienna: International Atomic Energy Agency, p. 171.

Lichtenberger, H. V., Thalgott, F. W., Kato, W. Y., Novick, M. (1955), "NaK Cooled Fast Reactor (EBR-I)," in: R. A. Charpie *et al.* (Eds.), *Proceedings of the International Conference on the Peaceful Uses of Atomic Energy*, Vol. 3, New York: United Nations, p. 345.

McCandless, R. J., Redding, J. R. (1989), "Simplicity: the Key to Improved Safety, Performance and Economics," *Nucl. Eng. Int.* **34** (424), 20–24.

McLane, V., Dunford, C. L., Rose, P. F. (1988), *Neutron Cross Sections*, Vol. 2, San Diego: Academic.

Mughabghab, S. F., Divadeenam, M., Holden, N. E. (1981), *Neutron Cross Sections*, Vol. 1, New York: Academic.

National Energy Software Center (1991), "Compilation of Program Abstracts," Report No. ANL-7411, Argonne, IL: Argonne National Laboratory.

National Research Council (1992), *Nuclear Power, Technical and Institutional Options for the Future*, Washington, DC: National Academy Press.

Peckinpaugh, C. L. (1986), "Fast Flux Test Facility: First Three Years of Operation," in: *Proceedings of a Symposium on Fast Breeder Reactors: Experience and Trends*, Vol. 1, Vienna: International Atomic Energy Agency, p. 203.

Peckinpaugh, C. L., Bennett, R. A., Wykoff, W. R. (1982), "Fast Flux Test Facility, Operational Results," in: *Proceedings of the International Conference on Nuclear Power Experience*, Vol. 5, Vienna: International Atomic Energy Agency, p. 155.

Radiation Shielding Information Center (1993), *RSIC Software Directory*, Oak Ridge, TN: Oak Ridge National Laboratory.

Ransom, V. H., Wagner, R. J., Trapp, J. A., Feinauer, L. R., Johnsen, G. W. (1985), *RELAP5/MOD2 Code Manual*, Vol. 1, "Code Structure, Systems Models, and Solution Methods," NTIS No. NUREG/CR-4312/HDM, Springfield, VA: National Technical Information Service.

Ransom, V. H., Wagner, R. J., Trapp, J. A., Johnsen, G. W., Miller, C. S. (1987), *RELAP5/MOD2 Code Manual*, Vol. 2, "Users Guide and Input Requirements," NTIS No. NUREG/CR-4312-REV-1/HDM, Springfield, VA: National Technical Information Service.

Rhodes, R. (1986), *The Making of the Atomic Bomb*, New York: Simon and Shuster.

Rogovin, M., Frampton, G. T., Jr., Cornell, E. K., DeYoung, R. C., Budnitz, R. (1980), *Three Mile Island: A Report to the Commissioners and to the Public*, Vol. 1, NTIS No. NUREG/CR-1250-V-1/HDM, Springfield, VA: National Technical Information Service.

Scheinman, L. (1987), *The International Atomic Energy Agency and World Nuclear Order*, Washington, DC: Resources for the Future.

Schnurr, N.M., Steinke, R. G., Spore, J. W., Martinez, V. (1992), *TRAC-PF1/MOD2 Code Manual*, NTIS No. NUREG/CR-5673-V2/HDM, Springfield, VA: National Technical Information Service.

Smyth, H. D. (1945), *Atomic Energy for Military Purposes*, Princeton, NJ: Princeton University Press.

Stahlkopf, K. E., DeVine, J. C., Sugnet, W. R. (1988), "US ALWR Programme sets out Utility Requirements for the Future," *Nucl. Eng. Int.* **33** (412) 16–19.

Stamm'ler, R. J. J., Abbate, M. J. (1983), *Methods of Steady-State Reactor Physics in Nuclear Design*, New York: Academic.

Summers R. M., Cole, R. K., Boucheron, E. A., Carmel, M. K., Dingman, S. E. (1991), *MELCOR 1.8.0: A Computer Code for Nuclear Reactor Severe Accident Source Term and Risk Assessment Analyses*, NTIS No. NUREG/CR-5531HDM, Springfield, VA: National Technical Information Service.

Swartout, J. A., Boch, A. L., Cole, T. E., Cheverton, R. D., Adamson, G. M., Winters, C. E. (1965), "High Flux Isotope Reactor (HFIR)," in: A. de Calmès *et al.* (Eds.), *Proceedings of the International Conference on the Peaceful Uses of Atomic Energy*, Vol. 7, New York: United Nations, p. 360.

U.S. Nuclear Regulatory Commission (1987), *Report on the Accident at the Chernobyl Nuclear Power Station*, NTIS No. NUREG-1250/HDM, Springfield, VA: National Technical Information Service.

Vijuk, R., Bruschi, H. (1988), "AP600 Offers a Simple Way to Greater Safety, Operability, and Maintainability," *Nucl. Eng. Int.* **33** (412), 22–28.

Waltar, A. E. Reynolds, A. B. (1981), *Fast Breeder Reactors*, New York: Pergamon, p. 772.

Further Reading

Duderstadt, J. J., Hamilton, L. J. (1976), *Nuclear Reactor Analysis*, New York: John Wiley & Sons.

Glasstone, S., Sesonske, A. (1987), *Nuclear Reactor Engineering*, New York: Van Nostrand Reinhold.

Greenspan, H., Kelber, C. N., Okrent, D. (1968), *Computing Methods in Reactor Physics*, New York: Gordon and Breach, 1968.

Henry, A. F. (1975), *Nuclear Reactor Analysis*, Cambridge, MA: MIT Press.

Hsu, Y. Y., Graham, R. E. (1976), *Transport Processes in Boiling and Two-Phase Systems*, Washington, DC: Hemisphere Publishing.

Ott, K. O., Bezella, W. A. (1989), *Introductory Nuclear Reactor Statics*, LaGrange Park, IL: American Nuclear Society.

Ott, K. O., Neuhold, R. J. (1985), *Introductory Nuclear Reactor Dynamics*, LaGrange Park, IL: American Nuclear Society.

Tong, L.S., Weisman, J. (1979), *Thermal Analysis of Pressurized Water Reactors*, 2nd ed., LaGrange Park, IL: American Nuclear Society.

NUCLEAR FUELS AND ISOTOPES

ALLEN G. CROFF, *Oak Ridge National Laboratory, Oak Ridge, Tennessee, U.S.A.*

3-527-28133-9/94/$5.00 + .50

INTRODUCTION

This article addresses the necessary or desirable operations that are performed to prepare the nuclear fuel required to sustain nuclear reactors and to manage the fuel constituents after removal from the reactor. Taken collectively, these operations are called a *nuclear fuel cycle*.

These operations include the following:

1. obtaining raw materials to make fuel;
2. processing the fuel materials to achieve the desired chemical and isotopic composition;
3. fuel manufacture;
4. processing of used fuel to recover valuable constituents, including optional recovery of radioactive isotopes for beneficial uses; and
5. transportation and storage of fuel materials at various facilities.

The design and operation of the nuclear reactors and the management of nuclear wastes are described in separate articles in this series entitled NUCLEAR ENERGY, FISSION, and NUCLEAR WASTE MANAGEMENT.

1. OVERVIEW

Since the practical implementation of nuclear power during World War II, many different types of nuclear reactors have been conceived, and many of these have been built. For most of these, multiple fuel cycles are possible, yielding a large number of reactor–fuel cycle combinations. However, for descriptive purposes, these can be grouped into three categories: once through, reprocessing and recycle, and continuous recycle. These three fuel cycle categories are depicted in Fig. 1.

A *once-through fuel cycle* involves the production of fissile material having a concentration and total mass adequate to sustain a nuclear chain reaction. This material is then fabricated into fuel assemblies that have optimal nuclear physics properties, provide for removal of the heat generated during fission, and keep the radioactive products of fission separated from the coolant. This fuel (called fresh fuel) is transported to the reactor, where it is inserted into a core containing many assemblies and used to generate heat, which then generates high-pressure steam that is used to generate electricity. At some point, the fissile material is sufficiently depleted so that fission will no longer occur (i.e., the reactor is subcritical). The oldest fuel in the reactor core is then discharged and stored in a manner that provides protection from its intense radiation and heat removal. At some point, the spent fuel will be packaged and sent to a deep-mined cavern (i.e., a repository) for permanent disposal.

The spent fuel from a once-through fuel cycle contains significant amounts of residual fissile material. In a fuel cycle involving *reprocessing and recycle*, the fuel is reprocessed to recover the fissile material, and this material is refabricated into recycle fuel for use in another reactor. The most promising of these cycles involves recovering the slightly enriched uranium and plutonium from the spent fuels. The uranium would be re-enriched to concentrate the fissile isotope ^{235}U. The plutonium (produced by irradiating ^{238}U with neutrons) is typically composed of about $\frac{2}{3}$ fissile isotopes and would be diluted with natural uranium to achieve the proper fissile enrichment, which can range from 3 to 25%. Systems have also been proposed in which ^{232}Th is irradiated to generate fissile ^{233}U, which is then diluted with ^{232}Th to form recycle fuel. The fuel assemblies made from these materials are then inserted into the reactor and used to generate electricity. Upon

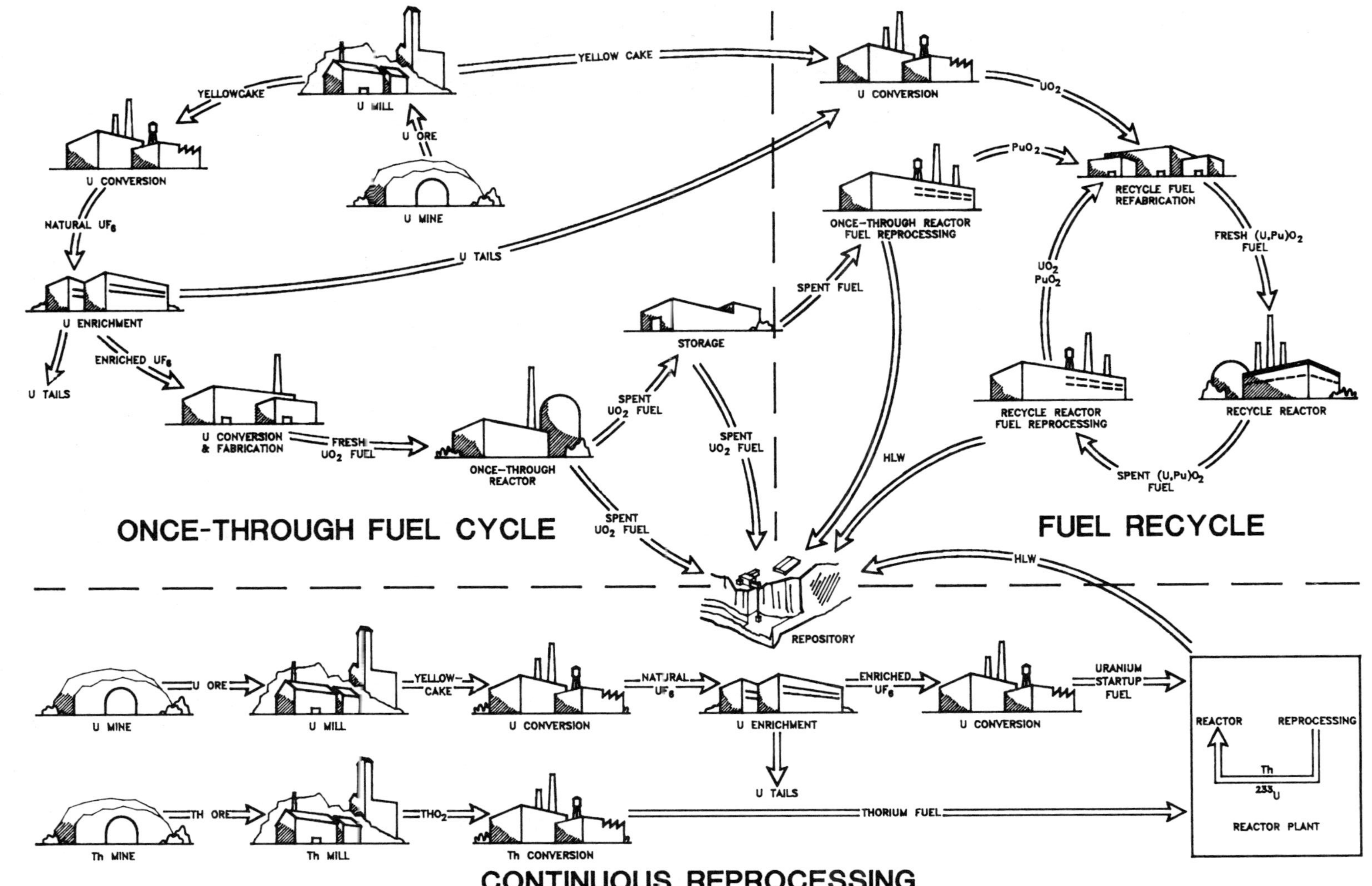

Fig. 1. Relationship of operations in once-through, recycle, and continuous reprocessing nuclear fuel cycles.

discharge, they are again reprocessed, and the cycle is repeated. Depending on the specific features of the reactor design, the overall recycle loop may be a net producer (i.e., a breeder) or net consumer (i.e., a converter) of fissile material.

In a fuel cycle involving *continuous recycle*, the fuel is dissolved or suspended in a liquid such as water, molten salt, or molten metal. The liquid is continuously circulated through a loop. At some point in the loop, its configuration is such that it is critical, generating the desired heat. At other points, the heat is removed to generate steam, and, at still other locations, the fuel is continuously reprocessed to remove unwanted by-products. Many such systems have been proposed since the 1940s, but they have not been well developed as compared to the batch systems described above.

The remainder of this article will more specifically address most of the steps depicted in Fig. 1. Exceptions to this are descriptions of specific types of nuclear reactors and management of radioactive wastes, both of which are the subject of separate articles (see NUCLEAR ENERGY, FISSION; NUCLEAR WASTE MANAGEMENT).

2. URANIUM MINING AND MILLING

2.1 Uranium Sources and Mining Techniques

The only naturally occurring fissile isotope is ^{235}U, which constitutes 0.71% of natural uranium. It is this material that is the wellspring of all nuclear fuel cycles. Uranium is a ubiquitous element, being widely distributed in the earth's crust at an average concentration of about 0.0003%. At this concentration, recovery is not economic. However, as a result of geologic and environmental processes, uranium has been concentrated in veins at many locations. A few high-grade veins containing the mineral pitchblende (U_3O_8) have been found to contain uranium at concentrations of 30–70% U_3O_8, mostly in Africa. Significant ore bodies having uranium concentrations up to 15% are now being developed in Canada. However, these are relatively uncommon, and most of the high-concentration veins were mined decades ago. Much more common are veins of carnotite ($K_2O \cdot 2UO_3 \cdot V_2O_5 \cdot 3H_2O$) ore that contain 0.1 to 0.2% uranium, typically in a sandstone matrix. These are widely distributed in the western United States and Canada in a band covering the eastern to middle portion of the Rocky Mountains, and in similar geologies in other countries. It is economically feasible to mine carnotite ores at uranium concentrations of 0.1% or above. Total known world resources of uranium outside the centrally planned economies are estimated to be 3.1 million metric tons (OECD, 1992).

The mining of carnotite ores is accomplished using a variety of methods. Standard tunnel mining has often been used, although less commonly so now because of the relatively high cost of such operations in combination with low uranium prices. Open-pit mining is often used for deposits relatively near the surface. A newer technique, called *in situ* leaching, involves injecting leaching solutions into a subsurface ore zone through a well and then drawing the uranium-bearing solution back to the surface through another well. The leaching chemicals are very similar to those used in uranium ore milling, described below.

Other dilute sources of uranium are also economic to mine when the uranium is a by-product of the production of other economic values. Significant amounts of uranium are generated as a by-product of phosphate mining. Most of this production occurs in the Gulf Coast states in the United States, and similar locations elsewhere. Uranium is also produced as a by-product of copper and gold mining at various locations throughout the world.

Recovery of uranium from seawater has been and continues to be studied, typically using ion-exchange technology. While the "ore" is easily accessible, the low concentration of the uranium and fouling of the ion exchange material continue to make this an uneconomic source.

2.2 Uranium Milling and Refining

Uranium milling is the concentration and purification of the uranium in ores. Ores are typically ground to 20 to 200 mesh to provide adequate contact with reagents. The ground ore is then contacted with an aqueous acid or alkaline leach, depending on the specific properties of the ore. Sulfuric acid is most commonly used, but all of the mineral acids are used to some extent. Alkaline leaching is performed with solutions of the carbonates and bicarbonates of sodium or ammonia. The acid-based process is generally more economic unless the ore contains large amounts of alkaline materials, which would require excessive acid consumption. In either case, the result is a selective dissolution of uranium from the ore.

The acid or alkaline slurry of uranium-bearing solution and solids is then filtered to remove the solids. These solids are somewhat radioactive, containing all of the decay products of the uranium that have achieved equilibrium levels during millions of years of decay. Most noteworthy of these are ^{226}Ra (with a half-life of about 1600 years) and its chemically inert gaseous decay product ^{222}Rn. These radionuclides are primarily responsible for the radiological effects of the tailings. The tailings are typically managed by simply creating small hills of the solids near the mill, along with engineered features to manage the runoff resulting from rain and to control the release of ^{222}Rn.

The uranium is recovered from the aqueous leach solution by precipitation using a variety of chemicals such as ammonia, magnesia, or caustic soda. The product of the precipitation is the diuranate of ammonia, magnesium, or sodium, with a typical chemical formula being $(NH_4)_3U_2O_7$. Depending on the specifics of the process, this product is then washed and dried to yield a 75-to 100%-pure U_3O_8 product. Although all of the diuranate intermediates and the impure product are commonly called *yellowcake*, only ammonium diuranate has the characteristic yellow color and these products are more properly called *uranium ore concentrates* (UOC)

2.3 Purification and Conversion of UOC to UF_6

The impure UOC product must be purified to be acceptable in subsequent processes. This is generally performed at a centralized facility separate from the mine and mill, and this same facility often includes the processes for conversion to UF_6. Purification is typically accomplished by dissolving the yellowcake in concentrated nitric acid and extracting the uranium from the acid by *solvent extraction*: contacting it with immiscible tributylphosphate (TBP) diluted with kerosene or *n*-dodecane and then back-extracting the uranium from the TBP solution using dilute nitric acid. Purification may also be accomplished using other solvent extraction reagents or ion exchange techniques. The uranium solution is then typically evaporated and calcined (to yield high-purity UO_3) and then reduced by hydrogen gas to yield uranium dioxide (UO_2) and nitric acid for internal recycle.

Uranium dioxide can be taken directly to fresh fuel fabrication for reactors that do not require isotopically enriched uranium. However, many reactors require uranium enriched so that it has a ^{235}U concentration of 1–5%, with the required concentration depending on the reactor design and operating parameters. All currently operating enrichment processes require the feed material to be in the form of uranium hexafluoride (UF_6), thus requiring *conversion* of the oxide to the hexafluoride. This conversion is accomplished in two steps. The first is to react the UO_2 with anhydrous hydrogen fluoride (HF) in a multiple-stage reactor, which yields uranium tetrafluoride, UF_4. The UF_4 is then reacted with elemental fluorine gas (F_2) to yield UF_6.

Purification and conversion can be carried out simultaneously by fluorinating the UOC directly by first using the HF-F_2 sequence described above. The result is that the uranium and a few other elements form fluorides that are easily volatilized, leaving most of the impurities in a solid residue. The gaseous fluoride stream is then further purified by fractional distillation to yield a pure UF_6 product.

3. ISOTOPE ENRICHMENT

The function of *isotope enrichment* processes is to increase the relative abundance of selected isotopes in an element. This also necessarily produces another output stream (typically called *tails*) that is depleted in the particular isotope. In some cases, the degree of enrichment is almost complete. For example, to produce "heavy water," the relative abundance of deuterium ($^2H_2$) is increased from the very low 0.015% natural abundance in water to about 99.75%. In other cases, the concentration of a particular isotope may be increased only a few percentage points. For example, uranium is enriched from its natural 0.71% ^{235}U concentration to less than 5% for most reactor applications.

3.1 Uranium

Uranium is the most extensively enriched material in the world. The first successful large-scale enrichment of uranium isotopes was achieved during World War II by the United States in its effort to obtain enough highly enriched (>90% ^{235}U) uranium to produce a nuclear weapon. A large number of uranium isotope separation methods were studied during this period, and most were able to enrich uranium to varying degrees. However, the cost of enrichment was very high for many of these processes, and less costly alternatives were subsequently developed. The following discussion describes uranium enrichment processes in their approximate order of large-scale development or use.

3.1.1 Electromagnetic Isotope Separation (EMIS) EMIS capitalizes on the fact that charged particles move in circular paths when subjected to a magnetic field perpendicular to the particle path. Different paths are followed by charged particles of different masses, with the path of a heavier particle having a slightly larger diameter. Thus, when ionized (charged) atoms such as those of uranium are injected into a strong magnetic field, the ^{235}U ions follow a different path than the ^{238}U ions, and the former may be collected in a different location from where the latter impinge (Fig. 2). Recovery of the separated material is then simply a matter of chemical processing.

The separation factor (a measure of the amount of enrichment attained with a single enrichment stage) is very large for the EMIS process. In theory, complete isotopic separation can be achieved in a single stage. In practice, the extent of enrichment attained is limited by ion beam dispersion, interactions with solid materials, and other second-order effects. However, highly enriched uranium can be produced with only two consecutive stages of EMIS.

The EMIS machines used during World War II in the United States to enrich uranium were called *calutrons*. Although the enrichment achieved in a single stage was high, the throughput of any one device was relatively small. As a result, a large number of calutrons were operated in parallel to produce the required amount of product (several tens of kilograms).

Although the calutrons were successful, they were expensive to operate because they used large amounts of electricity for the amount of enriched uranium produced and because they were very labor intensive. As a result, use of this technology for uranium enrichment ceased soon after the War, and most of the calutrons were dismantled.

3.1.2 Gaseous Diffusion Another method of uranium enrichment studied during World War II—*gaseous diffusion*—proved to be much less costly than EMIS, requiring far fewer people, operations, and units of electricity per unit of product. In the gaseous diffusion process, UF_6, which is gaseous at relatively low temperatures and pressures, is allowed to diffuse through very small openings in porous tubes called *barriers* contained within large cylindrical housings. The $^{235}UF_6$ molecules go through the pores a little faster than the heavier $^{238}UF_6$ molecules, and so the gas that passes through the barrier is slightly enriched in the lighter isotope. The slightly enriched product from this stage is then pumped to the next-higher stage, where the process is repeated. The gas that does not pass through the tubes is slightly depleted in $^{235}UF_6$ and is pumped to the next lower stage, where the process is repeated. In this way, the enriched uranium is moved continuously upward and repeatedly enriched in ^{235}U, while the depleted uranium is moved continuously downward and repeatedly depleted in ^{235}U. The assemblage of interconnected stages is called a *cascade*. Along with the housing and tubes (called a *convertor*), each stage also requires a compressor to recompress the gas received from other stages and a heat exchanger to remove the heat of compression. Such an arrangement is shown in Fig. 3.

As implied above, the amount of enrichment achieved by a single stage is small. In fact, the maximum theoretical amount by which the ^{235}U can be enriched in a single stage is given by the square root of the ratio of the masses of the heavier and lighter uranium molecules, which is $(352/349)^{0.5} = 1.004$. As a consequence, to enrich natural uranium containing 0.71% ^{235}U to 3% (an enrichment typically required by reactors) while produc-

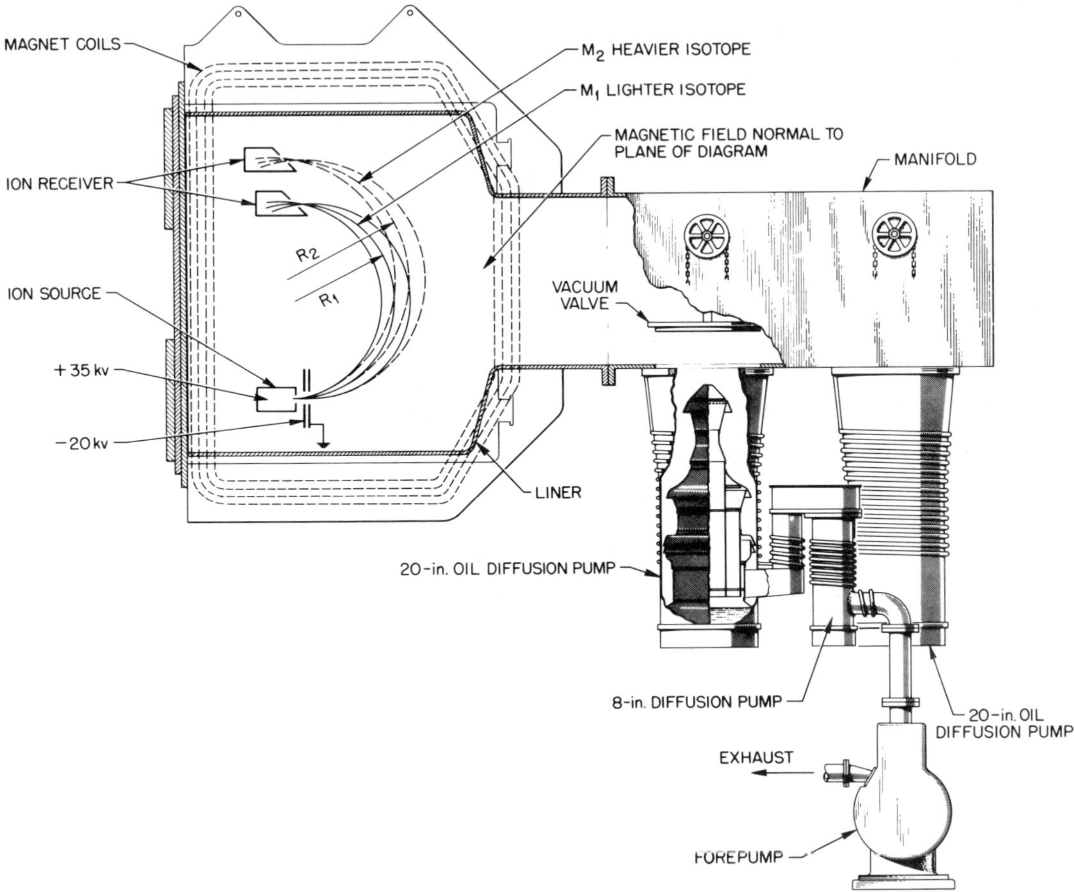

Fig. 2. Schematic diagram of electromagnetic isotope separation (a Calutron).

ing a tails stream containing 0.2% ^{235}U would require 1272 theoretical stages. This number is increased by the inevitable inefficiencies in any real process.

Despite the large number of stages required, the large physical size of the plants (which are multi-story buildings about 1 km long and 100 m wide), and the relatively large amounts of electricity required (a single plant requires on the order of 1000 MW), the large throughput of the plant and the low operation and maintenance cost resulted in the product cost being sufficiently low to make nuclear fuels economic. Gaseous diffusion plants for enriching uranium for both power reactors and military purposes have been built in the United States, Russia, France, Great Britain, and China. Construction of additional plants is not expected because the high electricity consumption makes the product relatively expensive compared to other technologies that have become available.

3.1.3 Gas Centrifuge A *gas centrifuge* separates uranium isotopes by placing UF_6 in an artificially created high-gravity environment, thus magnifying the importance of the small mass difference between ^{235}U and ^{238}U. The high-gravity environment is created by spinning a right-circular cylinder called a *rotor* around its axis at very high speeds, conceptually similar to the spin cycle on a washing machine. For a given rotor diameter, the separation factor is a function of the rate at which the centrifuge rotates. This rate is limited by the mechanical properties of the ro-

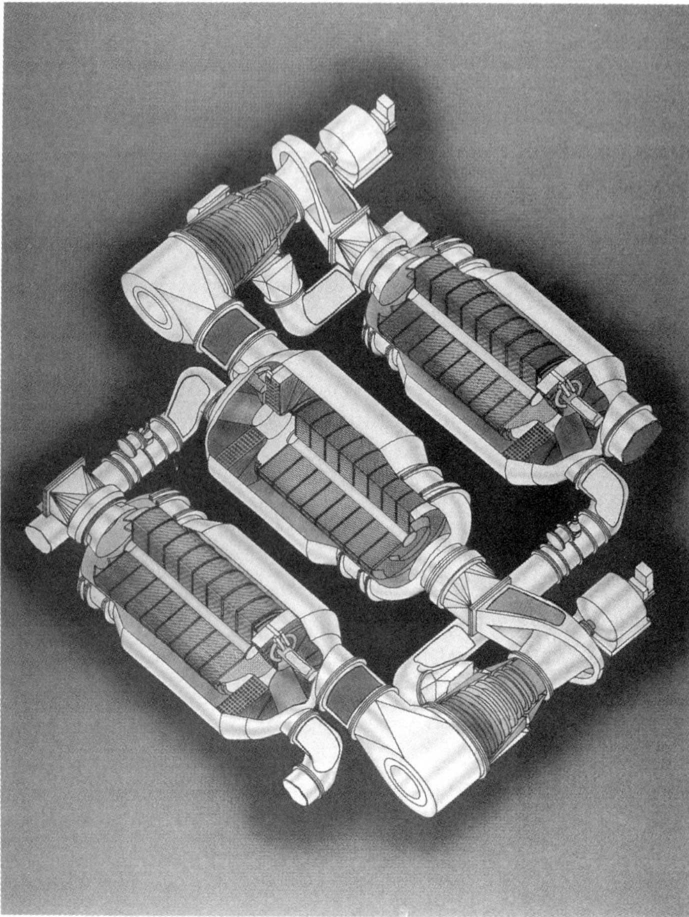

Fig. 3. Diagram of interconnected stages in a gaseous diffusion plant cascade.

tor materials, which may be advanced metal alloys or composites.

Because of the circulating flow within the centrifuge, each functions as a "miniature cascade," with multiple effective stages within each physical device. A schematic of one of the early centrifuge designs depicting this circulation is shown in Fig. 4. The net result is that the number of physical stages (i.e., centrifuge machines) required is far fewer than that required for gaseous diffusion, although a number of centrifuges may have to be operated in parallel to achieve the desired throughput. As a result of the processes within the centrifuge being reversible (the pressure drop in gaseous diffusion is irreversible), the electricity requirement for gas centrifuges is estimated to be about 4% of that for gaseous diffusion. However, the centrifuges are very sophisticated devices, requiring advanced materials to achieve economic operating speeds without destroying themselves. On balance, the cost of product from the gas centrifuge process appears to be less expensive than that from gaseous diffusion, but not greatly less.

Gas centrifuge plants for uranium enrichment are currently operating in the Netherlands, Great Britain, Germany, and Russia. Plants are under construction in other countries. A small pilot plant was built and operated in the United States, and construction started on a demonstration plant, but the project was cancelled in favor of a laser isotope separation process.

3.1.4 Laser Isotope Separation (LIS) A *laser isotope separation* (LIS) process separates uranium isotopes by capitalizing on slight differences in the wavelength of light

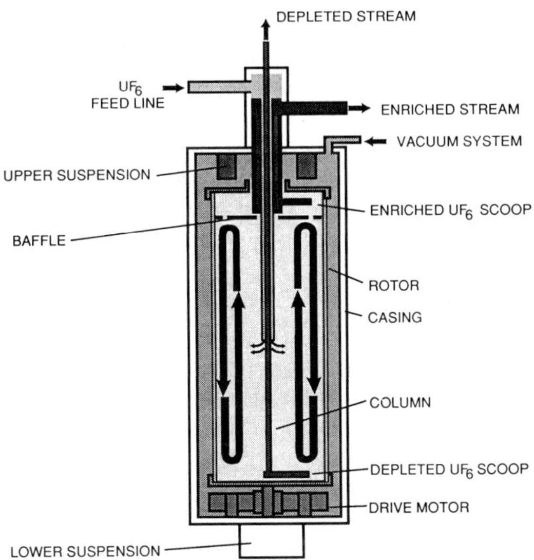

Fig. 4. Schematic diagram of a gas centrifuge for uranium enrichment.

that will excite uranium metal or uranium-bearing molecules to a species that can be readily separated. The variation in wavelengths is a result of the differing atomic properties of ^{235}U and ^{238}U that are a consequence of the difference in the masses of the two nuclei. The laser power requirements are significant (multikilowatt to megawatt). It was only with advanced developments in high-power lasers that the wavelengths could be adequately controlled to achieve the desired enrichment while producing the power required for the needed throughput.

As with EMIS, the selectivity of LIS processes is very high. In theory, only one stage would be required to separate the isotopes completely, while in practice, only two stages may be needed. The process requires much less electricity than EMIS. However, the cost of the equipment (especially the high-powered lasers) required for LIS and the precision with which it must be controlled is high. While this process is still under development, it is thought that the initial cost of product will be about the same as that from gas centrifuges.

The *atomic vapor laser isotope separation* (AVLIS) process (Fig. 5) is carried out by introducing uranium metal (or uranium metal alloy) vapor into a sealed vessel under low-pressure conditions. The uranium is exposed to laser light at one or more suitable wave-

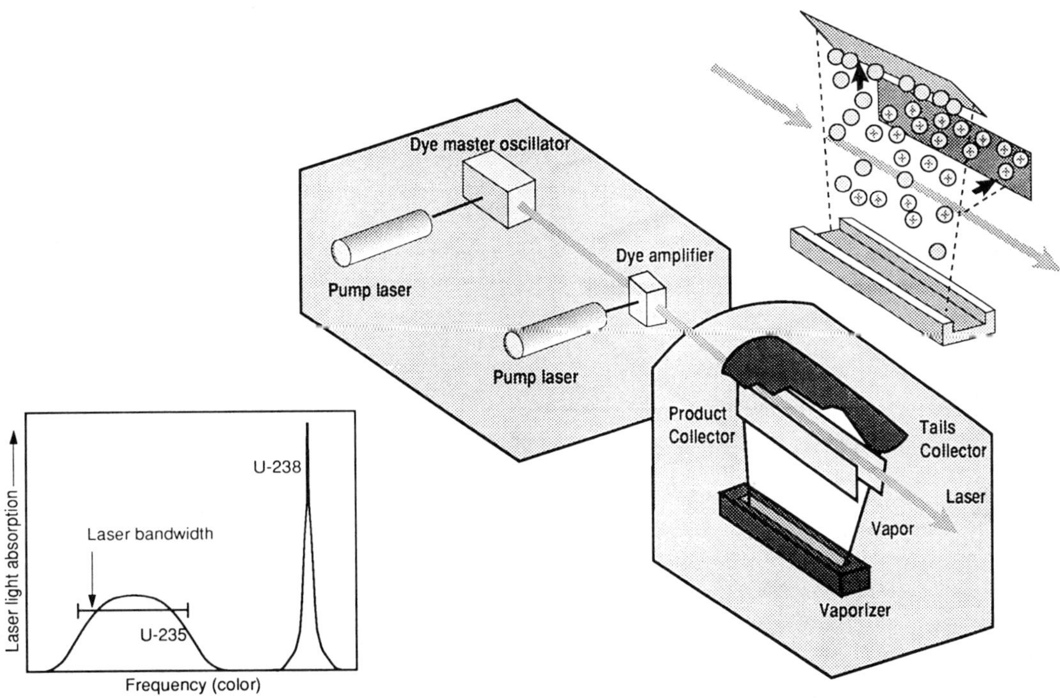

Fig. 5. Diagram of a laser isotope separation module.

lengths, resulting in the selective ionization of the ^{235}U fraction. The ionized material can then be diverted from the un-ionized (^{238}U) material with an electrostatic field and collected.

The *molecular laser isotope separation* (MLIS) process is conceptually similar to AVLIS. However, in this case, the feed material is the molecule UF_6. Upon exposure to laser light of suitable wavelengths, the UF_6 is dissociated by photolysis to form UF_5 and atomic fluorine (which may quickly combine to the more stable F_2 molecule). The UF_5 is a solid entrained within the flowing UF_6 gas, and can be separated from the gas stream by standard techniques such as by filtration with gaseous-diffusion–plant barrier material. Rapid separation of the UF_5 is required to prevent recombination with the atomic and molecular fluorine present in the gas stream.

3.1.5 Aerodynamic Processes Aerodynamic processes capitalize on the centrifugal forces resulting from forcing a gas (in this case, a mixture of UF_6 and hydrogen) through a small, curved nozzle shown in Fig. 6 (Becker *et al.* 1975) at high speed. The heavier fraction tends to be concentrated near the periphery of the nozzle, allowing a knife edge to separate it from the lighter fraction. The enriched and depleted streams from each stage are moved to other stages in a manner very similar to that used in gaseous diffusion.

As with gaseous diffusion, aerodynamic processes are inherently irreversible. As a consequence, the cost of enriched product appears to be at least as large as that of gaseous diffusion and higher than that of gas centrifuges. Small production plants based on these processes have been built and operated in Brazil and South Africa.

3.2 Stable Isotopes

A wide variety of separated *stable isotopes* (i.e., nonradioactive) are used for various aspects of nuclear power, in scientific experiments, and as precursors for production of radioactive isotopes (via irradiation in particle accelerators or small reactors) used in an expanding variety of medical procedures.

3.2.1 Deuterium Deuterium (^2H or D) is a heavy isotope of hydrogen containing one proton and one neutron in the nucleus as compared to the single proton in normal hydrogen (^1H). Deuterium occurs naturally (DHO at low concentrations, D_2O at high concentrations) in concentrations of about 0.015 at.% in all water. While it has some minor scientific uses, by far the predominant use of heavy water is as a neutron moderator in some types of nuclear reactors such as the Canadian CANDU. Deuterium's nuclear properties result in its having an exceedingly small chance of undergoing undesirable neutron absorp-

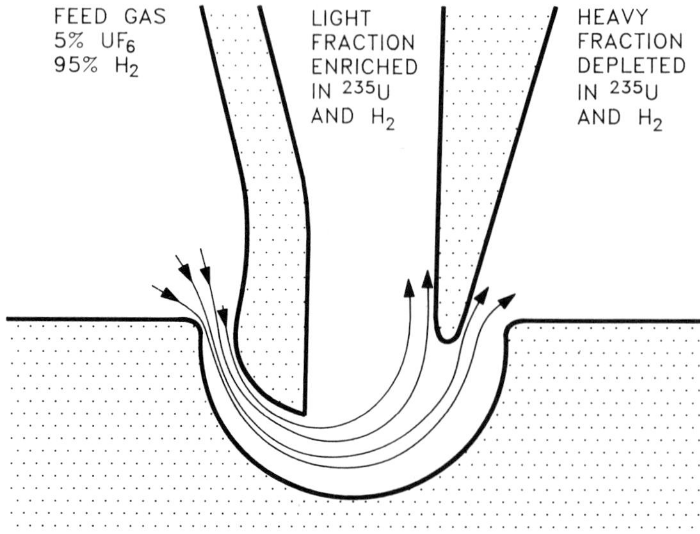

Fig. 6. Cross section of a basic separation nozzle used for uranium enrichment.

tion reactions, thus allowing a practical reactor to be operated using only natural (i.e., not enriched) uranium fuel. However, the savings in uranium enrichment costs are approximately balanced by the cost of having to enrich the deuterium to nearly 100% and the fact that a 600 MW_e CANDU reactor requires an inventory of about 450 metric tons of heavy water.

Many processes have been proposed and used since the first known heavy-water production plant was operated in 1934. Early plants were typically based on electrolysis of water, which, like the EMIS process for other stable isotopes, was effective but costly. Over the years, a wide variety of alternative processes have been developed and deployed to produce heavy water. The two general categories of processes that have been developed involve exchange of D and H between various chemicals such as ammonia, water, or hydrogen sulfide, and distillation of various chemicals such as water, ammonia, or methane.

After decades of experience, the preferred approach to producing essentially pure heavy water is the Girdler-sulfide, dual-temperature water–hydrogen-sulfide exchange process (GS) followed by water distillation. A flowsheet of the GS process is exceedingly complex because it involves extensive measures to conserve heat within the process. However, in conceptual terms, a tall vertical vessel is fed at the top with natural water at about 0° C. This is contacted with a gaseous stream of D_2S injected at the bottom of the vessel. The net result is that the deuterium preferentially moves to the aqueous phase (which becomes heavy water) and hydrogen to the gaseous phase. The D_2S is generated in a separate vessel by reacting part of the heavy water product with H_2S at a much higher temperature, where the formation of D_2S is favored, producing water depleted in deuterium. The sulfide compound is continuously recycled within the process. The product enrichment of this process is typically about 15 at.% deuterium.

The second step of this approach involves taking the partially enriched product from the GS process and fully enriching it in a simple water distillation column. The lighter H_2O favors the gaseous (steam) phase. Thus, injecting the partially enriched liquid product from GS into the top of a vessel which has H_2O steam rising from the bottom has the effect of stripping the rest of the H_2O from the liquid, yielding the desired high-purity product.

3.2.2 Lithium Lithium has two stable isotopes. Lithium-7 (abundance of 92.5 at.%) has a relatively low probability of undergoing undesirable interactions with neutrons, and its compounds have found limited use in controlling the water chemistry in nuclear reactors. ^{6}Li has a relatively large probability of undergoing interactions with neutrons, with the products being stable helium and radioactive ^{3}H (called tritium). Tritium is used in industry in a variety of applications related to radioluminescence. Tritium production does not require that the precursor material be enriched in ^{6}Li, and so the demand for enriched lithium products has been limited to the small demands for ^{7}Li and enrichment activities related to the production of ^{6}Li for nuclear weapons.

As with heavy water, a wide variety of processes have been investigated for the purpose of enriching lithium. In fact, the general categories of processes are the same as those for heavy water production (i.e., electrolysis, distillation, and chemical exchange), along with ion exchange processes. As a result of the limited demand for enriched lithium, the preferred process is not evident. However, there appears to be a tendency to favor a chemical exchange process involving contact of an aqueous solution of LiOH with a lithium amalgam (a solution of lithium in mercury).

3.2.3 Boron Boron has two stable isotopes. ^{10}B (abundance of 19.6 at.%) has an extremely high probability of undergoing reactions with neutrons, and it is used as a neutron absorber to control the chain reaction in nuclear reactors. Worldwide production is on the order of a few tons per year. ^{11}B has a relatively low probability of undergoing nuclear reactions. Although only a few small plants have been built, the preferred process involves a chemical exchange reaction between boron trifluoride (BF_3), which is gaseous at modest temperatures, and a complex of boron trifluoride and dimethyl ether [$(CH_3)_2O$], which acts as the liquid phase.

3.2.4 Other Light Isotopes There is also limited enrichment of the low-abundance, stable isotopes of carbon (^{13}C, abundance 1.11 at.%), nitrogen (^{15}N, abundance 0.365 at.%), and oxygen (^{17}O, abundance 0.037 at.%, and ^{18}O, abundance 0.204 at.%). All of these are used in scientific research to make biological compounds that can be traced through organisms and chemical reactions because of their enrichment. As with the other isotopes mentioned above, the preferred production methods involve distillation and chemical exchange. Worldwide production is estimated to be several kg/yr or less for these isotopes.

3.2.5 Heavier Stable Isotopes Beyond the stable isotopes listed above, demand for enriched products is typically small and depends on the identification of a specific use for the isotope. Most elements have been isotopically enriched in small amounts to provide relatively pure materials upon which to perform experiments concerning the unique nuclear properties of each. Other selected isotopes continue to be separated in larger quantities to provide precursor material to produce useful radioactive isotopes, especially for medical applications.

Although a few specialized enrichment processes have been used, most of the heavier elements are enriched using the EMIS process described earlier. While it is expensive to operate for the purpose of producing large quantities of enriched uranium, its flexibility and high enrichment per stage make it cost-effective for small scale separations.

4. CONVERSION OF UF_6 TO OTHER SPECIES

The fact that UF_6 is gaseous at modest temperatures makes it very useful for enrichment processes. However, this compound is not useful for the end applications of the uranium. As a result, the UF_6 must be converted to a chemical species that has properties appropriate to its end use.

4.1 Conversion of UF_6 to UO_2

Uranium dioxide (UO_2) is the species used to manufacture most nuclear reactor fuels, and it has few uses beyond this application. Three different processes have been developed to perform this function. All are used commercially.

The first method of production involves two steps. The first is gas-phase reduction of UF_6 by elemental hydrogen, yielding UF_4 and HF. The UF_4 is then reacted with steam, yielding UO_2 and a solution of HF.

The second method involves direct reaction of UF_6 with water to form UO_2F_2 and aqueous HF. This solution is then reacted with ammonium hydroxide (dissolved ammonia) to precipitate ammonium diuranate [$(NH_4)_2U_2O_7$) and ammonium fluoride. The ammonium diuranate is then filtered from the solution, dried, and reduced with hydrogen gas to yield ammonia and UO_2.

The third method involves reaction of the UF_6 with carbon dioxide and ammonia in aqueous solution to produce a precipitate of ammonium uranyl carbonate, $(NH_4)_4UO_2(CO_3)_3$. The ammonium uranyl carbonate is converted to UO_2 by reaction with steam and hydrogen gas at elevated temperatures.

In all cases, the characteristics of the product must be carefully controlled. Not only is the chemical composition of the product important, but its physical attributes such as grain size and sinterability are also very important in subsequent manufacture of fuel.

4.2 Conversion of UF_6 to Uranium Metal

Uranium metal has a variety of uses. It is used in nuclear fuels, typically alloyed with other metals (e.g., aluminum, zirconium) to provide acceptable mechanical properties. It is used to manufacture armor-piercing bullets for the military. The high density (19.1 g/cm^3) and pyrophoric nature of uranium metal when finely ground or molten make it preferable to lead (11.7 g/cm^3) or other materials in some applications. Uranium is also used to form radiation shields, especially for gamma rays. It has occasionally been used in other applications, such as counterweights in airplanes, where space is at a premium and its high density makes it cost-effective.

Only one process is used for the large-scale production of uranium metal. The first step is to reduce the UF_6 to UF_4 with hydrogen gas, as described in the previous section. The UF_4 is then mixed with a metal more reactive than uranium, typically magnesium or calcium, and an iodine catalyst. This mixture is placed in a sealed chemical bomb (a sturdy closed vessel), and the mixture is ignited electrically. The result is the reduction of the uranium to molten metal and the production of the molten fluoride of the reducing agent (e.g., CaF_2). The metal and fluoride salt are immiscible, and so, after cooling, the result is a solid layer of the fluoride salt atop a mass of uranium metal, often called a *button*.

5. FUEL FABRICATION AND REFABRICATION

The purpose of fuel fabrication and refabrication is to process raw nuclear materials into a form and configuration that are appropriate for the fuel of a particular reactor design. *Fabrication* involves processing unirradiated nuclear materials (i.e., uranium or thorium). *Refabrication* involves processing of materials recovered from spent fuels for recycle, with plutonium and ^{233}U being the most common examples.

Typically, the nuclear material begins in powder or metal form. This material is formed into larger shapes, generically called *fuel*, to withstand irradiation in the reactor better and to facilitate handling. The fuel is then enclosed in a metal container, generically designated as *cladding*, which serves the purpose of separating the radioactive fuel from the relatively clean fluid (e.g., water, liquid sodium, CO_2, gaseous helium) required to cool the fuel during irradiation. The units of clad fuel are called *fuel rods* or *fuel elements*. The fuel elements are then arranged into a fixed geometric array, called a *fuel assembly*. Dozens, if not hundreds, of fuel-cladding-assembly combinations have been developed over the last five decades of nuclear power. The following sections describe two of the more preferred combinations that have received some degree of large-scale use.

5.1 Oxide Pellet Fuels

By far the most popular fuel type is formed from uranium dioxide produced as described earlier. Most large reactors that produce electric power use this type of fuel, primarily because it is stable at relatively high temperatures. The following steps are typically followed to make oxide fuels:

1. If the fuel is to be composed of oxides of several elements (e.g., Pu-enriched uranium dioxide or $[Pu,U]O_2$), the oxides are thoroughly blended to achieve the desired enrichment, which would typically range from a few percent Pu (for thermal reactors) to 25% (for fast reactors).
2. The oxide powder is blended with binding agents and die lubricants.
3. The mixture is pressed into pellets that are the shape of a right circular cylinder, typically about the diameter of a pencil or somewhat larger.
4. The pellets are sintered at high temperatures to consolidate the powder into a solid pellet.
5. The pellets are inspected, ground to size, and placed in a vacuum to remove atmospheric gasses.
6. The pellets are then loaded into empty cladding tubes, which are typically 1–5 m long and made of Zircaloy (an alloy of zirconium with minor amounts of tin and iron) or stainless steel, and an end cap is attached by welding, yielding fuel rods.
7. The rods are inserted into *grid spacers* (metal squares subdivided into smaller squares sized to hold the rods in place with springs), and the entire assembly is bound together axially with end pieces that are connected with hollow tubes similar to those used for cladding.
8. The assembly is cleaned and readied for transport to the reactor.

A cutaway drawing of a typical fuel assembly is shown in Fig. 7.

5.2 Metal Fuels

A second common fuel type is made of metal and is commonly employed in test and research reactors where achieving high temperatures (which is required to produce electricity efficiently) is irrelevant. Following are the major steps typically required to make such fuel:

1. Uranium dioxide with an appropriate isotopic enrichment (which could range from natural to about 93% ^{235}U) is reduced from the oxide to the metal as described earlier.
2. The uranium metal is often alloyed with other metals (e.g., Al, Zr, Mo, Nb) to improve its physical characteristics.

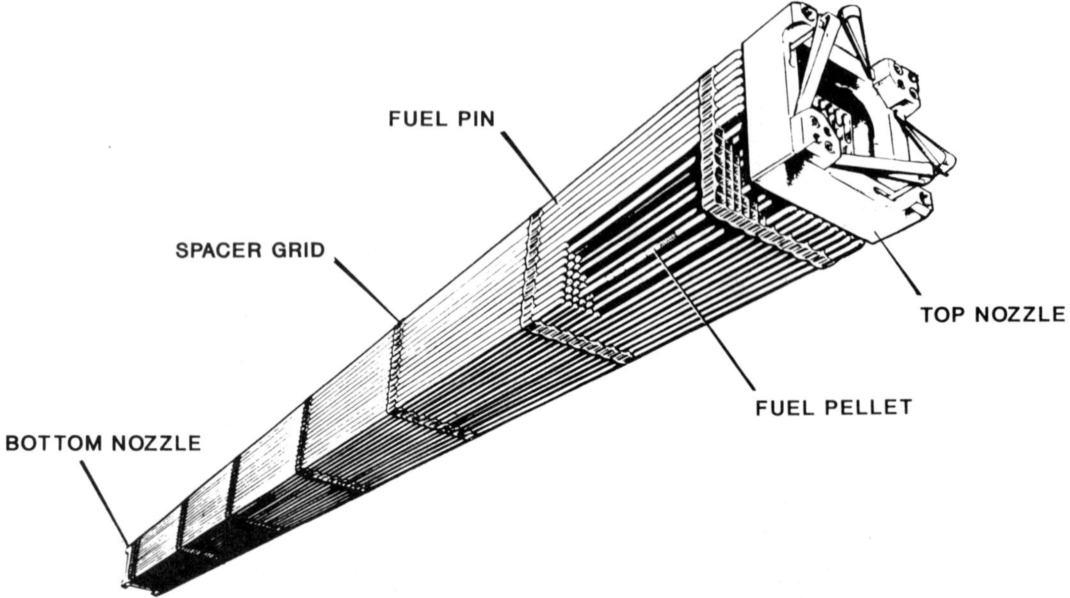

Fig. 7. Major components of a typical fuel assembly from a pressurized water reactor.

3. The alloy is then hot rolled and sized into plates in the range of 0.5–2 mm thick.
4. The plates are placed between two larger plates of cladding metal (typically Al or an Al alloy), which are then hot pressed to seal the edges into a fuel element.
5. The elements are then put into slots in two of four larger flat metal plates that form a box surrounding the plates but open at the ends to allow coolant to pass through. This assembly is typically about 1 m long and 10 cm on each side, although annular designs are also employed and size can vary widely.
6. The assembly is cleaned and prepared for transportation to the reactor.

There is substantial variability in these fuels, with fuel material made of uranium hydrides being relatively common and more advanced materials such as uranium silicides being developed.

6. EFFECTS OF FUEL IRRADIATION

6.1 Irradiation Processes and Products

As noted previously, there is a wide variety of fuel material, cladding material, coolants, and moderators in reactors throughout the world. Despite this diversity, the effects of irradiation are similar in all cases.

Conceptually, a nuclear reactor is a vessel in which a sufficient number of fuel assemblies are placed in close proximity in the *core* and controlled such that a stable fission rate is established in the fuel. Fission generates a significant amount of heat, and one purpose of the vessel is to contain the coolant required to remove the heat from the fuel. Each fission generates about 200 MeV of energy that is recovered as heat, and about 2.5 neutrons. As a consequence, a typical large nuclear power plant generating 1000 MW of electricity has a fission rate of about 9.4×10^{19} fissions/s and produces 2.3×10^{20} neutrons/s.

The 2.5 neutrons produced by each fission reaction are consumed by the nuclei of all elements in the reactor in the following ways:

1. Exactly 1.0 of the neutrons interacts with a nucleus and results in another fission, thus sustaining the chain reaction. This nucleus must be able to be fissioned by the neutrons produced by previous fissions. These neutrons have a spectrum of kinetic energies (i.e., speeds) ranging from fractions of an eV to about 3×10^6 eV. The only elements that can be fissioned by neutrons having this range of energies are

the actinides. By definition, this includes all elements with atomic numbers greater than 89 (that of actinium). In practice, only a few actinides occur naturally or can be produced in sufficient quantities to constitute a fuel material: Th, U, and Pu. Isotopes such as ^{235}U and Pu^{239} can be fissioned by neutrons of all energies and are called *fissile*. Isotopes such as ^{238}U and ^{240}Pu can be fissioned by only high-energy neutrons and are called *fissionable*. The sum total of the interaction of the 2.5 neutrons with the various actinide isotopes present in the reactor at a given instant results in 1.0 producing a fission that sustains the chain reaction. The other 1.5 neutrons interact as described below, with the proportions of these reactions varying significantly depending on the reactor design.

2. A significant portion of the neutrons interact with the actinides by *neutron capture*. In this case, the actinide is not fissioned. Instead, the neutron combines with the actinide nucleus to form the next heavier isotope of the same actinide element. For example, ^{238}U captures a neutron to form ^{239}U. Uranium-239 has a relatively short half-life and decays to ^{239}Np, which also has a relatively short half-life and decays to ^{239}Pu. During the residence of the fuel, dozens of actinide isotopes are produced by a complex chain of neutron captures and radioactive decays. Many of these are sufficiently long-lived so that they remain in the fuel when it reaches the end of its useful life. All are radioactive, and the dominant decay mechanism is alpha decay, although the decay of some radionuclides results in the emission of neutrons.
3. In the fission process, the actinide nucleus is split into two separate nuclei having much lower atomic numbers. These nuclei are called *fission products*. The split is usually not symmetrical (i.e., the two fission-product nuclei have very different masses). Further, an entire range of fission product elements and isotopes are produced. The distribution of these is shown in Fig. 8 (Katcoff, 1958). As produced, they are all radioactive, and the dominant decay mechanism is beta decay. Many have short half-lives, and these decay to long-lived or stable (i.e., nonradioactive) fission products that are present when the fuel reaches the end of its useful life.
4. Another fate of the 1.5 neutrons not involved in sustaining the chain reaction is to interact with nonactinide nuclei by neutron capture, resulting in *activation products*. The nonactinide nuclei are most often the other materials present in the reactor (e.g., cladding metal, the reactor vessel, or neutron absorbers used to control the chain reaction). The neutron capture process is the same as that described previously for the actinides. Most of the activation products are relatively short-lived and decay to innocuous levels during irradiation. However, a few have longer half-lives and are present in the fuel and reactor structural materials when it reaches the end of its useful life.

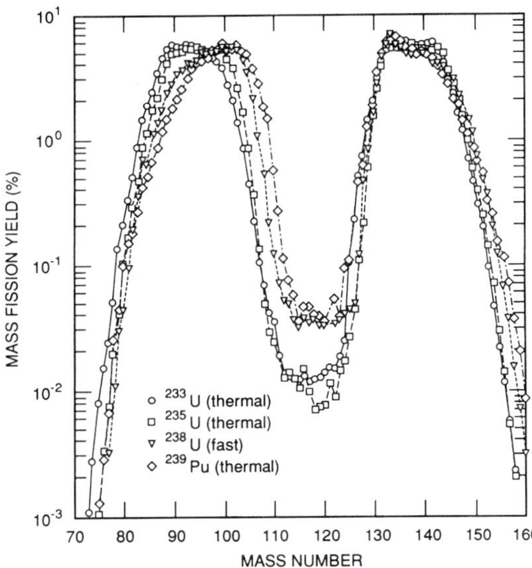

Fig. 8. Fission-product yield as a function of atomic mass number for 0.025 eV-neutron-induced fission of ^{233}U, ^{235}U, ^{239}Pu, and 14 MeV-neutron-induced fission of ^{238}U.

At the end of its useful life, the uranium dioxide fuel typically used to generate electricity contains about 1% actinides heavier than uranium (primarily Pu), 3% fission products, a variable but very small amount of activation products, and about 96% uranium isotopes (^{238}U with about 0.8% ^{235}U and 0.4% ^{236}U). The proportions of constituents other than uranium isotopes are increased as the amount of power extracted from a unit of fuel (called *burnup*) increases. The useful life of the fuel is limited by the consumption of fissile isotopes, production of fission product neutron absorbers, and physical degradation of the fuel assembly structural materials (e.g., cladding).

6.2 Distribution and Properties of Fuel Irradiation Products

This section discusses the characteristics of the fuel after it reaches the end of its useful life. At this point, it is removed from the reactor vessel (i.e., *discharged*) and becomes known as *spent fuel*. The characteristics to be discussed are the location of the various products within the fuel assembly and the salient properties of the various products.

6.2.1 Distribution of Irradiation Products The actinides in the spent fuel are contained almost exclusively within the fuel matrix. "Tramp" fuel on the cladding may result in a small amount being associated with it, but this is usually negligible.

The fission products are mostly contained within the fuel matrix. However, some of the fission product elements are gaseous (e.g., Kr, Xe) or are somewhat volatile at reactor operating conditions (e.g., Cs, I). Some of these elements will be present in nonfuel areas where it is relatively cool, such as in the top end of the fuel rod, which is designed as a gas plenum, or on the inside of cladding.

The activation products are distributed more widely. Impurities and the oxygen within the fuel are activated by irradiation and tend to remain there. The cladding and other metals used to make the fuel assembly and impurities within them also become activated. Most remain contained within the metals, although a small amount is dissolved by the coolant and finds its way to the reactor plant wastes. Although activation rates decrease as the distance from the reactor core increases, all of the other fixed internal parts of the reactor and the structure (typically steel and concrete) immediately surrounding the reactor become activated to varying degrees during their useful lives (typically 2–3 decades) and must be treated as radioactive waste when the plant is decommissioned. Finally, the coolant circulating through the core and the chemicals within it (if any) become activated to varying degrees, and the resulting radioactive isotopes find their way into reactor plant wastes during operation. Most of the activation products in the coolant are short lived, although a few are produced in significant quantities and are long lived (e.g., ^{3}H, ^{14}C).

6.2.2 Properties of Irradiation Products The irradiation products in spent fuel could be characterized in a variety of ways, including element concentrations, isotope concentrations, chemical form, etc. However, the most relevant property of these products is that they are radioactive. As such, they have emissions of varying energy that include photons (gamma rays, x rays, bremsstrahlung), electrons (beta particles, Auger electrons), alpha particles, neutrons, neutrinos, and (very seldom) positrons. These emissions are generally well characterized and documented in large volumes (Lederer and Shirley, 1978).

These emissions have impact in two important ways. The first is that some of them (photons and neutrons) are considered to be *penetrating radiation* because they are capable of passing through a significant thickness of shielding material before being absorbed. As a consequence, any significant source of these must be shielded to protect human health. Calculation of this impact, called the *radiation dose*, is very complex because of its dependence upon the specific mix of isotopes, the age of the source, the geometric relationship of the source to the target, and the composition and thickness of materials between the source and target. Results of a calculation giving the dose rates from an unshielded, 10-year-old spent-fuel assembly are shown in Fig. 9 (Croff *et al.*, 1979). As a basis for comparison, a dose of about 500 rem has been considered to be lethal to 50% of human recipients, although medical advances evident in treating the casualties of the accident at the Chernobyl reactor in the Ukraine now indicate that the dose required to cause a 50% fatality rate may be significantly higher.

The second important impact of the radioactive emissions is that all (except neutrinos) are eventually absorbed in the material surrounding the spent fuel. The energy of these emissions is manifested as energy deposited in the material and the resultant heating of the material. A typical fuel assembly from a modern power reactor generates

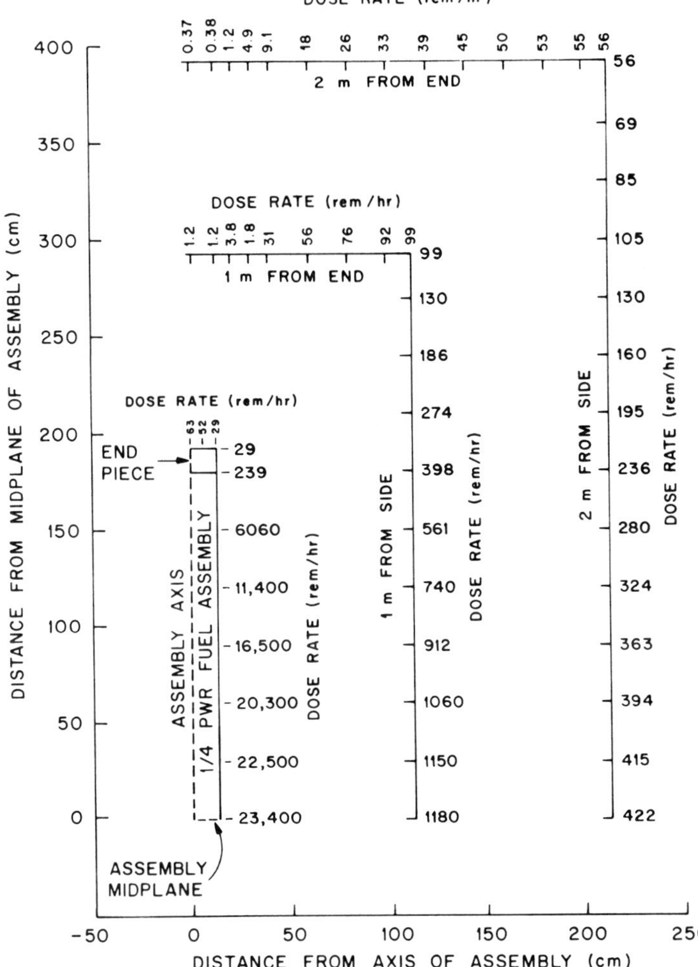

Fig. 9. Radiation dose from a 10-year-old spent pressurized water reactor fuel assembly that has been irradiated to 33 GWd per Mg of initial uranium.

heating of the material. A typical fuel assembly from a modern power reactor generates 17 300 kW of thermal power during irradiation. When the chain reaction ceases, the thermal power immediately declines to about 1200 kW. As shown in Fig. 10 (Croff and Alexander, 1980), the thermal power continues to decline at a decreasing rate thereafter until it finally attains negligible levels. However, for many years, the thermal power is sufficient to threaten the integrity of the fuel assembly if provisions are not made to remove the heat.

Radioactive isotopes can also impact living organisms if ingested or inhaled. For more discussion of these subjects, see ENVIRONMENTAL HEALTH AND SAFETY and NUCLEAR WASTE MANAGEMENT.

7. SPENT-FUEL REPROCESSING

After irradiation in a reactor and discharge from the reactor at the end of its useful life, spent fuel is stored for a significant amount of time (at least a couple of years, often many years) to allow shorter-lived irradiation products to decay. At this point, it becomes available for one of two possible dispositions. The first is that a determination is made that further chemical processing of the spent fuel is not worth while, and the spent

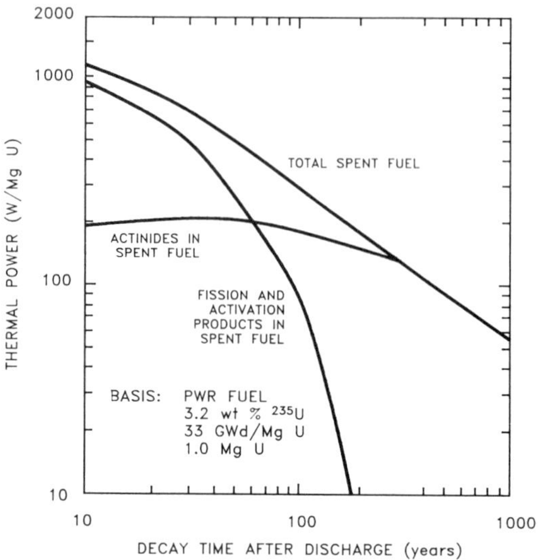

Fig. 10. Thermal power of major constituents in spent pressurized water reactor fuel that initially contained 1.0 Mg of uranium.

fuel should be treated as a waste (i.e., to have a once-through fuel cycle as described at the outset; see Fig. 1). In this case, the spent fuel would typically be stored for an extended time, sealed in a multiple-layer package, and sent to a permanent disposal site, where it would be buried deep underground (see article on NUCLEAR WASTE MANAGEMENT).

The second possible disposition is that further chemical processing of the spent fuel is worth while. This could be the case to recover various valuable constituents of the spent fuel, to change the form of the spent fuel to one more acceptable for waste disposal, or a combination of the two. The processing of the spent fuel is called *reprocessing*. The operations required for reuse of fissile and fertile constituents of the spent fuel are called *recycle*. Because of the intense radiation from the irradiation products, reprocessing must be carried out in remotely operated facilities that have thick, dense walls called *shielding* interposed between people and the nuclear material.

By far the most widely used reprocessing technology is the Plutonium-Uranium Extraction process, which is shortened to *Purex*. It has been used for decades to reprocess defense fuels to recover Pu for military purposes. More recently, large plants have been built in France and Great Britain to implement this process on civilian fuels, and more plants are under construction. The Purex process is described in some detail below as it is applied to oxide fuels clad in metals such as Zircaloy and stainless steel. This is followed by brief descriptions of variants of the Purex process and major alternative reprocessing technologies.

7.1 Purex Reprocessing

Reprocessing involves the application of a number of chemical processing steps to separate and purify the uranium and plutonium constituents of the spent fuel and to manage the residuum, which is considered to be waste. These are described in the following sections in the order in which they are performed.

7.1.1 Head End The *head end* of a reprocessing plant involves operations (generally physical) to expose the fuel material.

7.1.1.1 Remove Nonfuel Components. The first operation is to remove structural components that do not contain the fuel, such as end fittings and possibly the grid spacers. This may be accomplished by unscrewing threaded components or sawing and cutting. The removed metal pieces are treated as waste.

7.1.1.2 Segmentation of Fuel. The second operation is to expose the fuel material within the cladding in a process step called *segmentation*. This is typically accomplished with a mechanical shear, which is simply a mechanism to clamp a group of rods together with the ends projecting beyond a flat surface and chop off 2–4-cm segments with a hydraulically driven metal blade (much like chopping a bundle of celery stalks with a knife). It can also be accomplished with an abrasive saw or with a high-power laser.

7.1.1.3 Dissolution. The third operation is to dissolve the exposed fuel material in hot, concentrated nitric acid in an operation called *dissolution*. After being sheared off, the rod segments with the fuel exposed drop into a metal basket with small perforations. When the basket is sufficiently full, it is lowered into a vessel containing the acid. The acid is heated and sparged with air, and, over a period of time, most of the fuel matrix is dissolved. Fuel constituents that are not dissolved fall into three categories: cladding, insoluble metals, and volatile elements. The cladding does not dissolve in nitric acid, and remains in the basket in the form of *cladding hulls*. When fuel dissolution is complete, the cladding hulls are removed from the dissolver and treated as waste.

The insoluble metals typically comprise a small fraction of the fission products, which are themselves a small fraction of the fuel. It is the more noble fission products (i.e., Pd, Rh, Ru, Tc, Mo) that tend to be insoluble under dissolver conditions. These are typically removed from the nitric acid solution by centrifugation and may then be leached with more aggressive reagents (e.g., HNO_3-HF) to remove residual actinides before being treated as a waste or dissolved in aqua regia for further processing.

The volatile elements that evolve from the dissolver solution are called *off gas*. They may subsequently be captured to prevent their release to the environment and treated as waste. The inert gases Kr (which contains radioactive ^{85}Kr) and Xe are released quantitatively from the dissolver solution. Also released quantitatively under the highly acidic conditions are iodine (containing radioactive ^{129}I) and carbon as CO_2 (containing the radioactive activation product ^{14}C). Other species that are volatilized to some significant extent are activation- and fission-product tritium and Ru as RuO_4 (which contains radioactive ^{106}Ru). Nonradioactive oxides of nitrogen are also evolved from the dissolver solution.

7.1.2 Product Recovery The nitric acid solution containing most of the fuel constituents is then sent to two successive solvent extraction operations to recover impure uranium and plutonium in separate streams.

7.1.2.1 Primary Decontamination. In the first operation, the nitric acid solution undergoes multistage contact with a 30-vol% solution of tri-*n*-butyl phosphate (TBP) dissolved in an organic diluent such as *n*-dodecane. The aqueous and organic streams are immiscible and can be easily separated by gravity or centrifugation. The organic phase extracts virtually all (99+%) of the U and Pu (which are in the +6 and +4 valence states, respectively), a significant amount of the Np (which is in multiple valence states), and a significant amount of a few fission products such as Zr, Nb, and Ru. These leave the solvent extraction equipment in the organic stream. The nitric acid solution contains virtually all of the fission products, Am and Cm, and some of the Np. This solution is evaporated to concentrate it and recover the nitric acid. The remaining solution is called *high-level waste* and has little further value, except possibly for the recovery of small amounts of isotopes for specific beneficial uses.

The organic chemicals are recycled within the plant. However, because of the fission products, the organic extractant is exposed to intense radiation (primarily beta particles and gamma rays) while it is in contact with the aqueous solution. The result of this exposure is that some of the TBP is degraded to form undesirable impurities, with the levels increasing with increased exposure. It is necessary to remove the impurities from the solvent to make it acceptable for reuse. This can be accomplished by various means such as contacting it with an aqueous solution of sodium carbonate or passing it over a bed of silica gel.

7.1.2.2 Partitioning. The second product-recovery operation separates the U from the Pu. This is accomplished by contacting the organic stream laden with U and Pu with a dilute nitric acid stream containing a reductant such as ferrous sulfamate, hydrazine, or electrolytically generated U^{4+}. The reductant changes Pu to the +3 valence state, which is readily extractable by the nitric acid stream while the U remains in the organic stream.

7.1.2.3 Solvent Extraction Equipment. The key pieces of equipment in the above operations are the solvent extraction contactors. This is especially the case in the primary decontamination operation, where fast, efficient contact between immiscible phases followed by rapid, complete separation is important. The first equipment used was called a *mixer-settler*, which amounted to little more than mixing the organic and aqueous streams in a vessel using a powered agitator and then allowing the phases to separate by gravity in a baffled portion of the device. The two phases were then routed to neighboring mixer-settlers in counter-current fashion, much like the enriched and depleted streams in uranium enrichment described in Sec. 3.1.2. This equipment had relatively long contact times and low efficiencies, resulting in the development of more advanced equipment.

The successor to the mixer-settler in nuclear fuel reprocessing is the *pulse column*, which is presently in widespread use in most reprocessing plants. In this device, an aqueous (more dense) stream is introduced into the top of a vertical cylindrical column, and an organic (less dense) phase is introduced into the bottom. Spaced throughout the column are *sieve plates*, which are perforated horizontal metal plates. As the liquids pass through the plates, they are dispersed to promote intimate contact of the two phases. A *pulser* (much like a hydraulic cylinder) is periodically activated to force a small amount of liquid upward through the plates into the next stage. While this device was a distinct improvement over the mixer-settler, it still suffers from long contact times and requires large (and expensive) shielded process cells because the contactors are relatively tall.

The most recent advance in solvent extraction equipment is the *centrifugal contactor*. This is a piece of rotating equipment much like the gas centrifuge described in Sec. 3.1.3. The two streams are introduced, and, in one portion of the equipment, an agitator ensures good dispersion and contact between the phases, which quickly exit to another portion where they are quickly separated by centrifugal force. The equipment is relatively small and appears to meet the demands of fuel reprocessing quite well.

7.1.3 Product Purification At this point in the process, the separate U and Pu streams remain cross-contaminated (e.g., Pu in the U stream) as well as being contaminated with significant amounts of Np and fission products. The purpose of product purification is to remove these contaminants to acceptable levels.

7.1.3.1 Uranium Purification. The uranium is in the organic phase as it enters purification. It is then typically back extracted into dilute nitric acid and then subjected to one or two more cycles of TBP extraction and nitric acid back extraction to achieve the necessary purity. An aqueous solution of U may also be passed over a bed of silica gel for final purification. The purity requirements are higher if the uranium is being recycled for isotopic re-enrichment, as opposed to being simply used as feed material to be enriched with Pu.

7.1.3.2 Plutonium Purification. The plutonium is in the aqueous phase as it enters purification. It is typically subjected to one more cycle of TBP extraction and nitric acid back extraction. The final purification step consists of passing a nitric acid solution of the Pu through a column of anion-exchange resin to remove the residual contaminants.

7.1.4 Product Conversion The two product streams resulting from purification of U and Pu are aqueous acid solutions of the respective nitrates. The final forms of these products are generally UO_2 or UF_6 and PuO_2, respectively. Such conversion is usually done relatively promptly because the liquids are more vulnerable to leakage, criticality, and radiation effects and are expensive to store.

7.1.4.1 Uranium Conversion. The conversion of uranium nitrate solution to UO_2 generally proceeds as described in Sec. 2.2. Briefly, the uranium is precipitated as ammonium diuranate, which is then calcined to yield UO_2. This can be mixed with other fissile elements such as Pu to make mixed-oxide fuel, which can then be fabricated as described earlier and recycled to a reactor. Alternatively, the UO_2 can be further converted to UF_6 for re-enrichment, as described in Sec. 2.3.

7.1.4.2 Plutonium Conversion. The approaches used to convert plutonium nitrate to PuO_2, which is the only useful form in power production systems, are somewhat more diverse than for uranium. The plutonium can be precipitated with oxalic acid as plutonium oxalate, which is then filtered, dried, and calcined to PuO_2. The addition of hydrogen peroxide to the plutonium nitrate solution directly precipitates PuO_2, which can then be recovered through filtration and dried. Finally, there is some experience with direct evaporation and calcination of the nitrate solution to yield PuO_2, although this has not been used commercially.

7.1.4.3 Coconversion. Techniques have been developed and tested that form mixed $(U,Pu)O_2$ directly from a mixed-nitrate solution. These are called *coconversion processes*. They can involve direct calcination of the nitrate mixture to yield UO_3 and PuO_2, followed by hydrogen reduction to the mixed dioxide. Alternatively, the U and Pu can be coprecipitated and then calcined to the dioxide.

7.2 Reprocessing Variations

Numerous variations of reprocessing technology have been developed, and many have been deployed. Some of these variations represent the normal evolutionary improvement of any technology. Others include features necessary to produce somewhat different products or handle various types of spent fuel. Finally, fundamentally different approaches can be used for a specific fuel type.

7.2.1 Purex Variations A number of specific variations in the implementation of the Purex process are embodied in Sec. 7.1, and many more (primarily regarding chemical conditions in the processes and equipment differences) are possible, depending on cost estimates and the objectives of the Purex plant designer.

The only major variation of Purex that has been seriously considered (but not deployed) is to operate the process without the partitioning and purification operations (American Nuclear Society 1978). This yields a mixed U-Pu product that is contaminated with significant amounts of radioactive fission products for the purpose of making the material less desirable for diversion to undesirable uses. Such a material could only be recycled by designing facilities and equipment to carry out all steps of the fuel cycle with fully remote operations.

7.2.2 Aqueous Reprocessing Variations A number of aqueous reprocessing technologies other than Purex have been developed since the 1940s, when large-scale nuclear technology was born. Most of the early processes (e.g., bismuth phosphate precipitation, REDOX, Trigly, Butex) were displaced by the superior Purex process and others developed after Purex did not appear to offer enough advantages to justify the cost and risk of development. As a consequence, the Purex process continues to dominate the processing of uranium-based fuels to the exclusion of other aqueous processes.

Other aqueous processes are used in the processing of fuels that are not uranium based. The best known of these processes is *Thorex*, which is used for thorium-based fuels in which ^{233}U is produced by neutron capture. Thorex is similar to Purex in many respects, and the reader is referred to Sec. 7.1 for the operational sequence of the process. The primary difference in Thorex is that Th does not dissolve easily in nitric acid, and it is necessary to add HF to the dissolver solution to obtain an acceptable dissolution rate. However, the HF causes additional problems in the process, principally unacceptably high corrosion of metals. As a consequence, aluminum nitrate is added to the dissolver to complex the fluoride ion and keep corrosion to an acceptable level. The fluorides and aluminum report to the high-level waste and may result in problems or additional cost in its management. So far, Thorex has not been applied commercially.

7.2.3 Nonaqueous Reprocessing Variations Two major classes of nonaqueous processes have been developed: pyroprocesses and fluoride volatility. The potential advantages of these processes are that first, unlike organic solvents, radiation damage to the reagents does not create deleterious effects in the process, and second, the fundamentally different nature of the reagents offers unique possibilities concerning what is readily separated and what is not. The waste streams from these processes differ markedly from aqueous process wastes.

7.2.3.1 Pyroprocesses. *Pyroprocesses* are defined by the use of high-temperature (>500 °C) molten metals and salts as the reprocessing reagents. The feed material must typically be in the form of metal, and oxide fuels must be reduced to the metal (typically with Ca) before processing. Numerous pyroprocesses have been investigated over the last few decades, including volatilization, fractional crystallization, extraction with liquid metals, liquid extraction from fused salts, electrolysis, partial oxidation, skull reclamation, and salt transport. Most of these have not been developed to any appreciable extent because they manifested serious deficiencies compared to other processes, but those described below have been investigated to a significant extent.

A partial oxidation process (called *melt refining*) that involves melting uranium-based metallic spent fuel in a ZrO_2 crucible at 1400 °C has been applied to the recycle of fuel at the EBR-II test reactor in Idaho for almost 30 years. Many fission products are volatilized at this temperature, and other nonvolatile fission products that are more reactive than uranium and 5–10% of the feed uranium are oxidized by the ZrO_2 and form a separate layer (called a *skull*) in the crucible. The remaining metal is a mixture of fission products less reactive than uranium (e.g., Zr, Nb, Mo, Ru, Rh, Pd), which are collectively known as *fissium*, and uranium. This mixture is combined with fresh, enriched uranium metal, and the product is poured into glass molds to make recycle fuel. The fissium is highly radioactive, but its physical properties stabilize the uranium during irradiation in the reactor. The *skull reclamation* process uses a Mg-Zn alloy to reduce the valuable uranium in the skull selectively to the metal, which is soluble in the alloy. The Mg-Zn is then removed by distillation to recover the uranium.

An electrolysis process called *electrodeposition* is being investigated for processing metallic U-Pu fuels for advanced reactors cooled with liquid sodium. In this process, stainless-steel–clad fuel made of metallic U-Pu-Zr and a portion of the fission products are chopped and placed in a basket. The electrorefiner contains a bottom layer of molten Cd and a top layer of LiCl-KCl eutectic salt. The basket is lowered into the salt layer, and $CdCl_2$ is added to oxidize the alkali, alkaline-earth, and rare-earth metals in the spent fuel to their chlorides, which dissolve in the salt layer. Electrolysis occurs by making the basket and bottom Cd layer the anode. Two cathodes are immersed in the molten salt layer, with one part being steel and one part being molten Cd in a ceramic crucible. During electrolysis, U-Cd is deposited on the steel cathode, and a mix of U-Pu-Cd-lanthanides is deposited on the Cd cathode. The U-Pu product contains the other actinides heavier than U that were present in the spent fuel. Both products contain significant amounts of fission products and the Zr that was present in the initial fuel. When electrodeposition is complete, the two cathodes are removed to a retort where the Cd is volatilized. The remaining constituents are then removed to an injection melter, which melts the actinide products, adjusts the composition to the correct U-Pu-Zr proportions, and injects this mixture into fuel molds. The fuel is then recycled to the reactor.

7.2.3.2 Fluoride Volatility Processes. As might be evident from the section on uranium enrichment, fluorides of some elements are volatile at modest temperatures (a few hundred degrees Celsius or less). This property has been used as a basis for reprocessing spent fuel using a *fluoride volatility process*.

In this process, the fuel material is exposed by decladding operations much as in the Purex process. However, at this point, the fuel material is converted from the oxide to fluorides using a sequential combination of HF and F_2. Many of the fluorides are volatile but at varying temperatures. This permits a first separation of the U from other radionuclides by using simple distillation. The impure uranium stream from this operation is then further purified using techniques such as passing the gases through solid beds of NaF to remove undesirable elements by selective sorption of uranium. When this is used as a primary reprocessing technology, only the U is recovered, and the Pu follows the fission-product waste. As a consequence, this has been considered for only highly enriched uranium fuels where the Pu is mostly composed of ^{238}Pu, which is not worth recovering.

Fluoride volatility processes are capable of excellent decontamination from undesirable constituents and have been used to clean uranium feed to enrichment operations. A full-size reprocessing plant (Midwest Fuel Reprocessing Plant) in which the final cleanup operations were based on fluoride volatility processes was designed and built at Morris, Illinois, but never operated because of design inadequacies (especially inadequate surge capacity within the facility).

7.2.4 Continuous Reprocessing The reprocessing of spent fuel from an individual reactor, as described in Sec. 7.1, is performed intermittently on a group of discharged spent fuel assemblies sent to a separate facility. However, reactor systems have been developed and tested in which the fuel is circulated through a core region configured so that a nuclear chain reaction occurs and the fuel is heated. The fuel then flows to a heat exchanger, where the heat is removed by generating steam to make electricity in a turbine. The fuel (or a fraction thereof) is then circulated to a reprocessing operation that is an integral part of the fuel flow loop, where it is *continuously reprocessed* to remove deleterious fission products.

This concept has been tested in designs involving aqueous fuel [e.g., the Aqueous Homogeneous Reactor Experiment (Chapman *et al.*, 1963) and the Molten Salt Breeder Reactor (Weinberg *et al.*, 1970)]. In the aqueous case, reprocessing operations involve volatilization of the noble-gas fission products and use of the fact that actinide sulfates exhibit *retrograde solubility* (less soluble at higher temperatures) to precipitate them with some of the fission products and recover them with a cyclone. For the molten salt reactor, the reprocessing technology would be very similar to pyroprocessing as described in Sec. 7.2.3.1. These concepts have not been developed further than the test reactor phase, and there are no apparent plans to do so.

8. STORAGE AND TRANSPORTATION OF NUCLEAR MATERIALS

Because of economics of scale and other business factors, most facilities constituting the nuclear fuel cycle shown in Fig. 1 are generally not colocated. As a consequence, various radioactive product materials must be transported between facilities. This, in turn, dictates that the various facilities have significant capability for storing the feed and product materials in order to achieve economic transportation loads, to ensure uninterrupted facility operations, or to condition materials prior to processing.

Although there are radioactive materials transported between and stored at each of the facilities shown in Fig. 1, most of these operations are straightforward because the materials involved are only mildly radioactive. Considerations for transportation and storage of materials that have not been irradiated involve protecting them from dispersion and the natural elements (e.g., rain) and, in the case of precision-manufactured components such as fresh fuel assemblies, protecting them from physical damage during transport. These considerations, and the transportation and storage methods employed, are very similar to those used for mildly toxic industrial chemicals and other high-precision equipment and will not be detailed here. One exception is storage and transportation of highly enriched fissile material, which requires special considerations to provide adequate protection from illicit diversion and criticality control.

On the other hand, spent fuel and the high-level waste from the primary decontamination operation of fuel reprocessing are highly radioactive, emitting significant amounts of heat and penetrating radiation. The considerations in their transport and storage are unique and are described in some detail below.

8.1 Storage

8.1.1 Water Pools Spent fuel is most commonly stored in deep-water pools, which are usually rectangular and placed in a large open room with a high-capacity overhead crane. They are present at reactor plants, which typically have the discharged spent fuel from many years of operations stored in the pool since disposal facilities have not yet been built in any country. They are also present at fuel reprocessing plants and constitute the centerpiece of stand-alone facilities designed for long-term spent-fuel storage. Experience has shown that fuels clad in Zircaloy and stainless steel can be stored under water for decades with no significant degradation.

The pools have varying surface areas depending on the capacity of the facility. They are normally quite deep: 3–6 m plus the height of a fuel assembly. The water layer above the top of the fuel assemblies provides radiation shielding to the workers (e.g., crane operators) located around and above the pool, and one fuel-assembly height is required for storage. A portion of the pool is normally deeper yet by about an extra fuel-assembly length plus 1–2 m to accommodate a large, thick-walled vessel used to transport the fuel between sites (see Sec. 8.2 for details).

The pool is built with a grid structure of metal racks to hold the spent fuel in place in a configuration that eliminates the possibility of a critical reaction. The pools have a high mass and typically a high center of gravity, and extensive provisions are made to ensure integrity during potential seismic events. The water in the pool is circulated to maintain it at extremely high purity levels and to remove the decay heat from the spent fuel. Fuel assemblies in which the cladding of one or more elements has been penetrated are normally sealed in a metal can before being stored.

Other materials having significant radioactivity are often stored in the pools. Examples are contaminated or irradiated reactor components and metal hardware from dismantled fuel assemblies. In a reprocessing plant pool, the high-level waste packages and cladding waste packages are typically stored in the pool.

8.1.2 Dry Storage Storage of some materials under water may not be possible or desirable on the basis of technical or economic grounds. This may be the case because water will adversely affect the integrity of the fuel (e.g., fuels clad in Al and Mg or made of graphite), an existing pool is full and expansion is not possible, or the fuel characteristics are such that it is more economical to store it under dry conditions.

There are two common technologies for dry storage. One is to store the fuel in a container called a *storage cask* that is so massive that it provides adequate radiation protection. This cask is simply loaded with spent fuel or waste, sealed, and then placed outside on a concrete pad where heat is removed via conduction through the package and natural convection. This container is very similar to those used for transporting the spent fuel except that it may not be certified for off-site transportation. It is described in detail in Sec. 8.2.

The second common dry-storage technology is *dry wells*. These are typically an array of individual cells made of reinforced concrete and sized so that a fuel assembly can be lowered into them and a plug providing radiation shielding inserted into the top of the well. The wells are not normally ventilated, and heat removal occurs via conduction through the walls and then convection in a moving air stream. If the wells are ventilated and actively cooled, the gas stream must be purified before release. This type of design is also used in reprocessing plants for storage of high-level and other highly radioactive wastes.

8.2 Transportation

Spent fuel is transported between facilities via trucks, railroads, and ships in containers called *shipping casks* or *shipping packages*. This operation is challenging because the cask must simultaneously provide for radiation protection, heat removal, criticality protection, and barriers to the release of radionuclides from the spent fuel in the case of accidents.

8.2.1 Regulations Standards that casks must meet have been established by international organizations (IAEA, 1992) and national regulatory authorities and are remarkably uniform throughout the world. This is necessary because of the international nature of the nuclear fuel cycle and the need to move nuclear material between nations. The manner in which a particular designer meets these standards is flexible, depending on the specific characteristics of the fuel being transported. When a new shipping cask is to be built, or an existing cask significantly modified, it must be shown to meet the standards by actual testing, typically with a scale model, or by rigorous engineering analysis. A cask is subjected to the following tests in sequence:

1. a 9-m drop onto a flat, horizontal, unyielding surface so that it strikes in its most damaging orientation (typically an edge);
2. a 1-m drop onto a 0.15-m-diameter steel bar at least 0.2 m long, at its most vulnerable part;
3. a fire test in which the cask is totally engulfed in a 800 °C thermal environment for 30 min;
4. water immersion in which the cask is submerged under at least 0.9 m of water for 8 h, and a separate cask is completely immersed under 15 m of water for 8 h. In addition, for casks carrying over 1 MCi of radioactivity, the International Atomic Energy Agency requires that the immersion depth be increased to 200 m (a pressure chamber is used to simulate this depth).

To be acceptable, the cask must remain leak-tight after this test sequence.

Myriad additional regulations are often imposed on the transportation operation *per se*. These include how the cask is fastened to the transport vehicle, radiation level limits, shipping documentation, labeling, driver qualifications, emergency response, and routing.

8.2.2 Shipping Casks The above requirements, when coupled with the need for a significant payload and removal of significant amounts of decay heat from the spent fuel, result in the cask having to be made almost exclusively of metal components. A typical schematic of a spent-fuel shipping cask is shown in Fig. 11. The cask has an inner and outer shell usually made of stainless steel. Between these shells is a very thick layer of denser metal to provide shielding from penetrating photon radiation. Iron, lead, and uranium are often used to provide this

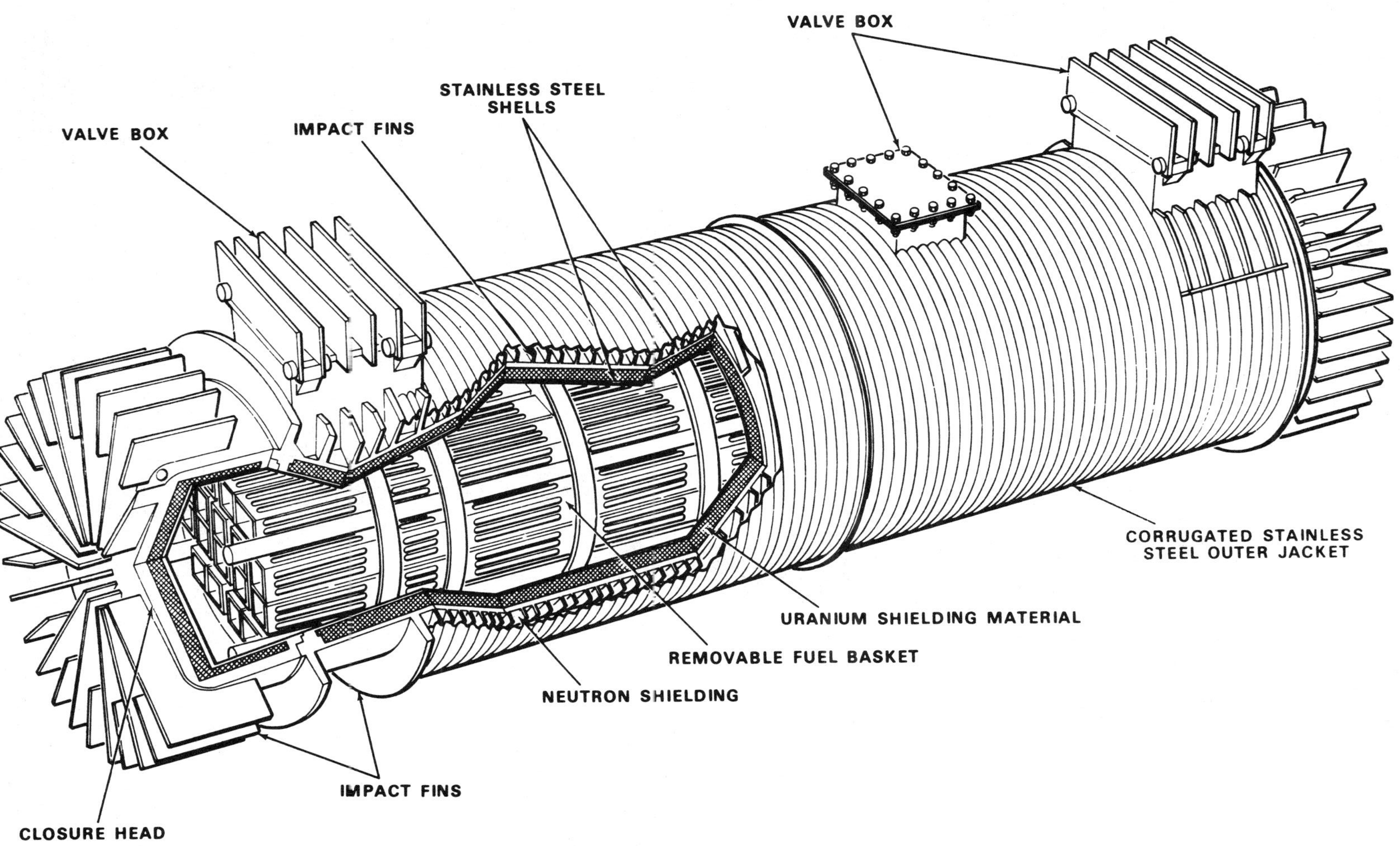

Fig. 11. Schematic diagram of a typical spent-fuel shipping cask.

shielding. In addition, if the spent fuel emits a significant amount of neutron radiation, there is often a layer of hydrogenous material (e.g, polyethylene) that contains a good neutron absorber (e.g., boron). Casks typically weigh between 25 and 100 Mg.

A grid called a *basket* that is appropriately sized for the spent fuel being transported is inserted into the hollow cavity in the middle of the cask. The basket holds the fuel in place to prevent damage during normal transport and minimize damage during accident conditions. It is also important in assuring that the array of fuel assemblies will remain subcritical during transport.

The cask is typically loaded by removing its lid and immersing it in the pool with the open end upward. The fuel is then loaded into the cask with a crane, the lid placed on the cask (to provide radiation shielding), and the cask removed from the pool. Water is drained from the cask interior, the lid bolted down, the inside dried with a vacuum pump, the cask leak tested, and the interior filled with an inert gas (typically nitrogen). The outside of the cask is decontaminated to remove the trace radioactivity present from the pool water, loaded onto the transport vehicle (trailer or rail car), tied down, and then transported to its destination, where the process is reversed to unload the fuel. The cask is cooled by conducting the heat through the walls; no active cooling measures are employed, although fins are often used to enhance convective heat transfer on the outer surface.

9. BENEFICIAL USE OF RADIOISOTOPES

The preceding sections of this article have been concerned primarily with the beneficial use of a very limited number of radioisotopes: ^{232}Th/^{233}U and ^{235,238}U/$^{239-242}$Pu, with the latter halves of the pairs being created during fuel irradiation. Thorium and ^{235,238}U occur naturally, although the latter often requires enrichment to achieve usable concentrations of ^{235}U. However, a large number of other radioisotopes have found beneficial uses in smaller, but nevertheless extremely important, applications. The purpose of this section is to describe the production and uses of radioisotopes other than those used in the manufacture of power reactor fuels.

9.1 Production and Recovery of Radioisotopes

Radioisotopes other than those that occur naturally are produced by two methods: irradiation of a target by nuclear particles and decay of other radioisotopes.

9.1.1 Production by Irradiation Radioisotope production by irradiation can be accomplished using nuclear reactors or particle accelerators (see NUCLEAR ENERGY: FISSION; ACCELERATORS, LINEAR). In the case of reactors, neutrons irradiate the target material and are captured or induce fission as a result. In the case of accelerators, the target is irradiated by high-energy charged particles such as protons, deuterons, or alpha particles. This results in the ejection of photons and/or other nuclear particles. In both cases, the target material, irradiating particle intensity, irradiation duration, and particle energy spectrum are selected to maximize the yield of the desired radioisotope or minimize contamination by undesirable isotopes.

Very often, the target material is a stable material enriched in a specific isotope (see Sec. 3.2.5) to maximize the yield and purity of the desired radioisotope. The targets are always small (micrograms to grams) in the case of accelerator irradiation, and in cases where expensive, enriched stable isotopes are used as targets in reactors. Targets are also small (ranging from 100 g for initial isotopes to micrograms for targets of the heaviest elements) when long-lived radioisotopes (the transuranic actinides in particular) are irradiated in reactors to produce heavier actinides. When the desired radioisotope is a fission product, the target may be nuclear fuel, which can be a relatively large assembly as described previously, or a target consisting of a few grams of enriched uranium.

9.1.2 Production by Decay In a few cases, very pure radioisotopes can be recovered by decay. Specifically, a pure, relatively long-lived isotope is isolated and allowed to decay. Periodically, its decay product is separated (or *milked*) from the parent for a particular use. This production method is significantly dependent upon serendipity, requiring the ability to isolate a suitable long-lived parent that does not produce competing decay products and that produces the desired product at an adequate rate. This method is used, for example, in the preparation of ^{99m}Tc from decay of fission product ^{99}Mo.

9.1.3 Recovery of Radioisotopes In general terms, radioisotopes are recovered by dissolving the target and applying appropriate chemical or physical processes to isolate a relatively pure product. When production is by decay, the parent is often maintained in dissolved form to facilitate separation. In the case of fission products, the feed material is most often the high-level waste produced after the primary decontamination operation in reprocessing. In other cases, the target may be dissolved in any of a number of reagents and the separation effected using a variety of techniques (e.g., solvent extraction, precipitation, ion exchange, vaporization). These techniques are specific to the nature of the target and product, as well as the product purity specifications. In most cases, the processing also involves recovery of the residual target material, which is often very valuable. The recovery operations take place in facilities ranging from standard ventilated laboratory hoods to elaborate cells with thick radiation shielding, depending on the characteristics and amount of the radioisotopes being processed. In very special cases, a few atoms of exotic radioisotopes such as some of the very heavy elements are separated by recoil from thin films that are bombarded with particles from accelerators.

9.2 Useful Radioisotopes and Their Applications

A wide variety of radioisotopes are regularly used in the modern world. Some of these applications are very specialized and esoteric, relating only to the frontiers of knowledge (e.g., the very heavy synthetic elements). Other applications occur every day in familiar settings and affect the lives of many people (e.g., medical diagnostics/treatment and the use of radioisotopes in smoke detectors). The following provides a framework of the applications of radioisotopes that is used here for the purposes of more detailed discussions: medical diagnostics and treatment, industrial applications, basic research in the physical sciences, radiotracer applications, power production, and military applications.

9.2.1 Medical Diagnostics and Treatment By far the most rapidly growing use of radioisotopes is in the medical community (see NUCLEAR MEDICINE). Within this community, radioisotopes are used for diagnosis (determining the cause of symptoms) and treatment (curing the problem that causes the symptoms). Approximately 300 radiopharmaceuticals are presently available, and this number is growing rapidly.

9.2.1.1 Diagnosis. The chemical properties of many isotopes result in their concentrating in particular organs when introduced into the body. The radioactive emanations from the decay of these isotopes can then be detected, providing an indication of the activity of the organ or even an image of the organ that can achieve remarkable clarity through computer enhancement. Although a wide variety of radioisotopes are used for this purpose (e.g., ^{131}I, ^{201}Tl, ^{47}Sc), the most commonly used is ^{99m}Tc, which is produced by decay from its longer-lived ^{99}Mo parent. By incorporating the ^{99m}Tc into various chemical compounds, it can be caused to concentrate in almost any organ of the human body. It is estimated that ^{99m}Tc is used in about 30 000 medical procedures per day.

9.2.1.2 Treatment. After a problem is diagnosed, radioisotopes are often used to effect a cure. In general, this involves use of the radioactive emanations of a radioisotope to destroy specific tissues such as tumors. For example, gamma rays from ^{131}I are often used in the treatment of the thyroid gland by introducing it into the body, and ^{60}Co is used as an external gamma-ray source for treating a variety of tumors. Californium-252 decays in part by the emission of neutrons and has been found to be effective in treating cancer of the cervix and prostate.

9.2.2 Industrial Applications

9.2.2.1 Sterilization. The quest to improve the quality of health and foodstuffs has resulted in an increasing need for sterilization. Regarding the former, sterilization of a wide variety of medical equipment is required. Because much of this material is now made of plastics and polymers, the previous practice of using high temperatures is unacceptable. Sterilization of foodstuffs is increasingly desirable to prevent spoilage while retaining the full nutritional content and taste, and, again, this is not possible by using high-temperature or drying technologies.

It has been found that irradiating the equipment or food with massive amounts of gamma radiation is very effective in sterilization without impairing other characteristics or making the target material radioactive. The medical equipment can be sterilized after being sealed in impermeable packages, and sterilized food can last for months without refrigeration. Although any radioisotope that emits high-energy gamma rays can be used in this application, by far the most common are ^{60}Co (produced by neutron irradiation of natural ^{59}Co), ^{192}Ir (produced by neutron irradiation of natural ^{191}Ir), and ^{137}Cs (a fission product). Gamma ray sterilization has also been tested on sewage sludges, although application has been very limited.

9.2.2.2 Radiation-Induced Reactions. Gamma rays (typically from the same sources as described immediately above) have also found use in creating desirable chemical reactions. One of the most common is the use of gamma rays to induce the polymerization of a monomer (e.g., linking ethylene monomer to form polyethylene). Gamma rays from radioisotopes can also be used to break chemical bonds when required, and this has been used to control the molecular weight of Teflon®.

9.2.2.3 Measurement and Testing. Radioactive emanations from radioisotopes can be used in a number of ways related to industrial measurement and testing.

1. A gamma-ray source and a number of detectors can determine the level or density of liquid in a tank as a means of process control.
2. Beta particles (e.g., from ^{90}Sr) can be used to measure the thickness of thin layers on a substrate such as metal plating or paint as a means of process or quality control.
3. Neutron and gamma radiation are used in petroleum exploration to determine the characteristics of the surrounding rock down-hole, which is called *well logging*.
4. Neutron and gamma radiation are used to radiograph industrial equipment to detect imbedded flaws (similar to an x ray of the human body).
5. Radioisotopes are used to make self-luminous materials (e.g., watch dials, egress lighting).
6. Americium-241 is used in many home smoke detectors, where the difference between radiation-induced ionization of the smoke and clean air results in activation of the alarm.

9.2.3 Basic Research in the Physical Sciences The properties of radioisotopes have made them very useful in research concerning the physical sciences. In some instances, the radioisotope is the subject of study. For example, the heavy actinide isotopes (e.g., einsteinium, fermium) are studied to learn more about the structure of the nucleus and the laws that govern it, as well as the chemical properties of these exotic elements. In most other cases, the properties of the radioisotope emanations are used as a tool to learn more about a specific phenomenon or object. One of the most common examples is *activation analysis*, in which an object is briefly exposed to neutrons and the induced radioactivity measured as a means to determine nondestructively the atomic composition of the object. Applications of activation analysis range from industry to criminology. The use of radioisotopes to determine the age of an object (e.g., ^{14}C dating) is also quite common in the archaeological and geological sciences.

9.2.4 Radiotracer Applications A *radiotracer* is a radioisotope that is used to follow the movement or reaction path of a material. For example, a biological compound into which radioactive tritium has been introduced can be used to study the behavior of the compound in the human body. Radiotracers are widely used in many applications described above, including medicine (e.g., examining blood circulation patterns), industry (e.g., tracking the movement of process fluids and leak detection), and the sciences (e.g., monitoring the movement of groundwater around waste-disposal sites). Radiotracers may be natural (e.g., ^{14}C, ^{226}Ra) or artificial (i.e., manufactured and introduced into the system being studied).

9.2.5 Power Production Some radioisotopes are used indirectly in the production of nuclear power. For example, a neutron source is needed to initiate the chain reaction in a nuclear reactor, especially a new reactor containing only unirradiated fuel. These sources are often made of an alpha-emitting radioisotope (e.g., ^{210}Po) and an element that releases neutrons when an energetic alpha particle is absorbed (e.g., beryllium) or of a radioisotope such as ^{252}Cf that decays in part by *spontaneous fission* and consequent neutron emission.

A much more common use of radioisotopes is as a source of power for specialized applications. Medium-lived radioisotopes such as ^{90}Sr, ^{238}Pu, and ^{244}Cm have been and continue to be used for this purpose. The heat from the radioisotope may be used without conversion (i.e., as a source of space or process heating). However, the heat is typically converted to electricity directly by using the heat to create a flow of electrons in a thermoelectric generator or indirectly by vaporizing a fluid which drives a turbine that turns an electric generator.

The major attribute of such power sources is a high power output per unit weight combined with a very long useful lifetime. The cost of radioisotopes makes their use as a power source impractical for normal applications. However, they have found extensive use in specialized applications such as powering equipment (e.g., military outposts, buoys, communications) where refueling would be more expensive, powering the instrumentation aboard spacecraft (especially interplanetary probes), and in early artificial hearts.

9.2.6 Military Applications There are specialized military applications for specific radioisotopes. One of the most common is the use of tritium in nuclear weapons, which can greatly increase the explosive power of the device. This occurs when the fissioning of the uranium or plutonium primary system provides sufficient energy to initiate a reaction in which two tritium nuclei are fused, with the resulting release of considerable additional energy. Other radioisotopes such as ^{252}Cf and ^{244}Pu have found use in diagnosing the sequence of events that occurs in underground nuclear weapons tests. Californium-252 is also used to radiograph intact military aircraft for flight-induced structural damage, an approach that has proven to be less costly than disassembling the aircraft and using more conventional radiographic techniques.

GLOSSARY

Activation Analysis: Analysis of the elemental content of a material by briefly irradiating it with neutrons and measuring the radioactive emissions from the resultant short-lived radioactive products.

Activation Product: The radioactive product of neutron capture by a material originally composed of nonradioactive isotopes.

Atomic Vapor Laser Isotope Separation: A physical method of uranium enrichment based on the selective adsorption of laser light by vaporized ^{235}U atoms to produce charged atoms (ions).

Burnup: A measure of fuel consumption (and power production) in a nuclear reactor. In the strict sense, burnup is the depletion of fissile material through fission or other processes and is expressed as a percentage. In this sense, fuel could have 75% burnup (in which case 75% of the initial ^{235}U nuclei have fissioned or been converted to another isotope). Burnup is more commonly used to express the amount of energy produced per metric ton of uranium metal (MWd/t). This is more a measure of the total number of fissions that occur, because heat is produced by fissioning of initial ^{235}U nuclei, fissioning of Pu produced in the reactor, and other nuclear processes. While it would be more correct to use the term "exposure" in this context, "burnup" continues in popular usage. Commercial power reactors attempt to maximize the amount of energy produced (i.e., fuel consumed) and thus typically have high burnups (greater than 10 000 MWd/t). "Production reactors" are operated to maximize production of weapons-grade ^{239}Pu. This is typically limited by the generation of ^{240}Pu that occurs at higher burnups. Production reactors thus typically have burnups of much less than 10 000 MWd/t.

Calutron: An electromagnetic isotope separation device used to enrich uranium during World War II and subsequently to produce a variety of enriched stable isotopes for research and medical applications.

Cascade: Isotope separation devices (gaseous diffusion or centrifuge) connected so as to multiply the amount of enrichment and/or depletion achievable by a single device.

Centrifugal Contactor: A solvent extraction device that mixes two immiscible liquids of differing density and then forces their rapid separation with the use of centrifugal force.

Cladding: The protective surrounding of reactor fuel materials to prevent their corrosion or erosion by the reactor coolant, to increase the strength of the fuel material, and to contain fission products that would contaminate the coolant.

Cladding Hulls: Pieces of nuclear fuel cladding remaining after fuel dissolution.

Coconversion: A process wherein uranium and plutonium concurrently and together undergo conversion from nitrate salts to oxide.

Continuous Recycle: A nuclear fuel cycle in which the fuel is dissolved or suspended in a liquid such as water, molten salts, or molten metals and continuously circulated through a loop involving fission, heat removal, waste product removal, and fuel renewal.

Continuous Reprocessing: Reprocessing technology used in continuous (as opposed to a discrete fuel batch) recycle, which involves on-line reprocessing of nuclear fuel.

Conversion: Changing the chemical form of uranium or plutonium to one desirable as feed to a subsequent step of the nuclear fuel cycle [e.g., yellowcake to UF_6, U/Pu nitrates to $(U,Pu)O_2$, UF_6 to UO_2].

Core: The power-producing heart of a nuclear reactor, containing the nuclear fuel that undergoes fission and releases energy. The core contains the fuel, the moderator (if required), and the control rods. The core is often surrounded by a reflecting material (e.g., beryllium) that bounces stray neutrons back into the fuel.

Dissolution: Immersion of fuel rod pieces resulting from segmentation in solvents, typically hot nitric acid, to dissolve the fuel.

Dry Well: A storage area containing air surrounded by thick radiation shielding in which spent fuel or nuclear waste is stored while being cooled by conduction of heat to an air stream outside the cell walls.

Electromagnetic Isotope Separation: An isotope separation or enrichment process that capitalizes on the different radii of the circular paths in a magnetic field of ions having different atomic masses.

Enrich: Increase the relative abundance of an isotope of an element, for example, to increase the ^{235}U concentration from its natural value of 0.72% to 3–5% for use in civilian power reactors.

Fertile Material: A material that is not fissionable by thermal (low-energy) neutrons, but that can be converted into a fissile material through neutron capture by irradiation in a reactor. There are two basic types of fertile materials: ^{238}U and ^{232}Th. When these fertile materials capture neutrons, they are converted into fissile ^{239}Pu and ^{233}U, respectively.

Fissile Isotope: An isotope that can fission by absorbing a thermal neutron and can support a thermal neutron chain reaction as well. Examples are ^{239}Pu and ^{235}U.

Fission Products: The complex mixture of elements produced as a result of nuclear fission. Fission products decay and form additional products, which may be radioactive or stable. The mixture of fission products so formed contains about 200 different isotopes of over 30 elements.

Fissionable Isotope: An isotope that can be fissioned by only high-energy neutrons.

Fluoride Volatility Process: A reprocessing technology based on the low-temperature volatility of fluorides of key elements (such as uranium).

Fuel Assembly: A number of fuel rods held in a fixed geometric array.

Fuel Fabrication: Processing of unirradiated nuclear and structural materials into a chemical and physical form and configuration that are appropriate for the fuel of a specific reactor design.

Fuel Refabrication: Processing of nuclear materials recovered from spent (irradiated) fuels and structural materials into a form and configuration that are appropriate for the fuel of a specific reactor design.

Fuel Rod (or Fuel Element): A container of nuclear fuel, usually a metal tube, suitable for insertion in the core of a reactor.

Gas Centrifuge: A device for enriching uranium (and other isotopes) in which gaseous UF_6 is contained in a hollow cylindrical rotor that is spun at high speed. The heavier $^{238}UF_6$ molecules tend to move closer to the rotor wall than the lighter $^{235}UF_6$ molecules, thus allowing enrichment to occur.

Gaseous Diffusion: A process for enriching uranium that uses gaseous UF_6 as the working fluid. The lighter $^{235}UF_6$ molecules flow through a porous barrier slightly faster than the heavier $^{238}UF_6$ molecules, thus allowing enrichment to occur.

Grid Spacers: Metal component of a fuel assembly used to hold fuel rods in place laterally.

Head End: The initial step of reprocessing, which includes spent fuel segmentation and dissolution.

High-Level Waste: The highly radioactive waste remaining after reprocessing.

Ion Exchange: A solid or liquid material with replaceable ions that can exchange with other ions in the surrounding solution.

Isotope: A nuclear species that is unique because of the number of neutrons and protons it contains. Most elements are composed of a mixture of isotopes, each with the same number of protons but differing in the number of neutrons. Isotopes may be nonradioactive or radioactive.

Isotope Enrichment: A process to increase the relative abundance of selected isotopes of an element.

Laser Isotope Separation: An enrichment process in which a selected isotope is excited by a laser beam tuned to a specific frequency (color). The excited isotope can then be removed by, for example, electromagnetic or chemical means. Important examples include Atomic Vapor Laser Isotope Separation (AVLIS) and Molecular Laser Isotope Separation (MLIS).

Metric Ton (t): 1000 kilograms or 1 Mg. Also written as tonne. Equal to 2200 pounds.

Milk: Periodically separate a desirable decay product being continuously produced by a long-lived radioactive parent.

Mixer-Settler: A solvent extraction device that mixes two immiscible liquids of differing densities and allows them to settle under normal gravitational force to effect the separation of components.

Molecular Laser Isotope Separation (MLIS): Laser isotope separation in which the feed is a uranium compound, typically UF_6.

Neutron Capture: A process in which an isotope absorbs a neutron but does not fission, thereby creating a heavier isotope of the same element.

Nuclear Fuel: The nuclear material that sustains a chain reaction in a nuclear reactor.

Nuclear Fuel Cycle: The sequence of operations involved in supplying fuel for nuclear power generation: irradiating the fuel in a nuclear reactor, handling and treating the fuel elements following discharge from the reactor, and disposing of the radioactive wastes. If spent fuel is to be reprocessed, the fuel cycle is called a "closed" cycle. If the fuel is not to be reprocessed, it is a "once-through" cycle.

Off Gas: Gases evolved during reprocessing of spent fuel.

Once-Through Nuclear Fuel Cycle: A nuclear fuel cycle in which the fuel constituents are not recovered for reuse, and the spent fuel is managed intact.

Penetrating Radiation: Particle or electromagnetic emissions, resulting from the decay of radioactive isotopes or devices such as an x-ray machine, that can penetrate significant thicknesses of solid materials.

Pulse Column: A vertical pipe in which a dissolved material (e.g., plutonium) can be extracted or exchanged by the contracting of two immiscible solutions of differing densities. The lighter solution is usually fed into the bottom of the column, while the heavier solution is fed near the top. The process is enhanced by filling the column with suitable materials to produce droplets of one solution (e.g., perforated horizontal plates) and periodically pulsing the flow.

Purex: A reprocessing technology involving aqueous nitric acid dissolution of spent fuel and recovery and separation of uranium and plutonium using solvent extraction from nitric acid solutions using tributyl phosphate.

Pyroprocessing: A reprocessing technology that employs high temperatures and typically uses molten metals and salts as the reprocessing reagents.

Radiation Dose: A measure of the radiation energy absorbed in materials, especially living organisms.

Radionuclide: Same as a radioactive isotope.

Radiotracer: A radioactive isotope used to follow the movement or reaction path of a material.

Recycle: Reuse of the fissile and fertile products from reprocessing. Recycle involves reprocessing and refabrication.

Reprocessing: Chemical processing of spent nuclear fuel to recover desired constituents, typically uranium and plutonium.

Reprocessing and Recycle: A nuclear fuel cycle in which fuel constituents are recovered for reuse.

Segmentation: A process used to reduce fuel rods to short pieces that are open at the ends, thus exposing the spent fuel.

Separation Factor: The ratio of isotopic concentration in the product (enriched) stream to the concentration in the waste (tails) stream of a single stage of an enrichment process.

Shipping Cask (or Shipping Package): A container with thick walls of radiation-absorbing material used to transport spent fuel while providing protection from radiation and preventing the release of nuclear material during an accident.

Solvent Extraction: Contacting two immiscible liquids to selectively move a dissolved constituent of one liquid to the other.

Spent Fuel: Fuel that has been removed from a nuclear reactor because it has reached the end of its useful life.

Spontaneous Fission: A mode of radioactive decay in which a nucleus fissions without being struck by a neutron. Observed only in heavy actinide isotopes.

Stable Isotopes: Nonradioactive isotopes of an element.

Stage: A single step in an isotope separation or chemical separation process. In an isotope enrichment cascade, a single stage consists of all those separators that receive feed material of the same enrichment. The number of stages in a cascade primarily depends on the desired cascade product enrichment and waste stream assays and on the separation factor of the process. The definition is the same for chemical separation, but stages are not necessarily physically distinct units as is common in isotope enrichment.

Tails: Material from isotope enrichment that is depleted in the isotopes of interest.

Thorex: A variant of the Purex process that is used to recover thorium and uranium from thorium-based fuels.

Tonne (abbreviation t): A metric ton, 1000 kg.

Tributyl Phosphate (TBP): A water-immiscible liquid used in reprocessing. It is used to extract uranium and plutonium from an acidic aqueous solution. TBP is usually diluted with an organic solvent such as kerosene.

Uranium Concentrates: The product of a uranium ore concentration plant. Uranium concentrates typically contain 65 to 85% uranium, expressed as U_3O_8. Concentrates are usually in the form of UO_3, U_3O_8, or yellowcake.

Uranium Milling: Concentration and purification of uranium in ores.

Yellowcake: Ammonium or sodium diuranate, chemical forms of uranium concentrates.

Works Cited

American Nuclear Society (1978), "CIVEX: Solution to Breeder/Diversion Dilemma?" *Nucl. News* **21** (5), 32–37.

Becker, E. W., Bier, W., Ehrfeld, W., Schubert, K., Schutte, R., Seidel, D. (1975), *Uranium Enrichment by the Separation Nozzle Process*, Gesellschaft für Kernforschung mbH, Karlsruhe, Germany, Report KfK-2235.

Chapman, R. H., Spiewak, I., Tobias, M. L., Vondy, D. R. (1963), "Design of Small Aqueous Homogeneous Breeder Reactor," *Nucl. Sci. Eng.* **15** 347–353.

Croff, A. G., Hermann, O. W., Alexander, C. W. (1979), *Calculated, Two-Dimensional Dose Rates from a PWR Fuel Assembly*, Oak Ridge National Laboratory Report No. ORNL/TM-6754 [available from the U.S. Department of Energy Office of Scientific and Technical Information, P.O. Box 62, Oak Ridge, TN 37831].

Croff, A. G., Alexander, C. W. (1980), *Decay Characteristics of Once-Through LWR and LMFBR Spent Fuels, High-Level Wastes, and Fuel-Assembly Structural Material Wastes*, Oak Ridge National Laboratory Report No. ORNL/TM-7431 [available from the U.S. Department of Energy Office of Scientific and Technical Information, P.O. Box 62, Oak Ridge, TN 37831].

International Atomic Energy Agency (1992), *Transport of Radioactive Materials*, IAEA/PI/A.2E 92–04899 [available from the International Atomic Energy Agency, Division of Public Information, P.O. Box 100, A-1400 Vienna, Austria].

Katcoff, S. (1958), "Fission Product Yields from U, Th, and Pu," *Nucleonics* **16** (4), 78–93.

Lederer, C. M. Shirley, V. S. (1978) *Table of Isotopes*, 7th ed., New York: Wiley.

Organization for Economic Cooperation and Development (OECD) and the International Atomic Energy Agency (1992), *Uranium 1991: Resources, Production, and Demand* [available from OECD Publications and Information Centre, Suite 700, 2001 L Street NW, Washington, D.C. 20036–4910, and numerous other locations worldwide].

Weinberg, A. M., Rosenthal, M. W., Briggs, R. B., Kasten, P. R., Haubenreich, P. N., Engel, J. R., Grimes, W. R., McCoy, H. E., Beatty, R. L., Cook, W. H., Gehlback, R. E., Kennedy, C. R., Koger, J. W., Litman, A. P., Sessions, C. E., Weir, J. R., Whatley, M. E., McNeese, L. E., Carter, W. L., Ferris, L. M., Nicholson, E. L., Scott, D., Eatherly, W. P., Bettis, E. S., Robertson, R. C., Perry, A. M., Bauman, H. F. (1970), series of papers on the molten salt breeder reactor, *Nucl. Appl. Tech.* **8** (2), 105–215.

Further Reading

Benedict, M., Pigford, T. H., Levi, H. W., (1981), *Nuclear Chemical Engineering*, 2nd ed., New York: McGraw-Hill. A comprehensive technical book on most of the processes involved in the nuclear fuel cycle.

Gregory, J. N., (1966), *The World of Radioisotopes*, Sydney: Halstead Press. (1966). Although a bit dated, still provides a reasonably comprehensive overview of the uses of radioisotopes.

Lederer, C. M., Shirley, V. S., (1978), *Table of Isotopes*, 7th ed., New York: Wiley. An extremely detailed compendium of all isotopes known at the time of publication and their nuclear characteristics, including half-lives and radioactive emanations.

Long, J. T., (1967), *Engineering for Nuclear Fuel Reprocessing*, New York: Gordon and Breach. A comprehensive technical book on the equipment used for nuclear fuel reprocessing.

National Research Council, (1982), *Separated Isotopes: Vital Tools for Science and Medicine*, Washington, DC: National Academy Press. Provides a good overview of the production and uses of enriched stable isotopes, which are often the precursors of radioisotopes.

Organization for Economic Cooperation and Development and the International Atomic Energy Agency, (1983), *Uranium Extraction Technology*, Vienna: OECD and IAEA. An overview of uranium mineralogy, extraction, and processing.

Wymer, R. G., Vondra, B. L., Jr. (Eds.), (1981), *Light-Water Reactor Nuclear Fuel Cycle*, Boca Raton, FL: CRC Press. A more descriptive book on the nuclear fuel cycle with emphasis on reprocessing and waste management technology.

NUCLEAR MEDICINE

THOMAS J. RUTH, *UBC/TRIUMF PET Program, University of British Columbia and TRIUMF, Vancouver, British Columbia, Canada*

INTRODUCTION

Nuclear medicine makes use of the fact that certain radioisotopes emit γ rays with sufficient energy that the γ rays can be detected outside of the body. These radioisotopes are attached to biologically active compounds, called radiopharmaceuticals, which either localize in certain bodily tissues or follow a particular biochemical pathway. Modern nuclear medicine encompasses many disciplines, including the use of *in vitro* and *in vivo* techniques. However, the following discussion will concentrate on the uses of radioactive substances for the diagnosis or therapeutic treatment of human disease.

1. HISTORICAL BACKGROUND

Nuclear medicine (NM) has its origins in the pioneering work of the Hungarian doctor G. Hevesy, who, in 1924, used radioactive isotopes of lead as tracers in bone studies. Shortly thereafter, R. H. Stevens made intravenous injections of radium chloride to study malignant lymphomas (1926).

However, it was not until the discovery of artificially produced radioactive isotopes that the number of available species suitable for use as tracers began to increase. The 1931 invention, by Ernest Lawrence, of the cyclotron made it possible to produce radioactive isotopes of a number of biologically important elements.

The use of these artificially produced radiotracers continued with J. G. Hamilton and R. Stone using radioactive sodium clinically in 1937. In 1938, S. Hertz, A. Roberts, and R. D. Evans used radioactive iodine in the study of thyroid physiology, followed, in 1939, by J. H. Lawrence, K. G. Scott, and L. W. Tuttle studying leukemia with radioactive phosphorus. By 1940, J. G. Hamilton and M. H. Soley were performing studies in iodine metabolism by the thyroid gland *in situ* by use of radioiodine in normal subjects and in patients with various types of goiters.

In 1941, the first medical cyclotron was installed at Washington University, St. Louis, Missouri, where radioactive isotopes of phosphorus, iron, arsenic, and sulfur were produced. With the development of the fission process, during the second World War, most radioisotopes of medical interest began to be

3-527-28133-9/94/$5.00 + .50

produced in nuclear reactors. After World War II, the wide use of radioactive materials in medicine established the new field of what was then called atomic medicine, and only later became known as nuclear medicine. Radioactive carbon, tritium, iodine, iron, and chromium found increasing use in the study of disease processes.

In the 1950s, B. Cassen developed the concept of the rectilinear scanner, which opened the way to obtaining in a short amount of time the distribution of radioactivity in a subject. This development was followed by production of the first γ camera by H. Anger in 1953 (see BIOMEDICAL ENGINEERING). The original design was modified later in the decade to what is now known as the Anger scintillation camera, thus heralding the modern era of γ cameras whose principles are still in use today.

D. Kuhl and colleagues introduced the concept of transaxial tomogram using a transverse imaging device. Extensions of this work led to almost exact tomographic sections. Hounsfield introduced the x-ray–transmission computed tomography by application of the reconstruction techniques used in radio astronomy. Its extension to emission computed tomography (ECT) is the basis of single-photon–emission tomography (SPECT) and positron-emission tomography (PET).

P. Richards was instrumental in the development of the ^{99}Mo/^{99m}Tc generator system following its invention at Brookhaven National Laboratory in 1966. Technetium-99*m* produced via this generator system has become the most widely used radionuclide in nuclear medicine today, accounting for as much as 85% of all *in vivo* diagnostic procedures.

The modern era of nuclear medicine may become known as molecular medicine as the field translates advances in molecular biology and biochemistry into treatment of humans and the diagnosis of disease and anatomical abnormalities.

2. RADIOISOTOPES; RADIONUCLIDE PRODUCTION

Radionuclide production is indeed true alchemy, that is, converting the atoms of one element into those of another. This conversion involves the altering of the number of protons and/or neutrons in the nucleus (target). If a neutron is added without the emission of particles, then the resulting nuclide will have the same chemical properties as the target nuclide. If, however, the target nucleus is bombarded by a charged particle, for example, a proton, the resulting nucleus will usually be that of a different element. The exact types of nuclear reactions a target undergoes depend on a number of parameters, including the type of bombarding particle and the energy of these projectiles.

The binding energy of nucleons in the nucleus is of the order of 8 MeV, on average. Therefore, if the incoming projectile has more than this amount of energy, the resulting reaction may cause other particles to be ejected from the target nucleus. By carefully selecting the target nucleus, the bombarding particle, and its energy, it is possible to produce a specific radionuclide.

Specific activity is a measure of the number of radioactive species (atoms or molecules) as compared with the total number of species present in the sample. The specific activity is usually expressed in terms of radiation units per mass unit. The traditional units have been Ci/mole (Ci/g) or fraction thereof. If the only atoms of the labeled element present in the sample are those of the radionuclide, then the sample is referred to as *carrier free*. For example, a compound labeled with ^{99m}Tc will be carrier free, since there are no stable isotopes of technetium. However, in most cases, there are small quantities of unlabeled compounds that have a similar chemical behavior and can act as *pseudocarrier*. The specific activity of an isotope or radiopharmaceutical is important in determining the chemical or biological effect the substance may have on the system under investigation.

As an example of specific activity, assume that glucose has been labeled with 10 mCi of ^{11}C with a half-life of 20.3 min in such a way that one of the glucose carbon atoms is the radioactive label and the remaining five carbons are natural nonradioactive carbon. Its carrier free specific activity (SA) would be as follows: atoms of $^{11}\mathrm{C} = N = dN/\lambda dT = 10 \times 3.7 \times 10^7/0.693(20.3 \times 60) = 6.5 \times 10^{11}$. For every radioactive carbon atom, there is one glucose molecule, and, using Avogadro's number, the number of moles is then 1.08×10^{-12}, and the $\mathrm{SA} = 9.3 \times 10^9$ Ci/mol.

If the radionuclide had been ^{14}C with its 5730-yr half-life, following the same process but using the λ for ^{14}C, the SA would be 62 Ci/mole. If the radiolabeled glucose had been prepared in a plant, the naturally occurring glucose would have lowered the SA because of the nonradioactive glucose molecules. Therefore, it is easy to see that the short lived radioisotopes potentially have a much higher specific activity.

The use of nuclear reactors for the production of radioisotopes relies on the fact that during the fission process in a reactor, there are large numbers of neutrons produced and thermalized by the surrounding media, usually water or graphite. These neutrons are ideal for initiating (n,γ) reactions. In some reactors, the higher-energy neutrons are used to produce radioisotopes via other reactions, for example, (n,p) or (n,α) reactions. Table 1 illustrates some neutron-induced reactions.

The fission process is a source of several widely used radioisotopes. For example, ^{90}Sr, ^{99}Mo, ^{131}I, and ^{133}Xe are all produced in reactors by fission and can be separated from the uranium fuel cells or from targets of enriched ^{235}U placed in the reactor for radioisotope production directly. The major drawback of using fission-produced materials is the production not only of isotopes of the desired species, but of large quantities of both radioactive waste material and other radionuclides as well. In producing ^{131}I from fission, the isotopes ^{127}I and ^{129}I also are formed, thus reducing the specific activity.

It is ironic that the first radionuclides artificially produced were produced on Lawrence's cyclotrons, but it was another 30 years before accelerator-produced radionuclides began to play a major role in the production of medically important radiopharmaceuticals. The principal advantage of accelerator-produced radioisotopes is the high specific activities that can be obtained. Another not insignificant advantage is that a smaller amount of radioactive waste is generated from charged-particle reactions, especially at low (30 MeV) bombarding energy.

Table 1. Reactor produced radionuclides.

Reactions
$^{14}N(n,p)^{14}C$
$^{31}P(n,\gamma)^{32}P$
$^{32}S(n,p)^{32}P$
$^{50}Cr(n,\gamma)^{51}Cr$
$^{124}Xe(n,\gamma)^{125}Xe \rightarrow {}^{125}I$

Cyclotrons designed for producing medical radioisotopes were initially capable of accelerating protons, deuterons, $^{3}He^{+2}$, and α particles. However, the primary radioisotopes used in medical applications can all be produced by protons, and the simplicity of design for proton-only cyclotrons has resulted in cyclotrons that accelerate H^{-} ions capable of two or more simultaneous beams of varying energies and intensities. The modern cyclotron is completely computer controlled and is capable of running for many days with minimal attention.

The production mode of the radionuclides used for diagnostic and therapeutic purposes uses both cyclotrons and reactors, as shown in Table 2. Table 3 shows the quantities produced for the common commercially available radionuclides and the change in demand over the last decade.

As can be seen from Table 4, the PET radionuclides are produced from either proton or deuteron reactions. Since most early PET research was performed at major research laboratories having accelerators capable of proton or deuteron production, these reactions were the standard. However, in the early 1980s, small, compact, proton-only cyclotrons became available, and cyclotrons specifically designed for producing PET radionuclides were installed in many hospitals. The major drawback of these proton-only cyclotrons is that, in some cases, an enriched target material must be used for sufficient product to be generated.

The success of the small low-energy cyclotron encouraged research into the design of even lower-energy accelerators, i.e., linear accelerators and cyclotrons of a few MeV extracted energy. To date, these accelerators are in the developmental stage, but there could be a proliferation of these devices in most medical centers if they prove inexpensive to purchase and operate.

The equation for the rate of production is

$$R = I\sigma t, \tag{1}$$

where R is the number of nuclei formed per second, I is the flux of bombarding particles per second, σ is the cross section (probability of the reaction occurring) in cm^2, and t is the target thickness expressed as the number of

Table 2. The production modes of some radionuclides widely used for diagnostic or therapeutic purposes. The principal mode of decay is given, along with the major γ-ray energies and their intensities (I) in percent of total decay rate.

Nuclide	Principal decay mode	E(I) keV (%)	Reaction
^{67}Ga	EC	93 (37)	^{68}Zn(p,2n)
		185 (20)	
^{99}Mo/^{99m}Tc	IT	140 (98)	^{98}Mo(n,γ), fission product
^{111}In	EC	171 (90)	^{112}Cd(p,2n)
^{123}I	EC	159 (83)	^{124}Xe(p,2n)^{123}Cs $\rightarrow$ ^{123}Xe $\rightarrow$
^{131}I	β^-	364 (81)	^{130}Te(n,γ)^{131}Te $\rightarrow$
^{127}Xe	EC	203 (68)	^{133}Cs(p,spall)
^{133}Xe	β^-	81 (36)	Fission product
^{201}Tl	EC	167 (11)	^{203}Tl(p,3n)^{201}Pb $\rightarrow$

nuclei per cm^2 (n/cm^2). It is of historical interest to note that the unit for cross section is the barn, which is defined as 10^{-24} cm^2. The expression barn comes from the fact that the probability of a neutron interacting with a target is proportional to the area of the nucleus, which, compared to the size of the neutron, is as big as a barn.

The rate of production is, of course, affected by the fact that the resulting nuclide is radioactive and, thus, undergoes radioactive decay. For short-lived nuclides, the competing reaction rates, production and decay, will come to equilibrium at sufficiently long bombardment times, since the rate of decay is proportional to the number of radionuclei present. The point where equilibrium has been reached is called saturation, meaning that, no matter how much longer the irradiation occurs, the production rate is equal to the rate of decay, and no additional product will be formed. At shorter irradiation times, the fraction of product produced is related to the saturation factor given by $1 - e^{-\lambda t}$, where λ is the decay constant of the decaying nuclide, and to the bombardment time t. It is evident that an irradiation equivalent to one half-life would result in 50% of saturation. For practical reasons, an irradiation rarely exceeds three half-lives (90% of saturation), except for the shortest-lived radioisotopes.

For long-lived species, the quantity produced is usually expressed in terms of integrated dose or total beam flux (μA-h). For example, with a long-lived radionuclide such as ^{82}Sr ($t_{1/2}$ = 25 d), the amount produced will be essentially the same whether it is produced from 100 μA in 1 h or 50 μA in 2 h (both represent 100 μA-h of beam).

Finally, the other source of radionuclides used in medicine is the generator. A radioactive generator takes advantage of the cases where one longer-lived (parent) radionuclide decays, usually by β^- emission, to a shorter-lived (daughter) radionuclide. The chemical differences in the two elements are exploited to separate the daughter product from the parent. The parent radionuclide is produced by one of the methods described above and then attached to an inert substance from which the desired product can be eluted. The product can be used directly, as in the case of ^{82}Rb$^+$ from the Sr/Rb generator, or after undergoing a chemical reaction, as in the case

Table 3. Some of the principal commercially available radionuclides and the quantities, in curies, used over the last decade. Note the rate of growth for ^{123}I and ^{201}Tl.

Nuclide	$t_{1/2}$	Retail consumption (1982)	Retail consumption (1987)	Retail consumption (1990)
^{67}Ga	78.3 d	800	800	820
^{99}Mo/^{99m}Tc	66 h/6 h	100 000 (^{99}Mo)	120 000	150 000
^{111}In	68 h	150	160	185
^{123}I	13.2 h	75	1250	3100
^{131}I	8.0 d	10 000	10 000	14 000
^{127}Xe	36.4 d	100	100	100
^{133}Xe	5.2 d	25 000	25 000	45 000
^{201}Tl	73 h	500	2500	6000

Table 4. The most commonly used positron emitters and typical reactions for their production; chemical forms of precursors for labeling more complex chemical species and typical radiopharmaceuticals.

Nuclide	$t_{1/2}$ (min)	Reaction	Precursor	Radiopharmaceutical
^{11}C	20.3	$^{14}N(p,\alpha)$	CO_2, CH_3I	Many
^{13}N	9.97	$^{16}O(p,\alpha)$	NH_3, NO_3^-	NH_3
^{15}O	2.03	$^{14}N(d,n)$ $^{15}N(p,n)$	O_2, H_2O, CO	same as precursor
^{18}F	109.7	$^{18}O(p,n)$ $^{20}Ne(d,\alpha)$	F^-, F_2	Fluorodeoxyglucose Fluorodopa

of ^{99m}Tc from the Mo/Tc generator (see below).

The equilibrium equations (Sorenson and Phelps, 1987) that reflect the relative radioactivity of parent and daughter are given by the general equation

$$A_d(t) = A_p(0)[\lambda_d/(\lambda_d - \lambda_p)] \times (e^{-\lambda_p t} - e^{\lambda_d t}) + A_d(0)e^{-\lambda_d t}, \quad (2)$$

where A is the radioactivity for daughter (d) and parent (p), respectively. The initial radioactivity present is expressed as $A(0)$ for both parent and daughter. The first term in Eq. (2) accounts for the growth of the daughter as a function of the decay of the parent and the disappearance of the daughter due to its decay, and the last term accounts for the presence of any daughter nuclei at zero time.

All generator systems used routinely in nuclear medicine form equilibrium between parent and daughter. For the situation where the parent has a half-life much longer than the daughter, for example, with ^{68}Ge/^{68}Ga and ^{82}Sr/^{82}Rb, the change in the amount of the parent during the time for steady state to be reached will be negligible; that is, $\lambda_p \ll \lambda_d$ and $e^{-\lambda_p t} \approx 1$. Thus, by using Eq. (2), the quantity of daughter radioactivity at any time can be expressed by

$$A_d(t) = A_p(0)\,(1 - e^{-\lambda_d t}). \quad (3)$$

Thus, after a very long time when $e^{-\lambda_d t} \approx 0$, the daughter and parent radioactivities are approximately equal. The steady-state condition is referred to as secular equilibrium.

In the case of the ^{99}Mo/^{99m}Tc generator, the parent (^{99}Mo) decays at a rate ($t_{1/2}$ = 66 h, where $t_{1/2} = \ln 2/\lambda$) long in comparison to its daughter (^{99m}Tc $t_{1/2}$ = 6 h), such that $\lambda_p < \lambda_d$, so that $e^{-\lambda_d t}$ is negligible in comparison to $e^{-\lambda_p t}$ when t is sufficiently long. Under these conditions, Eq. (2) can be solved so that the ratio of the daughter radioactivity to that of the parent is given by

$$A_d/A_p = T_p/(T_p - T_d), \quad (4)$$

where T is the half-life for each species, respectively. This condition is referred to as transient equilibrium, even though, in the strict sense, it is a steady state and not a true equilibrium.

3. RADIOACTIVE TRACERS

The term radiotracers refers to a radioactive species that is used to follow (trace) the uptake into or function of an organ system in a living plant or animal. Initially, the radiotracers used were radioactive isotopes of naturally occurring elements or chemical congeners of these elements, as in the case where de Hevesy used Pb to trace bone activity (see above). Other early uses of simple elements as radiotracers were the use of radioiodine to monitor iodine uptake in the thyroid and the use of radioactive phosphorus to monitor metabolism.

The characteristics of an ideal tracer include the following:

1. The behavior of the tracer should be identical or related in a known manner to the substance to be traced; and
2. the quantity of tracer used should not have an effect on the process to be studied.

Inherent in the first of these characteristics is that there should be negligible effects due to the different isotopes present, as in the case of ^{131}I substituted for the natural isotope of iodine, ^{127}I. The rates of movement of atoms or molecules vary inversely as the square roots of their masses. Thus, for very light nuclei, e.g., isotopes of hydrogen, there can be a 73% difference between the diffusion rates of ^{1}H and ^{3}H. The difference in the diffusion rates for more massive isotopes is much less, as in the case of the iodine isotopes, where the difference is about 1.5%. Fortunately, this effect is generally of no consequence in nuclear medicine.

Radioactive isotopes of the elements of sodium and potassium (^{24}Na, ^{42}K) have been used to measure the sodium and potassium content of the body, employing the *isotope dilution* technique. Analysis by isotope dilution involves adding a known mass and specific activity of a substance to a mixture containing an unknown quantity of the substance and taking an aliquot of the mixture and determining the new specific activity of the substance under investigation. It can be shown that the mass of the substance of interest in the unknown mixture can be expressed as

$$M_u = M_1(S_1/S_2 - 1), \tag{5}$$

where M_u is the mass of the substance in the system under investigation, S_1 and S_2 are the specific activities of the tracer before and after its addition to the system, respectively, and M_1 is the mass of the labeled substance added.

4. RADIOPHARMACEUTICALS

The term radiopharmaceutical is derived from the fact that a radionuclide has been attached to a biologically active compound and is used for the diagnosis or the therapeutic treatment of human diseases. Radiopharmaceuticals differ in one major aspect from regular pharmaceuticals in that they are given in such small concentrations that they do not elicit any pharmacologic response. Because of this difference, there have been several attempts to change the name used to describe these substances. Present-day radiopharmaceuticals are used for diagnostic purposes in about 95% of the cases, and the remainder are used in therapy. The therapeutic effect is in response to the radioactivity associated with the radiopharmaceutical and not to the chemical nature of the compound, which is administered in very small concentrations.

In order for a radiotracer (radiopharmaceutical) to be used in humans safely, it must meet quality standards that include chemical and radiochemical purity, that it be sterile and free from pyrogenic material. The ideal radiopharmaceutical should

1. be readily available at a low cost;
2. be a pure γ emitter, that is, have no emission of particles such as α's and β's that contribute radiation dose to the patient but do not provide any diagnostic information (see Sec. 8);
3. have a short biological and physical half-life so that it is eliminated from the body as quickly as possible;
4. have a high target to nontarget ratio so that the resulting image has a high contrast; that is, the background does not blur the image; and
5. possess the proper metabolic activity in that it follows or is trapped in the metabolic process of interest.

The ability to measure regional biochemical function requires a careful design process with these principles in mind. However, it is not possible in reality to meet all of these criteria. For example, all decay processes involve the emission of particles, as in the case of the pure γ emitters, which have Auger electrons emitted during some fraction of the decays. Thus, it is necessary to address the following steps (Eckelman, 1991) in the development of a biochemical probe:

1. develop a radiotracer that binds preferentially to a specific site;
2. determine the sensitivity of the radiotracer to a change in biochemistry; and
3. find a biochemical change as a function of a specific disease that matches that sensitivity.

Figure 1 illustrates the ability to image different functional systems using the same radionuclide by varying the compound to which it is attached.

A large number of radiotracers have been synthesized to probe metabolic turnover such as oxygen consumption, glucose utilization, and amino-acid synthesis. Enzymatic activity, neurotransmission, receptor density, and occupancy have all been measured via appropriately designed radiotracers. It should be pointed out that the development of radiotracers for PET fundamentally violates rule number 2 for the ideal tracer, since, by na-

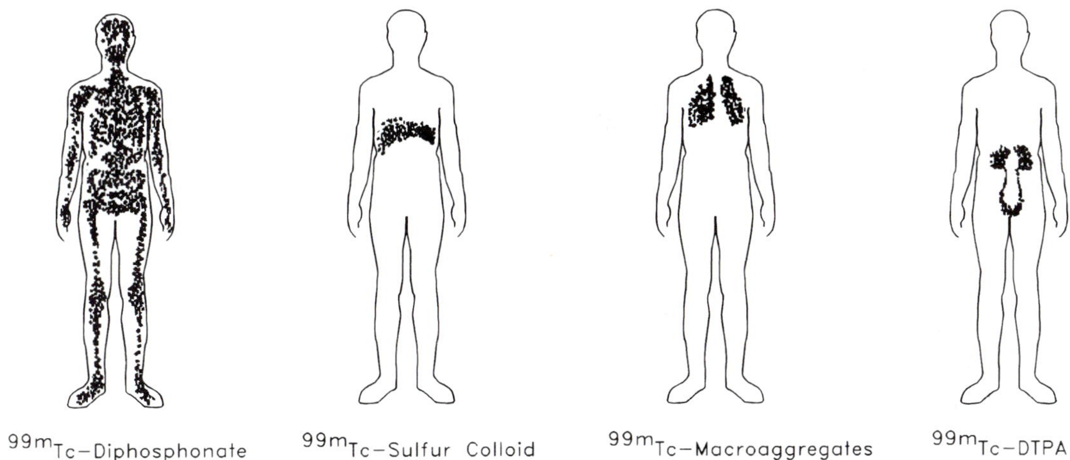

FIG. 1. Dependence of distribution of radioactivity on the radiopharmaceutical used. Left to right: the skeleton imaged by ^{99m}Tc-methylene diphosphonate, the liver and spleen with ^{99m}Tc sulfur colloid, the lungs with ^{99m}Tc albumin microspheres, and the kidney-ureter-bladder system with ^{99m}Tc DTPA (diethylenetriaminepenta-acetic acid). (Used with permission of J. S. Robertson, Mayo Clinic.)

ture, the PET radioisotopes emit β^+ particles. However, the resulting coincident γ rays from the β^+ annihilation form the basis for the technique, which will be described below.

In addition to these principles, the synthetic chemist must plan on how to insert the radioisotope into the molecule at a point in the synthesis so that there is minimal handling, and late enough in the synthesis process to minimize loss due to chemical yield and radioactive decay. For these reasons, the preparation of radiopharmaceuticals requires planning and techniques not encountered by traditional synthetic chemistry.

5. RADIOIMMUNOASSAYS

On the basis of the radioisotope techniques developed earlier in the 1950s, S. Berson and R. Yalow published in 1959 the first use of radioimmunoassay (RIA) to measure the concentration of insulin in unextracted human plasma. The RIA principle is simple and is illustrated by the schematic in Fig. 2.

The concentration of an unknown amount of unlabeled antigen can be determined by comparing its inhibitory effect on the binding of radioactively labeled antigen to specific antibody with the inhibitory effect of

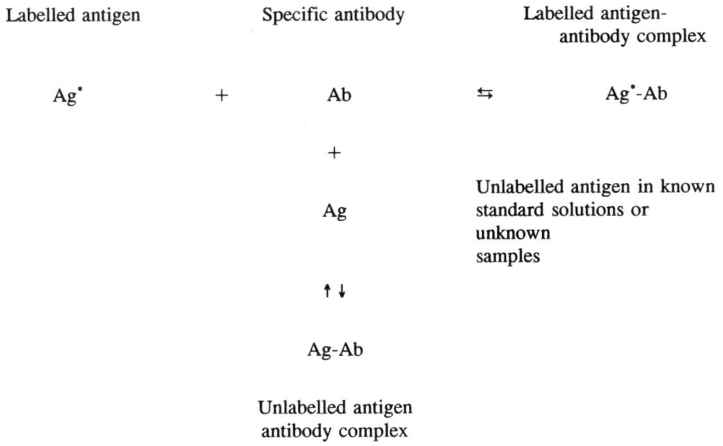

Labelled antigen — Specific antibody — Labelled antigen-antibody complex

$$Ag^* + Ab \leftrightarrows Ag^*\text{-}Ab$$

+

Ag — Unlabelled antigen in known standard solutions or unknown samples

↑↓

Ag-Ab

Unlabelled antigen antibody complex

FIG. 2. Schematic representation of the reactions used in radioimmunoassay.

known standards (Yalow, 1978). Because of the use of radioactive tracers, the method is highly sensitive for detecting very small quantities. Concentrations of less than 1 p*M* can be determined for some species. The technique involves the separation of the labeled antigen of interest into bound and unbound fractions after its interaction with an antibody in the presence of an unknown amount of unlabeled antigen. The ratio of the bound to free fractions of the labeled antigen is compared to the binding of known standards.

In order to use this method to measure the concentration of a substance, the antigen in the samples to be tested and the antigen in standard samples must have identical immunologic behavior. Therefore, it is not necessary for the unknown and the standards to be chemically identical or to have identical biological behavior. The original technique has been modified by making use of a variety of radiotracers (Harbert and da Roche, 1984) to measure concentrations of vitamins, enzymes, peptides, serum proteins, hormones, viruses, drugs, and tumor antigens, as shown in Table 5.

Commercially available *kits* provide the materials for performing RIA for most of these substances. A typical kit would consist of a series of standard samples, each containing a specific amount of an unlabeled antigen, a vial of labeled antigen, a vial of antibody solution, and a substance used to precipitate the antigen-antibody complex. In most cases, ^{125}I is the radioactive isotope employed.

6. IMAGING TECHNIQUES

The following sections describe the techniques used to obtain *in vivo* functional images following the injection of radiotracers. The underlying basis for the techniques is that the radioisotopes employed must emit photons of sufficient energy that they penetrate the body so that they may be detected. All modern nuclear-medicine imaging devices make use of a scintillator coupled to a device that converts the emitted light into an electrical pulse. The following sections describe the more common methods presently in use.

6.1 Planar Imaging

By far the most common imaging device in nuclear medicine is the planar camera or Anger camera, named after its inventor, H. Anger. The basic components of the camera include a thin crystal of NaI scintillator coupled to a cluster of photomultiplier tubes and an (x, y) positioning circuit, and a readout device that may be an oscilloscope or film. In between the radioactive source and the detector is a collimator constructed of lead metal. The collimator has one or more holes drilled through it to allow the γ rays through. Since γ rays cannot be bent or focused, the collimator's function is to absorb those γ rays that do not pass through the openings.

In its simplest form, the collimator has a single pinhole and acts like a camera lens. Other collimators are constructed with the holes converging, diverging, or parallel to the imaginary line connecting the object to be imaged and the camera face. The converging collimator has the effect of magnifying the imaged object, while the diverging collimator reduces the image. The parallel collimator is used for high resolution. Regardless of which collimator is used, they all absorb a large fraction of the photons emitted by the radiotracer in the patient.

Table 5. A few examples of the application of radioimmunoassay.

Test	Substrate	Radioisotope	Applications
Schilling test	Vitamin B_{12}	^{57}Co	Indication of absorption of vitamin B_{12} as in pernicious anemia.
Blood volume	Serum albumin	^{131}I	Determination of blood volume in pre- and postoperative patients
Red blood cell survival	Red cells	^{51}Cr	Hemolytic anemia patients have a very short red cell survival time.
Plasma iron turnover	Ferrous citrate	^{59}Fe	Hematologic disorders have abnormal iron turnover rates

The NaI scintillator has been designed to minimize multiple interactions with the incident γ rays so that the position of interaction can be accurately determined. Typical scintillation cameras have detectors 25–45 cm in diameter with thicknesses in the range of 0.64–1.27 cm.

The (x, y) positioning circuit relies on the light output from the many (19–91) photomultiplier tubes (PM tubes) mounted on the back of the scintillator. The photomultiplier tube closest to the position of the γ ray's interaction will receive the maximum amount of light, while the other PM tubes will receive light in proportion to the solid angle subtended by the tube at the point of interaction. The positioning circuit sums the output of the PM tubes and produces x and y pulses proportional to the (x, y) coordinates of the γ-ray interaction.

A more detailed discussion of the Anger camera can be found in BIOMEDICAL ENGINEERING.

6.2 Positron-Emission Tomography

Positron-emission tomography (PET) makes use of the unique characteristics of the radionuclides that decay by emitting positrons. When these positrons lose energy through their interactions with the electrons of the surrounding matter, each positron will annihilate with an electron, giving rise to two photons, or γ rays, which radiate at 180° from each other. This conversion of matter into energy is an example of the energy–mass relationship described by Einstein. With an appropriate array of detectors, these coincident annihilation γ rays can be detected. Thus, the annihilation event is self-collimating; that is, the back-to-back photons define the line of response between two detectors. By mathematically back-projecting the lines of response, it is possible to reconstruct a density map for the location of the radioactivity. If the positron radionuclide is attached to a biologically active compound (radiopharmaceutical), then the spatial and temporal distribution in an organ or a functional system can be determined.

The spatial resolution of a PET system is determined by the ability to measure the origin of the two annihilation photons. The theoretical limits arise from the range of the positrons in tissue and the lack of true collinearity of the emitted photons. The first limit is determined by the kinetic energy of the positron, which for the typical PET radionuclides have a range of 2–20 mm in tissue. The second limiting factor arises from the fact that the annihilation event occurs when the positron-electron pair still have a small amount of kinetic energy, and, in order to conserve momentum, the two photons are emitted at an angle slightly different from the 180° for the at-rest situation. The effect of this lack of collinearity depends on the detector separation, which means the larger the diameter of the tomograph, the greater the effect. Thus, the best resolution that can be expected for a PET scanner is of the order of 2 mm. Because positrons are emitted from the decaying nucleus over an energy spectrum and only a very small fraction have the maximum range, the broadening associated with positron range amounts to only a fraction of a millimeter for ^{18}F and ^{11}C, the two most widely used positron-emitting nuclides.

Initially, PET scanners were constructed with a single ring of detectors. However, the limited axial field of view made it difficult to perform meaningful studies. Subsequent designs employed multiple rings of detectors, separated by lead or tungsten septa to collimate the γ rays into planes or slices. Thus, the tomographs could acquire data over an extended field of view, and reconstructed images could be displayed in three dimensions.

The design of the detector system has the constraints of high efficiency to minimize the radioactive dose given to the patient and a small detector for finer resolution with the added complication and expense of more detectors and the associated electronics. The modern tomograph designers have solved this problem by using a block detector, which has many small detectors sharing a smaller subset of photomultiplier tubes (PMT); for example, a block of detector material is cut into an 8×8 array sharing four PMTs. Using Anger logic, the location (which crystal segment) of the incident photon can be accurately determined.

The strength of PET is the ability to determine the amount of attenuation caused by the surrounding material, e.g., the torso in a cardiac scan, by using an external positron source, usually ^{68}Ga in equilibrium with its parent ^{68}Ge. Since the count rate is determined by the coincidence rate, only those γ

rays that are not absorbed by the intervening material are registered. Thus, by comparing the count rate with and without the attenuating material (the patient), it is possible to correct for the attenuation that the γ rays originating internally will experience.

The sequence of measurements in a PET scan include a blank scan with a positron source without the patient, followed by the same measurement with the patient in place. This determines the attenuation factors. This is followed by removal of the external source and the injection of the PET radiotracer, and a third scan. The reconstructed image of this last measurement is corrected for attenuation.

The other factors affecting the quality of a PET scan are the amount of scatter of the γ rays and the number of random coincidences. Scatter is caused by one, or both, of the coincident γ rays scattering from the subject or from other material in the scanner, giving rise to an incorrectly encoded line of response. The γ rays are detected by different pairs of detectors than if they had not been scattered, which results in the blurring of the image. Scatter is corrected for by determining the scatter function at the edge of the field of view and extrapolating a profile across the field of view. The correction is applied before attenuation corrections and reconstruction are performed.

Random coincidences arise from the limitations of the detector design, which requires a finite timing window to allow the two γ rays to be registered in the opposing detectors and their electronics. The coincidence window for modern tomographs is usually set at between a couple of nanoseconds and 20 ns. The random rate is determined by

$$r = \tau N_1 N_2, \tag{6}$$

where τ is the coincidence resolving time and N_1 and N_2 are the count rates in the opposing detectors, respectively. Obviously, the shorter the resolving time, the lower the random coincidence rate; and conversely, the higher the individual detector counting rates, the higher the random coincidence rate. Because the random events have no directional character, the effect is to add noise uniformly throughout the image. Corrections for random events are made either by determining the single rates directly and applying Eq. (6) or, alternatively, by the use of a separate circuit with time delay to measure the random rate.

Another important factor in PET is that the radionuclides that decay by positron emission are the light elements associated with biological systems—carbon, oxygen, and nitrogen. Fluorine's atomic size makes it a suitable substitute for hydrogen. Thus, it is possible, in principle, to construct nearly any compound found in a biological system with a positron emitter. When using ^{18}F as the label, the resulting analog often has similar properties to the natural compound. The difficulty is, of course, the short half-lives of these isotopes, which makes the job of synthesizing them extremely difficult. The basic strategy is to insert the label as near the end of the synthesis as possible. There have been hundreds of compounds labeled with positron-emitting isotopes. However, simply having a labeled compound does not make it useful. What is required is a complete understanding of the metabolic processes the compound undergoes and how the label follows that process, because the PET image detects the location of the radionuclide, not its chemical form. Thus, modeling of a PET radiopharmaceutical is an important component in PET. Nevertheless, even with the complexity of synthesizing PET radiopharmaceuticals and the modeling required to understand the images, PET has proved to be an important tool in understanding basic *in vivo* biochemistry. There are a small number of compounds that are beginning to see use in clinical medicine and are mentioned in the relevant sections below. In particular, 2-[^{18}F]-fluoro-2-deoxyglucose (FDG) has been used to assess myocardium viability, as well as to detect hypermetabolic tissue associated with tumor growth.

6.3 Single-Photon–Emission Computed Tomography

As with PET, single-photon–emission computed tomography (SPECT) acquires views of the emitted photons from many different angles and reprojects these views to reconstruct the image or distribution of radioactivity in the object or patient. In SPECT, the radiopharmaceuticals used contain radioisotopes such as ^{99m}Tc that emit single photons (ones that are not in timed coincidence with one another). Directional information is achieved by collimating the photons incident on the detector. The collimator thus reduces the sensitivity of the camera, because all of the photons not parallel to the holes in the collimator are prevented from reaching the detector surface.

Since the reconstructed image contains the three-dimensional information on the distribution of radioactivity, SPECT also has the potential for quantification. The factors affecting this capability are similar to those in PET, for example, system sensitivity, dead time, spatial resolution, sampling interval, reconstruction filters, and the size of the object being imaged. Also, as in PET, the photons emerging from the subject are attenuated by the amount of matter between their origin and the detector and, of course, have a definite probability of being scattered along their path. These last two nonlinear effects are the most significant and difficult to correct for.

Because of the inherently lower energies (100–150 keV) of the photons emitted by radionuclides used in SPECT, the effect of attenuation can be quite dramatic, with reductions as great as a factor of 5 or more. Thus, with single-photon emitters, it is nearly impossible to determine whether data reflect a weak source near the surface or a stronger source located at some greater depth. In addition, the amount of scattering is strongly dependent on the energy of the photons, with the photons from ^{201}Tl having a scattering fraction of as much as 40–50% depending on the depth of the source.

Whereas scintillation camera images show the distribution of the radiopharmaceutical in defined regions in planar view, they suffer from the superimposition of organs and background contributions to the areas of interest. It is because of these shortcomings in planar imaging that SPECT has a major role to play in diagnostic imaging, regardless of whether SPECT can achieve the difficult task of providing quantitative information. The ability to view the distribution in three dimensions greatly affects the interpretation of the images. Figures 4 and 6 illustrate this point.

7. FUNCTIONAL IMAGING

Nuclear-medicine imaging differs from other radiological imaging in that the radiotracers used in nuclear medicine relate to the function of an organ system or metabolic pathway, and, thus, the tracing of these agents reveals the integrity of these systems or pathways. This is the basis for the unique information that a nuclear-medicine scan provides, as indicated in the variety of agents used to monitor the systems indicated in Table 6 and described in the following sections for various scanning procedures for the various organ or functional systems of the body. The structures of some of the ^{99m}Tc compounds are shown in Fig. 3.

7.1 Central Nervous System

Unlike other body tissues, the nervous system is highly discriminatory as to which solutes are transported into its tissues. This selective permeability is called the blood–brain barrier. The ability to measure changes in the blood–brain barrier (BBB) with radiophar-

Table 6. Standard nuclear-medicine procedures along with the radiopharmaceuticals commonly used for such procedures.

Standard radionuclide procedures	Radiopharmaceutical[a]
Brain perfusion	^{99m}Tc-HMPAO, ^{99m}Tc-ECD
Brain permeability	^{99m}Tc-DTPA
Blood-pool imaging	^{99m}Tc-RBC ^{99m}Tc-albumin
Bone imaging	^{99m}Tc-diphosphonates
Hepatobiliary	^{99m}Tc-DISIDA
Infection imaging	^{67}Ga-citrate Radiolabeled WBC
Liver–spleen	^{99m}Tc-colloid ^{99m}Tc-IDA
Lung	^{133}Xe (^{127}Xe) ^{99m}Tc-MAA
Myocardial perfusion imaging	^{201}Tl, ^{82}Rb, ^{18}FDG ^{99m}Tc-isonitriles ^{99m}Tc-teboroxines
Renal imaging	^{99m}Tc-DTPA ^{99m}Tc-MAG3
Thyroid imaging	^{123}I$^-$ ^{99m}Tc-pertechnetate
Tumor imaging	^{67}Ga-citrate Radiolabeled antibodies

[a]Abbreviations: HM-PAO, hexamethylpropylene amine oxime; ECD, ethyl cysteinate dimer; DTPA, diethylenediaminetetra-acetic acid; RBC, red blood cells; DISIDA, diisopropyliminodiacetic acid; WBC, white blood cells; IDA, iminodiacetic acid; MAA, macroaggregated albumin; FDG, fluorodeoxyglucose; and MAG3, mercaptoacetylglycylglycylglycine.

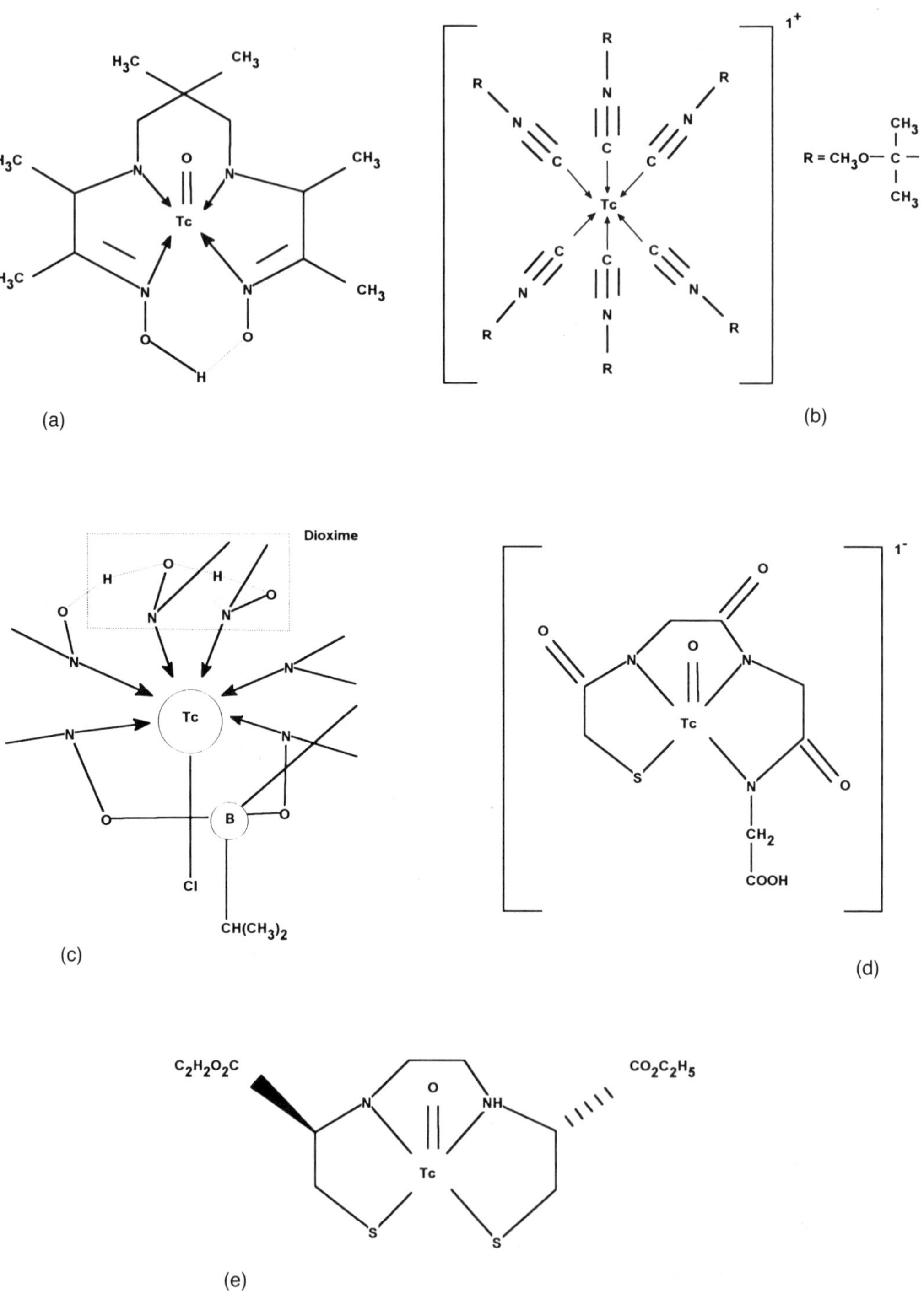

FIG. 3. Schematic of the structures of the ^{99m}Tc radiopharmaceuticals defined in the text. (a) Tc-HM-PAO, hexamethylpropylene amine oxime; (b) Tc-sestamibi, complexing with isonitriles; (c) Tc-teboroxime; (d) Tc-MAG3, mercaptoacetylglycylglycylglycine; and (e) Tc-ECD, ethyl cysteinate dimer.

maceuticals provided a major impetus for the growth of nuclear medicine over the last 25 years (Swanson *et al.*, 1990). In more recent times, the other radiological imaging techniques, such as computed tomography (CT) and, more recently, magnetic resonance imaging (MRI), have all but eliminated the use of radioactive agents to measure brain lesions.

However, recent developments in ligands that cross the BBB intact have opened the possibility of measuring functional changes in the brain associated with regional perfusion, metabolism, and the distribution and concentration of receptors. While very few of these new radiopharmaceuticals have achieved widespread clinical usage, the potential applications clearly indicate that brain imaging will again become an important aspect of nuclear medicine.

The brain is divided into two hemispheres along a fissure that runs from the anterior to the posterior. Each hemisphere has four lobes called the frontal, temporal, parietal, and occipital. The brain consumes up to 25% of the blood's oxygen supply and receives about 20% of all the blood that comes from the heart. The nervous system is bathed in a colorless fluid known as the cerebrospinal fluid.

The outer layer of the brain is composed of gray matter, which forms the cortex. Below this layer is the interconnecting tissue, made up of white matter interspersed with gray-matter areas. The gray matter is made up of neuronal endings, while the color of white matter is due to the myelin sheathes covering the axons.

Normal brain capillaries differ anatomically from capillaries found elsewhere in the body by their closed endothelial clefts and close approximation and continuity of the endothelial cells with the covering membrane. Molecules of glucose, water, and electrolytes apparently pass freely into the vascular space by passive diffusion and then into the brain cells themselves by active transport across its membrane, utilizing high-energy processes produced from the metabolism of glucose (the primary energy source of the brain).

Larger molecules are normally stopped at the endothelial lining, but, in disease, they may pass by pore filtration through the endothelial clefts widened by the disease process. These lesions that disturb the brain's integrity unselectively allow radiopharmaceuticals to permeate the nervous tissue.

Cerebral imaging normally consists of two phases (Harbert and da Roche, 1984), as follows:

1. a dynamic or angiographic study comprised of the arrival of the radioactive bolus in the cerebral hemisphere, which essentially constitutes a qualitative measure of brain perfusion, and
2. static images obtained approximately 1 h after the dynamic study, which provide a record of the distribution of the radiopharmaceutical in cerebral regions.

Scans at even later times can help delineate suspected lesions. The most commonly used agent for permeability studies is $^{99m}TcO_4^-$, pertechnetate, though the more expensive ^{99m}Tc-DTPA, diethylenetriamine-penta-acetic acid, is preferred because it clears more rapidly from the blood and does not accumulate in the blood vessels of the lateral ventricles (fluid-filled spaces in the brain), as does ^{99m}Tc-pertechnetate, resulting in a *hot spot* in the brain scan.

Single-photon–emission computed tomography (SPECT, see above) is gaining favor for three-dimensional perfusion studies of the brain. HM-PAO and ECD labeled with ^{99m}Tc are the agents of choice, hexamethylpropylene amine oxime and ethyl cysteinate dimer, respectively. These two compounds take advantage of the *p*H shift that occurs between the blood and the brain tissue (Saha, 1992). At the *p*H of blood (*p*H = 7.4), the agent is neutral and lipid soluble and can rapidly penetrate the intact blood-brain barrier. When it encounters the more acidic brain tissue (*p*H = 7.0), the agent acquires a charge, and, as an ion, it is incapable of reverse diffusion. Figure 4 depicts a blood flow image using SPECT and HM-PAO. A number of radioiodinated lipophilic agents, including IMP (isopropyl-*p*-iodoamphetamine) and HIPDM (*N*,*N*,*N*′-trimethyl-*N*-[2-hydroxy-3-methyl-5-iodobenzyl]-1,3-propanediamine), have shown promise but have been replaced by the technetium agents.

Although the concept of using positron emitters for biomedical investigations had been around since the 1951, positron-emission tomography (PET) did not become practical until the mid 1970s, based on the confluence of the PET scanner by Ter-Pogossian,

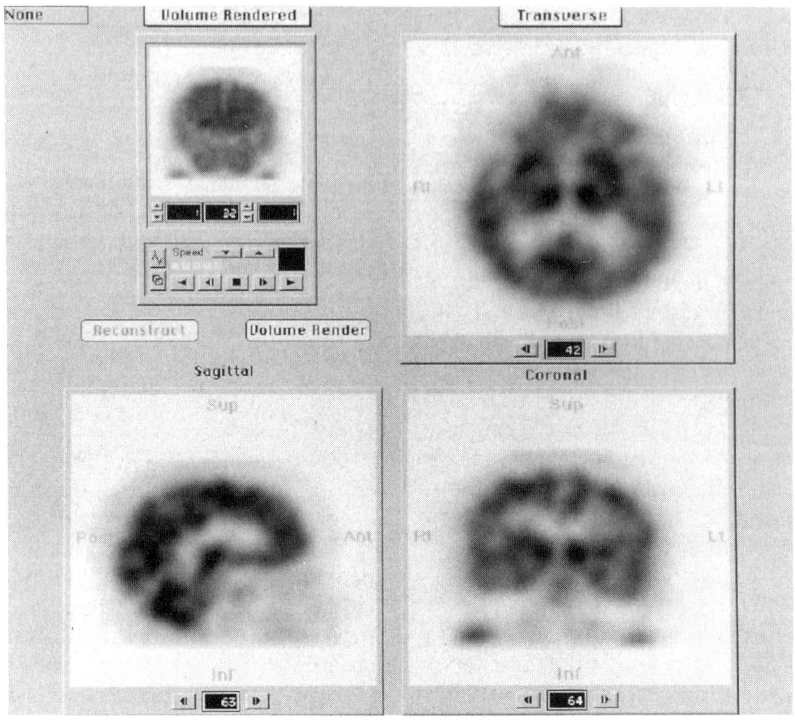

FIG. 4. SPECT image of the brain using ^{99m}Tc-HM-PAO, hexamethylpropylene amine oxime. The image is shown in a volume-rendered form, in the upper left, and three slice configurations: transverse, sagittal, and coronal. The abreviations are as follows: Ant—anterior (front); Post—posterior (back); Rt—right; Lt—left; Sup—superior (top); and Inf—inferior (bottom). (Courtesy I. Szaz, University Hospital, UBC Site.)

Phelps, and Hoffman (Ter-Pogossian *et al.*, 1975) in St. Louis and the development of the glucose analog ^{18}F-fluoro-deoxy-D-glucose (FDG) by Wolf and colleagues at Brookhaven National Laboratory (Reivich *et al.*, 1979). FDG is transported by plasma and crosses the BBB, where it is phosphorylated in a similar manner to glucose. The resulting FDG-6-phosphate is not a substrate for the other enzymes that are involved with glucose metabolism; thus, the labeled species is metabolically trapped. Once the relationship between FDG and glucose transport and uptake is taken into account, it is possible to obtain a quantitative measure of glucose metabolism on a region-specific basis. Mathematically, this relationship is achieved by determining the steady-state ratio of the net extraction of FDG to that of glucose at constant plasma levels of FDG and glucose and is referred to as the *lumped constant*. The FDG technique with PET has been used extensively in fundamental research of the functioning human brain to understand a number of normal brain functions, as well as disease processes. The clinical utility of this method has been employed to monitor tumor treatment and to locate the foci of epileptic seizures.

Oxygen consumption, blood flow, and blood volume have been measured using the ^{15}O-labeled molecules of O_2, H_2O, and CO, respectively. Neurotransmission and receptor density have also been measured for a number of neurological disorders, such as Parkinson's disease and schizophrenia. Figure 5 compares the uptake of ^{18}F-6-fluorodopa, FDOPA, in a normal subject and one affected by Parkinson's disease. The images in FDOPA scans reflect the uptake and conversion of fluorodopa to fluorodopamine and thus monitor the efficacy of the enzymatic systems involved in those processes. Research using PET is aimed not only at diagnosing disease but at an understanding of the disease process, as well as monitoring possible treatments.

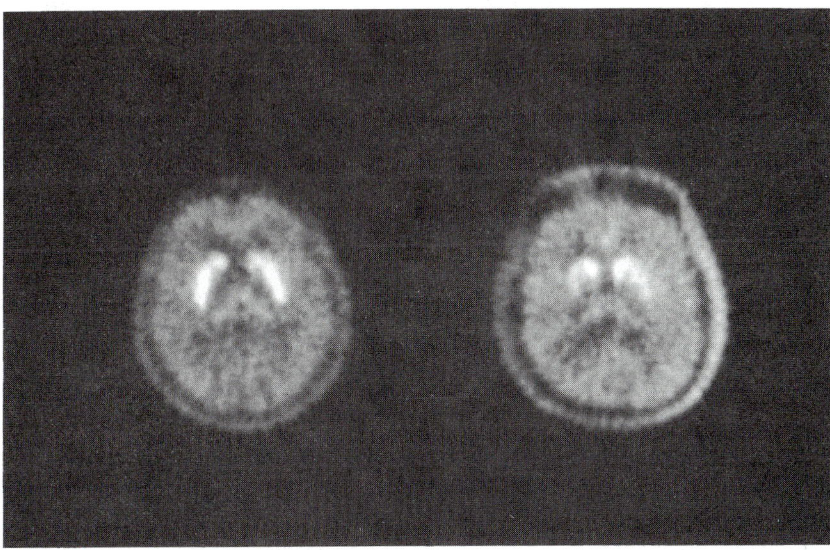

FIG. 5. PET image of ^{18}F-fluorodopa (FDOPA) uptake in (left) a normal patient and (right) a patient with Parkinson's Disease. Note the diminished uptake of FDOPA in the lower portion of the striatal region in the right image. (Courtesy UBC/TRIUMF PET Program.)

7.2 Cardiac Imaging

In general, nuclear medicine studies of the heart can be divided according to the clinical information sought (Swanson *et al.*, 1990). These studies include those that monitor the myocardium directly, such as perfusion assessment of metabolic activity, and determining whether the myocardium is healthy or infarcted. The imaging agents for perfusion use either potassium analogs or agents that are trapped in the myocardium in proportion to flow. The new technetium agents are included in the latter.

Among the perfusion agents, ^{201}Tl is the most widely used. Monovalent thallium has an ionic radius that is equivalent to that of K^+; thus, its uptake follows the active transport system for sodium and potassium. As much as 4% of the injected dose is taken up by the myocardial tissue, reaching its maximum in about 10–30 min after injection. The blood is cleared rapidly, with about 5% of the injected dose remaining in the blood pool after about 5 min.

^{201}Tl continuously redistributes between intracellular and interstitial compartments with differing rates for healthy and infarcted tissue. This behavior is typical for alkaline-like, singly charged cations. Thus, the equilibrium that is established upon redistribution can be used to distinguish between infarcted and ischemic tissue. In the case of ischemia, the diminished uptake disappears with redistribution, while persistent defects remain.

While the physiological behavior of ^{201}Tl is highly suited for perfusion imaging, its imaging characteristics are less than ideal. Principally, the energy of the emitted photons is extremely low (69–80 keV), which results in their being scattered, resulting in a loss in spatial resolution. Also, the half-life of ^{201}Tl (73 h) adds to the radiation burden for the patient. This radionuclide is cyclotron produced, making it expensive. Because of these negative attributes, there have been active attempts to find a ^{99m}Tc-based replacement.

Within the last decade, a series of ^{99m}Tc-labeled compounds have been proposed for myocardial perfusion studies (Saha, 1992). The

^{99m}Tc isonitriles are chemical complexes consisting of six monodentate CNR ligands surrounding a positively charged Tc^+ ion in the center, resulting in a stable cationic complex $Tc(CNR)_6{}^+$, where R represents one of the three functional groups tested thus far, tertiary butyl, carboxyisopropyl, and *hexakis*-2-methoxy-2-isobutyl groups. Of these agents, ^{99m}Tc-sestamibi (MIBI, where the complex is carboxyisopropyl isonitrile) has received the widest clinical usage. The myocardial uptake of this agent is proportional to regional myocardial blood flow. It does not redistribute with time, and, thus, a second injection must be performed the following day if a stress/rest protocol is desired. Figure 6 shows cardiac images acquired using ^{99m}Tc-MIBI.

Another class of technetium agents has been developed, consisting of boronic acid adducts of technetium dioxime (BATO). Unlike the isonitriles, the BATO complexes are neutral with respect to charge and are highly lipophilic. Lipid solubility is the uptake mechanism. ^{99m}Tc-teboroxime has undergone clinical trials. It consists of seven coordinate bonds coming from three dioxime molecules and one chlorine atom. ^{99m}Tc-teboroxime also does not redistribute and, thus, requires a second injection. BATO has a higher uptake in the myocardium than the isonitriles but does not clear as rapidly from the blood as does sestamibi and therefore does not have as high an image contrast.

$^{82}Rb^+$, as supplied from the ^{82}Sr generator, is the PET analog of K^+. Between 65 and 75% of the $^{82}Rb^+$ is extracted from plasma by the myocardium in the first pass at normal blood flow. Ammonia and water labeled with ^{13}N and ^{15}O, respectively, have also been used clinically to monitor cardiac blood flow. The use of these compounds is restricted by the fact that their short half-lives (^{13}N, $t_{1/2}$ = 10 min; ^{15}O, $t_{1/2}$ = 2 min) dictates production to be at or very near the site of use by an accelerator.

The metabolic function of the heart muscle can be monitored by a number of iodinated fatty acids (see below) or with the glu-

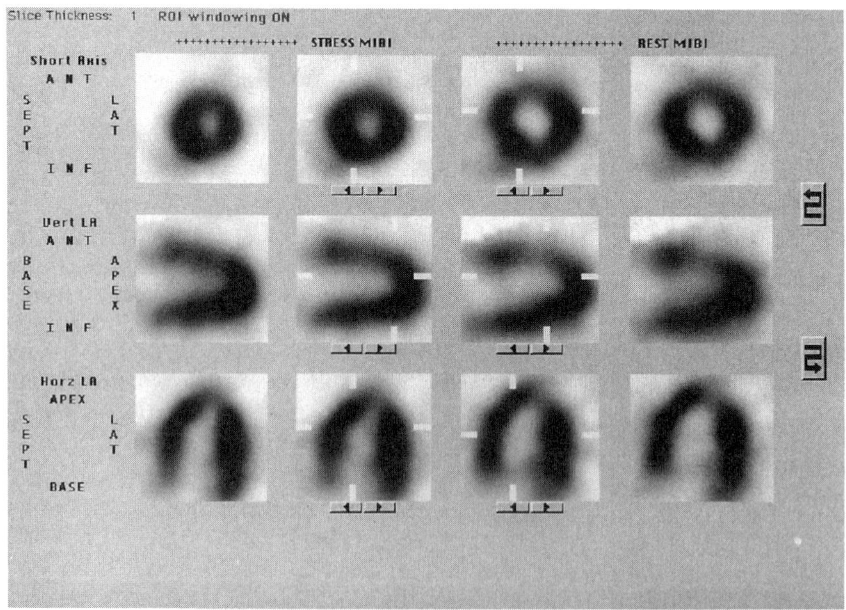

FIG. 6. SPECT image of the heart using ^{99m}Tc-MIBI, methoxy isobutyl isonitrile. The two columns on the left show the distribution in the heart at rest, and the two columns on the right show the distribution in the heart under stress. The top row has the images oriented in an oblique manner. This involves the reorientation of the heart images in such a way that slices at right angles to the long axis of the left ventricle become transverse or oblique images. The second row shows slices in a plane parallel to the long axis, from septum (the muscle that separates the the two chambers) to lateral wall (the thick outside wall of the heart), which are referred to as oblique/sagittal (Obl/Sag). The bottom row has the plane at right angles to oblique/sagittal and is the oblique/coronal (Obl/Cor). The abbreviations refer to ANT, anterior; INF, inferior; SEPT, septum; LAT, lateral; Base, base of the heart; APEX, the apex of the heart. (Courtesy I. Szaz, University Hospital, UBC Site.)

cose analog 2-fluoro(^{18}F)-2-deoxyglucose (FDG). FDG in conjunction with PET has the potential to measure the metabolic capacity of the myocardium quantitatively. Just as in the case for the brain, the metabolism of FDG essentially stops after phosphorylation, trapping the labeled analog in the cells.

The combination of ^{18}F-FDG and ^{13}N-ammonia with PET has received considerable interest in assessing the viability of the myocardium. Figure 7 compares a normal heart image with those that indicate ischemia. The ammonia scan measures blood flow and can detect obstructions in the blood supply to the myocardium, as shown in the figure. The FDG scan, however, indicates that the muscle is still capable of utilizing glucose and thus may return to normal function if reperfused via heart bypass surgery.

While FDG is an excellent compound for monitoring cardiac metabolism, its widespread use is hindered by the fact that these studies require a PET scanner, and, even though PET scanning is becoming more available, it is still only found in about 100 centers as of 1993. The primary energy source of the heart muscle under fasting conditions and rest is free fatty acids. A number of fatty acids have been labeled with ^{123}I. The most promising of these is ^{123}I-labeled *p*-iodophenylpentadecanoic acid (IPPA). The iodine atom is more strongly bound to aromatic carbons than to aliphatic carbons, thus minimizing the loss of the ^{123}I *in vivo*. With the favorable characteristics of ^{123}I and the rapid blood clearance of the fatty acids, IPPA holds promise for use in monitoring cardiac metabolism in the clinical setting, although the usefulness of such measurements has not been established.

Ventricular function is measured with blood-pool agents. These studies are performed by gating the data collection to the cardiac cycle. In this method, the radiopharmaceutical must reside in the vascular compartment for a sufficiently long time and at constant levels for the data acquisition.

7.3 Bone Scans

Although bone mineral is mainly composed of calcium, phosphate, and hydroxyl ions, living bone is a dynamic and physiologically active organ. The large adsorptive surface of bone provides chemically reactive sites that can be labeled with a wide variety of radionuclidic substances (e.g., Sr^{+2}, Pb^{+2}, Ra^{+2}, and Mg^{+2} for Ca^{+2}; F^{-} for OH^{-}; CO_3^{-2}, citrate, phosphate esters, diphosphonates, and pyrophosphate for phosphate).

In the formation of bone, amorphous calcium phosphate (ACP) precipitates on the surfaces of collagen fibers, where it grows over days and weeks to form hydroxyapatite [HA; $Ca_{10}(OH)_2(PO_4)_6$] (Swanson *et al.*, 1990). It has been shown that ^{99m}Tc-diphosphonate binds

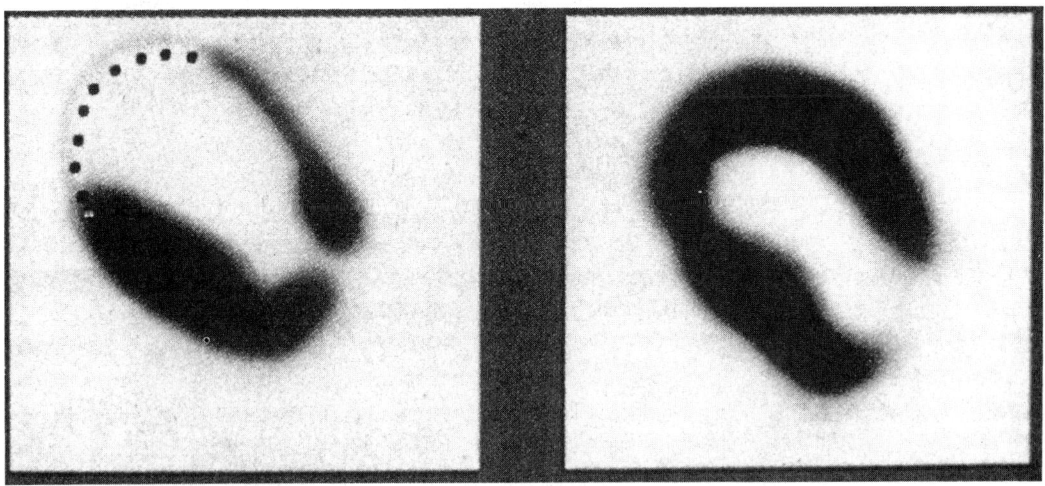

FIG. 7. PET image of the heart comparing blood flow with ^{13}N-ammonia on the left and metabolism with ^{18}F-FDG on the right. Note that there is a mismatch between flow and metabolism, indicating that the region of low flow is still capable of metabolizing glucose. (Courtesy M. Phelps, UCLA.)

to ACP to a higher degree than to crystalline HA. This selectivity forms the basis for detecting bone lesions, since areas of increased osteogenic activity contain higher concentrations of ACP than normal bone.

Shortly after its discovery in 1937, ^{18}F was discovered to have a high affinity for bone and teeth. This characteristic was exploited in the 1950s, when ^{18}F-fluoride was used extensively as a bone-scanning agent in nuclear medicine. With the advent of the ^{99m}Tc agents in the 1970s, the use of fluoride was replaced because of the better physical characteristics of half-life and photon energy, as well as the ease of delivery of ^{99m}Tc from a generator.

With the growing number of PET centers with their own cyclotrons, the interest in ^{18}F-fluoride as a bone-scanning agent has returned. The tomographic resolution and sensitivity for ^{18}F are superior to those of the single-photon systems. Thus, the resulting images can provide for accurate anatomical localization and improved lesion contrast. The first-pass extraction of fluoride from plasma is quantitative and in proportion to bone perfusion. Figure 8 illustrates the use of PET and whole-body ^{18}F-fluoride imaging.

7.4 Endocrine Systems

The endocrine systems regulate body functions through the secretion of various chemicals called hormones, from the Greek word meaning *arouse to activity*. The endocrine system includes the thyroid and parathyroid, pancreas, pituitary, gonads, and the adrenal glands. The thyroid gland helps control energy production, and the parathyroid provides for the distribution of calcium between the blood and bone. The adrenal glands regulate blood pressure, while the pancreas controls blood sugar levels. The pituitary controls growth, as well as regulating the activity of the gonads, the adrenal, and thyroid glands through the release of hormones. The gonads produce the germ cells of reproduction and a number of hormones related to the reproductive process.

The hormones generated by the thyroid that regulate the basal metabolic rate contain iodine. The iodine is accumulated in the thyroid from various food sources; thus, radioactive iodine becomes the simplest and nearly ideal tracer of metabolic function of the thyroid. With the production of ^{131}I from

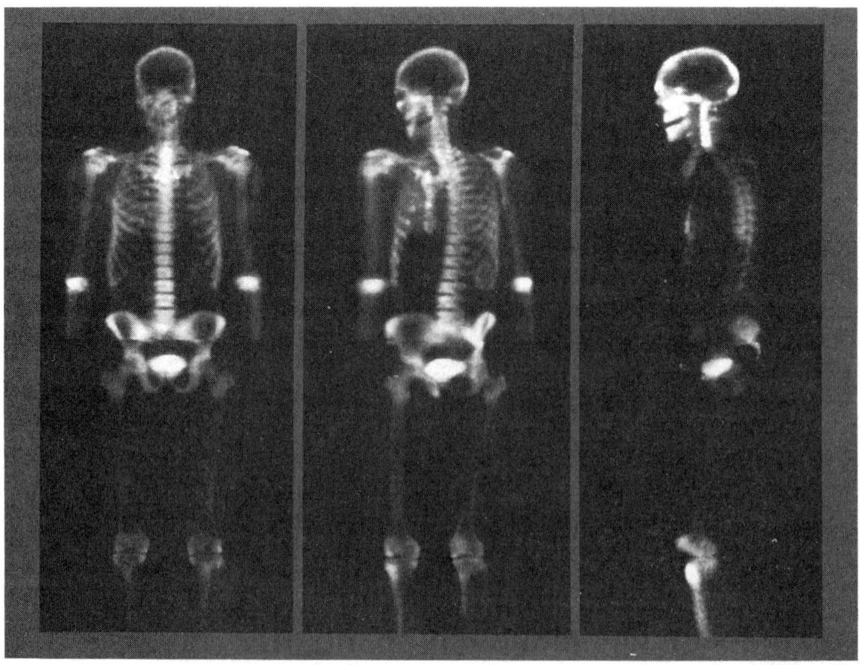

FIG. 8. Whole body PET scan of the skeleton using ^{18}F-fluoride in a normal male. The images from the left are the anterior projection, sagittal, and coronal images. (Courtesy C. Hoh, UCLA.)

reactors during and after the second World War, studies and treatment of thyroid diseases became widely available. There are a number of radioisotopes of iodine, but ^{123}I has the best physical properties for nuclear imaging, namely short half-life for low radiation exposure to patient, the energy of the principal γ ray matched to the present collimator systems on γ cameras, and no β-particle emission, which also reduces the dose to the patient in comparison to ^{131}I. Until recently, the production of ^{123}I required a high-energy accelerator. However, in the early 1980s, the use of a low-energy reaction on enriched ^{124}Xe to produce this radionuclide in high yield and extremely high purity increased the availability of this important radionuclide. The major drawback remains that its cost of production still makes the cheap and readily available ^{131}I competitive. As the reliability of ^{123}I becomes established, the universal acceptance of this radionuclide will be assured.

Radioiodine uptake studies utilizing sodium iodide are used to study hypothyroidism and hyperthyroidism, and to plan radioiodine treatment, the assessment of functional reserve, and the evaluation of thyroid gland autonomy. Radionuclide imaging of the thyroid is used to assess the size, position, and morphology of the thyroid gland, the detection of metastatic thyroid cancer and response to therapy, the evaluation of acute thyroid pain, the detection of accessory thyroid tissue, and the assessment of the functional status of nodules.

Because $^{99m}TcO_4^-$, pertechnetate, has a similar size and ionic characteristics to iodine (Swanson *et al.*, 1990), it is expected to be trapped in the thyroid. However, with the trapping, it is not incorporated into the functional hormonal system. Pertechnetate is used primarily for studying the morphologic structure of the thyroid.

Certain cells of the pancreas secrete enzymes that are used in the digestive system to react with proteins, carbohydrates, and lipids, while other cells secrete the hormones insulin and glucagon. Insulin controls the glucose level in blood. The amino acid methionine is incorporated in the digestive enzymes. Thus, an analog of methionine, selenomethionine, where the sulfur in methionine is replaced with radioactive ^{75}Se, has been used to image the pancreas to detect tumors, pancreatitis, and pseudocysts. However, because of high uptake in the liver and gastrointestinal areas, visualization of the pancreas is difficult.

The two adrenal glands positioned on top of the kidneys have two structures, an outer cortex and a central medulla. The adrenal cortical cells synthesize a number of steroid hormones from the circulating plasma cholesterol. Radioiodinated analogs of cholesterol, 19-iodocholesterol and 6β-iodomethyl-19-norcholesterol, have been used to image the cortical rim of the adrenal glands. Tumors and hyperfunction, Cushing's syndrome, can be diagnosed with increased uptake of these compounds.

7.5 Kidney

The urinary system is composed of the two kidneys, which are connected to the bladder via the ureters, and the bladder contents are discharged through the urethra. Just as in the case of the adrenal glands, morphologically the kidneys have an outer rim called the cortex and an inner area called the medulla. The nephron is the functional unit of the kidney and consists of tubules and ducts intermixed with the capillary bed of the blood. The nephrons perform three functions: the filtration of blood plasma, selective absorption of the materials the body requires, and the secretion of waste materials to the urine. The kidneys receive about 25% of the blood supply in each cycle, so that the entire blood volume passes through the kidneys every few minutes (Saha, 1992).

As part of the secretion process, there are two independent pathways for organic ions, and one for anions and one for cations. The pathway for anions has been exploited for imaging in nuclear medicine by the use of *o* iodohippuric acid (OIHA), initially labeled with ^{131}I and more recently with ^{123}I. A new agent that is replacing OIHA is ^{99m}Tc-mertiatide or MAG3, a chelating agent based upon amide nitrogen and thiolate sulfur groups. Since the excretion rate of a drug by the kidneys depends on the molecular size, lipid solubility, and pK_a of the drug, and the volume of distribution of the drug in the body as well as the rate of transport through the renal tubules, it is possible to quantify renal function.

After injection of ^{99m}Tc-MAG3, dynamic

images are acquired, followed by a delay to allow for plasma clearance, and a static image is then obtained. Renograms, or a display of the uptake and flow in the kidney, are obtained by placing a region of interest (ROI) on the temporal images of the kidneys and displaying the radioactivity present in the ROIs as a function of time.

^{99m}Tc-DTPA is often used to monitor the morphologic structure of the kidneys by acquiring a static image 30–60 minutes after injection (see Fig. 1).

7.6 Pulmonary System

The pulmonary system is imaged by the demonstration of pulmonary perfusion employing limited capillary blockade, as well as ventilation assessment with inspired inert gases, such as the noble gas xenon.

The airway of the respiratory system consists of the trachea, which divides into right and left main-stem bronchi, and these in turn divide to form lobar bronchi. The lobar divisions on the right are upper-, middle-, and lower-lobe bronchi, while, on the left, only upper- and lower-lobe bronchi exist. These lobes further divide into segments terminating in alveoli. In the adult, there are approximately 250 to 300 million alveoli, with an average diameter of 150 μm per alveolus. In addition to the direct pathway to the alveoli, there are pores and canals that interconnect them to allow collateral ventilation in the peripheral lung.

In the xenon ventilation studies, the patient breathes in the radioactive gas and holds his or her breath while the lungs are ventilated. Then the patient breathes while the air and xenon reach equilibrium, and, finally, the patient breathes in fresh air to wash out the xenon. Images are acquired in all three conditions. The ventilation studies can be used to determine if there is any airway obstruction, bronchitis, or emphysema.

For ventilation studies, ^{133}Xe is the most widely used noble gas, even though ^{127}Xe has better physical properties (higher photon yield, higher photon energy). The difference in availability has dictated the relative usage of these two isotopes of xenon (see Tables 2 and 3). Xenon-127 requires a high-energy accelerator to be produced, and the only accelerators producing ^{127}Xe at the present time are those associated with basic physics programs, thus making production and delivery unstable. Efforts to use the ^{126}Xe(n,γ) reaction may alter this in the future.

Perfusion studies employ the limited capillary blockade through the use of particles that become lodged in the capillary beds of the lung. The capillaries that surround the alveoli are between 7 and 10 μm in diameter. Thus, to achieve highest extraction efficiency, the ideal size for the particles is in the range of 10–20 μm. Macroaggregates (MAA) and human albumin microspheres (HAM) labeled with ^{99m}Tc have been used for 20 years to measure perfusion. More than 90% of the particles from either agent are extracted from the blood pool in the first passage through the lungs. Distribution in the lung is expected to be uniform, and, thus, the image appears "hot." Where blood flow is occluded or restricted because of emboli or vessel constriction, the particles will be prevented from passing these points of obstruction. The image thus appears with "cold spots" at these locations. Perfusion studies are used to diagnose obstructions caused by pulmonary embolism, tumor, infection, and tuberculosis.

7.7 The Liver, Spleen, and Bile

The lobules of the liver consist of two basic type of cells: hepatocytes and endothelial cells (Kupffer's cells). The hepatocytes make up about 85% of the cell population in the liver and are responsible for the major metabolic functions, while the endothelial cells make up the remaining 15% of all liver cells and serve to cleanse the blood. Each lobule consists of a central vein surrounded by a number of sinusoids, like spokes around the hub of a wheel. Sinusoids are vessels that transport blood from the portal vein to the central veins that eventually deliver blood to the hepatic vein (Saha, 1992).

The liver serves to metabolize the sugars and simple amino acids from digestion and to store the resulting peptides and proteins, as well as for the detoxification of harmful substances through methylation and conjugation. The liver excretes bile into the intestines through the hepatic duct.

The spleen consists of two main compartments. They are the white pulp, consisting of small lymphocytes and plasma cells, and red pulp, a swampy mass of vascular spaces whose supporting structure contains blood-cleansing cells. The two main functions of the spleen are antibody production in the white pulp and particle filtration in the red pulp. Blood entering the spleen through the splenic arterioles empties first into the white pulp, where plasma skimming occurs to remove soluble antigens. Passing through the white pulp, the blood enters the red pulp through an open meshwork of large oval cells at the junction of these two cell types. This cellular maze serves to remove abnormal cells and allows normal cells to proceed unhindered.

Labeled colloids are used to evaluate the blood-cleansing function of both the liver and the spleen. The most widely accepted agent today is ^{99m}Tc-sulfur colloid. After administering the ^{99m}Tc-sulfur colloid, the majority of the particles (0.3–1.0 μm) are accumulated by the reticuloendothelial cells of the liver (80–90%) and spleen (5–10%), and the remainder go to the bone marrow. In principle, the correlation between particle size and organ avidity for the colloid would have the spleen removing the larger particles, the smaller particles going to the liver and the smallest particles to the bone marrow. In the normal liver, a homogeneous distribution of radioactivity throughout the organ is expected. Any localized, space-occupying process in the liver may present as a focal area of decreased uptake referred to as a defect. Lesions as small as 8 mm in size may be identified, while those that are 20–25 mm in diameter are routinely imaged. Colloid scans are used to diagnose cirrhosis of the liver, abscesses, tumors, metastatic lesions, and hepatitis.

To evaluate the functional integrity of the hepatocytes, ^{99m}Tc-IDA has been developed. The ^{99m}Tc-IDA is given to fasting patients, where it rapidly appears in the liver and then progressively clears from the liver via the biliary tree, where major hepatic ducts and the gall bladder are able to be visualized.

7.8 Tumor and Inflammatory Processes

Although the precise mechanism by which tumor cells proliferate is not known, there are some common factors that can be exploited for imaging. With the increased metabolic activity and blood flow, along with high vascular permeability, agents that mark these events can be used to identify this change in function as compared to normal tissue. Tumors also have antigens associated with them that can be used to target a radiopharmaceutical specifically to all of the locations of the particular tumor.

The radiopharmaceuticals in routine clinical use are, in general, nonspecific with respect to tumor type and primarily reflect the altered functional state. In fact, ionic ^{67}Ga as gallium citrate was accidentally found to accumulate in soft-tissue tumors. It was quickly learned that ^{67}Ga localized in many nonmalignant areas, such as inflammatory lesions. It has thus become a common mode for evaluating patients with suspected infectious processes. Although the mechanism by which ^{67}Ga accumulates in these tissues is not understood, it is known that gallium quickly binds to transferrin in the blood and that this protein binding may be part of the process.

More recently, ^{111}In-labeled leukocytes have been used to detect abscesses in various parts of the body. Whereas ^{67}Ga is eliminated slowly through the gut, the labeled leukocytes are not excreted through the bowel and kidney. Thus, the abdominal background in this region is lower. The difficulty in labeling the leukocytes, however, is a major shortcoming.

As referred to in other sections, a number of other agents are used to monitor tumors associated with particular organs or functional systems. As the understanding of the use of monoclonal antibodies increases, the variety and specificity of these agents will increase their use.

8. RADIATION DOSIMETRY

The γ rays and particle emissions associated with the radionuclides used in nuclear medicine cause ionization to occur as they pass through matter. If the material these photons and particles are passing through is human tissue, these ionizations can break chemical bonds or kill cells if sufficient energy is deposited in these structures. Fortunately, in the case of nuclear-medicine imaging, the repair mechanisms of the human body usually prevent long-lasting damage from occurring. The restorative function in certain organs is more sensitive than others; thus, it becomes important to monitor the amount of energy deposited per unit mass in the various parts of the body by the decay of radioactive species. The energy absorbed from the ionizing radiation per unit mass of material is referred to as the absorbed dose. The *Système Internationale* (SI) unit for absorbed dose is the *gray* (Gy) and is equivalent to 1 J/kg. In spite of the efforts to make SI units the standard, the traditional units for radiation exposure and absorbed dose are commonly used. The term rad is used to express absorbed dose, and its relationship to the Gy is such that 100 rad = 1 Gy. The dose-equivalent unit (rem) was developed to account for the differences in the effects which various types of radiation have in causing biological damage. In radiation protection, the rem is defined as

$$\text{No. rems} = \text{No. rads} \times \text{QF} \times N$$

where QF is the quality factor and N the modifying factor of the radiation in question. The term QF is related to the amount of energy transfer to the surrounding medium the radiation traverses. The quality factor for photons is 1, while protons and neutrons have the value of 10 and heavy ions (α and other larger ions) have the value of 20, reflecting the greater loss of energy per unit length in matter for the heavier species. For practical purposes, the term N is assumed to be unity. The dose equivalent in SI units is the sievert (Sv), where 1 Sv is equal to 100 rem. Since it is difficult to determine the actual absorbed dose that radiation workers receive, it is common to refer to their radiation exposure in sieverts or rems and fractions thereof. However, for nuclear-medicine procedures, an effort is made to determine the patients' actual radiation exposure expressed in grays.

In order to estimate the absorbed dose due to a nuclear-medicine procedure, a method has been developed whereby dose estimates are calculated. These calculations require the knowledge of the biological distribution data and the physical properties of the radionuclide. The absorbed-dose method allows one to calculate the dose delivered to the tissue of interest, referred to as the target organ, from the radioactivity accumulated in a source organ, which may or may not be a number of sites within the body. The source organ can also be a target organ, which is generally the case when dealing with low-energy photons or particle emissions.

The actual process of calculating the dose (Loevinger *et al.*, 1988) includes the following steps:

1. The time integral of the amount of radioactivity (accumulated activity) that is in a source organ is determined by following the time course of radioactivity in the organ of interest.
2. The total amount of energy associated with the emissions of the radioactive species in the source organ is calculated. This is usually tabulated according to the type of radiation (penetrating or nonpenetrating), energy, and intensity (the frequency of the particular emission).
3. The fraction of the energy that is absorbed by the target organ is determined. The amount of energy absorbed depends on the type and energy of the emissions, as well as the physical relationship between the source and target organs, that is, the relative size, shape, and distance apart.

There are obvious difficulties in determining these quantities, especially in each human subject. Therefore, estimates and certain assumptions are made to simplify the calculations.

To calculate the cumulative activity A, the sum of all of the nuclear transitions in an organ is determined for the time interval of interest. Cumulative activity defines the time integral of the activity, which is proportional to the sum of all the nuclear transitions during a given time interval.

The rate of disappearance of the radioactive species from the body or organ within the body is due to the radioactive decay, which is dictated by the physical half-life of the radionuclide in question, and the biological processes that accumulate or eliminate the species. This latter process is measured as the biological half-life. Thus, the clearance from a source organ is measured by the effective half-life, a combination of the radioactive decay and biological elimination. Mathematically, this is represented by the sum of the decay constants for the physical and biological half-lives:

$$\lambda_e = \lambda_p + \lambda_b, \tag{7}$$

where the respective decay constants are equal to $(\ln 2)/t_{1/2}$ for each process, physical and biological. Residence time defines the average time that the administered radioactivity spends in the source organ. The residence time of a radiopharmaceutical in any particular organ is related to the fraction of the injected dose reaching the organ and to the chemical and biological processes that the radiopharmaceutical experiences.

Figure 1 illustrates how the distribution of different radiopharmaceuticals can affect the absorbed dose, not only to specific organs but also to the whole body. The compounds used in the figure are labeled with the same radionuclide, but the chemical form dramatically alters where they deposit in the body. The ^{99m}Tc-diphosphonate binds to the hydroxyapatite of bones and is distributed throughout the skeleton, while the spleen and liver have a high uptake of ^{99m}Tc-sulfur colloid. The lungs are clearly visible with ^{99m}Tc-albumin microspheres, and the kidney-ureter-bladder system concentrates the ^{99m}Tc-DTPA complex.

The distribution of the radiopharmaceutical will, of course, be dependent upon the functional integrity of the organ or functional system, which could be radically different in health and disease. Thus, to have as accurate an estimate of dose as possible, the actual distribution in the subject must be determined. Otherwise, the values used will only be population averages.

The data on the physical characteristics for the decay of the radionuclides used in nuclear medicine have, for the most part, been determined accurately. For a given radionuclide, each emission is characterized as to its energy and relative frequency. The product of these two values is the energy imparted per decay of the nuclide and referred to as the equilibrium absorbed-dose constant, symbolized by Δ. The value for this term in traditional units is g-rad/μCi-h, and Gy-kg/Bq-s in SI units.

The emissions associated with radioactive decay can be classified into two types of radiation, nonpenetrating and penetrating radiation. Nonpenetrating radiations are those particles emitted by the source that contribute a negligible dose to organs other than the source organ. These emissions include all of the particulate emissions, as well as very low-energy photons. Penetrating radiations are those particles emitted by the source that contribute a significant dose to organs other than the source organ. All of the emissions that can be detected outside of the body, as in the case for imaging, are obviously in this category.

The absorbed fraction is the fraction of energy emitted by the source organ that is absorbed by the target organ, and is given the symbol ϕ. The value of the absorbed fraction depends on the size, shape, and composition of the source organ, the attenuation factors between the two organs (distance and composition) of the intervening matter, along with the energy of the emission and its type. In the special case of nonpenetrating radiation (α and β particles, conversion and Auger electrons, and low-energy x rays or γ rays), where the source and target organs are one and the same, the value of ϕ is 1, and, for all other organs, ϕ equals 0. That is, all of the energy is absorbed in the source organ, while no energy is imparted to surrounding tissue.

Mathematical models have been derived to calculate the values of ϕ as a function of photon energy for pairs of organs in a standard adult of 70 kg. As an example, the fraction of absorbed energy in the heart (target organ) from a uniformly distributed source of 100-keV photons in the liver (source organ) equals 0.067. In establishing the values for ϕ, the reciprocity theorem has been used, which states that it does not matter which organ is the source and which the target, the relationship is the same. The energy absorbed per gram is the same for radiation traveling from organ A to organ B as it is in traveling from B to A.

To aid in the calculation of dose, a single parameter has been tabulated that is the product of the equilibrium absorbed-dose constant and the absorbed fraction per unit mass, and has been given the symbol S:

$$S = \Sigma \Delta \phi, \tag{8}$$

where $\Delta\phi$ is summed (Σ) over all emissions from the radionuclide.

The Medical Internal Radiation Dose Committee of the U.S. Society of Nuclear Medicine (MIRD Committee) has compiled tables of S values for organ pairs and all commonly used radionuclides in nuclear medicine. Thus, once the cumulated activity for a particular radiopharmaceutical is known for each organ, then the dose is determined from the product as in

$$D = AS, \tag{9}$$

for each organ in the body.

The S values are tabulated for about 20 organs acting both as source and target organs. The compiled values of the S are for a standard 70-kg man; therefore, in applying these values to other persons of differing size, the S values may not be correct.

Recently, the International Commission on Radiological Protection (ICRP) has adapted the concept of *effective dose equivalent* H_E for use in calculating dose to patients receiving radiopharmaceuticals (ICRP, 1988). The advantage of this approach is that a single figure can be used to express the radiation risk to patients undergoing different diagnostic procedures. The calculation is performed by summing the doses to the different organs in the body, weighted by a factor determined to reflect the relative sensitivity of that organ to radiation. For example, the weighting factors for gonads, red bone marrow, and skin are 0.20, 0.12, and 0.01, respectively. Tables of the effective dose per unit administered activity have been compiled by the ICRP for the standard radiopharmaceuticals. Table 7 gives some of these values. With this approach, it can be seen how the various procedures can be compared and, in the case of the same radiopharmaceutical, how the different radioisotopes affect the dose.

The calculation of absorbed dose is obviously a complex undertaking, but, as described here, a number of simplifications and assumptions have been made, so that useful values of estimated dose can be determined.

9. RADIOTHERAPY

The use of radionuclides for therapeutic purposes makes use of the modes of radioactive decay involving particulate emission. These modes include α and β decay (both positron and negatron), as well as any mode resulting in Auger electron and/or Coster–Kronig electron emission. The emission of γ rays may or may not accompany these particle emissions.

The use of particles emanating from the nucleus relies on the fact that these particles are stopped within a short distance. Alpha-emitting radionuclides are especially attractive in that α particles with 5–8 MeV energy have a range in tissue of about 40–80 μm. Figure 9 schematically illustrates the range of particles in tissue, along with a relative scale of tissue sizes to reflect the range along which the energy of the particles would be deposited. In choosing a therapeutic radionuclide, the precise site of localization of the radiopharmaceutical carrier in the tumor governs this choice. Thus, the Auger emitters must be transported to the nucleus if not incorporated into the DNA itself if *cell kill* is to be achieved. Table 8 gives some radionuclides that are considered to be candidates for use in radiotherapy.

Alpha particles with their high linear energy transfer (LET, the energy deposited per unit length by radiation as it passes through matter) are potentially lethal for cells if deposited on the cell surface or into the cytoplasm. There are very few acceptable α-emitting radionuclides because of the complex decay schemes associated with these isotopes. However, there are two candidates that have received special attention, ^{211}At and ^{212}Bi, both of which are cyclotron produced and have relatively short half-lives, which severely restrict their availability.

In addition to the types of particles emitted in their decay, there are a number of other considerations in assessing the suitability of radioisotopes for use in therapy. The physical half-life of a radionuclide used in therapy

Table 7. Radiation exposure due to selected radiopharmaceuticals. Note the factor of 100 difference between ^{123}I and ^{131}I in the thyroid uptake studies. This difference is due to the longer half-life of ^{131}I.

Radionuclide	Substance	E (mSv/MBq)
^{15}O	Oxygen gas, single breath, 20 s hold	4.2×10^{-4}
^{18}F	FDG	2.0×10^{-2}
^{99m}Tc	Tc-DPTA	5.2×10^{-3}
^{99m}Tc	Pertechnetate	1.2×10^{-2}
^{123}I	Iodide (thyroid uptake 35%)	2.2×10^{-1}
^{131}I	Iodide (thyroid uptake 35%)	2.4×10^{1}
^{201}Tl	Thallium chloride	2.3×10^{-1}

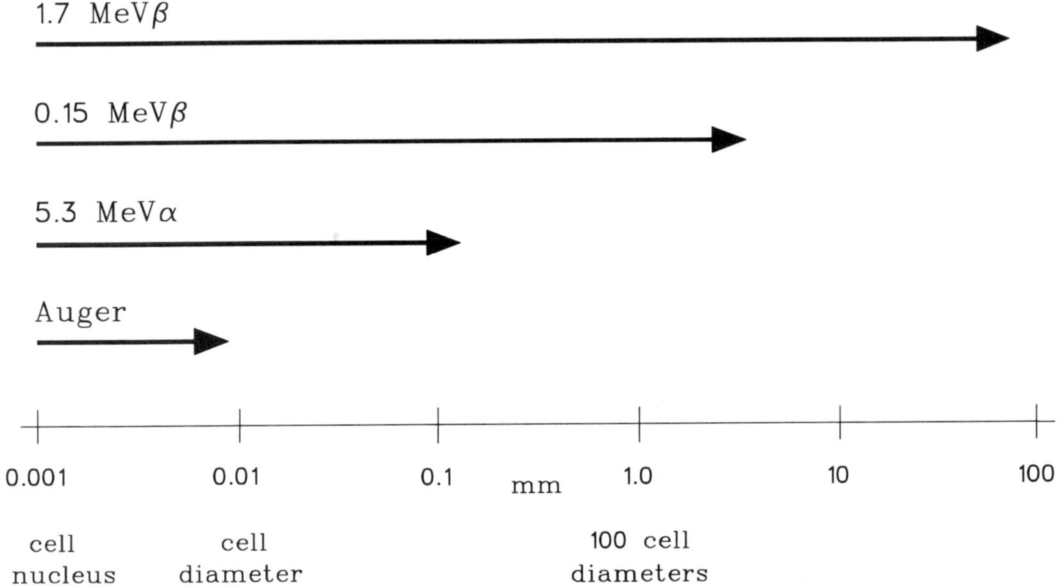

FIG. 9. Schematic representation of the ranges of particles in tissue in comparison to relative size of the cell structure.

must be long enough to allow the labeled species to clear the blood pool and concentrate in the region of interest but short enough to prevent the radionuclide from becoming a radiation burden to the rest of the body through metabolism of the primary agent. The physical half-life should be closely associated with the biological half-life of the radiopharmaceutical in the tumor. In this way, the bulk of the decays occur in the region where their damaging effects will be directed to the tumorous tissue and not to normal tissue.

One of the major difficulties in using pure β-emitting nuclides for therapy is the inability to detect their location accurately from outside of the body. However, a low-abundance γ ray with an energy of 100–200 keV can be used in imaging to assess localization and estimate delivered dose. Thus, the accompanying γ ray(s) should be in low abundance to remove radiation burden from surrounding healthy tissue.

The availability of a radionuclide and its associated chemistry determine its practical utility as far as receiving widespread clinical use is concerned. Just as in the case of the α emitters, even with their nearly ideal particulate energies, the difficulties in producing them have not only restricted the availability for clinical assessment but also basic research. The chemical form of the radionuclide has ramifications in determining how the radionuclide will be delivered to the site of interest. In this regard, ^{131}I has been con-

Table 8. Radionuclides with potential use as radiotoxic agents. The half-life of the radionuclide is included along with the particle emitted and its energy and the principal γ ray with its intensity (I) in percent.

Radionuclide	$t_{1/2}$	Particle	E_{max} (MeV)	E_γ (MeV) (I)
^{90}Y	64 h	β^-	2.3	None
^{111}In	2.83 d	Auger e^-	···	0.17 (90)
^{125}I	60.1 d	Auger e^-	···	0.035 (6)
^{131}I	8.1 d	β^-	0.6	0.36 (81)
^{153}Sm	46.7 h	β^-	0.81	0.103 (28)
^{186}Re	90.6 h	β^-	1.07	0.14 (9)
^{211}At	7.2 h	α	5.87 (7.45 from ^{211}Po)	0.69 (0.2)

sidered the nearly ideal radiotherapeutic agent for thyroid carcinoma because of the high natural uptake of iodine by the thyroid.

Radionuclides that are used in therapy are made in three different manners (see Sec. 2). The most common is from neutron capture, since neutron-rich radionuclides decay by β^- emission. The high fluxes of neutrons associated with nuclear reactors make it possible to produce large quantities of neutron-rich radionuclides by this method. However, the specific activity of reactor-produced radionuclides is generally low. An alternative source that is available indirectly from a reactor is from generators. Accelerator-produced radionuclides, as mentioned above, can, in principle, produce radionuclides with very high specific activity.

The interest in therapy with radionuclides has increased during the last decade, not only with the prospect of labeling site-specific compounds with these radiotoxic substances but also as palliative agents for treating metastatic tumors. Post-mortem studies indicate that a large majority of carcinomas develop bone metastases. While the exact mechanisms for causing pain in skeletal metastases are unknown, their management has been directed toward palliation. The use of targeted radionuclide therapy for this treatment is gaining favor. High specific uptake in bone lesions with rapid clearance from other tissue is an important consideration here, as in all tumor treatments using radiotoxic nuclides. Of the bone agents in use, ^{32}P orthophosphate and ionic ^{89}Sr have been used most extensively. More recently, ^{153}Sm-EDTMP (ethylenediaminetetramethylenephosphonic acid) and ^{186}Re-HEDP (1-hydroxymethylidene diphosphonate) show promise to be even more effective at rapid symptom relief.

The use of antibodies for direct treatment of tumors without damaging normal tissues has been of interest for at least a century. However, it was not until recently that the ability to isolate and label antibodies has made this approach possible on a large scale. Although a number of antibodies had been labeled with ^{131}I in the 1950s and 1960s, it was not until the mid 1970s, when the techniques for producing monoclonal antibodies were developed, that advances in radioimmunotherapy (RIT) became a reality.

There are five factors that determine the potential utility of radiolabeled antibodies for radioimmunotherapy:

1. the absolute quantity of radiopharmaceutical delivered to the tumor (target);
2. the amount of radioactivity in other organs (nontarget);
3. the radiosensitivity of the tumor;
4. the radiation dose delivery rate; and
5. the types and energies of radiations emitted.

There is still some dispute as to whether monoclonal antibodies are more efficacious than polyclonals for achieving an acceptable target to nontarget ratio. The isotopes ^{111}In, ^{90}Y, ^{125}I, and ^{186}Re have all been incorporated into antibodies and are the subject of clinical trials. While the results are promising, it is still too early to know in which direction the results will lead.

^{131}I is still widely accepted, even though its physical characteristics are less favorable than those of other radiotoxic nuclides. For example, its half-life of 8 d is, in general, longer than the biological half-life of most carriers. Also, its β energy ($E_{\max}$ = 0.6 MeV) is considered less than optimum for large tumors, as is the large number of γ rays that accompany the β decay. What remains attractive about this isotope is the wide variety of compounds into which iodine can be chemically inserted.

What is still a big question mark in the use of targeted radionuclide therapy is a complete understanding of the dose delivered, not only to the tumor but to surrounding healthy tissue. Dosimetric treatments that encompass situations where the microdosimetry on the cellular level is combined with biologic data could lead to more realistic dose estimates. This is especially true for the Auger emitters. When the decay of an Auger electron emitter occurs in the vicinity of DNA, the dose to microscopic volumes within the DNA may prove to be even more meaningful than the dose to the whole cell. The time-dependent distribution of the radiotoxic substance is required to gain a fuller understanding of the relationship between dose and effect.

10. FUTURE ASPECTS OF NUCLEAR MEDICINE

Probably the biggest advances in nuclear medicine in the future will be in the area of therapy. As the ability to deliver radionuclides to specific sites becomes a reality through research in monoclonal antibodies and their fragments, the use of therapy with radionuclides will become increasingly possible. As such, the rhenium isotopes ^{188}Re, from the ^{188}W/Re generator system, and ^{186}Re will be attractive choices because of rhenium's similar chemistry to ^{99m}Tc. Also, most experts feel that the growth area for PET will be in oncology for both diagnosis and treatment assessment. The ability to acquire quantitative metabolic and receptor density images throughout the whole body will provide a needed tool in the fight against cancer.

As the present-day PET tomographs approach theoretical spatial resolution, the remaining technological advances will be made in volume imaging, where the sensitivity of the tomograph is maximized by the removal of interplane septa. Work in this area has already begun with extremely promising results, not only in the increased sensitivity but in spatial sampling.

There is increasing interest in monitoring the effect on the neuronal system of the heart as a function of ischemia and infarction. One candidate is *m*-iodobenzylguanidine (MIBG), a norepinephrine analog. MIBG could monitor the depletion of the catecholamine stores of the heart in disease. Imaging of the muscarinic acetylcholine receptors in the heart using radioiodinated 3-quinuclidinyl-4-iodobenzylate (QNB) has been proposed. A ^{11}C-labeled QNB has also been proposed for use with PET. The availability of SPECT imaging has been a driving force in the development of ^{123}I-labeled compounds that are analogs of receptor antagonist used in PET. A number of these are undergoing clinical trials.

As the chemistry for attaching technetium becomes even more sophisticated, there will be more attempts at incorporating ^{99m}Tc into agents that have receptor sensitivity, so that receptor imaging will no longer be associated solely with PET radiopharmaceuticals or their iodinated analogs. However, in the meantime, more radiopharmaceuticals labeled with ^{123}I will find their way into widespread clinical use as they are validated in the assessment of disease and the availability of ^{123}I at a reasonable cost becomes assured.

There will be continuing efforts toward improving SPECT cameras as well as the software for correcting for scatter and attenuation in SPECT. To take advantage of the new radiopharmaceuticals for brain and heart, improved collimators will be designed.

Since radiopharmaceuticals generally are used to monitor function, there will be increasing use of these techniques to monitor treatment, for example, in determining the effectiveness of organ transplants. As an example of PET volume imaging, Fig. 10 illustrates the functional integrity following fetal

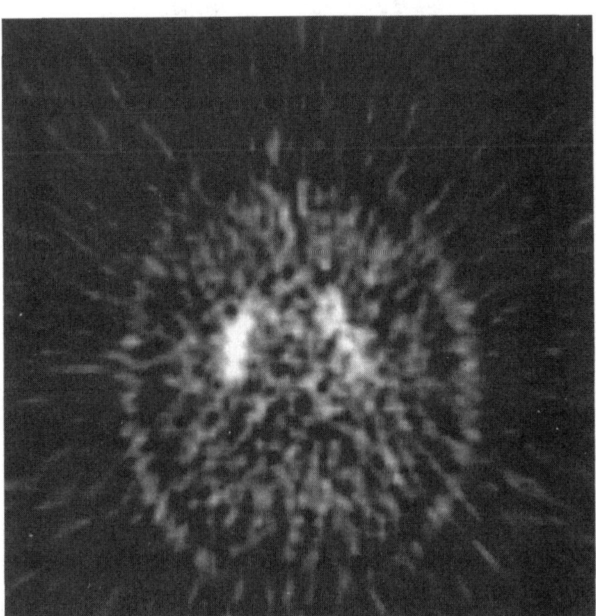

FIG. 10. PET image of ^{18}F-fluorodopa uptake in a patient who has received a fetal cell transplant. The arrows point to the four "hot spots" corresponding to the four surgical implants, indicating that the implanted cells are metabolically active with respect to the dopaminergic pathway. (Courtesy UBC/TRIUMF PET Program.)

cell transplantation. It can be expected that neuropsychiatric treatment will also be monitored by nuclear imaging processes as we gain a better understanding of the biochemistry of mental function.

As one of the pioneers of nuclear medicine, H. Wagner, recently said at the 1992 Annual Society of Nuclear Medicine meeting, nuclear medicine has begun the transformation of the art of diagnostic and therapeutic decision making into a quantitative science.

GLOSSARY

Alpha Decay: The mode of radioactive decay when the nucleus emits an α particle (the nucleus of ^{4}He), the product being two protons and two neutrons less.

Annihilation Radiation: The two 511-keV γ rays resulting from the annihilation of the β^+ with an electron in matter.

Antibody: A protein produced in the body in response to and as a counteragent to antigens.

Antigen: A substance that can induce the production of an antibody and bind with it specifically.

Auger Electron: The emission of an electron as an alternative to x-ray emission in an excited atom; thus, no x ray is emitted.

Becquerel: The SI unit of radioactivity, equivalent to one disintegration per second.

Beta Decay: The mode of radioactive decay when the nucleus emits an electron. In β^- decay, the atomic number increases by one, and, in β^+, decay the atomic number decreases by one.

Biological Half-Life: The time during which one-half of the amount of a substance is eliminated by a biological process such as urinary excretion.

Carrier Free: A term used in radiopharmaceutical sciences to indicate the absence of any stable atoms in a radionuclide sample. The term *no carrier added* is used to indicate the absence of purposely added nonradioactive atoms of the same element.

Curie: A measure of radioactivity equivalent to 3.7×10^{10} disintegrations per second.

ECD: The abbreviation for 1,1-ethylcysteinate dimer, which is used to monitor cerebral blood flow when labeled with ^{99m}Tc.

Electron Capture: The mode of radioactive decay when the nucleus captures an orbiting electron to convert a proton into a neutron; equivalent product to positron decay.

Electron Volt (eV): The kinetic energy required to move an electron through a potential difference of 1 V.

Half-Life ($t_{1/2}$): The time during which one-half the number of radioactive species decays. Same as physical half-life.

HIPDM: The abbreviation for *N*,*N*,*N*′-trimethyl-*N*-[2-hydroxy-3-methyl-5-iodobenzyl]-1,3-propanediamine, used as a blood-flow agent.

HM-PAO: The abbreviation for hexamethylpropylene amine oxime, which when labeled with ^{99m}Tc is used in cerebral blood-flow measurements.

IMP: The abbreviation for *N*-isopropyl-*p*-iodoamphetamine, which is used in assessing cerebral blood flow.

Infarct: An area of dead tissue resulting from a complete lack of blood circulation.

IPPA: The abbreviation for iodophenylpentadecanoic acid, a fatty acid used to monitor cardiac metabolism.

Ischemia: A condition where a tissue region has a deficiency in blood supply.

Kit: A package of the chemicals required to prepare a radiotracer.

Lumped Constant: The collection of mathematical terms used to express the relationship between the biological behavior of glucose and fluoro-deoxy-glucose.

MAG3: The abbreviation for triamide-monothiol-(N3M) complex ^{99m}Tc-mercaptoacetylglycylglycylglycine, which is used to monitor renal function.

MeV: Mega electron volts, a measure of energy equivalent to 1.6×10^{-19} joules.

RBC: The abbreviation for red blood cells.

Rectilinear Scanner: An imaging device whose detector is scanned mechanically in a rasterlike pattern over the area of interest.

Specific Activity: The measure of the amount of radioactivity per mass of sample.

Tritium: The only radioactive isotope of hydrogen, with 2 neutrons; it has a half-life of 12.3 years.

WBC: The abbreviation for white blood cells.

Works Cited

Eckelman, W. C. (1991), *Nucl. Med. Biol.* **18**, iii–vi.

Harbert, J., da Roche, A.F.G. (1984), *Textbook of*

Nuclear Medicine, 2nd ed., Vol. I, "Basic Science," Philadelphia: Lea & Febiger.

ICRP Publication 53 (1988), *Radiation Dose to Patients from Radiopharmaceuticals,* Oxford: Pergamon.

Loevinger, R., Budinger, T. F., Watson, E. E. (1988), *MIRD Primer for Absorbed Dose Calculations,* New York: Soc. Nuclear Medicine.

Reivich, M., Kuhl, D., Wolf, A., Greenberg, J., Phelps, M., Ido, T., Casella, V., Fowler, J., Hoffman, E., Alavi, A., Som, P., Sokoloff, L. (1979), *Circ. Res.* **44**, 127–137.

Saha, G. B. (1992), *Fundamentals of Nuclear Pharmacy,* 3rd ed., New York: Springer-Verlag.

Sorenson, J. A., Phelps, M. E. (1987), *Physics in Nuclear Medicine,* 2nd ed., New York: Grune & Stratton.

Swanson, D. P., Chilton, H. M., Thrall, J. H. (1990), *Pharmaceuticals in Medical Imaging,* New York: Macmillan.

Ter-Pogossian, M. M., Phelps, M. E., Hoffman, E. J., Mullani, N. A. (1975), *Radiology* **114**, 89–98.

Yalow, R. S. (1978), "Radioimmunoassay: A Probe for the Fine Structure of Biologic System," *Science* **200**, 1236–1245.

Further Reading

Burns, H. D., Gibson, R. E., Dannals, R. F., Siegl, P. K. S. (1993), *Nuclear Imaging in Drug Discovery, Development and Approval.* Boston: Birkhauser.

Diksic, M., Reba, R. C. (Eds.) (1990), *Radiopharmaceuticals and Brain Pathology Studied with PET and SPECT,* Boca Raton, FL: CRC Press.

Frost, J. J., Wagner, H. N. (Eds.) (1990), *Quantitative Imaging, Neuroreceptors, Neurotransmitters and Enzymes,* New York: Raven.

Harbert, J., da Roche, A.F.G. (1984), *Textbook of Nuclear Medicine,* 2nd ed., Vol. I, "Basic Science," Philadelphia: Lea & Febiger.

Loevinger, R., Budinger, T. F., Watson, E. E. (1988), *MIRD Primer for Absorbed Dose Calculations,* New York: Soc. Nuclear Medicine.

Sorenson, J. A., Phelps, M. E. (1987), *Physics in Nuclear Medicine,* 2nd ed., New York: Grune & Stratton.

Swanson, D. P., Chilton, H. M., Thrall, J. H. (1990), *Pharmaceuticals in Medical Imaging,* New York: Macmillan.

NUCLEAR REACTIONS

HERMAN FESHBACH, *Center for Theoretical Physics, Laboratory for Nuclear Science, and Department of Physics, Massachusetts Institute of Technology, Cambridge, Massachusetts, U.S.A.*

INTRODUCTION

In a nuclear reaction, a projectile collides with a target nucleus. The projectile may be one of the elementary particles, such as the proton, neutron, pion, kaon, antiproton, gamma ray, electron, neutrino, or muon (see PARTICLES, ELEMENTARY). The projectile might also be one of the atomic nuclei ranging from protons to uranium. The targets are the atomic nuclei (see PARTICLE TARGETS). The projectile energies available extend from very small values for neutrons to very large ones of the order of many GeV for which relativity is of special importance. The projectile beams are generally produced by accelerators, while the consequences of the collision are measured by detectors.

In view of the large number of experimental opportunities that are therefore accessible, it is not surprising that, through nuclear reactions, a rich set of phenomena has been uncovered. These phenomena permit the determination of the charge and mass distributions and corresponding currents of nuclei. They have led to the discovery of collective degrees of freedom characteristic not only of nuclei but of mesoscopic systems generally. They have provided insights into the nature of nuclear forces and nuclear dynamics. It is through nuclear reaction studies that most of our knowledge regarding nuclear systems is obtained. This has been true since the very beginning of nuclear physics. The nucleus itself was discovered through alpha-particle scattering by nuclei (Rutherford, 1911). The existence of a strong short-range nuclear force was exhibited by proton–proton scattering (White, 1936). Neutron-induced resonances (Amaldi *et al.*, 1935) and artificial radioactivity (Curie and Joliot, 1934; Fermi, 1934) are early examples of phenomena discovered by nuclear reaction studies.

In the post–World War II era, one saw a rapid expansion in experimental capability

3-527-28133-9/94/$5.00 + .50

with a corresponding increase in precision. This is governed by the quality of the beam produced by accelerators, such as its energy, current intensity, monoenergeticity, and beam divergence (see ACCELERATORS, LINEAR; ACCELERATORS, POTENTIAL DROP; CYCLOTRONS; SYNCHROTRONS). It requires that the detectors (see DETECTORS, SCINTILLATION; DETECTORS, SEMICONDUCTOR; SPECTROMETERS, ELECTRON AND ION; SPECTROMETERS, NEUTRON; SPECTROMETERS, GAMMA RAY) of the products of a reaction provide precision determination of their energies and directions of motion. Detectors measuring the masses and the charges of reaction products are needed when these consist of several different types of particles.

This enormous increase in experimental precision was accompanied by corresponding developments in the theory of nuclear reactions needed to interpret the data obtained in the laboratory. Before World War II, the emphasis was on the understanding of neutron resonances as provided by the Breit–Wigner formula (1936). In the postwar period, the discovery of important new phenomena led to the development of a comprehensive theory capable of interpreting the wide range of experimental studies and providing insights into nuclear dynamics.

A nuclear reaction is described by the symbol $X(a,b,c,\ldots)Y$, where X is the target nucleus, a is the incident projectile, Y is the residual nucleus, and b, c, ... are the reaction products. The final system may consist of two bodies $X(a,b)Y$, three bodies $X(a,bc)Y$, etc. The symbol (a,b) will sometimes be used to indicate a process in which the projectile a is transformed into b. *Baryon number* (equal to the number of neutrons and protons combined) and *charge* are conserved in a reaction. For example, in the reaction ${}^{51}_{23}\mathrm{V}(p,n){}^{51}_{24}\mathrm{Cr}$, the initial and final mass numbers are 52. The mass number of vanadium, V, is 51. Its atomic number is 23, while the initial charge is 23 + 1, equal to the atomic number of Cr. Conservation of energy and of linear and angular momentum as well as parity is rigorously obeyed in nuclear reactions (see SYMMETRIES AND CONSERVATION LAWS). Reactions involving the weak interactions, which violate parity conservation, will not be considered in this article (see WEAK INTERACTIONS).

Generally, two reference frames are used. In the *laboratory frame*, the target nucleus of mass M at rest (there is some thermal motion of no importance except at very low projectile energy) is struck by a projectile of mass m with energy E_L. In the second frame, the center-of-mass system, the target nucleons and projectile have equal and opposite momenta. The total energy is E, which is related to the energy E_L by the equation

$$E = \frac{M}{m+M} E_L. \tag{1}$$

It is the energy available for the reaction. The energy released in the reaction $X(a,b^*)Y^*$, the Q value, is given by

$$Q_{fi} = (m_x + m_a - m_Y - m_b)\, c^2 - \epsilon_b - \epsilon_Y. \tag{2}$$

In this equation, m_X is the mass of X, etc.; c is the velocity of light. The asterisks in b^* and Y^* indicate the possibility that internal energies ϵ_Y and ϵ_b have been transmitted to residual nucleus Y and reaction product b. We say in that case that b and Y are excited.

In nuclear reactions, energy, momentum, angular momentum, charge, and mass may be transferred from the projectile to the target and/or vice versa. When there is no change in total energy or in the character of the projectile and target, i.e., $X(a,a)X$, but there is a transfer of momentum, the process is referred to as *elastic scattering*. In elastic scattering, the colliding projectile and target nucleus change direction. None of the initial kinetic energy is used to excite either the target or projectile; that is, $Q_{fi} = 0$. The momentum transferred to the projectile, $\hbar q$, is given by $2p \sin \frac{1}{2}\vartheta$, where p is the momentum of the projectile in the center-of-mass frame and ϑ is the angle by which the projectile deviates from its initial direction.

If the collision results in the excitation of the target and/or projectile but with no other change, i.e., $X(a,a^*)X^*$, *inelastic scattering* has taken place.

Charge-exchange reactions involve the transfer of charge as exemplified by ${}^{51}_{23}\mathrm{V}(p,n){}^{51}_{24}\mathrm{Cr}$ or ${}^{A}_{Z}X(\pi^+,\pi^0){}^{A}_{Z+1}Y$ or ${}^{A}_{Z}X(\pi^+,\pi^-){}^{A}_{Z+2}Y$. Here, π^+, π^-, and π^0 refer to the three pions with charges as indicated.

Mass transfer reactions occur when a particle such as a nucleon or a cluster of nucleons is transferred from the projectile to the target nucleus or vice versa. The earliest example was the *stripping* reaction (d,p) or its inverse, the *pickup process* (p,d). In the latter case, the incident projectile is a proton, while the emergent one is a deuteron formed by picking up a neutron from the target nucleus. More massive pickups are possible. Such mass transfers lead to *transmutations*, that is, to the formation of a nucleus which differs in mass and/or charge from the original target nucleus. For example, the reaction $^{200}Hg(p,\alpha)^{197}Au$ (α = alpha particle) transmutes mercury (Hg) into gold (Au).

A very dramatic example of mass transfer is *radiative capture*, such as the (n,γ) reaction in which the incident neutron is captured by the target nucleus accompanied by the emission of a gamma ray. The inverse is called *photon absorption* in which the incident projectile is a gamma ray that ejects, for example, a neutron from the target nucleus in a (γ,n) reaction. Pions and kaons, like the photon, can also be absorbed. Because of their rest mass, absorption is usually accompanied by the emission of one or more nucleons.

Gamma rays and pions can be produced when a sufficiently energetic particle collides with a nucleus. Reactions like $(p,p\gamma)$ or (p,π^+) are examples of the production process. Production processes are used to produce secondary beams of unstable particles not found in nature, such as pions and neutrons.

When the incident projectile is an antiproton, $\bar{p}$, it will, upon colliding with a proton, annihilate, producing a number n of pions for the most part, $p + \bar{p} \rightarrow n\pi$. This process, referred to as *annihilation*, occurs whenever antiparticles, in this case the $\bar{p}$, are involved.

Finally, when the energy deposited in a nucleus is sufficiently large, the nucleus may break apart into several large fragments in a process called *fragmentation*. When two nuclei collide, they may combine to form a large nucleus with the emission of a few light particles if the nuclei are relatively massive. This process is called *fusion*, while its inverse is referred to as *fission*, in which a heavy nucleus splits into two or more fragments.

The consequence of a nuclear collision is expressed in terms of a cross section, which measures the probability that an event will occur. It is equal to that area which when placed orthogonal to the incident beam will intersect and remove from the beam a number of particles equal to that observed at a given angle, in a given energy range, of a given type, and so on. An incident beam of unit flux is assumed.

The total cross section σ_T equals the area that would remove a number of projectile particles involved in the reaction regardless of the momenta, energy, mass, etc. of the reaction products. It is related to the attrition of the beam traveling through a medium with a density ρ of scatterers. If the initial intensity of the beam is I_0, its intensity $I(x)$ after traveling a distance x is $I_0 \exp(-\rho\sigma_T x)$. The *differential cross section*, $d\sigma/d\Omega$, measures the probability of scattering into a solid angle $d\Omega$ located at an angle ϑ. The *double differential cross section* $d^2\sigma/d\Omega dE$ measures the probability of a reaction producing particles whose energies lie between E and $E + dE$ in a solid angle $d\Omega$ at ϑ. The energy dependence of this cross section at an angle ϑ (or integrated over ϑ) is the *energy spectrum*.

Theoretically, these cross sections are related to the *transition matrix* $\mathcal{T}_{fi}$ and the scattering amplitude f_{fi}, where the subscripts f and i refer to the final and initial states of the system, respectively. They are related as follows:

$$f_{fi} = -\frac{1}{2\pi}\frac{\mu}{\hbar^2}T_{fi} \tag{3}$$

and

$$\frac{d\sigma}{d\Omega} = \frac{1}{(2\pi)^2}\frac{\mu^2 k_f}{\hbar^4 k_i}|\mathcal{T}_{fi}|^2 = \frac{k_f}{k_i}|f_{fi}|^2. \tag{4}$$

In these equations, μ is the reduced mass:

$$\mu = \frac{m_a m_X}{m_a + m_X}, \tag{5}$$

while the initial projectile momentum in the center of the mass system is $\hbar k_i$ and the final observed particle momentum is $\hbar k_f$. The quantity f has the dimension of a length, while $\mathcal{T}$ has the dimensions of an energy times volume.

1. NUCLEAR REACTION MODELS

The theoretical analysis of nuclear reactions to be discussed is based on two general formalisms. These yield general formulas containing a number of parameters that can be adjusted to fit the experimental data. On the other hand, one can ask microscopic theory based on explicit models of the nucleus and nuclear force to predict the parameters so determined. The theory of nuclear reactions thus provides a framework for both the analysis and prediction of nuclear reactions.

1.1 Multiple Scattering

In the limit of high projectile energy, it is possible to assume that the projectile undergoes successive collisions with the nucleons that make up the target nucleus. Each of these collisions is considered to be a two-body collision, so that one can relate the many-body projectile–nucleus collision to the projectile–nucleon collision. The latter can be studied by using a nucleon as the target. Multiple scattering is valid when $kr_0 \gg 1$, where $2r_0$ is the distance between the nucleons in the nucleus and k is the projectile momentum/$\hbar$. Moreover, the nucleus must be sufficiently dilute that the collisions are two-body (rather than many-body, e.g., three-body collisions). This requirement is met if $r_0 \gg a$, where a is the correlation length, the distance over which two target nucleons are correlated. The *frozen approximation* is often used. Here, it is assumed that while the projectile passes through the nucleus, the nucleons in the nucleus are stationary. This approximation is valid if $kr_0\,(M/M_p) \gg 1$, where M is the nucleon mass and M_p is the projectile mass. When the frozen approximation holds, the transition matrix is a function of the coordinates and spin orientation of each of the target nucleons that prevail at the time the projectile passes through the nucleus:

$$\hat{\mathcal{T}} = \mathcal{T}(\mathbf{r}_1,\mathbf{r}_2,\ldots\mathbf{r}_A).$$

To compare with elastic scattering data, we average $\mathcal{T}$ by multiplying by the probability that the nucleons are at $\mathbf{r}_1$, $\mathbf{r}_2$, etc.

Using these approximations, one finds the elastic scattering transition amplitude describing the scattering of the projectile from an initial momentum of $\hbar\mathbf{k}_i$ to a final one of $\hbar\mathbf{k}_f$ to be

$$\mathcal{T}_{el}(\mathbf{k}_f,\mathbf{k}_i) = A\bar{t}(\mathbf{k}_f,\mathbf{k}_i)\,\bar{\rho}(\mathbf{k}_i - \mathbf{k}_f) \tag{6}$$

(Rayleigh–Lax potential), where A is the number of nucleons in the target nucleus, $\bar{t}(\mathbf{k}_f,\mathbf{k}_i)$ is the two-body projectile–nucleon scattering transition matrix, and $\bar{\rho}$ is the transform of the nuclear particle density $\rho(\mathbf{r})$;

$$\bar{\rho}(\mathbf{q}) = \int \exp\,(i\mathbf{q}\cdot\mathbf{r})\rho(\mathbf{r})\,d\mathbf{r}. \tag{7}$$

Generally, $\bar{\rho}(\mathbf{k}_i - \mathbf{k}_f)$ decreases rapidly with increasing angle ϑ between $\mathbf{k}_i$ and $\mathbf{k}_f$; the parameter measuring the rate of decrease is $kR \sin \frac{1}{2}\vartheta$, where R is the nuclear radius. The quantity $\rho(\mathbf{q})$ is referred to as the *form factor*.

On the other hand, $\bar{t}$ varies slowly with increasing angle. It is thus a good approximation to replace $\bar{t}(\mathbf{k}_f,\mathbf{k}_i)$ by its value at $\vartheta = 0$, which we symbolize by $t_E(0)$, where E is the projectile energy.

A similar treatment for inelastic scattering replaces the $\bar{\rho}(\mathbf{k}_i - \mathbf{k}_f)$ in Eq. (6) by $\bar{\rho}_{fi}(\mathbf{k}_i - \mathbf{k}_f)$, where $\bar{\rho}_{fi}$ is now the Fourier transform of the *transition density* or *transition form factor*:

$$\bar{\rho}_{fi}(\mathbf{q}) = \int \exp\,(i\mathbf{q}\cdot\mathbf{r})\rho_{fi}(\mathbf{r})\,d\mathbf{r}. \tag{8}$$

Here, ρ_{fi} measures the overlap of the initial and final wave functions. When there is little overlap, $\rho_{fi}(\mathbf{r}) \rightarrow 0$.

In a more accurate treatment, one finds that the scattering can be predicted using a potential that is complex in the Schrödinger equation (see QUANTUM MECHANICS):

$$\nabla^2\psi + \frac{2\mu}{\hbar^2}(E - V_{\text{opt}})\,\psi = 0. \tag{9}$$

This potential, V_{opt}, referred to as the *optical potential*, is approximately given by $V_{\text{opt}}(\mathbf{r}) = At_E(0)\rho(\mathbf{r})$. The quantity μ is defined in Eq. (5).

It can be seen from these equations that, once the elementary projectile–nucleon amplitude is known, one can use the projectile–nucleus cross section to determine the nuclear density $\rho(\mathbf{r})$. Indeed, by choosing appropriate projectiles, one can probe different aspects of $\rho(\mathbf{r})$. An example showing a comparison of experiment with the predictions is shown in Fig. 1. The uncertainties in the deduced neutron densities of Ni isotopes are shown in Fig. 2.

Improvements on the results given above include the effect of the motion of the nucleons in the nucleus as well as the effect of the correlations. For a discussion of these points, see the summary presented in Fesh-

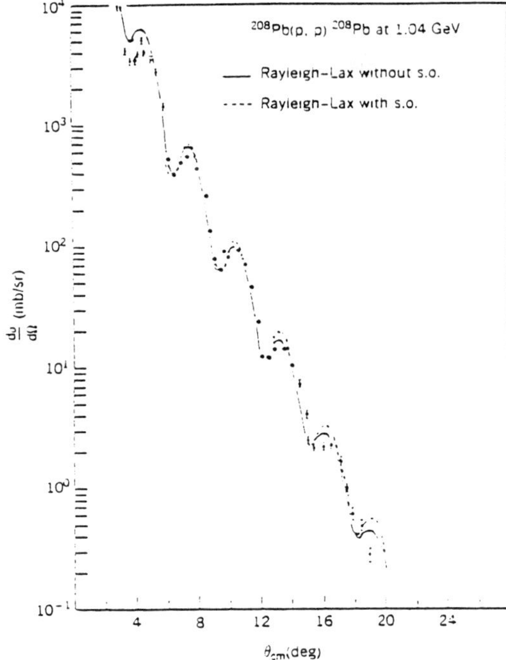

FIG. 1. Comparison of experimental angular distribution for the elastic scattering of 1.04-GeV protons by ^{206}Pb with the predictions employing the Rayleigh–Lax potential with and without spin-orbit (s.o) terms. The density-dependent Hartree–Fock densities are used. The Rayleigh–Lax potential is given by Eq. (6). (From Feshbach, 1992.)

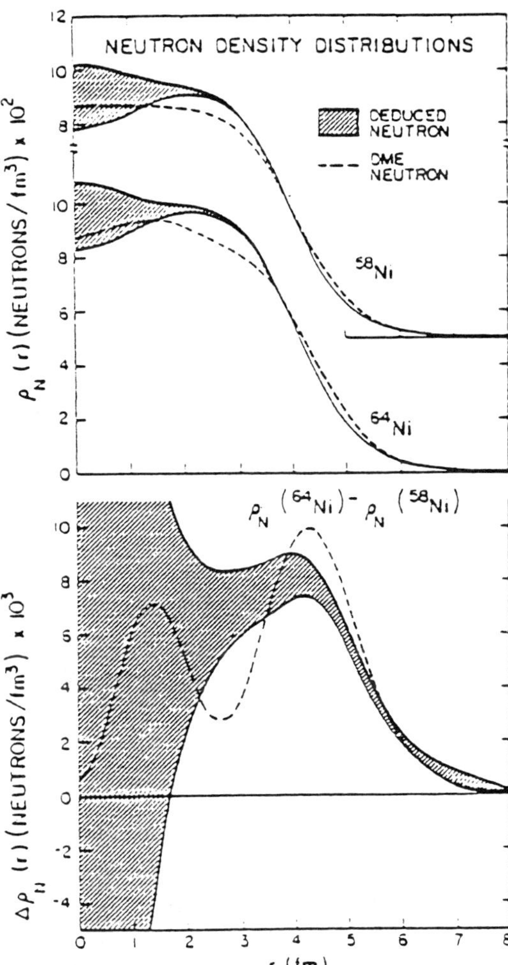

FIG. 2. Point-neutron density distributions for ^{58,64}Ni deduced from second-order Kerman *et al.* (1959) analysis (shaded bands) and predicted by the density-matrix expansion (DME) approach to Hartree–Fock theory (dashed curves). The difference between the ^{64}Ni-and ^{58}Ni-deduced neutron densities is compared with the DME prediction in the lower half of the figure. (From Ray, 1979.)

bach (1992) and the original paper of Kerman *et al.* (1959).

1.2 Interaction Time and General Properties of Nuclear Reactions

Another method for analyzing a reaction focuses on the interaction time, defined to be the time during which the reaction takes place. Very long interaction times permit the formation of very complex states from the initial simple one. These states are referred to as *compound nuclear states*, and the reactions of this type are called *compound nuclear reactions*. When the interaction time is small, that is, of the order of the time required for the incident projectile to traverse the nucleus, the states developed during that time cannot differ markedly from the initial state. These are called *direct reactions*. Between these two limiting situations, that is, for intermediate interaction times, the reactions are called *multistep reactions*.

The structure in the energy dependence of the observed cross section is related to the interaction time. If the width in energy of a structure in the cross section is Γ and the interaction time is τ, then $\Gamma = \hbar/\tau$. In the case of compound nuclear reactions, for which τ is very large, very narrow structures in the cross section become possible, as illustrated by Fig. 3. This is a *compound nuclear resonance*. These were discovered in the pre–World War II period using as projectiles slow neutrons, which do not have to overcome the

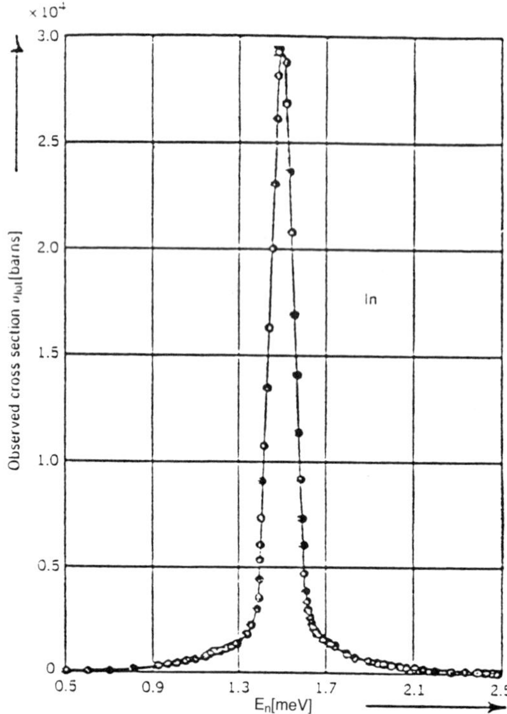

FIG. 3. Neutron total cross section for In (Landon and Sailor, 1955).

Coulomb electrostatic repulsion that charged projectiles such as a proton feel. In the post–World War II era, when strong sources of neutrons from reactors became available, such resonances were measured for all but the lightest nuclei. As the energy of the projectile increases, the resonances begin to overlap and the experimental energy resolution no longer permits the observation of individual resonances. The cross section will now be a *strongly fluctuating function* of the energy. Since the interaction time is long, memory of the incident direction is lost, the mechanism being the many collisions the projectile will have during its interaction with the nucleus. As a consequence, the angular distribution of the reaction products will be *isotropic*.

Contrast these results with those that are valid for the direct reactions. For these, more properly called the *prompt* reactions, τ is relatively short so that Γ is large. We may therefore expect the cross section for a prompt process to vary slowly with energy. This is illustrated by Fig. 4, where we observe very little variation of the cross section as the projectile energy changes by several MeV. More-

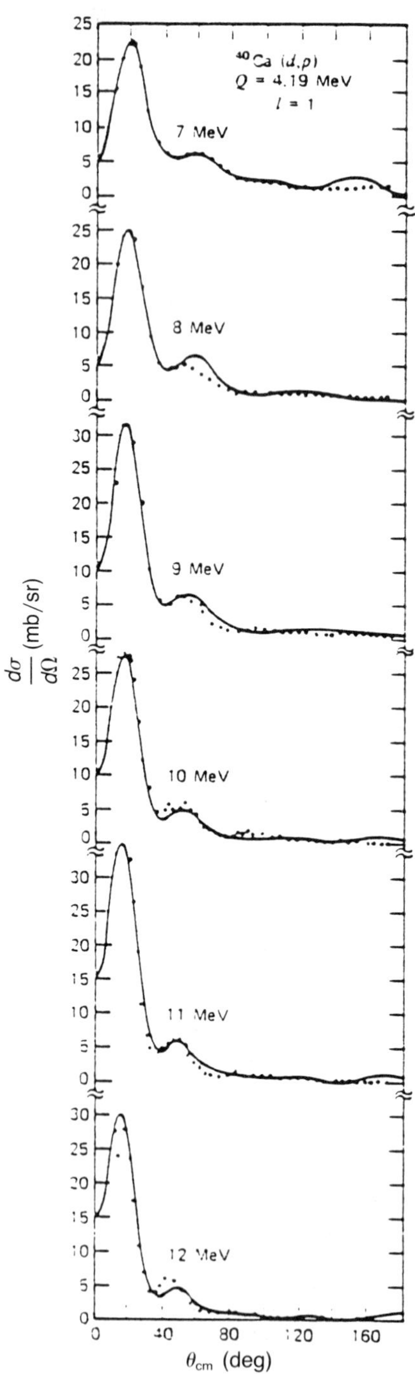

FIG. 4. The reaction $^{40}Ca(d,p)$ for bombarding energies in the range 4.19 to 12 MeV exciting a level in ^{41}Ca; the solid lines are theoretical (from Lee *et al.*, 1964).

over, memory of the incident direction is retained, since the angular distribution is sharply peaked near the forward direction.

Between the very short and very long interaction times, the multistep processes prevail. There are two kind, called the *multistep compound* and the *multistep direct*. The direct process involves just a single interaction. The multistep direct will then involve two or more interactions. The cross sections will still be peaked forward, but the peak will not be as sharp or as strong as it is for the direct process. The multistep compound reaction occurs when the reaction terminates before the compound-nucleus regime has been reached. Characteristically, the angular distribution after averaging over the fluctuations is symmetric about 90°.

This picture of a reaction is illustrated in Fig. 5. A (p,n) reaction is chosen as an example, but the shape of the spectrum is appropriate for any two-body final state. At the high–neutron-energy end, the process is direct, and low-lying levels of the residual nucleus are excited. As the energy of the neutrons produced in the reaction decreases, the reaction process passes from the direct to the multistep direct, to the multistep compound. Finally, the lowest-energy neutrons are emitted after forming a compound nuclear system. This correlation of regions in the energy spectrum with reaction types is qualitative. Indeed, at any energy, all reaction types contribute but the dominant ones are as indicated in the figure.

To isolate a given reaction mode, one theoretically makes use of the concept of *energy averaging*. Experimentally, one always averages, since the incident beam is never exactly monoenergetic. The energy average of the transition amplitude or cross section is taken over a domain ΔE. If ΔE is larger than Γ, a structure whose energy width is Γ will not be visible. If, on the other hand, ΔE is considerably smaller than Γ, then the structure is observable. One may therefore obtain information regarding a given reaction regime by averaging using the appropriate scale for ΔE.

Information regarding the reaction regime for interactions times τ greater than $\hbar/\Delta E$ is lost upon energy averaging. This is mirrored by the fact that reaction products that require a time greater than $\hbar/\Delta E$ are not included in the averaged transition amplitude.

Neutron elastic scattering was the first reaction in which energy averaging was introduced (Feshbach *et al.*, 1954). In this case, $\Delta E \sim 1$ MeV. It was found that total cross section as a function of energy, as well as the angular distributions for a variety of target nuclei, could be understood if the neutron were thought to be moving in a complex potential. The wave function satisfies Eq. (9). The imaginary part of the potential reflects absorption, which is a consequence of nonelastic processes and of energy averaging. This model is called the *optical model*. The use of a complex potential yields an equation similar to that satisfied by a light wave propagating in a medium with a complex index of refraction.

Long interaction times translate into complex, multicomponent wave functions. In qualitative terms, interaction time maps into complexity. An incident projectile plus a target nucleus forms the simplest, least complex system. Upon a single interaction, the pro-

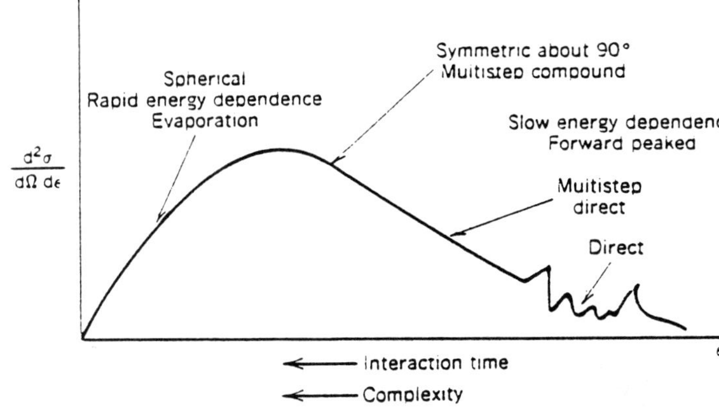

FIG. 5. Neutron energy (ϵ) spectrum at an angle $\theta = \theta_0$ (From Feshbach, 1992.)

jectile may lose energy, which is used to excite a nucleon to a new state forming thereby a particle-hole excitation. With successive interactions, projectile plus several particle-hole excitations are developed, the wave functions becoming more complex and the time during which interactions occur increasing correspondingly.

The *projection operator* method (Feshbach, 1992) is able to reflect directly the above analysis of reactions based on interaction times. Using projection operators, one resolves the wave function into those components that are of direct importance for the reaction and the rest. The latter may then be energy averaged. The interaction times thus selected will depend upon the complexity of the averaged components or equivalently on the complexity of those not averaged. If we eliminate all but the incident system, we are dealing with prompt reactions. For the case of compound nuclear systems, long interaction times, energy averaging is not performed. In these cases, the incident projectile interacts with the target, forming a nearly stable system with a long lifetime and with well-defined quantum numbers, such as angular momentum, energy, and parity. This can occur only at particular energies, the *resonance energies*. The resonance will have a half-width, the *resonance width*. For such *time-delayed* processes, the transition amplitude is, for an isolated resonance at E_s,

$$\mathcal{T}_{fi} = \mathcal{T}_{fi}^{(p)} + \frac{1}{2\pi} e^{i(\delta_i+\delta_f)} \frac{g_{fs}g_{si}}{E - E_s + \frac{1}{2}i\,\Gamma_s}, \tag{10}$$

where g_{si} are real and positive and are related to the partial widths Γ_{si} by $\Gamma_{si} = g_{si}^2$. The quantity $\mathcal{T}_{fi}^{(p)}$ is the transition matrix for the prompt component of the reaction amplitude. Γ_s, the total width, is equal to $\Sigma_k \Gamma_{sk}$. The partial width Γ_{sk} measures the probability that the resonant state with energy E_s will decay into a particular state designated by k. The quantities δ are *phase shifts* induced by the reaction in the absence of the resonance. The subscripts i and f refer to the initial and final states, respectively [see Eq. (19)]. For elastic scattering, the total cross section for a given value of the total angular momentum of the system, J, is

$$\sigma_J = \frac{4\pi}{k^2} \frac{2J+1}{(2I+1)(2i+1)} \times \left| e^{i\delta} \sin\delta - e^{2i\delta} \frac{\frac{1}{2}\Gamma_{\lambda\alpha}}{E - E_\lambda + \frac{1}{2}i\Gamma_\lambda} \right|^2. \tag{11}$$

For this case, $\delta_i = \delta_f \equiv \delta$. The quantity $\Gamma_{\lambda\alpha}$ measures the probability that elastic scattering will occur. If the scattering is not elastic, $\Gamma_{\lambda\alpha} < \Gamma_\lambda$. The quantities i and I are the spin of the projectile and the spin of the target nucleus, respectively. An example of the energy dependence is illustrated by Fig. 6. The minimum is the result of the interference between the prompt amplitude ($\sim e^{i\delta} \sin \delta$), sometimes called the *potential scattering*, and the resonant second term. Note also that

$$\sigma_J \le \frac{4\pi}{k^2} \frac{2J+1}{(2I+1)(2i+1)}. \tag{12}$$

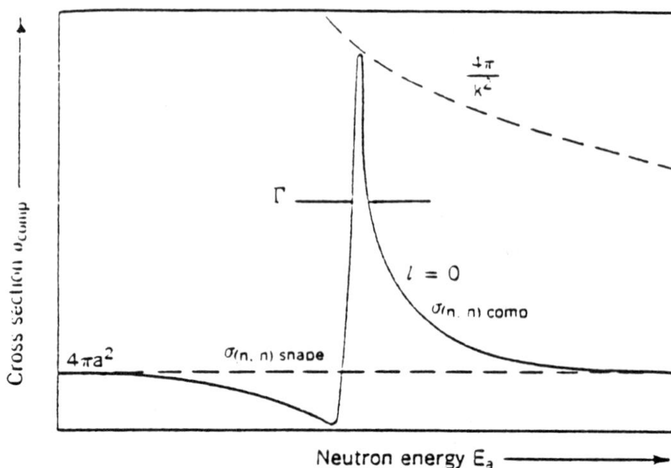

FIG. 6. Elastic-scattering cross sections for *S*-wave neutrons near a resonance in the compound nucleus. The figure shows the relationship between the shape-elastic and the compound-elastic cross sections for a spinless target nucleus. (From Marmier and Sheldon, 1970.)

When the resonances overlap, the *reactance matrix* description of the resonance term is used. This is to be added onto the prompt amplitude. The resonance term $\mathcal{T}_J^{(R)}$ for the case of a given J is

$$\mathcal{T}_J^{(R)} = \kappa/(1 + i\kappa), \tag{13}$$

where

$$\kappa = \frac{1}{2\pi} \sum_s \frac{\gamma_s^2}{E - e_s}, \tag{14}$$

where γ_s^2 and e_s are parameters.

The prompt processes are obtained by energy averaging using a relatively large ΔE. These include the phenomena described by the optical model and the distorted wave approximation (DWA) [see Eq. (18)]. Averaging yields a Schrödinger equation, Eq. (9), with a complex potential. Its solution predicts the average elastic scattering amplitude. In addition, it can be used to predict the probability that the projectile will penetrate the target nucleus. This probability is given by the *transmission factor* T_l, where the subscript denotes the orbital angular momentum in units of $\hbar$ involved. Qualitatively, there are two *barriers* to penetration. One is the *Coulomb barrier*, which acts when the projectile is positively charged because then there is an electrostatic repulsion. Another is the *angular momentum barrier*. When a particle has a momentum $\hbar k$, its maximum angular momentum is approximately kR in $\hbar$ units, where R is the nuclear radius. Therefore, the penetration of particles with angular momenta greater than kR will be reduced. Thus, an angular momentum barrier exists for $l > kR$. At low energies, $kR \ll 1$,

$$T_l \sim (kR)^{2l+1}\, C_l^2(\eta), \tag{15}$$

$$\eta = \frac{zZ}{137\beta}, \qquad \beta = \frac{v}{c} = \sqrt{\frac{2E}{\mu c^2}}. \tag{16}$$

The first factor in Eq. (15) gives the effect of the angular momentum barrier, and the second gives that of the Coulomb barrier:

$$\begin{aligned} C_l^2 &= \frac{2^{2l}}{[(2l)!]^2}(l^2 + \eta^2)[(l-1)^2 + \eta^2] \\ &\quad \times [(l-2)^2 + \eta^2]\cdots(1 + \eta^2)C_0^2(\eta) \\ C_0^2 &= 2\pi\eta/(e^{2\pi\eta} - 1). \end{aligned} \tag{17}$$

In Eq. (16), z and Z are the positive charges of the projectile and target nucleus, respectively.

If inelastic and/or transfer reactions as well as elastic scattering are included in the group of components that are not averaged, one obtains the *generalized optical model*, which consists of a set of *coupled* Schrödinger-type equations of the following form:

$$\nabla^2\psi_\alpha + (E - V_{\alpha\alpha})\psi\alpha = \sum_{\beta\neq\alpha} V_{\alpha\beta}\psi_\beta. \tag{18}$$

A perturbative solution of these equations is called the DWA.

Examples have been given of reactions with short interaction times, the optical model and the coupled channels leading to the DWA, and of long interaction times, the compound nucleus. In the large intermediate region, one again finds structures, known as *doorway state resonances*. These have widths that are not as narrow as that of the compound nuclear resonance nor as broad as the structures in the optical-model cross sections. Examples of these are the *isobar analog resonances*, the multipole *giant resonances*, and the *Gamow–Teller resonances*. These will be discussed in the next section in more detail. For the present, it will suffice to use the $^{27}Al(p,\gamma)^{28}Si$ resonance as an example. Figure 7 presents the results obtained using good energy resolution, while Fig. 8 presents the results obtained with successively poorer resolution. For the averaging interval of 300 keV, one observes four distinct resonances that are broken up into a series of fine-structure resonances in a good-resolution experiment. For the averaging interval of 1520 keV, only a broad maximum in the cross section remains. Thus, the full hierarchy of compound nucleus, doorway-state resonances, and the optical-model limit can be seen in this one experiment.

We recall that the optical model is a consequence of energy averaging over all configurations except that of the projectile plus target nucleus. The doorway-state resonance is a consequence of averaging over all configurations except that of the incident projectile and target nucleus and that of the system after the first interaction. The latter consists of the projectile at a reduced energy plus the excited target nucleus. The resulting resonance formula is similar to the com-

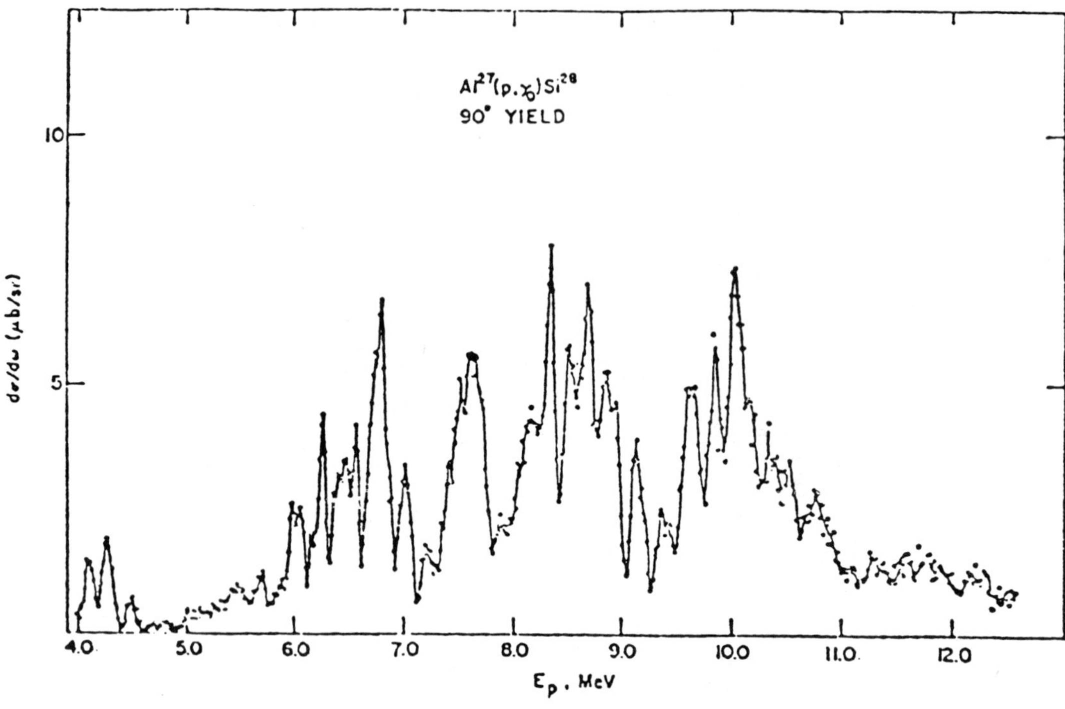

FIG. 7. Photocapture of protons by ^{27}Al to the ground state of ^{28}Si. Good energy resolution. (From Singh *et al.*, 1965.)

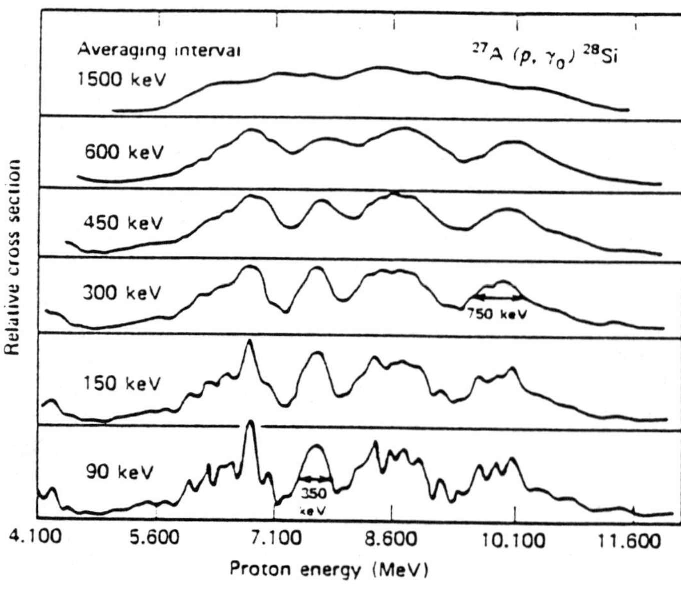

FIG. 8. Photocapture of protons by ^{27}Al to the ground state of ^{28}Si. The data are presented for various energy resolutions (taken from Singh *et al.*, 1965.)

pound-nuclear result, Eqs. (10) and (11). To obtain the expression replacing Eq. (11) for elastic scattering, replace Γ by $\Gamma_d^\uparrow + \Gamma_d^\downarrow$ and $\Gamma_{\lambda\alpha}$ by $\Gamma_d^\uparrow$. The width $\Gamma_d^\uparrow$ measures the probability of the transition from the doorway state to the final state. It is referred to as the *escape width*. The width $\Gamma_d^\downarrow$, the *spreading width*, measures the overall probability that the doorway state will make a transition to the more complex states that have been averaged. The spreading width thus acts like an absorption.

In the absence of a doorway-state resonance, reactions involving an intermediate reaction time can occur. These are often referred to as *pre-equilibrium* reactions. In the present discussion, we shall distinguish two types: the *multistep direct reaction* and the *multistep compound reaction*. Under certain statistical assumptions, the cross sections for the two processes combine additively. The multistep compound-nucleus reaction is important at relatively low energies. In the case of a nucleon projectile, the multistep direct process dominates above 15–20 MeV.

2. RESONANCES AND THE STATISTICAL THEORY OF REACTIONS

2.1 The Reaction Amplitude

Since energy, angular momentum, and parity are conserved, one can resolve the reaction amplitude (or $\mathcal{T}$ matrix) into a sum of terms, each of which has well-defined values of these quantum numbers. This sum is called a *partial-wave series*. In the simplest case, elastic scattering of a spinless projectile (e.g., an alpha particle) by a spinless target nucleus, the partial-wave series for the elastic scattering amplitude [see Eq. (6)] is

$$f(\vartheta) = \frac{1}{2ik} \sum_l (2l + 1) P_l(\cos \vartheta)(e^{2i\delta_l} - 1). \tag{19}$$

The angle ϑ is the angle made by the scattered particle with respect to the direction of the incident beam. The quantity δ_l is the phase shift, and P_l is the Legendre polynomial of order l defined by

$$P_l(x) = \frac{1}{2^l l!} \frac{d}{dx^l} (x^2 - 1)^l.$$

The radial dependence of the lth component of the wave function in the absence of any interaction approaches $\sin(kr - l\pi/2)$ in the limit of large r. In the presence of an interaction, it becomes $\sin(kr - l\pi/2 + \delta_l)$ in the same limit. The lth term in this series is a consequence of the scattering of a particle with angular momentum $l\hbar$.

When the projectile and target have spin, the partial-wave series is a series in J, the total angular momentum in units of $\hbar$. The possible values of J are determined by an angular-momentum coupling scheme. We shall make use of the *channel spin* coupling in which the spin of the target $\mathbf{I}$ and the spin of the projectile $\mathbf{i}$ are coupled to give the channel spin, $\mathbf{I} + \mathbf{i} = \mathbf{s}$. The vector notation implies the condition $|I - i| \leq s \leq I + i$. The channel spin is then coupled to the orbital angular momentum $l\hbar$ to obtain J, $\mathbf{s} + \mathbf{l} = \mathbf{J}$. Several values of s and l can combine to yield the same J. Moreover, the values of s and l for the incident wave can differ from those of the emerging wave, s' and l'. A *channel* is defined by the values of s, l, and J, as well as parity, energy, and the excitation energy of any of the complex systems in the initial or final states of the system. Let the subscript c be used to designate a particular channel. Then the reaction amplitude will be expressible as a partial-wave series: the individual terms are amplitudes for the transition from a channel c to a channel c'. The resulting general expression for the angular distribution can be expressed as a series in Legendre polynomials $P_L(\cos \vartheta)$ (see Feshbach, 1992). A general theorem states that the maximum value of L is the minimum value of the three quantities $2l_M$, $2l'_M$, and $2J_M$, where l_M, l'_M, and J_M are the maximum values of l, l', and J that appear in the partial-wave series. The values of l_M and l'_M are generally limited by the barrier penetrability, which decreases rapidly with increasing l. This is referred to as the *complexity theorem*.

For resonances, some simple results can be obtained. For the case when all the particles in the initial and final states have zero spin, for example, $^{12}O + {}^{16}O \rightarrow {}^{24}Mg + \alpha$,

$$\frac{d\sigma^{(R)}}{d\Omega} \sim [P_J(\cos\vartheta)]^2 \quad (J = l). \tag{20}$$

Finally, note that, in the DWA case, as we see from Fig. 4, the prompt cross section decreases rapidly with angle. A resonance with its characteristic angular distribution [see Eq. (20)] will be more easily observable at large angles.

2.2 Correlations and the Statistical Theory of Nuclear Reactions

At low excitation energies, the nuclear levels of the residual nucleus are well separated so that, with achievable energy resolution, it is possible to measure the cross section for the excitation of a particular nuclear level. However, as the excitation energy increases, the density of levels becomes so large that it is no longer possible to measure the excitation of individual levels. Instead of clean resonances, the cross section exhibits rapid fluctuations, the so-called *Ericson fluctuations*. Under these circumstances, one must resort to statistical measures, which must conform to experimentally achievable studies.

The first of these is the distribution of the cross section about its mean value. Under some simplifying assumptions, the probability density distribution $P(\sigma)$ is

$$P(\sigma) = \frac{1}{\sqrt{\langle\sigma\rangle}} \exp[-\sigma/\langle\sigma\rangle], \tag{21}$$

where $\langle\sigma\rangle$ is the mean value of σ.

Second, one can measure the energy correlations defined by

$$C(\epsilon,\vartheta) \equiv \frac{\langle\sigma(E+\epsilon)\,\sigma(E)\rangle - \langle\sigma\rangle^2}{\langle\sigma\rangle^2}, \tag{22}$$

where the brackets indicate averages. Then, the energy dependence of C is given by

$$C(\epsilon,\vartheta) = C(0,\vartheta)\frac{\Gamma^2}{\epsilon^2 + \Gamma^2}, \tag{23}$$

where Γ, the *coherence energy*, is the inverse of the interaction time. Correlations in angle are also expected from the complexity theorem [above Eq. (20)], which states that there is a maximum value of L in the angular distribution given by a sum over P_L. Correlations of the cross section between angles that differ by $1/L$ will then be expected. Angular correlations will also be induced by the energy correlations.

It is always possible to resolve the transition amplitude for a reaction into a fluctuating component and the average component. The latter is the prompt part. Thus, $\mathcal{T} = \mathcal{T}^{(P)} + \mathcal{T}^{(Fl)}$, where $\langle\mathcal{T}\rangle = \mathcal{T}^{(P)}$ and $\langle\mathcal{T}^{Fl}\rangle = 0$. Then the average cross section becomes $\sigma = \sigma^{(P)} + \sigma^{(Fl)}$, where $\sigma^{(Fl)} \sim \langle|\mathcal{T}^{Fl}|^2\rangle$. $\mathcal{T}^{Fl}$ consists of a sum of components that are assumed to be random. Thus, when the averages are carried out, there will be no interference between the individual terms because of the averaging. This has the immediate consequence that the angular distribution is symmetrical about 90°. This approximation is the *random-phase approximation*.

This is the result for the excitation of a given level in the residual nucleus. However, for sufficiently high excitations, it is more useful to consider the energy spectrum of the residual nucleus to be continuous with a level density of $\omega_f(U,I)$, where U is the excitation energy. Then,

$$\left\langle\frac{d\sigma^{(Fl)}}{dU}\right\rangle = \frac{\pi}{k^2}\sum\frac{(2J+1)}{(2i+1)(2I+1)}\,\omega_f(U,I_F) \times \frac{T_f T_i}{\sum_c\left(T_c + \int T_c\omega_c(U_c,I_f)\,dU_c\right)}, \tag{24}$$

where both a sum and integral are included in the denominator in order to account for both the discrete and the continuous energy spectrum. The quantity I_f is the spin of the residual nucleus. Equation (24) is referred to as the *Hauser–Feshbach* (1952) theory.

To conclude this discussion, we need $\omega_f(U,J)$. For the case of a noninteracting Fermi gas, one finds

$$\omega(U,J) = \frac{2J+1}{\sqrt{8\pi}\,\sigma^3}\,\omega(U)\exp\left[-\frac{(J+\frac{1}{2})^2}{2\sigma^2}\right], \tag{25}$$

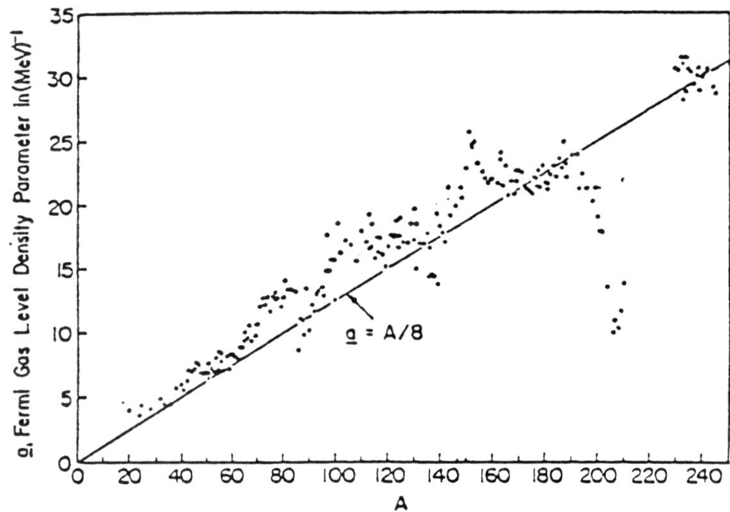

FIG. 9. Level-density parameter a vs A. The straight line corresponds to $a = A/8$ and dark points are experimental determinations. (From Lefort, 1976.)

where

$$\omega(U) = \frac{1}{\sqrt{48}\,(U - \Delta)} \exp[2\sqrt{a(U - \Delta)}] \quad (26)$$

and

$$\sum_{J} \frac{(2J + 1)}{\sqrt{8\pi}\,\sigma^3} \exp\left[-\frac{(J + \frac{1}{2})^2}{2\sigma^2}\right] \approx 1. \quad (27)$$

In these equations, a, σ, and Δ are semiempirical parameters. Figure 9 shows the experimental values of a, while Fig. 10 shows the values of σ at low excitation of the residual nucleus.

The function $\omega(U)$ can be expanded about the maximum excitation energy $U_M - \Delta$ of the residual nucleus. From energy conservation, $U = E_i + Q_{fi} - E_f$, where E_i and E_f are the energies of the emitted particles. Then, U_M equals $E_i + Q_{fi}$ and

$$\exp[2\sqrt{a(U - \Delta)}] \simeq \exp[2\sqrt{a(U_M - \Delta)}]\, e^{-E/T}, \quad (28)$$

where T, the *nuclear temperature*, in energy units is

$$T = \sqrt{\frac{U_M - \Delta}{a}}. \quad (29)$$

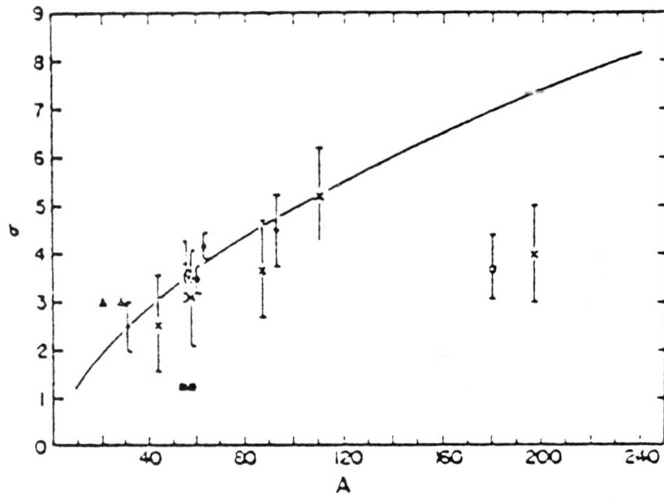

FIG. 10. Computation of values of σ. (From Huizenga, 1972.)

Statistical analysis can be used to derive the probability distribution $P(\Gamma)$ for widths. For a single channel, this is the *Porter–Thomas* distribution:

$$P(\Gamma)d\Gamma = \frac{1}{\sqrt{2\pi}}\left(\frac{\langle\Gamma\rangle}{\Gamma}\right)^{1/2} \exp\left[-\tfrac{1}{2}\frac{\Gamma}{\langle\Gamma\rangle}\right]\frac{d\Gamma}{\langle\Gamma\rangle}. \tag{30}$$

Here, (Γ) is the average of Γ. As the number of channels increases, the variance decreases so that, in the limit,

$$\langle f(\Gamma)\rangle \rightarrow f(\langle\Gamma\rangle), \tag{31}$$

where the angular brackets indicate that an average has been taken.

3. PROMPT REACTIONS

This section deals with those reactions with short interaction times. Characteristically, the variation of the cross section with energy is slow and the angular distribution is asymmetric (except at very low energies), thereby revealing the direction of the incident beam.

3.1 Elastic Scattering by Nuclei

Briefly, the angular distribution for prompt elastic scattering changes from isotropy at very low incident energy to a more complicated oscillatory angular dependence as the energy increases. At high energies, the angular distribution is peaked in the forward direction. The separation angle between peaks in the distribution is given approximately by $1/kR$, where R is the nuclear radius. The energy dependence of the elastic cross section at high energies is relatively slow, asymptotically tending toward πR^2, with an equal value for the absorption cross section so that the total cross section tends toward $2\pi R^2$.

The optical model (Feshbach *et al.*, 1953, 1954) was devised in order to provide a description of the prompt elastic scattering. Generally, a semiempirical approach is adopted. A form is chosen for the optical-model potential. The parameters in the form are chosen so as to achieve the best fit with experiment, that is, with the total cross section, with the angular distribution, and with polarization of the emergent particles when that is available. Regularity not only with energy but also with A, the mass number of the target nucleus, is required in conformity with the experimental results.

The optical model is obtained when the amplitude of the scattered particle is energy averaged. Experimentally, it is the cross section, σ_T, that is averaged. This difference does not affect the total cross section, sines σ_T is linearly related to the amplitude (*optical theorem*):

$$\sigma_T = (4\pi/k)\,\mathrm{Im}\, f(0°). \tag{32}$$

Thus, the experimental average indicated by a bar and the optical-model average indicated by angular brackets are equal, $\langle\sigma_T\rangle = \bar{\sigma}_T$. This is not the case for the angular distribution or other observables:

$$\langle|\mathcal{T}_{fi}|^2\rangle = |\mathcal{T}_{fi}^{\text{prompt}}|^2 + \langle|\mathcal{T}_{fi}^{(Fl)}|^2\rangle$$

so that to the optical-model result one must add the fluctuation cross section. For elastic scattering, the latter is referred to as *compound elastic scattering*. As the energy increases, it is expected that this term will become unimportant.

The prompt amplitude can be expressed in terms of a partial-wave series given by Eq. (19) for spinless-projectile, spinless-target collision. For a spin-$\frac{1}{2}$ projectile, spinless target nucleus (or vice versa), in addition to the differential cross section, other observables include the *polarization* and the *spin rotation*.

3.2 Low Energy

At low energies, only one term (the first one) in the partial-wave series is significant. Thus, low-energy scattering is characterized by two parameters, R and S_0. The elastic cross section is given by $4\pi R^2$; the reaction cross section, by $4\pi b/k$, where b is a constant. This is the $1/v$ law, since $k = mv/\hbar$. The constant b is related to the strength function S_0. The average width and the spacing can be measured by a good-resolution experiment in which each resonance is measured separately. Comparison between the optical model and experiment is made in terms of the *strength function*, S_0, defined as

$$S_0 = \frac{\langle\Gamma\rangle}{D}\left(\frac{E_0}{E}\right)^{1/2}, \tag{33}$$

where E_0 is taken to be 1 eV, $\langle \Gamma \rangle$ is average width, and D is the average distance in energy between levels. Substantial deviations from the predictions of the optical model exist near $A \simeq 110$. The microscopic explanation of this failure of the optical model is that in this mass-number region, the values of $\langle \Gamma \rangle / D$ are anomalously low because of the low number of doorway states through which the system must pass in order to form the compound nucleus.

3.3 Nucleon Optical-Model Potential for Low and Intermediate Energy

When the energy of the incident neutron or proton is less than about 100 MeV, the optical-model potential is composed of a *central* plus a *spin-orbit* potential:

$$V_{\text{opt}} = V_c + f(J,l)\, V_{\text{so}}. \tag{34}$$

V_c and V_{so} are functions only of r, the radial distance between the nucleon and the target nucleus center of mass. The function $f(J,l)$ is present because the potential depends upon the orientation of the spin of the nucleon and its orbital angular momentum. For the state in which the total angular momentum J is $l + \frac{1}{2}$, where l is the value of the orbital angular momentum, $f(J,l) = l$. When J is $l - \frac{1}{2}$, $f(J,l) = -(l+1)$. The central potential, V_c, is complex and given by

$$V_c(r) = V_{\text{Coul}} - Vf(x_0) - i\left[Wf(x_W) - 4W_D \frac{d}{dx_D} f(x_D) \right], \tag{35}$$

while

$$V_{\text{so}}(r) = \left(\frac{\hbar}{m_\pi c}\right)^2 v_{\text{so}} \frac{1}{r}\frac{d}{dr} f(x_{\text{so}}). \tag{36}$$

The quantities V, W, W_D, and v_{so} are constants. V_{Coul} is the Coulomb interaction of the proton with a uniformly charged sphere of radius R:

$$V_{\text{Coul}} = \begin{cases} \dfrac{Ze^2}{r}, & r \geq R, \\[2ex] \dfrac{Ze^2}{2R}\left(3 - \dfrac{r^2}{R^2}\right), & r \leq R. \end{cases} \tag{37}$$

It is absent for neutrons. The constant $\hbar/m_\pi c$ is approximately 1.4×10^{-13} cm. The function $f(x_i)$ is usually taken to be in the *Woods–Saxon* form,

$$f(x_i) = \frac{1}{1 + e^{x_i}}, \qquad x_i = \frac{r - R_i}{a_i}, \qquad R_i = r_i A^{1/3}. \tag{38}$$

This form is constant within the nucleus, $r < R$, and decays exponentially for $r > R$. The parameter R_i is the value of r at which $f(x_i)$ is $\frac{1}{2}$ of its value at $r = 0$. The function $-4df/dx$ peaks in the surface region $r = R_i \pm 1.5a_i$. There are therefore 12 parameters in the optical potential V for nucleon projectiles. Observe that the spin-orbit term is confined to the surface, while the absorptive term in V_c, the imaginary part of V_c, has a volume term proportional to W and a surface term.

By a fine tuning of the parameters, one can, in most cases, obtain a nearly perfect fit to the elastic-scattering data. Such precise fits are often necessary for the calculation of reaction processes. An example of a global fit is given by Becchetti and Greenless (1969) for nucleon energies less than 50 MeV for protons and 24 MeV for neutrons (Fig. 11). In the region 100–200 MeV, optical-model parameters have been obtained for protons incident on nuclei. There are considerable changes from the Becchetti–Greenless values. The central absorption is now a purely volume effect corresponding to the ability of the proton to penetrate more easily into the target nuclear volume. The energy dependence of the parameters differs sharply, in part because of pion production not present in the lower energy range, while the spin-orbit potential is now complex.

For 400-MeV protons, the Woods–Saxon form Eq. (38) fails and must be replaced by $f \to f - \gamma f^2$, where γ is an additional parameter. As a consequence of this replacement, the central potential is no longer negative everywhere. At small distance, it is repulsive.

Optical-model fits have been obtained for other projectiles. For deuterons, see Feshbach (1992), p. 490.

3.4 Inelastic Scattering

When an incident particle is scattered by a target nucleus, it can transfer both energy and angular momentum to the target nu-

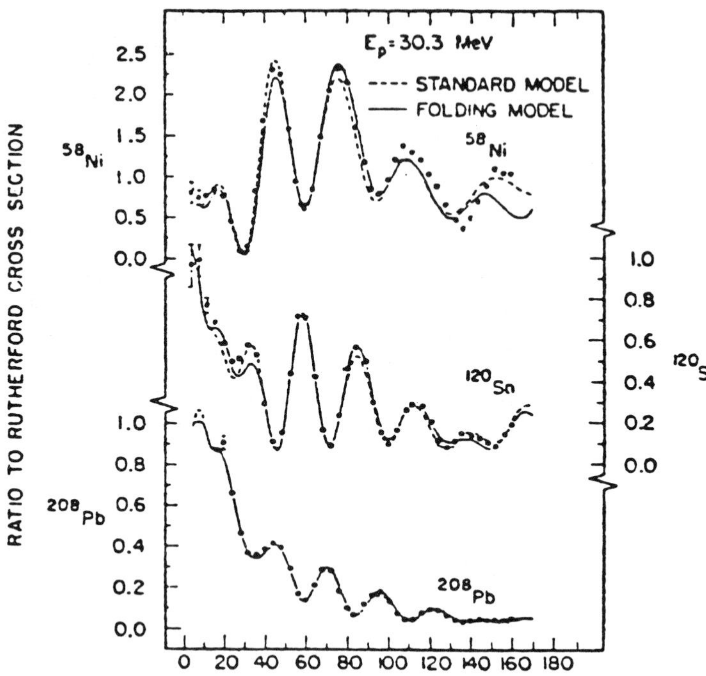

FIG. 11. Experimental differential cross-section data points, with errors, for the elastic scattering of 30.3-MeV protons, together with predictions. The two models illustrated are possible versions of the optical model. (From Greenlees, *et al.*, 1970.)

cleus. When the reaction is direct, that is, prompt, the energy variation of the cross section is slow, while the angular distribution peaks in the forward hemisphere. For reactions that occur mainly at the nucleon surface, the angular distribution will have a maximum at approximately

$$\theta = \hbar \Delta J / pR, \tag{39}$$

where ΔJ is the transferred angular momentum, R is the nuclear radius, and p is the magnitude of the incident momentum *at the nuclear surface*. Equation (39) assumes that the incident and final momenta are nearly equal. For angles less than θ, the cross section will decrease rather rapidly. For angles larger than θ, the cross section will oscillate, the amplitude decreasing as the angle increases. The *Blair phase rule* states that the oscillations in the angular distribution for the even orbital angular-momentum transfer will be 180° out of phase with those of the odd angular-momentum transfer. The elastic cross section will be in phase with the odd case. Moreover, approximately the transferred angular momentum is *perpendicular* to the *scattering plane* in which $\mathbf{k}_f$ and $\mathbf{k}_i$ lie.

The values of the initial and final orbital angular momenta, l_i and l_f, involved in a specific surface reaction extend over a small range. The maximum cross section is achieved for the *matching condition*

$$l_i/k_i = l_f/k_f = R. \tag{40}$$

Partial waves corresponding to values of l that do not satisfy Eq. (40) will have a reduced magnitude.

3.5 Transfer Reactions

Transfer reactions occur whenever mass is transferred from the projectile to the target nucleus or vice versa. The stripping reaction (d,p) is one in which the neutron originally in the deuteron is acquired by the target. The *pickup reaction* (p,d) is one in which the proton joins with a neutron originally in the target nucleus to form a deuteron. Proton transfer reactions include stripping, (He^3,d), and pickup, (d,He^3). Two-nucleon transfer occurs in the (t,p) and (p,t) reactions (t = triton, ^{3}H). When the projectile is a heavy ion, several nucleons can be transferred. The transfer of mass is accompanied by the transfer of energy, momentum, and angular momentum.

Transfer reactions are of major importance for the study of nuclear structure. The neutron in the (d,p) reaction was found to occupy a single-particle state in the residual nucleus, and from the analysis of the experiment, properties of that state could be determined.

This is possible because of the specificity of these reactions within a limited energy range above the Coulomb barrier energy. The reaction is a surface reaction since a weakly bound composite system, such as the deuteron, cannot penetrate deeply into the target nucleus, in contrast with the neutron and proton projectiles. The probability for a surface reaction is maximized for those partial waves of the projectile and the emergent particle whose amplitudes are a maximum at the nuclear surface. Thus, in the case of the (d,p) reaction,

$$k_{\mathrm{D}}R \sim l_i \quad \text{and} \quad \frac{A}{A+1} k_p R \sim l_f, \tag{41}$$

where $\hbar k_{\mathrm{D}}$ is the deuteron momentum and $\hbar k_p$ the proton momentum. A is the mass number of the target nucleus. The orbital angular momentum of the exchanged neutron L must lie between $l_f + l_i$ and $|l_f - l_i|$. Moreover, the momentum transfer is bounded by the momentum of the bound neutron, as given by the *momentum-matching* condition

$$\left| \frac{A}{A+1} k_p - k_D \right| < \sqrt{\frac{2m}{\hbar^2} |\epsilon|}, \tag{42}$$

where m is the neutron mass and $|\epsilon|$ its binding energy. Comparison with Eq. (41) yields the conclusion that L will usually be limited to relatively small values for the (d,p) reaction. In the case of heavy-ion reactions, for which large mass transfers involving several nucleons are possible, much larger values of L are possible.

The constraint on L is referred to as the *L window*. It has the consequences that the angular distribution is oscillatory. It is found that the angular distributions for a given L are nearly identical, permitting, as a consequence, the determination of the orbital angular momentum of the single-particle level occupied by the transferred nucleon. This is the most important insight into nuclear structure provided by these experiments. Beyond this, these experiments lead to information on the overlap of the initial wave function with the final wave function, which differs from the initial one by the addition or absence of one or more nucleons.

4. GIANT RESONANCES AND DOORWAY STATES

The first increase in complexity from the incident configuration of projectile plus target nucleus is called the *doorway state*. If the probability for coupling to states of even greater complexity is large, the system can continue to develop in complexity until the compound nucleus is formed. It is possible for the system to make a transition to the final state during this development. This is referred to as *precompound* emission, which will discussed in Sec. 5. If, however, the transition probability from the doorway state to a state of higher complexity is small, a *doorway-state resonance* becomes possible. The principal reasons for a small transition probability may lie in a symmetry, in the unusual geometric shape of the doorway state, or in the sparseness at the doorway-level energy of states with which it can couple.

The isobar analog state is an example of the effect of symmetry. It can be formed when, for example, the incident projectile is a proton. An example is shown in Fig. 12. The doorway state in this case has *isospin symmetry*, which is a consequence of *charge independence of nuclear forces*. This symmetry is broken by the electrostatic Coulomb interaction, which characteristically varies slowly over the interaction volume. As a consequence, the transition probability connecting this doorway state with states of differing isospin symmetry will be very small, and therefore a resonance becomes possible.

This resonance is not a compound nuclear resonance. As described earlier (see Sec. 1.2), a component of its width, the spreading width, originates in the possibility of transition to more complex states. Since there is a small but finite transition probability to such states, the system can form compound nuclear resonances. These are observable as fluctuations in experiments with good energy resolution, as seen in Fig. 13. When these are energy averaged, one obtains the doorway state resonance of Fig. 12.

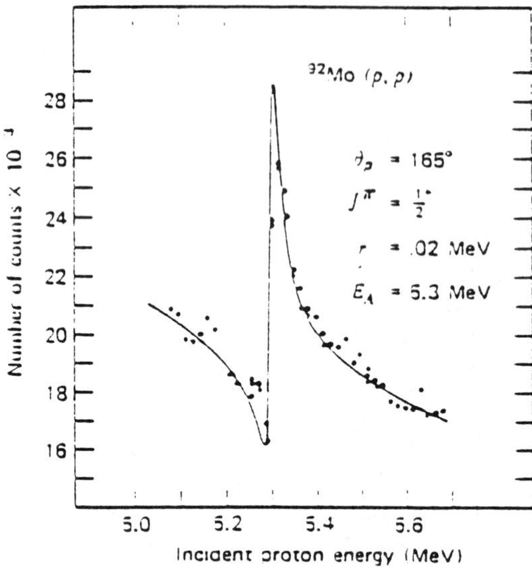

FIG. 12. Curve obtained by averaging the data of Fig. 13. (From Richard *et al.*, 1964.)

The isobar analog resonance might be called a Fermi resonance, because the interaction responsible for its formation is the same as the one that appears in Fermi β decay. There are also resonances that are a consequence of the Gamow–Teller interaction, which is used in the theory of Gamow–Teller β decay transitions; these are called *G–T resonances*. They are excited by charge-exchange reactions, such as (p,n), (He^3,t), etc. In the (p,n) reaction, they are seen at energies ranging above 150 MeV, where the spin- and charge-dependent interactions are an important component of the effective nucleon–nucleon interaction inside nuclei. The isobar analog resonances occur at relatively low energies, where the spin-independent component is significant.

The most thoroughly studied doorway states are the *giant resonances*, which are shape oscillations set into motion by the incident projectile. The earliest example is the *giant dipole resonance*, in which the neutrons oscillate out of phase with the proton oscillation creating thereby a dipole. This out-of-phase oscillation is called *isovector*. These oscillations are also characteristic of their multipole character. In addition to the dipole, the monopole or "breathing" mode, the quadrupole, the octupole, and even the hexadecapole modes have been observed. Finally, if spin is involved, magnetic multipole resonances must be included.

These resonances have been observed for nearly all nuclei. Their excitation energy var-

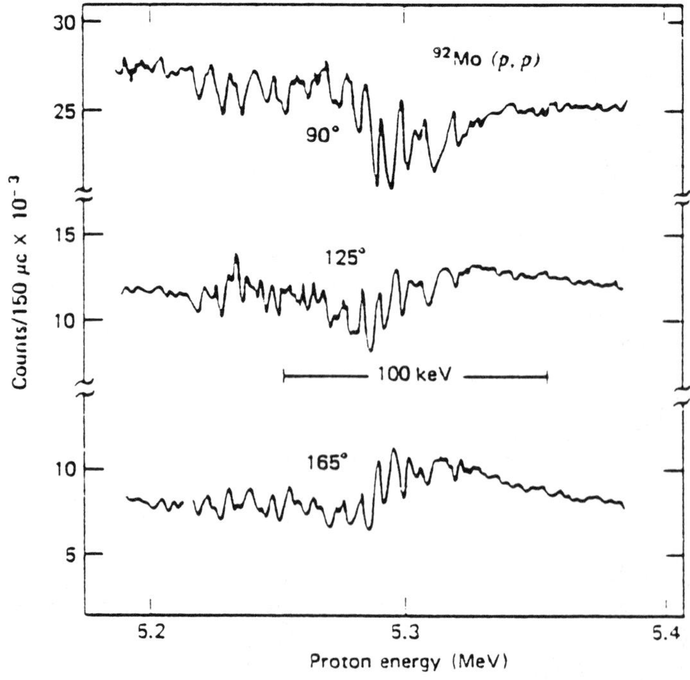

FIG. 13. Elastic scattering of protons by ^{92}Mo in the neighborhood of the *s*-wave isobar analog resonance at proton energy of 5.3 MeV. (From Richard *et al.*, 1964.)

ies smoothly with mass number A. One finds for the giant dipole $77A^{-1/3}$, and for the scalar quadrupole and monopole $63A^{-1/3}$ and $80A^{-1/3}$ MeV, respectively. Their widths are of the order of several MeV.

Giant resonances were first observed by measuring the absorption of photons. This was followed by experiments such as radiative capture reactions, e.g., (p,γ), and by the inelastic scattering of electrons and the inelastic scattering of the light ions, especially protons and alpha particles. Perhaps the most important addition to the experimental study is the use of heavy ions with energies of the order of several hundred MeV. The mechanism is the excitation of the resonances by the Coulomb field of the heavy ion, referred to as *Coulomb excitation*. Relatively recently, pions have been used to excite G–T resonances through a charge-exchange reaction in which the incident and emerging pion have different charges.

Shape resonances have been observed in fissioning nuclei. They are consequences of the fact that these nuclei can be bound for two very different shapes. For example, in one shape, the nucleus can be greatly deformed, and in another, nearly spherical. Thus, if the system assumes one of these shapes, it can only by tunneling make a transition to another shape. Thus, the spreading width will be correspondingly small.

Shape resonances have also been observed in the collision of heavy ions. Examples include $^{12}C + {}^{12}C$ and $^{12}C + {}^{16}O$. These resonances have been observed up to $^{28}Si + {}^{28}Si$. In these cases, the observed resonances are correlated with the greatly deformed states of, for example, ^{24}Mg for the $^{12}C + {}^{12}C$ case.

5. MULTISTEP REACTIONS

The reactions described in Secs. 2 and 3 present two extreme situations. Section 2, based on a statistical compound-nucleus model, is valid for relatively long interaction times. Section 3, for both inelastic and transfer reactions, assumes that the projectile interacts just once with the target nucleus, implying short interaction times. The statistical model works well for low-energy projectiles but fails, as the projectile energy increases, at back angles and for energetic emerging particles. The deviation from the statistical model can be as much as several orders of magnitude. The single-step, prompt-process prediction of the angular distribution is reliable for small angles but often fails at back angles. In both cases, the physics is inadequate. It thus becomes necessary to develop an understanding of nuclear reactions that in the limit of long and short interaction times yields the statistical and prompt descriptions, respectively, but, at the same time, permits the evaluation of contributions from intermediate interaction times.

When the reaction involves the excitation of low-lying levels, the single-step description is replaced by a set of coupled Schrödinger equations [see Eq. (18)], which take into account the significant coupling of the initial and final states to various other states and the coupling between these states. These equations must then be solved numerically. A particularly important case occurs when the nuclei are deformed, because then there are low-lying levels that are closely related. Taking these couplings into account alters the relationship among the various exit channels. General remarks about the results are not possible, because they depend critically upon idiosyncratic details as interference between the one-step and, for example, the two-step reactions becomes important. This method becomes rapidly unpractical as the number of states that are coupled increases. One then turns to a statistical theory, which presumes that the wave function of the system is made of very many components that can be assumed to be random. Averaging becomes a necessity.

The averaging distinguishes between two reaction types that can contribute. These are illustrated in Fig. 14. In the upper chain, the system progresses through a set of increasingly complex states characterized by the fact that at least one particle (or a cluster of particles) has a positive energy—that is, is not bound. At each step, the system can make a transition to the desired final state. This is referred to as the *statistical multistep direct reaction*; the word "statistical" implies that suitable averages have been taken. As is the case for single-step reactions, the cross section peaks in the forward hemisphere, but as the number of steps that need to be included increases, the peak broadens and the cross section at back angles increases. The nth step

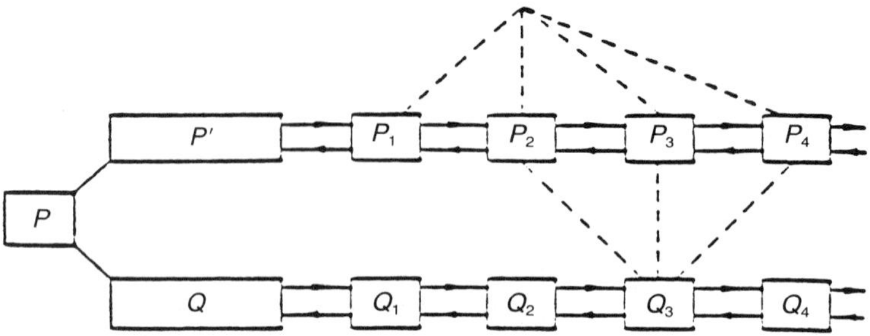

FIG. 14. Partition of Hilbert space into subspace of increasing complexity indicated by the subscript. (From Feshbach, 1992.)

will add a contribution whose peak is at a larger angle than that of the $(n - 1)$st step.

In the lower chain, the system passes through a set of increasingly complex states, but now all particles are bound. Eventually, as the number of steps increases, the compound nucleus is formed. Emission into the final state can, however, occur in the stages that precede the formation of the compound nucleus. For this reason, this contribution is referred to as precompound. The term *statistical multistep compound* refers to the entire process. It is clear that the precompound component contributes the higher-energy emerging particles needed for agreement with experiment. Importantly, the angular distribution remains symmetric about 90°.

The total cross section is the sum of the cross sections given by the two multistep processes and the single-step contribution. At the lowest energy of the emerging particles, the compound-nucleus cross section suffices. As the energy of the incident projectile increases, one must increasingly include the contribution from the precompound reaction. At some projectile energy at about 15–25 MeV, one must take the statistical multistep direct component into account. Above 25 MeV, the direct process dominates, the number of steps increasing with energy. At above 200 MeV, five steps are needed. However, at some stage the required number of steps will start to decrease, and eventually, at a high energy (~1 GeV), a single-step process will again suffice.

6. HEAVY IONS

Heavy-ion reactions are concerned with the collision of a nucleus, referred to as the heavy ion, with another nucleus. Energies of the heavy ions that have been produced extend from low energies of a few MeV per nucleon to 200 GeV per nucleon. These strike a stationary target. A collider in which two beams, each with an energy of 100 GeV per nucleon, move in opposite directions is now under construction at the Brookhaven National Laboratory (Upton, NY, U.S.A.). The high-energy beams are obtained using a number of accelerators working in tandem. As the beam passes from one accelerator to another, some of the atomic electrons are stripped from the heavy ions by passing the beam through a stripping foil. This process increases in efficiency with increasing energy. Before entering the last accelerator, most of the atomic electrons have been removed so that these ions have a net charge that is large and the electromagnetic fields used to accelerate the ion become very effective.

The rich set of phenomena produced in heavy-ion collisions has three fundamental sources: the strong electric field of the heavy ion, its large mass, and its compositeness.

Because of the strong and long-range electric field, the heavy ion can excite the target nucleus electromagnetically. This process is referred to as *Coulomb excitation* and has been most important for the determination of the energy spectrum of deformed nuclei. Excitation to high spin values becomes possible. Referring to Sec. 4, we recall that it is possible to Coulomb excite the giant resonances. Finally, the Coulomb field can disintegrate the target nucleus, an effect that grows approximately like Z^2, where Z is the atomic number of the projectile.

Because the mass and the radius of both the target and projectile are relatively large, the relative angular momentum is correspondingly large. For the same reason, the wavelength of the projectile is very small. This means that semiclassical methods may be used to compute the path of the projectile. If

the only force acting is the Coulomb force, the trajectory of the projectile relative to the target is a hyperbola. The distance of closest approach is given by

$$d = \frac{\eta}{k}\left(1 + \cos\frac{\vartheta}{2}\right), \tag{43}$$

where ϑ is the angle of scattering, k is the momentum in units of $\hbar$, and

$$\eta = Z_1 Z_2 e^2/\hbar v, \tag{44}$$

with Z_1 and Z_2, the atomic numbers of the colliding nuclei. The *grazing angle* is obtained from Eq. (43) by placing d equal to R, where R is the sum of the radii of the colliding. The cross section for Coulomb scattering is the *Rutherford cross section*:

$$\frac{d\sigma}{d\Omega} = \left(\frac{Z_1 Z_2}{2\mu v^2}\right)^2 \frac{1}{\sin^4 \vartheta/2}. \tag{45}$$

A wide variety of reactions becomes possible because of the complex structure of the nuclei. Energy, momentum, and relatively large amounts of mass can be exchanged. Much of the incident kinetic energy can be converted into internal nuclear energy. Much of the initial angular momentum can be converted into rotational angular momentum. Thus, new nuclear species, highly excited and with large values of the spin, can be formed in a heavy-ion collision.

The reaction type is governed by the ratio d/R. If the distance of closest approach d is very much larger than the sum of the two nuclear radii, $d/R \gg 1$, then elastic scattering and Coulomb excitation dominate. If $d/R \gtrsim 1$, one must add inelastic scattering as well as the exchange of nucleons and nuclear clusters. These reactions are referred to as *quasielastic scattering*. They can be understood by use of the DWA approximation discussed in Sec. 4.

At the other extreme, $d/R \ll 1$, the collision is approximately head on. There is a finite probability that the two nuclei will amalgamate, forming a compound system. The process is called *fusion*. The formation of the component system may be preceded by the emission of a cluster, such as the α particle. This facilitates the formation of the compound system, since it reduces its total angular momentum. There is a limiting angular momentum above which the compound system is not stable. An estimate of $100\hbar$ for a compound nucleus with $A \sim 130$ has been obtained.

After fusion, the resulting nucleus will decay. For light nuclei, α-particle and nucleon emission compete. For high angular-momentum states, α emission is favored. Gamma-ray emission becomes important near and below the threshold for particle emission. For medium-weight and heavy nuclei, fission will compete with neutron emission; charged-particle emission will become increasingly less important as the nuclear atomic number increases.

For intermediate values of $d/R \sim 1$, a new phenomenon appears called *deep inelastic scattering*. In this collision, nearly all of the initial kinetic energy is converted into internal energy, so that the observed kinetic energy of the final nuclei equals the Coulomb energy of the nuclei when touching. The final state of the reaction is primarily two-body, with the mass and charge of each final nucleus close to the initial mass and charge of the colliding nuclei. Two types of angular distribution are observed. In one, *strong focusing*, which prevails in collisions of very heavy nuclei and lower energies, the reaction products fall within a relatively narrow angular range peaked at the grazing angle. In the second, the *orbiting* distribution occurs for lighter systems and higher energy. In that case, with increasing energy loss, the angle at which the cross section is maximum decreases from the grazing angle, approaching zero, and then increases slowly for further increases in energy loss. An example is shown in Fig. 15.

The total reaction cross sections are of the order of several barns. At relatively low energies and for the lighter nuclei, the fusion cross section is roughly 0.4 of the total, the quasielastic about the same, and the deep inelastic the remainder. For a given laboratory energy, the ratio of the fusion cross section to the total reaction cross section decreases rapidly with increasing Z_1Z_2, where Z_1 and Z_2 are the atomic numbers of the colliding nuclei.

The simplest theory of heavy-ion collisions assumes that each of the colliding nuclei is a gas of nucleons with a momentum distribution given by Fermi–Dirac statistics.

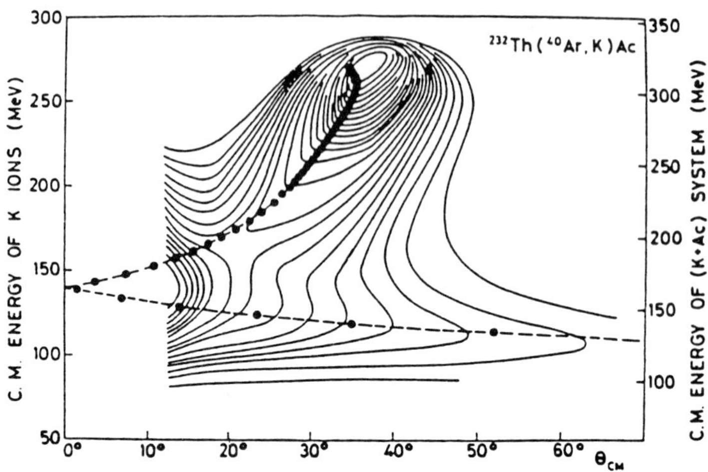

FIG. 15. Contour diagram of $d^2\sigma/dE\,d\theta$ for the reaction ^{232}Th(^{40}Ar,K) at E_{lab} = 388 MeV. The circles indicate the predicted correlation between scattering angle and final energy for different values of the angular momentum ranging from l = 180 to 250. (From Lefort and Ngô, 1978.)

The collision can be followed stepwise by calculating the effects of the collision of the individual nucleons, determining the resultant postcollision trajectories, and evaluating the ensuing collisions, repeating the process until the nuclei pass through each other. This classical (nonquantal) method is called the *internuclear cascade*. Another classical method makes use of the *Lorentz–Boltzmann* equation as modified by Uehling and Uhlenbeck (1933) to take into account the fermion nature of the nucleons. This classical equation yields the one-body distribution function in phase space. Using this method, one finds that the course of the reaction is determined by the relative importance of the mean field in which the nucleons move, the effect of the collisions, and the Pauli principle. At the lower energies, the mean field and the Pauli principle dominate. The nucleons of the incident projectile are more easily trapped but have a long mean free path. At the higher energies, the mean potential and the Pauli principle become less important and the effect of collisions becomes correspondingly more important.

7. HIGH-ENERGY NUCLEAR PHENOMENA

Projectiles such as electrons or protons whose energies are in the GeV or sub-GeV range are excellent probes of nuclear structure. Because of their short wavelength, they are sensitive to the spatial structure of nuclei. The electron accelerator can be thought of as an electron microscope designed to study the nucleus. The energetic electron beam probes the charge and current distributions inside the nucleus. Similarly, the proton beam or an energetic pion beam examines a combination of charge and mass distributions and currents. Inelastic scattering, (e,e') or (p,p'), localizes the spatial distributions in terms of the *transition densities*. In the $(e,e'p)$ and $(p,2p)$ reactions, a proton is ejected from the target nucleus. Qualitatively, this reaction should be thought of as a collision, of the projectile with the proton within the nuclear medium. It is referred to as quasielastic scattering. One can extract from the experimental data the momentum distribution of the proton hole formed in the collision for specific nuclear states.

At these energies and for the electron or proton projectile, the analysis of the experimental data is considerably simplified. In the electron case, this is because the interaction between the electron and the nucleons in the nucleus is very well known. For the protons, the interaction is more uncertain, but we can make use of multiple-scattering theory, which relates the proton-nucleus scattering with the proton-nucleon scattering that can be separately measured. In fact, one can determine the energy below and the angle above which the first order multiple-scattering theory fails, thus implying the importance of the effects of the medium: that is to say, when the scattering between nucleons within the nucleus differs from the scattering in free space.

The cross sections for elastic and inelastic scattering of electrons by nuclei are functions of $\mathbf{q}$, the momentum transfer to the target nucleus (divided by $\hbar$), and ω, the energy transfer (divided by $\hbar$). For elastic scattering, ω/c is just the recoil energy, $(q^2 + M^2c^4/\hbar^2)^{1/2}$, where M is the nuclear mass. In the limit of q equal to ω/c, the electron cross section reduces to the cross section for photon-induced reactions. For Coulomb *elastic scattering* and in a first-order approximation, the angular distribution for the scattering by the Coulomb field of the target nucleus is given by

$$\frac{d\sigma}{d\Omega} = \left(\frac{Ze^2}{2E}\right)^2 \frac{1}{1 + \dfrac{2E}{Mc^2}\sin^2 \frac{1}{2}\vartheta} \frac{\cos^2 \frac{1}{2}\vartheta}{\sin^4 \frac{1}{2}\vartheta} \rho(\mathbf{q}), \tag{46}$$

where, for $\rho(\mathbf{q})$, see Eq. (7). The quantity Z is the nuclear atomic number, M is the mass of the target, while e is the charge on a proton. For low q, one can determine the root mean square charge radius. Equation (46) holds only for the light nuclei. A more precise evaluation of the Coulomb scattering is necessary for the heavier nuclei.

In addition to Coulomb scattering, the electron can also be scattered by the charge current. The cross section including the effect of the charge scattering can be expressed in terms of the *Coulomb and transverse multipole moments*. The transverse moments dominate the cross section for large scattering angle, while the Coulomb multipole momenta make the major contribution at small angles. When momentum transfer (divided by $\hbar$) $q = \omega/c$, the transverse moments reduce to those which govern photon-induced transitions. More information is obtained with inelastic electron scattering, for then q can be different from ω/c. The investigation of elastic magnetic electron scattering is very informative. In contrast to charge scattering, magnetic scattering is sensitive to the properties of the valence nucleons since the contribution of the core nucleus is zero in the spherical shell-model description. One finds that several magnetic multipoles contribute; that the naive shell model is inadequate except for the case when the spin and orbital angular momentum of the valence nucleon combine to give a maximum value of the total angular momentum, *the stretched state*. When the interacting shell model—that is, a model in which the valence nucleons interact via a residual interaction—is used, agreement with experiment is restored. For all but the stretched state, significant reductions from the shell-model values in the multipole moments occur.

First-order multiple-scattering theory for a nucleon projectile collision provides a potential operator whose average with respect to the ground state of the target nucleus yields an optical-model potential. Using it, one can calculate the elastic scattering. Two elements enter into the calculation of the potential operator. One is the nuclear density, which is folded into the second element, given by the nucleon–nucleon scattering amplitude. It is found for even–even nuclear targets that the dominant contributions from the nucleon–nucleon amplitude are the central and spin-orbit terms. The resulting optical model potential is repulsive and absorptive. It can be weakly attractive for large nucleon–nucleus separations. Agreement with the experimental angular distribution above 0.5 GeV is excellent.

8. PION-INDUCED REACTIONS

The designation "pion" refers to a trio of unstable particles, π^+, π^0, π^-, of zero spin with about the same mass. The superscripts indicate the electric charge of each in units of e, the electron charge. The mass (mc^2) of the π^+ and π^- is 139.57 MeV, while that of the π^0 is 134.96 MeV. The $\pi^\pm$ lifetime is 2.6×10^{-8} s, while that of the π^0 is 0.83×10^{-16} s. Pions interact strongly with nucleons. Their exchange by nucleons is responsible for the long-range ($\gtrsim 1.4 \times 10^{-13}$ cm) component of the nucleon–nucleon interaction. They are produced in nucleon–nucleus collisions and are absorbed by nuclei in a pion–nucleus collision. Because of the relatively long lifetime of the $\pi^\pm$, it is possible to form beams of pions, which can interact with nuclei in such reactions as elastic and inelastic scattering as well as absorption. In addition, there are the *single* (SCX) and *double* (DCX) charge exchange reactions. The SCX reactions of interest are

$$\pi^- + {}_ZA \rightarrow \pi^0 + {}_{Z-1}A$$

$$\pi^+ + {}_ZA \rightarrow \pi^0 + {}_{Z+1}A.$$

The symbol $_{Z}A$ refers to a nucleus of mass number A and atomic number Z. These reactions lead to the same final nuclear states as the (n,p) and (p,n) reactions so that one can expect that they will excite isobar analog states (see Sec. 4). Because the pion–nucleus interaction differs from that of the nucleon with the nucleus, one can expect to obtain differing information regarding mass, charge, and spin densities of nuclei from scattering and SCX.

Double charge exchange results in the reactions

$$\pi^{-} + {}_{Z}A \rightarrow {}_{Z-2}A + \pi^{+}$$

$$\pi^{+} + {}_{Z}A \rightarrow {}_{Z+2}A + \pi^{-}.$$

These reactions will produce radioactive nuclei and will populate *double isobar analog states* as well as giant resonances.

A manifestation of the strong pion-nucleon interaction is the *pion–nucleon resonance*, the Δ (see Fig. 16). The Δ has a mass of about 1232 MeV and a width of about 115 MeV. It has a spin of $\frac{3}{2}$ and can exist in four charge states, Δ^{++}, Δ^{+}, Δ^{0}, and Δ^{-}, where the superscript indicates the charge state. Since the nucleon has a spin of $\frac{1}{2}$ and the pion has zero spin, the formation of the resonance involves a unit orbital angular momentum.

The pion–nucleon resonance dominates pion-induced reactions from low energies up to several hundred MeV. The pion upon striking a nucleon in the nucleus will form a Δ. The nuclear system is then in a Δ–hole (hole for the nucleon that was used to form the Δ) state, which now acts as a doorway state. The width of the doorway state has a spreading width, an escape width, and a Pauli-blocking contribution (Fig. 17). The Pauli-blocking term reduces the width from the free-space value for the Δ, but the contributions of the spreading and, especially, the escape widths more than compensate for this reduction. Indeed, just the spreading width restores the width to near its free-space value. Thus, a method has been developed for calculating the modification of the Δ by the nuclear medium. As

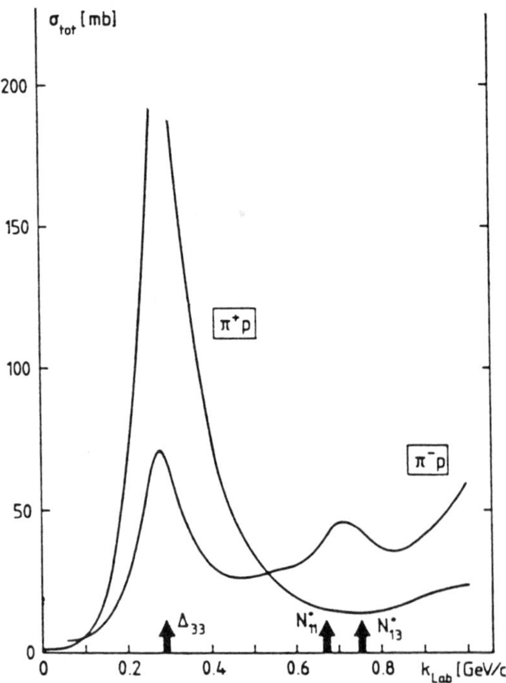

FIG. 16. Pion-nucleon total cross sections. (From Ericson and Weise, 1988.)

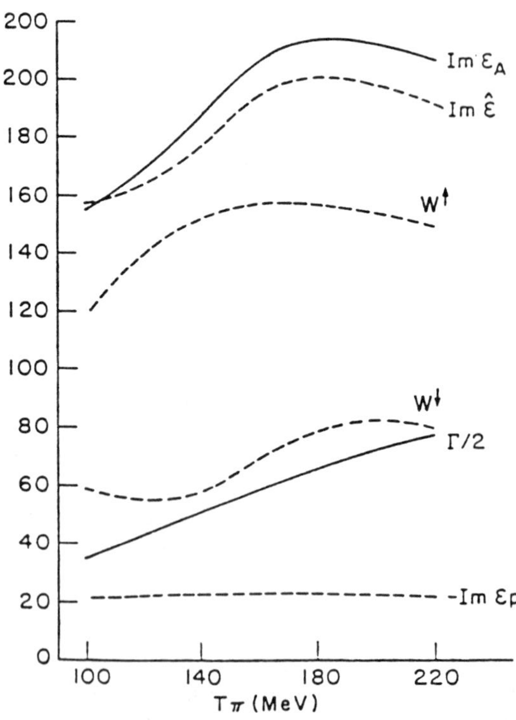

FIG. 17. Decomposition of the imaginary part of the doorway expectation value of the isobar-hole Hamiltonian. The eigenvalue of the leading eigenstate is denoted by $\hat{\epsilon}$, and Γ is the free-space isobar width. $W^{\uparrow}$ is the escape width, $W^{\downarrow}$ is the spreading width, and Im ϵ_p is the Pauli blocking width. (From Horikawa *et al.*, 1980.)

a corollary, the DWA method is not adequate. The resultant theory agrees quite well with elastic scattering, but some difficulties remain for inelastic scattering at the back angles for some cases.

The cross section for SCX at 0° is of the order of a few millibarns per steradian for pion energies of 165 MeV. At this energy, the DCX cross section at 0° is of the order of a microbarn/sr or less. It drops by an order of magnitude at about 20°, increasing somewhat at back angles. Theoretical studies are in agreement with the data at low energies but do deviate from experiment at back angles as the energy increases.

9. THE KAON–NUCLEUS INTERACTION

The kaon is an unstable particle of zero spin that possesses an internal attribute referred to as *strangeness*. The kaon K may have a positive charge, K^+, or it may be neutral, K^0. The mass of the K^+ is 493.7 MeV, with a lifetime of 1.24×10^{-8} s. The mass of the K^0 is 497.7 MeV. The antikaon ($\bar{K}$) doublet is then K^- and $\bar{K}^0$. The neutral kaon observed in nature is a linear combination of the K^0 and $\bar{K}^0$. There are two such combinations, the K_S and the K_L, where the lifetime of the K_S is 0.89×10^{-10} s while that of the K_L is 5.2×10^{-8} s. The *strangeness* of the K doublet is taken to be +1, while that of the $\bar{K}$ is −1. The strangenesses of the nucleon and pion are zero. Strangeness is conserved in a nuclear reaction. Thus, in the collision $\pi^- + p \rightarrow K^+ + \Sigma^-$, the Σ^- is referred to as a *hyperon* with strangeness −1. The other hyperons of strangeness −1 are the $\Sigma^{+,0}$ and the Λ, with masses of 1.189 and 1.115 GeV, respectively. Hyperons of strangeness −2 form a doublet Ξ^-, Ξ^0. The K_S decays principally into two pions with a very small admixture of three-pion decay, while for K_L, the opposite holds. Because of the relatively longer lifetime of K_L, a kaon beam will consist of K_L. However, by scattering the K_L from a nucleus, one can generate some K_S. The process is known as *regeneration*.

The interaction of the antikaon $\bar{K}$ with the nucleon is much stronger than that of the K. The cross section for the K^- collision with a proton is complex and is of the order of several tens of millibarns in the center-of-mass energy range 1.5 to 1.95 GeV. In part, this is because there is a release of 180 MeV in the reaction $K^-(p,\Lambda)\pi^0$ and 100 MeV from the reaction $K^-(p,\Sigma)\pi$. In addition, there are many

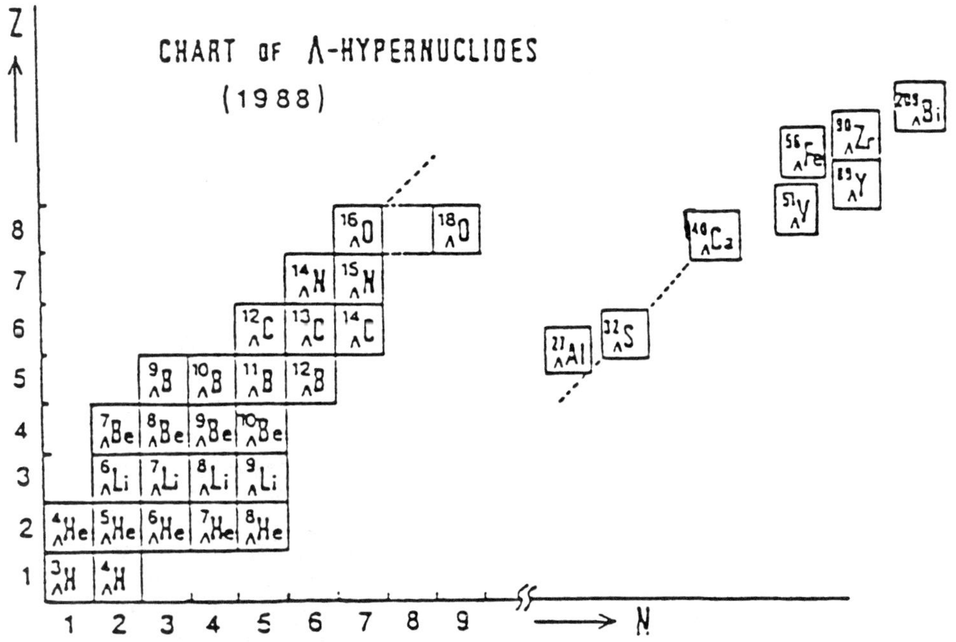

FIG. 18. Chart of observed Λ hypernuclei as of 1988. (From Bando, 1989.)

resonances beginning at 1.385 and 1.405 GeV center-of-mass energy.

Both the K^+p cross section is of the order 12 mb up to 0.8 GeV. It then rises to 17 mb and is constant up to 2.65 MeV laboratory energy. There are no confirmed resonances of this system. Turning to K-nucleus collisions, we observe that, because of the relatively weak K^+-nucleon interaction, the K^+ can penetrate much further into the nucleus than a nucleon. Hence, it would provide an important supplement to the information available from electron and proton scattering by nuclei.

Both the K and $\bar{K}$ projectiles can be scattered both inelastically and elastically by nuclei. The phenomena are similar to those that occur when the projectiles are nucleons. Strangeness exchange leads to a new class of reactions, since it permits the formation of nuclei with nonzero values of strangeness. An example of such reactions is the (K^-,π^-) reaction in which a neutron in the target nucleus is transformed into a Λ, producing thereby a *hypernucleus*. Hypernuclei are also produced by pions striking a nucleus—that is, by a (π^+,K^+) reaction. Over 30 hypernuclei have been made by these reactions (see Fig. 18).

ACKNOWLEDGMENTS

This work is supported in part by funds provided by the U.S. Department of Energy (DOE) under Contract No. DE-AC02-76ER03069.

GLOSSARY

Atomic Number: The number of protons in a nucleus, indicated by the letter Z.

Barn: A unit for cross sections, equal to 10^{-24} cm^2.

Baryon Number: The number of nucleons in a nucleus minus the number of antinucleons, indicated by the letter B.

Cross Section: The transition probability per unit time divided by the incident current density. It has the dimensions of an area.

Compound Nucleus: A well-defined nuclear state formed when a projectile fuses with a target nucleus.

Direct Reactions: Reactions that occur rapidly generally after just one interaction of the projectile with a nucleus.

Doorway State: A state through which a system must pass on the way to a more complicated final state.

Fission: The breakup of a nucleus into two or more fragments.

Fusion: The amalgamation of a projectile and a target nucleus to form another nucleus.

Mass Number: The number of nucleons in a nucleus, indicated by the letter A.

Multistep Reaction: A reaction that involves more than one interaction of the projectile with a nucleus.

Neutron Number: The number of neutrons in a nucleus, indicated by the letter N.

Pickup Reaction: A reaction in which one or more nucleons are transferred from the target nucleus to the projectile.

Pre-equilibrium Reaction: A reaction that terminates before a compound nucleus can be formed.

***Q* Value:** The energy released in a reaction [see Eq. (2)].

Resonance: A sharp change in the energy dependence of a cross section, usually a peak [see Eq. (10)].

Scattering: A reaction in which the projectile does not change. It is called *elastic* when there is no change of energy of either the projectile or target in the center-of-mass system. It is called *inelastic* when either the projectile or target acquires internal energy.

Stripping: A reaction in which one or more nucleons are transferred to the target nucleus.

Transfer Reaction: A reaction in which one or more nucleons is transferred between the projectile and the target nucleus.

Transmission Factor: A measure of the probability that a projectile will be able to interact with a nucleus [see Eq. (15)].

Transmutation: A reaction in which the mass and/or the atomic number of the projectile and the target nucleus change.

Works Cited

Amaldi, E., D'Agostino, R., Fermi, E., Pontecorvo, B., Rasetti, F., Segré, E. (1935), *Proc. R. Soc.* **149**, 522–558.

Bando, H. (1989), *J. Phys. Soc. Jpn., Suppl.* **58**, 379–393.

Becchetti, F. D., Jr., Greenlees, G. W. (1969), *Phys. Rev.* **182**, 1190.

Breit, G., Wigner, E. (1936), *Phys. Rev.* **49**, 519–531.

Curie, I., Joliot, F. (1934), *C. R. Acad. Sci.* **198**, 254–270.

Ericson, T., Weise, W. (1988), *Pions and Nuclei*, Oxford: Clarendon.

Fermi, E. (1934), *Ricerca Sci.* **5**, 283; *Nature* **133**, 757.

Feshbach, H. (1992), *Theoretical Physics: Nuclear Reactions*, New York: Wiley.

Feshbach, H., Porter, C. F., Weisskopf, V. F. (1953), *Phys. Rev.* **90**, 166–167.

Feshbach, H., Porter, C. F., Weisskopf, V. F. (1954), *Phys. Rev.* **96**, 448–464.

Greenlees, G. W., Hnizdo, V., Karban, O., Lowe, J., Makofake, W. (1970), *Phys. Rev. C* **2**, 1063–1071.

Hauser, W., Feshbach, H. (1952), *Phys. Rev.* **87**, 366–373.

Horikawa, V., Theis, N., Lens, F. (1980), *Nucl. Phys.* **A345**, 386–408.

Huizenga, J., Moretto, L. (1972), *Annu. Rev. Nucl. Sci.* **22**, 427–464.

Kerman, A. K., McManus, H., Thaler, R. (1959), *Ann. Phys. (N.Y.)* **8**, 551–635.

Landon, H. H., Sailor, V. L. (1955), *Phys. Rev.* **98**, 1267–1271.

Lee, L. L., Schiffer, J. P., Zeidman, B., Satchler, G. R., Drisko, G. R., Bassel, R. H. (1964), *Phys. Rev.*, **136**, B971–B993; Erratum (1965), *Phys. Rev.* **138**, AB6–AB7.

Lefort, M. (1976), in: H. Faraggi, R. A. Ricci (Eds.), *Nuclear Spectroscopy and Nuclear Reactions with Heavy Ions*, International School of Physics "Enrico Fermi," Corso LXII, Amsterdam: North-Holland, pp. 139–220.

Lefort, M., Ngô, Ch. (1978), *Ann. Phys. (Paris)* **3**, 5–114.

Marmier, P., Sheldon, E. (1969), *Physics of Nuclei and Particles*, Vol. 1; (1970), Vol. 2, New York: Academic.

Parmentola, J., Feshbach, H. (1982), *Ann. Phys. (N.Y.)* **139**, 314–342.

Ray, L. (1979), *Phys. Rev. C* **19**, 1855–1872.

Richard, P., Moore, C. F., Robson, D., Fox, J. D. (1964), *Phys. Rev. Lett.* **13**, 343–345.

Rutherford, E. (1911), *Philos. Mag.* **21**, 669–688.

Singh, P. P., Segal, R. E., Meyer-Schützmeister, L., Hanna, S. S., Allen, R. G. (1965), *Nucl. Phys.* **65**, 577–601.

Uehling, E., Uhlenbeck, G. (1933), *Phys. Rev.* **43**, 552–561.

White, M. G. (1936), *Phys. Rev.* **49**, 309–316.

Further Reading

The following are undergraduate texts.

Cohen, B. L. (1971), *Concepts of Nuclear Physics*, New York: McGraw-Hill.

Enge, H. (1966), *Introduction to Nuclear Physics*, Reading, MA: Addison-Wesley.

The following are graduate texts.

Austern, N. (1970), *Direct Nuclear Reaction Theories*, New York: Wiley.

Barrett, R. C., Jackson, D. F. (1977), *Nuclear Sizes and Structure*, Oxford: Clarendon.

Feshbach, H. (1992), *Theoretical Nuclear Physics: Nuclear Reactions*, New York: Wiley.

Marmier, P., Sheldon, E. (1969), *Physics of Nuclei and Particles*, Vol. 1; (1970), Vol. 2, New York: Academic.

Satchler, G. R. (1983), *Direct Nuclear Reactions*, Oxford: Clarendon.

The following deal with special topics.

Bromley, D. A. (Ed.) (1984), *Treatise on Heavy Ion Sources*, New York: Plenum.

Donnelly, T. W., Sick, I. (1984), "Electron Scattering," *Rev. Mod. Phys.* **56**, 461–560.

Dover, C. B., Walker, G. E. (1982), "Kaon Physics," *Phys. Rep.* **89**, 1–177.

Ericson, T., Weise, W. (1988), *Pions and Nuclei*, Oxford: Clarendon.

NUCLEAR STRUCTURE

J. P. DAVIDSON, *Department of Physics and Astronomy, University of Kansas, Lawrence, Kansas, U.S.A.*

INTRODUCTION

The study of nuclear structure today encompasses a vast territory from the study of simple, few-particle systems to systems with close to 300 particles, from ground-state properties to excitations of such energy that the nucleus disintegrates into substructures and individual constituents. Theories of the structure of the atomic nucleus begin with the fundamental paper by Ernest Rutherford (1911), in which he explained the large-angle alpha (α) particle scattering from gold that had been discovered earlier by Hans W. Geiger

3-527-28133-9/94/$5.00 + .50

and Ernest Marsden. Indeed, Rutherford, in this paper, proposed the first nuclear model (the α-particle model) with his statement "It may be remarked that the approximate value found for the central charge of the atom of gold ... is about that to be expected if the atom of gold consisted of 49 atoms of helium ..." In 1914, Henry Moseley (1913, 1914) showed that the nuclear charge number Z equaled the atomic number. The first attempt at understanding the relative stability of nuclear systems was made by Harkins and Madorsky in 1922. His model, like many others of the time, consisted of protons and electrons. In 1924, Wolfgang Pauli suggested that the optical hyperfine structure might be explained if the nucleus had a magnetic dipole moment, while, in 1931, Giulio Racah investigated the effect on the hyperfine structure if the nuclear charge were not spherically symmetric—that is, if it had an electric quadrupole moment.

All of these structure suggestions occurred before James Chadwick (1932) discovered the neutron, which not only explained certain difficulties of previous models (e.g., the problems of the confinement of the electron), but opened the way to a very rapid expansion of our knowledge of the structure of the nucleus. It may be difficult to believe today, 60 years after Chadwick's discovery, just how rapidly our knowledge of the nucleus increased in the mid-1930s. Hans A. Bethe's review articles (Bethe and Bacher, 1936; Bethe, 1937), one of the earliest and certainly the best known, discuss many of the areas that not only form the basis of our current knowledge but that are still being investigated, albeit with much more sophisticated methods.

The organization here will begin with general nuclear properties, such as size, charge, and mass for the stable nuclei, as well as half-lives and decay modes (alpha, beta, gamma, and fission) for unstable systems. Binding energies and the mass defect lead to a discussion of the stability of systems and the possibility of nuclear fusion. Then follow details of the charge and current distributions, which, in turn, lead to an understanding of static electromagnetic moments (magnetic dipole and octupole, electric quadrupole, etc.) and transitions. Next follows the discussion of single-particle and collective levels for the three classes of nuclei: even-even, odd-A, and odd-odd (i.e., odd Z and odd N). With these mainly experimental details in hand, a discussion of various major nuclear models follows. These attempt, in their own way, to categorize and explain the mass of experimental data. The article closes with an outline of an exciting new area of nuclear structure investigations: the hypernuclei. These are the usual garden-variety nuclei in which a strange particle (Λ, Σ, Ξ) is embedded.

1. GENERAL NUCLEAR PROPERTIES

1.1 Properties of Stable Nuclei

The discovery of the neutron allowed each nucleus to be assigned a number, A, the mass number, which is the sum of the number of protons (Z) and neutrons (N) in the particular nucleus. The atomic number of chemistry is identical to the proton number Z. The mass number A is the integer closest to the ratio between the mass of a nucleus and the fundamental mass unit. This mass unit, the unified atomic mass unit, has the value 1 u = 1.660 540 2 × 10^{-27} kg = 931.494 32 MeV/c^2. It has been picked so that the atomic mass of a ${}^{12}_{6}C_6$ atom is exactly equal to 12 u. The notation here is ${}^{A}_{Z}X_N$, where X is the chemical symbol for the given element—in this case, carbon. The number of electrons or protons, Z, of an element largely determines the gross chemical properties of the element. This notation contains some redundancy since Z and X are synonymous and $A = Z + N$. From this last expression, one can see that there may be several combinations of Z and N to yield the same A. These nuclides are called *isobars*. An example might be the pair ${}^{187}_{75}Re_{112}$ and ${}^{187}_{76}Pt_{111}$, both of which are stable. Furthermore, an examination of a table of nucleids shows many examples of nuclei with the same Z but different A's and N's. Such nuclei are said to be *isotopes* of the element. For example, oxygen (O) has three stable isotopes: ${}^{16}_{8}O_8$, ${}^{17}_{8}O_9$, and ${}^{18}_{8}O_{10}$. A group of nuclei that have the same number of neutrons, N, but different numbers of protons, Z (and, of course, A), are called *isotones*. An example might be ${}^{38}_{18}Ar_{20}$, ${}^{39}_{19}K_{20}$, and ${}^{40}_{20}Ca_{20}$, all three of which are stable. It should be noted that stability is not a requirement of these three definitions. Some elements have but one stable isotope (*e.g.*, ${}^{9}_{4}Be_5$, ${}^{19}_{9}F^{10}$, ${}^{197}_{79}Au_{118}$), others, two, three, or more. Tin ($Z = 50$) has the most at ten. A final definition of use for light nuclei is a *mirror pair,* which is a pair of nuclei with N and Z interchanged (i.e., ${}^{A}_{Z_1}X_{N_1}$ and ${}^{A}_{Z_2=N_1}X_{N_2=Z_1}$). An example of such a pair would be ${}^{23}_{11}Na_{12}$ and ${}^{23}_{12}Mg_{11}$.

The nuclear masses of stable isotopes are determined with a mass spectrometer, and we shall return to this fundamental property when we discuss the nuclear binding energy and the mass defect in Sec. 2.

After mass, the next property of interest is the size of a nucleus. The simplest assumption here is that the mass and charge form a uniform sphere whose size is determined by the radius. While not all nuclei are spherical nor of uniform density, the assumption of a uniform mass/charge density and spherical shape is an adequate starting assumption (more complicated charge distributions will be discussed in Sec. 3). The nuclear radius and, therefore, the nuclear volume or size are usually determined by electron-scattering experiments; the radius is given by the relation

$$R = r_0 A^{1/3}, \tag{1}$$

which with $r_0 = 1.25$ fm gives an adequate fit over the entire range of nuclear mass numbers. An expression like Eq. (1) implies that nuclei have a density independent of A, that is, that they are incompressible. A somewhat better fit to the nuclear sizes can be obtained from the Coulomb energy difference of mirror nuclei, which covers but a fifth of the total range of A. This yields $r_0 = 1.22$ fm. Even if the charge and/or mass distribution is neither spherical nor uniform, one can still define an equivalent radius as a size parameter.

Two important properties of a nuclide are the spin I and the parity π, often expressed jointly as I^{π}, of its ground state. These are usually listed in a table of isotopes and give important information about the structure of the nuclide of interest. An examination of such a table will show that the ground state and parity of all even-even nuclei is 0^+. The spin and parity assignments of the odd-A and odd-odd nuclei tell a great deal about the nature of the principal parts of their ground-state wave functions.

A final property of a given element is the relative abundance of its stable isotopes. These are determined again with a mass spectrograph and listed in various tables of the nuclides.

1.2 Properties of Radioactive Nuclei

A nucleus that is unstable, that is, can decay to a different or *daughter* nucleus, is characterized not only by its mass, size, spin, and parity but also by its half-life and decay mode or modes. (In fact, each level of a nucleus is characterized by its spin, parity, and half-life.) The half-life $t_{1/2}$ is the time in which half of the nuclei decay. The law of radioactive decay is simply

$$N(t) = N_0 e^{\lambda t}, \tag{2}$$

where N_0 is the number of nuclei initially present and λ is the decay constant. By setting $N = N_0/2$ in Eq. (2), one obtains the relation

$$t_{1/2} = 0.693/\lambda. \tag{3}$$

Another useful quantity is the mean lifetime τ, which is simply the reciprocal of the decay constant.

The decay mode can be alpha, beta (e^-), or positron (e^+) decay, electron (or K) capture, or spontaneous fission. In alpha decay, the parent nucleus emits an α particle (a nucleus of helium four—${}^{4}_{2}He_2$), leaving the daughter with two fewer neutrons and protons:

$${}^{A}_{Z}X_N \rightarrow {}^{4}_{2}He_2 + {}^{(A-4)}_{(Z-2)}X'_{(N-2)}. \tag{4}$$

The α particle has zero spin, but it can carry off angular momentum. In beta decay, the number of protons is increased by one, and the number of neutrons is decreased by one, while in positron decay, the opposite is true. In each of these decays, the charged particle (electron or positron) is accompanied by an antineutrino (β decay) or a neutrino (positron decay). All of these particles have spin $\frac{1}{2}$ and can also carry off angular momentum. Since the electron/positron is created from the decay energy at the instant of decay, it is possible that there is insufficient energy to produce a positron even though the parent is more massive than the daughter. Nevertheless, the decay can take place by the absorption of one of the atom's inner (K-shell) electrons. This process is called electron or K

capture. One very rare mode of decay is double β decay, in which a nucleus is unable to β decay to a $Z + 1$ daughter for energy reasons but can emit two electrons and make a transition to a $Z + 2$ daughter. An example is ${}^{238}_{92}U_{146} \rightarrow {}^{238}_{94}Pu_{144}$, with a half-life of $(2.0 \pm 0.6) \times 10^{21}$ yr ($\lambda = 1.1 \times 10^{-29}$ s^{-1}). All of these processes enable one to untangle portions of the structure of both the parent and daughter nuclei. In spontaneous fission, a very heavy nucleus simply breaks into two heavy pieces. For a given nuclide, the decay mode is not necessarily unique. If more than one mode occurs, then the branching ratio is also a characteristic of the radioactive nucleus in question.

An interesting example of a multimode radioactive nucleus is ${}^{242}_{95}Am_{147}$. Its ground state ($I^{\pi} = 1^-$, $t_{1/2} = 16.01$ h) can decay either by electron capture (17.3% of the time) to ${}^{242}_{94}Pu_{148}$ or by beta decay (82.7% of the time) to ${}^{242}_{96}Cm_{146}$. On the other hand, a low-lying excited state at 0.04863 MeV ($I^{\pi} = 5^-$, $t_{1/2} =$ 152 y) can decay either by emitting a gamma ray (99.52% of the time) and going to the ground state or by emitting an alpha particle (0.48% of the time) and going to ${}^{238}_{93}Np_{145}$. There is an excited state at 2.3 MeV with a half-life of 14.0 ms that undergoes spontaneous fission (Lederer and Shirley, 1978). The overall measured half-life of ${}^{242}Am$ is then determined by that of the 0.04863 MeV state. Such long-lived excited states are known as *isomeric states*. From this information on branching ratios, one easily finds the several decay constants for ${}^{242}Am$. For the ground state,

$$\lambda_{ec} = 2.080 \times 10^{-6}\ \mathrm{s}^{-1}, \tag{5}$$

$$\lambda_{\beta^-} = 9.944 \times 10^{-6}\ \mathrm{s}^{-1}, \tag{6}$$

while for the excited state at 0.04863 MeV,

$$\lambda_{\gamma} = 1.439 \times 10^{-10}\ \mathrm{s}^{-1}, \tag{7}$$

$$\lambda_{\alpha} = 6.639 \times 10^{-13}\ \mathrm{s}^{-1}, \tag{8}$$

and for the excited state at 2.3 MeV,

$$\lambda_{SF} = 49.5\ \mathrm{s}^{-1}. \tag{9}$$

2. NUCLEAR BINDING ENERGIES AND THE SEMIEMPIRICAL MASS FORMULA

2.1 Nuclear Binding Energies

One of the more important properties of any compound system, whether molecular, atomic, or nuclear, is the amount of energy needed to pull it apart, or, alternatively, the energy released in assembling it from it constituent parts. In the case of nuclei, these are protons and neutrons. The binding energy of a nucleus ${}^{A}_{Z}X_N$ can be defined as

$$B(A,Z) = ZM_H + NM_n - M_a(Z,A), \tag{10}$$

where M_H is the mass of a hydrogen atom, M_n the mass of a neutron, and $M_a(Z,A)$ the mass of a neutral atom of isotope A. Because the binding energy of atomic electrons is very much less than nuclear binding energies, they have been neglected in Eq. (10). The usual units are atomic mass units, u. Another quantity that contains essentially the same information as the binding energy is the *mass excess* or the *mass defect*, $\Delta = M(A) - A$. (Another useful quantity is the *packing fraction* $P = [M(A) - A]/A = \Delta/A$.)

The more interesting experimental quantity $B(A,Z)/A$ is the binding energy per nucleon, which varies from somewhat more than one MeV/nucleon (1.112 MeV/nucleon) for deuterium (${}^{1}_{1}H_0$) to a peak near ${}^{56}_{26}Fe_{30}$ of 8.790 MeV/nucleon and then falls slowly until, at ${}^{235}_{92}U_{143}$, it is 7.591 MeV/nucleon. Except for the very light nuclei, this quantity is roughly (within about 10%) 8 MeV/nucleon. Even for ${}^{4}_{2}He_2$, the value is 7.074 MeV/nucleon.

It is instructive to plot, for a given mass number, the packing fraction as a function of Z. These plots are quite accurately parabolas with the most β-stable nuclide at the bottom. The β emitters will occur on one side of the parabola (the left or lower-Z side) and the β^+ emitters on the other side. For odd-A nuclei, there is but one parabola, the β-unstable nuclei proceeding down each side of the parabola until the bottom or most stable nucleus is reached. For the even-A nuclei, there are two parabolas, with the odd-odd one lying above the even-even parabola. The fact that the odd-odd parabola is above the even-even one indicates that a pairing force exists that tends to increase the binding energy of the even-even nuclei. See Fig. 1 for the $A = 100$ mass chain. Another indication of the importance of this pairing force is the fact that only four stable odd-odd nuclei exist: ${}^{2}_{1}H_1$, ${}^{6}_{3}Li_3$, ${}^{10}_{5}B_5$, and ${}^{14}_{7}N_7$. For even-A nuclei, the β-unstable nuclei zig-zag between the odd-odd parabola and the even-even parabola until arriving at the most β-stable nuclide, usually an even-even one.

If these plots for each A are assembled into a three-dimensional plot (with A running along one long axis, Z along a perpendicular axis, and B mutually perpendicular to these two), one finds a "landscape" with a deep valley running from one end to the other. This valley is known as the *valley of stability*.

The immediate consequence of the behavior of $B(A,Z)/A$ is that a very large amount of energy per nucleon is to be gained from combining two neutrons and two protons to form a helium nucleus. This process is called *fusion*. The release of energy in the fission process follows from the fact that $B(A,Z)/A$ for uranium is less than for nuclei with more or less half the number of protons. Finally, the fact that the binding energy per nucleon peaks near iron is important to the understanding of those stellar explosions known as supernovae. In Fig. 2, the packing fraction, $P = \Delta/A$, is plotted against A for the most stable nuclei for a given mass number. Note that P has a broad minimum near iron ($A = 56$) and rises slowly until lawrencium ($A = 260$). This shows most clearly the energy gain from the fission of very heavy elements. (Data from Lederer and Shirley, 1978, who give a most extensive collection of the mass defect Δ.)

2.2 The Semiempirical Mass Formula

The semiempirical mass formula may be looked upon as simply the expansion of $B(A,Z)$ in terms of the mass number. Because $B(A,Z)/A$ is nearly constant, the most important term in this expansion must be the term in A. From Eq. (1) relating the nuclear radius to $A^{1/3}$, we see that a term proportional to A is a volume term. However, this term overbinds the sys-

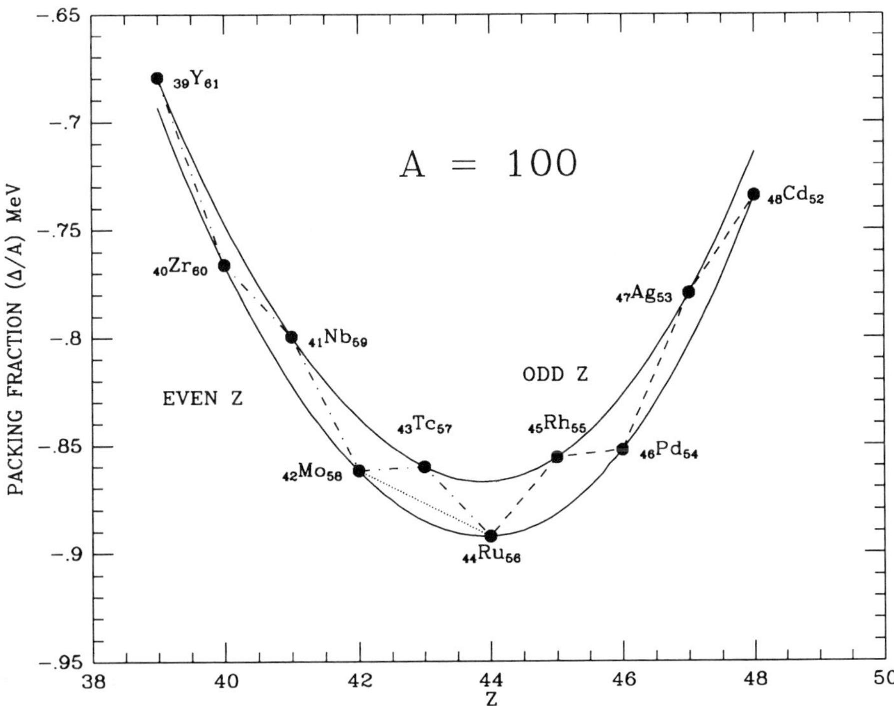

FIG. 1. The packing fraction Δ/A plotted against the nuclear charge Z for nuclei with mass number $A = 100$. Note that the odd-odd nuclei lie above the even-even ones. The β^- transitions are indicated by –·–, the β^+ transitions by –––, while the double-beta decay ${}^{100}_{42}Mo_{58} \rightarrow {}^{100}_{44}Ru_{56}$ is denoted by ···. Data from Lederer and Shirley (1978). The double-beta decay from Kudomi *et al.* (1992).

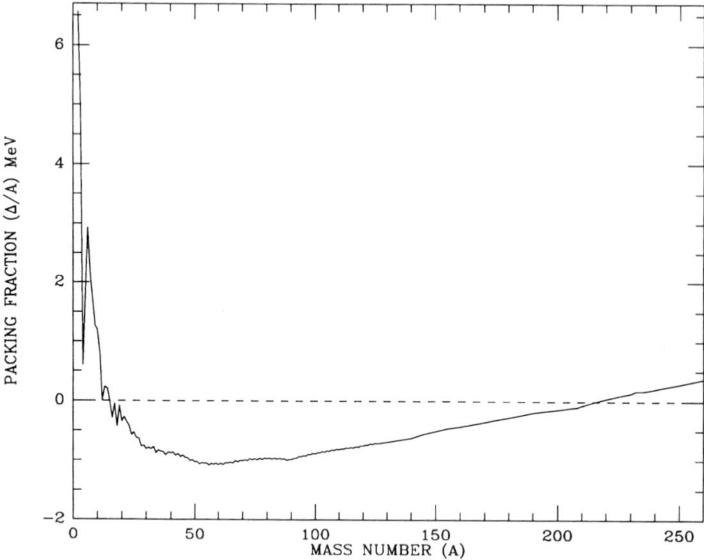

FIG. 2. The packing fraction Δ/A plotted against the mass number A for all nuclei from ${}^{2}_{1}D_{1}$ to ${}^{260}_{103}Lr_{157}$. Data from Lederer and Shirley (1978).

tem since it assumes that each nucleon is surrounded by the same number of neighbors. Clearly, this is not true for surface nucleons, and so a surface term proportional to $A^{2/3}$ must be subtracted from the volume term. (One might identify this with the surface tension found in a liquid drop.) Next, the repulsive Coulomb forces between protons must be included. Since this force is between pairs of protons, this term will be of the form $Z(Z - 1)/2$, the number of pairs of Z protons, divided by a characteristic nuclear length or $A^{1/3}$. Two other terms are necessary in this simple model. One term takes into account that, in general, $Z \approx A/2$, clearly true for stable light nuclei, and less so for heavier stable nuclei where more neutrons are needed to overcome the mutual repulsion of the protons. This term is generally taken to be of the form $\alpha_{\text{sym}}\,(N - Z)^2/A$. The other term takes into account the fact, noted in Sec. 2.1, that even-even nuclei are more tightly bound than odd-odd nuclei because all of the nucleons of the former are paired off. This is done by adding a term $\delta/2$ that is positive for even-even systems and negative for odd-odd systems and zero for odd-A nuclei. Thus, the two even-even parabolas are separated by δ. The semiempirical mass formula then becomes

$$M(A,Z) = ZM_H + NM_n - B(A,Z), \tag{11}$$

where

$$B(A,Z) = a_v A - a_s A^{2/3} - a_c Z(Z-1)A^{-1/3} - a_{\text{sym}}\frac{(N-Z)^2}{A} + \frac{\delta}{2}. \tag{12}$$

Originally, the constants a_v, a_s, a_c, a_{sym}, and δ were fixed by the measured binding energies and adjusted to give appropriate behavior with the mass number (Bohr and Mottelson, 1969). More recent developments by Myers and Swiatecki (1966; see also Myers, 1977) have included other terms to account for regions of nuclear deformation, as well as an exponential term of the form $-a_a A e^{-\gamma A^{1/3}}$ for which they provide no physical explanation beyond the fact that it reduces the deviation from experiment. Their formula, which they call the "finite-range droplet model," has nine adjustable parameters, which they have fixed using some 1500 pieces of data. Their formula fits binding energies quite well from about a 1% error near iron ($A \approx 50$) to 0.1% for very heavy nuclei like uranium ($A \approx 238$).

2.3 Some Consequences of the Semiempirical Mass Formula

A number of consequences flow from even a superficial examination of Eq. (12). We shall look at but two here. The first involves the saturation of the nuclear forces, while the second involves the stability of very heavy nuclei against spontaneous fission.

2.3.1 Saturation of Nuclear Forces The fact that the binding energy per nucleon, B/A, is essentially constant with A implies that the nuclear density is constant and, thus, the nuclear force saturates. That is, nucleons interact only with a small number of their neighbors. If this were not so, then each nucleon would interact with all others in the given nucleus (just as the protons interact with all other protons), and the leading term in $B(A,Z)$ would be proportional to the number of pairs of nucleons, which is $A(A-1)/2$ or roughly A^2. This would imply that B/A would go as A. Thus, not only does the nuclear force saturate (the Coulomb force does not); it is also of very short range (that of the Coulomb force is infinite) since the sizes of nuclei are of the order of 3.0 fm [recall Eq. (1)].

2.3.2 Stability against Spontaneous Fission In order to study the stability of a nucleus against fission or the breaking apart into two large pieces, one examines what happens to the binding energy as the nucleus, taken as a charged liquid drop, is distorted. Two effects and, thus, two terms of $B(A,Z)$ work against each other as the nucleus is distorted. Since a sphere has the smallest area of any shape encompassing a given volume, any distortion will increase the surface area and, thus, the surface energy. On the other hand, the Coulomb energy for a charged drop will be at a maximum for a sphere, so any distortion will reduce the Coulomb energy. Under small distortions from spherical, if the surface energy increases faster than the Coulomb energy decreases, the system will be stable.

To examine the energy change under distortion more analytically, call $E_s^0 = a_s A^{2/3}$ and $E_c^0 = a_c Z(Z-1)A^{-1/3}$ from Eq. (12). Now consider the ellipsoid given by

$$\eta x^2 + \kappa y^2 + z^2/\eta\kappa = R_0^2, \tag{13}$$

which has the volume of a sphere of radius R_0 for all values of η and κ. If we define the two eccentricities by

$$e_1 = (a-c)/R_0 \approx 2(1-\sqrt{\eta}) + (1-\sqrt{\kappa}), \tag{14}$$

$$e_2 = (b-c)/R_0 \approx (1-\sqrt{\eta}) + 2(1-\sqrt{\kappa}), \tag{15}$$

then the distorted surface and Coulomb energies become to first order in the eccentricities

$$E_s = E_s^0[1 + \tfrac{8}{45}(e_1^2 - e_1e_2 + e_2^2) + \cdots], \tag{16}$$

$$E_c = E_c^0[1 - \tfrac{4}{45}(e_1^2 - e_1e_2 + e_2^2) + \cdots]. \tag{17}$$

Calling ΔE the change in the surface plus Coulomb energies, one obtains

$$\Delta E = \tfrac{4}{45}(2E_s^0 - E_c^0)(e_1^2 - e_1e_2 + e_2^2). \tag{18}$$

Treating the eccentricities as variational parameters, one finds immediately that the spherical shape ($e_1 = e_2 = 0$) is the most stable. Also, the system will be stable against small deformations if $2E_s^0/E_c^0 > 1$, for then the deformation increases the energy of the system. Nominal values for a_s and a_c are 16.8 and 0.72 MeV, respectively, so the condition for a nucleus to be stable against spontaneous fission is

$$\frac{2E_s^0}{E_c^0} = \frac{2\times 16.8A}{0.72\times Z(Z-1)} = 46.7\frac{A}{Z(Z-1)} > 1, \tag{19}$$

or

$$\frac{Z(Z-1)}{A} \approx \frac{Z^2}{A} \leq 47. \tag{20}$$

No nucleus comes close to this value. From our previous example (Sec. 1.2) of $^{242}_{95}$Am, $Z^2/A = 37.3$, so it is quite stable against spontaneous fission. In fact, the ground state decays by internal conversion or beta decay. Only the excited state at 2.3 MeV undergoes spontaneous fission.

One final observation is that, for very heavy nuclei, the half-lives for spontaneous fission decrease as the ratio Z^2/A increases—not a surprising result.

3. NUCLEAR CHARGE DISTRIBUTIONS

3.1 General Comments

In his 1911 paper, Rutherford was able to conclude that the positive charge of the atom was concentrated within a sphere of radius less than 10^{-14} m (10 fm). This result came from α-particle scattering. However, for energetic enough α particles, the scattering result will contain a component due to nuclear interactions of the α particle, as well as the Coulomb interaction. For probing the structure of nuclei, electrons have the advantage that their scattering is purely Coulombic; however, to determine details of the internal nuclear structure, electron energies must be well over 100 MeV in order that their de Broglie wavelengths be less than nuclear dimensions. Well before the existence of such high-energy electron beams, nuclear structure effects were extracted from information provided by optical hyperfine spectra. In particular, nuclear charge distributions (electric quadrupole moments) and current distributions (magnetic dipole moments) were deduced from very accurate optical measurements (see the following section).

A result involving the innermost electrons of heavy atoms is the *isotope shift*, which can be observed in atomic x rays. This arises because the nuclear radii for two different isotopes of the same atom will produce slightly different binding energies of their K-shell electrons. Thus, the K x rays of these isotopes will be very slightly different in energy. As an example, the isotopic pair ${}^{203}_{81}\text{Tl}_{122}$ and ${}^{205}_{81}\text{Tl}_{124}$ have an isotope shift of about 0.05 eV.

Another early method to determine the charge radius is to take the difference between the binding energies of two mirror nuclei (cf. Sec. 1.1). This leads to an expression that only involves a_c and, thus, the nuclear radius. This is useful for light nuclei for which mirror pairs occur.

With the advent of copious beams of negative muons, much more accurate optical-type hyperfine spectrum studies could be made. The process is quite simple, and the advantages obvious. By stopping negative muons in a target, an exotic atom is formed in which the muon replaces an orbital electron (see MUONIC, MESONIC, AND BARYONIC ATOMS) and transitions to the muonic K shell follow. These transitions of the muon to the $1s_{1/2}$ state emit photons of the appropriate (but high) energies. (Since the muon is more than 200 times as massive as the electron, the radii of the muon orbits are reduced by that amount, so that electron-shielding problems are much reduced.) The energies of the photons are such that the $2p_{1/2}$-$2p_{3/2}$ splitting is easily measured (in ${}^{116}_{50}\text{Sn}_{66}$, it is 45.666 keV). Thus, both nuclear charge radii and isotope shifts are quite accurately determined. In some recent experiments, rms charge radii have been measured with a precision of 2×10^{-18} m (2 am).

Since electron scattering and muonic atoms are the two methods of measuring characteristics of the nuclear charge radius most susceptible of the greatest accuracy, they will be discussed in turn.

3.2 Nuclear Charge Distributions from Electron Scattering

In any scattering experiment, what is measured is the differential cross section ($d\sigma/d\Omega$). Rutherford developed an expression for alpha-particle scattering that can be used for low-energy, spinless particles incident on a spinless target. Both incident and target particles are assumed to be point particles. The differential cross section for scattering relativistic electrons off point charged particles leads to the expression for Mott scattering, while, if the target particle has nonzero spin (there is then a magnetic contribution), one obtains the Dirac scattering formula for $d\sigma/d\Omega$. However, real nuclei are not point particles, and so one needs to make use of the charge form factor $F(\mathbf{q})$, which is the Fourier transform of the charge density $\rho_{\text{ch}}(\mathbf{r})$,

$$F(\mathbf{q}) = \int \rho_{\text{ch}}(\mathbf{r}) e^{i\mathbf{q}\cdot\mathbf{r}} d\tau. \tag{21}$$

If one restricts the problem to spherically symmetric distributions, the angular integration follows at once, so that

$$F(q^2) = \frac{4\pi}{q} \int \rho_{\text{ch}}(r) \sin(qr) r \, dr, \tag{22}$$

where q^2 is the square of the momentum transfer and is a relativistic invariant. If the target nucleus has zero spin (applicable to all even-even nuclei), then the differential cross sections for a point target and a finite-sized target are related by

$$\left(\frac{d\sigma}{d\Omega}\right)_{\text{finite}} = \left(\frac{d\sigma}{d\Omega}\right)_{\text{point}} |F(q^2)|^2. \tag{23}$$

With the charge form factor determined experimentally, the inverse transform yields the radial charge density

$$\rho_{\text{ch}} = \frac{1}{2\pi^2 r}\int_0^\infty F(q^2)\sin(qr)q\,dq. \tag{24}$$

If the target nucleus is not of spin zero, then an additional term containing the so-called transverse form factor, $F_T(q^2)$, is needed. [The form factor defined in Eq. (22) is sometimes called the longitudinal form factor.] In any event, the charge distribution must be normalized to the number of protons (Z) in the target nucleus.

At this point, there are two ways to proceed. The first is a model-independent analysis of the form factor, or, second, one can assume a model with several parameters and fit these to the data. Limiting oneself to small momentum transfers, one can obtain the form factor as a power series in q^2 by expanding $\sin(qr)$ in Eq. (22) in a power series of its argument. Keeping only the lowest term of order q^2, one obtains

$$F(q^2) = Z(1 - \tfrac{1}{6}q^2\langle r^2\rangle), \tag{25}$$

with $\sqrt{\langle r^2\rangle} = R_{\text{rms}}$, the root mean square radius of the charge distribution. It should be noted that this is not the nuclear radius R, which is usually taken as the radius of the constant-density sphere. This yields

$$R = \sqrt{\tfrac{5}{3}\langle r^2\rangle}. \tag{26}$$

Recent data compilations (de Vries *et al.*, 1987) show that $R_{\text{rms}}/A^{1/3}$ is about constant with a value $R_{\text{rms}} = (0.97 \pm 0.04)A^{1/3}$ fm. Using this in Eq. (26) gives

$$R = 1.25A^{1/3}\ \text{fm}, \tag{27}$$

as was stated in Sec. 1.1.

The model-dependent approach requires using a parametrized charge distribution and fitting the parameters to the data. While there are several such distributions in use, only the two-parameter Fermi distribution, $\rho_{2pF}(r)$, and the three-parameter Fermi distribution, ρ_{3pF}, will be given here (see de Vries *et al.*, 1987, for several others):

$$\rho_{2pF} = \rho_0/\{1 + \exp[(r-c)/a]\}; \tag{28}$$

and

$$\rho_{3pF} = \rho_0\frac{1 + wr^2/c^2}{1 + \exp[(r-c)/a]}, \tag{29}$$

where c is the half-density radius and a is known as the diffuseness and is related to the skin thickness t by

$$t = 4a\ln 3. \tag{30}$$

The parameter w has been called the "wine-bottle" parameter. It can be either positive or negative, but, if the latter, the distribution must be modified so that it goes to zero and does not become negative. The three-parameter Fermi function with an appropriate choice of w can produce a central depression in the charge distribution (thus the appellation "wine-bottle"). This can be understood physically because of the mutual repulsion of the protons, which, in very heavy nuclei, would produce such a depression.

There are enough experimental electron-scattering data available throughout all regions of the stable nuclei that quite accurate charge parameters exist for almost all of the systems. The compendium by DeVries *et al.* (1987) lists these parameters fitted to the data for several distribution functions in addition to the two- and three-parameter Fermi functions.

3.3 Nuclear Charge Distributions from Muonic Atoms

In muonic atoms, the muon is much more tightly bound than is the electron in a similar atom. Not only is the radius of a given orbital reduced by the relative mass of the muon to the electron ($m_\mu/m_e = 206.8$); the muon binding energy is increased by a like amount. To obtain an estimate of the magnitude of these effects, one may use the results of the Bohr atom. This simple nonrelativistic calculation assumes a point nuclear charge and so is only useful for order-of-magnitude results. Taking as an example $^{238}_{92}U_{146}$, one gets for the radius of the muonic K shell 2.78 fm and a binding energy of 23.8 MeV. The nuclear radius from Eq. (27) for this nucleus is 7.75 fm. Thus, the muon orbit is well within the nuclear surface. Relativistic calculations show that, for such heavy nuclei, the muon spends about half of its time within the nucleus.

Since the value of $\langle r^2 \rangle$ for the lower muon orbitals is of the same order of magnitude as the nuclear radius, and because the transition energies are so high (in ${}^{238}_{92}U_{146}$, the measured $2p$–$1s$ transition is about 6.1 MeV), one must generate Dirac solutions for the muon moving in a Coulomb potential generated by a non-point-charge distribution. Also, since many nuclei are not spherical, several studies have used as the appropriate charge distribution a slightly modified form of Eq. (28), but where deformations can be taken into account:

$$\rho_{\mathrm{ch}}(\mathbf{r}) = \rho_0 \left\{ 1 + \exp\left[a^{-1} \left[r - c\left(1 + \sum_{n=1} \beta_{2n} \times Y_{2n0}(\theta,\phi) \right) \right] \right] \right\}^{-1}. \quad (31)$$

Here the β_{2n} are deformation parameters that determine the nuclear shape. If $\beta_{2n} \equiv 0$ for all n, the nucleus is spherical (e.g., ${}^{208}_{82}Pb_{126}$). Studies with β_2 and β_4 not zero have been done.

To these initial Dirac solutions one must add corrections, which, in order of size, are vacuum polarization, nuclear polarization, Lamb shift, and relativistic recoil. Electron-screening corrections are often included, but they are very tiny (for the $1s_{1/2}$ muonic state in ${}^{238}_{92}U_{146}$, this correction has the value of 11 eV).

Experiments to fit a, c, and, in deformed regions, β_{2n} have been made throughout the periodic table with results consistent with the electron data. However, to combine the results of electron-scattering experiments with those from muonic atoms, it is necessary to use the so-called "Barrett moment"

$$\langle r^k e^{-\alpha r} \rangle = \frac{4\pi}{Z} \int_0^\infty \rho(r) e^{-\alpha r} r^{k+2} dr, \quad (32)$$

where k and α are fitted to the experimental data. The muonic data are equivalent to data from electron-scattering experiments at low momentum transfer. The inclusion of the muonic Barrett moment improves the overall fit by reducing normalization errors. This then reduces the uncertainties over what would be obtained by fitting either the electron-scattering or the muonic atom data alone. Extensive tables of data fitted by various charge distribution models as well as model-independent analyses can be found in de Vries *et al.* (1987).

4. ELECTROMAGNETIC TRANSITIONS AND STATIC MOMENTS

4.1 Static Moments, First Measurements

The static electromagnetic nuclear moments played an important role in the unscrambling of the detailed measurements of atomic optical hyperfine structure well before the gross components of atomic nuclei were in hand. Almost a decade before the discovery of the neutron, Wolfgang Pauli (1924) suggested that the optical hyperfine splitting might be due, in part, to the interaction with a nuclear magnetic moment (μ). This suggestion lay fallow until 1930, when Goudsmit and Young, using the spectroscopic data of Schüler and of Granath, deduced the nuclear magnetic moment of ${}^7Li$ ($\mu = 3.29\mu_N$, where the nuclear magneton $\mu_N = e\hbar/2M_p = 5.050\,786 \times 10^{-27}$ J/T). Their value is quite close to the currently accepted value ($\mu = 2.327\mu_N$). Because of the existence, by then, of extensive hyperfine optical spectroscopic data, Goudsmit, in 1933, was able to publish a table of some 20 nuclear magnetic moments ranging from ${}^7Li$ to ${}^{209}Bi$ ($\mu = +4.0\mu_N$, modern value $\mu = +4.0802\mu_N$). In 1937, Schmidt published a simple, single-particle model of nuclear magnetic moments and supported it with the experimental moments of 32 odd-proton nuclei and 15 odd-neutron nuclei. This simple model yields what is now known as the "Schmidt limits," within which almost all nuclear magnetic moments lie (see below).

The suggestion that the nuclear electric quadrupole moments (Q) might also play an important role in optical hyperfine structure was again made before the discovery of the neutron. Racah (1931) was the first to work out the theory associated with "nuclear charge asymmetry" and the interaction with the atomic electrons. Casimir (1935), some time later, developed the theory of nuclear electric quadrupole hyperfine interaction and applied it to ${}^{151}Eu$ and ${}^{153}Eu$. In this paper, Casimir mentions work by Schüler and Schmidt, who determined Q for ^{175}Lu. A short time later, Gollnow (1936) obtained $Q = 5.9$ b for this nucleus, quite close to the currently accepted value of 5.68 b. This very large quadrupole moment (very much larger than can be accounted for by the single-particle shell model) was to provide, 20 years later, strong impetus for the development of the collective model of the nucleus.

In 1954, Schwartz (1955) extended the theory of nuclear hyperfine structure to examine the magnetic octupole hyperfine interaction and calculated the first four nuclear magnetic octupole moments (Ω) from data of the hyperfine structure of the nuclear ground states.

The next nuclear moment is the hexadecapole (H); however, no direct measurements of such static moments exist. What is known about these moments comes mainly from electromagnetic transitions of electrons and negative muons. For an in-depth theoretical study of all of these moments and how they can be used to test various nuclear models see, in particular, the text by Castel and Towner (1990).

4.2 Multipole-Moment Operators, Matrix Elements and Selection Rules

The generalized multipole-moment operator for the (λ, m)th electric moment is defined as

$$\mathscr{E}_{\lambda,m} = \sqrt{\frac{4\pi}{2\lambda+1}} \int \rho_e r^\lambda Y_{\lambda,m}(\theta,\phi)^* \, d\tau. \tag{33}$$

where ρ_e is the nuclear electric charge density, $Y_{\lambda,m}(\theta,\phi)$ a suitably normalized spherical harmonic, and $d\tau$ the differential volume element. The integral extends over the nuclear volume.

For the (λ, m)th magnetic multipole operator, we have:

$$\mathscr{M}_{\lambda,m} = \frac{i}{\lambda+1}\sqrt{\frac{4\pi}{2\lambda+1}} \times \int [\mathbf{L}^* \, r^\lambda Y_{\lambda,m}(\theta,\phi)^*]\mathbf{j}(\mathbf{r}) d\tau, \tag{34}$$

where $\mathbf{L}$ is the angular momentum operator ($\mathbf{L} = -i\mathbf{r} \times \boldsymbol{\nabla}$) and $\mathbf{j}(\mathbf{r})$ is the current density at $\mathbf{r}$.

Since these electric and magnetic multipole operators are λ-rank tensor operators, the selection rules can be determined by using the Wigner–Eckart theorem. If the initial and final states are $|N_i I_i M_i\rangle$ and $|N_f I_f M_f\rangle$, where $I_{i,f}$ are the initial and final angular momenta, $M_{i,f}$ their projections on the z-axis, and $N_{i,f}$ the non-angular-momentum quantum numbers, and calling $\mathscr{T}_{\lambda,m}$ either of $\mathscr{E}_{\lambda,m}$ or $\mathscr{M}_{\lambda,m}$, then

$$\langle N_f I_f M_f | \mathscr{T}_{\lambda,m} | N_i I_i M_i \rangle = C(I_i \lambda I_f;\, M_i m M_f) \times \langle N_f I_f \| \mathscr{T}_\lambda \| N_i I_i \rangle. \tag{35}$$

The quantity $\langle N_f I_f \| \mathscr{T}_\lambda \| N_i I_i \rangle$ is a reduced matrix element that does not depend on the projection quantum numbers but is model dependent, and $C(I_i \lambda I_f;\, M_i m M_f)$ is a Clebsch–Gordan coefficient that contains the selection rules. These coefficients vanish unless the triangle rule holds. That is, these coefficients vanish unless the three angular momenta, I_i, λ, and I_f, form the sides of a triangle. This then yields

$$I_f + I_i \geq \lambda \geq |I_f - I_i|. \tag{36}$$

For the static moments, the initial and final states will be the same, so that we have from Eq. (36) $I \geq \lambda/2$. Thus, spin-0 nuclei will have no static moments, spin-$\frac{1}{2}$ nuclei no quadrupole or higher moments, etc.

For electromagnetic transitions, the transition probability between initial and final states with energy difference $\hbar\omega$ is

$$T(\mathscr{T}_\lambda;\, I_i \rightarrow I_f) = \frac{8\pi(\lambda+1)}{\hbar\lambda[(2\lambda+1)!!]^2} (\omega/c)^{2\lambda+1} B(\mathscr{T}_\lambda;\, I_i \rightarrow I_f), \tag{37}$$

where B is a reduced transition matrix element defined as

$$B(\mathscr{T}_\lambda;\, I_i \rightarrow I_f) = \frac{2\lambda+1}{4\pi} \sum_{m,M_f} |\langle N_f I_f M_f \| \mathscr{T}_{\lambda,m} \| N_i I_i M_i \rangle|^2. \tag{38}$$

From these definitions, the electric moment operators have parity $(-1)^\lambda$, while the magnetic moment operators have parity $(-1)^{\lambda-I}$. Only those operators with even parity will have nonvanishing matrix elements for states with good parity, which include the nuclear ground and low-lying excited states.

Thus, we can see that such states will only have magnetic dipole ($\lambda = 1$), octupole ($\lambda = 3$), $\cdots$, and electric quadrupole ($\lambda = 2$), hexadecapole ($\lambda = 4$), $\cdots$ static moments.

It is customary to define the static operators in the following way:

$$\text{magnetic dipole, } \mu = \mathcal{M}_{1,0}, \tag{39}$$

$$\text{electric quadrupole, } Q = 2\mathcal{E}_{2,0}, \tag{40}$$

$$\text{magnetic octupole, } \Omega = -\mathcal{M}_{3,0}, \tag{41}$$

$$\text{electric hexadecapole, } H = \mathcal{E}_{4,0}. \tag{42}$$

Further development of the matrix elements depends upon model details. Results for simpler models will be given in the next section.

4.3 Magnetic Moments

If one assumes that the magnetic properties are associated with the individual nucleons, then Eq. (34) can be written as

$$\mathcal{M}_{\lambda,m} = \mu_N \sqrt{\frac{4\pi}{2\lambda + 1}} \sum_{k=1}^{A} [\nabla r_k^{\lambda} Y_{\lambda,m}(\theta_k,\phi_k)^*] \times \left[\frac{2g_l}{\lambda + 1} L_k + g_s^k S_k\right], \tag{43}$$

where the sum extends over all of the A nucleons, while L_k and S_k are the orbital and spin operators for the kth nucleon. The gyromagnetic ratios ($g_{l,s}^k$) are usually assigned the free-particle values: for protons

$$g_l^k = 1, \quad g_s^k = 5.587, \tag{44}$$

for neutrons

$$g_l^k = 0, \quad g_s^k = -3.826. \tag{45}$$

4.3.1 Dipole Moments To obtain the dipole moments for the single-particle model one finds from Eq. (43) for $\lambda = 1$

$$\mu = \frac{\mu_N}{2(j+1)} \{g_l[j(j+1) + l(l+1) - \tfrac{3}{4}] + g_s[j(j+1) - l(l+1) + \tfrac{3}{4}]\}. \tag{46}$$

Here $\mathbf{j}$ is the total angular momentum quantum number $\mathbf{j} = \mathbf{l} + \mathbf{s}$. The "Schmidt limits" occur then for the two cases $j = l \pm \frac{1}{2} \geq \frac{1}{2}$. It is interesting that, when plotted, almost all of these moments lie between the Schmidt limits. The moments that lie outside these limits occur mainly for some very light nuclei (^{3}H, ^{3}He, ^{13}C, ^{15}N). One may conclude that the single-particle model does possess some validity. Another set of limits, the Margenau–Wigner (M–W) limits, is obtained by replacing the free particle values for the orbital gyromagnetic ratios, g_l, by the uniform value Z/A. The justification for this is that one is in effect summing over all states that lead to the correct nuclear spin. This calculation represents an early attempt to account for core contributions to the dipole-moment operator. Figures 3 and 4 show plots of a number of the ground state magnetic moments for odd-Z (Fig. 3) and odd-N (Fig. 4) nuclei.

Since, in the single-particle model, one assumes that the odd nucleon's spin ($\mathbf{j}$) value is the nuclear spin (I), one can rewrite Eq. (46) as

$$\mu = \frac{\mu_N}{I+1} g_l I. \tag{47}$$

Then, for the axially symmetric core collective model, one has for the dipole moment

$$\mu = \frac{\mu_N}{I+1} (g_c + g_l) I, \tag{48}$$

where g_c is the gyromagnetic ratio of the core and either is a fitting parameter or depends upon calculations arising from more detailed nuclear models (cf. Sec. 6.4).

4.3.2 Octupole Moments To obtain the octupole moments for the single-particle model, one uses Eq. (43) but with $\lambda = 3$. Again, there are two cases:

$$j = l + \tfrac{1}{2} \geq \tfrac{3}{2},$$

$$\Omega = \mu_N \langle r^2 \rangle \frac{3(2j-1)}{8(j+1)} [g_l(j - \tfrac{3}{2}) + g_s]; \tag{49}$$

$$j = l - \tfrac{1}{2} \geq \tfrac{3}{2},$$

$$\Omega = \mu_N \langle r^2 \rangle \frac{3(j-1)(2j-1)}{8(j+1)(j+2)} [g_l(j + \tfrac{5}{2}) - g_s]. \tag{50}$$

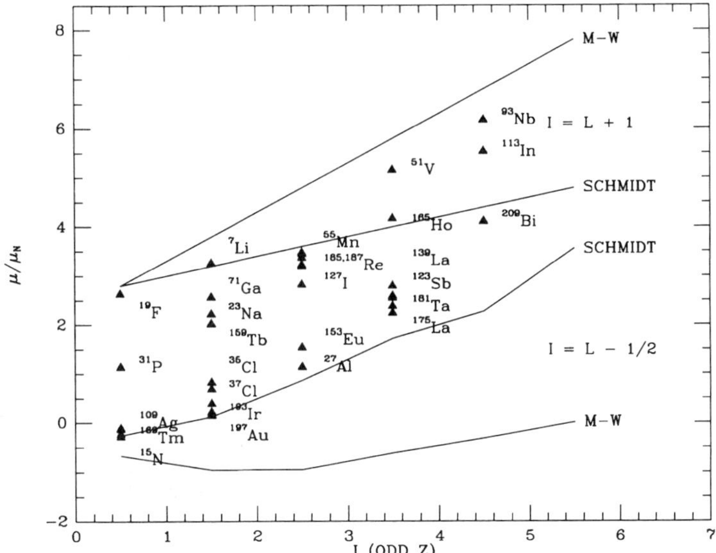

FIG. 3. Nuclear magnetic moments in units of the nuclear magneton (μ/μ_N) plotted against the nuclear spin (I) for a number of odd-Z nuclei. The Schmidt limits, as well as the Margenau-Wigner (M–W) limits, are shown as solid lines. Data from Lederer and Shirley (1978).

The data can be plotted in a manner similar to the dipole data. Again, one must assign values to the gyromagnetic ratios $g_{l,s}$ and to $\langle r^2\rangle$. If one uses the free-particle gyromagnetic ratios and takes $\langle r^2\rangle = \frac{3}{5}R_0^2$ (a uniformly charged sphere) with $R_0 = 1.2A^{1/3}$ fm, one obtains what has been called the "Schwartz" limits. As with the dipole moments, the few measured octupole moments lie within these limits.

4.4. Electric Moments

If one assumes that the nuclear core is symmetric, then the electric multipole properties will be associated with the extra-core nucleon. For such models, the reduced matrix elements of Eq. (35) are easily obtained:

$$\begin{aligned}&\langle N_f I_f \|\mathscr{E}_\lambda\| N_i I_i\rangle\\ &= eg_l\sqrt{\frac{2I_i+1}{2I_f+1}}\,C(I_i\lambda I_f;\,-\tfrac{1}{2}\,0)\langle N_f I_f\|r^\lambda\|N_i I_i\rangle,\\ &|l_i - l_f|,\ \lambda \quad \text{both odd or both even,}\end{aligned} \tag{51}$$

where $I = l + s$. Equation (48) displays all of the possible electric reduced matrix elements from which static moments or transition probabilities can be calculated for these models. For collective models in which the core may not be symmetric, one must determine the reduced matrix elements of the charge distribution $\langle N_f I_f\|\rho_e(r)\|N_i I_i\rangle$ from model parameters.

4.4.1 Quadrupole Moments

The quadrupole moment is obtained from Eq. (51) with $\lambda = 2$, which for an odd proton is

$$Q = -eg_l\,\tfrac{3}{5}\,R_0^2\,\frac{2I-1}{2(I+1)}, \tag{52}$$

where the reduced matrix element of r^2 has been taken to be that of a uniformly charged sphere, R_0 being the nuclear radius. This is easily extended to a closed shell minus a proton (or a proton-hole nucleus) if we replace $g_l^h = -g_l$ and assume that particles and holes have the same wave functions. This simple model would require odd-neutron nuclei to have zero quadrupole moment (recall $g_l = 0$ for neutrons). Not only is this not true, but the model fails to account for most measured quadrupole moments. Indeed, for ^{175}Lu, mentioned earlier, the model value is an order of magnitude too small.

This failure of the single-particle model simply means that all of the electric charge must be considered, especially if the core is not spherically symmetric. The axially symmetric core collective model has

$$Q = \frac{I(2I-1)}{(I+2)(2I+3)}\,Q_0, \tag{53}$$

with

$$Q_0 = \frac{3}{\sqrt{5\pi}}\,ZR_0^2\beta. \tag{54}$$

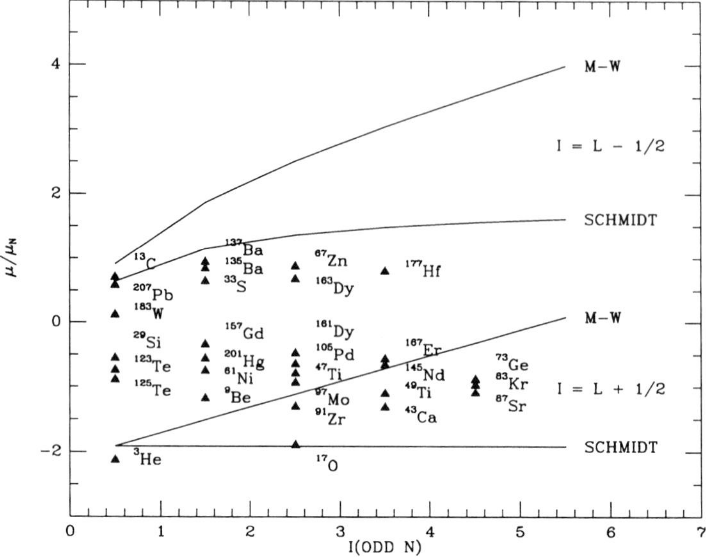

FIG. 4. Nuclear magnetic moments in units of the nuclear magneton (μ/μ_N) plotted against the nuclear spin (I) for a number of odd-N nuclei. The Schmidt limits, as well as the Margenau–Wigner (M–W) limits, are shown as solid lines. Data from Lederer and Shirley (1978).

Here β is a measure of the lack of spherical symmetry of the core.

Since electric quadrupole transitions play an important role in many collective nuclear models, we record here the $E2$ reduced transition matrix elements:

$$B(E2; I_i \to I_f) = \frac{5}{16\pi} e^2 Q_0^2 C^2(I_i 2 I_f; K, 0), \tag{55}$$

where K is an angular momentum projection quantum number for the core.

4.4.2 Hexadecapole Moments The expression for these static moments (with $\lambda = 4$) is easily obtained from Eqs. (33) and (35); however, since no direct measurements of these moments have been made, we shall not present it here. Hexadecapole moments would arise from a model of deformed nuclei in which such terms are kept in the surface expansion. The magnitude of their contribution depends on level systematics and Coulomb and other types of nuclear scattering.

5. LEVEL STRUCTURES

5.1 Regions of Different Level Structures

The most obvious characteristic of the various nuclei is that all of the even-even systems have ground-state spins and parities of 0^+. This not only categorizes one large group of nuclei but also indicates that the nuclear force is such that it couples, preferentially, pairs of like nucleons to angular momentum zero. This strong pairing of like nucleons is also supported by the ground-state spins and parities of the four stable odd-odd nuclei ${}^2_1H_1$ (1^+), ${}^6_3Li_3$ (1^+), ${}^{10}_5Be_5$ (3^+), and ${}^{14}_7N_7$ (1^+). What remains is the large set of odd-A nuclei, all having ground-state spins of half an odd integer.

Various sets of nuclear data indicate that certain numbers of nucleons, either neutrons or protons, correspond to the filling of angular momentum "shells." These are similar to the atomic shells often denoted as the K shell, L shell, etc. In nuclei, the shell filling is similar but not identical, and the principal shell closings occur at experimentally observed numbers. These are for either N or Z equal to 2, 8, 20, 28, 50, 82 and for neutrons only 126. These are the *magic numbers*. They occur where changes in neutron cross sections, nucleon separation energies, etc., change drastically. Because of these large gaps or changes in nuclear properties, they divide the low-lying excited states of nuclei into roughly three regions. These are, broadly, the nuclei in the neighborhood of the magic numbers, those nuclei far from the magic ones, which have a stable deformed shape, and ones found between these two.

An interesting, but small, group of nuclei

comprises the doubly magic ones—that is, nuclei with both neutron and proton numbers magic. The stable ones are ${}^{4}_{2}He_2$, ${}^{16}_{8}O_8$, ${}^{40}_{20}Ca_{20}$, ${}^{48}_{20}Ca_{28}$, and ${}^{208}_{82}Pb_{126}$. What is notable about these nuclei is that their first excited states are at a very high energy compared with their non-doubly magic neighbors. (The energy in MeV, spin, and parity of these first excited states is 20.1, 0^+; 6.05, 0^+; 3.35, 0^+; 3.83, 2^+; and 2.61, 3^-, respectively.)

Within major divisions (even-even, odd-A, and odd-odd), the excited levels are further divided into groups that are single-particle, vibrational, or rotational in nature. However, because of the strong pairing force, even-even nuclei do not show single-particle excited levels.

5.2 Even-Even Nuclei

5.2.1 Vibrational Levels The intermediate systems between the tightly bound magic nuclei and the deformed nuclei are the so-called vibrational nuclei. Here the number of nucleons outside of a deformable core is small, and the zero-point energy of the lowest oscillations is greater than the energy of deformation, so that the shape of the core is not stabilized. The motion then is of quantized surface oscillations, and one can introduce creation and destruction operators $b^*_{\lambda,\mu}$ and $b_{\lambda,\mu}$ where the quantum condition is just

$$[b_{\lambda,\mu},b^*_{\lambda',\mu'}] = \delta_{\lambda\lambda'}\,\delta_{\mu\mu'} \tag{56}$$

all other commutators zero. In this representation, the number operator is

$$n_{\lambda,\mu} = b^*_{\lambda,\mu}b_{\lambda,\mu}, \tag{57}$$

and the Hamiltonian is

$$H = \sum_{\lambda} \hbar\omega_\lambda[N_\lambda + \lambda + \tfrac{1}{2}], \tag{58}$$

where

$$N_\lambda = \sum_{\mu} n_{\lambda,\mu} = 0, 1, 2, \ldots. \tag{59}$$

This is just a quantized harmonic oscillator system with N_λ surfons. Finally, the z component of angular momentum is

$$L_z = \hbar \sum_{\lambda,\mu} \mu n_{\lambda,\mu} = \sum_{\lambda} L^\lambda_z. \tag{60}$$

Equation (60) shows that the "surfons" are quantized field particles with spin λ. Also, from the relation between angular momentum and parity, the λ-surfons will have parity $(-1)^\lambda$. Thus, positive-parity states will require λ to be even, and negative-parity states must have λ odd. The lowest vibrational surfons must have $\lambda = 2$ ($\lambda = 1$ surfons are forbidden since they represent center-of-mass motion, and the analysis is being carried out in the center-of-mass frame). The lowest states of either parity occur when no surfons are excited, so that $L = 0$. The first excited states are $L = 2$ or $L = 3$ for positive- and negative-parity states with one surfon.

Table 1 shows the angular momenta to which N surfons can couple for $\lambda = 2$ and 3. Note, in particular, the degeneracy occurring if $N \geq 2$.

Because an oscillator has equally spaced levels [i.e., $E(N = 2) = 2E(N = 1)$], an important requirement to judge whether or not a system is vibrational is if the ratio of the first $L = 4^+$ level to the first $L = 2^+$ level equals 2. This ratio is often denoted R_4.

A nice example of a "good" vibrational even-even nucleus is ${}^{110}_{48}Cd_{62}$. For $N_{\lambda=2}$ surfon states, one finds $E(N_2 = 1;\, L = 2) = 0.6578$ MeV, $E(N_2 = 2;\, L = 0, 2, 4) = 1.473$, 1.476, and 1.542 MeV, respectively, and $E(N_2 = 3;\, L = 0) = 2.0786$ MeV. The ratio $R_4 = 2.35$ is somewhat larger than the model gives.

5.2.2 Rotational Levels If the number of nucleons outside the deformable core is such that the zero-point oscillations are much less than the energy of deformation, then the system will have a stable but deformed shape, and one must quantize a rigid rotator. From classical considerations, we know that, if the system has a permanent, nonspherical shape,

Table 1. The angular momenta to which N surfons can couple.

	Surfons	
N	$\lambda = 2,\ \pi^+$ I	$\lambda = 3,\ \pi^-$ I
0	0	0
1	2	3
2	0, 2, 4	0, 2, 4, 6
3	0, 2, 3, 4, 6	1, 3, 3, 4, 5, 6, 7, 9
.	.	.
.	.	.
.	.	.

there exists a body-fixed system in which the inertial tensor $\mathfrak{I}$ is diagonal and is related to the laboratory-fixed system by an Euler transformation. We denote these two coordinate systems as (1,2,3) and (x,y,z), respectively. The inertial tensor then has components $\mathfrak{I}_1$, $\mathfrak{I}_2$, $\mathfrak{I}_3$, so that the Hamiltonian is just

$$H = \frac{\hbar^2}{2}\left[\frac{L_1^2}{\mathfrak{I}_1} + \frac{L_2^2}{\mathfrak{I}_2} + \frac{L_3^2}{\mathfrak{I}_3}\right], \tag{61}$$

where the body-fixed angular momentum operators satisfy

$$[L_1,L_2] = -iL_3, \quad \text{cyclically.} \tag{62}$$

The system specified by Eqs. (61) and (62) possesses the symmetry properties belonging to the point group $\mathbf{D}_2$ for which four representations exist. The Hamiltonian operator [Eq. (61)] does not mix different representations. The basis functions are those of a symmetric top, $|LMK\rangle$ being diagonal in L^2, L_z, and L_3 with the usual eigenvalues $L(L+1)$, M, and K, respectively. If the inertia tensor is such that $\mathfrak{I}_1 \neq \mathfrak{I}_2 \neq \mathfrak{I}_3$, an asymmetric-top problem, then the wave functions are to be expanded in terms of the symmetric-top functions

$$|LM\rangle = \sum_{K=-L}^{L} A_K|LMK\rangle. \tag{63}$$

The 1, 2, 3 labels of the momental ellipsoid's semiaxes are quite arbitrary and can be chosen in 24 different ways for right- (left-) handed systems. These 24 relabelings can be produced by three different relabeling transformations T_1, T_2, and T_3 (Bohr, 1952), which correspond to label interchanges and operate on the $|LM\rangle$ wave functions. They have the properties

$$T_1^2 = 1, \qquad T_2^4 = 1, \qquad \text{and } T_3^3 = 1. \tag{64}$$

From these relations, one can generate Table 2, relating the four representations and the values of angular momentum and parity allowed each. Note in particular that the A and B_2 representations are associated with positive-parity (π^+) states, while B_1 and B_3 representations are associated with negative-parity (π^-) states.

From this table, we note that only the rigid rotator systems belonging to the A representation have a zero angular momentum state. They must be associated with the lowest levels of even-even nuclei. Furthermore, one can easily show that the two $L = 2$ levels are related to the single $L = 3$ level by the simple relation

$$E(3) = E(2_1) + E(2_2). \tag{65}$$

Many other similar relations can be derived.

Negative-parity states in even-even nuclei could be associated with either the B_1 or B_3 representations. If one argues that, given two (or more) possibilities, the lowest set of states will be the most symmetric, then because the B_1 representation arises from sums over even K quantum numbers, one should associate this representation with the lowest-lying negative-parity states in even-even systems. Furthermore, in the deformed regions, the negative-parity bands in these nuclei not only lie above the ground-state rotational band, but they have as their band head a 1^- level. This is another indication that these negative-parity rotational bands belong to the B_1 representation.

The problem of solving the asymmetric, rigid rotor is far more involved than the symmetric rotor since all of the latter's eigenvalues can be given in closed form. To pro-

Table 2. Symmetry properties of the asymmetric rotator state functions, the representation of the point group $\mathbf{D}_2$ to which they belong, and the allowed values of the angular momentum and parity associated with them.

$T_1\|LM\rangle$	$T_2\|LM\rangle$	K	A_{-K}/A_K	Parity	Representation	Allowed L values
+	+	even	$(-1)^L$	+	A	0, 2, 2, 3, 4, 4, 4, 5, 5, ...
−	+	even	$(-1)^{L+1}$	−	B_1	
+	−	odd	$(-1)^{L+1}$	+	B_2	1, 2, 3, 3, 4, 4, 5, 5, 5, ...
−	−	odd	$(-1)^L$	−	B_3	

ceed along this line, one sets $\Im_1 = \Im_2 \neq \Im_3$, and we let $\Im_1 \equiv \Im_0$. The Hamiltonian of the system then becomes

$$H_{\text{sym}} = \frac{\hbar^2}{2}\left(\frac{L_1^2 + L_2^2}{\Im_0} + \frac{L_3^2}{\Im_3}\right) = \frac{\hbar^2}{2}\left(\frac{\mathbf{L}^2}{\Im_0} + \left(\frac{1}{\Im_3} - \frac{1}{\Im_0}\right)L_3^2\right), \tag{66}$$

and both of the operators appearing are diagonal. Thus, we obtain immediately the eigenvalues associated with H_{sym}:

$$E_{\text{sym}}(L,K) = \frac{\hbar^2}{2}\left[\frac{L(L+1)}{\Im_0} + \left(\frac{1}{\Im_3} - \frac{1}{\Im_0}\right)K^2\right]. \tag{67}$$

This expression is model independent and only depends upon the ratio of the moments of inertia $\Im_0/\Im_3$. Recent observations of "superdeformation" for nuclei with A near 200 yield spheroidal-axes ratios of 1.65:1. For symmetric rotational systems with $K = 0$, the ratio R_4 is 10/3 rather than 2 for the vibrational systems. Thus, this ratio differentiates well between rotational and vibrational systems. How well the R_4 ratio distinguishes between vibrational and rotational nuclear systems is to be found in the isotopes of samarium ($Z = 62$), which range from the magic-neutron nucleus ${}^{144}_{62}\text{Sm}_{82}$ to the rotational nucleus ${}^{154}_{62}\text{Sm}_{92}$. The following sequence is found for this ratio: $R_4(144) = 1.32$, $R_4(146) = 1.85$, $R_4(148) = 2.15$, $R_4(150) = 2.32$, $R_4(152) = 3.01$, and, finally, $R_4(154) = 3.25$.

There is no reason to believe that deformed nuclei are perfectly rigid, so that, as they rotate, they will tend to deform along an axis perpendicular to the rotational axis. This has the effect of increasing the $\Im_0$ moment of inertia and decreasing that of $\Im_3$. This effect can be easily taken into account for $K = 0$ bands, where a slight decrease in the rotational energy levels will occur. Since the $K = 0$ band forms the ground-state rotational band, one has

$$E_{\text{sym}}(L,K=0) = (\hbar^2/2\Im_0)L(L+1) - bL^2(L+1)^2. \tag{68}$$

Fitting of several low-lying levels by relation (68) shows that b, in general, is a small quantity. This relation is model independent as long as $\Im_0$ and b are simply fitting parameters. To determine them more exactly requires a model for the moments of inertia, and then, an explicit form can be given to b.

An interesting example of a rotational nucleus is ${}^{152}_{62}\text{Sm}_{90}$, for which $R_4 = 3.01$, so that one might apply Eq. (65), where one finds

$$E(3) = E(2_1) + E(2_2) = 121.7825 + 1085.897 = 1207.68\ \text{keV}, \tag{69}$$

while the lowest 3^+ level is at 1233.876 keV. Again, the value of R_4 is different from simple model predictions implying perhaps that $K \neq 0$. Interestingly enough, there is a negative-parity band starting at 963.376 keV with the sequence 1−, 3−, 5−, ... 13−, clearly a band belonging to the B_1 representation.

5.3 Odd-*A* Nuclei

5.3.1 Single-Particle Levels Near the magic numbers, the ground and lower excited states appear to be mainly single-particle states, that is, states that do not seem to possess collective properties. For instance, near the doubly magic nucleus ${}^{16}_{8}\text{O}_8$, one finds that both ${}^{15}_{7}\text{N}_8$ and ${}^{15}_{8}\text{O}_7$ have $\frac{1}{2}^-$ ground states and nearly identical lower excited states; $\frac{5}{2}^+$, $\frac{1}{2}^+$, $\frac{3}{2}^-$, $\frac{5}{2}^+$, $\frac{3}{2}^+$ (N) and $\frac{1}{2}^+$, $\frac{5}{2}^+$, $\frac{3}{2}^-$, $\frac{3}{2}^+$, $\frac{5}{2}^+$ (O). On the higher-A side, one finds that both ${}^{17}_{8}\text{O}_9$ and ${}^{17}_{9}\text{F}_8$ have $\frac{5}{2}^+$ ground states, but their lower excited states have spins $\frac{1}{2}^+$, $\frac{1}{2}^-$, $\frac{5}{2}^-$, $\frac{3}{2}^-$ (O) and $\frac{1}{2}^+$, $\frac{1}{2}^-$, $\frac{3}{2}^-$, $\frac{3}{2}^-$, (F). Again, near the doubly magic nucleus ${}^{40}_{20}\text{Ca}_{20}$, one has in ${}^{39}_{19}\text{K}_{20}$ and ${}^{39}_{20}\text{Ca}_{19}$ ground-state spins of $\frac{3}{2}^+$ and excited levels $\frac{1}{2}^+$, $\frac{7}{2}^-$, $\frac{3}{2}^-$, $\frac{9}{2}^-$, while ${}^{41}_{20}\text{Ca}_{21}$ and ${}^{41}_{21}\text{Sc}_{20}$ exhibit $\frac{7}{2}^-$ ground states and excited-state spins $\frac{3}{2}^-$, $\frac{3}{2}^+$, $\frac{3}{2}^-$, $\frac{5}{2}^+$, $\frac{1}{2}^+$. These data clearly support the notion that angular momentum shells are filling in a manner similar to the electronic shells of the elements.

5.3.2 Vibrational Levels The nucleons all possess intrinsic spin $\frac{1}{2}$ and angular momentum l, so that the simplest model for vibrational odd-A nuclei that one might construct is to add a single nucleon to an even-even core in which the nucleons of each kind pair to zero angular momentum. If the core is a good vibrational nucleus ($R_4 \approx 2$), then the ground state of the odd-A system will have the angular momentum properties of the odd nucleon. Furthermore, by coupling this nucleon to the excited surfon states of the core, one should expect to be able to determine the angular momenta of the odd-A nucleus's excited states. An example is the coupling of a neutron to the vibrational nucleus ${}^{110}_{48}Cd_{62}$ (see Sec. 5.2) to yield the odd-A nucleus ${}^{111}_{48}Cd_{63}$, which has a ground-state spin of $\frac{1}{2}^+$. One might expect that there would be two excited states built on the one-surfon state with spins $\frac{3}{2}^+$ $(2 - \frac{1}{2})$ and $\frac{5}{2}^+$ $(2 + \frac{1}{2})$. This nucleus does indeed have a $\frac{5}{2}^+$, $\frac{3}{2}^+$ pair as its first excited states, which are at 245.4 and 342.1 keV. Mixing of a $\frac{1}{2}^+$ nucleon with the two-surfon states would lead to excited-state spins from $\frac{1}{2}^+$ through $\frac{9}{2}^+$. Such a set of states does appear, except for a $\frac{1}{2}^+$ state. However, it is difficult to pick out which states might belong to the two-surfon core states.

5.3.3 Rotational Levels One might expect that a simple model for the odd-A rotational levels would follow that of the vibrational levels. However, the experimental data do not show such a similar structure. An example is the addition of a neutron to ${}^{152}_{62}Sm_{90}$ (see Sec. 5.3) to form ${}^{153}_{62}Sm_{91}$. The ground state is $\frac{3}{2}^+$, so that coupling a $\frac{3}{2}^+$ neutron to the first excited state of ${}^{152}_{62}Sm_{90}$ $(L - 2^-)$ would lead to states of spin $\frac{1}{2}^-$ through $\frac{7}{2}^+$. No $\frac{1}{2}^+$ state is seen, and the states appear in closely spaced doublets that apparently extend to $\frac{19}{2}^+$. Furthermore, the second excited state is $\frac{3}{2}^-$, which is much too low at 35.843 keV to be associated with the negative-parity band in ${}^{152}_{92}Sm_{60}$.

5.4 Odd-Odd Nuclei

As noted in Sec. 5.1, there are only four stable odd-odd nuclei; however, a great deal of data exist for all of this class of nuclei. Again, one might feel that these nuclei would be well represented by an even-even core plus a neutron-proton pair. Even if these are coupled to a spin-zero core, there is still considerable ambiguity as to how these couple to form the ground-state spin of the system. For instance, if the proton and neutron angular momenta are $\mathbf{j}_p$, $\mathbf{j}_n$, respectively, then their vector sum can lead to the range of J values

$$|j_p - j_n| \leq J \leq j_p + j_n. \tag{70}$$

One can obtain the values of j_p and j_n from the ground-state spins of the neighboring odd-A nuclei with one less neutron or one less proton, respectively. In order to determine more exactly the range of the possible ground-state spins of these nuclei, one must make use of Nordheim's rules, which are the following: strong rule, for $\eta = 0$,

$$J = |j_p - j_n|; \tag{71}$$

weak rule, for $\eta = \pm 1$,

$$J = \text{either } |j_p - j_n| \quad \text{or } j_p + j_n. \tag{72}$$

where $\eta = (j_p - l_p) + (j_n - l_n)$.

In general, these do not work all of the time. An example in light nuclei is ${}^{28}_{13}Al_{15}$ with ground-state spin 3^+ and excited-state spins 2^+, 0^+, 1^+. Taking the ground-state spin of ${}^{27}_{13}Al_{14}$, $\frac{5}{2}^+$, as the value of j_p and the ground-state spin of ${}^{27}_{12}Mg_{15}$, $\frac{1}{2}^+$, as the value of j_n, then Nordheim's weak rule applies. To predict the excited-states spins is even more difficult.

6. NUCLEAR MODELS

The nucleus, being a many-body collection of A particles interacting through a short-range but strong force, is not easily dealt with, from a theoretical point of view, by making use of an accurate many-body calculation since these are beyond the scope of current theoretical methods. Thus, calculations based upon various simple models are able to provide a good deal of physical insight as well as making predictions of measurable properties and providing a way to categorize the vast amounts of data available and flowing from nuclear physics laboratories. While these may seem to be organized in an historical order, all are to greater or less extent being refined and added to currently. We start with the oldest and simplest and proceed to the more complicated.

6.1 The Alpha-Particle Model

As mentioned in the Introduction, Rutherford, in 1911, made the first suggestion of a model to explain the nuclear charge of gold as being 49 atoms of helium. More serious attempts to use a model of n alpha particles for nuclei with $A = n(2Z + 2N)$ were made in the mid-1930s, soon after the discovery of the neutron. The initial problem of the model, that alpha particles could not be stable structures within a nucleus, was overcome when it was suggested that the stable lifetime would be longer than the periods of either rotational or vibrational modes. If this be true, then the nuclei under consideration might be looked upon as the chemist does polyatomic molecules. Alpha particles would be attracted to each other by a force similar to a van der Waals force having a long tail. For small values of the inter-alpha distance (less than the equilibrium distance), this force would be strongly repulsive. Furthermore, as in chemical binding, it is assumed that these forces are simply additive and depend upon the number of alpha-alpha bonds. Thus, ${}^{8}_{4}Be_4$ has a pair of alphas and one bond, ${}^{12}_{6}C_6$ has three alphas at the vertices of an equilateral triangle and so three bonds, ${}^{16}_{8}O_8$ has four alphas at the vertices of a tetrahedron with six bonds, etc. (the geometry gets a bit fancy beyond ${}^{16}O$, with various bipyramids). One can then define a quantity B_α, the inter-alpha binding energy, as

$$B_\alpha = nE(\alpha) - E(A), \tag{73}$$

where $E(\alpha)$ is the total energy of an alpha particle and $E(A)$ the total energy of the nucleus of mass number $A = n(2Z + 2N)$. Dividing B_α by the number of bonds yields the binding energy per bond, which, on the average, is about 2.5 MeV. An important exception is ${}^{8}_{4}B_4$, where $B_\alpha = -92$ keV, not surprising since this nucleus breaks up rapidly (70 as) into two alpha particles.

The model is reasonably successful in predicting ground-state energies. The excited states are another matter. One might expect these to be similar to molecular systems, with rotational and vibrational states with the latter more closely spaced. Unfortunately, this does not occur. The zero-point vibration of the alpha bonds is about equal to the inter-alpha distance, which in turn leads to the near equivalence of rotation and vibration energies, so that a strong rotation-vibration interaction would occur. This, in turn, would make a perturbation calculation impossible. Since this model deals only with even-even nuclei, it obviously predicts ground-state spins and parities of 0+ and, hence, zero electromagnetic moments.

Finally, one might make use of the analogy, again from chemistry, to the H_2^+ ion to investigate n-alpha + nucleon systems (i.e., ${}^{17}_{8}O_9$ or ${}^{17}_{9}F_8$). Since the mass of the nucleon is much closer to the mass of the n-alpha core than the electron is to the mass of two protons, the Born–Oppenheimer approximation, which works for the ion, will not be valid here.

Before leaving this subject, it is of interest to note that recent heavy-ion reactions indicate the fleeting existence of a highly excited state of ${}^{24}_{12}Mg_{12}$, which is best described as a linear chain of six alphas. For this arrangement, there are but five bonds, so that $B_\alpha \approx$ 12.5 MeV, whereas the ground state of this nucleus is in the form of an octahedron and has twelve bonds, so that $B_\alpha \approx 30$ MeV. This six-alpha–chain excited state is simply a longer version of the three-alpha chain found in ${}^{12}_{6}C_6$, which has been identified with the excited 0^+ state at 7.650 MeV and which has a half-life of 50 as. This system has two bonds with $B_\alpha \approx 5$ MeV, whereas the ground-state configuration has three bonds with $B_\alpha \approx 7.5$ MeV.

6.2 The Spherical-Shell Model

The alpha-particle model is conceptually very simple, replacing an A-particle system by an $(A/4)$-particle model, but it is useful only for a restricted type of nucleus and over a very restricted range of A ($8 \leq A \leq 40$). While it comes fairly close with simple calculations to estimating the ground-state energies, it says nothing about the large class of odd-A nuclei, even ones adjacent to the n-alpha systems. The independent-particle model, in which the single odd particle is taken to move in a potential well provided by the remaining even-even core, is used to address this problem. While it might seem improbable that nuclei made up of a large number of strongly interacting particles could in any way be represented by such a model, one must recall that the Pauli principle prevents most interactions of a nucleon in the nucleus. In general, scattering cannot take place, as most of the states into which a low-energy nucleon can scatter are filled. This implies that the mean free path for nucleon motion is quite long, and a single-particle model is worth exploring.

One obvious and important characteristic of odd-A nuclei is their ground-state spins and parities. Starting with the lightest nuclei, the general order of odd-A ground-state spins and parities is $\frac{1}{2}^+, \frac{3}{2}^-, \frac{1}{2}^-$; then, after ${}^{16}_{8}O_8$, it becomes $\frac{5}{2}^+, \frac{3}{2}^+, \frac{1}{2}^+, \frac{3}{2}^+$, and then, after ${}^{40}_{20}Ca_{20}$, $\frac{7}{2}^-, \frac{3}{2}^-, \frac{5}{2}^-$, etc., an order reminiscent of the atomic shells. To obtain a model with an ordering of angular momentum l values requires solving a single-particle Schrödinger equation with a reasonable potential function. The form of this potential is not too important since only minor changes in l-value order occur using one or another r dependence. The two simplest to calculate are the isotropic harmonic oscillator [$V(r) = \frac{1}{2} m\omega^2 r^2$] and the infinite square well. The latter is not realistic since the nucleon separation energy is quite finite and well known for most nuclei. [One might also consider a potential function related to the charge distributions of Eqs. (28) and (29); however, these would require extensive numerical computing to extract the eigenvalues.] The angular momentum order of the levels and the number of particles in each for the isotropic harmonic oscillator is 1s (2); 1p (6), 1d, 2s (12); 1f, 2p (20); 1g, 2d, 3s (30); etc. (The semicolons separate the levels with different quanta of energy $h\omega$, and most are degenerate.) The total number of levels thus becomes 2, 8, 20, 40, 70, 112, 168, ... Since the observed "magic numbers" are 2, 8, 20, 28, 50, 82 and, for neutrons, 126 (cf. Sec. 5.1), this is not adequate.

It took the genius of Maria Goeppert Mayer (1948) and others (Haxel *et al.*, 1949) to realize that by adding a spin-orbit term of the form

$$V_{ls}(r) = -V(r)\mathbf{l} \cdot \mathbf{s}, \tag{74}$$

the theoretical order could be brought into agreement with that observed. The minus sign guarantees that the $j = l + \frac{1}{2}$ level will lie below the $j = l - \frac{1}{2}$ level, as observed. This is just the opposite of the electron-level ordering in atoms.

The most obvious effect of adding a spin-orbit term is the change in the order of the levels, which now becomes 2, 8, 20, 28, 50, 82, 126, 184, ... This exactly what is necessary and explains in a straightforward manner the following observations:

1. Nuclei having Z or N a magic number have unusually large binding energies.
2. In nuclei having Z or N magic plus 1, this last nucleon is very weakly bound.
3. Neutron-capture cross sections decrease markedly at the magic numbers.
4. The number of stable isotopes with Z a magic number is greater than when Z is nearby and not magic. An example is the magic number 50. Tin (Z = 50) has ten stable isotopes, while indium (Z = 49) and antimony (Z = 51) have but two, and cadmium (Z = 48) and (Z = 52) each have eight.
5. The number of stable isotones with N a magic number is also greater than when it is not magic and nearby. Again, the N = 50 isotone series has five stable members, but the N = 49 and N = 51 each have but one stable member.
6. There is a sudden increase in the nuclear charge radius for Z magic when $\Delta N = +2$.
7. The first excited states of even-even nuclei with either Z or N or both magic are much higher in energy than nearby even-even nuclei. We have noted this for the doubly magic nuclei in Sec. 5.1.

Knowing the order of these shell-model levels, we at once can obtain the ground-state spins of the odd-A nuclei as being the spin of the last odd particle as we work up from a magic-number nucleus, or the spin of the last particle-hole as we work down from such a nucleus. This occurs since the spins of all of the particles in the core pair off to zero.

The model thus predicts that the odd-A hydrogen and helium nuclei should have ground states $\frac{1}{2}^+$ ($j = l + \frac{1}{2}$), then the odd-A nuclei of lithium and boron are $\frac{3}{2}^-$ ($j = l + \frac{1}{2}$), while for nitrogen they are $\frac{1}{2}^-$ ($j = l - \frac{1}{2}$). The even-Z nuclei all fit the model, except ${}^{11}_{4}\mathrm{Be}_7$, for which the ground state is $\frac{1}{2}^-$. The nucleus ${}^{15}_{6}\mathrm{C}_9$ also has a $\frac{1}{2}^+$ ground state, but this level could be a multiparticle state since ${}^{17}_{8}\mathrm{O}_9$ has a $\frac{5}{2}+$ ground state arising from the $1d_{5/2}$ level in the next shell. Between ${}^{17}_{8}\mathrm{O}_9$ and ${}^{40}_{20}\mathrm{Ca}_{20}$, the evidence is similar: first, $1d_{5/2^+}$ ground states, then $2s_{1/2^+}$ followed by $1d_{3/2^+}$, and then, for even-Z nuclei, $1f_{7/2}$ ground states. Similar results occur beyond ${}^{40}_{20}\mathrm{Ca}_{20}$, but the picture is not always as clear, for several reasons. One is that neutrons and protons are no longer filling the same shells. Also, while the ground-state magnetic moments fit into this scheme well (see Sec. 4.4.1), the electric quadrupole moments in the regions of the rare earths ($57 \le Z \le 71$) and the actinides ($Z \ge 90$) do not. In fact, quadrupole moments many times too large to be explained by the spherical-shell model require a model in which the charge distribution and, thus, the core, is no longer spherical.

6.3 The Deformed-Shell Model

Since the very large quadrupole moments of the nuclei in the middle of the higher shells cannot be explained by the simple spherical model, their explanation requires large contributions from the even-even core, which carries all (or almost all) of the charge. It can be shown that a single particle moving in a nonrigid potential well will have a lower energy if the well is not spherical, rather than if it is spherical. Since the lowering of the particle energy is proportional to the eccentricity, such a system will assume a deformed equilibrium shape. The nuclear Hamiltonian will now contain a term that produces deviations from spherical:

$$H = H_0 + H_d - V_{ls}(r) + cl^2, \tag{75}$$

where H_0 is the isotropic harmonic-oscillator Hamiltonian and $V_{ls}(r)$ is the spin-orbit term, both of which have proved so successful for the spherical-shell model. (The l^2 term helps give the proper level order in the isotropic limit.) Expanding the nuclear surface $S(\theta,\phi)$ in spherical harmonics gives

$$S(\theta,\phi) = R_0\left[a_0 + \sum_{\lambda>1,\mu} a_{\lambda\mu}Y_{\lambda\mu}(\theta,\phi)\right], \tag{76}$$

where R_0 is the radius of the spherical nucleus and a_0 is unity to quantities of second order and assures that the volume remains constant for small deformations. Taking the lowest order, one has that

$$S(\theta,\phi) = R_0\left[1 + \sum_{\mu=-2}^{2} a_{2\mu}Y_{2\mu}(\theta,\phi)\right]. \tag{77}$$

[This expression should be compared with the nuclear charge distribution of Eq. (31).] It is customary to take only the symmetric term a_{20} in this equation to be nonzero. This then leads to the deformation term in the Hamiltonian

$$H_d = -Ka_{20}\omega_0^2 r^2 Y_{20}(\theta,\phi). \tag{78}$$

The full Hamiltonian can now be diagonalized using a set of spherical shell-model basis states, and the resulting deformed, single-particle states are called "Nilsson" states after Sven Gösta Nilsson (1955), who first carried out this diagonalization.

The operators in Eq. (75) do not connect oscillator states differing by one principal oscillator quantum number, N, but do connect states differing by two. Nilsson neglected matrix elements not diagonal in N. This is a reasonably good assumption since oscillator states differing by $N = 2$ are rather far apart, except for states belonging to large N and large deformation (i.e., nuclei with large A and large quadrupole moments).

The Nilsson levels are labeled $\Omega[N,n_3,\Lambda]$. The symbol Ω is the projection of the particle angular momentum $\mathbf{j}$ on the 3 axis of the body-fixed axis system and is a good quantum number. In the limit of very large deformations, the single-particle energy is given by

$$E = (n_3 + \tfrac{1}{2})\hbar\omega_3 + (n_\perp + 1)\hbar\omega, \tag{79}$$

where the symbol ω_3 is the oscillator frequency along the symmetry axis and the symbol $\omega_\perp$ is the oscillator frequency in the perpendicular directions. Thus, in the limit of very large deformations, one has that $N = n_3 + n_\perp$. In this same limit, the projection of the orbital angular momentum on the symmetry axis becomes the last label,

$$\Lambda = \pm n_\perp, \pm(n_+ - 2), \ldots \pm 1 \text{ or } 0. \tag{80}$$

This scheme is used to label not only the ground states of deformed nuclei but also the excited-state rotational-band heads. The level order will now be I, $I + 1$, $I + 2$, . . ., as a result of core rotation, so that, when a level does not fit into this sequence, it must belong to a different single-particle band.

One now proceeds to assign quantum numbers to the states by determining the odd-nucleon number; then, since the Nilsson energy levels are plotted against the core deformation, one assigns the deformation by finding that place where a Nilsson level appears with the measured ground-state spin. The next Nilsson level up should give the excited-band head, etc. As an example, consider the nucleus ${}^{173}_{71}\mathrm{Lu}_{102}$, which has a $\frac{7}{2}^+$

ground state. Following the procedure above, the appropriate level is the $\frac{7}{2}[404]$, with excited members $\frac{9}{2}^+$, $\frac{11}{2}^+$, $\frac{13}{2}^+$. Just above the $\frac{9}{2}^+$ level is a close doublet with spins $\frac{5}{2}^-$, $\frac{1}{2}^-$, which has been assigned to the $\frac{1}{2}[541]$ Nilsson level. The $\frac{3}{2}^-$ level belonging to this band lies above the $\frac{9}{2}^+$ state. This nonuniform order of a $\frac{1}{2}$ band is an example of Coriolis coupling, which mixes the level order of $\Omega = \frac{1}{2}$ bands.

6.4 Collective Models of Even-Even Nuclei

In Sec. 5.2.2, where the rotational-level structures of even-even nuclei were discussed, it was pointed out that because of the nature of the rigid-body momental ellipsoid and the Euler-type transformations from space-fixed (laboratory) to body-fixed axes, numerous relations involving energy levels belonging to a given representation were totally model independent [cf. Eqs. (65) and (68)]. However, it was noted that more information could be extracted once a model of the momental ellipsoid was selected.

Å. Bohr (1952) proposed a model of the momental ellipsoid and thus of deformed even-even nuclei and the cores of odd-A and odd-odd nuclei. Several important assumptions were made in order to extract these moments of inertia. The first is that the nuclear core is incompressible, so that the nuclear density is constant. Second, the flow is assumed to be irrotational, so that a velocity potential exists and satisfies Laplace's equation. Finally, it must be assumed that the motion is such as to preserve the principal-axis system. The moments of inertia for quadrupole deformations then are

$$\mathfrak{I}_k^2 = 4B_2\beta^2 \sin^2[\gamma - (2\pi/3)k]. \tag{81}$$

Here B_2 and β are the quadrupole mass and deformation parameters, while γ is an asymmetry parameter whose range is $0 \le \gamma \le \pi/6$. There are similar relations for other deformations such as octupole (Y_3), hexadecapole (Y_4), etc.

The Hamiltonian for this system is now

$$H_{\text{quad}} = \tfrac{1}{2} B_2(\dot{\beta}^2 + \beta^2\dot{\gamma}) + \frac{1}{4B_2\beta^2} \sum_{k=1}^{3} \frac{L_k^2}{\sin^2(\gamma - 2\pi/k)} + \tfrac{1}{2} C_2\beta^2. \tag{82}$$

If the system is not rigid, then the deformation and asymmetry parameters are no longer fixed, and the system can execute either deformation (β) vibrations or asymmetry (γ) vibrations, or both. The beta vibrational bands have angular momentum 0—they are akin to molecular "breathing" modes—and the gamma vibrational bands have angular momentum 2. Thus, nuclei will exhibit ground-state bands where $n_\beta = 0$ and $n_\gamma = 0$, beta bands ($n_\beta \neq 0$, $n_\gamma = 0$), gamma bands ($n_\beta = 0, n_\gamma \neq 0$), and mixed bands ($n_\beta$ and $n_\gamma \neq 0$). Since beta vibrations are the simplest, beta bands will lie lower than gamma bands and the combined beta-gamma bands.

6.4.1 Models with Symmetric Deformation

In his original paper, Bohr (1952) argued that the nucleus would stabilize around $\gamma = 0$, so that the momental ellipsoid was symmetric (long and thin, needlelike), while the surface itself was spheroidal. Then $\mathfrak{I}_3 = 0$. This leads to difficulties since Eq. (67) shows that the energy eigenvalues become infinite unless $K = 0$. Thus, the ground-state bands of even-even nuclei have $K = 0$, and the angular momentum sequence becomes 0, 2, 4, 6, . . ., which is what is observed. These models automatically conform to the A representation.

Again, if the system is not rigid against beta vibrations, there will be one or more beta bands above the ground-state rotational band that mimic the ground-state band in spin sequence, but because $\langle\beta^2\rangle$ will be greater than for the ground state, the levels will be closer together.

Finally, in this model if gamma is not constant, gamma vibrations can arise with angular momentum 2. (One might think of these as equatorial bulges moving around the nuclear equator. They are akin to the well-known Jacobi ellipsoids of a self-gravitating, rotating fluid.) These gamma vibrations then produce a gamma band that is different in spin sequence from the beta and ground-state bands. This occurs because $K_\gamma = 2$, not 0. The gamma-band sequence is 2, 3, 4, 5, ... There can be several gamma bands with different K values and different spin sequences. The lowest one has $K_\gamma = 2$, while the next will have $K_\gamma = 4$, etc.

It must be made clear that the Bohr model is only for the low-lying positive-parity levels. The negative-parity levels belonging to the B_1 representation have moments of inertia that are considerably more complicated since they represent an octupole or pearlike shape.

An example of a well-developed rotational system is ${}^{154}_{62}\mathrm{Sm}_{92}$ where $R_4 = 3.25$. This nucleus has a well-developed ground-state rotational band with the spin sequence 0^+, 2^+, 4^+, . . ., 12^+ with the first excited state rather low in energy at 81.99 keV. It has a beta band with a 0^+ band head 1099.28 keV, followed by the first excited 2^+ state at 1177.78 keV (ΔE = 78.52 keV). This is a striking example of the increase in the moment of inertia due to the quantum of beta vibrational energy. Furthermore, the $K = 2$ gamma band, with sequence 2^+, 3^+, 4^+, . . ., starts at 1440.05 keV. Also to be seen is an octupole band starting with the 1^- level at 921.39 keV, followed by the 2^-, 3^-, 5^-, . . . in the expected order. Thus, one sees four well-developed rotational bands fitting quite nicely into the model scheme.

6.4.2 Models with Asymmetric Deformation If one does not follow the Bohr assumption that the nucleus will stabilize around $\gamma = 0$ but around some other value within its range, then one has to deal with an asymmetric rotator, which will have a considerably more complicated eigenvalue structure. Davydov and Filippov (1958) made an extensive investigation and showed that one obtained better results with such a model than with a symmetric one. In the first place, since K is no longer a good quantum number, the model, for the A representation, contains in a natural way the mixing of levels that arise in a symmetric model. Second, since the Bohr moments of inertia generate models belonging to the A representation, Davydov and Filippov were able to rely on energy sum rules like Eq. (65) and similar relations to obtain a quantitative determination of how well the model would fit. Returning again to the rotational nucleus ${}^{154}_{62}\mathrm{Sm}_{92}$, one finds using Eq. (65) that

$$E(2_1) + E(2_2) = 81.99 + 1440.05 = 1522.04\ \mathrm{keV}, \tag{83}$$

$$E(3) = 1539.3\ \mathrm{keV}, \tag{84}$$

which fits the model requirement (within 1.2%) and yields a value of gamma of about 10°. This discrepancy is well accounted for by the beta vibrational energy. The advantage to this asymmetric model is that what is called in the symmetric deformed model the gamma band is now placed exactly with respect to the other rotational levels. However, one gains the band placement for well-developed deformed rotational nuclei but loses the computational simplicity of the symmetric model.

6.5 Boson Models

The models discussed up until now and that have been used most extensively would seem to be almost mutually exclusive. The shell model and its modifications, such as Nilsson's extension to a deformed system, focus on the single-particle aspects of nuclear structure. On the other hand, the collective model of Bohr and Mottelson, and extensions, place the emphasis on the cooperative, or fluid, aspects of these systems. Thus, they seem to stand apart and be almost unrelated. This has led to the next step, the investigation of models that are called microscopic models or calculations. These include pairing models (similar to the Bardeen–Cooper–Schriefer theory of superconductivity) and the random-phase approximation, to name but two. A more recent and still developing set of microscopic models is called the interacting boson models or IBM. This is currently a rich field of investigation [Bonatsos (1988) indicates at least four different such models, IBM-1 through IBM-4], and only one or two of the salient features will be discussed here.

The basic idea is that, for even-even nuclei, the core is treated as being made up of the closed shells of protons and neutrons and is totally inert and is neglected. The so-called valence nucleons produce the properties of these nuclei. Furthermore, these valence nucleons are paired off to form integral-spin bosons of which two types seem sufficient to account for most of the low-energy properties. One type is called s bosons; the pair of like nucleons couple to $J = 0$ angular momentum—they are angular momentum scalars. In the other, called d bosons, the like-nucleon pair are coupled to $J = 2$ angular momentum and so are a five-component spherical tensor. The number of bosons, N, is the sum of the number of proton pairs, N_p, and the number of neutron pairs, N_n, each counted from the *nearest* appropriate closed shell. If the shell is less than half full, then N_i is that number of pairs beyond the closed shell. If it is more than half full, N_i is the number of hole pairs. Thus, the nucleus $^{154}_{62}Sm_{92}$ has $N_p = (62 - 50)/2 = 6$ and $N_n = (92 - 82)/2 = 5$; thus, $N = N_p + N_n = 11$. Now, in the model known as IBM-1, there is no distinction between neutrons and protons, which, for heavy nuclei, is realistic since these two types of nucleons are filling different shells. So their mutual interaction would be minimal.

The model now proceeds in a manner similar to the vibrational model discussed in Sec. 5.2.1. It is assumed that there are s- and d-boson creation operators $s^\dagger$ and $d_\mu^\dagger$ (six in all since $-2 \leq \mu \leq +2$), and corresponding annihilation operators s and d_μ satisfying [compare Eq. (56)]

$$[s, s^\dagger] = 1, \tag{85}$$

$$[d_\mu, d_\nu^\dagger] = \delta_{\mu\nu}. \tag{86}$$

all other commutators being zero. The number operators for s and d bosons are, as usual in such theories,

$$n_s = s^\dagger \bar{s}, \tag{87}$$

$$n_d = d^\dagger \bar{d} = \sum_{\mu=-2}^{2} (-1)^\mu d_\mu^\dagger \bar{d}_{-\mu} = \sum_{\mu=-2}^{2} d_\mu^\dagger d_\mu, \tag{88}$$

where the bar indicates the spherical tensor adjoint of the operator. Five more irreducible spherical tensors can be constructed of products of two of these boson operators. Then a linear combination of these five spherical tensors can be used to define a six-parameter Hamiltonian. These parameters can be fitted to known data. Setting some of these parameters equal to zero can explain the rotational structure; setting others equal to zero allows one to explain the quadrupole vibrational structure, etc.

This model describes only positive-parity levels of even-even nuclei. To describe negative-parity states, one must use negative-parity bosons constructed in a similar way to the s and d bosons. Finally, if one wants to describe odd-A nuclei, a model denoted IBFM (interacting boson-fermion model) must be used. These quite complicated types of investigations, often called algebraic calculations or models, have an extensive literature (cf. Bonatsos, 1988).

7. HYPERNUCLEI

7.1 The Exotic Nuclear Systems

An exotic nuclear system much studied recently is one in which a strange particle is embedded in an ordinary nucleus. These are the hypernuclei with either a Λ hyperon replacing a neutron, of which many have been observed, or a Σ hyperon bound in a nucleus (although the existence of Σ hypernuclei is considered "problematic"). These are the strangeness (S) -1 systems. A very few double-Λ hypernuclei have been reported from emulsion studies, and there exist several events that have been attributed to Ξ hypernuclei. These are $S = -2$ nuclear systems. Under the SU(3) particle classification system, these particles, Λ, Σ, Ξ, occupy the same octet representation with spin and parity $\frac{1}{2}^+$ as do the neutron and proton. The Λ and Σ have strangeness $S = -1$ and isospin $I = 0$ and 1, while the Ξ hyperon has $S = -2$, $I = \frac{1}{2}$.

7.2 Strangeness −1 Systems

There are two hypernuclear systems of interest here. The first involves the Λ hyperon whose mass is 1115.63 ± 0.05 MeV/c^2, $I(J^P) = 0(\frac{1}{2}^+)$, mean lifetime $(2.362 \pm 0.020) \times 10^{-10}$ s, and magnetic moment $-(0.613 \pm 0.004)\ \mu_N$. The second is the Σ hyperon with mass 1197.43 ± 0.06 MeV/c^2, $I(J^P) = 1(\frac{1}{2}^+)$, mean lifetime $(1.479 \pm 0.011) \times 10^{-10}$ s, and magnetic moment $-(1.157 \pm 0.025)\mu_N$. We deal with these in turn.

7.2.1 Λ Hypernuclei Certainly the best known and most studied of the strangeness -1 systems are the Λ hypernuclei. More than a dozen such systems, from $^3_\Lambda$H to $^{209}_\Lambda$Bi, are known and even several systems in which hypernuclear decay gamma rays have been observed.

Early studies were done exclusively with nuclear emulsions, and while such studies are tedious and time consuming, a good deal of information for many hypernuclei was determined, including the dependence of the Λ-binding energy on the nuclear mass number A. With the coming of accelerators with sufficient energy and intensity, inherently more accurate experiments could be carried out. The first of these used kaon beams, and the fundamental process of interest here is $K^- + n \rightarrow \Lambda + \pi^-$. It turns out that this process involves small momentum transfer, and, in fact, for certain experimental parameters there is a "magic momentum" in which the Λ is left at rest. Under these circumstances, the target nucleus will not break up, and the intended hypernucleus is formed. As the above reaction process shows, a neutron in the target nucleus is turned into a Λ, and, since the mass of the Λ is 1115.63 MeV/c^2, only 176.06 MeV/c^2 heavier than a free neutron, the nuclear wave function is essentially unchanged. These simplest reactions (for which $\Delta L = 0$) are called "substitutional" reactions, and the π^- is emitted in the forward direction ($\theta_{cm} \leq 4°$). If, however, the π^- is not emitted in the forward direction ($\theta_{cm} \geq 4°$), then transitions where $\Delta L = 1$ or 2 can take place. These reactions have a much lower cross section than do the $\Delta L = 0$ ones. Because the K^- and π^- are strongly interacting particles, all of these reactions will occur near or at the nuclear surface since there is little chance the former will get very far beyond the surface layer and the π^- still be able to exit.

Experimental difficulties with this process arise because of the low beam fluxes of the K^- particles, and the fact that most substitutional reactions are restricted to light nuclear targets. Another fundamental process that leads to Λ hypernuclei is $\pi^+ + n \rightarrow \Lambda + K^+$ which, while it does not have a "magic momentum," does have much higher π^+ fluxes available. Since momentum transfer and energy are larger, studies can be extended to medium- and heavy-mass nuclei. This process is termed "associated production."

Since the Λ is, from the point of view of the Pauli principle, a distinguishable particle for Λ hypernuclei in their ground states, it will be found in the Λ-1s state, so that one can examine several Λ-nucleon potentials from which the Λ binding energies (B_Λ) can be calculated and compared with experiment. If one plots the experimental binding energies against $A^{-2/3}$, one finds that the B_Λ fall along curves related to the nuclear shell-model single-particle states. Indeed, it has been said that "The Λ provides a superb example of single-particle structure in a many-body system" (cf. Chrien and Dover, 1989, and references therein cited).

The current experimental values of B_Λ are well fitted with a simple Woods–Saxon potential with a radius parameter of $r_0 = 1.128 + 0.439\,A^{-2/3}$ fm and a depth of 28 MeV. While numerous other potentials have been used, "From a practical point of view, this potential is as good as any for predicting Λ binding energies" (Millener *et al.*, 1988).

If the incident particle brings in enough angular momentum and energy, the Λ hypernucleus will be left in an excited state which can then undergo gamma decay. These decays are detected in coincidence experiments, so that one obtains very good determination of the level spacings. Targets of lithium and beryllium have led to the fixing of the excited states of the mirror pair $^4_\Lambda$H and $^4_\Lambda$He, as well as $^9_\Lambda$Be and $^7_\Lambda$Li. The information from the mirror pair shows that the 1^+ state in $^4_\Lambda$H is at 1.09 MeV and in $^4_\Lambda$He at 1.42 MeV above the 0^+ ground states. These are M1 spin-flip transitions and are important in fixing the spin dependence of the ΛN forces in the s shell. Recently, other experiments on ^{10}B and ^{16}O have looked for the M1 spin-flip transitions in the ground-state doublets with negative results for these p-shell nuclei. These doublets arise from the coupling of an s-shell Λ particle with a p-shell nuclear core yielding a $j = \frac{1}{2}$ doublet. The negative results over the gamma energy range $0.1 \leq E_\gamma \leq 0.5$ MeV, using germanium detectors, imply a doublet splitting of less than 100 keV.

7.2.2 Σ Hypernuclei There is considerable question as to whether or not any Σ hypernuclei exist, although numerous examples of possible such events occur in the literature. The problem is that the fundamental process $\Sigma + N \rightarrow \Lambda + N$ is very strong, so that narrow Σ excitations probably should not exist. In a recent review of the Σ hypernuclei, Dover *et al.* (1989) discuss the extant data from (K^-,$\pi^\pm$) reactions on ^{4}He, ^{6}Li, ^{7}Li, ^{9}Be, ^{12}C, and ^{16}O targets. These authors conclude that "The experimental status of narrow Σ-hypernuclear states remains controversial," while Feshbach (1990) states, "I remind you that the existence of these nuclei are [*sic*] problematic."

7.3 Strangeness −2 Systems

The $S = -2$ systems of interest are quite different in that the first is the double-Λ hypernuclei where two Λ hyperons are bound to the nuclear core. The other systems are

Table 3. Experimental parameters for the known ΛΛ systems.

System	$B_{\Lambda\Lambda}$ (MeV)	$\Delta B_{\Lambda\Lambda}$ (MeV)	ΛΛ Interaction	Reference
${}^{6}_{\Lambda\Lambda}$He	10.92 ± 0.6	4.68 ± 0.6	attractive	Chrien, 1989
${}^{10}_{\Lambda\Lambda}$Be	17.7 ± 0.08	4.29 ± 0.1	attractive	Chrien, 1989
${}^{13}_{\Lambda\Lambda}$B	27.5 ± 0.7	4.8 ± 0.7	attractive	Dover *et al.*, 1991

formed by Ξ^- or Ξ^0 hyperons bound to a nucleus. The properties of interest of the Ξ^0 are a mass of 1313.9 ± 0.6 MeV/c^2, $I(J^P) = \frac{1}{2}(\frac{1}{2}^+)$, mean lifetime (2.90 ± 0.09) 10^{-10} s, and magnetic moment $-(1.250 \pm 0.014)\mu_N$, while the similar quantities for the Ξ^- are 1321.32 ± 13 MeV/c^2, $\frac{1}{2}(\frac{1}{2}^+)$, (1.639 ± 0.015) × 10^{-10} s and $-(0.679 \pm 0.031)\mu_N$.

7.3.1 ΛΛ Hypernuclei These systems are important for the light they can throw on the ΛΛ interaction. From studies of the ΛΛ systems, one should obtain the separation energy of the two Λ's from the core nucleus, $B_{\Lambda\Lambda}$, as well as the quantity $\Delta B_{\Lambda\Lambda} = B_{\Lambda\Lambda} - 2B_{\Lambda}$. However, as recently stated, "The world supply of data on doubly strange (S = −2) ΛΛ hypernuclei is very small indeed" (Dover *et al.*, 1991). There are, to date, just three confirmed examples of this system, and all have been found in emulsion experiments. These are ${}^{6}_{\Lambda\Lambda}$He, ${}^{10}_{\Lambda\Lambda}$Be, and, most recently, ${}^{13}_{\Lambda\Lambda}$B. The fundamental processes involved are $K^- + p \rightarrow K^+ + \Xi^-$, followed by $\Xi^- + p \rightarrow \Lambda\Lambda$ where the protons are those found in the emulsion (the organic gelatine base in which is embedded the sensitive silver halides). Using the relation

$$B_{\Lambda\Lambda} = M({}^{A-2}Z) + 2M_{\Lambda} - M({}^{A}_{\Lambda\Lambda}Z), \tag{89}$$

analysis of these three ΛΛ systems leads to the experimental parameters given in Table 3.

These three examples, currently the entire "world supply," have led Dover *et al.* (1991) to the conclusion that "The meson-exchange potential for ΛΛ → ΛΛ in the ${}^1S_0$ channel is attractive in any reasonable model. . . ." While the results noted above are reasonably consistent, many more and different ΛΛ systems are needed to obtain a meaningful picture.

One other ΛΛ system has been looked for, and that is the dibaryon. This has been searched for in the reaction $p + p \rightarrow 2K^+ +$ H and in (K^-,K^+) reactions. However, no evidence for the the dibaryon has, as yet, been seen (Chrien and Dover, 1989; Dover *et al.*, 1991).

7.3.2 Ξ Hypernuclei Here again, the experimental data are sparse but not in quite as short supply as for ΛΛ systems. The fundamental processes are either $K^- + p \rightarrow K^+ + \Xi^-$ or $K^- + p \rightarrow K^0 + \Xi^0$. Currently, there are seven nuclear emulsion events whose analysis indicates the formation of a Ξ^- system (Dover *et al.*, 1989, and references cited therein, especially their Ref. 114). A number of other emulsion events have been analyzed, but the nuclear assignments are not as clear-cut.

The seven events have been fitted by a Woods–Saxon potential with the Ξ^- single-particle potential of the form

$$V_{\Xi}(r) = -V_{0\Xi}/(1 + e^{(r-R)/a}), \tag{90}$$

where $R = r_0 A^{1/3}$ fm, A the usual core mass number, and a = 0.65 fm the surface diffusivity parameter. The analysis of the emulsion data of the Ξ systems and their binding energies is then as shown in Table 4.

Taking the Ξ^- hyperon to be in its $1s_{1/2}$ state, then the well-depth parameter $V_{0\Xi}$ is found to depend upon the choice of r_0. The reported best-fit values for all but the magnesium systems are

$$V_{0\Xi} = 24 \pm 4 \text{ MeV}, \qquad r_0 = 1.1 \text{ fm}, \tag{91}$$

$$V_{0\Xi} = 21 \pm 4 \text{ MeV}, \qquad r_0 = 1.25 \text{ fm}. \tag{92}$$

Table 4. Binding energies of the known Ξ systems.

System	B_{Ξ} (MeV)
${}^{8}_{\Xi}$He	5.2 ± 1.2
${}^{11}_{\Xi}$B	9.2 ± 2.2
${}^{13}_{\Xi}$C	18.1 ± 3.2
${}^{15}_{\Xi}$C	16.0 ± 4.7
${}^{17}_{\Xi}$O	16.0 ± 5.5
${}^{28}_{\Xi}$Al	23.2 ± 6.8
${}^{29,30}_{\Xi}$Mg	2.4 ± 6.3

The magnesium systems are discussed in detail in the previously cited reference, where it is suggested that the observation might be of an excited state of the Mg hypernucleus with the spin $I(\mathrm{Mg}) = 4^+$. This might be either a $1d\ \Xi^-$ or a $2s\ \Xi^-$ single-particle level.

Clearly, much more experimental data will be necessary for further progress to be made with the $S = -2$ systems.

GLOSSARY

Bosons: Quantum particles obeying Bose–Einstein statistics and so having integral spin.

Daughter: In a radioactive decay in which the radioactive nucleus A decays into nucleus B, the nucleus B is said to be the daughter of nucleus A.

de Broglie Wavelength: The wavelength λ of a particle, given by the relation $\lambda = hc/E$, h Planck's constant, c the speed of light, and E the energy of the particle.

Decay Constant: If $\mathcal{N}_0$ nuclei are present at time t_0, then the decay constant λ is defined as

$$\lambda = -\frac{1}{\mathcal{N}_0}\frac{d\mathcal{N}_0}{dt}.$$

Electron Capture: A radioactive decay process, sometimes called K capture, where the decay energy Q is greater than the binding energy of one of the atom's cloud of electrons ($Q > BE_e$). The nucleus absorbs one of the atomic electrons so that $Z_{\text{new}} = Z_{\text{old}} - 1$. If $Q > 2\ m_ec^2$, then electron capture competes with positron decay.

Half-Life, $t_{1/2}$: In radioactive decay, the time in which half of the nuclei initially present decay.

Isobars: Nuclei with the same mass number A but with different numbers of protons (Z) and neutrons (N).

Isomeric State: An excited nuclear state whose half-life for gamma emission is quite long. Similar to a metastable state of an atomic system.

Isotones: Nuclei with the same number of neutrons (N) but different numbers of protons (Z) and, thus, different values for A.

Isotope Shift: Small changes in the wavelengths of x-ray, optical, and especially muonic transitions in going from one isotope to the next, which give a measure of the change in the nuclear radius as A changes by one unit.

Isotopes: Nuclei with the same number of protons (Z) but different numbers of neutrons (N) and, thus, different values of A.

Lamb Shift: A small quantum electrodynamic effect which is principally due to the radiative coupling of the orbiting electron (or muon) with the vacuum field.

Mass Defect: The difference between the mass of a nucleus of mass number A less A, or $\Delta = M(A) - A$.

Mass Excess: Negative of **Mass Defect**.

Mean Lifetime, τ: The reciprocal of the decay constant: $\tau = 1/\lambda$.

Mirror Pair: Two light nuclei with the same mass number A but with numbers of protons (Z) and neutrons (N) interchanged.

Muon: A member of the lepton family of elementary particles ($J = \frac{1}{2}$) next heavier than the electron, with mass $m_\mu = 206.77 m_e$, and mean life $\tau_\mu = 2.197 \times 10^{-6}$ s. They occur, as do electrons, with both positive and negative charges.

Nuclear Polarization: A small reduction in the Coulomb potential caused by the penetration of the nuclear volume by the bound, orbiting particle (an electron or muon).

Packing Fraction: The mass defect per unit mass number, $P = \Delta/A$.

Parity: The behavior of a state function upon reflection of the coordinates through the origin, $\mathbf{r} \to -\mathbf{r}$; then either $\psi(-\mathbf{r}) = +\psi(+\mathbf{r})$ and the state is said to have positive parity, or $\psi(-\mathbf{r}) = -\psi(+\mathbf{r})$ and the state is said to have negative parity.

Pauli Principle: The requirement that, in a system of like particles obeying Fermi–Dirac statistics (spin-$\frac{1}{2}$ particles), no two particles can have the same set of quantum numbers.

Strangeness: An additional quantum number S, assigned to strongly interacting elementary particles. It is conserved in all strong and electromagnetic processes but not in weak processes.

Surfon: A quantized surface vibration of a quantum fluid.

Vacuum Polarization: A small radiative correction to the Coulomb potential of an atom arising from the emission and reabsorption of virtual positron-electron pairs.

Valley of Stability: In a three-dimensional plot of the binding energy $B(Z,N)$ with the proton number Z as abscissa, neutron number N as ordinate and $B(Z,N)$ normal to the

ZN-plane, the stable nuclei will be found along a region where $N \approx Z$, which appears to form a deep valley—the Valley of Stability.

Woods–Saxon Potential: For a real potential function, this is written as

$$V_{\mathrm{Re}} = -\frac{V_0}{1 + e^{(r - r_0 A^{1/3})/a}},$$

where a is an adjustable surface diffuseness parameter. For a complex potential, one adds a similar term, V_{Im}.

Works Cited

Bethe, H. A., Bacher, R. F. (1936), *Rev. Mod. Phys.* **8**, 82–229.

Bethe, H. A. (1937), *Rev. Mod. Phys.* **9**, 69–244.

Bohr, Å. (1952), *K. Danske Vidensk. Selsk., Mat. Fys. Medd.* **26**, No. 14.

Bohr, Å., Mottelson, B. (1969), *Nuclear Structure*, Vol. I, New York: Benjamin.

Bohr, Å., Mottelson, B. (1975), *Nuclear Structure*, Vol. II, Reading, MA.: Benjamin.

Bonatsos, D. (1988), *Interacting Boson Models of Nuclear Structure*, Oxford: Clarendon Press.

Casimir, H. B. G. (1935), *Physica* **2**, 719–723.

Castel, B., Towner, I. S. (1990), *Modern Theories of Nuclear Moments*, Oxford: Clarendon Press, and references therein.

Chadwick, J. (1932), *Nature* **129**, 312.

Chrien, R. E., Dover, C. B. (1989), *Annu. Rev. Nucl. Part. Sci.* **39**, 113–150.

Davydov, A. S., Filippov, G. F. (1958), *Nucl. Phys.* **8**, 237–249.

de Vries, H., de Jager, C. W., de Vries, C. (1987), *At. Data Nucl. Data Tables* **36**, 495–536.

Dover, C. B., Millener, D. J., Gal, A. (1989), *Phys. Rep.* **184**, 1–97.

Dover, C. B., Millener, D. J., Gal, A., Davis, D. H. (1991), *Phys. Rev. C* **44**, 1905–1909.

Feshbach, H. (1990), *Nucl. Phys.* **A507**, 219–238.

Gollnow, G. (1936), *Z. Phys.* **103**, 443–453.

Goudsmit, S., Young, L. A. (1930), *Nature* **125**, 461–462.

Harkins, W. D., Majorsky, S. L. (1922), *Phys. Rev.* **19**, 135–156.

Haxel, O., Jensen, J. H. D., Suess, H. E. (1949), *Phys. Rev.* **75**, 1766.

Kudomi, N., Ejiri, H., Nagata, K., Okada, K., Shibata, T., Shima, T., Tanaka, J. (1992), *Phys. Rev.* **46**, R2132–R2135.

Lederer, C. M., Shirley, V. S. (Eds.) (1978), *Table of Isotopes*, 7th Ed., New York: Wiley.

Mayer, M. G. (1948), *Phys. Rev.* **74**, 235–239; (1949), *Phys. Rev.* **75**, 1969–1970. See also Mayer, M. G., Jensen, J. H. D. (1955), *Elementary Theory of Nuclear Shell Structure*, New York: Wiley. Mayer and Jensen shared the 1963 Nobel Prize (with E. P. Wigner) for this work.

Millener, D. J., Dover, C. B., Gal, A. (1988), *Phys. Rev. C* **38**, 2700–2708.

Moseley, H. G. J. (1913), *Philos. Mag.* **26**, 1024–1034.

Moseley, H. G. J. (1914), *Philos. Mag.* **27**, 703–713.

Myers, W. D., Swiatecki, W. J. (1966), *Nucl. Phys.* **81**, 1–60.

Myers, W. D. (1977), *Droplet Model of Atomic Nuclei*, IFI, New York: Plenum, and references therein.

Nilsson, Sven Gösta (1955), *Kgl. Danske Videnskab. Selskab, Mat. Fys. Medd.* **29**, No. 16.

Pauli, W. (1924), *Naturwiss.* **12**, 741–743; reprinted in Kronig, R., Weisskopf, V. F. (Eds.) (1964), *Collected Scientific Papers*, Vol. 2, New York: Wiley, pp. 198–200.

Racah, T. (1931), *Z. Phys.* **71**, 431–434.

Rutherford, E. (1911), *Philos. Mag.* **21**, 669–688.

Schmidt, T. (1937), *Z. Phys.* **106**, 358–361.

Schwartz, C. (1955), *Phys. Rev.* **97**, 380–395.

Further Reading

In this list, the symbols E, I, and A following the references indicate the level of difficulty: elementary, intermediate, and advanced, respectively.

Bethe, H. A. (1936), *Rev. Mod. Phys.* **8**, 82–229 (I).

Bethe, H. A., Bacher, R. F. (1937), *Rev. Mod. Phys.* **9**, 69–244 (I). This and the preceding form the best review of the early state of the subject.

Bohr, Å., Mottelson, B. (1969), *Nuclear Structure*, Vol. 1, New York: Benjamin (A).

Bohr, Å., Mottelson, B. (1975), *Nuclear Structure*, Vol. 2, New York: Benjamin (A). This and the preceding, while old, are comprehensive and still useful.

Bonatsos, D. (1988), *Interacting Boson Models of Nuclear Structure*, Oxford: Clarendon (A).

Castel, B., Towner, I. S. (1990), *Modern Theories of Nuclear Moments*, Oxford: Clarendon (A).

Casten, R. F. (1990), *Nuclear Structure from a Simple Perspective*, New York: Oxford Univ. Press (I). An excellent text, more advanced than Pal (1983).

Janssens, R. V. F., Khoo, T. L. (1991), "Superdeformed Nuclei," *Annu. Rev. Part. Sci.* **41**, 321–355 (A). Discusses many aspects, both theoretical and experimental, of this fascinating area of nuclear structure.

Lederer, C. M., Shirley, V. S. (Eds.) (1978), *Table of Isotopes*, 7th ed., New York: Wiley. A compendium of nuclear data, of great value.

Pal, M. K. (1983), *Theory of Nuclear Structure*,

New York: Van Nostrand Reinhold (E). Nice treatments of several of the older microscopic models.

Wong, S. S. M. (1990), *Introductory Nuclear Physics*, Englewood Cliffs, NJ: Prentice-Hall (I). Another excellent, more advanced text.

Two serials of importance:

Annu. Rev. Nucl. Part. Sci. includes many reviews of current works in nuclear structure.

Nucl. Data Tables is another compendium of nuclear data.

ERRATUM

INTERCALATION COMPOUNDS (Vol. 8, pp. 133–155). M. S. DRESSELHAUS AND G. DRESSELHAUS.

The figure shown as Fig. 6 is incorrect. The correct figure is shown here.

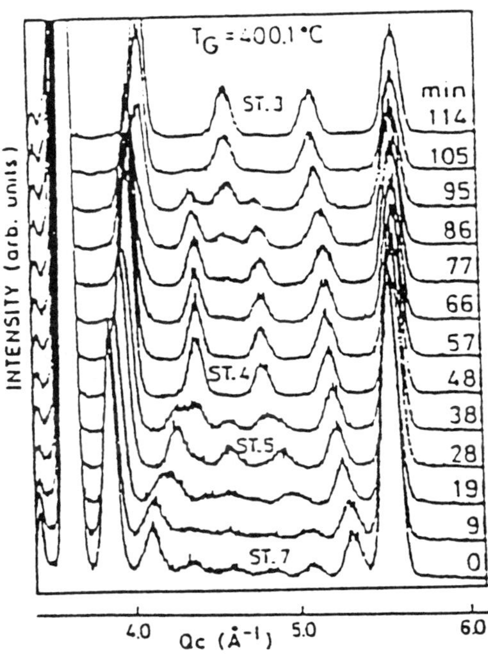

FIG. 6. *In situ* x-ray study of staging transitions (where T_G, the graphite temperature, is 400.1 °C). The kinetics of the staging process in K-GICs is obtained from the time dependence of (00*l*) reflections for transitions from a stage-7 compound to a stage-3 compound (Suematsu *et al.*, 1988).

LIST OF RECOMMENDED UNITS AND SYMBOLS

RECOMMENDED UNITS AND CONVERSION FACTORS

The SI system provides six basic units: meter m, kilogram kg, second s, ampere A, kelvin K, candela cd, and mole mol.

Some important derived units are also allowed and bear special names, e.g.:

1 N (newton) = 1 kg m s^{-2}
1 J (joule) = 1 N m = 1 kg m^2 s^{-2}
1 W (watt) = 1 J s^{-1} = 1 kg m^2 s^{-3}
1 Pa (pascal) = 1 N m^{-2} = 1 kg m^{-1} s^{-2}

For mass, the gram g, or the metric ton t which equals 1000 kg, may be used instead of kilogram kg.

The so-called "long ton" (UK) and "short ton" (US) have been abandoned and will not be used in the *Encyclopedia of Applied Physics*.

For pressure, the bar (name and symbol alike) may be used, and for temperature, the degree celsius °C.

From all of these, decimal multiples or fractions can be derived:

Power of ten	Prefix	Symbol	Power of ten	Prefix	Symbol
10	deca	da	10^{-1}	deci	d
10^2	hecto	h	10^{-2}	centi	c
10^3	kilo	k	10^{-3}	milli	m
10^6	mega	M	10^{-6}	micro	μ
10^9	giga	G	10^{-9}	nano	n
10^{12}	tera	T	10^{-12}	pico	p
10^{15}	peta	P	10^{-15}	femto	f
10^{18}	exa	E	10^{-18}	atto	a

For mass, multiples or fractions are derived from g, not from kg (since the latter already contains a prefix), e.g., mg.

Units with the prefixes are considered as one entity and can, therefore, be raised to any power, e.g., cm^3.

The liter is now considered as synonymous with dm^3 (which does not hold in the older literature!). Please use the capital letter L as unit symbol (following a recent IUPAC recommendation). The use of the Ångström unit is discouraged; it should be replaced by fractional meters (1 Å = 100 pm = 0.1 nm).

SELECTED QUANTITIES, UNITS, AND SYMBOLS

Name of quantity[a]	Symbol[b]	SI unit[c]	Name of unit	Other units
Space and time				
length*	l	m	meter[d]	
breadth, width	b	m		
height	h	m		
radius	r	m		
thickness	d	m		
area	A,S	m^2		a (are) h (hectare)
volume	V	m^3		L, l(liter)[e]
plane angle	$\alpha,\beta,\gamma,$ ϑ,φ	1, rad	radian	° (degree) ′ (minute) ″ (second)
solid angle	ω,Ω	1, sr	steradian	
wavelength	λ	m		
wave number	σ,ν	m^{-1}		
time*	t	s	second	min (minute) h (hour) d (day)
frequency	ν, f	s^{-1}		Hz (hertz)[f]
relaxation time	τ	s		
velocity	u,v	$m\ s^{-1}$		km/h
acceleration	a	$m\ s^{-2}$		
Mechanics				
mass*	m	kg	kilogram	g (gram) t (tonne)[g]
(mass) density	ρ	$kg\ m^{-3}$		g/cm^3
momentum	$\mathbf{p}$	$kg\ m\ s^{-1}$		
angular momentum	$\mathbf{L}$	$kg\ m^2\ s^{-1}$		
force	$\mathbf{F}$	N	newton[f]	
moment of force	$\mathbf{M}$	N m		
weight	G, W	N		
pressure	p	Pa	pascal	bar (bar)[h]
energy	E, W	J	joule	W h (watt hour)[i] eV (electron volt)[j]
work	W,A	J		
power	P	W	watt	J/s, V A[k]
Molecular Physics and Thermodynamics				
thermodynamic temperature*	T	K	kelvin	°C (degrees Celsius)
Celsius temperature	ϑ,t			°C
number of entities	N			
Avogadro constant[l]	N_A,L	mol^{-1}	(particles) per mole	
Boltzmann constant[m]	k,k_B	$J\ K^{-1}$		
Planck constant[n]	h	J s		
(molar) gas constant[o]	R	$J\ mol^{-1}\ K^{-1}$		
(quantity of) heat	Q	J		
entropy[p]	S	$J\ K^{-1}$		
internal energy[p]	U	J		
Helmholtz function,[p] (Helmholtz) free energy, Helmholtz energy	F,A	J		
enthalpy[p]	H	J		
Gibbs function,[p] (Gibbs) free energy, Gibbs energy	G	J		
heat capacity[p]	C_p,C_V	$J\ K^{-1}$		

Name of quantity[a]	Symbol[b]	SI unit[c]	Name of unit	Other units
Chemical Physics				
amount of substance*	n	mol	mole	
relative atomic mass	A_r	1		
relative molecular mass	M_r	1		
atomic mass constant	m_u	kg		u (atomic mass unit)[q]
mass of a portion (of substance B)	m_B,m(B)	kg		g (gram)
molar mass (of substance B)	M_B,M(B)	kg mol^{-1}		
concentration (of substance B)	c_B,c(B)	mol m^{-3}		mol/L
mole fraction[r] (of substance B)	κ_B,κ(B)	1		
mass fraction (of substance B)	ω_B,ω(B)	1		%,‰,ppm,ppb
volume fraction (of substance B)	φ_B,φ(B) ϕ_B,ϕ(B)	1		%,‰,ppm,ppb
mass concentration	ρ	kg m^{-3}		g/L
molality		mol kg^{-1}		mmol/kg
volume concentration[s]	σ	1		
molar volume	V_m	m^3 mol^{-1}		L/mol
molar heat capacity	C_m	J mol^{-1} K^{-1}		
molar conductivity	Λ_m	S m^2 mol^{-1}	(S: siemens)	
Faraday constant[t]	F	C mol^{-1}		
Electricity and Magnetism				
quantity of electricity[u]	Q	C	coulomb	
charge density	ρ	C m^{-3}		
electric potential	ϕ,V	V	volt	
electric potential difference, voltage	U,$\Delta\phi$,ΔV	V		
electric dipole moment	$\mathbf{p}$, $\mathbf{p}_c$	C m		
electric current*	I	A	ampere	
electric current density	j	A m^{-2}		
electric field strength	$\mathbf{E}$	V m^{-1}		
electric displacement	$\mathbf{D}$	C m^{-2}		
capacitance	C	F	farad	
permittivity	ε	F m^{-1}		
relative permittivity	ε	1		
dielectric polarization	$\mathbf{P}$	C m^{-2}		
electric susceptibility	χ_c	1		
polarization (of a particle)	α	m^2 C V^{-1}		
magnetic flux	Φ	Wb	weber	
magnetic flux density	$\mathbf{B}$	T	tesla	
magnetic field strength	$\mathbf{H}$	A m^{-1}		
permeability	μ	H m^{-1}, N A^{-2}	(H: henry)	
relative permeability	μ_r	1		
magnetization	M	A m^{-1}		
magnetic susceptibility	χ	1		
molar magnetic susceptibility	χ_m	m^3 mol^{-1}		
(electrical) resistance	R	Ω	ohm	
(electrical) conductance	G	S	siemens	
(electrical) resistivity	ρ	Ω m		
(electrical) conductivity	κ,σ	S m^{-1}		
self-inductance	L	H	henry	
Radiation				
radiant energy	Q,W,Q_c	J		
luminous intensity*	I	cd	candela	
radiant intensity	I_c	W sr^{-1}, W		
emissivity, emittance	ε	1		

Name of quantity[a]	Symbol[b]	SI unit[c]	Name of unit	Other units
absorptance	α	1		
reflectance	ρ, R	1		
transmittance	τ	1		
absorption coefficient:				
linear (decadic)	a	m^{-1}		
molar (decadic)	ε	$m^2\ mol^{-1}$		
refractive index	n	1		
molar refraction	R_m	$m^3\ mol^{-1}$		
angle of optical rotation	α	1, rad		
Transport Properties				
flux of quantity X	J_X, J	(varies)		
mass flow rate	$q_m, \dot{m}$	$kg\ s^{-1}$		
volume flow rate	$q_V, \dot{V}$	$m^3\ s^{-1}$		
heat flow rate	Φ	W		
thermal conductivity	κ, k, λ	$W\ m^{-1}\ K^{-1}$		
coefficient of heat transfer	h	$W\ m^{-2}\ K^{-1}$		
thermal diffusivity	a	$m^2\ s^{-1}$		
diffusion coefficient	D	$m^2\ s^{-1}$		
thermal diffusion coefficient	D_T	$m^2\ s^{-1}$		
viscosity	η, μ	Pa s		
kinematic viscosity	ν	$m^2\ s^{-1}$		

[a]SI base quantities are marked by asterisks (*)
[b]Recommended by IUPAC
[c]SI base units as well as derived and supplementary units are listed; all are to be used with prefixes as needed
[d]do not use "metre"
[e]do not use "litre"; $1\ L = 10^{-3}\ m^3$
[f]$1\ Hz = 1\ s^{-1}$; $1\ N = 1\ kg\ m\ s^{-2}$
[g]Formerly metric ton; $1\ t = 10^3\ kg$
[h]$1\ bar = 10^5\ Pa$
[i]$1\ W\ h = 3.6 \times 10^3\ J$
[j]$1\ eV = 1.602\ 189 \times 10^{-19}\ J$
[k]$1\ W = 1\ J/s = 1\ VA$
[l]$N_A = 6.022\ 136\ 7 \times 10^{23}\ mol^{-1}$
[m]$k = 1.380\ 658 \times 10^{-23}\ J\ K^{-1}$
[n]$h = 6.626\ 075\ 5 \times 10^{-34}\ J\ s$
[o]$R = 8.314\ 510\ J\ mol^{-1}\ K^{-1}$
[p]Molar quantities can be distinguished from the quantity of a system by adding the subscript m; e.g., molar internal energy U_m, in $J\ mol^{-1}$
[q]$1\ u = 1.660\ 565\ 5 \times 10^{-27}\ kg$
[r]A more accurate, but rather uncommon, name is "amount-of-substance fraction"
[s]σ refers to the total volume of a mixture, whereas the volume fraction φ relates the volume of a substance to the volume of several components before mixing
[t]$F = 9.648\ 530\ 9 \times 10^4\ C\ mol^{-1}$
[u]Also called electric charge